Nanotechnology in the Life Sciences

Series Editor

Ram Prasad
Department of Botany
Mahatma Gandhi Central University
Motihari, Bihar, India

Nano and biotechnology are two of the 21st century's most promising technologies. Nanotechnology is demarcated as the design, development, and application of materials and devices whose least functional make up is on a nanometer scale (1 to 100 nm). Meanwhile, biotechnology deals with metabolic and other physiological developments of biological subjects including microorganisms. These microbial processes have opened up new opportunities to explore novel applications, for example, the biosynthesis of metal nanomaterials, with the implication that these two technologies (i.e., thus nanobiotechnology) can play a vital role in developing and executing many valuable tools in the study of life. Nanotechnology is very diverse, ranging from extensions of conventional device physics to completely new approaches based upon molecular self-assembly, from developing new materials with dimensions on the nanoscale, to investigating whether we can directly control matters on/in the atomic scale level. This idea entails its application to diverse fields of science such as plant biology, organic chemistry, agriculture, the food industry, and more.

Nanobiotechnology offers a wide range of uses in medicine, agriculture, and the environment. Many diseases that do not have cures today may be cured by nanotechnology in the future. Use of nanotechnology in medical therapeutics needs adequate evaluation of its risk and safety factors. Scientists who are against the use of nanotechnology also agree that advancement in nanotechnology should continue because this field promises great benefits, but testing should be carried out to ensure its safety in people. It is possible that nanomedicine in the future will play a crucial role in the treatment of human and plant diseases, and also in the enhancement of normal human physiology and plant systems, respectively. If everything proceeds as expected, nanobiotechnology will, one day, become an inevitable part of our everyday life and will help save many lives.

More information about this series at http://www.springer.com/series/15921

Anand Krishnan
Balasubramani Ravindran
Balamuralikrishnan Balasubramanian
Hendrik C. Swart • Sarojini Jeeva Panchu
Ram Prasad
Editors

Emerging Nanomaterials for Advanced Technologies

Editors
Anand Krishnan
Department of Chemical Pathology
University of the Free State
Bloemfontein, Free State, South Africa

Balamuralikrishnan Balasubramanian
Department of Food Science and
Biotechnology
Sejong University
Seoul, Republic of Korea

Sarojini Jeeva Panchu
School of Chemistry and Physics
University of KwaZulu Natal
Durban, South Africa

Balasubramani Ravindran
Department of Environment Energy
and Engineering
Kyonggi University
Suwon-si, Republic of Korea

Hendrik C. Swart
Department of Physics
University of the Free State
Bloemfontein, South Africa

Ram Prasad
Department of Botany
Mahatma Gandhi Central University/ Bihar
Motihari, Bihar, India

ISSN 2523-8027   ISSN 2523-8035 (electronic)
Nanotechnology in the Life Sciences
ISBN 978-3-030-80373-5   ISBN 978-3-030-80371-1 (eBook)
https://doi.org/10.1007/978-3-030-80371-1

© The Editor(s) (if applicable) and The Author(s), under exclusive license to Springer Nature Switzerland AG 2022
This work is subject to copyright. All rights are solely and exclusively licensed by the Publisher, whether the whole or part of the material is concerned, specifically the rights of translation, reprinting, reuse of illustrations, recitation, broadcasting, reproduction on microfilms or in any other physical way, and transmission or information storage and retrieval, electronic adaptation, computer software, or by similar or dissimilar methodology now known or hereafter developed.
The use of general descriptive names, registered names, trademarks, service marks, etc. in this publication does not imply, even in the absence of a specific statement, that such names are exempt from the relevant protective laws and regulations and therefore free for general use.
The publisher, the authors, and the editors are safe to assume that the advice and information in this book are believed to be true and accurate at the date of publication. Neither the publisher nor the authors or the editors give a warranty, expressed or implied, with respect to the material contained herein or for any errors or omissions that may have been made. The publisher remains neutral with regard to jurisdictional claims in published maps and institutional affiliations.

This Springer imprint is published by the registered company Springer Nature Switzerland AG
The registered company address is: Gewerbestrasse 11, 6330 Cham, Switzerland

Preface

Recent years, the world has perceived the rise of nanotechnology, a fascinating field that creates numerous divisions of the scientific community. The book covers a broad spectrum of the scientific fields such as synthesis techniques, various innovative characterization techniques, and growth mechanisms of nanomaterials, physics and chemistry of nanomaterials, diverse functionalization methods as well as their applications in biological, therapeutic, energy, food and environmental science. Also, it focuses on the applications of nanostructured materials, integrative applications such as nano and microelectronic sensor devices, as well as agriculture and environmental remediation applications. This book comprises a collection of chapters on advances in functionalized nanomaterials and discusses the early stages of development of functionalized nanostructures, including a look at the future of 2D nanomaterials and 3D objects. Further, it includes a chapter on nanomaterial research developments, highlighting work on the life-cycle analysis of nanostructured materials and toxicity aspects. The contents of this book will prove useful for researchers and professionals working in the field of nanomaterials and green technology. Researchers, in the field of nanotechnology from entrants to specialized researchers, in a number of disciplines ranging from biology, chemistry and materials science to engineering and manufacturing in both of academia and industry sectors. The book targets scientists, researchers, academicians, graduates and doctoral students working in biological sciences and waste management.

Our sincere gratitude goes to the contributors for their insights on applications of various nanomaterials in industrial and medical sector.

We sincerely thank Dr. Eric Stannard, Senior Editor Botany, Springer, and the production editor for their generous assistance, constant support and patience in finalizing this book.

Bloemfontein, South Africa	Anand Krishnan
Suwon, South Korea	Balasubramani Ravindran
Seoul, South Korea	Balamuralikrishnan Balasubramanian
Bloemfontein, South Africa	Hendrik C. Swart
Durban, South Africa	Sarojini Jeeva Panchu
Bihar, India	Ram Prasad

Contents

1 **An Insight on Emerging Nanomaterials for the Delivery of Various Nutraceutical Applications for the Betterment of Heath**............ 1
T. Karpagam, Balasubramanian Balamuralikrishnan,
B. Varalakshmi, A. Vijaya Anand, and J. Sugunabai

2 **Nanoscale Smart Drug Delivery Systems and Techniques of Drug Loading to Nanoarchitectures**............................... 29
B. Varalakshmi, T. Karpagam, A. Vijaya Anand, and
B. Balamuralikrishnan

3 **Recent Advances in Nanomaterials-Based Drug Delivery System for Cancer Treatment**... 83
Prakash Ramalingam, D. S. Prabakaran, Kalaiselvi Sivalingam,
V. Uma Maheshwari Nallal, M. Razia, Mayurkumar Patel, Tanvi Kanekar, and Dineshkumar Krishnamoorthy

4 **Novel Organic and Inorganic Nanoparticles as a Targeted Drug Delivery Vehicle in Cancer Treatment**..................... 117
Saradhadevi Muthukrishnan, A. Vijaya Anand, Kiruthiga Palanisamy, Gayathiri Gunasangkaran, Anjali K. Ravi, and Balamuralikrishnan Balasubramanian

5 **Potential of Metal Oxide Nanoparticles and Nanocomposites as Antibiofilm Agents: Leverages and Limitations**................. 163
P. Sriyutha Murthy, V. Pandiyan, and Arindam Das

6 **Nanomaterials for A431 Epidermoid Carcinoma Treatment**........ 211
S. Christobher, P. Kalitha Parveen, Murugesh Easwaran, Haripriya Kuchi Bhotla, Durairaj Kaliannan, Balamuralikrishnan Balasubramanian, and Arun Meyyazhagan

7	Efficacy of Nanomaterials and Its Impact on Nosocomial Infections.................................... 237
	P. Kalitha Parveen, S. Christobher, Balamuralikrishnan Balasubramanian, Durairaj Kaliannan, Manikantan Pappusamy, and Arun Meyyazhagan
8	Nanonutraceuticals in Chemotherapy of Infectious Diseases and Cancer..................................... 261
	C. Sumathi Jones, V. Uma Maheshwari Nallal, and M. Razia
9	Trends of Biogenic Nanoparticles in Lung Cancer Theranostics.. 301
	V. Uma Maheshwari Nallal, C. Sumathi Jones, M. Razia, D. S. Prabakaran, and Prakash Ramalingam
10	Therapeutic Applications of Nanotechnology in the Prevention of Infectious Diseases......................... 323
	Rajkumari Mazumdar and Debajit Thakur
11	Nanotechnology's Promising Role in the Control of Mosquito-Borne Disease................................. 345
	Gopalan Rajagopal, Shenbagamoorthy Sundarraj, Krishnan Anand, and Sakkanan Ilango
12	Phytosynthesized Metal Nanomaterials as an Effective Mosquitocidal Agent................................. 369
	M. Suresh, Satheeshkumar Balu, S. Cathy Jose, and Jaison Jeevanandam
13	Perspectives of Metals and Metal Oxide Nanoparticles for Antimicrobial Consequence – An Overview............ 397
	R. L. Rengarajan, A. Rathinam, N. Suganthy, B. Balamuralikrishnan, A. Vijaya Anand, and S. Velayuthaprabhu
14	Advancement in Nanomaterial Synthesis and its Biomedical Applications.. 419
	Benil P. Bharathan, Rajakrishnan Rajagopal, Ahmed Alfarhan, Mariadhas Valan Arasu, and Naif Abdullah Al-Dhabi
15	Perspectives of Nanotechnology in Aquaculture: Fish Nutrition, Disease, and Water Treatment................. 463
	Ndakalimwe Naftal Gabriel, Habte-Michael Habte-Tsion, and Mayday Haulofu
16	Nanomaterials in Electrochemical Biosensors and Their Applications.. 487
	J. R. Anusha, Mariadhas Valan Arasu, Naif Abdullah Al-Dhabi, and C. Justin Raj

17	**Nano-Adsorbents and Nano-Catalysts for Wastewater Treatment** Zeenat Sheerazi and Maqsood Ahmed	517
18	**Nano-Bioremediation Using Biologically Synthesized Intelligent Nanomaterials** S. Sakthinarendran, M. Ravi, and G. Mirunalini	541
19	**Recent Developments in Nanotechnological Interventions for Pesticide Remediation** Rictika Das and Debajit Thakur	553
20	**Potential Applications of Nanomaterials in Agronomy: An African Insight** Hupenyu A. Mupambwa, Adornis D. Nciizah, Patrick Nyambo, Ernest Dube, Binganidzo Muchara, Morris Fanadzo, and Martha K. Hausiku	581
21	**Nanomaterials for Wastewater Remediation: Resolving Huge Problems with Tiny Particles** Ambikapathi Ramya, Periyasamy Dhevagi, and S. S. Rakesh	601
22	**Impact of Nanomaterials on Waste Management: An Insight to the Modern Concept of Waste Abatement** Ram Kumar Ganguly, Susanta Kumar Chakraborty, Sujoy Midya, and Balasubramani Ravindran	621
23	**Applicability of Emerging Nanomaterials in Microbial Fuel Cells as Cathode Catalysts** Vikash Kumar, Prasanta Pattanayak, and Subrata Hait	643
24	**Metal Oxide Nanostructured Materials for Photocatalytic Hydrogen Generation** Bishal Kumar Nahak, Lucky Kumar Pradhan, T. Suraj Kumar Subudhi, Arveen Panigrahi, Biranchi Narayan Patra, Satya Sopan Mahato, and Shrabani Mahata	665
25	**Recent Advances in the Synthesis of Heterocycles Over Heterogeneous Cerium-Based Nanocatalysts** Cong Chien Truong, Dinesh Kumar Mishra, and Hoang Long Ngo	709
Index		761

Contributors

Maqsood Ahmed Indian Institute of Chemical Technology, Hyderabad, India

Naif Abdullah Al-Dhabi Department of Botany and Microbiology, College of Science, King Saud University, Riyadh, Saudi Arabia

Ahmed Alfarhan Department of Botany and Microbiology, College of Science, King Saud University, Riyadh, Saudi Arabia

J. R. Anusha Department of Chemistry, Dongguk University, Jung-gu, Seoul, Republic of Korea

Mariadhas Valan Arasu Department of Botany and Microbiology, College of Science, King Saud University, Riyadh, Saudi Arabia

A. Vijaya Anand Department of Human Genetics and Molecular Genetics, Bharathiar University, Coimbatore, Tamil Nadu, India

Balamuralikrishnan Balasubramanian Department of Food Science and Biotechnology, College of Life Science, Sejong University, Seoul, South Korea

Satheeshkumar Balu Department of Ceramic Technology, A.C. Tech Campus, Anna University, Chennai, Tamil Nadu, India

Benil P. Bharathan Department of Agadatantra, Vaidyaratnam P.S Varier Ayurveda College, Kottakkal, Kerala, India

Haripriya Kuchi Bhotla Bioknowl Insights Private Limited, Coimbatore, Tamil Nadu, India

Susanta Kumar Chakraborty Department of Zoology, Vidyasagar University, Midnapore, WB, India

S. Christobher Department of Zoology, Nallamuthu Gounder Mahalingam College, Pollachi, Tamil Nadu, India

Arindam Das Homi Bhabha National Institute, Anushaktinagar, Mumbai, India

Surface and Nanoscience Division, Material Science Group, Indira Gandhi Center for Atomic Research, Kalpakkam, Tamil Nadu, India

Rictika Das Microbial Biotechnology Laboratory, Life Sciences Division, Institute of Advanced Study in Science and Technology (IASST), Guwahati, Assam, India

Department of Molecular Biology and Biotechnology, Cotton University, Guwahati, Assam, India

Periyasamy Dhevagi Department of Environmental Sciences, Tamil Nadu Agricultural University, Coimbatore, Tamil Nadu, India

Ernest Dube School of Natural Resources Management, Nelson Mandela Metropolitan University, George Campus, George, South Africa

Murugesh Easwaran Bioknowl Insights Private Limited, Coimbatore, Tamil Nadu, India

Morris Fanadzo Department of Agriculture, Faculty of Applied Sciences, Cape Peninsula University of Technology, Wellington, South Africa

Ndakalimwe Naftal Gabriel Department of Fisheries and Aquatic Sciences, Sam Nujoma Campus, University of Namibia, Henties Bay, Namibia

Ram Kumar Ganguly Department of Zoology, Vidyasagar University, Midnapore, WB, India

Gayathiri Gunasangkaran Department of Biochemistry, Bharathiar University, Coimbatore, Tamil Nadu, India

Habte-Michael Habte-Tsion Cooperative Extension-Aquaculture Research Institute, University of Maine, Orono, Maine, United States

Subrata Hait Department of Civil and Environmental Engineering, Indian Institute of Technology Patna, Bihar, India

Mayday Haulofu Sam Nujoma Coastal and Marine Resources Research Center, University of Namibia, Henties Bay, Namibia

Martha K. Hausiku Sam Nujoma Marine and Coastal Resources Research Centre, Sam Nujoma Campus, University of Namibia, Henties Bay, Namibia

Sakkanan Ilango Department of Zoology, Ayya Nadar Janaki Ammal College, Sivakasi, India

Jaison Jeevanandam CQM – Centro de Química da Madeira, MMRG, Universidade da Madeira, Campus da Penteada, Funchal, Portugal

C. Sumathi Jones Department of Pharmacology and Environmental Toxicology, University of Madras, Chennai, India

Department of Pharmacology, Asan Memorial Dental College and Hospital, Chengalpattu, India

S. Cathy Jose Postgraduate and Research Department of Advanced Zoology and Biotechnology, Loyola College, Chennai, Tamil Nadu, India

Durairaj Kaliannan Zoonosis Research Center, Department of Infection Biology, School of Medicine, Wonkwang University, Iksan, Republic of Korea

P. Kalitha Parveen PG Department of Zoology, Hajee Karutha Rowther Howdia College of Arts and Science, Uuthamapalayam, Tamil Nadu, India

Tanvi Kanekar Product Development, Genus Lifesciences Inc, Allentown, PA, USA

T. Karpagam Department of Biochemistry, Shrimati Indira Gandhi College, Tiruchirappalli, Tamil Nadu, India

Dineshkumar Krishnamoorthy Department of Plant Science, School of Biological Sciences, Central University of Kerala, Kasaragod, Kerala, India

Anand Krishnan Department of Chemical Pathology, School of Pathology, Faculty of Health Sciences and National Health Laboratory Service, University of the Free State, Bloemfontein, South Africa

Vikash Kumar Department of Civil and Environmental Engineering, Indian Institute of Technology Patna, Bihar, India

Shrabani Mahata Department of Chemistry, National Institute of Science and Technology, Brahmapur, India

Satya Sopan Mahato Department of Electronics and Communication Engineering, National Institute of Science and Technology, Brahmapur, India

Rajkumari Mazumdar Microbial Biotechnology Laboratory, Life Sciences Division, Institute of Advanced Study in Science and Technology (IASST), Department of Science and Technology, Ministry of Science and Technology (India), Guwahati, Assam, India

Department of Molecular Biology & Biotechnology, Cotton University, Guwahati, Assam, India

Arun Meyyazhagan Department of Life Sciences, CHRIST (Deemed to be University), Bangalore, Karnataka, India

Sujoy Midya Department of Zoology, Vidyasagar University, Midnapore, WB, India

G. Mirunalini DST-FIST Sponsored Centre, ESTC Cell – Marine Biotechnology, Sathyabama Institute of Science and Technology, Chennai, India

Dinesh Kumar Mishra Korea Institute of Industrial Technology (KITECH), Cheonan-si, Republic of Korea

Binganidzo Muchara Graduate School of Business Leadership, University of South Africa, Midrand, South Africa

Hupenyu A. Mupambwa Sam Nujoma Marine and Coastal Resources Research Centre, Sam Nujoma Campus, University of Namibia, Henties Bay, Namibia

P. Sriyutha Murthy Water and Steam Chemistry Division, Bhabha Atomic Research Centre, Kalpakkam, Tamil Nadu, India

Homi Bhabha National Institute, Anushaktinagar, Mumbai, India

Saradhadevi Muthukrishnan Department of Biochemistry, Bharathiar University, Coimbatore, Tamil Nadu, India

Bishal Kumar Nahak Department of Electronics and Communication Engineering, National Institute of Science and Technology, Brahmapur, India

Adornis D. Nciizah Agricultural Research Council – Institute for Soil, Climate and Water, Pretoria, South Africa

Hoang Long Ngo NTT Hi-Tech Institute, Nguyen Tat Thanh University, Ho Chi Minh City, Vietnam

Ambikapathi Nivetha Department of chemistry, Bharathiar University, Coimbatore, Tamil Nadu, India

Patrick Nyambo Department of Agronomy, University of Fort Hare, Alice Campus, Alice, South Africa

Kiruthiga Palanisamy Department of Biochemistry, Bharathiar University, Coimbatore, Tamil Nadu, India

V. Pandiyan Nehru Memorial College, Puthanampatti, Tiruchirapalli, Tamil Nadu, India

Arveen Panigrahi Department of Electronics and Communication Engineering, National Institute of Science and Technology, Brahmapur, India

Manikantan Pappusamy Department of Life Sciences, CHRIST (Deemed to be University), Bangalore, Karnataka, India

Mayurkumar Patel Product Development, Genus Lifesciences Inc, Allentown, PA, USA

Biranchi Narayan Patra Department of Electrical and Electronics Engineering, National Institute of Science and Technology, Brahmapur, India

Prasanta Pattanayak Advanced Polymer Laboratory, Department of Polymer Science & Technology, University of Calcutta, Kolkata, West Bengal, India

D. S. Prabakaran Department of Radiation Oncology, Chungbuk National University College of Medicine, Cheongju, Republic of Korea

Department of Biotechnology, Ayya Nadar Janaki Ammal College, Sivakasi, Sivakasi, Tamil Nadu, India

Lucky Kumar Pradhan Department of Electronics and Communication Engineering, National Institute of Science and Technology, Brahmapur, India

Mohan Uma Priya Department of Biotechnology, Kalasalingam Academy of Research and Education, Krishnankoil, India

C. Justin Raj Department of Chemistry, Dongguk University, Jung-gu, Seoul, Republic of Korea

Gopalan Rajagopal Department of Zoology, Ayya Nadar Janaki Ammal College, Sivakasi, India

Rajakrishnan Rajagopal Department of Botany and Microbiology, College of Science, King Saud University, Riyadh, Saudi Arabia

S. S. Rakesh Department of Environmental Sciences, Tamil Nadu Agricultural University, Coimbatore, Tamil Nadu, India

Prakash Ramalingam Product Development, Genus Lifesciences Inc, Allentown, PA, USA

Ambikapathi Ramya Department of Environmental Sciences, Tamil Nadu Agricultural University, Coimbatore, Tamil Nadu, India

A. Rathinam Department of Animal Science, Bharathidasan University, Tiruchirappalli, Tamil Nadu, India

Key Laboratory for Genome Stability and Disease Prevention, Guangdong Province, Shenzhen University, Shenzhen, China

Anjali K. Ravi Department of Biochemistry, Bharathiar University, Coimbatore, Tamil Nadu, India

M. Ravi DST-FIST Sponsored Centre, ESTC Cell – Marine Biotechnology, Sathyabama Institute of Science and Technology, Chennai, India

Balasubramani Ravindran Department of Environmental Energy and Engineering, Kyonggi University, Suwon, Republic of Korea

M. Razia Department of Biotechnology, Mother Teresa Women's University, Kodaikanal, Tamil Nadu, India

R. L. Rengarajan Department of Animal Science, Bharathidasan University, Tiruchirappalli, Tamil Nadu, India

S. Sakthinarendran DST-FIST Sponsored Centre, ESTC Cell – Marine Biotechnology, Sathyabama Institute of Science and Technology, Chennai, India

Zeenat Sheerazi Department of Chemistry, Jamia Millia Islamia, New Delhi, India

Kalaiselvi Sivalingam Department of Pharmaceutical Sciences, Irma Lerma Rangel College of Pharmacy, Texas A&M University, Kingsville, TX, USA

Shenbagamoorthy Sundarraj Department of Zoology, Ayya Nadar Janaki Ammal College, Sivakasi, India

T. Suraj Kumar Subudhi Department of Electronics and Communication Engineering, National Institute of Science and Technology, Brahmapur, India

N. Suganthy Department of Nanoscience and Technology, Alagappa University, Karaikudi, Tamil Nadu India

J. Sugunabai Department of Biochemistry, Seethalakshmi Ramaswamy College, Tiruchirappalli, Tamil Nadu, India

M. Suresh Loyola Institute of Frontier Energy (LIFE), School of Environmental Toxicology and Biotechnology, Postgraduate and Research Department of Advanced Zoology and Biotechnology, Loyola College, Chennai, Tamil Nadu, India

Debajit Thakur Microbial Biotechnology Laboratory, Life Sciences Division, Institute of Advanced Study in Science and Technology (IASST), Department of Science and Technology, Ministry of Science and Technology (India), Guwahati, Assam, India

Cong Chien Truong Department of Bio-functional Molecular Engineering, Graduate School of Science and Engineering, University of Toyama, Toyama, Japan

V. Uma Maheshwari Nallal Department of Biotechnology, Mother Teresa Women's University, Kodaikanal, Tamil Nadu, India

B. Varalakshmi Department of Biochemistry, Shrimati Indira Gandhi College, Tiruchirappalli, Tamil Nadu, India

S. Velayuthaprabhu Department of Biotechnology, Bharathiar University, Coimbatore, Tamil Nadu, India

About the Editors

Anand Krishnan, PrChemSA, MRSC has expertise in organic chemistry/medical biochemistry/integrative medicine/nano(bio)technology/drug discovery. He received his doctoral degree in organic chemistry from the Department of Chemistry, Durban University of Technology, in collaboration with the Department of Medical Biochemistry, University of KwaZulu-Natal, in 2014. He completed his master's degree in organic chemistry from Bharathiar University, India, and bachelor's degree in chemistry from Madurai Kamaraj University, India. He was a postdoctoral researcher at Durban University of Technology, South Africa, from November 2014 to November 2016. Later, he worked as a senior researcher in the disciplines of medical biochemistry and chemical pathology, School of Laboratory Medicine and Medical Sciences, University of KwaZulu-Natal, Durban, South Africa, from January 2017 to June 2019. Recently, he received the prestigious Innovation Postdoctoral Research Fellowship from the Department of Science and Innovation (DSI) and the National Research Foundation (NRF), South Africa, and is conducting research in the Department of Chemical Pathology, School of Pathology, Faculty of Health Sciences and National Health Laboratory Service (NHLS), University of the Free State, Bloemfontein, South Africa. He has published many scientific articles in international peer-reviewed journals and has authored many chapters as well as review articles. He is recognized for his contributions and has received awards from national and international organizations. He has been awarded Best Postdoctoral Researcher Award for 2016 and 2017 by Durban University of Technology and Young Scientist Researcher Award 2016 by Pearl Foundation.Dr. Krishnan was evaluated by the National Research Foundation and was awarded a Y1 rating, which is given to promising young researchers. He is a member of various editorial boards of internationally reputed journals. His research interests include organic chemistry, heterocyclic chemistry, medicinal biochemistry, drug discovery and delivery, extracellular vesicles, nanotoxicology, clinical biochemistry, and chemical pathology.

Balasubramani Ravindran, PhD is an assistant professor in the Department of Environmental Energy & Engineering, Kyonggi University, Suwon, South Korea. He obtained his doctorate from the Environmental Science and Engineering Division, Council of Scientific and Industrial Research (CSIR), Central Leather Research Institute (CLRI), which is an affiliated with the University of Madras, Tamil Nadu, India. His primary research focuses on the development and evaluation of treatment technologies for solid waste and wastewater from domestic and industrial outlets through aerobic and anaerobic fermentation of industrial waste, and composting/vermicomposting of fermented metabolites/solid waste/nanotechnology applications and phytotoxic/plant growth studies. Dr. Ravindran has also developed an improved process technology for the fast production of nutrient-rich vermicompost from different waste sources using different types of earthworms. He co-developed a process for the separation of enzymes from the aqueous phase using surface modified mesoporous activated carbon. Besides, he has filed patents based on the above novel process. His current research focus is on biochar/black carbon amendment and the strategies for carbon sequestration, and reducing greenhouse gas (GHG) emissions from compost and vermicompost production systems while degrading the different waste and animal manures. He has published more than seventy research papers in peer-reviewed journals, edited a book, published notable book chapters, and has three patents to his credit. Dr. Ravindran has received national and international funds for his research project. He is serving as guest editor in several international journals and is acting as a potential reviewer in top international journals, and has also received Outstanding Reviewer Award from Elsevier and Springer Journals. He has also received the prestigious "Best Researcher – IBET 2017" award (in waste management research) for exceptional performance and contributions to international bioenergy technology/ecoprotection/organic food/green businesses given by the International Centre for Biogas & Bioenergy Technology, India.

Balamuralikrishnan Balasubramanian, PhD is an assistant professor in the Department of Food Science and Biotechnology, Sejong University, Seoul, South Korea. His area of research is multidisciplinary and focuses on biological science, molecular genetics, food microbiology, animal science, nutrition science and food biotechnology, especially food resources and microbiological science. His interests include synthesis of nanomaterials/bioactive compounds from natural by-products and their nutraceutical applications; isolation and characterization of probiotic for food/feed supplementation and biological field; and nutrition in aquaculture. To his credit, he has participated in various international/symposia/conferences in the USA, Canada, Japan, Austria, Italy, Czech Republic, Thailand and South Korea and has published more than seventy-five research papers in international journals of repute. He has been serving as guest editor in *Animals* [MDPI] and is acting as potential reviewer in many highly reputed journals. He has been awarded life membership of various scientific societies such as The Korean Society of Food Science and Technology, Poultry Science Association, Korean Society of Animal Science and Technology, and Animal Nutrition Society of India (ANSI). Dr. B. Balasubramanian worked as a postdoctoral researcher in Department of Animal Science, Dankook University, Cheonan, South Korea.

Hendrik C. Swart is an internationally acclaimed researcher and currently a senior professor in the Department of Physics at the University of the Free State. He brought luminescence materials to South Africa in the beginning of 1996 after a highly productive sabbatical spent in the lab of Paul Holloway, Florida University, Gainesville. This laid the foundation for his subsequent research at the UFS and was one of the most exhilarating times of his academic career. Since then, he has led research in the area of the degradation of phosphors for field emission displays, as well as developing materials for nano solid state lighting. He has been key in the development of processes to synthesize and deposit thin films of various types of semiconductor nanoparticles, which will enhance the colour, luminescent intensity and life of such displays. His research led to the establishment of a strong group working on luminescent materials and also to the establishment of several smaller groups all over South Africa. He has more than 690 publications in international peer-reviewed journals as well as 112 peer-reviewed conference proceedings, and is editor/author or co-editor/author of 25 book chapters or books with more than 12,360 cited author references, H-index of 47, i10 index of 372 on google scholar (40 and 308 since 2015) and more than 660 national and international conference contributions (authored and co-authored). He has an ISI H-and Scopus index of 41 (rid=g-2696-2012) and 42, respectively. He is a reviewer for more than 100 international and national professional journals in his field (or in related fields), and a member of the editorial board of the high impact factor journal *Critical Reviews in Solid State and Materials Sciences* (IF-8.344). He is on the editorial board of *Applied Surface Science* (IF- 6.182). Hendrik has received the South African National Science and Technology Forum (NSTF) Award in 2009 for research capacity development of students in the niche area of nanophysics. His commitment to the next generation of scientists is also reflected by the awards he received from the Faculty of Natural and Agricultural Sciences at the University of the Free State, South Africa, for excellence (deans medal) (2012), research (2014), mentorship (2008), academic entrepreneurship (2009) and best researcher (2018). He received honorary membership of the Golden Key Association

(2012). The Radio Rosestad Award for outstanding research and postgraduate teaching in 2017. Hendrik was chair of national and international conferences. He has supervised more than 80 PhD and MSc students successfully in the past with another 20 in progress and has established a National Nano Surface Characterization Facility (NNSCF) containing state-of-the-art surface characterization equipment. A research chair in solid state luminescent and advanced materials was awarded to him from the South African Research Chairs Initiative (SARChI) at the end of 2012, which was renewed for another 5 years at the end of 2017. The main focus of his research group is the improvement of luminescent materials for applications in flat panel displays, solar cells, solid state lighting, dosimetry and thermometry.

Sarojini Jeeva Panchu, PhD is an postdoctoral researcher in the Department of Physics, University of the Free State, Bloemfontein, South Africa. She has expertise in functional nanomaterials for energy and biomedical applications. She received her doctoral degree in physics from the Discipline of Physics, School of Chemistry and Physics, University of KwaZulu-Natal, South Africa. She has worked as a research assistant in central electrochemical research institute (CSIR-CECRI), Karaikudi, India, from 2012 to 2016. She has received National Research Foundation (NRF) Fellowship in South Africa for her doctoral research in the Discipline of Physics, School of Chemistry and Physics, University of KwaZulu-Natal, Durban, South Africa. She has published scientific articles in international peer-reviewed journals and is a member of the Energy Science Society of India (ESSI). Her research interest includes nanomaterials synthesis by using hydrothermal, solvothermal, physical vapour deposition and chemical vapour deposition methods, solar cells, electrochemistry, fabrication of energy storage energy conversion, and sensor devices systems for multianalyte analysis.

Ram Prasad, PhD is a Associate Professor in the Department of Botany, Mahatma Gandhi Central University, Motihari, Bihar, India. His research interest includes applied and environmental microbiology, plant–microbe interactions, sustainable agriculture and nanobiotechnology. Dr. Prasad has more than two hundred twenty-five publications to his credit, including research papers, review articles and book chapters, and six patents issued or pending, and has edited or authored several books. Dr. Prasad has 12 years of teaching experience and has been awarded the Young Scientist Award and Prof. J.S. Datta Munshi Gold Medal by the International Society for Ecological Communications; FSAB fellowship by the Society for Applied Biotechnology; the American Cancer Society UICC

International Fellowship for Beginning Investigators, USA; Outstanding Scientist Award in the field of microbiology by Venus International Foundation; and BRICPL Science Investigator Award and Research Excellence Award. He has been serving as editorial board member of *BMC Microbiology, BMC Biotechnology, IET Nanobiotechnology, Journal of Nanomaterials, Current Microbiology, Annals of Microbiology, Archives of Microbiology, Archives of Phytopathology and Plant Protection, Journal of Renewable Materials, Journal of Agriculture and Food Research*; including series editor of Nanotechnology in the Life Sciences, Springer Nature, USA. Previously, Dr. Prasad served as assistant professor Amity University, Uttar Pradesh, India; visiting assistant professor, Whiting School of Engineering, Department of Mechanical Engineering at Johns Hopkins University, Baltimore, USA; and research associate professor at the School of Environmental Science and Engineering, Sun Yat-sen University, Guangzhou, China.

Chapter 1
An Insight on Emerging Nanomaterials for the Delivery of Various Nutraceutical Applications for the Betterment of Heath

T. Karpagam, Balasubramanian Balamuralikrishnan, B. Varalakshmi, A. Vijaya Anand, and J. Sugunabai

Contents

1.1	Introduction...	2
	1.1.1 Role of Nutraceuticals..	3
	1.1.2 Role of Phytonutrients..	4
1.2	Classification of Nutraceuticals...	4
	1.2.1 Nutraceuticals with Nutrients...	5
	1.2.2 Nutraceuticals with Phytochemicals...	5
	1.2.3 Nutraceuticals with Dietary Supplements..	5
1.3	Sources of Nutraceuticals..	5
1.4	Recent Trends of Nutraceuticals in the Global Market.......................................	6
1.5	Bioavailability of Bioactive Compounds of Nutraceuticals................................	6
	1.5.1 Enhancing the Bioavailability of Nutraceuticals by Nanotechnology.....	7
1.6	Nanotechnology...	7
	1.6.1 Nanocarriers as Nano Delivery System...	7
	1.6.2 Synthesis of Nanoparticles...	8
	1.6.3 Scenarios of Nanotechnology in Nutraceuticals......................................	9

B. Balamuralikrishnan contributed equally with all other contributors.

T. Karpagam (✉) · B. Varalakshmi
Department of Biochemistry, Shrimati Indira Gandhi College,
Tiruchirappalli, Tamil Nadu, India

B. Balamuralikrishnan
Department of Food Science and Biotechnology, College of Life Science, Sejong University,
Seoul, South Korea

A. Vijaya Anand
Department of Human Genetics and Molecular Genetics, Bharathiar University,
Coimbatore, Tamil Nadu, India

J. Sugunabai
Department of Biochemistry, Seethalakshmi Ramaswamy College,
Tiruchirappalli, Tamil Nadu, India

© The Author(s), under exclusive license to Springer Nature Switzerland AG 2022
A. Krishnan et al. (eds.), *Emerging Nanomaterials for Advanced Technologies*, Nanotechnology in the Life Sciences, https://doi.org/10.1007/978-3-030-80371-1_1

1.7 Food-Grade Nanomaterials.. 10
  1.7.1 Lipid as Nanomaterials... 10
  1.7.2 Polymer as Nanomaterials.. 17
  1.7.3 Cellulose as Nanomaterials... 19
  1.7.4 Protein as Nanomaterials.. 20
  1.7.5 Polysaccharide as Nanomaterials... 21
1.8 Conclusion.. 22
References... 23

1.1 Introduction

An ancient Siddha medicine aphorism "unave marundhu, marundhe unavu" which means food as medicine, medicine as food well emphasizes the importance of diet for disease-free living. The role of food components in inhibiting disease and health enhancement is becoming more apparent to the researchers as well as consumers (Palzer 2009). The relationship between diet and health, and awareness on foods, from a wide range of sources, that either inherently contains health-enhancing active ingredients or included via fortification is important (Taneja and Singh 2012). Thus, there is a demand for a balanced diet and functional food products that focus on specific health benefits. Consumption of functional ingredients enriched foods keeps the body in good health. Diets rich in essential nutrients along with regular exercise decrease threats related to many diseases and maintain body weight. The case of age-related diseases, such as malignancy (e.g., gastrointestinal cancer), cardiovascular diseases, and diabetes, is more prevalent with the increase in life expectancy of humans. Escalating the use of food products from plants delays the development of these chronic diseases, and various health organizations all over the world recommend this. Plant-derived food products show a positive effect on the reduction of chronic diseases due to the presence of phytoconstituents. These are nonnutritive secondary metabolites with widespread biological functions. As bioactive metabolites, these phytochemicals ensure low effectiveness in comparison with pharmaceutical products. However, if regularly ingested in the diet, a perceptible long-term physiological effect can result without side effects (Shampa Sen and Yashwant Pathak 2016). These bioactive components are beneficial to our health and exploited for nutraceutical application. De Felice (1995) defined nutraceuticals as "food or a part of food that provides medical or health benefits, which include prevention and treatment of diseases." The term nutraceutical exists between foods and drugs, and it arose from the combination of nutrition and pharmaceuticals and was framed by Stephen L. DeFelice (Fig. 1.1), in the year 1989 (Kalra 2003). This concept is in modern food science, and the area is beyond the diet, but before the drugs (El Sohaimy 2012).

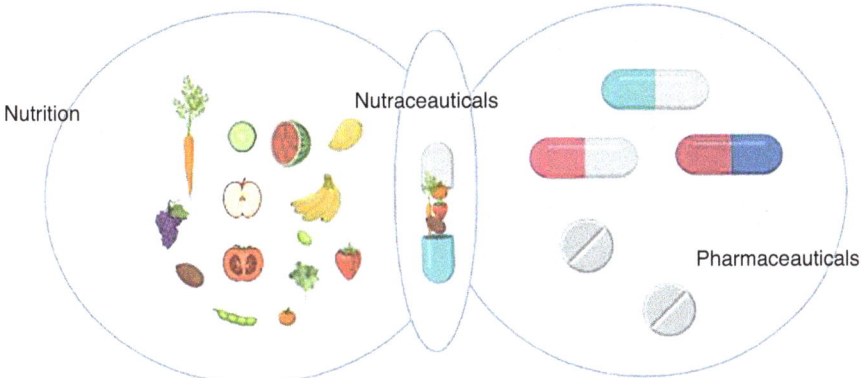

Fig. 1.1 Stephen L. DeFelice definition of nutraceuticals

1.1.1 Role of Nutraceuticals

Nutraceuticals are dietary supplements which deliver nutrients to the body. They are made accessible within a nonfood medium by supplementing phytoconstituents in the health/food product. Nutraceuticals enhance human health by introducing different dosages of active compounds from food in a higher quantity than the amount that can be by the consumption of regular food. Nutraceuticals provide resistance against several diseases and thus contribute significantly to the therapeutic performances. According to the theory of nutritional therapy, the nutraceuticals function by cleansing the body, evading deficiencies due to lack of protective food, reestablishing healthy food practices, and thus restoring healthy absorption of nutrients.

1.1.1.1 The Functional Role of Nutraceuticals

The functional role of nutraceuticals is the enhancement of nourishment for specific groups of people with impaired metabolism, in a specific physical state, who reduce the consumption of certain substances in food.

The following are the categories of nutraceuticals which include foods for special medical purposes (FSMPs):

- Nutrition for newborn and subsequent nourishment for the next stage
- Foods with cereal and foods devoid of cereal for infants and preschool children
- Low-calorie foods for decreasing obesity
- Foods deprived of or lacking phenylalanine
- Foods without gluten
- Nutrition designed for carbohydrate metabolic disorder

- Lactose-lacking foods
- Foods composed of decreased protein
- Foods for sportspersons and persons with augmented physical activity

FSMPs are recommended to consume under the direction of medical persons, and it must hold information regarding their use.

1.1.2 Role of Phytonutrients

Phytonutrients have a biological function which enhances health. They generally help plants to thwart competitors, predators, or pathogens – phytochemicals in food work by acting as a substrate and cofactors for biochemical reactions. As inhibitors of biochemical reaction, it removes the unwanted component in the intestine and augments the absorption and stability of vital nutrients. Also, they function as a selective growth factor for good flora and as inhibitors of infectious bacteria. They also deactivate harmful chemicals and act as ligands to antagonize the receptors that are present either extracellular or intracellularly (Baby Chauhan et al. 2013).

1.2 Classification of Nutraceuticals

Nutraceuticals are dietary component with health benefits. Examples include diets rich in fiber, probiotics, prebiotics, PUFA, antioxidants, vitamins, minerals, polyphenols, and spices. Natural sources, chemical nature of components, nature of the reaction, and pharmacological functions form the basis of nutraceutical classification.

The broad classifications of nutraceuticals are as follows:

(a) Potential nutraceuticals
(b) Established nutraceuticals

Clinical data on the medical benefits are needed to make the potential nutraceutical to established nutraceuticals (Pandey et al. 2010).

Classification of nutraceuticals based on its composition/functions (Akobundu et al. 2004):

1.2.1 Nutraceuticals with nutrients
1.2.2 Nutraceuticals with herbals or phytochemicals
1.2.3 Nutraceuticals with dietary supplements

1.2.1 Nutraceuticals with Nutrients

Nutraceuticals with nutrients are foods rich in nutrients and do the nutritive function, e.g., carbohydrates, fatty acids, amino acids, vitamins, and minerals. Foods comprising of carbohydrates, proteins, and lipids are necessary for the appropriate functioning of the body and its calorific requirements. Vitamins and minerals are not synthesized within the human body and play the role as protective foods. Hence they must be supplied via diet for the appropriate functioning of the body. Nutraceuticals play the role to improve health by combating against some chronic diseases (Mc Clements 2012).

1.2.2 Nutraceuticals with Phytochemicals

Nutraceuticals with phytochemicals possess herbs or botanical products. Intake of plant-based food offers enormous benefits to human health. Plants contain various phytochemical compounds, mostly polyphenols. These polyphenols are responsible for the beneficial activity. Pharmaceutical products with active ingredient have appeared in health products. These active ingredients are phytoconstituents with bioactivity (Espín et al. 2007). Important phytochemicals are anthocyanins, resveratrol, isoflavones, and polyphenols like ellagic acid, proanthocyanins, and flavanones.

1.2.3 Nutraceuticals with Dietary Supplements

Nutraceuticals act as dietary supplements composed of probiotics, prebiotics, antioxidants, and enzymes. They deliver bioactive constituents to the body, which is a mixture of several ingredients, metabolites, or constituents in the form of liquid, capsule, or tablet. The subclassification includes extracts from plants, supplements from herbs, proteins, vitamins, and minerals.

1.3 Sources of Nutraceuticals

Sources of nutraceuticals include plants, animals, and microbial.

Plant sources: Some significant plant sources include beta-glucan, ascorbic acid, gamma-tocotrienol, luteolin, cellulose, quercetin, gallic acid, indole3carbinol, pectin, perillyl alcohol, glutathione, potassium, allicin, D-limonene, daidzein, genistein, lycopene, hemicellulose, lignin, capsaicin, alpha-tocopherol, and zeaxanthin.

Animal sources: Some significant animal sources include conjugated linoleic acid (CLA), eicosapentaenoic acid (EPA), docosahexaenoic acid (DHA), selenium, and zinc.

Microbials: Some significant microbial sources include yeast, *Bifidobacterium*, *Lactobacillus*, and *Streptococcus*. Nowadays, the intake of foods is not only to satisfy the energy needs but also combat diseases and augment both physical and mental health (Menrad 2003).

1.4 Recent Trends of Nutraceuticals in the Global Market

Due to the enormous change in living style, the incidence of lifestyle-associated diseases is also growing. Life expectancy with these diseases is also growing. After gaining knowledge regarding the underlying cause of disease, now the consumers across the world are shifting from disease-causing chemical products to preventive healthcare products like nutraceuticals. Hence, the growth of the nutraceutical market is also rising. The value of global nutraceutical market was 205.39 billion USD in 2016 and is foreseen to elevate to around 294.79 billion USD by 2022, at 6.3% compound annual growth rate (CAGR) from 2017 to 2022 (Mordointelligence.com, 2017). Last year (2019), the global nutraceutical market was at 382.51 billion USD, at 8.3% CAGR which was higher than the estimation (https://www.grandviewresearch.com/industry-analysis/nutraceuticals-market, 2020). In forthcoming years, the market is expected to reach 423.2 billion USD by 2025, at a CAGR of 6.8% during 2020–2025. The tremendous increase in the global nutraceutical market is due to the increased utility of nutraceuticals by the people around the world. Improved delivery of nutraceuticals is may be one cause.

1.5 Bioavailability of Bioactive Compounds of Nutraceuticals

The efficiency of nutraceuticals depends on its bioavailability. With the increasing knowledge and understanding of the protective role of nutraceuticals, its availability into the body has become a significant concern. This essential factor must be considered by the manufacturers when producing nutraceuticals (Rapaka and Coates 2006). The quantity of a functional/bioactive compound that enters the blood circulation is called bioavailability (Esfanjani et al. 2018). Decreased bioavailability lowers the benefits of nutraceuticals inside the body. The route of administration/intake of nutraceuticals is oral, and they pass through GIT to reach the blood. Various constrain and obstructions like inadequate gastric retention, reduced permeability, solubility in the gut, and unstable conditions encountered in the GI tract reduce the activity and hence availability of nutraceuticals (Bell 2001). Apart from decreased bioavailability, it is necessary to regard other challenges like poor solubility, instability, and crystallization before incorporating these bioactive molecules into products while manufacturing (Augustin and Sanguansri 2012). To overcome the physiological limitations of the body and to enhance the bioavailability and absorption of nutraceutical compounds as nutritionally and pharmacologically important one, the bioactive compound needs to be conjugated with the suitable delivery system, thus enhancing its bioavailability.

1.5.1 Enhancing the Bioavailability of Nutraceuticals by Nanotechnology

Many proposed targeted delivery systems have not been considered globally as a pertinent system for the delivery of bioactive compounds since each bioactive compound has its typical molecular structure requiring different systems. Nanotechnology-based targeted delivery systems are now booming to evade problem raised in the bioavailability of bioactive compounds. Development of delivery systems with nanomaterials is to boost the biological availability of functional components present in nutraceuticals. The encapsulation of bioactive compounds combats the acidity and breakdown by enzymes present in the GIT. These delivery systems offer an increased surface area and enhance the bioavailability. Reducing the particle size of delivery systems improves its efficiency, solubility, and biological activity of the compounds due to the availability of greater surface area per unit molecule (Prasad et al. 2019).

1.6 Nanotechnology

Employing and handling materials, at the nanometer scale, is called nanotechnology (Fathi et al. 2012). A nanometer is one-millionth of a millimeter. The dimensions of nanoparticles (NPs) are roughly between 1 and 100 nm (Prasad et al. 2016, 2017a, 2018). Nanotechnology shows the excellent possibility of enhancing the delivery of nutraceuticals. It facilitates controlled release, improves bioavailability, and protects the nutraceuticals during processing, storage, and distribution.

1.6.1 Nanocarriers as Nano Delivery System

The material used for the synthesis of nanoparticles (NPs) or nanocarriers is of organic or inorganic material (Bhushan et al. 2014) or a mixture of both (Fig. 1.2). Nanocarriers synthesized from organic material are comprised of polymeric and lipid-based nanoparticles, and those that are synthesized from inorganic materials are from metallic nanostructures such as quantum dots (Prasad et al. 2017b).

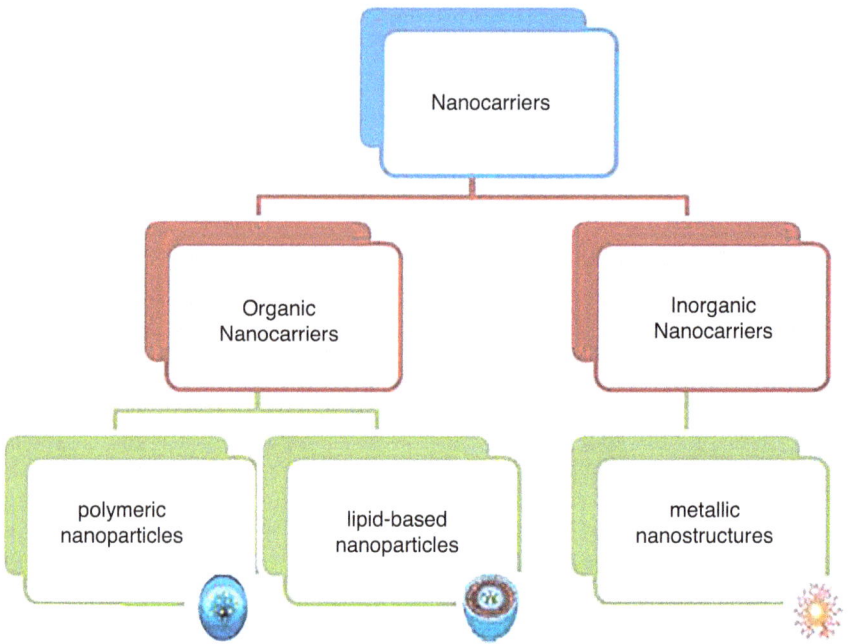

Fig. 1.2 Nanocarriers

1.6.2 Synthesis of Nanoparticles

Nanoparticles can be produced either by top-down methods (fluidization, dispergation, emulsifying technologies, or homogenization methods) or by bottom-up methods (precipitation or condensation, controlled sol-gel syntheses, evaporation) (Fig. 1.3) (Brayner et al. 2013). The size of the mechanical method of production of nanoparticles yields 100–1000 nm, while the size of nanoparticles from the chemical and bottom-up method yields 10–100 nm (Acosta 2009). Various biodegradable natural biopolymers for nanoencapsulation are frequently used (Jampilek and Kralova 2017).

Application of nanotechnology in nutraceuticals has enhanced the properties of nutraceuticals. They alter the difficulties encountered during delivery inside the body. The increased surface area of nutraceuticals is due to the decrease in size of a prepared nutraceutical, which in turn enhanced the desired delivery process inside the body. In a study by Javed et al. (2011), reformulation of silymarin into nanoliposomes increased its bio-absorption. The unique properties, such as size, shape, and internal structure, are essential while designing and fabricating nanomaterials. The fabrication monitors not only the incorporation of bioactive compounds but also on their stability, entrapment and release behaviors, and biological function (de Souza Simo͂es et al. 2017). Thus, nanotechnology offers an essential role in the delivery systems of a nutraceutical (Acevedo-Fani et al. 2017). Some of the

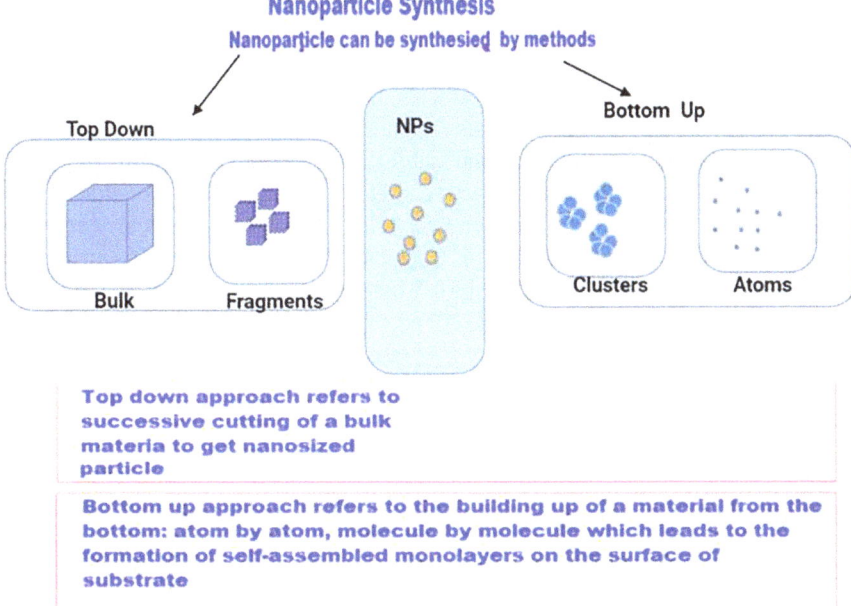

Fig. 1.3 Synthesis of nanoparticles

fabrication techniques of nanomaterials are emulsification, solid dispersion, spray drying, spray freeze drying, electro-spraying process, anti-solvent precipitation, complex coacervation, and layer-by-layer (LbL) deposition. The microscopically developed instruments that characterize nanomaterials help in better understanding of structures in nano-size and interpretation of their role (Tolles and Rath 2003). The nano-architecture endows protection and bioavailability of nutraceuticals by modifying the extent of solubility and its release (Ezhilarasi et al. 2013). Liquid crystalline mesophase is known as lyotropic formed by amphiphilic substances such as lipids along with surfactants and copolymers dispersed in a polar solvent. Under appropriate conditions, hydrophobic interactions cause them to self-assemble into stable crystals in nanometer size. Examples include micelles, hexagonal structures, lamellar structures, and cubosomes, which retain a high degree of molecular orientation in a liquid state.

1.6.3 Scenarios of Nanotechnology in Nutraceuticals

In the present scenario, nutraceutical delivery systems need to be familiar with the location, the load needed to be delivered, a decision on release, and feedback control. Advances in nanotechnologies revolved this by aiding its delivery. Aforementioned can be compared to the "smart drug delivery" presently under medical research. Thus, nanoscience, engineering, and technology guaranteed the proper delivery of nutraceuticals. Innovation in encapsulated formulations and bioavailability of nutraceuticals are considerably augmented. Even though nanotechnology aids in the delivery of nutraceuticals, however, the essential step necessary is the selection of the loading materials for the delivery systems, which plays a vital role in the encapsulation efficiency and stability, and the available option of materials is also limited. Loading of bioactive ingredients into nanotechnology-based delivery systems is essential. The nanotechnology-based nutraceuticals can attain commercialization by overcoming the boundaries related to them (Augustin and Hemar 2009), by using food grade-based nanomaterials which offer uniformity, flavor, taste, and texture (Prasad et al. 2017c; Chausal et al. 2021).

1.7 Food-Grade Nanomaterials

Various food-grade nanomaterials for delivery of nutraceuticals are as follows:

1.7.1 Lipids as nanomaterials
1.7.2 Polymer as nanomaterials
1.7.3 Cellulose as nanomaterials
1.7.4 Protein as nanomaterials
1.7.5 Polysaccharide as nanomaterials

1.7.1 Lipid as Nanomaterials

Lipids have a tremendous capability for encapsulating nutraceuticals. Some of the examples of lipid-based nanoparticles (Fig. 1.4) are solid-lipid nanoparticles (SLNs), nanostructured lipid carriers (NLCs), liposomes, nanoemulsions, and nanoliposomes systems. They encapsulate majority of the natural bioactive compounds. A high dose of different molecules can be encapsulated and used on distinct sites using these lipid-based nanomaterials.

Solid-lipid nanoparticles (SLNs): The first-generation lipid-based nanocarriers are solid-lipid nanoparticles (SLNs). The lipids used in the synthesis are fatty acids, monoglycerides, diglycerides, and triglycerides. They remained in the solid state at body temperature and are stabilized by emulsifying agents (Müller et al. 2000). The SLNs are highly lipophilic, which enables the transportation of core material. Synthesis of SLNs is shown in Fig. 1.5, and list of constituents necessary for the synthesis of nutraceuticals encapsulated in SLNs is shown in Table 1.1.

Advantages of solid-lipid nanoparticles: Solid-lipid nanoparticles have a combination of advantages of liposomes, nanoemulsions, and polymeric nanoparticles. SLNs have good tolerance, high bioavailability, and biodegradability. There is no formation of toxic breakdown products in SLNs (Cacciatore et al. 2016). Therapeutic

1 An Insight on Emerging Nanomaterials for the Delivery of Various Nutraceutical... 11

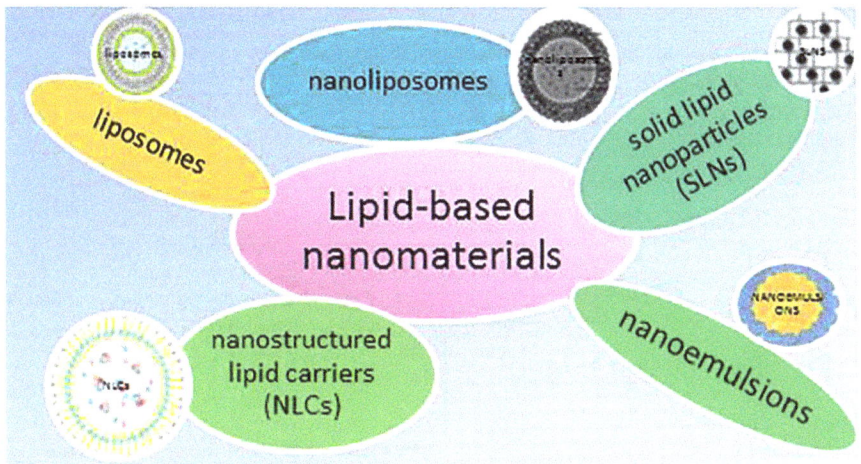

Fig. 1.4 Lipid-based nanomaterials

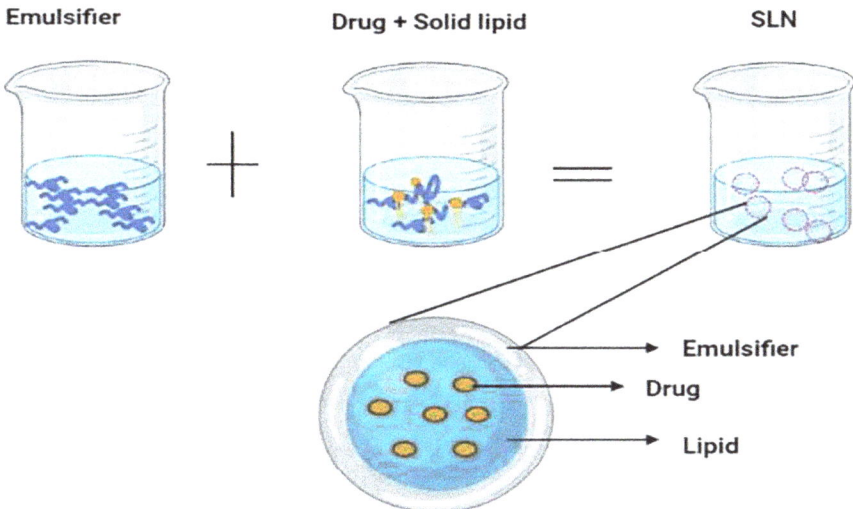

Fig. 1.5 Synthesis of solid-lipid nanoparticles

application of SLNs in different ailments is shown in Table 1.2. The significant benefits of SLNs are the following:

- Organic solvents are not necessary for their preparation.
- The physical stability is high.
- The drug release and drug targeting is well organized.
- The active substances are stable.
- Both lipophilic and hydrophilic compounds are encapsulated with SLNs.

Table 1.1 Constituents used for the synthesis of solid-lipid nanoparticles

Types of lipids used in the synthesis of SLNs	The emulsifier used in the synthesis of SLNs	Techniques for the synthesis of SLNs
Triglycerides Partial glycerides Fatty acids Steroids Waxes (cetyl palmitate)	Pluronic F 68 Pluronic F 127 Polysorbate Lecithin Sodium cholate Tyloxapol Taurodeoxychocolic acid sodium Sodium glycocholate Butyric acid and butanol (Campos et al. 2014; Severino et al. 2012)	High shear homogenization Hot homogenization Cold homogenization Ultrasonication Solvent emulsification Microemulsion Supercritical fluid Spray drying method and double emulsion method (Mukherjee et al. 2009)

Table 1.2 Therapeutic application of solid-lipid nanoparticles

Bioactive compound encapsulated in SLNs	Therapeutic uses of SLNs

Rosmarinic acid (RA)	Protects against
	  Elevated blood pressure
	  Myocardial infarction (MI)
	  Aging
	  Atherosclerosis
	  Diabetes
	  Neurodegenerative disorders like Huntington's disease
	  Parkinson's and Alzheimer's diseases (Klyachko et al. 2012)
	  Autoimmune disorders (Caccamo et al. 2012)
	  Malignancy (Riemann et al. 2011; Tasset et al. 2011)
	  Inflammatory disease (Lee et al. 2011) and excess O_2 in tissues and organs (hyperoxia) (Gao et al. 2012)

β-Carotene	Can suppress the production of free radical Offers protection against lipid peroxidation (Souyoul et al. 2018)
Nicotinamide (NA)	Has antioxidant and anti-inflammatory activity (Takechi et al. 2013) Protects the blood-brain barrier (BBB) Protects against neurodegenerative disorders
Quercetin	Can decrease serum lipids in hyperlipidemia Can widen the narrowed coronary arteries Acts as anticarcinogenic agent Stops platelet aggregation Provides antioxidant and anti-inflammatory activity Suppresses anaphylactic effects and elevates the level of hemoglobin in anemic condition (Wang et al. 2007)

- Simple in preparation hence scale up is also simple, so large-scale production is possible.
- Deliverance systems with safe ingredients.
- Sterilization is possible (Weber et al. 2014).

Limitation of solid-lipid nanoparticles: Although SLNs have many advantages, there are some potential disadvantages also. The main disadvantage is its drug-loading capacity which is low (Yoon et al. 2013), expelling of the drug from the carrier as a result of the polymorphic transitions, which depends on the physical and chemical structure of the active substances (e.g., hydrophilic molecules), and this mostly happens during storage. Another drawback is initial burst release and its dispersion (Makwana et al. 2015), with a large amount of water. Nanostructured lipid carriers (NLCs): Nanostructured lipid carriers (NLC) are the second-generation delivery system. In this system, partly crystallized lipids are present in the aqueous layer along with emulsifiers. It is one of the appropriate systems for nutraceutical delivery.

The main advantages of NLCs are as follows:

- The drug-loading efficiency is high.
- They have stable bioactive compounds with increased bioavailability (Muller and Keck 2004).
- More cost-effective to fabricate.
- Has low crystallinity and less dense lipid packaging.
- They provide augmented shelf life and controlled release of encapsulated materials.

Organic solvents are not necessary since it is water-based and are easy to scale up (Doktorovova et al. 2014). NLCs have the capability of transporting both lipophilic and hydrophilic drugs at the same time. NLCs can be added directly to transparent/opaque pasteurized products (e.g., fruit juice and milk). However, they are supplemented to foods prior to pasteurization. They can be spray-dried or freeze-dried. NLC has all the necessary characteristics for the use as nanocarriers for food and nutraceuticals. A combination of solid lipid and liquid lipid (or oil) is used to circumvent lipid crystallinity.

Limitation of nanostructured lipid carriers: Even though NLCs have many advantages as a nano delivery vehicle, they also have certain limitations:

- Toxic effects on the cells which depend on the matrix
- Irritating and sensitizing action of surfactants

Types of NLCs: The types of NLCs depend on the composition of the lipids mixed and fabrication techniques used. The principal mechanism involved is to arrange for a nanostructure for the lipid matrix, in turn, and escalate the load of bioactive compounds and decrease the discharge of these compounds during storage. The three different types of NLCs (Westesen et al. 1997) are imperfect, amorphous, and multiple. Production of NLCs using high-pressure homogenization (HPH) is used for quantitative production of SLNs and NLCs. It is a beneficial, trustworthy, and powerful technique. The procedure involves top-down methodology wherein a very high shear stress pushes the lipid with high pressure (100–2000 bars), which disrupts large particles to submicron or nanometer size. Five to ten percent of lipids is used for the production. When compared to other techniques used for preparation, HPH does not show scaling up problems. Homogenization can be performed either at a higher temperature (hot homogenization) or at a lower temperature (cold homogenization) (Schwarz et al. 1994). Different approaches used for the preparation and its constituents necessary are depicted in Table 1.3. Nanoemulsions: Emulsions are combination of oil, water or aqueous buffer, surfactant, and cosurfactant. Nanoemulsions are colloids with oil droplets dispersed in an aqueous medium. Hence they can solubilize drugs or bioactive compounds which have an affinity for lipid environment, thus enhancing their delivery. The isotropic systems of nanoemulsions are kinetically stable systems of oil and water (both the liquids do not mix) and become stable by forming an interfacial film using surfactant and cosurfactant. They can also encapsulate hydrophilic drugs (Tshweu et al. 2013) and can enhance the bioavailability of encapsulated compounds (Tshweu et al. 2014). The sizes of nanoemulsions are in the range of 5–200 nm and are transparent (Solans et al. 2005). They give protection from hydrolysis (Ozturk 2017) and can increase the therapeutic activity of essential oil (Katata-Seru et al. 2017).

The different types of nanoemulsions are water-in-oil (W/O) type, oil-in-water (O/W) type, and bi-continuous water-in-oil-in-water (W/O/W) type. The different techniques for the production of nanoemulsions are low and high energy type (Fig. 1.6).

Low energy type involves spontaneous emulsification and phase inversion temperature with the change in solubility, and high energy type are microfluidics, high-pressure homogenizers, or ultrasound equipment methods. O/W nanoemulsions can encapsulate lipophilic nutraceuticals and form stable systems for oral delivery. The food-grade nanoemulsions have been prepared by homogenization, mixing, and shearing (Aditya et al. 2017). The factors that prevent the oral availability of lipophilic bioactive agents are absorption, accessibility inside the body, and transformation (Aboalnaja et al. 2016). The encapsulation of lipophilic bioactive agents to nanoemulsions could help in diffusion across the membrane that will improve oral availability (Sivakumar et al. 2014). The size of nanoemulsions is small and globular with augmented surface area resulting in improved digestion rate. It also protects the bioactive compounds from oxidation and decreases their breakdown in the GIT (Frede et al. 2014). Thus, nanoemulsions can protect and deliver both lipid-soluble and water-soluble bioactive agents through the oral route and dermal route, similar

Table 1.3 Constituents used for the synthesis of nanostructured lipid carriers (NLCs)

Types of lipids used in NLC	The emulsifier used in NLC	Techniques used in NLC
Glycerides	Sodium dodecyl sulfate	High-speed homogenizer
Fatty acids	Sodium oleate	Hot homogenization
Waxes (cetyl palmitate)	Span 20, 80, and 85	Cold homogenization
	Tween 20 and 80	Ultrasonication
	Egg phospholipid	Phase inversion
	Sodium taurodeoxycholate	Microemulsion
	Sodium glycocholate	Evaporation solvent injection
	Phospholipon 80 H	Membrane contractor technique (Sanap and Mohanta 2013)

1 An Insight on Emerging Nanomaterials for the Delivery of Various Nutraceutical... 15

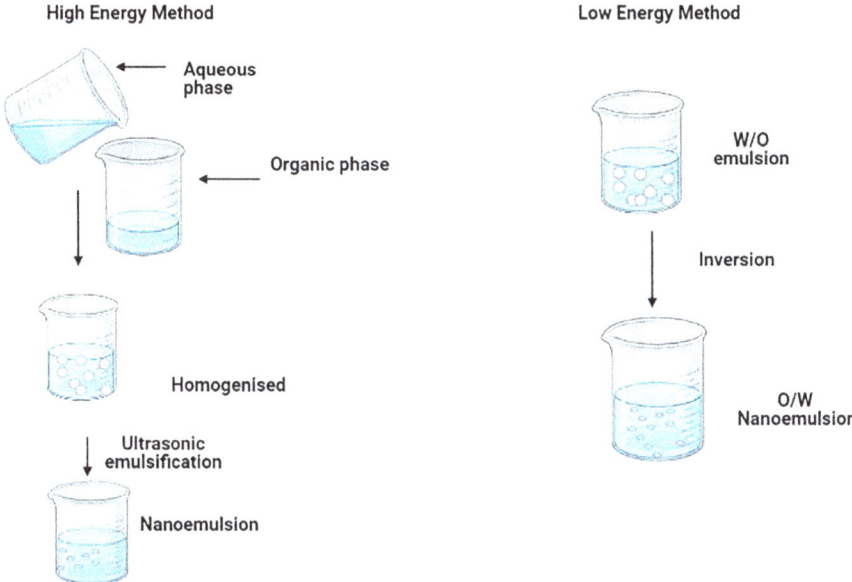

Fig. 1.6 Preparation of nanoemulsions (Typical low-energy (bottom-up) methods and high-energy (top-down) methods for producing nanoemulsions)

to that of pharmaceuticals. Although nanoemulsions have many advantages, they also have limits. The main disadvantage is the stability of nanoemulsions. Although nanoemulsions are stable for years, due to their minimal size, the Oswald ripening could disrupt them, making their role inadequate. Most of the time, they are made only at the time of usage. Liposomes and nanoliposomes: Liposomes are used as a carrier for the transportation of active components inside the body by oral absorption or preventing breakdown in the acidic environment present in GIT. The phospholipids arranged with their heads in the aqueous environment and their tails toward each other (Fig. 1.7).

These structures have trapped aqueous-soluble components and protect them from aggressive environments. Interactions between the membrane of liposome and cell membranes enhanced cellular uptake through endosomal mechanisms. Because of small size, liposomes provide improved bioavailability of entrapped components by their large surface area to the biological tissues. Examples of antioxidants encapsulated using liposomes are shown in Fig. 1.8. Resveratrol-loaded liposomal formulations showed improved bioavailability and are used as therapeutic agents to treat neurodegenerative diseases like Parkinson's disease (Wang et al. 2011).

Liposomes are the well-designed optimal drug carrier system for a broad range of bioactive substances. The biological advantage of liposome is due to the amphipathic nature of phospholipids used for encapsulation. They are small in size, compatible, and degradable within the body and also lack toxicity.

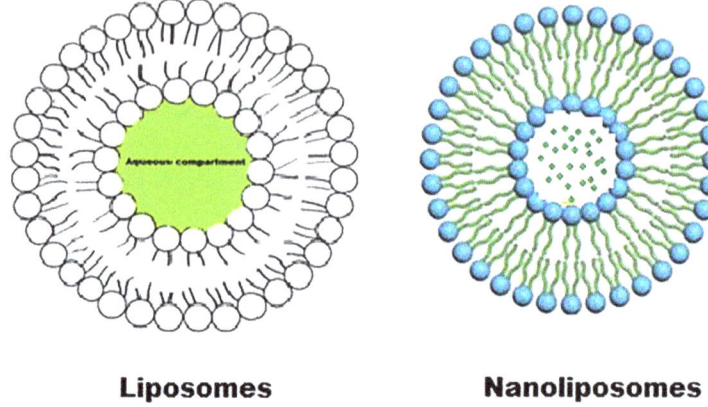

Fig. 1.7 Liposomes and nanoliposomes

Fig. 1.8 Examples of antioxidants encapsulated using liposomes

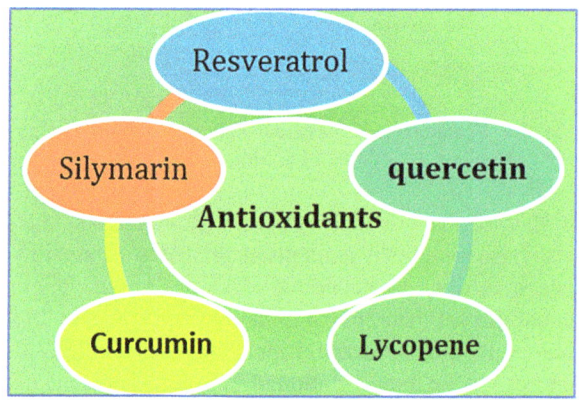

Some of the advantages of liposomal delivery are as follows:

- Increased biological availability and bio-absorption.
- Liposomal delivery vehicle protects the bio-component against the acidic environment of the GIT and enhances mucosal absorption.
- It augments intracellular delivery.
- Can entrap hydrophilic and lipophilic components.
- The dose can be lowered with the same effect by liposomal delivery and thus cost-effective.

Limitations of liposomal delivery are as follows:

- Possibility of large particle size with decreased core material during the manufacturing
- Lack of stability

Nanoliposomes: Nanostructured delivery systems are promising candidates that permit efficient and targeted delivery of bioactive compounds. Lipid carrier system includes liposomes and nanoliposomes (Fig. 1.7). Nanoliposomes are liposomes at nanoscale lipids as encapsulating agents. Both liposomes and nanoliposomes have common physicochemical structure, with kinetically stable characteristics. However, nanoliposomes offer additional surface area because they are nanometer in size (the smaller the size, the higher the surface area) and the solubility is also more with enhanced biological availability, sustained release, and site-directed targeting of the bioactive agents at an increased level than liposomes. Nanoliposomes are colloidal structures formed by a mixture of phospholipids which have an excellent emulsifying property along with the other constituents in water (Danaei et al. 2018). Dispersion of lipid and phospholipids for the synthesis of liposomes and nanoliposomes consumes energy (Mozafari 2005). Nanoliposomes are exploited as a useful drug carrier in the delivery of nutraceuticals. The routes of delivery of nanoliposomes are oral as well as parenteral (Shoji and Nakashima 2004). Stable polymers are used to protect the nanoliposomes.

Advantages of nanoliposomes are as follows:

- Nanoliposomes provide benefits in delivering and targeting of bioactive compounds by encapsulation (Chaudhry et al. 2017).
- Nanoliposomes are metastable. This stability resists the change in size in an aqueous medium when compared with other lipid encapsulating agents.
- Another benefit of nanoliposomes is that they can encapsulate lipophilic, hydrophilic, and amphiphilic molecules.
- Another added benefit is that they can encapsulate concurrently two different bioactive components differing in their solubility and be used as bifunctional nanosomes with synergistic function (Gowda et al. 2017).

1.7.2 Polymer as Nanomaterials

Biodegradable polymers are used for the transportation of nutraceuticals and synthesized in the form of nanoparticles (Vorhies and Nemunaitis 2009). By encapsulating the nutraceuticals, they are protected from the acidic environment of GIT. Thus, polymers are used as the means to deliver nutraceuticals. Biodegradable polymers are broken down either by biological processes or by hydrolysis. Biodegradable polymers are composed of natural and synthetic polymers (Fig. 1.9) (Nicolas et al. 2013). They are designed to break down within the body by biochemical processes. The breakdown products are also biologically endurable. These polymers deliver nutraceuticals. They are also biocompatible. Natural biodegradable polymers are biologically originated molecules. They can undergo chemical modifications to break down and are biocompatible. Within the body, the polymer can metabolize and clear them. Collagen, gelatin, albumin, gliadin, zein, and casein are examples of polymers of proteins, and carrageenan, alginate, chitin, chitosan,

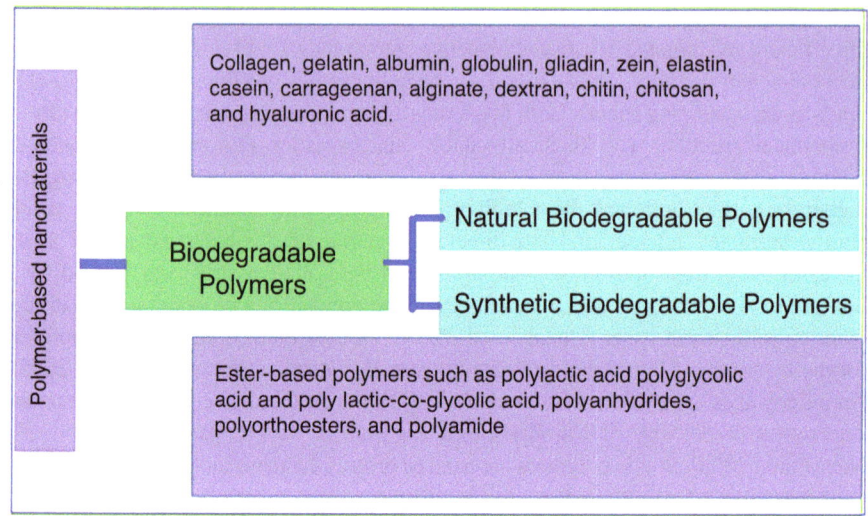

Fig. 1.9 Polymer as nanomaterials

and hyaluronic acid are examples of polymers of polysaccharides. These natural polymers have some limitation like poor mechanical properties and have complex structure. Natural polymers are extensively used most of the time because they are commercially available. The synthetic biodegradable polymers enhance the release of nutraceuticals. Synthetic polymers have additional benefits when compared to natural polymers in the sustained release of encapsulated active compounds for many days. The limitations of synthetic polymers are toxicity and chronic inflammation (Coelho et al. 2010). Toxicity and immunogenicity must be taken into account when using synthetic polymers for the delivery of nutraceuticals.

Examples of synthetic polymers are polylactic acid (PLA), polyglycolic acid (PGA), poly (lactic-co-glycolic) acid (PLGA), polyanhydrides, polyorthoesters, and polyamide. These have many substantial applications in medicine (Makadia and Siegel 2011). PLA, PGA, and PLGA are polymers of esters and are also extensively used. The composition of PLA is lactic acid. It is biodegradable by hydrolysis. The composition of PGA is glycolic acid. It is a sturdy polymer with fiber-forming properties. The limitation of PGA is that it cannot be made into films or rods since it is not soluble in polymer solvents (Makadia and Siegel 2011). The polymer properties of PLGA are due to the composition of two different monomers (lactic and glycolic acid) (Makadia and Siegel 2011). Polyanhydrides are made into microspheres and tubes for sustained release of nutraceuticals. Polyorthoesters under dry conditions are stable. The drug release rates of polyorthoester are from days to months (Engesaeter et al. 1992). The polyamide application is limited due to its ability to induce an immune response.

1.7.3 Cellulose as Nanomaterials

The nanomaterials made from cellulose are synthesized from renewable sources. They are light in weight and biocompatible and possess excellent strength. The physicochemical properties of cellulose have made these nanomaterials attractive for its role in nutraceutical. Examples of nanomaterials made from cellulose used for various applications are enzyme immobilization, active compounds, microorganisms, stabilizers, and additives.

The application of nanomaterials made from cellulose depends on its origin and extraction conditions (Grishkewich et al. 2017). Precise acid hydrolysis of cellulose produces rod-shaped crystalline nanoparticles. During this process, hydrogen bonds break down, thus dissolving the amorphous region of the fiber. The dimension also varies depending on the resources. For example, nanomaterial obtained from hardwood has 3–5 nm width and 100–300 nm length, while those that are obtained from a marine animal have a 15–30 nm width and length of 1000–1500 nm (Grishkewich et al. 2017). Incorporation of chemical and mechanical methods like microfluidization, high-pressure homogenization, high ultrasonic treatment, and cryo-crushing is involved in the processes of formation of cellulose nanomaterial (Khan et al. 2017). Cellulose nanomaterials are made up of mixtures of amorphous and crystalline cellulose chains having width range of 4–20 nm and length in micrometers. They are capable of forming intertwined linkage and semicrystalline structure, with excellent flexibility and mechanical strength. *Lactobacillus* species and *Bifidobacteria* are beneficial bacteria. These organisms are present naturally in GIT, which are beneficial to the body. They are exploited as nutraceuticals for health benefits. The main disadvantage of it is its decreased shelf life. The shelf life can be enhanced by using cellulose nanomaterials as encapsulating material (Huq et al. 2017). Vitamins and antioxidants are encapsulated with nanofibers and nanocrystals of cellulose acetate and cellulose. Appropriate delivery of antioxidants prevents long-term complications of metabolic diseases such as nephropathy and neuropathy by shielding the ill effects of oxidative damage. Vitamin C is an excellent nonenzymatic antioxidant which suppresses reactive oxygen species. It can be encapsulated using chitosan oligosaccharide (Akhlaghi et al. 2015). The encapsulated vitamin C showed stability and potent scavenger of reactive oxygen species when compared with nonencapsulated vitamin C. Cellulose nanocrystals can also be used for preparing nutraceuticals for other scavengers of reactive oxygen species. The oxidation-reduction reaction of ascorbic acid and hydrogen peroxide is brought out using cellulose nanocrystals as an encapsulant. This was in turn enhanced by γ-irradiation of cellulose nanocrystal suspension, and then the suspension can be made to react with ascorbic acid and hydrogen peroxide, and then gallic acid was added. The enhanced features of cellulose nanocrystal derivative showed enhanced antioxidant activity compared to nonencapsulated substances (Criado et al. 2016).

1.7.4 Protein as Nanomaterials

Proteins are biomolecules with one or more polypeptide chains of amino acids, and they provide essential amino acid to the human body. They are a part of natural ingredients, e.g., milk, eggs, gelatin, silk protein, whey protein, and casein. Even though proteins are one of the main dietary components, they are used for encapsulating nutraceuticals. Proteins play an outstanding role as encapsulant. They are flexible biopolymers. They are used as nanocarrier system due to their biocompatibility, encapsulation efficiency (Fathi et al. 2018), low cytotoxicity, biodegradability, controlled release of bioactive agents, and site-specific drug delivery (Chakraborty and Dhar 2017). As delivery vehicles, various kinds of animal proteins (Fig. 1.10) are used including gelatin (Payne et al. 2002), collagen (Swatschek et al. 2002), casein, albumin (Tomlinson and Burger 1985), and whey protein (Picot 2004), and plant proteins used are glycinin from soy (Lazko et al. 2004), zein from corn (Liu et al. 2005), and gliadin from wheat (Ezpeleta et al. 1996).

Of the above protein-based nanomaterial for nutraceutical deliveries, collagen and gelatin are widely used since they are biodegradable and non-immunogenic. The articulation of collagen into sponge, particles, gels, and films facilitated the delivery of nutraceuticals (Friess 1998). Next to collagen, albumin is a widely used polymer because it is nontoxic, non-immunogenic, biologically compatible, and degradable (Bae et al. 2012).

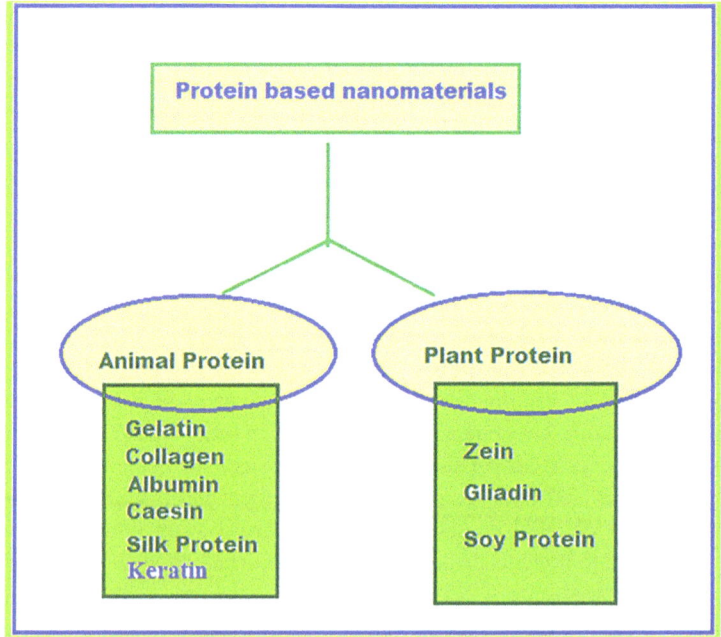

Fig. 1.10 Protein-based nanomaterial

Gliadin and zein are prolamins, and these plant proteins are abundant in proline. They are present in cereal grains (wheat (gliadin), corn (zein)) and used as an oral delivery system. SOD and catalase are antioxidant proteins. Within the body, they scavenge free radicle and thus have a therapeutic role. These antioxidants are encapsulated with gliadin or zein, to protect them from the acidity of GIT (Lee et al. 2013). Protein as nanoparticles can be prepared with ease, and the size distribution can also be supervised (Mac Adam et al. 1997). They can be modified, forming complexes with polysaccharides, lipids, or polymers. Since they have distinct primary structure, a wide variety of nutrients can be incorporated by modifying the surface. To the primary amino groups or sulfhydryl groups, the bioactive can be covalently added (Weyermanna et al. 2005). Nanoscale phenomena of proteins are also due to denaturation and aggregation properties, which pave the way to design them. These related properties make them promising agent for entrapping nutraceuticals. Nanostructured proteins have improved properties for the incorporation of nutraceuticals (Ramos et al. 2017). Submicron sizes are obtained by strong bonds and weak forces (Clark and Ross-Murphy 1987). The possible mechanism leading to the formation of these nanoaggregates involves electrostatic interactions, hydrophobic interactions, and intermolecular disulfide bonds. To increase the bioavailability of nutraceuticals, the surfaces of nanocarriers are coated with proteins which can modify the adhesive properties of it and their behavior in the GI tract. Coated proteins can bind to sugar-bearing sites on epithelial cells in the GIT (Goldstein et al. 1980). Coating of protein on nanoparticles provides added protection to the nutraceutical. Broadly proteins are used as nanocarriers for the delivery of nutraceuticals (Chakraborty and Dhar 2017). The whey proteins are utilized mainly as functional components in prepared foods as they are cost-effective and generally safe and ensure significant function (Gunasekaran et al. 2006). Furthermore, proteins possess beneficial biological properties against diseases such as cancer, viral infection, and indigestibility. It also modulates the immune system (Dissanayake and Vasiljevic 2009). These essential properties make them promising encapsulating agents.

1.7.5 *Polysaccharide as Nanomaterials*

Polysaccharides are polymer of monosaccharide, and they do both structural and functional role. The polysaccharides obtained from plants are pectin, inulin, fiber, and starch and from animals are chitosan, glycogen, and chondroitin sulfate. The microorganism present in the colon breaks these polysaccharides into monomers. When polysaccharides are used as a nanomaterial, they protect the nutraceuticals from the harsh environment of the GI tract. When they reach the colon, they get hydrolyzed and release nutraceuticals present within them. The primary therapeutic role of these systems is to deliver probiotics such as bifidobacteria and lactobacilli. Polysaccharides are one of the main classes of biological polymers. They are bioactive, hydrophilic, biologically degradable, biologically compatible, cost-effective, and without toxicity and possess a wide range of properties. These can be altered to enhance the stability of bioactive compounds (Sinha and Kumria 2001). Polyelectrolytes and non-polyelectrolytes are the types of nanomaterials made from polysaccharides. The polyelectrolytes are classified based on their intrinsic charge into cationic, anionic, and neutral. Examples of cationic polyelectrolytes are chitosan; examples of anionic polyelectrolytes are alginate, pectin, hyaluronic acid, and heparin; and examples of neutral polyelectrolytes are pullulan and dextran subgroups. Based on sources of origin, polysaccharides are mainly classified as (a) algal origin (alginate); plant origin such as guar gum, and pectin; (b) microbial origin, e.g., xanthan gum, dextran; and (c) animal origin, e.g., chondroitin, chitosan (Zheng et al. 2015; Augustin and Hemar 2009). The majority of the natural polysaccharides are used for nanoencapsulation of different types of bioactive. The types of polysaccharides used for nanoencapsulation depend on safety and cost. Different methods are utilized for the synthesis of nanoencapsulation of bioactive components depending upon the physical and chemical characteristics of both bioactive and polysaccharides. They can encapsulate both hydrophilic and hydrophobic bioactives. Their structural flexibility and site-specific targeting make them as suitable carriers for the controlled and targeted delivery of nutraceuticals for GIT (Sinha and Kumria 2001).

1.8 Conclusion

The significance of nutrition containing all necessary nutrients, along with antioxidants, vitamins, and minerals in appreciable amount, is necessary for health. Intake of nutraceuticals with bioactive ingredients will prevent the people from ailments and improve their health. For this purpose, nanoencapsulation of active compounds synthesized by using various biodegradable natural/semisynthetic-based nanocarriers that include polymeric nanoparticles, micelles, liposomes, nanoliposomes, nanoemulsions, solid-lipid nanoparticles (SLNs), nanostructured lipid carriers (NLCs), nanoemulsions, nanomaterials made from cellulose, protein, and polysaccharides is carried out. These nanomaterials have improved stability and sustained release of nutrients. Many nano delivery systems have not been considered as a universally appropriate system because the bioactive compounds have individual characteristics which necessitate new types of diverse systems. To conclude, this chapter gives an understanding of nutraceuticals, its types, nanocarriers and nanomaterials, and their encapsulation efficiency. In the upcoming years, nanomaterial generation for nutraceutical applications can benefit them in the global market.

References

Aboalnaja KO, Yaghmoor S, Kumosani TA, McClements DJ (2016) Utilization of nanoemulsions to enhance bioactivity of pharmaceuticals, supplements, and nutraceuticals: Nanoemulsion delivery systems and nanoemulsion excipient systems. Expert Opin Drug Deliv 13(9):1327–1336

Acevedo-Fani A, Soliva-Fortuny R, Martı'n-Belloso O (2017) Nanostructured emulsions and nanolaminates for delivery of active ingredients: improving food safety and functionality. Trends Food Sci Technol 60:12–22

Acosta E (2009) Bioavailability of nanoparticles in nutrient and nutraceutical delivery. Curr Opin Colloid Interface Sci 14:3–15

Aditya NP, Espinosa YG, Norton IT (2017) Encapsulation systems for the delivery of hydrophilic nutraceuticals: food application. Biotechnol Adv 35:450–457

Akhlaghi SP, Berry RM, Tam KC (2015) Modified cellulose nanocrystal for vitamin C delivery. AAPS PharmSciTech 16(2):306–314

Akobundu UO, Cohen NL, Laus MJ, Schulte MJ, Soussloff MN (2004) Vitamins A and C, calcium, fruit, and dairy products are limited in food pantries. J Am Diet Assoc 104(5):811–813

Augustin MA, Sanguansri L (2012) Challenges in developing delivery systems for food additives, nutraceuticals, and dietary supplements. In: Garti N, Mc Clements DJ (eds) Encapsulation technologies and delivery systems for food ingredients and nutraceuticals. Woodhead Publishing, Cambridge UK, pp 19–48

Augustin MA, Hemar Y (2009) Nano-and micro-structured assemblies for encapsulation of food ingredients. Chem Soc Rev 38(4):902–912

Chauhan B, Kumar G, Kalam N, Ansari SH (2013) Current concepts and prospects of herbal nutraceutical: a review. J Adv Pharm Technol Res 4(1):4–8

Bae S, Ma K, Kim TH, Lee ES, Oh KT, Park ES, Lee KC, Youn YS (2012) Doxorubicin-loaded human serum albumin nanoparticles surface-modified with TNF-related apoptosis-inducing ligand and transferrin for targeting multiple tumor types. Biomaterials 33:1536–1546

Bell LN (2001) Stability testing of nutraceuticals and functional foods. In: Wildman REC (ed) Handbook of nutraceuticals and functional foods. CRC Press, New York, pp 501–516

Bhushan B, Luo D, Schricker SR, Sigmund W, Zauscher S (2014) Handbook of nanomaterials properties. Springer, Berlin/Heidelberg

Brayner R, Fievet F, Coradin T (2013) Nanomaterials: a danger or a promise? A chemical and biological perspective. Springer, London, UK

Caccamo D, Curro M, Ferlazzo N, Condello S, Ientile R (2012) Monitoring of transglutaminase 2 under different oxidative stress conditions. Amino Acids 42:1037–1043

Cacciatore I et al (2016) Solid lipid nanoparticles as a drug delivery system for the treatment of neurodegenerative diseases. Expert Opin Drug Deliv 13:1–11

Campos DA, Madureira AR, Gomes AM, Sarmento B, Pintado MM (2014) Optimization of the production of solid Witepsol nanoparticles loaded with rosmarinic acid. Colloids Surf B: Biointerfaces 115:109–117

Chakraborty A, Dhar P (2017) A review on potential of proteins as an excipient for developing a nano-carrier delivery system. Crit Rev Ther Drug Carrier Syst 34(5):453–488

Chaudhry Q, Castle L, Watkins R (eds) (2017) Nanotechnologies in food. Royal Society of Chemistry

Chausali N, Jyoti Saxena J, Prasad R (2021) Recent trends in nanotechnology applications of bio-based packaging. Journal of Agriculture and Food Research, https://doi.org/10.1016/j.jafr.2021.100257

Clark AH, Ross-Murphy SB (1987) Structural and mechanical properties of biopolymer gels. Adv Polym Sci 83:57–192

Coelho JF, Ferreira PC, Alves P, Cordeiro R, Fonseca AC, Gois JR, Gil MH (2010) Drug delivery systems: advanced technologies potentially applicable in personalized treatments. EPMA J 1:164–209

Criado P, Fraschini C, Salmieri S, Becher D, Safrany A, Lacroix M (2016) Free radical grafting of gallic acid (GA) on cellulose nanocrystals (CNCS) and evaluation of antioxidant reinforced gellan gum films. Radiat Phys Chem 118:61–69

Danaei M, Dehghankhold M, Ataei S, Hasanzadeh Davarani F, Javanmard R, Dokhani A et al (2018) Impact of particle size and polydispersity index on the clinical applications of lipidic nanocarrier systems. Pharmaceutics 10(2):57

De Felice SL (1995) The nutraceutical revolution: its impact on food industry R&D. Trends Food Sci Technol 6:59–61

de Souza Simoˉes LDA, Madalena AC, Pinheiro JA, Teixeira AA, Vicente OL, Ramos (2017) Micro- and nano bio-based delivery systems for food applications: in vitro behavior. Adv Colloid Interface Sci 243:23–45

Dissanayake M, Vasiljevic T (2009) Functional properties of whey proteins affected by heat treatment and hydrodynamic high-pressure shearing. J Dairy Sci 92:1387–1397

Doktorovova S, Souto EB, Silva AM (2014) Nanotoxicology applied to solid lipid nanoparticles and nanostructured lipid carriers—a systematic review of in vitro data. Eur J Pharm Biopharm 87(1):1–18

El Sohaimy SA (2012) Functional foods and nutraceuticals-modern approach to food science. World Appl Sci J 20:691–708

Engesaeter LB, Sudmann B, Sudmann E (1992) Fracture healing in rats inhibited by locally administered indomethacin. Acta Orthop Scand 63:330–333

Esfanjani AF, Assadpour E, Jafari SM (2018) Improving the bioavailability of phenolic compounds by loading them within lipid-based nanocarriers. Trends Food Sci Technol 76:56–66

Espín JC, GarcíaConesa MT, TomásBarberán FA (2007) Nutraceuticals: facts and fiction. Phytochemistry 68(22–24):2986–3008

Ezhilarasi PN, Karthik P, Chhanwal N, Anandharamakrishnan C (2013) Nanoencapsulation techniques for food bioactive components: a review. Food Bioprocess Technol 6:628–647

Ezpeleta I, Irache JM, Stainmesse S, Chabenat C, Gueguen J, Popineau Y, Orecchioni A (1996) Gliadin nanoparticles for the controlled release of all-trans-retinoic acid. Int J Pharm 131:191–200

Fathi M, Donsi F, Julian D, McClements (2018) Protein-based delivery Systems for the nanoencapsulation of food ingredients. Compr Rev Food Sci Food Saf 17(4):920–936

Fathi M, Mozafari MR, Mohebbi M (2012) Nanoencapsulation of food ingredients using lipid based delivery systems. Trends Food Sci Technol 23(1):13–27

Frede K, Henze A, Khalil M, Baldermann S, Schweigert FJ, Rawel H (2014) Stability and cellular uptake of lutein-loaded emulsions. J Funct Foods 8:118–127

Friess W (1998) Collagenebiomaterial for drug delivery. Eur J Pharm Biopharm 45:113–136

Gao Z, Spilk S, Momen A, Muller MD, Leuenberger UA, Sinoway LI (2012) Vitamin C prevents hyperoxia-mediated coronary vasoconstriction and impairment ofmyocardial function in healthy subjects. Eur J Appl Physiol 112:483–492

Goldstein IJ, Hughes RC, Monsigny M, Osawa T, Sharon N (1980) What should be called lectin? Nature 285:66–69

Gowda R, Kardos G, Sharma A, Singh S, Robertson GP (2017) Nanoparticlebasedcelecoxib and plumbagin for the synergistic treatment of melanoma. Mol Cancer Ther 16(3):440–452

Grishkewich N, Mohammed N, Tang J, Tam KC (2017) Recent advances in the application of cellulose nanocrystals. Curr Opin Colloid Interface Sci 29:32–45

Gunasekaran S, Xiao L, OuldEleya MM (2006) Whey protein concentrate hydrogels as bioactive carriers. J Appl Polym Sci 99:2470–2476

Huq T, Fraschini C, Khan A, Riedl B, Bouchard J, Lacroix M (2017) Alginate based nanocomposite for microencapsulation of probiotic: effect of cellulose nanocrystal (CNC) and lecithin. Carbohydr Polym 168:61–69

Jampilek J, Kralova K (2017) Nanomaterials for delivery of nutrients and growth-promoting compounds to plants. In: Prasad R, Kumar M, Kumar V (eds) Nanotechnology: an agricultural paradigm. Springer, Singapore, pp 177–226

Javed S, Kohli K, Ali M (2011) Reassessing bioavailability of silymarin. Altern Med Rev 16(3):239–249
Kalra EK (2003) Nutraceutical-definition and introduction. AAPS Pharm Sci 5:27–28
Katata-Seru L, Lebepe TC, Aremu OS, Bahadur I (2017) Application of Taguchi method to optimize garlic essential oil nanoemulsions. J Mol Liq 244:279–284
Khan A, Wen Y, Huq T, Ni Y (2017) Cellulosic nanomaterials in food and nutraceutical applications: a review. J Agric Food Chem 66(1):8–19
Klyachko NL, Manickam DS, Brynskikh AM, Uglanova SV, Li S, Higginbotham SM et al (2012) Crosslinked antioxidant nanozymes for improved delivery to CNS. Nanomedicine 8:119–129
Lazko J, Popineau Y, Legrand J (2004) Soy glycinin microcapsules by simple coacervation method. Colloids Surf B Biointerfaces 37:1–8
Lee DM, Jackson KW, Knowlton N, Wages J, Alaupovic P, Samuelsson O et al (2011) Oxidative stress and inflammation in renal patients and healthy subjects. PLoS One 6:e22360
Lee S, Alwahab NS, Moazzam ZM (2013) Zein-based oral drug delivery system targeting activated macrophages. Int J Pharm 454:388–393
Liu X, Sun Q, Wang H, Zhang L, Wang JY (2005) Microspheres of corn protein, zein, for an ivermectin drug delivery system. Biomaterials 26:109–115
MacAdam AB, Shafi ZB, James SL, Marriott C, Martin GP (1997) Preparation of hydrophobic and hydrophilic albumin microspheres and determination of surface carboxylic acid and amino residues. Int J Pharm 151:47–55
Makadia HK, Siegel SJ (2011) Poly lactic-co-glycolic acid (PLGA) as biodegradable controlled drug delivery carrier. Polymers (Basel) 3:1377–1397
Makwana V, Jain R, Patel K, Nivsarkar M, Joshi A (2015) Solid lipid nanoparticles (SLN) of Efavirenz as lymph targeting drug delivery system: elucidation of mechanism of uptake using chylomicron flow blocking approach. Int J Pharm 495(1):439–446
Mc Clements DJ (2012) Requirements for food ingredient and nutraceutical delivery systems. In: Encapsulation technologies and delivery systems for food ingredients and nutraceuticals. Wood head Publishing, Cambridge, pp 3–18
Menrad K (2003) Market and marketing of functional food in Europe. J Food Eng 56:181–188
Mozafari MR (2005) Liposomes: an overview of manufacturing techniques. Cell Mol Biol Lett 10(4):711–719
Mukherjee S, Ray S, Thakur RS (2009) Solid lipid nanoparticles: a modern formulation approach in drug delivery system. Indian J Pharm Sci 71(4):349–358
Müller RH, Mäder K, Gohla S (2000) Solid lipid nanoparticles (SLN) for controlled drug delivery - a review of the state of the art. Eur J Pharm Biopharm 50(1):161–177
Muller RH, Keck CM (2004) Challenges and solutions for the delivery of biotech drugs—a review of drug nanocrystal technology and lipid nanoparticles. J Biotechnol 113:151–170
Nicolas J, Mura S, Brambilla D, Mackiewicz N, Couvreur P (2013) Design, functionalization strategies and biomedical applications of targeted biodegradable/biocompatible polymer-based nanocarriers for drug delivery. Chem Soc Rev 42:1147–1235
Ozturk B (2017) Nanoemulsions for food fortification with lipophilic vitamins: production challenges, stability, and bioavailability. Eur J Lipid Sci Technol 119:1–18
Palzer S (2009) Food structures for nutrition, health and wellness. Trends Food Sci Technol 20(5):194–200
Pandey M, Verma RK, Saraf SA (2010) Nutraceuticals: new era of medicine and health. Asian J Pharm Clin Res 3:11–15
Payne RG, Yaszemski MJ, Yasko AW, Mikos AG (2002) Development of an injectable, in situ crosslinkable, degradable polymeric carrier for osteogenic cell populations. Part 1. Encapsulation of marrow stromal osteoblasts in surface crosslinked gelatin microparticles. Biomaterials 23:4359–4371
Picot ALC (2004) Encapsulation of bifidobacteria in whey protein-based microcapsules and survival in simulated gastrointestinal conditions and in yoghurt. Int Dairy J 14:505–515

Prasad R, Kumar V, Kumar M, Choudhary D (2019) Nanobiotechnology in Bioformulations. Springer International Publishing (ISBN 978-3-030-17061-5) https://www.springer.com/gp/book/9783030170608

Prasad R, Pandey R, Barman I (2016) Engineering tailored nanoparticles with microbes: quo vadis. WIREs Nanomed Nanobiotechnol 8:316–330. https://doi.org/10.1002/wnan.1363

Prasad R, Bhattacharyya A, Nguyen QD (2017a) Nanotechnology in sustainable agriculture: Recent developments, challenges, and perspectives. Front Microbiol 8:1014. https://doi.org/10.3389/fmicb.2017.01014

Prasad R, Pandey R, Varma A, Barman I (2017b) Polymer based nanoparticles for drug delivery systems and cancer therapeutics. In: Natural Polymers for Drug Delivery (eds. Kharkwal H and Janaswamy S), CAB International, UK 53–70

Prasad R, Kumar V, Kumar M (2017c) Nanotechnology: Food and Environmental Paradigm. Springer Nature Singapore Pte Ltd. (ISBN 978-981-10-4678-0)

Prasad R, Jha A, Prasad K (2018) Exploring the Realms of Nature for Nanosynthesis. Springer International Publishing (ISBN 978-3-319-99570-0) https://www.springer.com/978-3-319-99570-0

Ramos OL, Pereira RN, Martins A, Rodrigues R, Fucinos C, Teixeira JA, Pastrana L, Malcata FX, Vicente AA (2017) Design of whey protein nanostructures for incorporation and release of nutraceutical compounds in food. Crit Rev Food Sci Nutr 57:1377–1393

Rapaka RS, Coates PM (2006) Dietary supplements and related products: a brief summary. Life Sci 78:2026–2032

Riemann A, Schneider B, Ihling A, Nowak M, Sauvant C, Thews O, Gekle M (2011) Acidic environment leads to ROS-inducedMAPKsignaling in cancer cells. PLoS One 6:e22445

Sanap GS, Mohanta GP (2013) Design and evaluation of miconazole nitrate loaded nanostructured lipid carriers (NLC) for improving the antifungal therapy. J Appl Pharm Sci 3:46–54

Schwarz C, Mehnert W, Lucks JS, Müller RH (1994) Solid lipid nanoparticles (SLN) for controlled drug delivery. I. Production, characterization and sterilization. J Control Release 30:83–96

Severino P, Andreani T, Macedo AS, Fangueiro JF, Santana MHA, Silva AM, Souto EB (2012) Current state-of-art and new trends on lipid nanoparticles (SLN and NLC) for oral drug delivery. J Drug Deliv 12:1–10

Sen S, Pathak Y (2016) Nanotechnology in nutraceuticals: production to consumption. CRC Press. 465 pages

Shoji Y, Nakashima H (2004) Nutraceutics and delivery systems. J Drug Target 12:385–391

Sinha VR, Kumria R (2001) Polysaccharides in colon-specific drug delivery. Int J Pharm 224:19–38

Sivakumar M, Tang SY, Tan KW (2014) Cavitation technology—a greener processing technique for the generation of pharmaceutical nanoemulsions. Ultrason Sonochem 21:2069–2083

Solans C, Izquierdo P, Nolla J, Azemar N, Garciacelma M (2005) Nano-emulsions. Curr Opin Colloid Interface Sci 10(3–4):102–110

Souyoul SA, Saussy KP, Lupo MP (2018) Nutraceuticals: a review. Dermatol Ther:1–12

Swatschek D, Schatton W, Muller W, Kreuter J (2002) Microparticles derived from marine sponge collagen (SCMPs): preparation, characterization and suitability for dermal delivery of all-trans retinol. Eur J Pharm Biopharm 54:125–133

Takechi R, Pallebage-Gamarallage MM, Lam V, Giles C, Mamo JC (2013) Nutraceutical agents with anti-inflammatory properties prevent dietary saturated-fat induced disturbances in blood–brain barrier function in wild-type mice. J Neuroinflammation 10(1):842

Taneja A, Singh H (2012) Challenges for the delivery of long-chain n-3 fatty acids in functional foods. Annu Rev Food Sci Technol 3:105–123

Tasset I, Pontes AJ, Hinojosa AJ, de la Torre R, Tunez I (2011) Olive oil reduces oxidative damage in a 3-nitropropionic acid-induced Huntington's disease-like rat model. Nutr Neurosci 14:106–111

Tolles WM, Rath BB (2003) Nanotechnology, a stimulus for innovation. Curr Sci 85:1746–1759

Tomlinson E, Burger JJ (1985) Incorporation of water-soluble drugs in albumin microspheres. Methods Enzymol 112:27–43

Tshweu L, Katata L, Kalombo L, Chiappetta DA, Hocht C, Sosnik A et al (2014) Enhanced oral bioavailability of the antiretroviral efavirenz encapsulated in poly(epsilon-caprolactone) nanoparticles by a spray-drying method. Nanomedicine (Lond) 9(12):1821–1833

Tshweu L, Katata L, Kalombo L, Swai H (2013) Nanoencapsulation of water-soluble drug, lamivudine, using a double emulsion spray-drying technique for improving HIV treatment. J Nanopart Res 15:1–11

Vorhies JS, Nemunaitis JJ (2009) Synthetic vs. natural/biodegradable polymers for delivery of shRNA-based cancer therapies. Methods Mol Biol 480:11–29

Wang D, Zhao P, Cuia F, Li X (2007) Preparation and characterization of solid lipid nanoparticles loaded with total flavones of Hippophae rhamnoides (TFH). PDA J Pharm Sci Technol 61:110–120

Wang Y, Xu H, Fu Q, Ma R, Xiang J (2011) Protective effect of resveratrol derived from Polygonum cuspidatum and its liposomal form on nigral cells in Parkinsonian rats. J Neurol Sci 304:29–34

Weber S, Zimmer A, Pardeike J (2014) Solid Lipid Nanoparticles (SLN) and Nanostructured Lipid Carriers (NLC) for pulmonary application: a review of the state of the art. Eur J Pharm Biopharm 86:722

Westesen K, Bunjes H, Koch MHJ (1997) Physicochemical characterisation of lipid nanoparticles and evaluation of their drug loading capacity and sustained release potential. J Control Release 48:223–236

Weyermanna J, Lochmanna D, Georgensa C, Zimmer A (2005) Albumin–protamine–oligonucleotide-nanoparticles as a new antisense delivery system. Part 2: cellular uptake and effect. Eur J Pharm Biopharm 59:431–438

Yoon G, Park JW, Yoon IS (2013) Solid lipid nanoparticles (SLNs) and nanostructured lipid carriers (NLCs): recent advances in drug delivery. J Pharm Investig 43(5):353–362

Zheng Y, Monty J, Linhardt RJ (2015) Polysaccharide-based nanocomposites and their Sapplications. Carbohydr Res 405:23–32

Web Reference

https://www.grandviewresearch.com/industry-analysis/nutraceuticals-market, 2020
https://www.Mordointelligence.com

Chapter 2
Nanoscale Smart Drug Delivery Systems and Techniques of Drug Loading to Nanoarchitectures

B. Varalakshmi, T. Karpagam, A. Vijaya Anand, and B. Balamuralikrishnan

Contents

2.1	Introduction.	30
2.2	Nanoscale Drug Delivery Strategies.	31
	2.2.1 Passive and Active Targeting.	32
	2.2.2 Cellular Internalization of Drug Nanocarriers by Endocytosis.	32
	2.2.3 Release of Drug from Nanoparticles.	34
2.3	Nanoparticle Drug Delivery Systems.	35
	2.3.1 Classification of Nanoparticles Based on Composition.	35
2.4	Organic Nanoparticles as Drug Carriers.	36
	2.4.1 Lipid-Based Amphiphilic Drug Delivery Systems.	37
	2.4.2 Polymer-Based Drug Delivery Systems.	40
	2.4.3 Carbon-Based Drug Delivery Systems.	46
2.5	Inorganic Nanoparticles as Drug Carriers.	50
	2.5.1 Gold Nanoparticles (Au NPs).	50
	2.5.2 Nanoshells.	52
	2.5.3 Quantum Dots (QDs).	53
	2.5.4 Superparamagnetic Iron-Oxide Nanoparticles (SPIONs).	56
	2.5.5 Mesoporous Silica Nanoparticles (MSNs).	56
2.6	Techniques for Drug Loading to Nanoparticles.	58
	2.6.1 Drug Loading to Polymeric Nanoparticles.	58
	2.6.2 Drug Loading Techniques to Various Nanostructures.	67
	2.6.3 Drug Loading Efficiency.	75
	2.6.4 Stability and Storage of Nanoparticles.	75
	2.6.5 Conclusion.	76
References.		76

B. Varalakshmi (✉) · T. Karpagam
Department of Biochemistry, Shrimati Indira Gandhi College,
Tiruchirappalli, Tamil Nadu, India

A. Vijaya Anand
Department of Human Genetics and Molecular Genetics, Bharathiar University,
Coimbatore, Tamil Nadu, India

B. Balamuralikrishnan
Department of Food Science and Biotechnology, Sejong University, Seoul, South Korea

2.1 Introduction

The term "nanomedicine" refers to a wide variety of nanomaterials and structures used for treatment and diagnosis. Nanoparticles are colloidal in nature with size ranges from 10 to 1000 nm. It can include nanosized polymers, dendrites, micelles, liposomes, particles, and capsules. They are constructed from organic or inorganic (or both) compounds and perfectly apt for optimized drug delivery with negligible adverse effects (Kim et al. 2010).

Requirements for a good drug delivery system are as follows:

1. The half-life in blood circulation should be considerably long enough for the accumulation of drug at the target site.
2. They should be able to carry significant quantity of drug compound.
3. Either the drug delivery system or its degradation products should not have adverse toxic outcomes.
4. The shelf life must be significantly long to permit storage and transport.
5. It should be economic, or the cost can be directly proportional to effectiveness of the drug.

Disadvantages of conventional drug delivery system are as follows:

1. The drug is quickly degraded or inactivated in the body prior to reach its target site.
2. The drugs lack the capacity to cross physiological barriers such as cell membranes, placenta, blood-brain barrier, etc.
3. Poor specificity and hence distribution of drugs to all types of tissues and organs. All these lead to decreased therapeutic efficiency or undesirable side effects.

Nanomedicines have more advantages over conventional medicine. Nanoparticles can be used for both treatment and diagnosis simultaneously. Theranostic nanoparticles carry both drugs for targeted therapy and imaging molecules for simultaneous diagnosis of diseases. The physiochemical characteristics of nanoparticles can directly influence the function and make them advantageous over many conventional medicines. Physiochemical properties of nanocarriers depend on compositions (organic, inorganic, or hybrid), size (nanoscale in dimension ranging from 10 nm to 1000 nm), shape (sphere, rod, highly branched, multilayer), and surface characters (active groups, electrostatic charges, shelling methods, or binding of ligands) (Tran et al. 2017).

The size of nanoparticles is basically very tiny, ranging from 1 to 1000 nm. They have typical structural, optical, and electrical characters which many larger molecules do not have. They have better solubility, so that they are able to change insoluble or sparingly soluble therapeutic agents into soluble aqueous suspensions; thus, the need for toxic organic solvents for the drug formulation is eliminated. The small nanosize of these particles increases circulation time and bioavailability. Hence, the circulation time of nanoparticles is definitely longer than any conventional drugs. Generally, nanoparticles coated with hydrophilic polymers show prolonged circulation time in the blood, so they are a better choice for increasing the efficiency of drugs having shorter half-life periods. The dissolution kinetics of drug is improved; commencement of therapeutic function is enhanced so the drug dosage can be reduced. Coating of nanoparticle with suitable material can make them to escape from the recognition and elimination by the immune system. Nanoparticles are available in different shapes (rods, cylindrical, spherical, hemispheres, discs, tubes, cones, and wires). Nanoparticles can also be porous, hollow, or solid (Husain 2017). These features influence the loading, interactive, and transport efficacy of nanoparticles. For instance, a hollow nanoparticle may be a smart carrier for both therapeutic drug and diagnostic imaging compounds. Nanoparticles have large surface area due to nanosize. The sum of surface area increases exponentially as particle size diminishes. The meaning of increase in surface area is that a larger fraction of atoms are positioned on the surface relative to the inner core. This makes nanoparticles more reactive compared with traditional bulky drugs. Amplified surface area also leads to better solubility in aqueous environment and better bioavailability of nanoparticles. The larger surface area of nanoparticles permits them to possess wide varieties of surface characters, including binding with biomolecules or charged groups.

Due to smaller dimension, nano drug carriers can cross biological barriers (blood-brain barrier, etc.) easily by enhanced permeability and retention effect (EPR) and deliver drugs to sites that are not normally reachable by conventional therapeutic agents. The increased permeability of nanoparticles allows delivery of drugs to inflamed tissue and delivery of cancer drugs into tumors by passing via neovessel pores of diameter less than 1 μm (Jong and Borm 2008). The use of novel nanotechnology-based drug delivery systems has paved the way to enhance the pharmacokinetic characters of drugs than conventional drugs. Pharmacokinetics deals with the proportion of dosage of a drug and its concentrations in biological fluids over a period. The factors influencing pharmacokinetics are rate and extent of drug absorption, distribution, metabolism, and excretion processes (ADME studies). The measureable pharmacokinetic factors of nanoparticles are maximum serum concentration (C_{max}), volume of distribution (V_d) area under the curve (AUC) of serum concentration-time profile, time to maximum serum concentration (T_{max}), half-life ($t_{1/2}$), and clearance (CL). Plant-derived compounds or synthetic compounds encapsulated or entrapped into nanoparticles gave good results to overcome their undesirable pharmacokinetic characters and improved stability of the loaded drugs (Abdifetah and Na Bangchang 2019).

2.2 Nanoscale Drug Delivery Strategies

Nanoparticles/nanocarriers are nanosized, colloidal particles, normally ranging below 1000 nm in size, characterized by having larger surface area to size ratio. These particles are made of natural or synthetic biodegradable polymers, and the therapeutic drug is loaded by encapsulation or entrapped/embedded within polymeric matrix or adsorbed or conjugated onto the surface. These drug-loaded nanoparticles are better drug delivery system than conventional carriers. Nanoparticle-based drug delivery in nanomedicine depends on two factors, enhanced encapsulation/entrapment/adsorption of the therapeutic compounds and delivery and release of drugs to the target site.

2.2.1 Passive and Active Targeting

Nanotechnology-based drug delivery strategy follows two main techniques: *passive* and *active targeting*. The fundamental principle for accumulation of nanoparticle into target site is the same for both approaches. According to EPR (enhanced permeability and retention) effect, nanocarriers extravasate from blood capillaries via interendothelial membrane openings by passive diffusion and reach target tissues. This kind of passive transfer mechanism is referred to as paracellular transport pathway. This pathway is based on nanoparticle concentration gradient between the blood and target tissue interstitium which drives nanoparticles to diffuse passively across the endothelium via interendothelial gaps/openings and assists accumulation in target tissue (Fig. 2.1a).

To perform active targeting mechanism, nanoparticles are attached with specific ligands on particle surface and these ligands are known as targeting ligands. These targeting ligands may be biomolecules such as RNA/DNA, antibodies, or peptides which can selectively bind to distinct cell membrane receptors on target cells with good attraction. Due to very high specific cellular interaction by the nanoparticles, the active targeting approach is widely employed in nanomedicine, thus reducing the chance of nonspecific cellular interaction by nanoparticles (Fig.2.1b) (Rosenblum et al. 2018).

2.2.2 Cellular Internalization of Drug Nanocarriers by Endocytosis

Polymer-based nanoparticles bind to target cellular surface and deliver therapeutic compounds which enter into the cell by simple diffusion. Also, intact polymeric nanoparticles can penetrate into the target cell by endocytosis process. These nanoparticles attach to the cell surface receptor and form endosomes. Once reaching the cytoplasm, these endosomes are lysed by enzymes or by the applied external stimulus; thereby, the drug-loaded nanoparticles and the drug are released in the cytoplasm. Thus, cellular internalization of drug nanocarriers takes place (Fig. 2.2) (Wang et al. 2012).

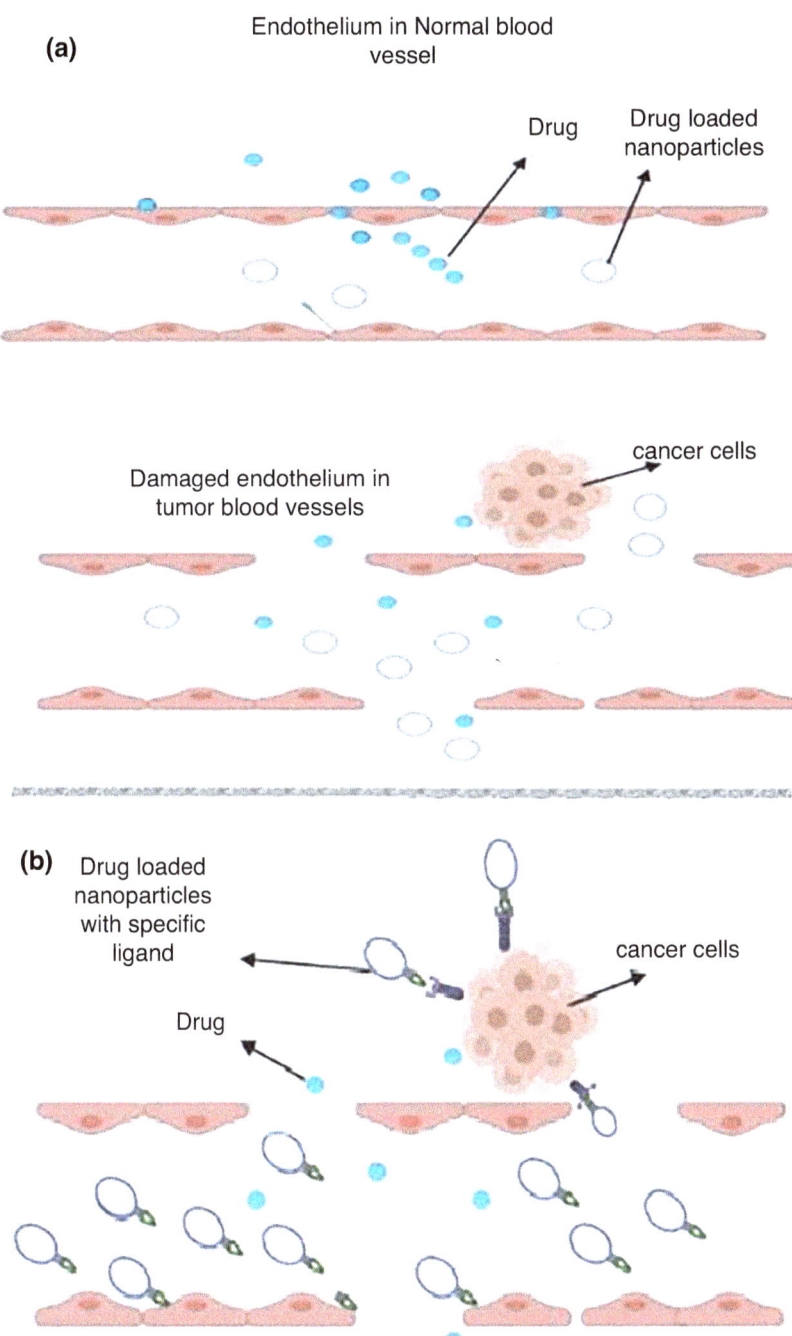

Fig. 2.1 (**a**) Mechanism of passive targeting. (**b**) Mechanism of active targeting

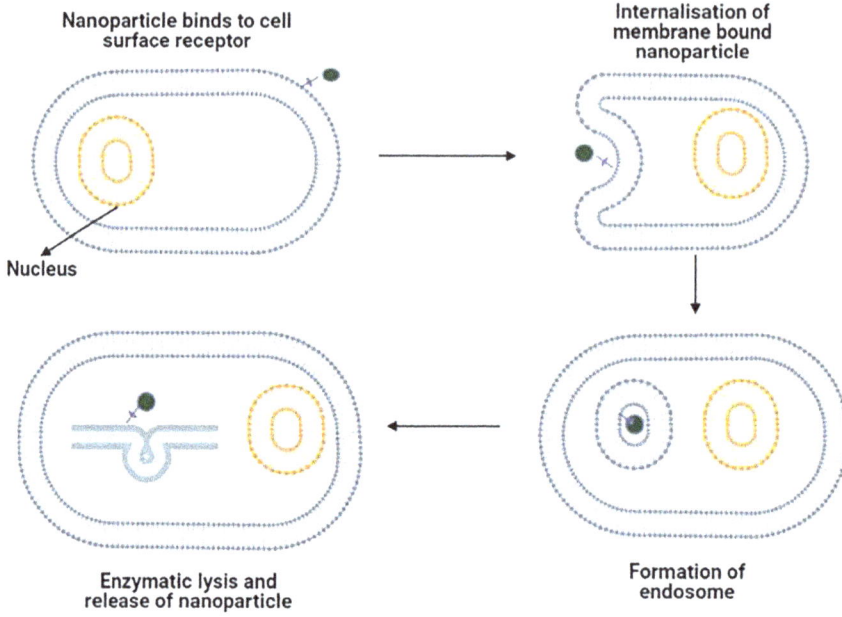

Fig. 2.2 Cellular internalization of drug nanocarriers by endocytosis

2.2.3 Release of Drug from Nanoparticles

The molecular nature of the nanoparticles (e.g., organic, inorganic, or hybrid compounds), the site where drugs are loaded to nanoparticles (shell region or matrix region), and the strength (strong covalent or weak non-covalent interactions) of binding of drug with nanoparticle are the three main parameters which dictate drug delivery profile (Mattos et al. 2017). The most prominent advantage of nano drug delivery system is the accurate rate of drug release at the target site. When the drug-loaded nanoparticle reaches the target site, the drug molecule has to be released from the nanocarrier at definite controlled rate. This can be achieved either *spontaneously* or by applying proper *stimulus*. The spontaneous release involves few mechanisms like gradual diffusion along with the degradation of delivery system or by solvent extraction or by chemical process. For stimulus-responsive nanosystems, the stimulus applied or used may be internal or external stimulus. The stimulus-responsive prototypes involve the application of *physical stimuli* like temperature, light, ultrasonic vibrations, magnetic field, *and* ionic strength; *chemical stimuli* like redox and pH; or *biological stimuli* like enzymes, etc. (Fig. 2.3) (Ding and Li 2017).

Smart stimuli-responsive drug delivery systems (SDDS) are the basis for fabricating several controlled drug delivery systems (DDSs). SDDS have drawn much focus and attraction because of controlled drug release with an effective

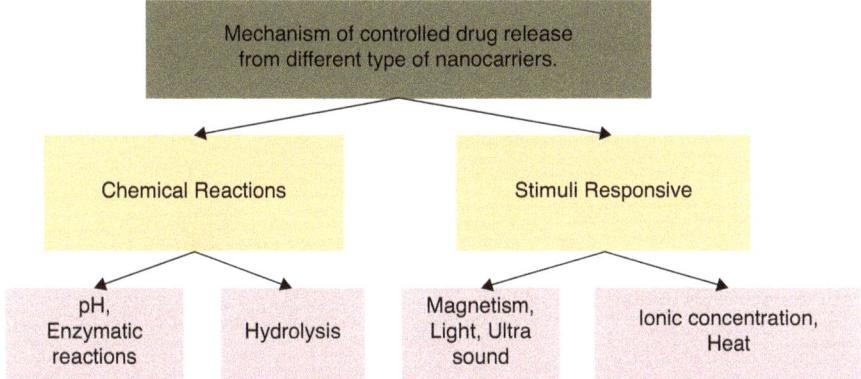

Fig. 2.3 Mechanism of controlled drug release from different types of nanocarriers

concentration for an extended duration, thus avoiding the need for repeated dosing and thus reducing the side effects. Also, SDDS prevent premature degradation of chemotherapeutic agents and induce their absorption into solid tumors (Hu et al. 2017; Ma et al. 2017).

2.3 Nanoparticle Drug Delivery Systems

Nanocarriers like nanotubes, nanoshells, nanopores, nanoliposomes, solid lipid nanoparticles, nanocapsule, dendrimers, fullerene, quantum dots, nanosphere, nanocrystals, gold nanoparticles, mesoporous silica nanoparticles, etc. are reliable novel drug delivery systems. Moreover, nano robotics and nanochips are modern systems developed for drug delivery (Couvreur 2013).

2.3.1 Classification of Nanoparticles Based on Composition

Generally nanoparticles are classified based on their composition, namely, *organic*, *inorganic*, and *hybrid* (Fig. 2.4). The organic or polymeric nanoparticles are not toxic and biodegradable. Dendrimers, solid lipid nanoparticles, nanogels, and hollow spheres like micelles and nanoliposomes are examples for organic nanocarriers. The carbon-based organic nanoparticles are fullerenes, graphene, and carbon nanotubes which are entirely made of carbon only.

The inorganic nanocarriers are principally composed of metals (silver, gold, iron, copper, aluminum, cadmium, cobalt, and zinc) or metal oxides (titanium oxide, iron oxide, magnetite, silicon dioxide, cerium oxide, and zinc oxide). Metal oxide-based nanocarriers show advanced properties than their metallic counterparts. Examples

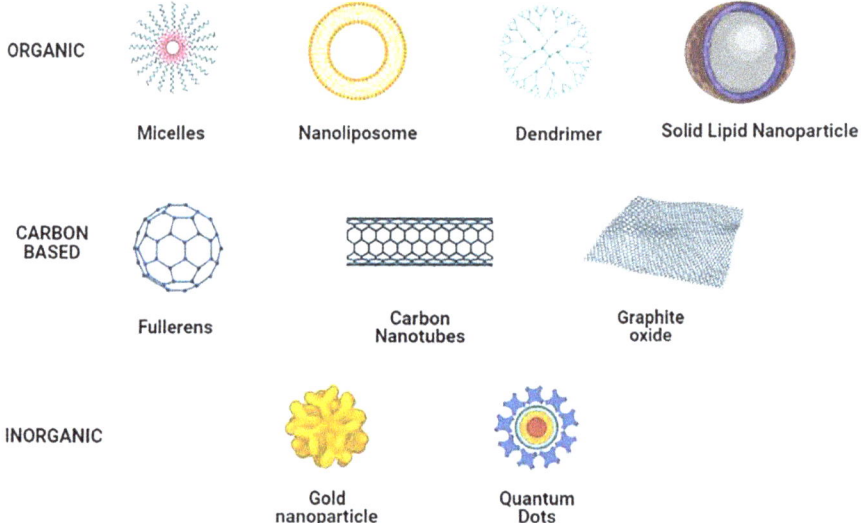

Fig. 2.4 Classification of nanoparticles for drug delivery applications

of metal oxide-based nanocarriers include quantum dots (QDs), mesoporous silica nanoparticles (MSNs), superparamagnetic iron-oxide nanoparticles (SPION), etc. (Domenico et al. 2019).

2.4 Organic Nanoparticles as Drug Carriers

The nanocarriers composed of organic compounds are endowed with excellent biological compatibility, enhanced drug carrying ability, high colloidal stability, and optimum size. Organic nanoparticles are well suited to carry a range of hydrophilic or lipophilic/hydrophobic drugs. Based on method of synthesis, nanocarriers are classified into two main types: amphiphilic systems obtained by self-assembly processes (such as nanovesicles and nanomicelles) and those nanocarriers prepared by *specific synthesis processes* (e.g., dendrimer nanoparticles, chemical cross-linked nanogels, CNTs, fullerenes, graphenes, etc.,) (Lombardo et al. 2015). Amphiphilic macromolecules contain both *hydrophilic* and *hydrophobic (lipophilic) r*egions. Hydrophilic regions that interact with surrounding aqueous medium may be negatively or positively charged or neutral. *Lipophilic* regions are usually made of long hydrocarbon chains. Self-assembly interactions involved are non-covalent mild interactions, namely, van der Waals forces, hydrogen bonding, hydrophobic attractions, hydration and electrostatic interactions, *pi-pi* bonding, coordination bonding, steric interactions, depletion, and restoration reactions (Degiorgio and Corti 1985).

2.4.1 Lipid-Based Amphiphilic Drug Delivery Systems

2.4.1.1 Nanoliposomes

Vesicles are made of natural or synthetic lipids known as liposomes. They are self-assembled amphiphilic systems. In water, they form a flexible bilayer vesicle with the hydrophilic regions facing the aqueous phase while hydrophobic regions are embedded inside by self-assembly process (Khosa et al. 2018). Generally, the degree fluidity of a lipid bilayer is based on two factors: composition and temperature. As the temperature increases, a bilayer composed of phospholipids transform from a rigid crystalline gel form to more mobile liquid form. Hence, desired fluidity of vesicle can be attained by adjusting composition and temperature. Fluidity plays an important role in uptake and discharge actions of cells (Kiselev and Lombardo 2017). Nanosized liposomes are called nanoliposomes. They contain hydrophilic head regions and lipophilic tail regions. Nanoliposomes are engineered to carry and deliver small drug molecules, imaging compounds, peptides, proteins, DNA and RNA, etc. Usually hydrophilic medicaments are loaded in hydrophilic regions of nanoliposomes, and lipophilic medicaments are loaded in hydrophobic tail regions. Nanoliposomes can easily escape from decomposition and clearance by macrophages present in the liver. They can execute both passive targeting and active targeting strategy for the delivery of therapeutic compounds (Kumar et al. 2010). Nanoliposomes principally get accumulated in the target tissue by passive targeting and discharge loaded drugs for longer duration. Active targeting is attained by attaching antibodies, proteins, peptides, and ligands on the outer region of nanoliposomes. In active targeting, nanoliposomes specifically reach the target diseased organs or tissues and enhance the sustained release of drug compound for longer duration. Active targeting is very specific in action since healthy cells are unaffected and diseased cells alone exposed to the drug (Riaz et al. 2018). This is the advantage of active targeting over passive targeting. For example, nanoliposomes engineered with C6-ceremide ligand are specifically targeted toward overexpressed leukemic cells and hence can be used as therapeutics for leukemia (blood cancer) (Kumar et al. 2012). Nanoliposomes composed of polyethylene glycol modify the pharmacokinetic characters of drug compounds resulting in prolonged half-life time for drug clearance (Dadashzadeh et al. 2008).

Nanoliposomes are able to present slow and steady release of an encapsulated drug at the target tissue resulting in enhanced efficacy. Many stimulus-responsive drug release nanoliposome models have been developed. Shi et al. (2017) developed light or photo-stimulated nanoliposomes (PNLs) against drug-resistant human breast cancer (MCF-7/MDR) cell lines and MCF-7/MDR mice tumor models. The PNLs are designed to carry two compounds, a photosensitizer and an anticancer drug molecule. The photosensitizer (hematoporphyrin monomethyl ether) molecule was incorporated in the outer lipid bilayer, and an anticancer medicine doxorubicin was encapsulated in the inner region of the nanoliposome. When photosensitizer molecule is activated by light, it leads to enhanced cytotoxicity and decreases drug resistance in tumor models. This reaction is carried out in synchrony with a photo-initiation reaction and fast release of anticancer medicine. Thus, combining effective chemotherapy and photodynamic therapy showed much improved antitumor activity and attained notable tumor suppression in drug-resistant tumor models.

2.4.1.2 Solid Lipid Nanoparticles (SLNs)

SLNs are innovative colloidal lipospheres of size less than 1000 nm. They exist in the form of nanoscale lipid emulsions, in which the lipid in oily (liquid) state is replaced by a solid lipid. SLN has exclusive biophysical characteristics such as tiny size with large surface area, excellent drug loading capacity, and the significant interaction of phases at the interface, all of which makes it an efficient drug delivery device. SLN lipospheres are basically solid at human normal body temperature (98.6 °F) (Pardeike et al. 2009). SLNs are composed of a variety of lipids, including mono-, di-, and triacylglycerols like tristearoyl glycerol, partial acylglycerols like Imwitor, long chain fatty acids like stearate and palmitate, and steroids like cholesterol and waxes like cetyl palmitic acid. They have outer shell and inner core regions. The lipids present in SLN are mostly physiological lipids which makes it advantageous in reducing the danger of acute as well as chronic toxic consequences. The lipid dispersions are generally stabilized by using a wide range of emulsifiers or combination of emulsifiers (Cavalli et al. 1993). The inner solid hydrophobic lipid core is loaded with drug compound which is solubilized in solid lipid matrix of high melting point. This core is enclosed by single layer of phospholipid. The loaded drug can be triggered easily for successive release from SLNs. Stabilizing surfactants are mixed to SLNs to convert it to an administrable emulsion.

Release of loaded drug from SLNs is carried out by disruption or degradation of solid matrix of SLN and diffusion of drug. Basically, three drug loading models are applicable for SLNs: homogenous matrix model, drug enriched shell in core shell model, and drug enriched core in core shell model (Fig. 2.5) (Pardeshi et al. 2012). Many kinds of stimulus-responsive SLNs have been designed to attain efficient drug release at target site. The stimulus applied can be either external or internal type. The main factor which affects the release of the drug from SLNs is particle size. Small size particles with larger surface area showed fast drug delivery profile when compared with large size SLNs. Another factor which affects the rate of drug release from SLNs is the site of drug loading on SLNs, for instance, rapid release of drug could be noticed when drug is loaded on outer shell (drug enriched shell in core shell model) than in inner core (drug enriched core in core shell model) (Rabinarayan and Padilama 2010).

Drugs in SLN are inserted in both regions: the matrix and shell (homogenous matrix model). This type of drug incorporation enables versatile dual drug release (immediate and sustained release) from SLN. Drugs loaded on the shell region of SLN will release first followed by the release of drugs from matrix region. The drug released from the shell is rapid and results in immediate drug effect. Afterward, depending on the composition of the matrix lipid, the matrix degrades or erodes and

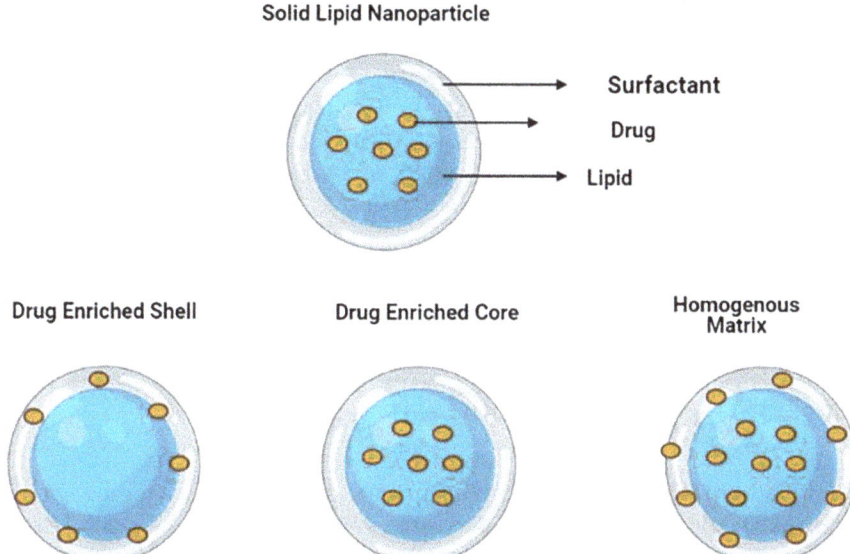

Fig. 2.5 Models of drug loading into SLN

leads to sustained or controlled drug release, thus enabling gradual drug effect. Thus dual drug release profile is obtained with SLNs. The release of drug from SLNs is based on the diffusion of drug compounds through lipid matrix and in vivo breakdown of lipid matrix by the action of lipase enzymes present in cells (Westesen and Siekmann 1996). Muller et al. (1993) studied on the in vitro degradation of SLNs in solutions of pancreatic lipase. The rate of lipid degradation was analyzed by quantifying the concentration of liberated free fatty acids by turbidimetry technique. The rate of lipid degradation was depending on the composition of the lipid matrix. The kinetics of breakdown is very high for triacylmyristic acid (triglyceride) moderate for acetylpalmitic acid (wax) and comparatively slower for triglycerides with very long chain fatty acid, for example, tribehenin.

High temperature or high concentration of surfactant can lead to burst discharge of drug from SLNs. So, the synthesis of SLN is generally carried out at room temperature to prevent burst discharge and partitioning of drug compound in aqueous phase but helps the partitioning of large amount of drug in lipid phase. Thus, at low temperatures, slow and controlled drug release is obtained from SLN without immediate or burst release of drug. Based on solid-liquid transition character of SLN upon applying heat, thermoresponsive SLNs have been developed by Rehman et al. (2017). Different combinations of fatty acids (laurate and oleate, laurate, and linoleate) were used to develop SLNs. The drug release profile was studied, and the results proved that rapid or immediate discharge of loaded 5-fluorouracil drug was >90% observed at 39 °C featured to the liquefaction of lipid core, whereas 22–34% of drug discharge was observed at 37 °C where solid core is retained.

In another research work, pH-responsive cholesterol-polyethylene glycol-coated SLNs were prepared and loaded with anticancer drug doxorubicin. The electric charge attraction between the positively charged doxorubicin and anionic (negatively charged) lipid core laurate (by protonation) plays an important role on drug release pattern. The release pattern was investigated, and it was reported that these stimulus-responsive particles showed rapid discharge of doxorubicin at pH 4.7 (intracellular pH) compared to slow drug discharge at pH 7.4 (extracellular pH). The rapid release of doxorubicin from SLN at low intracellular pH is due to the weakening of electrostatic attraction between the lipid core and the drug doxorubicin (Chen et al. 2015).

2.4.2 Polymer-Based Drug Delivery Systems

Many biodegradable synthetic and biological polymers are utilized for the synthesis of polymeric nanoparticles. Widely used synthetic polymers are polylactide (PLA), poly (D, L-lactide-co- glycolide) (PLGA), poly (glycolic acid) (PGA), and polyethylene glycol (PEG). These polymers have characteristic functions in drug delivery processes. They can provide controlled discharge of therapeutic agents by crossing the biological and also pathology barriers of the tissues, hydrolysis under in vivo conditions, and having high biocompatibility and less immunogenic and negligible toxicity. These properties make them good drug delivery system. PLGA is highly degradable and nontoxic because the hydrolytic products of this polymer are lactic acid and glycolic acid, which are degraded and eliminated naturally from the body (Tyler et al. 2016). Alginate is a natural polysaccharide polymer made of mannuronic and glucuronic acid monomers. It is an anionic mucoadhesive biopolymer used in nanoparticles to carry drugs (Maitra and Shukla 2014). Chitosan polymer is another natural, mucoadhesive, degradable, and biocompatible molecule with positively charged functional groups to which drugs can be bound (Yoo and Park 2001).

2.4.2.1 Polymer-Based Micelle and Vesicles

Micelle-like nanocarriers are made of amphiphilic polymers formed by the *self-assembly* process. The micelles contain inner hydrophobic (lipophilic) core and outer hydrophilic shell regions. The inner region forms a nanoenvironment for the loading of hydrophobic drugs and considerably improves the solubility and hence bioavailability of lipid-based drugs. Meanwhile, the outer shell forms a stabilizing interface between the lipophilic core and the aqueous phase; this helps in improving colloidal stability and suppressing aggregation and decreases unnecessary interactions between other components in the blood. The micelles of dimension less than 200 nm can escape from detection and clearance by reticuloendothelial cells and exhibit increased EPR effect at target tissues (passive targeting) (Yin et al. 2016).

In many polymeric micelles used for efficient drug delivery purpose, the core contains hydrophobic PLA or PLGA polymers, and the hydrophilic shell is made with the most hydrophilic polymer polyethylene glycol. For example, in PEG-PLGA copolymeric micelles, lipophilic PLGA core can proficiently encapsulate several kinds of drugs, while the hydrophilic PEG shell inhibits the binding of proteins and phagocytic cells, thus prolonging exist time in the blood (Cho et al. 2016). *Chitosan*-derived self-assembled *amphiphilic micelles* exhibit mucoadhesive characteristics, suitable to target at the epithelial tight junctions. Chitosan-derived nanomicelles are extensively used for continued drug delivery to different kinds of epithelia, such as buccal, intestinal, nasal, eye, and pulmonary. Chitosan-based nanocarriers can allow either parental or nonparental routes of administrations and be used for the treatment of skin and gastrointestinal tract diseases, lung diseases, and eye infections (Yoo and Park 2001).

Polymersomes are *vesicles* made of amphiphilic polymers, characterized by the presence of bilayer arrangement with an aqueous interior core. The water-soluble therapeutic agents can be encapsulated in the aqueous core, and lipophilic drugs can be integrated into the internal region of the bilayer structure. *Critical micelle concentration (CMC)* is defined as the concentration of surfactants above which micellar structure is formed by self-assembly process. Polymer-based vesicles and micelles retain their structural integrity above the CMC value. Below the CMC, the assembly of nanoarchitectures dissociate into component polymeric chains, which leads to loss of their drug carrying ability (Mikhail and Allen 2009). Stimulus-responsive polymersomes are engineered which are enabled to elicit controlled drug release and improved imaging sensitivity at tumor and intracellular microenvironment. Weak acidic pH, range of temperatures, many specifically overexpressed enzymes, and redox species are the most used key endogenous stimuli for the triggered drug release at the target tissues.

For example, temperature-sensitive polymeric nanocarriers are used in drug delivery applications for cancer treatment. Lower critical solution temperature (LCST) is the temperature below which the polymers are dissolved incompletely in water phase. Below LCST, polymer holds water through hydrogen bonds. However, at temperatures above LCST, the hydrogen bonds are disrupted rendering the polymer hydrophobic to precipitate out. Poly (N-isopropylacrylamide) (PNIPAm) is the most frequently used thermoresponsive polymer with LCST at 33 °C (Wei et al. 2009). Thermosensitive micelles are made of thermoresponsive polymers, which can undergo a sharp variation in their solubility in water with change in temperature. When the solubility is disrupted by temperature, the micellar structure is destabilized. Thus, the micellar structure can be destabilized by altering the surrounding temperature slightly above or below the LCST. This phase change is utilized for the controlled release of encapsulated drug by applying heat or cooling the environment for a definite period (Ward and Georgiou 2011).

2.4.2.2 Polymer-Based Nanogels

Polymeric nanogels are three-dimensional networks of size ranging from 100 to 200 nm. The cross-link between polymer chains may be chemical or physical in nature. The very peculiar character of nanogels is swelling in water. When diffused in water medium, they form semisolid states by swallow of large amount of water (hydrogels).

To synthesize chemically cross-linked hydrogels (Fig. 2.6), the hydrophilic polymer solution is treated with a bifunctional cross-linking agent. Few examples of chemical cross-linking agents used are glutaraldehyde and mono- and polycarboxylic acids. These agents can cross-link both proteins and polysaccharides (Sosnik and Seremeta 2017; Reddy et al. 2015),

while the cross-linking in physically cross-linked hydrogels is accomplished by the formation of non-covalent bonds (hydrogen bond, ionic, van der Waals forces, hydrophobic interactions) between polymers. Complex of polyanionic polymer with polycationic polymer has been used in many drug delivery approaches. The degree of cross-linking interactions (hydrogen bonding or ionic boning) between polymer chains is based on a variety of parameters such as polymer concentration, nature of solvent used, and solution temperature and pH. By using optimum conditions, these cross-linking interactions can be disrupted, and the controlled delivery of entrapped drug at target tissues is performed by these hydrogels (Hennink and van Nostrum 2002).

The small size of nanogels enables them to undergo rapid structural modifications in response to surrounding changes. Stimulus-responsive nanogels show

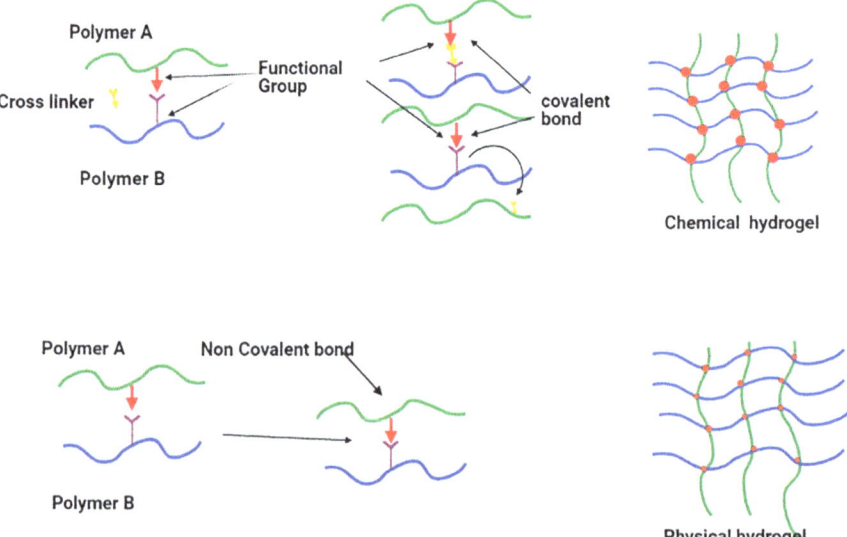

Fig. 2.6 Chemical and physical hydrogel

structural transitions in response to internal/external stimuli such as temperature, pH, electric field, and light (Soni and Yadav 2016). In disulfide-based cross-linked nanogels, the disulfide bond is stimuli-sensitive and biodegradable by biochemical reductants such glutathione or thioredoxin I/II. This system was used for the controlled release of doxorubicin (Chacko et al. 2012). Nanogels containing doxorubicin are studied for the treatment of cancer in the form of pH- and temperature-sensitive nanogels made of maleic acid poly-(N-isopropylacrylamide) polymers. Doxorubicin is liberated from these nanogels following a slight decrease in pH or through temperature stimulus (Sharma et al. 2016). Nanogel carriers are used to deliver large proteins and small peptides to the target tissues. Various types of nanogels used in the controlled delivery of drugs, protein, and peptides are listed in Table 2.1 (Kousalová and Etrych 2018).

Alginate is an anionic mucoadhesive biopolymer, used in nanogel formulations to encapsulate cells and proteins. It is biocompatible and nonimmunogenic in nature. Alginate is a naturally occurring polysaccharide consisting of mannuronic acid and glucuronic acid monomers. At room temperature and biological pH, alginate polymers produce a cross-linked gel complex in the presence of calcium ions. The gel structure can be distorted by removal of calcium ions from alginate gel by using chelating agent. Upon destabilization, the encapsulated drug is released from calcium alginate hydrogels and delivered to target tissue. This kind of stimulus-responsive controlled drug release strategy is used in various medicinal applications like scaffolding for cell cultures, drug delivery, and animal tissue engineering (like wound dressing) (Maitra and Shukla 2014). Nanogels of size 10–100 nm is used for systemic drug administration, since they can pass from small blood vessels via gaps in the endothelial membrane and reach tissues (passive targeting). Passively targeting nanogels are used for neoplastic disease treatment (Rigogliuso et al. 2012). Encapsulated paclitaxel in nanogels was actively targeted to the liver, breast, or prostate tumors using galactosamine, transferrin, anti-HER2, or parts of mAbs (anti-HER$_2$ scFv F5) as ligands selective for those cancer cells (Kousalová and Etrych 2018). Nanogels are an especially attractive tool for delivery of nucleic acid

Table 2.1 Stimulus-responsive nanogels for controlled delivery of drugs

Nanogel composition	Type of stimulus	Uses
Poly(N-isopropylacrylamide-copolyethylenimine-co-N,N′-methylenebisacrylamide)	Temperature and pH	5- Fluorouracil-loaded nanogel used for mastocarcinoma therapy
Poly(N-isopropylacrylamide-cobutylacrylate-co-N,N′-methylenebisacrylamide)	pH	Methotrexate-loaded nanogels used for therapy for breast and lung cancer, leukemia, and lymphoma
Poly(methylacrylic acid-co- N,N′-ethylenebisacrylamide)-coated Fe3O4 nanoparticles	pH and temperature	Controlled delivery of α-chymotrypsin using a magnetic field
Self-assembled cholesterol-bearing pullulan	Heat and light	Controlled delivery of bone anabolic agents, e.g., recombinant hormones and cytokines

drugs (plasmid DNA) and are hence widely used for gene therapy-based treatment such as in cancers, hemophilia, and viral infections (Ginn et al. 2013). DNA delivery using nanogels is advantageous over nonencapsulated DNA because of their enhanced cellular uptake and extended period of circulation in the blood (Peer et al. 2007). Cationic nanogels composed of PEO and poly (ethylenimine) are used to increase the transport of oligonucleotides across the GI tract epithelial layer and blood-brain barrier. Nanogels composed of polymer-protein conjugates have desirable drug carrier characteristics such as longer plasma half-life and improved protein stability.

2.4.2.3 Dendrimer-Polymeric Nanocarriers

Dendrimers are three-dimensional, treelike highly branched polymeric nanostructures of diameter ranging from 2.5 to 10 nm. Dendrimers exist in various shapes like spheres and flattened spheroids (disks) and ameba-like and starfish-like structures. The polymeric branch points are covalently bonded to middle core. The branching units are arranged into typical symmetric concentric layers around the core; these layers are known as generations. Each branching unit terminates with different kinds of functional groups exposed on the surface of the nanoparticle (Fig. 2.7) (Duncan and Izzo 2005). Dendrimers are prepared by *specific synthesis* process. The dendrimers are made of synthetic or biopolymers.

For medical applications, polyamidoamine (PAMAM) class and polypropyleneimine class dendrimers are generally used. A few other kinds of dendrimers are peptide dendrimers, glycodendrimers, polyethyleneimine (PEI) dendrimers, etc. Dendrimers are applied for the transport of DNA in cancer treatment or viral infections in various tissues and organs. Dendrimers display desirable drug delivery characteristics like monodispersity, presence of numerous functional groups, definite structure, and multivalency which makes them versatile nano drug carriers for various diseases especially for cancer treatment.

Controlled drug release from dendrimers depends on two factors.

(a) Chemical modification of dendrimer: For example, the drug indomethacin loaded on G4-NH2 dendrimer showed slow and controlled drug release when compared to the rapid discharge noted with G4-COOH dendrimer. Thus, the chemical modification of functional groups determines the strength of interaction of drugs with dendrimers which in turn determines the drug release kinetics from dendrimers (Chauhan 2018).
(b) Physical loading: The second parameter which determines the rate of release of therapeutic agent from dendrimers is physical loading which means dendrimer-to-drug molar ratio. For example, the drug release kinetics of cisplatin from dendrimer was observed to be directly related to the cisplatin/dendrimer molar ratio (Kulhari et al. 2015).

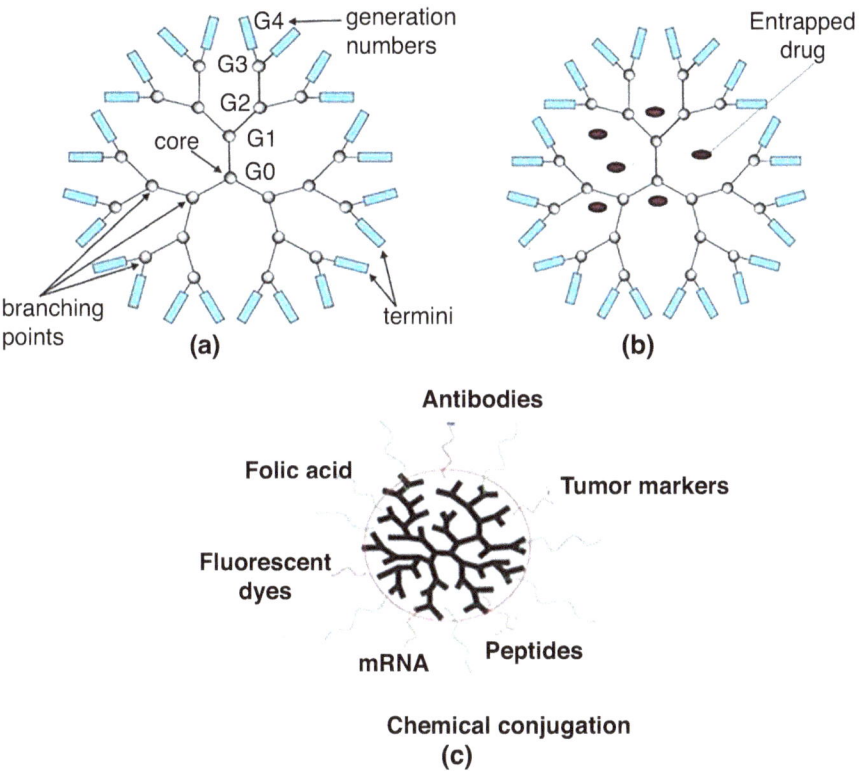

Fig. 2.7 (**a**) Structure of dendrimer nanocarrier. (**b**) Physical entrapment of drugs in dendrimer internal gaps. (**c**) Chemical conjugation of drugs through surface functional groups

Stimulus-responsive controlled drug release strategy depends on the in vivo breakdown of covalent bond linking drug and dendrimer by suitable enzymes or favorable physical environment like pH, temperature, etc.

Figure 2.8 depicts the use of dendrimers as gene delivery vectors. For gene therapy purpose, polyamidoamine (PAMAM)-based dendrimer is widely used. The possible mechanism for gene therapy by dendrimer is as follows: At first, the plasmid DNA (drug) is loaded on to dendrimer to form dendriplex (dendrimer and DNA complex) under in vitro conditions. Then, the dendriplex is mixed with cells under in vitro condition or is applied directly into animals in vivo or ex vivo by which dendriplex can reach the target cells by the blood circulation. On interaction with cell membrane, the dendriplex gets internalized into cytoplasm by endocytosis process. Now, the dendriplex experiences the change in pH from 7.4 (extracellular pH) to 5.5 (intracellular pH); this variation in pH triggers deprotonation of dendrimer functional groups which leads the dissociation of dendrimer and DNA from dendriplex and also results in diffusion of H+ and Cl− and water into the endosomes, leading to osmotic swelling and burst of endosome. This pathway of release of free nucleic acid into cytoplasm by dendrimer destruction is known as endosome escape

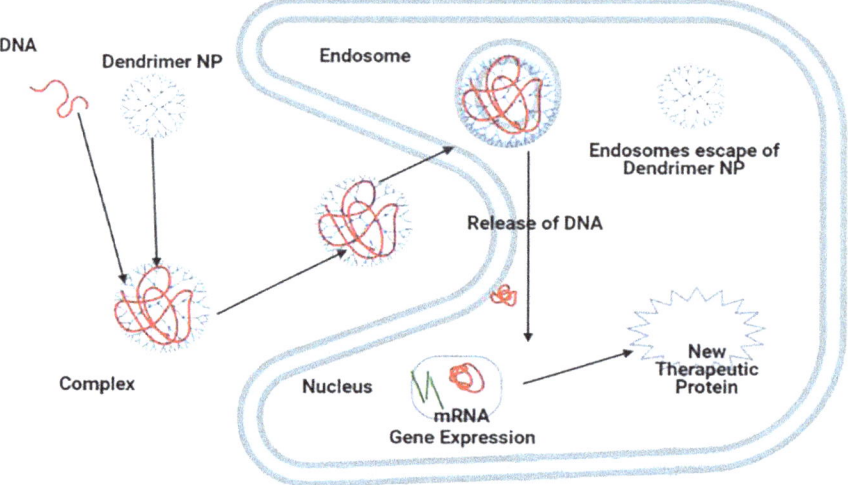

Fig. 2.8 Schematic presentation of application of dendrimers for gene therapy

pathway. In this pathway, the cargo of endosome is released into cytoplasm prior to the fusion of endosome with lysosome. Finally, the DNA from cytoplasm reaches the nucleus resulting in expression of gene. Thus, pH acts as internal stimulus for controlled targeted drug release (Palmerston Mendes et al. 2017).

2.4.3 Carbon-Based Drug Delivery Systems

These nanoparticles are made with carbons only which include carbon nanotubes, graphene, fullerenes, etc. Carbon-based nanodevices have distinctive, electric, optical, thermal, and mechanical characteristic, finding application in medicine as carriers of drug and imaging agents (Tonelli et al. 2015). Carbon-based nanomaterials have potential toxicity, but by suitable surface modifications, these nanomaterials are made susceptible to enzymatic degradation, thus rendering them nontoxic and more biocompatible (Bhattacharya et al. 2013).

2.4.3.1 Carbon Nanotubes (CNTs)

The CNTs are nanometer-scale tubelike structures made of carbon atoms in sp2 hybridization. The tube wall in CNTs consists of single (SWCNTs) or multiple (MWCNTs) layer of graphene sheets. CNTs comprise versatile drug delivery systems. CNTs show strong absorption in the near IR light and Raman scattering, photo-acoustic characteristics which render them as versatile device for in vivo biomedical applications like diagnosis and therapy (Kushwaha et al. 2013). The surface of raw CNTs is highly hydrophobic in nature and is insoluble in aqueous medium. To overcome this dilemma, functionalization of CNTs is carried out to increase its solubility. Functionalization is the process of binding required functional groups onto the walls of CNTs by chemical synthesis method to produce functionalized CNTs (f-CNT). To enhance the solubility, generally carboxylic or ammonium groups are incorporated to carbon nanotubes. f-CNTs are used for various biomedical applications like transport of small proteins, DNA/RNA, and drug molecules. The functionalization of CNTs not only improves solubility but also increases biocompatibility of loaded drugs and encapsulation tendency hence used for multimodal drug delivery and imaging. CNT-drug composites are easily eliminated from the body (Spitalsky et al. 2010).

Drugs can be incorporated onto CNTs either by mild non-covalent bonds (pi-pi bonding, van der Waals attractions, and hydrophobic interactions) or by covalent bonds. CNTs covalently attached cisplatin (drug) and with epidermal growth factor (ligand) are designed and actively targeted toward overexpressing EGF receptors present on head and neck squamous cancer cells. Thus, they are used for the treatment of malignancy (Bhirde et al. 2009). The CNTs can easily transport DNA molecules across plasma membrane, thus finding application in gene therapy. Genes can be loaded either at the tips or inside of CNTs. Al-Jamal et al. (2011) reported successful targeting of siRNA to the CNS by stereotactic administration of MWCNTs, resulting in the protection of neurons in experimental animals. Sacchetto et al. (2014) provided evidence that when PEG-treated SWCNTs were loaded with antisense oligonucleotides and administered to intra-particular region of chondrocytes in mice, they showed undisturbed cartilage homeostasis and no systemic side effects. Carbon nanotubes are used for the delivery of prodrugs (precursors of active drug). The prodrug can be converted to active form once delivered inside the cancer cells, for example, drug cisplatin. By using carbon nanotubes, the prodrug is delivered in the inactive form of platinum (oxidation IV state) which is then reduced to the active form of platinum (oxidation II state) selectively inside the cancerous cell only. Thus, the active drug does not interact with noncancerous cells, thus leading to targeted drug delivery (Hirsch et al. 2006). CNTs can even act as anticancer agents by themselves. Once CNTs entered cancerous cells, external electromagnetic radiation such as radiofrequency or NIR radiation were applied to heat up the carbon nanotubes and thus kill the cancer cells by the generated heat (Tomalia et al. 2007).

Heister et al. (2012) studied drug loading, dispersion stability, and stimulus-responsive targeted drug release of cancer drugs with oxidized SWCNTs (oxSWCNTs) and reported that pH 8 for doxorubicin and pH 9 for mitoxantrone were optimum for binding to oxSWCNTs by non-covalent interactions at 4 °C in the dark. Further, dispersion stability was improved by PEGylation of oxSWCNTs with various PEG formulations. It was also standardized that drug/oxSWCNTsPEG weight ratio of 1:2 was optimum for quantitative drug binding. The drug/oxSWCNTsPEG was tested for the density of drug loaded by suitable techniques, for example, by UV-VIS spectrophotometric analysis (at 479 nm for doxorubicin or at 550 nm for mitoxantrone). It was also reported that drug liberation of both doxorubicin and mitoxantrone from oxSWCNTsPEG and drug uptake by HeLa cells were significant at pH 5.5, which is the estimated pH for any drug delivery system that would involve in the endosomal pathway. The endosomal pathway is an endocytotic, energy-requiring engulf mechanism, by which the drug/oxSW-CNTsPEG complexes were engulfed into endosomes and subsequently the drug was released from carrier due to the low endo-lysosomal pH. The drug freed from carrier would then be translocated to the nucleus to carry out its cytotoxicity function by interfering with DNA synthesis (Heister et al. 2012).

2.4.3.2 Fullerenes

Fullerenes are another promising nanomaterial for drug delivery and imaging. Fullerenes are made of carbon atoms joined to three other carbon atoms by covalent bonding resulting in hollow sphere or ellipsoid tube structures. Fullerenes are also named as "buckyballs" (Fig. 2.9). The size of fullerenes ranges from those containing 20 carbon atoms to 100 carbon atoms. However, the most widespread fullerene is C_{60} made with 60 carbon atoms. C_{60} are hydrophobic in nature and aggregate very quickly in aqueous phase (Prato 1997). To overcome this problem, several methods are used to make C_{60} more hydrophilic and make it as an effective drug delivery system. Fullerenes are converted to highly hydrophilic functionalized nanostructures by treating with amino acid, carboxylic acid, polyhydroxyl group, amphiphilic polymers, etc. (chemical functionalization process) (Chen et al. 2001). Fullerenes and C_{60} derivatives lack immunogenetic action which supports their wide use in biomedical field.

Fullerenes are used for transport of antiviral, antibacterial, and chemotherapeutic drugs. C_{60} derivatives have potential antiviral activity. Dendrofullerene 1, a derivative of C_{60}, has strong antiprotease activity. It can precisely fit into the hydrophobic pocket of HIV protease enzyme and hinder the binding substrate molecules at the active site of enzyme (Schuster et al. 2000). Amino acid derivatives of fullerene C_{60} interfere with multiplication of HIV and human cytomegalovirus (Kotelnikova et al.

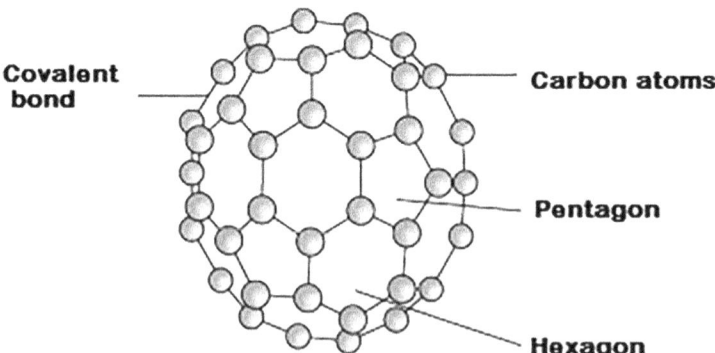

Fig. 2.9 Fullerenes

2003). Monoclonal antibodies targeted against melanoma tumors are conjugated with multiple C_{60} buckyballs to develop a new system of active targeting and simultaneous multiple drug delivery to tumors (Darshana Nagda et al. 2010). Buckyballs do not break down in the body and are excreted intact. This property has significant use to deliver anticancer drugs that are harmful to normal cells. For example, buckyball drug delivery nanoparticles that carry hazardous radioactive elements would allow for the complete excretion of radiation from the body following cancer treatment (Chan 2007). The conventional drug delivery methods have disadvantage of uncontrolled drug release in circulation and the slow release of drug at target tissue. Many stimulus-responsive drug release models are developed with C_{60} derivatives. Shi et al. (2016) engineered an "on- off" model of controlled drug release system with C_{60} derivative. In this, doxorubicin was covalently bonded to C_{60} through a reactive oxygen species (ROS)-sensitive thioketal linker (C_{60}-DOX), and then the hydrophilic shell was fixed to the external surface of C_{60}-DOX, to get C_{60}-DOX-NGR. The hydrophilic shell can impart enhanced stability in body fluids and also provides efficient tumor targeting. C_{60}-DOX-NGR was able to strongly hold doxorubicin (off state) even at weak acidic pH (pH 5.5). C_{60}-DOX-NGR can be switched to "on state" when large amounts of reactive oxygen species were generated by C_{60}, resulting in breaking of ROS-sensitive thioketal linker, thereby leading to burst release of doxorubicin. The "off-on" state of C_{60}-DOX-NGR could be accurately remote-controlled by irradiating with a 532 nm laser beam (at a low power density) with a high spatial/temporal resolution (combined phototherapy with chemotherapy). This new C_{60}-based drug delivery system with "off-on" switch showed efficient anticancer activity and a low toxicityunder in vivo and in vitro conditions.

A biocompatible, hydrophilic fluorescent fullerene derivative, C_{60}-TEG-COOH, was prepared by treating C_{60} with tetraethylene glycol. C_{60}-TEG-COOH was then coated on MSN which had been treated with 3-aminopropyltriethoxysilane to obtain an amino-modified MSN (MSN-NH_2). The anticancer drug doxorubicin hydrochloride (DOX) was loaded on the surface of the MSN-NH_2 at pH = 7.4 to get MSN@C60-DOX complex. This nanostructure can act as pH-sensitive drug delivery and fluorescence cell imaging system. This complex showed high cytotoxicity to HeLa cancer cell lines. The release of doxorubicin hydrochloride is stimulus responsive and could be encountered under a mild acidic environment (lysosomal pH = 5.0) due to the protonation of C_{60}-TEG-COO−, which disrupts the association between C_{60}-TEG-COOH and MSN and leads to controlled drug release (Tan et al. 2016).

2.4.3.3 Graphenes

Graphenes/graphene derivatives can be functionalized easily and possess large surface area, and delocalized pi electrons hence provided new openings for new drug delivery applications (Tonelli et al. 2015). Graphenes exhibited adequate drug loading, efficient in vivo drug distribution, and drug release (Novoselov et al. 2012). Graphene-based biological stimulus-responsive controlled drug release models were designed. The drug doxorubicin was chemically attached to "PEI-PEG polymer-graphene oxide" by means of a matrix metalloproteinase 2 (MMP2)-cleavable peptide linker. Matrix metalloproteinases are a family of enzymes largely secreted by cancer cells. Normally, the intrinsic fluorescence property of doxorubicin is quenched by graphene oxide. On reaching cancer cells, doxorubicin-loaded graphenes are acted upon by MMP2 enzymes, and then the peptide linker is cleaved and releases doxorubicin precisely at tumor cells. The unbound doxorubicin emits fluorescence for tumor cell imaging, thus making a versatile dual purpose theranostic system (Qin et al. 2014).

2.5 Inorganic Nanoparticles as Drug Carriers

Inorganic nanocarriers consist of two parts: core and shell. The core is generally made of inorganic component like gold, quantum dots, silica, iron oxide, etc., and the shell is composed mostly of organic polymers (or metals) to which biomacromolecules can be attached.

2.5.1 Gold Nanoparticles (Au NPs)

Gold nanoparticles are metallic nanostructures having size ranges from 1 to 8 μm and exist in diverse shapes such as nanosphere, nanorod, nanocage, and nanoshells (Fig. 2.10). AuNPs are nontoxic drug carriers. Optical and electrical characteristics of AuNPs are dictated by its shapes and sizes. A combined resonance oscillation of electrons on the surface of gold nanoparticle stimulated by incident light at given wavelength is called surface plasmon resonance (SPR) effect. SPR effect converts light energy to heat energy. AuNPs are attractive nano drug carriers because of the presence of the surface plasmon shapes and sizes, which allow them to convert the incident light to heat and spread the generated heat to destroy the tumor cells (Sreejivungsa et al. 2016).

AuNPs generally use three main pathways for cellular internalization which includes receptor-mediated endocytosis, phagocytosis, and fluid-phase endocytosis (Nalawade et al. 2012). Gold NPs for drug delivery are depicted in Fig. 2.11.

Many AuNP-based stimulus-responsive smart drug delivery systems have been developed. Drugs are linked to AuNP surfaces by ionic or covalent bonding or physical absorption. The controlled drug release at target site can be achieved through internal biological stimulus (pH or glutathione) or applied external stimulus (light) (Kong et al. 2017). The controlled discharge of therapeutic agent at target cells is based on the strength of conjugation between drug and AuNPs and methods of drug discharge inside the cells. Weak non-covalent interactions are used for binding hydrophobic drugs, which does not need any more alteration in order to be released. Some prodrug compounds are linked to AuNPs by strong covalent interactions requiring exposure to internal or external stimulus for the release. The fine-tunable

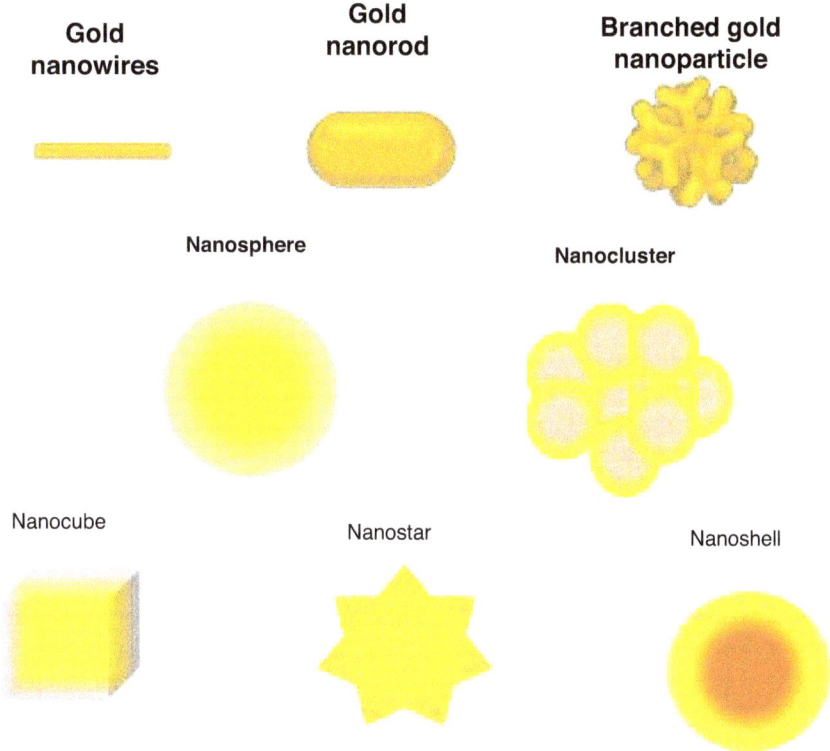

Fig. 2.10 Morphology of synthesized gold NPs

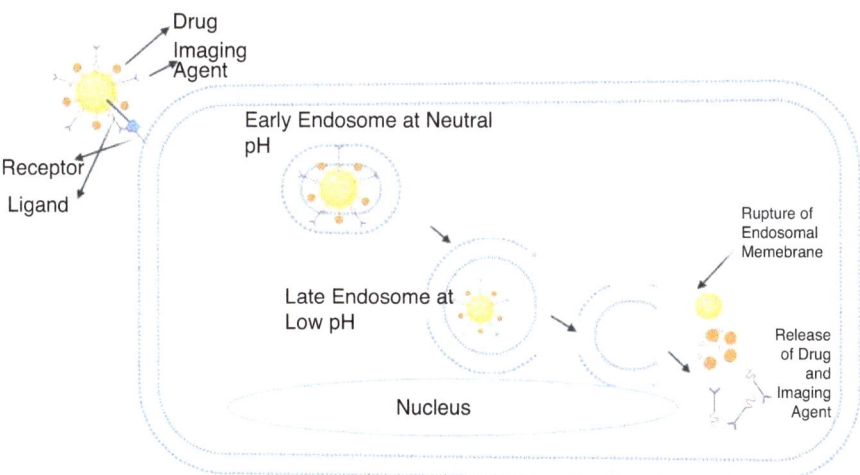

Fig. 2.11 Theragnostic gold NP-based systems for tumor treatment and imaging

optical properties of gold nanoparticle surface are utilized for the release of drug by applying internal or external stimuli.

Photo-regulated release of the prodrug depends on the absorption of light by AuNP-prodrug complex which leads to photo-cleavage of AuNP-prodrug complex with simultaneous release and activation of prodrug (Li et al. 2013). You et al. 2010 administered near IR light on novel AuNPs and achieved targeted controlled drug release by SPR effect. AuNano cages coated with thermosensitive polymers were designed to carry drugs, and by applying near IR light, controlled drug release was attained (Gou et al. 2010). In another approach, pH-dependent controlled release of doxorubicin (DOX) was studied. Doxorubicin was linked to 30 nm gold nanoparticles via a pH-sensitive linker. This kind of DOX-AuNP conjugation permits the intracellular controlled discharge of doxorubicin from gold nanoparticles once AuNPs accumulated inside acidic cell organelles. This approach gave better therapeutic benefits in drug-resistant tumor cells because of rapid intracellular accumulation of doxorubicin (Wang et al. 2011).

Another internal stimulus for controlled drug release is the higher intracellular concentration of glutathione. It is a nonenzymatic approach adopted for the effective delivery of therapeutic agents into target cells. The basic concept involved is the existence of difference between intracellular glutathione concentrations (high) compared with intercellular glutathione concentrations (low). Drug delivery nanoparticles designed with disulfide linkages can sense difference in glutathione concentrations and selectively release the drug molecule inside the target cell (cancer cell) where the intracellular concentration of glutathione is high. In another study, smart theranostic functionalized AuNPs were designed, i.e., the DOX was linked to gold nanoparticles via Au-S bond by using an octapeptide which can be selectively cleaved by overexpressed protease enzyme present in tumor tissues. Once the bond is cleaved, DOX can be precisely released at the tumor site, and thus normal cells are not exposed to the drug. The animal studies proved that after injection of this kind of smart theranostic AuNPs to the tumor mice, the overexpressed protease in tumor tissue and high intracellular glutathione concentration have led to the fast liberation of doxorubicin from the gold nanoparticles. This showed not only inhibition of growth of tumor but also simultaneous fluorescence imaging of tumor (Chen et al. 2013).

2.5.2 Nanoshells

The size of nanoshells ranges from 100 to 200 nm and finds wide applications in theranosis. Two types of nanoshells are commonly used: metallic nanoshells and E-LbL nanoshells. Metallic nanoshells contain silica core and a thin metallic shell (e.g., gold). Nanoshells with silica core and gold shell are particularly used in whole-blood immune analysis. For immune analysis purpose, nanoshells are linked with specific antibodies that function as recognition molecules for specific target compounds (Hirsch et al. 2003a). Silica-gold nanoshells of size 120 nm are conjugated with antibodies or peptides on surface used to kill cancer cells. The antibodies on the nanoshell are bound to the cancer cell receptor. The location of the tumor is then exposed to infrared laser radiations, which heat up the gold shell adequately, and the tumor cells are selectively killed by the generated heat (Hirsch et al. 2003b). E-LbL nanoshells are synthesized by electrostatic layer-by-layer molecular self-assembly method. Nanoshells with silica core encapsulated by E-LbL method with electrolyte compounds like gelatine B or carboxymethylcellulose were prepared. This nanoshell was loaded with the peptide drug, Phor21-bCG (ala). The drug release profile and its efficiency to kill breast tumor cells were found to be improved in vitro (Hirsch et al. 2003a). The drug release profile and its efficiency to kill breast tumor cells were found to be improved in vitro (Hirsch et al. 2003a).

Nanoshells are also implanted in a polymeric hydrogel to develop nanoshell-hydrogel composite material. This composite nanodevice is designed for pulsatile drug release at the target tissues when stimulated with external stimulus. For example, gold nanoshells are conjugated into temperature-responsive hydrogels to prepare a novel kind of drug delivery device that disintegrates on exposure to laser radiation (Sershen et al. 2000). Upon irradiation with laser light, SPR effect takes place which results in photo-to-heat energy transition in embedded gold nanoshells. The liberated heat then shrinks the volume of the hydrogel and releases the drug at target site. Thus, the nanodevice is collapsed by external remote stimulus (laser light). This mechanism is utilized for efficient pulsatile drug release by nanoshell-hydrogel composite material. If the total quantity of drug load is not liberated during the first irradiation cycle, successive burst release of the drug can be obtained by applying next round of irradiations. Once the laser irradiation is terminated, the drug ejection is carried out by diffusion, and the quantity of discharged drug is lesser than that caused by laser radiation. Also, once the laser light is switched off, the hydrogel will begin to enlarge again and resume to its equilibrium state. Next, irradiation given at this moment will cause the hydrogel to disrupt again, resulting once again in burst release of the "drug" moiety. This type of pulsatile discharge of a therapeutic agent is absolutely used in insulin therapy (De Villiers and Lvov 2007).

2.5.3 Quantum Dots (QDs)

QDs are tiny semiconductor crystal particles of size less than 10 nm, made with atoms of group II and VI of periodic table (CdS, CdTe, and ZnS) that can be stimulated to emit different colors of fluorescence upon irradiated with light. The biomedical applications include drug delivery and cellular imaging. Mostly QDs consist of three regions: very tiny central core of diameter from 2 to10 nm made of semiconductor compounds (e.g., CdSe), enclosed by another semiconductor compound (ZnS); next, this double-layer structure is encapsulated by a cap made up of various kinds of compounds (Fig. 2.12). QDs made with CdSe core and ZnS shell are the most common nanoplatform used for biomedical applications. Due to

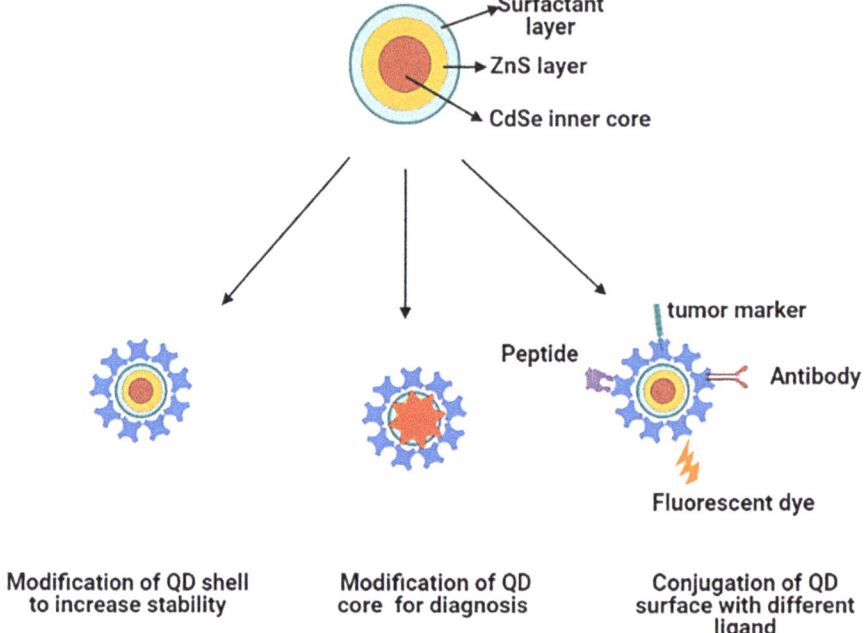

Fig. 2.12 Designs of QD nanocarriers for nanomedicine applications

their nanosize and quantum effects, they elicit peculiar optical (photophysical) characteristics that permit visualizing and monitoring the malignant cells in real time, during the drug transport and drug delivery at the target tissues (Matea et al. 2017).

QDs are labeled with biocompounds and used as extremely sensitive probes. QDs are used for imaging of sentinel node in cancer patients for diagnosis of stage of tumor and scheduling of therapy. CdSe-based QDs can be used for detection of malignant tissues because when irradiated with UV rays, they glow. This phenomenon helps in selective surgical removal of tumor. The surgeon administers these QDs into cancerous tumors and can visualize the glowing tumor; thus, the tumor can be precisely removed with ease, for example, used in the diagnosis and therapy of malignant breast tumors.

Another application of quantum dot is in the diagnosis of viral infection. Fast and accurate diagnosis of respiratory syncytial virus is essential for control of infection and development of antiviral agents. QD nanoparticles conjugated with antibodies are used for rapid and sensitive detection of this virus and also quantify the relative level of expression of viral surface protein. Many controlled drug release models have been developed with QDs. The intact QD-drug complex is transported to the target organs or tissues, and the therapeutic agents are either liberated when the polymeric particle is collapsed when it encounters a low pH or just diffuse out from polymeric particle. For instance, Bagalkot et al. (2007)

engineered and experimented a new QD-aptamer(Apt)-doxorubicin (Dox) conjugate (QD-Apt(Dox)) as a theranostic tool for prostate tumor imaging and treatment. The ideology is that the QD was functionalized with an RNA aptamer (A10 PSMA aptamer) which is then followed by intercalation of the drug DOX. This intercalation quenches the fluorescence emission from both QD and DOX via a fluorescence resonance energy transfer (FRET) mechanism. FRET usually monitors the transfer of energy between two light-sensitive molecules, in this case between QD and doxorubicin. The target site for the aptamer is cell surface domain of the prostate-specific antigen (PSA). Selective engulfment of QD-Apt(Dox) complex into prostate tumor cells takes place by endocytosis process. The release of doxorubicin from the QD-Apt(Dox) complex takes place due to low pH in cancer cells, and once released, DOX and QD both regain fluorescence which is then monitored by FRET mechanism. This system enables sensing, imaging, and killing of prostate cancer cells (Fig. 2.13).

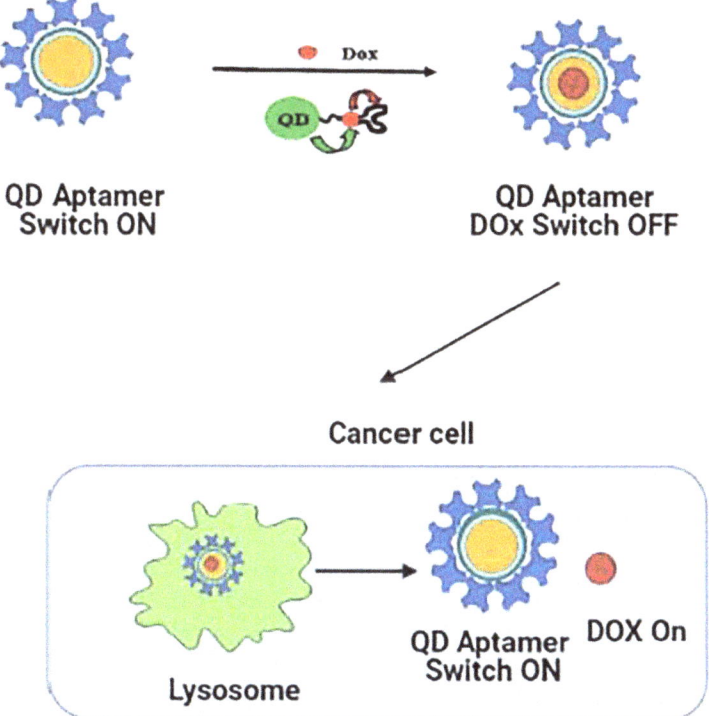

Fig. 2.13 Drug delivery by QD-Apt(Dox) FRET system

2.5.4 Superparamagnetic Iron-Oxide Nanoparticles (SPIONs)

Magnetic particles of dimension 10–20 nm exhibit a very high level of the magnetization up to saturation, and this phenomenon is known as a super paramagnetism effect (Wahajuddin and Arora 2012). SPIONs, like magnetite (Fe_3O_4) and maghemite (Fe_2O_3), are successfully proven for targeted drug release by applying magnetic energy (external stimulus). Modifying SPION surface with different ligand molecules like proteins, antibodies, peptides, and anticancer agents permits specific binding to their target receptors that are expressed only on tumor surface (Watermann and Brieger 2017). For example, SPIONs conjugated with polymers or lipids have been proven to induce a controlled drug delivery by applying external magnetic energy (Alonso et al. 2016).

A novel 100-nm-long nanoparticle chain was prepared by chemical conjugation of three magnetic iron-oxide nanospheres. One of the spheres was attached with DOX-loaded liposome. Once the nanochain enters into the malignant cells, magnetic nanospheres were vibrated by the applying radiofrequency field which leads to the burst release of drug from liposome. Then the free form of the drug spreads all over the malignant tissues. Thus, the drug release is controlled by applying field of radiofrequency as external stimulus (Peiris et al. 2012).

2.5.5 Mesoporous Silica Nanoparticles (MSNs)

Silica (SiO2)-based nanomaterials find wide applications in medicine due to their undemanding synthesis techniques and porous architecture attributes. MSNs are considered as perfect nanoparticles for drug carrier applications due to their firm architecture, adjustable pore size and volume, larger surface area, definite surface characters, and excellent biocompatibility (Tan et al. 2016). Comparatively a sufficient quantity of drug can be incorporated into MSNs, thus aiding in drug accumulation at target tissues by passive targeting. Hence, MSNs are well suited for theranosis (Angelova et al. 2015).

A variety of MSN-based systems have been designed for stimulus-responsive controlled drug release. The stimuli generally used for triggered drug release include physical (temperature, light, pH, magnetic, electrical, and mechanical), biological (enzymes), or chemical (chemical reactions) (Hu et al. 2016). The stimulus-responsive controlled delivery of drugs from MSNs is effort by smart capping agents known as gatekeepers that modify the surface of the mesopore. A wide variety of gatekeepers in use are polymers, proteins, supramolecular assemblies, and inorganic nanoparticles. Upon exposure to stimulus (pH, temperature, light, and redox), the smart capping agent and the gatekeeper material are displaced, which leads to the discharge of therapeutic agents from the mesopores (Tan et al. 2016). This kind of release ensures "Nil premature delivery" and drug release is under the control of specified stimulus only. Moreover, the framework of silica in MSN can readily dissolve to silicic acid, under physiological conditions, which is a nontoxic compound (Croissant et al. 2017).

Sun et al. (2018) studied targeted drug delivery and cytotoxicity of a new core-shell-type nanoparticle (CSNP) on MCF-7 human breast tumor cell lines. This core-shell NP has been developed to deliver doxorubicin (drug) and indocyanine green (photosensitizer compound) simultaneously to cancer tissues. Irradiation of core-shell NP with near IR rays stimulates photothermal conversion effect. This system elicits a collective chemo- and photothermal therapy for cancer treatment. In another system, electrostatically self-joined core-shell NPs have been synthesized by amino-functionalized MSNs (MSN-NH2) as the positively charged internal core and DSPE-PEG2000-COOH and DSPE-PEG2000-FA modified phosphatidyl choline as the negatively charged external shell. This self-assembled core-shell NPs upon irradiation with near IR stimulus at 808 nm evidence the controlled drug release under in vitro condition on breast cancer MCF-7 cell lines.

A novel kind of redox-sensitive controlled drug delivery nanodevice with mesoporous carbon nanoparticles (MCNs) with custom-made fluorescent carbon dots (CDs) as gatekeeper was designed (Zhang et al. 2016). The mesoporous carbon nanoparticles were modified with a disulfide group. This makes the nanodevice responsive to higher intracellular glutathione concentration. The fluorescent carbon dots are attached to outside of the mesoporous carbon nanoparticles through an electrostatic attraction, act as gatekeeper or cap, and close the mesopores. This capping of mesopore prevents the outflow of doxorubicin which is filled inside the mesopore channel. When this nanodevice is exposed to high concentration of glutathione at the biological conditions, the integrity of the nanodevice is collapsed by the cleavage of the disulfide linkage; at the same time, stripping the carbon dots to open the gate thus facilitates the fast discharge of the encapsulated anticancer drug. The fluorescence of CDs is quenched (switched off) when attached to the surface of mesoporous carbon NPs, and it restores fluorescence (switched on) when disconnected from the surface of the mesoporous carbon NPs. Thus, the fluorescent carbon dots serve as both a gatekeeper to ensure controlled drug discharge and a fluorescent probe for the monitoring of the drug release profile. Thus, by combining therapy and imaging, this kind of drug delivery nanodevice can be hopefully used for controlled drug delivery under the control of in situ stimulus in the cells.

A thiolated hydrophobic surface of mesoporous NP was made by treating with octadecanethiol via disulfide bond. On this modified mesoporous NP, Pluronic P123 was coated via hydrophobic interactions. Pluronic 123 is a triblock copolymer made with hydrophilic poly (ethylene oxide) blocks and lipophilic poly (propylene oxide) blocks. Thus, a nano assembly consisting of P123 and octadecyl group-modified mesoporous NP was constructed. Before coating mesoporous NP, the drug was loaded into the mesopores. The disulfide linkages function as "gatekeeper control switch" to confer the redox stimuli-sensitive drug release system. Almost all cancer cells have 100–1000 times greater intracellular glutathione concentration than extracellular concentration. When this nano assembly is entered into the malignant cells, disulfide linkage can be broken by excess glutathione concentration inside tumor cells, and then the hybrid coating would collapse opening the pores and fast release of therapeutic agents at the specific target cell. The benefit of this kind of drug carrier systems is that they could inhibit tumor growth and at the same time suppress tumor metastasis (Sha et al. 2018). Neetu et al. (2011) designed and experimented a polymeric mesoporous NP to deliver doxorubicin (DOX). The controlled drug release from these nanoparticles is stimulated by specific proteases present in cancer tissues.

2.6 Techniques for Drug Loading to Nanoparticles

2.6.1 Drug Loading to Polymeric Nanoparticles

Nanoencapsulation/entrapment techniques are widely adopted for the synthesis of drug-incorporated polymeric NPs of size 1–1000 nm. Based on drug loading, nanoparticles are categorized into two types, viz., nanocapsules and nanospheres. Nanocapsules are vesicular type in which the therapeutic agent is limited to a cavity containing an internal liquid core enclosed by a polymeric membrane, but sometimes the drug may also be adsorbed to the outer surface of capsule. Nanospheres are provided with matrix-type architecture. Therapeutic compounds may be either adsorbed on the surface of sphere or embedded in the polymeric matrix (Fig. 2.14).

Different kinds of techniques have been devised to synthesize drug-loaded nanoparticles; these techniques are categorized into two classes based on if the nanoparticle synthesis needs a polymerization reaction or synthesis is carried out with readymade or pre-synthesized polymers. Further, polymerization techniques

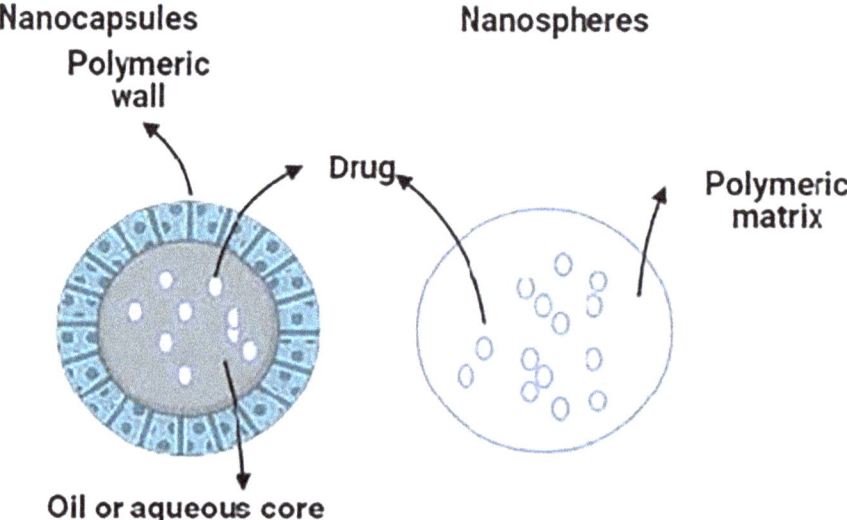

Fig. 2.14 Drug loading to polymeric nanoparticles

can be categorized into two: emulsion and interfacial polymerization, and the emulsion polymerization is further classified in to two types – organic and aqueous – based on the type of continuous phase used. Nanoparticles are synthesized directly from synthetic or biopolymers and by desolation of polymer (Couvreur et al. 1995).

2.6.1.1 Nanoparticles Synthesized by Polymerization Method

In this method, either organic solvent or water can be used as continuous phase for the synthesis of polymeric nanoparticles. The outline of *continuous organic phase* technique is shown in Fig. 2.15 (Lowe and Temple 1994; Harmia-Pulkkinen et al. 1989).

The outline of continuous aqueous phase technique is shown in Fig. 2.16 (Kreuter et al. 1979). Polymethylmethacrylate (PMMA)-based polymeric nanospheres can be used as proper adjuvants for immunizing agents, and also the synthesis of this nanoparticle is easy. The monomeric MMA are polymerized by radical emulsion polymerization method usually in the absence of emulsifiers to obtain PMMA nanospheres (Kreuter et al. 1979). A variety of therapeutic agents and imaging compounds can be efficiently entrapped into PMMA nanospheres (e.g., DOX) (Rolland et al. 1986).

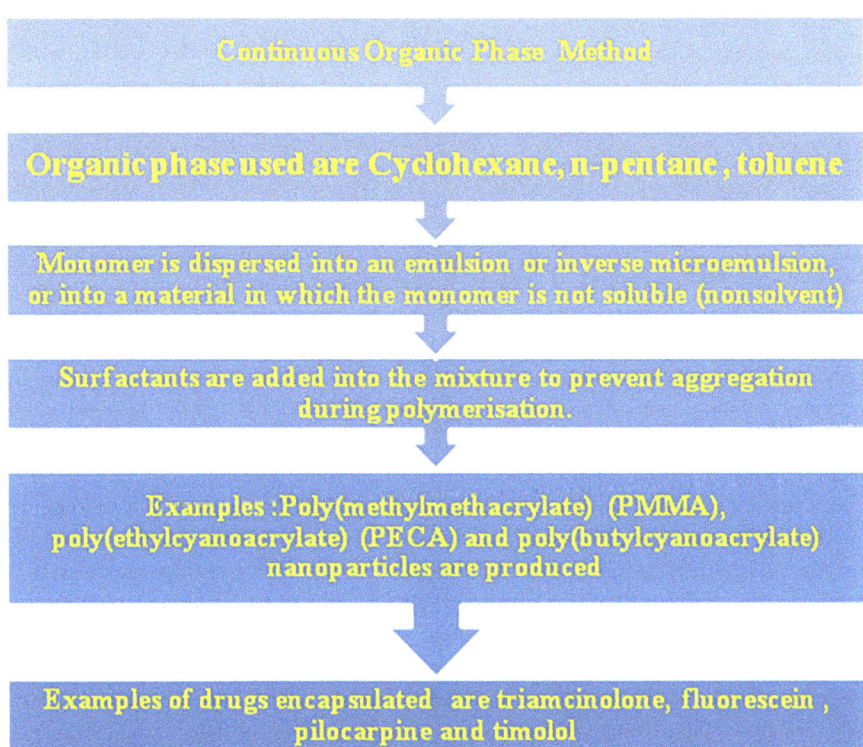

Fig. 2.15 Emulsion polymerization-continuous organic phase technique

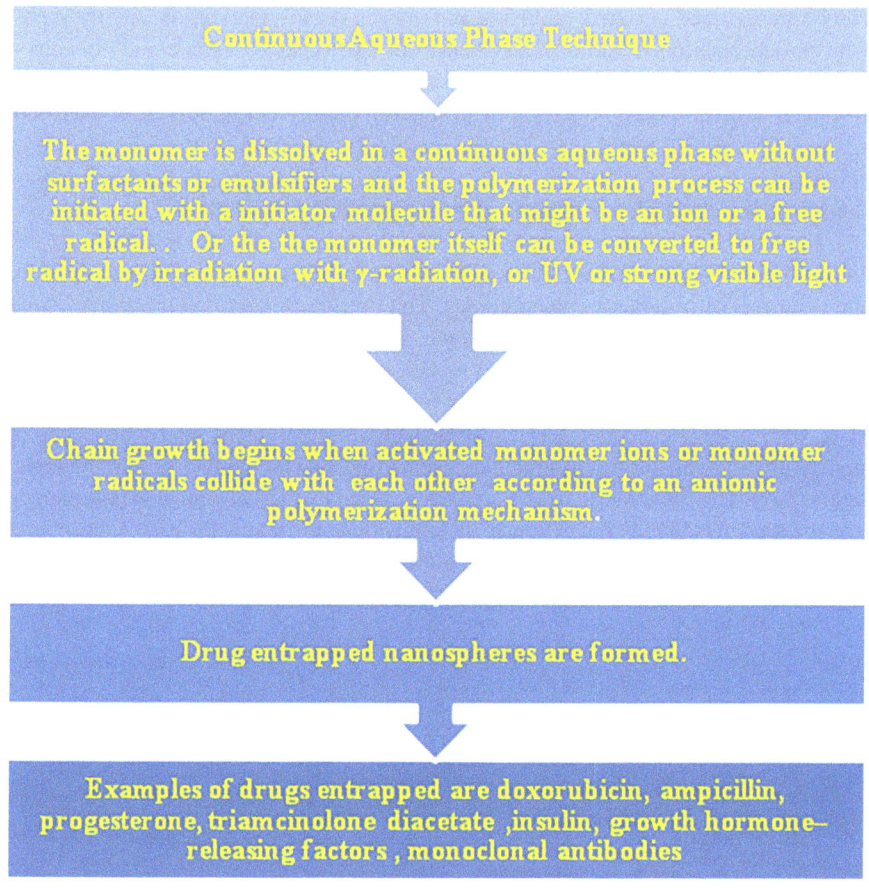

Fig. 2.16 Emulsion polymerization-continuous aqueous phase technique

Synthesis of poly (alkylcyanoacrylate) (PACA) nanoparticles has the advantage that polymerization can be initiated by anionic process at room temperature, and these polymers are highly biodegradable. Hydrophilic drugs, for example, ampicillin and doxorubicin, are encapsulated with good efficiency (Seijo et al. 1990). Sparingly water-soluble drugs, e.g., progesterone, triamcinolone diacetate, are successfully entrapped into PACA nanoparticles by dissolving the drug in a suitable solvent or surface-acting agent prior to mixing to the aqueous polymerization medium. Insulin, (Michel et al. 1991), GHRF, and monoclonal Abs have been entrapped into PACA nanospheres successfully (Grangier et al. 1991; Kubiak et al. 1988).

By interfacial polymerization technique, poly (alkylcyanoacrylate), poly (ethylcyanoacrylate) (PECA), poly(isobutylcyanoacrylate), and poly (isohexylcyanoacrylate) nanoparticles are successfully produced. The protocol for the synthesis of poly (alkylcyanoacrylate) nanoparticle is shown in Fig. 2.17 (Ammoury et al. 1991). The drug encapsulation efficiency obtained by this technique is significantly high, for example, 95% for insulin (Couvreur et al. 2002).

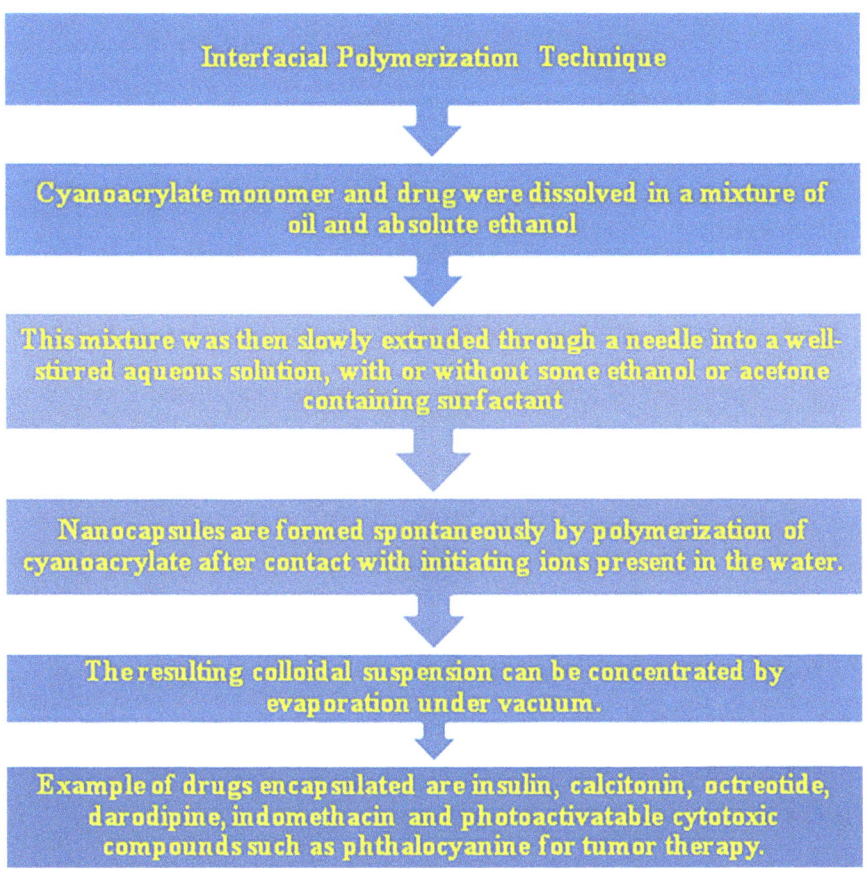

Fig. 2.17 Interfacial polymerization technique – synthesis of poly (alkylcyanoacrylate) nanoparticles

By interfacial polycondensation method, polymeric nanoparticles of size smaller than 500 nm can be synthesized. The nanoparticles were prepared by polycondensation of hydrophobic (phthaloyldichloride) and hydrophilic (diethylenetriamine) monomers with or without surface-acting agents. Urethane and ether urethane monomers were subjected to interfacial polycondensation to prepare nanocapsules to encapsulate α-tocopherol (Bouchemal et al. 2004).

2.6.1.2 Nanoparticles Synthesized from Synthetic Polymers

Emulsification/solvent evaporation technique takes place in two stages. In initial stage, the polymeric solution is emulsified into an aqueous phase. In the next stage, polymer solvent is evaporated which stimulates the precipitation of polymeric nanospheres (Figs. 2.18 and 2.19). Liposoluble drugs alone can be entrapped

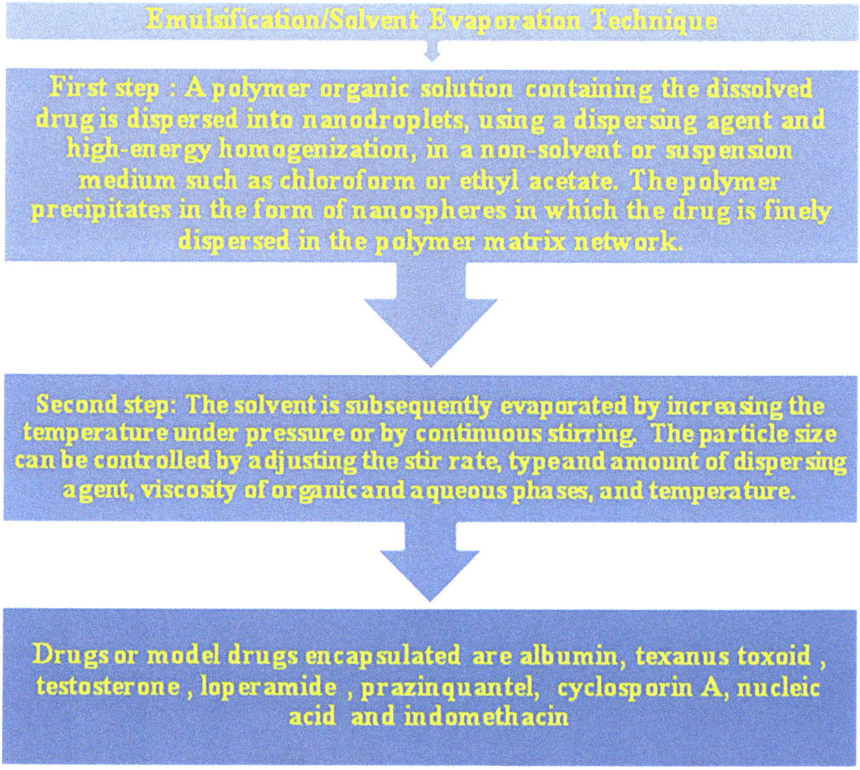

Emulsification/Solvent Evaporation Technique
First step : A polymer organic solution containing the dissolved drug is dispersed into nanodroplets, using a dispersing agent and high-energy homogenization, in a non-solvent or suspension medium such as chloroform or ethyl acetate. The polymer precipitates in the form of nanospheres in which the drug is finely dispersed in the polymer matrix network.
Second step: The solvent is subsequently evaporated by increasing the temperature under pressure or by continuous stirring. The particle size can be controlled by adjusting the stir rate, type and amount of dispersing agent, viscosity of organic and aqueous phases, and temperature.
Drugs or model drugs encapsulated are albumin, texanus toxoid , testosterone , loperamide , prazinquantel, cyclosporin A, nucleic acid and indomethacin

Fig. 2.18 Emulsification/solvent evaporation method

Organic Solution :
Polymer + Drug in aqueous non miscible solvent
Aqueous Solution Stabilized in Water
Step I
Solvent Evaporation
Step II

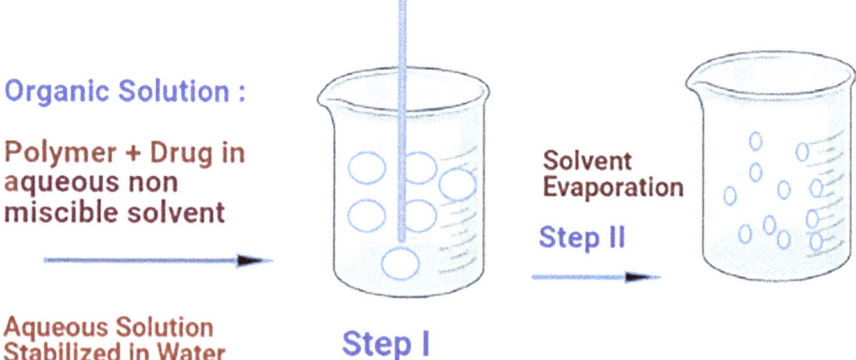

Fig. 2.19 Emulsification-evaporation method

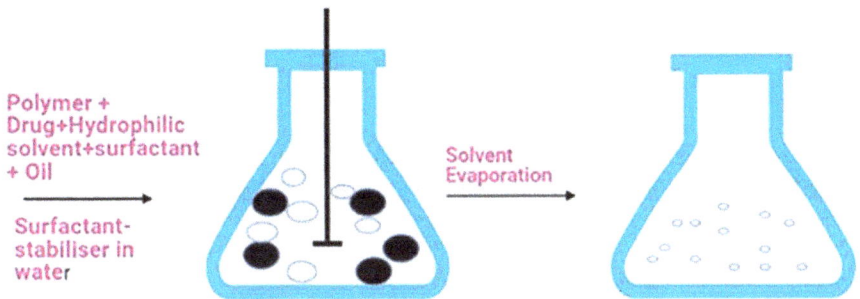

Fig. 2.20 SD technique

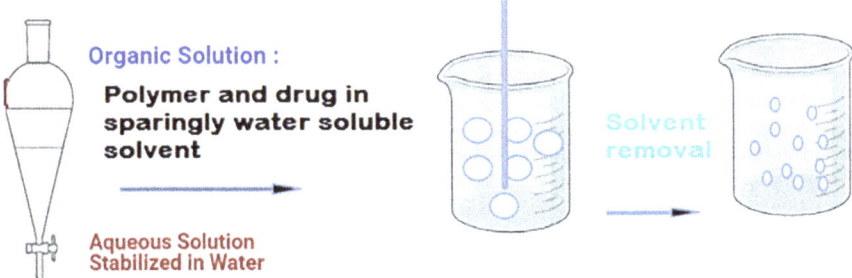

Fig. 2.21 ESD method

into nanoparticles by this method. Another disadvantage is the high energy requirement in homogenization. Often used polymers are poly lactic acid, poly(lactic-co-glycolic acid), poly(E-caprolactone), and poly(h-hydroxybutyrate) (Pinto Reis et al. 2006).

Solvent displacement (SD) and interfacial deposition (ID) methods both have nearly same principle. Both drug loaded nanospheres and nanocapsules are formed by solvent displacement method (Fig. 2.20) but by interfacial deposition method nanocapsules are only formed. SD precipitation of polymer in organic solution takes place in the presence or absence of surfactants (Ganachaud and Katz 2005). This technique is essentially adopted for loading of hydrophobic drugs into nanoparticles because of the use of solvents miscible with aqueous phase. Even though drug entrapment efficiency is high, this method is inefficient to encapsulate hydrophilic drugs (Barichello et al. 1999). Nanoparticles prepared with amphiphilic cyclodextrins are suitable for the parenteral injection of the sparingly soluble fungicidal agents like bifonazole and clotrimazole (Memisoglu et al. 2003). ID is an emulsification/solidification technique resulting in the synthesis of nanocapsules. A compound of oily nature, miscible with the solvent of the polymer but not miscible with the mixture is used. The polymer precipitates at the interface between the finely dispersed oil drops and the aqueous phase, resulting in the synthesis of nanocapsules of size 230 nm (Couvreur et al. 1995).

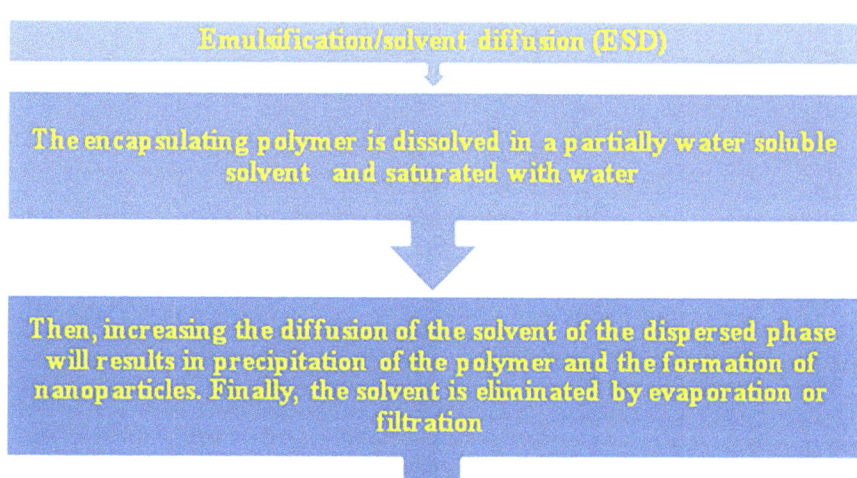

Fig. 2.22 ESD method

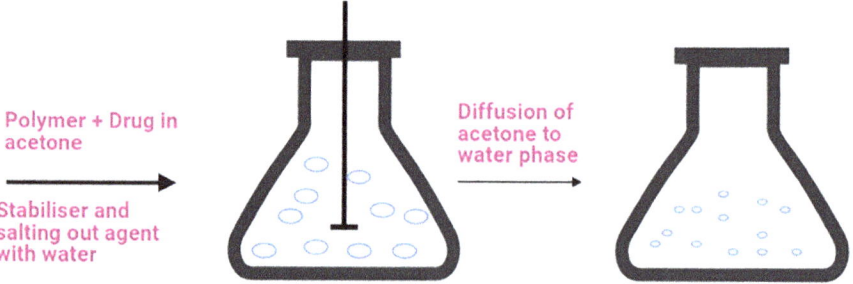

Fig. 2.23 Salting-out method

The protocol for *emulsification/solvent diffusion (ESD) technique* is illustrated in Figs. 2.21 and 2.22. This method is suitable to encapsulate lipophilic drugs (Pinto Reis et al. 2006).

The salting-out technique is similar to emulsification method with few modifications. Polymers along with drugs are first dissolved in a solvent (acetone), which is quickly emulsified into water-based gel consisting of a salting-out agent and a colloidal stabilizer. The salting-out agent may be an electrolyte or nonelectrolyte material. Electrolytes generally used are $MgCl_2$, $CaCl_2$, and $Mg(CH_3COO)_2$, and nonelectrolyte used is sucrose. The chief stabilizers used are PVP (polyvinyl pyrrolidine) or HEC (hydroxyethylcellulose). The emulsion (oil/water) is diluted with an adequate quantity of water to increase the distribution of acetone into the aqueous layer which stimulates the synthesis of nanospheres (Fig. 2.23). This technique is useful for the loading of lipophilic drugs (Quintanar-Guerrero et al. 1998).

2.6.1.3 Nanoparticles Synthesized from Natural Polymer

Albumin nanospheres can be synthesized by homogenizing the oil with albumin droplets and stabilized by heat at 175–180 °C for 10 minutes. It was then treated with ethyl ether to decrease the viscosity of the oil phase, which was then separated by centrifugal force. The thermal-stable drug molecules only can be entrapped by the heat treatment of albumin. This disadvantage is obviated by emulsifying albumin in cottonseed oil at 25 °C, subsequently denaturing the albumin protein by suspending it again in ether along with the cross-linking agents 2,3-butadiene. Centrifugal force was applied to separate, followed by lyophilization to get dried particles. The drug release profile of doxorubicin was rapid than particles synthesized by thermal treatment (Patil 2003).

The gelatin nanoparticles are synthesized as follows: Gelatin protein was emulsified to get gelatin droplets. Then they were hardened by refrigeration which resulted in gelation. By filtration, gelatine droplets were recovered and then cross-linked with formaldehyde. Thus, gelatine nanoparticles of dimension from 100 to 600 nm with an average of 280 nm were synthesized. This is well suited for the entrapment of heat-sensitive drugs into gelatine nanoparticles (Yoshioka et al. 1981). Similarly, gliadin nanoparticles were synthesized from vegetable protein gliadins from wheat gluten. This is used to competently encapsulate lipophilic drugs, for example, alpha-tocopherol (Duclairoir et al. 2002).

Alginate nanoparticles were synthesized as follows: Sodium alginate, a natural hydrophilic polymer, can be converted to gel form by treating with multivalent cations like calcium. Alginate nanoparticles of size 1 to 5 µm are generally prepared by dropwise expulsion of sodium alginate into $CaCl_2$ solution by using an air atomizer (Fig. 2.24) (Reis et al. 2005).

Chitosan nanoparticles (CNPs) are suitable nanocarriers to encapsulate many protein drugs like BSA, toxoid vaccines (tetanus and diphtheria), antimutagenic agents insulin and DNA/RNA. Two methods are adapted to produce CNPs. In the first method, CNPs are synthesized by treating chitosan with tripolyphosphate (polyanion) to form a chitosan-polyanion complex which showed a quasi-spherical shape with size ranging from 200 to 500 nm. The second method is based on stimulating gelation in an emulsification-based technique. Gel transition of a chitosan

Fig. 2.24 Synthesis of alginate nanoparticles by emulsification-internal gelation method

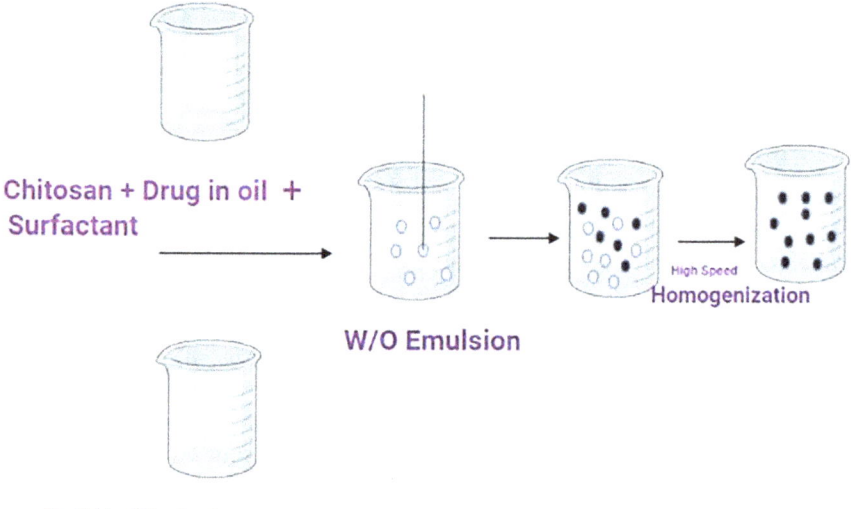

Fig. 2.25 Synthesis of chitosan nanoparticles by the emulsification method

solution dispersed in oil emulsion resulted in the synthesis of CNPs of size around 400 nm (Fig. 2.25) (Vauthier and Couvreur 2000).

Chitosan is a mucoadhesive polymer. So, chitosan-coated nanoparticles interacted much with mucus to extend the duration of drug release at these sites and save the entrapped drugs from enzyme action, thus enhancing transmucosal drug release (Sailaja and Amareshwar 2011). For example, chitosan significantly increases the absorption of insulin across the nasal epithelium, hence used in insulin aerosol therapy (Illum et al. 1994).

Agarose nanoparticles (ANPs) are suitable for therapeutic delivery of proteins and peptides. Agarose in water can be converted to hydrogels when cooled at a temperature less than the gel forming temperature (318–368 °C). Temperature-based gel formation process forms helicoid structures in three-dimensional meshes which

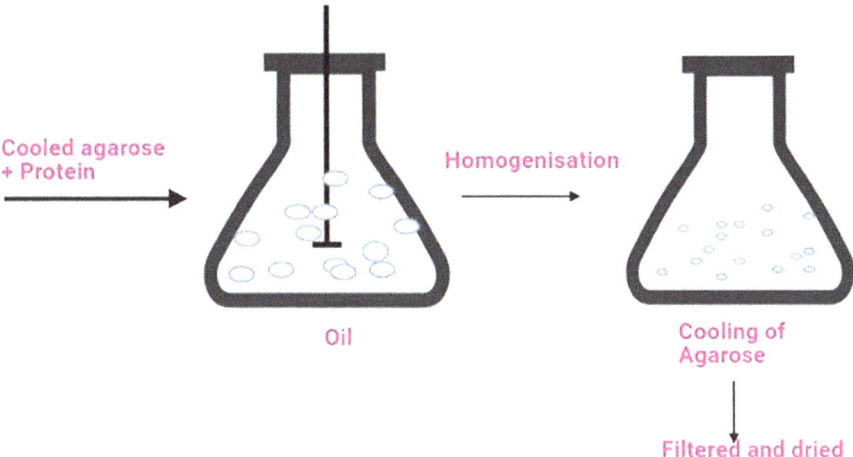

Fig. 2.26 Synthesis of agarose nanoparticles by the emulsification method

can hold large quantity of water. These gels are water loving, inert, and biologically compatible and can form matrix for proteins and peptides to entrap them during synthesis. ANPs formed by an emulsion-based technique are shown in Fig. 2.26. This technique needs the synthesis of agarose solution in corn oil emulsion at 408 °C.

Peptides and proteins are first added to the agarose solution. Dispersed nanodroplets can be obtained by homogenization. The gel formation of agarose is next stimulated by mixing the emulsion in corn oil at low temperature and stirred. The liquid nanodroplets are then gelled and entrap the protein drug. Thus, protein-loaded agarose hydrogel nanoparticles are synthesized (Vauthier and Couvreur 2000).

2.6.2 Drug Loading Techniques to Various Nanostructures

Let us discuss few drug loading techniques used to incorporate drugs into specific nanocarrier structures.

2.6.2.1 Drug Loading to Micelles

Drugs can be encapsulated inside the polymeric micelles by three main methods: 1. direct dissolution, 2. solvent evaporation, and 3. dialysis. By direct dissolution method, the copolymer and the drugs are kept in an aqueous medium and they are combined by self-assembly process to form drug-loaded micelles. In the solvent evaporation method, the copolymer and the chosen drug are kept in a volatile solvent, and the solvent evaporation forms the drug-loaded micelles. In the dialysis method, drug-loaded micelles are formed when both the drug and the copolymer in organic solvent are mixed and placed inside dialysis membrane bag (Mourya et al. 2011).

2.6.2.2 Drug Loading to SLN

Basically three drug loading models are applicable for SLNs which are Homogenous matrix model, Drug enriched shell in core shell model and Drug enriched core in core shell model (Pardeshi et al. 2012) as shown in Fig. 2.5. In the first prototype, the core contains drug as unstructured clusters. It is applicable for incorporating extreme hydrophobic drugs. This kind of drug loading can be achieved either by hot or cold homogenization method.

The hot homogenization process takes place at temperatures higher than the melting temperature of the lipid. By using a high shear mixing equipment, pre-emulsion which has drug-loaded lipid melt and the aqueous emulsifier phase (temperature is maintained in the same level) is prepared. Desired nanosize of SLN can be attained by lowering the viscosity of the lipid. To lower the viscosity of the lipid, a high processing temperature can be used, but high temperature may lead to degradation of both drug and lipid molecules (Lander et al. 2000). Better results are achieved after numerous repetitions (three to five times) through the high-pressure homogenizer.

The colloidal hot oil in aqueous emulsion is prepared after many rounds of homogenization; on cooling, it crystallizes the lipids to nanospheres and thus solid lipid nanoparticles are formed (Fig. 2.27). Degradation, partitioning, and loss of drug into aqueous medium are the disadvantages of hot homogenization technique. To obviate drawbacks, cold homogenization was developed. The initial step is

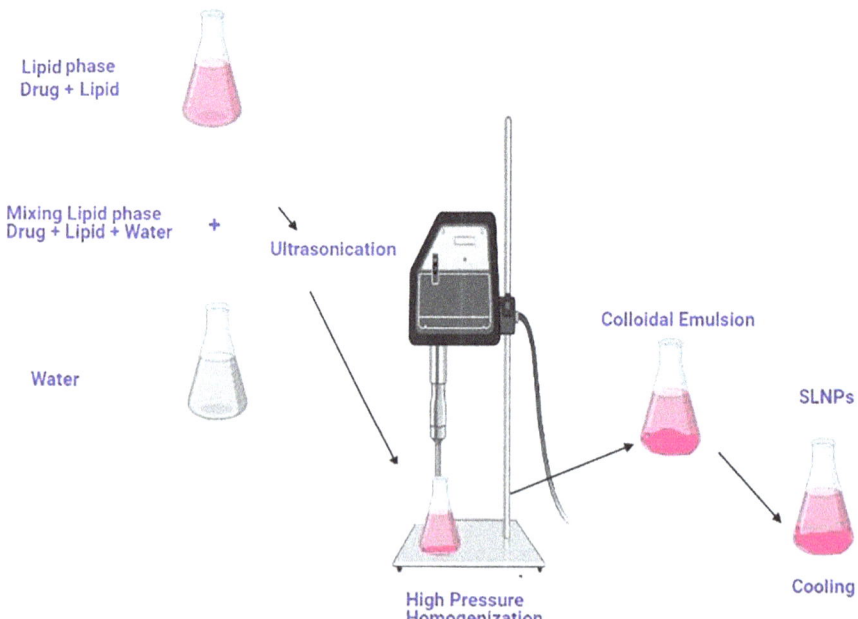

Fig. 2.27 Proposed protocol of hot homogenization technique

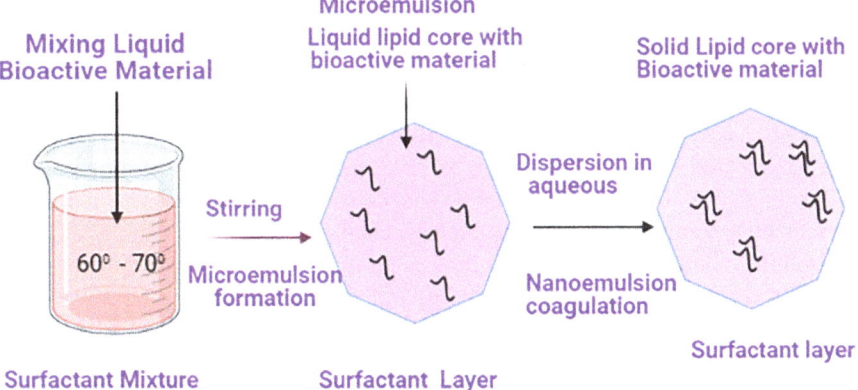

Fig. 2.28 Synthesis of SLN by microemulsion method

related to that of hot homogenization procedure, but further steps are altered. The drug containing melted lipid is cooled quickly (by applying liquid nitrogen) to attain consistent drug distribution. Then, the drug-loaded solid lipid is pulverized to microparticles of size 50–100 microns by ball/mortar milling (Jahnke 1998). In drug enriched shell prototype, the drug is loaded inside the shell, and the lipid core is free of the drug. Phase separation takes place when the preparation is cooled and precipitates resulting in lipid core free of the drug. Simultaneously, the drug repartitions into liquid-lipid phase, thus steadily increasing drug concentration in the outermost shell of the lipid core. A drug enriched core prototype nano emulsion is formed which is accomplished by melting the drug in the lipid at its supersaturating concentration. When this supersaturated melted lipid is cooled, it precipitates the drug before lipid. Additional cooling will precipitate lipid around the drug precipitate, thus acting as a membrane to loaded drugs (Fig. 2.28) (Esposito et al. 2018).

2.6.2.3 Drug Loading to Nanogels

Nanogels are versatile drug delivery vehicles for various drugs from hydrophobic to hydrophilic ones. There are several methods for the encapsulation or attachment of drugs. The commonly used approaches are the covalent conjugation method, direct addition method, and soaking method.

Covalent conjugation method forms a covalent bond between suitable compounds of the drug and nanogel. The most widely used covalent bonds are stimuli-sensitive, thus enabling the liberation of the drug at the place of therapeutic function inside the body. A pH-responsive hydrazone bond between doxorubicin and the methacrylamide polymeric nanogel was used recently for solid tumor drug delivery and initiation of reaction (Chen et al. 2017). Furthermore, biomacromolecules can be also covalently bound to nanogels. For example, enzymes are attached through a two-step reaction. The initial step is the reaction of the enzyme with

N-hydroxysuccinimidoacrylate under mild conditions. This reaction generates double bonds on the surface of enzyme. The second step is in situ polymerization with acrylamide monomer, N,N′-methylenebisacrylamide, as the cross-linking agent and N,N,N′,N′-tetramethylethylenediamine as the initiator of reaction (Yan et al. 2006). Otherwise, polyacrylamide nanogels with incorporated modified α-chymotrypsin can be prepared by copolymerization in an inverse micro polymerization reaction (Khmelnitsky et al. 1992). Nanogels with proteins with covalent bond can enhance their stability at high temperature and plasma half-life.

In direct addition technique, the drug is mixed with the monomer in the water phase of the emulsion before the synthesis of the nanogel. The drug is therefore encapsulated in the nanogel structure during its formation by hydrophobic or electrostatic interaction. Using this procedure, aspirin-containing nanogels were prepared by photoisomerization using a solution of the aspirin salt dispersed in a solution of linear dextran containing N-(6-aminohexyl)-4-[4- hydroxyphenylazo]-benzamide substituent attached via an amide linkage. The primary step in the synthesis was preparation of the hydrophobic substituent, while the second step was the reaction of the substituent with dextran. The nanogel was then formed through non-covalent self-aggregation induced by photoisomerization (Patnaik et al. 2007).

Soaking method is useful in the case of amphiphilic nanogels containing hydrophobic moieties such as cholesterol. The drugs are introduced by dipping the nanogels in a supersaturated solution of the drug. For example, this method was used to synthesize indomethacin-carrying nanogels (Sahiner et al. 2007).

2.6.2.4 Drug Loading to Dendrimers

Dendrimers are synthetic, branched structures with core at the center. The dendrimers can be synthesized by divergent or by convergent methods. In the first method, dendrimers extend outward away from core compound. The components in the core act in response with monomer sharing one reactive group, and two inactive groups give rise to the formation of first-generation dendrimer. On other hand, in the second method, the dendrimers are built from the periphery toward the center. Once the chains are sufficiently large, they bind to the core compound. The second method is advantageous because of the minimal defects in the final dendrimer structure. The degree of branching depends on the synthesis processes. The sizes are controlled while in the process of synthesis of dendrimers. The structure of dendrimers in the solution is based on many factors, like generation, spacer size, ionic strength, surface adjustment, pH, and temperature. Factors influencing drug delivery process are charge effect and electrostatic forces (Ballauff and Likos 2004).

Drugs of low molecular weight are entrapped inside the dendrimers and the drugs are immobilised by hydrophobic interactions or hydrogen or covalent bonds or linked to the functional compounds present on the exterior of dendrimer for a short period. Dendrimers enhance the solubility and bioavailability of lipophilic drug molecules (Lombardo 2009). Dendrimer nanodevices are developed to carry both targeting ligands and imaging molecules. The drug moieties conjugated to the dendrimers for delivery follow two ways: formulation and nano-construction methods. Drugs are physically entrapped into the cavities of dendrimer branches via non-covalent interactions (formulation approach) or by covalent interactions (nanoconstruct approach) (Chauhan 2018).

Poly(amidoamine) PAMAM dendrimers are the most frequently used drug delivery systems. The dendrimer construction has three regions for entrapping the drug by using different binding forces: (i) void space entrapping, (ii) branching point bonding, and (iii) outside surface group interactions. The site where drug is entrapped and the type of binding force depend on the structure of both (dendrimers and drug). A dendrimer is a versatile nano drug delivery device that can perform many functions starting from enhancement of solubility to drug targeting as described below.

Solubility enhancers: PAMAM dendrimers enhance the solubility of entrapped hydrophobic molecules (Svenson and Chauhan 2008). The drug indomethacin was entrapped into G4-NH2 dendrimers by ionic interactions between negatively charged carboxylate groups of indomethacin and positively charged amine groups of dendrimers. These interactions are pH dependent (Chauhan et al. 2004).

Stability Dendrimers enhance the stability of the drug, for example, entrapped resveratrol into dendrimer shows more stability and solubility.

Dissolution Dendrimer-drug composite confirms faster drug dissolution than hydrophobic drug alone (Chauhan et al. 2018).

Drug release Controlled drug release from dendrimers depends on two factors.

(a) Chemical modification of dendrimer: For example, the drug indomethacin loaded on G4-NH2 dendrimer showed slow and controlled drug release when compared to the more rapid discharge noticed with G4-COOH dendrimer. Thus, chemical modification of functional groups determines the strength of interaction of drugs with dendrimers which in turn determines the kinetics of drug release by dendrimers (Chauhan 2018).
(b) Physical loading: The second parameter which plays an important role in the rate of drug release from dendrimers is physical loading which means dendrimer-to-drug molar ratio. For example, the rate of release of cisplatin from dendrimer was directly related to the cisplatin by dendrimer molar ratio (Kulhari et al. 2015).

2.6.2.5 Drug Loading to CNTs

Drug loading efficiency of carbon nanotubes is remarkably high. The higher drug packing efficiency is due to more surface area which offers sufficient room by loading more drugs and imaging molecules, both inside (cavity) and on the surface. CNTs provide easiness in cell uptake and thermal ablation, and these properties make it a unique drug delivery system.

Carbon nanotubes are capped or end-closed structures. So, in order to load the drug into the inner cavity, two approaches are used to fill carbon nanotubes in situ. They are either filling of drug during synthesis of CNTs or post-synthesis of CNTs. Filling in situ while producing CNTs is less efficient than post-synthesis of CNTs. The appropriate method depends on the melting point, surface tension, reactivity, and sensitivity of the drug to be incorporated (Monthioux 2002). Filling of CNTs by post-synthesis requires the ends must be opened by passing electricity, attacking the CNTs with acid or oxidizing by using CO_2 (Tsang et al. 1994; Ajayan et al. 1993). After opening the ends, drug molecules can be filled by two methods: decoration or capillarity. In the less efficient decoration method, functional groups are bonded to the inner wall or outer wall of CNTs (Ebbesen 1996). The most frequent mechanism followed for loading CNTs is capillarity. The drug loading by capillarity is based on two factors, such as the width of the CNT and the surface tension which is around 200mN/m. After filling, they are rinsed using a solution with partial solubility to the impregnating fluid so that the deposits left outside the CNT alone are removed. Then, the CNTs are closed by passing an electricity which closes the ends (Fu et al. 2008). Heister et al. (2012) studied drug filling, dispersion stability, and site directed drug release of cancer drugs with oxidized SWCNTs (oxSWCNTs) and reported that pH of 8 for doxorubicin and pH 9 for mitoxantrone are optimum for binding to oxSWCNTs by non-covalent bond at 4 °C in the dark.

2.6.2.6 Drug Loading to Fullerenes

Fullerenes can be produced when gaseous carbon is condensed in the presence helium gas. First, the vaporized carbon is produced when a strong beam of laser light is irradiated on the carbon surface. The liberated vaporized carbon is admixed with stream of helium gas, and the carbon atoms are joined to form clusters which are composed of few atoms to hundreds of atoms. When this condensed carbon is passed into a vacuum chamber, it expands and encountered a low temperature treatment just above absolute zero degrees. Also, by this technique, some drug molecules can be entrapped inside fullerenes (Szoka 1980).

2.6.2.7 Drug Loading to Gold Nanoparticles (AuNPs)

The surface modification of gold NPs can be carried out very easily, and this feature makes its use as versatile drug delivery system. After synthesis, the gold NPs are subjected to surface modification by using stabilizing agents; this imparts a net surface charge on gold NPs. The negative charge on gold core nanoparticles is utilized to conjugate a variety of small or large drug agents (antibiotics, proteins, nucleic acids targeting ligands) easily on gold NPs through physical absorption and ionic or covalent interactions.

Ligands binding to gold NPs through covalent bond enhance the drug stability and protect the drug from extreme conditions (e.g., elevated ionic strength, high serum concentrations); otherwise, it would lead to undesirable clumping and insolubility. Generally covalent binding allows for easy binding to a range of ligands like biomolecules and biopolymeric compounds through thiol, amine, and carboxylate groups to gold NPs.

The widely used covalent bond and attaching group is a thiol linker. The gold-thiol bond is strong (45 kcal/mol). Generally gold NPs are synthesized by colloidal synthesis method. In this method, $AuCL_4$ salts are reduced by treating with sodium tetrahydroborate, in the presence of thiol group donating compounds which form a layer around the core gold atom and depend on gold to thiol ratio (Kong et al. 2017). A maximum drug load (up to 60%) of doxorubicin to gold NPs was attained since the drug forms a thin layer both on the internal and the external surfaces of the gold NPs through electrostatic interaction (Dreaden et al. 2012). Modified polymers and biomolecules contain a-sulfhydryl group which improves their conjugation on the surface of the AuNp.

Poly(ethylene glycol) (PEG) conjugated to 30 nm AuNPs via sulfhydryl group was found to impart improved stability of NPs inside the body and allows further conjugation of a NLS peptide (importins) with thiol groups of cysteine on the peptide (Kang et al. 2010). In addition, linking molecules are modified to possess a thiol group at one end, by which it can attach itself to gold surface and either amino or carboxylic group on the other end. Thus, the amino or carboxylic group allows drug to bind to the gold NPs.

McIntosh et al. (2001) prepared gold NPs coated with a cationic stabilizing agent, which binds with anionic phosphate groups of DNAs to the surface of the nanoparticle through non-covalent interactions, for gene therapy purpose. Anionic, citrate-stabilized Au nanospheres were prepared and conjugated efficiently with antibodies for prostate-specific antigen via electrostatic interactions. A 20 nm increase in size of Au nanosphere was observed after joining with antibody. Thus, the binding of nanosphere with antibodies was proven by measuring the size of hydrodynamic diameter of the Au nanosphere. This Au-antibody nanocomposite is used for the diagnosis of prostate cancer (Liu et al. 2008).

2.6.2.8 Drug Loading to Quantum Dots

The QDs are minute luminescent crystals of dimension lesser than 10 nm and emit fluoresce light of diverse colors when irradiated with visible light. They are prepared with exceptional optical properties, thus functioning as probes or tracers and drug carrier for theragnostic uses. QDs are designed for active targeting to specified cells or tissues by attaching with targeting ligands.

QDs with colloidal core-shell are used for medical treatments. They are synthesized by hot colloidal method. For example, core CdSe QDs are produced by treating CdO in oleic acid and Se in trioctylphosphine at very high temperature (up to 300 °C). By this process, monodispersed QDs are synthesized. For biomedical applications, the core of QDs must be passivated with a thin coating of nontoxic bandgap material like ZnS or ZnSe, which forms the shell (Peng and

Peng 2002). ZnS shell of CdSe QDs leads to red shifts (~10 nm) in the absorption and photoluminescence peaks (Akerman et al. 2002). Two techniques are in vogue for the formulation of drug-loaded QDs: (a) directly binding drug moieties to QD surface and upon reaching the target tissues and releasing the drug from drug-QD conjugate in response to internal biological stimulus like enzymes or pH and (b) loading the drug in a polymeric nano-delivery system that has either lipophilic or hydrophilic QDs, which depends on the kind of polymer used for its encapsulation (Bagalkot et al. 2007). Generally, hydrophilic QDs are made by terminating their surface with groups like COOH, NH_2, and SH to which targeting ligands are attached by traditional conjugation methods. Avidin-biotin cross-linking is an additional well-known system for binding biocompounds on the surface of QDs (Medintz et al. 2005).

2.6.2.9 Drug Loading to MSNs

MSNs are efficient drug carrier system utilized for the targeted delivery of different varieties of medicinal compounds from micro- to macromolecules. Since there is no chemical change during drug loading, the structure of drugs is not altered after loading and following its release. These systems are more suitable for hydrophobic drugs like proteins, since they resolve the troubles related to these drugs. The volume and pore diameter of MSNs are features which determine the drug loading efficiency. MSNs have inner cylindrical mesopores (internal surface) and an outer polymeric surface (external particle surface). Inner and outer surface can be selectively functionalized with diverse conjugation molecules. Drugs are loaded to either inner inorganic core or outer organic polymer shell (Lodha et al. 2012).

Zhang et al. (2010) synthesized and functionalized a new MSN-based drug carrier nanodevice to deliver the hydrophobic, antihypertensive medicine telmisartan (TEL) to the target tissues. In brief, MSNs were developed by an organic template approach using tetraethyl orthosilicate (TEOS) and cetyltrimethyl ammonium bromide (CTAB) as surface-acting agents. The functionalization was done by treating with aminopropyl groups by a post-synthesis process. MSNs were soaked in acetic acid solution of telmisartan. Thus, the drug was absorbed into MSN. During this process, MSN nanocarrier/drug ratio was kept at 2:3 (w:w). By applying ultrasonic waves, the mixture was vibrated for some time followed by vortexing for 10 hours which enhances the drug incorporation into the mesopores of MSNs. The final traces of acetic acid were removed completely by drying the mixture at 55 °C for a day. The dried drug-loaded MSN composite can be used as therapeutic agent. Similarly, Lodha et al. (2012) synthesized cyclosporine A-loaded mesoporous silica nanoparticles. In brief, they prepared MSN using TEOS and CTAB. The MSN nanocarrier/drug ratio was kept at 1:1 (w:w). Then, the sample and MSN mixture was stirred at 300 rpm for 24 hours. Thus, by stirring and centrifugation, the drug was loaded into pores of MSN.

2.6.3 Drug Loading Efficiency

The drug loading (entrapment/encapsulation) efficiency of synthesized nanoparticles is identified by finding the amount of free drug in the dispersion solution. The percentage of entrapment/encapsulation efficiency of the nanoparticle is defined as the ratio of mass of drug loaded into nanoparticle to the mass of initial drug. The entrapment/encapsulation efficiency is calculated using the following formula:

$$\text{Entrapment / encapsulation efficiency} (\%) = \frac{\text{Wt of initial drug} - \text{Wt of free drug}}{\text{Wt of initial drug}} \times 100$$

where "Wt of initial drug" = mass of initial drug and "Wt of free drug" = mass of free drug measured in dispersion medium.

2.6.4 Stability and Storage of Nanoparticles

Synthesized NPs must be stored in the active form till administered. Many factors can influence the stability of these NPs. Commonly, the colloidal suspension of nanoparticles is very stable and will not deposit because of continuous mixing by diffusion and convection. But sometimes, clumping may result in deposition and precipitation of nanoparticle in suspension. To overcome this, some additives are added. Chemical integrity of drug is another factor which is very important for the stability assessments. Other factors to maintain the stability of the loaded drug are the (i) length of contact period with aqueous phase (e.g., if the drug is hydrophilic, then the length contact period in water must be long), (ii) pH (for pH-sensitive drugs), and (iii) light exposure (for light-sensitive drugs). Hence, the pH of the medium and exposure to light must be taken care while in storage. Stability studies are more crucial and must be carried out in accordance to the characteristics of drugs and polymers. Few techniques are used to maintain the stability of the NPs. Lyophilization is the chief and economical stabilizing process. After lyophilization process, the desiccated powder form of NPs is obtained and packed in vials. The powder form is advantageous than colloidal suspension form since it is easy to handle, easy for transportation, and easy for storage. NP vials can be stored in vacuum-packed containers at appropriate temperatures, particularly for temperature-liable drugs. Before use, the freeze-dried powder can readily be converted to suspension form by simply dispersing in aqueous solutions, or sometimes ultrasonication is needed for redispersion (Esquisabel et al. 1997).

2.6.5 Conclusion

Over the past few decades, proactive and continuous research has been performed in the area of nanomedicine for theranostic purpose. Site-specific or targeted drug delivery is the primary aim in any therapeutic investigations to enhance the therapeutic effects of drugs while lowering drug toxicity. Conservative drug carrier systems forever suffer from the unpredicted drug discharge in circulation and the slow discharge of drug at the target tissues. But recently developed nanoscale drug delivery systems provide novel opportunities for precisely targeted and controlled drug delivery. However, several problems remain to be solved and need further intense research. This chapter provides insight into various types of nanocarriers, mechanism of drug targeting, stimulus-responsive drug delivery systems, and drug loading techniques to nanostructures.

References

Abdifetah O, Na-Bangchang K (2019) Pharmacokinetic studies of nanoparticles as a delivery system for conventional drugs and herb-derived compounds for cancer therapy: a systematic review. Int J Nanomedicine 14:5659–5677

Ajayan PM, Ebbesen TW, Ichihashi T, Iijima S, Tanigaki K, Hiura H (1993) Opening carbon nanotubes with oxygen and implications for filling. Nature 362(6420):522–525

Akerman ME, Chan WCW, Laakkonen P, Bhatia SN, Ruoslahti E (2002) Nanocrystaltargeting in vivo. Proc Natl Acad Sci 99:12617–12621

Al-Jamal KT, Gherardini L, Bardi G, Nunes A, Guo C, Bussy C et al (2011) Functional motor recovery from brain ischemic insult by carbon nanotube-mediated siRNA silencing. Proc Natl Acad Sci USA 108:10952–10957

Alonso J, Khurshid H, Devkota J, Nemati Z, Khadka NK, Srikanth H, Pan JJ, Phan MH (2016) Superparamagnetic nanoparticles encapsulated in lipid vesicles for advanced magnetic hyperthermia and biodetection. J Appl Phys 119:083904

Ammoury N, Fessi H, Devissaguet JP, Dubrasquet M, Benita S (1991) Jejunal absorption, pharmacological activity and pharmacokinetic evaluation of indomethacin-loaded poly(d,l- lactide) and poly(isobutylcyanoacrylate) nanocapsules in rats. Pharm Res 8:101–105

Angelova A, Angelov B, Mutafchieva R, Lesieur S (2015) Biocompatible mesoporous and soft nanoarchitectures. J Inorganic Organometal Polym Mater 25(2):214–232

Bagalkot V, Zhang L, Levy-Nissenbaum E, Jon S, Kantoff PW, Langer R et al (2007) Quantum DotâAptamer conjugates for synchronous cancer imaging, therapy, and sensing of drug delivery based on Bi-fluorescence resonance energy transfer. Nano Lett 7:3065–3070

Ballauff M, Likos CN (2004) Dendrimers in solution: insight from theory and simulation. Angewandte Chemie 43(23):2998–3020

Barichello JM, Morishita M, Takayama K, Nagai T (1999) Encapsulation of hydrophilic and lipophilic drugs in PLGA nanoparticles by the nanoprecipitation method. Drug Dev Ind Pharm 25:471–476

Bhattacharya K, Andón FT, El-Sayed R, Fadeel B (2013) Mechanisms of carbon nanotube-induced toxicity: focus on pulmonary inflammation. Adv Drug Deliv Rev 65:2087–2097

Bhirde AA, Patel V, Gavard J, Zhang G, Sousa AA, Masedunskas A et al (2009) Targeted killing of cancer cells in vivo and in vitro with EGF directed carbon nanotube-based drug delivery. ACS Nano 3:307–316

Bouchemal K, Briancon S, Perrier E, Fessi H, Bonnet I, Zydowicz N (2004) Synthesis and characterization of polyurethane and poly(ether urethane) nanocapsules using a new technique of interfacial polycondensation combined to spontaneous emulsification. Int J Pharm 269:89–100

Cavalli R, Caputo O, Gasco MR (1993) Solid liposphered of doxorubicin and idarubicin. Int J Pharm 89:R9–R12

Chacko RT, Ventura J, Zhuang J, Thayumanavan S (2012) Polymer nanogels: a versatile nanoscopic drug delivery platform. Adv Drug Deliv Rev 64:836–851

Chan WCW (2007) Bio-applications of nanoparticles. Landes Bioscience, Austin, TX

Chauhan AS (2018) Dendrimers for Drug Delivery. Molecules 23:938, 9 pages

Chauhan AS, Jain NK, Diwan PV, Khopade AJ (2004) Solubility enhancement of indomethacin with poly(amidoamine) dendrimers and targeting to inflammatory regions of arthritic rats. J Drug Target 12:575–583

Chauhan A, Newenhouse E, Gerhardt A (2018) Compositions comprising a Dendrimer-resveratrol complex and methods for making and using the same. U.S. Patent 9,855,223 B2, 2 January

Chen Y, Cai RF, Chen S et al (2001) Synthesis and characterization of fullerol derived from C60n-precursors. J Phys Chem Solids 62:999–1001

Chen H, Zhang X, Dai S (2013) Multifunctional gold nano star conjugates for tumor imaging and combined photothermal and chemo-therapy. Theranostics 3(9):633–649

Chen HH, Huang WC, Chiang WH, Liu TI, Shen MY, Hsu YH, Lin SC, Chiu HC (2015) pH-responsive therapeutic solid lipid nanoparticles for reducing P-glycoprotein-mediated drug efflux of multidrug resistant cancer cells. Int J Nanomedicine 10:5035–5048

Chen Y, Tezcan O, Li D, Beztsinna N, Lou B, Etrych T, Ulbrich K, Metselaar JM, Lammers T, Hennink WE (2017) Overcoming multidrug resistance using folate receptor-targeted and pH-responsive polymeric nanogels containing covalently entrapped doxorubicin. Nanoscale 9:10404–10419

Cho H, Gao J, Kwon GS (2016) PEG-b-PLA micelles and PLGA-b-PEG-b-PLGA sol–gels for drug delivery. J Control Release 240(28):191–201

Couvreur P (2013) Nanoparticles in drug delivery: past, present and future. Adv Drug Deliv Rev 65(1):21–23

Couvreur P, Dubernet C, Puisieux F (1995) Controlled drug delivery with nanoparticles: current possibilities and future trends. Eur J Pharm Biopharm 41:2–13

Couvreur P, Barrat G, Fattal E, Legrand P, Vauthier C (2002) Nanocapsule technology. Crit Rev Ther Drug Carrier Syst 19:99–134

Croissant JG, Fatieiev Y, Khashab NM (2017) Degradability and clearance of silicon, Organosilica, Silsesquioxane, silica mixed oxide, and mesoporous silica nanoparticles. Adv Mater 29(9):51

Dadashzadeh S, Vali AM, Rezaie M (2008) The effect of PEG coating on in vitro cytotoxicity and in vivo disposition of topotecan loaded liposomes in rats. Pharm Nanotechnol 353:251–259

DarshanaNagda KS, Rathore M, Bharkatiya S, Sisodia S, Nema RK (2010) Bucky balls: a novel drug delivery system. J Chem Pharm 2(2):240–248

De Villiers MM, Lvov YM (2007) Nanoshells for drug delivery (Chapter 12). In: Kumar CSSR (ed) Nanotechnologies for the life sciences Vol. 10 Nanomaterials for medical diagnosis and therapy. WILEY-VCH Verlag GmbH & Co. KGaA, Weinheim. ISBN: 978-3-527-31390-7

Degiorgio V, Corti M (eds) (1985) Physics of Amphiphiles: micelles, vesicles and microemulsions. North-Holland, Amsterdam

Ding CZ, Li ZB (2017) A review of drug release mechanisms from nanocarrier systems. Mater Sci Eng 76:1440–1453

Domenico Lombardo, Kiselev MA, Caccamo MT (2019) Smart nanoparticles for drug delivery application: development of versatile nanocarrier platforms in biotechnology and nanomedicine. Hindawi J Nanomater Article ID 3702518:26 pages

Dreaden EC, Austin LA, Mackey MA, El-Sayed MA (2012) Size matters: gold nanoparticles in targeted cancer drug delivery. Ther Deliv 3(4):457–478

Duclairoir C, Orecchioni AM, Depraetere P, Nakache E (2002) α-Tocopherol encapsulation and in vitro release from wheat gliadin nanoparticles. J Microencapsul 19:53–60

Duncan R, Izzo L (2005) Dendrimer biocompatibility and toxicity. Adv Drug Deliv Rev 57(15):2215–2237

Ebbesen TW (1996) Wetting, filling and decorating carbon nanotubes. J Phys Chem Solids 57(6-8):951–955

Esposito E, Pecorelli A, Sguizzato M, Drechsler M, Mariani P, Carducci F, Valacchi G (2018) Production and characterization of nanoparticle based hyaluronate gel containing retinylpalmitate for wound healing. Curr Drug Deliv 15(8):1172–1182

Esquisabel A, Herna'ndez RM, Igartua M, Gasco'n AR, Calvo B, Pedraz JL (1997) Production of BCG alginate-PLL microcapsules by emulsification/internal gelation. J Microencapsul 14:627–638

Fu Q, Weinberg G, Su DS (2008) Selective filling of carbon nanotubes with metals by selective washing. New Carbon Mater 23(1):17–20

Ganachaud F, Katz JL (2005) Nanoparticles and nanocapsules created using the ouzo effect: spontaneous emulsification as an alternative to ultrasonic and high-shear devices. ChemPhysChem 6:209–216

Ginn SL, Alexander IE, Edelstein ML, Abedi MR, Wixon J (2013) Gene therapy clinical trials worldwide to 2012–an update. J Gene Med 15:65–77

Gou XC, Liu J, Zhang HL (2010) Monitoring human telomere DNA hybridization and G-quadruplex formation using gold nanorods. Anal Chimacta 668:208–214

Grangier JL, Puygrenier M, Gauthier JC, Couvreur P (1991) Nanoparticles as carriers for growth hormone releasing factors (GRF). J Control Release 15:3–13

Harmia-Pulkkinen T, Tuomi A, Kristoffersson E (1989) Manufacture of polyalkylcyanoacrylate nanoparticles with pilocarpine and timolol by micelle polymerization: factors influencing particle formation. J Microencapsul 6:87–93

Heister E, Neves V, Lamprecht C, Ravi S, Silva P, Helen M, Coley A, McFadden J (2012) Drug loading, dispersion stability, and therapeutic efficacy in targeted drug delivery with carbon nanotubes. Carbon 50:622–632

Hennink WE, van Nostrum VF (2002) Novel crosslinking methods to design hydrogels. Adv Drug Deliv Rev 54(1):13–36

Hirsch LR, Jackson A, Lee A, Halas NJ, West JL (2003a) A whole blood immunoassay usinggold nanoshells. Anal Chem 75:2377–2381

Hirsch LR, Stafford RJ, Bankson JA, Sershen SR, Rivera B, Price RE, Hazle JD, Halas NJ, West JL (2003b) Nanoshell-mediated nearinfrared thermal therapy of tumors under magnetic resonance guidance. Proc Natl Acad Sci U S A 100:13549–13554

Hirsch LR, Gobin AM, Lowery AR, Tam F, Drezek RA, Halas NJ, West JL (2006) Metal Nanoshells. Ann Biomed Eng 34(1):15–22

Hu JJ, Liu LH, Li ZY, Zhuo RX, Zhang XZ (2016) MMP-responsive theranosticnanoplatform based on mesoporous silica nanoparticles for tumor imaging and targeted drug delivery. J Mater Chem B 4(11):1932–1940

Hu X, Zhang Y, Xie Z, Jing X, Bellotti A, Gu Z (2017) Stimuli-responsive polymersomes for biomedical applications. Biomacromolecules 18(3):649–673

Husain Q (2017) Nanosupport bound lipases their stability and applications. Biointerface Res Appl Chem 7:2194–2216

Illum L, Farraj NF, Davis SS (1994) Chitosan as novel nasal delivery system for peptide drugs. Pharm Res 11:1186–1189

Jahnke S (1998) The theory of high pressure homogenization. In: Muller RH, Benita S, Bohm B (eds) Emulsions and nanosuspensions for the formulation of poorly soluble drugs. Medpharm Scientific Publishers, Stuttgart, pp 177–200

Jong WHD, Borm PJA (2008) Drug delivery and nanoparticle applications and hazards. Int J Nanomedicine 3(2):133–149

Kang B, Mackey MA, El-Sayed MA (2010) Nuclear targeting of gold nanoparticles in cancer cells induces DNA damage, causing cytokinesis arrest and apoptosis. J Am Chem Soc 132(5):1517–1519

Khmelnitsky YL, Neverova IN, Gedrovich AV, Polyakov VA, Levashov AV, Martinek K (1992) Catalysis by α-chymotrypsin entrapped into surface-modified polymeric nanogranules in organic solvent. Eur J Biochem 210:751–757

Khosa A, Reddi S, Saha RN (2018) Nanostructured lipid carriers for site-specific drug delivery. Biomed Pharmacother 103:598–613

Kim B, Rutka JT, Chan W (2010) Nanomedicine. New Engl J Med 363:2434–2443

Kiselev MA, Lombardo D (2017) Structural characterization in mixed lipid membrane systems by neutron and X-ray scattering. Biochimicaet Biophysica Acta (BBA) - General Subjects 1861(1):3700–3717

Kong FY, Zhang JW, Li RF, Wang ZX, Wang WJ, Wang W (2017) Unique roles of gold nanoparticles in drug delivery, targeting and imaging applications. Molecules 22(9):article 1445

Kotelnikova RA, Bogdanov GN, Frog EC et al (2003) Nanobionics of pharmacologically active derivatives of fullerene C60. J Nanopart Res 5:561–566

Kousalová J, Etrych T (2018) Polymeric Nanogels as drug delivery systems. Physiol Res 67(Suppl. 2):S305–S317

Kreuter J, Tauber U, Illi V (1979) Distribution and elimination of poly (methyl-2-^{14}C-methacrylate) nanoparticle radioactivity after injection in rats and mice. J Pharm Sci 68:1443–1447

Kubiak C, Manil L, Couvreur P (1988) Sorptive properties of antibodies onto cyanoacrylic nanoparticles. Int J Pharm 41(181-18):7

Kulhari H, Pooja D, Singh MK, Chauhan AS (2015) Optimization of carboxylate-terminated poly(amidoamine) dendrimer-mediated cisplatin formulation. Drug Dev Ind Pharm 41:232–238

Kumar A, Badde S, Kamble R, Pokharkar VB (2010) Development and characterization of liposomal drug delivery system for nimesulide. Int J Pharm Pharm Sci 2(4):87–89

Kumar KPS, Bhowmik D, Deb L (2012) Recent trends in liposomes used as novel drug delivery system. J Pharm Innov 1(1):26–34

Kushwaha SKS, Ghoshal S, Rai AK, Singh S (2013) Carbon nanotubes as a novel drug delivery system for anticancer therapy: a review. Braz J Pharm Sci 49(4):629–643

Lander R, Manger W, Scouloudis M, Ku A, Davis C, Lee A (2000) Gaulin homogenization: a mechanistic study. Biotechnol Prog 16:80–85

Li J, Gupta S, Li C (2013) Research perspectives: gold nanoparticles in cancer theranostics. Quant Imag Med Surg 3(6):284–291

Liu X, Dai Q, Austin L et al (2008) A one-step homogeneous immunoassay for cancer biomarker detection using gold nanoparticle probes coupled with dynamic light scattering. J Am Chem Soc 130(9):2780–2782

Lodha A, Lodha M, Patel A, Chaudhuri J, Dalal J, Edwards M, Douroumis D (2012) Synthesis of mesoporous silica nanoparticles and drug loading of poorly water soluble drug cyclosporin A. J Pharm Bioallied Sci 4(Suppl 1):S92–S94

Lombardo D (2009) Liquid-like ordering of negatively charged poly (amidoamine) (PAMAM) dendrimers in solution. Langmuir 25(5):3271–3275

Lombardo D, Kiselev MA, Magazù S, Calandra P (2015) Amphiphiles self-assembly: basic concepts and future perspectives of supramolecular approaches. Adv Condens Matter Phys:Article ID 151683, 22 pages

Lowe PJ, Temple CS (1994) Calcitonin and insulin in isobutylcyanoacrylatenanocapsules: protection against proteases and effect on intestinal absorption in rats. J Pharm Pharmacol 46:547–552

Ma GL, Lin WF, Yuan ZF, Wu J, Qian HF, Xua LB, Chen SF (2017) Development of ionic strength/pH/enzyme triple-responsive zwitterionic hydrogel of the mixed l-glutamic acid and l-lysine polypeptide for site-specific drug delivery. J Mater Chem B 5:935–943

Maitra J, Shukla VK (2014) Cross-linking in hydrogels – a review. Am J Polym Sci 4(2):25–31

Matea CT, Mocan T, Tabaran F, Pop T, Mosteanu O, Puia C, Iancu C, Mocan L (2017) Quantum dots in imaging, drug delivery and sensor applications. Int J Nanomedicine 12:5421–5431

Mattos BD, Tardy BL, Magalhaes WLE, Rojas OJ (2017) Controlled release for crop and wood protection: recent progress toward sustainable and safe nanostructured biocidal systems. J Control Release 262:139–150

McIntosh CM, Esposito EA, Boal AK, Simard JM, Martin CT, Rotello VM (2001) Inhibition of DNA transcription using cationic mixed monolayer protected gold clusters. J Am Chem Soc 123(31):7626–7629

Medintz IL, Uyeda HT, Goldman ER, Mattoussi H (2005) Quantum dot bioconjugates for imaging, labelling and sensing. Nat Mater 4:435–446

Memisoglu E, Bochot A, O¨ zalp M, Sen M, Ducheˆne D, Hincal A (2003) Direct formation of nanospheres from amphiphilic beta-cyclodextrin inclusion complexes. Pharm Res 20:117–125

Michel C, Roques M, Couvreur P, Vranchx H, Baldschmidt P (1991) Isobutylcyanoacrylatenanoparticles as drug carrier for oral administration of insulin. Proc Int Symp Control Release Bioact Mater 18:97–98

Mikhail AS, Allen C (2009) Block copolymer micelles for delivery of cancer therapy: transport at the whole body, tissue and cellular levels. J Control Release 138(3):214–223

Monthioux M (2002) Filling single-wall carbon nanotubes. Carbon 40(10):1809–1823

Mourya V, Inamdar N, Nawale R, Kulthe S (2011) Polymeric micelles: general considerations and their applications. Ind J Pharm Educ Res 45:128–138

Muller RH, Schwarz C, Mehenert W, Lucks JS (1993) Production of solid lipid nanoparticles for controlled drug delivery. Proc Int Symp Control Release Bioact Mater 20(1993):480–481

Nalawade P, Mukherjee T, Kapoor S (2012) High-yield synthesis of multispiked gold nanoparticles: characterization and catalytic reactions. Colloid Surfac A: Physicochem Eng Asp 396:336–340

Neetu S, Amrita K, Luo G, Kevin L, Jordan SM, Christopher SC, Michael JS, Sangeeta NB (2011) BioresponsiveMesoporous silica nanoparticles for triggered drug release. J Am Chem Soc 133:19582–19585

Novoselov KS, Fal'ko VI, Colombo L, Gellert PR, Schwab MG, Kim K (2012) A roadmap for graphene. Nature 490:192–200

Palmerston Mendes L, Pan J, Torchilin V (2017) Dendrimers as nanocarriers for nucleic acid and drug delivery in cancer therapy. Molecules 22(9):article 1401

Pardeike J, Hommoss A, Müller RH (2009) Lipid nanoparticles (SLN, NLC) in cosmetic and pharmaceutical dermal products. Int J Pharm 366(1-2):170–184

Pardeshi C, Rajput P, Belgamwar V, Tekade A, Patil G, Chaudhary K, Sonje A (2012) Solid lipid based nanocarriers: an overview. Acta Pharma 62:433–472

Patil GV (2003) Biopolymer albumin for diagnosis and in drug delivery. Drug Dev Res 58:219–247

Patnaik S, Sharma AK, Garg BS, Gandhi RP, Gupta KC (2007) Photoregulation of drug release in azo-dextran nanogels. Int J Pharm 342:184–193

Peer D, Karp JM, Hong S, Farokhzad OC, Margalit R, Lange R (2007) Nanocarriers as an emerging platform for cancer therapy. Nat Nanotechol 2:751–760

Peiris PM, Bauer L, Toy R, Tran E, Pansky J, Doolittle E, Schmidt E, Hayden E, Mayer A, Keri RA, Griswold MA, Karathanasis E (2012) Enhanced delivery of chemotherapy to tumors using a multicomponent nanochain with radio-frequency-tunable drug release. ACS Nano 6(5):4157–4168

Peng ZA, Peng X (2002) Nearly monodisperse and shape-controlled CdSe nanocrystals via alternative routes: nucleation and growth. J Am Chem Soc 124:3343–3353

Pinto Reis C, Neufeld RJ, Ribeiro AJ, Veiga F (2006) Nanoencapsulation-I methods for preparation of drug-loaded polymeric nanoparticles. Nanomedicine 2:8–21

Prato M (1997) [60] Fullerene chemistry for materials science applications. J Mater Chem 7:1097–1109

Qin SY, Feng J, Rong L, Jia HZ, Chen S, Liu XJ et al (2014) Theranostic GO based nanohybrid for tumor induced imaging and potential combinational tumor therapy. Small 10:599–608

Quintanar-Guerrero D, Alle'mann E, Fessi H, Doelker E (1998) Preparation techniques and mechanism of formation of biodegradable nanoparticles from preformed polymers. Drug Dev Ind Pharm 24:1113–1128

Rabinarayan P, Padilama S (2010) Production of solid lipid nanoparticles-drug loading and release mechanism. J Chem Pharm Res 2:211–227

Reddy N, Reddy R, Jiang Q (2015) Crosslinking biopolymers for biomedical applications. Trends Biotechnol 33(6):362–369

Rehman M, Ihsan A, Madni A, Bajwa SZ, Shi D, Webster TJ, Khan WS (2017) Solid lipid nanoparticles for thermoresponsive targeting: evidence from spectrophotometry, electrochemical, and cytotoxicity studies. Int J Nanomedicine 12:8325–8336

Reis CP, Neufeld RJ, Ribeiro AJ, Viega F (2005) Insulin-alginate nanospheres: influence of calcium on polymer matrix properties. Proceedings of the 13th International Workshop on Bioencapsulation. Kingston, Ontario, Canada: QueenTs University

Riaz M, Riaz M, Zhang X et al (2018) Surface functionalization and targeting strategies of liposomes in solid tumor therapy: a review. Int J Mol Sci 19(1):195

Rigogliuso S, Sabatino MA, Adamo G, Grimaldi N, Dispenza C, Ghersi G (2012) Polymeric nanogels: Nanocarriers for drug delivery application. Chem Eng Trans 27:247–252

Rolland A, Gibassier D, Sado P, Le Verge R (1986) Purification et proprie'te'sphysico-chimiques des suspensions de nanoparticulesdepolyme're. J Pharm Belg 41:94–105

Rosenblum D, Joshi N, Tao W, Karp JM, Peer D (2018) Progress and challenges towards targeted delivery of cancer therapeutics. Nat Commun 9:1410

Sacchetto C, Liu-Bryan R, Magrini A, Rosato N, Bottini N, Bottini M (2014) Glycol-modified single-walled carbon nanotubes for intra-articular delivery to chondrocytes. ACS Nano 8:12280–12291

Sahiner N, Alb AM, Graves R, Mandal T, McPherson GL, Reed WF, John VT (2007) Core-shell nanohydrogel structures as tunable delivery systems. Polymer (Guildf) 48:704–711

Sailaja A, Amareshwar P (2011) Preparation of chitosan coated nanoparticles by emulsionpolymerization technique. Asian J Pharm Clin Res 4(Suppl 1):73–74

Schuster DI, Wilson SR, Kirschner AN et al (2000) Evaluation of the anti-HIV potency of a water-soluble dendrimeric fullerene. Proc Electrochem Soc 9:267–270

Seijo B, Fattal E, Roblot-Treupel L, Couvreur P (1990) Design of nanoparticles of less than 50 nm diameter: preparation, characterization and drug loading. Int J Pharm 62:1–7

Sershen SR, Westcott SL, Halas NJ, West JL (2000) Temperature sensitive polymer–nanoshell composites for photothermally modulated delivery. J Biomed Mater Res 51:293–298

Sha L, Wang D, Mao Y, Shi W, Gao T, Zhao Q, Wang S (2018) Hydrophobic interaction mediated coating of pluronics on mesoporous silica nanoparticle with stimuli responsiveness for cancer therapy. Nanotechnology 29(34):345101

Sharma A, Garg T, Aman A, Panchal K, Sharma R, Kumar S, Markandeywar T (2016) Nanogel-an advanced drug delivery tool: current and future. Artif Cells Nanomed Biotechnol 44:165–177

Shi J, Wang B, Wang L, Lu T, Fu Y, Zhang H, Zhang Z (2016) Fullerene (C 60)-based tumor-targeting nanoparticles with "off-on" state for enhanced treatment of cancer. J Control Release 235:245–258

Shi J, Su Y, Liu W, Chang J, Zhang Z (2017) A nanoliposome based photoactivable drug delivery system for enhanced cancer therapy and overcoming treatment resistance. Int J Nanomedicine 12:8257–8275

Soni G, Yadav KS (2016) Nanogels as potential nanomedicine carrier for treatment of cancer: a mini review of the state of the art. Saudi Pharm J 24:133–139

Sosnik A, Seremeta K (2017) Polymeric hydrogels as technology platform for drug delivery applications. Gels 3(3):25

Spitalsky Z, Tasis D, Papagelis K, Galiotis C (2010) Carbon nanotubepolymer composites: chemistry, processing, mechanical and electrical properties. Prog Polym Sc 35(3):357–401

Sreejivungsa K, Suchaichit N, Moosophon P, Chompoosor A (2016) Light-regulated release of entrapped drugs from photoresponsive gold nanoparticles. J Nanomater 2016:7 pages

Sun K, You C, Wang S, Gao Z, Wu H, Tao WA, Sun B (2018) NIR stimulus-responsive Core-shell type nanoparticles based on photothermal conversion for enhanced antitumor efficacy through chemo-photothermal therapy. Nanotechnology 29(28):285302

Svenson S, Chauhan AS (2008) Dendrimers for enhanced drug solubilization. Nanomedicine 3:679–702

Szoka F Jr (1980) Annu Rev Biophys Bioeng 9:467–508

Tan L, Wu T, Tang ZW, Xiao JY, Zhuo RX, Shi B, Liu CJ (2016) Water-soluble photoluminescent fullerene capped mesoporous silica for pH-responsive drug delivery and bioimaging. Nanotechnology 27(31):315104

Tomalia DA, Reyna LA, Svenson S (2007) Biochem Soc Trans 35:61–67

Tonelli FM, Goulart VA, Gomes KN, Ladeira MS, Santos AK, Lorençon E et al (2015) Graphene-based nanomaterials: biological and medical applications and toxicity. Nanomedicine (Lond) 10:2423–2450

Tran S, DeGiovanni PJ, Piel B, Rai P (2017) Cancer nanomedicine: a review of recent success in drug delivery. Clin Transl Med 6(1):44

Tsang SC, Chen YK, Harris PJF, Green MLH (1994) A simple chemical method of opening and filling carbon nanotubes. Nature 372(6502):159–162

Tyler B, Gullotti D, Mangraviti A, Utsuki T, Brem H (2016) Polylactic acid (PLA) controlled delivery carriers for biomedical applications. Adv Drug Deliv Rev 107:163–175

Vauthier C, Couvreur P (2000) Development of polysaccharide nanoparticles as novel drug carrier systems. In: Wise DL (ed) Handbook of pharmaceutical controlled release technology. Marcel Dekker, New York7, pp 413–429

Wahajuddin, Arora (2012) Superparamagnetic iron oxide nanoparticles: magnetic nanoplatforms as drug carriers. Int J Nanomedicine 2012(7):3445–34712

Wang F, Wang YC, Dou S, Xiong MH, Sun TM, Wang J (2011) Doxorubicin-tethered responsive gold nanoparticles facilitate intracellular drug delivery for overcoming multidrug resistance in cancer cells. ACS Nano 5:3679–3692

Wang H, Wu L, Reinhard BM (2012) Scavenger receptor mediated endocytosis of silver nanoparticles into J774A.1 macrophages is heterogeneous. ACS Nano 6:7122–7132

Ward MA, Georgiou TK (2011) Thermoresponsive polymers for biomedical applications. Polymers 3(3):1215–1242

Watermann A, Brieger J (2017) Mesoporous silica nanoparticles as drug delivery vehicles in cancer. Nano 7(7):189

Wei H, Cheng SX, Zhang XZ, Zhuo RX (2009) Thermosensitive polymeric micelles based on poly(N-isopropylacrylamide) as drug carriers. Prog Polym Sci 34(9):893–910

Westesen K, Siekmann B (1996) Biodegradable colloidal drug carrier systems based on solid lipids. In: Benita S (ed) Microencapsulation methods and industrial applications. Marcel Dekkar, Inc., New York

Yan M, Ge J, Liu Z, Ouyang P (2006) Encapsulation of single enzyme in nanogel with enhanced biocatalytic activity and stability. J Am Chem Soc 128:11008–11009

Yin J, Chen Y, Zhang ZH, Han X (2016) Stimuli-responsive block copolymer-based assemblies for cargo delivery and theranostic applications. Polymers 8(7):268

Yoo HS, Park TG (2001) Biodegradable polymeric micelles composed of doxorubicin conjugated PLGA–PEG block copolymer. J Control Release 70(1-2):63–70

Yoshioka T, Hashida M, Muranishi S, Sezaki H (1981) Specific delivery of mitomycin C to the liver, spleen, and lung: nano- and microspherical carriers of gelatin. Int J Pharm 8:131–141

You J, Zhang G, Li C (2010) Exceptionally high payload of doxorubicin in hollow gold nanospheres for near-infrared light-triggered drug release. ACS Nano 4(2):1033–1041

Zhang Y, Zhi Z, Jiang T, Zhang J, Wang Z, Wang S (2010) Spherical mesoporous silica nanoparticles for loading and release of the poorly water-soluble drug telmisartan. J Control Release 145(3):257–263

Zhang Y, Han L, Zhang Y, Chang YQ, Chen XW, He RH, Wang JH (2016) Glutathione-mediated mesoporous carbon as a drug delivery nanocarrier with carbon dots as a cap and fluorescent tracer. Nanotechnology 27(35):355102

Chapter 3
Recent Advances in Nanomaterials-Based Drug Delivery System for Cancer Treatment

Prakash Ramalingam, D. S. Prabakaran, Kalaiselvi Sivalingam,
V. Uma Maheshwari Nallal, M. Razia, Mayurkumar Patel, Tanvi Kanekar,
and Dineshkumar Krishnamoorthy

Contents

3.1	Introduction	85
3.2	Limitations of Conventional Cancer Treatment	86
3.3	Nanomaterials as Drug Delivery System for Cancer Treatment	87
3.4	Unique Advantages of Nano DDS	88
	3.4.1 Particle Size	88
	3.4.2 High Drug Payload	88
	3.4.3 Controlled Drug Release	89
	3.4.4 Surface Modification	89

P. Ramalingam (✉) · M. Patel · T. Kanekar
Product Development, Genus Lifesciences Inc, Allentown, PA, USA

D. S. Prabakaran
Department of Radiation Oncology, Chungbuk National University College of Medicine,
Cheongju, Republic of Korea

Department of Biotechnology, Ayya Nadar Janaki Ammal College, Sivakasi,
Sivakasi, Tamil Nadu, India

K. Sivalingam
Department of Pharmaceutical Sciences, Irma Lerma Rangel College of Pharmacy, Texas
A&M University, Kingsville, TX, USA

V. U. M. Nallal · M. Razia
Department of Biotechnology, Mother Teresa Women's University,
Kodaikanal, Tamil Nadu, India

D. Krishnamoorthy
Department of Plant Science, School of Biological Sciences, Central University of Kerala,
Kasargode, Kerala, India

3.5	Physiology of Tumor and Tumor Targeting Using Nano DDS...........................	90
	3.5.1 Angiogenesis and Tumor Vasculatures...	90
	3.5.2 Mechanisms of Tumor Targeting by Nano DDS...................................	90
3.6	Nano DDS for Cancer Treatment...	92
	3.6.1 Organic Nanomaterials for Cancer Treatment.....................................	92
	3.6.2 Inorganic Nanomaterials for Cancer Treatment...................................	98
3.7	Challenges and Future Perspectives...	103
3.8	Conclusion...	104
References..		105

Abbreviations

BCS	Biopharmaceutical Classification System
CNTs	Carbon nanotubes
CS	Chitosan
DOX	Doxorubicin
DTX	Docetaxel
EPR	Enhanced permeation and retention effect
FA	Folic acid
FDA	Food and Drug Administration
GEM	Gemcitabine
GNPs	Gold nanoparticles
HA	Hyaluronic acid
MNPs	Magnetic nanoparticles
MPS	Mononuclear phagocytic system
MRI	Magnetic resonance imaging
MSNs	Mesoporous silica nanoparticles
Nano DDS	Nano drug delivery system
PAMAM	Polyamidoamine
PCL	Poly (ε-caprolactone)
PDCs	Polymer-drug conjugates
PET	Positron emission tomography
PEG	Polyethylene (glycol)
PLA	Polylactic acid
PLGA	Poly (D, L-lactide-*co*-glycolide)
PLL	Poly-l-lysine
PPI	Poly (propylamine)
PTT	Photothermal therapy
PTX	Paclitaxel
QDs	Quantum dots
RES	Reticuloendothelial system

SLNs	Solid lipid nanoparticles
TPGS	D-Tocopherol polyethylene glycol1000 succinate
WHO	World Health Organization

3.1 Introduction

Cancer is the second most severe lethal disease in the current world and spreading further with continuance and growing incidence in the twenty-first century. According to the estimates from the GLOBOCAN cancer statistics 2018 (International Agency for Research on Cancer, WHO), there are 9.6 million cancer cases deaths in 2018. More than 18.1 million cancer cases are diagnosed, and this rate has been estimated to rise to 29.5 million by the year 2040 (Faisca Phillips 2019; Bray et al. 2018). The condition is so alarming that every fourth person is having a lifetime cancer risk. Is cancer treatable? The short reply to this question is "yes." Cancer mortality rates can be decreased if cancer cases are detected and treated early with better treatment strategies (Siegel et al. 2019; Wild 2019). Cancer begins from transforming healthy normal tissues into tumor tissues in a multistage development that usually progresses from a precancerous to a malignant tumor. Many types of cancers affect the people, and the cancer cells show no symptoms at an initial stage of development (Papaccio et al. 2017; Kulikov et al. 2017). Cancer cells proliferate and continue to increase unless one of three things occur: (i) The tumor tissues are removed surgically, (ii) using radiation therapy, or (iii) using chemotherapy.

There are different methods of cancer treatment. Current cancer treatment options can be surgical intervention, radiation therapy, chemotherapy, immunotherapy and hormone therapy, or a combination of these options (Miller et al. 2019; Chowdhury et al. 2016). The types of cancer treatment that patients receive depend on the type of cancer patients have and what stage advanced it is. The treatment of cancer by surgery works best for small size solid tumors that are localized in one area (Tyson II et al. 2018; Derks et al. 2017). The surgery to remove the entire tumorous mass should not harm the surrounding normal healthy cells or tissues. Nonsurgical cancer treatment commonly followed is radiation therapy or chemotherapy medication. Radiation therapy practices with high ionizing radiation dose to eradicate cancer cells and slow tumor growth by damaging the DNA (Liu et al. 2016a; Baskar and Itahana 2017). The radiation therapy is commonly used in combination with the surgery to reduce the tumor size, so the tumor can be easily removed by surgical treatment (Bishop et al. 2018a, b). The body can safely receive a limited amount of radiation over the course of the treatment. The radiation dose to be delivered to the cancer site depends upon various factors such as the cancer type, tumor size and location in the body, age of the person, general health and medical history, and possible side effects on the nearby normal tissues (Ghahremani et al. 2018; Cabrera et al. 2016). Immunotherapy is a biological cancer therapy that supports the immune system battle against cancer, and it is not yet as extensively used as surgery, radiation therapy, and chemotherapy (Zaidi and Jaffee 2019; Ishihara et al. 2017). Hormone therapy uses hormones to stop the growth of cancers (Axelrad et al. 2020; Eeles et al. 2016).

Chemotherapy or combined chemotherapy, a very common cancer treatment, uses anticancer drugs to kill or destroy the uncontrolled proliferation of cancerous cells. Conventional chemotherapy works principally by interfering with the synthesis of DNA and mitosis, leading to the death of rapidly proliferating and dividing cancer cells (Senapati et al. 2018; Wang et al. 2016). Unfortunately, due to nonspecific drug targeting by anticancer medicines, conventional chemotherapy fails to target the tumor specifically without interacting with the normal healthy cells (Kumari et al. 2016; Wakaskar 2017; Raza et al. 2019).

This chapter aims to present the limitations of conventional cancer treatment and principal concepts of nanomaterials for cancer treatment, to emphasize the distinguished advantage of nanomaterials-based drug delivery systems (nano DDS) and the mechanism of action underlying their selective targeted drug delivery effects, and to introduce successful recent nano drug delivery system for cancer treatment and diagnosis.

3.2 Limitations of Conventional Cancer Treatment

The conventional cancer treatments effectively destroy the cancer cells, but they are also harmful to the normal healthy cells and tissues (Johnson et al. 2018; Kalyanaraman 2017). Cancer cells cannot be entirely removed by the surgery, and even the existence of a single cancer cell that is unseen can redevelop into a new tumor and metastasize to other parts of the body. The cancer treatment by the surgical procedure is not used for hematological cancers or cancers that have metastasized to other tissues or parts of the body. The radiation therapy administered both internally or externally can also destroy the normal healthy cells and induce the side effects due to the ionizing radiation. The radiation therapy is not used if the tumor is located at extremely vulnerable locations or if the cancer is at the advanced stages. Immunotherapy and hormone therapy cause side effects in the body, and hormone therapy blocks the ability to produce hormones in the body system.

Chemotherapy is considered as an effective type of cancer treatment for all types of cancers, but it damages either normal healthy tissues or cells that divide rapidly, such as cells in the macrophages, digestive tract, bone marrow, and hair follicles. The notable drawback of conventional chemotherapy is that it cannot provide specific target action only to the cancer cells. The nonspecific delivery of chemotherapeutic drugs causes severe side effects such as mucositis, myelosuppression, organ dysfunction, alopecia, and thrombocytopenia, and these side effects impose treatment delay, dose reduction, and therapy discontinuation. Furthermore, most of the available chemotherapeutic drugs often cannot penetrate the outer membranes of solid tumors and reach the inside core of solid tumors, failing to destroy the cancer cells. Also, the repeated administration of nonselective chemotherapeutic drugs can influence drug resistance.

Chemotherapeutic drugs are often eliminated from the plasma circulation engulfed by macrophages and P-glycoprotein, acting as the efflux pump, which is overexpressed on the cancer cells surface and prevents the accumulation of drugs inside the tumor. Thus, chemotherapeutic drugs stay in the plasma circulation for a very short and limited time and cannot interact with the cancer cells resulting in the chemotherapy entirely unsuccessful. The low drug solubility, large particle size, low specificity, and high toxicity of chemotherapeutic drugs are also important issues in conventional chemotherapy, making them unable to improve the bioavailability and reach the chemotherapeutic drugs at the tumor sites.

To circumvent the pitfalls as mentioned above and the limitations of conventional cancer treatments, chemotherapeutic drugs need to reformulate with various types of nanomaterials and drug delivery systems.

3.3 Nanomaterials as Drug Delivery System for Cancer Treatment

Since innovative researches and understanding of biological mechanisms of cancer tissues are emerging regularly, novel cancer treatment procedures are being developed to have improved effectiveness of the treatment, thereby enabling the patient's survivability and improving their quality of life. With the recent technological advances in medical sciences, different types of cancer treatment have been practiced in the past, and many new therapies, such as targeted therapy, are currently being practiced. There have been significant successes in the nanotechnology medical applications (nanomedicine) in recent years, particularly in the drug delivery system (Wolfram and Ferrari 2019; Salvioni et al. 2019; van der Meel et al. 2019; Tran et al. 2017; Prasad et al. 2017).

Treating cancer cells using a nanoparticulate drug delivery system (nano DDS) approach plays a pivotal role in circumventing the limitations of conventional cancer treatment methods by providing simultaneous diagnosis and treatment. The application of nano DDS to cancer treatment could extend beyond the drug delivery system into the making of new therapeutics capable of killing the cancer cells with negligible damage to normal healthy cells and tissues. Various types of organic and inorganic nanomaterials are used to formulate chemotherapeutic drug-loaded nano DDS for cancer diagnosis and treatment. Most of the organic nanomaterials (liposomes, solid lipid nanoparticles, dendrimers, polymeric micelles, polymeric (natural or synthetic) nanoparticles, and polymer-drug conjugates) and inorganic nanomaterials (mesoporous silica nanoparticles, gold nanoparticles, magnetic nanoparticles, carbon nanotubes, and quantum dots) were developed as a vehicle in nano DDS for cancer treatment.

3.4 Unique Advantages of Nano DDS

3.4.1 Particle Size (Kumar et al. 2017; Arms et al. 2018; Ghasemiyeh and Mohammadi-Samani 2018; Tiruwa 2016; Ghasemiyeh and Mohammadi-Samani 2020; Sarcan et al. 2018)

Particle size distribution and small size with high surface area characteristics of nanoparticles are the most important key factors for drug delivery applications. The great advantage of nano DDS is that the particle size and size distributions are tunable. Several types of research have reported that nanoparticulate systems have plenty of advantages over other microparticulate systems. Nanoparticles can improve drug loading, stability, controlled drug release, high cellular uptake, in vivo pharmacokinetics, plasma circulation half-life, biodistribution, targeted drug delivery, tumor accumulation, and ability to cross the blood-brain barrier and transport the drugs to the brain due to their smaller size and flexibility (Prasad et al. 2019). Nanoparticles can also be coated with different types of polymers or surface-functionalized with targeting moieties, peptides, and nucleic acids that bind to specific cancer target sites. The nanoparticles used in a nano DDS should be small size enough to escape or avoid capture by macrophages in the circulation system. Systemically administered nano DDS should have a particle size ranging from 10 to 200 nm, particle size less than 200 nm to avoid sequestration by the liver and spleen, and particle size larger than 10 nm to avoid first-pass metabolism or elimination through the kidneys, benefiting accumulation/clearance and biodistribution behavior. The particle size of nano DDS has been shown to influence the surface functionalization and targeted drug delivery applications for cancer treatment.

3.4.2 High Drug Payload (Ghasemiyeh and Mohammadi-Samani 2018; Meunier et al. 2017; Liu et al. 2020; Qu et al. 2016; Huang et al. 2016)

An effective nanoparticulate system should load and hold a higher amount of drugs, thereby decreasing the frequent dose of uptake and increasing drug plasma concentration after administration in the body. Drug loading in the nano DDS can be done by adsorption/absorption and incorporation techniques. A high drug loading capacity and encapsulation efficiency mainly depend on the classification of drugs (e.g., biopharmaceutical classification systems (BCS) Class I–IV) and drug solubility in the nano DDS, which is related to the drug-polymer interactions, compositions of excipients, and the presence of active functional groups from drug and excipients. For instance, the solid lipid core of solid lipid nanoparticles can accommodate a higher amount of hydrophobic chemotherapeutic drugs, and liposomes can load and hold both hydrophobic and hydrophilic chemotherapeutic drugs due to their unique characteristics.

3.4.3 Controlled Drug Release (Li et al. 2016a; Kamaly et al. 2016; Deodhar et al. 2017; Liu et al. 2019a; Paris et al. 2018)

It is crucial to take consideration of both polymer biodegradation and drug release kinetics in simulated body conditions when formulating a nano DDS. The drug release behavior from nano DDS mainly depends on (i) solubility of active pharmaceutical ingredient, (ii) nano DDS degradation or erosion, (iii) desorption from the surface-attached drug or incorporated drug from the inside polymer core, (iv) drug diffusion through the nano DDS, and (v) the combination of diffusion and erosion processes. For example, the drug release of uniformly drug distributed nanospheres occurs by diffusion or matric erosion. If the active drug diffusion is more rapid than matrix erosion, then the drug release mechanism is mostly maintained by diffusion. The burst drug release from nanoparticles at the early stage is primarily attributed to surface-attached drug molecules to the large surface of nano DDS. It is indicated that the method of drug loading has a pivotal role in the drug release profile from nanoparticles. If the active pharmaceutical ingredient is entrapped in the nano DDS by the incorporation technique, then the nano DDS has a negligible amount of burst drug release and controlled drug release profile. If the nano DDS is surface-modified or coated by other synthetic or natural polymers, the drug release profile is then controlled by drug diffusion from the surface polymeric membrane.

3.4.4 Surface Modification (Ahmad et al. 2018a; Choi and Meghani 2016; Ahmad et al. 2018b; Ganesan et al. 2018; Ramalingam and Ko 2016; Ramalingam and Ko 2015; Ramalingam et al. 2016)

Surface modification or coating on the nano DDS can improve drug biodistribution, pharmacokinetics, and oral and brain drug delivery. To enhance drug targeting, it is crucial to prolong the nanoparticle circulation and minimize the opsonization in vivo, and it can be accomplished by coating or surface modification of nano DDS with biodegradable hydrophilic polymers, e.g., natural polymers such as chitosan and their derivatives, PEG, polysorbate 80, poloxamer, and polyethylene oxide. Several researches publish that PEG surface modification on nano DDS avoids opsonization and reduces phagocytosis.

3.5 Physiology of Tumor and Tumor Targeting Using Nano DDS

3.5.1 Angiogenesis and Tumor Vasculatures

A well understanding and knowledge of the angiogenesis and tumor vasculature characteristics have facilitated effective cancer treatment against various types of cancers. To develop the nano DDS, it is essential to find the biomarkers of the tumor microenvironment and the important differences in normal healthy cells (Liu et al. 2021). The process of angiogenesis in tumor sites promotes new blood vessels with discontinuous epithelium from preexisting vascular systems. The irregular blood vessels present in tumor regions have unusual morphological and physiological conditions dissimilar from normal vasculatures. The discontinuities between epithelial cells or vascular gap openings of tumors are remarkably 10 and 100 times larger in tumor models than in normal tissues. Lack of lymphatic drainage with leakiness favors the passive accumulation of long-circulating macromolecules and into the tumor (Li et al. 2016b; Park et al. 2016; Yang and Gao 2017; Wong et al. 2016). These findings suggest that the nano DDS of certain sizes can penetrate leaky tumor vasculatures and selectively carry the chemotherapeutic drugs to the tumor regions.

3.5.2 Mechanisms of Tumor Targeting by Nano DDS

Tumor-targeted drug delivery can be attained by inherent passive targeting and adopted active targeting strategies. Active drug targeting of chemotherapeutic drugs can be accomplished by conjugating the targeting moiety on the nano DDS. Passive drug targeting is achieved by loading chemotherapeutic drugs into a nano DDS that passively reaches the cancer target site or tissue through the EPR effect. For example, several studies reported that liposomes surface-modified with targeting moiety influenced the drug targeting and it can work as a drug reservoir exhibiting controlled drug release profile and drug accumulation at the tumor site (Kanamala et al. 2016; Masood 2016; Anarjan 2019; Derakhshandeh and Azandaryani 2016; Dai et al. 2016).

3.5.2.1 Passive Tumor Targeting

The EPR effect-mediated chemotherapeutic drug deliveries of nano DDS have been considered one of the strategies to accumulate the drug at the tumor sites. Compared to blood vessels in normal tissues, angiogenic blood vessels at the tumor sites have bigger size openings between nearby vascular endothelial cells. This can help the nano DDS to accumulate at the tumor tissues and then release a higher concentration of the drugs specifically into the tumor cells, thus permitting effective cancer treatment with least systemic side effects. Various studies have demonstrated that EPR plays a pivotal part in passive drug targeting. The EPR effect mainly depends on many factors, such as the nano DDS surface properties, tumor types, and immunogenicity. Passive drug targeting is due to the faulty leaky tumor vasculature with irregular epithelium, reduced level of lymphatic drainage, and lowered uptake of the interstitial fluid, supporting passive targeting of nano DDS in tumors (Kumari et al. 2016; Wakaskar 2017; Masood 2016; Mahato 2017).

3.5.2.2 Active Tumor Targeting

Passive tumor targeting can help the localization of nano DDS at the tumor sites, but it is not able to encourage cellular uptake by tumor cells. This can be accomplished by active tumor targeting. Compared to passive tumor targeting, active tumor targeting strategy relies on a biological communication between targeting ligand on the surface of nano DDS and the receptor on the target tumor cell surface. Active tumor targeting strategy can easily differentiate the normal healthy cells and tumor cells. A large number of targeting ligands and targets have been identified and evaluated for facilitating active drug targeting of nano DDS for various types of cancers (Table 3.1). Such ligands on the surface of nano DDS often actively attach to specific receptors on the tumor cell surface, increasing the drug-containing nano DDS internalization by receptor-mediated endocytosis, improving the therapeutic efficacy, controlling the delivery of chemotherapeutic drugs to healthy tissues, and also decreasing the systemic adverse effects. Hence, active tumor targeting has displayed promising outcomes in circumventing different pitfalls, such as multidrug resistance in tumors and bypassing the blood-brain barrier (Anarjan 2019; He et al. 2020; Lin et al. 2016; Nag and Delehanty 2019).

Table 3.1 Targeting moiety and targets for active targeting of nano DDS

Targets	Targeting moiety	Type of cancer treatment
CD44 receptor	Hyaluronic acid	Human hepatocellular carcinoma, human lung adenocarcinoma, breast cancer (Yang et al. 2018a; Liu et al. 2016b; Song et al. 2017)
CD13	NGR motif peptide	Liver cancer, non-small cell lung cancer (Zheng et al. 2017; Schmidt et al. 2017; Corti et al. 2017)
FA receptor	Folic acid	Breast cancer, liver cancer (Vinothini et al. 2019; Zhang et al. 2018a)
Integrin $\alpha_v\beta_3$	RGD peptide	Prostate tumor, breast cancer (Kim et al. 2017; Wu et al. 2017a)
Prostate-specific membrane antigen	Aptamer	Prostate cancer (Ptacek et al. 2020; Pan et al. 2017)
Transferrin receptor	Transferrin	Breast cancer, lung cancer (Li et al. 2019a; Zhang et al. 2017; Xu et al. 2018)

3.6 Nano DDS for Cancer Treatment

3.6.1 Organic Nanomaterials for Cancer Treatment

Most of the organic nanomaterials (liposomes, solid lipid nanoparticles, polymeric micelles, dendrimers, polymeric nanoparticles, and polymer-drug conjugates) are used as a carrier and targeting system for cancer treatment (Fig. 3.1).

3.6.1.1 Liposomes

Liposomes are described as phospholipid vesicles comprising of one or more concentric bilayer vesicles surrounding the discrete aqueous phase. Because liposome composition is identical to that of cellular membranes, liposomes are safer and biocompatible than other synthetic polymers. Because of the unique structure of liposomes, both hydrophobic and hydrophilic drugs can be incorporated in liposomes. Liposomes can load and hold hydrophobic drugs in the lipid bilayers and hydrophilic drugs in the aqueous core. Liposomes have several advantages than other drug delivery systems, and it is administrated as a potential nanocarrier for drug delivery of chemotherapeutic drugs (Mishra et al. 2018; Ahmed et al. 2019). Currently, there are many liposomal products in the market (Table 3.2) and clinical development (Table 3.3) for cancer treatment.

The types of phospholipids, targeting ligand, PEGylation, and stimuli-sensitive materials determined the charge of the surface of the liposomes. In addition, liposomes with surface modification protect the incorporated drug from degradation, increase the targeting, improve the pharmacokinetic and pharmacodynamics properties, and reduce the toxic side effect of the chemotherapeutic drugs (Patel 2020; Mohamed et al. 2019). PEG conjugation has been identified as a unique strategy for the evasion of RES uptake. The targeting ligands, peptides, and nucleic acid-functionalized liposomes can specifically deliver the chemotherapeutic drugs to the tumor sites. The use of liposome targeted delivery systems in combination therapies of chemotherapy and phototherapy to transport anticancer drugs and photosensitizer can reduce the side effects, significantly enhance the drug accumulation at the target site, and improve the effectiveness of chemotherapy and photodynamic therapy (Cao et al. 2018). Different types of liposomes for targeted anticancer drug delivery are summarized in Table 3.4.

3.6.1.2 Solid Lipid Nanoparticles

Solid lipid nanoparticles (SLNs) are made from biological and safe grade lipids, and it is biocompatible and less toxic compared to polymeric or inorganic nanomaterials. SLNs promote the high drug upload of multiple hydrophobic and hydrophilic drugs. SLNs are a versatile drug delivery system that has been applied to enhance

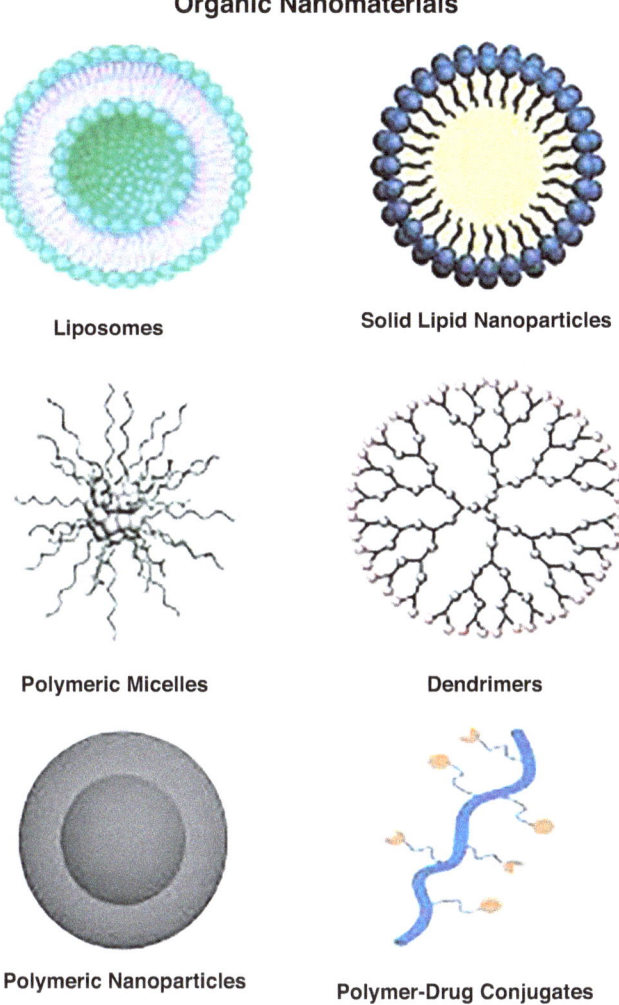

Fig. 3.1 Different types of organic nanomaterials for cancer treatment

the therapeutic effect of chemotherapeutic drugs. Targeted delivery of chemotherapeutic drugs from SLNs reduces the systemic side effects and improves the therapeutic action. SLNs can enhance the chemotherapeutic drug delivery applications for cancer treatment by tumor targeting mechanisms of actions such as passive, active, and codelivery mechanisms (Ganesan et al. 2018; Ramalingam and Ko 2016; Lingayat et al. 2017; Patel et al. 2018). Several studies have reported that SLNs are used as a targeted drug delivery vehicle for different types of tumors. The outcomes of SLNs as carriers of chemotherapeutic drugs are summarized in Table 3.5.

Table 3.2 Approved and marketed liposome-based drugs for cancer treatment

Product name	Drug name	Type of cancer treatment
Abraxane	PTX PTX + gemcitabine	Various cancers (Bobo et al. 2016) Metastatic pancreatic cancer (Saif 2013)
DaunoXome®	Daunorubicin	AIDS-related Kaposi's sarcoma (Dawidczyk et al. 2014)
Doxil®/Caelyx®	DOX	Ovarian cancer, AIDS-related Kaposi's sarcoma and multiple myeloma (Barenholz 2012, 2016)
DepoCyt	Cytarabine	Lymphomatous meningitis (Bobo et al. 2016)
Lipusu®	PTX	Solid tumors (Barkat et al. 2019)
Lipo-dox®	DOX	Kaposi's sarcoma, breast and ovarian cancer (Chou et al. 2015)
Marqibo®	Vincristine	Acute lymphoblastic leukemia (Silverman and Deitcher 2013)
Myocet®	DOX	Metastatic breast cancer (Anselmo and Mitragotri 2016)
Oncaspar	PEGasparaginase	Acute lymphocytic leukemia (Alconcel et al. 2011)

Table 3.3 Liposome-based drugs in clinical development for cancer treatment

Product name	Drug name	Type of cancer treatment
Atragen™	Tretinoin	Acute promyelocytic leukemia, prostate cancer (Nayak et al. 2019)
CPX-1	Irinotecan HCl	Colorectal cancer (Pandey et al. 2016)
EndoTAG®-1	Paclitaxel	Breast cancer, pancreatic cancer (Sofias et al. 2017)
INX-0125	Vinorelbine	Advanced solid tumors (Rahman et al. 2017)
Lipoplatin™	Cisplatin	Pancreatic cancer, lung cancer, breast cancer (Serinan et al. 2018)
L-Annamycin	Annamycin	Acute lymphocytic leukemia (Eryılmaz and Canpolat 2017)
SPI-077	Cisplatin	Head and neck cancer, lung cancer (Zahednezhad et al. 2020)
ThermoDox®	Doxorubicin	Primary hepatocellular carcinoma, breast cancer (Lyon et al. 2017)

3.6.1.3 Polymeric Micelles

Polymeric micelles composed of amphiphilic block copolymers with a hydrophilic corona and hydrophobic core are colloidal nanoparticulate drug delivery systems for chemotherapeutic drugs. Polymeric micelles form a self-assembled structure spontaneously in an aqueous environment. The hydrophobic core of the polymeric micelles possesses a high drug loading of water insoluble chemotherapeutic drugs, and hydrophilic corona provides steric stability to avoid rapid uptake by the RES, resulting in extended drug circulation in the body. In addition to passive drug targeting, polymeric micelles can be surface-modified with targeting ligands for active tumor targeting to enhance the selectivity for cancer cells and improve intracellular delivery of anticancer drugs by receptor-mediated endocytosis while reducing systemic toxicity and severe side effects compared to systemic chemotherapy (Marzbali

Table 3.4 Liposome-based targeted drug delivery systems for cancer treatment

Liposome type	Drug	Ligand	Type of cancer treatment
Plain liposomes	PTX	Aspartic acid	Bone metastasis (Zhao et al. 2020)
	Resveratrol	Transferrin	Glioblastoma (Jhaveri et al. 2018)
	5-Fluorouracil	Transferrin	Colon cancer (Moghimipour et al. 2018)
Cationic liposomes	Daunorubicin and Honokiol	Hyaluronic acid	Breast cancer (Ju et al. 2018)
	Sorafenib	Hyaluronic acid	Cancer (Mo et al. 2018)
	DOX	Asparagine glycine Arginine (NGR) peptide	Breast adenocarcinoma (Yang et al. 2015)
pH-sensitive liposomes	Losartan	TH peptides	Cancer (Jain and Jain 2018)
	DTX	Eph A10	Cancer (Zhang et al. 2018b)
Photothermal therapy	Rapamycin and polypyrrole	Trastuzumab	Breast cancer (Nguyen et al. 2017)
Thermosensitive liposomes	DOX	iRGD	Cancer (Deng et al. 2016)
Thermoresponsive magnetic liposomes	DOX	Magnetic targeting	Cancer (Dai et al. 2017)
Magnetic liposomes	Curcumin	Magnetic targeting	Cancer (Hardiansyah et al. 2017)

and Khosroushahi 2017; Gothwal et al. 2016; Biswas et al. 2016). Currently, many chemotherapeutic drug-loaded polymeric micelles are evaluated for effective cancer treatment (Table 3.6).

3.6.1.4 Dendrimers

Dendrimers are highly branched globular macromolecules with their 3D nonpolymeric architectures: a central core, a corona with functional groups, and a hyperbranched mantle. Dendrimers' unique properties like polyvalency, well-defined molecular weight, nanosize, the high degree of branching, water solubility, and simple synthesis procedure make them promising drug carrier systems for anticancer drugs. The dendrimers' biological effect is initiated by terminal moieties, and the dendrimers seem to be excellent candidates for carriers of anticancer drugs. A variety of dendrimers, including PAMAM, PEG, PPI, and PLL, have been successfully developed for drug delivery applications, and the PAMAM is most widely employed for targeted cancer therapy. Surface modification or conjugation of

Table 3.5 Application of SLNs against different types of cancers

Drug/formulations	Ligand	Type of cancer treatment
PTX/SLNs	Tyr-3-octreotide	Antiangiogenic and anti-glioma (Banerjee et al. 2016)
	Folate-grafted chitosan	Lung cancer (Rosiere et al. 2018)
	TAT	Cervical cancer (Liu et al. 2017a)
Methotrexate/SLNs	Protein functionalization	Brain cancer (Muntoni et al. 2019)
	Fucose	Brain cancer (Garg et al. 2016)
Curcumin/SLNs		Breast cancer (Wang et al. 2018a)
Resveratrol/SLNs		Breast cancer (Wang et al. 2017a)
Erlotinib/SLNs		Non-small lung cancer (Bakhtiary et al. 2017)
Omega-3 PUFA/SLNs		Colorectal cancer (Serini et al. 2018)
Linalool/SLNs		Liver cancer (Rodenak-Kladniew et al. 2017)
DOX/SLNs	cRGD	Breast cancer (Zheng et al. 2019)
IR-780 dye/SLNs	cRGD	Photothermal therapy (Kuang et al. 2017)

Table 3.6 Application of polymeric micelles against different types of cancers

Drug	Polymeric micelles	Ligand	Type of cancer treatment
DOX	Poloxamer 407 and vitamin TPGS	pH-responsive FA	Ovarian carcinoma (Butt et al. 2015)
	PLA-PEG	Aptamer	Prostate cancer (Xu et al. 2013)
	Cholic acid – PE	–	Colorectal cancer (Amjad et al. 2012)
	Succinylated gelatin micelles	Folic acid	Breast cancer (Wang et al. 2018b)
	PLGA-PEG	–	Cancer (Ma et al. 2016)
PTX	Redox-responsive micelles	Albumin	Breast cancer (Zhang et al. 2018c)
	Pluronic F87-PLA/TPGS	Folate	Cancer (Xiong et al. 2017)
	Pluronic F127-PEG	–	Ovarian cancer (Zhai et al. 2018)

dendrimers with PEG and other ligands can help reduce the cytotoxicity of dendrimers and enhance plasma circulation time and accumulation of tumor through the EPR effect (Kaur et al. 2016; Augustus et al. 2017; Munir et al. 2016; Parajapati et al. 2016; Abedi-Gaballu et al. 2018; Sherje et al. 2018). Numerous researches that have been conducted to study the application of dendrimers in cancer treatment are presented in Table 3.7.

3.6.1.5 Polymeric Nanoparticles

The polymeric nanoparticulate system from natural and synthetic biodegradable polymers has earned more attention due to their biodegradability, biocompatibility, tailorability and stability, ease of coating or surface modification, and low cost. Polymeric nanoparticles, in general, can be used to improve solubility, controlled

Table 3.7 Dendrimer-based nano DDS for cancer treatment

Polymer	Drug	Modification	Type of cancer treatment
PAMAM	DOX	–	Breast cancer (Khodadust et al. 2014)
	DTX	Trastuzumab	Breast cancer (Kulhari et al. 2016)
	Camptothecin	N-acetyl-D-glucosamine	Lung cancer (Pooja et al. 2020)
	pDNA/siRNA	–	Cancer (Li et al. 2018a)
PLL	DOX	PEG	Cancer (Mehta et al. 2018)

release, and bioavailability for systemic delivery of anticancer drugs. Drug-loaded polymeric nanoparticles can be developed to actively or passively accumulate in sites of the tumor by controlling their particle size or surface functionalizing with targeting moieties. Polymers like hyaluronic acid and pullulan are used to activate nanoparticles for active targeted drug delivery. These polymers degrade in physiological body conditions, and by-products of the polymers are not harmful to the body. Various natural and synthetic polymers-based nanoparticles were developed and reported for cancer treatment and diagnosis (Masood 2016; Prasad et al. 2017; Conte et al. 2016; Wong et al. 2020; Espinosa-Cano et al. 2018; Taghipour-Sabzevar et al. 2019). Natural and synthetic polymers-based nano DDS for cancer treatment are summarized in Tables 3.8 and 3.9.

3.6.1.6 Polymer-Drug Conjugates

Polymer-drug conjugates (PDCs) can be prepared as nano DDS by covalently conjugating one or more drugs to a polymer backbone before the synthesis of nanoparticles. PDCs are identified as the most examined type of nano DDS, and currently, many PDs in clinical trials and several polymer-drug conjugates are successfully transformed into clinical practice. For example, N-(2-hydroxypropyl) methacrylamide-DOX was the first chemotherapeutic PDC to reach clinical trial studies about 22 years ago. The conjugation of therapeutic drugs to polymers provides many benefits, including improved drug solubilization, stability, controlled drug delivery, enhanced efficacy and improved pharmacokinetics, biodistribution, as well as reduced toxicity and immunogenicity. The main advantage of using PDCs is that the physical and chemical characteristics of polymers can be modified to reduce the toxicity and improve the therapeutic efficacy of the loaded chemotherapeutics. In addition, PDCs have displayed increased accumulation of tumors, improved therapeutic index, prolonged circulation, controlled release of the anticancer drugs, and active tumor uptake by active targeting (Ekladious et al. 2019; Thanou and Duncan 2003; Vicent and Duncan 2006; Li and Wallace 2008) (Table 3.10).

Table 3.8 Natural polymers-based nano DDS for cancer treatment

Polymer	Conjugation	Drug	Type of cancer treatment
Chitosan	N-acetyl histidine and arginine	DOX	Breast (Raja et al. 2017)
	Trimethyl and folic acid	PTX	Hepatoma and colon (He and Yin 2017)
	TPGS and transferrin	DTX	Brain (Agrawal et al. 2017)
Alginate	PEI and FA	Curcumin	Cervical (Anirudhan et al. 2017)
	Glycyrrhetinic acid	Tetravalent platinum	Liver and lung (Wang et al. 2019)
	Chitosan	DOX	Breast (Katuwavila et al. 2016)
Pullulan	Arabinogalactan	DOX	Liver (Pranatharthiharan et al. 2017)
	PEI and MSA	DOX	Glioma (Priya and Rekha 2017)
	Folic acid	PTX	Liver (Huang et al. 2018)
Dextran	Folic acid	DOX	Breast and lymphoma (Tang et al. 2018a)
	Albumin	PTX	Colorectal (Zhang et al. 2019)
	Folic acid	Resveratrol	Lung (Zhao et al. 2017)
HA	Chitosan	5-Fluorouracil	Lung and liver (Wang et al. 2017b)
	PLGA	PTX	Breast (Cerqueira et al. 2017)
	PLGA	DTX	Lung (Wu et al. 2017b)

Table 3.9 Synthetic polymers-based nano DDS for cancer treatment

Polymer	Conjugation	Drug	Type of cancer treatment
PLGA	Folate-PEG	GEM and DTX	Ovarian (Li et al. 2019b)
	Transferrin	PTX	Breast and brain (Cui et al. 2017)
	Chondroitin sulfate	DOX	Glioma (Liu et al. 2019b)
PLA	Hydroxyethyl starch	DOX	Liver (Yu et al. 2017)
	PEG	DTX	Ovarian (Qi et al. 2017)
	FA-PEG	PTX	Ovarian (Yao et al. 2018)
PCL	PEG	Curcumin	Liver (Guo et al. 2017)
	PEG	Artemisinin	Breast (Manjili et al. 2018)
	TPGS	Sorafenib	Liver (Tang et al. 2018b)
PEG	Glycyrrhetinic acid-PCL	Curcumin	Liver (Feng et al. 2017)
	Lactoferrin-PLGA	Shikonin	Glioma (Li et al. 2018b)

3.6.2 Inorganic Nanomaterials for Cancer Treatment

Inorganic nanomaterials have been intensively studied for cancer therapy and diagnostic imaging due to their great advantages, such as high drug loading, large surface area, improved bioavailability, reduced toxic side effects and controlled release of anticancer drugs, and their tolerance to most organic solvents. Mesoporous silica nanoparticles, gold nanoparticles, magnetic nanoparticles, carbon nanotubes, and

Table 3.10 Polymer-drug conjugates for cancer treatment

Drug	Polymer	Conjugates	Type of cancer treatment
Dihydroartemisinin	HA	HA-dihydroartemisinin	Lung cancer (Kumar et al. 2019)
DOX	N-(2-hydroxypropyl) methacrylamide	N-(2-hydroxypropyl) methacrylamide-DOX	Breast cancer (Bobde et al. 2020)
	PEG	PEG-DOX	Breast cancer (Gu et al. 2018)
	Poly-l-glutamic acid	Poly-l-glutamic acid-DOX	Non-small cell lung cancer (Li et al. 2013)
DOX and GEM	HA	HA-DOX-GEM	Breast and lung cancer (Alven et al. 2020)
FA and trastuzumab	PEG	PEG-FA-trastuzumab	Breast and lung cancer (Alven et al. 2020)
PTX	N-(2-hydroxypropyl methyl) acrylamide	N-(2-hydroxypropyl methyl) acrylamide copolymer-gadolinium-PTX	Breast and lung cancer (Alven et al. 2020)
	HA	HA-PTX	Cancer (Wang et al. 2017c)
	PEG	PEG-PTX	Lung cancer (Luo et al. 2016)
GEM	Poly (l-glutamic acid)-g-methoxy PEG	Poly (l-glutamic acid)-g-methoxy PEG-GEM	Cancer (Yang et al. 2018b)

quantum dots are commonly used in cancer treatment and diagnosis in various ways (Fig. 3.2) (Khafaji et al. 2019; Veeranarayanan and Maekawa 2019; Liu et al. 2017b).

3.6.2.1 Mesoporous Silica Nanoparticles (Senapati et al. 2018; Ahmadi Nasab et al. 2018; Moreira et al. 2016; de Oliveira Freitas et al. 2017; Yang and Yu 2016; Saini and Bandyopadhyaya 2019)

Silica nanoparticles are extensively used nanoparticle systems in cancer treatment due to its various benefits such as easy synthesis, well-controlled diameter, adjustable pore volume, and potential surface modification. There are two types of silica nanoparticles (core or shell silica nanoparticles and mesoporous silica nanoparticles (MSNs)) established for cancer treatment. Of the two types, MSNs are mostly used as a nano DDS in cancer treatment. One study demonstrated that gemcitabine-loaded MSNs are used to treat pancreatic cancer. One research group developed the rod-shaped magnetic MSNs for suicide gene therapy. The shapes of the MSNs also

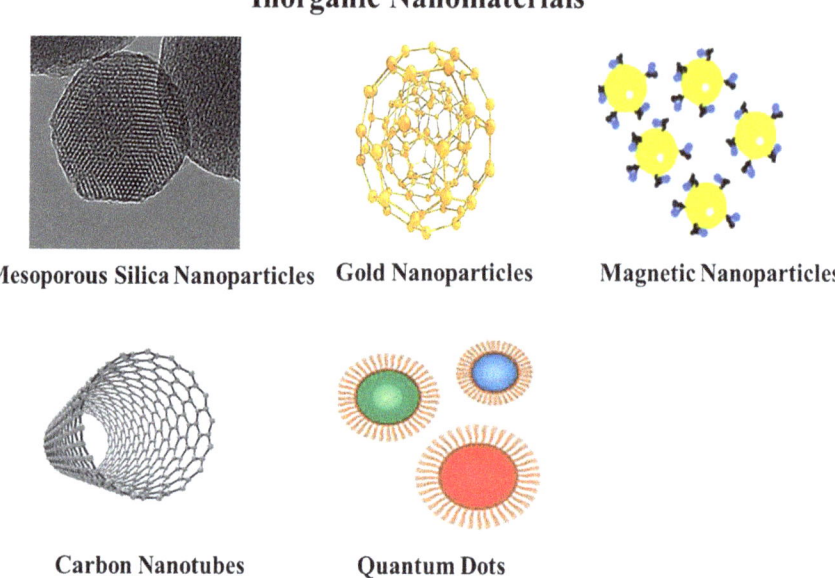

Fig. 3.2 Different types of inorganic nanomaterials for cancer treatment

play a vital role in drug delivery applications. Compared to spherical MSNs, rod shape-like MSNs displayed higher drug loading, better drug release, and gene delivery.

The research carried out by Lee et al. showed how MSNs decorated with doxorubicin-loaded multiple magnetite nanocrystals promoted effective cell death in a melanoma model, confirming passive targeting and nanoparticle accumulation in the tumor site. Huan et al. used MSNs modified with polyethyleneimine/PEG to deliver doxorubicin jointly with P-glycoprotein siRNA. This research explained that nanoparticles were efficiently biodistributed, resulting in 8% of the EPR effect at the tumor site. MSNs can also be surface-functionalized with various types of ligand molecules such as aptamers, growth factors, peptides, and vitamins to actively target tumors via receptor-mediated endocytosis. In the study carried out by Kayuan et al., DOX-loaded HB5 aptamer-functionalized MSNs were used for combined chemo-photothermal therapies. This study verified that combination therapies promote cancer cell killing compared to chemo-photothermal therapy alone. MSNs achieve a satisfactory level of active targeting and reduce toxic side effects in the healthy normal cells.

3.6.2.2 Gold Nanoparticles (Sztandera et al. 2018; Peng and Liang 2019; Kumar et al. 2012; Singh et al. 2018)

Gold nanoparticles (GNPs) have been investigated for its potential application in cancer treatment, diagnostics, and targeted drug delivery. Current researches confirm numerous advantages of GNPs for cancer treatment, primarily due to enabling the control of preparation of GNPs with multiple sizes and shapes and the possibility of surface functionalization on GNPs with various functional and targeting agents. Many features of GNPs are related to their shape and size. The size of spherical GNPs influenced plasma concentration, circulation time, and cellular uptake. It was also reported that the smaller particles of GNPs permeated into the blood-brain barrier, deep layers of skin, and placental barrier. Surface functionalization of GNPs provides significant effects on plasma half-life, protection against aggregation, biocompatibility, preventing the removal by the MPS and RES, targeted transport and drug accumulation at the desired site. For the GNP-based drug delivery system, passive targeting, active targeting, or a combination of both strategies can improve tumor accumulation. A remarkable approach confirming the intracellular delivery of chemotherapeutic drugs involves their conjugation to the surface of GNPs through thiol functional groups. The examples of chemotherapeutic drugs conjugated with GNPs are listed in Table 3.11.

Due to their exceptional properties of absorption and scattering of electromagnetic radiation, GNPs are of specific interest for the PTT in cancer treatment. This PTT treatment procedure involves the utilization of electromagnetic radiation or laser radiation to generate local heating and hyperthermia for the thermal destruction of cancerous cells. The PTT efficacy may be additionally improved by the application of photothermal compounds such as transition metal oxide/sulfide nanomaterials and nanocarbons, enabling an improved transformation of light into heat.

Table 3.11 Gold nanoparticles for cancer treatment

Nanomaterials	Targeting agents	Drug
Gold nanoparticles	PEG	Tamoxifen
	PEG, tumor necrosis alpha	PTX
	3-Mercaptopropionic acid	Daunorubicin
	PEG, folate	DOX
	Poly(L-aspartate), PEG, folate	DOX
	–	Methotrexate
	–	Gemcitabine
	Photocleavable and zwitterionic thiol ligands	5-Fluorouracil

3.6.2.3 Magnetic Nanoparticles (Zhang et al. 2018d; Kolosnjaj-Tabi and Wilhelm 2017; Fathi Karkan et al. 2017; Fathi et al. 2020; Lungu et al. 2016)

Magnetic nanoparticles (MNPs) have been discovered as a potential carrier system to modify the pharmacokinetics of loaded drugs, decrease the cytotoxicity, improve the controlled release, and increase the half-life. Due to the unique properties of higher magnetic moments and surface to volume ratios, it can be used for hyperthermia therapy of cancer treatment and targeted delivery. MNPs are in magnetic resonance imaging to enhance the image contrast of targeted tumor tissues. MNPs can be functionalized with high affinity ligands such as peptides and antibodies to enhance the selectivity further and localize MNPs at the tumor sites. Recently, the MNP application in biosensors has been extensively studied for rapid cancer diagnosis and prevention of cancer metastasis. Various types of MNPs employed in cancer treatment and diagnosis are summarized in Table 3.12.

3.6.2.4 Carbon Nanotubes (Chen et al. 2017; Son et al. 2016; Pardo et al. 2018)

Carbon nanotubes (CNTs) are very popular systems for cancer treatment and diagnosis due to their many unique properties such as structure and high specific surface area to volume. CNTs are classified into single-walled carbon nanotubes and multi-walled carbon nanotubes based on the number of graphene sheets used for the preparation. CNTs have been investigated in all the cancer treatment modalities, including thermal, photodynamic, and gene therapy, drug delivery, lymphatic targeted chemotherapy, and diagnostic techniques. Recently developed single-walled carbon nanotube-based drug delivery systems for cancer treatment are summarized in Table 3.13. CNTs may help the attached chemotherapeutic drugs to penetrate through the target cell to treat cancer.

The CNTs are used as a photosensitizer for photodynamic therapy. CNTs are used as a contrast medium for diagnostic imaging techniques, and it can be used in

Table 3.12 Magnetic nanoparticles for cancer treatment

Drug	Magnetic nanoparticles	Type of cancer treatment
Methotrexate	Chitosan grafted pH and thermoresponsive	Ovarian cancer (Fathi et al. 2020)
Doxorubicin	FA conjugated Fe3O4	Cancer (Rana et al. 2016)
	PEG coated	Hyperthermia therapy (Dabbagh et al. 2019)
	Dual stimuli responsive polymer modified	MR imaging (Bhattacharya et al. 2016)
	pH-sensitive polymer coating	Cancer, pH-sensitive release (Lungu et al. 2016)

Table 3.13 Carbon nanotube-based systems for cancer treatment

Drug	Surface functionalization	Type of cancer treatment
DOX	FA	Chemo-photothermal (Wang et al. 2017d)
PTX	Riboflavin and thiamine	Cancer (Singh et al. 2016a)
DOX	Polyphosphazene coated	Redox responsive and photothermal (Wang et al. 2017e)
DOX	Polyampholyte	Cervical cancer (Phan et al. 2020)
Temozolomide	Vitamin B6 and PEG	Cancer (Saberinasab et al. 2019)
DOX	Hyaluronic acid coated	Breast cancer (Liu et al. 2019c)
DOX	pH-sensitive nanogels	Glioblastoma (Seyfoori et al. 2019)
DTX	Vitamin E TPGS	Lung cancer (Singh et al. 2016b)

ultrasonography, photoacoustic imaging, PET, and MRM for cancer diagnostic applications.

3.6.2.5 Quantum Dots (Zhao et al. 2016; Fang et al. 2017; Lee et al. 2017)

Quantum dots (QDs) are nanosized crystals comprised of a semiconductor core within a shell composed of second semiconductor material. QDs have outstanding optical properties, such as high brightness, tunable emission spectra, and resistance to photo-bleaching. Quantum dots have been used in targeting and localizing tumors and sentinel lymph node mapping in vivo. New imaging techniques like quantum dots resolve the limitations of sensitivity and specificity from current imaging techniques like X-ray, ultrasound, radionuclide imaging, computed tomography, and MRI. Recent studies in surface functionalization of QDs improve their potential application in imaging of cancer. Bioconjugation of QDs with peptides and antibodies can be used for tumor-targeted drug delivery, nanodiagnostics, imaging, and photodynamic therapy. The application of quantum dot conjugates is listed in Table 3.14.

3.7 Challenges and Future Perspectives

Despite numerous advanced technologies in the production of safe biopolymers and nanomaterials, there remain controversies regarding the safety of nanoformulations. Although the benefits of some biopolymers, dendrimers, and metal-based inorganic nanomaterials are remarkable, toxicity remains a serious problem. It has been proven, for example, that PEI and excessive positive charges of dendrimers destabilize the cell membrane. Thus, advancements in biopolymer synthesis and purification techniques promise to reduce side effects and enhance treatment efficacy. The instability, immune response, potential toxicity, and chronic inflammation

Table 3.14 Quantum dots for cancer treatment and diagnosis

Conjugates	Application
DOX-D-glucosamine-folate-QD conjugates	Cancer cell imaging and treatment (Ranjbar-Navazi et al. 2018)
Antibody-QD conjugates	In vitro and in vivo molecular imaging (Tsuboi et al. 2017)
Titanium nitride MXene QDs	Phototheranostics in both NIR-I/II bio windows (Shao et al. 2020)
Polydopamine-black phosphorus QDs	Cancer theranostics (Li et al. 2019c)
pH-responsive fluorescent graphene QDs	Fluorescence-guided cancer surgery and diagnosis (Fan et al. 2017)
Aptamer conjugated graphene QDs	Photothermal therapy and photodynamic therapy (Cao et al. 2017)
Graphitic-C3N4 QDs	Photodynamic therapy (Chu et al. 2017)

challenges for micelles and inorganic nanomaterials need to be focused so that more effective cancer treatment strategies can be developed. Combination therapy with nanomaterials for different types of cancers remains a challenge because of the distinct cancer development mechanisms. For targeted drug therapy, inorganic nanomaterials and micelles can be surface-functionalized with target agents such as magnetic, light, and pH imaging contrast agents; the major limitation of these clinical treatment methods is the poor tissue penetration. All the nanomaterials are not biodegradable so that it can be retained and circulated in the body system for a more extended period after administration. Various research and strategies aimed at overcoming all these challenges will facilitate nanomaterial usage as a drug delivery system and eventually enhance patient survival.

The future perspective of stimuli-responsive nanomaterials can be obtained by various strategies, including enzymatic activation, pH variants, magnetic fields, ultrasound, light, redox potential, and thermal gradients for efficient cancer treatment and diagnosis. Further advancements in the nanomaterials system can improve their application in localizing metastasis, quantitative measurement of molecular targets, and monitoring the efficacy and tracking of drug delivery.

3.8 Conclusion

This chapter has summarized a variety of nanomaterials that are either being used or have the potential to be used as nano drug delivery systems for cancer treatment. Nanomaterials-based cancer treatment has shown significant advantages and new strategies over conventional cancer treatment. Passive or active targeting can significantly remove the systemic side effects of conventional chemotherapies. Targeted drug delivery has made a considerable impact on selective recognizing of the tumor tissues, controlled drug delivery, and overcoming limitations of the conventional chemotherapies. Numerous nanomedicines have been approved by the FDA and indicated satisfactory performance in clinical practice. Although some nanomaterials have not been approved upon their clinical translation, new strategies and promising nanomaterials that are under progress show great assurance, thus providing hope for innovative cancer treatment choices in the near future.

References

Abedi-Gaballu F, Dehghan G, Ghaffari M, Yekta R, Abbaspour-Ravasjani S, Baradaran B et al (2018) PAMAM dendrimers as efficient drug and gene delivery nanosystems for cancer therapy. Appl Mater Today 12:177–190

Agrawal P, Singh RP, Kumari L, Sharma G, Koch B, Rajesh CV et al (2017) TPGS-chitosan cross-linked targeted nanoparticles for effective brain cancer therapy. Mater Sci Eng C 74:167–176

Ahmad N, Alam MA, Ahmad R, Naqvi AA, Ahmad FJ (2018a) Preparation and characterization of surface-modified PLGA-polymeric nanoparticles used to target treatment of intestinal cancer. Artif Cells Nanomed Biotechnol 46(2):432–446

Ahmad N, Ahmad R, Alam MA, Ahmad FJ (2018b) Enhancement of oral bioavailability of doxorubicin through surface modified biodegradable polymeric nanoparticles. Chem Cent J 12(1):65

Ahmadi Nasab N, Hassani Kumleh H, Beygzadeh M, Teimourian S, Kazemzad M (2018) Delivery of curcumin by a pH-responsive chitosan mesoporous silica nanoparticles for cancer treatment. Artif Cells Nanomed Biotechnol 46(1):75–81

Ahmed KS, Hussein SA, Ali AH, Korma SA, Lipeng Q, Jinghua C (2019) Liposome: composition, characterisation, preparation, and recent innovation in clinical applications. J Drug Target 27(7):742–761

Alconcel SN, Baas AS, Maynard HD (2011) FDA-approved poly (ethylene glycol)–protein conjugate drugs. Polym Chem 2(7):1442–1448

Alven S, Nqoro X, Buyana B, Aderibigbe BA (2020) Polymer-drug conjugate, a potential therapeutic to combat breast and lung cancer. Pharmaceutics 12(5):406

Amjad MW, Amin MCIM, Katas H, Butt AM (2012) Doxorubicin-loaded cholic acid-polyethyleneimine micelles for targeted delivery of antitumor drugs: synthesis, characterization, and evaluation of their in vitro cytotoxicity. Nanoscale Res Lett 7(1):687

Anarjan FS (2019) Active targeting drug delivery nanocarriers: ligands. Nano-Struct Nano-Objects 19:100370

Anirudhan TS, Anila MM, Franklin S (2017) Synthesis characterization and biological evaluation of alginate nanoparticle for the targeted delivery of curcumin. Mater Sci Eng C 78:1125–1134

Anselmo AC, Mitragotri S (2016) Nanoparticles in the clinic. Bioeng Transl Med 1(1):10–29

Arms L, Smith DW, Flynn J, Palmer W, Martin A, Woldu A et al (2018) Advantages and limitations of current techniques for analyzing the biodistribution of nanoparticles. Front Pharmacol 9:802

Augustus EN, Allen ET, Nimibofa A, Donbebe W (2017) A review of synthesis, characterization and applications of functionalized dendrimers. Am J Polym Sci 7(1):8–14

Axelrad JE, Bazarbashi A, Zhou J, Castañeda D, Gujral A, Sperling D et al (2020) Hormone therapy for cancer is a risk factor for relapse of inflammatory bowel diseases. Clin Gastroenterol Hepatol 18(4):872–80.e1

Bakhtiary Z, Barar J, Aghanejad A, Saei AA, Nemati E, Ezzati Nazhad Dolatabadi J et al (2017) Microparticles containing erlotinib-loaded solid lipid nanoparticles for treatment of non-small cell lung cancer. Drug Dev Ind Pharm 43(8):1244–1253

Banerjee I, De K, Mukherjee D, Dey G, Chattopadhyay S, Mukherjee M et al (2016) Paclitaxel-loaded solid lipid nanoparticles modified with Tyr-3-octreotide for enhanced anti-angiogenic and anti-glioma therapy. Acta Biomater 38:69–81

Barenholz YC (2012) Doxil®—the first FDA-approved nano-drug: lessons learned. J Control Release 160(2):117–134

Barenholz YC (2016) Doxil®–the first FDA-approved Nano-drug: from basics via CMC, cell culture and animal studies to clinical use. Nanomed Design Deliv Detect 51:315–345

Barkat MA, Beg S, Pottoo FH, Ahmad FJ (2019) Nanopaclitaxel therapy: an evidence based review on the battle for next-generation formulation challenges. Nanomedicine 14(10):1323–1341

Baskar R, Itahana K (2017) Radiation therapy and cancer control in developing countries: can we save more lives? Int J Med Sci 14(1):13

Bhattacharya D, Behera B, Sahu SK, Ananthakrishnan R, Maiti TK, Pramanik P (2016) Design of dual stimuli responsive polymer modified magnetic nanoparticles for targeted anti-cancer drug delivery and enhanced MR imaging. New J Chem 40(1):545–557

Bishop AJ, Zagars GK, Torres KE, Bird JE, Feig BW, Guadagnolo BA (2018a) Malignant peripheral nerve sheath tumors: a single institution's experience using combined surgery and radiation therapy. Am J Clin Oncol 41(5):465

Bishop AJ, Zagars GK, Demicco EG, Wang W-L, Feig BW, Guadagnolo BA (2018b) Soft tissue solitary fibrous tumor: combined surgery and radiation therapy results in excellent local control. Am J Clin Oncol 41(1):81–85

Biswas S, Kumari P, Lakhani PM, Ghosh B (2016) Recent advances in polymeric micelles for anticancer drug delivery. Eur J Pharm Sci 83:184–202

Bobde Y, Biswas S, Ghosh B (2020) PEGylated N-(2 hydroxypropyl) methacrylamide-doxorubicin conjugate as pH-responsive polymeric nanoparticles for cancer therapy. React Funct Polym 151:104561

Bobo D, Robinson KJ, Islam J, Thurecht KJ, Corrie SR (2016) Nanoparticle-based medicines: a review of FDA-approved materials and clinical trials to date. Pharm Res 33(10):2373–2387

Bray F, Ferlay J, Soerjomataram I, Siegel RL, Torre LA, Jemal A (2018) Global cancer statistics 2018: GLOBOCAN estimates of incidence and mortality worldwide for 36 cancers in 185 countries. CA Cancer J Clin 68(6):394–424

Butt AM, Amin MCIM, Katas H (2015) Synergistic effect of pH-responsive folate-functionalized poloxamer 407-TPGS-mixed micelles on targeted delivery of anticancer drugs. Int J Nanomedicine 10:1321

Cabrera AR, Kirkpatrick JP, Fiveash JB, Shih HA, Koay EJ, Lutz S et al (2016) Radiation therapy for glioblastoma: executive summary of an American Society for Radiation Oncology evidence-based clinical practice guideline. Pract Radiat Oncol 6(4):217–225

Cao Y, Dong H, Yang Z, Zhong X, Chen Y, Dai W et al (2017) Aptamer-conjugated graphene quantum dots/porphyrin derivative theranostic agent for intracellular cancer-related microRNA detection and fluorescence-guided photothermal/photodynamic synergetic therapy. ACS Appl Mater Interfaces 9(1):159–166

Cao J, Chen Z, Chi J, Sun Y, Sun Y (2018) Recent progress in synergistic chemotherapy and phototherapy by targeted drug delivery systems for cancer treatment. Artif Cells Nanomed Biotechnol 46(sup1):817–830

Cerqueira BBS, Lasham A, Shelling AN, Al-Kassas R (2017) Development of biodegradable PLGA nanoparticles surface engineered with hyaluronic acid for targeted delivery of paclitaxel to triple negative breast cancer cells. Mater Sci Eng C 76:593–600

Chen Z, Zhang A, Wang X, Zhu J, Fan Y, Yu H et al (2017) The advances of carbon nanotubes in cancer diagnostics and therapeutics. J Nanomater 2017:1–13

Choi J-S, Meghani N (2016) Impact of surface modification in BSA nanoparticles for uptake in cancer cells. Colloids Surf B Biointerfaces 145:653–661

Chou H, Lin H, Liu JM (2015) A tale of the two PEGylated liposomal doxorubicins. Onco Targets Ther 8:1719

Chowdhury S, Yusof F, Salim WWAW, Sulaiman N, Faruck MO (2016) An overview of drug delivery vehicles for cancer treatment: nanocarriers and nanoparticles including photovoltaic nanoparticles. J Photochem Photobiol B Biol 164:151–159

Chu X, Li K, Guo H, Zheng H, Shuda S, Wang X et al (2017) Exploration of graphitic-C3N4 quantum dots for microwave-induced photodynamic therapy. ACS Biomater Sci Eng 3(8):1836–1844

Conte C, Maiolino S, Pellosi DS, Miro A, Ungaro F, Quaglia F (2016) Polymeric nanoparticles for cancer photodynamic therapy. Light-responsive nanostructured systems for applications in nanomedicine, 61–112

Corti A, Fiocchi M, Curnis F (2017) Targeting CD13 with Asn-Gly-Arg (NGR) peptide-drug conjugates. In Next-generation therapies and technologies for immune-mediated inflammatory diseases. Springer, Cham , pp 101–122

Cui Y-N, Xu Q-X, Davoodi P, Wang D-P, Wang C-H (2017) Enhanced intracellular delivery and controlled drug release of magnetic PLGA nanoparticles modified with transferrin. Acta Pharmacol Sin 38(6):943–953

Dabbagh A, Hedayatnasab Z, Karimian H, Sarraf M, Yeong CH, Madaah Hosseini HR et al (2019) Polyethylene glycol-coated porous magnetic nanoparticles for targeted delivery of chemotherapeutics under magnetic hyperthermia condition. Int J Hyperthermia 36(1):104–114

Dai L, Liu J, Luo Z, Li M, Cai K (2016) Tumor therapy: targeted drug delivery systems. J Mater Chem B 4(42):6758–6772

Dai M, Wu C, Fang H-M, Li L, Yan J-B, Zeng D-L et al (2017) Thermo-responsive magnetic liposomes for hyperthermia-triggered local drug delivery. J Microencapsul 34(4):408–415

Dawidczyk CM, Kim C, Park JH, Russell LM, Lee KH, Pomper MG et al (2014) State-of-the-art in design rules for drug delivery platforms: lessons learned from FDA-approved nanomedicines. J Control Release 187:133–144

de Oliveira Freitas LB, de Melo Corgosinho L, Faria JAQA, dos Santos VM, Resende JM, Leal AS et al (2017) Multifunctional mesoporous silica nanoparticles for cancer-targeted, controlled drug delivery and imaging. Microporous Mesoporous Mater 242:271–283

Deng Z, Xiao Y, Pan M, Li F, Duan W, Meng L et al (2016) Hyperthermia-triggered drug delivery from iRGD-modified temperature-sensitive liposomes enhances the anti-tumor efficacy using high intensity focused ultrasound. J Control Release 243:333–341

Deodhar GV, Adams ML, Trewyn BG (2017) Controlled release and intracellular protein delivery from mesoporous silica nanoparticles. Biotechnol J 12(1):1600408

Derakhshandeh K, Azandaryani AH (2016) Active-targeted nanotherapy as smart cancer treatment. Smart Drug Delivery System:91–116

Derks M, van Lonkhuijzen LR, Bakker RM, Stiggelbout AM, de Kroon CD, Westerveld H et al (2017) Long-term morbidity and quality of life in cervical cancer survivors: a multicenter comparison between surgery and radiotherapy as primary treatment. Int J Gynecol Cancer. 27(2):350–356

Eeles RA, Morden JP, Gore M, Mansi J, Glees J, Wenczl M et al (2016) Adjuvant hormone therapy may improve survival in epithelial ovarian cancer: results of the AHT randomized trial. Obstet Gynecol Surv 71(4):223–224

Ekladious I, Colson YL, Grinstaff MW (2019) Polymer–drug conjugate therapeutics: advances, insights and prospects. Nat Rev Drug Discov 18(4):273–294

Eryılmaz E, Canpolat C (2017) Novel agents for the treatment of childhood leukemia: an update. Onco Targets Ther 10:3299

Espinosa-Cano E, Palao-Suay R, Aguilar MR, Vázquez B, San Román J (2018) Polymeric nanoparticles for cancer therapy and bioimaging. In: Nanooncology. Springer, pp 137–172.

Faisca Phillips AM (2019) Recent developments in anti-cancer drug research. Curr Med Chem 26(41):7282–7284

Fan Z, Zhou S, Garcia C, Fan L, Zhou J (2017) pH-Responsive fluorescent graphene quantum dots for fluorescence-guided cancer surgery and diagnosis. Nanoscale 9(15):4928–4933

Fang M, Chen M, Liu L, Li Y (2017) Applications of quantum dots in cancer detection and diagnosis: a review. J Biomed Nanotechnol 13(1):1–16

Fathi Karkan S, Mohammadhosseini M, Panahi Y, Milani M, Zarghami N, Akbarzadeh A et al (2017) Magnetic nanoparticles in cancer diagnosis and treatment: a review. Artif Cells Nanomed Biotechnol 45(1):1–5

Fathi M, Barar J, Erfan-Niya H, Omidi Y (2020) Methotrexate-conjugated chitosan-grafted pH- and thermo-responsive magnetic nanoparticles for targeted therapy of ovarian cancer. Int J Biol Macromol 154:1175–1184

Feng R, Deng P, Song Z, Chu W, Zhu W, Teng F et al (2017) Glycyrrhetinic acid-modified PEG-PCL copolymeric micelles for the delivery of curcumin. React Funct Polym 111:30–37

Ganesan P, Ramalingam P, Karthivashan G, Ko YT, Choi D-K (2018) Recent developments in solid lipid nanoparticle and surface-modified solid lipid nanoparticle delivery systems for oral delivery of phyto-bioactive compounds in various chronic diseases. Int J Nanomed 13:1569

Garg NK, Singh B, Jain A, Nirbhavane P, Sharma R, Tyagi RK et al (2016) Fucose decorated solid-lipid nanocarriers mediate efficient delivery of methotrexate in breast cancer therapeutics. Colloids Surf B Biointerfaces 146:114–126

Ghahremani F, Shahbazi-Gahrouei D, Kefayat A, Motaghi H, Mehrgardi MA, Javanmard SH (2018) AS1411 aptamer conjugated gold nanoclusters as a targeted radiosensitizer for megavoltage radiation therapy of 4T1 breast cancer cells. RSC Adv 8(8):4249–4258

Ghasemiyeh P, Mohammadi-Samani S (2018) Solid lipid nanoparticles and nanostructured lipid carriers as novel drug delivery systems: applications, advantages and disadvantages. Res Pharm Sci 13(4):288

Ghasemiyeh P, Mohammadi-Samani S (2020) Potential of nanoparticles as permeation enhancers and targeted delivery options for skin: advantages and disadvantages. Drug Des Devel Ther. 14:3271

Gothwal A, Khan I, Gupta U (2016) Polymeric micelles: recent advancements in the delivery of anticancer drugs. Pharm Res 33(1):18–39

Gu Z, Gao D, Al-Zubaydi F, Li S, Singh Y, Rivera K et al (2018) The effect of size and polymer architecture of doxorubicin–poly (ethylene) glycol conjugate nanocarriers on breast duct retention, potency and toxicity. Eur J Pharm Sci 121:118–125

Guo F, Guo D, Zhang W, Yan Q, Yang Y, Hong W et al (2017) Preparation of curcumin-loaded PCL-PEG-PCL triblock copolymeric nanoparticles by a microchannel technology. Eur J Pharm Sci 99:328–336

Hardiansyah A, Yang M-C, Liu T-Y, Kuo C-Y, Huang L-Y, Chan T-Y (2017) Hydrophobic drug-loaded PEGylated magnetic liposomes for drug-controlled release. Nanoscale Res Lett 12(1):1–11

He R, Yin C (2017) Trimethyl chitosan based conjugates for oral and intravenous delivery of paclitaxel. Acta Biomater 53:355–366

He Z, Zhang Y, Feng N (2020) Cell membrane-coated nanosized active targeted drug delivery systems homing to tumor cells: a review. Mater Sci Eng C 106:110298

Huang J, Li Y, Orza A, Lu Q, Guo P, Wang L et al (2016) Magnetic nanoparticle facilitated drug delivery for cancer therapy with targeted and image-guided approaches. Adv Funct Mater 26(22):3818–3836

Huang L, Chaurasiya B, Wu D, Wang H, Du Y, Tu J et al (2018) Versatile redox-sensitive pullulan nanoparticles for enhanced liver targeting and efficient cancer therapy. Nanomed Nanotechnol Biol Med 14(3):1005–1017

Ishihara D, Pop L, Takeshima T, Iyengar P, Hannan R (2017) Rationale and evidence to combine radiation therapy and immunotherapy for cancer treatment. Cancer Immunol Immunother 66(3):281–298

Jain A, Jain SK (2018) Stimuli-responsive smart liposomes in cancer targeting. Curr Drug Targets 19(3):259–270

Jhaveri A, Deshpande P, Pattni B, Torchilin V (2018) Transferrin-targeted, resveratrol-loaded liposomes for the treatment of glioblastoma. J Control Release 277:89–101

Johnson SB, Park HS, Gross CP, Yu JB (2018) Use of alternative medicine for cancer and its impact on survival. JNCI 110(1):121–124

Ju R-J, Cheng L, Qiu X, Liu S, Song X-L, Peng X-M et al (2018) Hyaluronic acid modified daunorubicin plus honokiol cationic liposomes for the treatment of breast cancer along with the elimination vasculogenic mimicry channels. J Drug Target 26(9):793–805

Kalyanaraman B (2017) Teaching the basics of cancer metabolism: developing antitumor strategies by exploiting the differences between normal and cancer cell metabolism. Redox Biol 12:833–842

Kamaly N, Yameen B, Wu J, Farokhzad OC (2016) Degradable controlled-release polymers and polymeric nanoparticles: mechanisms of controlling drug release. Chem Rev 116(4):2602–2663

Kanamala M, Wilson WR, Yang M, Palmer BD, Wu Z (2016) Mechanisms and biomaterials in pH-responsive tumour targeted drug delivery: a review. Biomaterials 85:152–167

Katuwavila NP, Perera A, Samarakoon SR, Soysa P, Karunaratne V, Amaratunga GA et al (2016) Chitosan-alginate nanoparticle system efficiently delivers doxorubicin to MCF-7 cells. J Nanomater 2016:1–12

Kaur D, Jain K, Mehra NK, Kesharwani P, Jain NK (2016) A review on comparative study of PPI and PAMAM dendrimers. J Nanopart Res 18(6):146

Khafaji M, Zamani M, Golizadeh M, Bavi O (2019) Inorganic nanomaterials for chemo/photothermal therapy: a promising horizon on effective cancer treatment. Biophys Rev 11:335–352

Khodadust R, Unsoy G, Gunduz U (2014) Development of poly (I: C) modified doxorubicin loaded magnetic dendrimer nanoparticles for targeted combination therapy. Biomed Pharmacother 68(8):979–987

Kim Y-M, Park S-C, Jang M-K (2017) Targeted gene delivery of polyethyleneimine-grafted chitosan with RGD dendrimer peptide in αvβ3 integrin-overexpressing tumor cells. Carbohydr Polym 174:1059–1068

Kolosnjaj-Tabi J, Wilhelm C (2017) Magnetic nanoparticles in cancer therapy: how can thermal approaches help? Nanomedicine (Lond) (Future Medicine) 12(6):573–575

Kuang Y, Zhang K, Cao Y, Chen X, Wang K, Liu M et al (2017) Hydrophobic IR-780 dye encapsulated in cRGD-conjugated solid lipid nanoparticles for NIR imaging-guided photothermal therapy. ACS Appl Mater Interfaces 9(14):12217–12226

Kulhari H, Pooja D, Shrivastava S, Kuncha M, Naidu V, Bansal V et al (2016) Trastuzumab-grafted PAMAM dendrimers for the selective delivery of anticancer drugs to HER2-positive breast cancer. Sci Rep 6(1):1–13

Kulikov AV, Luchkina EA, Gogvadze V, Zhivotovsky B (2017) Mitophagy: link to cancer development and therapy. Biochem Biophys Res Commun 482(3):432–439

Kumar A, Ma H, Zhang X, Huang K, Jin S, Liu J et al (2012) Gold nanoparticles functionalized with therapeutic and targeted peptides for cancer treatment. Biomaterials 33(4):1180–1189

Kumar B, Jalodia K, Kumar P, Gautam HK (2017) Recent advances in nanoparticle-mediated drug delivery. J Drug Deliv Sci Technol 41:260–268

Kumar R, Singh M, Meena J, Singhvi P, Thiyagarajan D, Saneja A et al (2019) Hyaluronic acid-dihydroartemisinin conjugate: synthesis, characterization and in vitro evaluation in lung cancer cells. Int J Biol Macromol 133:495–502

Kumari P, Ghosh B, Biswas S (2016) Nanocarriers for cancer-targeted drug delivery. J Drug Target 24(3):179–191

Lee JJ, Yazan LS, Abdullah CAC (2017) A review on current nanomaterials and their drug conjugate for targeted breast cancer treatment. Int J Nanomed 12:2373

Li C, Wallace S (2008) Polymer-drug conjugates: recent development in clinical oncology. Adv Drug Deliv Rev 60(8):886–898

Li M, Song W, Tang Z, Lv S, Lin L, Sun H et al (2013) Nanoscaled poly (L glutamic acid)/doxorubicin-amphiphile complex as pH-responsive drug delivery system for effective treatment of nonsmall cell lung cancer. ACS Appl Mater Interfaces 5(5):1781–1792

Li J, Wang H, Yang B, Xu L, Zheng N, Chen H et al (2016a) Control-release microcapsule of famotidine loaded biomimetic synthesized mesoporous silica nanoparticles: controlled release effect and enhanced stomach adhesion in vitro. Mater Sci Eng C 58:273–277

Li X, Wu M, Pan L, Shi J (2016b) Tumor vascular-targeted co-delivery of anti-angiogenesis and chemotherapeutic agents by mesoporous silica nanoparticle-based drug delivery system for synergetic therapy of tumor. Int J Nanomed 11:93

Li J, Liang H, Liu J, Wang Z (2018a) Poly (amidoamine)(PAMAM) dendrimer mediated delivery of drug and pDNA/siRNA for cancer therapy. Int J Pharm 546(1-2):215–225

Li H, Tong Y, Bai L, Ye L, Zhong L, Duan X et al (2018b) Lactoferrin functionalized PEG-PLGA nanoparticles of shikonin for brain targeting therapy of glioma. Int J Biol Macromol 107:204–211

Li Y, Chen M, Yao B, Lu X, Zhang X, He P et al (2019a) Transferrin receptor-targeted redox/pH-sensitive podophyllotoxin prodrug micelles for multidrug-resistant breast cancer therapy. J Mater Chem B 7(38):5814–5824

Li S, Li X, Ding J, Han L, Guo X (2019b) Anti-tumor efficacy of folate modified PLGA-based nanoparticles for the co-delivery of drugs in ovarian cancer. Drug Des Devel Ther. 13:1271

Li Z, Xu H, Shao J, Jiang C, Zhang F, Lin J et al (2019c) Polydopamine-functionalized black phosphorus quantum dots for cancer theranostics. Appl Mater Today 15:297–304

Lin W, Ma G, Kampf N, Yuan Z, Chen S (2016) Development of long-circulating zwitterionic cross-linked micelles for active-targeted drug delivery. Biomacromolecules 17(6):2010–2018

Lingayat VJ, Zarekar NS, Shendge RS (2017) Solid lipid nanoparticles: a review. Nanosci Nanotechnol Res 2:67–72

Liu J, Yang Y, Zhu W, Yi X, Dong Z, Xu X et al (2016a) Nanoscale metal− organic frameworks for combined photodynamic & radiation therapy in cancer treatment. Biomaterials 97:1–9

Liu Y, Zhou C, Wang W, Yang J, Wang H, Hong W et al (2016b) CD44 receptor targeting and endosomal pH-sensitive dual functional hyaluronic acid micelles for intracellular paclitaxel delivery. Mol Pharm 13(12):4209–4221

Liu B, Han L, Liu J, Han S, Chen Z, Jiang L (2017a) Co-delivery of paclitaxel and TOS-cisplatin via TAT-targeted solid lipid nanoparticles with synergistic antitumor activity against cervical cancer. Int J Nanomedicine 12:955

Liu Y, Zhang G, Guo Q, Ma L, Jia Q, Liu L et al (2017b) Artificially controlled degradable inorganic nanomaterial for cancer theranostics. Biomaterials 112:204–217

Liu Q, Jing Y, Han C, Zhang H, Tian Y (2019a) Encapsulation of curcumin in zein/caseinate/sodium alginate nanoparticles with improved physicochemical and controlled release properties. Food Hydrocoll 93:432–442

Liu P, Chen N, Yan L, Gao F, Ji D, Zhang S et al (2019b) Preparation, characterisation and in vitro and in vivo evaluation of CD44-targeted chondroitin sulphate-conjugated doxorubicin PLGA nanoparticles. Carbohydr Polym 213:17–26

Liu D, Zhang Q, Wang J, Fan L, Zhu W, Cai D (2019c) Hyaluronic acid-coated single-walled carbon nanotubes loaded with doxorubicin for the treatment of breast cancer. Die Pharmazie 74(2):83–90

Liu Y, Yang G, Jin S, Xu L, Zhao CX (2020) Development of high-drug-loading nanoparticles. ChemPlusChem 85(9):2143–2157.

Liu Z, Parida S, Prasad R, Pandey R, Sharma D, Barman I (2021) Vibrational spectroscopy for decoding cancer microbiome interactions: Current evidence and future Perspective. Semin Cancer Biol. https://doi.org/10.1016/j.semcancer.2021.07.004.

Lungu II, Radulescu M, Mogosanu GD, Grumezescu AM (2016) pH sensitive core-shell magnetic nanoparticles for targeted drug delivery in cancer therapy. Rom J Morphol Embryol 57(1):23–32

Luo T, Magnusson J, Préat V, Frédérick R, Alexander C, Bosquillon C et al (2016) Synthesis and in vitro evaluation of polyethylene glycol-paclitaxel conjugates for lung cancer therapy. Pharm Res 33(7):1671–1681

Lyon PC, Griffiths LF, Lee J, Chung D, Carlisle R, Wu F et al (2017) Clinical trial protocol for TARDOX: a phase I study to investigate the feasibility of targeted release of lyso-thermosensitive liposomal doxorubicin (ThermoDox®) using focused ultrasound in patients with liver tumours. J Therap Ultrasound 5(1):1–8

Ma G, Zhang C, Zhang L, Sun H, Song C, Wang C et al (2016) Doxorubicin-loaded micelles based on multiarm star-shaped PLGA–PEG block copolymers: influence of arm numbers on drug delivery. J Mater Sci Mater Med 27(1):17

Mahato R (2017) Nanoemulsion as targeted drug delivery system for cancer therapeutics. J Pharm Sci Pharm 3(2):83–97

Manjili HK, Malvandi H, Mousavi MS, Attari E, Danafar H (2018) In vitro and in vivo delivery of artemisinin loaded PCL–PEG–PCL micelles and its pharmacokinetic study. Artif Cells Nanomed Biotechnol 46(5):926–936

Marzbali MY, Khosroushahi AY (2017) Polymeric micelles as mighty nanocarriers for cancer gene therapy: a review. Cancer Chemother Pharmacol 79(4):637–649

Masood F (2016) Polymeric nanoparticles for targeted drug delivery system for cancer therapy. Mater Sci Eng C 60:569–578

Mehta D, Leong N, McLeod VM, Kelly BD, Pathak R, Owen DJ et al (2018) Reducing dendrimer generation and PEG chain length increases drug release and promotes anticancer activity of PEGylated polylysine dendrimers conjugated with doxorubicin via a cathepsin-cleavable peptide linker. Mol Pharm 15(10):4568–4576

Meunier M, Goupil A, Lienard P (2017) Predicting drug loading in PLA-PEG nanoparticles. Int J Pharm 526(1-2):157–166

Miller KD, Nogueira L, Mariotto AB, Rowland JH, Yabroff KR, Alfano CM et al (2019) Cancer treatment and survivorship statistics, 2019. CA Cancer J Clin 69(5):363–385

Mishra H, Chauhan V, Kumar K, Teotia D (2018) A comprehensive review on Liposomes: a novel drug delivery system. J Drug Deliv Therap 8(6):400–404

Mo L, Song JG, Lee H, Zhao M, Kim HY, Lee YJ et al (2018) PEGylated hyaluronic acid-coated liposome for enhanced in vivo efficacy of sorafenib via active tumor cell targeting and prolonged systemic exposure. Nanomed Nanotechnol Biol Med 14(2):557–567

Moghimipour E, Rezaei M, Kouchak M, Ramezani Z, Amini M, Ahmadi Angali K et al (2018) A mechanistic study of the effect of transferrin conjugation on cytotoxicity of targeted liposomes. J Microencapsul 35(6):548–558

Mohamed M, Abu Lila AS, Shimizu T, Alaaeldin E, Hussein A, Sarhan HA et al (2019) PEGylated liposomes: immunological responses. Sci Technol Adv Mater 20(1):710–724

Moreira AF, Dias DR, Correia IJ (2016) Stimuli-responsive mesoporous silica nanoparticles for cancer therapy: a review. Microporous Mesoporous Mater 236:141–157

Munir M, Hanif M, Ranjha NM (2016) Dendrimers and their applications: a review article. Pakistan J Pharm Res 2(1):55–66

Muntoni E, Martina K, Marini E, Giorgis M, Lazzarato L, Salaroglio IC et al (2019) Methotrexate-loaded solid lipid nanoparticles: protein functionalization to improve brain biodistribution. Pharmaceutics 11(2):65

Nag OK, Delehanty JB (2019) Active cellular and subcellular targeting of nanoparticles for drug delivery. Pharmaceutics 11(10):543

Nayak R, Meerovich I, Dash AK (2019) Translational multi-disciplinary approach for the drug and gene delivery systems for cancer treatment. AAPS PharmSciTech 20(4):160

Nguyen HT, Tran TH, Thapa RK, Dai Phung C, Shin BS, Jeong J-H et al (2017) Targeted co delivery of polypyrrole and rapamycin by trastuzumab-conjugated liposomes for combined chemo-photothermal therapy. Int J Pharm 527(1-2):61–71

Pan M, Li W, Yang J, Li Z, Zhao J, Xiao Y et al (2017) Plumbagin-loaded aptamer-targeted poly D, L-lactic-co-glycolic acid-b-polyethylene glycol nanoparticles for prostate cancer therapy. Medicine 96(30):e7405

Pandey H, Rani R, Agarwal V (2016) Liposome and their applications in cancer therapy. Brazilian Arch Biol Technol 59

Papaccio F, Paino F, Regad T, Papaccio G, Desiderio V, Tirino V (2017) Concise review: cancer cells, cancer stem cells, and mesenchymal stem cells: influence in cancer development. Stem Cells Transl Med 6(12):2115–2125

Parajapati SK, Maurya SD, Das MK, Tilak VK, Verma KK, Dhakar RC (2016) Potential application of dendrimers in drug delivery: a concise review and update. J Drug Deliv Therap 6(2):71–88

Pardo J, Peng Z, Leblanc RM (2018) Cancer targeting and drug delivery using carbon-based quantum dots and nanotubes. Molecules 23(2):378

Paris JL, Mannaris C, Cabañas MV, Carlisle R, Manzano M, Vallet-Regí M et al (2018) Ultrasound-mediated cavitation-enhanced extravasation of mesoporous silica nanoparticles for controlled-release drug delivery. Chem Eng J 340:2–8

Park J-S, Kim I-K, Han S, Park I, Kim C, Bae J et al (2016) Normalization of tumor vessels by Tie2 activation and Ang2 inhibition enhances drug delivery and produces a favorable tumor microenvironment. Cancer Cell 30(6):953–967

Patel V (2020) Liposome: a novel carrier for targeting drug delivery system. Asian J Pharm Res Develop 8(4):67–76

Patel SG., Patel MD, Patel AJ, Chougule MB, Choudhury H (2018) Solid lipid nanoparticles for targeted brain drug delivery. In Nanotechnology-based targeted drug delivery systems for brain tumors. Academic Press, pp 191–244.

Peng J, Liang X (2019) Progress in research on gold nanoparticles in cancer management. Medicine 98(18):e15311

Phan QT, Patil MP, Tu TT, Le CM, Kim G-D, Lim KT (2020) Polyampholyte-grafted single walled carbon nanotubes prepared via a green process for anticancer drug delivery application. Polymer 193:122340

Pooja D, Reddy TS. Kulhari H, Kadari A, Adams DJ, Bansal V, Sistla R (2020). N-acetyl-d-glucosamine-conjugated PAMAM dendrimers as dual receptor-targeting nanocarriers for anticancer drug delivery. Eur J Pharm Biopharm 154:377–386.

Pranatharthiharan S, Patel MD, Malshe VC, Pujari V, Gorakshakar A, Madkaikar M et al (2017) Asialoglycoprotein receptor targeted delivery of doxorubicin nanoparticles for hepatocellular carcinoma. Drug Deliv 24(1):20–29

Prasad R, Pandey R, Varma A, Barman I (2017) Polymer based nanoparticles for drug delivery systems and cancer therapeutics. In: Natural Polymers for Drug Delivery (eds. Kharkwal H and Janaswamy S), CAB International, UK 53–70

Prasad R, Kumar V, Kumar M, Choudhary D (2019) Nanobiotechnology in Bioformulations. Springer International Publishing (ISBN 978-3-030-17061-5) https://www.springer.com/gp/book/9783030170608

Priya S, Rekha M (2017) Redox sensitive cationic pullulan for efficient gene transfection and drug retention in C6 glioma cells. Int J Pharm 530(1–2):401–414

Ptacek J, Zhang D, Qiu L, Kruspe S, Motlova L, Kolenko P et al (2020) Structural basis of prostate-specific membrane antigen recognition by the A9g RNA aptamer. Nucleic Acids Res

Qi D, Gong F, Teng X, Ma M, Wen H, Yuan W et al (2017) Design and evaluation of mPEG-PLA micelles functionalized with drug-interactive domains as improved drug carriers for docetaxel delivery. J Biomater Sci Polym Ed 28(14):1538–1555

Qu J, Zhang L, Chen Z, Mao G, Gao Z, Lai X et al (2016) Nanostructured lipid carriers, solid lipid nanoparticles, and polymeric nanoparticles: which kind of drug delivery system is better for glioblastoma chemotherapy? Drug Deliv 23(9):3408–3416

Rahman M, Beg S, Anwar F, Kumar V, Ubale R, Addo RT et al (2017) Liposome-based nanomedicine therapeutics for rheumatoid arthritis. Crit Rev Ther Drug Carrier Syst. 34(4):283–316

Raja MA, Arif M, Feng C, Zeenat S, Liu C-G (2017) Synthesis and evaluation of pH-sensitive, self-assembled chitosan-based nanoparticles as efficient doxorubicin carriers. J Biomater Appl 31(8):1182–1195

Ramalingam P, Ko YT (2015) Enhanced oral delivery of curcumin from N-trimethyl chitosan surface-modified solid lipid nanoparticles: pharmacokinetic and brain distribution evaluations. Pharm Res 32(2):389–402

Ramalingam P, Ko YT (2016) Improved oral delivery of resveratrol from N-trimethyl chitosan-g-palmitic acid surface-modified solid lipid nanoparticles. Colloids Surf B Biointerfaces 139:52–61

Ramalingam P, Yoo SW, Ko YT (2016) Nanodelivery systems based on mucoadhesive polymer coated solid lipid nanoparticles to improve the oral intake of food curcumin. Food Res Int 84:113–119

Rana S, Shetake NG, Barick K, Pandey B, Salunke H, Hassan P (2016) Folic acid conjugated Fe 3 O 4 magnetic nanoparticles for targeted delivery of doxorubicin. Dalton Trans 45(43):17401–17408

Ranjbar-Navazi Z, Eskandani M, Johari-Ahar M, Nemati A, Akbari H, Davaran S et al (2018) Doxorubicin-conjugated D-glucosamine-and folate-bi-functionalised InP/ZnS quantum dots for cancer cells imaging and therapy. J Drug Target 26(3):267–277

Raza A, Hayat U, Rasheed T, Bilal M, Iqbal HM (2019) "Smart" materials-based near-infrared light-responsive drug delivery systems for cancer treatment: a review. J Mater Res Technol 8(1):1497–1509

Rodenak-Kladniew B, Islan GA, de Bravo MG, Durán N, Castro GR (2017) Design, characterization and in vitro evaluation of linalool-loaded solid lipid nanoparticles as potent tool in cancer therapy. Colloids Surf B Biointerfaces 154:123–132

Rosiere R, Van Woensel M, Gelbcke M, Mathieu V, Hecq J, Mathivet T et al (2018) New folate-grafted chitosan derivative to improve delivery of paclitaxel-loaded solid lipid nanoparticles for lung tumor therapy by inhalation. Mol Pharm 15(3):899–910

Saberinasab A, Raissi H, Hashemzadeh H (2019) Understanding the effect of vitamin B6 and PEG functionalization on improving the performance of carbon nanotubes in temozolomide anticancer drug transportation. J Phys D Appl Phys 52(39):395402

Saif MW (2013) US Food and Drug Administration approves paclitaxel protein-bound particles (Abraxane®) in combination with gemcitabine as first-line treatment of patients with metastatic pancreatic cancer. JOP 14(6):686–688

Saini K, Bandyopadhyaya R (2019) Transferrin-conjugated polymer-coated mesoporous silica nanoparticles loaded with gemcitabine for killing pancreatic cancer cells. ACS Appl Nano Mater 3(1):229–240

Salvioni L, Rizzuto MA, Bertolini JA, Pandolfi L, Colombo M, Prosperi D (2019) Thirty years of cancer nanomedicine: success, frustration, and hope. Cancers 11(12):1855

Sarcan ET, Silindir-Gunay M, Ozer AY (2018) Theranostic polymeric nanoparticles for NIR imaging and photodynamic therapy. Int J Pharm 551(1-2):329–338

Schmidt LH, Brand C, Stucke-Ring J, Schliemann C, Kessler T, Harrach S et al (2017) Potential therapeutic impact of CD13 expression in non-small cell lung cancer. PLoS One 12(6):e0177146

Senapati S, Mahanta AK, Kumar S, Maiti P (2018) Controlled drug delivery vehicles for cancer treatment and their performance. Signal Transduct Target Ther 3(1):1–19

Serinan E, Altun Z, Aktaş S, Çeçen E, Olgun N (2018) Comparison of cisplatin with lipoplatin in terms of ototoxicity. J Int Adv Otol 14(2):211

Serini S, Cassano R, Corsetto PA, Rizzo AM, Calviello G, Trombino S (2018) Omega-3 PUFA loaded in resveratrol-based solid lipid nanoparticles: Physicochemical properties and antineoplastic activities in human colorectal cancer cells in vitro. Int J Mol Sci 19(2):586

Seyfoori A, Sarfarazijami S, Seyyed Ebrahimi S (2019) pH-responsive carbon nanotube-based hybrid nanogels as the smart anticancer drug carrier. Artif Cells Nanomed Biotechnol 47(1):1437–1443

Shao J, Zhang J, Jiang C, Lin J, Huang P (2020) Biodegradable titanium nitride MXene quantum dots for cancer phototheranostics in NIR-I/II biowindows. Chem Eng J 400:126009

Sherje AP, Jadhav M, Dravyakar BR, Kadam D (2018) Dendrimers: a versatile nanocarrier for drug delivery and targeting. Int J Pharm 548(1):707–720

Siegel RL, Miller KD, Jemal A (2019) Cancer statistics, 2019. CA Cancer J Clin 69(1):7–34

Silverman JA, Deitcher SR (2013) Marqibo®(vincristine sulfate liposome injection) improves the pharmacokinetics and pharmacodynamics of vincristine. Cancer Chemother Pharmacol 71(3):555–564

Singh S, Mehra NK, Jain N (2016a) Development and characterization of the paclitaxel loaded riboflavin and thiamine conjugated carbon nanotubes for cancer treatment. Pharm Res 33(7):1769–1781

Singh RP, Sharma G, Singh S, Kumar M, Pandey BL, Koch B et al (2016b) Vitamin E TPGS conjugated carbon nanotubes improved efficacy of docetaxel with safety for lung cancer treatment. Colloids Surf B Biointerfaces 141:429–442

Singh P, Pandit S, Mokkapati V, Garg A, Ravikumar V, Mijakovic I (2018) Gold nanoparticles in diagnostics and therapeutics for human cancer. Int J Mol Sci 19(7):1979

Sofias AM, Dunne M, Storm G, Allen C (2017) The battle of "nano" paclitaxel. Adv Drug Deliv Rev 122:20–30

Son KH, Hong JH, Lee JW (2016) Carbon nanotubes as cancer therapeutic carriers and mediators. Int J Nanomedicine 11:5163

Song L, Pan Z, Zhang H, Li Y, Zhang Y, Lin J et al (2017) Dually folate/CD44 receptor-targeted self-assembled hyaluronic acid nanoparticles for dual-drug delivery and combination cancer therapy. J Mater Chem B 5(33):6835–6846

Sztandera K, Gorzkiewicz M, Klajnert-Maculewicz B (2018) Gold nanoparticles in cancer treatment. Mol Pharm 16(1):1–23

Taghipour-Sabzevar V, Sharifi T, Moghaddam MM (2019) Polymeric nanoparticles as carrier for targeted and controlled delivery of anticancer agents. Ther Deliv 10(8):527–550

Tang Y, Li Y, Xu R, Li S, Hu H, Xiao C et al (2018a) Self-assembly of folic acid dextran conjugates for cancer chemotherapy. Nanoscale 10(36):17265–17274

Tang X, Lyu Y, Xie D, Li A, Liang Y, Zheng D (2018b) Therapeutic effect of sorafenib-loaded TPGS-b-PCL nanoparticles on liver cancer. J Biomed Nanotechnol 14(2):396–403

Thanou M, Duncan R (2003) Polymer-protein and polymer-drug conjugates in cancer therapy. Curr Opin Investig Drugs (London, England: 2000) 4(6):701–709

Tiruwa R (2016) A review on nanoparticles–preparation and evaluation parameters. Indian J Pharm Biol Res 4(2):27–31

Tran S, DeGiovanni P-J, Piel B, Rai P (2017) Cancer nanomedicine: a review of recent success in drug delivery. Clin Transl Med 6(1):44

Tsuboi S, Sasaki A, Sakata T, Yasuda H, Jin T (2017) Immunoglobulin binding (B1) domain mediated antibody conjugation to quantum dots for in vitro and in vivo molecular imaging. Chem Commun 53(68):9450–9453

Tyson MD II, Koyama T, Lee D, Hoffman KE, Resnick MJ, Wu X-C et al (2018) Effect of prostate cancer severity on functional outcomes after localized treatment: comparative effectiveness analysis of surgery and radiation study results. Eur Urol 74(1):26–33

van der Meel R, Sulheim E, Shi Y, Kiessling F, Mulder WJ, Lammers T (2019) Smart cancer nanomedicine. Nat Nanotechnol 14(11):1007–1017

Veeranarayanan S, Maekawa T (2019) External stimulus responsive inorganic nanomaterials for cancer theranostics. Adv Drug Deliv Rev 138:18–40

Vicent MJ, Duncan R (2006) Polymer conjugates: nanosized medicines for treating cancer. Trends Biotechnol 24(1):39–47

Vinothini K, Rajendran NK, Ramu A, Elumalai N, Rajan M (2019) Folate receptor targeted delivery of paclitaxel to breast cancer cells via folic acid conjugated graphene oxide grafted methyl acrylate nanocarrier. Biomed Pharmacother 110:906–917

Wakaskar RR (2017) Passive and active targeting in tumor microenvironment. Int J Drug Develop Res 9(2):37–41

Wang H, Yu J, Lu X, He X (2016) Nanoparticle systems reduce systemic toxicity in cancer treatment. Nanomedicine (Lond) (Future Medicine) 11(2):103–106

Wang W, Zhang L, Chen T, Guo W, Bao X, Wang D et al (2017a) Anticancer effects of resveratrol-loaded solid lipid nanoparticles on human breast cancer cells. Molecules 22(11):1814

Wang T, Hou J, Su C, Zhao L, Shi Y (2017b) Hyaluronic acid-coated chitosan nanoparticles induce ROS-mediated tumor cell apoptosis and enhance antitumor efficiency by targeted drug delivery via CD44. J Nanobiotechnol 15(1):7

Wang W, Li M, Zhang Z, Cui C, Zhou J, Yin L et al (2017c) Design, synthesis and evaluation of multi-functional tLyP-1-hyaluronic acid-paclitaxel conjugate endowed with broad anticancer scope. Carbohydr Polym 156:97–107

Wang D, Ren Y, Shao Y, Yu D, Meng L (2017d) Facile preparation of doxorubicin-loaded and folic acid-conjugated carbon nanotubes@ poly (N-vinyl pyrrole) for targeted synergistic chemo–Photothermal Cancer treatment. Bioconjug Chem 28(11):2815–2822

Wang D, Ren Y, Shao Y, Meng L (2017e) Multifunctional polyphosphazene-coated multi-walled carbon nanotubes for the synergistic treatment of redox-responsive chemotherapy and effective photothermal therapy. Polym Chem 8(45):6938–6942

Wang W, Chen T, Xu H, Ren B, Cheng X, Qi R et al (2018a) Curcumin-loaded solid lipid nanoparticles enhanced anticancer efficiency in breast cancer. Molecules 23(7):1578

Wang Y, Ren J, Liu Y, Liu R, Wang L, Yuan Q et al (2018b) Preparation and evaluation of folic acid modified succinylated gelatin micelles for targeted delivery of doxorubicin. J Drug Deliv Sci Technol 46:400–407

Wang X, Chang Z, Nie X, Li Y, Hu Z, Ma J et al (2019) A conveniently synthesized Pt (IV) conjugated alginate nanoparticle with ligand self-shielded property for targeting treatment of hepatic carcinoma. Nanomed Nanotechnol Biol Med 15(1):153–163

Wild CP (2019) The global cancer burden: Necessity is the mother of prevention. Nat Rev Cancer 19(3):123–124

Wolfram J, Ferrari M (2019) Clinical cancer nanomedicine. Nano Today 25:85–98

Wong P-P, Bodrug N, Hodivala-Dilke KM (2016) Exploring novel methods for modulating tumor blood vessels in cancer treatment. Curr Biol 26(21):R1161–R1166

Wong KH, Lu A, Chen X, Yang Z (2020) Natural ingredient-based polymeric nanoparticles for cancer treatment. Molecules 25(16):3620

Wu P-H, Onodera Y, Ichikawa Y, Rankin EB, Giaccia AJ, Watanabe Y et al (2017a) Targeting integrins with RGD-conjugated gold nanoparticles in radiotherapy decreases the invasive activity of breast cancer cells. Int J Nanomedicine 12:5069

Wu J, Deng C, Meng F, Zhang J, Sun H, Zhong Z (2017b) Hyaluronic acid coated PLGA nanoparticulate docetaxel effectively targets and suppresses orthotopic human lung cancer. J Control Release 259:76–82

Xiong XY, Pan X, Tao L, Cheng F, Li ZL, Gong YC et al (2017) Enhanced effect of folated pluronic F87-PLA/TPGS mixed micelles on targeted delivery of paclitaxel. Int J Biol Macromol 103:1011–1018

Xu W, Siddiqui IA, Nihal M, Pilla S, Rosenthal K, Mukhtar H et al (2013) Aptamer-conjugated and doxorubicin-loaded unimolecular micelles for targeted therapy of prostate cancer. Biomaterials 34(21):5244–5253

Xu G, Chen Y, Shan R, Wu X, Chen L (2018) Transferrin and tocopheryl-polyethylene glycol-succinate dual ligands decorated, cisplatin loaded nano-sized system for the treatment of lung cancer. Biomed Pharmacother 99:354–362

Yang S, Gao H (2017) Nanoparticles for modulating tumor microenvironment to improve drug delivery and tumor therapy. Pharmacol Res 126:97–108

Yang Y, Yu C (2016) Advances in silica based nanoparticles for targeted cancer therapy. Nanomed Nanotechnol Biol Med 12(2):317–332

Yang Y, Yang Y, Xie X, Wang Z, Gong W, Zhang H et al (2015) Dual-modified liposomes with a two-photon-sensitive cell penetrating peptide and NGR ligand for siRNA targeting delivery. Biomaterials 48:84–96

Yang Y, Zhao Y, Lan J, Kang Y, Zhang T, Ding Y et al (2018a) Reduction-sensitive CD44 receptor-targeted hyaluronic acid derivative micelles for doxorubicin delivery. Int J Nanomedicine 13:4361

Yang C, Song W, Zhang D, Yu H, Yin L, Shen N et al (2018b) Poly (L-glutamic acid)-g-methoxy poly (ethylene glycol)-gemcitabine conjugate improves the anticancer efficacy of gemcitabine. Int J Pharm 550(1-2):79–88

Yao S, Li L, Su X-T, Wang K, Lu Z-J, Yuan C-Z et al (2018) Development and evaluation of novel tumor-targeting paclitaxel-loaded nano-carriers for ovarian cancer treatment: in vitro and in vivo. J Exp Clin Cancer Res 37(1):29

Yu C, Zhou Q, Xiao F, Li Y, Hu H, Wan Y et al (2017) Enhancing doxorubicin delivery toward tumor by hydroxyethyl starch-g-polylactide partner nanocarriers. ACS Appl Mater Interfaces 9(12):10481–10493

Zahednezhad F, Zakeri-Milani P, Shahbazi Mojarrad J, Valizadeh H (2020) The latest advances of cisplatin liposomal formulations: essentials for preparation and analysis. Expert Opin Drug Deliv 17(4):523–541

Zaidi N, Jaffee EM (2019) Immunotherapy transforms cancer treatment. J Clin Invest 129(1):46–47

Zhai J, Luwor RB, Ahmed N, Escalona R, Tan FH, Fong C et al (2018) Paclitaxel-loaded self-assembled lipid nanoparticles as targeted drug delivery systems for the treatment of aggressive ovarian cancer. ACS Appl Mater Interfaces 10(30):25174–25185

Zhang B, Zhang Y, Yu D (2017) Lung cancer gene therapy: transferrin and hyaluronic acid dual ligand-decorated novel lipid carriers for targeted gene delivery. Oncol Rep 37(2):937–944

Zhang J, Zhao X, Xian M, Dong C, Shuang S (2018a) Folic acid-conjugated green luminescent carbon dots as a nanoprobe for identifying folate receptor-positive cancer cells. Talanta 183:39–47

Zhang J, Yang C, Pan S, Shi M, Li J, Hu H et al (2018b) Eph A10-modified pH-sensitive liposomes loaded with novel triphenylphosphine–docetaxel conjugate possess hierarchical targetability and sufficient antitumor effect both in vitro and in vivo. Drug Deliv 25(1):723–737

Zhang Y, Guo Z, Cao Z, Zhou W, Zhang Y, Chen Q et al (2018c) Endogenous albumin-mediated delivery of redox-responsive paclitaxel-loaded micelles for targeted cancer therapy. Biomaterials 183:243–257

Zhang H, Liu XL, Zhang YF, Gao F, Li GL, He Y et al (2018d) Magnetic nanoparticles based cancer therapy: current status and applications. Sci China Life Sci 61(4):400–414

Zhang X, Zhang R, Huang J, Luo M, Chen X, Kang Y et al (2019) Albumin enhances PTX delivery ability of dextran NPs and therapeutic efficacy of PTX for colorectal cancer. J Mater Chem B 7(22):3537–3545

Zhao M-X, Zhu B-J, Yao W-J, Chen D-F (2016) Therapeutic effect of quantum dots for cancer treatment. RSC Adv 6(114):113791–113795

Zhao Q-S, Hu L-L, Wang Z-D, Li Z-P, Wang A-W, Liu J (2017) Resveratrol-loaded folic acid-grafted dextran stearate submicron particles exhibits enhanced antitumor efficacy in non-small cell lung cancers. Mater Sci Eng C 72:185–191

Zhao Z, Chen C, Xie C, Zhao Y (2020) Design, synthesis and evaluation of liposomes modified with dendritic aspartic acid for bone-specific targeting. Chem Phys Lipids 226:104832

Zheng YB, Gong JH, Liu XJ, Li Y, Zhen YS (2017) A CD13-targeting peptide integrated protein inhibits human liver cancer growth by killing cancer stem cells and suppressing angiogenesis. Mol Carcinog 56(5):1395–1404

Zheng G, Zheng M, Yang B, Fu H, Li Y (2019) Improving breast cancer therapy using doxorubicin loaded solid lipid nanoparticles: synthesis of a novel arginine-glycine-aspartic tripeptide conjugated, pH sensitive lipid and evaluation of the nanomedicine in vitro and in vivo. Biomed Pharmacother 116:109006

Chapter 4
Novel Organic and Inorganic Nanoparticles as a Targeted Drug Delivery Vehicle in Cancer Treatment

Saradhadevi Muthukrishnan, A. Vijaya Anand, Kiruthiga Palanisamy, Gayathiri Gunasangkaran, Anjali K. Ravi, and Balamuralikrishnan Balasubramanian

Contents

4.1	Introduction	119
4.2	Polymeric Micelles	121
4.3	Polymeric Nanoparticles	123
	4.3.1 Role of Polymeric Nanoparticles in Cancer Treatments	124
4.4	Liposomes	126
	4.4.1 Liposome-Encapsulated Drugs in Cancer Treatment	127
4.5	Dendrimers	128
4.6	Polymer Drug Conjugates	130
	4.6.1 Polymer Drug Conjugates Against Cancer	132
4.7	Silica Nanoparticles	133
	4.7.1 Silica Nanoparticles toward Cancer Therapy	133
4.8	Gold Nanoparticles	135
	4.8.1 Applications of Gold Nanoparticles	136
	4.8.2 Gold Nanoparticle Toward Cancer Therapy	136
	4.8.3 Combinational Therapy of Phytochemicals with Gold Nanoparticle in Cancer Cells	138
4.9	Carbon Nanotubes	139
	4.9.1 Carbon Nanotubes in Cancer Therapy	139

Kiruthiga Palanisamy, Gayathiri Gunasangkaran and Anjali K. Ravi contributed equally with all other contributors.

S. Muthukrishnan (✉) · K. Palanisamy · G. Gunasangkaran · A. K. Ravi
Department of Biochemistry, Bharathiar University, Coimbatore, Tamil Nadu, India
e-mail: saradhadevi@buc.edu.in

A. Vijaya Anand
Department of Human Genetics and Molecular Genetics, Bharathiar University, Coimbatore, Tamil Nadu, India

B. Balasubramanian
Department of Food Science and Biotechnology, Sejong University, Seoul, South Korea

4.10	Quantum Dots	141
	4.10.1 Quantum Dot-Conjugated Inhibitors in Cancer Treatment	143
4.11	Nanographene	145
	4.11.1 Nanographene-Conjugated Inhibitors in Cancer Treatment	145
4.12	Magnetic Nanoparticles	147
	4.12.1 Role of Magnetic Nanoparticles in Cancer Treatments	147
4.13	Conclusion	149
References		149

Abbreviations

5FU	5-Fluorouracil
AITC-SiQDs	Allyl isothiocyanate-silicon QDs
AKT/mTOR	Serine/threonine-specific protein kinase/mechanistic target of rapamycin kinase
APAF1	Apoptotic protease activating factor 1
AuNPs	Gold nanoparticle
Bcl-2	B-cell lymphoma 2
CNTs	Carbon nanotubes
DSC	Differential scanning calorimetry
EGF	Epidermal growth factor
FA-NGO-PVP	Folic acid-graphene oxide-polyvinylpyrrolidone
FDA	Food and Drug Administration
GA-CdTe	Gambogic acid-cadmium-tellurium
GO-PVCL	Graphene oxide-poly N-vinyl caprolactam
HA	Hyaluronic acid
HA-DOX-GQD@MSN	Hyaluronic acid-doxorubicin-N-graphene quantum dots-mesoporous silica nanoparticles
HCC	Hepatocellular carcinoma
HER2 genes	Human epidermal growth factor receptor-2 genes
HPMA	N-2-hydroxypropyl methacrylamide
IGF1R	Type I insulin-like growth factor receptor
MALDI-MS	Matrix-assisted laser desorption ionization-mass spectrometry
MMP	Matrix Metallo Proteinases
NF-kB	Nuclear factor-κB
NGO-HA	Nanographene oxide-hyaluronic acid
NIPAAM	N-isopropylacrylamide
NMR	Nuclear magnetic resonance
PAMAM	Polyamidoamine
PEDOT:PSS	Poly 3,4-ethylenedioxythiophene-poly styrenesulfonate
PEG	Polyethylene glycol
PEG-A	Polyethylene glycol monoacrylate

PEG-PE	Polyethylene glycol-phosphatidyl ethanolamine
PEO-PCL	Polyethylene oxide-poly epsilon-caprolactone
PLA-PEG-PLL- DTPA	Poly lactic acid-polyethylene glycol-poly L-lysine-diethylene Triamine pentaacetic acid
PLGA	Poly D,L-lactic acid-co-glycolic acid
PLGA	Polylactide-co-glycolide
PLH-PEG-biotin	Poly L-histidine-polyethylene glycol-biotin
PPI	Polypropylene imine
PTT	Photothermal therapy
QDs	Quantum dots
ROS	Reactive oxygen species
SEM	Scanning electron microscope
TEM	Transmission electron microscopy
TPTN	Theranostic polymeric nanoparticle
TRC-NPs	Thermo-responsive chitosan-g-poly (N-vinylcaprolactam) nanoparticles
VP	N-vinyl-2-pyrrolidone
γ-PGA	Polyvinyl pyrrolidone and poly-γ-glutamic acid

4.1 Introduction

Free radicals are the toxic substances produced in our body due to our modern lifestyle patterns like food habits, lack of exercise, stress, and work pressure. Free radicals can multiply through a chain reaction mechanism resulting in the release of thousands of cellular oxidants which cause DNA damage and mutation, alter DNA codes, and compromise immunity, resulting in disease like cancer (Lobo et al. 2010). Cancer is medically known as neoplasm. It arises when the homeostatic balance between cell growth and death is disturbed. Current standard treatments for cancer like chemotherapy and radiation and with synthetic compounds affect not only the cancer cells but also normal cells causing organ dysfunction, reduced production of white blood cells, hair loss, inflammation of the mucous membranes lining the digestive tract, etc. (Labi and Erlacher 2015). But the nanoparticles or nanocarriers have diversiform structures, multiple targets, diversified pharmacological potential, targeted drug releasing capacity, sustainability, and solubility which could act as a better choice to treat cancer. Another advantage is the high permeability of nanoparticles along with the anticancer agents at tumor site, compared with healthy cells (ud Din et al. 2017). Homeostatic imbalance in cancer cells can be readily reverted by nanocarriers or nanoparticles (organic or inorganic) by enhancing the immune response and suppressing the cancer activity in our body. Our aim is to couple the anticancer drugs like 5-FU, irinotecan, oxaliplatin, bevacizumab, and capecitabine with nano-based encapsulation technology, for treating primary or advanced metastatic tumors. Although anticancer drugs have potent anticancer activity, sometimes it causes poor water solubility, due to its smaller size easily excreted in urine, less availability, and less target specificity (Ungari et al. 2017). Nanotechnology is the most advanced in the field of medicine to overcome the limitations of conventional low molecular weight drugs. And it acts as a better anticancer drug delivery vehicle by using both organic and inorganic nanoparticles like polymeric micelles, liposomes, dendrimers, silica, gold, silver, carbon nanotubes, quantum dots, nanographene, and magnetic nanoparticles (Fig. 4.1). This nanocarriers have different biophysicochemical properties like different shapes, size, surface area, and material which may be hard (organic) or soft (inorganic) (Chenthamara et al. 2019). Smaller-size nanoparticles can easily pass through the circulation and release anticancer drugs at the specific tumor tissues. The lifespan of nanoparticles was particularly determined by the hydrophilic surfaces; this surfaces help to evade the nanoparticles from the macrophages when passing through the circulatory mechanisms (Patra et al. 2018). Coating the nanoparticles with a hydrophilic polymer (PEG) will provide the hydrophilic surface. Green synthesis of nanoparticles can be achieved by using the plant extracts or bioactive compounds of anticancer medicinal plants which are environment friendly with no toxic side effects and are the better choice to fight against cancer (Wei et al. 2019; Roy et al. 2021). Nanoparticles are regarded to be an ideal vehicle for antitumor drugs because of their hydrophobic inner core and hydrophilic outer shell, which enhanced the permeation and retention effect in tumor tissue (Bae et al. 2011). This review paper aims to cover the advantages of organic and inorganic nanoparticles as a novel drug carrier vehicle in cancer treatment in order to overcome the limitations of conventional drugs.

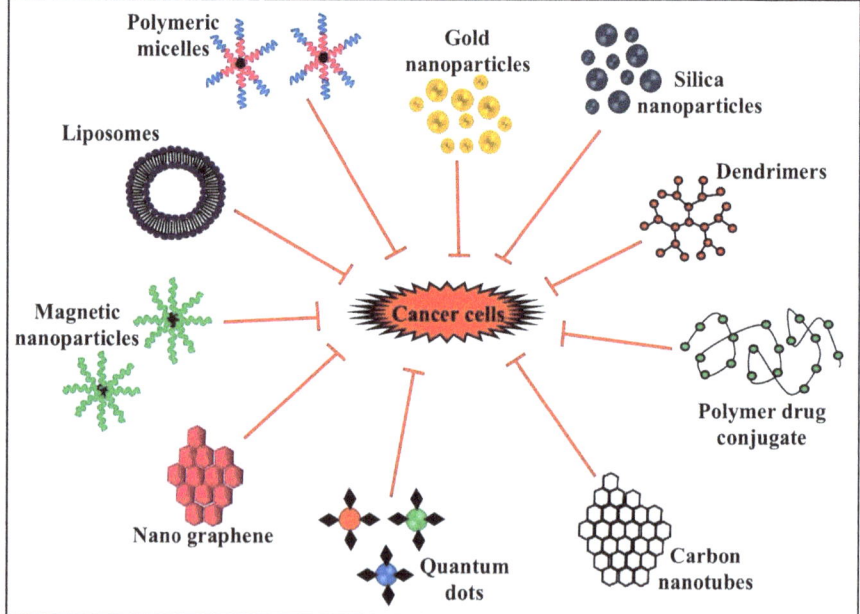

Fig. 4.1 Organic and inorganic nanoparticles as an anticancer drug delivery vehicle

4.2 Polymeric Micelles

A number of anticancer delivery vehicles are there, but polymeric micelles gained importance due its small sizes (10–100 nm) and its efficacy to solubilize the insoluble drugs and because they are more efficient in slow release of drug in the tumor tissue. Polymeric micelles consist of core shell with inner hydrophobic domains suitable for water-insoluble drugs and hydrophilic outer shell (Zhang et al. 2014a, Zhang et al. 2014b). Polymeric micelles act as an efficient drug carrier due to its prolonged and stable drug release at the specific site of cancer tissue (Rijcken et al. 2007) (Table 4.1). The core shell structure of polymeric micelles makes it more efficient to load the hydrophobic drugs and protect the drugs from protein degradation (Gaucher et al. 2004). Polymeric micelle-conjugated Genexol proves to be more efficient against breast cancer approved by the US FDA (Oerlemans et al. 2010). Polymeric micelles of naringin with a better anticancer and anti-ulcer activity by decreasing the level of mucosal damage, gastric expression of malondialdehyde, tumor necrosis factor-alpha, caspase-3, nuclear factor kappa-light-chain enhancer of activated B cells, and interleukin-6 with the elevation of gastric reduced glutathione and superoxide dismutase (Mohamed et al. 2018). 7-Ethyl-10-hydroxy-camptothecin (SN-38)-incorporating polymeric micelles have a better anticancer activity by inhibiting vascular endothelial growth factor in human colorectal cancer (Koizumi et al. 2006). Combinational anticancer effect of polymeric micelles along with the 5-fluorouracil showed the cell cycle arrest at S phase in colorectal cancer (Nakajima et al. 2008). Polymeric micelles incorporating cisplatin show complete cancer inhibition in solid tumor and reduced the nephrotoxic and neurotoxic effects (Nishiyama et al. 2003; Uchino et al. 2005). Non-covalent interactions of polymeric micelles with doxorubicin showed a potent anticancer activity and kinetic stability against human liver carcinoma cells (Yang et al. 2012a, Yang et al. 2012b, Yang et al. 2012c). Adriamycin as physically entrapped in the inner core of polymeric micelles showed antitumor activity against solid tumor (Yokoyama et al. 1998). Cisplatin-conjugated polymeric micelles (NC-6004) showed more efficient cytotoxic effect in solid tumor patients and contribute to low toxicity of NC-6004 (Plummer et al. 2011). Polyethylene glycol (5 K)-embelin nanomicellar delivered the paclitaxel in the tumor tissues of breast and prostate cancer and acts as a best anticancer drug vehicle (Lu et al. 2013). Polyethylene glycol-phosphatidyl ethanolamine (PEG-PE) and vitamin E tumor-targeted immunomicelles act as best anticancer drug carriers to carry water-insoluble camptothecin which showed improved cytotoxic activity in various cancer cells (Sawant et al. 2008). Doxorubicin-loaded polylactide-poly(ethylene glycol) aptamer micelles act as a best nanocarrier to treat prostate cancer (Xu et al. 2013a, Xu et al. 2013b). Polymeric micelles co-conjugated with gemcitabine and miR-205 inhibited tumor growth and increase apoptosis in pancreatic cancer (Mondal et al. 2017). Charge reversed polymer micelles conjugated with paclitaxel/disulfiram overcome multidrug resistance in breast cancer cell (Huo et al. 2017). Overexpression of glucose transporter-1 and glutathione in cancer cells was exploited to assemble aminoglucose-conjugated, redox-responsive

Table 4.1 Polymeric micelles as a drug delivery system

Sl. no.	Polymeric micelles	Cancer	References
1.	Polymeric micelle-conjugated Genexol	Brest cancer	Oerlemans et al. (2010)
2.	7-Ethyl-10-hydroxy-camptothecin (SN-38)-incorporating polymeric micelles	Colorectal cancer	Koizumi et al. (2006)
3.	Polymeric micelle with 5-fluorouracil	Colorectal cancer	Nakajima et al. (2008)
4.	Polymeric micelles incorporating cisplatin	Reduced solid tumors, nephrotoxic, neurotoxic effects	Nishiyama et al. (2003) and Uchino et al. (2005)
5.	Polymeric micelles with doxorubicin	Liver cancer	Yang et al. (2012a), Yang et al. (2012b), Yang et al. (2012c)
6.	Adriamycin with polymeric micelles	Solid tumor	Yokoyama et al. (1998)
7.	Cisplatin-conjugated polymeric micelles	Solid tumor	Plummer et al. (2011)
8.	Polyethylene glycol (5 K)-embelin nanomicellar with paclitaxel	Breast cancer, prostate cancer	Lu et al. (2013)
9.	Polyethylene glycol-phosphatidyl ethanolamine (PEG-PE) and vitamin E tumor-targeted immune micelles with camptothecin	Various cancer	Sawant et al. (2008)
10.	Doxorubicin-loaded polylactide-poly(ethylene glycol) aptamer micelles	Prostate cancer	Xu et al. (2013a), Xu et al. (2013b)
11.	Polymeric micelles co-conjugated with gemcitabine and miR-205	Pancreatic cancer	Mondal et al. (2017)
12.	Polymer micelles conjugated with paclitaxel/disulfiram	Breast cancer	Huo et al. (2017)
13.	Aminoglucose-conjugated nanomicelles with polyethylene glycol and polylactic acid	Lung cancer	Zhou et al. (2017)
14.	Polymeric micelles with mitoxantrone	Breast cancer	Li et al. (2017)
15.	Cyclic-Arg-Gly-Asp (cRGD) peptide-conjugated polymeric micelles with epirubicin	Brain tumor	Quader et al. (2017)
16.	Polymeric micelles with paclitaxel and tariquidar	Ovarian carcinoma	Zou et al. (2017)
17.	Polymeric micelles with paclitaxel and honokiol	Breast cancer	Wang et al. (2017)
18.	Polymer nanomicelles with doxorubicin and P-gp siRNA	Breast cancer	Babu et al. (2017)

(continued)

Table 4.1 (continued)

Sl. no.	Polymeric micelles	Cancer	References
19.	PEGylated cholesterol/α-tocopheryl succinate conjugated to polymer with curcumin	Murine melanoma, breast cancer	Muddineti et al. (2017)
20.	Blood-stable polymeric micelles with curcumin	Erythroleukemia	Gong et al. (2017)

nanomicelles, polyethylene glycol, and polylactic acid to overcome chemoresistance in lung cancer cells (Zhou et al. 2017). Polymeric micelles loaded with mitoxantrone have reversed multidrug resistance in breast cancer cells via photodynamic therapy (Li et al. 2017). Cyclic-Arg-Gly-Asp (cRGD) peptide-conjugated polymeric micelles loaded with epirubicin inhibit the growth of an orthotopic glioblastoma multiforme in brain tumor tissue (Quader et al. 2017). Transferrin-modified polymeric micelles loaded with paclitaxel and tariquidar reverse the multidrug resistance in ovarian carcinoma cells (Zou et al. 2017). pH-responsive polymeric micelles loaded with paclitaxel and honokiol inhibited multidrug resistance and metastasis in breast cancer (Wang et al. 2017). Penta block polymeric micelles grafted with folic acid loaded with doxorubicin are pronounced for targeting anticancer drug delivery and control release (Chen et al. 2018). Polymer nanomicelles co-conjugated with doxorubicin and P-gp siRNA showed cytotoxic effect in breast cancer cells (Babu et al. 2017).

Curcumin is the most important anticancer agent in many cancer cells against multidrug resistance. But curcumin alone has some disadvantages with lower bioavailability in the tumor sites due to the properties of water insolubility and more excretion through the kidneys. In order to overcome its disadvantages, it should be conjugated to some nanoparticles to increase its bioavailability and efficacy against cancer. A self-assembled PEGylated cholesterol/α-tocopheryl succinate conjugated to polymer loaded with curcumin to form a micellar system inhibits the cell proliferation and cancer progression in murine melanoma cell line and human breast cancer, MDA-MB-231 cell lines (Muddineti et al. 2017). Blood-stable polymeric micelles loaded with curcumin increases higher uptake and slower elimination of curcumin into the erythroleukemia with better cytotoxic activity (Gong et al. 2017).

4.3 Polymeric Nanoparticles

Polymeric nanoparticles are widely used as carriers in the pharmaceutical sector for targeted drug delivery mechanism (Prasad et al. 2017). Top-down and bottom-up approaches are used for the preparation of polymeric nanoparticles (Krishnaswamy and Orsat 2017). They are solid in nature and colloidal particles of size range 10 nm to 1 μm with two possible structures: nanosphere and nanocapsule (Sharma 2019). Polymeric nanoparticles can be made from synthetic polymers that are biodegradable and nonbiodegradable in nature (Zhang et al. 2013). The potent applications of polymeric nanoparticles are the biocompatibility in nature which assists in drug release at specific target sites (Rana and Sharma 2019) (Table 4.2).

4.3.1 Role of Polymeric Nanoparticles in Cancer Treatments

van Vlerken et al. (2007) investigated the therapeutic efficiency of a combinational chemotherapeutic agent of ceramide with paclitaxel, which is encapsulated by poly (ethylene oxide)-modified poly (epsilon-caprolactone) (PEO-PCL) nanoparticles used as a drug-delivering vehicle to regain apoptotic signaling in ovarian cancer cell line SKOV3. The nanoparticle drug delivery method shows 100-fold increase in chemosensitization due to restoration of apoptotic signaling. Rejinold et al. (2011) studied the curcumin nanoparticle formulation for cancer drug delivery combined with biodegradable thermo-responsive chitosan-g-poly (N-vinylcaprolactam) nanoparticles (TRC-NPs). Curcumin-loaded TRC-NP treatment increased mitochondrial-mediated apoptosis on PC3. Bisht et al. (2007) developed a polymeric nanoparticle, N-isopropylacrylamide (NIPAAM), with N-vinyl-2-pyrrolidone (VP) and poly (ethylene glycol)monoacrylate (PEG-A) encapsulated nanocurcumin. The treatment enhances cellular apoptosis, downregulates pro-inflammatory cytokines (IL-6, IL-8, and TNFα), and inhibits nuclear factor kappa B (NFκB) activation. Katiyar et al. (2016) studied the formulation of nanoparticles of rapamycin combined with a chemosensitizer piperine, and they are loaded in poly (D,L-lactide-co-glycolide) nanoparticles against breast cancer. The uptake of rapamycin and its efficacy are increased by the presence of chemosensitizer piperine in killing breast cancer cells. The combined drug formulation along with poly (D,L-lactide-co-glycolide) nanoparticles results in improved bioavailability and long-term therapeutic action with less dosing frequency. Gong et al. (2013) investigated the NIR-absorbing conjugated polymer PEDOT:PSS with PEGylation surface coating to serve as a drug loading platform in cancer therapy for various types of aromatic therapeutic molecules . They found that PEDOT:PSS-PEG nanoparticles can act as a multifunctional drug carrier that has great applications in combined chemo-photodynamic and photothermal therapy of cancer. Aryal et al. (2011) examined the combination chemotherapy by loading doxorubicin and camptothecin into a single polymeric nanoparticle to obtain a drug-polymer conjugate, which is then encapsulated into the lipid-coated polymeric nanoparticles. This study points out that the dual drug approach offers a solution to overcome the challenge in ratiometric control over the loading of different types of drugs onto the same drug delivery vehicle. Jain et al. (2011) studied the therapeutic effects of tamoxifen-loaded PLGA nanoparticles (Tmx-NPs) against breast cancer. The results showed that when compared to untreated groups, oral Tmx-NP-treated group tumor size was reduced up to 41.56% along with reduced hepatotoxicity. Xiao et al. (2015) developed hyaluronic acid (HA)-functionalized polymeric nanoparticle (HA-CPT/CUR-NPs) to act as carriers to co-deliver camptothecin (CPT)/curcumin (CUR) in various weight ratios for colon cancer-targeted combination chemotherapy. HA-CPT/CUR-NPs have high

Table 4.2 Polymeric nanoparticle for the targeted drug delivery system

Sl. no.	Polymeric nanoparticle	Biological applications	Authors
1.	PEO-PCL nanoparticles	It shows 100-fold increase in chemosensitization due to restoration in apoptotic signaling	van Vlerken et al. (2007)
2.	TRC-NP nanoparticles	The treatment increased mitochondrial-mediated apoptosis on PC3	Rejinold et al. (2011)
3.	N-isopropylacrylamide (NIPAAM), with N-vinyl-2-pyrrolidone (VP) and poly (ethylene glycol) monoacrylate (PEG-A) nanoparticle	Enhanced cellular apoptosis, downregulates pro-inflammatory cytokines (IL-6, IL-8, and TNFα) and inhibits nuclear factor kappa B (NFκB) activation	Bisht et al. (2007)
4.	Poly (D,L-lactide-co-glycolide) nanoparticles	Improved bioavailability and long-term therapeutic action with less dosing frequency	Katiyar et al. (2016)
5.	PEDOT:PSS-PEG nanoparticles	Acts as a multifunctional drug carrier	Gong et al. (2013)
6.	Lipid-coated polymeric nanoparticles	It can load different types of drugs onto the same drug delivery vehicle	Aryal et al. (2011)
7.	Tmx-NP nanoparticles	Tumor size was reduced up to 41.56% along with reduced hepatotoxicity	Jain et al. (2011)
8.	HA-CPT/CUR-NP nanoparticles	High colon cancer cell-targeting ability	Xiao et al. (2015)
9.	Lactic-co-glycolic acid nanoparticles	Improved cellular uptake and highest cytotoxicity against CD44+ cells	Muntimadugu et al. (2016)
10.	Hypoxia-responsive nanoparticles	Drug showed higher toxicity toward hypoxic cells, and it has the ability to deliver doxorubicin into tumor cells under hypoxic conditions	Thambi et al. (2014)
11.	Temozolomide-loaded nanoparticles	Nanoparticle vector can act as an image-guided treatment of malignant glioma	Bernal et al. (2014)
12.	TPTN nanoparticle	Higher antitumor effect in H22 and HepG2 tumor cells Higher resolution and longer imaging time (>90 min) in the MRI diagnosis	Liu et al. (2014)
13.	PEO-PCL nanoparticles	Improved antitumor efficacy without any acute toxicity	Devalapally et al. (2008)
14.	HER-DMPNP nanoparticles	Great cancer cell affinity and ultrasensitivity via magnetic resonance imaging	Yang et al. (2007)

colon cancer cell-targeting ability. The extremely low 1:1 weight ratio shows highest antitumor efficiency against Colon-26 cells. Muntimadugu et al. (2016) formulated an efficient drug delivery system for the co-delivery of salinomycin and paclitaxel in the form of poly (lactic-co-glycolic acid) nanoparticles for targeting both cancer cells and cancer stem cells. In vitro cytotoxicity studies show that the combination therapy has great specificity toward the targeted cells, improved cellular uptake, and highest cytotoxicity against CD44+ cells. Thambi et al. (2014) examined the efficiency of self-assembled hypoxia-responsive nanoparticles loaded with doxorubicin as a potent drug carrier against hypoxic tumor tissues. The conjugated drug showed higher toxicity toward hypoxic cells, and it has the ability to deliver doxorubicin into tumor cells under hypoxic conditions. Bernal et al. (2014) reported a polymeric nanoparticle vector which has the ability to deliver viable therapeutic drugs and also be tracked in vivo using MRI. Convection-enhanced delivery of temozolomide-loaded nanoparticles improved the survival ratio of animals. And the nanoparticle vector can act as an image-guided treatment of malignant glioma. Liu et al. (2014) formulated a self-assembled target pH-sensitive theranostic polymeric nanoparticle (TPTN) concerning the cancer cell examination and also treatment. They loaded sorafenib, an anti-hepatocellular carcinoma drug, inside the multi-block polymer, poly (lactic acid)-poly (ethylene glycol)-poly (L-lysine)-diethylenetriamine pentaacetic acid and poly(L-histidine)-poly(ethylene glycol)-biotin. The TPTN nanoparticle showed antitumor effect against HepG2 cells, higher antitumor effect in H22 tumor cells, and higher resolution and longer imaging time (>90 min) in the MRI diagnosis. Overall it has the capacity for drug loading, imaging agents, precise targeting, pH-triggered drug-delivering qualities, and also exceptional biocompatibility. Devalapally et al. (2008) studied the potency and preceding protection of tamoxifen with paclitaxel in the biodegradable polymeric nanoparticle. Nanoparticle consists of poly (ethylene oxide) revised poly (epsilon-caprolactone), (PEO-PCL) nanoparticles, toward ovarian adenocarcinoma. The results showed improved antitumor efficacy without any acute toxicity. Yang et al. (2007) developed antibody-conjugated doxorubicin magnetic poly (D,L-lactide-co-glycolide) (PLGA) nanoparticles (HER-DMPNP) for diagnostic and treatment of cancer. The nanoparticle has high cancer cell affinity and ultrasensitivity via magnetic resonance imaging.

4.4 Liposomes

Liposomes are small spherical shape drug distributing vesicles created from nontoxic substances like phospholipids and cholesterol with varying sizes ranging from 0.025 µm to 2.5 µm. There are different types of liposomes like multilamellar vesicles; small unilamellar vesicle and large unilamellar vesicle were used for drug encapsulation to treat different types of cancers. It is widely used as an auxiliary in vaccination, signal carriers, or inducers in disease diagnosis and also in encapsulation technology for the targeted drug delivery against different types of cancer cells

Table 4.3 Liposomal-based drug delivery system

Sl. no.	Type of the liposomal encapsulates	Biological applications	References
1.	Doxorubicin-loaded anti-HER2 immunoliposomes	Suppressed the activity of overexpressing HER2 genes	Park et al. (2002)
2.	Liposome conjugated paclitaxel and gemcitabine	Induce endothelial apoptotic activity, inhibit the vasculature system	Eichhorn et al. (2010)
3.	Liposome-coloaded paclitaxel/epigallocatechin gallate	Arrest the MMP-2 and MMP-9 expression	Ramadass et al. (2015)
4.	Liposomal encapsulate curcumin and resveratrol	Enhance the apoptotic activity, cyclin D1 activity, and androgen receptor proteins and arrest AKT/mTOR signaling pathways	Narayanan et al. (2009)
5.	Paclitaxel- and curcumin-loaded liposomes	Arrest the NF-κB and Akt pathways, G2/M cell cycle phase	Ruttala and Ko (2015)
6.	Vincristine-/quercetin-loaded liposomes	Upregulate the activity of trastuzumab and arrest JIMT-1 cell-insensitive activity	Wong and Chiu (2011)
7.	PEGylated liposomal doxorubicin	Arrests the activity of HER2 genes and decreases the cardiac toxicity	O'Shaughnessy (2003)
8.	Magnetic liposomes with doxorubicin	Enhance cytotoxic effects	Hardiansyah et al. (2014)
9.	Liposomal encapsulated curcumin nanoparticles	Induce apoptotic activity	Saengkrit et al. (2014)

(Daraee et al. 2016) (Table 4.3). These liposomal drug formulations were used in several disease treatments based on their capacity of self-assembly, to carry large drug payloads, and biocompatibility (Sercombe et al. 2015). The liposomal encapsulating drugs offer a safe platform which have potent cytotoxic effect toward cancer cells with no toxic side effects in normal cells (Akbarzadeh et al. 2013).

4.4.1 Liposome-Encapsulated Drugs in Cancer Treatment

Park et al. (2002) developed doxorubicin-loaded anti-HER2immunoliposomes that suppressed the activity of overexpressing HER2 genes and decrease the cardiotoxicity in breast cancer tumor-induced xenograft models, compared to the doxorubicin treated group alone. Eichhorn et al. (2010) demonstrate that liposome conjugated paclitaxel and gemcitabine were targeted against specific tumor sites to induce endothelial apoptotic activity and inhibit the vasculature (reduce blood circulation in tumor area) system in lung and pancreatic cancer. Ramadass et al. (2015) investigated paclitaxel-/epigallocatechin gallate-coloaded liposomes induced apoptotic activity via caspase enzymes and arrests the MMP-2 (Matrix Metallo Proteinases are zinc-dependent family proteases) and MMP-9 expressions in breast cancer cells.

Narayanan et al. (2009) investigated the liposomal encapsulated curcumin and resveratrol drug to enhance the apoptotic activity, cyclin D1 activity, androgen receptor proteins and arrest AKT/mTOR signaling pathways for decreasing the tumor progression in prostate cancer cells. Ruttala and Ko (2015) studied paclitaxel- and curcumin-loaded liposomes to inhibit NF-κB and Akt pathways and G2/M cell cycle phase via increasing the apoptotic activity and subG1 cell population in skin and breast cancer cells. Wong and Chiu (2011) demonstrated vincristine-/quercetin-loaded liposomes inhibit drug-resistant activity and downregulate cell proliferation in breast cancer cells. O'Shaughnessy (2003) investigated PEGylated liposomal (liposomes enclosed with in polyethylene glycol (PEG) layer) doxorubicin drug shows an efficient progress compared with the different dosage/other combination of drugs in breast cancer cells against HER2 gene expression. These PLD drugs decrease the tumor progression as well as cardiac toxicity, myelosuppression, and alopecia. Hardiansyah et al. (2014) developed magnetic liposome (citric acid-coated magnetic nanoparticles encapsulated by liposomes) encapsulated with doxorubicin that is a best antineoplastic agent to arrest the colorectal cancer cell progression. Saengkrit et al. (2014) developed liposomal encapsulated curcumin nanoparticles that caused cell membrane damage by inducing the apoptotic activity in cervical cancer cells.

4.5 Dendrimers

Dendrimers are a molecule consisting of a central core and branches. It is a polymer classified as hyperbranched polymers or brush polymers (Morikawa 2016). It acts as a best antitumor vehicle by conjugating with anticancer drugs, monoclonal antibodies, and plant-based bioactive compounds through by its branches. Interactions between the branches of dendrimers and drugs are hydrophobic/hydrogen-bond and electrostatic. It also has encapsulating ability by encapsulating the drug using the central cavity or between the multiple channels of dendrons (Mignani and Majoral 2013) (Table 4.4). It has attracted much attention of the researchers due to its highly branched symmetrical architecture and its center core to encapsulate the molecular drugs. The dendrimer molecular structure was characterized using nuclear magnetic resonance (NMR) of proton (1 H) and carbon (13C), differential scanning calorimetry (DSC), matrix-assisted laser-desorption ionization-mass spectrometry (MALDI-MS), transmission electron microscopy (TEM), scanning electron microscope (SEM), rheology, and scattering techniques (Valdés Lizama et al. 2016).

PAMAM-doxorubicin conjugate is a best anticancer drug-delivering vehicle, which shows cytotoxic effect by inhibiting GLUT1 transporter in human breast adenocarcinoma cell lines (Sztandera et al. 2019). Polyamidoamine-docetaxel-trastuzumab and polyamidoamine-paclitaxel-trastuzumab conjugates exhibit antiproliferative impact on HER-2-positive breast cancer cells (Marcinkowska et al. 2019). Polyamidoamine, astramol dendrimers, and maltose-modified PPI dendrimers exhibit cytotoxic effect toward human ovarian carcinoma cell lines and

Table 4.4 Dendrimers as drug delivery agents

Sl. no.	Dendrimers	Cancer	References
1.	PAMAM-doxorubicin	Breast cancer	Sztandera et al. (2019)
2.	PAMAM-doc-trastuzumab and PAMAM-ptx-trastuzumab	Breast cancer	Marcinkowska et al. (2019)
3.	PPI dendrimers	Ovarian cancer	Janaszewska et al. (2012)
4.	Trastuzumab-grafted PAMAM	Breast cancer	Kulhari et al. (2016)
5.	Dendrimer-doxorubicin conjugates	Hepatocellular carcinoma	Kuruvilla et al. (2017)
6.	DOX conjugated with EGF receptor-binding peptide	Colon cancer	Ai et al. (2013)
7.	Lactobionic acid dendrimers with polyethylene glycol	Liver cancer	Fu et al. (2014)
8.	Dendrimer-curcumin	Breast cancer	Debnath et al. (2013)
9.	Cationic chlorambucil-dendrimer	Colon cancer	Seixas et al. (2019)
10.	G3 PAMAM-NH$_2$ dendrimer-chlorambucil	Breast cancer	Bielawski et al. (2011)
11.	Methotrexate-loaded polyether-copolyester dendrimers	Gliomas	Dhanikula et al. (2008)
12.	PAMAM-paclitaxel-conjugated omega-3 fatty acid	Gastrointestinal cancer	Dichwalkar et al. (2017)
13.	Doxorubicin-conjugated dendrimer	Lung cancer	Kaminskas et al. (2014)
14.	Polyamidoamine dendrimers with hyaluronic acid	Pancreatic cancer	Kesharwani et al. (2015)
15.	Dendrimer-camptothecins	Breast cancer	Morgan et al. (2006)
16.	Cisplatin-dendrimer	Lung cancer	Nguyen et al. (2015)
17.	J591 antibody-dendrimer	Prostate cancer	Patri et al. (2004)

Chinese hamster ovary cell lines (Janaszewska et al. 2012). PAMAM dendrimers stimulate the mitochondria of dermal cell line to synthesize reactive oxygen production, apoptosis, and DNA damage that shows a better cytotoxic and antiproliferative effect (Mukherjee et al. 2010). Trastuzumab-grafted PAMAM dendrimers have better intra-tumoral delivery and apoptosis stimulation in HER2-positive breast cancer cells; thus, it acts as a better candidate to deliver anticancer drugs (Kulhari et al. 2016). Dendrimer-doxorubicin conjugates showed efficient anticancer activity in murine hepatocellular carcinoma cells and the conjugation of dendrimer to doxorubicin which reduce the cardiotoxicity induced by doxorubicin (Kuruvilla et al. 2017). DOX conjugated with an EGF receptor-binding peptide shows targeted drug delivery and decreases the drug resistance and anticancer activity in human colon cancer cells (Ai et al. 2013). Lactobionic acid-modified dendrimers conjugated with polyethylene glycol spacer showed the better cytotoxic effect in liver cancer cells (Fu et al. 2014). Dendrimer-surcumin conjugate shows cytotoxic activity in breast cancer cell lines by inducing apoptosis via caspase-3 activation (Debnath et al.

2013). Dendrimer-curcumin conjugate dissolves in water and thus is used as a potent cytotoxic factor against breast cancer cell lines. This conjugate is efficient in inducing cytotoxicity, as estimated by the MTT assay, and it also efficiently induced cellular apoptosis measured by caspase-3 activation. Cationic chlorambucil dendrimer will exhibit the cell cycle arrest at the G2 phase of mitosis and stimulate the caspase-independent apoptosis activity in prostate and colon cancer cell lines (Seixas et al. 2019). G3 PAMAM-NH_2 dendrimer-chlorambucil conjugate induces apoptotic effect and cytotoxic activity in human breast cancer cells (Bielawski et al. 2011). The anti-HER2 mAb conjugated to polyamidoamine dendrimer generation 5 was labeled with alexaFluor 488 and showed cytotoxic activity against HER2-expressing tumors (Shukla et al. 2006). Resorcinarene-PAMAM-dendrimer conjugates of flutamide have more apoptotic and anticancer activity than flutamide derivatives (Pedro-Hernández et al. 2018).

Methotrexate-loaded polyether-copolyester dendrimers act as a potential delivery system in the treatment of gliomas with enhanced permeability across the blood-brain barrier (Dhanikula et al. 2008). DHATX is significantly more potent than PTX or PAX at inhibiting cellular proliferation, suppressing long-term survival, and inducing cell death in UGC cells. PAMAM-paclitaxel-conjugated omega-3 fatty acid inhibits the cellular proliferation and induced cell death in upper gastrointestinal cancer cells (Dichwalkar et al. 2017). Doxorubicin-conjugated dendrimer inhibits the drug-resistant activity in lung cancer and improves the anticancer activity (Kaminskas et al. 2014). Polyamidoamine dendrimers conjugated with hyaluronic acid were an efficient drug delivery vehicle of curcumin to target and inhibit CD44 in pancreatic cancer cells (Kesharwani et al. 2015). Dendrimer-encapsulated camptothecins increased cellular uptake, increased intracellular drug retention, and afford enhanced anticancer activity in human breast adenocarcinoma (Morgan et al. 2006). Cisplatin-dendrimer nanocomplex has sufficient antiproliferative activity against lung cancer (Nguyen et al. 2015). J591 antibody-dendrimer conjugate is used as a targeted drug delivery vehicle in prostate cancer (Patri et al. 2004).

4.6 Polymer Drug Conjugates

Polymer is a chemical compound which is essentially combined together to form repeating subunits. Polymers are classified in to two types: natural polymers (naturally found in plants and animals) or synthetic polymers (synthetically prepared). There is an increased use of polymers in several fields like aerospace, sports, 3D printing, holography, water purification, molecular recognition, and drug distribution. Naturally polymers had a unique property like being hard, lightweight, and strong and having thermal and electrical stability, low cost, high specificity, and adaptability. Based on these qualities, polymers are used in biological-related researches, treatments, and drug distributions. Polymers were conjugated with drugs to form polymer drug conjugates to treat different types of diseases including cancer (Table 4.5).

Table 4.5 Polymeric conjugates for drug delivery

Sl. no.	Polymer drug conjugate	Cancer	References
1.	Poly L-glutamic acid with paclitaxel	Breast cancer, ovarian cancer, non-small cell lung cancer	Oldham et al. (2000), Li et al. (1999), Auzenne et al. (2002), and Zou et al. (2004)
2.	Poly1,2-glycerol carbonate with paclitaxel	Peritoneal cancer, lung cancer	Ekladious et al. (2017)
3.	N-2-hydroxypropylmethacrylamide copolymer with paclitaxel and gemcitabine	Ovarian cancer	Zhang et al. (2014a), Zhang et al. (2014b)
4.	Polyl-γ-glutamyl glutamine-conjugated docetaxel	Breast cancer	Tavassolian et al. (2014)
5.	γ-Polyglutamic acid with docetaxel	Colon cancer, gastric cancer	Maya et al. (2014) and Sreeranganathan et al. (2017)
6.	N-2-Hydroxypropylmethacrylamide copolymer/docetaxel	Prostate cancer	Liu et al. (2012a), Liu et al. (2012b)
7.	Poly D,L-lactide-co-glycolide/hyaluronic acid with docetaxel	Breast cancer	Huang et al. (2014)
8.	Poly-γ-glutamic acid-coated doxorubicin	Liver cancer, non-small cell lung cancer	Qi et al. (2017) and Li et al. (2013)
9.	Polyethylene glycol-conjugated doxorubicin and paclitaxel	Non-small cell lung cancer	Lv et al. (2014)
10.	N-2-hydroxypropylmethacrylamide copolymer-doxorubicin	Ovarian cancer	Shiah et al. (2001)
11.	Poly-l-glutamic acid-gemcitabine	Breast cancer	Kiew et al. (2010) and Voon et al. (2012)
12.	Poly N-2-hydroxypropylmethacrylamide gemcitabine and paclitaxel	Ovarian cancer	Larson et al. (2013)
13.	Gemcitabine-polyethylene glycol	Lung cancer	Garg et al. (2012)
14.	Gemcitabine-loaded polylactide-*co*-glycolide	Pancreatic cancer	Jaidev et al. (2015)
15.	Irinotecan with poly 2-ethyl 2-oxazoline-*b*-poly L-glutamic acid	Colorectal carcinoma	Salmanpour et al. (2019)
16.	Poly-L-glutamic acid-camptothecin	Melanoma, lung cancer	Singer et al. (2001), Bhatt et al. (2003), and Zou et al. (2001)
17.	Camptothecin-polyethylene glycol	Ovarian cancer	Minko et al. (2002)

Polymer drug conjugates (PDCs) are nano medicines that conjugated with specific drugs to act as a carrier molecule. The polymer drug-conjugated technology enhances the solubility of anticancerous drugs and controls the target site-specific drug distributions in cancer cells by enhanced permeability and retention (Girase et al. 2019). Different types of polymers like poly (L-glutamic acid), polyethylene glycol (PEG), N-(2-hydroxypropyl) methacrylamide (HPMA), polylactide-co-glycolide (PLGA), polyvinyl pyrrolidone, and poly-γ-glutamic acid (γ-PGA) were used for drug conjugation to treat different types of cancers. Familiar cancer drugs like paclitaxel, docetaxel, doxorubicin, gemcitabine, irinotecan, and camptothecin were commonly used in polymer drug conjugation process in cancer treatment (Duncan et al. 2005).

4.6.1 Polymer Drug Conjugates Against Cancer

Poly L-glutamic acid with paclitaxel conjugate induces apoptotic, tumor suppressor p53 activity and G2/M cell cycle arrest, downregulated HER2/neu expression in breast cancer cells (Oldham et al. 2000; Li et al. 1999), and decreases the tumor progression in ovarian cancer (Auzenne et al. 2002) and non-small cell lung cancer (Zou et al. 2004). Ekladious et al. (2017) reported that high dose of poly 1,2-glycerol carbonate with paclitaxel conjugates is exhibiting greater efficiency to inhibit cell proliferation of cancer cells when compared to normal multi-dosage of paclitaxel drug in mesothelioma cancer cells, lung cancer cells, and pancreatic cancer cells. N-2-hydroxypropylmethacrylamide copolymer with paclitaxel and gemcitabine induces apoptotic activity and replication arrest in ovarian cancer. Paclitaxel and gemcitabine inhibit the micro tubulin formation and decrease the tumor cell proliferation in ovarian cancer cell (Zhang et al. 2014a, Zhang et al. 2014b). Tavassolian et al. (2014) investigated the anticancer efficacy of poly l-γ-glutamyl glutamine-conjugated docetaxel in mice model and MCF7 cancer cells; compared to chemotherapeutic drub Taxotere, the poly l-γ-glutamyl glutamine-conjugated docetaxel has better efficacy to inhibit tumor growth in breast cancer. γ-Poly glutamic acid-docetaxel inhibits the cell cycle progression of G2/M phase by inducing apoptotic function in colon cancer cells and downregulating EGFR, in gastric cancer cells (Maya et al. 2014; Sreeranganathan et al. 2017). N-2-Hydroxypropylmethacrylamide copolymer/docetaxel conjugate increases the enhanced permeability and retention effect to downregulate the tumor progression in prostate cancer cells (Liu et al. 2012a, Liu et al. 2012b). Poly D,L-lactide-co-glycolide/hyaluronic acid copolymers with docetaxel conjugate target the overexpressing CD44 and inhibit cell proliferation in breast cancer (Huang et al. 2014). Poly-γ-glutamic acid-coated doxorubicin arrests the cell cycle at S phase and enhances apoptotic activity in liver cancer and non-small cell lung cancer (Qi et al. 2017; Li et al. 2013).

Lv et al. (2014) developed poly-ethylene glycol-conjugated doxorubicin and paclitaxel co-drug delivery inducing apoptotic activity was monitored and visualized by fluorescence images in A549 xenograft models. The co-drug delivery exhibits high level of antitumor activity and reduces the tumor size compared to single drug-loaded polymers and free drug combinations. N-2-hydroxypropylmethacrylamide copolymer with doxorubicin conjugates cleaved the oligopeptides of cancer cells and incorporated antibodies into specific antigens to induce antitumor efficacy in ovarian cancer cells (Shiah et al. 2001). Poly-l-glutamic acid-gemcitabine conjugate enhances the plasma stability and decreases the tumor growth in breast cancer cells (Kiew et al. 2010; Voon et al. 2012). Poly N-2-hydroxypropylmethacrylamide gemcitabine and paclitaxel conjugates are found to be with potent anticancer activity to inhibit cell proliferation in ovarian cancer cells (Larson et al. 2013). Garg et al. (2012) developed gemcitabine with polyethylene glycol conjugate tightly bound to sigma receptors of cancer cells to effectively transport the drug inside the cancer cells and to bring high level of cytotoxic response in lung cancer cells. Jaidev et al. (2015) demonstrate that gemcitabine-loaded polylactide-*co*-glycolide was to increase the antiproliferative and apoptotic activity in pancreatic cancer. The polymer polylactide-*co*-glycolide is an efficient drug carrier to release the drugs within 3 hours in pancreatic cancer cells. Irinotecan with poly 2-ethyl 2-oxazoline and *b*-poly L-glutamic acid double copolymers decreases tumor proliferation rate in colorectal cancer (Salmanpour et al. 2019). Poly-L-glutamic acid-camptothecin conjugates decrease the tumor progression in melanoma and lung cancer cells (Singer et al. 2001; Bhatt et al. 2003; Zou et al. 2001). Camptothecin-polyethylene glycol conjugates to induce apoptotic genes APAF-1 and caspases 3 and 9 by downregulating the BCL-2 gene in ovarian cancer cells (Minko et al. 2002).

4.7 Silica Nanoparticles

Silica nanoparticle also termed as silicon dioxide nanoparticles or nano silica is the promising tool in biomedical research and also widely used in various applications due to their excellent thermal stability, biocompatibility, low toxicity, and large-scale synthetic availability. The particle with its high surface area has the chances for the possibility of chemical modification which could help in drug delivery, gene therapy, and site-specific target therapy (Slowing et al. 2007) (Table 4.6). Stober method is one of the most widely used and accepted methods in the synthesis of silica nanoparticle. It has the property of uniform synthesis and the particle size ranges from 5 to 2000 nm. Based on the structure and particle size, it could be divided into P-type and S-types. The synthesized particle can be characterized by SEM, TEM, X-ray diffraction, absorption spectroscopy, and EPR spectroscopy studies (Tan et al. 2004).

4.7.1 Silica Nanoparticles toward Cancer Therapy

Nano medicine against cancer as an upcoming field has major impact in the field of health and medicine. Previous studies have claimed that silica nanoparticle has the potent anticancer activity to induce apoptosis in cancer cells. Conjugated mesoporous silica nanoparticles with anticancer drug of paclitaxel have significant effect to arrest cell cycle and inhibit cell proliferation in human pancreatic cancer cell lines

Table 4.6 Silica-based nanoparticles as drug delivery systems

Sl. no.	Compound	Type of cancer	Action	References
1.	Paclitaxel-conjugated mesoporous silica nanoparticles	Pancreatic cancer	Inhibit cell proliferation	Lu et al. (2007)
2.	Doxorubicin-loaded silica nanoparticles	Various types of cancer	Inhibit oncogenes	Chen et al. (2009)
3.	Hyaluronic acid-loaded mesoporous silica nanoparticles	Colon cancer	Downregulate CD44 expression	Yu et al. (2013)
4.	Doxorubicin-loaded silica nanoparticles	Cervical cancer	Decrease cancer cell viability	Qiu et al. (2013)
5.	Curcumin-loaded silica nanoparticles	Breast Cancer	Induce cytotoxic effect	Ma'mani et al. (2014)
6.	Quercetin-loaded mesoporous silica nanoparticles	Breast cancer	Reduce overexpressed signaling pathway	Sarkar et al. (2016)
7.	Lectin-conjugated silica nanoparticles	Bone cancer	Reduce cancer cell viability	Martínez-Carmona et al. (2018)
8.	Chemo drug-loaded silica nanoparticles	Lung cancer	Induce apoptosis	Liu et al. (2012a), Liu et al. (2012b)
9.	Epirubicin-loaded silica nanoparticles	Colon cancer	Arrest targeted tumor cells growth	Xiong et al. (2015)
10.	Snake venom (*Walterinnesia aegyptia*)-conjugated silica nanoparticles	Breast and prostate cancer	Alter mitochondrial function	Badr et al. (2013)
11.	Glycosylated cytochrome C-conjugated silica nanoparticles	Cervical cancer	Induce apoptosis	Méndez et al. (2014)
12.	Cisplatin- and nitric oxide-loaded mesoporous silica nanoparticles	Lung cancer	Induce C toxicity effect	Munaweera et al. (2015)
13.	Curcumin-loaded silver nanoparticles	Head and neck cancer	Inhibits NF-κB pathway	Singh et al. (2014)

(Lu et al. 2007). Loading of doxorubicin drug to silica nanoparticles enhances the anticancer activity by suppressing the tumor gene and induces apoptosis in multidrug-resistant cancer cell lines (Chen et al. 2009). Modified hyaluronic acid capped with mesoporous silica nanoparticles acts as an efficient drug delivery vehicle and induces cellular uptake to react against CD44 overexpressed human colon cancer cells (Yu et al. 2013). Doxorubicin-loaded silica nanoparticles successfully act toward cancer cells and reduced the tumor volume in cervical cancer (Qiu et al. 2013). Conjugation of cetuximab with silica nanoparticle was an efficient drug delivery vehicle to target tumor cell function in human colon cancer cells (Cho et al. 2010). Curcumin-loaded silica nanoparticle efficiently defeats cancer cell propagation and generates apoptosis in breast cancer cell lines (Ma'mani et al. 2014).

Quercetin-loaded mesoporous silica nanoparticles strongly inhibit cell cycle progression and stimulate cancer cell death by inhibiting Akt/Bax pathway. The developed compound also increases the expression of caspase protein and decreases cancer cell viability in breast cancer cells (Sarkar et al. 2016). Lectin-conjugated silica nanoparticle reduces the cancer cell viability and inhibits the cancer inducing proteins in bone cancer (Martínez-Carmona et al. 2018). Combinational loading of both hydrophilic and hydrophobic types of chemo drugs with silica nanoparticle was efficiently internalized into cancer cells to induce apoptosis effect in human pulmonary adenocarcinoma cell lines (Liu et al. 2012a, Liu et al. 2012b). Dual targeting ligands such as folic acid and dexamethasone are modified with mesoporous silica nanoparticle which significantly inhibit only the cancer growth and prevent the non-cancer cell from the toxic side effects by means of receptor-mediated cellular uptake in cervical cancer cell line (Xiong et al. 2015). Epirubicin-loaded silica nanoparticle shows the anticancer activity in C-26 colon cancer-induced mice (Hanafi-Bojd et al. 2015). Loading of silica nanoparticle with snake venom (*Walterinnesia aegyptia*) raises the reactive oxygen species level and modifies the mitochondrial membrane potential and then subsequently arrests tumor cell spread and provokes apoptosis in breast and prostate cancer cell lines (Badr et al. 2013). Administration of glycosylated cytochrome C-conjugated silica nanoparticle induces cytotoxic and apoptotic effect toward cervical cancer cell lines (Méndez et al. 2014). Cisplatin- and nitric oxide-loaded mesoporous silica nanoparticles exhibit toxicity effect and induce tumor cell death in lung cancer (Munaweera et al. 2015). Curcumin-loaded silver nanoparticles suppress NF-κB proteins and induce necrotic cell death in human squamous cell carcinoma cell line (Singh et al. 2014).

4.8 Gold Nanoparticles

Gold nanoparticles (AuNPs) have the smallest nanoparticle size ranging from 1 to 100 nm in diameter. It could effectively act as a carrier vehicle for gene and drug delivery in therapeutic applications. It was differentiated into various subtypes based on their shape, size, and morphological characters with greater potential in fighting against various diseases including cancer. It also extends its applications in molecular diagnosis, molecular therapy, and molecular profiling (Arvizo et al. 2010). Synthesis of gold nanoparticles by way of chemical reduction method by using gold salts of citrate and hydrogen tetrachloroaurate (HAuCl4) as reducing agents is one of the novel methods (Salcedo and Sevilla III 2013). Synthesized nanoparticles hold novel chemical and physical traits to deliver antibiotics and drugs to the aspired targeted molecules accurately (Wilczewska et al. 2012). Covalent attachments and supramolecular assembly are the two primary strategies accomplished to incorporate the gold nanoparticles in gene therapy. The nanoparticle appears in different colors like blue, red, etc., based on the morphology, assemblage, and their local environment (Vigderman and Zubarev 2013).

4.8.1 Applications of Gold Nanoparticles

Gold nanoparticles exhibit a wide range of application in the field of medicine, bioimaging, biological research, gene therapy, photothermal therapy, etc. and as catalyst, drug delivery vehicle, and biosensors (Ghosh et al. 2008) (Table 4.7). As a biosensor device, gold nanoparticles are extensively used for the purpose of detecting the enzyme. The major advantages of using gold nanomaterial include easy scaling up, increased bioavailability, and enhanced time-resistant and targeted drug delivery at the specific target (Sperling et al. 2008). Gold nanoparticles have been utilized to increase the sensitivity of magnetic resonance imaging (MRI). One of the newly developed technologies of DNA labeled gold nanoparticle has gained much attention in the field of bio-nanotechnology research (Cai et al. 2008). Gold nanoparticle has numerous biomedical applications in the development of potential therapies for serious human diseases like HIV, cancer, etc. (Dykman and Khlebtsov 2012).

4.8.2 Gold Nanoparticle Toward Cancer Therapy

Administration of gold nanoparticles into cancer cell selectively damages the DNA double strand and inhibits the mitotic division of binucleate formation, leading to the complete arrest of cancer cell division which induces cell death (Kang et al. 2010). Conjugation of anticancer drug doxorubicin with gold nanoparticles provides the efficient drug releasing capacity with greater solubility and stability in cancer cells. This developed compound significantly reduces the cancer cell proliferation and provides efficient cytotoxic effect at the metastatic levels of cancer cells (Aryal et al. 2009). Dendrimer encapsulated gold nanoparticle loaded with thiol-containing anticancer drug shows efficient cytotoxic effect and induces mitochondrial enzyme function to activate apoptosis in cervical cancer cell lines (Wang et al. 2013). Chloroquine-loaded gold nanoparticle alters oncogenes and inhibits cancer cell proliferation in breast cancer cells (Joshi et al. 2012). Methotrexate conjugate gold nanoparticle shows rapid drug release and inhibits tumor development in syngeneic lung cancer model (Chen et al. 2007). Porphyran encapsulated gold nanoparticles stimulate the tumor suppressor gene to resist cancer cell initiation and arrest the cell cycle progression in human glioma cell lines (Venkatpurwar et al. 2011). Conjugation of gallic acid with gold nanoparticle significantly increases the caspase enzymes activity to induce apoptosis by both the extrinsic and intrinsic pathway in cervical cancer cell lines (Daduang et al. 2015). Fucoidan mimetic glycopolymer-coated gold nanoparticle increases the level of caspase enzyme through death receptor and mitochondrial apoptotic pathway. This decreases the cancer cell migration and invasion in human colon cancer cell lines (Tengdelius et al. 2015). Exposure to 1.4 nm range of gold nanoparticle increases oxygen consumption in cervical cancer cells to repair the mitochondrial dysfunction and to induce cell death (Pan et al. 2009). Conjugated gold nanoparticle with galactoxyloglucan decreases cancer cell

Table 4.7 Gold nanoparticles in targeted cancer drug delivery

Sl. no.	Compound name	Cancer type	Action of compound	Type of compound	References
1.	Doxorubicin-conjugated gold nanoparticle	Metastatic cancer	Inhibits cancer cell proliferation	Synthesized drug	Aryal et al. (2009)
2.	Thiol-containing anticancer drug-loaded gold nanoparticle	Cervical cancer	Apoptosis activity	Synthesized drug	Wang et al. (2013)
3.	Chloroquine-loaded gold nanoparticle	Breast cancer	Decreases oncogene expression	Synthesized drug	Joshi et al. (2012)
4.	Methotrexate-loaded gold nanoparticle	Lung cancer	Inhibits tumor growth	Synthesized drug	Chen et al. (2007)
5.	Porphyran-synthesized gold nanoparticle	Brain cancer	Induces tumor suppressor gene	Synthesized drug	Venkatpurwar et al. (2011)
6.	Gallic acid-synthesized gold nanoparticle	Cervical cancer	Increases apoptotic enzyme function	Synthesized drug	Daduang et al. (2015)
7.	Fucoidan mimetic glycopolymer-loaded gold nanoparticle	Colon cancer	Induces cell death	Synthesized drug	Tengdelius et al. (2015)
8.	1.4 nm gold nanoparticle	Cervical cancer	Increases ROS level	Synthesized drug	Pan et al. (2009)
9.	Galactoxyloglucan-loaded gold nanoparticle	Various types of cancer	Increases life span	Synthesized drug	Joseph et al. (2014)
10.	*C. guianensis*-loaded gold nanoparticle	Blood cancer	Decreases cell viability	Phytomedicine	Geetha et al. (2013)
11.	*Enterococcus* sp.-loaded gold nanoparticle	Lung cancer	Regulates signaling molecule	Phytomedicine	Rajeshkumar (2016)
12.	*Abelmoschus esculentus* (L.)-synthesized gold nanoparticle	Blood cancer	Increases ROS level	Phytomedicine	Mollick et al. (2014)
13.	*Indigofera tinctoria*-synthesized gold nanoparticle	Lung cancer	Cytotoxicity	Phytomedicine	Vijayan et al. (2018)
14.	*Gymnema sylvestre*-synthesized gold nanoparticle	Epithelial cancer	Increases ROS level	Phytomedicine	Nakkala et al. (2015)
15.	*Cassia tora*-synthesized gold nanoparticle	Colon cancer	Inhibits cell proliferation	Phytomedicine	Abel et al. (2016)
16.	*Nerium oleander*-synthesized gold nanoparticle	Breast cancer	Cell death	Phytomedicine	Barai et al. (2018)
17.	*Bauhinia purpurea*-synthesized gold nanoparticle	Various types of cancer	Reduces cancer cell volume	Phytomedicine	Vijayan et al. (2019)

(continued)

Table 4.7 (continued)

Sl. no.	Compound name	Cancer type	Action of compound	Type of compound	References
18.	*Mimosa pudica*-synthesized gold nanoparticle	Breast cancer	Cell cycle arrest	Phytomedicine	Uma Suganya et al. (2016)
19.	*Abies spectabilis*-synthesized gold nanoparticle	Bladder cancer	Alters apoptotic enzymes	Phytomedicine	Wu et al. (2019)
20.	*B. citriodora*-synthesized gold nanoparticle	Liver cancer	Induces cytotoxicity	Phytomedicine	Khandanlou et al. (2018)
21.	*Solanum xanthocarpum*-synthesized gold nanoparticle	Head and neck cancer	Induces autophagy	Phytomedicine	Zhang et al. (2018)
22.	*Sargassum swartzii*-synthesized gold nanoparticle	Cervical cancer	Alters mitochondrial function	Phytomedicine	Dhas et al. (2014)

viability and increased the life span of murine cancer-induced mouse model (Joseph et al. 2014).

4.8.3 Combinational Therapy of Phytochemicals with Gold Nanoparticle in Cancer Cells

Aqueous flower extract of *C. guianensis*-loaded gold nanoparticle efficiently acts against cancer cells and induces apoptotic activity in human leukemia cancer cells (Geetha et al. 2013). *Enterococcus* sp. marine bacteria-loaded gold nanoparticle downregulates overexpressed signaling molecule of Akt/Ras/m-TOR pathway in lung cancer cells (Rajeshkumar 2016). Synthesis of gold nanoparticles using *Abelmoschus esculentus* (L.) pulp extracts significantly increases apoptotic activity by regulating mitochondrial enzyme. This compound also elevates intracellular oxygen level to prevent cancer cell proliferation in human blood cancer cells (Mollick et al. 2014). *Indigofera tinctoria* leaf extract-synthesized gold nanoparticle exhibits more toxicity toward lung cancer cells (Vijayan et al. 2018). *Gymnema sylvestre* leaf extract-loaded gold nanoparticles increase intracellular oxygen concentration and induce antiproliferative effect in human epithelial cancer cell line (Nakkala et al. 2015). Gold nanoparticles capped with *Cassia tora* arrest cancer cell proliferation in colon cancer cells (Abel et al. 2016). Gold nanoparticle conjugates with *Nerium oleander* extract significantly increase cellular oxygen level and induce apoptosis in breast cancer cells (Barai et al. 2018). *Bauhinia purpurea* leaf extract-synthesized gold nanoparticle acts as an excellent anticancer agent on different types of cancer cells (Vijayan et al. 2019). Synthesized gold nanoparticles using *Mimosa pudica* extract on cancer cell reveals apoptotic effect and induced cell cycle arrest at early G1/S phase in breast cancer cell lines (Uma Suganya et al. 2016). *Abies spectabilis* plant extract-synthesized gold nanoparticle upregulates the Beclin-1, Bax, and caspase-3 enzyme and downregulates Bcl-2 and BidT24 in bladder cancer cells (Wu et al. 2019). *Backhousia citriodora* leaf extract-synthesized Au-NPs induce cytotoxicity effect and apoptotic activity in breast and liver cancer cells (Khandanlou et al. 2018). *Solanum xanthocarpum*-synthesized AuNPs increase ROS level, autophagy, and apoptosis through activation of caspase-3 and caspase-9 and nuclear fragmentation in nasopharyngeal carcinoma cell lines (Zhang et al. 2018). Biosynthesis of AuNPs using *Sargassum swartzii* induces cytotoxic effect and reverts mitochondrial function in HeLa cell lines (Dhas et al. 2014).

4.9 Carbon Nanotubes

Carbon nanotubes (CNTs) are cylindrical molecules made up of hexagonal arrangements of hybridized carbon atoms. It appears in tubular shape that was coated with graphite carbon atoms in the form of layer. Carbon nanotubes are classified in two types: single-walled carbon nanotubes with diameter ranging from 1 to 10 nm and multiwalled carbon nanotubes with diameters ranging approximately 5 to 30 nm. The nanotubes consist of 1–100. There are two major methods to synthesize carbon nanotubes using chemical vapor deposition (CVD) method (fossil-based hydrocarbon method) and plant-based hydrocarbon method. It has unique properties such as high thermal conductivity, highly flexible property, and strong tensile strength. The inner and outer surfaces of carbon nanotubes are well modified with a variety of functional groups which helps in conjugating with targeting ligands and drug molecules. The strength and flexibility of carbon nanotubes are used to control the other nanoscale structures which play significant role in nano engineering field. Carbon nanotubes have a wide range of applications as tiny sensors, electronic devices, optical devices, catalyst process, batteries, fuel cells, solar cells, and drug delivery vehicles (Prasek et al. 2011) (Table 4.8).

4.9.1 Carbon Nanotubes in Cancer Therapy

Carbon nanotubes stand as a popular tool in cancer therapy due to their unique physiochemical properties. It is also considered as one of the most promising nanomaterials with capability of both detecting tumor cells and delivering drug therapy to cancer cells. Single-walled carbon nanotube selectively activates caspase-3 and cytochrome C enzyme to induce mitochondrial function which inhibits tumor growth in breast cancer cells (Zhou et al. 2011). Methotrexate-conjugated multiwalled carbon nanotube possesses rapid drug release and exhibits higher cytotoxicity effect toward breast cancer cells (Samorì et al. 2010). Single-walled carbon nanotube conjugate with anticancer drug of SN-38 carries good biocompatibility

Table 4.8 Carbon nanotubes in cancer therapy and drug delivery

Sl. no.	Compound name	Types of cancer	Mode of action	Reference
1.	Single-walled carbon nanotube	Breast cancer	Induces mitochondrial function	Zhou et al. (2011)
2.	Methotrexate-conjugated multiwalled carbon nanotube	Breast cancer	Cytotoxic effect	Samorì et al. (2010)
3.	SN-38-loaded single-walled carbon nanotube	Colon cancer	Induces apoptosis	Lee et al. (2013)
4.	Paclitaxel-loaded single-walled carbon nanotube	Breast cancer	Inhibits cancer cell proliferation	Liu et al. (2008a), Liu et al. (2008b)
5.	Curcumin-loaded carbon nanotubes	Prostate cancer	Downregulate Wnt signaling pathway	Li et al. (2014)
6.	Betulinic acid-conjugated multiwalled carbon nanotubes	Lung cancer	Inhibit Warburg pathway	Tan et al. (2014)
7.	Polyethylene glycol-coated carbon tubes	Lung cancer	Induce Bcl-2-mediated apoptosis	Kim et al. (2017)
8.	Doxorubicin-loaded multiwalled carbon nanotube	Breast cancer	Decreases cancer cell viability	Ali-Boucetta et al. (2008)
9.	Cisplatin-loaded carbon nanotubes	Solid tumor	Inhibit neuroblastoma cell growth	Vittorio et al. (2014)
10.	Doxorubicin-conjugated single-walled carbon nanotubes	Cervical cancer	Induce apoptotic activity	Meng et al. (2012)
11.	Cisplatin-loaded carbon nanotubes	Cervical cancer	Enhanced antiproliferative effect	Peng et al. (2010)
12.	Doxorubicin-loaded carbon nanotubes	Cervical cancer	Autophagy reaction	Zhang et al. (2009a), Zhang et al. (2009b)
13.	Oxaliplatin encapsulated single-walled carbon nanotubes	Various types of cancer	Inhibit tumor growth	Rezvani et al. (2016)
14.	Indole-3-carbinol cyclic-loaded single-walled carbon nanotube	Breast cancer cells	Promotes antiproliferative and cytopathic effect	De Santi et al. (2013)
15.	Polyethylene glycol-coated multiwalled carbon nanotubes	Pancreatic cancer cell line	Enhance mitochondrial function	Mocan et al. (2014)

and efficient cellular uptake and induces cancer cell death in colon carcinoma cell lines (Lee et al. 2013). Paclitaxel-loaded single-walled carbon nanotube inhibits cancer cell proliferation and induces apoptotic activity in murine breast cancer model (Liu et al. 2008a, Liu et al. 2008b). Curcumin-loaded carbon nanotubes downregulate Wnt signaling pathway in human prostate cancer (Li et al. 2014).

Oxidized multiwalled carbon nanotube conjugates with betulinic acid suppress Warburg pathway by inhibiting lactate dehydrogenase enzyme secretion and enhance anticancer activity in lung cancer cell line (Tan et al. 2014). Polyethylene glycol-coated carbon tubes generate intracellular oxygen level and induce Bcl-2-mediated apoptosis in lung cancer cells (Kim et al. 2017). Doxorubicin-loaded multiwalled carbon nanotube enhances cytotoxic effect and reduces cancer cell viability in breast cancer cell line (Ali-Boucetta et al. 2008). Cisplatin-loaded carbon nanotubes significantly inhibit neuroblastoma cell growth in nerve tissues and enhance cytotoxic effect against human neuroblastoma cell lines (Vittorio et al. 2014). Doxorubicin-conjugated single-walled carbon nanotubes increase ROS level and induce apoptotic activity in cervical cancer (Meng et al. 2012). Cisplatin-loaded carbon nanotubes promote extracellular drug release and enhance the antiproliferative effect in cervical cancer cell line (Peng et al. 2010). Doxorubicin-loaded carbon nanotubes accumulate inside the cancer cell lysosomes and inhibit transcription activity which ultimately leads to autophagy reaction in cervical cancer cell line (Zhang et al. 2009a, Zhang et al. 2009b). Oxaliplatin encapsulated single-walled carbon nanotubes provide efficient drug releasing ability and inhibit cancer cell growth in various types of cancer cells (Rezvani et al. 2016). Anticancer drug of indole-3-carbinol cyclic-loaded single-walled carbon nanotube induces cell death and promotes antiproliferative and cytotoxic effect in breast cancer cells (De Santi et al. 2013). Polyethylene glycol-coated multiwalled carbon nanotubes stimulate mitochondrial function which in turn activates free radicals in the cells and induces oxidative state in the pancreatic cancer cell lines (Mocan et al. 2014).

4.10 Quantum Dots

Quantum dots (QDs) are nanoparticles ranging from 2 to 10nm semiconductor crystals which have unique optical properties including intensive fluorescence and tunable wavelength (Zhang et al. 2008). QDs differ from natural biogenic nanoparticles by its crystalline metalloid structure and quantum confinement effect (Hardman 2006). These small probes can penetrate cells and organelles (Jovin 2003). QDs have a semiconductor core made up of CdS, CdSe, CdTe, ZnS, and PbS metals, and the core region is overcoated by a shell to enhance its optical activity and an outermost cap for increasing solubility in aqueous buffers (Ghasemi et al. 2009). QDs can be used for bio labeling and biosensing by conjugating them with the appropriate biomolecules (Chan and Nie 1998; Bruchez et al. 1998). Just like organic fluorophores, quantum dots can perform as photosensitizers because they have energy levels in the range of 1–5 eV (Bakalova et al. 2004; Samia et al. 2003). QDs can act as high-energy photons like gamma and X-rays, so they can be used as radiosensitizers for the targeted therapy of cancer cells (Carter et al. 2007) (Table 4.9).

Table 4.9 Quantum dot nanoparticles for drug delivery and diagnostic systems

Sl. no.	Quantum dots	Biological applications	Authors
1.	Cadmium and tellurium conjugated with gambogic acid	It promotes drug accumulation in HepG2 cells It repressed cancer cell propagation It influenced the G2 cell cycle phase arrest of cancer cell lines, raising cell apoptosis	Xu et al. (2013a), Xu et al. (2013b)
2.	Anti-GRP78 scFv	The antitumor activity is detected by the enhanced pyrophosphate-AKT-ser473 Inhibit the breast cancer growth	Xu et al. (2012)
3.	AVE-1642	Downregulates IGF1R levels in MCF-7 cells. Cell proliferation is inhibited	Zhang et al. (2009a), Zhang et al. (2009b)
4.	AITC-SiQDs	Lower and long-lasting activation of Nrf2 translocation into the nucleus is detected ROS production is triggered	Liu et al. (2018)
5.	QD-labeled monoclonal anti-HER2 antibody	Quantitatively analyzed the six processes of drug delivery Identified the rate-limiting constraints on QD-antibody delivery	Tada et al. (2007)
6.	HER2-RQDs	Target the gastric cancer MGC803 cells Inhibit the growth of gastric cancer tissues. Prevent protein synthesis and induce cell apoptosis	Ruan et al. (2012)
7.	ZnO QDs-conjugated gold nanoparticles	Nanocarriers have anticancer activity against HeLa cells	Chen et al. (2013)
8.	Mn-doped ZnS QDs	Great thermal stability (20–80 °C) Determined the glucose in real serum samples	Wu et al. (2010)
9.	CdS QDs	Anticancer activity by arresting the A549 cell growth at the S phase	Shivaji et al. (2018)
10.	Peptide-conjugated graphene Qds	Target and image tumor cells simultaneously	Su et al. (2015)
11.	Graphene QDs	Cytotoxic effect on cervical cancer cell lines and breast cancer cell lines	Thakur et al. (2016)
12.	Carbon QDs	Green fluorescent nature Possess imaging and phototherapy on cancer cell lines	Meena et al. (2019)
13.	Peptide-PEGylated lipid QDs	Inhibited the prostate cancer cell growth	Yeh et al. (2016)
14.	HA-DOX-GQD@MSN	Cytotoxicity effect against HeLa cell lines. Fluorescent monitoring ability	Gui et al. (2018)
15.	Cadmium telluride QDs	Inhibited the p53 inhibitor pifithrin-α Decrease cell viability	Choi et al. (2008)

(continued)

Table 4.9 (continued)

Sl. no.	Quantum dots	Biological applications	Authors
16.	Carbon QDs	Cytotoxic efficiency against MCF-7 and PC-3 cell lines by the activation of caspase-3 protein	Arkan et al. (2018)
17.	CdSe/ZnS QDs loaded with curcumin	Enhance apoptosis, cell death, ROS generation, and single/double DNA strand breaks in HL-60 cells	Belletti et al. (2017)
18.	Curcumin-coated QDs	Enhance the cellular internalization of curcumin. Tracked drug release and nanocarrier's destiny into cells	

4.10.1 Quantum Dot-Conjugated Inhibitors in Cancer Treatment

Various types of semiconductor QDs are nowadays studied for labeling, imaging, targeted drug delivery, and photodynamic therapy (Wang et al. 2009). Xu et al. (2013a) and Xu et al. (2013b) examined the fluorescent nanocomposites gambogic acid (GA) and cadmium tellurium QDs. They adjusted it with cysteamine for cancer cell labeling and merged treatment in HepG2 cell lines. GA-CdTe treatment promotes drug accumulation in HepG2 cell lines. GA-CdTe nanocomposites also inhibited cancer cell proliferation and enhance the drug action of GA molecules in HepG2 cells and induced the G2/M phase arrest of the cancer cell cycle, promoting cell apoptosis. Xu et al. (2012) described a conjugating QD-anti-GRP78 scFv, by conjugating QDs with tiny antibody components targeting membrane-bound proteins (GRP78). And its pattern is observed by visualization of multicolor fluorescence imaging in both in vitro and in vivo techniques. It can efficiently internalize cancer cells. The antitumor activity is evaluated from the elevated pyrophosphate-AKT-ser473 levels and shows hindrance of breast cancer growth in a xenograft model. Zhang et al. (2009a) and Zhang et al. (2009b) developed AVE-1642 (IGF1R antibody, AVE-1642) QDs to measure type I insulin-like growth factor (IGF) receptor (IGF1R) levels in breast cancer cells. AVE-1642 QDs can downregulate IGF1R levels in MCF-7 cells and rendered cells refractory to IGF-I stimulation; thus, cell proliferation is inhibited by AVE-1642 QDs. Liu et al. (2018) investigated the anticancer activities of allyl isothiocyanate, a dietary phytochemical, conjugated with silicon QDs (AITC-SiQDs). Lower and long-lasting activation of Nrf2 translocation into the nucleus is shown by AITC-SiQDs by the fluorescence detection of their cellular uptake. They concluded that ROS production may trigger the anticancer effect of AITC-SiQDs. Tada et al. (2007) studied the tracking of a quantum dot labeled monoclonal anti-HER2 antibody in mice with HER2-overexpressing breast cancer to analyze the molecular processes of its mechanistic delivery to the tumor. They used a dorsal skinfold chamber and a high-speed confocal microscope with a high-sensitivity camera for tracking process. They quantitatively analyzed the six processes of drug delivery and identified the rate-limiting constraints on QD-antibody delivery. Ruan et al. (2012) developed HER2-RQDs (HER2 monoclonal antibody-conjugated RNase A-associated CdTe quantum dot cluster) for the evaluation of its cytotoxicity, bio-distribution, and therapeutic effects. HER2-RQD nanoprobes can effectively target the gastric cancer MGC803 cells and can inhibit the growth of gastric cancer tissues by destroying functional RNAs in the cytoplasm by RNase A released from HER2-RQDs nanoprobes, preventing protein synthesis and inducing cell apoptosis. Chen et al. (2013) developed core-shell structured nanocarriers for targeted anticancer drug delivery using ZnO quantum dot-conjugated gold nanoparticles and amphiphilic hyperbranched block copolymer as core and shell, respectively, against HeLa cells. Wu et al. (2010) examined the conjugation of glucose oxidase onto phosphorescent Mn-doped ZnS quantum dots for effective glucose biosensing. The conjugate showed greater thermal stability in the range of 20–80 °C and effectively determines the glucose in real serum samples. Shivaji et al. (2018) demonstrated a CdS QDs (2–5nm) from tea leaf extract. The Qds show anticancer activity by arresting the A549 cell growth at the S phase for inhibiting the growth of lung cancer cell. Su et al. (2015) demonstrated a peptide with trifunctional motifs and conjugated with graphene Qds. This nano hybrid has the capacity to target and image tumor cells simultaneously. Thakur et al. (2016) developed aqueous soluble graphene QDs for drug delivery and imaging in cancer, and it is synthesized from cow milk by microwave-assisted heating. The QDs were then loaded with cysteamine hydrochloride-berberine hydrochloride complex, which showed a potent cytotoxic effect on cervical cancer and breast cancer cell lines. Meena et al. (2019) reported a simple and cost-effective method for the synthesis of fluorescent carbon QDs from the leaves of *Azadirachta Indica*, *Ocimum tenuiflorum*, and *Tridax procumbens*. These QDs are green fluorescent in nature and have homogeneous size distribution (~6–12 nm). They possess cancer cell imaging property and phototherapy against cancer cell lines. Yeh et al. (2016) developed conjugated QDs using peptide-PEGylated lipids for targeting prostate cancer cells. This peptide conjugate inhibited the tumor growth. Gui et al. (2018) illustrated a bifunctional mesoporous silica nanoparticle coated with blue fluorescent N-graphene quantum dots. This conjugate is bound with the drug doxorubicin and then covered with hyaluronic acid (HA-DOX-GQD@MSN). The conjugated drug has cytotoxicity effect against HeLa cell lines and also has fluorescent monitoring ability. Choi et al. (2008) investigated the epigenomic and genotoxic response of cadmium telluride QDs in breast cancer cell lines. QD treatment inhibited the p53 inhibitor pifithrin-α and it simultaneously decreases cell viability. Epigenetic changes have more effects on gene expression. Arkan et al. (2018) reported a simple hydrothermal method for the synthesis of carbon quantum dots from walnut oil, and also its cytotoxic and apoptogenic properties are analyzed. QDs possessed potent cytotoxic efficiency toward PC-3 and MCF-7 cell lines by the activation of the caspase-3 protein, which prompts apoptosis. Goo et al. (2013) reported a conjugated CdSe/ZnS QDs with curcumin to reduce QD-induced cytotoxicity on the cancer treatment using HL-60 cells and normal lymphocytes. The combined treatment enhances apoptosis, cell death, ROS generation, and single/double DNA strand breaks in HL-60 cells, and at the same time curcumin protects the normal lymphocyte cell viability, apoptosis, and ROS generation. Belletti et al. (2017) described a single-emulsion procedure for pure curcumin-loaded nanoparticle coated with QDs against primary effusion lymphoma. The QD conjugate can enhance the cellular internalization of curcumin and also track drug release and nanocarrier's destiny into cells.

4.11 Nanographene

Nanographene, a class of two-dimensional carbon nanomaterial, is widely used for biosensing, drug/gene delivery, different types of cancer therapies, as well as tissue engineering and imaging in the area of biomedicine (Geim and Novoselov 2010; Yang et al. 2016a, Yang et al. 2016b). Nanographene and its derivatives could act as photothermal agents for efficient photothermal therapy (PTT) of cancer if it has an absorbance at near-infrared rays (Yang et al. 2012a, Yang et al. 2012b, Yang et al. 2012c). Various types of biomolecules can be loaded in nanographene due to its high specific surface area (Liu et al. 2008a, Liu et al. 2008b). Various inorganic nanoparticles and radionuclides such as 64Cu, 66Ga, 125I, and 131I can be loaded with nanographene to offer additional optical, magnetic properties and to find applications in nuclear imaging and radiotherapy of cancer, respectively (Yang et al. 2012a, Yang et al. 2012b, Yang et al. 2012c; Hu et al. 2012; Hong et al. 2012a, Hong et al. 2012b) (Table 4.10).

4.11.1 Nanographene-Conjugated Inhibitors in Cancer Treatment

Yang et al. (2016a) and Yang et al.(2016b) demonstrated the efficient targeting of breast cancer metastasis using conjugated nanographene oxide with a monoclonal antibody against follicle-stimulating hormone receptor. This method can be used as an early detector of metastasis in breast cancer. Jung et al. (2014) designed a nanographene oxide using hyaluronic acid conjugate. It can be used for the photothermal ablation treatment of melanoma skin cancer utilizing a near-infrared laser. It can be transdermally given to the tumor tissues in the skin of mice to enable the enhanced penetration and retention of nanoparticles. The near-infrared radiation produced a complete ablation of tumor tissues with no account of tumorigenesis. Kavitha et al. (2014) examined an efficient nano-cargo vehicle for the distribution of drugs into cells using graphene oxide. The nanocarrier has polyvinyl caprolactam conjugate and the drug camptothecin is loaded inside it. It can be used as an efficient drug delivery vector with high biocompatibility, solubility, and stability in physiological solutions. Camptothecin-loaded polyvinyl caprolactam conjugate exhibited immense potency in killing cancer cells. Zeng et al. (2017) developed multifunctional graphene, folic acid (FA)-conjugated polyethylenimine-modified PEGylated

Table 4.10 Nanographene conjugate for target-specific delivery of an anticancer drug

Sl. no.	Type of the nanographene conjugate	Biological applications	Authors
1.	Nanographene oxide with a monoclonal antibody (64Cu-NOTA-GO)	Used as an early detector of metastasis in breast cancer	Yang et al. (2016a), Yang et al. (2016b)
2.	Nanographene oxide- hyaluronic acid (NGO-HA)	Exhibits the total ablation of tumor tissues with no appearance of tumorigenesis	Jung et al. (2014)
3.	Graphene oxide-poly N-vinyl caprolactam (GO-PVCL) loaded with camptothecin	Used as an effective drug transportation vector Camptothecin-loaded polyvinyl caprolactam conjugate has high potency in killing cancer cell lines	Kavitha et al. (2014)
4.	Folic acid (FA)-conjugated polyethylenimine-modified PEGylated nanographene (PPG-FA/siRNA/Dox)	It has high drug and siRNA loading ability, gene silencing effect, and efficient intracellular delivery of doxorubicin	Zeng et al. (2017)
5.	Nanographene oxide conjugated with D-α-tocopheryl polyethylene glycol-1000 succinate	Reduced the cell viability of breast cancer cell lines Enhanced drug stability	de Melo-Diogo et al. (2017)
6.	Nano-conjugate of artesunate with PEGylated nanographene oxide (nGO-PEG-ARS)	It showed the complete tumor cure ability within 15 days without causing any apparent histological lesion	Pang et al. (2017)
7.	FA-NGO-PVP	Targeted chemo-photothermal therapy showed high anticancer efficiency	Qin et al. (2013)
8.	^{66}Ga-NOTA-GO-TRC105	It has excellent stability and tumor target specificity	Hong et al. (2012a), Hong et al. (2012b)

nanographene (PPG-FA/siRNA/Dox), for the delivery system for siRNA and doxorubicin. This nano-conjugate exhibited high drug/siRNA loading ability and satisfactory gene silencing effect as well as efficient intracellular delivery of doxorubicin. de Melo-Diogo et al. (2017) investigated the therapeutic efficiency of D-α-tocopheryl polyethylene glycol-1000 succinate conjugated with nanographene oxide in breast cancer cells. It enhanced the drug stability and also reduced the cell viability of breast cancer cell lines. Pang et al. (2017) developed a nano-conjugate of artesunate with PEGylated nanographene oxide (nGO-PEG-ARS) for obtaining photothermal effect. The nano-conjugate nGO-PEG-ARS with near-infrared irradiation resulted in complete tumor cure within 15 days without causing any apparent histological lesion. Qin et al. (2013) designed a nanographene oxide (NGO) conjugated drug using polyvinylpyrrolidone functionalized NGO with folic acid (FA-NGO-PVP), for testing the combination chemotherapy and near-infrared photothermal therapy. The drug conjugate targeted chemo-photothermal therapy showed high anticancer efficiency. Hong et al. (2012a) and Hong et al. (2012b) designed a nanographene conjugate by covalently linking with polyethylene glycol, NOTA, and TRC105 ([66]Ga-NOTA-GO-TRC105) for tumor targeting. The nanographene conjugate has excellent stability and tumor target specificity which was vasculature specific with little extravasation.

4.12 Magnetic Nanoparticles

Magnetic nanoparticles consist of magnetic elements, such as iron, nickel, cobalt, chromium, manganese, gadolinium, and their chemical compounds. Chemo- and radiotherapy are used for the treatment of cancer cells. However, none of them is cancer cell-specific. Nanotechnology can bring a useful alternative to current cancer therapies, which can target the cancer-specific cells from normal cells (Pautler and Brenner 2010). Superparamagnetic iron oxide nanoparticle size can vary within 10-100nm and used extensively in biomedical treatments. Magnetic hyperthermia is a nonintrusive strategy for tumor excision (Pankhurst et al. 2003). The magnetic nanoparticle consists of a magnetic core with a specific coating. Superparamagnetic nanoparticles have a single magnetic domain because of their small magnetic core and are magnetize under an externally applied magnetic field (Yigit et al. 2012) (Table 4.11).

4.12.1 Role of Magnetic Nanoparticles in Cancer Treatments

Yallapu et al. (2011) invented a water-soluble superparamagnetic iron oxide nanoparticle. It can be used for hyperthermia, magnetic resonance imaging, and drug distribution purposes. Drug-loaded formulation of F127250 has good stability, enhanced cellular uptake, and multilayer imaging contrast properties. Curcumin packed drug formulation (F127250-CUR) exhibited equal growth inhibition impacts on (A2780CP) ovarian cancer cell line (MDA-MB-231), breast cancer cell line, and (PC-3) prostate cancer cell lines. Li et al. (2015) examined a hollow magnetic nanoparticle (HMNPs) as an actively established drug delivery scheme which performs both infrared thermal imaging and magnetic resonance imaging characteristics. Under the presence of an alternating magnetic field, this system showed the potential for thermo-chemo combination to induce cancer cell apoptosis via the in vivo method. Chu et al. (2013) studied the photothermal impact of surface-functionalized Fe3O4 magnetic nanoparticles in both in vitro and in vivo approaches. Fe_3O_4 nanoparticles are taken up by esophageal cancer cells, and upon irradiation at 808 nm and incubation with near-infrared radiation, the cell viability and cellular organelles are damaged. Mouse esophageal tumor growth is reduced by the photothermal effect of Fe3O4 nanoparticles. Sadhukha et al. (2013) synthesized an epidermal growth factor receptor-targeted, inhalable superparamagnetic iron oxide

Table 4.11 Magnetic nanoparticles for drug delivery

Sl. no.	Type of magnetic nanoparticles	Biological applications	Authors
1.	Superparamagnetic iron oxide (F127250)	It has good stability, enhanced cellular uptake, and multilayer imaging contrast properties. The curcumin-loaded formulation exhibited equal tumor repression influences on ovarian, breast, and prostate cancer cell lines	Yallapu et al. (2011)
2.	Hollow magnetic nanoparticles (HMNPs)	The system exhibits both infrared thermal imaging and magnetic resonance imaging characteristics. Induce cancer cell apoptosis	Li et al. (2015)
3.	Fe3O4 magnetic nanoparticles	Cancer cell viability and cellular organelles are damaged. Mouse esophageal tumor growth is reduced	Chu et al. (2013)
4.	Superparamagnetic iron oxide nanoparticles (SPIO)	Magnetic hyperthermia causes a greater reduction in lung tumor growth	Sadhukha et al. (2013)
5.	Magnetic nanoparticles embedded in polylactide-co-glycolide	It can deliver both hydrophobic and hydrophilic cancer therapeutic drugs	Singh et al. (2011)
6.	ODN-dendrimer-MNP-magnetic nanoparticle composites	Inhibit cancer cell growth in a dose- and time-dependent manner by downregulating the survivin gene and protein	Pan et al. (2007)
7.	Fe3O4-curcumin conjugate	It showed a high loading cellular uptake. Antitumor efficiency of curcumin is delivered to the cells through macrophages	Dai Tran et al. (2010)

nanoparticles against magnetic hyperthermia of lung cancer cells. Magnetic hyperthermia treatment using the inhalable superparamagnetic iron oxide nanoparticles showed a greater reduction in lung tumor growth. Singh et al. (2011) developed and characterized the magnetic nanoparticles embedded in polylactide-co-glycolide (PLGA-MNPs) matrixes as dual drug delivery and imaging system capable of encapsulating both hydrophilic and hydrophobic drugs by examining the biocompatibility, cellular uptake, cytotoxicity, membrane potential, and apoptosis in MCF-7 and PANC-1 cell lines. Magnetic nanoparticles embedded in polylactide-co-glycolide matrixes showed an enhanced contrast effect due to higher T2 relaxivity with a blood circulation half-life ~47 min. Thus it can deliver both hydrophobic and hydrophilic cancer therapeutic drugs. Pan et al. (2007) developed a modified magnetic nanoparticle with a diameter of 8nm using different polyamidoamine dendrimers and mixed with antisense survivin oligodeoxynucleotide and incubated in (MCF-7) breast cancer cell line, (MDA-MB-435) melanoma cell line, and (HepG2) liver cancer cell lines. The modified magnetic nanoparticle (ODN-dendrimer-MNP composites) can enter into tumor cells within 15 min which appears in the hindrance of cancer cell growth in a shot- and the time-dependent manner by downregulating the survivin gene and protein. Dai Tran et al. (2010) investigated the magnetic drug targeting using Fe_3O_4-curcumin conjugate, which has magnetic nano Fe3O4 core and chitosan as outer shell and entrapped curcumin, and also studied its efficiency to label, target, and treat tumor cells. It showed a high loading cellular uptake. Antitumor efficiency of curcumin is delivered to the cells through macrophages for enhanced phagocytosis process.

4.13 Conclusion

Conventional anticancer drugs against proliferating cells with high toxicity show diverse severe dilemmas upon the gastrointestinal tract, hematopoietic system, immune system, and nervous systems. This drawback can be reduced by using the organic and inorganic nanoparticles. These nanocarriers are the ideal candidate to carry anticancer drugs to the specific cancer sites. Due to its smaller size, it potentially reacts to solubilize the insoluble drugs and in sustainable release of drug in the tumor tissue. It acts as a best antitumor vehicle by conjugating with anticancer drugs, monoclonal antibodies, and plant-based bioactive compounds due to its inner hydrophobic domains and hydrophilic outer shell. Inorganic nanoparticles are due to the uniform size, shape, and optoelectronic properties used in the field of medicine. Particularly, the metal nanoparticles are highly utilized in imaging, optical, sensors, cancer therapy, and drug delivery. Poly(lactic-*co*-glycolic acid) (PLGA) acid is a biodegradable polymer which is extensively accepted as a matrix to fuse a broad range of therapeutic tools. That comprises hydrophobic and hydrophilic small particles, nucleic acids, and proteins. PLGA is licensed by the Food and Drug Administration (FDA) for aid in pharmaceutical commodities. Association of poorly soluble cytostatic agents and drugs into polymeric nanoparticles is presumably the most efficient and simplest way to enhance their therapeutic effectiveness. Nanoparticles including polymeric nanoparticles, polymeric micelles, gold, dendrimers, or quantum dots are effectively used in cancer treatments by enhancing apoptotic activity, cell death, ROS generation, and single/double DNA strand breaks and inhibit cell growth in cancer cells.

References

Abel EE, Poonga PRJ, Panicker SG (2016) Characterization and in vitro studies on anticancer, antioxidant activity against colon cancer cell line of gold nanoparticles capped with Cassia tora SM leaf extract. Appl Nanosci 6(1):121–129

Ai S, Jia T, Ai W, Duan J, Liu Y, Chen J et al (2013) Targeted delivery of doxorubicin through conjugation with EGF receptor–binding peptide overcomes drug resistance in human colon cancer cells. Br J Pharmacol 168(7):1719–1735

Akbarzadeh A, Rezaei-Sadabady R, Davaran S, Joo SW, Zarghami N, Hanifehpour Y et al (2013) Liposome: classification, preparation, and applications. Nanoscale Res Lett 8(1):102

Ali-Boucetta H, Al-Jamal KT, McCarthy D, Prato M, Bianco A, Kostarelos K (2008) Multiwalled carbon nanotube–doxorubicin supramolecular complexes for cancer therapeutics. Chem Commun 4:459–461

Arkan E, Barati A, Rahmanpanah M, Hosseinzadeh L, Moradi S, Hajialyani M (2018) Green synthesis of carbon dots derived from walnut oil and an investigation of their cytotoxic and apoptogenic activities toward cancer cells. Adv Pharm Bull 8(1):149

Arvizo R, Bhattacharya R, Mukherjee P (2010) Gold nanoparticles: opportunities and challenges in nanomedicine. Expert Opin Drug Deliv 7(6):753–763

Aryal S, Grailer JJ, Pilla S, Steeber DA, Gong S (2009) Doxorubicin conjugated gold nanoparticles as water-soluble and pH-responsive anticancer drug nanocarriers. J Mater Chem 19(42):7879–7884

Aryal S, Hu CMJ, Zhang L (2011) Polymeric nanoparticles with precise ratiometric control over drug loading for combination therapy. Mol Pharm 8(4):1401–1407

Auzenne E, Donato NJ, Li C, Leroux E, Price RE, Farquhar D, Klostergaard J (2002) Superior therapeutic profile of poly-L-glutamic acid-paclitaxel copolymer compared with taxol in xenogeneic compartmental models of human ovarian carcinoma. Clin Cancer Res 8(2):573–581

Babu A, Munshi A, Ramesh R (2017) Combinatorial therapeutic approaches with RNAi and anticancer drugs using nanodrug delivery systems. Drug Dev Ind Pharm 43(9):1391–1401

Badr G, Al-Sadoon MK, Rabah DM (2013) Therapeutic efficacy and molecular mechanisms of snake (Walterinnesiaaegyptia) venom-loaded silica nanoparticles in the treatment of breast cancer-and prostate cancer-bearing experimental mouse models. Free Radic Biol Med 65:175–189

Bae KH, Chung HJ, Park TG (2011) Nanomaterials for cancer therapy and imaging. Mol Cells 31(4):295–302

Bakalova R, Ohba H, Zhelev Z, Ishikawa M, Baba Y (2004) Quantum dots as photosensitizers? Nat Biotechnol 22(11):1360

Barai AC, Paul K, Dey A, Manna S, Roy S, Bag BG, Mukhopadhyay C (2018) Green synthesis of Nerium oleander-conjugated gold nanoparticles and study of its in vitro anticancer activity on MCF-7 cell lines and catalytic activity. Nano Conv 5(1):1–9

Belletti D, Riva G, Luppi M, Tosi G, Forni F, Vandelli MA et al (2017) Anticancer drug-loaded quantum dots engineered polymeric nanoparticles: diagnosis/therapy combined approach. Eur J Pharm Sci 107:230–239

Bernal GM, LaRiviere MJ, Mansour N, Pytel P, Cahill KE, Voce DJ et al (2014) Convection-enhanced delivery and in vivo imaging of polymeric nanoparticles for the treatment of malignant glioma. Nanomedicine 10(1):149–157

Bhatt R, de Vries P, Tulinsky J, Bellamy G, Baker B, Singer JW, Klein P (2003) Synthesis and in vivo antitumor activity of poly (L-glutamic acid) conjugates of 20 (S)-camptothecin. J Med Chem 46(1):190–193

Bielawski K, Bielawska A, Muszyńska A, Popławska B, Czarnomysy R (2011) Cytotoxic activity of G3 PAMAM-NH2 dendrimer-chlorambucil conjugate in human breast cancer cells. Environ Toxicol Pharmacol 32(3):364–372

Bisht S, Feldmann G, Soni S, Ravi R, Karikar C, Maitra A, Maitra A (2007) Polymeric nanoparticle-encapsulated curcumin ("nanocurcumin"): a novel strategy for human cancer therapy. J Nanobiotechnol 5(1):3

Bruchez M, Moronne M, Gin P, Weiss S, Alivisatos AP (1998) Semiconductor nanocrystals as fluorescent biological labels. Science 281:2013–2016

Cai W, Gao T, Hong H, Sun J (2008) Applications of gold nanoparticles in cancer nanotechnology. Nanotechnol Sci Appl 1:17

Carter JD, Cheng NN, Qu Y, Suarez GD, Guo T (2007) Nanoscale energy deposition by X-ray absorbing nanostructures. J Phys Chem B 111(40):11622–11625

Chan WCW, Nie SM (1998) Quantum dot bioconjugates for ultrasensitive nonisotopic detection. Science 281:2016–2018

Chen YH, Tsai CY, Huang PY, Chang MY, Cheng PC, Chou CH, Wu CL (2007) Methotrexate conjugated to gold nanoparticles inhibits tumor growth in a syngeneic lung tumor model. Mol Pharm 4(5):713–722

Chen AM, Zhang M, Wei D, Stueber D, Taratula O, Minko T, He H (2009) Co-delivery of doxorubicin and Bcl-2 siRNA by mesoporous silica nanoparticles enhances the efficacy of chemotherapy in multidrugresistant cancer cells. Small 5(23):2673–2677

Chen T, Zhao T, Wei D, Wei Y, Li Y, Zhang H (2013) Core–shell nanocarriers with ZnO quantum dots-conjugated Au nanoparticle for tumor-targeted drug delivery. Carbohydr Polym 92(2):1124–1132

Chen Q, Zheng J, Yuan X, Wang J, Zhang L (2018) Folic acid grafted and tertiary amino based pH-responsive pentablock polymeric micelles for targeting anticancer drug delivery. Mater Sci Eng C 82:1–9

Chenthamara D, Subramaniam S, Ramakrishnan SG, Krishnaswamy S, Essa MM, Lin FH, Qoronfleh MW (2019) Therapeutic efficacy of nanoparticles and routes of administration. Biomater Res 23(1):1–29

Cho YS, Yoon TJ, Jang ES, Hong KS, Lee SY, Kim OR, Park C, Kim YJ, Yi GC, Chang K (2010) Cetuximab-conjugated magneto-fluorescent silica nanoparticles for in vivo colon cancer targeting and imaging. Cancer letters 299(1): 63–71

Choi AO, Brown SE, Szyf M, Maysinger D (2008) Quantum dot-induced epigenetic and genotoxic changes in human breast cancer cells. J Mol Med 86(3):291–302

Chu M, Shao Y, Peng J, Dai X, Li H, Wu Q, Shi D (2013) Near-infrared laser light mediated cancer therapy by photothermal effect of Fe3O4 magnetic nanoparticles. Biomaterials 34(16):4078–4088

Daduang J, Palasap A, Daduang S, Boonsiri P, Suwannalert P, Limpaiboon T (2015) Gallic acid enhancement of gold nanoparticle anticancer activity in cervical cancer cells. Asian Pac J Cancer Prev 16(1):169–174

Dai Tran L, Hoang NMT, Mai TT, Tran HV, Nguyen NT, Tran TD et al (2010) Nanosized magnetofluorescent Fe3O4–curcumin conjugate for multimodal monitoring and drug targeting. Colloids Surf A Physicochem Eng Asp 371(1–3):104–112

Daraee H, Etemadi A, Kouhi M, Alimirzalu S, Akbarzadeh A (2016) Application of liposomes in medicine and drug delivery. Artif Cells Nanomed Biotechnol 44(1):381–391

Debnath S, Saloum D, Dolai S, Sun C, Averick S, Raja K, Fata JE (2013) Dendrimer-curcumin conjugate: a water soluble and effective cytotoxic agent against breast cancer cell lines. Anti Cancer Agents Med Chem (Formerly Current Medicinal Chemistry-Anti-Cancer Agents) 13(10):1531–1539

de Melo-Diogo D, Pais-Silva C, Costa EC, Louro RO, Correia IJ (2017) D-α-tocopheryl polyethylene glycol 1000 succinate functionalized nanographene oxide for cancer therapy. Nanomedicine 12(5):443–456

De Santi M, Antonelli A, Menotta M, Sfara C, Serafini S, Lucarini S, Magnani M (2013) Singlewalled carbon nanotubes functionalization for the delivery of the water insoluble anticancer agent indole-3-carbinol cyclic. J Nanopharm Drug Deliv 1(1):45–51

Devalapally H, Duan Z, Seiden MV, Amiji MM (2008) Modulation of drug resistance in ovarian adenocarcinoma by enhancing intracellular ceramide using tamoxifen-loaded biodegradable polymeric nanoparticles. Clin Cancer Res 14(10):3193–3203

Dhanikula RS, Argaw A, Bouchard JF, Hildgen P (2008) Methotrexate loaded polyethercopolyester dendrimers for the treatment of gliomas: enhanced efficacy and intratumoral transport capability. Mol Pharm 5(1):105–116

Dhas TS, Kumar VG, Karthick V, Govindaraju K, Narayana TS (2014) Biosynthesis of gold nanoparticles using Sargassumswartzii and its cytotoxicity effect on HeLa cells. Spectrochim Acta A Mol Biomol Spectrosc 133:102–106

Dichwalkar T, Patel S, Bapat S, Pancholi P, Jasani N, Desai B et al (2017) Omega-3 fatty acid grafted PAMAM-Paclitaxel conjugate exhibits enhanced anticancer activity in upper gastrointestinal cancer cells. Macromol Biosci 17(8):1600457

Duncan R, Vicent MJ, Greco F, Nicholson RI (2005) Polymer–drug conjugates: towards a novel approach for the treatment of endrocine-related cancer. Endocr Relat Cancer 12(Supplement_1):S189–S199

Dykman L, Khlebtsov N (2012) Gold nanoparticles in biomedical applications: recent advances and perspectives. Chem Soc Rev 41(6):2256–2282

Eichhorn ME, Ischenko I, Luedemann S, Strieth S, Papyan A, Werner A et al (2010) Vascular targeting by EndoTAG™-1 enhances therapeutic efficacy of conventional chemotherapy in lung and pancreatic cancer. Int J Cancer 126(5):1235–1245

Ekladious I, Liu R, Zhang H, Foil DH, Todd DA, Graf TN et al (2017) Synthesis of poly (1, 2-glycerol carbonate)–paclitaxel conjugates and their utility as a single high-dose replacement for multi-dose treatment regimens in peritoneal cancer. Chem Sci 8(12):8443–8450

Fu F, Wu Y, Zhu J, Wen S, Shen M, Shi X (2014) Multifunctional lactobionic acid-modified dendrimers for targeted drug delivery to liver cancer cells: investigating the role played by PEG spacer. ACS Appl Mater Interfaces 6(18):16416–16425

Garg NK, Dwivedi P, Campbell C, Tyagi RK (2012) Site specific/targeted delivery of gemcitabine through anisamide anchored chitosan/poly ethylene glycol nanoparticles: an improved understanding of lung cancer therapeutic intervention. Eur J Pharm Sci 47(5):1006–1014

Gaucher G, Fournier E, Le Garrec D, Khalid MN, Hoarau D, Sant V, Leroux J (2004) Delivery of hydrophobic drugs through self-assembling nanostructures. In: 2004 international conference on MEMS, NANO and smart systems (ICMENS'04), August. IEEE, pp 56–57

Geetha R, Ashokkumar T, Tamilselvan S, Govindaraju K, Sadiq M, Singaravelu G (2013) Green synthesis of gold nanoparticles and their anticancer activity. Cancer Nanotechnol 4(4–5):91–98

Geim AK, Novoselov KS (2010) The rise of graphene. Nanosci Technol Collect Rev Nat J:11–19

Ghasemi Y, Peymani P, Afifi S (2009) Quantum dot: magic nanoparticle for imaging, detection and targeting. Acta Biomed 80(2):156–165

Ghosh P, Han G, De M, Kim CK, Rotello VM (2008) Gold nanoparticles in delivery applications. Adv Drug Deliv Rev 60(11):1307–1315

Girase ML, Patil PG, Ige PP (2019) Polymer-drug conjugates as nanomedicine: a review. Int J Polym Mater Polym Biomater:1–25

Gong H, Cheng L, Xiang J, Xu H, Feng L, Shi X, Liu Z (2013) Near-infrared absorbing polymeric nanoparticles as a versatile drug carrier for cancer combination therapy. Adv Funct Mater 23(48):6059–6067

Gong F, Chen D, Teng X, Ge J, Ning X, Shen YL, Wang S (2017) Curcumin-loaded blood-stable polymeric micelles for enhancing therapeutic effect on erythroleukemia. Mol Pharm 14(8):2585–2594

Goo S, Choi YJ, Lee Y, Lee S, Chung HW (2013) Selective effects of curcumin on CdSe/ZnS quantum-dot-induced phototoxicity using UVA irradiation in normal human lymphocytes and leukemia cells. Toxicol Res 29(1):35–42

Gui W, Zhang J, Chen X, Yu D, Ma Q (2018) N-doped graphene quantum dot@ mesoporous silica nanoparticles modified with hyaluronic acid for fluorescent imaging of tumor cells and drug delivery. Microchim Acta 185(1):66

Hanafi-Bojd MY, Jaafari MR, Ramezanian N, Xue M, Amin M, Shahtahmassebi N, Malaekeh-Nikouei B (2015) Surface functionalized mesoporous silica nanoparticles as an effective carrier for epirubicin delivery to cancer cells. Eur J Pharm Biopharm 89:248–258

Hardiansyah A, Huang LY, Yang MC, Liu TY, Tsai SC, Yang CY et al (2014) Magnetic liposomes for colorectal cancer cells therapy by high-frequency magnetic field treatment. Nanoscale Res Lett 9(1):1–13

Hardman R (2006) A toxicologic review of quantum dots: toxicity depends on physicochemical and environmental factors. Environ Health Perspect 114(2):165–172

Hong H, Yang K, Zhang Y, Engle JW, Feng L, Yang Y, Barnhart TE (2012a) In vivo targeting and imaging of tumor vasculature with radiolabeled, antibody-conjugated nanographene. ACS Nano 6(3):2361–2370

Hong H, Zhang Y, Engle JW, Nayak TR, Theuer CP, Nickles RJ et al (2012b) In vivo targeting and positron emission tomography imaging of tumor vasculature with 66Ga-labeled nanographene. Biomaterials 33(16):4147–4156

Hu SH, Chen YW, Hung WT, Chen IW, Chen SY (2012) Quantum-dot-tagged reduced graphene oxide nanocomposites for bright fluorescence bioimaging and photothermal therapy monitored in situ. Adv Mater 24(13):1748–1754

Huang J, Zhang H, Yu Y, Chen Y, Wang D, Zhang G et al (2014) Biodegradable self-assembled nanoparticles of poly (d, l-lactide-co-glycolide)/hyaluronic acid block copolymers for target delivery of docetaxel to breast cancer. Biomaterials 35(1):550–566

Huo Q, Zhu J, Niu Y, Shi H, Gong Y, Li Y et al (2017) pH-triggered surface charge-switchable polymer micelles for the co-delivery of paclitaxel/disulfiram and overcoming multidrug resistance in cancer. Int J Nanomedicine 12:8631

Jaidev LR, Krishnan UM, Sethuraman S (2015) Gemcitabine loaded biodegradable PLGA nanospheres for in vitro pancreatic cancer therapy. Mater Sci Eng C 47:40–47

Jain AK, Swarnakar NK, Godugu C, Singh RP, Jain S (2011) The effect of the oral administration of polymeric nanoparticles on the efficacy and toxicity of tamoxifen. Biomaterials 32(2):503–515

Janaszewska A, Mączyńska K, Matuszko G, Appelhans D, Voit B, Klajnert B, Bryszewska M (2012) Cytotoxicity of PAMAM, PPI and maltose modified PPI dendrimers in Chinese hamster ovary (CHO) and human ovarian carcinoma (SKOV3) cells. New J Chem 36(2):428–437

Joseph MM, Aravind SR, George SK, Pillai KR, Mini S, Sreelekha TT (2014) Antitumor activity of galactoxyloglucan-gold nanoparticles against murine ascites and solid carcinoma. Colloids Surf B: Biointerfaces 116:219–227

Joshi P, Chakraborti S, Ramirez-Vick JE, Ansari ZA, Shanker V, Chakrabarti P, Singh SP (2012) The anticancer activity of chloroquine-gold nanoparticles against MCF-7 breast cancer cells. Colloids Surf B: Biointerfaces 95:195–200

Jovin TM (2003) Quantum dots finally come of age. Nat Biotechnol 21:32–33

Jung HS, Kong WH, Sung DK, Lee MY, Beack SE, Keum DH et al (2014) Nanographene oxide–hyaluronic acid conjugate for photothermal ablation therapy of skin cancer. ACS Nano 8(1):260–268

Kaminskas LM, McLeod VM, Ryan GM, Kelly BD, Haynes JM, Williamson M et al (2014) Pulmonary administration of a doxorubicin-conjugated dendrimer enhances drug exposure to lung metastases and improves cancer therapy. J Control Release 183:18–26

Kang B, Mackey MA, El-Sayed MA (2010) Nuclear targeting of gold nanoparticles in cancer cells induces DNA damage, causing cytokinesis arrest and apoptosis. J Am Chem Soc 132(5):1517–1519

Katiyar SS, Muntimadugu E, Rafeeqi TA, Domb AJ, Khan W (2016) Co-delivery of rapamycin- and piperine-loaded polymeric nanoparticles for breast cancer treatment. Drug Deliv 23(7):2608–2616

Kavitha T, Kang IK, Park SY (2014) Poly (N-vinyl caprolactam) grown on nanographene oxide as an effective nanocargo for drug delivery. Colloids Surf B: Biointerfaces 115:37–45.4

Kesharwani P, Xie L, Banerjee S, Mao G, Padhye S, Sarkar FH, Iyer AK (2015) Hyaluronic acid-conjugated polyamidoamine dendrimers for targeted delivery of 3, 4-difluorobenzylidene curcumin to CD44 overexpressing pancreatic cancer cells. Colloids Surf B: Biointerfaces 136:413–423

Khandanlou R, Murthy V, Saranath D, Damani H (2018) Synthesis and characterization of gold-conjugated Backhousiacitriodora nanoparticles and their anticancer activity against MCF-7 breast and HepG2 liver cancer cell lines. J Mater Sci 53(5):3106–3118

Kiew LV, Cheong SK, Sidik K, Chung LY (2010) Improved plasma stability and sustained release profile of gemcitabine via polypeptide conjugation. Int J Pharm 391(1–2):212–220

Kim SW, Lee YK, Lee JY, Hong JH, Khang D (2017) PEGylated anticancer-carbon nanotubes complex targeting mitochondria of lung cancer cells. Nanotechnology 28(46):465102

Koizumi F, Kitagawa M, Negishi T, Onda T, Matsumoto SI, Hamaguchi T, Matsumura Y (2006) Novel SN-38–incorporating polymeric micelles, NK012, eradicate vascular endothelial growth factor–secreting bulky tumors. Cancer Res 66(20):10048–10056

Krishnaswamy K, Orsat V (2017) Sustainable delivery systems through green nanotechnology. In: Nano-and microscale drug delivery systems. Elsevier, Amsterdam, pp 17–32

Kulhari H, Pooja D, Shrivastava S, Kuncha M, Naidu VGM, Bansal V et al (2016) Trastuzumab-grafted PAMAM dendrimers for the selective delivery of anticancer drugs to HER2-positive breast cancer. Sci Rep 6(1):1–13

Kuruvilla SP, Tiruchinapally G, Crouch AC, ElSayed ME, Greve JM (2017) Dendrimer-doxorubicin conjugates exhibit improved anticancer activity and reduce doxorubicin-induced cardiotoxicity in a murine hepatocellular carcinoma model. PLoS One 12(8):e0181944

Labi V, Erlacher M (2015) How cell death shapes cancer. Cell Death Dis 6(3):e1675–e1675

Larson N, Yang J, Ray A, Cheney DL, Ghandehari H, Kopeček J (2013) Biodegradable multiblock poly (N-2-hydroxypropyl) methacrylamide gemcitabine and paclitaxel conjugates for ovarian cancer cell combination treatment. Int J Pharm 454(1):435–443

Lee PC, Chiou YC, Wong JM, Peng CL, Shieh MJ (2013) Targeting colorectal cancer cells with single-walled carbon nanotubes conjugated to anticancer agent SN-38 and EGFR antibody. Biomaterials 34(34):8756–8765

Li C, Price JE, Milas L, Hunter NR, Ke S, Yu DF et al (1999) Antitumor activity of poly (L-glutamic acid)-paclitaxel on syngeneic and xenografted tumors. Clin Cancer Res 5(4):891–897

Li M, Song W, Tang Z, Lv S, Lin L, Sun H et al (2013) Nanoscaled poly (L-glutamic acid)/doxorubicin-amphiphile complex as pH-responsive drug delivery system for effective treatment of nonsmall cell lung cancer. ACS Appl Mater Interfaces 5(5):1781–1792

Li H, Zhang N, Hao Y, Wang Y, Jia S, Zhang H, Zhang Z (2014) Formulation of curcumin delivery with functionalized single-walled carbon nanotubes: characteristics and anticancer effects in vitro. Drug Deliv 21(5):379–387

Li J, Hu Y, Hou Y, Shen X, Xu G, Dai L et al (2015) Phase-change material filled hollow magnetic nanoparticles for cancer therapy and dual modal bioimaging. Nanoscale 7(19):9004–9012

Li Z, Cai Y, Zhao Y, Yu H, Zhou H, Chen M (2017) Polymeric mixed micelles loaded mitoxantrone for overcoming multidrug resistance in breast cancer via photodynamic therapy. Int J Nanomedicine 12:6595

Liu Z, Chen K, Davis C, Sherlock S, Cao Q, Chen X, Dai H (2008a) Drug delivery with carbon nanotubes for in vivo cancer treatment. Cancer Res 68(16):6652–6660

Liu Z, Robinson JT, Sun X, Dai H (2008b) PEGylated nanographene oxide for delivery of water-insoluble cancer drugs. J Am Chem Soc 130(33):10876–10877

Liu J, Kopečková P, Pan H, Sima M, Bühler P, Wolf P et al (2012a) Prostate-cancer-targeted N-(2-hydroxypropyl) methacrylamide copolymer/docetaxel conjugates. Macromol Biosci 12(3):412–422

Liu Q, Zhang J, Sun W, Xie QR, Xia W, Gu H (2012b) Delivering hydrophilic and hydrophobic chemotherapeutics simultaneously by magnetic mesoporous silica nanoparticles to inhibit cancer cells. Int J Nanomedicine 7:999

Liu Y, Feng L, Liu T, Zhang L, Yao Y, Yu D, Zhang N (2014) Multifunctional pH-sensitive polymeric nanoparticles for theranostics evaluated experimentally in cancer. Nanoscale 6(6):3231–3242

Liu P, Behray M, Wang Q, Wang W, Zhou Z, Chao Y, Bao Y (2018) Anti-cancer activities of allyl isothiocyanate and its conjugated silicon quantum dots. Sci Rep 8(1):1–11

Lobo V, Patil A, Phatak A, Chandra N (2010) Free radicals, antioxidants and functional foods: impact on human health. Pharmacogn Rev 4(8):118

Lu J, Liong M, Sherman S, Xia T, Kovochich M, Nel AE, Tamanoi F (2007) Mesoporous silica nanoparticles for cancer therapy: energy-dependent cellular uptake and delivery of paclitaxel to cancer cells. NanoBiotechnology 3(2):89–95

Lu J, Huang Y, Zhao W, Marquez RT, Meng X, Li J et al (2013) PEG-derivatizedembelin as a nanomicellar carrier for delivery of paclitaxel to breast and prostate cancers. Biomaterials 34(5):1591–1600

Lv S, Tang Z, Li M, Lin J, Song W, Liu H et al (2014) Co-delivery of doxorubicin and paclitaxel by PEG-polypeptide nanovehicle for the treatment of non-small cell lung cancer. Biomaterials 35(23):6118–6129

Ma'mani L, Nikzad S, Kheiri-Manjili H, al-Musawi S, Saeedi M, Askarlou S, Shafiee A (2014) Curcumin-loaded guanidine functionalized PEGylated I3ad mesoporous silica nanoparticles KIT-6: practical strategy for the breast cancer therapy. Eur J Med Chem 83:646–654

Marcinkowska M, Stanczyk M, Janaszewska A, Sobierajska E, Chworos A, Klajnert-Maculewicz B (2019) Multicomponent conjugates of anticancer drugs and monoclonal antibody with PAMAM dendrimers to increase efficacy of HER-2 positive breast cancer therapy. Pharm Res 36(11):154

Martínez-Carmona M, Lozano D, Colilla M, Vallet-Regí M (2018) Lectin-conjugated pH-responsive mesoporous silica nanoparticles for targeted bone cancer treatment. Acta Biomater 65:393–404

Maya S, Sarmento B, Lakshmanan VK, Menon D, Jayakumar R (2014) Actively targeted cetuximab conjugated γ-poly (glutamic acid)-docetaxel nanomedicines for epidermal growth factor receptor over expressing colon cancer cells. J Biomed Nanotechnol 10(8):1416–1428

Meena R, Singh R, Marappan G, Kushwaha G, Gupta N, Meena R et al (2019) Fluorescent carbon dots driven from ayurvedic medicinal plants for cancer cell imaging and phototherapy. Heliyon 5(9):e02483

Méndez J, Morales Cruz M, Delgado Y, Figueroa CM, Orellano EA, Morales M, Griebenow K (2014) Delivery of chemically glycosylated cytochrome c immobilized in mesoporous silica nanoparticles induces apoptosis in HeLa cancer cells. Mol Pharm 11(1):102–111

Meng L, Zhang X, Lu Q, Fei Z, Dyson PJ (2012) Single walled carbon nanotubes as drug delivery vehicles: targeting doxorubicin to tumors. Biomaterials 33(6):1689–1698

Mignani S, Majoral JP (2013) Dendrimers as macromolecular tools to tackle from colon to brain tumor types: a concise overview. New J Chem 37(11):3337–3357

Minko T, Paranjpe PV, Qiu B, Lalloo A, Won R, Stein S, Sinko PJ (2002) Enhancing the anticancer efficacy of camptothecin using biotinylated poly (ethyleneglycol) conjugates in sensitive and multidrug-resistant human ovarian carcinoma cells. Cancer Chemother Pharmacol 50(2):143–150

Mocan T, Matea CT, Cojocaru I, Ilie I, Tabaran FA, Zaharie F, Mocan L (2014) Photothermal treatment of human pancreatic cancer using PEGylated multi-walled carbon nanotubes induces apoptosis by triggering mitochondrial membrane depolarization mechanism. J Cancer 5(8):679

Mohamed EA, Hashim IIA, Yusif RM, Shaaban AAA, El-Sheakh AR, Hamed MF, Badria FAE (2018) Polymeric micelles for potentiated antiulcer and anticancer activities of naringin. Int J Nanomedicine 13:1009

Mollick MMR, Bhowmick B, Mondal D, Maity D, Rana D, Dash SK, Chakraborty M (2014) Anticancer (in vitro) and antimicrobial effect of gold nanoparticles synthesized using Abelmoschusesculentus (L.) pulp extract via a green route. RSC Adv 4(71):37838–37848

Mondal G, Almawash S, Chaudhary AK, Mahato RI (2017) EGFR-targeted cationic polymeric mixed micelles for codelivery of gemcitabine and miR-205 for treating advanced pancreatic cancer. Mol Pharm 14(9):3121–3133

Morgan MT, Nakanishi Y, Kroll DJ, Griset AP, Carnahan MA, Wathier M et al (2006) Dendrimer-encapsulated camptothecins: increased solubility, cellular uptake, and cellular retention affords enhanced anticancer activity in vitro. Cancer Res 66(24):11913–11921

Morikawa A (2016) Comparison of properties among dendritic and hyperbranched poly (ether ether ketone) s and linear poly (ether ketone) s. Molecules 21(2):219

Muddineti OS, Kumari P, Ray E, Ghosh B, Biswas S (2017) Curcumin-loaded chitosan–cholesterol micelles: evaluation in monolayers and 3D cancer spheroid model. Nanomedicine 12(12):1435–1453

Mukherjee SP, Lyng FM, Garcia A, Davoren M, Byrne HJ (2010) Mechanistic studies of in vitro cytotoxicity of poly (amidoamine) dendrimers in mammalian cells. Toxicol Appl Pharmacol 248(3):259–268

Munaweera I, Shi Y, Koneru B, Patel A, Dang MH, Di Pasqua AJ, Balkus KJ Jr (2015) Nitric oxide-and cisplatin-releasing silica nanoparticles for use against non-small cell lung cancer. J Inorg Biochem 153:23–31

Muntimadugu E, Kumar R, Saladi S, Rafeeqi TA, Khan W (2016) CD44 targeted chemotherapy for co-eradication of breast cancer stem cells and cancer cells using polymeric nanoparticles of salinomycin and paclitaxel. Colloids Surf B: Biointerfaces 143:532–546

Nakajima TE, Yasunaga M, Kano Y, Koizumi F, Kato K, Hamaguchi T et al (2008) Synergistic antitumor activity of the novel SN-38-incorporating polymeric micelles, NK012, combined with 5-fluorouracil in a mouse model of colorectal cancer, as compared with that of irinotecan plus 5-fluorouracil. Int J Cancer 122(9):2148–2153

Nakkala JR, Mata R, Bhagat E, Sadras SR (2015) Green synthesis of silver and gold nanoparticles from Gymnemasylvestre leaf extract: study of antioxidant and anticancer activities. J Nanopart Res 17(3):151

Narayanan NK, Nargi D, Randolph C, Narayanan BA (2009) Liposome encapsulation of curcumin and resveratrol in combination reduces prostate cancer incidence in PTEN knockout mice. Int J Cancer 125(1):1–8

Nguyen H, Nguyen NH, Tran NQ, Nguyen CK (2015) Improved method for preparing cisplatin-dendrimer nanocomplex and its behavior against NCI-H460 lung cancer cell. J Nanosci Nanotechnol 15(6):4106–4110

Nishiyama N, Okazaki S, Cabral H, Miyamoto M, Kato Y, Sugiyama Y et al (2003) Novel cisplatin-incorporated polymeric micelles can eradicate solid tumors in mice. Cancer Res 63(24):8977–8983

Oerlemans C, Bult W, Bos M, Storm G, Nijsen JFW, Hennink WE (2010) Polymeric micelles in anticancer therapy: targeting, imaging and triggered release. Pharm Res 27(12):2569–2589

Oldham EA, Li CHUN, Ke S, Wallace SIDNEY, Huang PENG (2000) Comparison of action of paclitaxel and poly (L-glutamic acid)-paclitaxel conjugate in human breast cancer cells. Int J Oncol 16(1):125–157

O'Shaughnessy JA (2003) Pegylated liposomal doxorubicin in the treatment of breast cancer. Clin Breast Cancer 4(5):318–328

Pan B, Cui D, Sheng Y, Ozkan C, Gao F, He R et al (2007) Dendrimer-modified magnetic nanoparticles enhance efficiency of gene delivery system. Cancer Res 67(17):8156–8163

Pan Y, Leifert A, Ruau D, Neuss S, Bornemann J, Schmid G, Jahnen-Dechent W (2009) Gold nanoparticles of diameter 1.4 nm trigger necrosis by oxidative stress and mitochondrial damage. Small 5(18):2067–2076

Pang Y, Mai Z, Wang B, Wang L, Wu L, Wang X, Chen T (2017) Artesunate-modified nano-graphene oxide for chemo-photothermal cancer therapy. Oncotarget 8(55):93800

Pankhurst QA, Connolly J, Jones SK, Dobson JJ (2003) Applications of magnetic nanoparticles in biomedicine. J Phys D Appl Phys 36(13):R167

Park JW, Hong K, Kirpotin DB, Colbern G, Shalaby R, Baselga J et al (2002) Anti-HER2 immunoliposomes: enhanced efficacy attributable to targeted delivery. Clin Cancer Res 8(4):1172–1181

Patra JK, Das G, Fraceto LF, Campos EVR, del Pilar Rodriguez-Torres M, Acosta-Torres LS et al (2018) Nano based drug delivery systems: recent developments and future prospects. J Nanobiotechnol 16(1):71

Patri AK, Myc A, Beals J, Thomas TP, Bander NH, Baker JR (2004) Synthesis and in vitro testing of J591 antibody– dendrimer conjugates for targeted prostate cancer therapy. Bioconjug Chem 15(6):1174–1181

Pautler M, Brenner S (2010) Nanomedicine: promises and challenges for the future of public health. Int J Nanomedicine 5:803

Pedro-Hernández LD, Martínez-Klimova E, Martínez-Klimov ME, Cortez-Maya S, Vargas-Medina AC, Ramírez-Ápan T et al (2018) Anticancer activity of resorcinarene-PAMAM-dendrimer conjugates of flutamide. Anti-Cancer Agents Med Chem (Formerly Current Medicinal Chemistry-Anti-Cancer Agents) 18(7):993–1000

Peng YH, Jin Y, Wang XX, Sun SB (2010) Study on the enhancement of anticancer effects of cisplatin by single-walled carbon nanotubes with larger diameters [J]. Chin J Clin Pharmacol 26(12):920–922

Plummer R, Wilson RH, Calvert H, Boddy AV, Griffin M, Sludden J et al (2011) A Phase I clinical study of cisplatin-incorporated polymeric micelles (NC-6004) in patients with solid tumours. Br J Cancer 104(4):593–598

Prasad R, Pandey R, Varma A, Barman I (2017) Polymer based nanoparticles for drug delivery systems and cancer therapeutics. In: Natural Polymers for Drug Delivery (eds. Kharkwal H and Janaswamy S), CAB International, UK 53–70

Prasek J, Drbohlavova J, Chomoucka J, Hubalek J, Jasek O, Adam V, Kizek R (2011) Methods for carbon nanotubes synthesis. J Mater Chem 21(40):15872–15884

Qi N, Tang B, Liu G, Liang X (2017) Poly (γ-glutamic acid)-coated lipoplexes loaded with doxorubicin for enhancing the antitumor activity against liver tumors. Nanoscale Res Lett 12(1):1–9

Qin XC, Guo ZY, Liu ZM, Zhang W, Wan MM, Yang BW (2013) Folic acid-conjugated graphene oxide for cancer targeted chemo-photothermal therapy. J Photochem Photobiol B Biol 120:156–162

Qiu K, He C, Feng W, Wang W, Zhou X, Yin Z, Mo X (2013) Doxorubicin-loaded electrospun poly (L-lactic acid)/mesoporous silica nanoparticles composite nanofibers for potential postsurgical cancer treatment. J Mater Chem B 1(36):4601–4611

Quader S, Liu X, Chen Y, Mi P, Chida T, Ishii T et al (2017) cRGD peptide-installed epirubicin-loaded polymeric micelles for effective targeted therapy against brain tumors. J Control Release 258:56–66

Rajeshkumar S (2016) Anticancer activity of eco-friendly gold nanoparticles against lung and liver cancer cells. J Genet Eng Biotechnol 14(1):195–202

Ramadass SK, Anantharaman NV, Subramanian S, Sivasubramanian S, Madhan B (2015) Paclitaxel/epigallocatechin gallate coloaded liposome: a synergistic delivery to control the invasiveness of MDA-MB-231 breast cancer cells. Colloids Surf B: Biointerfaces 125:65–72

Rana V, Sharma R (2019) Recent advances in development of nano drug delivery. In: Applications of targeted nano drugs and delivery systems. Elsevier, Amsterdam, pp 93–131

Rejinold NS, Muthunarayanan M, Divyarani VV, Sreerekha PR, Chennazhi KP, Nair SV, Jayakumar R (2011) Curcumin-loaded biocompatible thermoresponsive polymeric nanoparticles for cancer drug delivery. J Colloid Interface Sci 360(1):39–51

Rezvani M, Ahmadnezhad I, Darvish Ganji M, Fotukian M (2016) Theoretical insights into the encapsulation of anticancer Oxaliplatin drug into single walled carbon nanotubes. J Nanoanal 3(3):69–75

Rijcken CJF, Soga O, Hennink WE, Van Nostrum CF (2007) Triggered destabilisation of polymeric micelles and vesicles by changing polymers polarity: an attractive tool for drug delivery. J Control Release 120(3):131–148

Roy A, Datta S, Bhatia KS, Bhumika, Jha P, Prasad R (2021) Role of plant derived bioactive compounds against cancer. South African Journal of Botany. https://doi.org/10.1016/j.sajb.2021.10.015

Ruan J, Song H, Qian Q, Li C, Wang K, Bao C, Cui D (2012) HER2 monoclonal antibody conjugated RNase-A-associated CdTe quantum dots for targeted imaging and therapy of gastric cancer. Biomaterials 33(29):7093–7102

Ruttala HB, Ko YT (2015) Liposomal co-delivery of curcumin and albumin/paclitaxel nanoparticle for enhanced synergistic antitumor efficacy. Colloids Surf B: Biointerfaces 128:419–426

Sadhukha T, Wiedmann TS, Panyam J (2013) Inhalable magnetic nanoparticles for targeted hyperthermia in lung cancer therapy. Biomaterials 34(21):5163–5171

Saengkrit N, Saesoo S, Srinuanchai W, Phunpee S, Ruktanonchai UR (2014) Influence of curcumin-loaded cationic liposome on anticancer activity for cervical cancer therapy. Colloids Surf B: Biointerfaces 114:349–356

Salcedo ARM, Sevilla FB III (2013) Citrate-capped gold nanoparticles as colorimetric reagent for copper (II) ions. Philipp Sci Lett 6:90–96

Salmanpour M, Yousefi G, Samani SM, Mohammadi S, Anbardar MH, Tamaddon A (2019) Nanoparticulate delivery of irinotecan active metabolite (SN38) in murine colorectal carcinoma through conjugation to poly (2-ethyl 2-oxazoline)-b-poly (L-glutamic acid) double hydrophilic copolymer. Eur J Pharm Sci 136:104941

Samia AC, Chen X, Burda C (2003) Semiconductor quantum dots for photodynamic therapy. J Am Chem Soc 125(51):15736–15737

Samorì C, Ali-Boucetta H, Sainz R, Guo C, Toma FM, Fabbro C, Bianco A (2010) Enhanced anticancer activity of multi-walled carbon nanotube–methotrexate conjugates using cleavable linkers. Chem Commun 46(9):1494–1496

Sarkar A, Ghosh S, Chowdhury S, Pandey B, Sil PC (2016) Targeted delivery of quercetin loaded mesoporous silica nanoparticles to the breast cancer cells. Biochim Biophys Acta (BBA)-Gen Subj 1860(10):2065–2075

Sawant RR, Sawant RM, Torchilin VP (2008) Mixed PEG–PE/vitamin E tumor-targeted immunomicelles as carriers for poorly soluble anti-cancer drugs: improved drug solubilization and enhanced in vitro cytotoxicity. Eur J Pharm Biopharm 70(1):51–57

Seixas N, Ravanello BB, Morgan I, Kaluđerović GN, Wessjohann LA (2019) Chlorambucil conjugated Ugi dendrimers with PAMAM-NH2 core and evaluation of their anticancer activity. Pharmaceutics 11(2):59

Sercombe L, Veerati T, Moheimani F, Wu SY, Sood AK, Hua S (2015) Advances and challenges of liposome assisted drug delivery. Front Pharmacol 6:286

Sharma M (2019) Transdermal and intravenous nano drug delivery systems: present and future. In: Applications of targeted nano drugs and delivery systems. Elsevier, Amsterdam, pp 499–550

Shiah JG, Dvořák M, Kopečková P, Sun Y, Peterson CM, Kopeček J (2001) Biodistribution and antitumour efficacy of long-circulating N-(2-hydroxypropyl) methacrylamide copolymer–doxorubicin conjugates in nude mice. Eur J Cancer 37(1):131–139

Shivaji K, Mani S, Ponmurugan P, De Castro CS, Lloyd Davies M, Balasubramanian MG, Pitchaimuthu S (2018) Green-synthesis-derived CdS quantum dots using tea leaf extract: antimicrobial, bioimaging, and therapeutic applications in lung cancer cells. ACS Appl Nano Mater 1(4):1683–1693

Shukla R, Thomas TP, Peters JL, Desai AM, Kukowska-Latallo J, Patri AK et al (2006) HER2 specific tumor targeting with dendrimer conjugated anti-HER2 mAb. Bioconjug Chem 17(5):1109–1115

Singer JW, Bhatt R, Tulinsky J, Buhler KR, Heasley E, Klein P, de Vries P (2001) Water-soluble poly-(l-glutamic acid)–Gly-camptothecin conjugates enhance camptothecin stability and efficacy in vivo. J Control Release 74(1–3):243–247

Singh A, Dilnawaz F, Mewar S, Sharma U, Jagannathan NR, Sahoo SK (2011) Composite polymeric magnetic nanoparticles for co-delivery of hydrophobic and hydrophilic anticancer drugs and MRI imaging for cancer therapy. ACS Appl Mater Interfaces 3(3):842–856

Singh SP, Sharma M, Gupta PK (2014) Enhancement of phototoxicity of curcumin in human oral cancer cells using silica nanoparticles as delivery vehicle. Lasers Med Sci 29(2):645–652

Slowing II, Trewyn BG, Giri S, Lin VY (2007) Mesoporous silica nanoparticles for drug delivery and biosensing applications. Adv Funct Mater 17(8):1225–1236

Sperling RA, Gil PR, Zhang F, Zanella M, Parak WJ (2008) Biological applications of gold nanoparticles. Chem Soc Rev 37(9):1896–1908

Sreeranganathan M, Uthaman S, Sarmento B, Mohan CG, Park IK, Jayakumar R (2017) In vivo evaluation of cetuximab-conjugated poly (γ-glutamic acid)-docetaxel nanomedicines in EGFR-overexpressing gastric cancer xenografts. Int J Nanomedicine 12:7165

Su Z, Shen H, Wang H, Wang J, Li J, Nienhaus GU et al (2015) Motif-designed peptide nanofibers decorated with graphene quantum dots for simultaneous targeting and imaging of tumor cells. Adv Funct Mater 25(34):5472–5478

Sztandera K, Działak P, Marcinkowska M, Stańczyk M, Gorzkiewicz M, Janaszewska A, Klajnert-Maculewicz B (2019) Sugar modification enhances cytotoxic activity of PAMAM-doxorubicin conjugate in glucose-deprived MCF-7 cells–possible role of GLUT1 transporter. Pharm Res 36(10):140

Tada H, Higuchi H, Wanatabe TM, Ohuchi N (2007) In vivo real-time tracking of single quantum dots conjugated with monoclonal anti-HER2 antibody in tumors of mice. Cancer Res 67(3):1138–1144

Tan W, Wang K, He X, Zhao XJ, Drake T, Wang L, Bagwe RP (2004) Bionanotechnology based on silica nanoparticles. Med Res Rev 24(5):621–638

Tan JM, Karthivashan G, Arulselvan P, Fakurazi S, Hussein MZ (2014) Characterization and in vitro studies of the anticancer effect of oxidized carbon nanotubes functionalized with betulinic acid. Drug Des Devel Ther 8:2333

Tavassolian F, Kamalinia G, Rouhani H, Amini M, Ostad SN, Khoshayand MR et al (2014) Targeted poly (l-γ-glutamyl glutamine) nanoparticles of docetaxel against folate over-expressed breast cancer cells. Int J Pharm 467(1–2):123–138

Tengdelius M, Gurav D, Konradsson P, Påhlsson P, Griffith M, Oommen OP (2015) Synthesis and anticancer properties of fucoidan-mimetic glycopolymer coated gold nanoparticles. Chem Commun 51(40):8532–8535

Thakur M, Mewada A, Pandey S, Bhori M, Singh K, Sharon M, Sharon M (2016) Milk-derived multi-fluorescent graphene quantum dot-based cancer theranostic system. Mater Sci Eng C 67:468–477

Thambi T, Deepagan VG, Yoon HY, Han HS, Kim SH, Son S et al (2014) Hypoxia-responsive polymeric nanoparticles for tumor-targeted drug delivery. Biomaterials 35(5):1735–1743

Uchino H, Matsumura Y, Negishi T, Koizumi F, Hayashi T, Honda T et al (2005) Cisplatin-incorporating polymeric micelles (NC-6004) can reduce nephrotoxicity and neurotoxicity of cisplatin in rats. Br J Cancer 93(6):678–687

ud Din F, Aman W, Ullah I, Qureshi OS, Mustapha O, Shafique S, Zeb A (2017) Effective use of nanocarriers as drug delivery systems for the treatment of selected tumors. Int J Nanomedicine 12:7291

Uma Suganya KS, Govindaraju K, Prabhu D, Arulvasu C, Karthick V, Changmai N (2016) Anti-proliferative effect of biogenic gold nanoparticles against breast cancer cell lines (MDA-MB-231 & MCF-7). Appl Surf Sci 371:415–424

Ungari AQ, Pereira LRL, Nunes AA, Peria FM (2017) Cost-effectiveness analysis of XELOX versus XELOX plus bevacizumab for metastatic colorectal cancer in a public hospital school. BMC Cancer 17(1):691

Valdés Lizama O, Vilos C, Durán-Lara E (2016) Techniques of structural characterization of dendrimers. Curr Org Chem 20(24):2591–2605

van Vlerken LE, Duan Z, Seiden MV, Amiji MM (2007) Modulation of intracellular ceramide using polymeric nanoparticles to overcome multidrug resistance in cancer. Cancer Res 67(10):4843–4850

Venkatpurwar V, Shiras A, Pokharkar V (2011) Porphyran capped gold nanoparticles as a novel carrier for delivery of anticancer drug: in vitro cytotoxicity study. Int J Pharm 409(1–2):314–320

Vigderman L, Zubarev ER (2013) Therapeutic platforms based on gold nanoparticles and their covalent conjugates with drug molecules. Adv Drug Deliv Rev 65(5):663–676

Vijayan R, Joseph S, Mathew B (2018) Indigoferatinctoria leaf extract mediated green synthesis of silver and gold nanoparticles and assessment of their anticancer, antimicrobial, antioxidant and catalytic properties. Artif Cells Nanomed Biotechnol 46(4):861–871

Vijayan R, Joseph S, Mathew B (2019) Anticancer, antimicrobial, antioxidant, and catalytic activities of green-synthesized silver and gold nanoparticles using Bauhinia purpurea leaf extract. Bioprocess Biosyst Eng 42(2):305–319

Vittorio O, Brandl M, Cirillo G, Spizzirri UG, Picci N, Kavallaris M, Hampel S (2014) Novel functional cisplatin carrier based on carbon nanotubes–quercetinnanohybrid induces synergistic anticancer activity against neuroblastoma in vitro. RSC Adv 4(59):31378–31384

Voon KL, Keng CS, Ramli E, Sidik K, Lim TM (2012) Efficacy of a poly-L-glutamic acid-gemcitabine conjugate in tumor-bearing mice. Drug Dev Res 73(3):120–129

Wang S, Li Y, Bai J, Yang Q, Song Y, Zhang C (2009) Characterization and photoluminescence studies of CdTe nanoparticles before and after transfer from liquid phase to polystyrene. Bull Mater Sci 32(5):487

Wang X, Cai X, Hu J, Shao N, Wang F, Zhang Q, Cheng Y (2013) Glutathione-triggered "off–on" release of anticancer drugs from dendrimer-encapsulated gold nanoparticles. J Am Chem Soc 135(26):9805–9810

Wang Z, Li X, Wang D, Zou Y, Qu X, He C et al (2017) Concurrently suppressing multidrug resistance and metastasis of breast cancer by co-delivery of paclitaxel and honokiol with pH-sensitive polymeric micelles. Actabiomaterialia 62:144–156

Wei QY, He KM, Chen JL, Xu YM, Lau AT (2019) Phytofabrication of nanoparticles as novel drugs for anticancer applications. Molecules 24(23):4246

Wilczewska AZ, Niemirowicz K, Markiewicz KH, Car H (2012) Nanoparticles as drug delivery systems. Pharmacol Rep 64(5):1020–1037

Wong MY, Chiu GN (2011) Liposome formulation of co-encapsulated vincristine and quercetin enhanced antitumor activity in a trastuzumab-insensitive breast tumor xenograft model. Nanomedicine 7(6):834–840

Wu P, He Y, Wang HF, Yan XP (2010) Conjugation of glucose oxidase onto Mn-doped ZnS quantum dots for phosphorescent sensing of glucose in biological fluids. Anal Chem 82(4):1427–1433

Wu T, Duan X, Hu C, Wu C, Chen X, Huang J, Cui S (2019) Synthesis and characterization of gold nanoparticles from Abiesspectabilis extract and its anticancer activity on bladder cancer T24 cells. Artif Cells Nanomed Biotechnol 47(1):512–523

Xiao B, Han MK, Viennois E, Wang L, Zhang M, Si X, Merlin D (2015) Hyaluronic acid-functionalized polymeric nanoparticles for colon cancer-targeted combination chemotherapy. Nanoscale 7(42):17745–17755

Xiong L, Du X, Kleitz F, Qiao SZ (2015) Cancer-cell-specific nuclear-targeted drug delivery by dual ligand modified mesoporous silica nanoparticles. Small 11(44):5919–5926

Xu W, Liu L, Brown NJ, Christian S, Hornby D (2012) Quantum dot-conjugated anti-GRP78 scFv inhibits cancer growth in mice. Molecules 17(1):796–808

Xu W, Siddiqui IA, Nihal M, Pilla S, Rosenthal K, Mukhtar H, Gong S (2013a) Aptamer-conjugated and doxorubicin-loaded unimolecular micelles for targeted therapy of prostate cancer. Biomaterials 34(21):5244–5253

Xu P, Li J, Shi L, Selke M, Chen B, Wang X (2013b) Synergetic effect of functional cadmium–tellurium quantum dots conjugated with gambogic acid for HepG2 cell-labeling and proliferation inhibition. Int J Nanomedicine 8:3729

Yallapu MM, Othman SF, Curtis ET, Gupta BK, Jaggi M, Chauhan SC (2011) Multi-functional magnetic nanoparticles for magnetic resonance imaging and cancer therapy. Biomaterials 32(7):1890–1905

Yang J, Lee CH, Park J, Seo S, Lim EK, Song YJ, Haam S (2007) Antibody conjugated magnetic PLGA nanoparticles for diagnosis and treatment of breast cancer. J Mater Chem 17(26):2695–2699

Yang C, Attia ABE, Tan JP, Ke X, Gao S, Hedrick JL, Yang YY (2012a) The role of non-covalent interactions in anticancer drug loading and kinetic stability of polymeric micelles. Biomaterials 33(10):2971–2979

Yang K, Hu L, Ma X, Ye S, Cheng L, Shi X et al (2012b) Multimodal imaging guided photothermal therapy using functionalized graphene nanosheets anchored with magnetic nanoparticles. Adv Mater 24(14):1868–1872

Yang K, Wan J, Zhang S, Tian B, Zhang Y, Liu Z (2012c) The influence of surface chemistry and size of nanoscale graphene oxide on photothermal therapy of cancer using ultra-low laser power. Biomaterials 33(7):2206–2214

Yang D, Feng L, Dougherty CA, Luker KE, Chen D, Cauble MA et al (2016a) In vivo targeting of metastatic breast cancer via tumor vasculature-specific nano-graphene oxide. Biomaterials 104:361–371

Yang K, Feng L, Liu Z (2016b) Stimuli responsive drug delivery systems based on nano-graphene for cancer therapy. Adv Drug Deliv Rev 105:228–241

Yeh CY, Hsiao JK, Wang YP, Lan CH, Wu HC (2016) Peptide-conjugated nanoparticles for targeted imaging and therapy of prostate cancer. Biomaterials 99:1–15

Yigit MV, Moore A, Medarova Z (2012) Magnetic nanoparticles for cancer diagnosis and therapy. Pharm Res 29(5):1180–1188

Yokoyama M, Fukushima S, Uehara R, Okamoto K, Kataoka K, Sakurai Y, Okano T (1998) Characterization of physical entrapment and chemical conjugation of adriamycin in polymeric micelles and their design for in vivo delivery to a solid tumor. J Control Release 50(1–3):79–92

Yu M, Jambhrunkar S, Thorn P, Chen J, Gu W, Yu C (2013) Hyaluronic acid modified mesoporous silica nanoparticles for targeted drug delivery to CD44-overexpressing cancer cells. Nanoscale 5(1):178–183

Zeng Y, Yang Z, Li H, Hao Y, Liu C, Zhu L et al (2017) Multifunctional nanographene oxide for targeted gene-mediated thermochemotherapy of drug-resistant tumour. Sci Rep 7:43506

Zhang H, Yee D, Wang C (2008) Quantum dots for cancer diagnosis and therapy: biological and clinical perspectives. Nanomedicine (Lond) 3:83–91

Zhang X, Meng L, Lu Q, Fei Z, Dyson PJ (2009a) Targeted delivery and controlled release of doxorubicin to cancer cells using modified single wall carbon nanotubes. Biomaterials 30(30):6041–6047

Zhang H, Sachdev D, Wang C, Hubel A, Gaillard-Kelly M, Yee D (2009b) Detection and downregulation of type I IGF receptor expression by antibody-conjugated quantum dots in breast cancer cells. Breast Cancer Res Treat 114(2):277–285

Zhang Z, Tsai PC, Ramezanli T, Michniak-Kohn BB (2013) Polymeric nanoparticles-based topical delivery systems for the treatment of dermatological diseases. Wiley Interdiscip Rev Nanomed Nanobiotechnol 5(3):205–218

Zhang X, Huang Y, Li S (2014a) Nanomicellar carriers for targeted delivery of anticancer agents. Ther Deliv 5(1):53–68

Zhang R, Yang J, Sima M, Zhou Y, Kopeček J (2014b) Sequential combination therapy of ovarian cancer with degradable N-(2-hydroxypropyl) methacrylamide copolymer paclitaxel and gemcitabine conjugates. Proc Natl Acad Sci 111(33):12181–12186

Zhang P, Wang P, Yan L, Liu L (2018) Synthesis of gold nanoparticles with Solanumxanthocarpum extract and their in vitro anticancer potential on nasopharyngeal carcinoma cells. Int J Nanomedicine 13:7047

Zhou F, Wu S, Wu B, Chen WR, Xing D (2011) Mitochondriatargeting singlewalled carbon nanotubes for cancer photothermal therapy. Small 7(19):2727–2735

Zhou Y, Wen H, Gu L, Fu J, Guo J, Du L et al (2017) Aminoglucose-functionalized, redox-responsive polymer nanomicelles for overcoming chemoresistance in lung cancer cells. J Nanobiotechnol 15(1):87

Zou YIYU, Wu QP, Tansey W, Chow D, Hung MC, Charnsangavej C et al (2001) Effectiveness of water soluble poly (L-glutamic acid)-camptothecin conjugate against resistant human lung cancer xenografted in nude mice. Int J Oncol 18(2):331–336

Zou Y, Fu H, Ghosh S, Farquhar D, Klostergaard J (2004) Antitumor activity of hydrophilic paclitaxel copolymer prodrug using locoregional delivery in human orthotopic non–small cell lung cancer xenograft models. Clin Cancer Res 10(21):7382–7391

Zou W, Sarisozen C, Torchilin VP (2017) The reversal of multidrug resistance in ovarian carcinoma cells by co-application of tariquidar and paclitaxel in transferrin-targeted polymeric micelles. J Drug Target 25(3):225–234

Chapter 5
Potential of Metal Oxide Nanoparticles and Nanocomposites as Antibiofilm Agents: Leverages and Limitations

P. Sriyutha Murthy, V. Pandiyan, and Arindam Das

Contents

5.1	Introduction	164
5.2	Biofilms: The Way of Life of Microbial Cells	166
	5.2.1 Stages in Biofilm Development	167
	5.2.2 Detrimental Effects of Biofilms	168
	5.2.3 Antibiotic and Biocide Tolerance/Resistance Mechanisms	168
5.3	Routes of Synthesis on Physicochemical Characteristics of MONs Influencing Antibiofilm Efficacy	170
	5.3.1 Chemical and Hydrothermal Synthesis of MONs	170
	5.3.2 Sol-Gel Synthesis of MONs	170
	5.3.3 Sonochemical Synthesis of MONs	170
	5.3.4 Coprecipitation Synthesis of MONs	171
	5.3.5 Wet Chemical Synthesis of MONs	171
	5.3.6 Electrochemical Synthesis of MONs	171
	5.3.7 Biosynthesis of MONs	171
	5.3.8 Growth of Metal Oxide Nanomaterials	172
	5.3.9 Influence of Growth of MONs on Morphology Vis-a-vis Synthesis Methods	174
5.4	Applications of Metal Oxide Nanoparticles	179
5.5	Polymer Nanocomposites as Antibiofilm Agents	181
	5.5.1 Silver Nanoparticles	182
	5.5.2 Copper Oxide Nanoparticles	184
	5.5.3 Zinc Oxide Nanoparticles	185

P. S. Murthy (✉)
Water and Steam Chemistry Division, Bhabha Atomic Research Centre, Kalpakkam, Tamil Nadu, India

Homi Bhabha National Institute, Anushaktinagar, Mumba, India

V. Pandiyan
Nehru Memorial College, Puthanampatti, Tiruchirapalli, Tamil Nadu, India

A. Das
Homi Bhabha National Institute, Anushaktinagar, Mumba, India

Surface and Nanoscience Division, Material Science Group, Indira Gandhi Center for Atomic Research, Kalpakkam, Tamil Nadu, India
e-mail: adas@igcar.gov.in

	5.5.4	Titanium Dioxide Nanoparticles...	187
	5.5.5	Control of Biofilm in Biomedical Settings and Implant Surfaces................	188
	5.5.6	Control of Oral Biofilms by Polymer Nanocomposites...........................	191
5.6	Mechanism of Antibacterial Action of MONs...		191
5.7	Conclusions...		193
References...			194

5.1 Introduction

The transformation of atoms from one state to a smaller form was first proposed by Richard Feynman in 1959 who is termed as the "father of nanotechnology" (Feynman 1959) with the term "nanotechnology" coined later in 1974 by Norio Taniguchi (Taniguchii 1974). It was these findings which laid the foundations leading to the discovery that properties of materials changed drastically compared to conventional solids when their dimensions were reduced to less than 1–100 nm in one dimension leading to the concept of "nanoparticles." Metal oxide nanoparticles (MONs) find application as semiconductors, catalyst, sensors, and solid oxide fuel cells and as antifouling/antimicrobials, with metallic, insulator, and semiconductor properties.

Metal oxide NPs have been used to combat biofilms and biofouling ranging from *Ship Hulls to Bandages* (Kurtz and Schiffman 2018). The use of MONs was initiated as an alternate antimicrobial to the emergence of antibiotic and multidrug-resistant strains (MDR). MONs offered an alternate viable method to control these bacterial strains as resistance to metal ions seemed to be a nonviable proposition (Negi et al. 2012). Compared to their planktonic counterparts, bacterial biofilms are more resilient to the penetration of biocides and antibiotics. This was also an area of concern, where the use of MONs became prominent with the development of antimicrobial coatings for marine biofouling control, public hygiene, and water treatment. Mortality due to antibiotic resistance in the USA has been reported to be 23K and projected to be around 10 million/year worldwide by 2050 (O'Neill 2014; CDCP 2013) which also has been instrumental in developing alternate antimicrobial agents.

The science of nanotechnology advanced with the fruits of these small-sized particles, due to their low surface area/mass ratio and high reactivity making them efficient from their bulk counterparts (Nair et al. 2009). A reduction in particle size from 10 μm to 10 nm increases the surface contact area hypothetically by 10^9 which enhance the probability of surface contact toward a bacterial cell imparting toxicity (Hamouda 2012). Apart from surface area, novel physicochemical properties of nanoparticles also play a significant role in antimicrobial activity (Pal et al. 2007). An inverse relationship exists between nanoparticle size and antimicrobial activity, and it is understood that most nanoparticles in the size range of 1–10 nm elicit the highest antimicrobial activity (Morones et al. 2005). Several properties of MONs like size (Azam et al. 2012), shape (Gold et al. 2018), crystallinity (Espetia et al.

2012; Cha et al. 2015), solubility (Zhong et al. 2017), dispersion (Guan et al. 2019), agglomeration (Zhong et al. 2017), surface charge (Van et al. 2013), concentration (Jones et al. 2008; Pandiyarajan et al. 2013), chemical composition, and physicochemical characteristics (Lemire et al. 2013) have a say on antimicrobial activity. In this context, studies have tried to improvise on synthesis methods to yield nanoparticles of reduced sizes for superior antimicrobial activity.

Ideally, microbial cells require trace amounts of Cu, Zn, Ni, and the divalent cations Ca^{2+}, Mg^{2+} $Fe^{2+,3+}$ for their metabolic process, whereas excess quantities of the same are toxic (Lemire et al. 2013). Metal oxide nanoparticle surfaces are highly reactive due to the presence of increased number of unsaturated atoms/ions at the surface as a result of nano size features enabling easy release of metal ions from MONs. Surface charge of MONs is another important factor influencing antibacterial activity (AB) which is represented by the zeta potential. In general, metal oxide nanoparticles with zeta potential values between +30 and -30 mV are stable in suspension (Kadu et al. 2011). Stanic and Tanaskovic (2020) outlined the importance of "point of zero charge" (PZC) of MONs as a function of pH wherein the total number of positive and negative charges becomes zero or neutral which is vital for antimicrobial activity of MONs at a given pH. PZC of MONs is again dependent on impurities, crystallinity, and type of electrolyte. MONs with positive charge showed high antimicrobial activity followed by neutral and negative charge. Another significant factor involving MONs is with regard to their solubility, with particle size, pH of media, and temperature playing a major role during antimicrobial assays evaluating MON efficacy. Solubility was found to increase with decreasing particle size which was in turn affected by their aggregation property (Zhong et al. 2017).

Concentration of MONs is another vital factor involved in antimicrobial activity. This is usually represented by the minimum inhibitory concentration (MIC) which is the concentration which inhibits visible growth of microorganisms, and for biofilms, it is termed as biofilm inhibitory concentration (BIC) (Jones et al. 2008). Antimicrobial activity and environmental toxicity (to nontarget organisms by release from products) of nanoparticles are two sides of a coin. For assessing toxicity to aquatic organisms and human cell lines, the protocol of lethal concentration 50 (LC 50) is followed which is the concentration at which 50% mortality of an organism is observed. Interestingly, MONs were less toxic to bacteria than to higher-level organisms. MIC of Ag (7.1 mg/L), CuO (200 mg/L), and ZnO (500 mg/L) were recorded for bacteria, whereas LC50 values were around 0.01, 2.1, and 2.3 for aquatic organisms and 1.36, 100, and 3.0 for fish and 11.3, 25, and 43 mg/L for mammalian cells for the three nanoparticles, respectively (Bondarenko et al. 2013). Hence, environmental release of MONs is still an issue hindering their extensive application.

Several classes and combination of metallic nanoparticles have been synthesized and have been classified into (1) monometallic NP, viz., copper (Cu), copper oxide (CuO), cuprous oxide (Cu_2O), gold (Au), silver (Ag), iron oxide (FeO), lead oxide (PbO), aluminum oxide (Al_2O), calcium oxide (CaO), magnesium oxide (MgO), zinc oxide (ZnO), nickel (Ni), nickel oxide (NiO), platinum (Pt), palladium (Pd), selenium (Se), silver chloride ($AgCl_2$), silver sulfide (AgS), tellurium (Te), tin oxide

(SnO_2), and titanium dioxide (TiO_2); (2) bimetallic NP, viz., cobalt ferrites ($CoFe_2O_4$), lead and selenium (Pb/Se), copper and platinum (Cu/Pt), iron oxide and silver (Fe/Ag), zirconium oxide (ZrO_2), silver and gold (Ag/Au), silver and iron (Ag/Fe), silver and nickel (Ag/Ni), silver and palladium (Ag/Pd), and silver and ZnO (Ag/ZnO); and (3) trimetallic nanoparticles, viz., cerium oxide/copper oxide/zinc oxide (CeO/CuO/ZnO) and copper/chromium/nickel (Cu/Cr/Ni). Another significant development is the quantum dots (cadmium sulfide and cadmium selenide) which have also demonstrated antimicrobial properties (Ju-Nam and Lead 2008). Monometallic nanoparticles have been synthesized using a single metallic salt solution, whereas bimetallic nanoparticles have been synthesized with reduction of two metal salt solutions, with trimetallic nanoparticles being a recent development. The antibacterial activity of bimetallic nanoparticles has been observed to be higher than monometallic nanoparticles due to their higher reactivity and synergistic activity of both metallic NP (Sumbal et al. 2019). Bimetallic nanoparticles have been very effective against strong biofilm producers like *C. albicans*, *S. aureus*, and *P. aeruginosa* (Yallappa et al. 2015). Trimetallic nanoparticles have been shown to be more effective antibacterial agents compared to mono- and bimetallic nanoparticles (Yadav et al. 2018). Crystalline structure of MONs also influences antimicrobial activity, and different synthesis methods and precursors have yielded 0D (nanoparticles, nanocube), 1D (nanorods, nanotubes, nanowires and nanofibers), 2D (nanosheets, nanoplates), and 3D (nanoflowers, nanopillars) (Nikalova and Chavali 2019).

Several advancements of MONs with respect to their synthesis methods, physicochemical characteristics like shape, size, charge, solubility, dispersion, aggregation, binding with polymers, antibacterial activity against planktonic and biofilm forms, mechanism of action on different organisms viz: bacteria, fungi and virus, incorporation into polymer matrix to develop nanocomposites, role in medical implants, prosthesis, drug delivery agents, imaging and detection of pathogens have all been extensively reviewed. For detailed reviews on these aspects, refer to (Raghunath and Perumal 2017), (Negi et al. 2012), (Seabra and Duran 2015), (Abozeid and Williams 2019), (Kobayashi et al. 2019), (Malaekeh-Nikouei et al. 2020), (Peng et al. 2020), (Kawish et al. 2020), (Jeyaprakashvel et al. 2020), (Jandt et al. 2009), (Dhanalekshmi et al. 2020), and (Ahamadbadi et al. 2020).

5.2 Biofilms: The Way of Life of Microbial Cells

In general, bacterial cells were thought to be existent in a planktonic mode of life. It was revealed by Claude E. Zobell (1943) that bacterial cells adhered to surfaces. The concept of "biofilm theory" was proposed later on by Costerton et al. (1978). Biofilm formation is ubiquitous on all surfaces whether biotic or abiotic and is influenced by environmental conditions. The switching from a planktonic to a sessile (biofilm) mode involves environmental factors, species composition, signaling, and genetic factors. The developmental process of biofilms on surfaces does not follow a common cycle and varies depending on local environment conditions, viz., multispecies biofilms in the case of natural aquatic environments and unispecies biofilms growing on implant materials (Wille and Coenye 2020).

5.2.1 Stages in Biofilm Development

A general biofilm development/maturation cycle (Fig. 5.1) involves (1) changes to an immersed surface by formation of a conditioning film comprising of dissolved organics and biomolecules; (2) attraction of bacterial cells to the surfaces, held together by weak electrostatic forces of attraction; (3) sensing of substratum suitability resulting in secretion of exopolymeric substances (EPS) and firm adhesion of microbial cells resulting in biofilm maturation and microcolony formation with increase in density and diversity (O'Toole et al. 2000); and (4) dispersion of cells from mature biofilms again to a planktonic mode of life (Davies 2011; Flemming and Rumbaugh 2017). Adhesion of bacteria to surfaces is a complex phenomenon involving substratum properties as well as bacterial cell surface properties to influence the adhesion process. Initial adhesion is mediated by hydrophilic, hydrophobic cell surface interactions and hydrogen bonding to substratum. Bacterial cell surface charge plays a role with a predominantly net negative charge on their hydrophilic surfaces and coexistence of positively charged hydrophobic areas which are involved

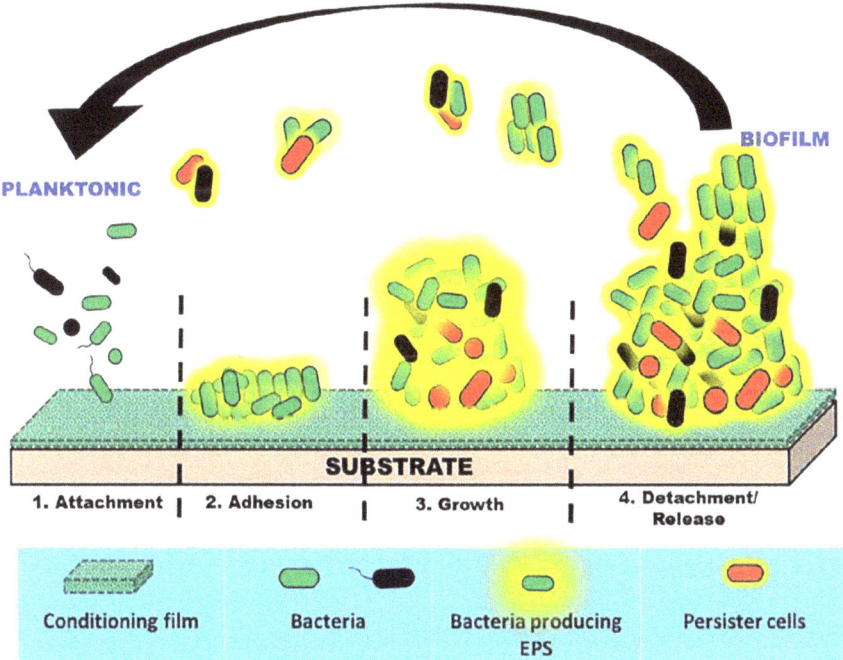

Fig. 5.1 Stages in biofilm development on biotic and abiotic surfaces

in the adhesion of cells to the substratum. The firm adhesion is brought about by secretion of EPS.

5.2.2 Detrimental Effects of Biofilms

Biofilms constitute 80% of the microbial infections as outlined by the National Institute of Health (Davies 2003). Biofilms have been shown to be physiologically heterogeneous (Stewart and Franklin 2008). Biofilms offer several advantages to resident microbes by providing increased nutrient availability and dissolved oxygen (Xu et al. 2000), quenching of antimicrobials and biocides, and gene transfer within communities and serve as a reservoir for microbes. Biofilm-induced infections have been revealed to be the cause of pathogenesis in several cases and are very difficult to eradicate (Costerton et al. 1999; Stewart 2015; Hall-Stoodley et al. 2004). Biofilm matrix/mode offers protection to microorganisms with a 1000-fold increase in concentration of antibiotics/biocides required to eliminate them compared to their planktonic counterparts (Hall and Mah 2017). Similarly, Nickel et al. (1985) also reported that biofilm cells of *P. aeruginosa* were ~1000-fold resistant to the antibiotic tobramycin compared to planktonic cells. Tetz et al. (2009) have reviewed and reported studies wherein antibiotics with 10^2 and 10^4 times their MIC concentration had no killing effect on bacteria in biofilms. Involvement of extracellular DNA in the tolerance process was observed, and when neutralized by DNase, the response/sensitivity to antibiotics increased.

Biofilm infections, development on medical devices, pathogenesis, and intervention strategies have been reviewed extensively (Donlan and Costerton 2002; Prasad et al. 2020). Donlan and Costerton (2002) also have reported that the common biofilm forming bacterial species *Streptococcus viridans*, *S. epidermidis*, *S. aureus*, *Enterococcus faecalis*, *E. coli*, *Moraxella catarrhalis*, *K. pneumoniae*, *Proteus mirabilis*, and *P. aeruginosa* are potential infection-causing agents. Common chronic biofilm infections caused by biofilm bacteria are cystic fibrosis, endocarditis, meningitis, periodontitis, dental caries, rhinosinusitis, otitis, osteomyelitis, chronic wounds, prosthesis and implantable devices. Biofilm infections in humans are resistant to the host immune defense systems and have to be systemically treated with antibiotics. In addition, implant devices like orthopedic prostheses, heart valves, venous and urinary catheters, surgical sutures, contact lens, and arteriovenous shunts are prone to biofilm development and need protective strategy.

5.2.3 Antibiotic and Biocide Tolerance/Resistance Mechanisms

Resistance or tolerance to antibiotics and biocides by microbes is dependent on several variables like bacterial strains, diffusion, or penetrating power of antimicrobials and adaptive phenotypic response of organisms in biofilm matrix. Hall and

Mah (2017) have reviewed several studies where different antibiotics were able to penetrate into biofilm matrix of different strains, without affecting the viability of cells inside the biofilm matrix (Tseng et al. 2013) which is attributed to the slow penetration of antibiotics resulting in adaptive tolerance of the cells to antimicrobial compounds. To a certain extent, this effect can be explained by the decreased diffusion limitation of biofilms, with their high cell densities and extracellular polymer matrix (EPS) resulting in throttling or impeding convective transport (Stewart 2003). Mechanism of antibiotic resistance in biofilms has been known to be mediated by diffusion barrier and quenching of antibiotics by components of the extracellular polymer matrix (EPS) and the existence of persister cells (Percival et al. 2011). Another aspect of action of antibiotics is that most of the antibiotics are active against growing cells and exhibit growth-dependent killing activity (Olsen 2015). Alteration of pH within the biofilm matrix and depletion of substrate result in reduced metabolic activity which influences the cells growing in biofilm. The switching of the toxin-antitoxin (TA) system by starvation and DNA damage is also known for the development of persister cells (Sharma et al. 2015). Stewart (2015) after reviewing several studies came up with the conclusion that the antibiotic resistance is mainly due to the physiological status of the cells in the biofilm. However, the factor responsible/involved in full biofilm developing tolerance/resistance mechanism is still elusive, and how physiologically different cells in a biofilm render this property on the entire biofilm is still a question mark.

Biofilms also have an altered chemical microenvironment which is due to the presence of aerobic species at the surface layers and anaerobic species deep down near the substratum as the biofilm matures (deBeer et al. 1994). The presence of both aerobic and anaerobic strains in biofilms is another matter of concern with conventional aminoglycoside antibiotics shown to be ineffective against anaerobes (Gupta et al. 2016). In addition, production of enzymes like the β-lactamase enzyme which cleaves/inactivates the β-lactam rings of antibiotics neutralizing it (Kawai et al. 2018) has necessitated the search for alternative antimicrobials. AmpC-type cephalosporinase enzyme isolated from cystic fibrosis patients infected with *P. aeruginosa* revealed degradation of the drug cephalosporin (Chalhoub et al. 2018) which is also indicative of the resistance mechanisms operating for the major known antibiotics in use. Other factors like local antibiotic sequestration by electrostatic interactions, negative charge of EPS matrix interacting with the positively charged regions of antibiotics, and gene transfer confer antibiotic resistance to the microbes are also responsible for the observed tolerance mechanisms (Molin and Nielsen 2003). Due to increased tolerance of biofilms, higher doses of antibiotics need to be administered which also leads to systemic toxicity. Similar to efflux of metal ions from cytosol of bacterial cells to overcome metal toxicity, antibiotic efflux from cells through multidrug efflux system of bacterial cells and transposons has also been involved in antibiotic tolerance/resistance (Levy 2002).

This has triggered the research into alternative biocidal agents wherein metal oxide nanoparticles (MONs) have filled the void as antibiofilm agents/coatings, assisted as carriers for antibiotics (clinical settings) in enhancing antibiofilm and antimicrobial activity. However, environmental biofilms constitute a heterogeneous proposition with respect to coexistence of multiple phenotypes, differential gene expression, and metabolic status which all pose new challenges to the development of a biofilm-resistant surface/coating. Nano functionalized surfaces and polymer nanocomposites which can inhibit, kill, and prevent biofilm formation are at a nascent stage even though voluminous literature on their antimicrobial activity is available, and as a much sought after requirement, research efforts for the development of a broad-spectrum antibiofilm surface are a priority.

5.3 Routes of Synthesis on Physicochemical Characteristics of MONs Influencing Antibiofilm Efficacy

5.3.1 Chemical and Hydrothermal Synthesis of MONs

The chemical method involves reaction of salt solutions using precursors and surfactants at high temperature. Copper oxides, zinc oxide, and magnesium oxide NP have been synthesized using this method. This route of synthesis is used wherein aqueous salt solutions are difficult to dissolve at normal conditions. The process involves the use of high temperature and high pressure autoclaves with temperatures ranging from 120 to 200 °C and pressures ranging from 15 to 200 psi (Byrappa and Haber 2001).

5.3.2 Sol-Gel Synthesis of MONs

Sol-gel method involves condensation and hydroxylation reactions of reactants, and this method offers a scope for variation of pH and temperature of gels to regulate hydrolysis rates as well as condensation reactions (Ennas et al. 1998). Example novel MgO nanoparticles have been synthesized by this process which has demonstrated ~98% biofilm inhibition of *S. aureus*, *E. coli*, and *C. albicans* (Wong et al. 2020). A characteristic feature observed was that MgO particles synthesized by sol-gel process had minimum aggregation tendency.

5.3.3 Sonochemical Synthesis of MONs

Most of the metal oxide nanoparticles have been synthesized using this method wherein the metal salt solution is sonicated which relies on acceleration and collision of particles in solution which impart different particle size, composition, and morphologies. This method is usually applied for preparing polymer nanocomposites where NPs are incorporated into polymers (Malka et al. 2013).

5.3.4 Coprecipitation Synthesis of MONs

This is a simple facile method for large-scale production of MONs with the addition of either ammonium hydroxide or sodium hydroxide to a metal salt solution under constant stirring and temperature conditions. The process involves initial nucleation of the metals and then nuclei growth onto the crystal surface through diffusion of the solutes (Sugimoto 2003).

5.3.5 Wet Chemical Synthesis of MONs

This route of synthesis offers for large-scale production of nanoparticles and is a simple and inexpensive method. The process involves mixing of reactants by stirring and use of mild heating in some cases (Wu et al. 2005).

5.3.6 Electrochemical Synthesis of MONs

This involves the electrolysis process under inert conditions using a suitable electrolyte with the bulk material as the anode. In the process, the cations move toward the cathode, and the bulk material is oxidized at the anode resulting in release of small metal clusters which are stabilized by stabilizers. During the process, the residual oxygen present in the electrolyte oxidizes the metals into respective metal oxide nanoparticles (Reetz and Helbig 1994).

5.3.7 Biosynthesis of MONs

Physical and chemical methods of synthesis of nanoparticles involve the use of high temperature and vacuum conditions with a time- and energy-consuming process. However, in the last decade, the advent of green synthesis using biological agents like plants, bacteria, fungi, and different cell cultures has added advantages compared to physical and chemical methods in that the biological metabolites also act as capping agents which aid in better dispersion of nanoparticles (Zhang et al. 2016; Prasad et al. 2016, 2018; Srivastava et al. 2021) in solution and also offer large-scale synthesis and production of these NP. A major disadvantage of this approach is that the growth of biological organisms like bacteria, algae, and fungi is dependent to a great extent on different synthetic media, varying nutrient requirements, and wide range of environmental conditions which trigger production of different metabolites during their growth phase. These metabolites reduce metal ions resulting in synthesis of green nanoparticles. The active metabolic moiety involved in the process of reduction of salt solution has however not been characterized in majority of studies using the bio-route for synthesis. The reproducibility of size, shape, and charge properties of such green synthesized nanoparticles is still a big question mark compared to those synthesized by physical and chemical methods where there is more control over the synthesis parameters.

5.3.8 Growth of Metal Oxide Nanomaterials

The advent of nanotechnology as an outstanding versatile technology has paved the way for superior functionality of products in various sectors (D'Souza and Richards 2007; Fernandez-Garcia et al. 2004; Jeevanandam et al. 2018; Pradeep 2007). The characteristic of nanomaterials such as large specific surface area with higher interfacial nature has given them unique fascinating features and functionalities. The synthesis of a material is a backbone for novel applications, and it is truly a challenging demand to synthesis nanomaterials for antifouling applications (Abioye et al. 2019; Wang and Chen 2019). Depending on the application of a material, the requirements need to be met. For instance, self-cleaning coatings should have efficacy in the specific environment against the relevant microbes with toxicity to the microbes. In this respect, the use of nanomaterials has yielded promising results for antimicrobial and antifouling coatings (Sathya et al. 2019; Salazar-Hernandez et al. 2019). Biofouling and biofilm formation are a major concern in the marine industry due to the impact it has on the maintenance of system for functioning and the cost for prevention (Scardino et al. 2009; Sankar et al. 2015). Methods such as mechanical and chemical processes related to the application of biocides currently used to control biofouling are not always effective and environment friendly. The need for alternative methods for the prevention of biofouling therefore exists. In this respect, nanomaterials have yielded promising results. There is widespread use of nanomaterials in the form of metal or metal compounds, mostly as oxides which are chemically stable even in ambient and moist environments (Scardino et al. 2009; Sathya et al. 2019; Ruiz-Sanchez et al. 2020).

In defining nanomaterials, it is usually described as control of matter at dimensions between approximately 1 nm and 100 nm. Importantly, these materials at the nanoscale reveal new physiochemical and biological properties compared to their bulk counterparts. Noble metals, like silver, gold, and copper nanoparticles, have, for example, shown higher antimicrobial activities against certain bacterial species (Yael et al. 2017; Bankier et al. 2019). Even pure carbons in the forms of nanotube and graphene do show antimicrobial activities (Wiaraja et al. 2018; Mohammed et al. 2020). Metal oxide nanomaterials are used for biological applications, and these oxides have to cope with the process-related demands and challenges. Thus, essential methods may have to be tuned wisely to fit the needs of an application. For instance, one-dimensional (1D) metal oxides are efficient for biomedical applications (Yah et al. 2011; Bonu et al. 2019), while the nanoparticles (0D) suit better for the antifouling treatment (D'Souza and Richards 2007; Sankar et al. 2015; Yael et al. 2017; Pradeep 2007). Some well-known adopted approaches are classified such as "bottom-up" – starting from atoms or molecules to reach to the desired goal with shape, size, and morphology – and another approach is the "top-down" which deals with larger sizes to reach to smaller ones (Fig. 5.2). Considering the focus of applications on antifouling treatments, bottom-up approach allowing large-scale synthesis with ease processable methods is likely to be adopted. Top-down process such as ball milling can be of use for large-scale process; however, possible contamination-free uniformity nanoparticles are a formidable task (Basnet et al. 2019).

A few methods which are well utilized for synthesis of metal oxides, such as CuO, ZnO, TiO_2, SnO_2, etc., are precipitation, sol-gel, hydrothermal, solvothermal, and template-assisted techniques (Fernandez-Garcia et al. 2004; D'Souza and Richards 2007; Rao et al. 2007; Lee and Soltis 2014). In coprecipitation, a soluble metal salt in the form of chloride or nitrate is used in a solvent to form a precipitate mostly with the help of a base such as $NaOH$ and NH_4OH. The simplicity of the technique attends attention for the preparation of coprecipitation of mixed-metal oxides. Furthermore, with control on precursor concentration and pH solvent medium along with the use of capping/surfactants, the method is able to provide uniform nanoparticles including quantum dots (Deshmukh and Niederberger 2017; Juine and Das 2020).

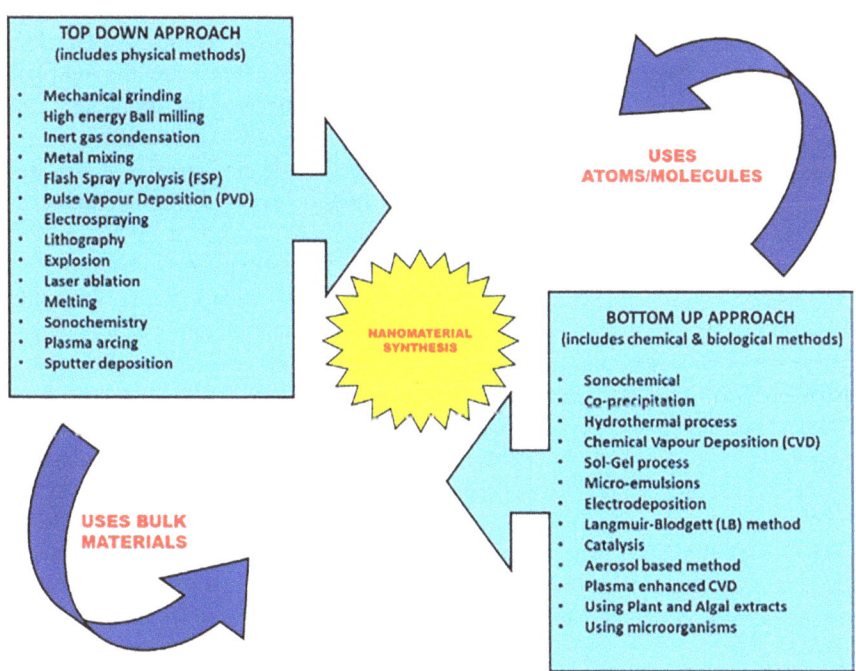

Fig. 5.2 Approaches toward synthesis of metal oxide nanoparticles

Similarly, sol-gel, a highly economic and simple method, is of high interest for the preparation of metal oxides (Li et al. 2015; Deshmukh and Niederberger 2017; Juine and Das 2020). In this process, hydrolysis of precursors results in the corresponding oxo-hydroxide, followed by the condensation of molecules to release water molecules to provide a gel-like formation, via polymerized hydroxides. Further drying process and sometime high temperature calcinations offer stable metal oxides. The process can be aquatic or non aquatic friendly environment. Highly porous metal oxides are obtained by the process. A popular method is solvo-thermal/hydrothermal process where a solvent (water) is raised to the boiling point in a closed vessel to impart pressure (Li et al. 2015, 2016). An appropriate chemical molecule is used as a capping agent for controlling the size and shape. There are several methods like template, light assisted and electrochemistry for creating varied sizes of nanoparticles (Rao et al. 2007; Li et al. 2016).

5.3.9 Influence of Growth of MONs on Morphology Vis-a-vis Synthesis Methods

In most of the above methods, the formation of crystal may not follow a traditional nucleation and growth mechanism. Undoubtedly, understanding of the growth process is helpful for the desired outcome and further improvement in a predictive way. In general, it includes pre-nucleation, nucleation and growth processes, assembly, and agglomeration of nanomaterials. However, unique mechanistic explanations with the predictive character are strongly limited. In fact, foreseeing the morphology and the composition of the crystal structure considering the influence of all chemical species remains a demanding challenge. There are several concepts on the growth, oriented attachment, cluster-mediated pre-nucleation, particle-based and mesocrystals which do not conform the classical view of crystallization (Ludi et al. 2012; Lee and Soltis 2014; Li et al. 2015; Li et al. 2016; Zhang et al. 2009; Liu et al. 2020; Weigiang et al. 2014). With the presence of further liquid/gas phase as seen, for instance, in solvothermal methods for final metal oxide crystals, steps like Ostwald ripening may also take part. Crystallization is an important phenomenon to define the size, shape, and crystalline orientation of the final nanoparticles. In a broad view, the basic crystallization process may be categorized into two types – (1) classical and (2) nonclassical crystallization – depending on whether an atom-/ion-mediated growth or a particle-mediated growth mechanism takes place. In classical crystallization model, small units arising from precursors, like atoms, ions, or molecules, undergo nucleation from a supersaturation point and thus its nucleation leads to a cluster. Total energy including the surface and crystal lattice energies of the system determine the growth or disintegration of it. In contrast, particles or clusters participate in nonclassical crystallization process such as oriented attachment growth (Lee and Soltis 2014). The smaller units have same orientation and get together to form crystal structures. TiO_2 nanoparticles can be synthesized mechanically for self-supporting networks of macroscopic size, e.g., aerogels. Studies with HRTEM reveal the formation the anatase nanocrystals which leads to the oriented attachments (Dalmaschio and Leite 2012). Typical surface dominance of {101} faces is also reported. Such mechanism may proceed in various directions to result in 3D morphologies and provides various shapes, like spindle and flowers (Jianfeng et al. 2011). Precipitation methods are utilized for the growth of SnO_2 nanoparticles where a selective dopant is also incorporated (Das and Jayaraman 2014; Das and Panda 2019). In general, chloride salt of Sn is used to react with base, NaOH or NH_4OH. The precipitated gel is washed and calcinated for obtaining nanomaterials. Further, in gel with suitable surfactant, a hydrothermal method is used to produce improved quality of SnO_2 crystals. Similarly, oriented attachment is also reported, while $SnCl_4$ and benzyl alcohol are used with anisotropic growth along the <110> directions, a low energy facet of rutile SnO_2 phase (Daniel et al. 2011). Another important metal oxide is ZnO which is well utilized for the antimicrobial studies (Sathya et al. 2019).

The most commonly used semiconducting metal oxide, ZnO, has attracted large attention as a promising material for a wide range of technological applications mainly wastewater treatment via photocatalysis and antibacterial activity (Bhuyan et al. 2015; Enas et al. 2020). Various methods, such as precipitation; solvothermal approaches with metal salts like zinc acetate and zinc nitrate with base; and capping agents are well utilized. Various morphologies resulting from a sol-gel process with zinc acetylacetonate hydrate in benzyl alcohol are well documented (Li et al. 2015). Here, authors discuss the critical role of concentration for supersaturation to lead a nucleation and further growth through agglomeration into different shapes. Synthesis methods and precursors (Fig. 5.3) were found to influence morphology and also influence catalytic activity.

ZnO NPs obtained from zinc nitrate and oxalic acid via precipitation-decomposition method were found to have a heterogeneous morphology with some of the particles spherical in shape and that also appeared as nano-bundles (Fig. 5.1a, b) due to the formation of intermediate chemical moieties of zinc oxalate. The precursors zinc nitrate and oxalic acid aqueous solutions were brought to their boiling points separately. After reaching their boiling points, both were mixed rapidly and stirred for a time, and the obtained zinc oxalate precipitate was filtered and dried. In order to get the ZnO, the formed zinc oxalate was decomposed at 450 °C. In comparison with ZnO, NP obtained from zinc acetate and sodium hydroxide via simple physical grinding method using pestle and mortar offered a rodlike structure (Fig. 5.1c, d). The length of the nanorods ranged from 100 to 300 nm, and the width of the rods was approximately 20–40 nm (Krishnakumar and Imae 2014; Krishnakumar et al. 2014). In physical grinding method using zinc acetate and NaOH precursors, formation of zinc hydroxide intermediate was observed which was washed with water and ethanol, filtered and dried at room temperature. The dry crystals were calcined for 3 h at 200 °C in a muffle furnace to obtain rodlike ZnO. The method of preparation and precursors slightly influenced the bandgap energy of the materials. From DRS analysis, the bandgap energies of the prepared

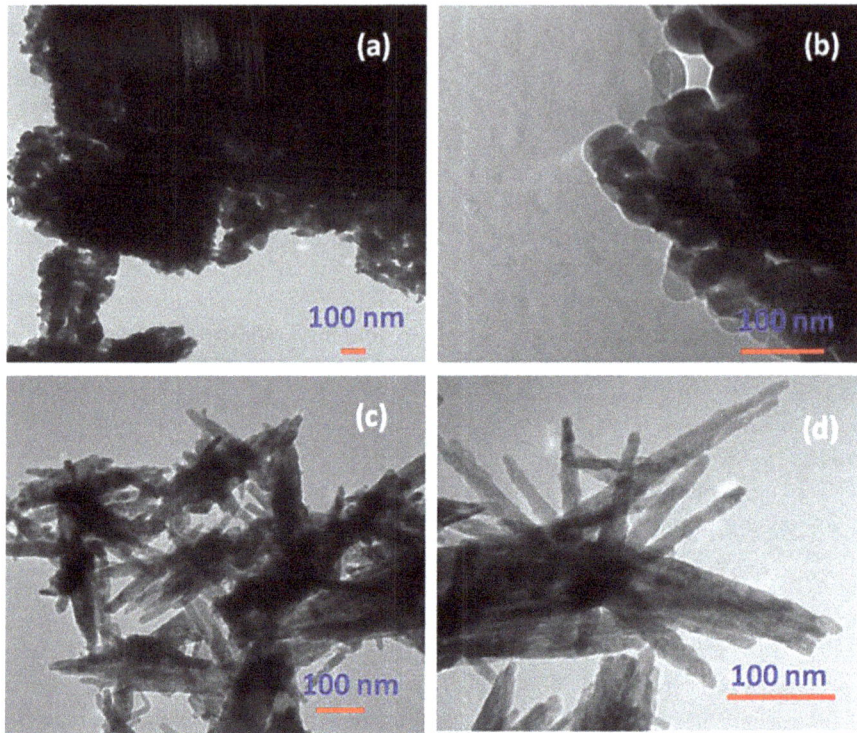

Fig. 5.3 TEM images indicating morphological variation of ZnO synthesized from precipitation-decomposition method (**a, b**) and physical grinding method (**c, d**) with different precursors

ZnOs were found to be 3.02 and 3.13, respectively, for ZnO obtained from precipitation-decomposition and physical grinding methods.

TiO_2 NP synthesized via sol-gel method and gelatin incorporation influenced the morphology as observed by corresponding FE-SEM images (Fig. 5.4a {sol-gel} and Fig. 5.4b {gelatin-assisted}). Morphology of TiO_2 was not much affected with addition of gelatin during synthesis. Both sol-gel-derived and gelatin-assisted TiO_2 yielded spherical-shaped particles. However, the variation in the BET surface area and pore size was observed (Krishnakumar et al. 2018).

SiO_2 prepared by a simple sol-gel method and gelatin-assisted Ni/SiO_2, ZnO-loaded SiO_2, and CdS-loaded SiO_2 were synthesized which revealed bare SiO_2 had spherical morphology from FESEM images (Fig. 5.5a–f) whose diameter was found to be between 400 and 500 nm. The particles were well separated and did not show aggregation (Fig. 5.5a) (Krishnakumar et al. 2017, 2019, 2020). Incorporation of Ni was carried out to the particles by the sol-gel method where the gelatin was dissolved in hot liquid and $Ni(NO_3)_2 \cdot 6H_2O$ solutions were added resulting in the formation of the g-Ni/SiO_2 NP which was found to be aggregated (Fig. 5.5b). However, ZnO and CdS synthesized by similar route did not affect the shape of the SiO_2 particles, and the spherical-shaped were retained without aggregation (Figs. 5.5c–f).

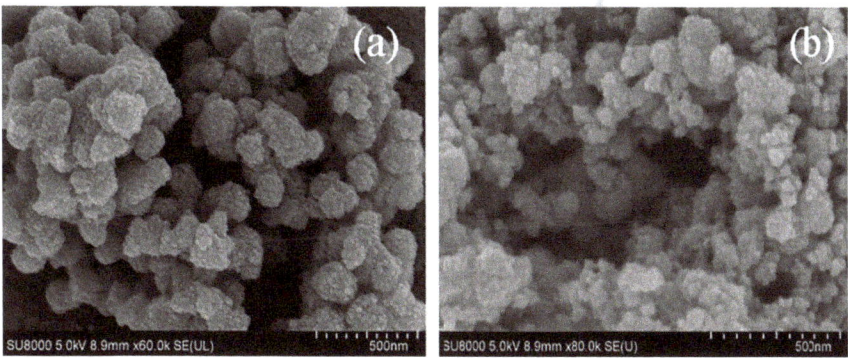

Fig. 5.4 FE-SEM images of (**a**) TiO$_2$ and (**b**) gelatin assisted TiO$_2$

Fig. 5.5 FE-SEM images of (**a**) SiO$_2$, (**b**) g/Ni-SiO$_2$, (**c**, **d**) ZnO/SiO$_2$, and (**e**, **f**) CdS/SiO$_2$

Copper oxides (Fig. 5.6) are used as antimicrobial agents. The growth of CuO nanoparticles commencing from its salts, like nitrate and acetates, with ammoniacal solution via Cu-OH formation and decomposition is well described (Zhang et al. 2006). Different kinds of CuO structures, i.e., urchin-like and sheetlike, are obtained (Vaseem et al. 2008). The growth of urchin-like CuO structures is observed with the high alkaline solution of copper nitrate where the critical role of surfaces of the copper powder and a layer of copper hydroxide are correlated according to a chemical reaction Cu(OH)$_2$ → CuO + H$_2$O.

Likewise, in the method of preparation as shown in Fig. 5.6, precursors and surfactant have tremendous influence on the morphology. Typical CuO nanomaterials grown with various precursors and surfactant are shown in Fig. 5.7. As precursor changes from Cu(NO$_3$)$_2$ in Fig. 5.7a to Cu(CH$_3$COO)$_2$ as in Fig. 5.7c, a visible change in the morphology was found. With the use of hexamine, an organic base, bitter gourd-type morphology with large small nanoparticles (Fig. 5.7b) develops

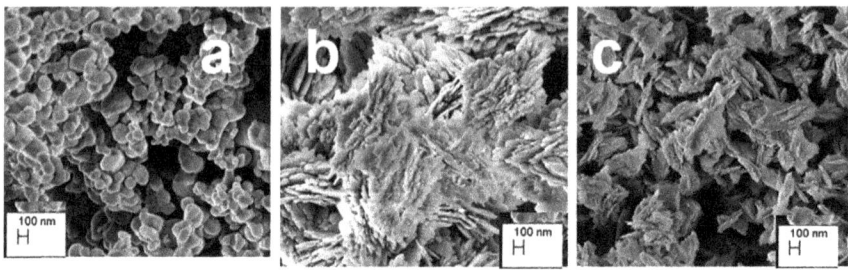

Fig. 5.6 FESEM of CuO nanomaterials. Clear demonstration of morphological variation from bulk (**a**) is found when a hydrothermal process at 100 °C (**b**) or a surfactant CTAB (**c**) is used

Fig. 5.7 FESEM images of CuO nanostructures arising due to change of precursor and surfactant

compared to the use of inorganic base, ammonia solution as shown in Fig. 5.7a where the morphology is more like snowflakes. In both cases, Cu $(NO_3)_2$ is used as starting material. Furthermore, the use of CTAB as a surfactant was found to impact morphology as seen in Fig. 5.7d, e where 1.0 M CTAB was added to starting materials for Fig. 5.7a, c, respectively. The above morphology links to growth of oriented attachments (Zhang et al. 2006; Vaseem et al. 2008).

In addition to size and shape, the growth processes influence crystal phase and can lead to different polymorphs. For instance, TiO_2 exhibits anatase and rutile crystal phases, and they differ strongly in physiochemical properties. Thus, controlled synthesis of particular phase is an important aspect to consider for an application. By using various solvents and surfactants, formation of crystals can be controlled. For instance, the lowest surface energy crystallographic facet {111} of Fe_2O_3 is predominantly obtained with no energy stabilizer, whereas surfactant-mediated growth provides other facets (Guo et al. 2015). Simple solvothermal reaction is found to yield α−Fe_2O_3 rhombohedra with {104} facet by variation of water and 1−propanol solvent (Wang et al. 2010). Thus, it shows potential control on the growth direction and crystallographic phase and hence the shape of the nanomaterials.

It is obvious to have large surfaces with decreasing size of nanoparticles which may facilitate improved applications. The physical and chemical properties are also drastically influenced when the size approaches below 10 nm. Such a quantum confinement size effect may take place as the particle size becomes equivalent to the Bohr's radius to cause a blue shift in the optical bandgap of the nanomaterials (Das and Jayaraman 2014; Deshmukh and Niederberger 2017). Moreover, large surfaces result in larger surface energy and surface defects rather than the counter bulk material. Prevalent defect states, such as oxygen or metal vacancies in SnO_2 or in CuO, offer new properties by influencing electronic structures (Das and Jayaraman 2014). Importantly, defects in SnO_2, TiO_2, or ZnO make them *n*-type semiconductors whereas CuO becomes a p-type semiconductor. At nanoscale, large defects inflict strongly a non-stoichiometric layer and influence the surface chemistry including possible leaching of ions from nanoparticles. Furthermore, shapes and crystallographic orientation with different surface energies can also affect surface reactivity. To preserve the surfaces for an application like antifouling becomes a significant step, as fabricating a coating may need other supporting material, like polymers. In the above context, material synthesis and their critical physical and chemical properties are of profound interest to resolve and correlate to a specific application. Most importantly, the morphology covering the size, shape, and crystallographic nature is grossly an important aspect to deduce. Very often, phase purity is done by X-ray diffraction study. In addition, simple Scherer's formula (Patterson 1939; Das and Jayaraman 2014; Bonu et al. 2019) allows determination of size which may further be supported by TEM or SEM images. Both later techniques are direct method to map the morphology, shape, and size of the nanoscale materials.

The precipitation method with 1 M $SnCl_4$ and NH_4OH allows the formation of ultrafine SnO_2 nanoparticles as shown in Fig. 5.3. These particles are equivalent to the Bohr exciton radius and show a blue shift to 4.4 eV from the bulk optical bandgap of 3.67 eV (Pradeep 2007; Bonu et al. 2019). Further, annealing those as-prepared quantum dots (QDs) at varied temperatures in the air atmosphere improves crystallite sizes as depicted in TEM images in Fig. 5.8. High-resolution TEM as an inset shown in Fig. 5.8 provides a d spacing value which supports the Rutile structure of the SnO_2 NPs.

5.4 Applications of Metal Oxide Nanoparticles

Among a plethora of nanoparticles (NPs), copper, silver, zinc, and titanium dioxide NP have been extensively used/applied due to their broad-spectrum antimicrobial activity (Brayner et al. 2006; Jones et al. 2008) and photocatalytic property (Norman et al. 2008; Aziz et al. 2015). As biofilm formation is a surface-associated phenomenon, protective coatings offer promise. This review limits to four oxide nanoparticles, viz., silver, copper, zinc, and titanium. Metal oxide nanoparticle (MON)

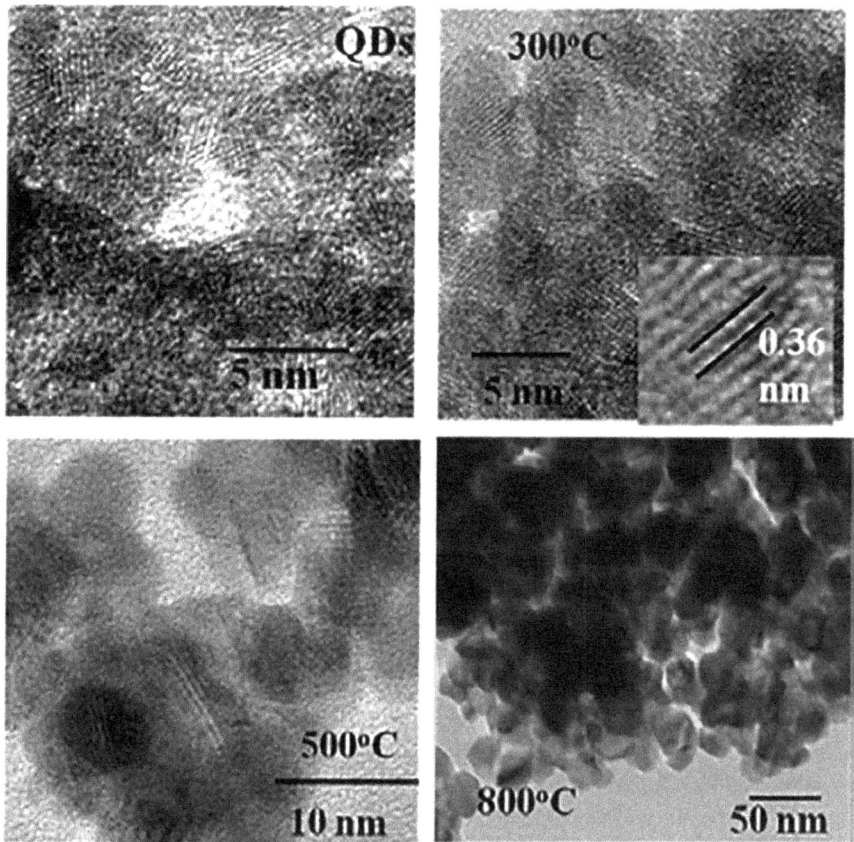

Fig. 5.8 Growth of SnO_2 from QDs to larger nanoparticles by annealing process

coatings find application for the protection of materials in biomedical devices (Stoica et al. 2017), reverse osmosis desalination membranes (Zunita et al. 2018), hospital equipment and biomedical implants (Khatoon et al. 2018), cancer treatment (Ren et al. 2015), biomedicine (Augustine and Hasan 2020) and public hygiene surfaces (Nikalova and Chavali 2019), coatings for food processing and storage equipment (Ogunsona et al. 2020), marine antifouling coatings (Natalia et al. 2012), and exterior protective coatings (Kaegi et al. 2008). In spite of significant improvements in the development of advanced functional coatings, challenges still exist due to the instability of metal oxide nanoparticles with respect to their size, shape, dispersion, aggregation, binding, interaction with polymer matrix, and release of metal ions (Rong et al. 2006; Anyaogu et al. 2008). Among the metal nanoparticles, silver (Ag), gold (Au), and zinc oxide (ZnO) have been extensively used in biomedical settings (Wong and Liu 2010). Qayyum and Khan (2016) have reviewed the antibiofilm activity of different MONs and have outlined the limitations in the application of MONs due to their toxicity concerns as a result of chemicals used during synthesis which are retained in the NP and have urged for developing more eco-friendly methods for the synthesis of MONs. Silicon dioxide (SiO_2) and nanoclay (zeolites) have also demonstrated excellent antibiofilm properties and have been incorporated into polymer matrices. Single-walled carbon nanotube (SWCNT)-coated surfaces were found to inhibit initial bacterial growth and prevent biofilm maturation of the *E. coli* strain. However, as the biofilm matured, cells became less sensitive to the presence of SWCNT indicating a response similar to the antibiotic resistance which was attributed to the production of soluble exopolymeric substances (EPS) which mitigated the toxic effects of SWCNT (Rodrigues and Elimelech 2010). Interestingly cells without soluble EPS were susceptible and detached from biofilms compared to cells which produced soluble EPS which required a tenfold increase in concentration of SWCNT for inhibition.

5.5 Polymer Nanocomposites as Antibiofilm Agents

The concept of developing antimicrobial surfaces is paramount as bacteria live in biofilms. Metal oxide nanoparticles embedded in polymer matrixes offer release of toxic metal ions which have been demonstrated to inhibit either adhesion of microbial cells or contact killing of adsorbed cells on surfaces. In general ceramics, metal oxide-polymer and of late metal oxide-carbon (graphene oxide) nanocomposites have been developed with sol-gel blending, hydrogels, and in situ polymerization being the most preferred methods of synthesis of polymer NCs. Starting from simple biopolymers like starch, alginate, cellulose, carboxymethylcellulose, guar gum, gelatin, chitosan, polyhydroxyalkanoates, polyhydroxybutrate, poly-caprolactone, and poly-lactic acid to synthetic polymers like polyvinyl chloride (PVC), polyvinyl alcohol (PVA), polydimethylsiloxane (PDMS), polyurethane (PU), polyvinyl pyrrolidine (PVP), polymethylmethacrylate (PMMA), poly-sulfobetaine, poly(amide-imide), poly-vinylidenefluoride (PVDF), polyamide, polysulfone, polystyrene, polyethylene glycol (PEG), polyethylene, and polyisopropanol have all be used to develop metal oxide-polymer nanocomposites for antifouling and antibiofilm applications (Mallakpour et al. 2020). Polymers act as releasing agents which deliver biocides at the surface, and latest developments have resulted in smart antibacterial surfaces which are triggered with surface response. Nanoparticles act as nanofillers in a polymer matrix and are known to improve physical and mechanical strength as well (Fig. 5.9). Moreover, metal oxides or metals for antifouling application need a medium to deliver it suitably. In general, biocides including metal particles are embedded or loaded in a selective polymer by physical and chemical reactions (Sankar et al. 2015; Sathya et al. 2019; Salazar-Hernandez et al. 2019; Ruiz-Sanchez et al. 2020).

An ideal polymer nanocomposite matrix should not aggregate NP, have optimum loading of nanofillers (NPs), and retain the properties of nanoparticles (Chen and Gonsalves 1997). However, due to their high surface free energy, NPs bind strongly to other materials and to each other resulting in agglomeration (Klaus and Sigusch 2009). The strong binding features of NPs are advantageous in a polymer matrix;

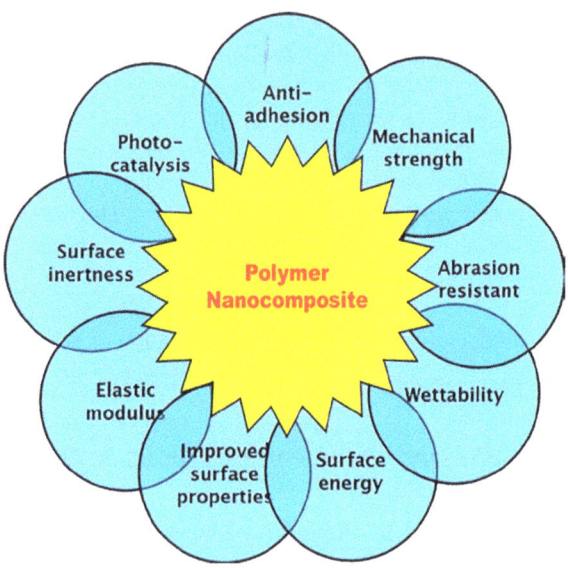

Fig. 5.9 Properties of polymer nanocomposites involved in antimicrobial/antifouling coatings

however, the agglomeration phenomenon affects its uniform dispersion in the polymer matrix. Extensive research has gone into nanoparticle dispersion in polymer matrix as NPs with high surface energy tend to agglomerate and the interaction of hydrophilic nanoparticles in a hydrophobic matrix results in weak interfacial interactions (Rong et al. 2006). Immobilization of nanoparticles into polymer matrix to develop polymer nanocomposites offers the best suitable surface protection strategy as antimicrobial surfaces.

In general, approaches toward developing antibiofilm/antifouling coatings involve (Fig. 5.10) (1) delivery of biocides (oxidizing, quaternammonium compounds-QAC), (2) delivery of antibiotics, (3) metal oxide nanoparticle MON-incorporated coatings for metal leaching and toxicity, (4) phytochemicals and essential oil based (5) polycation based (6) antimicrobial peptide based (AMP) (7) low surface energy & foul release polymeric coatings (PDMS and fluoropolymers) (8) micro / nano structured surface coatings (9) biomimetic coatings (10) superhydrophobic coatings (11) coatings based on quorum sensing inhibitors (12) enzyme based coatings (13) nitric oxide (NO) release coatings (14) sol-gel, hydrogel, xerogel, amphiphilic, zwitterionic coatings (15) photoactive coatings (16) slippery liquid infused surfaces (SLIPS) (Unal 2018).

5.5.1 Silver Nanoparticles

Silver nanoparticle (SNP/Ag_2O) with a bandgap of 1.46 eV has wide application and is the most investigated nanomaterial due to its broad biocompatibility (at low concentrations of Ag ions) and broad-spectrum antimicrobial activity ranging from bacteria, fungi, to virus and even multidrug-resistant strains such as

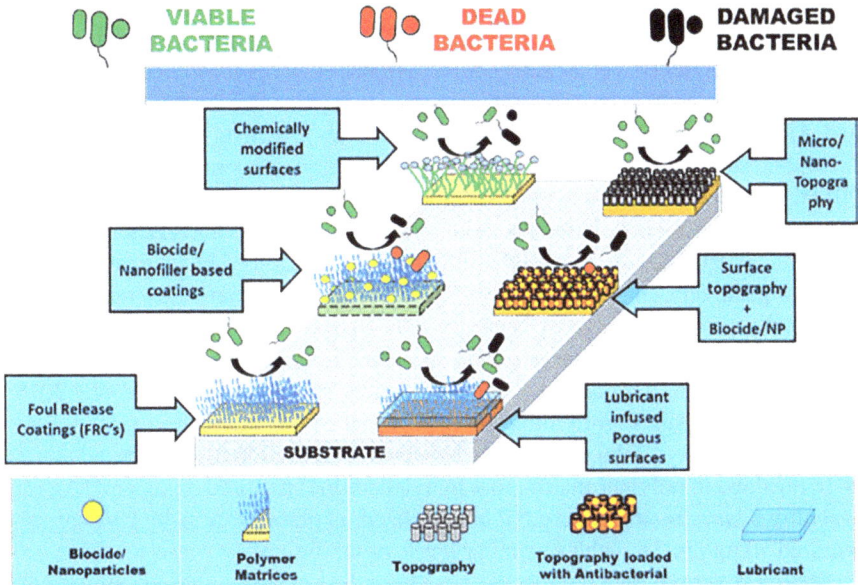

Fig. 5.10 Advancements in approaches toward antimicrobial/antifouling coatings

methicillin-resistant *Streptococcus aureus* (Pal et al. 2007). Silver nanoparticles with less than 100 nm are known to possess 10,000 to 15,000 atoms (Oberdoster et al. 2005). However, the extensive use of silver is still limited by its toxicity to humans, exerted by silver ions (Hamouda 2012). Very low concentrations of silver ions are not toxic, but high concentrations have been reported to exert cytotoxic effects. Nano silver toxicity is still to be clearly understood/established, and in vivo research is a priority area of concern (Chaloupka et al. 2010) for wider application.

Self-assembled monolayers of AgNP on glass surfaces prepared by aminosilanization exhibited strong antibiofilm activity against *S. epidermidis* RP62A which offers prolonged release and increased local release of Ag ions at the surface (Tagletti et al. 2014). A silver-tolerant strain of *Bacillus cereus* was able to produce SNP of ~17.51 nm by biosynthetic pathway which exhibited good antibacterial activity against MDR strains (Khan et al. 2020). Silver NP-loaded polyethylenimine and polyethersulfone membranes developed by electrospinning showed excellent antibiofilm as well as growth inhibition properties (Maziya et al. 2020).

Regarding the effective mechanism, for the well-studied Ag nanoparticles, it is postulated that the binding of Ag nanoparticles onto microbial cells alters their membranes and causes damage to cellular organ after penetrating the microbial cells and the dissolution of Ag nanoparticles releasing Ag+ ions for antimicrobial effects (Yael et al. 2017; Aziz et al. 2014, 2016, 2019). Obviously, release of ions, cell penetration, and toxicity of it will highly depend on the nature of nanomaterials which may have various shapes, crystallographic orientations, and sizes (D'Souza and Richards 2007; Fernandez-Garcia et al. 2004; Jeevanandam et al. 2018; Pradeep 2007; Abioye et al. 2019; Rao et al. 2007).

Small-sized/nano silver of 25 nm has found topical application in biomedical field. In addition, low concentrations as low as 1.69 µg/mL of SNP elicited good antibiofilm activity (Pal et al. 2007). Silver NP has been demonstrated to act against HIV virus at an early stage inhibiting its replication postentry into systems (Al-Jabri and Alenzi 2009) which makes it all the more an effective antimicrobial agent. Incorporation of silver in bone implants offers prophylactic properties (Stanic et al. 2011). The use of bimetallic NP like Ag-incorporated TiO_2 coatings showed excellent AB activity against dental bacterial strains of *S. mutans* and *Lactobacillus* sp. (Lv et al. 2019). The superior action of SNP is also influenced by its surface charge, wherein positively charged NPs have better electrostatic interaction with the predominantly negative charge of bacterial cells (Kim et al. 2007). Silver ions from SNPs are also known to generate ROS in bacterial species and interfere with cell respiration, DNA replication, and translation process (Joshi et al. 2018). Incorporation of silk fibroin into Ag_2O NP enhanced wound healing and improved antibiofilm activity (Babu et al. 2018). However, Tiller (2006) demonstrated that Ag NP embedded in polymer matrix poly (ethyleneimine) required 32 µg cm^{-2} Ag ions to inhibit *S. aureus* whereas Ag NP immobilized on glass surfaces by CVD process required 10 µg cm^{-2} to achieve similar levels of inhibition.

Several silver polymer composites have been developed exhibiting antimicrobial activity. Incorporation of Ag_2O into chitosan, a natural biopolymer, to develop a composite film has increased its antimicrobial properties and finds application in food packaging industries (Tripathi et al. 2011; Chausali et al. 2022). Photocatalytic antibacterial paper has been developed using Ag_2O incorporation into cellulose and graphite fibers (Chen and Liu 2016). Glass ionomer cements have been developed with silver nanoparticles with very low concentrations of 10 µg Ag capsules, reducing biofilm by 99% (Porter et al. 2020). Synthesis of Ag_2O by reducing/stabilizing agents like starch, dextran, polyvinylpyrrolidone, and β-cyclodextrin has been effective in inhibiting biofilms of common pathogenic strains (Bryaskova et al. 2011; Habash et al. 2014; Mohanty et al. 2012). A unique mechanism of action of SNP has been reported by Shrivastava et al. (2007) which involves modulating of cellular signaling activity by dephosphorylating tyrosine residues of peptides, inhibiting bacterial growth. Tyrosine phosphorylation of proteins has been shown to be involved in capsular and extracellular production of polysaccharides in both Gram-negative and Gram-positive bacterial strains (Grangeasse et al. 2003). The superior antibacterial activity of SNP may be due to the inactivation/interference with this signaling pathway.

5.5.2 Copper Oxide Nanoparticles

CuO NPs have wide application in environmental settings where it has been used as biocide for more than half a century. Copper oxide exists in two forms: as (1) copper oxide (CuO with bandgap of 1.21–1.55 eV) and (2) cuprous oxide (Cu_2O with bandgap of 2.2–2.25 eV), like silver depends on release of metal ions to impart toxicity towards microbial cells. CuO is relatively a more stable MON with respect to its physical and chemical properties as well as cheap compared to silver and gold and blends well with most of the polymer matrices. In general, CuO NPs have been shown to develop as needles, nanoflowers, nanorods, and nanowires. CuO NP has been shown to possess broad-spectrum antibacterial activity with Cu_2O NP showing better antibacterial activity than CuO nanoparticles (Kumar et al. 2019). However, CuO have been shown to produce more hydroxyl radicals compared to Cu_2O nanoparticles (Meghana et al. 2015). CuO NP has also been shown to generate ROS with its electron donor nature. Copper oxide NP has been combined with antifungal agents like fluconazole for medication. Copper nanoparticles functionalized with acrylic monomers were found to be stabilized in the polymer backbone which exhibited better release rate of biocidal ions. Copper acrylate surfaces showed good antimicrobial activity which was not dependent on the concentration of the NP; viz., 1.0 wt% showed similar activity as 10% and 25%; however, higher loading resulted in enhanced release rates (Anyaogu et al. 2008).

Mallakpour et al. (2020) reviewed extensively about benign synthesis of CuO and applications of different CuO nanocomposites. Cuprous oxide nanocube on graphene oxide (0.5–2 nm) composite sheets synthesized by facile method was effective in inhibiting biofilms with low MIC values of 5.9, 2.9, and 2.929 μg/mL for *E. coli*, *P. aeruginosa*, and *B. subtilis*, respectively (Selim et al. 2020). Silver-copper-graphene oxide (Ag/Cu/GO) nanocomposites synthesized were effective against *Methylobacterium* sp., *Sphingomonas* sp., and *P. aeruginosa* at concentrations which were harmless to humans (Jang et al. 2020). Copper oxide-titanium dioxide nanocomposites synthesized by pulsed laser ablation technique were effective in inhibiting methicillin-resistant *S. aureus* and *P. aeruginosa* by damaging the cell membrane (Baig et al. 2020). Erci et al. (2020) demonstrated that biosynthesized CuO NP from leaf extracts also had higher cytotoxicity to mouse fibroblast L929, with antibiofilm activity against *S. aureus*. A limitation in the study was CuO NPs were effective only against Gram-positive bacteria. This again demonstrates the intrinsic mechanisms involved in action of NP vis-a-vis their synthesis methods and that cytotoxicity of biosynthesized NP also occurs.

5.5.3 Zinc Oxide Nanoparticles

Similar to copper, zinc oxide (a wide bandgap material 3.3 eV; binding energy of 60 meV) nanoparticles possess broad-spectrum antibacterial activity with less toxicity to humans as it is an essential trace element and is widely used in medicine, cosmetics, and wound healing (Antonijevic et al. 2019). Shape and size of ZnO NP influence the photocatalytic activity and hence extensive research into synthesis methods. Zinc oxide exists as wurtzite, cubic zinc, and cubic rock salt. Comparison of toxicity of different nanoparticles revealed high toxicity by $ZnO < CuO < TiO_2 < Co_3O_4$ in normal conditions and $ZnO < CuO < Co_3O_4 < TiO_2$ under light and dark conditions (Dasari et al. 2013). Mechanism of toxicity of ZnO nanoparticles has been through contact-mediated killing as well as release of toxic metal ions (Li et al. 2011a, b) resulting in generation of ROS, oxidative stress (Dwivedi et al. 2014), cell membrane damage, alteration of membrane permeability (Sirelkhatim et al. 2015), and inactivation of biomolecules and enzymes resulting in cell death (Li et al. 2012). ZnO-PAM.Nc (polyacrylamide) developed by in situ emulsion polymerization technique showed broad-spectrum activity against bacteria and fungi (Morsi et al. 2016). Comparison of activity of different NPs, viz., MgO, TiO_2, CuO, and CeO_2, with ZnO on inhibition of biofilms of *S. aureus* strain RN6390 revealed high activity with ZnO NP (>50%) (Jones et al. 2008).

Zinc oxide is soluble and releases zinc ions, and the net positive charge on ZnO NPs acts on proteases of bacterial cell membrane affecting metabolism. $Zinc^{2+}$ ions have been found to inhibit biofilm formation of the bacterium *Bacillus amyloliquefaciens* FZB42 by suppression of the response regulator SPo0F (Huang et al. 2020). Photoactivation of ZnO releases e- during its transition from VB to CB and results in the formation of hole (hb) (He et al. 2014). The hb reacts with electrons and hydroxyl ions from water resulting in release of oxygen. Singlet oxygen is also a strong oxidant, and the formed OH ions react with nucleic acid, amino acids, proteins, and lipids of bacterial cell membrane. Similar to copper, zinc NP has also shown to exhibit a size-dependent activity with maximum activity observed with smaller-sized particles (3–10 nm). However, wire and rod-shaped ZnO NP has been shown to exert higher activity than spherical nanoparticles. Bare ZnO NP has been shown to exhibit higher antibacterial activity compared to capped nanoparticles (Datta et al. 2012). Carbon-stabilized ZnO showed 100% inhibition of biofilm formation of *S. aureus* and *P. aeruginosa* with loading of carbon to influence photocatalytic activity (Janani et al. 2020).

Enhanced antibiofilm activity was observed with incorporation of Ag into ZnO (Lu et al. 2008) which is attributed to the high positive charge on electroneutral Ag particles adhered on ZnO. The strong electrostatic interaction caused by exchange of electrons between Ag and ZnO as well as between positively charged Ag and negatively charged bacterial cells has been attributed to the enhanced antibiofilm activity. Similar mechanism has been observed with gold (Au) NP incorporation with ZnO which showed good antibacterial activity against *S. aureus*, whereas reduced toxicity to mouse fibroblast cells was observed (Khan et al. 2018). Oxidative stress has been attributed to cell death with Au-ZnO nanocomposites. Changing of bandgap of Au-ZnO composites by UV irradiation increases electron transfer and increases ROS generation (He et al. 2013) leading to increased antimicrobial activity. Low concentration of 0.05 mg/mL of photoactivated Au-ZnO was lethal to *S. aureus* and *E. coli* compared to plain ZnO which required three times the concentration for similar effect (Li et al. 2011a). Combining Cu NP with ZnO decreased survival rate of *E. coli* (3.12%) compared to plain NP, viz., ZnO (58.02%) and Cu (76.12%) (Nithya et al. 2015), which has been attributed to the increase of mesoporous property.

Doping is another mechanism to increase antimicrobial activity which is cost-effective and alters the material property. Mn doping of ZnO (Khan et al. 2017), Fe doping of ZnO (Basith et al. 2014), Ag doping of ZnO (Dutta et al. 2010), and Cu and CuO doping of ZnO/polyaniline (Liang et al. 2012) have all shown good antimicrobial activity against different bacterial strains. Incorporation of organic molecules is another approach to increase AB property of ZnO. Chitosan with reactive OH- and –NH$_2$ groups binds with ZnO to form composites which have been used as bandage material (Kumar et al. 2012). Di(octyl)phosphinic acid-capped ZnO NP has been incorporated into silicone to reduce implant-related infection (Stefan et al. 2013). ZnO-chitosan with polyaniline and montmorillonite (Trivedi et al. 2014) and nano-ZnO bacterial cellulose (BC) (Dinca et al. 2018) have also shown good AB activity. ZnO-GO (Wang et al. 2014), ZnO-starch (Vigneshwaran et al. 2006), ZnO-SiO$_2$ (Barani and Hossein 2014), chitosan-ZnO (Malini 2015), PA6/ZnO (Erem et al. 2011), Ag-ZnO (Matai 2014), alginate-silica-ZnO (Ahmed et al. 2018), and ZnO-PMMA (Anzlovar et al. 2011) nanocomposites have all shown good antibacterial activity.

5.5.4 Titanium Dioxide Nanoparticles

Since the demonstration of water splitting ability of titanium dioxide (Fujishima and Honda 1972), TiO$_2$ has found applications in nanomedicine and nanobiotechnology, magnetic resonance imaging (Zeng et al. 2013); black TiO$_2$ as cancer photothermal therapeutic agents (Ren et al. 2015); photocatalytic, biocompatibility and low cytotoxicity (Mou et al. 2016); solar and electrochemical cells, wastewater treatment, food packaging technology, gas sensing, paints, cosmetics, paper production, hydrogen fuel generation, ink for printing, plastic manufacturing, self cleaning surfaces, antiseptics and antibacterial creams (Akakuru et al. 2020). Clearances by the US FDA for incorporation into domestic products like toothpastes, fillers in medicines – tablets and capsules – and dental pastes (Gomes et al. 2018; Skocaj et al. 2011) have increased TiO$_2$ applications. TiO$_2$ NP exists in nature in three forms, viz., rutile, anatase, and brookite, with maximum photocatalytic activity observed with the anatase form (Su et al. 2011; Pantaroto et al. 2018). Some of the features of TiO$_2$ like hydrophilicity, chemical inertness and stability, film transparency in the visible region (decrease in size from 200 to 10 nm and changes from opaque to transparency), and low oxygen absorption capability offering good corrosion resistance have made TiO$_2$ a more attractive material. In addition, TiO$_2$ has also been used as a coloring agent in paints (Manesh et al. 2018).

Titanium dioxide nanoparticles also exhibit a broad-spectrum antibacterial activity due to its photocatalytic properties which was first described by Matusunga et al. (1985). Titanium dioxide NP has been more effective against the MDR nosocomial pathogen *P. aeruginosa* isolated from endotracheal tract and bronchoalveolar regions at concentrations of 350 mg/mL by free radicals (Arora et al. 2015). TiO$_2$ NP has been effective against another nosocomial pathogen methicillin-resistant strain *S. aureus* which is known to form prolific biofilms on catheters causing bacteremia and pneumonia in the lungs (Jesline et al. 2015). Biofilms of these MDR strains produce EPS-containing nucleic acids and lipids which withstand the host immune system as well as decrease the penetrating power of antibiotics resulting in less susceptibility of biofilm cells.

TiO_2 films have shown excellent antibiofilm properties against the three oral pathogens – *Streptococcus sanguinis*, *Actinomyces naeslundii*, and *Fusobacterium nucleatum* – with 99% inhibition on surfaces (Pantaroto et al. 2018). TiO_2 has been extensively applied in polymer matrix, doped with other MONs to improve its photocatalytic and antimicrobial properties (Gupta and Tripathi 2011). Synthesis of TiO_2 nanorods using near-infrared (NIR) source (808 nm) activation has been demonstrated as an effective strategy for killing of single species biofilms using the principle of photothermal therapy and photodynamic therapy. The mechanism of action of TiO_2 nanoparticles is found to be inactivation of coenzyme A; DNA damage induced by hydroxyl radicals; oxidative stress induced in cell membranes; reactive oxygen species generation, hyperthermia; singlet oxygen production; affect iron homeostasis; affect expression of genes involved in growth (Zhang et al. 2021; Akakuru et al. 2020).

5.5.5 Control of Biofilm in Biomedical Settings and Implant Surfaces

A prerequisite for an implant material apart from its functionality is to resist bacterial adhesion/colonization to prevent pathogenesis. Even with technological advancements, currently infection of implantable biomaterials still results in significant patient morbidity/mortality (Shah et al. 2013). Biofilm-associated infection of implant devices was reported as early as 1972 (Johanson 2013). Antibiotic resistance in implant-related infections is still a major problem to be circumvented, and this has triggered the search for alternate new materials to prevent biofilm-/biomaterial-associated infection (BAI) (Saldarriaga 2010). Orthopedic implants, nonvalvular cardiovascular stents, and urinary catheters are all prone to BAI. Surface modification of biomaterials through coatings offers protection. Different coating approaches have been tried, viz., use of low surface energy hydrophobic coatings (Everaert et al. 1998), polymer brushes (Norde and Gage 2004), zwitterionic coatings (Cheng et al. 2008), and positively charged coatings (Gottenbos et al. 2001; Roosjen et al. 2006) and use of quaternary ammonium compounds (QAC) (Tiller et al. 2001).

Orthopedic implant materials generally involve polymethyl methacrylate (PMMA), titanium and stainless steel, hydroxyapatite for a range of applications like bone cement, fixation devices, screws, osseointegrated implants for limb prosthesis. Gadd et al. (2012) reported PMMA to be highly fouled polymers followed by stainless steel and then titanium. PMMA implants have been shown to develop biofilms causing acute, chronic, and delayed infections (Trampuz and Zimmerli 2006; Minelli et al. 2011). Even antibiotic (vancomycin)-loaded PMMA has been observed to harbor biofilms with PMMA beads recovered from patients harboring drug-resistant bacterial strains (Neut et al. 2001). Biofilm development on implant materials has been attributed to local bacterial flora along with genetic factors involved in attachment (Gadd et al. 2012). These findings have led to the development of concept of synthetic delivery vehicles for controlled and sustained release of antibiotics in implant materials. Titanium dioxide-nanostructured films have been effective in inhibiting biofilm formation *Listeria monocytogenes* a common oral/environmental pathogen by about 3 log reduction (Chorianopoulous et al. 2011).

Another osteoconductive material is hydroxyapatite (HAP), $(Ca_{10}(PO_4)_6(OH)_2)$, similar in composition to the teeth and bone. *Staphylococcus aureus* is the most widespread pathogen on implants forming biofilms with other groups like *Streptococci*, *E. coli*, *Pseudomonas* sp., and *Enterobacter* sp. Zinc has been demonstrated to increase osteogenesis by promoting osteoblast cell proliferation and differentiation (Mourino et al. 2011). Zinc oxide nanoparticles have been classified as *generally recognized as safe* by the US FDA, which is a broad-spectrum antimicrobial compound (Espitia et al. 2016). ZnO-loaded hydroxyapatite composites were evaluated for antibiofilm efficacy against *S. aureus* and *E. coli* which exhibited ~52–54% inhibition (Beyene and Ghosh 2019).

For over two decades, antibiotic-loaded carrier biodegradable polymers like poly(lactic-co-glycolic acid) (PLGA), poly(e-caprolactone) (PCL), and poly(DL-lactic acid) (PLA) (Shi et al. 2010) have been the workhorse for surface modification of implant materials and antimicrobial coatings. Modification of PCL polymers as fine fibers by electrospinning, microcapsules and microspheres to improve their ability to deliver antibiotics in a controlled and sustained manner; preserve the bioactivity of the antibiotic; and release of the total volume of the antibiotic yielding better results (Shi et al. 2010; Huang et al. 2006). Alteration to morphology of PMMA as PMMA beads showed better antibiotic release lasting in time scale of several weeks (Shi et al. 2010). To this effect, a coaxial spun collagen with PLA carrier gentamicin was able to release antibiotics over a 2-week period (Torres-Giner et al. 2011). Synthesis of novel quaternized chitosan derivate (hydroxypropyltrimethyl ammonium chloride chitosan, HACC) demonstrated strong antibiofilm properties against *S. epidermidis* (Tan et al. 2012). Similarly polyethylene oxide coatings developed on glass surfaces using silyl ether bonds were also effective in preventing adhesion of *S. epidermidis* (Roosjen et al. 2005).

Adsorption of cationic molecules on biomaterial surfaces offers favorable sites for bacterial colonization (Cheng et al. 2008). Hence surface modification of implant material with polyethylene glycol (PEG) has demonstrated extensive anti-adhesive property to biomaterials by inhibiting deposition of plasma proteins (Park et al. 1998). Further improvements to PEG coatings are zwitterionic coatings which possess both positive and negative charges with an overall neutral charge balance. Zwitterionic surfaces created by grafting silver nanoparticles into poly (sulfobetaine methacrylate) (pSBMA) and poly (carboxybetaine methacrylate) (pCBMA) polymer matrix demonstrated killing of bacteria upon contact and releasing dead bacteria (Hu et al. 2013). Further enhancement of killing and release of microorganisms upon contact (smart switching) was achieved by poly (N, N-dimethylN-(ethoxycarbonylmethyl)-N-[2'-(methacryloyloxy) ethyl] – ammonium bromide) (pCBMA-1 C2, cationic precursor) which were able to kill 99.8% of *E. coli* strains upon contact, and 98% of the cells were released by hydrolysis of cationic derivatives to nonfouling zwitterionic polymers (Cheng et al. 2008).

In comparison with MONs, magnesium fluoride (MgF_2) nanoparticle (25 nm)-coated surfaces were effective in inhibiting adhesion of *S. aureus* and *E. coli* strains. The nanoparticles were also reported to penetrate cell membranes and alter membrane potential, reactive oxygen species (ROS) generation, and binding to DNA (Lellouche et al. 2009) parallel to mechanisms reported for MONs. Antimicrobial property of fluorides is well known as well as their mechanistic activity on microbial cells. Metal fluoride complexes of aluminum and beryllium cations have been demonstrated to interact with F-ATPases and nitrogenase enzymes (Sturr and Marquis 1990), disrupt proton movement across the cell due to formation of HF (Guha-Chowdhury et al. 1997a), and inhibit enzymes like enolase involved in the glycolytic pathway by a complex form of F^- and Mg^{2+} (Guha-chowdhury et al. 1997b).

Novel antibiotic-free hydrogels have been developed as wound dressing materials (Zhong et al. 2020). Antibiofilm fabrics have been developed using small molecules and Au nanoparticles using sonochemical techniques which have displayed efficient biofilm inhibition of multidrug-resistant bacteria (Wang et al. 2020). Novel nano-enabled surfaces coated with ZnO or ZnO/Ag composites have exhibited enhanced antibiofilm activity against *S. aureus*, *E. coli*, and *C. albicans* (Rosenberg et al. 2020). Such surfaces find application in high touch surfaces in hospital and medical settings. Sputter coating of catheter materials with silver-copper films reduced biofilm formation by *P. aeruginosa* (McLean et al. 1993). Coating of silver on polyurethane catheters inhibited biofilm formation (Jansen et al. 1994). Similarly ion beam implantation of silver on catheters reduced their susceptibility to biofilm formation (Davenas et al. 2002). Smart polymer surfaces of polypropylene-grafted polyacrylic acid (PP-gPAAc) loaded with antibiotic vancomycin inhibited MRSA biofilm formation (Munoz-Munoz et al. 2009). A significant development is the self-sterilizing materials for intravenous catheters using NIR carbon dots (CD) exhibiting broad-spectrum antimicrobial activity (Mauro et al. 2020). Novel chitosan hydrogels incorporating Ag NP and antibiotic ampicillin have inhibited the formation of biofilms of *Enterococcus faecium* and *S. epidermidis* with a tenfold increased concentration required to inhibit beta-lactamase-positive *E. cloacae* (Lopez-Carrizales et al. 2020).

Magnesium oxide (MgO) and calcium oxide (CaO) nanoparticles comprising of essential elements of magnesium and calcium have extensive application in repair of dental tissues and in dental care (Milosevic 1991). Both these nanoparticles hydrolyze to their hydroxide forms in the presence of water and have strong antibacterial activity due to strong alkaline nature, generation of ROS, and adsorption to bacterial cell membranes. The high pH values of 10.4 generated by $Mg(OH)_2$ and 12.5 by $Ca(OH)_2$ have a cascading effect on microbial cells. This was demonstrated by Dong et al. (2010) wherein *E. coli* cells were not affected by pH of 10 in LB medium whereas in the presence of both these hydroxide particles tested individually at the same pH of 10, the organisms succumbed. High alkaline conditions disrupt structural proteins in bacterial cell membranes, and in contrast to other MONs, CaO and MgO do release Mg^{2+} and Ca^{2+} ions, which do not impart any toxicity. In fact low concentration of Mg^{2+} and Ca^{2+} ions promote biofilm. An important feature is that CaO NPs have been used as drug delivery agents and chemotherapeutic agents due to their unique structural and optical properties (Meghana et al. 2015). Aluminum oxide NP has been shown to inhibit EPS production and reduced biofilm formation by 50% of MDR *Acinetobacter baumannii* (Muzammil et al. 2020). Combination of selenium and iron oxide NP incorporated into chitosan coatings with pentasodium triphosphate as cross-linker yielded excellent antibiofilm activity as well as less toxicity to human dermal fibroblast cells (Li et al. 2020).

5.5.6 Control of Oral Biofilms by Polymer Nanocomposites

Biocompatibility is an issue in the use of nanoparticles in biomaterials and human medicine. Primitively silver nitrate was the most common disinfectant used in dental surgery for the prevention of dental caries. However, silver and titanium NPs have found wide applications in control of biofilms within the oral cavity and are now used as topically applied agents in dental materials (Hamouda 2012). Ag NP demonstrated strong antibacterial activity with 25-fold lower growths compared to the mouthwash chlorhexidine with little influence of silica and titanium dioxide nanoparticles on the oral pathogen *Streptococcus mutans* (Besinis et al. 2014). *S. mutans* and *S. lactobacilli* are known to colonize dentine, enamel and soft tissues, and prosthetic materials, producing acids resulting in dental carries and damage to the hard tissue. This has resulted in incorporating antimicrobial property to dental materials, viz., filling materials, cements, and sealants orthodontic adhesives (Ahn et al. 2009). Lee et al. (2008) demonstrated superior antimicrobial activity of low concentrations of silver-zinc-zeolite incorporation into polymethyl methacrylate (PMMA), used for tissue conditioners, acrylic resins, denture bases, and orthodontic appliances. Silver-zeolite, powdered zinc citrate/acetate, and titanium dioxide have all been incorporated into mouth rinses and toothpastes as whiteners and plaque control agents (Allaker 2010; Boldyryera et al. 2005; Giertsen 2004).

5.6 Mechanism of Antibacterial Action of MONs

The primary mechanism of action of MONs (Fig. 5.11) is by (1) adsorption to the cell membrane and mechanical damage of cell wall, (2) internalization into cytoplasm, (3) release of metal ions, (4) generation of reactive nitrogen species, (5) oxidative stress by ROS generation, and (6) binding with DNA (Ren et al. 2020). Each of these mechanisms may act independently or in synergy to bring about toxicity. MONs have been effective in penetrating thick and simple cell wall of Gram$^+$ (20–80 nm) as well as thin and multilayered (peptidoglycan and

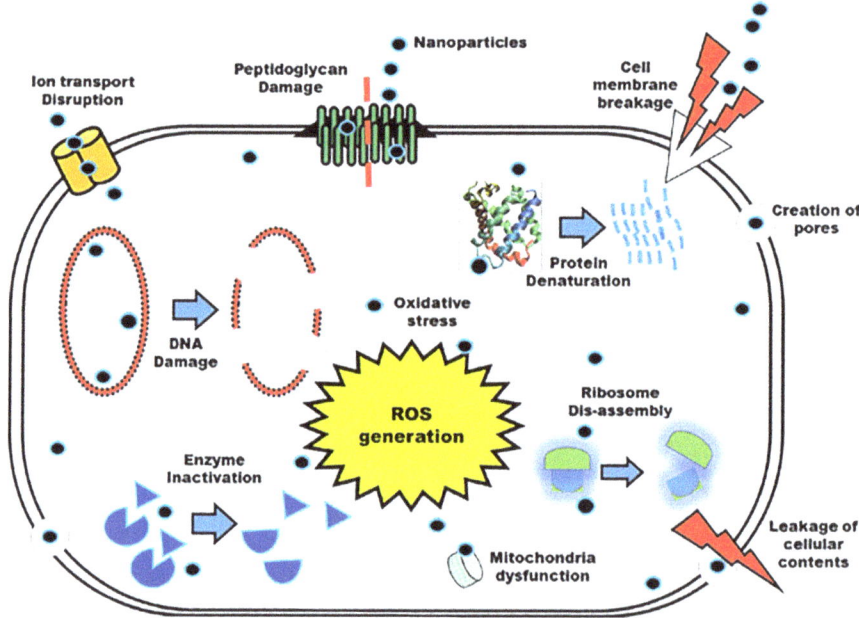

Fig. 5.11 Mechanism of action and toxicity of metal oxide nanoparticles to bacterial cells

lipopolysaccharides) cell walls of Gram$^-$ (10–15 nm) bacteria. Mechanism of adsorption of MONs on bacterial cell membrane and its lysis occurs by electrostatic interactions of positively charged MONs with the predominantly negatively charged bacterial (due to teichoic acids in Gram$^+$ and lipopolysaccharides in Gram$^-$) cell surface (Silhavy et al. 2020). MONs have other advantages; along with adsorption to cell membranes, they also release cationic ions (Beyth et al. 2015) into the media which get further internalized into the cytosol increasing their toxicity.

Bacterial cell membrane provides structural integrity apart from performing functions like transport of solutes, ions, and stimuli and transduction of signals (Gold et al. 2018). Lysis of cell membrane results in leaking of intracellular water from cytosol which actively triggers the proton efflux pumps and electron transport system. Imbalances in ionic composition due to cell lysis impair respiration and gradually result in cell death (Pelgrift and Friedman 2013). Metal ions released from CuO, Ag$_2$O, TiO$_2$, and ZnO NP have all been found to bind to thiol groups (-SH), amino groups (-NH), carboxylic groups (-COOH) of cell wall proteins, change in membrane permeability, degradation of lipopolysacchrides, denaturation of proteins, destruction of lipids, cytochrome enzymes, DNA in the cytosol causing homeostatic imbalances. Some bacterial strains have exhibited tolerance to high levels of metal ions like Cu^{++} which was demonstrated by Padmavathi et al. (2017) for a marine strain, *Staphylococcus lentus*, which was not inhibited in the presence of 1000 μg/mL of Cu. Interestingly, concentration of 100 μg/mL of copper ions enhanced biofilm formation, compared to controls. The increased metal ion tolerance of these bacteria can be attributed to Cu ion efflux pumps, repression of genes, adaptation to Cu ion stress, and plasmid-mediated resistance. Alternatively Cu NPs have also been demonstrated to inhibit antibiotic efflux pumps imparting resistance to bacteria in biofilms. Concentrations as low as 0.065 mM NP showed efflux inhibition of wild-type *S. aureus* and *P. aeruginosa* with less inhibition of efflux pumps in MRSA drug-resistant strain of *S. aureus* (Christena et al. 2015).

In general exogenous events (Bogdan et al. 2015) like binding of MONs to cell membranes and endogenous influx of metal ions have both been associated with the production of reactive oxygen species (ROS) by a Fenton reaction (Sperandio et al. 2013) attributing to antimicrobial activity (Ezraty et al. 2017) and no respite from resistance mechanism (Rout et al. 2017). ROS moieties include hydrogen peroxide (H_2O_2), free radicals such as hydroxyl (OH) and singlet oxygen (O_2), and superoxide ions (O_2^-) among which hydroxyl and singlet oxygen are the most toxic and not neutralized by intracellular enzymes (Ezraty et al. 2017). Hydroxyl radicals with a high oxidation potential of 2.8 V act on proteins, carbohydrates, lipids, and nucleic acids (Stanic and Tanaskovic 2020). Photosensitive MONs like TiO_2, ZnO, CuO, SiO_2, MgO, and Fe_2O_3 have been demonstrated to produce ROS by photo irradiation and UV spectrum absorption due to recombination of electron-hole pairs (Chen et al. 2014, 2018). The mechanism of photocatalytic action is by excitation of electrons from the valence band (VB) to the conduction band (CB) to produce photoexcited electrons (e^-) and an electronic hole (h^+) resulting in an electronic hole pair (e^-/h^+). Smaller-sized particles have increased bandgap energies with enhanced redox potential with photogenerated electrons and holes resulting in enhanced antibacterial activity compared to larger-sized particles.

ROS moieties impact cell membrane, protein, DNA, and electron transport chain and cause oxidative damage to membrane lipids, viz., polyunsaturated fatty acids and phospholipids of membranes (lipid peroxidation). Lipid peroxidation occurs by removal of hydrogen atom from the lipid by the hydroxyl radical resulting in the formation of a lipid radical. The lipid radical reacts with oxygen to form lipid peroxyl radical. The peroxyl radical interacts with biomolecules to form lipid hydroperoxides which in the presence of Fe^{2+} ions results in the formation of alkoxy radicals (Stanic and Tanaskovic 2020). Lipid peroxidation increases membrane fluidity and damages cell integrity. ROS moieties act on membrane proteins oxidizing amino acids, impairing membrane permeability leading to cell death (Cabiscol et al. 2000). ROS acts on nucleic acids by oxidation of double-stranded DNA breaking their backbone and adduction of base pairs and sugar groups (Kim et al. 2013).

5.7 Conclusions

The surge in synthesis and testing of different metal oxide nanoparticles (MONs) arose due to inactivity of conventional antibiotics toward MDR strains and need for combating biofilms in different applications. Significant improvements have been achieved in synthesis protocols, control over morphology, techniques to increase solubility, and dispersal and understanding of mechanism of action on microbes. Nanoparticle size and dispersion are important parameters influencing activity. The primary reason behind the use of MONs is their nano configuration which aids in mechanical damage to cells, release of ions imparting toxicity, and oxidative stress. Further Cu, Ag, and Zn ions at trace levels are required for organisms and are toxic at higher concentrations. Silver, zinc, and titanium dioxide NPs have transformed from in vitro experimentation to in situ application in biomedical field as implant coatings, antibiotic carriers, and wound dressing agents. Silver has been extensively used in cosmetics and equipment/high touch surfaces in public hygiene due to its certain degree of biocompatibility. CuO finds wide application for biofilm control in industrial settings like reverse osmosis membranes and antifouling coatings. In comparison, TiO_2 finds application as fillers in paint coatings for exterior protection. Fe_2O_3 has applications in biomedical imaging with MgO and CaO finding application in dental settings. However, the cytotoxicity of MONs is a cause of concern which hinders its wide application. However encapsulation and incorporation into polymer matrices have been used as a delivery system wherein the leaching of toxic metal ions at the polymer surface has been exploited and achieved success. However, release of metal ions from NP and depletion of NP with time in polymer matrix render the surfaces benign to microbial colonization. MONs have become a success in marine antifouling coatings, desalination membranes, and antibacterial coatings for public hygiene; however, their in situ use in human medicine is hindered by their residual toxicity caused by long-term accumulation (chronic toxicity) in the skin and internal organs. In general, qualification of new drugs/modification of existing certified devices/disinfectant formulations/antimicrobial nanocomposites depends on two criteria, viz., 1) demonstrating higher efficacy than existing technology for a particular application and 2) that the new product is safer. It is here that research on MONs needs to focus on developing MONs with low MIC values (1–5 μg/L) to minimize environmental toxicity and cytotoxicity. Currently MIC concentration range from 10 μg/mL to 300 μg/mL is reported in the literature for various NPs. To achieve this, synthesis methods should be fine-tuned to obtain NP in the size range of 1–10 nm. Biosynthesized NP in this range is a boon. Lower sizes have been achieved by quantum dots which have immediately found application in imaging and drug delivery. Next clinical safety-related studies should be initiated with potential MONs for successful realization as products.

References

Abioye OP, Loto CA, Fayomi OSI (2019) Evaluation of anti-biofouling progresses in marine application. J Bio Tribo-Corrosion 5:22. https://doi.org/10.1007/s40735-018-0213-5

Abo-zeid Y, Williams GR (2019) The potential anti-infective applications of metal oxide nanoparticles: a systematic review. WIREs Nanomed Nanobiotechnol 2019:e1592019. https://doi.org/10.1002/wnan.1592

Ahmadabadi HY, Yua K, Kizhakkedathu JN (2020) Surface modification approaches for prevention of implant associated infections. Colloids Surf B: Biointerfaces 193:111–116. https://doi.org/10.1016/j.colsurfb.2020.111116

Ahmed S, Diaba MA, Abou-Zeida RE, Aljohani HA, Kamel Rizq Shoueir KR (2018) Crosslinked alginate/silica/zinc oxide nanocomposite: a sustainable material with antibacterial properties. Compos Commun 7:7–11

Ahn SJ, Lee SJ, Kook JK, Lim BS (2009) Experimental antimicrobial orthodontic adhesives using nanofillers and silver nanoparticles. Dent Mater 25(2):206–213. https://doi.org/10.1016/j.dental.2008.06.002

Akakuru OU, Iqbal ZM, Wu A (2020) TiO_2 nanoparticles: properties and applications. In: Wu A, Ren W (eds) Nanoparticles: applications in nanobiotechnology and nanomedicine, 1st edn. Wiley-VCH Verlag GmbH & Co, pp 1–66

Al-Jabri AA, Alenzi FQ (2009) Vaccines, virucides and drugs against HIV/AIDS: hopes and optimisms for the future. Open AIDS J 3:1–3. https://doi.org/10.2174/1874613600903010001

Allaker RP (2010) The use of nanoparticles to control oral biofilms. J Dent Res 89(11):1175–1186. https://doi.org/10.1177/0022034510377794

Antonijevic MN, Drazic B, Antic J, Stankovic S, Tanaskovic S (2019) New mixed-ligand Ni(II) and Zn(II) macrocyclic complexes with bridged bicyclo-[2,2,1]-hept-5-en-endo-2,3-cis-dicarboxylate: synthesis, characterization, antimicrobial and cytotoxic activity. J Serb Chem Soc 84(9):28. https://doi.org/10.2298/JSC181216028A

Anyaogu KC, Fedorov AV, Neckers DC (2008) Synthesis, characterization, and antifouling potential of functionalized copper nanoparticles. Langmuir 24:4340–4346. https://doi.org/10.1021/la800102f

Anžlovar A, Kogej K, Orel ZC, Žigon M (2011) Polyol mediated nano size zinc oxide and nanocomposites with poly (methyl methacrylate). Express Polym Lett 5(7):604–619. https://doi.org/10.3144/expresspolymlett.2011.59

Arora B, Murar M, Dhumale V (2015) Antimicrobial potential of TiO_2 nanoparticles against MDR *Pseudomonas aeruginosa*. J Exp Nanosci 10:819–827. https://doi.org/10.1080/17458080.2014.902544

Augustine R, Hasan A (2020) Emerging applications of biocompatible phytosynthesized metal/metal oxide nanoparticles in healthcare. J Drug Deliv Sci Technol 56:101516. https://doi.org/10.1016/j.jddst.2020.101516

Azam A, Ahmed AS, Oves M, Khan MS, Habib SS, Memic A (2012) Antimicrobial activity of metal oxide ´nanoparticles against gram-positive and gram-negative bacteria: a comparative study. Int J Nanomedicine 7:6003–6009. https://doi.org/10.2147/IJN.S35347

Aziz N, Fatma T, Varma A, Prasad R (2014) Biogenic synthesis of silver nanoparticles using *Scenedesmus abundan*s and evaluation of their antibacterial activity. Journal of Nanoparticles, Article ID 689419, https://doi.org/10.1155/2014/689419

Aziz N, Pandey R, Barman I, Prasad R (2016) Leveraging the attributes of *Mucor hiemalis*-derived silver nanoparticles for a synergistic broad-spectrum antimicrobial platform. Front Microbiol 7:1984. https://doi.org/10.3389/fmicb.2016.01984

Aziz N, Faraz M, Sherwani MA, Fatma T, Prasad R (2019) Illuminating the anticancerous efficacy of a new fungal chassis for silver nanoparticle synthesis. Front Chem 7:65. https://doi.org/10.3389/fchem.2019.00065

Aziz N, Faraz M, Pandey R, Sakir M, Fatma T, Varma A, Barman I, Prasad R (2015) Facile algae-derived route to biogenic silver nanoparticles: Synthesis, antibacterial and photocatalytic properties. Langmuir 31:11605–11612. https://doi.org/10.1021/acs.langmuir.5b03081

Babu PJ, Doble M, Raichur AM (2018) Silver oxide nanoparticles embedded silk fibroin spuns: microwave mediated preparation, characterization and their synergistic wound healing and antibacterial activity. J Colloid Interface Sci 513:62–71. https://doi.org/10.1016/j.jcis.2017.11.001

Baig U, Ansari MA, Gondal MA, Akhtar S, Khane FA, Falatha WS (2020) Single step production of high-purity copper oxide-titanium dioxide nanocomposites and their effective antibacterial and anti-biofilm activity against drug-resistant bacteria. Mater Sci Eng C 113:110992. https://doi.org/10.1016/j.msec.2020.110992

Bankier C, Matharu RK, Cheong YK, Ren GG, Cloutman-Green E, Ciric L (2019) Synergistic antibacterial effects of metallic nanoparticle combinations. Sci Rep 9:16074. https://doi.org/10.1038/s41598-019-52473-2

Barani H, Hossein (2014) Preparation of antibacterial coating based on in situ synthesis of ZnO/SiO$_2$ hybrid nanocomposite on cotton fabric. Appl Surf Sci 320:429–434. https://doi.org/10.1016/j.apsusc.2014.09.102

Basith NM, Vijaya JJ, Kennedy LJ, Bououdina M, Jenefar S, Kaviyarasan V (2014) Co-doped ZnO nanoparticles: structural, morphological, optical, magnetic and antibacterial studies. J Mater Sci Technol 30(11):1108–1117. https://doi.org/10.1016/j.jmst.2014.07.013

Basnet P, Anderson E, Zhao Y (2019) Hybrid Cu$_x$O–TiO$_2$ Nanopowders prepared by ball milling for solar energy conversion and visible light induced wastewater treatment. ACS Appl Nano Mater 2(4):2446–2455. https://doi.org/10.1021/acsanm.9b00325

Besinis A, Peralta TC, Handy RD (2014) The antibacterial effects of silver, titanium dioxide and silica dioxide nanoparticles compared to the dental disinfectant chlorhexidine on *Streptococcus mutans* using a suite of bioassays. Nanotoxicology 8(1):1–16. https://doi.org/10.3109/17435390.2012.742935

Beyene Z, Ghosh R (2019) Effect of zinc oxide addition on antimicrobial and antibiofilm activity of hydroxyapatite: a potential nanocomposite for biomedical applications. Mater Today Commun 21:100612. https://doi.org/10.1016/j.mtcomm.2019.100612

Beyth N, Houri-Haddad Y, Domb A, Khan W, Hazan R, Based E (2015) Alternative antimicrobial approach: Nano-antimicrobial materials. Complem Alterna Med 246012. https://doi.org/10.1155/2015/246012

Bhuyan T, Mishra K, Khanuja M, Prasad R, Varma A (2015) Biosynthesis of zinc oxide nanoparticles from *Azadirachta indica* for antibacterial and photocatalytic applications. Mater Sci Semicond Process 32: 55–61

Bogdan J, Zarzynska J, Plawinska-Czarnak J (2015) Comparison of infectious agent's susceptibility to photocatalytic effects of nanosized titanium and zinc oxides: a practical approach. Nanoscale Res Lett 10:1023. https://doi.org/10.1186/s11671-015-1023-z

Boldyreva H, Umeda N, Plaskin OA, Takeda Y, Kishimoto N (2005) High fluency implantation of negative metal ions into polymers for surface modification and nanoparticle formation. Surf Coat Tech 196:373–377. https://doi.org/10.1016/j.surfcoat.2004.08.159

Bondarenko O, Juganson K, Ivask A, Kasemets K, Mortimer M, Kahru A (2013) Toxicity of Ag, CuO and ZnO nanoparticles to selected environmentally relevant test organisms and mammalian cells in vitro: a critical review. Arch Toxicol 87:1181–1200. https://doi.org/10.1007/s00204-013-1079-4

Bonu V, Sahu BK, Das A, Amirthapandian S, Dhara S, Barshilia HC (2019) Sub-wavelength waveguide properties of different morphological 1 D and surface functionalized SnO$_2$ nanostructures. Beilstein J Nanotechnol 10:379. https://doi.org/10.3762/bjnano.10.37

Brayner R, Ferrari-Iliou R, Brivois N, Djediat S, Benedetti MF, Fievet F (2006) Toxicological impact studies based on *Escherichia coli* bacteria in ultrafine ZnO nanoparticles colloidal medium. Nano Lett 6(4):866–870. https://doi.org/10.1021/nl052326h

Bryaskova R, Pencheva D, Nikolov S, Kantardjiev T (2011) Synthesis and comparative study on the antimicrobial activity of hybrid materials based on silver nanoparticles (AgNps) stabilized by polyvi- nylpyrrolidone (PVP). J Chem Biol 4:185. https://doi.org/10.1007/s12154-011-0063-9

Byrappa K, Haber M (2001) Handbook of hydrothermal technology. William Andrews Publishing, Noyes Publications, Park Ridge, NJ

Cabiscol E, Tamarit J, Ros J (2000) Oxidative stress in bacteria and protein damage by reactive oxygen species. Int Microbiol 3:38. http://hdl.handle.net/10459.1/56751

Center for Disease Control and Prevention (2013) Antibiotic resistance threats in the United States, (2013); CS239559-B. Centres for Disease Control and Prevention, US Department of Health and Human Services, Washington, DC

Cha SH, Hong J, McGuffie M, Yeom B, VanEpps JS, Kotov NA (2015) Shape-dependent biomimetic inhibition of enzyme by nanoparticles and their antibacterial activity. ACS Nano 9(9):9097–9105. https://doi.org/10.1021/acsnano.5b03247

Chalhoub H, Sáenz Y, Nichols WW, Tulkens PM, Van Bambeke F (2018) Loss of activity of ceftazidime-avibactam due to MexAB-OprM efflux and overproduction of AmpC cephalosporinase in *Pseudomonas aeruginosa* isolated from patients suffering from cystic fibrosis. Int J Antimicrob Agents 52:697e701. https://doi.org/10.1016/j.ijantimicag.2018.07.027

Chaloupka K, Malam Y, Seifalian AS (2010) Nanosilver as a new generation of nanoproduct in biomedical applications. Trends Biotechnol 28:580–588. https://doi.org/10.1016/j.2010.07.006

Chausali N, Jyoti Saxena J, Prasad R (2022) Recent trends in nanotechnology applications of bio-based packaging. Journal of Agriculture and Food Research, https://doi.org/10.1016/j.jafr.2021.100257

Chen X, Gonsalves KE (1997) Synthesis and properties of an aluminium nitride/polyimide nanocomposite prepared by a nonaqueous suspension process. 12(5):1274–1286. https://doi.org/10.1557/JMR.1997.0176

Chen H, Liu W (2016) Cellulose-based photocatalytic paper with Ag_2O nanoparticles loaded on graphite fibers. J Bioresour Bioprod 1:192198

Chen S, Guo Y, Zhong H, Chen S, Li J, Ge Z, Tang J (2014) Synergistic antibacterial mechanism and coating application of copper/titanium dioxide nanoparticles. Chem Eng J 256(15):238–246. https://doi.org/10.1016/j.cej.2014.07.006

Chen Y, Chen M, Zhang Y, Lee JH, Escajadillo T, Gong H, Fang RH, Gao W, Nizet V, Zhang L (2018) Antivirulence therapy: Broad-spectrum neutralization of pore-forming toxins with human erythrocyte membrane-coated nanosponges. Adv Healthc Mater 7:1701366. https://doi.org/10.1002/adhm.201870049

Cheng G, Xue H, Zhang Z, Chen S, Jiang S (2008) A switchable biocompatible polymer surface with self-sterilizing and nonfouling capabilities. Angenwadte Chemie 47(46):8831–8834. https://doi.org/10.1002/ange.200803570

Chorianopoulos NG, Tsoukleris DS, Panagou EZ, Falaras P, Nychas GJE (2011) Use of titanium dioxide (TiO_2) photocatalysts as alternative means for *Listeria monocytogenes* biofilm disinfection in food processing. Food Microbiol 28:164e170. https://doi.org/10.1016/j.fm.2010.07.025

Christena LR, Mangalagowri V, Pradheeb P, Ahmed KBA, BIS S, Vidyalakshmi M, Anbazhagan V, Subramanian NS (2015) Copper nanoparticles as an efflux pump inhibitor to tackle drug resistant bacteria. RSC Adv 5:12899–12909. https://doi.org/10.1039/c4ra15382k

Costerton JW, Geesey GG, Cheng KJ (1978) How bacteria stick. Sci Am 238:86–95. https://doi.org/10.1038/scientificamerican0178-86

Costerton JW, Stewart PS, Greenberg EP (1999) Bacterial biofilms: a common cause of persistent infections. Science 284:1318–1322. https://doi.org/10.1126/science.284.5418.1318

D'Souza L, Richards R (2007) Synthesis of metal-oxide nanoparticles: liquid-solid transformations. In: Rodríguez JA, Fernández-García M (eds) Synthesis, properties and applications of oxide nanoparticles. Wiley, New Jersey, pp 81–117

Dalmaschio CJ, Leite ER (2012) Detachment induced by Rayleigh-instability in metal oxide nanorods: insights from TiO_2. Cryst Growth Des 12:3668–3674. https://doi.org/10.1021/cg300473u

Daniel GS, Montoro LA, Beltrán A, Conti TG, Silva ROD, Andrés J, Leite ER, Ramirez AJ (2011) Anomalous oriented attachment growth behavior on SnO_2 nanocrystals. Chem Commun 47:3117. https://doi.org/10.1039/C0CC04570E

Das S, Jayaraman V (2014) SnO_2: a comprehensive review on structures and gas sensors. Prog Mater Sci 66:112–255. https://doi.org/10.1016/j.pmatsci.2014.06.003

Das A, Panda D (2019) SnO_2 tailored by CuO for improved CH_4 sensing at low temperature. Phys Stat Solid: B 256:1800296. https://doi.org/10.1002/pssb.201800296

Dasari TP, Pathakoti K, Hwang HM (2013) Determination of the mechanism of photoinduced toxicity of selected metal oxide nanoparticles (ZnO, CuO, Co_3O_4 and TiO_2) to *E. coli* bacteria. J Environ Sci 25(5):882–888. https://doi.org/10.1016/S1001-0742(12)60152-1

Davenas J, Thevenard P, Philippe F, Arnaud MN (2002) Surface implantation treatments to prevent infection complications in short term devices. Biomol Eng 19(2-6):263–268. https://doi.org/10.1016/S1389-0344(02)00037-0

Davies D (2003) Understanding biofilm resistance to antibacterial agents. Nat Rev Drug Dis 2:114–122. https://doi.org/10.1038/nrd1008

Davies DG (2011) Biofilm dispersion. In: Flemming HC, Wingender J, Szewzyk U (eds) Biofilm highlights, vol 5. Springer Berlin Heidelberg, Berlin, Heidelberg, pp 1–28. https://doi.org/10.1007/978-3-642-19940-0_1

de Beer D, Stoodley P, Roe F, Lewandowski Z (1994) Effects of biofilm structures on oxygen distribution and mass transport. Biotechnol Bioeng 43:1131–1138. https://doi.org/10.1002/bit.260431118

Deshmukh R, Niederberger MC (2017) Mechanistic aspects in the formation, growth and surface functionalization of metal oxide nanoparticles in organic solvents. Chem Eur J 23(36):8542–8570. https://doi.org/10.1002/chem.201605957

Dhanalekshmi KI, Nguyen VY, Magesan P (2020) Nanosilver loaded oxide nanoparticles for antibacterial application. Smart Nanocontainers. https://doi.org/10.1016/B978-0-12-816770-0.00026-5

Dincă V, Mocanu A, Isopencu G, Busuioc C, Brajnicov S, Vlad A, Icriverzi M (2018) Biocompatible pure ZnO nanoparticles-3D bacterial cellulose biointerfaces with antibacterial properties. Arab J Chem 13(1):3521–3533. https://doi.org/10.1016/j.arabjc.2018.12.003

Dong C, Cairney J, Sun Q, Maddan OL, He G, Deng Y (2010) Investigation of Mg(OH)$_2$ nanoparticles as an antibacterial agent. J Nanopart Res 12:2101–2109. https://doi.org/10.1007/s11051-009-9769-9

Donlan RM, Costerton JW (2002) Biofilms: survival mechanisms of clinically relevant microorganisms. Clin Microbiol Rev 15:167–193. https://doi.org/10.1128/CMR.15.2.167-193.2002

Dutta RK, Sharma P, Bhargava R, Kumar N, Pandey A (2010) Differential susceptibility of *Escherichia coli* cells toward transition metal-doped and matrix embedded ZnO nanoparticles. J Phys Chem B 29:114(16):5594–5599. https://doi.org/10.1021/jp1004488

Dutta RK, Nenavathu BP, Gangishetty MK, Reddy AVR (2012) Studies on antibacterial activity of ZnO nanoparticles by ROS induced lipid peroxidation. Colloids Surf B Biointerfaces 94:143–150. https://doi.org/10.1016/j.colsurfb.2012.01.046

Dwivedi S, Wahab R, Khan F, Mishra YK, Musarrat J, Al-Khedhairy AA (2014) Reactive oxygen species mediated bacterial biofilm inhibition via zinc oxide nanoparticles and their statistical determination. PLoS One 9(11):e111289. https://doi.org/10.1371/journal.pone.0111289

Enas ND, Hjiri M, Abdel-wahab M.Sh, Alonizan NH, Mir LE, Aida MS (2020) Antibacterial activity of In-doped ZnO nanoparticles. Inorg Chem Commun 122:108281. https://doi.org/10.1016/j.inoche.2020.108281

Ennas G, Musinu A, Piccaluga G, Zedda D, Gatteschi D, Sangregorio C (1998) Characterization of iron oxide nanoparticles in an Fe2O3–SiO2 composite prepared by a sol-gel method. Chem Mater 10:495–502. https://doi.org/10.1021/cm970400u

Erci F, Cakir-Koc R, Yontem M, Torlak E (2020) Synthesis of biologically active copper oxide nanoparticles as promising novel antibacterial-antibiofilm agents. Prep Biochem Biotechnol. https://doi.org/10.1080/10826068.2019.1711393

Erem AD, Ozcan G, Skrifvars M (2011) Antibacterial activity of PA6/ZnO nanocomposite fibers. Text Res J 81(16):1638–1646. https://doi.org/10.1177/0040517511407380

Espitia PJP, Soares NFF, Coimbra JSR, Andrade NJ, Cruz RS, Medeiros EAA (2012) Zinc oxide nanoparticles: synthesis, antimicrobial activity and food packaging applications. Food Bioprocess Technol 5:1447–1464. https://doi.org/10.1007/s11947-012-0797-6

Espitia P, Otoni C, Soares N (2016) Zinc oxide nanoparticles for food packaging applications. Antimicrob Food Pack:425–431. https://doi.org/10.1016/B978-0-12-800723-5.00034-6

Everaert EPJM, Van der Mei HC, Busscher HJ (1998) Adhesion of yeasts and bacteria to fluoro-alkylsiloxane layers chemisorbed on silicone rubber. Coll Surfaces B: Biointerfaces 10(4):179–190. https://doi.org/10.1016/S0927-7765(98)00003-4

Ezraty B, Gennaris A, Barras F, Collet JF (2017) Oxidative stress, protein damage and repair in bacteria. Nat Rev Microbiol 15(7):385–396. https://doi.org/10.1038/nrmicro.2017.26

Fernández-García M, Martínez-Arias A, Hanson JC, Rodriguez JA (2004) Nanostructured oxides in chemistry: characterization and properties. Chem Rev 104:4063. https://doi.org/10.1021/cr030032f

Feynman RP (1959) There's plenty of room at the bottom. Eng Sci:22e36

Fleming D, Rumbaugh K (2017) Approaches to dispersing medical biofilms. Microorganisms 5:15. https://doi.org/10.3390/microorganisms5020015

Fujishima A, Honda K (1972) Electrochemical photolysis of water at a semiconductor electrode. Nature 238:37–38. https://doi.org/10.1038/238037a0

Gad GFM, Aziz AAA, Ibrahem RA (2012) In-vitro adhesion of *Staphylococcus* spp. to certain orthopedic biomaterials and expression of adhesion genes. J Appl Pharm Sci 02(06):145–149. https://doi.org/10.7324/JAPS.2012.2634

Giertsen E (2004) Effects of mouth rinses with triclosan, zinc ions, copolymer, and sodium lauryl sulphate combined with fluoride on acid formation by dental plaque in vivo. Caries Res 38:430–435. https://doi.org/10.1159/000079623

Gold K, Slay B, Knackstedt M, Gaharwar V (2018) Antimicrobial activity of metal and metal-oxide based nanoparticles. Adv Therap 1:1700033 r. https://doi.org/10.1002/adtp.201700033

Gomes SIL, Roca CP, Kammer FVD, Scott-Fordsman JJ, Amorim MJB (2018) Mechanisms of (photo) toxicity of TiO_2 nanomaterials (NM103, NM104, NM105): using high-throughput gene expression in *Enchytraeus crypticus*. Nanoscale 10:21960–21970. https://doi.org/10.1039/C8NR03251C

Gottenbos B, Grijpma DW, Van der Mei HC, Feijen J, Busscher HJ (2001) Antimicrobial effects of positively charged surfaces on adhering Gram-positive and Gram-negative bacteria. J Antimicrob Chemother 48:7–13. https://doi.org/10.1093/jac/48.1.7

Grangeasse C, Obadia B, Mijakovic I, Deutscher J, Cozzone AJ, Doublet P (2003) Autophosphorylation of the *Escherichia coli* protein kinase Wzc regulates tyrosine phosphorylation of Ugd, a UDP-glucose dehydrogenase 1. J Biol Chem 41(39323):39329

Guan R, Zhai H, Sun D, Zhang J, Wang Y, Li Y (2019) Effects of Ag doping content and dispersion on the photocatalytic and antibacterial properties in ZnO nanoparticles. Chem Res Chin Univ 35:271276. https://doi.org/10.1007/s40242-019-8275-6

Guha-Chowdhury N, Iwami Y, Yamada T (1997a) Effect of low levels of fluoride on proton excretion and intracellular pH in glycolysing *Streptococcal* cells under strictly anaerobic conditions. Caries Res 31:373–378. https://doi.org/10.1159/000262421

Guha-Chowdhury N, Clark AG, Sissons CH (1997b) Inhibition of purified enolases from oral bacteria by fluoride. Oral Microbiol Immunol 12:91–97. https://doi.org/10.1111/j.1399-302X.1997.tb00623.x

Guo T, Yao M, Lin YH, Nan C (2015) A comprehensive review on synthesis methods for transition−metal oxides nanostructures. Cryst Eng Comm 17:3551–3585. https://doi.org/10.1039/C5CE00034C

Gupta SM, Tripathi M (2011) A review of TiO_2 nanoparticles. Chin Sci Bull 56:1639–1657. https://doi.org/10.1007/s11434-011-4476-1

Gupta P, Sarkar S, Das B, Bhattacharjee S, Tribedi P (2016) Biofilm, pathogenesis and prevention—a journey to break the wall: a review. Arch Microbiol 198:1–15. https://doi.org/10.1007/s00203-015-1148-6

Habash MB, Park AJ, Vis EC, Harris RJ, Khursigara CM (2014) Synergy of silver nanoparticles and aztreonam against *Pseudomonas aeruginosa* PAO1 biofilms. Antimicrob Agents Chemother 58:5818–5830. https://doi.org/10.1128/AAC.03170-14

Hall CW, Mah T (2017) Molecular mechanisms of biofilm-based antibiotic resistance and tolerance in pathogenic bacteria. FEMS Microbiol Rev fux010 41:276–301. https://doi.org/10.1093/femsre/fux010

Hall-Stoodley L, Costerton JW, Stoodley P (2004) Bacterial biofilms: from the natural environment to infectious diseases. Nat Rev Microbiol 2:95–108. https://doi.org/10.1038/nrmicro821

Hamouda IM (2012) Current perspectives of nanoparticles in medical and dental biomaterials. J Biomed Res 26(3):143–151. https://doi.org/10.7555/JBR.26.20120027

He (2013) Photogenerated charge carriers and reactive oxygen species in ZnO/Au hybrid nanostructures with enhanced photocatalytic and antibacterial activity. J Am Chem Soc 136(2):750–757. https://doi.org/10.1021/ja410800y

He W, Jia H, Zheng Z, Wu H, Wamer W, Yin J, Kim H, Zheng J (2014) Unraveling the enhanced photocatalytic activity and phototoxicity of zNO/metal hybrid nanostructures from generation of reactive oxygen species and charge carriers. ACS Appl Mater Interfaces 6(17):15527–15535. https://doi.org/10.1021/am5043005

Hu R, Li G, Jiang Y, Zhang Y, Zou JJ, Wang L, Zhang XW (2013) Silver zwitterion organic-inorganic nanocomposite with antimicrobial and antiadhesive capabilities. Langmuir 29(11):3773–3779. https://doi.org/10.1021/la304708b

Huang ZM, He CL, Yang A, Zhang Y, Han X, Yin J, Wu Q (2006) Encapsulating drugs in biodegradable ultrafine fibers through co-axial electro spinning. J Biomed Mater Res 77A:169–179. https://doi.org/10.1002/jbm.a.30564

Huang Z, Wu L, Li X, Ma L, Borriss R, Gao X (2020) Zn(II) suppresses biofilm formation in *Bacillus amyloliquefaciens* by inactivation of the Mn(II) uptake. Environ Microbiol 22(4):1547–1558. https://doi.org/10.1111/1462-2920.14859

Janani B, Syed A, Raju LL, Al Harthi HF, Thomas AM, Das A, Khan SS (2020) Synthesis of carbon stabilized zinc oxide nanoparticles and evaluation of its photocatalytic, antibacterial and anti-biofilm activities. J Inorg Organomet Polym Mater 30:2279–2288. https://doi.org/10.1007/s10904-019-01404-9

Jandt KD, Sigusch BW (2009) Future perspectives of resin-based dental materials. Dent Mater 25:1001–1006. https://doi.org/10.1016/j.dental.2009.02.009

Jang J, Lee JM, Oh SB, Choi YH, Jung HS, Choi J (2020) Development of antibiofilm nanocomposites: Ag/Cu bimetallic nanoparticles synthesized on the surface of graphene oxide nanosheets. ACS Appl Mater Interfaces 12:35826–35834. https://doi.org/10.1021/acsami.0c06054

Jansen B, Rinck M, Wolbring P, Strohmeier A, Jahns T (1994) In vitro evaluation of the antimicrobial efficacy and biocompatibility of a silver-coated central venous catheter. J Biomater Appl 9:55–70. https://doi.org/10.1177/088532829400900103

Jayaprakashvel M, Sami M, Subramani R (2020) Antibiofilm, antifouling, and anticorrosive biomaterials and nanomaterials for marine applications. In: Prasad R (ed) Nanostructures for antimicrobial and antibiofilm applications. Nanotechnology in the life sciences. Springer Nature Switzerland AG, p 233. https://doi.org/10.1007/978-3-030-40337-9_10

Jeevanandam J, Barhoum A, Chan YS, Dufresne A, Danquah MK, Beilstein (2018) Review on nanoparticles and nanostructured materials: history, sources, toxicity and regulations. J Nanotechnol 9:1050–1074. https://doi.org/10.3762/bjnano.9.98

Jesline A, John NP, Narayanan PM, Vani C, Murugan S (2015) Antimicrobial activity of zinc and titanium dioxide nanoparticles against biofilm-producing methicillin-resistant *Staphylococcus aureus*. Appl Nanosci 5:157–162. https://doi.org/10.1007/s13204-014-0301-x

Jianfeng Ye, Liu W, Cai JG, Chen SA, Zhao XW, Zhou HH, Qi LM (2011) Nanoporous Anatase TiO_2 Mesocrystals: additive-free synthesis, remarkable crystalline-phase stability, and improved lithium insertion behavior. J Am Chem Soc 133(4):933–940. https://doi.org/10.1021/ja108205q

Johanson G (2013) Nosocomial respiratory infections with gram-negative *Bacilli*. Ann Intern Med 77(5):701. https://doi.org/10.7326/0003-4819-77-5-701

Jones N, Ray B, Ranjit KT, Manna AC (2008) Antibacterial activity of ZnO nanoparticle suspensions on a broad spectrum of microorganisms. FEMS Microbiol Lett 279(1):71–76. https://doi.org/10.1111/j.1574-6968.2007.01012.x

Joshi N, Jain N, Pathak A, Singh J, Prasad R, Upadhyaya CP (2018) Biosynthesis of silver nanoparticles using *Carissa carandas* berries and its potential antibacterial activities. J Sol-Gel Sci Techn 86(3):682–689. https://doi.org/10.1007/s10971-018-4666-2

Juine RN, Das A (2020) Surfactant-free green synthesis of ZnS QDs with active surface defects for selective nanomolar oxalic acid colorimetric sensors at room temperature. ACS Sustain Chem Eng 8(31):11579–11587

Ju-Nam Y, Lead JR (2008) Manufactured nanoparticles: an overview of their chemistry, interactions and potential environmental implications. Sci Total Environ 400(1-3):396–414. https://doi.org/10.1016/j.scitotenv.2008.06.042

Kadu PJ, Kushare SS, Thacker DD, Gattani SG (2011) Enhancement of oral bioavailability of atorvastatin calcium by self-emulsifying drug delivery systems (SEDDS). Pharm Dev Technol 16:65–74. https://doi.org/10.3109/10837450903499333

Kaegi R, Ulrich A, Sinnet B, Vonbank R, Wichser A, Zuleeg S, Simmler H, Brunner S, Vonmont H, Burkhardt M, Boller M (2008) Synthetic TiO_2 nanoparticle emission from exterior facades into the aquatic environment. Environ Pollut 156:233–239. https://doi.org/10.1016/j.envpol.2008.08.004

Kawai Y, Mickiewicz K, Errington J (2018) Lysozyme counteracts β-lactam antibiotics by promoting the emergence of L-form bacteria. Cell 172(5):1038–1049.e10

Kawish M, Ullah F, Ali FS, Saifullah S, Ali I, Rehman JU, Imran M (2020) Bactericidal potentials of silver nanoparticles: novel aspects against multidrug resistance bacteria. In: Metal nanoparticles for drug delivery and diagnostic applications. https://doi.org/10.1016/B978-0-12-816960-5.00010-0

Khan S, Shahid S, Bashir W, Kanwal S, Iqbal A (2017) Synthesis, characterization and evaluation of biological activities of manganese-doped zinc oxide nanoparticles. Trop J Pharm Res 16(10):2331–2339. https://doi.org/10.4314/tjpr.v16i10.4

Khan M, Behera S, Paul P, Das B, Suar M, Jayabalan R, Fawcett D, Poinern G, Tripathy S, Mishra A (2018) Biogenic Au@ZnO Core–Shell nanocomposites kill Staphylococcus aureus without provoking nuclear damage and cytotoxicity in mouse fibroblasts cells under hyperglycaemic condition with enhanced wound healing proficiency. Med Microbiol Immunol 208(5):609–629. https://doi.org/10.1007/s00430-018-0564-z

Khan MH, Sneha U, Ramalingam K (2020) Bactericidal potential of silver-tolerant bacteria derived silver nanoparticles against multi drug resistant ESKAPE pathogens. Biocatal Agric Biotechnol S1878-8181(18):30941–1. https://doi.org/10.1016/j.bcab.2018.12.004

Khatoon Z, McTiernan CD, Suuronen EJ, Mah TF, Alarcon EI (2018) Bacterial biofilm formation on implantable devices and approaches to its treatment and prevention. Heliyon 4:e01067. https://doi.org/10.1016/j.heliyon.2018.e01067

Kim JS, Kuk E, Yu KN, Kim JH, Park SJ, Lee HJ (2007) Antimicrobial effects of silver nanoparticles. Nanomed Nanotechnol Biol Med 3:95–101

Kim S, Ghafoor K, Lee J, Feng M, Hong J, Lee D (2013) Bacterial inactivation in water, DNA strand breaking, and membrane damage induced by ultraviolet-assisted titanium dioxide photocatalysis. Water Res 47:44034411

Klaus DJ, Sigusch BW (2009) Future perspectives of resin-based dental materials. Dent Mater 25:1001–1006. https://doi.org/10.1016/j.dental.2009.02.009

Kobayashi KRT, Nishio EK, Scandorieiro S, Saikawa GIA, Rocha SPDD, Nakazato G (2019) Metallic nanoparticles as a potential antimicrobial for catheters and prostheses. In: Materials for biomedical engineering: bioactive materials for antimicrobial, anticancer and gene therapy. Elsevier Publications. https://doi.org/10.1016/B978-0-12-818435-6.00006-2

Krishnakumar B, Imae T (2014) Chemically modified novel PAMAM-ZnO nanocomposite: synthesis, characterization and photocatalytic activity. Appl Catal A 486:170–175. https://doi.org/10.1016/j.apcata.2014.08.010

Krishnakumar B, Imae T, Miras J, Esquena J (2014) Synthesis and Azo dye photodegradation activity of ZrS_2-ZnO nanocomposites. Sep Purif Technol 132:281–288. https://doi.org/10.1016/j.seppur.2014.05.018

Krishnakumar B, Balakrishna A, Nawabjan SA, Pandiyan V, Aguiar A, Sobral AJFN (2017) Solar and visible active amino porphyrin/SiO_2-ZnO for the degradation of naphthol blue black. J Phys Chem Solids 111:364–371. https://doi.org/10.1016/j.jpcs.2017.08.012

Krishnakumar B, Hariharan R, Pandiyan V, Abilio A, Sobral JFN (2018) Gelatin-assisted g-TiO_2/BiOI heterostructure nanocomposites for Azo dye degradation under visible light. J Environ Chem Eng 6:4282–4288. https://doi.org/10.1016/j.jece.2018.06.035

Krishnakumar B, Santosh Kumar S, Gil SM, Mani D, Arivanandhan M, Sobral AJFN (2019) Synthesis and characterization of g/Ni-SiO$_2$ composite for enhanced hydrogen storage applications. Int J Hydrog Energy 44:23249–23256. https://doi.org/10.1016/j.ijhydene.2019.07.073

Krishnakumar B, Ravikumar S, Pandiyan V, Nithya V, Sylvestre S, Sivakumar P, Surya C, Agnel Arul John N, Sobral AJFN (2020) Synthesis, characterization of porphyrin and CdS modified spherical shaped SiO$_2$ for Reactive Red 120 degradation under direct sunlight. J Mol Struct 1210:128021. https://doi.org/10.1016/j.molstruc.2020.128021

Kumar P, Lakshmanan V, Anilkumar T, Ramya C, Reshmi P, Unnikrishnan AG, Nair S, Jayakumar R (2012) Flexible and microporous chitosan hydrogel/nano ZnO composite bandages for wound dressing: in vitro and in vivo evaluation. ACS Appl Mater Interfaces 4(5):2618–2629. https://doi.org/10.1021/am300292v

Kumar S, Ojha AK, Bhorolua D, Das J, Kumar A, Hazarika A (2019) Facile synthesis of CuO nanowires and Cu$_2$O nanospheres grown on rGO surface and exploiting its photocatalytic, antibacterial and super capacitive properties. Physica B 558:7481

Kurtz IS, Schiffman JD (2018) Current and emerging approaches to engineer antibacterial and antifouling electrospun nanofibers. Materials 11:1059. https://doi.org/10.3390/ma11071059

Lee RP, Soltis JA (2014) Characterizing crystal growth by oriented aggregation. Cryst Eng Comm 16:1409–1418. https://doi.org/10.1039/C3CE41773E

Lee C, Lee M, Nam K (2008) Inhibitory effect of PMMA denture acrylic impregnated by silver nitrate and silver nano-particles for *Candida albicans*. J Korean Chem Soc 52:380–386. https://doi.org/10.5012/jkcs.2008.52.4.380

Lellouche J, Kahana E, Elias S, Gedanken A, Banin E (2009) Antibiofilm activity of nano-sized magnesium fluoride. Biomaterials 30:5969–5978. https://doi.org/10.1016/j.biomaterials.2009.07.037

Lemire JA, Harrison JJ, Turner RJ (2013) Antimicrobial activity of metals: mechanisms, molecular targets and applications. Nat Rev Microbiol 11:371–384. https://doi.org/10.1038/nrmicro3028

Levy SB (2002) Factors impacting on the problem of antibiotic resistance. J Antimicrob Ther 49(1):25–30. https://doi.org/10.1093/jac/49.1.25

Li M, Zhu L, Lin D (2011a) Toxicity of ZnO nanoparticles to *Escherichia coli*: mechanism and the influence of medium components. Environ Sci Technol 45(5):1977–1983. https://doi.org/10.1021/es102624t

Li P, Wei Z, Wu T, Peng Q, Li Y (2011b) Au−ZnO hybrid nanopyramids and their photocatalytic properties. J Am Chem Soc 133(15):5660–5663. https://doi.org/10.1021/ja111102u

Li Y, Zhang W, Niu J, Chen Y (2012) Mechanism of photogenerated reactive oxygen species and correlation with the antibacterial properties of engineered metal-oxide nanoparticles. ACS Nano 6(6):5164–5173. https://doi.org/10.1021/nn300934k

Li ZQ, Chen WC, Guo FL, Mo LE, Hu LH, Dai SY (2015) Mesoporous TiO$_2$ Yolk-Shell microspheres for dye-sensitized solar cells with a high efficiency exceeding 11%. Sci Rep 5:14178. https://doi.org/10.1038/srep14178

Li J, Wu Q, Wu J (2016) Synthesis of nanoparticles via Solvothermal and hydrothermal methods. Chapter −12. In: Aliofkhazraei M (ed) Handbook of nanoparticles. Springer International Publishing Switzerland. https://doi.org/10.1007/978-3-319-15338-4_17

Li S, Chang R, Chen J, Mi G, Xie Z, Webster TJ (2020) Novel magnetic nanocomposites combining selenium and iron oxide with excellent anti-biofilm properties. J Mater Sci 55:1012–1022. https://doi.org/10.1007/s10853-019-04019-0

Liang X, Sun M, Qiao LL, Chen K, Xiao Q, Feng X (2012) Preparation and antibacterial activities of polyaniline/Cu 0.05Zn 0.95O nanocomposites. Dalton Trans 41(9):2804–2811. https://doi.org/10.1039/c2dt11823h

Liu L, Nakouzi E, Sushko ML, Schenter GK, Mundy CJ, Chun J, Yoreo JED (2020) Connecting energetics to dynamics in particle growth by oriented attachment using real-time observations. Nat Commun 11:1045. https://doi.org/10.1038/s41467-020-14719-w

Lopez-Carrizales M, Mendoza-Mendoza E, Peralta-Rodriguez RD, Perez-Dıaz MA, Portales-Perez D, Magana-Aquino M, Aragon-Pina A, Infante-Martınez R, Barriga-Castro ED, Sanchez-Sanchez R, Martinez-Castanon GA, Martinez-Gutierrez F (2020) Characterization, antibiofilm and biocompatibility properties of chitosan hydrogels loaded with silver nanoparticles and ampicillin: an alternative protection to central venous catheters. Colloids Surf B: Biointerfaces. https://doi.org/10.1016/j.colsurfb.2020.111292

Lu W, Liu G, Gao S, Xing S, Wang J (2008) Tyrosine-assisted preparation of Ag/ZnO nanocomposites with enhanced photocatalytic performance and synergistic antibacterial activities. Nanotechnology 19(44):445711. https://doi.org/10.1088/0957-4484/19/44/445711

Ludi B, Sueess MJ, Werner IA, Niederberger M (2012) Mechanistic aspects of molecular formation and crystallization of zinc oxide nanoparticles in benzyl alcohol. Nanoscale 4:1982–1995. https://doi.org/10.1039/C1NR11557J

Lv Y, Wu Y, Lu X, Yu X, Fu S, Yang L (2019) Microstructure, bio-corrosion and biological property of Ag-incorporated TiO_2 coatings: influence of Ag_2O contents. Ceram Int 45:2235722367. https://doi.org/10.1016/j.ceramint.2019.07.265

Malaekeh-Nikouei B, Fazly Bazzaz BS, Mirhadi E, Tajani AS, Khameneh B (2020) The role of nanotechnology in combating biofilm-based antibiotic resistance. J Drug Deliv Sci Technol. https://doi.org/10.1016/j.jddst.2020.101880

Malini (2015) A versatile Chitosan/ZnO nanocomposite with enhanced antimicrobial properties. Int J Biol Macromol 80:21–29. https://doi.org/10.1016/j.ijbiomac.2015.06.036

Malka E, Peralshtein I, Lipovsky A, Shalom Y, Naparstek L, Perkas N (2013) Eradication of multidrug resistant bacteria by a novel Zn-doped CuO nanocomposite. Small 9(23):4069–76. https://doi.org/10.1002/smll.201301081

Mallakpour S, Azadi E, Chaudhery Mustansar Hussain CM (2020) Environmentally benign production of cupric oxide nanoparticles and various utilizations of their polymeric hybrids in different technologies. Coord Chem Rev 419:213378. https://doi.org/10.1016/j.ccr.2020.213378

Manesh RR, Grassi G, Bergami E, Marques-Santos LF, Faleri C, Liberatori G, Corsi I (2018) Co-exposure to titanium dioxide nanoparticles does not affect cadmium toxicity in radish seeds (*Raphanus sativus*). Ecotoxicol Environ Saf 148:359–366. https://doi.org/10.1016/j.ecoenv.2017.10.051

Matai (2014) Antibacterial activity and mechanism of Ag-ZnO nanocomposite on *S. aureus* and GFP-expressing antibiotic resistant *E. coli*. Colloids Surf B: Biointerfaces 115:359–367. https://doi.org/10.1016/j.colsurfb.2013.12.005

Matusunga T (1985) Sterilization with particulate photosemiconductor. J Antibacterial Antifungal Agents 13:211–220

Mauro N, Fiorica C, Giuffrè M, Calà C, Maida MC, Giammon G (2020) A self-sterilizing fluorescent nanocomposite as versatile material with broad spectrum antibiofilm features. Mater Sci Eng C 117:111308. https://doi.org/10.1016/j.msec.2020.111308

Maziyaa K, Dlamini BC, Malinga SP (2020) Hyperbranched polymer nanofibrous membrane grafted with silver nanoparticles for dual antifouling and antibacterial properties against *Escherichia coli*, *Staphylococcus aureus* and *Pseudomonas aeruginosa*. React Funct Polym 148:104494. https://doi.org/10.1016/j.reactfunctpolym.2020.104494

McLean RJ, Hussain AA, Sayer M, Vincent PJ, Hughes DJ, Smith TJ (1993) Antibacterial activity of multilayer silver-copper surface films on catheter material. Can J Microbiol 39(9):895–899. https://doi.org/10.1139/m93-134

Meghana S, Kabra P, Chakraborty S, Padmavathy N (2015) Understanding the pathway of antibacterial activity of copper oxide nanoparticles. RSC Adv 5:12293–12299. https://doi.org/10.1039/C4RA12163E

Milosevic A (1991) Calcium hydroxide in restorative dentistry. J Dent 19:313. https://doi.org/10.1016/0300-5712(91)90028-W

Minelli EB, Bora TD, Benini A (2011) Different microbial biofilm formation on polymethylmethacrylate (PMMA) bone cement loaded with gentamicin and vancomycin. Anaerobe 17(6):380–383. https://doi.org/10.1016/j.anaerobe.2011.03.013

Mohammed H, Bekyarova AKE, Deethi YAH, Zhang Z, Chen M, Ansari MS, Cochis A, Rimondini L (2020) Antimicrobial mechanisms and effectiveness of graphene and graphene-functionalized biomaterial: a scope review. Front Bioeng Biotechnol 8:465. https://doi.org/10.3389/fbioe.2020.00465

Mohanty S, Mishra S, Jena P, Jacob B, Sarkar B, Sonawane A (2012) An investigation on the antibacterial, cytotoxic, and antibiofilm efficacy of starch-stabilized silver nanoparticles. Nanomed Nanotechnol Biol Med 8(6):916–924. https://doi.org/10.1016/j.nano.2011.11.007

Molin S, Nielsen TT (2003) Gene transfer occurs with enhanced efficiency in biofilms and induces enhanced stabilisation of the biofilm structure. Curr Opin Biotechnol 14(3):255–261. https://doi.org/10.1016/S0958-1669(03)00036-3

Morones JR, Elechiguerra JL, Camacho A, Ramirez JT (2005) The bactericidal effect of silver nanoparticles. Nanotechnology 16:2346–2353

Morsi RE, Labena A, Khamis EA (2016) Core/shell (ZnO/polyacrylamide) nanocomposite: in-situ emulsion polymerization, corrosion inhibition, anti-microbial and anti-biofilm characteristics. J Taiwan Inst Chem Eng 63:512–522. https://doi.org/10.1016/j.jtice.2016.03.037

Mou J, Lin T, Huang F, Chen H, Shi J (2016) Black titania-based theranostics nanoplatform for single NIR laser induced dual-modal imaging-guided PTT/PDT. Biomaterials 84:13–24. https://doi.org/10.1016/j.biomaterials.2016.01.009

Mourino V, Cattalini JP, Boccaccini AR (2011) Metallic ions as therapeutic agents in tissue engineering scaffolds: an overview of their biological applications and strategies for new developments. J R Soc Interface 9(68):401–419. https://doi.org/10.1098/rsif.2011.0611

Muñoz-Muñoz F, Ruiz J-C, Alvarez-Lorenzo C, Concheiro A, Bucio E (2009) Novel interpenetrating smart polymer networks grafted onto polypropylene by gamma radiation for loading and delivery of vancomycin. Eur Polym J 45:1859–1867. https://doi.org/10.1016/j.eurpolymj.2009.04.023

Muzammil S, Khurshid M, Nawaz I, Hussnain Siddique M, Zubair M, Atif Nisar M, Imran M, Hayat S (2020) Aluminium oxide nanoparticles inhibit EPS production, adhesion and biofilm formation by multidrug resistant *Acinetobacter baumannii*. Biofouling 36(4):492–504. https://doi.org/10.1080/08927014.2020.1776856

Nair S, Sasidharan A, Rani VVD, Menon D, Nair S, Manzoor K (2009) Role of size scale of ZnO nanoparticles and microparticles on toxicity toward bacteria and osteoblast cancer cells. J Mater Sci Mater Med 20:S235–S241. https://doi.org/10.1007/s10856-008-3548-5

Natalio F, Andre R, Hartog AF, Stoll B, Jochum KP, Wever R, Tremel W (2012) Vanadium pentoxide nanoparticles mimic vanadium haloperoxidases and thwart biofilm formation. Nat Nanotechnol 7:530–535. https://doi.org/10.1038/nnano.2012.91

Negi H, Agarwal T, Zaidi MGH, Goel R (2012) Comparative antibacterial efficacy of metal oxide nanoparticles against Gram negative bacteria. Ann Microbiol 62:765–772. https://doi.org/10.1007/s13213-011-0317-3

Neut D, Belt HVD, Stokroos I, Horn JRV, Mei HCVD, Busscher HJ (2001) Biomaterial-associated infection of gentamicin-loaded PMMA beads in orthopaedic revision surgery. J Antimicrob Chemother 47(6):885–891. https://doi.org/10.1093/jac/47.6.885

Nickel JC, Ruseska I, Wright JB (1985) Tobramycin resistance of Pseudomonas aeruginosa cells growing as a biofilm on urinary catheter material. Antimicrob Agents Chemother 27:619–624. https://doi.org/10.1128/AAC.27.4.619

Nikolova MP, Chavali MS (2019) Metal oxide nanoparticles as biomedical materials. Biomimetics 5:27. https://doi.org/10.3390/biomimetics5020027

Nithiya P, Chakra C, Ashok C (2015) Synthesis of TiO_2 and ZnO nanoparticles by facile polyol method for the assessment of possible agents for seed germination. Mater Today Proc 2(9):4483–4488. https://doi.org/10.1016/j.matpr.2015.10.056

Norde W, Gage D (2004) Interaction of bovine serum albumin and human blood plasma with PEO-tethered surfaces: influence of PEO chain length, grafting density, and temperature. Langmuir 20 (10):4162–4167. https://doi.org/10.1021/la030417t

Norman RS, Stone JW, Gole A, Murphy CJ, Sabo-Attwood TL (2008) Targeted photothermal lysis of the pathogenic bacteria, *Pseudomonas aeruginosa*, with gold nanorods. Nano Lett 8:302–306. https://doi.org/10.1021/nl0727056

O'Neill J (2014) Antimicrobial resistance: tackling a crisis for the health and wealth of nations; the review on antimicrobial resistance, London, UK

O'Toole G, Kaplan HB, Kolter R (2000) Biofilm formation as microbial development. Annu Rev Microbiol 54:49–79. http://ehp.niehs.nih.gov/members/2005/7339/

Oberdörster G, Oberdörster E, Oberdörster J (2005) Nanotoxicology: an emerging discipline evolving from studies of ultrafine particles. Environ Health Perspect 113:823

Ogunsona EO, Muthuraj R, Ojogbo E, Valerio O, Mekonnen TH (2020) Engineered nanomaterials for antimicrobial applications: a review. Appl Mater Today 18:100473. https://doi.org/10.1016/j.apmt.2019.100473

Olsen I (2015) Biofilm-specific antibiotic tolerance and resistance. Eur J Clin Microbiol Infect Dis 34:877–886. https://doi.org/10.1007/s10096-015-2323-z

Padmavathi AR, Sriyutha Murthy P, Das A, Nishad PA, Pandian R, Subba Rao T (2017) Copper oxide nanoparticles as an effective anti-biofilm agent against a copper tolerant marine bacterium, *Staphylococcus lentus*. Biofouling 35(9):1007–1025. https://doi.org/10.1080/08927014.2019.1687689

Pal S, Tak YK, Song JM (2007) Does the antibacterial activity of silver nanoparticles depend on the shape of the nanoparticle? A study of the Gram negative bacterium *Escherichia coli*. Appl Environ Microbiol 73:1712–1720. https://doi.org/10.1128/AEM.02218-06

Pandiyarajan T, Udayabhaskar R, Vignesh S, James RA, Karthikeyan B (2013) Synthesis and concentration dependent antibacterial activities of CuO nanoflakes. Mater Sci Eng C 33(4):2020–2024. https://doi.org/10.1016/j.msec.2013.01.021

Pantaroto HN, Ricomini-Filho AP, Bertolini MM, da Silvad JDH, Neto NFA, Sukotjo C (2018) Antibacterial photocatalytic activity of different crystalline TiO_2 phases in oral multispecies biofilm. Dent Mater 34:e182–e195. https://doi.org/10.1016/j.dental.2018.03.011

Park KD, Kim YS, Han DK, Kim YH, Lee EUB, Suh H, Choi KS (1998) Bacterial adhesion on PEG modified Polyurethane surfaces. Biomaterials 19(7-9):851–859. https://doi.org/10.1016/S0142-9612(97)00245-7

Patterson A (1939) The Scherrer formula for X-Ray particle size determination. Phys Rev 56:978–982. https://doi.org/10.1103/PhysRev.56.978

Pelgrift RY, Friedman AJ (2013) Nanotechnology as a therapeutic tool to combat microbial resistance. Adv Drug Deliv Rev 65(13-14):1803–1815. https://doi.org/10.1016/j.addr.2013.07.011

Peng G, Zil-e-Huma, Umair M, Hussain I, Javed I (2020) Nanosilver at the interface of biomedical applications, toxicology, and synthetic strategies. In: Metal nanoparticles for drug delivery and diagnostic applications. https://doi.org/10.1016/B978-0-12-816960-5.00008-2

Percival SL, Hill KE, Malic S, Thomas DW, Williams DW (2011) Antimicrobial tolerance and the significance of persister cells in recalcitrant chronic wound biofilms. Wound Repair Regen 19(1):1–9. https://doi.org/10.1111/j.1524-475X.2010.00651.x

Porter GC, Tompkins GR, Schwass DR, Li KC, Waddell JN, Meledandri CJ (2020) Anti-biofilm activity of silver nanoparticle-containing glass ionomer cements. Dent Mater 36(8):1096–1107. https://doi.org/10.1016/j.dental.2020.05.001

Pradeep T (2007) Nano: the essential–understanding nanoscience and nanotechnology. Tata McGraw-Hill Publishing Company Limited, New Delhi

Prasad R, Jha A, Prasad K (2018) Exploring the Realms of Nature for Nanosynthesis. Springer International Publishing (ISBN 978-3-319-99570-0). https://www.springer.com/978-3-319-99570-0

Prasad R, Pandey R, Barman I (2016) Engineering tailored nanoparticles with microbes: quo vadis. WIREs Nanomed Nanobiotechnol 8:316–330. https://doi.org/10.1002/wnan.1363

Prasad R, Siddhardha B, Dyavaiah M (2020) Nanostructures for Antimicrobial and Antibiofilm Applications. Springer International Publishing (ISBN 978-3-030-40336-2) https://www.springer.com/gp/book/9783030403362

Qayyum S, Khan AU (2016) Nanoparticles vs. biofilms: a battle against another paradigm of antibiotic resistance. Med Chem Commun 7:1479. https://doi.org/10.1039/c6md00124f

Raghunath A, Perumal E (2017) Metal oxide nanoparticles as antimicrobial agents: a promise for the future. Indian J Antimicrob Agents 49(2):137–152. https://doi.org/10.1016/j.ijantimicag.2016.11.011

Rao CNR, John TP, Kulkarni GU (2007) Nanocrystals: synthesis, properties and applications, Springer series in Materials Science, vol 95. Springer publisher

Reetz MT, Helbig W (1994) Size-selective synthesis of nanostructured transition metalclusters. J Am Chem Soc 116:7401–7402

Ren W, Yan Y, Zeng L, Shi Z, An G, Schaaf P, Wang D, Zhao J, Zou B, Yu H, Ge C, Michael E, Brown B, Wu A (2015) A near infrared light triggered hydrogenated black TiO_2 for cancer photothermal therapy. Adv Healthc Mater 4(10):1526–1536. https://doi.org/10.1002/adhm.201500273

Ren En, Zhang C, Li D, Pang X, Liu G (2020) Leveraging metal oxide nanoparticles for bacteria tracing and eradicating. Willey open access Journal VIEW 2020;20200052. https://doi.org/10.1002/VIEW.20200052

Rodrigues DF, Elimelech M (2010) Toxic effects of single-walled carbon nanotubes in the development of *E. coli*. Biofilm Environ Sci Technol 44:4583–4589. https://doi.org/10.1021/es1005785

Rong MZ, Zhang MQ, Ruan WH (2006) Surface modification of nanoscale fillers for improving properties of polymer nanocomposites: a review. Mater Sci Technol 22:787. https://doi.org/10.1179/174328406X101247

Roosjen A, Vries JD, Mei HCVD, Norde W, Busscher HJ (2005) Stability and effectiveness against bacterial adhesion of poly(ethylene oxide) coatings in biological fluids. J Biomed Mater Res Part B Appl Biomater 73B(2):347–354. https://doi.org/10.1002/jbm.b.30227

Roosjen A, Norde W, Van der Mei HC, Busscher HJ (2006) The use of positively charged or low surface free energy coatings versus polymer brushes in controlling biofilm formation. Prog Coll Polym Sci 132:138–144. https://doi.org/10.1007/2882_026

Rosenberg M, Visnapuu M, Vija H, Kisand V, Kasemets K, Kahru A, Ivask A (2020) Selective antibiofilm properties and biocompatibility of nano-ZnO and nano-ZnO/Ag coated surfaces. 10:13478. https://doi.org/10.1038/s41598-020-70169-w

Rout B, Liu CH, Wu WC (2017) Photosensitizer in lipid nanoparticle: a nano-scaled approach to antibacterial function. Sci Rep 7:7892. https://doi.org/10.1038/s41598-017-07444-w

Ruiz-Sanchez AJ, Guerin AJ, El-Zubira O, Dura G, Ventura C, Dixon LI, Houlton A, Horrocks BR, Jakubovics NS, Guard PA, Simeone G, Clare AS, Fulton DA (2020) Preparation and evaluation of fouling-release properties of amphiphilic perfluoropolyether-zwitterion cross-linked polymer films. Prog Org Coat 140:105524. https://doi.org/10.1016/j.porgcoat.2019.105524

Salazar-Hernández C, Salazar-Hernández M, Carrera-Cerritos R, Manuel Mendoza-Miranda J, Elorza-Rodríguez E, Miranda-Avilés R, Mocada-Sánchez CD (2019) Anticorrosive properties of PDMS-Silica coatings: effect of methyl, phenyl and amino groups. Prog Org Coat 136:105220. https://doi.org/10.1016/j.porgcoat.2019.105220

Saldarriaga Fernández Isabel Cristina (2010) Crosslinked poly (ethylene glycol) based polymer coatings to prevent biomaterial-associated infections Thesis submitted to University of Groningen: s.n. p 133

Sankar GG, Sathya S, Murthy PS, Das A, Pandiyan R, Venugopalan VP, Doble M (2015) Polydimethyl siloxane nanocomposites: their antifouling efficacy in vitro and in marine conditions. Int Biodeterior Biodegradation 104:307–314. https://doi.org/10.1016/j.ibiod.2015.05.022

Sathya S, Murthy PS, Devi VG, Das A, Anandkumar B, Sathyaseelan VS, Doble M, Venugopalan VP (2019) Antibacterial and cytotoxic assessment of poly (methyl methacrylate) based hybrid nanocomposites. Mater Sci Eng C 100:886–896. https://doi.org/10.1016/j.msec.2019.03.053

Scardino AJ, Zhang H, Cookson DJ, Lamb RN, de Nys R (2009) The role of nano-roughness in antifouling. Biofouling 25(8):757–767. https://doi.org/10.1080/08927010903165936

Seabra AB, Duran N (2015) Nanotoxicology of metal oxide nanoparticles. Metals 5:934–975. https://doi.org/10.3390/met5020934

Selim MS, Samak NA, Hao Z, Jianmin Xing (2020) Facile design of reduced graphene oxide decorated with Cu_2O nanocube composite as antibiofilm active material. Mater Chem Phys 239:122300. https://doi.org/10.1016/j.matchemphys.2019.122300

Shah SR, Tatara AM, D'Souza RN, Mikos AG, Kasper FK (2013) Evolving strategies for preventing biofilm on implantable materials. Mater Today 16(5):177–182. https://doi.org/10.1016/j.mattod.2013.05.003

Sharma B, AV, Brown NE, Matluck LT, Hu K, Lewis (2015) *Borrelia burgdorferi*, the causative agent of Lyme disease, forms drug-tolerant persister cells. Antimicrob Agents Chemother 59:4616–4624. https://doi.org/10.1128/aac.00864-15

Shi M, Kretlow JD, Nguyen A, Young S, Baggett LS, Wong ME, Kasper FK, Mikos AG (2010) Antibiotic-releasing porous polymethylmethacrylate constructs for osseous space maintenance and infection control. Biomaterials 31(14):4146–4156 https://doi.org/10.1016/j.biomaterials.2010.01.112

Srivastava S, Usmani Z, Atanasov AG, Singh VK, Singh NP, Abdel-Azeem AM, Prasad R, Gupta G, Sharma M, Bhargava A (2021) Biological nanofactories: Using living forms for metal nanoparticle synthesis. Mini-Reviews in Medicinal Chemistry 21(2): 245–265.

Shrivastava S, Bera T, Roy A, Singh G, Ramachandrarao P, Dash D (2007) Characterization of enhanced antibacterial effects of novel silver nanoparticles. Nanotechnology 18:225103. (9pp). https://doi.org/10.1088/0957-4484/18/22/225103

Silhavy TJ, Kahne D, Walker S (2020) The bacterial cell envelope. Cold Spring Harb Perspect Biol 2:a000414. https://doi.org/10.1101/cshperspect.a000414

Sirelkhatim A, Mahmud S, Seeni A, Kaus NHM, Ann LC, Bakhori SKM (2015) Review on zinc oxide nanoparticles: antibacterial activity and toxicity mechanism. Nano-Micro Lett 7:219242. https://doi.org/10.1007/s40820-015-0040-x

Skocaj M, Filipic M, Petkovic J, Novak S (2011) Titanium dioxide in our everyday life; is it safe? Radiol Oncol 45:227–247. https://doi.org/10.2478/v10019-011-0037-0

Sperandio FF, Huang YY, Hamblin MR (2013) Antimicrobial photodynamic therapy to kill Gram-negative bacteria. Recent Patents on Anti-Infective Drug Discovery 8(2):108–120 CODEN: RPADCX; ISSN: 1574-891X

Stanic V, Tanaskovi SB (2020) Antibacterial activity of metal oxide nanoparticles. Nano 241–274. https://doi.org/10.1016/B978-0-12-819943-5.00011-7

Stanic V, Janackovic D, Dimitrijevic S, Tanaskovic SB, Mitric M, Pavlovic MS (2011) Synthesis of antimicrobial monophase silver-doped hydroxyapatite nanopowders for bone tissue engineering. Appl Surf Sci 257(9):4510–4518. https://doi.org/10.1016/j.apsusc.2010.12.113

Stefan B, Ovidiu O, Georgeta V, Anton F, Ecaterina A, Andrei T, Alina H (2013) Synthesis and characterization of a novel controlled release zinc oxide/gentamicin–chitosan composite with potential applications in wounds care. Int J Pharm. https://doi.org/10.1016/j.ijpharm.2013.11.035

Stewart P (2003) Diffusion in biofilms. J Bacteriol 185(5):1485–1491. https://doi.org/10.1128/JB.185.5.1485-1491.2003

Stewart PS (2015) Antimicrobial tolerance in biofilms. Microbiol Spectr 3. https://doi.org/10.1128/microbiolspec.MB-0010-2014

Stewart PS, Franklin MJ (2008) Physiological heterogeneity in biofilms. Nat Rev Microbiol 6:199–210. https://doi.org/10.1038/nrmicro1838

Stoica P, Chifiriuc MC, Rapa M, Lazăr V (2017) Overview of biofilm-related problems in medical devices biofilms and implantable medical devices. https://doi.org/10.1016/B978-0-08-100382-4.00001-0

Sturr MG, Marquis RE (1990) Inhibition of proton-translocating ATPases of Streptococcus mutans and *Lactobacillus casei* by fluoride and aluminum. Arch Microbiol 155:22–27. https://doi.org/10.1007/BF00291269

Su R, Bechstein R, Sø L, Vang RT, Sillassen M, Esbjornsson M (2011) How the anatase-to-rutile ratio influences the photo reactivity of TiO_2. J Phys Chem C 115(49):24287–24292. https://doi.org/10.1021/jp2086768

Sugimoto T (2003) Formation of monodispersed nano and micro particles controlled in size, shape and internal structure. Chem Eng Technol 26:313–21. https://doi.org/10.1002/ceat.200390048

Sumbal, Nadeem A, Naz S, Ali JS, Abdul M, Zia M (2019) Synthesis, characterization and biological activities of monometallic and bimetallic nanoparticles using *Mirabilis jalapa* leaf extract. Biotechnol Rep 22:e00338. https://doi.org/10.1016/j.btre.2019.e00338

Taglietti A, Arciola CE, D'Agostino A, Dacarro G, Montanaro L, Campoccia D, Cucca L, Vercellino M, Poggi A, Pallavicini P, Visai L (2014) Antibiofilm activity of a monolayer of silver nanoparticles anchored to an amino-silanized glass surface. Biomaterials 35:1779e1788. https://doi.org/10.1016/j.biomaterials.2013.11.047

Tan H, Penga Z, Li Q, Xu X, Guo S, Tan T (2012) The use of quaternised chitosan-loaded PMMA to inhibit biofilm formation and downregulate the virulence-associated gene expression of antibiotic-resistant *Staphylococcus*. Biomaterials 33(2):365–377. https://doi.org/10.1016/j.biomaterials.2011.09.084

Taniguchii N (1974) On the basic concept of nanotechnology. Proc Intl Conf Prod Eng Tokyo II 1974:18e23

Tetz GV, Artemenko NK, Tetz VV (2009) Effect of DNase and antibiotics on biofilm characteristics. Antimicrob Agents Chemother 53:1204–9. https://doi.org/10.1128/AAC.00471-08

Tiller JC (2006) Silver-based antimicrobial coatings. In: Polymeric drug delivery II: Svenson S, ACS symposium series. American Chemical Society, Washington DC. https://doi.org/10.1021/bk-2006-0924.ch014

Tiller JC, Liao CJ, Lewis K, Klibanov AM (2001) Designing surfaces that kill bacteria on contact. Proc Natl Acad Sci U S A 98:5981–5985. https://doi.org/10.1073/pnas.111143098

Torres-Giner S, Martinez-Abad A, Gimeno-Alcañiz JV, Ocio MJ, Lagaron JM (2011) Controlled delivery of gentamicin antibiotic from bioactive electrospun polylactide-based ultrathin fibers. Adv Eng Mater 14(4):B112–B122. https://doi.org/10.1002/adem.201180006

Trampuz A, Zimmerli W (2006) Diagnosis and treatment of infections associated with fracture-fixation devices injury. Int J Care Injur 37:S59—S66. https://doi.org/10.1016/j.injury.2006.04.010

Tripathi S, Mehrotra GK, Dutta PK (2011) Chitosansilver oxide nanocomposite film: preparation and antimicrobial activity. Bull Mater Sci 34:2935

Trivedi J, Rao S, Kumar A (2014) Facile preparation of Agarose-chitosan hybrid materials and nanocomposite Ionogels using an ionic liquid via dissolution, regeneration and sol-gel transition. Green Chem 16(1):320–330. https://doi.org/10.1039/c3gc41317a

Tseng BS, Zhang W, Harrison JJ (2013) The extracellular matrix protects *Pseudomonas aeruginosa* biofilms by limiting the penetration of tobramycin. Environ Microbiol 15:2865–2878. https://doi.org/10.1111/1462-2920.12155

Unal H (2018) Antibiofilm coatings. In: Tiwari A (ed) Handbook of antimicrobial coatings, pp 301–313. https://doi.org/10.1016/B978-0-12-811982-2.00015-9

Van OM, Schäfer T, Gazendam JA, Ohlsen K, Tsompanidou, de Goffau MC, Harmsen HJ, Crane LM, Lim E, Francis KP (2013) Real-time in vivo imaging of invasive- and biomaterial-associated bacterial infections using fluorescently labelled vancomycin. Nat Commun 4:2584. https://doi.org/10.1038/ncomms3584

Vaseem M, Umar A, Kim S, Al-Hajry A, Hahn Y (2008) Growth and structural properties of CuO urchin-like and sheet-like structures prepared by simple solution process. Mater Lett 62(10-11):1659–1662. https://doi.org/10.1016/j.matlet.2007.09.054

Vigneshwaran N, Kumar S, Kathe AA, Varadarajan PV, Prasad V (2006) Functional finishing of cotton fabrics using zinc oxide-soluble starch nanocomposites. Nanotechnology 17(20):5087–5095. https://doi.org/10.1088/0957-4484/17/20/008

Wang X, Chen X (2019) Novel nanomaterials for biomedical, environmental and energy applications, 1st edn. Elsevier

Wang X, Zhao Z, Qu J, Wang Z, Qiu J (2010) Shape-control and characterization of magnetite prepared *via* a one-step solvothermal route. Cryst Growth Des 10(7):2863–2869. https://doi.org/10.1021/cg900472d

Wang YW, Cao AN, Jiang Y, Zhang J, Liu H, Liu YF, Wang HF (2014) Superior antibacterial activity of zinc oxide/graphene oxide composites localized around bacteria. ACS Appl Mater Interfaces 6(4):2790–2797. https://doi.org/10.1021/Am4053317

Wang L, Natan M, Zheng W, Zheng W, Liu S, Jacobi G, Perclshtein I, Gedanken A, Banin E, Jiang X (2020) Small molecule-decorated gold nanoparticles for preparing antibiofilm fabrics. Nanoscale Adv 2:2293–2302. https://doi.org/10.1039/D0NA00179A

Weiqiang Lv, He W, Wang X, Niu Y, Cao H, Dickerson JH, Wang Z (2014) Understanding the oriented-attachment growth of nanocrystals from an energy point of view: a review. Nanoscale 6:2531–2547. https://doi.org/10.1039/C3NR04717B

Wille J, Coenye T (2020) Biofilm dispersion: the key to biofilm eradication or opening Pandora's box? Biofilms 2:100027. https://doi.org/10.1016/j.bioflm.2020.100027

Wiraja C, Xu C, Wei J, Wang Y, Wang L, Liu F, Che Y (2018) Graphene materials in antimicrobial nanomedicine: current status and future perspectives. Adv Healthc Mater 7:170–140. https://doi.org/10.1002/adhm.201701406

Wong KK, Liu X (2010) Silver nanoparticles – the real "silver bullet" in clinical medicine. Med Chem Comm 1:125–131. https://doi.org/10.1039/C0MD00069H

Wong CW, Chan YS, Jeevanandam J, Pal K, Bechelany M, Elkodous MA, El-Sayyad GS (2020) Response surface methodology optimization of mono-dispersed MgO nanoparticles fabricated by ultrasonic-assisted Sol–Gel method for outstanding antimicrobial and antibiofilm activities. J Clust Sci 31:367–389. https://doi.org/10.1007/s10876-019-01651-3

Wu X, Zheng L, Wu D (2005) Fabrication of superhydrophobic surfaces from microstructured ZnO-based surfaces via a wet-chemical route. Langmuir 21(7):2665–2667. https://doi.org/10.1021/la050275y

Xu KD, McFeters GA, Stewart PS (2000) Biofilm resistance to antimicrobial agents. Microbiology 146:547–549. https://doi.org/10.1099/00221287-146-3-547

Yadav N, Jaiswal AK, Dey KK, Yadav VB, Nath G, Srivastava AK, Yadav RR (2018) Trimetallic Au/Pt/Ag based nanofluid for enhanced antibacterial response. Mat Chem Phys 218:10–17. https://doi.org/10.1016/j.matchemphys.2018.07.016

Yael N, Jason S, Asnis, Häfeli UO, Bach H (2017) Metal nanoparticles: understanding the mechanisms behind antibacterial activity. J Nanobiotechnol 15:65. https://doi.org/10.1186/s12951-017-0308-z

Yallappa S, Manjanna J, Dhananjaya BL (2015) Phytosynthesis of stable Au, Ag and Au–Ag alloy nanoparticles using *J. Sambac* leaves extract, and their enhanced antimicrobial activity in presence of organic antimicrobials. Spectrochim Acta Part A: Mol Biomol Spec 137:236–243. https://doi.org/10.1016/j.saa.2014.08.030

Yan R, Park JH, Choi Y, Heo CJ, Yang SM, Lee LP, Yang P (2011) Nanowire-based single-cell endoscopy. Nat Nanotechnol 7:191–196. https://doi.org/10.1038/nnano.2011.226

Zeng L, Ren W, Xiang L, Zheng J, Chen B, Wu A (2013) Multifunctional Fe_3O_4–TiO_2 nanocomposites for magnetic resonance imaging and potential photodynamic therapy. Nanoscale 5:2107–2113. https://doi.org/10.1039/C3NR33978E

Zhang Y, Wang S, Li X, Chen L, Qian Y, Zhang Z (2006) CuO shuttle-like nanocrystals synthesized by oriented attachment. J Cryst Growth 291:196–201. https://doi.org/10.1016/j.jcrysgro.2006.02.044

Zhang Q, Liu SJ, Yu SH (2009) Recent advances in oriented attachment growth and synthesis of functional materials: concept, evidence, mechanism, and future. J Mater Chem 19:191–207. https://doi.org/10.1039/B807760F

Zhang XF, Liu ZG, Shen W, Gurunathan S (2016) Silver nanoparticles: synthesis, characterization, properties, applications and therapeutic approaches. Int J Mol Sci 17:1534. https://doi.org/10.3390/ijms17091534

Zhang X, Zhang G, Chai M, Yao X, Chen W, Chu PK (2021) Synergistic antibacterial activity of physical-chemical multi-mechanism by TiO_2 nanorod arrays for safe biofilm eradication on implant. Bioactive Mater 6:12–25. https://doi.org/10.1016/j.bioactmat.2020.07.017

Zhong L, Yu Y, Lian H, Hu X, Fu H, Chen Y (2017) Solubility of nano-sized metal oxides evaluated by using in vitro simulated lung and gastrointestinal fluids: implication for health risks. J Nanopart Res 19:375. https://doi.org/10.1007/s11051-017-4064-7

Zhong Y, Xiao H, Seidi F, Jin Y (2020) Natural polymer-based antimicrobial hydrogels without synthetic antibiotics as wound dressings. Biomacromolecules 21(8):2983–3006. https://doi.org/10.1021/acs.biomac.0c00760

Zobell CE (1943) The effect of solid surfaces upon bacterial activity. J Bacteriol 46:39–56 PMC373789

Zunita M, Makertihartha IGBN, Saputra FA, Syaifi YS, Wenten IG (2018) Metal oxide based antibacterial membrane. IOP Conf Series Mater Sci Eng 395:012021. https://doi.org/10.1088/1757-899X/395/1/012021

Chapter 6
Nanomaterials for A431 Epidermoid Carcinoma Treatment

S. Christobher, P. Kalitha Parveen, Murugesh Easwaran,
Haripriya Kuchi Bhotla, Durairaj Kaliannan,
Balamuralikrishnan Balasubramanian, and Arun Meyyazhagan

Contents

6.1	Introduction	212
	6.1.1 Grouping of Nanoparticles	214
	6.1.2 Types of Nanoparticles	216
	6.1.3 Properties of Nanoparticles	216
6.2	Cancer and Its History	217
	6.2.1 Epidemiology	218
	6.2.2 Cancer and Its Classifications	218
6.3	Nanotherapuetics to Overcome Cancer	220
6.4	Epidermoid Carcinoma (Skin Cancer)	221
	6.4.1 Sorts of Skin Growth	222
	6.4.2 Basal Cell Carcinoma	222
	6.4.3 Squamous Cell Carcinoma (Epidermoid Carcinoma)	223
	6.4.4 Melanoma	223
	6.4.5 Signs and Side Effects of Skin Disease	223

S. Christobher (✉)
Department of Zoology, Nallamuthu Gounder Mahalingam College,
Pollachi, Tamil Nadu, India

P. Kalitha Parveen
PG Department of Zoology, Hajee Karutha Rowther Howdia College of Arts and Science,
Uthamapalayam, Tamil Nadu, India

M. Easwaran · H. K. Bhotla
Bioknowl Insights Private Limited, Coimbatore, Tamil Nadu, India

D. Kaliannan
Zoonosis Research Center, Department of Infection Biology, School of Medicine, Wonkwang University, Iksan, Republic of Korea

B. Balasubramanian
Department of Food Science and Biotechnology, College of Life Sciences, Sejong University, Seoul, South Korea

A. Meyyazhagan (✉)
Department of Life Sciences, CHRIST (Deemed to be University), Bangalore, India

© The Author(s), under exclusive license to Springer Nature Switzerland AG 2022
A. Krishnan et al. (eds.), *Emerging Nanomaterials for Advanced Technologies*,
Nanotechnology in the Life Sciences, https://doi.org/10.1007/978-3-030-80371-1_6

	6.4.6 Chance Elements Prompting Skin Growth.	224
	6.4.7 Treatment of Skin Tumor.	224
	6.4.8 Curettage and Drying Up.	224
	6.4.9 Radiation Treatment.	225
	6.4.10 Cryosurgery.	225
	6.4.11 Therapeutic Treatment.	225
	6.4.12 MTT Assay for Skin cancer Cell Lines Using Green Synthesis of Nanoparticles.	225
6.5	Nano Drug Delivery in Skin Cancer.	228
	6.5.1 Liposomes.	228
	6.5.2 Solid Lipid Nanoparticles (SLNs).	229
	6.5.3 Dendimers.	229
	6.5.4 Quantum Dots.	230
	6.5.5 Nanotubes.	230
6.6	Efficacy of Nano Treatment in Future for Skin Cancer.	230
6.7	Conclusion.	232
References.		232

6.1 Introduction

One of the most unique fields in today's cutting-edge science is nanotechnology. Nanoparticles show proof of inventive or upgraded properties dependent on their exact uniqueness, for example, measurement, distribution, and surface. Nanotechnology involves blends, delineation, investigation, and apparatus of nano-sized (1–100 nm) materials (Ayyanar and Ignacimuthu 2009). It includes the materials whose structures reveal significantly novel and upgraded physical, substance, and organic properties, wonders, and usefulness due to their nano-scale size. Nanotechnology has an interesting part in contemporary history, with plentiful strategies to join nanoparticles of circumspect shape and size contingent upon explicit necessities. New uses of nanoparticles and nonmaterials are expanding quickly. The idea of nanotechnology was introduced in 1959 by Richard Feynman, who gave a discussion on the idea of nanotechnology. He explained how sub-atomic machines worked with nuclear accuracy and talked about nanoparticles in a lecture titled "There's plenty of room at the bottom." The clinical condition was completely changed in the 1960s, considering the truth that nanoparticles could go about as medication and in addition be utilized as antibodies. The primary paper distributed in 1980 by K. Eric Drexler of Space Systems Laboratory, Massachusetts Institute of Technology was titled "A way to deal with the improvement of general abilities for atomic control." The expression "nanotechnology" was first time utilized as a logical field by Nario Tanigushi in his 1974 paper "Nanotechnology" that for the most part comprises the handling of, partition, union, and distortion of materials by one iota or one atom (Zhang and Zhang 2013).

Nanoscience is the science dependent on nanoparticles whose size is 1–100 nanometers similar to iota's and atoms. Polymers and their subunits together with nanoscale go about as bio-macromolecules. Nanoscience can be characterized in a progressively consistent way as:

The future holds incredible breadth for organizations fabricating "nanoproducts," which are to beutilized for humanity advances. Nanoparticles are nano sized, and some are easily created and amicable. Nanoparticles are incorporated in three ways: physical, compound, and organic (Jemal et al. 2006). In synthetic blends, an extraordinary measure of nanoparticles are relied upon and the compound substances utilized for the combination and the steady condition of nanoparticles are harmful and non-eco agreeable. The requirement for ecological non-harmful manufactured conventions for nanoparticle combinations prompts enthusiasm for organic methodologies, which eliminate the utilization of toxic synthetic substances as a result. In this way, there is an expanding interest for "green nanotechnology" (Zhao et al. 2015). There are a few reports where a wide assortment of organic substances ranging from amino acids to living microorganisms have been utilized to accomplish this goal.

Plants are the best wellspring of creating nanoparticles. Moreover, utilizing plant extracts likewise decreases the sticker price of confinement of microorganisms and culture media affecting the cost of nanoparticles amalgamation by microorganisms. Part of the esteemed methodology is utilized to save microbial societies. There are different verdures used by specialists and botanists to orchestrate nanoparticles in the organic way, and their concentrates are used as medication. Some plants as well as some microbial species are likewise used to blend nanoparticles, and they are *Fusarium oxysporum, Penicillium* sp. also, microbes, for example, *Bacillus subtilis*. The amalgamation of nanoparticles by methods using plant extracts are the largest part of the strategy for green, biological production of nanoparticles, and furthermore has the uncommon advantage that the plants are widely scattered, easily accessible, a lot more secure to hold and proceed as a premise of various metabolites (Suriyavathana and Kumar 2010). Some of the restorative plants used to execute the blend of nanoparticles are *Oryza sativa, Helianthus annus, Saccharum officinarum, Sorghum bicolour, Zea mays, Basella alba, Aloe Vera, Capsicum annuum, Magnolia kobus, Medicago sativa (Alfalfa), Cinnamomum camphora,* and *Geranium sp.* Pharmaceutical and natural businesses utilize these nanoparticles.

Currently, plants assume an effective job in incorporating silver nanoparticles. In spite of the fact that there are numerous courses accessible for the combination of silver nanoparticles, the natural blend utilizing plant sources offers a few favorable circumstances, for example, best in cost-viability, non-harmful, and eco-accommodating specialist. Silver nanoparticles have a variety of utilizations in vitro and in vivo. Historically, silver has been proven as an antimicrobial operator that destroys pathogenic microorganisms. A plethora of vegetation has been utilized to integrate metal nanoparticles. Because silver is a sensitive white, glossy part, a basic use of silver nanoparticles is to give a thing a silver coating. In any case, the strikingly strong antimicrobial development is the genuine bearing for progression of nano-silver things. Cases are sustenance packaging materials and food supplements, smell safe materials, equipment, family machines, magnificence care items and remedial advices, water disinfectants, and room showers. Silver is one of the fundamental parts that make up our planet. It is a remarkable, yet naturally occurring part, to some degree harder than gold and outstandingly adaptable and malleable. Unadulterated silver has the high electrical and warm conductivity and has the least contact obstruction. Silver can be accessible in three differing oxidation states: AgO, Ag^{2+}, Ag^+. Metallic silver it is insoluble in water, but metallic salts, for instance, $AgNO_3$ and AgCl, are dissolvable in water. Metallic silver is used for prosthesis and supports, fungicides, and coinage. Dissolvable silver blends, for instance, silver salts, have been used to treat afflictions, such as epilepsy, nicotine development, gastroenteritis, and diseases, including syphilis and gonorrhea. Assessments have shown that these groupings of Ag^+ particles are too low to have a harmful quality. Metallic silver appears to pose an inconsequential danger to prosperity, while dissolvable silver could cause troublesome effects. Owing to the wide arrangement of uses, silver can have presentation through various courses of area into the body. Ingestion is the basic course of entry for silver blends and colloidal silver proteins. Dietary confirmation of silver is surveyed at 70–90 μg/day. Silver in any form is not believed to be perilous to the vulnerable, cardiovascular, nervous, or regenerative structure and it is not believed to be malignancy causing, as such silver is reasonably non-toxic. Silver solicitations will most likely ascend as silver finds new uses, particularly in materials, plastics, and remedial endeavors, changing the case of silver release as these progressions and things diffuse through the overall economy.

6.1.1 Grouping of Nanoparticles

Nanoparticles are grouped by their temperament, measurement, and viability. They are used as medication carriers or imaging specialists in biomedical applications. The characterization of nanoparticles is important to determine their possible applications, and to this end procedures from materials science are used, such as X-beam photoelectron spectroscopy (XPS), electron microscopy (TEM, SEM), nuclear constrain microscopy (AFM), powder X-beam diffractrometry (XRD), bright obvious spectroscopy, Fourier change infrared spectroscopy (FTIR), dynamic light dispersing (DLS), grid helped laser desorption time of light mass spectrometry (MALDI-TOF), and particular surface range and a high division of surface iota's are the qualities of metallic nanoparticles. Because of the extraordinary physicochemical characteristics of nanoparticles, including synergist activity, optical properties, electronic properties, antibacterial properties, and alluring properties, they are receiving attention from scientist to use in novel procedures (Sukirtha et al. 2012a).

The amalgamation of metal and semiconductor nanoparticles is an incredible region of exploration because of its potential applications, which were realized in the progression of novel advances. Nano-crystalline silver particles have found enormous applications in the fields of high influence capacity biomolecular acknowledgment, diagnostics, antimicrobials, therapeutics, catalysis, and scaled down scale devices. Regardless, there is still a necessity for a fiscally viable and nonpolluting course to coordinate the silver nanoparticles. Nanotechnology is one of the exceptional domains of examination in the current field of materials science. Nanoparticles demonstrate absolutely new or improved properties, for instance, size, transport, and morphology of the particles. Novel uses of nanoparticles and nonmaterials are rising rapidly in various fields. Silver is prominent for having an inhibitory effect toward various bacterial strains and microorganisms commonly shown in restorative and mechanical systems. In drugs, silver and silver nanoparticles have a bountiful application, including skin medicines and creams containing silver to hinder defilement of burns and open wounds, restorative devices, and supplements masterminded with silver-impregnated polymers. In the material business, silver-embedded surfaces are used as a piece of shaking gear. Nanoparticles can be joined using various systems, including creation, physical, and natural. However, an engineered procedure for amalgamation requires a short period of time for association of a broad measure of nanoparticles; this methodology requires beating agents for size alteration of the nanoparticles. Synthetic concoctions used for nanoparticles amalgamation and alteration are unsafe and sometimes non eco-friendly. The necessity for normal non-noxious builds for nanoparticles prompts the use of natural techniques that avoid the use of deadly synthetic substances. As such, there is an extending enthusiasm for green nanotechnology. Moreover, using plants lessens the expense of microorganism's disengagement and culture media, improving the cost centered feasibility over nanoparticles mix by microorganisms (Alam et al. 2011). Various natural procedures for both extracellular and intracellular nanoparticles mixtures have been represented to date using microorganisms, including minuscule creatures, parasites, and plants. To a great extent, the amalgamation of nanoparticles using various plants and their concentrates can be good over other common association techniques that incorporate the amazingly mind-boggling procedure of maintaining microbial social orders. A mix of nanoparticles using plant concentrates is the most understood procedure for green, eco-obliging formation of nanoparticles and has a remarkably favored point of view because the plants are comprehensively cycled, easily available, secure to manage and a wellspring of a couple of metabolites.

There has also been a couple of examinations performed on the mix of silver nanoparticles using restorative plants, for instance, *Oryza sativa, Helianthus annus, Saccharum officinarum,* and *Sorghum bicolour, Zea mays* (Sukirtha et al. 2012b), in the field of pharmaceutical applications and natural organizations. In addition, a green association of silver nanoparticles using a methanolic concentrate of *Eucalyptus hybrida* was also investigated. Silver nanoparticles have been incorporated from naturally occurring sources, including green tea (*Camellia sinensis*), Neem (*A. indica*), leguminous bush (*Sesbania drummondii*), distinctive leaf stock, trademark versatile, starch, Aloe Vera plant extract, lemongrass leaves extract, etc. (Jung et al. 2014). With respect to the microorganisms, the silver nanoparticles become joined to the cell divider, as such disturbing the permeability of cell divider and cell breath. The nanoparticles may in like manner invade some place inside the cell divider, inducing cell injury by interfacing with phosphorus and sulfur containing blends, for instance, DNA and protein, shown inside the cell. In addition, the intensity of the antibacterial effects is dependent on the proportion of the nanoparticles. The smaller particles have higher antibacterial activities because of the practically identical silver mass substance (Panhwar and Abro 2007).

6.1.2 Types of Nanoparticles

Nanoparticles can be widely collected into categories, normal nanoparticles, which fuse carbon nanoparticles (fullerenes), while a part of the inorganic nanoparticles fuse appealing nanoparticles, fair metal nanoparticles (such as gold and silver), and semi-transmitter nanoparticles (such as titanium oxide and zinc oxide). There is a demand for inorganic nanoparticles, for example, of noteworthy metal nanoparticles (gold and silver), as they give overwhelming material properties utilitarian adaptability. As a result of their size, segments, and focal points over available compound imaging drug administrators and prescriptions, inorganic particles have been assessed as likely gadgets for helpful imaging and furthermore to treat ailments. Inorganic nanomaterials have been comprehensively used for cell movement in light of their adaptable components such as wide openness, rich convenience, extraordinary similitude, and capacity of centered drug transport and controlled appearance of medications (Bernal et al. 2010).

6.1.3 Properties of Nanoparticles

The closeness of a high segment of iotas establishing the nanoparticles on the particle surface rather than in the atom and the huge surface range available per unit volume of the material is the best result for the nanoparticles. Both of these properties increase in significance with a decrease in particle size. Consequently, the unique physical, manufactured, and natural properties of nanoparticles start from these two segments. Quantum effects of some nanoscale materials occur, considering different captivating applications (Liu 2004). The colossal specific surface zone of nanoparticles includes the beginning stage of a portion of their novel applications. Catalysis is updated by the high surface range per unit volume and the homogenous allotment of nanoparticles. Regardless, high surface regions provide a strong relationship between the nanoparticles and the solid system wherein they may be united. The platelet morphology and gigantic specific surface areas of silicate nanoparticles improve the limit properties of polymer films by massively growing the pathway for a nuclear vehicle of invading substances (Tang et al. 2006).

6.2 Cancer and Its History

Malignant growth is a class of sicknesses caused by uncontrolled division of cells and the capacity of these cells to attack different tissues, either by direct development into nearby tissue through intrusion or by implantation into remote locales by metastasis. Metastasis is characterized as the phase where malignant growth cells are moved through the circulatory system or lymphatic framework. Cancer may influence individuals at all ages; however, chance will, in general, increase with age because DNA damage is progressively evident in maturing DNA.

The oldest depictions of malignant growth are found in the Edwin Smith Papyrus (referenced a breast tumor), composed in roughly 3000 BC, and the Ebers Papyrus, dating to 1500 BC (referenced a delicate tissue tumor and diseases of the stomach, uterus, skin, and rectum). Until the nineteenth century, blades, salts, searing, and arsenic glue were utilized by the Egyptians to treat tumors. The Indians and Chinese relied upon homegrown medication, including metals such as iron, mercury, and copper in different structures for over 3000 years. Hippocrates (460–375 BC), the father of medicine utilized the expressions "carcinoma" and "carcinos," which in Greek signified "crab." He and his followers had immense information about the various sorts of shallow and deep carcinomas and utilized different moisturizers, searing methods, and deep tumors were extracted with a blade. He thought about a range of treatments from palliative medicines to removal of breast and colon malignancies (Akhani et al. 2008).

Aulus Celsus (25 BC–AD 50), a Roman doctor, in his book De Medicina wrote about diseases of the spleen, liver, colon, and stomach and suggested early medical procedures. He made an interpretation of carcinos to the Latin word disease. Pliny the Roman (AD 23–79), in Materia Medica gathered different solutions for inner use. His solution of a boiled blend of ocean crabs, egg white, nectar, and powdered hawk defecation was generally valued. Aretaeus (AD 81–138) additionally gave a detailed depiction about different diseases. Galen (AD 130–200) is notable for the humoral hypothesis and numerous unyielding speculations (Ribeiro et al. 2010).

Lanfranc (1252–1315) established French medical procedure. Henri de Mondeville (1260–1320) and Guy de Chauliac (1300–1368), two French doctors, dismissed the old Galen hypotheses, prompting headways in treatment of malignant growth. Paracelsus (1493–1541) presented chemotherapy by directing different metals such as lead, zinc, copper, arsenic, and so forth for ingestion. The mid-1900s demonstrated advances in chemotherapy, distinguishing proof of natural cancer-causing agents, impact of hormones in disease, performing mammography, ID of DNA, and so forth. From 1970 onwards, much accentuation was given to malignant growth research, which prompted distinguishing proof of oncogenes, tumor silencers, and infections causing disease. Different screening techniques were created such as MRI (attractive reverberation imaging), tomography, immunohistochemistry, and so on. The campaign against neoplasia has served to effectively treat and decrease malignancy mortality, which has been conceivable because of the discoveries and commitments made by doctors, specialists, and a multidisciplinary setting, including scientists, oncologists, radiologists, and so forth (Ankanna et al. 2010).

6.2.1 Epidemiology

For malignant growth, the study of disease transmission is the investigation of the frequency of malignancy as an approach to gather potential patterns and causes. The main such reason for malignant growth was distinguished by British specialist Percivall Pott, who found in 1775 that malignancy of the scrotum was a typical ailment among smokestack cleaners. In some Western nations, for example, the USA and the UK, malignant growth is overwhelming cardiovascular disease as the main source of death. In numerous developing nations, malignancy rate (to the extent that this can be estimated) shows up lower, in all probability as a result of the higher mortality rates because of irresistible sickness or injury, with the expanded command over jungle fever and tuberculosis in some developing nations, frequency of disease is expected to rise; this is termed epidemiologic progress in epidemiological phrasing. For malignant growth, the study of disease transmission intently reflects the chance to calculate spread in different nations. Hepatocellular carcinoma (liver cancer) is uncommon in the West; however, it is the fundamental malignant growth in China and neighboring nations, in all likelihood because of the endemic nearness of hepatitis B and aflatoxin in that populace. Also, with tobacco smoking becoming progressively normal in different developing nations, lungs malignant growth rates have expanded in an equal manner (Ankanna et al. 2010).

6.2.2 Cancer and Its Classifications

Malignant growths are grouped by the sort of cells that look like the tumor, and hence the tissue that was the starting point of the tumor. The accompanying general classes are normally acknowledged. (1) Carcinoma – Malignant tumors formed in epithelial cells. This gathering speaks to the most widely recognized diseases, including the basic types of breast, prostate, lung, and colon malignancies. (2) Lymphoma and Leukemia – Malignant tumors formed in blood and bone marrow cells. (3) Sarcoma – Malignant tumors formed in connective tissue or mesenchymal cells. (4) Mesothelioma – Tumors formed in the mesothelial cells coating the peritoneum and the pleura. (5) Glioma – Tumors formed in glia, the most widely recognized sort of synapse. (6) Germinoma – Tumors formed in germ cells, regularly found in the gonad and ovary. (7) Choriocarcinoma – Malignant tumors formed in the placenta. Carcinogenesis, which implies the commencement or age of malignancy, is the procedure of derangement of the pace of cell division because of damage to DNA. A wide range of malignant growth shares these characteristics (Hanahan and Weinberg 2000a).

Proto-oncogenes are qualities that advance cells development and mitosis, a procedure of cell division, and tumor silencer qualities demoralize cells development, or incidentally end cell division and thus DNA changes to these qualities are required before a typical cell changes into a malignant growth cell. Proto-oncogenes advance cell development through an assortment of ways. Many can create hormones, a chemical messenger between cells that empowers mitosis, the impact of which relies upon the sign transduction of the accepting tissue or cells. Some are liable for the sign transduction framework and sign receptors in cells and tissues themselves, in this way controlling the affectability toward such hormones (Burkill 2004; Hanahan and Weinberg 2000b). They frequently produce mitogens, or they are associated with interpretation of DNA in protein blend, which makes the proteins and catalysts responsible for creating the components advanced biochemical cells utilize and interface with. Changes in proto-oncogenes can alter their appearances and capacity, expanding the sum or action of the component protein. At the point when this occurs, they become oncogenes, and in this manner, cells have a higher opportunity to partition exorbitantly and wildly. The possibility of malignancy cannot be diminished by expelling proto-oncogenes from the genome because they are essential for development and homeostasis of the body (Hanahan and Weinberg 2000b; Suffiness and Pezzuto 1990). Tumor silencer qualities code for hostile to expansion singles and proteins that stifle mitosis and cell development. For the most part, tumor silencers are translation factors that are actuated by cell stress or DNA damage. Frequently, DNA damage will cause the nearness of free-drifting hereditary material just as different signals and will trigger proteins and pathways that lead to the initiation of tumor silencer qualities. The elements of such qualities are to capture the movement of the cell cycle to complete the DNA fix, keeping transformations from being given to daughter cells. Accepted tumor silencers incorporate the p53 protein, which is an interpretation factor initiated by numerous cell stressors, including hypoxia and UV radiation damage. The Warburg impact is the particular utilization of glycolysis for vitality to support disease development. p53 has been shown to control movement from the respiratory to the glycolytic pathway (Emerich and Thanos 2006). Amalgamation of cytochrome c oxidase 2 (SCO2) has been perceived as the downstream middleman of this impact. SCO2 is essential for directing the cytochrome c oxidase complex inside the mitochondria, and p53 can disturb the SCO2 quality p53 guideline of SCO2 and mitochondrial breath may give a potential clarification to the Warburg impact (Brooks et al. 1998). Through complex correspondence between the tumor cells, stromal cells and its microenvironment, tumor cells can attack and metastasize (Chandran et al. 2006).

6.3 Nanotherapuetics to Overcome Cancer

A significant reason for mortality overall is malignant growth. Malignant growth is one of the most widely recognized issues and genuine medical problems worldwide. It has been seen that more than one out of three individuals will develop some type of malignancy during their lifetime. An assortment of starting points for disease exist, for example, thyroid, prostate, bladder disease, kidney malignancy, pancreatic, breast malignant growth, melanoma, leukemia, oral, colon rectal malignancy, and so on. When cells duplicate and develop rapidly, shaping dangerous tumors and attacking nearby parts of the body, it is called malignant growth. A total remedy for this malady is yet to be found to date. Various techniques have been utilized in the past three decades by scientists and clinicians, but often they only forestall or hinder the development of disease (Chandran et al. 2006). The most widely recognized sorts to treat malignant growth are radiation, chemotherapy, medical procedure, immunotherapy, and photodynamic treatment. Medical procedure is for the most part (Altankov and Groth 1994) used to deal with and to break down the disease if there should arise an occurrence of any neoplasm overwhelmingly when the malignancy has not metastasized to lymph nodes or different parts of the body. Elimination of malignant growth cells by damaging their DNA with vitality radiation is a commonly used technique in radiation treatment. This treatment utilizes the host's own resistant framework to help battle disease and is called immunotherapy (Fabricant and Farnsworth 2001).

In photodynamic treatment, extraordinary medications, called photosensitizing agents, are utilized alongside light to destroy disease cells. According to the substance structure and instrument by which they act, they can be separated into four gatherings: (i) Alkalyting specialists, which damage DNA to forestall the development of malignant growth cells. (ii) Anti-metabolites, which meddle with the replication of DNA or translation of RNA by subbing the ordinary structure squares of RNA and DNA and from this time forward can cause standard cell cycle capture. (iii) Antitumor anti-infection agents, which incorporate anthracycline sactinomycin–D. (iv) Plant alkaloids obtained from regular items are mitotic inhibitors that stop mitosis and thereafter hinder the development of the cell cycle. To avoid the issues related to the abovementioned treatments, as of late, researchers have turned to nanoparticles, which can be utilized as targeted medication conveyance for disease treatment where the malignant growth cells will be annihilated instead of normal cells. More significant than fixing is the counteraction of malignancy. We realize that 80% of all tumors are related with lifestyle causes, such as smoking, biting tobacco, dietary substances, alcohol use, radiation, working environment exposure, drugs, and so forth. Cancer is an extensive issue that can involve early discovery, remedial measures, recovery of the patient, and mental issues monitored by a patient and their family members (Cho et al. 2005).

A superior method to wipe out an issue is to wipe out the reason. Nanotechnology is as an impressive means to treat malignant growth. Nanotechnology incorporates promising outcomes to decimate disease cells with insignificant harm to ordinary cells and organs, and furthermore to wipe out the malignant growth cells before they move up to the tumor. Bismuth nanoparticles have been researched and shown to amass radiation utilized in radiation treatment to treat malignant growth. The specific properties of diseases can be targeted by nanoparticles. An overabundant admission of folic corrosive may likewise secure the malignant growth development. The study of nanotechnology relies upon the way capacities change in an unexpected way when things are at nano-scale (Brooks et al. 1998).

6.4 Epidermoid Carcinoma (Skin Cancer)

Skin disease is a dangerous development on the skin that for the most part forms in the epidermis of the skin. Because of its development in the peripheral layer of the skin, it is effectively discernible in the beginning phases. Skin malignant growths are the fastest developing kind of tumors in the United States. More than one million Americans are determined to have skin malignancy yearly. There are three sorts of skin malignant growth, each of which is named after the kind of skin cell it emerges from. The first is basal cell carcinoma, which emerges from the basal cell. The subsequent one is squamous cell carcinoma also called epidermoid carcinoma, which are thin, flat cells that look like fish scales under a magnifying lens. The third kind is referred to as melanoma, which is the most dangerous type of skin malignant growth since it spreads rapidly all through the body (Habbal et al. 2011a). The word squamous is derived from the Latin word squama, which implies the size of a serpent or fish. They are likewise found in the lining of the hollow organs of the body, the respiratory and digestive tracts. The beginning phases of this carcinoma are named actinic keratosis. They show up as red unpleasant bumps on the face, ears, scalp, and rear of the hands. Men are more consistently affected than women. The manifestation causes changes of the skin that do not heal, staining, ulcers in the skin, and changes in existing moles. Exposure to UV light is a typical reason for skin damage, which is a significant reason for skin malignant growth in people. Constant non-healing wounds, particularly burns or ulcers dependent on their appearance, form into epidermoid carcinoma (Brooks et al. 1998; Dai and Mumper 2010).

The biomimetic properties of nanoparticles are utilized in location, counteraction, and in the treatment of oncologic sicknesses. Nanoparticles can also help to improve the ineffectively water-solvent medications, bioavailability, decreasing immunogenicity, and to lessen sedate digestion. The use of nanoparticles as medications and their objective toward malignant growth cells expands the viability of the particular treatment of disease cells while leaving the solid cells inert. Metastasized (epidermoid carcinoma) usually leads to a 90% chance of death within 5 years. In epidermoid carcinoma treatment, nano medication conveyance is a help. Because of their typical metabolic exercises, all plants can be used to produce synthetic substances. The phytochemicals obtained from plants are separated into two essential metabolites, incorporating sugars and fats, and optional metabolites are mixes that are found in small scale in plants but have high explicit capacities. They are helpful and financially inexpensive. The bioactive substances present in therapeutic plants are utilized in the treatment of numerous human ailments. When contrasting with advanced medications, the conventional or homegrown medication is generally less expensive (Zhang et al. 2008; Wang and Chen 2011; Roy et al. 2021).

Skin tumors usually occur in the epidermis (the uttermost layer of skin), and appear as clearly unquestionable areas. This makes most skin developments distinguishable in the beginning phases, which is in contrast to that of other tumors such as those beginning in the lung, pancreas, and stomach, where only a small minority of those affected can eradicate the disease. Skin malignancies are the fastest developing kind of tumors in the United States. Skin tumors are the most common danger, beating lung, chest, colorectal, and prostate development. More than one million Americans are determined to have skin sickness yearly. The yearly rates of a wide range of skin tumors are growing each year, creating an open concern. It has been evaluated that almost half of all Americans who live to age 65 will have a skin malady at least once.

6.4.1 Sorts of Skin Growth

Skin growth is of three sorts, each of which is named after the kind of skin cell emerges from. The first being basal cell carcinoma, which emerges from the basal cell. Second is squamous cell carcinoma (as also called epidermoid carcinoma) that emerges from the squamous cells, which are thin, level cells that look like fish scales under a magnifying instrument. These two are the most widely recognized types of skin disease. Together, these two are additionally alluded to as non-melanoma skin growths. The third sort is called melanoma, which is the most dangerous type of skin malignancy since it tends to spread (metastasize) all through the body rapidly (Daniel 2006).

6.4.2 Basal Cell Carcinoma

Basal cell carcinoma is the most frequent tumor among white people. Progression of basal cell carcinoma is due to the interaction between genes and the environment, especially uv irradiation. Several genes have been associated with basal cell carcinoma development. A basal cell carcinoma patient has increased risk of developing further skin cancers (Eg: malignant melanoma, squamous cell carcinoma as well as non-cutaneous malignancies). Increasing and repeated occurrence of basal cell carcinoma in affected individuals make them an imperative community health predicament. Treatment of basal cell carcinoma includes different forms of surgery, radiotherapy, photodynamic therapy, topical fluorouracil, and imiquimod.

6.4.3 Squamous Cell Carcinoma (Epidermoid Carcinoma)

It begins in the squamous cells, which are thin, flat cells that look like fish scales under a magnifying lens. The word squamous is derived from the Latin squama, signifying "the size of a fish or snake" because of the nearness of the cells. Squamous cells are found in the tissue that structures the outside of the skin, the lining of the hollow organs of the body, the respiratory and digestive tracts. Therefore, squamous cell carcinomas can develop in any of these tissues. Men are affected more consistently than women. The timeliest sort of squamous cell carcinoma is called actinic (or sun-based) keratosis. Actinic keratoses appear as unpleasant red bumps on the scalp, face, ears, and back of the hands. They often appear on mottled, sun-damaged skin. They can be sore and sensitive, in relation to the degree of their appearance. In a patient with actinic keratoses, the rate at which one such keratosis may further develop in the skin into a squamous cell carcinoma is found to be approximately 10–20% over 10 years, but it may take less time. An actinic keratosis that quickly becomes obviously thicker raises the concern that it may have changed into a prominent squamous cell carcinoma (Bernal et al. 2010).

6.4.4 Melanoma

Melanoma is the most dangerous kind of skin tumors. If it is seen and treated early, it is approximately 100% reparable. If it is not, the danger can advance and spread to various parts of the body, where it ends up being hard to treat and can be deadly. While it is not the most common of the skin developments, it causes the highest mortality. Melanoma is an undermining tumor that starts in melanocytes, the cells which make the pigment melanin that tone our skin, hair, and eyes. Regardless, some melanomas are skin-concealed, pink, red, purple, and blue or white (Hu et al. 2009).

6.4.5 Signs and Side Effects of Skin Disease

There is a collection of different skin ailment reactions. These include changes in the skin that do not retouch, ulcers in the skin, recoloring, and changes in existing moles.
- Basal cell carcinoma typically looks like a raised, smooth, eminent bump on the sun-exposed skin of the head, neck, or shoulders. Rarely little veins can be seen inside the tumor. Crusting and leaking in the point of convergence of the tumor routinely occurs. It is often stirred up by an irritated area that does not recover.

- Squamous cell carcinoma (SCC) is typically a red, scaling, thickened spot on sun-exposed skin. Ulceration and depleting may occur. When SCC is not dealt with, it may form into an immense mass.
- Most melanomas are darker to dull looking injuries. Signs that may indicate a perilous melanoma include change in size, shape, concealing, or tallness of a mole, the nearness of another mole during adulthood, or new pain, shivering, ulceration or death.

6.4.6 Chance Elements Prompting Skin Growth

Exposure to UV light is common cause of skin damage and the main reason behind skin threat in individuals. Effects of UV light include burns from the sun, aggravation, erythema, immunosupression, DNA damage, and apoptosis; in any case, the epidermal damage caused by extreme UV exposure will disappear within a few weeks. Never-ending or repetitive prolonged exposure to UV light on the other hand, can provoke photoaging and skin threat. In the scope of UV light, UVB and less significantly UVA are involved in the improvement of skin infection (Altankov and Groth 1994). Chronic non-repairing wounds, especially burns, are called Marjolin's ulcers considering their appearance, and can form into squamous cell carcinoma.

Genetic tendency, including "Inherent Melanocytic Nevi Syndrome" (CMNS) is depicted by the proximity of "nevi" or moles of fluctuating size that either appear at or inside a half year of birth. Nevi greater than (3/4″) in size are likely to become harmful. Skin tumor is one of the feasible dangers of significant germicidal light.

6.4.7 Treatment of Skin Tumor

Most skin malignances can be dealt with by excision of the injury, ensuring that the (edges) are free of the tumor cells. These surgical extractions give the best cure to both early and high-chance illness (Djeridane et al. 2006).

6.4.8 Curettage and Drying Up

Dermatologists as often as possible use this technique, which contains scooping out the basal cell carcinoma by using a spoon-like instrument, called a curette. Evaporating is the additional use of an electric flow to control depleting and eliminate the remainder of the development cells. The skin repairs without sewing. This framework is generally suitable for small tumors in non-basic reaches, for instance, the capacity compartment and cutoff points (Matsuo et al. 2005).

6.4.9 Radiation Treatment

Specialists regularly utilize radiation medicines for skin disease occurring in regions that are hard to treat with surgery. A decent restorative outcome often requires 25 to 30 treatment sessions.

6.4.10 Cryosurgery

A few specialists can accomplish great outcomes by solidifying basal cell carcinomas. Commonly, liquid nitrogen is connected to the development to stop and eliminate the anomalous cells.

6.4.11 Therapeutic Treatment

It uses creams that attack tumor cells (5-Fluorouracil- – 5-FU, Efudex, Fluoroplex) or brace the invulnerable system. These are used a couple of times every week for a brief period. They can cause irritation. The advantages of this procedure are that it avoids medical procedure, allows the patient to perform treatment at home, and may give an unrivaled helpful result, weighed against inconvenience and a lower fix rate, which makes clinical treatment unsuitable for treating most skin malignancies on the face. If disease has spread (metastasized) further, chemotherapy may be required. Analysts have recently been coordinating preliminaries on what they have named "safe getting ready." This treatment is still in its beginning phases, but has been shown to suitably deter dangers such as contaminations and lock onto and attack skin malignancies. Even more starting, recently researchers have focused their undertakings on fortifying the body's own regularly delivered "assistant T cells" that recognize and jolt onto damaged cells and help control the killer cells to the disease. Researchers pervaded patients with approximately 5 billion of the associate T cells with no harsh meds or chemotherapy. This sort of treatment whenever demonstrated to be convincing has no side effects and could change the manner in which malignant growth patients are managed. The probability of metastasis makes it essential to break down squamous cell carcinomas early and treat them thoroughly (Singh and Lillard Jr. 2009).

6.4.12 MTT Assay for Skin cancer Cell Lines Using Green Synthesis of Nanoparticles

In the MTT test, test samples were enhanced as 2X convergence of the cell in 100 µl volume and the amounts were: 1000, 100, 10, 1.0, 0.1 µg/ml. Because the mixes are solvent in the medium, the mixes are diminished a reasonable amount and broke up in DMSO and extra weakening were in the media when 100 µl of stock was added to the cell. The dishes were helper brooded for 48 h in the CO_2 incubator. MTT arrangement was 3-(4,5-dimethylthiazol-2-yl)- 2,5-diphenyl tetrazolium bromide (MTT) at 5 mg/ml in phosphate supported saline (1.5 mM KH_2PO_4, 6.5 mM Na_2HPO_4, 137 mM NaCl, 2.7 mM KCl; pH 7.4), from this arrangement 50 µl was pipette out into each well to achieve 1 mg/ml. The dish was also incubated for 2.30 h in an incubator and the medium was deliberately tapped. The formazan gems were air dried in a desolate spot and broke up in 100 µl DMSO, and the plates were gently shaken at room temperature and the OD was determined utilizing Synergy HT smaller scale plate pore at 570 nm. From the optical densities, the rate developments were expected with the accompanying recipe:

$$\text{Rate growth} = 100 \times \left[(T - T_0)/(C - T_0)\right]$$

On the off chance that T is more noteworthy than or equivalent to T_0, and if T is under T_0,

$$\text{Rate development} = 100 \times \left[(T - T_0)/T_0\right],$$

where T is the optical thickness of test,
  C is the optical thickness of control, and
  T_0 is the optical thickness at time zero.

From the rate developments a portion reaction bend was produced and GI50 values were introduced as of the development bends.

In cell imaging, the preliminary compound of every fixation was made in quadruplicate and combined variety was maintained under 20% among the information targets. Three arrangements of the cell lines were tried in a 96 well plate as portrayed in the under 96 well organization. The test mixes displayed unrivaled cytotoxicity/anticancer activity in both cell lines used. Test sample extract showed high cell development restraint displayed great hindrance in A431 cell lines. RPM indicated the GI50 territory of 9.2 µg/ml as a target in the A431 cell line. The result delineates with the end goal of the cytotoxic impacts of biosynthesized AgNPs more noteworthy than before within the sight of the concentrate on malignancy cell line. These deeds can be applied to the mixes in the concentrate that improved the activity of the AgNPs.

Immovability of AgNPs can be applied to preserver these mixes in the wrapping of nanoparticles. The biosynthesized AgNPs were separated from the concentrate, and their anticancer impacts were inspected. In this investigation, the anticancer exercises of biosynthesized AgNPs were contrasted. As seen in the convergence of AgNPs in blepharis maderaspatensis, cytotoxicity in A431 cells was more

6 Nanomaterials for A431 Epidermoid Carcinoma Treatment

noteworthy than previously. The IG50 estimation of disengaged AgNPs in the concentrate was 64.4 and 9.2 μg/ml. Biosynthesized AgNPs could partake in a basic errand in cultivating their bioavailability similar to remedial applications in illnesses such as malignant growth. This investigation exhibits the chance of utilizing AgNPs to slow the development of malignant growth cells and their cytotoxicity for pending remedial medicines, and offers another strategy to fight different ailments, for example, disease, arthritis, and neovascularization. The test samples have been tried in the MTT test for the cytotoxic potential in A431 cell line. The concentrate displayed superb anticancer action comparable to SNP. The test samples in SNP and concentrates showed a G150 of 64.4 and 9.2 μg in A431 cell line. This concentrate restraint is powerful and groundbreaking; furthermore, the test samples can be taken for other anticancer examinations, including apoptosis and cell cycle examination to find the component of cell development hindrance/cytotoxicity (Aziz et al. 2019). Silver nanoparticles are synthesized by using three routes physical, chemical, and biological, but the biological route plays a vital role because it is eco-friendly in nature (Prasad 2014; Prasad et al. 2016, 2018, 2020; Srivastava et al. 2021), as shown in Fig. 6.1.

The skin cancer detection has the following steps (as shown in Fig. 6.2): the image input, preprocessing by median filter, segmentation of cluster ring cells, extraction of features, detection of them, and then post processing.

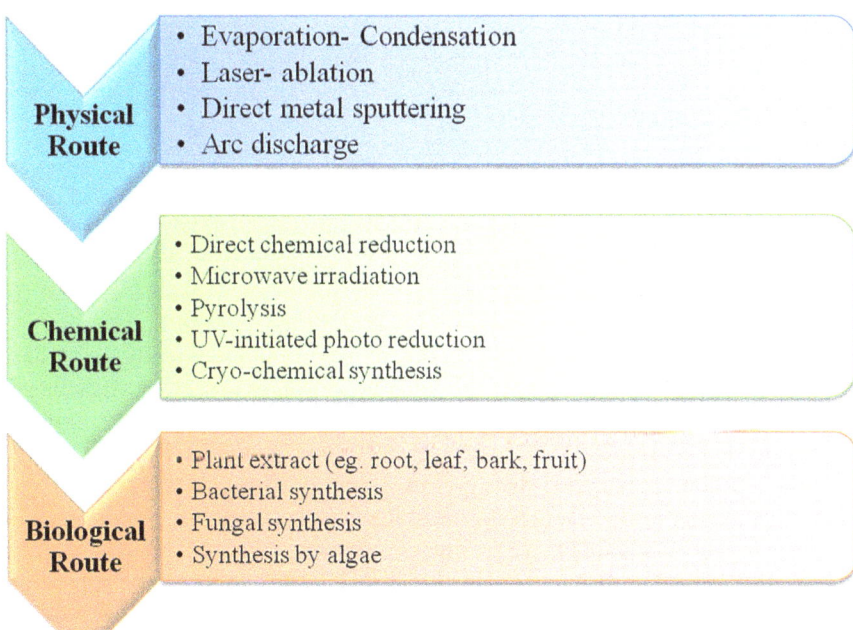

Fig. 6.1 Different routes of AgNps synthesis

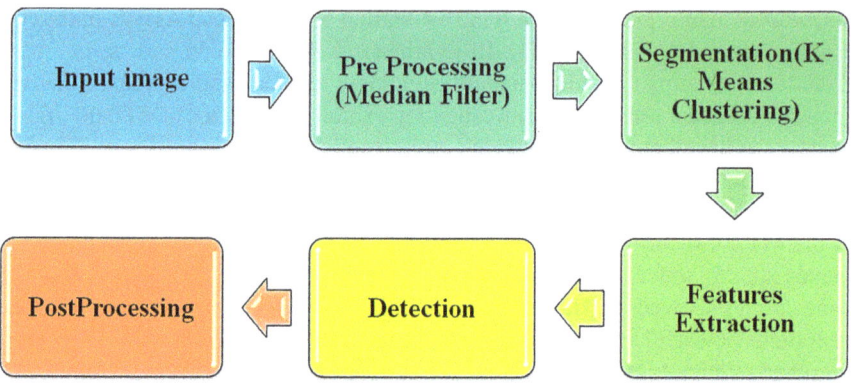

Fig. 6.2 Flow chart of skin cancer detection

6.5 Nano Drug Delivery in Skin Cancer

Numerous nanoparticles have been studied for the treatment of skin diseases, particularly in melanoma treatment, including liposomes, dendrimers, polymersomes, carbon-based nanoparticles, inorganic nanoparticles, and protein-based nanoparticles. In the accompanying sections, the qualities of the regular nanoparticles utilized in skin malignant growth treatment are portrayed.

6.5.1 Liposomes

Liposomes are phospholipid vesicles (measurement of 50–100 nm and much larger) that have a bilayered film structure, similar to that of organic layers, along with a neutral watery stage. Liposomes are grouped by size and number of layers into multi-, ligo, or unilamellar. The fluid center can be utilized for embodiment of water-soluble medications, though the lipid bilayers may hold hydrophobic or amphiphilic mixes. To escape from reticulo endothelial system (RES) take-up after i.v. infusion, PEGylated liposomes, "secrecy liposomes," were produced for diminishing leeway and drawing out course half-life (Sanvicens and Marco 2008).

Early exploration exhibited that liposomes stay in the tumor interstitial liquid close to the tumor vessels. At present, a few liposomal plans in the clinical practice contain a few medications for the treatment of various sorts of malignant growth, including melanoma (Allen and Cullis 2004). A few other liposomal chemotherapeutic medications are at the different phases of clinical trials. In addition, progress with cationic liposomes prompted the fruitful conveyance of small interfering RNA (siRNA). Liposomes can likewise be altered to join amagnetic components for use in checking their development inside the body utilizing MRI or to capture gases and medications for ultrasound-controlled medication conveyance (Caldecott and Tierra 2006).

6.5.2 Solid Lipid Nanoparticles (SLNs)

SLNs were presented toward the start of the 1990s as an elective conveyance framework to liposomes, emulsion, and polymeric NP. SLNs present a high physical strength; that is, they can ensure the medications against corruption, and they permit a simple control for the medication discharge. The planning of SLNs does not require the utilization of natural solvents. They are biodegradable and biocompatible and have low toxicity. In addition, the production and sterilization on a large scale are rather easy. Solid lipid nanoparticles (SLNs) containing docetaxel improve the efficacy of this chemotherapeutic agent in colorectal (C-26) and malignant melanoma (A-375) cell lines in "in vitro" and "in vivo" experiments (Jabr-Milane et al. 2008).

6.5.3 Dendimers

Dendrimers are unimolecular, monodisperse, synthetic polymers (<15 nm) with layered architectures that are composed of a central core, an internal region consisting of repeating units, and various terminal groups that determine the three-dimensional dendrimer characteristics structures. Dendrimers can be prepared for the delivery of both hydrophobic and hydrophilic drugs, nucleic acids, and imaging agents due to their attractive properties such as well-defined size and molecular weight, monodispersed, multivalence, number of available internal cavities, high degree of branching, and high number of surface functional groups (Torchilin 2005). Several literature sources demonstrate the ability of dendrimer targeting ligands to induce the specific targeting and destruction of tumors. They include oligosaccharides, polysaccharides, oligopeptides, and polyunsaturated fatty acids as well as folate and tumor associated antigens (Yuan et al. 1994). New developments in polymer and dendrimer chemistry have provided a new class of molecules called dendronized polymers, which are linear polymers that bear dendrons at each repeat unit, obtaining drug delivery advantages because of their enhanced circulation time. Another approach is to synthesize or conjugate the drug to the dendrimers so that incorporating a degradable link can be further used to control the release of the drug. They have also found applications in the diagnostic imaging of cancer cells, such as MRI. Gadolinium-conjugates dendrimers have allowed the selective comprehensive targeting and imaging of tumors (Jerant et al. 2000).

6.5.4 Quantum Dots

Quantum dots (QDs) are colloidal fluorescent semiconductor nanocrystals (2–10 nm). They possess a broad absorption band and a symmetric, narrow emission band, typically in the visible to near infrared (NIR) spectral range (Ling et al. 2011). Quantum dots are photostable; therefore, the optical properties of QDs make them suitable for highly sensitive, long term, and multi target bio-imaging applications (Tsai et al. 2009). Indeed, in order for QDs to be used for melanoma detection, the surface must be treated to increase hydrophilicity and the desired tumor-targeting ligand must be attached. Possible ligands include antibodies, peptides, and small-molecule drugs/inhibitors. New methodologies, for example, the expansion of a silica covering or a biocompatible polymer covering, have additionally expanded the biocompatibility and diminished their toxicity. In fact, even though quantum dots offer many points of interest in detecting and imaging and as complexity specialists in different procedures such as MRI, PET, IR fluorescent imaging, and processed tomography, there is uncertainty surrounding the toxicity of the materials used (Amstad et al. 2011).

6.5.5 Nanotubes

Carbon nanotubes have a place with the group of fullerenes and are made of coaxial graphite sheets (<100 nm) folded up into chambers. These structures can be obtained either as single- (one graphite sheet) or multi-walled nanotubes (a few concentric graphite sheets). They show incredible physical, photochemical, and electrochemical properties. Attributable to their metallic or semiconductor conduct, nanotubes are frequently utilized as biosensors. Carbon nanotubes can be additionally utilized as medication bearers and tissue repair frameworks (Bañobre-López et al. 2013). Tumor targeting single-walled carbon nanotubes (SWCNT) have been integrated by covalent connection of various duplicates of tumor-explicit monoclonal antibodies, radiation particle chelates, and fluorescent tests (Hu et al. 2016). This conveyance framework can be stacked with a few atoms of an anticancer medication, in light of the fact that no covalent bonds are required, so the expanded payload does not fundamentally change the targeting capacity of the counter acting agent. They have likewise been rebuilt to convey gadolinium molecules for MRI of tumors and have been surface functionalized with receptor agonists and foes for tumor targeting (Vannucci et al. 2012).

6.6 Efficacy of Nano Treatment in Future for Skin Cancer

Novel treatments are fast approaching for almost any ailment, and they are intended to be progressively productive, less expensive, and with no penalty to patient security, sometimes even improving it. To accomplish the last mentioned, a perfect treatment ought to be formed considering acceptable patient consistence, a superior treatment proficiency, a low conceivable harmfulness, and extremely exceptional returns of arriving at the target hand in the body per unit mass of the medication. Nanotechnology based treatments can be progressively effective, since they can be adorned with target moieties (e.g., antibodies) (Sahoo et al. 2011). Both referenced also lead to a decreased toxicity, since lower dosages are important to accomplish a similar impact, as the target and controlled discharge at the target in the vicinity, and renders the harmfulness exceptionally limited. The last is likewise in direct connection to the significant returns of the conveyance (Ma et al. 2012). Knowing these likely points of interest of such plans, it is not unexpected that these are intensely explored and that there is an appeal for their take-up into clinical practice once demonstrated safe. Furthermore, without a doubt, there are at any rate two previously endorsed nanotechnology-based plans utilized in malignancy treatment; Doxil (Janssen Biotech, Horsham, PA, USA), a doxorubicin containing liposome infusion, and Nab-paclitaxel (Abraxane), which contains paclitaxel bound to egg whites' nanoparticles (nm).

Considering all referenced skin malignant growth treatments, there is some encouraging research progress that could mean huge upgrades in the potential treatment result for skin disease patients. A few unique types of novel nano-sized plans have been recently created and are in various phases of testing (Maver et al. 2009). Among others, different lipid-based particles (e.g., micelles, strong lipid particles) are especially mainstream. These not only frequently have the capacity to outmaneuver the host defenses of the living being, which for itself prompts improved yields of the payload conveyance (Navarro-Pardo et al. 2016), but also provide other diverse targeting ways to further improve this yield significantly. Carbon nanostructures remain among the most explored kinds of nanostructures for biomedical applications (Makovec et al. 2009) particularly because of their different applications (e.g., sub-atomic gadgets), and countless adjustment approaches are accessible for them making them intriguing contenders for future viable malignant growth treatment (Siu et al. 2014).

In addition, the most encouraging future treatment approaches lie in the field of theranostics, which joins all the right "fixings" to meet all the prerequisites of the perfect treatment approach. A few novel treatment arrangements in skin malignant growth have been examined (Huber et al. 2015).

It is important that, from our viewpoint, MNPs appear to be among the most encouraging nanoparticulate specialists, particularly in the treatment of "difficult-to-treat" disease structures (e.g., MM). Among MNPs, the most regularly utilized particles are unquestionably based on iron oxides (Lu 2014). The last were joined with different medications (e.g., epirubicin) and are on their own equipped for being utilized to initiate attractive hyperthermia, and since they are generally functionalized so as to forestall agglomeration, their surface is reasonable for securing distinctive targeting moieties (e.g., antibodies) (Sharma et al. 2016).

6.7 Conclusion

Nanomedicine, as a promising instrument for disease treatment, has received increased attention and yielded an incredible number of health advantages. Recently, different sorts of nanomaterials, which could display better cell take-up, tumor site explicitness, and delayed course time after surface change, have been explored for imaging, locating, and treatment of malignant growths. Regarding these points of interest, nanomaterial-based therapeutics have shown equivalent or even better anticancer viability than other industry standards, while showing decreased symptoms, giving new methodologies to battle against carcinomas. An assortment of nanomaterials have been introduced, including polymeric NPs, liposomes, dendrimers, polymeric micelles, polymer-medicate conjugates, silica NPs, carbon nanotubes, nanographene, attractive NPs, AuNPs, and quantum specks. The examination of these nanomaterials through improving their structure and cell targeting capacity has given an increasingly useful restorative conveyance. A large portion of these works have indicated extraordinary potential to make life changing nanomedicine treatments for clinical malignancy treatment. The most significant concern would be nanomaterial toxicity and potential wellbeing dangers, as little is known about how they carry on in people. Liposomes are the most evolved, are clinically affirmed, and currently have the best number of clinical preliminaries with certain plans effectively accessible in the commercial center, while numerous other nanomaterial-based devices, particularly inorganic ones, have not received such endorsement and achievement. Some nanomaterials are not biodegradable and might be held inside the body for extensive periods of time after insertion. The advancement of biocompatible and biodegradable nanomaterials for malignant growth treatment could, along these lines, have an a lot higher clinical worth. Considering everything, tranquilize conveyance has all the earmarks of being a promising methodology for a superior viable melanoma treatment.

References

Akhani H, Ghasemkhani M, Chuong SDX, Edwards GE (2008) Occurrence and forms of Kranz anatomy in photosynthetic organs and characterization of NAD-ME subtype C4 photosynthesis in *Blepharis ciliaris* (L.) B. L. Burtt(Acanthaceae). J Exp Bot 59(7):1755–1765

Alam MB, Hossain S, Haque ME (2011) Antioxidant and anti-inflammatory activities of the leaf extract of *Brassica nigra*. Int J Pharm Sci Res 2:303–310

Allen TM, Cullis PR (2004) Drug delivery systems: entering the mainstream. Science 303(5665):1818–1822

Altankov G, Groth A (1994) Reorganization of substratum-bound fibronectin on hydrophilic and hydrophobic materials is related to biocompatibility. J Mater Sci Mater Med 5:732

Amstad E, Textor M, Reimhult E (2011) Stabilization and functionalization of iron oxide nanoparticles for biomedical applications. Nanoscale 3(7):2819–2843

Ankanna S, Prasad TNVKV, Elumalai EK, Savithramma N (2010) Production of biogenic silver nanoparticles using boswellia ovalifoliolata stem bark. Digest J Nanomater Biostruct 5:369–372

Ayyanar M, Ignacimuthu S (2009) Herbal medicines for wound healing among tribal people in Southern India: Ethnobotanical and Scientific evidences. Int J App Res Nat Prod 3:29–42

Aziz N, Faraz M, Sherwani MA, Fatma T, Prasad R (2019) Illuminating the anticancerous efficacy of a new fungal chassis for silver nanoparticle synthesis. Front Chem 7:65. https://doi.org/10.3389/fchem.2019.00065

Bañobre-López M, Teijeiro A, Rivas J (2013) Magnetic nanoparticle-based hyperthermia for cancer treatment. Rep Pract Oncol Radiother 18(6):397–400

Bernal J, Mendiola JA, Ibanez E, Cifuentes A (2010) Advanced analysis of nutraceuticals. J Pharm Biomed Anal 55:S758–S774

Brooks AC, Whelan CJ, Purcell WM (1998) Reactive oxygen species generation and histamine release by activated mast cells: modulation by nitric oxide synthase inhibition. Br J Pharmacol 128:585–590

Burkill HM (2004) The useful plants of west tropical Africa. Royal Botanic Gardens, Kew. Burg, K. J. L.; Porter, S.; Kellam, J. F, 2000. Biomaterials 21:2347

Caldecott T, Tierra M (2006) Ayurveda: the divine science of life. Elsevier Health Sciences, New York, pp 161–162

Chandran SP, Chaudhary M, Pasricha R, Ahmad A, Sastry M (2006) Synthesis of gold nanotriangles and silver nanoparticles using Aloe vera plant extract. Biotechnol Prog 22:577–583

Cho K, Park J, Osaka T, Park S (2005) The study of antimicrobial activity and preservative effects of nanosilver ingredient. Electrochim Acta 51:956–960

Dai J, Mumper RJ (2010) Plant Phenolics: extraction, analysis and their antioxidant and anticancer properties. Molecules 15:7313–7352

Daniel M (2006) Medicinal plants: chemistry and properties. Science Publishers, New York, p 53

Djeridane A, Yousfi M, Nadjemi B, Boutassouna D, Stocker P, Vidal N (2006) Antioxidant activity of some Algerian medicinal plants extracts containing phenolic compounds. Food Chem 97:654–660

Emerich DF, Thanos CG (2006) The pinpoint promise of nanoparticle-based drug delivery and molecular diagnosis. Biomol Eng 23(4):171–184

Fabricant DS, Farnsworth NR (2001) The value of plants used in traditional medicine for drug discovery. Environ Health Perspect 109(1):69–75

Habbal O, Hasson SS, El-Hag AHZ, Al-Hashmi N, Al-Bimani Z, Al-Baluschi MS (2011a) Antibacterial activity of *Lawsonia inermis* Linn (Henna) against *Pseudomonas eruginosa*. Asian Pac J Trop Biomed 1:173–176

Habbal O, Hasson SS, El-Hag AHZ, Al-Hashmi N, Al-Bimani Z, Al-Baluschi MS (2011b) Antibacterial activity of *Lawsonia inermis* Linn (Henna) against *Pseudomonas aeruginosa*. Asian Pac J Trop Biomed 1:173–176

Hanahan D, Weinberg RA (2000a) The hallmarks of cancer. Cell 100:57–70

Hanahan D, Weinberg RA (2000b) The hallmarks of cancer. Cell 100:57–70

Hu R, Yong K-T, Roy I, Ding H, He S, Prasad PN (2009) Metallic nanostructures as localized plasmon resonance enhanced scattering probes for multiplex dark-field targeted imaging of cancer cells. J Phys Chem C 113(7):2676–2684

Hu R, Ma S, Ke X, Jiang H, Wei D, Wang W (2016) Effect of interleukin-2 treatment combined with magnetic fluid hyperthermia on Lewis lung cancer-bearing mice. Biomedi Rep 4(1):59–62

Huber LA, Pereira TA, Ramos DN et al (2015) Topical skin cancer therapy using doxorubicin-loaded cationic lipid nanoparticles and iontophoresis. J Biomed Nanotechnol 11(11):1975–1988

Jabr-Milane LS, vanVlerken LE, Yadav S, Amiji MM (2008) Multi-functional nanocarriers to overcome tumor drug resistance. Cancer Treat Rev 34(7):592–602

Jemal A, Siegel R, Ward E, Murray T, Xu J, Smigal C, Thun MJ (2006) Cancer statistics CA cancer. J Clin 56:106–130

Jerant F, Johnson JT, Sheridan CD, Caffrey TJ (2000) Early detection and treatment of skin cancer. Am Fam Physician 62(2):357–382

Jung HS, Kong WH, Sung DK et al (2014) Nanographene oxide-hyaluronic acid conjugate for photothermal ablation therapy of skin cancer. ACS Nano 8(1):260–268

Ling Y, Wei K, Luo Y, Gao X, Zhong S (2011) Dual docetaxel/superparamagnetic iron oxide loaded nanoparticles for both targeting magnetic resonance imaging and cancer therapy. Biomaterials 32(29):7139–7150

Liu RH (2004) Potential synergy of phytochemicals in cancer prevention: mechanism of action. Supplement to the international research conference on food, nutrition, and cancer. J Nutr 134:3479–3485S

Lu Z-R (2014) Theranostics: fusion of therapeutics and diagnostics. Pharm Res 31(6):1355–1357

Ma M, Hao Y, Liu N et al (2012) A novel lipid-based nanomicelle of docetaxel: evaluation of anti-tumor activity and biodistribution. Int J Nanomedicine 7:3389–3398

Makovec D, Čampelj S, Bele M et al (2009) Nanocomposites containing embedded superparamagnetic iron oxide nanoparticles and rhodamine 6G. Colloids Surf A Physicochem Eng Asp 334(1–3):74–79

Matsuo M, Sasaki N, Saga K, Kaneko T (2005) Cytotoxicity of flavonoids toward cultured normal human cells. Biol Pharmacol Bull 28(2):253–259

Maver U, Bele M, Makovec D, Čampelj S, Jamnik J, Gaberšček M (2009) Incorporation and release of drug into/from superparamagnetic iron oxide nanoparticles. J Magn Magn Mater 321(19):3187–3192

Navarro-Pardo F, Martinez-Hernandez AL, Velasco-Santos C (2016) Carbon nanotube and graphene based polyamide electrospun nanocomposites: a review. J Nanomater 2016:16. Article ID 3182761

Panhwar AQ, Abro H (2007) Ethnobotanical studies of MahalKohistan. Pak J Bot 39:2301–2315

Prasad R (2014) Synthesis of silver nanoparticles in photosynthetic plants. Journal of Nanoparticles, Article ID 963961, 2014, https://doi.org/10.1155/2014/963961

Prasad R, Pandey R, Barman I (2016) Engineering tailored nanoparticles with microbes: quo vadis. WIREs Nanomed Nanobiotechnol 8:316–330. https://doi.org/10.1002/wnan.1363

Prasad R, Jha A, Prasad K (2018) Exploring the Realms of Nature for Nanosynthesis. Springer International Publishing (ISBN 978-3-319-99570-0) https://www.springer.com/978-3-319-99570-0

Prasad R, Siddhardha B, Dyavaiah M (2020) Nanostructures for Antimicrobial and Antibiofilm Applications. Springer International Publishing (ISBN 978-3-030-40336-2) https://www.springer.com/gp/book/9783030403362

Ribeiro A, Romeiras MM, Tavares J, Faria MT (2010) Ethnobotanical survey in Canhane village, district of Massingir, Mozambique: medicinal plants and traditional knowledge. J Ethnobiol Ethnomed 6:33

Roy A, Datta S, Bhatia KS, Bhumika, Jha P, Prasad R (2021) Role of plant derived bioactive compounds against cancer. South African Journal of Botany. https://doi.org/10.1016/j.sajb.2021.10.015

Sahoo NG, Bao H, Pan Y et al (2011) Functionalized carbon nanomaterials as nanocarriers for loading and delivery of a poorly water-soluble anticancer drug: a comparative study. Chem Commun 47(18):5235–5237

Sanvicens N, Marco MP (2008) Multifunctional nanoparticles—properties and prospects for their use in human medicine. Trends Biotechnol 26(8):425–433

Sharma P, Mehra NK, Jain K, Jain NK (2016) Biomedical applications of carbon nanotubes: a critical review. Curr Drug Deliv 13(6):796–817

Singh R, Lillard JW Jr (2009) Nanoparticle-based targeted drug delivery. Exp Mol Pathol 86(3):215–223

Siu KS, Chen D, Zheng X et al (2014) Non-covalently functionalized single-walled carbon nanotube for topical siRNA delivery into melanoma. Biomaterials 35(10):3435–3442

Srivastava S, Usmani Z, Atanasov AG, Singh VK, Singh NP, Abdel-Azeem AM, Prasad R, Gupta G, Sharma M, Bhargava A (2021) Biological nanofactories: Using living forms for metal nanoparticle synthesis. Mini-Reviews in Medicinal Chemistry 21(2):245–265

Suffiness M, Pezzuto JM (1990) Assays related to cancer drug discovery. In: Hostettmann K (ed) Methods in plant biochemistry: assays for bio-activity. Academic Press, London

Sukirtha R, Priyanka KM, Antony JJ, Kamalakkannan S, Thangam R, Gunasekaran P, Krishnan M, Achiraman S (2012a) Cytotoxic effect of Green synthesized silver nanoparticles using Melia azedarach against in vitro HeLa cell lines and lymphoma mice model. Process Biochem 47:273–279

Sukirtha R, Priyanka KM, Antony JJ, Kamalakkannan S, Thangam R, Gunasekaran P, Krishnan M, Achiraman S (2012b) Cytotoxic effect of Green synthesized silver nanoparticles using Melia azedarach against in vitro HeLa cell lines and lymphoma mice model. Process Biochem 47:273–279

Suriyavathana M, Kumar M (2010) Int J Appl Biol Pharm Technol 3:1098–1100

Tang Z, Wang Y, Podsiadlo P, Kotov NA (2006) Biomedical Applications of Layer-by-Layer Assembly: From Biomimetics to Tissue Engineering. Adv Mater 18:3203–3224

Torchilin VP (2005) Recent advances with liposomes as pharmaceutical carriers. Nat Rev Drug Discov 4(2):145–160

Tsai C, Chen C, Hung Y, Chang F, Mou C (2009) Monoclonal antibody-functionalized mesoporous silica nanoparticles (MSN) for selective targeting breast cancer cells. J Mater Chem 19(32):5737–5743

Vannucci L, Falvo E, Fornara M et al (2012) Selective targeting of melanoma by PEG-masked protein-based multifunctional nanoparticles. Int J Nanomedicine 7:1489–1509

Wang Y, Chen L (2011) Quantum dots, lighting up the research and development of nanomedicine. Nanomedicine 7(4):385–402

Yuan F, Leunig M, Huang SK, Berk DA, Papahadjopoulos D, Jain RK (1994) Microvascular permeability and interstitial penetration of sterically stabilized (stealth) liposomes in a human tumor xenograft. Cancer Res 54(13):3352–3356

Zhang L, Zhang N (2013) How nanotechnology can enhance docetaxel therapy. Int J Nanomedicine 8(1):2927–2941

Zhang L, Gu FX, Chan JM, Wang AZ, Langer RS, Farokhzad OC (2008) Nanoparticles in medicine: therapeutic applications and developments. Clin Pharmacol Therap 83(5):761–769

Zhao H, Li Z, Yang B, Wang J, Li Y (2015) Synthesis of dual-functional targeting probes for cancer theranostics based on iron oxide nanoparticles coated by centipede-like polymer connected with pH-responsive anticancer drug. J Biomater Sci Polym Ed 26(16):1178–1189

Chapter 7
Efficacy of Nanomaterials and Its Impact on Nosocomial Infections

P. Kalitha Parveen, S. Christobher, Balamuralikrishnan Balasubramanian, Durairaj Kaliannan, Manikantan Pappusamy, and Arun Meyyazhagan

Contents

7.1	Introduction..	238
	7.1.1 Miniature Things with Marvelous Impact.............................	240
	7.1.2 Infectious Disease...	241
7.2	Infectious Agent..	242
	7.2.1 Types of Infectious Agent...	243
7.3	Antimicrobial Agents..	244
7.4	Antifungal Agents..	245
7.5	Nano War against Infectious Disease..	245
	7.5.1 Nanomaterials in Bacterial Detection.................................	246
7.6	Nanomaterials in Viral Detection...	251
	7.6.1 SERS..	252
	7.6.2 Electrochemical Biosensing...	252
	7.6.3 Other Biosensing Methods..	253
7.7	Advanced Nano Biomaterials to Treat Infectious Disease..................	253
	7.7.1 Nano Vaccine...	254
	7.7.2 Nano Adjuvant..	254

P. Kalitha Parveen (✉)
PG Department of Zoology, Hajee Karutha Rowther Howdia College of Arts and Science, Uthamapalayam, Tamil Nadu, India

S. Christobher
Department of Zoology, Nallamuthu Gounder Mahalingam College,
Pollachi, Tamil Nadu, India

B. Balasubramanian
Department of Food Science and Biotechnology, College of Life Sciences, Sejong University, Seoul, South Korea

D. Kaliannan
Zoonosis Research Center, Department of Infection Biology, School of Medicine, Wonkwang University, Iksan, Republic of Korea

M. Pappusamy · A. Meyyazhagan (✉)
Department of Life Sciences, CHRIST (Deemed to be University),
Bangalore, Karnataka, India

7.7.3	Quorum Sensing	255
7.8	Conclusion	256
References		257

7.1 Introduction

Nanotechnology is a multidisciplinary area that covers a tremendous scope from designing, material science and various other science disciplines. Nanotechnology has enabled tremendous advances in science and innovation, opening doors for advances in the fields of medicine, gadgets, nutrition, and the earth. It comprises the group of tiny structures and the prefix "nano" is a Greek expression defined as "diminutive person or scaled down" (Feynman 1959). Nanotechnology gives the ability to design the properties of assets by using their size, and this has led to exploration towards an enormous area of likely uses for nanomaterials. The advantages in edifying available treatments propel established researchers to continue searching for inventive ways to fight contaminations (Kannan et al. 2014).

Nanotechnology is empowering innovation that manages nanometer-sized things. Nanotechnology involves a few levels: materials, gadgets, and frameworks. Nanomaterials are the principal components of the quickly growing field of nanomedicine and bionanotechnology. Nanoscale structures and materials (nanoparticles, nanowires, nanofibers, and nanotubes) have been investigated in many natural applications (biosensing, organic severance, sub-atomic imaging, and anticancer treatment) because their novel properties and capacities vary significantly from their large counterparts. Nanotechnology can be used in medication and medical procedures as designers find novel approaches to apply these particles. There is abundant room for up and coming endeavors with tranquilize conveyance frameworks and cell targeting to create more efficient applications. As shown in Fig. 7.1, the size of

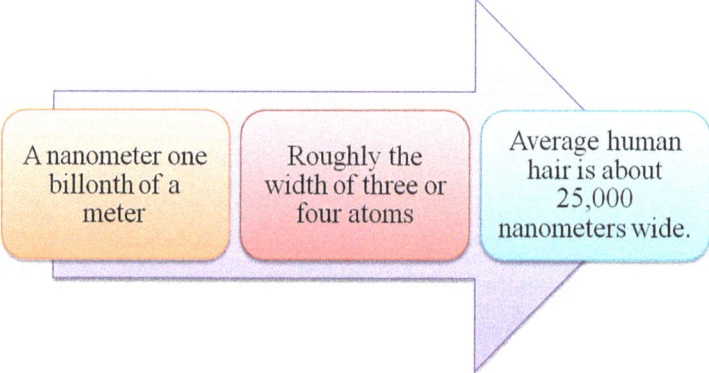

Fig. 7.1 Size of a nanometer

a nanometer is one billionth of a meter, they are roughly the width of three or four atoms, and the human hair is about 25,000 nanometers wide.

The nanoparticles (NPs) have become a fixture because of their use in industry frameworks, along with client items, pharmaceuticals, beauty care products, transportation, influence, cultivating, and so forth, and are continuously implemented in new modern applications. A captivating application of NPs in the field of life sciences is their use in rich freedom frameworks. Metal NPs are of enormous specialized consideration because they connect the gap between the massive and nuclear structures. NPs have supreme physicochemical properties, i.e., raised surface territory, transcending reactivity, tunable pore size, and molecule morphology. Current progress in nanotechnology includes the amalgamation of metallic NPs into changed industrialized, healing, and household items. The size comparison of nanotechnology is shown in (Fig. 7.2) as milliliter, micrometer and nanometer; the examples are five million red blood cells in a drop of blood for millimeter, the blood cells micrometers, and the strand of DNA present in the whole blood cells are 2 nm wide.

Moreover, it is possible that these engaged methodologies could become multiuseful with various procedures and health advantages. The universe of nanotechnology in medicine is currently available, and there is significantly more to learn. In science and medicine, nanotechnology includes the materials, gadgets, and frameworks whose structure and capacity are connected for small length scales, from nanometers (10–9 m) through microns (10–6 m). Various aspects of nanotechnology rely upon the way that it is likely to adjust the structures of assets at extremely

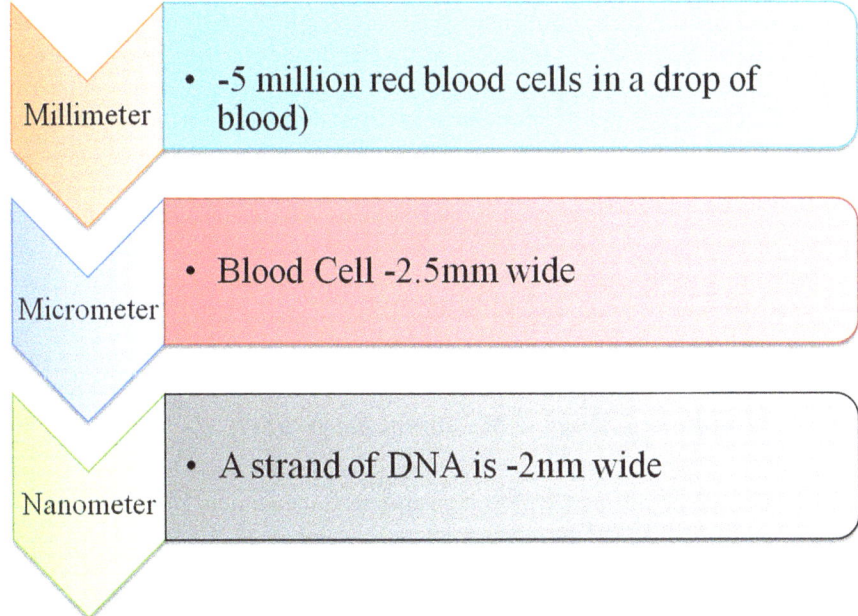

Fig. 7.2 Size comparisons of nanotechnology

small scales to realize specific properties (Buzea et al. 2007). Nano biotechnology is a field that concerns the natural framework streamlined through movement, for example, cells, cell instrument, nucleic corrosive, and proteins to smooth the advancement of utilitarian nano organized and mesoscopic engineering contained natural and inorganic materials. Bio functionalization of nanoparticles is a basic job of current day nano biotechnology. Then again, bio nanotechnology generally refers to how the objectives of nanotechnology can be guided by concentrating on how organic "machines" work and adjust these natural themes into improving available nanotechnologies or making novel ones (Fang et al. 2014).

We can characterize a nanoparticle as a molecule with at least one measurements under 100 nm. The history of nanoparticles dates to the ninth century in Mesopotamia where craftsmen used a few mixes to cover pots. These "paints" molded a glittering impact on the outside of the pots. Because of its size, a nanoparticle shows matchless optical, physical, and compound properties, for example, colossal electrical and warm conductivity, photoemission, and amazing synergist action, among others. Maybe the most common use of nanoparticles in medicine today is in tranquilize conveyance frameworks. The points of interest are numerous over customary conveyance frameworks.

Nanosystems might be utilized for the diagnosis and treatment of viral and contagious diseases. Gainful expository tests dependent on nanosystems are available (Kannan et al. 2013). Differing techniques dependent on nanoparticles (NPs) have been created to recognize unambiguous agents or to separate Gram-positive and Gram-negative microorganisms. Biosensors dependent on nanoparticles have been helpful in viral location to advance reachable basic strategies (Zazo et al. 2017). A few purpose-of-care (POC) tests have been foreseen that can give earlier results, simpler, and at lesser expense than common methods and can even be used in inaccessible locales for viral analysis. Quorum sensing is an upgraded strategy interrelated with population density that microorganisms utilize to authorize biofilms creation. Nanostructured materials that hamper signal particles concerned in biofilm growth have been expanded for the intensity of contaminations involved with biofilm-related diseases. In summary, nanoparticles make an engaging platform for theranostic applications, and frameworks that consolidate drugs and specific sorts of nanoparticles dispense helpful specialist delivery such as the imaging of an objective organ or tissue (Burlage and Tillmann 2017).

7.1.1 Miniature Things with Marvelous Impact

Nanoscience is a promising territory of science that includes the group of materials on an ultra-small scale and the novel properties that these materials possess. One of the most elating fundamentals of administration in the nanoworld is that effects act in an alternate manner when things are ultra-small. At the point when molecule sizes are dense to the nanoscale, the extent of surface area to amount increases extensively. The ability to adjust the center structures of materials at the nanoscale to achieve exact properties is at the heart of nanotechnology. A couple of instances of contemporary nanotechnology include the following. Nanosensors in wrapping can see salmonella and different contaminants in food. Other blending improvements grasp the chance of utilizing nanotechnology to intensify the extension of nerve cells (for instance, in a harmed cerebrum or spinal string), and by methods for nano strands to encourage fortification of crushed spinal nerves (currently being tried on mice).

Nanotechnology is being used in a progression of energy territories—to recuperate the ability and cost-adequacy of sun-based boards, produce modern sorts of batteries, advance the skill of fuel making by methods for upgraded catalysis, and produce improved enlightenment frameworks. Nanoengineered materials are in a variety of items, including high-power batteries, fuel added substances and energy units, and upgraded exhaust systems, which produce cleaner exhaust for longer periods. Nanostructured channels that can kill infection cells and different flotsam and jetsam from water may in the long run encourage production of soil free, economical, and plentiful drinking water. A nanofabric paper towel, which can sop up multiple times its weight in oil, can be used for oil slick clean up tasks. Nanoscale added substances in textures help resist recoloring, wrinkling, and bacterial development. There is additionally the possibility that nanomaterials may venture out from life form to life form, or completely through natural pecking orders. Despite these worries, most researchers expect that nanoscience will be instrumental in mammoth advances in cures, biotechnology, industry, data innovation, and other territories. Nanoscience is about the ultra-small; however, it has the likelihood to have an epic impact on our lives.

7.1.2 Infectious Disease

Pathogenic diseases have been consistently prosperous. Plants and creatures as well as people are habitually infected by such pathogenic elements causing fierce illnesses; some are even basic and some lead to raised recuperating cost, other prosperity payment, and high mortality hazard. These inconvenient microorganisms can cause high death rates, incapacity, and ailments in plants and creatures. Thinking back to the former times of human maladies, during the nineteenth century, it was thought that microorganisms were liable for a variety of irresistible ailments that had been afflicting mankind from old days. Some bacterial illnesses, for example, tuberculosis, typhus, plague, diphtheria, typhoid fever, cholera, loose bowels, and pneumonia have negatively affected humankind. In 1997, coronary illness and malignancies represented 55% all things considered, with 4.5% owing to pneumonia, flu, and human immunodeficiency infection (HIV) disease (Hoyert et al. 1999). Organisms are likewise the source of various sicknesses in plants, which, if crop plants or woods assets, may have basic practical or social results.

Plant sicknesses have forever been opposed to plant development and yield creation in a few parts of the planet. Plant ailments can influence plants in various ways, for example, the absorbance and translocation of stream and supplements, photosynthesis, bloom and natural product improvement, plant growth and augmentation, and cell division and amplification. Plant illnesses can be brought about by various sorts of organisms, microbes, phytoplasma, infections, viroids, nematodes, and different agents. Plant maladies are notable to decrease the food available to people by reducing crop yields. This can result in deficient food to people or lead to starvation and death in the most shocking cases. For instance, late blight of potato, which is brought about by *Phytophthora infestans*, destroyed potatoes, which were the principle crop in Ireland during 1845–1850. This brought about the Great Famine (or Great Hunger), where approximately one million individuals died and another million moved to Canada, the USA, and different nations (Nowicki et al. 2011). One of the most widely recognized ways by which plant maladies can trouble people is through the release of harmful metabolites "mycotoxins" by parasites contaminating plant parts. In spite of the fact that the parasites creating these mycotoxins defile vegetation but not people, these mycotoxins can have direct effects on people and creatures, following in maladies and death. Instances of contagious species creating mycotoxinsinclude *Aspergillus flavus, Fusarium* spp., and *Penicillium* sps. (Schaafsma and Hooker 2007).

Aflatoxin B1 is one of the gravest mycotoxins, since it is risky at high fixation and is cancer causing to people in small dosages and can result in condensed liver capacity, retching, and stomach torment. Yearly deaths in certain parts of Africa because of the impact of Aflatoxin have been recorded at 250,000 every year (Hong et al. 2013). In the present situation, it has been seen that the pathogenic organisms, such as microbes, infection, growths, protozoans, and so forth, are battling with antipathogenic substances. The rise of multi-tranquilize resistant (MDR) microscopic organisms has become a thorough hazard to general wellbeing (Tanwar et al. 2014). There are various sought after procedures, including testing for new antimicrobials from common items, change of open anti-infection classes, and the advancement of antimicrobial peptides. Nanoparticles are currently carefully being investigated as anti-microbials and seem to have a high potential to translate the hitch of the surfacing of microbial multidrug obstruction (O'Connell et al. 2013). To start nanoparticles in the field of drugs and to have quality as a matter of first importance, we must be aware of the microorganisms and their impact on living creatures.

7.2 Infectious Agent

A pathogen or infectious agent is a biological agent that causes disease or sickness to its host. The idiom is most often used for agents that interrupt the usual physiology of a multicellular animal or plant. Some pathogens have been shown to be responsible for immense numbers of afflicted groups. Today, while countless remedial advances have been ready to treat illness caused by pathogens, through the use of vaccination, antibiotics and fungicide, pathogens continue to threaten human life. Pathogens are usually divergent from the ordinary flora. Our ordinary microbial populaces only cause trouble if our immune systems are destabilized or if they gain access to a normally sterile part of the body (Alberts et al. 2002).

7.2.1 Types of Infectious Agent

7.2.1.1 Bacteria

Microscopic organisms are innocuous or gainful; a couple of pathogenic microbes can cause irresistible illnesses. They are *Staphylococcus aureus, Pseudomonas aeruginosa, Xanthomonas compestris.*

Staphylococcus aureus – It is Gram-positive microscopic organisms, of the family *Staphylococcaceae*. These microorganisms are fundamentally responsible for causing skin virus in people; however, they may also transmit an infection to different pieces of the body, for example, respiratory system, cerebrum, and can likewise be destructive for plants. *S. aureus* causes serious sicknesses, for example, pneumonia, meningitis, osteomyelitis, endocarditis, poisonous stun condition, bacteremia, and sepsis. It is as yet one of the five most regular reasons for clinic obtained contaminations and is frequently the reason for wound diseases following medical procedure (Masalha et al. 2001). *Pseudomonas aeruginosa* – It is a Gram-negative bacterium that is in the family Pseudomonadaceae. These microorganisms are found broadly in soil, water, plants, and creatures. It is an entrepreneurial microorganism and only sometimes causes affliction in strong individuals yet can expand effectively in immunocompromised patients. It can cause serious nosocomial contaminations (Itah and Essien 2005). *Xanthomonas compestris* – It is responsible for the dark decay in crucifers such as bacterial wither of turfgrass. It is known as the most horrible microbe, which obliterates the entire vegetation of *Brassica* (Slusarenko et al. 2000).

7.2.1.2 Fungi

Growths include the eukaryotic realm of microorganisms that are generally saprophytes (absorb dead and decaying things); however, it can establish illnesses in people, creatures, and plants. Organisms are the most well-known reason for maladies in crops and different plants. The average contagious spore size is 1–40 micrometer long (Chauhan et al. 2014).

Fusarium graminearum – *Fusarium graminearum* is in the phylum Ascomycota of family nectriaceae. It is a pathogenic organism causing *Fusarium* head curse, which happens in wheat and different grains. The illness has the ability to obliterate a possibly high yield inside half a month of reap. It causes despair and greatness misfortunes because of sterility of the floret and arrangement of stained, contracted, and light test weight pieces. In people, *F. graminearum* has been linked to nutritious discharging and contact dermatitis poison levels and seizures (Schmale III and Bergstrom 2003). *Colletotrichum gloeosporioidis* – It is one of the most basic contagious microbes of the phylum Ascomycota family phyllochoraceae. It is chiefly known for causing anthracnose, a plant sickness occurring on different hosts going from trees to grass. Side effects of this ailment are shown by shaded withered spots on practically all the airborne pieces of the host plant. Skin break out might be engorged prompting shrink, shrivel, and hang from the tainted plant. It requires muggy and clingy environmental factors to root infection on a plant. Consequently, this microorganism is significant for plant pathologists as it could impact the money related framework tumbling crop production worldwide (Waller 1992). *Mycosphaerella pinodes* – It is a hemibiotrophic contagious plant microbe in the family Didymellaceae. It causes curse on pea, it likewise defiles an assortment of species, for example, *Lathyrus sativus, Lupinus albus, Medicago spp., Trifolium spp., Vicia sativa* (Khan et al. 2013).

7.2.1.3 Virus

Infections typically are approximately 20–300 nanometers lengthwise. Pathogenic viral maladies are generally brought about by the groups of Picornaviridae, Herpesviridae, Togaviridae, Adenoviridae, Orthomyxoviridae, Paramyxoviridae, Papovaviridae, Flaviviridae, Polyomavirus, Hepadnaviridae, Rhabdoviridae, and Retroviridae.

7.2.1.4 Prion

As per the prion hypothesis, prions are infectious microorganisms that do not hold nucleic acids. These exceptionally collapsed proteins are occur in a number of ailments, for example, scrapie, cow-like spongiform encephalopathy (mad cow disease), and Creutzfeldt–Jakob ailment. Even though prions do not adequately to meet Koch's hypothesizes, their recognition as a modern class of microorganism drove Stanley B. Prusiner to get the Nobel Prize in Physiology or Medicine in 1997.

7.3 Antimicrobial Agents

There are numerous antimicrobial agents present in the commercial market that are utilized to treat a group of microbial maladies, for example, bacterial, contagious, viral ailments such as ailment brought about by protozoan's and helminthes. Remembering these conditions together with the expanding responsiveness of medication security, we are as of now confronting conditions of partially altered microbial agents. The opposition of organisms lined up with common antimicrobial agents is one of the significant dangers to human wellbeing. Anti-toxins are the broadly utilized antibacterial agents; they are chiefly used to treat bacterial contaminations and numerous infections brought about by the spread of microscopic organisms in the human body. There are two kinds of antibacterial agents. Bactericide: These are the most routinely utilized antibacterial agents; they execute the bacterial strains effectively, for instance, cephalosporins, amino glycosides, fluoroquinollines, vancomycin, daptomycin, and metronidazole. Bacteriostatic: They fundamentally slow the acceleration of microorganisms; however, never executes them, for instance: macrolides, antibiotic medications, trimethoprim, and sulfonamides (Webster 2005).

7.4 Antifungal Agents

These are fungicide or fungi static depending upon the method of activity that is utilized close by contagious contaminations on plants, creatures, or individuals. Amphotericin is the best antifungal agent accessible; however, it caries considerable danger of toxicity and mortality. Fluconazole is an imidazole that is being used as an alternative to amphotericin for grievous contagious diseases. Fungicides, herbicides, and bug sprays are on the whole pesticides utilized in plant assurance. A fungicide is an exact sort of pesticide that controls parasitic illness by explicitly repressing or killing the organism causing the infection (Bhattacharyya et al. 2016). Fungicides have been utilized to lessen mycotoxins virus in wheat influenced by Fusarium head curse, but most fungicides used widely so far have not been sufficiently viable to be valuable for working on mycotoxins related with different sicknesses (Roco 2011).

A difficult microbe is not always affected by the fungicide, which results in the fungicide being less adequate or even futile. Fungicides that are designed for specific catalysts or proteins arranged by growths do not harm plant tissue, in this way they can puncture and move in the inside of leaves empowering helpful properties and expanding the measure of plant tissue shielded to a larger area than where the fungicide was applied. Because the method of activity of these fungicides is so explicit, small hereditary changes in organisms can beat the viability of these fungicides and microorganism populaces can form resistance to future applications. Although ordinary antimicrobial agents have been significant against numerous irresistible infections from old occasions, recently, they have increasingly been used against numerous bacterial and parasitic strains; thus, because of the increase in the quantity of different anti-toxin safe microorganisms and the standing accentuation on social insurance costs, numerous researchers have explored techniques to broaden new productive antimicrobial agents that overcome the protections of these microorganisms and are also cost effective. Nanoscale materials are presently considered to be an adequate alternative to regular substance antimicrobial agents and have a high plausibility to take care of the issue of the bacterial multidrug obstruction.

7.5 Nano War against Infectious Disease

Subsequently, the use of nanotechnology in pharmaceuticals and microbiology is a way to forestall destructive outcomes. A clear and unfortunate case of the necessity for the capacities of nanotechnology to increase and accelerate microorganism uncovering, and to initiate at the purpose of need, is the ongoing Ebola infection flare-up in

West Africa. Nano-empowered targeting discharge offers promising treatment of jungle fever and other intracellular contaminations. Liposomes, nano emulsions, dendrimers, and chitosan nano carriers outline huge success, by improving defense and targeting one of the most persuasive anti-malarial drugs, artemisinin. Nosocomial diseases (NI), otherwise called Hospital Associated/Acquired Infections (HAI), are those contaminations that occur during a patient's stay in an emergency clinic or other kind of clinical offices, which were absent at the hour of admittance. A wide range of microscopic organisms, infections, growths, and parasites may cause nosocomial diseases. Diseases might be caused by a microorganism procured from someone else in the clinic (cross-contamination) or may be by the patient's own already present infectious agent (endogenous disease) (Ducel et al. 2002).

7.5.1 Nanomaterials in Bacterial Detection

Nanotechnology is being expanded to check, break down, and treat transmittable maladies, with some in or approaching the clinical preliminary stage. Irresistible illnesses brought about by irresistible microorganisms are a question of spreading from either a weak host or vector to a strong host. Rapid, vulnerable, and exact understanding of microbes is essential for recognizing the wellspring of contamination, edifying patient consideration with legitimate treatment, and plotting the expansion of sickness (Wilson et al. 2011). Regular techniques utilized for the acknowledgment of microscopic organisms depend on the way of life of the microorganisms on agar plates and the portrayal of their phenotypic properties (Buszewski et al. 2017). Systems dependent on gold or silver nanoparticles, glass nanospheres, or quantum spots, among others, have been created to recognize specific agents or to recognize Gram-positive and Gram-negative microorganisms. Generally, unique physicochemical and immunological techniques have been created for bacterial identification, for example, fluorescence spectroscopy, mass spectrometry, catalyst connected immunosorbent test, and so forth (Lin et al. 1998). Various sorts of nanoparticles, for example, gold, silver, silica and functionalized nanoparticles, among others, have permitted the improvement of specific and sensitive techniques for the finding and eliminating of microorganisms, with various applications in biomedicine and different fields. Fast and careful discovery of pathogenic microscopic organisms is a significant research area for medicinal services, the earth, food production, and so on.

7.5.1.1 Magnetic Nanoparticles

Tuberculosis is a significant medical issue worldwide, *Mycobacterium tuberculosis* strains are aligned with increased depression and shorter lifespan in afflicted patients. A magneto resistive biosensor to recognize *Mycobacterium bovis* (BCG) microorganisms for tuberculosis decision dependent on the use of attractive

7 Efficacy of Nanomaterials and Its Impact on Nosocomial Infections

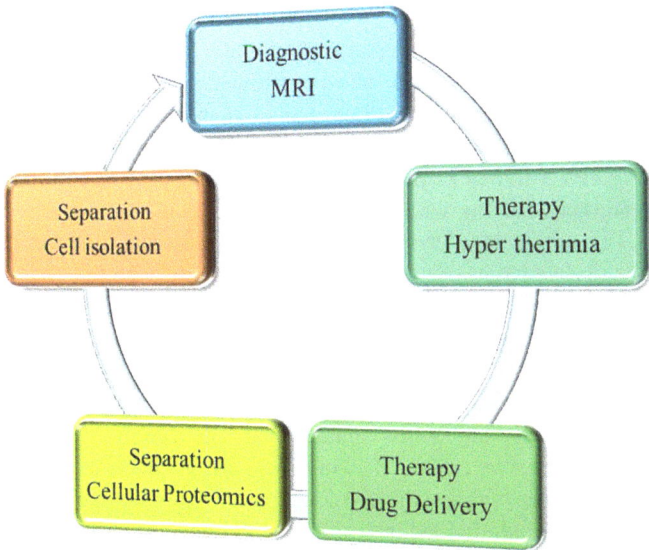

Fig. 7.3 Functionalization of magnetic nanoparticles

nanoparticles has recently been created (Barroso et al. 2018). In addition, a magnetophoretic immunoassay sensor for unfavorable determination of tuberculosis from sputum tests has been created (Kim et al. 2017). Attractive nanoparticles adjusted with a manufactured ligand bis-Zn-DPA can expel *Escherichia coli* (*E. coli*) from ox-like whole blood with practically 100% freedom at streams as high as 60 mL/h. In one ongoing examination, Lowery and associates built up a T2 attractive reverberation (T2MR)-based SPION symptomatic stage that can quickly and reproducibly recognize five Candida species in whole blood inside 3 h. Furthermore, ligand-altered attractive nanoparticles have additionally been joined with attractive microfluidic gadgets for clearing microscopic organisms and endotoxins from the circulation system (Lee et al. 2014). Use of this remarkable profile of attractive nanoparticles related to novel discovery procedures offers boundless potential in delicate and multiplex identification of microbes (Bizzini et al. 2010). The functionalization of magnetic nanoparticles are shown in (Fig. 7.3) as MRI diagnosis, hyperthermia therapy, drug delivery, cellular proteomics, and cell isolation.

7.5.1.2 Silver Nanoparticles

Silver nanoparticles (AgNPs) are considered an antibacterial agent and are used to alter orthopedic inserts to forestall disease. Silver (Ag) has been determined to have a significant antibacterial impact and has been widely utilized in medicine. Ag can be imagined into silver nanoparticles (AgNPs) through nanotechnology to have improved physical, synthetic, and natural properties. Many studies have examined the antimicrobial action of AgNPs, but the promising anti-toxin components and planned weakness remain unclear. Planning to increase the biocompatibility of AgNPs, biosynthesis procedure can be useful to alter the morphology and surface qualities of AgNPs. Strategies, for example, biosynthesis, modifications of physical properties, and consolidating with biomolecules to expand the similarity of AgNPs are featured. Two antibacterial systems are broadly recognized: contact slaughtering and particle intervened murdering. It has been shown that AgNPs can append to the bacterial cell divider and subsequently invade it. It was additionally determined that the antibacterial impact of AgNPs on Gram-negative microorganisms was stronger than Gram-positive microscopic organisms. Furthermore, it has been demonstrated that the cell film of microorganisms has a negative charge because of the nearness of carboxyl, phosphate, and amino gatherings. The functionalization of silver nanoparticles are shown as a flow chart in (Fig. 7.4). While tetracycline and silver nanoparticles are under functionalization, it produces tetracycline silver nano complex, which is further sub divided into kinetic of tetracycline binding, instrumental method, and antimicrobial test.

The positive charge presents electrostatic attraction among AgNPs and adversely charges cell layers of the microorganisms, consequently encouraging AgNP connection onto cell films. After attachment to the bacterial divider, AgNPs can likewise puncture the layer and infiltrate the microorganisms and

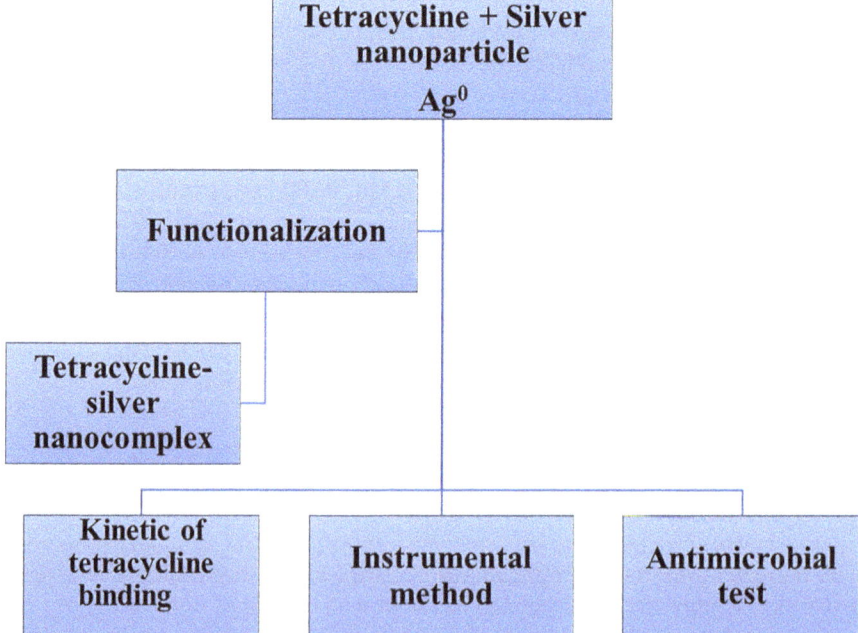

Fig. 7.4 Functionalization of silver nanoparticles

can arrive at the cytoplasm. AgNPs enter inside the microbial cell; it might interact with cell structures and biomolecules, for example, proteins, lipids, and DNA. Communication among AgNPs and cell structures or biomolecules will accompany to bacterial decline and ultimately death (Prasad 2014; Prasad and Swamy 2013; Aziz et al. 2014, 2015, 2016). One of the basic boundaries of AgNPs lined up with microorganisms is the surface zone of the nanomaterials. AgNPs can economically free Ag^+ all through microscopic organisms. In an ongoing report, it is being shown that AgNPs improved bacterial protection from anti-microbials by advancing pressure avoidance through direction of intracellular ROS. Gram-negative microbes *E. coli* 013, *Pseudomonas aeruginosa* CCM 3955 and *E. coli* CCM 3954 can create protection from AgNPs after repeating introduction. The antibiofilm disturbance of AgNPs has been checked in various investigations. One spearheading study was performed to dissect the collaborations of AgNPs with *Pseudomonas putida* biofilms. The outcomes proposed that biofilms are affected by the treatment with AgNPs.

7.5.1.3 Gold Nanoparticles

Au nanoparticles have optical and electrochemical properties that stirred great enthusiasm for their application as detecting materials (Uehara 2010). Au nanoparticles have been broadly utilized as tests for quick distinguishing proof of microbes whose genome arrangement is known to contain remarkable nucleic corrosive marks. Storhoff and colleagues additionally built up a "spot-and-read" colorimetric discovery technique for recognizing the mec A quality found in MRSA strains (Mirkin et al. 1996). Au nanoparticle tests marked with oligonucleotides and Raman-dynamic colors have been used for multiplexed acknowledgment of oligonucleotide targets with increasing affectability and selectivity. Six unique DNA targets are well known with six Raman-named Au nanoparticle tests with a recognition breaking point of 20 femtomolar. Mirkin et al. built a bio-standardized identification test for exceptionally discerning nucleic corrosive and protein targets (Hill and Mirkin 2006). Sandwich structure with Au nanoparticles and attractive microparticles for attractive detachment and dithiothreitol (DTT) intervened in the arrival of scanner tag strands, which are in this manner recognized and evaluated on a microarray. Au nanoclusters embedded inside lysozymes that can tie with peptidoglycans on bacterial cell dividers were created to target pathogenic microbes for MALDI-MS-based recognizable proof (Chan and Don 2013). Human serum egg whites or its coupling peptide with Au nanoclusters also settled unequivocal partiality with *S. aureus* and MRSA for their separation detection. Au nanoparticles can likewise be utilized to set up antimicrobial weakness by estimating the movements in the surface Plasmon band, upon the Con An instigated grouping of dextran-covered Au nanoparticles near starch in bacterial suspension (Nath et al. 2008). The applications of gold nanoparticles are shown in (Fig. 7.5). They are controlled drug delivery, catalyst, used in cancer therapy & diagnostics, and molecular detection.

Fig. 7.5 Applications of gold nanoparticles

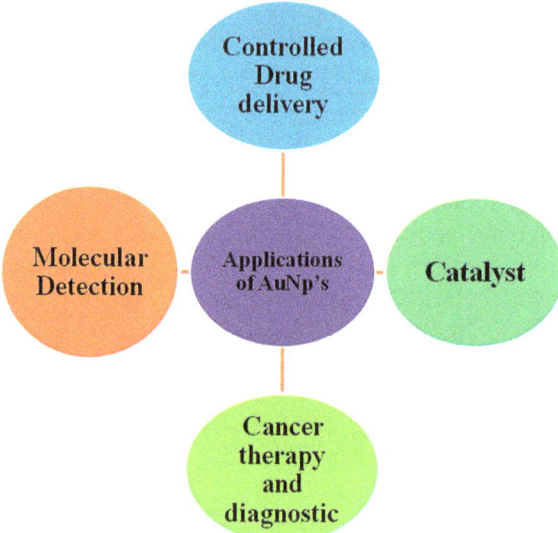

7.5.1.4 Localized Surface Plasmon's

A limited surface plasmon (LSP) is the outcome of the constrainment of a surface plasmon in a nanoparticle of size equal to or smaller than the frequency of light used to invigorate the plasmon. At the point when a small circular metallic nanoparticle is illuminated by light, the wavering electric field influences the conduction electrons. A solitary nanohole in a metal layer is equipped for supporting an LSP (Spackova et al. 2016). Innovations dependent on nanocavity-formed photonic precious stones with solid plasmonic signals have been created with expected applications in bacterial identification. Thue–Morse (T–M) exhibit nanoholes in a polymeric film to procure metallic gold nanocavities that allows the declaration of surface plasmons (Rippa et al. 2016). These sorts of structures, together with SERS, grant fast and detailed bacteriophage understanding of pathogenic microscopic organisms, for example, *Brucella* sp. (Zhang et al. 2001).

7.5.1.5 Fluorescent Nanoparticles

When microscopic organisms breathe, they produce acids. The acids decrease the pH and oxidize the carbon atoms, causing changes to arrangement of the particles. As a result, they exude a more splendid fluorescent gleam. The specialists put a couple of *E. coli* microscopic organisms and the fluorescent nanoparticles in a little gel microsphere that holds them. When the microscopic organisms begin to partition the carbon atoms, they started to shine more brilliantly. Nanomaterials with fluorescent properties or nanoparticles named/typified with fluorescent colors have been helpful for microbial identification. They also created diverse fluorescence reverberation vitality transfer (FRET) silica nanoparticles by co-embodying three colors that radiate interesting hues upon excitation with a solitary frequency (Wang et al. 2005). Fluorescence imaging is a non-obtrusive, sensitive strategy that permits examining natural life forms with high tridimensional achievement continuously, by utilizing reasonable fluorescent differentiation agents. From the perspective of fluorescence splendor, the capacity of NP to create an exceptional fluorescence signal, even in the low force excitation system, results from the co-nearness of a high number of MF in every NP. These nanoparticle combination demonstrative systems depend on knowing the bacterial genome groupings/biomarkers by focusing on tests, and in this way may not classify transformed or potentially new microorganisms strains. As medication safe strains are slowly determined, another noteworthy way is the advancement of symptomatic nanotechnology able to detect the nearness of microbes, with the ability to decide the defenselessness of the microorganisms to antimicrobial medications simultaneously.

7.6 Nanomaterials in Viral Detection

Pathogens are the least known organisms, yet they cause the most significant misfortunes to human wellbeing. More often than not, the most popular remedy for infections is the intuitive immunological resistance system of the host; in any case, the starter counteraction of viral disease is the main substitute. Infections are caused by amazing microorganisms called pathogens that cause an extraordinarily number of maladies and mortality worldwide. Currently, popular contaminations and associated infections are significant reasons for death in humankind, and under the current setting of industrialization and migration, they occur and spread at a quick pace, causing huge human, social, and budgetary expenses. Unfavorable analysis is consistently favorable for the control of irresistible sicknesses. Various methodologies have been established for nanoparticles to liven up explanatory procedure qualities, in any event, permitting the extension of simple and quick purpose-of-care (POC) measures to analyze in situ in remote locales.

Various types of nanoparticles have increased their approaches for the determination, recuperation, and expectation of viral diseases in numerous applications, primarily those nanoparticles with viral material or frameworks that copy infection qualities. This portion centers around the indicative procedures expanded for the understanding and evaluation of infections themselves, and specifically for the infections that have been deliberately researched and are more essential (Hassanpour et al. 2018). It is interesting to see that in therapeutics against infections, some special nanoparticles have emerged: the "virus-like particles" (VLPs). They are nanoparticles formed from viral proteins that aggregate in structures similar to authentic infection particles even though they require irresistible nucleic corrosive groupings (Lee et al. 2016). Early conclusions have been consistently positive for the control of irresistible illnesses. Various procedures have been actualized utilizing nanoparticles to improve systematic strategy qualities, including permitting the advancement of straightforward and quick purpose-of-care (POC) tests to analyze in situ in remote locales. The exceptional and adaptable properties of nanoparticles themselves and the atoms that can be related with them empower quick, complex, and savvy analyses (Jorquera and Tripp 2016).

7.6.1 SERS

The use of nanoparticles in Raman spectroscopy intensifies the signs, prompting SERS that has been applied for various types of infection (Tanwar et al. 2021; Liu et al. 2021). The nanoplasmonic properties of gold nanoparticles have been helpful in human immunodeficiency virus (HIV) load measurement from whole blood tests. Along these lines, a knowledge platform with specific antiviral antibodies preset to the biosensing surface has been built that has the option to identify and measure various HIV subtypes and could be changed for different microorganisms that have known biomarkers (Halfpenny and Wright 2010). Some of the strategies dependent on these intrinsic properties of gold nanoparticles have been used in respiratory infections, taking into consideration the differentiation among various flu infections and hepatitis viruses (Park et al. 2012). An optofluidic-nanoplasmonic sensor that could be utilized as a POC for Ebola investigation, even in bio barrier settings, has been structured. This nano opening-based detecting raised zone has increased its capacity to identify unblemished infections from naturally pertinent media with simple model preparation and the creators prescribe that it could be extrapolated to different infections (Yanik et al. 2010).

7.6.2 Electrochemical Biosensing

Distinctive biosensors dependent on nanoparticles have been used in flu infection identification, among others. Because they have ideal attributes for biosensors for POC examines that can suggest insightful result quicker, simpler, at lower cost than traditional strategies and with amazing selectivity and sensitivity (Tepeli and Ülkü 2018). An anode incorporating graphene and polyaniline nanowires has additionally been anticipated as an approach to advance its DNA discovery affectability (Diba et al. 2015). In addition, immunoassays dependent on complementary metal–oxide–semiconductors (CMOS) that have sensor innovation utilizing indium nanoparticle (InNP) substrates have been utilized for hepatitis infection recognition (Devadhasan and Kim 2015).

In the examination field of Ebola analysis, specialists have explicitly highlighted the need for a biosensor that permits the identification of Ebola infections at the point of care using the relationship of nanoparticles to symptomatic methods created for different infections and on the premise that scaled down chips with immobilized antibodies have built up their ability to recognize pM levels of different biomarkers (Vasudev et al. 2013). A few creators suggest that scaling down the electrochemical insusceptible detecting ability would be a sensible way to create gadgets for quick and in situ Ebola screening (Kaushik et al. 2016).

7.6.3 Other Biosensing Methods

Regarding infection, soluble phosphatase (ALP) has been utilized as a sign tag for immunoreactions. Shading change was seen within the sight of the infection because of silver expression on the outside of gold nanoparticles initiated by the catalyst. Combined with attractive advancement, this strategy has been exhibited to be basic, quick, and profoundly sensitive, permitting H9N2 infection identification straightforwardly in complex samples (Chin et al. 2011). Nano arrays obtained by nanolithography have demonstrated an upgrade such as quicker discovery than the ordinary colorimetric enzyme-linked immunosorbent assay (ELISA). Double luminophore-doped silica nanoparticles with various surface changes have been used for multiplexed investigation. Together with stream cytometry, it has been proposed that these frameworks have fascinating worthwhile properties with regard to the identification of microorganisms, particularly for those that have issues with typical colors because of their negligible specific antigens. Results uncovered that these nanoparticles have high sign intensification, superb photostability, and simple surface bioconjugation for biomarker location, which marks this framework as a perfect biolabeling reagent in antigens and nucleic acids identification (Wang et al. 2009).

Different techniques dependent on colorimetric discovery have been applied to infection identification. Through the relationship of gold nanoparticles with switch translation circle intervened isothermal intensification, a straightforward test for hepatitis E was created, whose outcomes can be assessed with the unaided eye because of shading changes. It has been proposed as an option in contrast to other costly and tedious techniques normally utilized (Chen et al. 2014). Fe_3O_4 attractive nanoparticles have been additionally applied as nanozyme tests, tackling their regular inborn peroxidase-like action that can be outwardly distinguished because of the undeniable shading response. By marking them with specific antibodies and close by peroxide substrates, they have been used for immuno attractive Ebola infection discovery (Duan et al. 2015).

7.7 Advanced Nano Biomaterials to Treat Infectious Disease

Clinical gadgets assume a significant job in current medicinal services practice, but their application may increase the dangers of nosocomial disease. The microbes most generally found in contaminated gadgets include *S. epidermidis, S. aureus,* and *P. aeruginosa.* These microorganisms can be amazingly impervious to anti-toxin treatment because of the development of biofilms, and foundational organization of anti-infection agents as a rule does not show agreeable outcomes (Krishnasami et al. 2002). Medical gadgets with innate antimicrobial properties have been used for quite a long time, with the objective that a good mix with have cells while forestalling any bacterial bond or biofilm development.

7.7.1 Nano Vaccine

The host's immune system response has been exhibited to be extremely powerful in securing them against microbial disease. Different existing immunizations for organisms show a significant variety in immunogenicity and wellbeing. Worries with the utilization of live constricted bacterial antibodies include the conceivable inversion of pathogenicity and the prior insusceptibility to the vector, such as the risk to reward traded off for people (Smith et al. 2013). Advances in biotechnology empower the creation of cutting-edge bacterial immunizations, including disengaged proteins, polysaccharides, and exposed DNA. Novel antibodies are regularly less immunogenic than conventional immunizations, for example, those utilizing live weakened organisms. To address this test, the use of nanotechnologies to upgrade the resistant reactions of these antibodies has pulled in extraordinary intrigue (Reddy et al. 2011). Nanoparticles have likewise been demonstrated to be viable conveyance frameworks for mucosal immunization. A defensive, dependable mucosal insusceptible reaction is imperative to shield the host from likely bacterial contamination. Accordingly, mucosal organization through intranasal, inhalational, or gastrointestinal courses is becoming a supported course of immunization. Distinctive nanoparticle conveyance vehicles have been proposed to improve mucosal immunization through their immunostimulatory exercises (Kammona and Kiparissides 2012).

7.7.2 Nano Adjuvant

Nanoemulsions, or oil-in-water emulsions framed by isotropic blends of oil and surfactant with bead distance across in the nanometer scale, are compelling non-provocative mucosal adjuvants. The adjuvanticity of nanoemulsions has been proposed to add to expanded cell take-up of antigens, enlistment of monocytes and granulocytes, and upgraded arrival of cytokines and chemokines (Hamouda et al. 2001). Intranasally controlled recombinant *Bacillus anthracis* defensive antigen blended in nanoemulsion prompted both serum IgG and bronchial IgA and IgG antibodies after a couple of mucosal organizations in mice and guinea pigs (Bielinska et al. 2007). In correlation, industrially accessible human *Bacillus anthracis* immunization requires six subcutaneous infusions more than year and a half and yearly promoter. Cationic liposomes complexed with non-coding plasmid DNA were additionally answered to be compelling as parenteral and mucosal immunization adjuvants (Makidon et al. 2010).

7.7.3 Quorum Sensing

Quorum sensing is an upgraded process connected with population density that microorganisms use to manage biofilm development. The drawback to Quorum sensing is the need for a methodology to battling its pathogenicity. Common or manufactured Quorum sensing inhibitors may be hostile to biofilm agents and be helpful in rewarding multi-tranquilize safe microscopic organisms. Microscopic organisms can speak with one another through discharged flagging elements, named autoinducers. These compound signs are combined intra cellularly and discharged to the extracellular medium where they are perceived by the nearby cells enacting the statement of related qualities (Lazdunski et al. 2004). The autoinducer movement and the conduct changes are possibly activated when an edge level is reached (Turan et al. 2017; Sintim et al. 2010; Galloway et al. 2012). These occasions require, at that point, high cell densities (to collect adequate sign).

The base conduct unit has been portrayed as many microbes and, in this manner, this method of bacterial correspondence has been named majority sensing (Mukherjee et al. 2008). The QS procedure among cells was first found to control bioluminescence in the marine microscopic organisms *Vibrio fischeri*, where for low cell densities a homoserine lactone is discharged to the medium, while for high cell densities, it is aggregated inside when it triggers the interpretation of radiance qualities (Stevens and Greenberg 1997). In *Pseudomonas aeruginosa*, whose Quorum sensing framework has been the most considered, it manages the creation of a few intensifies that assume significant jobs in biofilm arrangement. This includes rhamnolipids, lectin A (LecA)/LecB, and pyochelin and pyoverdine siderophores. The least complex Gram-positive quorum sensing framework was first found in *Lactococcus lactis* and *Streptococcus pneumoniae* (Tielker et al. 2005; Diggle et al. 2006).

Currently, the expansion in safe bacterial strains and the absence of new antitoxins make it important to scan for new techniques to battle contaminations. Because of the significant job that quorum sensing plays in bacterial harmfulness, the interruption of this bacterial correspondence framework is drawing in a lot of enthusiasm as another antimicrobial methodology (Defoirdt 2017). The mediation methodology is named "quorum quenching" (QQ), a term used to incorporate any methodology that meddles with legitimate microbial QS flagging. This should be possible at various focuses: restraint of autoinducer amalgamation, corruption of the autoinducer, and capture of its collaboration with the receptor (Brooks and Brooks 2014). Numerous restorative plant species, for example, garlic, ginger, basic oils of cinnamon and clove are additionally known to have QSI uses. Carrot, chamomile, garlic, and numerous peppers have been demonstrated to have hostile to QS action, even though the systems for a significant number of them have not yet been recognized. Additionally, flavonoids, for example, baicalin, quercetin, naringenin, kempferol, and apigenin, have all been found to be effective in threatening bacterial QS. In many examinations, a prophylactic use has been concentrated by overseeing the counter QS agents simultaneously as the microorganisms inoculum, and a significant improvement in the contamination result has been found (Khajanchi et al. 2011; Musthafa et al. 2012).

7.8 Conclusion

Various advances in nanoparticle-based frameworks for the demonstrative and treatment of bacterial contaminations have been distributed with possible applications in the battle against multidrug resistant strains and bacterial biofilms, among other areas. The possible effect of nanotechnology on microbial irresistible sicknesses has just been exhibited by the clinical endorsement of numerous nanotechnology-based items for the location of bacterial contamination, the conveyance of anti-toxins, and the improvement of clinical gadgets with antimicrobial coatings. Nanoparticles with elite physiochemical properties have empowered the detection of microbial sickness with high affectability, selectivity, and quick readout. The attributes of particular kinds of nanoparticles and extra functionalization present perfect properties for application in indicative examination, permitting scaling down and improvement of some customary procedures of microbe location. Progressed explanatory strategies, for example, SERS, joined with the utilization of metallic nanoparticles are magnificent apparatuses for the location of microbes and infections. Regardless of these enchanting accomplishments, the full potential of nanotechnology in running microbial contamination, especially in the territories of antimicrobial treatment and antibodies, is far from being reached. The epic field of theranostic is very much perceived as possible for specially crafted malignancy treatment, including the hang-up of bacterial quorum sensing systems by the utilization of metallic and different kinds of nanoparticles comprises a promising methodology in the battle against bacterial contaminations. The consolidation of QS inhibitors into these nanosystems expands their effectiveness for biofilm treatment. Antimicrobial nanotechnologies can be encouraged by developing more clinically relevant creature models, recognizing the instruments of microbial pathogenesis and new biomarkers, tolerating the microenvironment of bacterial pollution destinations, and overcoming the authoritarian boundaries. With ceaseless progress in antimicrobial nanomedicine, we can expect that many more nanotechnology-based items will be developed to manage each bit of microbial disease.

References

Alberts B, Johnson A, Lewis J et al (2002) Introduction to pathogens. In: Molecular biology of the cell, 4th edn. Garland Science, USA. p 1. Retrieved 26 April 2016

Aziz N, Fatma T, Varma A, Prasad R (2014) Biogenic synthesis of silver nanoparticles using Scenedesmus abundans and evaluation of their antibacterial activity. Journal of Nanoparticles, Article ID 689419, http://dx.doi.org/10.1155/2014/689419

Aziz N, Faraz M, Pandey R, Sakir M, Fatma T, Varma A, Barman I, Prasad R (2015) Facile algae-derived route to biogenic silver nanoparticles: Synthesis, antibacterial and photocatalytic properties. Langmuir 31:11605–11612. https://doi.org/10.1021/acs.langmuir.5b03081

Aziz N, Pandey R, Barman I, Prasad R (2016) Leveraging the attributes of Mucor hiemalis-derived silver nanoparticles for a synergistic broad-spectrum antimicrobial platform. Front Microbiol 7:1984. https://doi.org/10.3389/fmicb.2016.01984

Barroso TG, Martins RC, Fernandes E, Cardoso S, Rivas J, Freitas PP (2018) Detection of BCG bacteria using a magnetoresistive biosensor: a step towards a fully electronic platform for tuberculosis point-of-care detection. Biosens Bioelectron 100:259–265

Bhattacharyya A, Duraisamy P, Govindarajan M, Buhroo AA, Prasad R (2016) Nano-biofungicides: Emerging trend in insect pest control. In: Advances and Applications through Fungal Nanobiotechnology (ed. Prasad R), Springer International Publishing Switzerland 307–319

Bielinska AU, Janczak KW, Landers JJ, Makidon P, Sower LE, Peterson JW, Baker JR (2007) Mucosal Immunization with a Novel Nanoemulsion-Based Recombinant Anthrax Protective Antigen Vaccine Protects against *Bacillus anthracis* Spore Challenge. Infect Immun 75:4020–4029

Bizzini A, Durussel C, Bille J, Greub G, Prod'hom G (2010) Performance of matrix-assisted laser desorption ionization-time of flight mass spectrometry for identification of bacterial strains routinely isolated in a clinical microbiology laboratory. J Clin Microbiol 48:1549–1554

Brooks BD, Brooks AE (2014) Therapeutic strategies to combat antibiotic resistance. Adv Drug Deliv Rev 78:14–27

Burlage RS, Tillmann J (2017) Biosensors of bacterial cells. J Microbiol Methods 138:2–11

Buszewski B, Rogowska A, Pomastowski P, Zloch M, Railean-Plugaru V (2017) Identification of microorganisms by modern analytical techniques. J AOAC Int 100:1607–1623

Buzea C, Pacheco II, Robbie K (2007) Nanomaterials and nanoparticles: sources and toxicity. Biointerphases 2:17–71

Chan YS, Don MM (2013) Optimization of process variables for the synthesis of silver nanoparticles by Pycnoporus sanguineus using statistical experimental design. J Korean Soc Appl Bio Chem 56:11–20

Chauhan P, Mishra M, Gupta D (2014) Potential application of nanoparticles as Antipathogens. In: Tiwari A, Syväjärvi M (eds) Advanced materials for agriculture, food, and environmental safety. © Scrivener Publishing LLC, USA. pp 333–368

Chen Q, Yuan L, Wan J, Chen Y, Du C (2014) Colorimetric detection of hepatitis E virus based on reverse transcription loop mediated isothermal amplification (RT-LAMP) assay. J Virol Methods 197:29–33

Chin CD, Laksanasopin T, Cheung YK, Steinmiller D, Linder V, Parsa H, Wang J, Moore H, Rouse R, Umviligihozo G et al (2011) Microfluidics-based diagnostics of infectious diseases in the developing world. Nat Med 17:1015–1019

Defoirdt T (2017) Quorum-sensing systems as targets for Antivirulence therapy. Trends Microbiol 26(4):313–328

Devadhasan JP, Kim S (2015) Label free quantitative immunoassay for hepatitis B. J Nanosci Nanotechnol 15:85–92

Diba FS, Kim S, Lee HJ (2015) Amperometric bioaffinity sensing platform for avian influenza virus proteins with aptamer modified gold nanoparticles on carbon chips. Biosens Bioelectron 72:355–361

Diggle SP, Stacey RE, Dodd C, Camara M, Williams P, Winzer K (2006) The galactophilic lectin, LecA, contributes to biofilm development in Pseudomonas aeruginosa. Environ Microbiol 8:1095–1104

Duan D, Fan K, Zhang D, Tan S, Liang M, Liu Y, Zhang J, Zhang P, Liu W, Qiu X et al (2015) Nanozyme-strip for rapid local diagnosis of Ebola. Biosens Bioelectron 74:134–141

Ducel G, Fabry J, Nicolle L, Girard R, Perraud M, Pruss A, Savey A (2002) Prevention of hospital-acquired infections, A practical guide, Department of Communicable Disease, Surveillance and Response, Editors; 2nd edn, Available at WHO/CDS/CSR/EPH/2002.12

Fang Y-S, Wang H-Y, Wang L-S, Wang J-F (2014) Electrochemical immunoassay for procalcitonin antigen detection based on signal amplification strategy of multiple nanocomposites. Biosens Bioelectron 51:310–316

Feynman RP (1959) Plenty of room at the bottom. Am Phy Soc. Available online: http://www.pa.msu.edu/~yang/RFeynman_plentySpace.pdf. Accessed on 30 June 2016

Galloway WR, Hodgkinson JT, Bowden S, Welch M, Spring DR (2012) Applications of small molecule activators and inhibitors of quorum sensing in Gram-negative bacteria. Trends Microbiol 20:449–458

Halfpenny KC, Wright DW (2010) Nanoparticle detection of respiratory infection. Wiley Interdiscip Rev Nanomed Nanobiotechnol 2:277–290

Hamouda T, Myc A, Donovan B, Shih AY, Reuter JD, Baker JR (2001) A novel surfactant nanoemulsion with a unique non-irritant topical antimicrobial activity against bacteria, enveloped viruses and fungi. Microbiol Res 156:1–7

Hassanpour S, Baradaran B, Hejazi M, Hasanzadeh M, Mokhtarzadeh A, de la Guardia M (2018) Recent trends in rapid detection of influenza infections by bio and nanobiosensor. TrAC Trends Anal Chem 98:201–215

Hill HD, Mirkin CA (2006) The bio-barcode assay for the detection of protein and nucleic acid targets using DTT-induced ligand exchange. Nat Protoc 1:324–336

Hong KW, Koh CL, Sam CK, Yin WF, Chan KG (2013) Quorum quenching revisited–from signal decays to humoral and cellular immune responses. J Control Release 168:271–279

Hoyert DL, Kochanek KD, Murphy SL (1999) Deaths: final data for 1997. Hyattsville, Maryland: US Department of Health and Human Services, Public Health Service, CDC, National Center for Health Statistics. (National vital statistics reports, vol 47, no. 19)

Itah A, Essien J (2005) Growth profile and Hydrocarbonoclastic potential of microorganisms isolated from Tarballs in the bight of bonny, Nigeria. World J Microbiol Biotechnol 21: 1317–1322

Jorquera PA, Tripp RA (2016) Synthetic biodegradable microparticle and nanoparticle vaccines against the respiratory syncytial virus. Vaccine 4:45

Kammona O, Kiparissides C (2012) Recent advances in nanocarrier-based mucosal delivery of biomolecules. J Control Release 161:781–794

Kannan RRR, Arumugam R, Ramya D, Manivannan K, Anantharaman P (2013) Green synthesis of silver nanoparticles using marine macroalga Chaetomorpha linum. Appl Nanosci 3:229–233

Kannan RM, Nance E, Kannan S, Tomalia DA (2014) Emerging concepts in dendrimer-based nanomedicine: from design principles to clinical applications. J Intern Med 276:579–617

Kaushik A, Tiwari S, Dev Jayant R, Marty A, Nair M (2016) Towards detection and diagnosis of Ebola virus disease at point-of-care. Biosens Bioelectron 75:254–272

Khajanchi BK, Kirtley ML, Brackman SM, Chopra AK (2011) Immunomodulatory and protective roles of quorum-sensing signaling molecules N-acyl homoserine lactones during infection of mice with Aeromonas hydrophila. Infect Immun 79:2646–2657

Khan TN, Timmerman-Vaughan GM, Rubiales D, Warkentin TD, Siddique KHM, Erskine W, Barbetti MJ (2013) *Didymella pinodes* and its management in field pea: challenges and opportunities Field Crop Res 148:61–77

Kim J, Lee KS, Kim EB, Paik S, Chang CL, Park TJ, Kim HJ, Lee J (2017) Early detection of the growth of Mycobacterium tuberculosis using magnetophoretic immunoassay in liquid culture. Biosens Bioelectron 96:68–76

Krishnasami Z, Carlton D, Bimbo L, Taylor ME, Balkovetz DF, Barker J, Allon M (2002) Management of hemodialysis catheter-related bacteremia with an adjunctive antibiotic lock solution. Kidney Int 61:1136–1142

Lazdunski AM, Ventre I, Sturgis JN (2004) Regulatory circuits and communication in Gram-negative bacteria. Nat Rev Microbiol 2:581

Lee JJ, Jeong KJ, Hashimoto M, Kwon AH, Rwei A, Shankarappa SA, Tsui JH, Kohane DS (2014) Synthetic Ligand-Coated Magnetic Nanoparticles for Microfluidic Bacterial Separation from Blood. Nano Lett 14:1–5

Lee KL, Twyman RM, Fiering S, Steinmetz NF (2016) Virus-based nanoparticles as platform technologies for modern vaccines. Wiley Interdiscip Rev Nanomed Nanobiotechnol 8:554–578

Lin HY, Huang CH, Hsieh WH, Liu LH, Lin YC, Chu CC, Wang ST, Kuo IT, Chau LK, Lowy FD (1998) *Staphylococcus aureus* infections. N Engl J Med 339(8):520–532

Liu Z, Parida S, Prasad R, Pandey R, Sharma D, Barman I (2021) Vibrational spectroscopy for decoding cancer microbiome interactions: Current evidence and future Perspective. Seminars in Cancer Biology. https://doi.org/10.1016/j.semcancer.2021.07.004

Makidon PE, Knowlton J, Groom JV, Blanco LP, LiPuma JJ, Bielinska AU, Baker JR Jr (2010) Induction of immune response to the 17 kDa OMPA Burkholderia cenocepacia polypeptide and protection against pulmonary infection in mice after nasal vaccination with an OMP nanoemulsion-based vaccine. Med Microbiol Immunol 199:81–92

Masalha M, Borovok I, Schreiber R, Aharonowitz Y, Cohen G (2001) Analysis of transcription of the *Staphylococcus aureus* aerobic class Ib and anaerobic class III ribonucleotide reductase genes in response to oxygen. J Bacteriol 183(24):7260–7272

Mirkin CA, Letsinger RL, Mucic RC, Storhoff JJ (1996) A DNA-based method for rationally assembling nanoparticles into macroscopic materials. Nature 382:607–609

Mukherjee P, Roy M, Mandal BP, Dey GK, Mukherjee PK, Ghatak J et al (2008) Green synthesis of highly stabilized nanocrystalline silver particles by a non-pathogenic and agriculturally important fungus T. Asperellum Nanotechnol 19:075103

Musthafa KS, Balamurugan K, Pandian SK, Ravi AV (2012) 2,5-Piperazinedione inhibits quorum sensing-dependent factor production in Pseudomonas aeruginosa PAO1. J Basic Microbiol 52:679–686

Nath S, Kaittanis C, Tinkham A, Perez JM (2008) Rapid Nanoparticle-Mediated Monitoring of Bacterial Metabolic Activity and Assessment of Antimicrobial Susceptibility in Blood with Magnetic Relaxation. Anal Chem 80:1033–1038

Nowicki M et al (2011) Potato and tomato late blight caused by Phytophthora infestans: an overview of pathology and resistance breeding. Plant Dis 96:4–17

O'Connell KMG, Hodgkinson JT, Sore HF, Welch M, Salmond GPC, Spring DR (2013) Combating multidrug- resistant Bacteria: current strategies for the discovery of novel Antibacterials. Angew Chem Int Ed 52:10706–10733

Park TJ, Lee SJ, Kim DK, Heo NS, Park JY, Lee SY (2012) Development of label-free optical diagnosis for sensitive detection of influenza virus with genetically engineered fusion protein. Talanta 89:246–252

Prasad R, Swamy VS (2013) Antibacterial activity of silver nanoparticles synthesized by bark extract of Syzygium cumini. Journal of Nanoparticles 2013, http://dx.doi.org/10.1155/2013/431218

Prasad R (2014) Synthesis of silver nanoparticles in photosynthetic plants. Journal of Nanoparticles, Article ID 963961, 2014, http://dx.doi.org/10.1155/2014/963961

Reddy ST, van der Vlies AJ, Simeoni E, Angeli V, Randolph GJ, O'Neil CP, Lee LK, Swartz Roy V, Adams BL, Bentley WE (2011) Developing next generation antimicrobials by intercepting AI-2 mediated quorum sensing. Enzym Microb Technol 49:113–123

Rippa M, Castagna R, Pannico M, Musto P, Bobeico E, Zhou J, Petti L (2016) High-performance nanocavities-based meta-crystals for enhanced plasmonic sensing. Opt Data Process Storage 2:22–26

Roco MC (2011) The long view of nanotechnology development: the national nanotechnology initiative at 10 years. Nanotechnology Research Directions for Societal Needs in 2020, Volume 1 of the series Science Policy Reports, pp 1–28

Schaafsma AW, Hooker DC (2007) Climatic models to predict occurrence of *fusarium* toxins in wheat and maize. Int J Food Microbiol 119(1–2):116–125

Schmale DG III, Bergstrom GC (2003) Fusarium head blight in wheat. Plant Health Instructor. https://doi.org/10.1094/PHI-I-2003-0612-01

Sintim HO, Smith JA, Wang J, Nakayama S, Yan L (2010) Paradigm shift in discovering next-generation anti-infective agents: targeting quorum sensing, c-di-GMP signaling and biofilm formation in bacteria with small molecules. Future Med Chem 2:1005–1035

Slusarenko AJ, Fraser Alberts, van Loon LC (eds) (2000) Mechanisms of resistance to plant diseases. Kluwer Academic Publishers, Dordrecht, pp 21–52

Smith DM, Simon JK, Baker JR Jr (2013) Applications of nanotechnology for immunology. Nat Rev Immunol 13:592–605

Spackova B, Wrobel P, Bockova M, Homola J (2016) Optical biosensors based on Plasmonic nanostructures: a review. J Proc IEEE 104:2380–2408

Stevens AM, Greenberg EP (1997) Quorum sensing in Vibrio fischeri: essential elements for activation of the luminescence genes. J Bacteriol 179:557–562

Tanwar S, Paidi SK, Prasad R, Pandey R, Barman I (2021) Advancing Raman spectroscopy from research to clinic: Translational potential and challenges. Spectrochimica Acta Part A: Molecular and Biomolecular Spectroscopy. https://doi.org/10.1016/j.saa.2021.119957

Tanwar J, Das S, Fatima Z, Hameed S (2014) Multidrug resistance: an emerging crisis. Interdiscip Perspect Infect Dis 2014:541340. 7 pages

Tepeli Y, Ülkü A (2018) Electrochemical biosensors for influenza virus a detection: the potential of adaptation of these devices to POC systems. Sens Actuators B Chem 254:377–384

Tielker D, Hacker S, Loris R, Strathmann M, Wingender J, Wilhelm S, Rosenau F, Jaeger KE (2005) Pseudomonas aeruginosa lectin LecB is located in the outer membrane and is involved in biofilm formation. Microbiology 151:1313–1323

Turan NB, Chormey DS, Büyükpınar Ç, Engin GO, Bakirdere S (2017) Quorum sensing: little talks for an effective bacterial coordination. TrAC Trends Anal Chem 91:1–11

Uehara N (2010) Polymer-functionalized Gold Nanoparticles as Versatile Sensing Materials. Anal Sci 26:1219–1228

Vasudev A, Kaushik A, Tomizawa Y, Norena N, Bhansali S (2013) An LTCC-based microfluidic system for label-free, electrochemical detection of cortisol. Sens Actuators B Chem 182:139–146

Waller JM (1992) *Colletotrichum* diseases of perennial and other cash crops. In: Bailey JA, Jeger MJ (eds) Colletotrichum: biology, pathology and control. CABI, Wallingford. ISBN 978-0851987569

Wang L, Yang C, Tan W (2005) Dual-luminophore-doped silica nanoparticles for multiplexed signaling. Nano Lett 5:37–43

Wang L, Chen W, Xu D, Shim BS, Zhu Y, Sun F, Kotov NA (2009) Simple, rapid, sensitive, and versatile SWNT_ paper sensor for environmental toxin detection competitive with ELISA. Nano Lett 9(12):4147–4152

Webster P (2005) World nanotechnology market frost and Sullivan. Nanomedicine 1(2):140–142

Wilson BA, Salyers AA, Whitt DD, Winkler ME (2011) Bacterial pathogenesis: a molecular approach. American Society for Microbiology, Washington, DC

Yanik AA, Huang M, Kamohara O, Artar A, Geisbert TW, Connor JH, Altug H (2010) An Optofluidic Nanoplasmonic biosensor for direct detection of live viruses from biological media. Nano Lett 10:4962–4969

Zazo H, Millán CG, Colino CI, Lanao JM (2017) Chapter 15—applications of metallic nanoparticles in antimicrobial therapy. In: Grumezescu AM (ed) Antimicrobial nanoarchitectonics. Elsevier, New York, pp 411–444

Zhang B, Zhang ZJ, Wang B, Yan J, Li JJ, Cai SM (2001) A study of designed current oscillations of Fe in H2SO4 solution. Acta Chim Sin 59:1932

Chapter 8
Nanonutraceuticals in Chemotherapy of Infectious Diseases and Cancer

C. Sumathi Jones, V. Uma Maheshwari Nallal, and M. Razia

Contents

8.1	Introduction to Nanotechnology and Nutraceuticals.	262
	8.1.1 Nano Fabrication.	262
	8.1.2 Types of Nanocarriers.	263
8.2	Nutraceuticals in Biomedical Applications.	266
8.3	Nanonutraceutical Drug Delivery System.	267
8.4	Nanosize Nutraceutical Formulations in Biomedical Applications.	270
	8.4.1 Nanovitamins and Nanominerals.	270
	8.4.2 Nanoencapsulation of Probiotics.	272
	8.4.3 Nanophytochemicals.	273
8.5	Nanonutraceuticals in the Chemotherapy of Cancer.	274
8.6	Nanonutraceuticals and Anti-inflammatory Activity.	278
8.7	Nanonutraceuticals in Medical Imaging.	279
8.8	Nanonutraceuticals in Prophylaxis, Diagnosis, and Treatment of Infectious Diseases.	280
8.9	Nanonutraceuticals as Antibacterial Agents.	282
8.10	Nanonutraceuticals in Antiviral Therapy.	283
8.11	Utilization of Nanonutraceutical in COVID-19 Therapy: Pharmacological and Toxicological Aspects.	284
	8.11.1 Toxicities of Nanonutraceuticals.	287
8.12	Challenges and Future Perspectives.	288
8.13	Conclusion.	289
Bibliography.		289

C. S. Jones (✉)
Department of Pharmacology and Environmental Toxicology,
DR ALM PG Institute of Basic Medical Sciences (IBMS),
University of Madras, Chennai, India

Department of Pharmacology, Asan Memorial Dental College and Hospital,
Chengalpattu, India

V. U. M. Nallal · M. Razia
Department of Biotechnology, Mother Teresa Women's University,
Kodaikanal, TamilNadu, India

© The Author(s), under exclusive license to Springer Nature Switzerland AG 2022
A. Krishnan et al. (eds.), *Emerging Nanomaterials for Advanced Technologies*,
Nanotechnology in the Life Sciences, https://doi.org/10.1007/978-3-030-80371-1_8

8.1 Introduction to Nanotechnology and Nutraceuticals

Nanotechnology, an upcoming field of nanoscience and engineering research, is an empowering technology leading to transformation in food, pharmaceutical, cosmetics, and nutraceutical industry (Prasad et al. 2017a). Nanotechnology involves the ability to control and manipulate the atomic scale, creating and utilizing the small and intermediate size structures, devices, and systems that have novel properties and functions and research and technology development at the atomic, molecular, and macromolecular levels (1 to 100 nanometer). Nanotechnology remains a key research area in the science and engineering field with the enhanced utilization of nanoparticle-based medicines (Godwin et al. 2009; Rostamabadi et al. 2019; Saglam et al. 2021).

Nanoparticles are synthesized using organic or inorganic materials either by top-down methods starting with bulk material and removing the material to attain nanomaterial through milling, fluidization, homogenization processes, or emulsification or bottom-up methods starting with atomic or molecular precursors and combining them to form nanomaterials through precipitation/condensation, evaporation, or controlled sol-gel synthesis methods (Bhushan et al. 2014; Singh and Colonna 2015). Nowadays, encapsulation with the natural biodegradable polymer is the frequently used technique (Zhao et al. 2008; Shukla & Iravani 2019).

Nanomaterials that occurs naturally in food include protein, fats, carbohydrates, and nanostructures are cell membranes, hormones, etc (Powell 2008). As nanoparticles have major advantages and worthy of its value, billions of dollars are invested in the nano research, and scientists are exploring all the possible nanomechanisms to encapsulate nanomaterials in nutraceuticals for its safe and wise use to achieve maximum health benefits.

However, a great challenge in nanonutraceutical research is the efficiency in nanomanufacturing which is dependent (Bernhardt et al. 2010; McClements et al. 2015) on the identification and understanding of naturally occurring therapeutically useful material at the molecular level and large-scale production synthesizing at the nanoscale level (Roco and Bainbridge 2001).

8.1.1 Nano Fabrication

Nanomaterials are categorized into the following:

- Category 1: zero-dimensional structures are synthesized using vaporization (Physical /Chemical), aerosol processing, condensation of inert gas, precipitation etc. Eg; Quantum dots, nanoclusters, nanoparticles, nanocrystallites etc
- Category 2: nanometer-sized thin material processed by PVD/CVD/laser ablation/ion implantation (Geng et al. 2006; Singh and Colonna 2015)

8.1.2 Types of Nanocarriers

Nanoparticles are synthesized from natural (lipids, proteins, and polysaccharides) and synthetic (polymers) sources which are usually well-tolerated, biodegradable, and biocompatible (Nalwa 2004; Andreas Wicki et al. 2015; Doktorovova et al. 2016) (Fig. 8.1).

8.1.2.1 Lipid-Based and Surfactant-Based Nanocarriers

Emulsion types which include nanoemulsion/microemulsion/lipid concentrate (SMEDDS and SEDDS) are prepared from liquid lipid (oils) SLNs (solid lipid nanoparticles and NLCs). Nanostructured lipid carriers are synthesized from solid or semisolid fats (Doktorovová et al. 2016; Jafari et al. 2017).

(i) *Liquid Lipid Preparations*

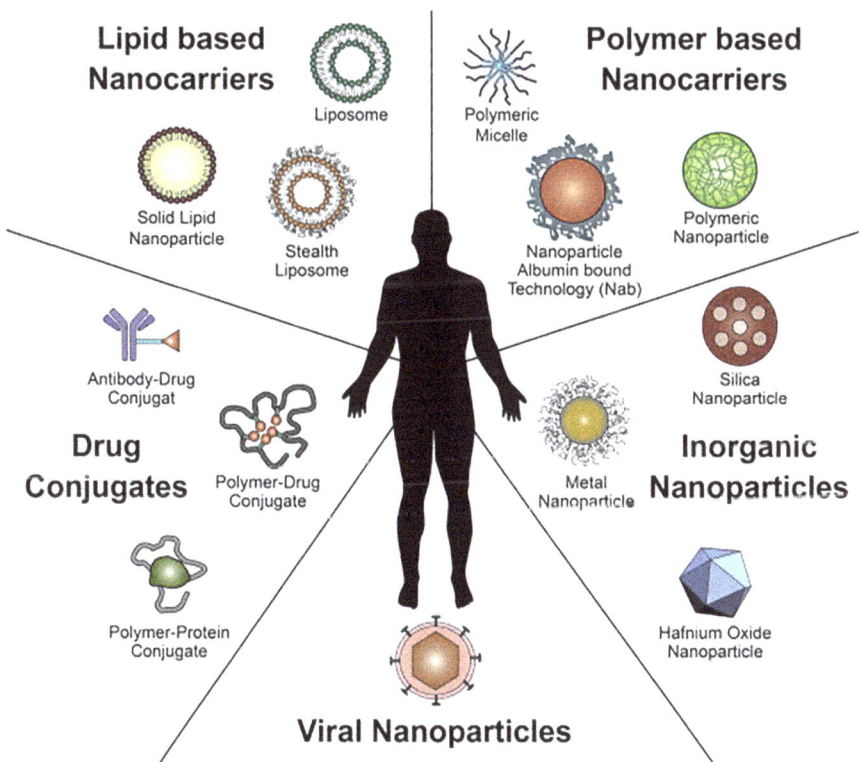

Fig. 8.1 Various types of nanocarriers. (Adapted from Andreas Wicki et al. 2015)

(a) *Liposomes* consist of small (~100–400 nm) amphiphilic artificial vesicles spherical in shape lipid bilayers that are used in the delivery of drugs, nutraceuticals, nutrients, enzymes, vitamins, antimicrobials, and additives (Godwin et al. 2009). Liposomes are an efficacious and safe drug delivery system with increased biocompatibility and are easily biodegradable. Moreover, it is devoid of toxicity or immunogenicity (Lungu et al. 2019), e.g., encapsulation of gallic acid with Zein fibers (Thomas and Sayre 2005).

(b) *Nanoemulsions* (5–200 nm) are kinetically stable, colloidal dispersion of oil droplets in an aqueous medium (O/W, W/O, bi-continuous type). The preparation techniques include the utilization of different energy levels. Low energy techniques include phase inverse and spontaneous emulsification and the techniques that utilize high energy are microfluids and ultrasound-homogenizers. The recent studies include GEON (garlic essential oil nanoemulsion) and Taguchi method(vitamins A, D, and E, encapsulated in O/w nanoemulsion, carotenoids, polyphenols lycopene, β-carotene, and curcumin (Castro et al. 2018; Saura-Calixto and Pérez-Jiménez 2018; Yuan et al. 2019).

(c) *Microemulsion* is a system that is thermodynamically stable comprising of oil, surfactant, cosurfactant, and water. The synthesis includes the self-assembling of surfactant molecules into core micelles and a cosurfactant that minimizes the interfacial energy, e.g., nanoemulsion droplet size (10 to 100 nm) of lycopene (Jafari et al. 2017; Amiri-Rigi and Abbasi 2019).

(d) *SEDDS and SMEDDS* include isotropic oil solution with the lipophilic component and a surfactant that will form microemulsion or emulsion with water on agitation, e.g., d-α-tocopherol polyethylene glycol 1000 succinate (Kuentz 2012).

(e) *Micelles* can encapsulate the electrolyte stabilized lipophilic or lipophobic drugs (~10–100 nm) usually formed by self-assembled amphiphilic particles. The solubilization capacity is directly proportional to the volume of the hydrophobic domain of the micelle core, e.g., casein (Yoksan et al. 2010; Abbasi et al. 2014).

(f) *Cubosomes* and *hexosomes* are viscous and formed from a dispersion of inversed cubic or hexagonal mesophases in water. They are synthesized by hydrating the unsaturated monoacylglycerols like monoolein, monovaccenin, and monopalmitolein or phytantriol and used preferably in encapsulation of hydrophilic, lipophilic, and amphiphilic food and drug components (Spicer 2004; Meikle et al. 2017) (Fig. 8.2).

(g) *SLNs* and *NLCs* (30 to 1000 nm) are lipid-based nanoparticles produced from lipids (stearic acid) with a high melting point (>40 c) (Ganesan et al. 2018). The advantages of SLNs and NLCS are nontoxic and have maximum encapsulation efficiency (Rostamabadi et al. 2019).

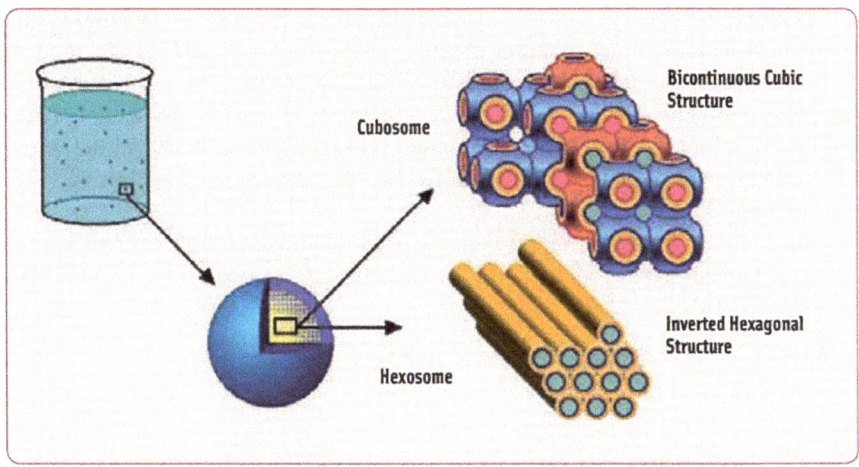

Fig. 8.2 The structure of cubosome and hexosome. (Adapted from Spicer 2004)

8.1.2.2 Biopolymeric Nanocarriers

GRAS (generally recognized as safe) biodegradable and natural biopolymers are used as nanocarriers and classified as follows:

(a) *Polysaccharide-based nanocarriers*

Homopolysachharides (cellulose containing glucose) or hetero polysaccharides (alginates made up of mannuroninc and glucuronic acid) are used as it forms gel by cold, heat set, or isotropic mechanism (McClements 2012).

(b) *Protein-based nanocarriers*

The proteins used for encapsulation are zein, gelatin, casein, myofibrillar, soy, egg, and BSA (bovine serum albumin) proteins (DeFrates et al. 2018).

(c) *Polymeric micelles*

Polymeric micelles (PMs) (10–100 nm) are amphiphilic copolymers containing poly amino acid and propylene oxide (Garti and Aserin 2012; Cho et al. 2015).

(d) *Dendrimers*

Dendrimers (~3–20 nm), a covalently conjugated branchlike structure formed through polymerization reactions (Castro et al. 2018), act as a vehicle to transport hydrophilic nucleic acid-based chemotherapies that cannot pass through the cell membrane (Baker 2009; Bharali et al. 2017; Mendes et al. 2017; Chakraborty 2019).

(e) *Nanocapsules* (~10–1000 nm): Nucleic acid (DNA and RNA) and drugs are restricted in a core-shell-like vesicles with a polymeric coating (Mao et al. 2009; Mendes et al. 2017).

(f) *Nanoparticles* (~20–200 nm): Polymeric nanoparticles include solid carriers that are capable of adsorption, dispersion, and entrapping the active ingredients into the matrix and are synthesized using preformed polymers through solvent evaporation, dialysis, salting, supercritical fluid technology, and nanoprecipitation. Silicon dioxide, titanium dioxide, and silver nanoparticles are used for sustained release (Arshak et al. 2007; Acosta 2009; Varela-Moreira et al. 2017).

(g) *Nanoconjugate drug molecules are covalently bound to the polymer.* Nanocochleates composed of soy-based phospholipids coil around the micronutrients and enhance the quality of the processed food (Liang and Subirade 2016).

(h) *Virosomes* (~150) are used as carriers in chemotherapy of cancer and vaccine development against viral infections (hepatitis, *H. influenza*) (Krishnamachari et al. 2010).

Nanotechnology based Nutraceutical applications

8.2 Nutraceuticals in Biomedical Applications

Nutraceutical, coined by DeFelice (1989), is an umbrella term used for the intersection of nutrition and pharmaceutical and a food or food product that provides a physiological benefit or provides some protection against chronic disease (DeFelice 1995). About 50%–70% of nutraceuticals are used in developed countries, and this number is increasing by age (Gupta and Prakash 2015). Women use more nutraceuticals than men. Nutraceuticals include functional foods and beverages and dietary supplements.

Classification of Nutraceuticals

A dietary supplement is a nutrient-containing drug obtained from food or food products that are prescribed in the form of powder, capsule or pills, and liquids (tonics, syrups). Functional food may be whole or fortified food with beneficial health effects (Chaudhry et al. 2008; Isidoro 2020) (Table 8.1).

The bioactive phytochemicals derived from plants have become a greater ingredient (Charu et al. 2007). Out of these phytochemicals, several groups of polyphenols (flavanones, isoflavones, anthocyanins, proanthocyanidins, ellagic acid, and resveratrol) are currently found useful in the nutraceutical industry. Nutraceuticals as dietary supplements provide incredible health and medicinal value for the prevention and treatment of various ailments (Prankash et al. 2012; Charu et al. 2014; Prakash and Sharma 2014).

Primary food elements including carbohydrates, proteins, and lipids necessary for basic life provide energy and normal the function of the body. However, secondary elements like vitamins and minerals are essential but not produced endogenously; hence, they must be administered in the diet or in the form of medicine (Bernela et al. 2018; Assadpour et al. 2020).

Nutraceuticals are minor food elements that regulate normal physiology and provide the defense against invading pathogens and cancer cells by immunomodulating effect (McClements 2012).

Table 8.1 Classification of nutraceuticals

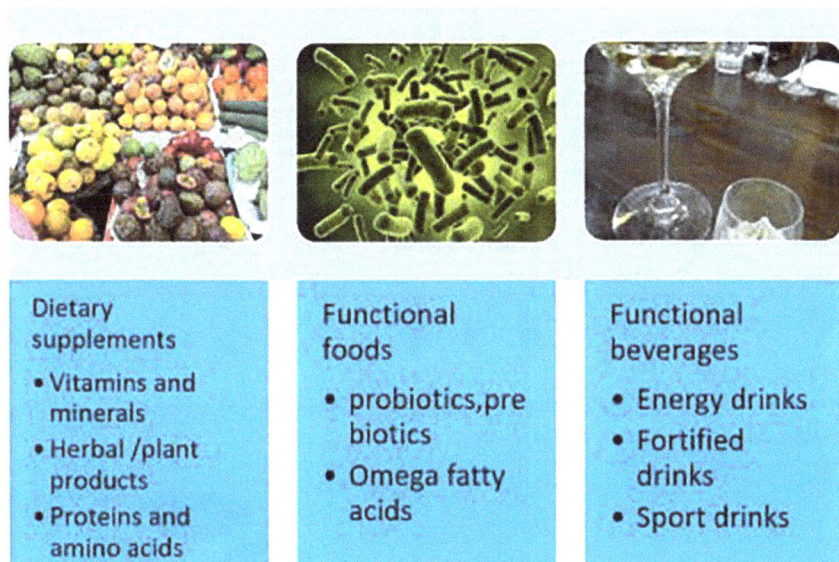

However, when nutraceuticals are ingested orally, poor water solubility, low permeability, reduced absorption, insufficient gastric residence time, and decreased bioavailability are the major factors that limit the efficacy of the nutraceuticals (Augustin and Hemar 2009).

8.3 Nanonutraceutical Drug Delivery System

Nanoencapsulation of food and nutrients with GRAS and ecofriendly biodegradable material facilitate sustained and steady release of the active ingredient (Sen and Pathak 2016; Dima et al. 2020).

Materials used for encapsulation of nutraceuticals are as follows:
- Polysaccharides: plant starch, carrageenan, pectin, etc.
- Microorganism: dextran, xanthan gum, etc.
- Protein food: egg, milk, oat, soy, gelatin, etc.
- Emulsifiers: Tween, sugar esters, monoglycerides, lecithin, etc.

Nutraceuticals are encapsulated and formulated based on shape, particle size, zeta potential, etc. of the nanomaterial to improve the stability, bioavailability, and biocompatibility of the nutraceutical (Fig. 8.3). The basic steps to be considered during the selection of nanocarriers are identification, isolation, purification, and characterization of the properties of incorporated food components (nutritional and

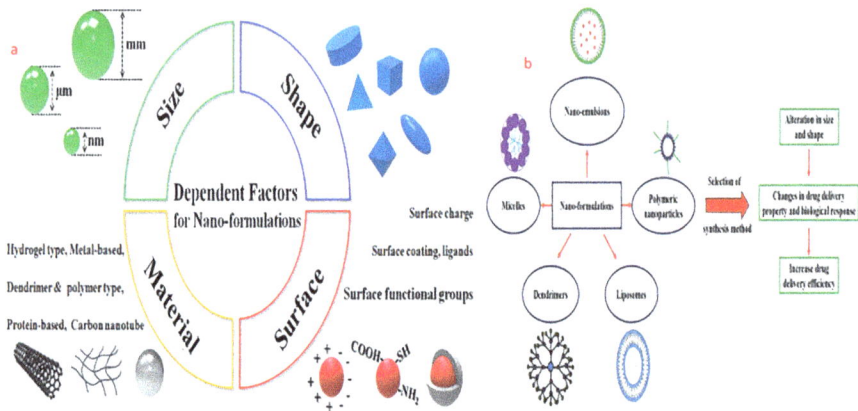

Fig. 8.3 Factors determining synthesis of nanocarriers and biosynthetic process. (Adapted from Jeevanandam et al. 2016; Bajpai et al. 2018)

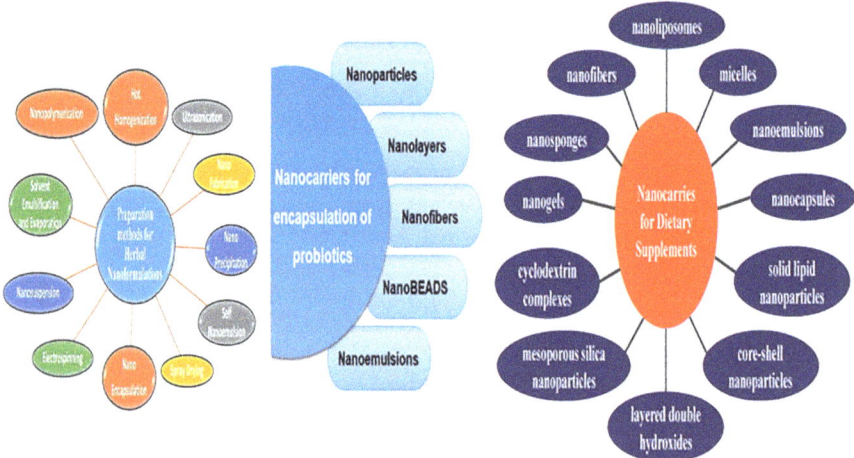

Fig. 8.4 Nanonutraceutical carriers. (Adapted from Dhawan et al. 2018; Jampilek et al. 2019; Machado et al. 2020)

medicinal value) (Jeevanandam et al. 2016; Bajpai et al. 2018). Nanoformulations of vitamins include spray cooling, spray drying, phase separation, liposome, solid lipid nanoparticles (SLN), and inclusion complexation. Figure 8.4 depicts nanoencapsulation of herbal products, probiotics, and dietary supplements (Dhawan et al. 2018; Jampilek et al. 2019; Machado et al. 2020). Archeosome is a nanoencapsulated delivery system for antioxidants pp obtained from archaebacterial membrane lipids. The degradation of milk is prevented by nanoencapsulation of α- tocopherol in fat droplets and canola active oil for the nanoencapsulation of fortified phytosterols with canola oil and Shelf-life of vitamins B9, B12, and vitamin C were increased using chitosan biopolymer (Chaudhry et al. 2008; Ali et al. 2019; Czech et al. 2019).

8 Nanonutraceuticals in Chemotherapy of Infectious Diseases and Cancer

The instrumental techniques used to characterize the physicochemical parameter of the nanomaterial are the following:

The particle size in nanometer is determined using (TEM, SEM, dynamic light scattering, laser diffraction, static image analysis, etc), crystal morphology (powder x-ray diffraction), magnetic properties (vibrating sample magnetometer), chemical parameters (FTIR, NMR, Raman spectroscopy, organic (LC-MS), inorganic(inductively coupled plasma spectroscopy, Electron microscopy with energy-dispersive X-ray spectroscopy).surface chemistry(Xray photon electron microscopy, scanning probe microscopy) etc.

Limitations of Nutraceuticals

However, physicochemical properties of nutraceuticals have a number of limitations.

All these factors may contribute to reduced bioavailability and hence reduced therapeutic effect.

Nanoformulation of nutraceuticals offers improved pharmacokinetic and pharmacodynamic properties that facilitate target-specific treatment and diagnosis, low systemic side effects, and accurate therapeutic monitoring (Chau et al. 2007) (Table 8.2). Therefore, more innovative research are being carried out for the transition of nutraceuticals into nanonutraceuticals, and most of them are patented (Razak et al. 2018).

The factors affecting pharmacokinetics of nanonutraceuticals include complete absorption, targeted delivery to the site of action, metabolized in the liver and excreted mainly via the kidney in the urine (Fig. 8.5).

The main advantages of nanocarrier-mediated nutraceutical delivery is enhancement of solubility, rate of dissolution and oral bioavailability, and reduction in dosage, time of onset, and systemic toxicity (Fig. 8.6) (Bae and Park 2011; Klajnert et al. 2013; Shende and Mallick 2020).

Major advantages of nanosizing of nutraceuticals are depicted in Fig. 8.7.

Table 8.2 Comparison between nutraceutical and nanonutraceuticals

Nutraceuticals	Nanonutraceuticals
Poor solubility	Chemically and physically stable to environmental stresses while preserving its functional characteristics
Instability of the compound	
Impermeability across the biological membranes	Improve gastric stability of labile bioactive nutrients
	Maintain constant dosage level within systemic circulation
Disintegration in the stomach due to the presence of an acid	Capable of facilitating lymphatic transport
	Extend the gastric retention time
Widespread distribution	Target-specific delivery
Nontargeted delivery	
High dose required	
Systemic toxicity	

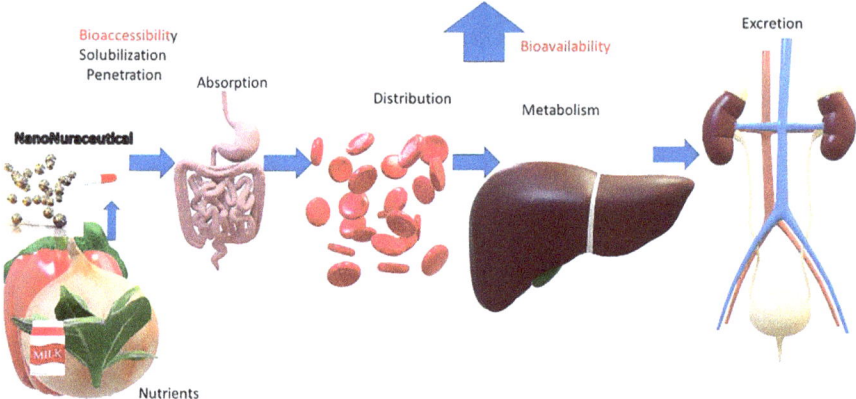

Fig. 8.5 Pharmacokinetics of nanonutraceuticals

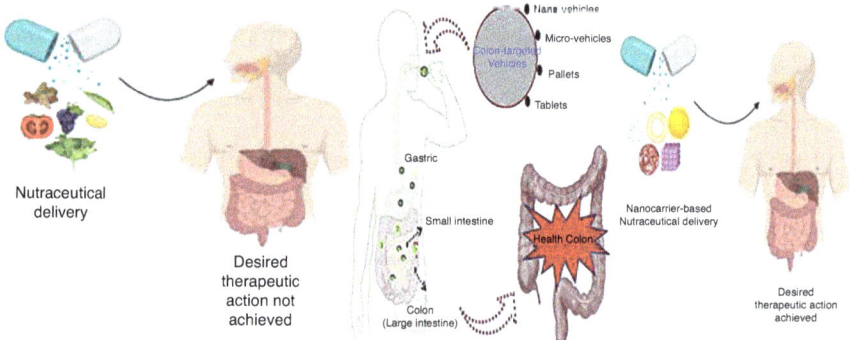

Fig. 8.6 Target-based nanonutraceutical drug delivery. (Adapted from Shende and Mallick 2020)

8.4 Nanosize Nutraceutical Formulations in Biomedical Applications

Nanonutraceuticals in medicine include screening of drugs, gene and drug delivery, and diagnosis, detection, and monitoring of diseases (Fig. 8.8).

Nanoformulated drugs enhance effectiveness in different dosage forms.

Nanonutraceuticals in healthcare: nanovitamins, nano-calcium, nano-magnesium, nano-iron, nanoprobiotics, and nanophytochemicals

8.4.1 Nanovitamins and Nanominerals

Nanoformulation of vitamins and minerals is of foremost importance as it aids in target-specific delivery of vitamins/minerals to the tissues and organs (Iman Katouzian and Seid Mahdi Jafari 2016). Therefore, wastage and adverse effect due

8 Nanonutraceuticals in Chemotherapy of Infectious Diseases and Cancer

Fig. 8.7 Major advantages of nanonutraceuticals

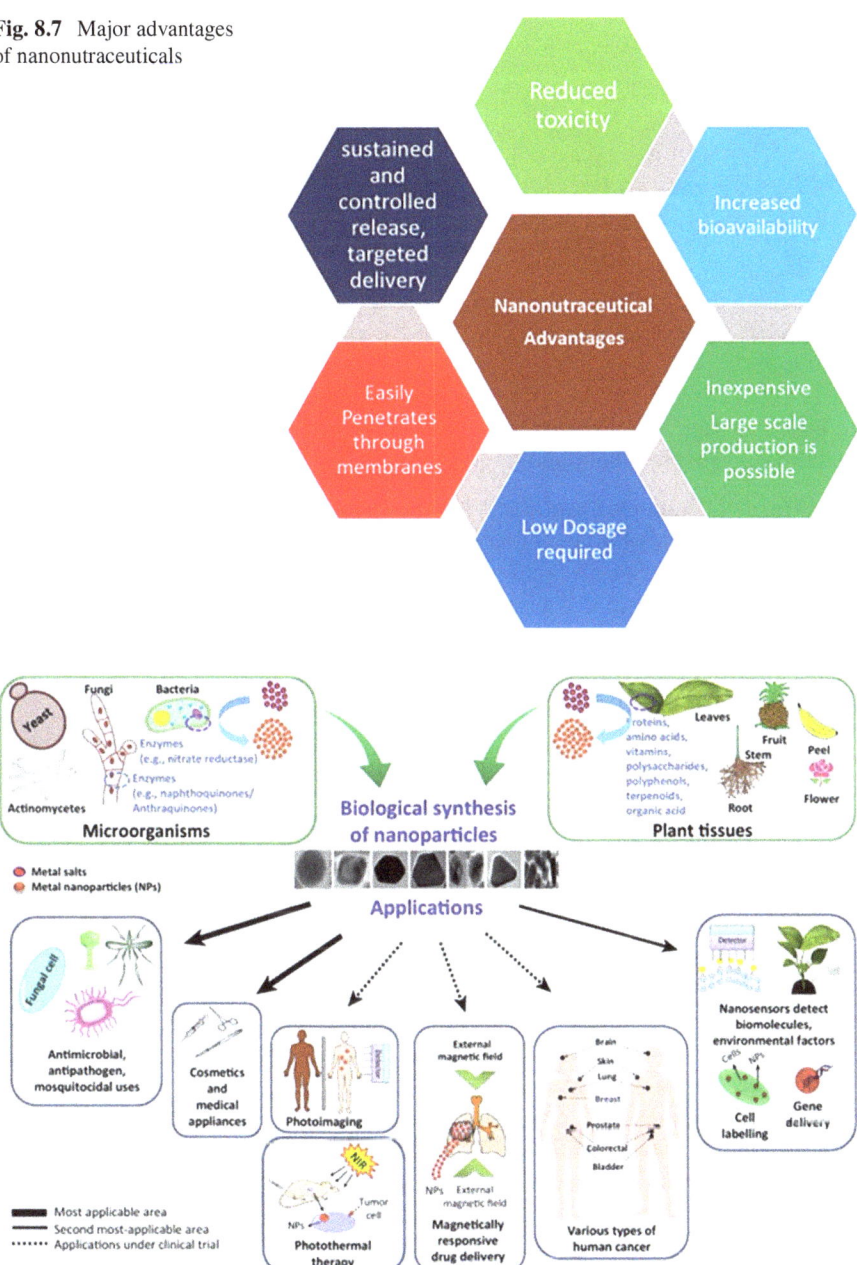

Fig. 8.8 Biomedical applications of nanotechnology. (Adapted from He et al. 2019)

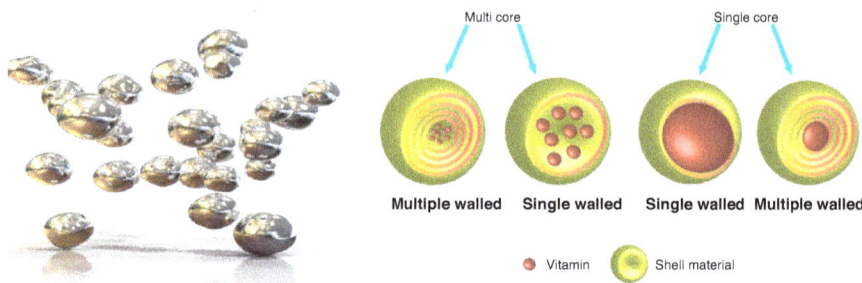

Fig. 8.9 Nanoencapsulation of vitamins. (Adapted from Iman Katouzian and Seid Mahdi Jafari 2016)

to higher dosage is minimized, e.g., nanoencapsulation of vitamins and minerals with liposomes (phospholipids) (Fig. 8.9).

The Merits of Nanoencapsulated Vitamins and Minerals
The dissolution rate of drugs is quick comfortable to swallow, reaches the intestine directly hence complete absorption of vitamins and minerals due to the prevention of first-pass metabolism and destruction by gastric juice, distribution and bioavailability is high and enhanced therapeutic action due to target-specific delivery. Additionally, nanotechnology enables to reduce the usage of additives, fillers, and binders during the manufacturing processes.

8.4.2 Nanoencapsulation of Probiotics

The imbalance or disequilibrium in normal gut flora results in pathological conditions due to disturbances in cellular components. Numerous nutraceutical probiotics counteract the effect of ROS production and thereby can prevent major ailments that include cancer and cardiac and cerebrovascular diseases (MCquade et al. 2019) (Fig. 8.10). Recently, nanoencapsulation of probiotics has been studied with gold and selenium (10–1000 nm) nanoparticles (Pathak and Akhtar 2018; Machado et al. 2020).

The probiotic metabolites produced in the form of secretory proteins (extracellular proteins), short-chain fatty acids, enzymes, extracellular vesicles, bacteriocins, indoles, prodigiosin, and menaquinones enhance the mucus secretion and intervene with the receptor function as a defensive mechanism to afford the protection to the epithelial layer of the intestine (Biersack and Schobert 2012; Kumar et al. 2012; Sumathi et al. 2014).

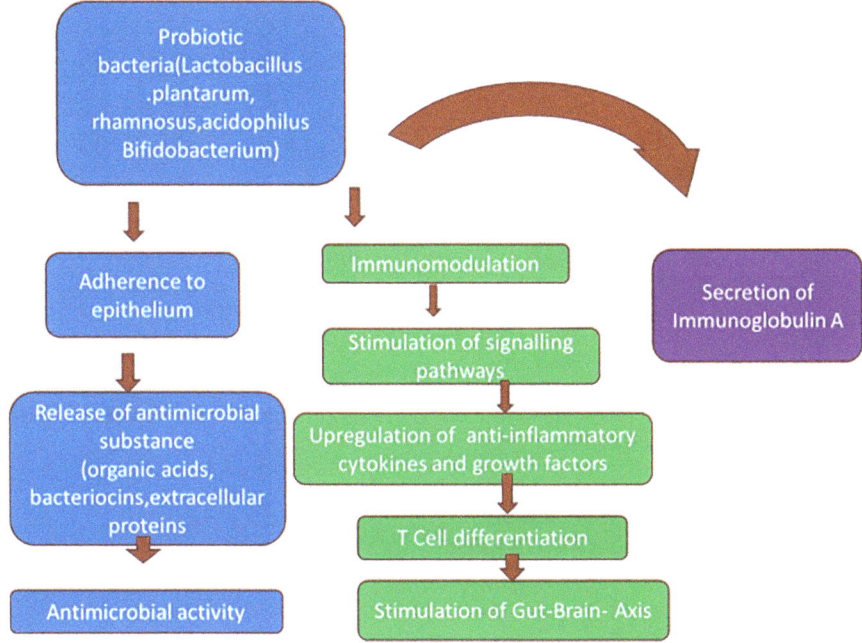

Fig. 8.10 Pharmacodynamics of probiotics

8.4.3 Nanophytochemicals

Similarly, phytochemicals including polyphenols, flavonoids, carotenoids, and glycosides can be effectively utilized in the treatment of hypertension, diabetes, cancer, and GIT disorders and as an immunostimulant owing to their antioxidant potential (Bayir et al. 2019). Nevertheless, the main drawback of the phytochemicals was low bioavailability. Nanoencapsulation of the phytochemicals with small size lipids increases the oral bioavailability and improved therapeutic efficacy due to targeted delivery (Roy et al. 2021; Traitler et al. 2015; Üner 2016; Zachariah et al. 2020) (Fig. 8.11).

Pathophysiology of Cancer, Infectious Diseases, and Inflammation
The imbalance between the ROS production and antioxidant defense mechanism creates oxidative stress that leads to various pathological conditions. ROS interacts with the biomolecules (lipids, a cellular protein, DNA, and RNA) and causes physiological and biochemical alterations in the cell leading to various diseases (Lee et al. 2004; Nosrati et al. 2017; Zewen et al 2018) (Fig. 8.12a).

Nutraceuticals induce the antioxidant mechanism through stimulation of NrF2, ARE and other genes that are involved in antagonizing the adverse effect due to ROS and RNS production leading to deleterious effects (Fig. 8.12b) (Seifalian et al. 2014; Li et al. 2017b; Prasad et al. 2017; Calvani et al. 2020; Godugu et al. 2020).

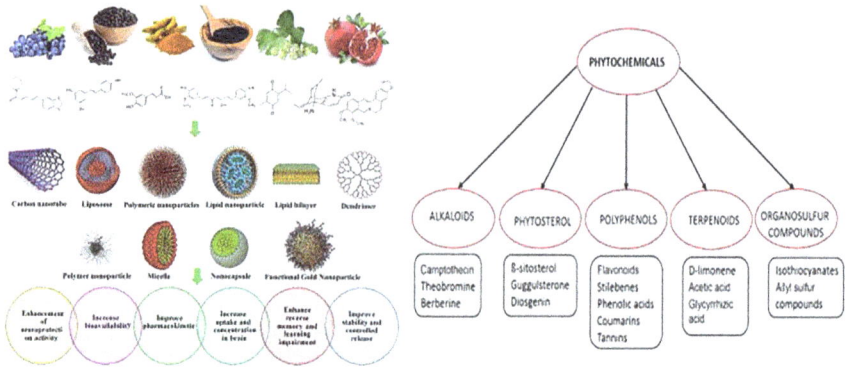

Fig. 8.11 The different forms of phytochemicals and its nanoencapsulation. (Adapted from Bayir et al. 2019; Zachariah et al. 2020)

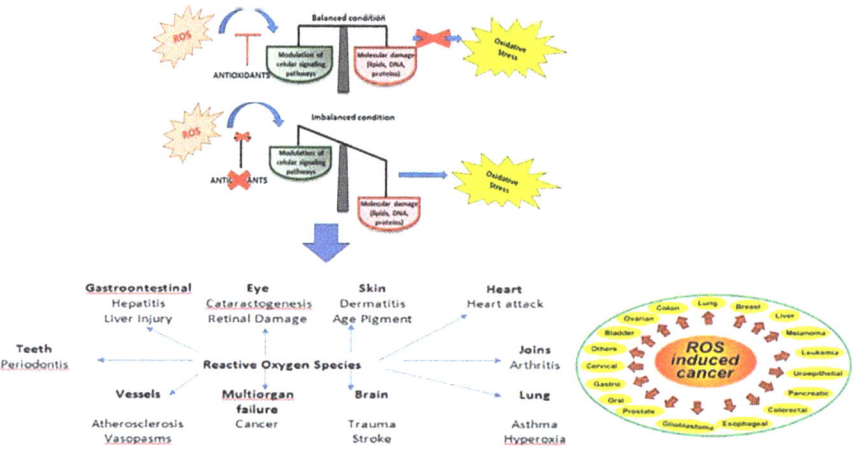

Fig. 8.12 (**a**) Pathophysiological mechanism involved in cancer, infectious diseases, and inflammation. (Adapted from Lee et al. 2004; Prasad et al. 2017; Calvani et al. 2020). (**b**) Nanonutraceuticals in cancer, infectious diseases, and inflammation

8.5 Nanonutraceuticals in the Chemotherapy of Cancer

The mechanism of action of nutraceutical are mainly by affecting epigenetics (up- or downregulating DNA methylation and acetylation of protein, stimulating specific miRNAS, downregulation of B-Cell lymphoma-2(Bcl-2)/ cyclin 01, angiogenic factors, metalloproteinases-9, and IL-6 pathways by blocking inflammatory pathways and NF-kB inhibition (Hegazy et al. 2019). Nutraceuticals reset the normal epigenetic marks and control the metastasis condition including hematological malignancies (Deshantri et al. 2018).

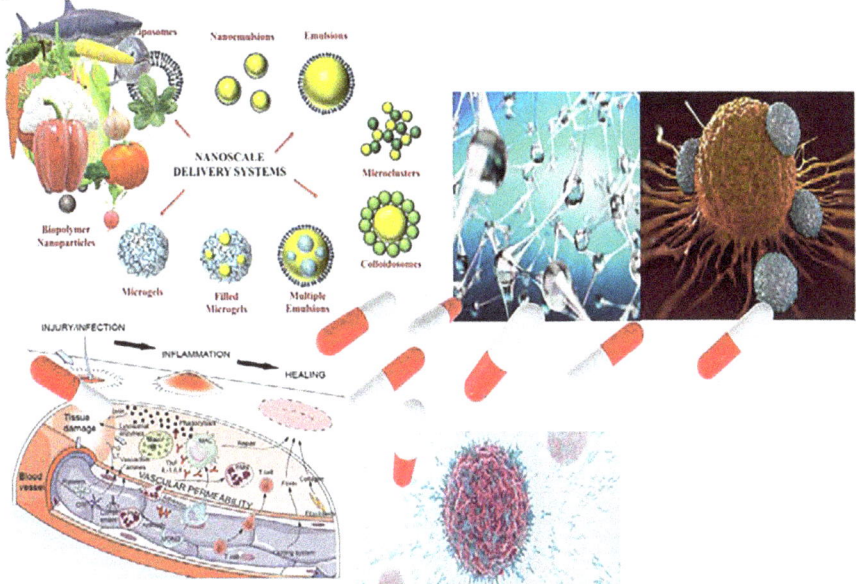

Fig. 8.12 (continued)

Phytochemicals possess antioxidant activity that plays a pivotal role in cancer therapy by preventing the oxidative stress-induced DNA damage, e.g., indoles in cabbage decrease estrogen and thereby the risk of breast cancer (Biersack and Schobert 2012; Singhal et al. 2017). Capsaicin, in chili pepper, protects DNA from carcinogens (Clark et al. 2015). Nanoformulation of these nutraceuticals similar to nanomedicines will facilitate specific and targeted effects with enhanced benefits (Wolfram and Ferrari 2019; Czech et al. 2019; Podsednik et al. 2020).

The probiotic action is mediated through the production of antimicrobial substances, multi-pathogen competition, protection of epithelial layer through secretion of mucus, and immunomodulation. The probiotics adhere to the epithelium and prevent the entry of pathogens. Probiotics affect the immune system by augmenting and upregulating the signaling pathway leading to anti-inflammatory responses via the gut-brain axis (Iacono et al. 2011; Sumathi et al. 2012; Dasari et al. 2017; Langella et al. 2019) (Fig. 8.13h).

Moreover, bioactive nutraceuticals interfere in the electron transport chain mechanism by augmenting phosphorylation and AMPK production. Simultaneously, nutraceuticals will cease the spread of cancerous cells and will provide resistance to ROS-induced cell damage during chemotherapy through synergistic prooxidant activity with anticancer drugs (Hegazy et al. 2019; Pratheeshkumar and Kuttan 2011).

The biotransformation or detoxification process involves phase I (non-synthetic) and phase II (synthetic) pathway. Phase I biotransformation may lead to the accumulation of toxic metabolites that can induce cancer. Hence, activation of phase II synthetic pathway will lead to excretion of these tumorigenic potential compounds.

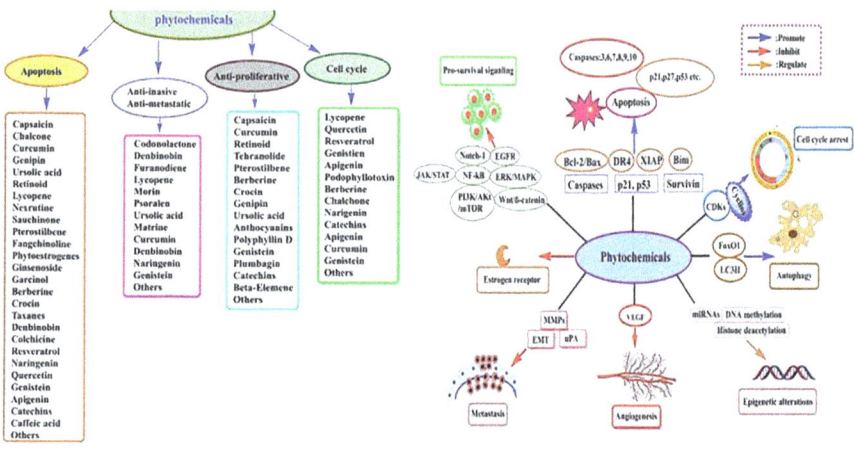

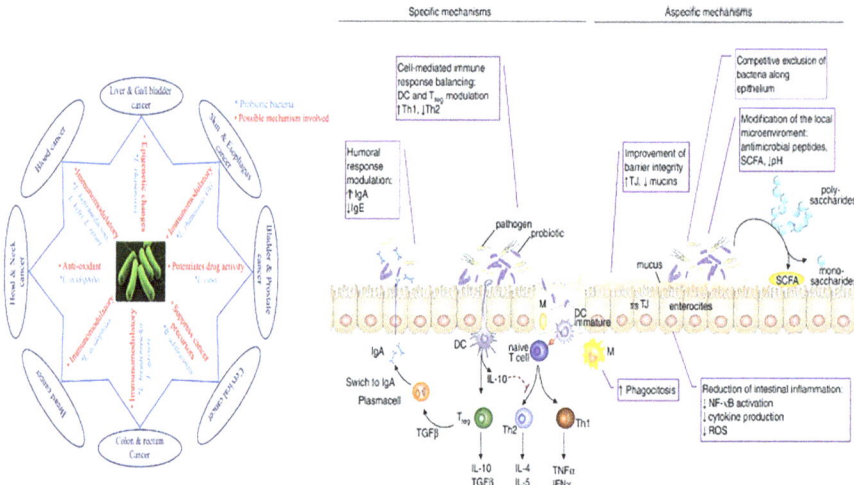

Fig. 8.13 (**a**) The anticancer activity of a. phytochemicals. (Adapted from Iqbal et al. 2018). (**b**) probiotics. (Adapted from Iacono et al. 2011; Dasari et al. 2017)

Nanonutraceuticals upregulate synthetic phase II detoxifying enzymes and lower the excess production of toxic intermediates that can lead to tumorigenic potential and inflammatory responses (Zhao and Agarwal 1999: Zhu et al. 2020).

Nanoformulations of nutraceuticals allow the specific target of tumor cell that surpasses hepatic metabolism and improves efficacy and safety due to dosage reduction and prevention of toxic effects (Flores et al. 2017; Li et al. 2019b).

The phytochemicals possess medicinal value in cancer chemotherapy due to its antioxidant activity and suppression of oxidative stress-induced DNA damage (flavonoids, capsaicin, etc.) and reduce hormone (vinca alkaloids, cabbage in breast cancer). Nano targeting of tumor requires active high-affinity ligand for interacting with the target moiety which is overexpressed in cancerous cells (Clark et al. 2015; Iqbal et al. 2018; Jampilek et al. 2019; Lungu et al. 2019). Phytochemicals so far studied are noscapine, berberine, brucine, eugenol, homoharringtonne, sanguinarine, taxifolin, quercetin, dihydroartemisinin, gambogic acid, ursolic acid, nobiletin, ellagic acid, curcumin, etc. that acts by inhibiting STAT-3 signaling pathway and vascular endothelial growth factor (VEGF) and regulation of miRNA expression and platelet-derived growth factor (PDGF) receptors (Eun and Koh 2004; Li et al. 2012; Srivastava et al. 2016; Luo et al. 2017; Abdullah et al. 2018; Soto-Quintero et al. 2019; Madkour 2020) (Fig. 8.13).

Vitamins that are used in the therapy of various cancer like colon, rectal, ovarian, and breast include folic acid, B6, B12, and vitamin D that act by interfering with DNA synthesis, receptor-mediated activity, and cell signaling pathways (Liu 2011; Kuppusamy et al. 2014; Pludowski et al. 2019; Singh et al. 2020).

Subsequently, the limitations of physicochemical properties of phytochemicals and probiotics that include polyphenols, phytosterols, carotenoids, vitamins, and minerals are solubility, stability, and permeability that can be overcome by utilizing natural bioactive nanocarriers (PLGA, chitosan, and natural that can improve bioaccessibility and bioavailability) (Boik 2001; Haidar et al. 2008; Sundraraman and Jayakumari 2019). Recent research with nanoencapsulated nutraceuticals has proven effective against various types of cancer cells as it increases the bioavailability of up to 80% (Sahoo et al. 2017; Loutfy et al. 2019; Nayak et al. 2019; Mousa et al. 2020).

Nanocarriers which are proven effective in the encapsulation of nutraceuticals in cancer chemotherapy with highly target-specific delivery and bioavailability are SLN (Date et al. 2019) Biodegradable TPGS-b-PCL NP, Gold NP Triptolide-loaded cationic liposomes, thermosensitive polymer NP hybrid nanomaterial, PEGylated silica NP, dithiodiglycolic acid, Silk fibroin NP, naringenin loaded PCL NP,, naringenin-loaded PLGA NP PLGA, PLA-vitamin E TPGS copolymer, multiwalled carbon nanotubes alginate NPs, PVP conjugate micelle, soy protein NPs α-CD derivatives, liposomal formulation, magnetic nanoprecipitation, NP, hollow capsules, albumin nanosuspension in Brain, cardiovascular system, leukemia, hepatic, colon, prostrate, lung, renal, neuroblastoma, colorectal, breast, prostate cancers (Yoksan et al. 2010; Teleanu et al. 2018; Zhang 2018; Calvani et al. 2020; Fuster et al. 2020; Montalbán et al. 2018; Salama et al. 2020) (Fig. 8.14.)

Similarly, resistance develops to chemotherapy due to lack of specific target and presence of efflux pumps in cancer cells. Nanotechnology can be applied in diagnosis, treatment, and prophylaxis (Divya Arora and Sundeep Jaglan 2016; Zhang et al. 2018; Sarwar et al. 2019). Nanocarrier-mediated drug or nutraceutical delivery aims at the specific target, i.e., tumor cell, and scientists are still exploring in this recent research area.

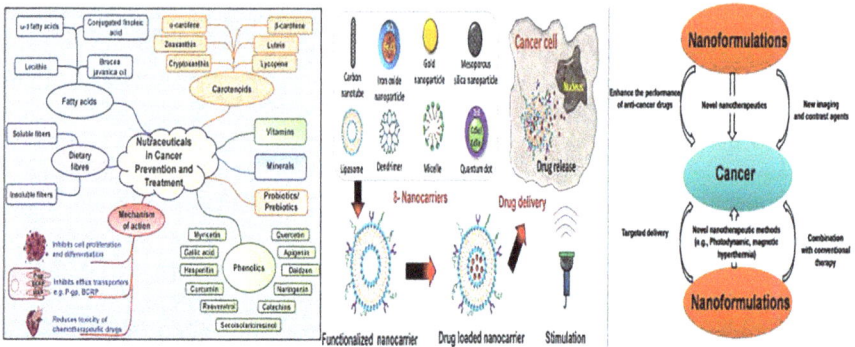

Fig. 8.14 Nanonutraceuticals in cancer therapy. (Adapted from Divya Arora and Sundeep Jaglan 2016; Sarwar et al. 2019; Sahoo et al. 2017)

8.6 Nanonutraceuticals and Anti-inflammatory Activity

Nutraceuticals reduce the reactive oxygen species (ROS) production by either upregulation of anti-inflammatory process or downregulation of inflammatory responses through the alteration of signaling pathways that include NF-κB, MAPK, STAT, etc. or mediators (prostaglandins; TNF-α; interleukins 1, 6, and 7; IFN-γ; etc.) (Kim et al. 2013a; Gupta et al. 2016; Afonina et al. 2017) (Fig. 8.15a, b).

Nutraceuticals possess nutrigenomic potential that can alter gene expressions. Utilizing the nutrigenomic principles in the prevention and therapy of major diseases by recovering the normal homeostasis is the novel approach. However, owing to its higher molecular weight and structure and lower bioavailability, it is not recommended for routine use in cancer therapy (Dandawate et al. 2016; Khoder et al. 2016).

Nutraceuticals from plants and animals are extensively studied so as to utilize in chemotherapy and prevention of major diseases owing to their ability to modify inflammatory response and interfere with cellular functions (Chikara et al. 2018; Al-Mssallem et al. 2019).

There exists a cross talk between NF-κB and Nrf2 that regulates the cellular responses to oxidative stress. The anti-inflammatory activity of nutraceutical involves the regulation of multiple gene expression which is influenced by the NF-κB family of transcription factors in coordination with immune responses and inflammatory responses. The most potent activators of NF-κB involved in cell signaling pathways (canonical and alternate) leading differentiation, proliferation, and apoptosis are tumor necrosis factor (TNF-α), bacterial lipopolysaccharide (LPS), and interleukin (IL)-1β (Jesus and Yamamoto-Furusho. 2007; Maithili Karpaga Selvi 2015; Chen et al. 2018; Darwish et al. 2019; Deng 2020).

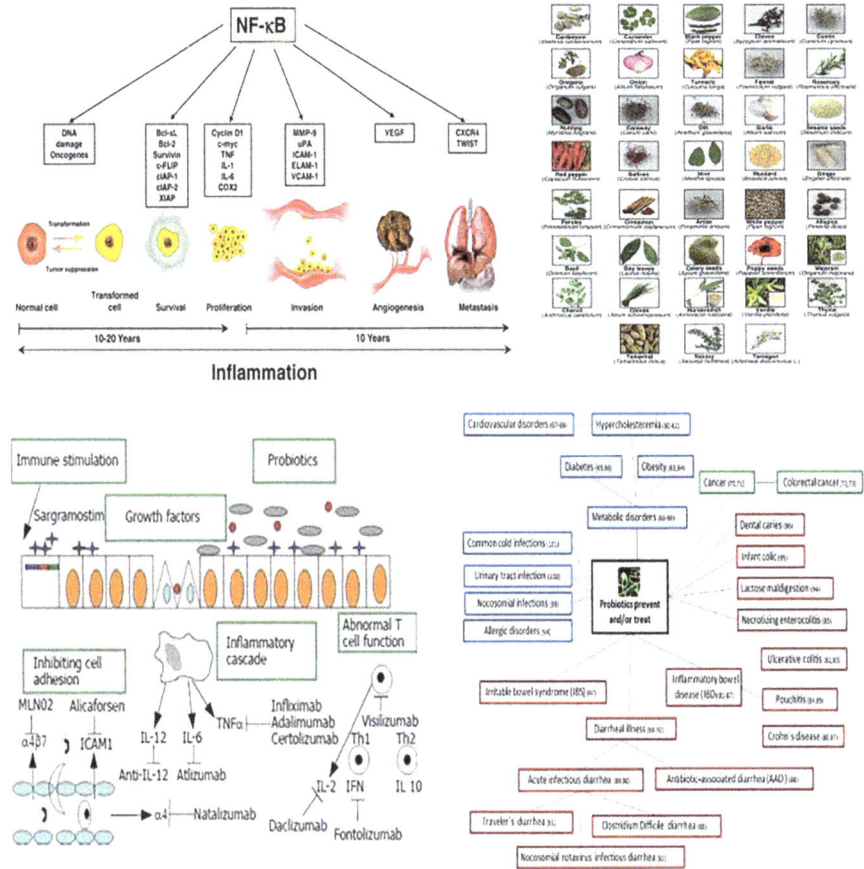

Fig. 8.15 The inflammatory process and anti-inflammatory activity of (**a**) phytochemicals. (Adapted from Agarwal et al. 2009). (**b**) probiotics in inflammatory conditions. (Adapted from Jesus and Yamamoto-Furusho 2007; Khoder et al. 2016)

8.7 Nanonutraceuticals in Medical Imaging

Nanomedicine has been successfully utilized in targeting the specific site and identifying the cancer cells and other serious clinical conditions (Salama et al. 2017).

Nanoparticles that are useful in medical imaging are silica in brain tumor, carbon nanoparticles in colorectal cancer, and ferumoxytol- ironoxide nanoparticles in recognizing the metastasis condition, nanosensors in diseases, nanochips in B- cell lymphomas (Bawarski et al. 2008; Etheridge et al. 2013; Xu et al. 2015; Surendran et al. 2018; Pellico et al. 2019) (Fig. 8.16).

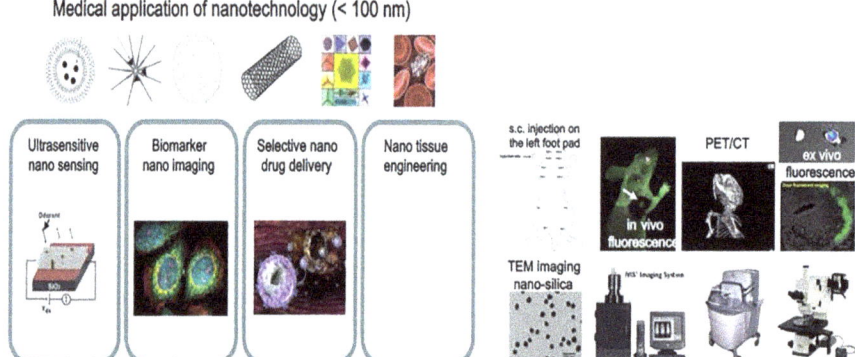

Fig. 8.16 Nanotechnology in medicine and biomedical engineering

8.8 Nanonutraceuticals in Prophylaxis, Diagnosis, and Treatment of Infectious Diseases

Epidemiological studies have indicated that dietary constituents have a major impact on the health of the individual (Chiu et al. 2020). The daily consumption of nutraceuticals in a balanced proportion helps to prevent chronic cardiovascular and neurodegenerative diseases, gallstone, and diabetes. The nutraceuticals regulate the microRNA expression which is involved in the maintenance of normal cellular processes and health conditions (Watson and Preedy 2014; Cameron and Chrubasik 2014; Brüll et al. 2016) (Fig. 8.17).

Furthermore, nutraceuticals cause induction of:

- Glutathione, a major contributor of cellular redox status which gets depleted during oxidative stress and downregulated glutamate-cysteine ligase, a rate-limiting enzyme associated with NF-κB (Zhang et al. 1992).
- Thioredoxin is a disulfide-linked protein involved in thiol-dependent cellular oxidative defense mechanism – signal transduction, cellular growth, and proliferation – and that influences hormones like insulin and glucocorticoid receptors and nitric oxide synthase a transcriptional factor.
- NAD(P)H quinone dehydrogenase, an inducible enzyme that protects cells against the cell damage due to redox cycling of quinones and glutathione depletion.
- Heme oxygenase-1 (HO-1) inducible cytoprotective isoform, a first and rate-controlling enzyme of the degradation of heme into iron, carbon monoxide, and biliverdin that causes direct inhibition of NADPH oxidase activity, hence the SOD generation. HO-1 consists of a binding site for transcription factors of inflammation (NF-κB and activator protein-1 (AP1) and a number of cytokines) (McElvee et al. 2007) (Fig. 8.18).
- The major cardiovascular, cerebrospinal, respiratory, gastrointestinal, and renal diseases and cancer are mainly due to the imbalance between Nrf2 and NF-κB

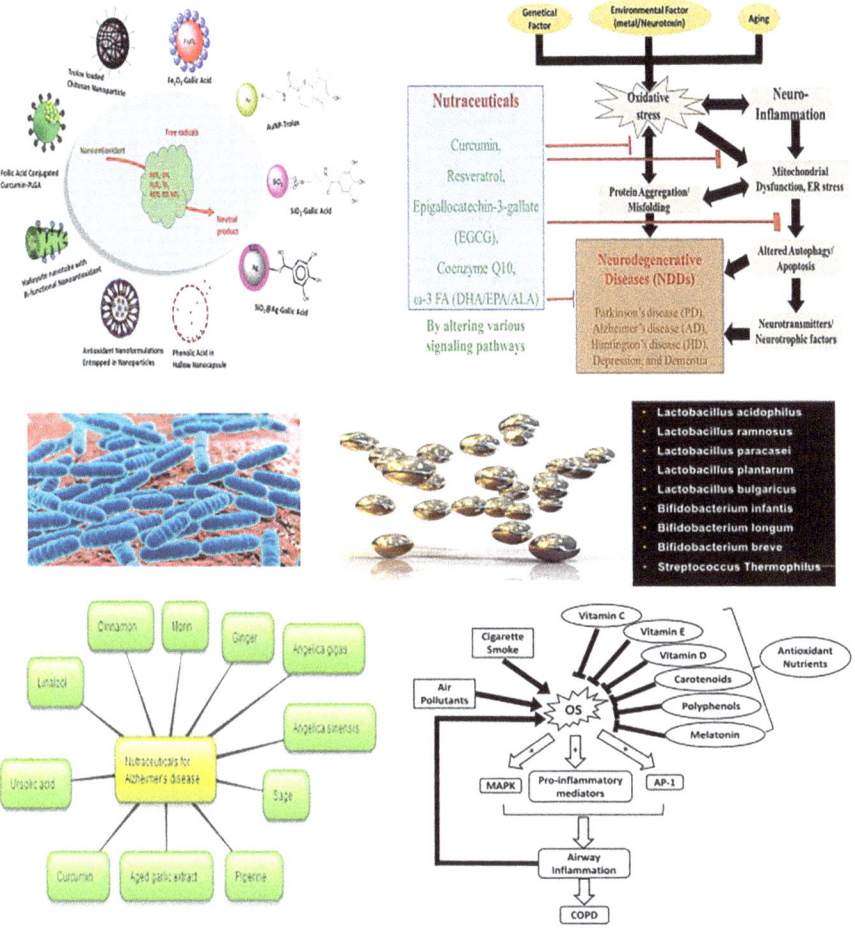

Fig. 8.17 Nanoencapsulation of nutraceuticals in the treatment of diseases. (Adapted from Liu et al. 2018; Chiu et al. 2020)

(Davis et al. 2013; Acevedo-Murillo et al. 2019; Nagaprashantha et al. 2019; Fuster et al. 2020).

Nanoparticles like ω-liposomes can actively penetrate through the immune cells and hence can effectively deliver the nutraceutical (Haidar et al. 2008; Subramani and Ganapathyswamy 2020).

Recent research indicates that fish oil encapsulated with ω-liposomes has a synergistic effect that blocks the production and release of pro-inflammatory cytokines and inflammatory mediators from neutrophils and macrophages and prevents tumor-cell proliferation in head and neck squamous carcinoma (Jampilek et al. 2019; Hamsa and Kuttan 2010).

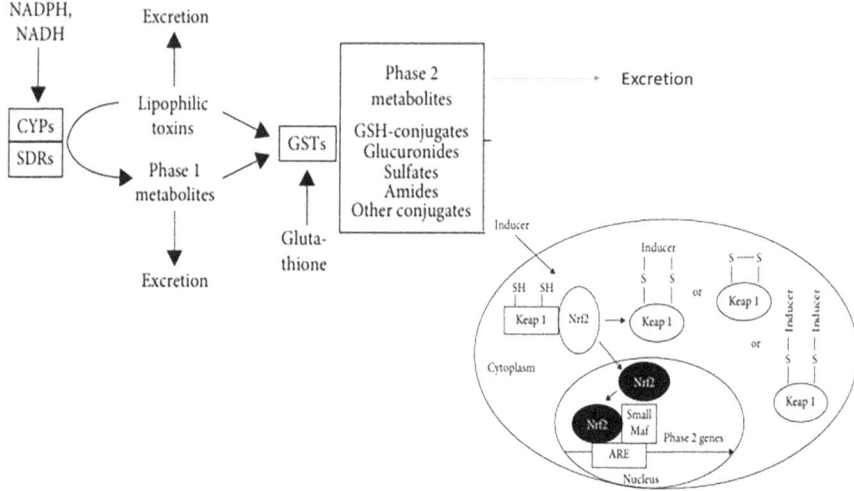

Fig. 8.18 Induction of metabolism by neutraceuticals. (Adapted from Zhang et al. 1992; McElvee et al. 2007)

Similarly, antimalarial herbals include *Ocimum gratissimum*, *Mangifera indica*, *Citrus sinensis*, *Carica papaya*, etc. (Lawal et al. 2015). Peppermint oil and silymarin are effective in the treatment of irritable bowel syndrome. *Artemisia afra*, *Clausena anisata*, and *Haemanthus albiflos* are antitubercular herbals. *Alpinia* and *Acanthopanax* are useful as antiparkinsonian drugs, and *Gardenoia*, *Rubia*, etc. are used in the treatment of gout and arthritis due to its uricosuric effect. *Ficus* have an anthelminthic, antioxidant, and antidiarrheal effect. *Zinger officinalis* is effective against *Leishmania amazonensis*. *Aloe vera* was effective in reducing chronic ulceration condition that include aphthous ulcer (Pawar et al. 2010; Bhalang et al. 2013; Lage et al. 2015; Duarte et al. 2016). The therapeutic effect of these plant extracts is due to the presence of tannins, alkaloids, terpenes, glycosides, acemannan, brucine, matrine, etc. (Cameron and Chrubasik 2014; Heber et al. 2014; Liu et al. 2016; Saura-Calixto and Pérez-Jiménez 2018; Sudha et al. 2020).

8.9 Nanonutraceuticals as Antibacterial Agents

TLR2 and TLR4 act as innate sensors in the identification of the distinct molecular patterns on the cell wall of invading pathogens and reciprocate with the innate and adaptive immune responses and consequent activation of NF-κB and inflammatory cytokines. Nanonutraceuticals suppress TLR4 oligomerization and reduces hs-CRP, IL-6, and TNF-α. Another gut pathogen *H. pylori*, a pathobiont, is a cancer promoter, causing cancer due to impairment of protein and mineral metabolism due to a rise in pH of gastric content. Nanonutraceuticals afford gastroprotective effect and modulate the immune response by simultaneous Nrf2 upregulation and NF-Kb downregulation (Grierson and Afolayan 1999; Sahin et al. 2010; Wang et al. 2011; Voukeng et al. 2016; Swanson et al. 2019).

Nanonutraceuticals provide perspective in antimicrobial, anticancer, and anti-inflammatory therapy due to its relatively nontoxic and multiple mechanisms through interference with the activity of phase I and phase II detoxifying enzymes, anti-inflammatory mediators, cancerous cell cycle arrest, apoptosis, and the epigenetic regulation on cyclins, Nrf2-Keap1, CDK, etc. (Sikka and Sethi 2015; Liu et al. 2014; Liu et al. 2016; Mathur et al. 2018; Shende and Mallick 2020).

Subsequently, nanoencapsulation with phytochemicals and probiotics produces a synergistic effect with conventional anticancer antibiotics (nano-TQ with doxorubicin causes upregulation of p53 and downregulation of Bcl2). Nano-TQ offers additional protection against inflammation, diabetes, and cardiovascular and central nervous system diseases due to its antioxidant activity (Alam et al. 2012; Chen et al. 2018). Vitamin E induces apoptosis and enhances the efficiency of chemotherapy (Joshi et al. 2013; Li et al. 2017a; Jampílek and Králová 2018; Menditto et al. 2018).

Probiotic produces secondary metabolites with a broad antimicrobial spectrum (*L. salivarius Bacillus megaterium, Pontibacter sps, Lactococcus lactis, L.reutri, Streptococcus lactis antagonizes S. aureus, Listeria monocytogenes, S.hemolyticus, E. faecalis(Ent V)* inhibits *C.albicans*(Boneca et al. 2007; Siezen et al. 2014; Liong 2015; Pathak and Akhtar 2018).

The bioavailability of nutraceuticals depends upon the absorption, bioaccessibility, and biotransformation. Nanocarriers effective in anticancer and antimicrobial therapy include alginic acid, chitosan, titanium dioxide, silver, etc. (Slavin et al. 2017). These nanoparticle acts as a bactericidal agent, alters the metabolic enzyme activities, prevents microbial contamination, and improves the shelf life of the nutraceuticals (Acosta 2009; Alam et al. 2012; Liang and Subirade 2016; Li et al. 2019a).

8.10 Nanonutraceuticals in Antiviral Therapy

Phytochemicals released by *Erythrina abyssinica, Mangifera indica, Aaloe vera,* and *Warburgia salutaris* are proven effective in the treatment of AIDS patients (Kim et al. 2013b; Konur 2016).

Probiotics that are are commonly used in the treatment of viral infections include *Lactobacillus rhamnosus, Lactobacillus gasseri, Bifidobacterium bifidum, Lactobacillus casei, Bacillus coagulans, Lactobacillus acidophilus, Saccharomyces boulardii,* and *Enterococcus faecium* with or without prebiotic inclusion(galacto oligosaccharide, poly dextrose, hi-maize,rice bran, inulin, pectin are effective in the elimination of viral pathogens by macrophage in HSV-1 and reduced the incidence of (RTIs) through metabolite the production which interferes with the regulation of ca2+ and NSP4 protein involved in reteroviral diarrhea.(Shim et al. 2017; Kanauchi et al. 2018; Liu et al. 2020; Loutfy et al. 2020; Shen 2020).

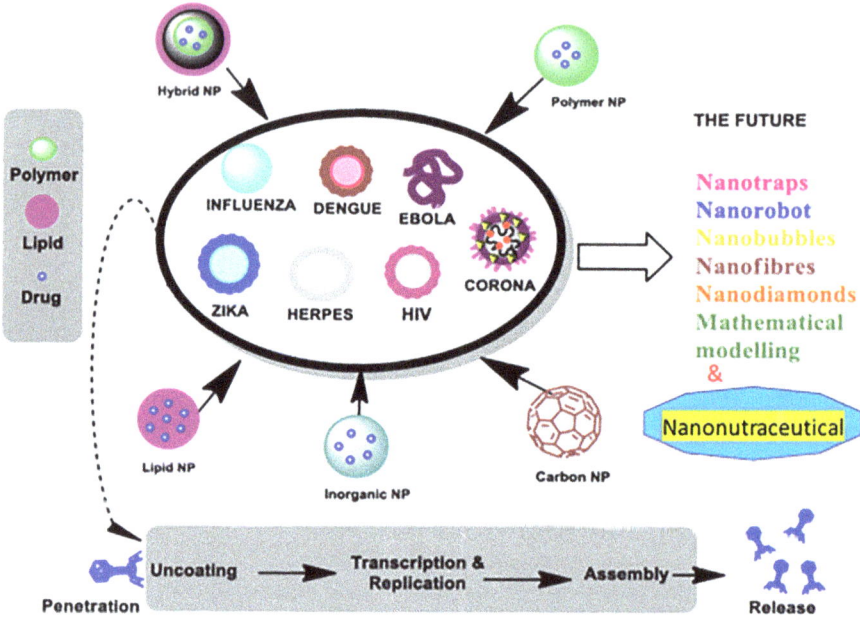

Fig. 8.19 Nanonutraceuticals in antiviral therapy. (Adapted from Chakravarty and Vora 2020)

However, in order to maintain the health benefits of these probiotics and prebiotics, processing, storage, and bioavailability play a crucial role in treatment. Nanoencapsulation with polymers, micelles, and metals has proven to be effective to withstand the adverse conditions of light, temperature, pH, etc. and to regulate the function and viability and also imparts texture, biocompatibility, odor, color, etc. (Fig. 8.19) (Ying et al. 2016; Raddatz, et al. 2020; Rodrigues et al. 2020). The nanocarriers used are nanocomposite, alginate-chitosan capsules, microcapsules, alginate-pectin gels, pectin microparticles, etc. (Chen and Subirad 2006; Ying et al. 2013; Müller et al. 2016; McCarty and DiNicolantonio 2020).

8.11 Utilization of Nanonutraceutical in COVID-19 Therapy: Pharmacological and Toxicological Aspects

The novel pandemic, corona virus disease (COVID-19), a crucial global health calamity, is posing a great threat to the human population as preexisting immune responses may be limited against this new virus (nCoV) strain from Wuhan since December 2019. The new emerging and reemerging animal and human coronaviruses (CoVs) are a major menace and remain challenging to the second world. Most countries around the world have been on lockdown to halt the spread of the virus. Coronaviruses (CoVs) primarily infect birds and mammals but were shown capable of species crossover by infecting humans in 2002 with SARS and in 2012 with MERS. In spite of intensive efforts taken since the SARS epidemic in 2003, the antiviral drugs to treat CoV infections have not been approved by the FDA so far.

The innate immune system is vital in 2019-nCoV replication. The proteins involved are spike, envelope, membrane, and nucleocapsid protein. The S protein binds to the ACE2 receptor that is most prevalent in the respiratory tract and also present in the heart, kidney, blood vessels, and intestine. ACE2 is protective to the respiratory epithelium (surfactant), and this is one of the reasons for respiratory failure and impact on other organs. The infection causes the overproduction and release of pro-inflammatory cytokines. Most important are IL-1, TNF-a, and IFN. The mostly bind to the Toll-like receptor in macrophages which causes the production of IL-1 beta, destruction and fibrosis of lung tissue, and stimulation of the inflammasome Inflammasomes are formed by the lipopolysaccharide (LPS) that includes pathogen-associated and damage-associated molecular patterns (PAMPs, DAMPs) and pro-inflammatory cytokines (IL1-B, TNF-a). Viruses, bacteria, and fungi mostly contain PAMPs (Bosch et al. 2003; Wu et al. 2020).

The cytokine storm or hypercytokinemia-induced pathological responses play a pivotal role in COVID-19 viral infection. The immune and inflammatory responses include disseminated vascular permeability and coagulation, severe respiratory infections, and natural M protease inhibitors that can be considered as an approach to interfering with viral replication (Conti et al. 2019; Monteil et al. 2020; Fisher and Emdad 2020).

Nutraceuticals have immunomodulating effect and are capable of reducing symptoms of encapsulated RNA viruses that include corona and influenza viruses (Fig. 8.20). Phytochemicals, vitamins, minerals, and probiotics are proven to be involved in host defense mechanism (Pathak and Akhtar 2018; Sander et al. 2019; Adem et al. 2020). Curcumin, quercetin, melatonin, EGCG, resveratrol, vitamins, thymoquinone, and minerals are nutraceuticals that can inhibit M protease and corona viral infection (Alam et al. 2012; Rafiee et al. 2019. Swanson et al. 2019; Jin et al. 2020).

Vitamin D, a fat-soluble vitamin, modulates both innate and adaptive immune responses to respiratory viral infections including *H. influenzae* A and B, respiratory syncytial virus (RSV), parainfluenza 1 and 2, and HCV by production of peptides (cathelicidin) with antimicrobial activity (Liu 2011; Pludowski et al. 2019; Zdrenghea et al. 2016).

Vitamin A stimulates immune function and regulates both cellular and humoral immune responses; vitamins have been effective in chronic hepatitis B and deficiency leading to impairment of cellular and humoral immunity (Stephensen 2017). Vitamin C deficiency may cause dysregulation of host defense mechanism and affects T cell-mediated immune responses and adaptive immunity. Vitamin B deficiency results in immunosuppression due to reduced blood cell count (Liu 2011; Huang et al. 2018; Skrajnowska and Bobrowska-Korcza 2019).

The combined therapy with a multivitamin and multimineral supplement improves the immune responses to corona viral infections (Hemilä and Chalker 2013).

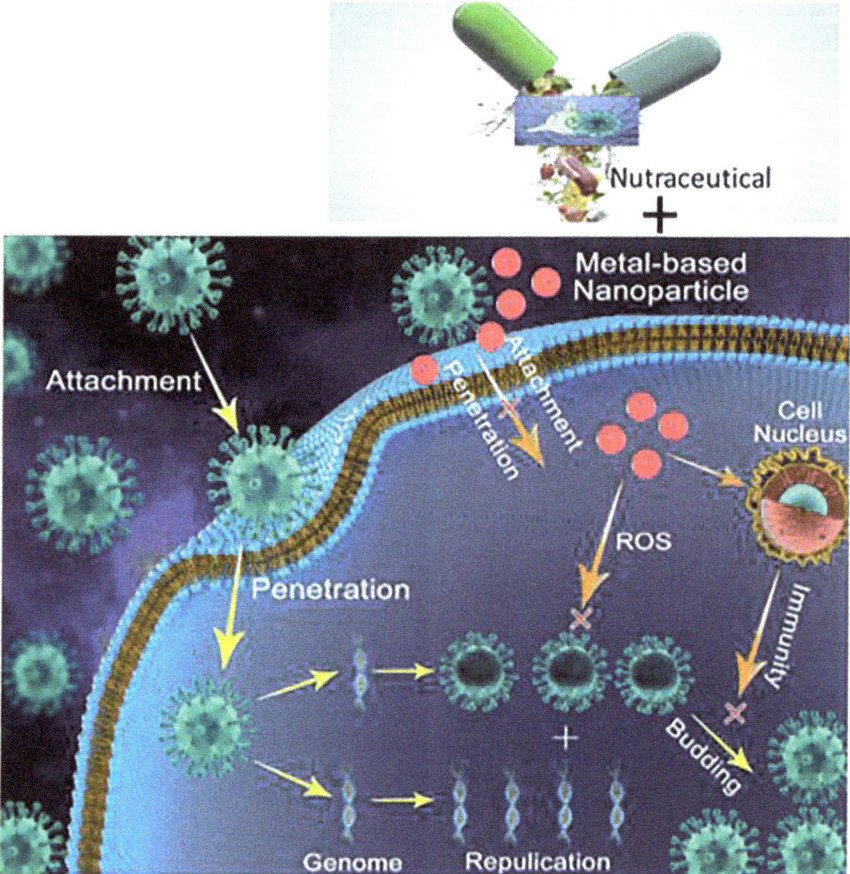

Fig. 8.20 (**a**) Possible utilization of nanonutraceutical in COVID therapy. (**b**) Mechanism of action. (Adapted from Zhou et al. 2020)

Zinc is a crucial trace element involved in immunity, growth, and development. Zinc deficiency leads to a disrupted zinc homeostasis that alters immune mechanism and causes an alteration in intercellular communication, defective lymphopoiesis, and production of reactive oxygen species that results in reduced resistance to viral infections (HIV, HCV) (Acevedo-Murillo et al. 2019; Hemilä Chalker 2019).

Selenium is an antioxidant with anti-inflammatory properties. Selenium in reduced concentration will increase mortality due to lack of immune response and when the concentration is gradually increased offers protection against viral infections in a dose-dependent manner with an increase in IL-10 and T-cell proliferation (poliovirus) (Rayman 2012; Limaye et al. 2018; Mousa et al. 2020).

Magnesium controls immune function through immunoglobulin synthesis and adherence of T helper cells and B cell, affecting the binding of IgM and macrophage response to lymphokines. Iron is essential for the cells that make up our immune defenses (Liang et al. 2012). Copper is important in the differentiation and development of immune cells. Copper has antioxidant activity and its supplements increased the activity of plasma ceruloplasmin, benzylamine oxidase, and superoxide dismutase activity (Potter et al. 2011; Rosato et al. 2017; Devi and Ahmaruzzaman 2018; Li et al. 2019c; Sevanian 2019).

Omega-3 fatty acids, beta 1,3/1,6 glucan have been recently used as an immune booster for the infection-fighting antibodies and cytokines in viral infections (upper respiratory tract infections, flu, hepatitis, and HPV the phytochemical act as immune boosters and augments the functions of macrophages, white blood cells. Similarly, a Chinese herb *Astragalus* has antiviral property and shortens the duration and intensity of *H. influenza* (Brüll et al. 2016; Müller et al. 2016).

Probiotics are beneficial in COVID therapy. Previous studies indicate that those gut microbes.

Bacteroides ferment dietary fiber into short-chain fatty acids (acetate, butyrate, and propionate that facilitate the formation of the gut barrier and stimulate mucus production that acts as a protective layer against invading pathogens) (VanHook 2015). The researchers found that dietary fiber or butyrate was able to reduce the influx of immune cells called neutrophils into the airways and another mechanism that may be involved is the PPAR pathway. The microbiome competes for binding sites on the intestinal epithelial cells by blocking (i) pathogenic attachment and entry in the host cell, (ii) nutrients in the GI tract to preserve its habitat, and reduces the flourishing of the pathogen by producing antiviral substances. (Sanders 2008; Mackie 2012; Schwabe and Jobin 2013; Rook et al. 2017; Shen et al. 2018a, b; Kanauchi et al. 2018; McQuade et al. 2019; Song et al. 2019).

Recently phytochemical constituents are preferred as immunostimulants for the protection against viral infections. Green tea (polyphenols, catechin epigallocatechin, epicatechin gallate) (Sahin et al. 2010) and lycopene (non-provitamin A carotenoid) provide protection against cellular disturbances due to free radicals (Yang et al. 2017). Curcumin reduces reactive oxygen species and causes degradation of antioxidant enzymes by modulating Nrf2 (He 2012; Wang et al. 2011; Farooqui 2016) bound oligonucleotides that offer stability to DNA triplexes or G-quadruplexes and tumors inhibiting telomerase and topoisomerase (Ortiz et al. 2014) Vanillin used worldwide could reduce free radical cancer promotion (Bezerra et al. 2016). Taurine and curcumin in cactus pear is used for its anticancer, antiviral, and antidiabetic properties (Sabella 2014; Jiang et al. 2018; Meng et al. 2018; Loutfy et al. 2019. Shao et al. 2020).

8.11.1 Toxicities of Nanonutraceuticals

The adverse effects are prominently observed with nanonutraceuticals as there are fewer in vivo studies to assess the efficiency of nanoparticle encapsulation. There are controversial results wherein a set of research have proven that antioxidants, such as vitamins C and E, and carotenoids enhance the efficacy of radiation or

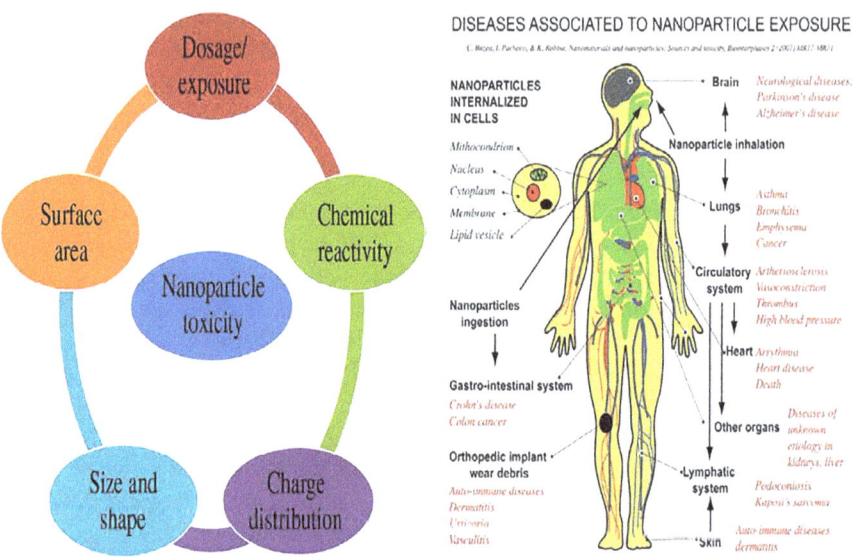

Fig. 8.21 (**a**) Physicochemical parameters of nanotoxicity. (Adapted from Naseer et al. 2018). (**b**) Toxic symptoms of nanoparticles. (Adapted from Cristina Buzea 2007)

chemotherapy by increasing tumor response and minimizing systemic toxicity. Contrastingly, another study reveals that the antioxidants protect the cancer cells from chemotherapy or radiation induced damage (Lam et al. 2004) washed 3 kinds of carbon nanotubes into the lungs of mice;all caused lung granulomas), (Dupont injected nanotubes into rat lungs; 15% died -highest death rate seen in such studies) (Rice University studies show nanoparticles bioaccumulate in living tissues). The toxicities of nanonutraceuticals mainly depend on the dosage of the nutraceuticals and nanoparticles and physicochemical chemical properties of the nanoencapsulation (Naseer et al. 2018) (Fig. 8.21a). The nanocarriers can induce mild to severe toxicities in tissues and organs (Fig. 8.21b) (Cristina Buzea 2007:Andreas Elsaesser and Howard. 2012; Zhao et al. 2015; Limaye et al. 2018; Sukhanova et al. 2018; Marcinowska-Suchowierska et al. 2018).

8.12 Challenges and Future Perspectives

Nanonutraceutical formulations remain a great challenge due to the cost involved in the production. Nutraceuticals are highly complex molecules; hence, the choice of nanoparticles for encapsulation according to size, shape, vehicles, and optimization of dosage for therapeutic purpose is the new area of research. Further, large-scale production requires highly efficient equipment, more space, and time which in turn leads to a higher cost of production and acquisition. Proper planning and implementation of a cost-effective method is the essential criteria for the formulation of nanonutraceuticals. For therapeutic use, efficacy, potency, and toxicity should be carefully monitored through pharmacokinetic and pharmacodynamic profiling of nanonutraceutical studies based on specific targeting, bioavailability, and toxicity studies. However, nanonutraceuticals seem beneficial in the prevention of infections and inflammation and in cancer therapy. Hence, nanonutraceutical formulations with vitamins, minerals, herbal supplements, and probiotics represent a valuable and promising strategy to prevent diseases and maintain health status in the individual. However, the encapsulation of nanocarriers with nutraceuticals should be strictly monitored with guidelines from regulatory authorities for the safe and efficient use of these nanonutraceuticals.

8.13 Conclusion

Nanonutraceuticals are safe for administration with better bioavailability; however, utilization of nanomaterials in the food and the pharmaceutical industry remains a conflict of the modern world due to inadequate invasive studies. The nanotechnology can increase the shelf life of food and drug, reduce the dosage and toxicity, prevent microbial contamination, and improve food and drug packing using nanosensors and production of nanonutraceutical. However, the major concern is toward the toxicity of engineered nanomaterials in nanofood or nano-drug on human health and environment. The one way that can solve this concern is proper research, information, and regulation of nanofood. Several entities are now in place to govern every aspect of nanofood, starting from the synthesis of nanomaterials to their usage in food industries. Laws and regulation implemented by entities such as the US-FDA, ESFA, etc. need to be followed to have a nanofood that is safe for human consumption. There is an urgent need for natural products as a medicament to overcome pandemic diseases. Recently, standardization of dosage of herbal products after identification and screening with human cell line studies has thrown light in the nanonutraceutical research area. The future work will be to explore and develop novel nanonutraceutical with valid in vivo studies for the utilization in severe pathological conditions leading to life-threatening infections and inflammatory diseases and cancer.

Bibliography

Abbasi E, Aval S, Akbarzadeh A, Milani M, Nasrabadi H, Joo S, Hanifehpour Y, Nejati-Koshki K, Pashaei-Asl (2014) Dendrimers: synthesis, applications, and properties. Nanoscale Res Lett 9(1):247

Abdullah ML, Hafez MM, Al-Hoshani A, Al-Shabanah O (2018) Anti-metastatic and antiproliferative activity of eugenol against triple negative and HER2 positive breast cancer cells. BMC Complement Altern Med 18(1):321

Acevedo-Murillo JA, García León ML, Firo-Reyes V, Santiago-Cordova JL, Gonzalez-Rodriguez AP, Wong-Chew RM (2019) Zinc supplementation promotes a Th1 response and improves clinical symptoms in fewer hours in children with pneumonia younger than 5 years old a randomized controlled clinical trial. Front Pediatr 7:431

Acosta E (2009) Bioavailability of nanoparticles in nutrient and nutraceutical delivery. Curr Opin Colloid Interf Sci 14(1):3–15

Adem S, Eyupoglu V, Sarfraz I, Rasul A, Ali M (2020) Identification of potent COVID-19 main protease (Mpro) inhibitors from natural polyphenols: an in-Silico strategy unveils a hope against CORONA. Preprints.org; 2020. https://doi.org/10.20944/preprints202003.0333.v1

Afonina IS, Zhong Z, Karin M, Beyaert R (2017) Limiting inflammation—the negative regulation of NF-κb and the NLRP3 inflammasome. Nat Immunol 18(8):861–869

Aggarwal BB, Van Kuiken M, Iyer LH, Harikumar KB, Sung B (2009) Molecular targets of nutraceuticals derived from dietary spices: potential role in suppression of inflammation and tumorigenesis. Exp Biol Med 234(8):825–849

Alam S, Mustafa G, Khan ZI, Islam F, Bhatnagar A, Ahmad F, Kumar (2012) Development and evaluation of thymoquinone-encapsulated chitosan nanoparticles for nose-to-brain targeting: a pharmacoscintigraphic study. Int J Nanomedicine 7:5705

Ali A, Ahmad U, Akhtar J, Badruddeen KMM (2019) Engineered nano scale formulation strategies to augment efficiency of nutraceuticals. J Funct Foods 62:103554

Al-Mssallem MQ, Alqurashi RM, Al-Khayri JM (2019) Bioactive compounds of date palm (Phoenix dactylifera L). In: Reference series in phytochemistry, pp 1–15

Amiri-Rigi A, Abbasi S (2019) Extraction of lycopene using a lecithin-based olive oil microemulsion. Food Chem 272:568–573

Andreas Elsaesser C, Howard V (2012) Toxicology of nanoparticles. Adv Drug Deliv Rev 64:129–137

Arora D, Jaglan S (2016) Nanocarriers based delivery of nutraceuticals for cancer prevention and treatment: a review of recent research developments. Trends Food Sci Technol 54:114–126

Arshak K, Adley C, Moore E, Cunniffe C, Campion M, Harris J (2007) Characterisation of polymer nanocomposite sensors for quantification of bacterial cultures. Sensors Actuators B Chem 126(1):226–231

Assadpour E, Dima C, Jafari SM (2020) Fundamentals of food nanotechnology. In: Handbook of food nanotechnology, pp 1–35

Augustin MA, Hemar Y (2009) Nano- and micro-structured assemblies for encapsulation of food ingredients. Chem Soc Rev 38(4):902–912

Bae YH, Park K (2011) Targeted drug delivery to tumors: myths, reality and possibility. J Control Release 153(3):198–205

Bajpai VK, Shukla S, Kang S-M, Hwang SK, Song X, Huh YS, Han Y-K (2018) Developments of cyanobacteria for nano-marine drugs: relevance of nanoformulations in cancer therapies. Mar Drugs 16:179

Baker JR (2009) Dendrimer-based nanoparticles for cancer therapy. Hematology 2009(1):708–719

Bawarski WE, Chidlowsky E, Bharali DJ, Mousa SA (2008) Emerging nanopharmaceuticals. Nanomedicine 4(4):273–282

Bayir AG, Kiziltan HS, Kocyigit A (2019) Chapter 1 - Plant family, Carvacrol, and putative protection in gastric cancer. In: Watson RR, Preedy VR (eds) Dietary interventions in gastrointestinal diseases. Academic Press, pp 3–18

Bernela M, Kaur P, Ahuja M, Thakur R (2018) Nano-based delivery system for nutraceuticals: the potential future. In: Advances in animal biotechnology and its applications, pp 103–117

Bernhardt ES, Colman BP, Hochella MF, Cardinale BJ, Nisbet RM, Richardson CJ, Yin L (2010) An ecological perspective on nanomaterial impacts in the environment. J Environ Qual 39(6):1954–1965

Bezerra DP, Soares AK, De Sousa DP (2016) Overview of the role of vanillin on redox status and cancer development. Oxidative Med Cell Longev 2016:1–9

Bhalang K, Thunyakitpisal P, Rungsirisatean N (2013) Acemannan, a polysaccharide extracted from Aloe vera, is effective in the treatment of oral aphthous ulceration. J Altern Complement Med 19(5):429–434

Bharali DJ, Bawarski WE, Chidlowsky E, Mousa SA (2017) Emerging Nanopharmaceuticals. In: Nanomedicine in Cancer, pp 99–127

Bhushan B, Luo D, Schricker SR, Sigmund W, Zauscher S (2014) Handbook of nanomaterials properties. Springer-Verlag, Berlin Heidelberg, pp 1199–1226

Biersack Z, Schobert R (2012) Indole compounds against breast cancer: recent developments. Curr Drug Targets 13(14):1705–1719

Boik J (2001) Natural compounds in cancer therapy: promising nontoxic antitumor agents from plants and natural sources. Oregan Medical Press, Princeton

Boneca IG, Dussurget O, Cabanes D, Nahori M, Sousa S, Lecuit M, Psylinakis E, Bouriotis V, Hugot J, Giovannini M, Coyle A, Bertin J, Namane A, Rousselle J, Cayet N, Prévost M, Balloy V, Chignard M, Philpott DJ, Girardin SE (2007) A critical role for peptidoglycan N-deacetylation inListeriaevasion from the host innate immune system. Proc Natl Acad Sci 104(3):997–1002

Bosch BJ, Van der Zee R, De Haan CA, Rottier PJ (2003) The coronavirus spike protein is a class I virus fusion protein: structural and functional characterization of the fusion core complex. J Virol 77(16):8801–8811

Brüll F, De Smet E, Mensink RP, Vreugdenhil A, Kerksiek A, Lütjohann D, Wesseling G, Plat J (2016) Dietary plant stanol ester consumption improves immune function in asthma patients: results of a randomized, double-blind clinical trial1. Am J Clin Nutr 103(2):444–453

Cristina Buzea (2007) Pathways of exposure to nanoparticles and associated diseases as suggested by epidemiological, in vivo and in vitro studies

Calvani M, Pasha A, Favre C (2020) Nutraceutical boom in cancer: inside the labyrinth of reactive oxygen species. Int J Mol Sci 21(6):1936

Cameron M, Chrubasik S (2014) Oral herbal therapies for treating osteoarthritis. Cochrane Database Syst Rev 5:CD002947

Castro RI, Forero-Doria O, Soto-Cerda L, Peña-Neira A, Guzmán L (2018) Protective effect of Pitao (*Pitavia punctata* (RP) Molina) polyphenols against the red blood cells Lipoperoxidation and the in vitro LDL oxidation. In: Evidence-based complementary and alternative medicine, pp 1–9

Chakraborty AK (2019) Nucleic acid-based. In: Nanocarriers nanocarriers for drug delivery, pp 155–172

Chakravarty M, Vora A (2020) Nanotechnology-based antiviral therapeutics. Drug Deliv Transl Res 3:1–40

Charu G, Dhan P, Garg AP, Sneh G (nd) (2014) Nutraceuticals from microbes Phytochemicals of nutraceutical importance. pp 79–102

Chau C, Wu S, Yen G (2007) The development of regulations for food nanotechnology. Trends Food Sci Technol 18(5):269–280

Chaudhry Q, Scotter M, Blackburn J, Ross B, Boxall A, Castle L, Aitken R, Watkins R (2008) Applications and implications of nanotechnologies for the food sector. Food Addit Contam Part A 25(3):241–258

Chen L, Subirad EM (2006) Alginate–whey protein granular microspheres as oral delivery vehicles for bioactive compounds. Biomaterials 27(26):4646–4654

Chen Y, Wang B, Zhao H (2018) Thymoquinone reduces spinal cord injury by inhibiting inflammatory response, oxidative stress and apoptosis via PPAR-γ and PI3K/Akt pathways. Exp Ther Med 15(6):4987–4994

Chikara S, Nagaprashantha LD, Singhal J, Horne D, Awasthi S, Singhal SS (2018) Oxidative stress and dietary phytochemicals: role in cancer chemoprevention and treatment. Cancer Lett 413:122–134

Chiu H-F, Venkatakrishnan K, Wang C-K (2020) The role of nutraceuticals as a complementary therapy against various neurodegenerative diseases: a mini-review. J Tradit Complement Med 10(5):434–439

Cho H, Lai TC, Tomoda K, Kwon GS (2015) Polymeric micelles for multi-drug delivery in cancer. AAPS Pharm Sci Tech 16:10–20

Clark R, Lee J, Lee S (2015) Synergistic anticancer activity of capsaicin and 3,3'-Diindolylmethane in human colorectal cancer. J Agric Food Chem 63(17):4297–4304

Conti P, D'Ovidio C, Conti C, Gallenga CE, Lauritano D, Caraffa A, Kritas S, Ronconi G (2019) Progression in migraine: role of mast cells and pro-inflammatory and anti-inflammatory cytokines. Eur J Pharmacol 844:87–94

Czech T, Lalani R, Oyewumi MO (2019) Delivery systems as vital tools in drug repurposing. AAPS Pharm SciTech 20(3):116

Dandawate PR, Subramaniam D, Jensen RA, Anant S (2016) Targeting cancer stem cells and signaling pathways by phytochemicals: novel approach for breast cancer therapy. Semin Cancer Biol 40-41:192–208

Darwish NH, Sudha T, Godugu K, Bharali DJ, Elbaz O, El-ghaffar HA, Azmy E, Anber N, Mousa SA (2019) Novel targeted nano-parthenolide molecule against NF-kb in acute myeloid leukemia. Molecules 24(11):2103

Dasari S, Kathera C, Janardhan A, Kumar AP, Viswanath B (2017) Surfacing role of probiotics in cancer prophylaxis and therapy: a systematic review. Nutrition 36(6):1465–1472

Date T, Paul K, Singh N, Jain S (2019) Drug–lipid conjugates for enhanced oral drug delivery. AAPS Pharm SciTech 20(2):41

Davis PJ, Davis FB, Luidens MK, Lin H, Mousa SA (2013) Tetraiodothyroacetic acid (Tetrac). In: Nanotetrac and anti-angiogenesis angiogenesis modulations in health and disease, pp 107–117

DeFelice SL. (1989) FIM Rationale and Proposed Guidelines for the Nutraceutical Research & Education Act - NREA, November 10, 2002. Foundation for Innovation in Medicine. Available at: http://www.fimdefelice.org/archives/arc.researchact.html.

DeFelice SL (1995) The nutraceutical revolution: its impact on food industry R D. Trend Food Sci Technol 6(2):59–61

DeFrates K, Markiewicz T, Gallo P, Rack A, Weyhmiller A, Jarmusik B, Hu X (2018) Protein polymer-based nanoparticles: fabrication and medical applications. Int J Mol Sci 19(6):1717

Deng H (2020) Nrf2 and its modulation in inflammation. Springer Nature

Deshantri AK, Varela Moreira A, Ecker V, Mandhane SN, Schiffelers RM, Buchner M, Fens MH (2018) Nanomedicines for the treatment of hematological malignancies. J Control Release 287:194–215

Devi TB, Ahmaruzzaman M (2018) Green synthesis of silver, copper and iron nanoparticles: synthesis, characterization and their applications in wastewater treatment. In: Green metal nanoparticles, pp 441–466

Dhawan S, Hooda P, Nanda S (2018) Herbal Nano formulations: patent and regulatory overview applied clinical research. Clin Trials Regul Affairs 5(3):159

Dima C, Assadpour E, Dima S, Jafari SM (2020) Bioavailability of nutraceuticals: role of the food matrix, processing conditions, the gastrointestinal tract, and nanodelivery systems comprehensive. Rev Food Sci Food Safety 19(3):954–994

Doktorovová S, Kovačević AB, Garcia ML, Souto EB (2016) Preclinical safety of solid lipid nanoparticles and nanostructured lipid carriers: current evidence from in vitro and in vivo evaluation. Eur J Pharm Biopharm 108:235–252

Duarte MC, Tavares GS, Valadares DG, Lage DP, Ribeiro TG, Lage LM, Rodrigues MR, Faraco AA, Soto M, Da Silva ES, Chávez Fumagalli MA, Tavares CA, Leite JP, Oliveira JS, Castilho RO, Coelho EA (2016) Antileishmanial activity and mechanism of action from a purified fraction of *zingiber officinalis* Roscoe against leishmania amazonensis. Exp Parasitol 166:21–28

Etheridge ML, Campbell SA, Erdman AG, Hayne CL, Wolf SM, McCullough J (2013) The big picture on nanomedicine: the state of investigational and approved nanomedicine products *Nanomedicine: Nanotechnology*. Biol Med 9(1):1–14

Eun J, Koh GY (2004) Suppression of angiogenesis by the plant alkaloid, sanguinarine. Biochem Biophys Res Commun 317(2):618–624

Farooqui AA (2016) Therapeutic potentials of curcumin for Alzheimer disease. Springer

Fisher PB, Emdad L (2020) Faculty opinions recommendation of inhibition of SARS-Cov-2 infections in engineered human tissues using clinical-grade soluble human ACE2. Faculty Opinions – Post-Publication Peer Review of the Biomedical Literature

Flores O, Santra S, Kaittanis C, Bassiouni R, Khaled AS, Khaled AR, Grimm J, Perez JM (2017) PSMA-targeted Theranostic Nanocarrier for prostate cancer. Theranostics 7(9):2477–2494

Fuster MG, Carissimi G, Montalbán MG, Víllora G (2020) Improving anticancer therapy with naringenin-loaded silk fibroin nanoparticles. Nano 10(4):718

Ganesan P, Ramalingam P, Karthivashan G, Ko YT, Choi D (2018) Recent developments in solid lipid nanoparticle and surface-modified solid lipid nanoparticle delivery systems for oral delivery of phyto-bioactive compounds in various chronic diseases. Int J Nanomedicine 13:1569–1583

Garti N, Aserin A (2012) Micelles and microemulsions as food ingredient and nutraceutical delivery systems. In: Encapsulation technologies and delivery systems for food ingredients and nutraceuticals, pp 211–251

Geng J, Lv Y, Lu D, Zhu J (2006) Sonochemical synthesis of PbWO4crystals with dendritic, flowery and star-like structures. Nanotechnology 17(10):2614–2620

Godugu K, El-Far AH, Al Jaouni S, Mousa SA (2020) Nanoformulated Ajwa (Phoenix Dactylifera) bioactive compounds improve the safety of Doxorubicin without compromising its anticancer efficacy in breast cancer. Molecules 25(11):2597

Godwin HA, Chopra K, Bradley KA, Cohen Y, Harthorn BH, Hoek EM, Holden P, Keller AA, Lenihan HS, Nisbet RM, Nel AE (2009) The University of California center for the environmental implications of nanotechnology. Environ Sci Technol 43(17):6453–6457

Grierson D, Afolayan A (1999) Antibacterial activity of some indigenous plants used for the treatment of wounds in the Eastern Cape, South Africa. J Ethnopharmacol 66(1):103–106

Guamán Ortiz L, Lombardi P, Tillhon M, Scovassi A (2014) Berberine, an epiphany against cancer. Molecules 19(8):12349–12367

Gupta C, Prakash D (2015) Nutraceuticals for geriatrics. J Tradit Complement Med 5(1):5–14

Gupta C, Prasad S, Aggarwal B (2016) Anti-inflammatory nutraceuticals and chronic diseases. Springer

Haidar ZS, Hamdy RC, Tabrizian M (2008) Protein release kinetics for core–shell hybrid nanoparticles based on the layer-by-layer assembly of alginate and chitosan on liposomes. Biomaterials 29(9):1207–1215

Hamsa T, Kuttan G (2010) Harmine inhibits tumour specific Neo-vessel formation by regulating vegf, mmp, timp and pro-inflammatory mediators both in vivo and in vitro. Eur J Pharmacol 649(1–3):64–73

He H (2012) Curcumin attenuates Nrf2 signaling defect, oxidative stress in muscle and glucose intolerance in high fat diet-fed mice. World J Diab 3(5):94

He X, Deng H, Hwang H-m (2019) The current application of nanotechnology in food and agriculture. J Food Drug Anal 27(1):1–21

Heber D, Li Z, Garcia-Lloret M, Wong AM, Lee TY, Thames G, Krak M, Zhang Y, Nel A (2014) Sulforaphane-rich broccoli sprout extract attenuates nasal allergic response to diesel exhaust particles. Food Funct 5(1):35–41

Hegazy AM, El-Sayed EM, Ibrahim KS, Abdel-Azeem AS (2019) Dietary antioxidant for disease prevention corroborated by the Nrf2 pathway. J Complem Integr Med 16(3)

Hemilä H, Chalker E (2013) Vitamin C for preventing and treating the common cold. Cochrane Database Syst Rev 2013:CD000980

Hemilä H, Chalker E (2019) Zinc for preventing and treating the common cold. Cochrane Database Syst Rev 2017:CD012808

Huang Z, Liu Y, Qi G, Brand D, Zheng S (2018) Role of vitamin A in the immune system. J Clin Med 7(9):258

Iacono A, Raso GM, Canani RB, Calignano A, Meli R (2011) Probiotics as an emerging therapeutic strategy to treat NAFLD: focus on molecular and biochemical mechanisms. J Nutr Biochem 22(8):699–711

Iqbal J, Abbasi BA, Batool R, Mahmood T, Ali B, Khalil AT, Kanwal S, Shah SA, Ahmad R (2018) Potential phytocompounds for developing breast cancer therapeutics: Nature's healing touch. Eur J Pharmacol 827:125–148

Isidoro C (2020) Nutraceuticals and diet in human health and disease the special issue at a glance. J Tradit Complement Med 10(3):175–179

Jafari SM, Paximada P, Mandala I, Assadpour E, Mehrnia MA (2017) Encapsulation by nanoemulsions. In: Nanoencapsulation technologies for the food and nutraceutical industries, pp 36–73

Jampílek J, Kráľová K (2018) Nanomaterials applicable in food protection. In: Nanotechnology applications in the food industry, pp 75–96

Jampilek J, Kos J, Kralova K (2019) Potential of nanomaterial applications in dietary supplements and foods for special medical purposes. Nano 9(2):296

Jeevanandam J, Chan YS, Michael K (2016) Danquah Nano-formulations of drugs: recent developments, impact and challenges. Biochimie 128–129:99–112

Jesus K, Yamamoto-Furusho (2007) Innovative therapeutics for inflammatory bowel disease. World J Gastroenterol 7 13(13):1893–1896

Jiang L, Zhou S, Zhang X, Wu W, Jiang X (2018) Dendrimer-based nanoparticles in cancer chemotherapy and gene therapy. Sci China Mater 61(11):1404–1419

Jin Z, Du X, Xu Y et al (2020) Structure of M^{pro} from SARS-CoV-2 and discovery of its inhibitors. Nature 582:289–293

Joshi VB, Geary SM, Salem AK (2013) Biodegradable particles as vaccine antigen delivery systems for stimulating cellular immune responses. Human Vaccin Immunother 9(12):2584–2590

Kanauchi O, Andoh A, AbuBakar S, Yamamoto N (2018) Probiotics and Paraprobiotics in viral infection: clinical application and effects on the innate and acquired immune systems. Curr Pharm Des 24(6):710–717

Katouzian I, Jafari SM (2016) Nano-encapsulation as a promising approach for targeted delivery and controlled release of vitamins. Trends Food Sci Technol 53:34–48

Khoder G, Asma A, Al-Menhali, Al-Yassir F, Karum SM (2016) Potential role of probiotics in management of gastric ulcer. Exp Ther Med 12:3–17

Kim Y, Kim H, Bae S, Choi J, Lim SY, Lee N, Kong JM, Hwang Y, Kang JS, Lee WJ (2013a) Vitamin C is an essential factor on the anti-viral immune responses through the production of interferon-α/β at the initial stage of influenza A virus (H3N2) infection. Immune Netw 13(2):70

Kim S, Oh S, Lee J, Han J, Jeon M, Jung T, Lee SK et al (2013b) Berberine suppresses TPA-induced Fibronectin expression through the inhibition of VEGF secretion in breast cancer cells. Cell Physiol Biochem 32(5):1541–1550

Klajnert B, Peng L, Cena V (2013) Dendrimers in biomedical applications. Royal Society of Chemistry

Konur O (2016) Scientometric overview regarding the nanobiomaterials in antimicrobial therapy. In: Nanobiomaterials in antimicrobial therapy, pp 511–535

Krishnamachari Y, Geary SM, Lemke CD, Salem AK (2010) Nanoparticle delivery systems in cancer vaccines. Pharm Res 28(2):215–236

Kuentz M (2012) Lipid-based formulations for oral delivery of lipophilic drugs. Drug Discov Today Technol 9(2):97–104

Kumar M, Nagpal R, Verma V, Kumar A, Kaur N, Hemalatha R, Gautam SK, Singh B (2012) Probiotic metabolites as epigenetic targets in the prevention of colon cancer. Nutr Rev 71(1):23–34

Kuppusamy P, Yusoff MM, Maniam GP, Ichwan SJ, Soundharrajan I, Govindan N (2014) Nutraceuticals as potential therapeutic agents for colon cancer: a review. Acta Pharm Sin B 4(3):173–181

Lage PS, Chávez-Fumagalli MA, Mesquita JT, Mata LM et al (2015) Antileishmanial activity and evaluation of the mechanism of action of strychnobiflavone flavonoid isolated from *Strychnos pseudoquina* against *Leishmania infantum*. Parasitol Res 114(12):4625–4635

Lam C-W, James JT, McCluskey R, Hunter RL (2004) Pulmonary toxicity of single-wall carbon nanotubes in Mice 7 and 90 days after intratracheal instillation. Toxicol Sci 77(1):126–134

Langella P, Guarner F, Martín R (2019) Next-generation probiotics: from commensal bacteria to novel drugs and food supplements. Front Media SA 10:1973

Lawal I, Grierson D, Afolayan (2015) The antibacterial activity of *Clausena anisata hook*, a South African medicinal plant. Afr J Tradit Complement Altern Med 12(1):23

Lee J, Koo N, Min DB (2004) Reactive oxygen species, aging, and antioxidant nutraceuticals. Compr Rev Food Sci Food Saf 3:21–33

Li J, Liang X, Yang X (2012) Ursolic acid inhibits growth and induces apoptosis in gemcitabine-resistant human pancreatic cancer via the JNK and PI3K/Akt/NF-κb pathways. Oncol Rep 28(2):501–510

Li T, Chen X, Liu Y, Fan L, Lin L, Xu Y, Chen S, Shao J (2017a) PH-sensitive mesoporous silica nanoparticles anticancer prodrugs for sustained release of ursolic acid and the enhanced anticancer efficacy for hepatocellular carcinoma cancer. Eur J Pharm Sci 96:456–463

Li Y, Li S, Meng X, Gan R, Zhang J, Li H (2017b) Dietary natural products for prevention and treatment of breast cancer. Nutrients 9(7):728

Li C, Li Y, Ding C (2019a) The role of copper homeostasis at the host-pathogen Axis: from bacteria to fungi. Int J Mol Sci 20(1):175

Li W, Yalcin M, Bharali DJ, Lin Q, Godugu K, Fujioka K, Keating KA, Mousa SA (2019b) Pharmacokinetics, biodistribution, and anti-angiogenesis efficacy of Diamino propaneTetraiodothyroacetic acid-conjugated biodegradable polymeric nanoparticle. Sci Rep 9(1):9006

Liang L, Subirade M (2016) β-lactoglobulin: bioactive nutrients delivery. In: Encyclopedia of biomedical polymers and polymeric biomaterials, pp 421–428

Liang R, Wu W, Huang J, Jiang S, Lin Y (2012) Magnesium affects the cytokine secretion of CD4+T lymphocytes in acute asthma. J Asthma 49(10):1012–1015

Limaye A, Yu R, Chou C, Liu J, Cheng K (2018) Protective and detoxifying effects conferred by dietary selenium and curcumin against AFB1-mediated toxicity in livestock: a review. Toxins 10(1):25

Liong M (2015) Beneficial microorganisms in food and nutraceuticals. Springer

Liu PT (2011) The role of vitamin D in innate immunity *vitamin D*, pp 1811–1823

Liu L, Cheng S, Shieh P, Lee J, Chen J, Ho C, Kuo S, Kuo D, Huang L, Way T (2014) The methanol extract of euonymus laxiflorus, rubia lanceolata and gardenia jasminoides inhibits xanthine oxidase and reduce serum uric acid level in rats. Food Chem Toxicol 70:179–184

Liu JP, Zhang J, Lu H (2016) Chinese herbal medicines for treating diabetic foot ulcers. Cochrane Database Syst Rev

Liu Q, Yu Z, Tian F, Zhao J, Zhang H, Zhai Q, Chen W (2020) Surface components and metabolites of probiotics for regulation of intestinal epithelial barrier. Microb Cell Factories:19(1)

Loutfy S, Eldin HA, Elberry MH, Farroh K, Mohamed B, Faraag AH, Dawood R, Awady ME (2019) THU-155-Polymeric, metallic nanoparticles, and curcumin as inhibitors of hepatitis C virus genotype 4a replication in vitro. J Hepatol 70(1):230

Loutfy SA, Elberry MH, Farroh KY, Mohamed HT, Mohamed AA, Mohamed EB, Faraag AH, Mousa SA (2020) Antiviral activity of Chitosan nanoparticles encapsulating curcumin against hepatitis C virus genotype 4a in human Hepatoma cell lines. Int J Nanomedicine 15:2699–2715

Lungu II, Grumezescu AM, Volceanov A, Andronescu E (2019) Nanobiomaterials used in cancer therapy: an up-to-Date overview. Molecules 24(19):3547

Luo J, Hu Y, Wang H (2017) Ursolic acid inhibits breast cancer growth by inhibiting proliferation, inducing autophagy and apoptosis, and suppressing inflammatory responses via the PI3K/AKT and NF-κb signaling pathways in vitro. Exp Ther Med 14(4):3623–3631

Machado D, Almeida D, Seabra CL, Andrade JC, Gomes AM, Freitas AC (2020) Nanoprobiotics: when technology meets gut health. In: Thangadurai D, Sangeetha J, Prasad R (eds) Functional bionanomaterials. Nanotechnology in the life sciences. Springer, Cham, pp 389–425

Mackie A (2012) Interaction of food ingredient and nutraceutical delivery systems with the human gastrointestinal tract. In: Encapsulation technologies and delivery systems for food ingredients and nutraceuticals, pp 49–70

Madkour LH (2020) Cellular signaling pathways with reactive oxygen species (ROS). In: Reactive Oxygen Species (ROS), nanoparticles, and Endoplasmic Reticulum (ER) stress-induced cell death mechanisms, pp 37–79

Maithili Karpaga Selvi N (2015) Curcumin attenuates oxidative stress and activation of redox-sensitive kinases in high fructose- and high-fat-Fed male Wistar rats. Sci Pharm 83(1):159–175

Mao L, Xu D, Yang J, Yuan F, Gao Y, Zhao J (2009) Effects of small and large molecule emulsifiers on the characteristics of b carotene nanoemulsions prepared by high pressure homogenization. Food Technol Biotechnol 47:336–342

Marcinowska-Suchowierska E, Kupisz-Urbańska M, Łukaszkiewicz J, Płudowski P, Jones G (2018) Vitamin D toxicity–a clinical perspective. Front Endocrinol 9:550

Mathur H, Field D, Rea MC, Cotter PD, Hill C, Ross RP (2018) Fighting biofilms with lantibiotics and other groups of bacteriocins. Biofilms Microbiomes 4:9

McCarty MF, DiNicolantonio JJ (2020) Nutraceuticals have potential for boosting the type 1 interferon response to RNA viruses including influenza and coronavirus. Prog Cardiovasc Dis 63(3):383–385

McClements D (2012) Requirements for food ingredient and nutraceutical delivery systems. In: Encapsulation technologies and delivery systems for food ingredients and nutraceuticals, pp 3–18

McClements DJ, Li F, Xiao H (2015) The nutraceutical bioavailability classification scheme: classifying nutraceuticals according to factors limiting their oral bioavailability. Annu Rev Food Sci Technol 6(1):299–327

McElwee JJ, Schuster E, Blanc et al (2007) Evolutionary conservation of regulated longevity assurance mechanisms. Genome Biol 8(7):132

McQuade JL, Daniel CR, Helmink BA, Wargo JA (2019) Modulating the microbiome to improve therapeutic response in cancer. Lancet Oncol 20(2):77–91

Meikle TG, Zabara A, Waddington LJ, Separovic F, Drummond CJ, Conn CE (2017) Incorporation of antimicrobial peptides in nanostructured lipid membrane mimetic bilayer cubosomes. Colloids Surf B: Biointerfaces 152:143–151

Menditto E, Cahir C, Aza-Pascual-Salcedo M, Bruzzese D, Poblador-Plou B, Malo S, Costa E, González Rubio F, Gimeno-Miguel A, Orlando V, Kardas P, Prados-Torres A (2018) Adherence to chronic medication in older populations: application of a common protocol among three European cohorts. Patient Prefer Adherence 12:1975–1987

Meng X, Li S, Li Y, Gan R, Li H (2018) Gut microbiota's relationship with liver disease and role in Hepatoprotection by dietary natural products and probiotics. Nutrients 10(10):1457

Montalbán M, Coburn J, Lozano-Pérez A, Cenis J, Víllora G, Kaplan D (2018) Production of curcumin-loaded silk fibroin nanoparticles for cancer therapy. Nano 8(2):126

Monteil V, Kwon H, Prado PL (2020) Inhibition of SARS-CoV-2 infections in engineered human tissues using clinical-grade soluble human ACE2. Cell. https://doi.org/10.1016/j.cell.2020.04.004

Mousa DS, El-Far AH, Saddiq AA, Sudha T, Mousa SA (2020) Nanoformulated bioactive compounds derived from different natural products combat pancreatic cancer cell proliferation. Int J Nanomedicine 15:2259–2268

Müller L, Meyer M, Bauer RN, Zhou H, Zhang H, Jones S, Robinette C, Noah TL, Jaspers I (2016) Effect of broccoli sprouts and live attenuated influenza virus on peripheral blood natural killer cells: a randomized, double-blind study. PLoS One 11(1):e0147742

Nagaprashantha LD, Singhal J, Chikara S, Gugiu G, Horne D, Awasthi S, Salgia R, Singhal SS (2019) 2′-Hydroxyflavanone induced changes in the proteomic profile of breast cancer cells. J Proteome 192:233–245

Nalwa HS (2004) Encyclopedia of nanoscience and nanotechnology. American Scientific Publishers

Naseer B, Srivastava, Qadri O, Faridi S, Islam R, Younis K (2018) Importance and health hazards of nanoparticles used in the food industry. Nanotechnol Rev 7(6):623–641

Nayak R, Meerovich I, Dash AK (2019) Translational multi-disciplinary approach for the drug and gene delivery systems for cancer treatment. AAPS Pharm SciTech 20(4):160

Nosrati N, Bakovic M, Paliyath G (2017) Molecular mechanisms and pathways as targets for cancer prevention and progression with dietary compounds. Int J Mol Sci 18(10):2050

Palmerston Mendes L, Pan J, Torchilin V (2017) Dendrimers as Nanocarriers for nucleic acid and drug delivery in cancer therapy. Molecules 22(9):1401

Pathak K, Akhtar (2018) Nanoprobiotics: Progress and issues. In: Nano Nutraceuticals. Taylor & Francis, pp 147–164

Pawar R, Patil U, Gadekar R, Singour P, Chaurasiya P (2010) A potential of some medicinal plants as an antiulcer agents. Pharmacogn Rev 4(8):136

Pellico J, Ellis CM, Davis JJ (2019) Nanoparticle-based paramagnetic contrast agents for magnetic resonnce imaging Contrast Media. Mol Imaging 2019:1–13

Pludowski P, Grant WB, Konstantynowicz J, Holick MF (2019) Classic and pleiotropic actions of vitamin D. Front Media SA 10:341

Podsednik A, Jacob A, Li L, Xu H (2020) Relationship between optical redox status and reactive oxygen species in cancer cells. React Oxyg Species 9:95

Potter AS, Foroudi S, Stamatikos A, Patil BS, Deyhim F (2011) Drinking carrot juice increases total antioxidant status and decreases lipid peroxidation in adults. Nutr J 10(1):96

Powell M (2008) Nanotechnology and food safety: potential benefits, possible risks? CAB Rev: Perspectives in Agriculture, Veterinary Science Nutrition and Natural Resources 3(038)

Prakash D, Sharma G (2014) Phytochemicals of nutraceutical importance CABI

Prankash D, Gupta C, Sharma G (2012) Importance of phytochemicals in nutraceuticals. JCMRD 1:70–78

Prasad S, Gupta SC, Tyagi AK (2017) Reactive oxygen species (ROS) and cancer: role of antioxidative nutraceuticals. Cancer Lett 387:95–105

Prasad R, Kumar V and Kumar M (2017a) Nanotechnology: Food and Environmental Paradigm. Springer Nature Singapore Pte Ltd. (ISBN 978-981-10-4678-0)

Pratheeshkumar P, Kuttan G (2011) Nomilin inhibits tumor-specific angiogenesis by downregulating VEGF, NO and proinflammatory cytokine profile and also by inhibiting the activation of MMP-2 and MMP-9. Eur J Pharmacol 668(3):450–458

Raddatz GC, de Souza da Fonseca B, Poletto G, Jacob-Lopes E, Cichoski AJ (2020) Influence of the prebiotics hi-maize, inulin and rice bran on the viability of pectin microparticles containing Lactobacillus acidophilus LA-5 obtained by internal gelation/emulsification. Powder Technol 362:409–415

Rafiee Z, Nejatian M, Daeihamed M, Jafari SM (2019) Application of curcumin-loaded nanocarriers for food, drug and cosmetic purposes. Trends Food Sci Technol 88:445–458

Rayman MP (2012) Selenium and human health. Lancet 379(9822):1256–1268

Razak S, Afsar T, Ullah A, Almajwal A, Alkholief M, Alshamsan A, Jahan S (2018) Taxifolin, a natural flavonoid interacts with cell cycle regulators causes cell cycle arrest and causes tumor regression by activating Wnt/ β -catenin signaling pathway. BMC Cancer 18(1):1043

Roco MC, Bainbridge WS (2001) Nanotechnology and societal interactions. In: Societal implications of nanoscience and nanotechnology, pp 12–17

Rodrigues FJ, Cedran MF, Bicas JL, Sato HH (2020) Encapsulated probiotic cells: relevant techniques, natural sources as encapsulating materials and food applications – a narrative review. Food Res Int 137:109682

Rook G, Bäckhed F, Levin BR, McFall-Ngai MJ, McLean AR (2017) Evolution, human-microbe interactions, and life history plasticity. Lancet 390(10093):521–530

Rosato A, Natile G, Arnesano F (2017) Copper homeostasis in humans and bacteria. In: Reference module in chemistry, molecular sciences and chemical engineering

Rostamabadi H, Falsaf SR, Jafari SM (2019) Starch-based nanocarriers as cutting-edge natural cargos for nutraceutical delivery. Trends Food Sci Technol 88:397–415

Roy A, Datta S, Bhatia KS, Bhumika, Jha P, Prasad R (2021) Role of plant derived bioactive compounds against cancer. South African Journal of Botany https://doi.org/10.1016/j.sajb.2021.10.015

Sabella S (2014) Impact of Bionanointeractions of engineered nanoparticles for nanomedicine. In: Nanotoxicology, pp 21–36

Saglam N, Korkusuz, F, Prasad R (2021) Nanotechnology Applications in Health and Environmental Sciences. Springer International Publishing (ISBN: 978-3-030-64410-9) https://www.springer.com/gp/book/9783030644093

Sahin K, Orhan C, Tuzcu M, Ali S, Sahin N, Hayirli A (2010) Epigallocatechin gallate inhibits hepatic glucose production in primary hepatocytes via downregulating PKA signaling pathways and transcriptional factor FoxO1. J Agric Food Chem 67:3651

Sahoo AK, Verma A, Pant P (2017) Nanoformulations for cancer therapy. In: Rai M, Alves dos Santos C (eds) Nanotechnology applied to pharmaceutical technology. Springer, Cham

Salama L, Pastor ER, Stone T, Mousa SA (2017) Chapter 4: Emerging nanopharmaceuticals nanomedicine in cancer, pp 99–128

Sanders ME (2008) Probiotics: definition, sources, selection, and uses. Clin Infect Dis 46(S2):S58–S61

Sanders ME, Merenstein DJ, Reid G et al (2019) Probiotics and prebiotics in intestinal health and disease: from biology to the clinic. Nat Rev Gastroenterol Hepatol 16:605–616

Sarwar H, Khalid Hossain M, Basher MK, Mia MNH, Rahman MT, Uddin MJ, Smart (2019) Nanocarrier-based drug delivery systems for cancer therapy and toxicity studies: a review. J Adv Res 15:1–18

Saura-Calixto F, Pérez-Jiménezn J (2018) Non-extractable polyphenols and carotenoids: importance in human nutrition and health. Royal Society of Chemistry

Schwabe RF, Jobin C (2013) The microbiome and cancer. Nat Rev Cancer 13(11):800–812

Seifalian A, Rizvi S, Rouhi S, Taniguchi S, Yang SY, Green M, Keshtgar M (2014) Near-infrared quantum dots for HER2 localization and imaging of cancer cells. Int J Nanomedicine 9:1323

Sen S, Pathak Y (2016) Nanotechnology in nutraceuticals: production to consumption. CRC Press

Sevanian A (2019) Lipid peroxidation, membrane damage, and phospholipase A2 action. In: Cellular Antioxidant Defense Mechanisms, pp 77–96

Shao J, Fang Y, Zhao R, Chen F, Yang M, Jiang J, Chen Z, Yuan X, Jia L (2020) Evolution from small molecule to nano-drug delivery systems: an emerging approach for cancer therapy of ursolic acid. Asian J Pharm Sci 15:685

Shen L (2020) Gut, oral and nasal microbiota and Parkinson's disease. Microb Cell Factories 19(1)

Shen X, Liu L, Peek RM, Acra SA, Moore DJ, Wilson KT, He F, Polk DB, Yan F (2018a) Supplementation of p40, a *Lactobacillus rhamnosus* GG-derived protein, in early life promotes epidermal growth factor receptor-dependent intestinal development and long-term health outcomes. Mucosal Immunol 11(5):1316–1328

Shen X, Liu L, Peek RM, Acra SA, He F, Polk DB, Yan F (2018b) P079 neonatal supplementation of a *Lactobacillus rhamnosus* GG (lgg)-derived protein, p40, promotes intestinal development and prevents colitis in adulthood. Inflamm Bowel Dis 24(s1):S29–S29

Shende P, Mallick C (2020) Nanonutraceuticals: a way towards modern therapeutics in healthcare. J Drug Deliv Sci Technol 58:101838

Shim J, Kim J, Tenson T, Min J, Kainov D (2017) Influenza virus infection, interferon response, viral counter-response, and Apoptosis. Viruses 9(8):223

Shukla AK, Iravani S (2019) Preface green synthesis, characterization and applications of nanoparticles, xxv

Siezen RJ, Kok J, Abee T (2014) Lactic acid bacteria: genetics, metabolism and applications

Sikka S, Sethi G (2015) Anticancer effects of agents derived from fruits and vegetables against stomach cancer. In: Anticancer properties of fruits and vegetables, pp 309–335

Singh OV, Colonna T (2015) Nanotechnology. Bio-Nanoparticles:303–329

Singh AK, Verma A, Singh A, Arya RK, Maheshwari S, Chaturvedi P et al (2020) Salinomycin inhibits epigenetic modulator EZH2 to enhance death receptors in colon cancer stem cells. bioRxiv. https://doi.org/10.1101/2020.02.03.932269

Singhal J, Nagaprashantha L, Chikara S, Awasthi S, Horne D, Singhal SS (2017) 2'-Hydroxyflavanone: a novel strategy for targeting breast cancer. Oncotarget 8(43):75025–75037

Skrajnowska D, Bobrowska-Korczak B (2019) Potential molecular mechanisms of the anti-cancer activity of vitamin D. Anticancer Res 39(7):3353–3363

Slavin YN, Asnis J, Häfeli UO, Bach H (2017) Metal nanoparticles: understanding the mechanisms behind antibacterial activity. J Nanobiotechnol 15:6

Song W, Anselmo AC, Huang L (2019) Nanotechnology intervention of the microbiome for cancer therapy. Nat Nanotechnol 14(12):1093–1103

Soto-Quintero A, Guarrotxena N, García O, Quijada-Garrido I (2019) Curcumin to promote the synthesis of silver NPs and their self-assembly with a Thermoresponsive polymer in core-shell Nanohybrids. Sci Rep 9(1):18187

Spicer P (2004) Cubosomes: Bicontinuous cubic liquid crystalline nanostructured particles. In: Schwarz JA, Contescu C, Putyera K (eds) Encyclopaedia of nanoscience and nanotechnology. Marcel Dekker, New York

Srivastava S, Somasagara RR, Hegde M, Nishana M, Tadi SK, Srivastava M, Choudhary B, Raghavan SC (2016) Quercetin, a natural flavonoid interacts with DNA, arrests cell cycle and causes tumor regression by activating mitochondrial pathway of Apoptosis. Sci Rep 6(1):24049

Stephensen CB (2017) Vitamin A, immunity, and o: 181–196

Subramani T, Ganapathyswamy H (2020) An overview of liposomal nano-encapsulation techniques and its applications in food and nutraceutical. J Food Sci Technol 57:3545–3555

Sudha T, Mousa DS, El-Far AH, Mousa SA (2020) Pomegranate (Punica granatum) fruit extract suppresses cancer progression and tumor angiogenesis of pancreatic and colon cancer in chick Chorioallantoic membrane model. Nutr Cancer:1–7

Sukhanova A, Bozrova S, Sokolov P (2018) Dependence of nanoparticle toxicity on their physical and chemical properties. Nanoscale Res Lett 13:44

Sumathi C, Priya DM, Mandal AB, Sekaran G (2012) Production of different proteases from fish gut micro flora utilizing Tannery fleshing. Eng Lifesci 12(3):1–15

Sumathi C, Priya DM, Somasundaram S, Sekaran G (2014) Production of prodigiosin from tannery fleshing through microbial fermentation processes. Sci World J:Article ID 290327

Sundraraman G, Jayakumari LS (2019) Meticulous Taxifolin releasing performance by the zinc oxide nanoparticles: as a short road to drug delivery system for cancer therapeutics. J Clust Sci 31(1):241–255

Surendran SP, Moon MJ, Park R, Jeong YY (2018) Bioactive nanoparticles for cancer immunotherapy. Int J Mol Sci 19(12):3877

Swanson KV, Deng M, Ting JP (2019) The NLRP3 inflammasome: molecular activation and regulation to therapeutics. Nat Rev Immunol 19(8):477–489

Teleanu D, Chircov C, Grumezescu A, Volceanov A, Teleanu R (2018) Impact of nanoparticles on brain health: an up to date overview. J Clin Med 7(12):490

Thomas K, Sayre P (2005) Research strategies for safety evaluation of nanomaterials, part I: evaluating the human health implications of exposure to nanoscale materials. Toxicol Sci 87(2):316–321

Traitler H, Coleman B, Hofmann K (2015) Food industry design, technology and innovation

Üner M (2016) Characterization and imaging of solid lipid nanoparticles and nanostructured lipid carriers. In: Handbook of nanoparticles, pp 117–141

VanHook AM (2015) Butyrate benefits the intestinal barrier. Sci Signal 8(378):135

Varela-MoreiraA SY, Fens MH, Lammers T, Hennink WE, Schiffelers RM (2017) Clinical application of polymeric micelles for the treatment of cancer. Mater Chem Front 1(8):1485–1501

Voukeng IK, Beng VP, Kuete V (2016) Antibacterial activity of six medicinal Cameroonian plants against Gram-positive and Gram-negative multidrug resistant phenotypes. BMC Compl Altern Med 16(1):388

Wang X, Zhang F, Yang L, Mei Y, Long H, Zhang X, Zhang J, Qimuge-Suyila SX (2011) Ursolic acid inhibits proliferation and induces Apoptosis of cancer cells in vitro and in vivo. J Biomed Biotechnol 2011:1–8

Watson RR, Preedy VR (2014) Bioactive nutraceuticals and dietary supplements in neurological and brain disease: prevention and therapy. Academic Press

Wicki A, Witzigmann D, Balasubramanian V, Huwyler J (2015) Nanomedicine in cancer therapy: challenges, opportunities, and clinical applications. J Control Release 200:138–157

Wolfram J, Ferrari M (2019) Clinical cancer nanomedicine. Nano Today 25:85–98

Wu C, Liu Y, Yang Y, Zhang P, Zhong W, Wang Y et al (2020) Analysis of therapeutic targets for SARS-Cov-2 and discovery of potential drugs by computational methods. Acta Pharm Sin B 10(5):766–788

Xu B, Watkins R, Wu L, Zhang C, Davis R (2015) Natural product-based nanomedicine: recent advances and issues. Int J Nanomedicine 10:6055

Yang P, Chen H, Huang Y, Hsieh C, Wung B (2017) Lycopene inhibits NF-κb activation and adhesion molecule expression through nrf2-mediated heme oxygenase-1 in endothelial cells. Int J Mol Med 39(6):1533–1540

Ying D, Schwander S, Weerakkody R, Sanguansri L, Gantenbein Demarchi C, Augustin MA (2013) Microencapsulated *Lactobacillus rhamnosus* GG in whey protein and resistant starch matrices: probiotic survival in fruit juice. J Funct Foods 5(1):98–105

Ying D, Sanguansri L, Weerakkody R, Bull M, Singh TK, Augustin MA (2016) Effect of encapsulant matrix on stability of microencapsulated probiotics. J Funct Foods 25:447–458

Yoksan R, Jirawutthiwongchai J, Arpo K (2010) Encapsulation of ascorbyl palmitate in chitosan nanoparticles by oil-in-water emulsion and Ionic gelation processes. Colloids Surf B: Biointerfaces 76(1):292–297

Yuan X, Xiao J, Liu X, McClements DJ, Cao Y, Xiao H (2019) The gastrointestinal behavior of emulsifiers used to formulate excipient emulsions impact the bioavailability of β-carotene from spinach. Food Chem 278:811–819

Zachariah MS, Saeideh M, Zahra B, Hosein FM, Mohammad A (2020) Nanoformulations of herbal extracts in treatment of neurodegenerative disorders. Front Bioeng Biotechnol 8:238

Zdrenghea MT, Makrinioti H, Bagacean C, Bush A, Johnston S, Stanciu LA (2016) Vitamin D modulation of innate immune responses to respiratory viral infections. Rev Med Virol 27(1):1909

Zewen L, Zhangpin R, Jun Z, Chia-Chen C, Eswar K, Tingyang Z, Li Z (2018) Role of ROS and nutritional antioxidants in human diseases. Front Physiol 9:477

Zhang Y (2018) Preparation of silk fibroin nanoparticles and enzyme-entrapped silk fibroin nanoparticles. Bio-Protocol 8(24):e3113

Zhang Y, Talalay P, Cho CG, Posner GH (1992) A major inducer of anticarcinogenic protective enzymes from broccoli: isolation and elucidation of structure. Proc Natl Acad Sci U S A 89(6):2399–2403

Zhang X, Xu X, Gu Z (2018) Biomimetic dendritic peptides as precise nanodrugs for efficient tumor therapy. Nanomedicine 14(5):1760–1761

Zhao J, Agarwal R (1999) Tissue distribution of silibinin, the major active constituent of silymarin, in mice and its association with enhancement of phase II enzymes: implications in cancer chemoprevention. Carcinogenesis 20(11):2101–2108

Zhao R, Torley P, Halley PJ (2008) Emerging biodegradable materials: Starch- and protein-based bio-nanocomposites. J Mater Sci 43(9):3058–3071

Zhao T, Liu Y, Gao Z, Gao D, Li N, Bian Y, Dai K, Liu Z (2015) Self-assembly and cytotoxicity study of PEG-modified ursolic acid liposomes. Mater Sci Eng C53:196–203

Zhou J, Hu Z, Zabihi F et al (2020) Progress and perspective of antiviral protective material. Adv Fiber Mater 2:123–139

Zhu C, Sawrey-Kubicek L, Beals E, Rhodes CH, Houts HE, Sacchi R, Zivkovic AM (2020) Human gut microbiome composition and tryptophan metabolites were changed differently by fast food and Mediterranean diet in 4 days: a pilot study. Nutr Res 77:62–72

Chapter 9
Trends of Biogenic Nanoparticles in Lung Cancer Theranostics

V. Uma Maheshwari Nallal, C. Sumathi Jones, M. Razia, D. S. Prabakaran, and Prakash Ramalingam

Contents

9.1	Introduction..	302
9.2	Classification and Molecular Biology of Lung Cancer................................	304
	9.2.1 Small Cell Lung Cancer (SCLC)...	304
	9.2.2 Non-small Cell Lung Cancer (NSCLC)...	305
	9.2.3 Malignant Pleural Mesothelioma (MPM).......................................	306
9.3	Limitations and Challenges of Conventional Lung Cancer Theranostics.....	307
9.4	Nanotechnology: A Promising Theranostic Tool for Lung Cancer...............	309
9.5	Role of Biogenic Nanoparticles in Lung Cancer Theranostics.....................	311
	9.5.1 Synthesis and Mode of Action of Gold Nanoparticles (Au NPs)...	311
	9.5.2 Fabrication and Application of Silver Nanoparticles (Ag NPs).....	311
	9.5.3 Biogenic Synthesis and Use of Magnetic Nanoparticles................	312
	9.5.4 Green Synthesis and Theranostic Use of Copper Oxide Nanoparticles (CuONPs)...	313
	9.5.5 Bio-Fabrication and Mechanism of Action of Titanium Dioxide Nanoparticles (TiO_2NPs)..	314
9.6	Trends of Semisynthetic Nanoparticles in Lung Cancer Theranostics.........	315
	9.6.1 Chitosan Nanoparticles..	315

V. U. M. Nallal · M. Razia (✉)
Department of Biotechnology, Mother Teresa Women's University,
Kodaikanal, Tamil Nadu, India

C. S. Jones
Department of Pharmacology and Environmental Toxicology, University of Madras,
Chennai, Tamil Nadu, India

Asan Memorial Dental College and Hospital, Chengalpattu, Tamil Nadu, India

D. S. Prabakaran
Department of Radiation Oncology, Chungbuk National University College of Medicine,
Cheongju, Republic of Korea

Department of Biotechnology, Ayya Nadar Janaki Ammal College, Sivakasi,
Sivakasi, Tamil Nadu, India

P. Ramalingam
Product Development, Genus LifesciencesInc, Allentown, PA, USA

    9.6.2 Polyherbal Nanoparticles.. 315
9.7 Recent Trends, Challenges, and Future Prospects of Biogenic Nanoparticles in Lung
    Cancer.. 316
9.8 Conclusion.. 317
References.. 318

9.1 Introduction

Cancer is a fatal disease that has significant impact on mental health of the diagnosed patients. It imposes emotional, physical, and economic burden on individuals and also on the society. Being noncommunicable in nature, cancer and mental illness are closely interdependent. Cancer patients undergo stress, anxiety, and depression that tend to have a negative effect on the physical stability of the patient and their mortality (Purushotham et al. 2013). Cancer deaths increase each year with the simultaneous increase in the number of new cases annually. According to the estimation of International Agency for Research on Cancer, 22.2 million new cancer cases would be recorded by 2030 with an average of 13.2 million deaths worldwide (Cryer and Thorley 2019). Cancer is certainly an insidious disease that takes away millions of life per year, yet cancer is curable. Lack of knowledge on the possibilities of treatment and exaggeration on the painful deaths are the major reasons that affect the mental health of the patients. Since the past, new treatment strategies are being identified and implemented to treat and cure cancer in a painless manner. However, the hesitation and anxiety among patients to try new treatment strategies remain as a hindrance in the development of diagnostic tools as well as efficient therapies for cancer (Bandyopadhyay et al. 2015).

Lung carcinoma is the most commonly reported cancer among men and the leading cause of cancer deaths. It has been suggested that lung cancer might overtake the incidence of breast cancer among European women by 2050. Globally, 18.4% cancer-related deaths are due to lung cancer and are estimated to have poor survival rate of 15% with a survival extension of 5 years (Mukherjee et al. 2019; Babu et al. 2013). Long-term tobacco smoking and exposure to mutagenic and carcinogenic agents are considered as the major cancer causing agents. Interaction of genetic and environmental factors can aggravate the risk of developing lung carcinoma. Nevertheless, exposure to secondhand smoke; air pollution; and inhalation of toxins such as asbestos, arsenic, etc. also tend to elevate the threat of lung cancer (Que et al. 2019). Histological data suggest that lung cancer can be classified into two main types: small cell lung cancer (SCLC) and non-small cell lung cancer (NSCLC). NSCLC accounts for more than 85% of cases, and patients are reported to have a life expectancy of 6–18%. Though extreme and innovative measures have been taken in the last five decades to lower the mortality rates of cancer, it has been possible only to an extent. Lack of early theranostics in about 50% of the cases and the confirmation of the disease in stage IV lead to low survival rates (Padmini et al. 2016; Parvanian et al. 2017). Diagnosis technologies that include magnetic resonance imaging (MRI) and computed tomography (CT) and treatment options such as chemotherapy, hormone therapy, radiotherapy, and surgery fail to treat lung cancer in several patients. Inaccessibility of the deep tissues and barrier penetration within the lungs further complicate the treatment process of conventional therapies (Mao and Liu 2020; Moorthi et al. 2011).

High-quality theranostics for the complete eradication of cancer requires the implementation of new approaches that are safe, eco-friendly, and cost-effective with fewer or nil side effects (Zugazagoitia et al. 2016). Nanoscale materials have the greatest advantage of being smaller in size which ensures their accessibility across the bronchial epithelial barrier to the deeper lung tissues (Fig. 9.1). Nanotechnology has become a part of basic cancer research in recent times. Semiconductor nanocrystals and eco-friendly organic and inorganic nanomaterials have gained the attention of researchers worldwide (Sukumar et al. 2013). These nanoparticles (NPs) are widely employed for cancer theranostics as imagining tools and drug delivery systems. Detection of various cancer metastases that were undetectable in earlier times has now become possible due to the amalgamation of magnetic NPs with MRI scans that help in increasing the contrast of the images (Karkan et al. 2016). Nanoscale formulations such nanogels and nanosprays are efficaciously administered into the respiratory tract by intratracheal routes and showed better delivery results when compared to parental routes. Clinical detections of lung tumors have become possible owing to the efficient use of nanoparticles. For instance, quantum dot-doped polystyrene NPs are utilized to identify the presence of CEA and CYFRA21-1 in lung cancer patient serum (Mottaghitalab et al. 2019). Nevertheless, bioinspired NPs have secured special attention due to their amicable physicochemical properties (Fig. 9.2). Encouraging results have been obtained from bacterial NPs, plant NPs, viral NPs, aptamers, solid-lipid NPs, etc. (Madamsetty et al. 2019). Thus, this chapter addresses the emerging trends of biogenic nanoparticles in lung cancer theranostics due to their manifold applications and promising results in lung cancer diagnosis and therapy.

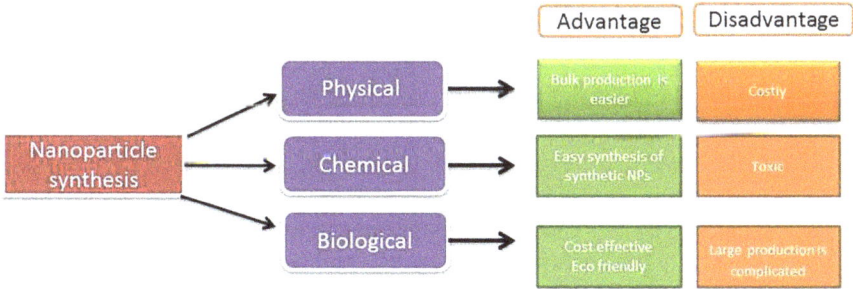

Fig. 9.1 Advantages and disadvantages of different approaches used in NP synthesis

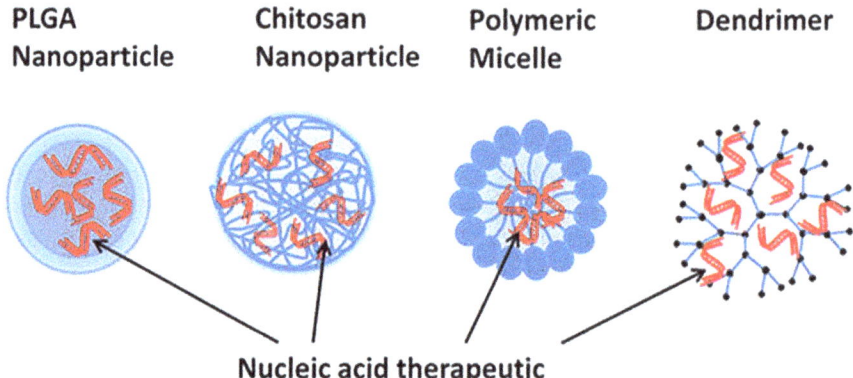

Fig. 9.2 Nanoparticle-mediated gene delivery for lung cancer. (Adapted from Narsireddy et al. 2017)

9.2 Classification and Molecular Biology of Lung Cancer

Neoplastic metamorphosis of the lung epithelial cells results in lung cancer. Progression of the disease is influenced by multiple factors such as molecular aberrations and epigenetic and genetic factors. These factors also have a profound influence in the theranostics and predictive outcomes of lung cancer. In order to identify an unfailing and reliable treatment strategy, it is necessary to understand the underlying molecular and genetic information of lung cancer. Lung cancer is divided into three types based on their histopathological characteristics. SCLC and NSCLC are the major classifications of lung cancer along with malignant pleural mesothelioma (Cooper et al. 2013) (Fig. 9.3).

9.2.1 Small Cell Lung Cancer (SCLC)

SCLC is commonly addressed as the smoker's disease since the incidence of the disease is higher in long-term smokers. The risk of SCLC increases with the exposure to smoke with over 90% of the patients falling into this category. Globally 14% of the reported lung cancer cases are under SCLC category, and more than one million cases are diagnosed and treated each year. According to the World Health Organization (WHO), this category of lung carcinoma is classified as a neuroendocrine malignancy with pathological features characterized by nuclear molding, mitoses, and necrosis (Travis 2012; Padmini et al. 2020). CD56 and thyroid transcription factor-1 aid as biomarkers for the confirmation of lung cancer. Diagnostic exclusion tools that are additionally employed are chromogranin and synaptophysin staining techniques, which act as strong neuroendocrine markers. The least progression is perceived in SCLC genetic studies when compared to other lung cancer

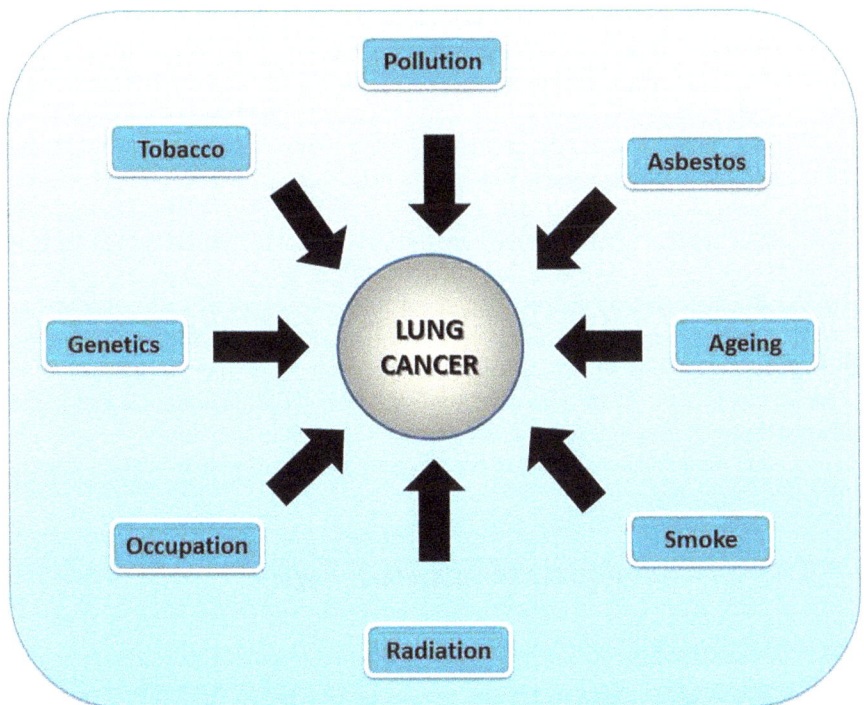

Fig. 9.3 Causes of lung cancer

classifications. The paucity of recognized targets unambiguous to SCLC remains as the plausible reason for the impediment in such cases. Tumor suppressor genes such as TP53 and RB1 are also lost in majority of SCLC patients which leaves them in higher risk (Byers and Rudin 2014).

9.2.2 Non-small Cell Lung Cancer (NSCLC)

Eighty-five percent of the total lung cancer cases are identified as NSCLC that tags the type as the predominant form of lung cancer. Metastasis or local advancement of the disease is identified in more than 50% of the cases. Like SCLC, cigarette smoking is a profound risk factor in NSCLC, yet other factors such as exposure to secondhand cigarette smoke, unhealthy food habits, insalubrious lifestyle, and exposure to carcinogens and mutagens along with genetic variation can trigger the disease (Cryer and Thorley 2019). Active mutations in the epidermal growth factor receptor are commonly diagnosed in NSCLC patients. Squamous cell carcinoma, adenocarcinoma, and large-cell carcinoma are clustered under the umbrella term NSCLC though they are exemplified with different epigenetic, genetic, molecular, and cellular features. Adenocarcinoma is the most prevalent NSCLC with incidence rates adversely increasing amid women. The major distinction between adenocarcinoma and squamous cell carcinoma patients is the positive expression of mucin, cytokeratin-7, and thyroid transcription factor-1 (Gridelli et al. 2015). Twenty to 30% of NSCLC cases account for the squamous cell carcinoma subtype, and incidence is accounted commonly in men. Familial predispositions, polymorphisms, and cigarette smoking are the major factors that lead to the development of squamous cell carcinoma (Stinchcombe 2014). Keratinization and pearl formation with dense cytoplasm and irregular nuclei are the characteristic features of this subtype. Respiratory malignancies that do not fall under the category of adenocarcinoma and squamous cell carcinoma are grouped under the term large-cell carcinoma and are devoid of differentiating characteristics when viewed under a light microscope. Difference in the characteristics gives rise to various but specific microenvironments that contribute to arduous treatment strategies. Histopathological and immunohistochemical staining, genetic analysis, and imaging techniques are currently employed to understand the disease progression in NSCLC patients (Sholl 2014).

9.2.3 Malignant Pleural Mesothelioma (MPM)

The incidence of MPM is very less when compared to other types of lung cancer. Approximately 2000–3000 new cases are reported each year worldwide, but the incidence rate is considered to increase rapidly in the coming years (Cryer and Thorley 2019). The primary factors that increase the risk of MPM are exposure to ionizing radiations, viral infection due to simian virus, and acquaintance to asbestos. Dyspnea due to pleural effusion is the common symptom noticed in patients diagnosed with MPM. Additionally, patients experience chest pain due to thoracic wall invasion. Other symptoms such as hyperhidrosis, fatigue, weight loss, and tiredness are common in patients. Intrinsic heterogeneity makes it difficult to obtain a genetic footprint; however, TP53 and RB1 mutations occur rarely in MPM (Papp et al. 2001). Cellular proliferation in MPM is not limited to physiological restrictions, since deregulated metabolic pathways play an influential role in determining the disease pathology. Epigenetic alterations and promoter hypermethylation have been remarked at several loci on MPM. Histone deacetylases (HDAC) are inhibited which allows the tumor suppressor genes to remain unfolded and available for transcription (Vandermeers et al. 2009).

Lung cancer is demarcated with the growth of abnormally dividing cells in the lungs that can later infiltrate into different cells of the body. A human body is generally tailored with a remarkable cell cycle assessment system that constraints the growth of new cells as per requirement of the body structure. Additionally, the system is equipped with the ability to eliminate unnecessary and harmful cells naturally by a process renowned as apoptosis. Tumor repressor genes also assist the body systems to control the growth of tumor and proliferation of redundant cells (Mukherjee et al. 2019). However mutations can cause an imbalance in the normal functioning of the cell cycle, for instance, Ras gene (growth promoting gene) mutations can lead to the uncontrolled growth and multiplication of cells to form tumor- or neoplasm-like structures. Not only oncogenes are altered due to mutations, but the rate of mutations in tumor suppressor genes and genes involved in DNA repair mechanism is also high. It is certain that a fast-growing tumor can consume copious amounts of oxygen and key nutrients that are required for the growth and multiplication of normal cells compelling them to behave abnormally. New territories are gained by invasive cancer cells by vitiating the extracellular matrix with the help of enzyme proteases and penetrate into the normal adjacent cells (Burstein and Schwartz 2008; Muthuraj et al. 2016).

The molecular basis of lung cancer is intricate and diverse. Basic understanding of molecular shifts and genetic, epigenetic, or protein expression along with their functions have an influential role in lung cancer theranostics. Progression and development of lung cancer is a multistep process epitomized by genetic variations and recurrent mutations at high frequency (Larsen and Minna 2011). Insights of genetic variations that affect a common cluster of oncogenic signaling pathways in lung cancer have led to the major improvements in cancer theranostics. New strategies that are being developed effectively target the types and subtypes of lung cancer. The major gene mutations that are correlated with lung cancer have been identified as v-Ki-ras2 Kirsten rat sarcoma viral oncogene (KRAS), epidermal growth factor receptor (EGFR), and phosphatidylinositol 3-kinase (P13K). Recent studies focus on genes such as MEK, HER2, ALK, and ROS1. The major concerns in adenocarcinoma and squamous cell carcinoma are MET and FGFR1, respectively (Cooper et al. 2013; Padmini et al. 2017). Race, gender, and smoking behaviors are commonly associated with the genetic alterations of carcinogenesis. Nevertheless, mutations and alterations associated with host factors such as tumor suppressor genes (TP53, LKB11, RBI, PTEN, and p16) have become a huge concern to researchers in lung cancer development and progression (Lynch et al. 2004; Yip et al. 2013).

9.3 Limitations and Challenges of Conventional Lung Cancer Theranostics

Systematic therapy (chemotherapy, hormonal therapy, targeted therapy, and biological therapy) and local therapy (surgery, radiation) are the conventional methods used to treat lung cancer. At various stages of lung cancer, systematic and local therapy are administered together to reduce the progress of abnormal cell growth in lung cancer patients. Health and age of the patient along with the stage, type, and size of the tumor in the lung determine the type of treatment that has to be administered to the patient. For instance, eradication of the cancerous cells may be possible (curative) or only measure to reduce the pain caused by the cancer can be possible (palliative). Primarily administered therapy is prescribed with an additional therapy in order to enhance the effect of the primary therapy. For example, surgical removal of the cancerous tissues in the lung is followed by chemotherapy or radiation therapy (Bandyopadhyay et al. 2015).

Surgery is prescribed to patients initially at stage I and stage II where small size tumors present in the lung are excised surgically, or sometimes lobectomy is advised to patients who are exposed to the risk of metastasis. The limitations of surgery in lung cancer are remarked by fewer possibilities to remove large size tumors and the inefficiency to remove cancerous cell completely (Rani et al. 2012). Chemotherapy is the administration of chemicals in the form of drugs into the patient's body either after surgery or as the primary treatment. The major inadequacies of chemotherapy are permeability and retention of the drug within the cells. Chemotherapy may be unsuccessful in patients with poor vasculature and limited interstitial pressure. Moreover, chemotherapeutic drugs can cause cardiotoxicity and nephrotoxicity along with other acute side effects (Mukherjee et al. 2019). Radiation therapy involves the use of X-rays or high energy ionizing radiations to kill the cancerous cells. The possibility of developing radiation-resistant tumors due to the production of reactive oxygen species during the process of radiation therapy is high. Side effects of radiation therapy include nausea, vomiting, fatigue, irritation of the digestive tract, and a sudden drop in the count of white blood cells followed by reduced platelet count (Moorthi et al. 2011) (Fig. 9.4). However, the abovementioned conventional therapies have a prominent effect only during the early stages of lung cancer. Thus, certainly a novel approach to diagnose and treat lung cancer at its various stages effectually without limitations is necessary (Zugazagoitia et al. 2016).

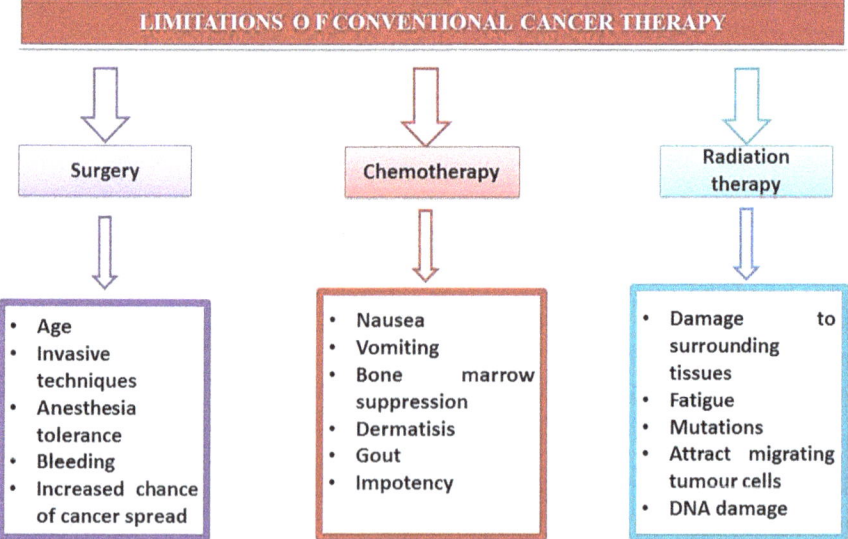

Fig. 9.4 Limtations of conventional cancer therapy

9.4 Nanotechnology: A Promising Theranostic Tool for Lung Cancer

Nanoparticles have enticed substantial attention in the field of nano-medical technology due to their unique and interesting characteristics. Particles that have dimension less than 100 nm are acknowledged as nanoparticles. Nanotechnology has created a cutting edge in cellular imaging, experiment therapies, and theranostics (Fig. 9.5). There is no uncertainty in the appreciable outcomes of nanoparticles in the medicinal field since nanoparticles have claimed their applications in various other fields such as electronics, cosmetics, solar cells, textiles, optical devices, and even more (Bandyopadhyay et al. 2015; Parvathi et al. 2015). Detection at the right stage and proficiency of the prescribed treatment can aid in new prospects of lung cancer theranostics. Interestingly, nature has fabricated the human system with numerous nanoscale biological entities which emphasizes the need of theranostic notions in the same scale. The major advantage of nanoparticles lies in their small size with large surface area. This enables them to absorb or bind micro- and macromolecules that are either hydrophobic or hydrophilic in nature. Nanoparticles can easily carry molecules such as DNA, RNA, drugs, and probes through the intricate barriers of the human body (Ahmad et al. 2011).

Drug targeting and drug delivery of sparingly water soluble drugs have become possible through the use of NPs. Interaction between the NPs and drug ligand to be

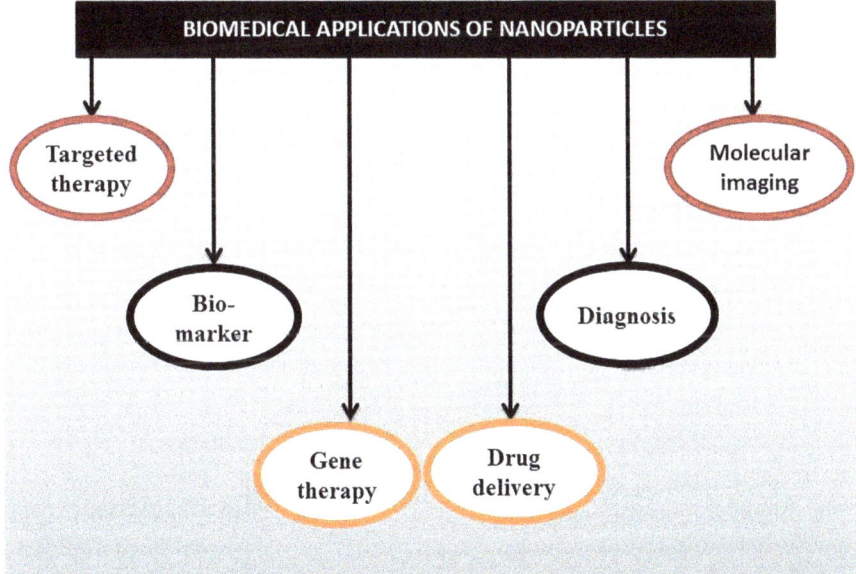

Fig. 9.5 Biomedical applications of nanoparticles

delivered determines the efficiency and the success of the experimental therapy. Diagnosis has reached better apices due to the opportunity of integrating poorly soluble probes to NPs. Attachment of the drug or the probe to the NPs that have smaller size and large surface area improves their dissolving ability in the bloodstream. This in turn increases the bioavailability and retention of the drug in the human system, thus delivering the drugs to specific tissues at the right time (Sattler 2010; Jong and Borm 2008). Degradation of the drugs or the probes is a major concern when it comes to oral and intravenous administration since the acids present in the gastrointestinal tract as well as the stomach can have adverse effects on them. On the other hand, the drugs might be easily metabolized by the hepatic portal system and eliminated from the body (Fig. 9.6). These shortcomings have been overcome by the introduction of NPs that are stable over a wide range of pH and temperature. This characteristic of the NPs also helps in effective targeting of the cancerous cells, since the pH of the cancerous lung cells are acidic in nature when compared to the normal cells. Drugs that have failed in previous experiments due to low solubility and toxicity can be retried by conjugating them with NPs that can improve their solubility and reduce the toxicity (Mohanraj and Chen 2006; Sivasankar and Kumar 2010).

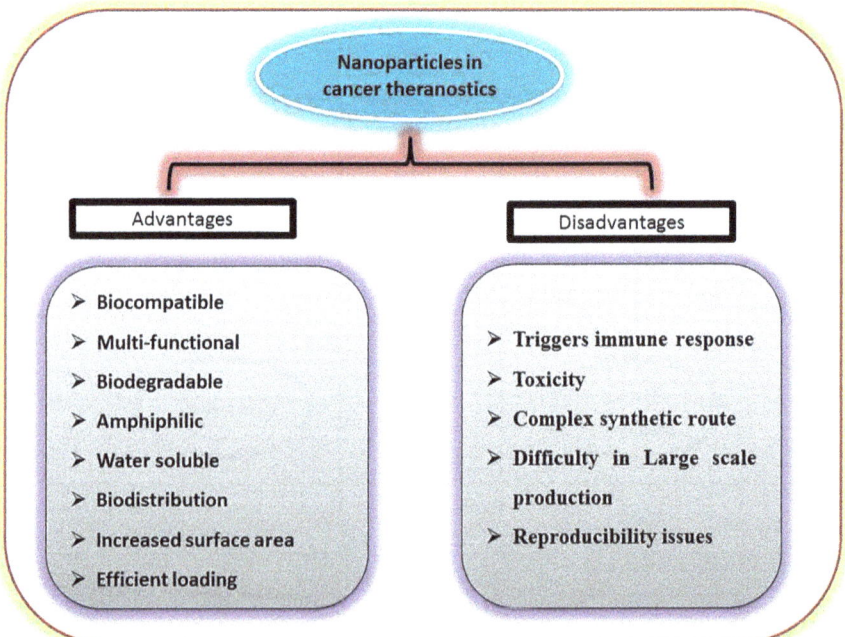

Fig. 9.6 Nanoparticles in cancer theranostics

9.5 Role of Biogenic Nanoparticles in Lung Cancer Theranostics

9.5.1 Synthesis and Mode of Action of Gold Nanoparticles (Au NPs)

Synthesis of gold nanoparticles from plant extracts has been an easy process, and Au NPs are harmless when compared to other NPs (Kumar et al. 2011). Au NPs can produce reactive oxygen species that creates oxidative stress within the tumor environment leading to cell apoptosis. Au NPs synthesized from Dendropanax morbifera leaf extract exhibited potential anticancer activity against HaCaT cells by inducing the production of reactive oxygen species and later persuading the cells to apoptosis process (Wang et al. 2016). Elevated ROS production provoked by Au NPs synthesized from Magnolia officinalis in A549 cells showed nuclear fragmentation and mitochondrial depolarization within the cells (Zheng et al. 2019). Sun et al. (2019) showed that green synthesized Au NPs from Marsdenia tenacissima extract A549 cells serve as effective cancer agents by controlling the proliferation of the cancerous cells through apoptotic pathway. Au NPs downregulated the expressions of Bax, caspase 3, caspase 8, and caspase 9 in lung cancer pathway. Moreover, the modulation of Bax/Bcl 2 induced apoptosis in lung cancer cells (Zheng et al. 2019).

9.5.2 Fabrication and Application of Silver Nanoparticles (Ag NPs)

Ag NPs are widely being recognized as promising candidates in the field of lung cancer theranostics. They not only act as alluring drug delivery vehicles and probes but also as an assuring therapeutic agent (Ratan et al. 2020) (Fig. 9.7). Dried and cleaned powders of various herbal plants are used as reducing, stabilizing, and capping agents in the synthesis of Ag NPs. Purified particles are subjected to characterization techniques to confirm the reduction of the particles to nanosize. He et al. (2016) demonstrated that Ag NPs synthesized using herbal extracts of Dimocarpus longan showed effective chemopreventive activity against human lung cancer H1299 cells. Studies have demonstrated that nanoparticle internalization allows the slow and sustained release of Ag NPs in A549 cells and degradation of the NPs is promoted by the acid environment of the lysozymes. Ag NPs are assumed to regulate the apoptosis process and NF-κB activity. They can effectively slow down the growth of tumors from 28 to 36 days (He et al. 2016; Matsushita et al. 2000). Ag NPs synthesized from *Gossypium hirsutum* leaf extracts, *Syzygium aromaticum*

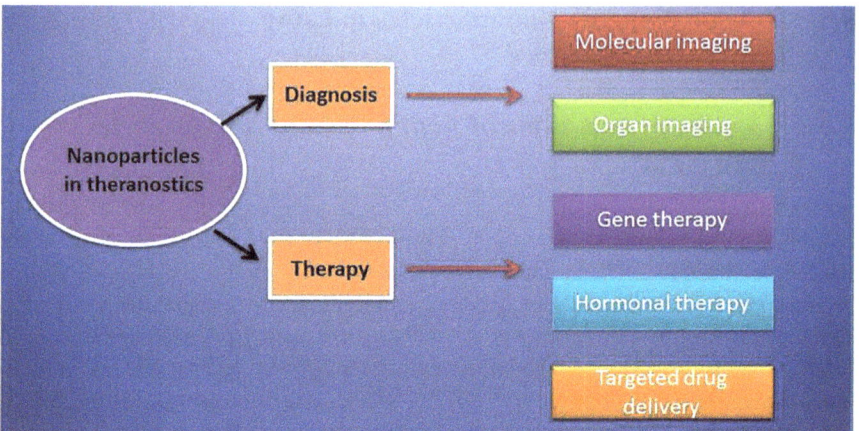

Fig. 9.7 Application of nanoparticles in cancer theranostics

fruit extracts, *Rosa damascene* petal extracts, and *Penicillium decumbens* microbe were effective in inhibiting the growth of A549 lung cancer cells in a dose-dependent manner (Ratan et al. 2020).

9.5.3 Biogenic Synthesis and Use of Magnetic Nanoparticles

Magnetic NPs are progressively gaining consideration as potential lung cancer theranostic agents. Magnetic NPs are advantageous for diagnosis as well as therapy since they evade cytotoxic effects and side effects of conventional cancer therapy. Iron oxide NPs are highly stable and can be fabricated in different shapes through environmentally safe processes. The super magnetic property and the ability to modify their surfaces are an important characteristic that aids in theranostics. Magnetic NPs have superior magnetic signal strength that can be used accurately for cancer detection in the lungs. Most of the magnetic NPs are low toxic in nature, and they increase the level of the contrasts while imaging the lung anatomy in cancer patients. Tumor margins can be precisely identified with the help of NPs as diagnostic tools (Rafique et al. 2017). Magnetic NPs are highly recommended for imaging, targeted drug delivery, slow and subsequent drug release, and site-specific targeting and therapy (Fig. 9.8). Iron oxide NPs are easily engulfed by the macrophages, thus aiding in the clear and accurate imaging of macrophage-rich organs such as the bone marrow, liver, and spleen by targeting the reticular endothelial system. Grape seeds, brown algae, peel extracts of plantain, eucalyptus leaves, sorghum, alfalfa, and *Aloe vera* extract have been successfully used to synthesize magnetic NPs (Karmous et al. 2019). Yusefi et al. (2019) have successfully synthesized iron oxide NPs using *Punica granatum* fruit peel extract and tested its cytotoxic effects on A549 lung cancer cells for which positive anticancer results were obtained.

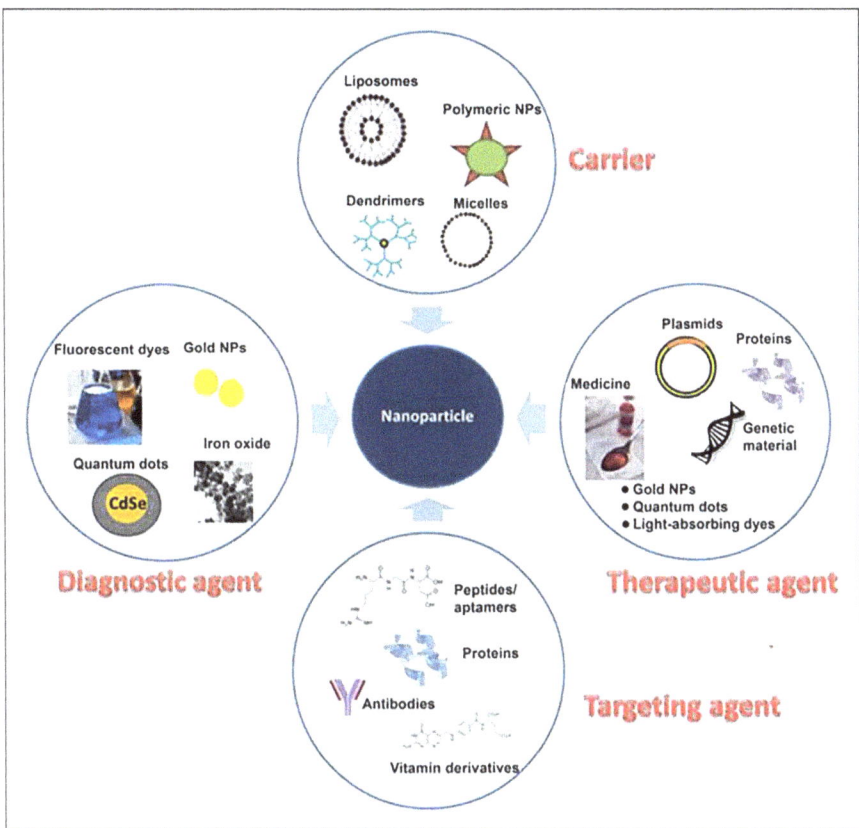

Fig. 9.8 Applications of magnetic nanoparticles. (Adapted from Menon et al. 2013)

9.5.4 Green Synthesis and Theranostic Use of Copper Oxide Nanoparticles (CuONPs)

Copper oxide nanoparticles are transitional metal oxide NPs that have been used as theranostic agents in recent years. Performance and applications of CuONPs depend on the strong absorption capacity that they exhibit. They also show improved and efficient biological photocatalytic activities when compared to other metal oxide NPs. Various approaches such as physical and chemical are being applied for the synthesis of CuONPs, yet biological approach is considered safe and environmental friendly (Fig. 9.9). Plant extracts as well as bioactive metabolites such as flavonoids, tannins, phenolic acids, terpenoids, and proteins are also employed in the synthesis of CuONPs (Rezaie et al. 2017). CuONPs that were synthesized from dried and purified leaf extracts of *Coleus aromaticus*, *Hibiscus rosa-sinensis*, *Murray akoenigii*, and *Azadirachta indica* showed effective anticancer activity against lung cancer cells (Akintelu et al. 2020).

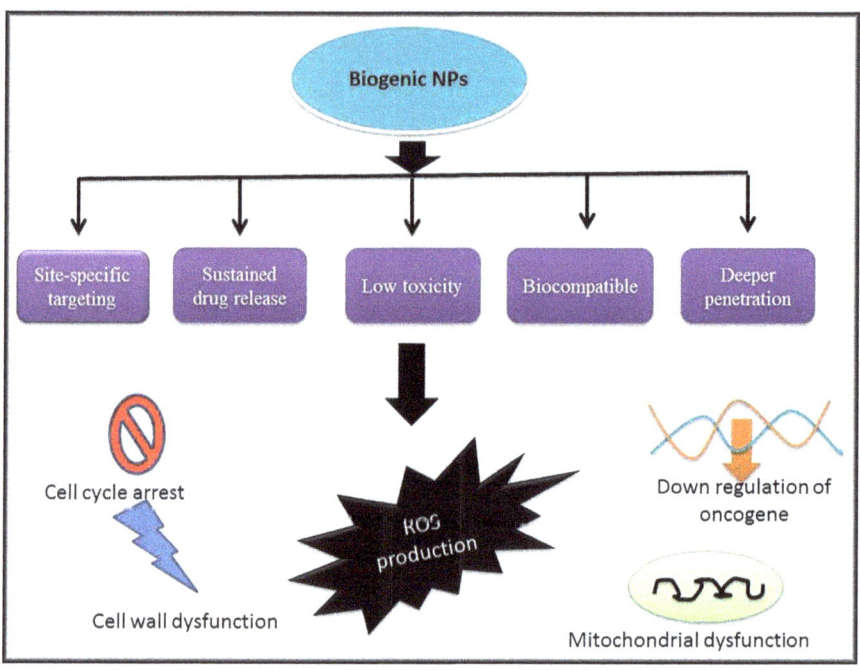

Fig. 9.9 Characteristics of biogenic NPs and their mechanism in lung cancer therapy

9.5.5 Bio-Fabrication and Mechanism of Action of Titanium Dioxide Nanoparticles (TiO$_2$NPs)

Titanium dioxide nanoparticles have plentiful applications in the field of cosmetics, medicine, energy, and bio sensing (Youssef et al. 2017). TiO$_2$NPs are being designed successfully to deliver drugs at specific sites without longer retention time in the tissues. Stimuli-trigger drug release and tumor cell targeting are possible with the use of TiO$_2$NPs. NPs have photocatalytic activity and large surface area, that is, highly functional remarkable cancer theranostic applications can be achieved by TiO$_2$NPs (Wang et al. 2015). These NPs can be engineered with different structure and size within the nanoscale range to make them highly compatible and bioavailable within the human system. The major limitations of conventional cancer therapies such as drug release, retention, and site-specific targeting are possible when TiO$_2$NPs are used as theranostic agents. Additionally TiO$_2$NPs possess low toxicity to the normal cells and are completely environmental friendly when synthesized in biological routes (Skocaj et al. 2011). TiO$_2$NPs synthesized from C*ynodon dactylon* leaf extract were shown to inhibit A549 lungs effectively at a concentration of 140 µg/ml. The biomolecules present in the plant extract can donate excess electrons to TiO$_2$ which provides additional opportunity for the TiO$_2$NPs to produce reactive oxygen species. ROS can generate stress within the cells that induce the cancerous cells to undergo apoptosis. Moreover, ROS is used to break the cell walls of cancerous cells. The physiochemical properties of TiO$_2$NPs favor their use as efficient lung cancer theranostic candidates (Hariharan et al. 2017).

9.6 Trends of Semisynthetic Nanoparticles in Lung Cancer Theranostics

Semisynthetic NPs are synthesized by combining natural moieties with synthetic molecules or polymers. The efficacies of the biogenic nanoparticles are improved with the integration of synthetic particles to treat cancer effectively. Further use of natural agents permits environmental-friendly techniques to synthesize semisynthetic nanoparticles

9.6.1 Chitosan Nanoparticles

Chitosan is made up of a nitrogenous polysaccharide with various amino acid groups attached to them. They are biodegradable as well as biocompatible in nature. Positive charge of chitosan molecules and negative charge sites on the cell lines favor strong electrostatic attraction that can promote the penetration of the drugs into the cells. The major advantage of chitosan molecules is their degrading ability when exposed to the cellular lysozyme. This helps in eliminating the chitosan NPs completely from the lungs after the diagnosis process or drug delivery (Bandyopadhyay et al. 2015). A mitotic cell cycle arresting drug PTX which is insoluble in water was loaded with chitosan chloride NPs and was treated to xenografted mouse with Lewis lung cancer cells. This experiment showed successful decrease in the tumor volume with the increase in drug accumulation in the tumor site (Lv et al. 2011). Chitosan microspheres loaded with 2′ 2′ difluorodeoxycytidine and adhered to dextran sulfate exhibited potential anticancer activity when experimented on human lung cancer A549 cells. Chitosan NPs loaded with antisense oligonucleotide that prevents telomere reduction during cell proliferation in NSCLC cells had potent inhibitory effect on A549 and Calu-3 cells. Chitosan-modified semisynthetic nanoparticles also show efficient theranostic properties (Galbiati et al. 2011; Okamoto et al. 2003).

9.6.2 Polyherbal Nanoparticles

Herbal medications are preferred due to their nontoxic property and absence of side effects. A major limitation in the use of NPs for drug delivery has been their accumulation in the target site when they are required to deliver large amounts of drugs.

In order to overcome this limitation, herbal formulations such as ethanolic extract of rattlesnake root and ginger that are reported to have anti-lung cancer activity are encapsulated in polymeric NPs that were biodegradable. It was observed that the cellular uptake of the NPs and sustained release of the polyherbal formulation effectively inhibited the lung cancer cells and were completely eliminated without accumulation in the tissues (Jadhav 2012).

9.7 Recent Trends, Challenges, and Future Prospects of Biogenic Nanoparticles in Lung Cancer

Nanoparticles have refurbished the use of drugs and probes for lung cancer theranostics (Tran et al. 2017). Recent trends of nanoparticles in lung cancer theranostics include the use of hybrid NPs. Hybrid NPs are made from organic/inorganic moieties from multiple origins and are acknowledged to have different functional operations. Hybrid super magnetic iron oxide NPs, lipid-polymer hybrid NPs, stimuli-responsive lipid-polymer hybrid NPs, gene-coupled hybrid nanoparticles, and gene-loaded hybrid NPs that might reduce drug resistance are the recently popular advancements in the use of NPs in lung cancer theranostics (Mottaghitalab et al. 2019). Separate NPs were designed for the purpose of diagnosis and lung cancer therapy in the last decades; recently, researchers are interested in developing single nano-system that can work as a theranostic tool; i.e., the same system is employed to diagnose the location of the tumor within the lung as well as deliver the drug required for the therapy at the specific diagnosed spot. Furthermore, the side effects of the chemotherapeutic agents have been considerably reduced with the use of nano vehicles. Thus, hybrid NPs are promising candidates in the field of cancer theranostics at present (Parvanian et al. 2017).

Nothing is completely perfect on the planet. NPs have their own shortcomings and challenges that have to be further amended to revolutionize the use of NPs in lung cancer theranostics (Table 9.1). The physiochemical properties of NPs must be additionally improvised to meet astonishing theranostic standards. Homogeneity and large-scale production of hybrid NPs are still at a stake due to technical concerns. NPs have the advantage of easily penetrating the tissues yet can get accumulated in the cell and cell organelles. Inorganic nanoparticles at times exert poisonous effects due to their accumulation in the tissues (Bhattacharyya et al. 2011). Further research must also be carried out in the aspects of nano-toxicology and environment safety (Ahmad et al. 2011). Another major hindrance observed in the use of lung cancer therapy is hemolysis. Hemoglobin tends to aggregate on the surface of the NPs that causes its degradation and at worse can lead to anemia resulting in patient's mortality. Opsonization is yet another challenge in the use of NPs, since opsonized NPs are targeted by white blood cells of the system's immune system and easily cleared from the body, thus reducing the bioavailability of the NPs. Moreover, the possibilities of NPs to create oxidative stress and release of reactive oxygen species are also a vital challenge (Sattler 2010).

Table 9.1 Advantages and disadvantages of nanoparticles in biological applications

Contrast agent (NPs)	Advantages	Disadvantages
Metal oxide NPs	FDA approved for clinical use; examples: Combidex, Ferridex Gadodiamide, LumenHance. SPIONs exhibit intrinsic magnetic properties	Toxicity and biocompatibility concerns Responsible for inflammatory responses in respiratory system
Gold NPs	Flexible surface modification with different coatings. Able to scatter and absorb light with strong excitation peaks in the visible and near-infrared region wavelengths depending on nanoparticle size	Toxicity concern
Quantum dots	Narrow wavelength emission. High fluorescent efficiency and photostability with wide range of emission spectra (400 nm–2000 nm) covering both the visible and near-infrared wavelengths	Toxicity and biocompatibility concerns specially heavy metal-containing QDs
Carbon nanotubes	Intrinsic properties, enabling imaging modalities. Ultrahigh surface area and hollow inside space available for efficient drug loading	Toxicity and biocompatibility concerns
Radioisotopes and fluorescent molecules	Flexible incorporation into a variety of drug delivering nanoparticle. Ability to perform as noninvasive monitoring and early diagnosis capabilities	Lack of targeting specificity, short half-life. Toxicity and biocompatibility concerns

Adapted from Howell et al. (2013)

The optimistic future of biogenic NPs in the field of lung cancer theranostics lies in addressing the aforementioned challenges. In vivo toxicity, bioavailability, bio distribution, and retention of hybrid NPs must be evaluated to determine their clinical efficacy. In the near future, hybrid NPs can be used for molecular imaging instead of anatomical imaging which is currently in use. Poor therapeutic consequences, low or poor solubility of drugs or probes, nonspecific targeting, and side effects of conventional cancer theranostics can be convalesced (Yu et al. 2012). Platforms in NP usage must be modified to encourage larger reproducibility, one-pot synthesis and simple hybrid NP synthesis, cost-effective production, environmental safety, higher functional properties, and superior structural properties for appreciable outcomes. Precise and robust research is required to confront the challenges in employing NPs for lung cancer theranostics (Mottaghitalab et al. 2019).

9.8 Conclusion

Every phase of consumer products has seen the eminent contribution of nanoparticles, especially the field of diagnosis and therapy. Extensive research is being carried out worldwide to create nanoparticles for theranostic use that are eco-friendly, less toxic, cost-effective, safe, and effective at the same time. Despite the overwhelming positive results obtained from laboratory experiments, NP formulations for lung cancer theranostics do not have enough success in clinical trials. The hesitation of public to try new therapies and the producers to invest in novel medications acts as barrier in the development of biogenic nanoparticles as promising therapeutic candidates. NPs can considerably overcome the limitations of conventional cancer therapies such as nonspecific targeting, low solubility, drug retention, bioavailability, toxicity, and side effects. Though NP synthesis is an easy and eco-friendly process when fabricated using biological approach, lack of large reproducibility is considered as the major shortcoming. Fortunately, hybrid NPs and biogenic NPs are being designed to overcome these limitations and will efficiently enter the pharmaceutical markets in the near future.

References

Ahmad MB, Tay MY, Shameli K, Hussein MZ, Lim JJ (2011) Green synthesis and characterization of silver/chitosan/polyethylene glycol nanocomposites without any reducing agent. Int J Mol Sci 12:4872–4884

Akintelu SA, Folorunso AS, Folorunso FA, Oyebamiji AK (2020) Green synthesis of copper oxide nanoparticles for biomedical application and environmental remediation. Heliyon 6(7):e04508

Babu A, Templeton AK, Munshi A, Ramesh R (2013) Nanoparticle-based drug delivery for therapy of lung cancer: progress and challenges. J Nanomater 2013:1–11

Bandyopadhyay A, Das T, Yeasmin S (2015) Nanoparticles in lung cancer therapy. In: Recent trends Springer Briefs in molecular science. ISSN No:978-81-322-2175-3, pp. 17–26

Bhattacharyya S, Kudgus R, Bhattacharya R, Mukherjee P (2011) Inorganic nanoparticles in cancer therapy. Pharm Res 28:237–259

Burstein HJ, Schwartz RS (2008) Molecular origins of cancer. N Engl J Med 358(5):527–527

Byers LA, Rudin CM (2014) Small cell lung cancer: where do we go from here? Cancer 121(5):664–672

Cooper WA, Lam DCL, O'Toole SA, Minna JD (2013) Molecular biology of lung cancer. J Thorac Dis 5(S5):S479–S490

Cryer AM, Thorley AJ (2019) Nanotechnology in the diagnosis and treatment of lung cancer. Pharmacol Ther 198:189–205

Galbiati A, Tabolacci C, Morozzo Della Rocca B, Mattioli P, Mattioli S, Beninati G, Paradossi G, Desideri A (2011) Targeting tumor cells through chitosan-folate modified microcapsules loaded with camptothecin. Bioconjug Chem 22:1066–1072

Gridelli C, Rossi A, Carbone DP, Guarize J, Karachaliou N, Mok T, Rosell R (2015) Non-small-cell lung cancer. Nat Rev Dis Primers 1:15009

Hariharan D, Srinivasan K, Nehru LC (2017) Synthesis and characterization of Tio 2 nanoparticles using CynodonDactylon leaf extract for antibacterial and anticancer (A549 cell lines) activity. J Nanomed Res 5(6):00138

He Y, Du Z, Ma S, Liu Y, Li D, Huang H, Zheng X (2016) Effects of green-synthesized silver nanoparticles on lung cancer cells in vitro and grown as xenografttumors in vivo. Int J Nanomedicine 11:1879

Howell M, Wang C, Mahmoud A, Hellermann G, Mohapatra SS, Mohapatra S (2013) Dual-function theranostic nanoparticles for drug delivery and medical imaging contrast: perspectives and challenges for use in lung diseases. Drug Deliv Transl Res 3(4):352–363

Jadhav UG (2012) Fabrication of polyherbal nanoparticles—a targetted herbal drug delivery. Pharm Anal Acta 3:121

Jong WHD, Borm PJ (2008) Drug delivery and nanoparticles: applications and hazards. Int J Nanomedicine 3:133–149

Karkan FS, Mohammadhosseini M, Panahi Y, Milani M, Zarghami N, Akbarzadeh A, Davaran S (2016) Magnetic nanoparticles in cancer diagnosis and treatment: a review artificial cells. Nanomed Biotechnol 45(1):1–5

Karmous I, Pandey A, Haj KB, Chaoui A (2019) Efficiency of the green synthesized nanoparticles as new tools in cancer therapy: insights on plant-based bioengineered nanoparticles, biophysical properties, and anticancer roles. Biol Trace Elem Res 196:330–342

Kumar A, MazinderBoruah B, Liang XJ (2011) Gold nanoparticles: promising nanomaterials for the diagnosis of cancer and HIV/AIDS. J Nanomater 2011:1–17

Larsen JE, Minna JD (2011) Molecular biology of lung cancer: clinical implications. Clin Chest Med 32:703–740

Lv PP, Wei W, Yue H, Yang TY, Wang LY, Ma GH (2011) Porous quaternized chitosan nanoparticles containing paclitaxel nanocrystals improved therapeutic efficacy in non-small-cell lung cancer after oral administration. Biomacromolecules 12:4230–4239

Lynch TJ, Bell DW, Sordella R (2004) Activating mutations in the epidermal growth factor receptor underlying responsiveness of non–small-cell lung cancer to gefitinib. N Engl J Med 350:2129–2139

Madamsetty VS, Mukherjee A, Mukherjee S (2019) Recent trends of the bio-inspired nanoparticles in cancer theranostics. Front Pharmacol 10:1264

Mao Y, Liu X (2020) Bioresponsive Nanomedicine: the next step of deadliest cancers' theranostics. Front Chem 8:257

Matsushita H, Morishita R, Nata T (2000) Hypoxia-induced endothelial apoptosis through nuclear factor-kappaB (NF-kappaB)-mediated bcl-2 suppression: in vivo evidence of the importance of NF-kappaB in endothelial cell regulation. Circ Res 86(9):974–981

Menon JU, Jadeja P, Tambe P, Vu K, Yuan B, Nguyen KT (2013) Nanomaterials for photo-based diagnostic and therapeutic applications. Theranostics 3(3):152–166

Mohanraj VJ, Chen Y (2006) Nanoparticles—a review. Trop J Pharm Res 5:561–573

Moorthi C, Manavalan R, Kathiresan K (2011) Nanotherapeutics to overcome conventional cancer chemotherapy limitations. J Pharm Pharm Sci 14(1):67–77

Mottaghitalab F, Farokhi M, Fatahi Y, Atyabi F, Dinarvand R (2019) New insights into designing hybrid nanoparticles for lung cancer: diagnosis and treatment. J Control Release 295:250–267

Mukherjee A, Paul M, Mukherjee S (2019) Recent progress in the theranostics application of nanomedicine in lung cancer. Cancers 11(5):597

Muthuraj B, Mukherjee S, Patra CR, Iyer PK (2016) Amplified fluorescence from Polyfluorene nanoparticles with dual state emission and aggregation caused red shifted emission for live cell imaging and cancer theranostics ACS. Appl Mater Interf 8:32220–32229

Narsireddy A, Anish B, Ranganayaki M, Anupama M, Rajagopal R (2017) Polymeric nanoparticle-mediated gene delivery for lung cancer treatment. Top Curr Chem 375(35):1–23

Okamoto H, Nishida S, Todo H, Sakakura Y, Iida K, Danjo K (2003) Pulmonary gene delivery by chitosan–pDNA complex powder prepared by a supercritical carbon dioxide process. J Pharm Sci 92:371–380

Padmini R, Razia M, Roopavahini R, Aruna S (2016) Molecular docking studies of Allylsulfur compounds from Allium sativum against EGFR receptor. Int J Curr Res Acad Rev 3:207–218

Padmini R, Razia M, Sitrarasi R (2017) Molecular docking studies of bioactive compounds from Allium sativum against EML4-ALK receptor. Res J Pharm Technol 10(11):3741–3747

Padmini R, Uma Maheshwari V, Arun M, Razia M, Anand K, Ravindran B, Chung WJ (2020) Myricetin: versatile plant based flavonoid for cancer treatment by inducing cell cycle arrest and ROS–reliant mitochondria-facilitated apoptosis in A549 lung cancer cells and in silico prediction. Mol Cell Biochem 476(1):57–68

Papp T, Schipper H, Pemsel H, Bastrop R, Muller KM, Wiethege T, Weiss DG, Dopp E, Schiffmann D, Rahman Q (2001) Mutational analysis of N-ras, p53, p16INK4a, p14ARF and CDK4 genes in primary human malignant mesotheliomas. Int J Oncol 18(2):425–433

Parvanian S, Mostafavi SM, Aghashiri M (2017) Multifunctional nanoparticle developments in cancer diagnosis and treatment. Sens Bio-Sens Res 13:81–87

Parvathi PV, Mahalingam U, Raj BR (2015) Improved waste water treatment by biosynthesized graphene sand composite. J Environ Manag 162:299e305

Purushotham A, Bains S, Lewison G, Szmukler G, Sullivan R (2013) Cancer and mental health—a clinical and research unmet need. Ann Oncol 24(9):2274–2278

Que YM, Fan XQ, Lin XJ, Jiang XL, Hu PP, Tong XY, Tan QY (2019) Size dependent anti-invasiveness of silver nanoparticles in lung cancer cells. RSC Adv 9(37):21134–21138

Rafique M, Sadaf I, Rafique M, Tahir MB (2017) A review on green synthesis of silver nanoparticles and their applications. Artif Cells 7:1–20

Rani D, Somasundaram VH, Nair S, Koyakutty M (2012) Advances in cancer nanomedicine. J Indian Inst Sci 92:187–218

Ratan ZA, Haidere MF, Nurunnabi M, Shahriar SM, Ahammad AJS, Shim YY, Cho JY (2020) Green chemistry synthesis of silver nanoparticles and their potential anticancer effects. Cancers 12(4):855

Rezaie AB, Montazer M, Rad MM (2017) Photo and biocatalytic activities along with UV protection properties on polyester fabric through green in-situ synthesis of cauliflower-like CuO nanoparticles. J Photochem Photobiol B Biol 176:100–111

Sattler KD (2010) Handbook of nanophysics: nanomedicine and nanorobotics. CRC Press, Boca Raton

Sholl LM (2014) Large-cell carcinoma of the lung: a diagnostic category redefined by immunohistochemistry and genomics. Curr Opin Pulm Med 20:324–331

Sivasankar M, Kumar BP (2010) Role of nanoparticles in drug delivery system. Int J Pharm Biomed Res 1:41–66

Skocaj M, Filipic M, Petkovic J, Novak S (2011) Titanium dioxide in our everyday life; is it safe? Radiol Oncol 45(4):227–247

Stinchcombe TE (2014) Unmet needs in squamous cell carcinoma of the lung: potential role for immunotherapy. Med Oncol 31:960

Sukumar UK, Bhushan B, Dubey P, Matai I, Sachdev A, Packirisamy G (2013) Emerging applications of nanoparticles for lung cancer diagnosis and therapy. Int Nano Lett 3:45

Sun B, Hu N, Han L, Pi Y, Gao Y, Chen K (2019) Anticancer activity of green synthesised gold nanoparticles from Marsdeniatenacissima inhibits A549 cell proliferation through the apoptotic pathway. Artif Cells Nanomed Biotechnol 47:4012–4019

Tran S, DeGiovanni PJ, Piel B, Rai P (2017) Cancer nanomedicine: a review of recent success in drug delivery. Clin Transl Med 6(1):44

Travis WD (2012) Update on small cell carcinoma and its differentiation from squamous cell carcinoma and other non-small cell carcinomas. Mod Pathol 25:S18–S30

Vandermeers F, Hubert P, Delvenne P, Mascaux C, Grigoriu B, Burny A, Willems L (2009) Valproate, in combination with pemetrexed andcisplatin, provides additional efficacy to the treatment of malignant mesothelioma. Clin Cancer Res 15:2818–2828

Wang T, Jiang H, Wan L (2015) Potential application of functional porous tio2 nanoparticles in light-controlled drug release and targeted drug delivery. Acta Biomater 13:354–363

Wang C, Mathiyalagan R, Kim YJ, Castro-Aceituno V, Singh P, Sungeun A, Wang D, Yang DC (2016) Rapid green synthesis of silver and gold nanoparticles using Dendropanaxmorbifera leaf extract and their anticancer activities. Int J Nanomedicine 11:3691–3701

Yip PY, Yu B, Cooper WA (2013) Patterns of DNA mutations and ALK rearrangement in resected node negative lung adenocarcinoma. J Thorac Oncol 8:408–414

Yu Z, Pestell TG, Lisanti MP, Pestell RG (2012) Cancer stem cells. Int J Biochem Cell Biol 44:2144–2151

Youssef Z, Vanderesse R, Colombeau L, Baros F, Roques-Carmes T, Frochot C, Gazzali AM (2017) The application of titanium dioxide, zinc oxide, fullerene, and graphene nanoparticles in photodynamic therapy. Cancer Nanotechnol 8(1):6

Yusefi M, Shameli K, Ali RR, Pang SW, Teow SY (2019) Evaluating anticancer activity of plant-mediated synthesized Iron oxide nanoparticles using PunicaGranatum fruit Peel extract. J Mol Struct 1204:127539

Zheng Y, Zhang J, Zhang R, Luo Z, Wang C, Shi S (2019) Gold nano particles synthesized from Magnolia officinalis and anticancer activity in A549 lung cancer cells. Artif Cells Nanomed Biotechnol 47:3101–3109

Zugazagoitia J, Guedes C, Ponce S, Ferrer I, Molina-Pinelo S, Paz-Ares L (2016) Current challenges in cancer treatment. Clin Ther 38(7):1551–1566

Chapter 10
Therapeutic Applications of Nanotechnology in the Prevention of Infectious Diseases

Rajkumari Mazumdar and Debajit Thakur

Contents

10.1	Introduction	324
10.2	Role of Nanotechnology in the Treatment of Infectious Diseases	325
10.3	Nanomaterials in Nanomedicine	325
10.4	Synthetic Nanomedicines	326
	10.4.1 Silver Nanoparticles	326
	10.4.2 Carbon-Based Nanoparticles	328
10.5	Gold Nanoparticles	331
10.6	Biological-Based Nanomedicines	332
	10.6.1 Chitosan-Based Nanoparticles	332
	10.6.2 Poly-L-Lactide Nanoparticles	333
10.7	Nanoparticle Vaccines Against Infectious Diseases	334
10.8	Types of Nano-Immuno Activators	335
	10.8.1 Liposomes	335
	10.8.2 VLPs (Virus like Particles)	336
	10.8.3 Dendrimers	336
10.9	Conclusion	338
References		339

R. Mazumdar
Microbial Biotechnology Laboratory, Life Sciences Division, Institute of Advanced Study in Science and Technology (IASST), Department of Science and Technology, Ministry of Science and Technology (India), Guwahati, Assam, India

Department of Molecular Biology & Biotechnology, Cotton University, Guwahati, Assam, India

D. Thakur (✉)
Microbial Biotechnology Laboratory, Life Sciences Division, Institute of Advanced Study in Science and Technology (IASST), Department of Science and Technology, Ministry of Science and Technology (India), Guwahati, Assam, India

10.1 Introduction

The word nanotechnology originates from the Greek word *nano*, meaning dwarf. It applies the ethics of engineering and manufacturing at the infinitesimal level. Nanoscience is an interdisciplinary field of research about the study of minute particles on a molecular scale, where the dimension is measured in nanometers. A nanometer means one billionth (i.e., 10–9 m) of a meter. Nanotechnology focuses mainly on how materials respond and function at the atomic, subatomic, or molecular levels (i.e., nanoscale). Nanotechnology involves the handling of the matter in the 1–100 nanometer scale range where atoms are approximately one-third of a nanometer. Thus, particles with a range and dimensions of this size are nanoparticles. The field of medicine, in which nanoscaled particles with dimensions ranging from 1 to 100 nm are built as biomedical tools for research, has been transformed by the use of nanotechnology in numerous therapeutic sectors. In recent years, the widespread and improper use of antimicrobial agents by the residential population has inevitably culminated in the production of pathogenic microorganisms with antibiotic resistance as an alternative to survival strategies. The commercially available antibiotics which are widely used against pathogens demonstrate huge changes in their resistivity. It is therefore well understood that the identification and discovery of new antimicrobial agents are essential to fight against these highly drug-resistant pathogens. The concern regarding the antibiotic resistance of pathogenic microorganisms requires the development of new antimicrobial agents to combat infectious diseases. The evolution of infectious diseases and the emergence of drug-resistant pathogenic microorganisms at an unprecedented pace are a subject of huge concern. Despite the substantial understanding of the pathogenesis of microorganisms and the use of modern therapies, morbidity and fatality connected with microbial infections remain high (Kolar et al. 2001). A significant number of microorganisms causing infectious diseases are known to become multidrug-resistant (MDR), including *Mycobacterium tuberculosis, A. baumannii, Burkholderia cepacia, Escherichia coli, Enterobacter* spp., *Campylobacter jejuni, Enterococcus faecium, Enterococcus faecalis, Haemophilus influenzae, Proteus mirabilis, Klebsiella pneumoniae, Pseudomonas aeruginosa, Salmonella* spp., *Staphylococcus epidermidis, Staphylococcus aureus*, and *Streptococcus pneumoniae*. All these microorganisms are also called "superbugs" because of their high mutation rate, and they cause a high degree of antimicrobial resistance, amplified virulence, and low susceptibility (Davies and Davies 2010). Hence, there is an urgent need to identify novel methods and new antimicrobial agents and formulations that contemplate the use of nanotechnology to create the next wave of drugs to control microbial infection (Prasad et al. 2019). Nanoparticles are being studied extensively because of their chemical and physical properties. In the field of medicine, nanoparticles are widely studied because of their size-dependent physicochemical properties. Nanoparticles are comparable in scale to those of most biomolecule structures. This makes them a convincing candidate for application in biomedical research. The findings of their implementation in the branch of medicine have contributed their application mainly in guided drug delivery, bioimaging, and biosensing. Their function as potent antimicrobial agents to target multidrug-resistant pathogens is another fascinating avenue for their medical research (Saglam et al. 2021; Maddela et al. 2021). But, for the application of nanoparticles in the biological study, biocompatibility is the most desirable trait. Biocompatibility is the ability of the substance to function therapeutically without unintended local or systemic consequences (Samia et al. 2006).

10.2 Role of Nanotechnology in the Treatment of Infectious Diseases

Infectious diseases caused by microbes such as bacteria, fungi, viruses, and parasites are responsible for over 15 million deaths worldwide (Panacek et al. 2009). In the latest days, the extensive use of antimicrobial agents by the resident community has undoubtedly led to the emergence of antibiotic-resistant pathogens as an alternative to survival mechanisms and restricts the effectiveness of existing infection treatments and thus faces a significant challenge in the fight against infectious diseases. Recent advancements in nanotechnology allow us to tackle this problem at two levels: diagnostics and therapy. Many new antimicrobial possibilities have been created with the advent of nanotechnology. Nanomaterials present a potential forum for alternate prevention strategies to control infectious diseases. However, relative to other typical antimicrobial agents with short-term action, they deliver sustained antimicrobial activity with minimal toxicity. Nanomaterials provide additional advantages due to their small size, allowing for an improved ability to overcome any physiological hurdles to meet their specific target. The high ratio of surface area to volume allows them for the boosted potential to communicate with the pathogen membranes and cell walls. The most important prerequisite for any successful therapeutic agent is the delivery of medications to the required target at the appropriate doses for the appropriate time. Hence, the nanomaterials serve as one of the most exciting smart drug delivery systems by using the outstanding size and surface properties.

10.3 Nanomaterials in Nanomedicine

Nanomaterials are enormously smaller in scale (dimensions from 1.0 to 100.0 nm), having 100 nm or less in at least one dimension. Nanomaterials may be one-dimensional (such as surface films), two-dimensional (such as wires or fibers), or three-dimensional (such as particles). Nanotechnology uses nanomaterials (nanoparticles, nanofibers, nanotubes, etc.) with novel physical and chemical properties such as their small scale, low toxicity, high biocompatibility, and ability to selectively penetrate the cell membrane of pathogenic microorganisms. These exceptional properties of nanoparticles could offer significant advantages to nanomedicine. For example, the small size of a nanoparticle can encourage them to cross biological barriers; various nanoparticle architectures can improve the bioavailability of non-soluble or unstable drugs; the configurable surface of nanoparticles can enable the required targeting potential for either imaging or precise delivery of drugs to the diseased region. Nanomedicine is a field of science that uses nanoscale materials for the diagnosis and treatment of human disease. It has arisen as an integral part of therapeutics. By developing new drugs for both diagnosis and care, nanomedicine has made promising advances in clinical practice, allowing it to resolve unmet patient needs through diagnosis, controlled drug release, surveillance, and control; increasing bioavailability, prevention and treatment of diseases, and drug targeting; identifying and managing diseases and analgesic effects; and improving human health. Many drug delivery carriers have been built over time, based on various nanomaterials. Specifically, antimicrobial agents are being placed into polymer-coated crystalline nanoparticles, PLGA, composite hydrogel/glass particles, liposomes, homogenized particulate suspensions, cationic, dendrimers, gold nanoparticles, and silver nanoparticles and are developed for the treatment of a wide variety of microbial infections along with drug-resistant tuberculosis. Nanocarriers have developed a novel vehicle for the delivery of therapeutic agents specifically for the target. Recently, these nanoparticles have arisen as an appealing candidate for the delivery of medications as well as the diagnosis and treatment of many infectious diseases.

Nanomedicine can be divided into two main groups: synthetic nanomedicines and biological-based nanomedicines. The first group consists of nanomedicine derived from synthetic or inorganic materials, and the second group consists of materials based on nanoscale-engineered biological or organic components (shown in Fig. 10.1).

10.4 Synthetic Nanomedicines

10.4.1 Silver Nanoparticles

Silver was most often studied as an antimicrobial agent, and it is used since ancient times to counter microbial infections. Silver nanoparticles (AgNPs) are frequently used as an antifungal (Medda et al. 2015), anti-inflammatory (Hebeish et al. 2014), and antiviral agents (Bekele et al. 2016). Researchers have shown that AgNPs are used as carriers to deliver small drug molecules or large biomolecules to particular targets. They are very target specific. Once the AgNPs have reached their specific target, the release of the drug could potentially be triggered by an internal or external stimulus. Targeted and stored nanoparticles could provide high concentrations of drugs at particular target sites and minimize side effects. Due to the emergence of drug-resistant pathogens which are crass to traditional drugs, the exceptional

10 Therapeutic Applications of Nanotechnology in the Prevention of Infectious Diseases

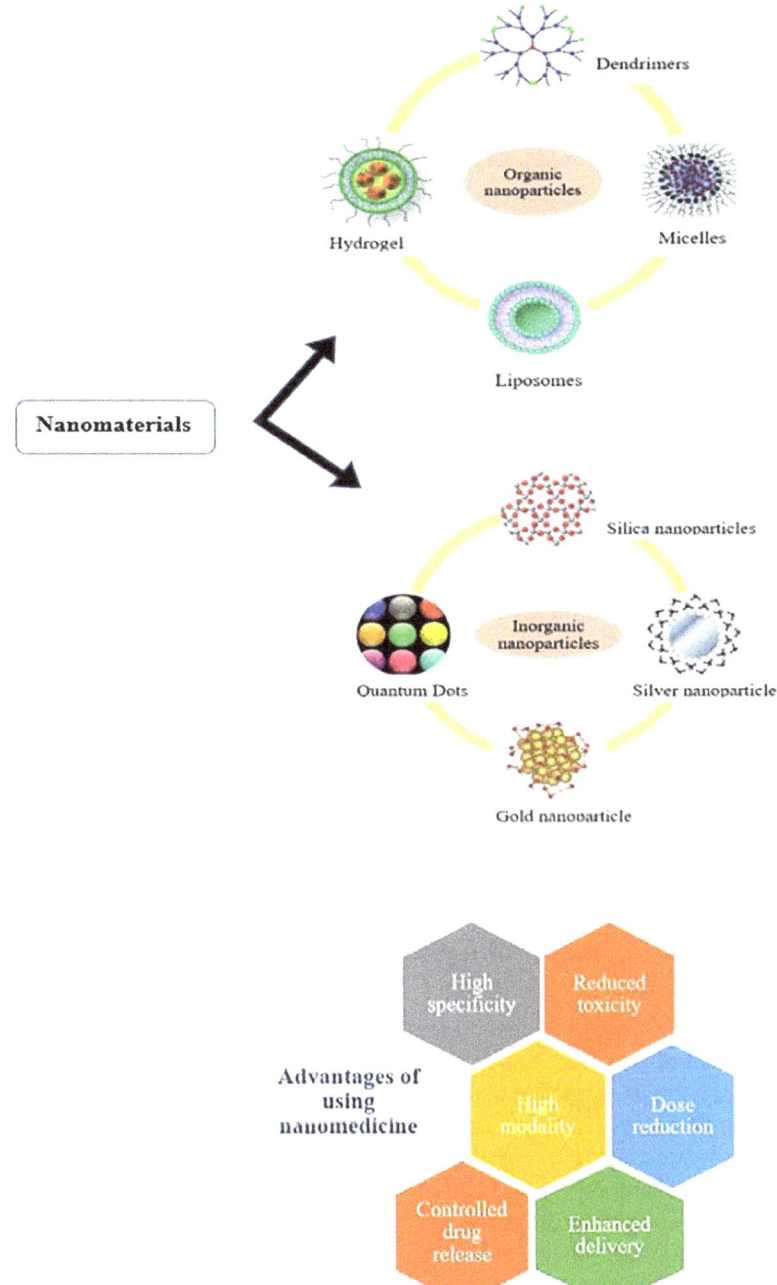

Fig. 10.1 Various types of nanoparticles and their advantages in nanomedicine

properties and broad function of silver and their compounds have prompted scientists from all over the world to investigate their potent antimicrobial behavior. AgNPs showed promising antimicrobial activity against various microorganisms because of their large surface area to volume ratio. These properties are used as bactericides in burned lesions and fillers in dental cavities to deter infection. AgNPs alone or with the combination of other typical antimicrobial agents have been shown to have antimicrobial efficacy against a broad variety of microorganisms. Nano formulated drugs are smart drugs. They are engineered to distribute medications directly to the intended infection sites and in areas of the body that are often impossible to access through available therapies. AgNPs have been reported to display antifungal activity by a mechanism called apoptosis (Hwang et al. 2012). Aggregation of intracellular reactive oxygen species (ROS) was commonly documented as a key feature in cell death and has been related with many of the recognized apoptotic pathways in yeast (Carmona-Gutierrez et al. 2010). AgNP-induced apoptosis cells undergo unique phenomena including ROS production, DNA fragmentation, caspase activation, phosphatidylserine exposure, cytochrome c release, mitochondrial membrane depolarization, nuclear condensation, and cell cycle arrest (Aziz et al. 2014, 2015, 2016, 2019; Bortner and Cidlowski 2007; Prasad and Swamy 2013; Prasad 2014). The injection of silver into bacterial cells triggers a high level of structural and morphological modifications that can contribute to the death of cells. Till now, three main mechanisms for its antibacterial mode of action have been described: (i) dissociation of cell wall and membrane, (ii) intercellular diffusion and damage, and (iii) oxidative stress (Dakal et al. 2016; Duran et al. 2016) (shown in Fig. 10.2). The synergy of AgNPs with bacterial cells allowed the cell wall and cell membrane to disturb and demolish, resulting in the shrinkage of the cell membrane and the leakage of the intercellular products and eventually the death of the bacteria (Abalkhil et al. 2017).

10.4.2 Carbon-Based Nanoparticles

Invariably, the excessive and inappropriate use of antimicrobial agents has led to the establishment of antibiotic resistance. This has been a big concern in recent years. Antibiotics currently available that are widely used against microbial infection display dramatic changes in their resistivity. Thus, the discovery of novel antimicrobial agents for the treatment of these infections is urgently required. In biological studies related to antimicrobial agents, drug delivery, and biosensing, carbon nanostructures are successfully implemented (Bitounis et al. 2013). In particular, carbon-based nanomaterials have shown significant antibacterial properties. Carbon nanotubes and fullerenes are carbon-based nanoparticles. Many biomedical applications of carbon-based nanoparticles have resulted in the development of a new field in diagnostics and therapeutics. The fact that carbon-based nanoparticles exhibit good antimicrobial properties has been well known in the last few years. The size, shape, and surface area are the most important parameters affecting their antimicrobial activity. The mechanism for the activity of interaction with bacterial cells can

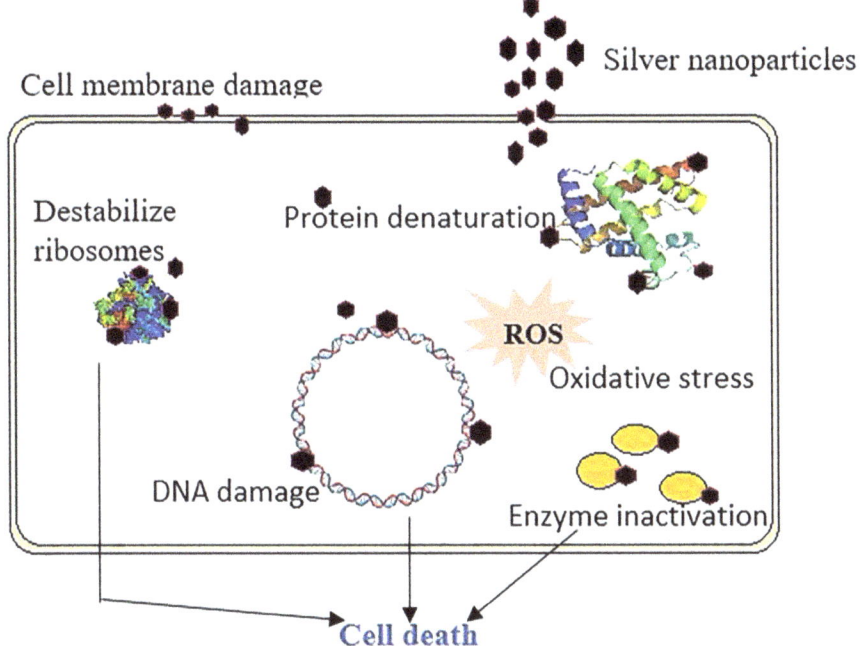

Fig. 10.2 Mechanism of antimicrobial action of silver nanoparticle. (Described by Dakal et al. 2016 and Duran et al. 2016)

be enhanced by increasing the nanoparticle's surface area by decreasing its size (Buzea et al. 2007). In the beginning, it has been revealed that the interaction of carbon-based nanoparticles with bacterial causes damage to the cell membrane due to oxidative stress (Shvedova et al. 2012; Vecitis et al. 2010). But, according to some recent studies, the main cause of the antimicrobial behavior of these nanostructures is the functional association of carbon-based nanomaterials with bacteria instead of oxidative stress (Pacurari et al. 2012; Manke et al. 2013).

10.4.2.1 Carbon Nanotubes

Carbon nanotubes (CNTs) are hollow barrel-shaped nanotubes. They are classified into single-walled carbon nanotubes (SWCNTs) and multiwalled carbon nanotubes (MWCNTs) based on the number of concentric panels. SWCNTs vary in diameter from 0.4 to 2 nm, and MWCNTs vary from 10 to 100 nm. Lijima and Lchihashi, in 1991, synthesized SWCNTs. The antimicrobial activity of SWCNTs on *E. coli* was first described by Kang et al. (2008). In another study (2008), they stated that the size of carbon nanotubes was a significant aspect of exhibiting antibacterial property. In particular, the length of nanotubes during interactions with the cell membrane is crucially significant. In contrast to longer tubes, the shorter tube has better

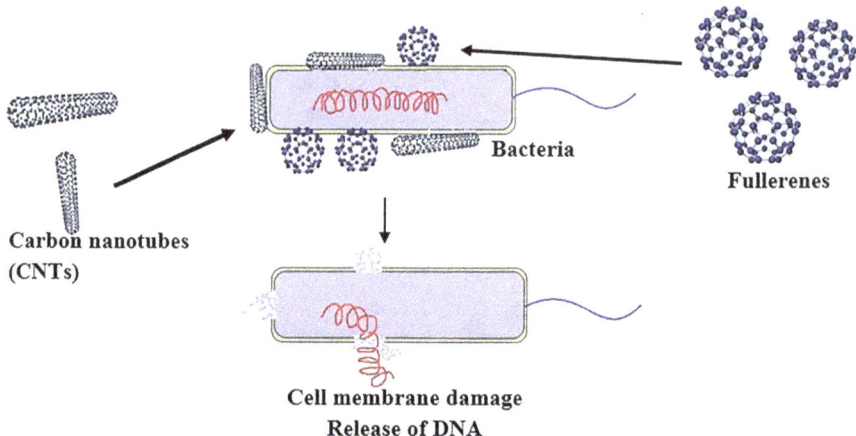

Fig. 10.3 Mechanism of antagonistic activity of carbon nanotubes and fullerenes. (Mechanism described by Kang et al. (2008), Tegos et al. (2005), and Cataldo and Da Ros (2008))

bactericidal efficiency. They have the potential to communicate best with open ends of nanotubes and a microbe, resulting in increased damage to the cell membrane (Aslan et al. 2010). They prepared and tested the antimicrobial effect of SWCNTs and MWCNTs against *E. coli*. Their findings confirmed that SWCNTs were more noxious than MWCNTs to bacteria. Thus, they reported that SWCNTs could be a better choice of antibacterial agents based on their findings. Arias and Yang (2009) tested the antimicrobial efficacy of SWCNTs and MWCNTs against rod-shaped or round-shaped Gram-negative and Gram-positive bacteria with different surface groups. They reported that SWCNTs showed potent antibacterial activity for both Gram-positive and Gram-negative bacteria with surface groups of -OH and -COOH, whereas MWCNTs with the same surface groups did not display any convincing antibacterial activity. Their findings revealed that CNT binding to bacterial cells triggers the cell membrane to disrupt and then release its DNA material, eventually contributing to the death of the bacteria (shown in Fig. 10.3). Moreover, the surface area of SWCNTs proposed improved synergy with the microbial cell wall.

When the bacterial cells come in direct contact with SWCNTs, it influenced the morphology, metabolism processes, and cellular membrane integrity of *E. coli*. SWCNTs easily penetrate the bacterial cell wall due to their smaller nanotube diameter causing severe damage to the cell membrane.

10.4.2.2 Fullerenes

Fullerenes is an allotrope of carbon, composed of carbon atoms connected by single and double bonds. Their design is much like a soccer ball. In discovering new antimicrobial agents for different infectious diseases, the antimicrobial role of fullerenes and their derivatives received considerable interest. They are also known for demonstrating antagonistic action against various bacteria, such as *E. coli*, *Streptococcus* spp., and *Salmonella* (Tegos et al. 2005). In another study, they mentioned the antimicrobial potency of fulleropyrrolidinium salts after photoirradiation. Their reports revealed that 99.9 percent of bacterial and fungal cells were dead after treatment. After white light irradiation, the cationic-substituted fullerene derivative is extremely efficient in destroying a broad range of microbial cells (Mizuno et al. 2011). They investigated a new category of synthetic fullerene derivatives with prominent antimicrobial activity, carrying either basic or quaternary amino groups. They proposed that in the treatment of superficial infections, for example, in burns and cuts, where light infiltration into tissue is not difficult, quaternized fullerenes can be used efficiently. A bactericidal mechanism that explains the induction of cell membrane disruptions was stated by Cataldo and Da Ros (2008). The hydrophobic surface of the fullerenes can conveniently unite with membrane lipids and intercalate into them. Among three separate groups of fullerene compounds (cationic, acidic, and anionic), cationic derivatives displayed the highest antimicrobial activity on E. coli and Shewanella oneidensis, while the anionic derivatives were nearly inactive. This is due to the robust synergy of cationic fullerenes with the negatively charged bacteria (Nakamura and Mashino 2009).

10.5 Gold Nanoparticles

In recent years, gold nanoparticles (AuNPs) have appeared as exclusive noninvasive drug carriers to target drugs at their site of action. Their site sensitivity has tended to improve the potency of medications at a reduced dosage and reduce their side effects (Kumari et al. 2019). In the field of biomedical research and diagnostics, AuNPs have become an important factor. The unique physicochemical properties of AuNPs make them a superior candidate for an antimicrobial agent (Allaker et al. 2012). The antimicrobial action of gold nanoparticles can be due to the formation of reactive oxygen species (ROS) that increases the oxidative stress of microbial cells. Photothermal therapy (PTT) can be used to improve the antimicrobial efficacy of AuNPs. Gold nanoparticles can produce heat because of the excitation of electrons when irradiated with a laser. This makes it easier to use them as anticancer or antimicrobial agents. For example, AuNPs have boosted the bactericidal effect on *S. aureus* when exposed to laser energy (Riley and Day 2017). The antifungal activity of AuNPs against *Candida* isolates was also reported by some researchers (Ahmad et al. 2013) (Table 10.1).

Table 10.1 Antimicrobial nanomedicines

Nanomedicine	Mechanism of action	Antimicrobial activity
Synthetic antibacterial nanomedicine		
Carbon nanotubes and fullerenes	Cell membrane damage, the release of DNA content, fullerenes can make radical-oxygen species	*E. coli* DH5α, *Vibrio fischeri*, *Bacillus subtilis*
Silver nanoparticles	Cell membrane damage by accumulation, DNA damage, protein denaturation	*E. coli, P. aeruginosa, S. aureus*
Gold nanoparticles	Disruption of the cell membrane	*E. coli, Salmonella typhimurium*
Bioactive glasses	Production of alkaline species	*E. faecalis*
Metal oxide nanoparticles	Electrostatic synergy between nanoparticle and bacteria	*E. coli, B. subtilis, S. aureus*
Magnesium oxide nanoparticles	Formation of superoxide anions	*Bacillus subtilis, S. aureus*
Zinc oxide nanoparticles	Introduction of oxidative stress	*C. jejuni, Salmonella enterica serovar Enteritidis, E. coli* O157:H7
Silicon dioxide nanoparticle	Discharge of nitric oxide	*S. aureus*
Biological-based antibacterial Nanomedicines		
Chitosan nanofiber	Loss of membrane permeability	*E. coli, S aureus*
Targeted drug-carrying phage medicines	Delivery of antimicrobial agents in the target pathogen	*S aureus, Streptococcus pyogenes, E. coli*
Poly-L-lactide nanoparticles	Release of antimicrobial protein nisin	*Lactobacillus delbrueckii*

Yacoby and Benhar 2008; Matthews et al. 2010; Hajipour et al. 2012

10.6 Biological-Based Nanomedicines

10.6.1 Chitosan-Based Nanoparticles

Chitosan (C) is a semisynthetic linear polysaccharide comprising of randomly assigned β-(1 → 4)-linked D-glucosamine (deacetylated unit) and *N*-acetyl-D-glucosamine (acetylated unit) (Prabaharan 2008). They are developed commercially by chitin deacetylation, which is the principal component in the exoskeleton of crustaceans. The techniques used for the synthesis of chitosan-based nanoparticles are ionotropic gelation, emulsification solvent diffusion, microemulsion, and emulsion-based solvent evaporation (Mohammed et al. 2017), as shown in Fig. 10.4. Chitosan has fascinated numerous biomedical and pharmaceutical industries because of its biodegradability and biocompatibility, in addition to its mucoadhesive and bacteriostatic activity (Nagpal et al. 2010). Chitosan nanoparticles (NPs) are widely studied as carriers of medications, proteins, and genes and have been used as a drug delivery vehicle in polymeric nanoparticle via multiple routes of

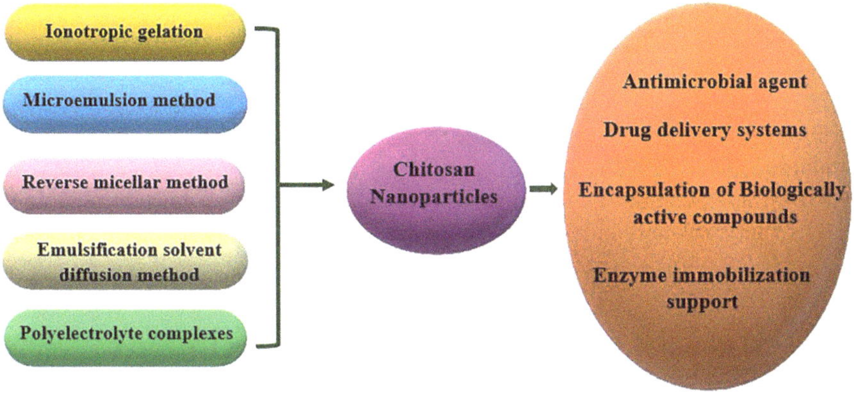

Fig. 10.4 Different methods of preparation and biomedical applications of chitosan nanoparticles

administration (Rampino et al. 2013). Chitosan and chitosan derivative NPs hold a cationic charge on their surface and mucoadhesive features that can bind to mucus membranes and release the drug consignment, in a continuous release fashion (Mohammed et al. 2017). It has an antibacterial and antifungal function as well (Ducheyne et al. 2011). The antibacterial activity of the electrospun nanofiber mats developed from chitosan was reported by Ignatova et al. (2006). The association of protonated quaternized chitosan (QCh) mats with the negatively charged bacterial surface results in the lack of bacterial membrane permeability, cell leakage, and eventually cell death. The QCh mats showed bactericidal characteristic on Staphylococcus aureus and Escherichia coli. Electrospun mats are useful as wound dressings because they can reduce the risk of bacterial infection. Chitosan-based NPs have all the required properties for quick, nontoxic wound healing, hemostatic activity, biodegradability, and the power to affect the function of macrophages.

10.6.2 Poly-L-Lactide Nanoparticles

Poly (L-lactic acid) or (PLLA) is a linear aliphatic thermoplastic polyester produced from sustainable resources (Wang et al. 2012). PLLA has attracted considerable attention because of the combination of its biodegradable property, bioresorbability, biocompatibility, and shape memory effect. It has been broadly practiced in biomedical fields such as bone screws, surgical sutures, tissue engineering, and controlled drug delivery (varkey 2019). Also, its low consumption of moisture and great wicking may benefit the appropriate exudates from the wound (Gupta et al. 2007; Davachi and Kaffashi 2015). In the medicine world, antimicrobial products and biocides are commonly used to kill or inhibit the development of pathogenic and other harmful microbes. Therefore, anti-infection alterations of polymers are introduced to suppress the development of certain microorganisms. Three approaches can accomplish the anti-infective properties of polymers, specifically anti-infective factors combined with polymers, copolymerization of anti-infective factors with monomers, and adequate surface treatment of medical polymers (Popelka et al. 2012; Kugel et al. 2012). A variety of studies have been performed on the antimicrobial efficacy of PLLA in different formulations, including neat, blended, and composite, and also using various types of antimicrobial agents, along with nisin, Nisaplin, and nano-silver, under a broad dimension of test conditions (Liu et al. 2009; Prapruddivongs and Sombatsompop 2012).

10.7 Nanoparticle Vaccines Against Infectious Diseases

Recently, preventing and immunizing infectious diseases has become a big problem due to the proliferation of numerous novel variants of pathogenic and drug-resistant microbes. As reported by the World Health Organization (WHO) in 2016, there were nearly 3.2 million deaths globally from lower respiratory or lung infections and 1.4 million deaths from tuberculosis alone. No effective medicines against these infectious diseases are available which has become a major obstacle in combating those diseases. The production of the desired vaccines against these diseases is urgent. In vaccine production, nanotechnology plays a major role. Vaccines based on nanotechnology are more effective than traditional vaccines. Any of the essential facets of any regular vaccine include (i) protection, (ii) security, and (iii) ability to provide a long-lasting and sufficient immune response at a minimum dose level (Atkins et al. 2006; Beverley 2002). For the production of new-generation vaccines, nanotechnology-based formulations provide various benefits. In addition to shielding the vaccines from premature degradation, the nanocarrier-based delivery mechanism also offers increased antigen stability, good adjuvant properties, targeted delivery, and gradual release of the drug and helps to deliver an immunogen to the antigen-presenting cells (APCs) (Pati et al. 2018). Several nanoparticle-based therapeutic methods have been introduced recently to monitor the function of T cells against viral, bacterial, or fungal infections. Effective transmission of antigens to APCs, particularly in dendritic cells (DCs), and APC activation are some important factors in the production of successful vaccines. It is now conceivable to target delivery to DCs, APCs activation, and monitor the release of the antigen by nanoparticle-based vaccine delivery systems. In the presence of co-stimulatory compounds and cytokines, the synergy between MHC I and T-cell receptors (TCR) kills the infected cells by triggering cytotoxicity. It provides a cellular immune response. Antigens are also expressed on the surface of the APC by class II MHC molecules to the (CD4+). Subsequently, helper T CD4 + cells stimulate B cells and thus produce antibodies (Pati et al. 2018) (shown in Fig. 10.5). It is responsible for humoral immunity.

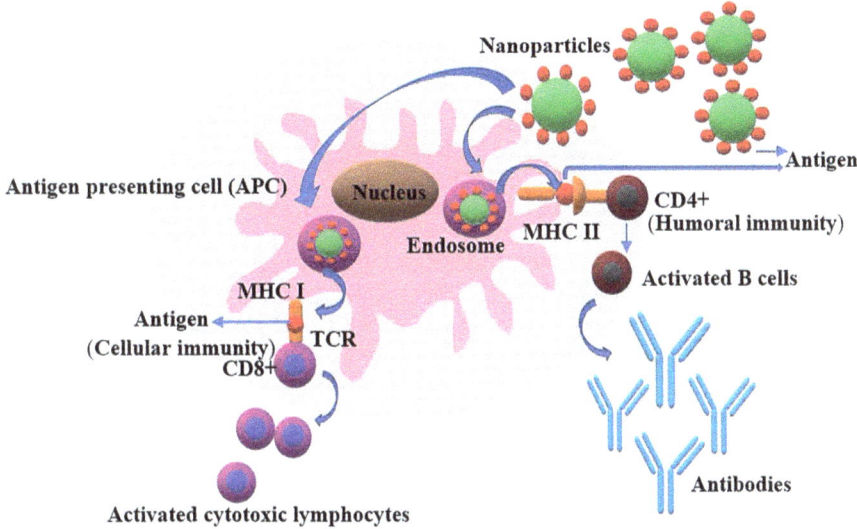

Fig. 10.5 Targeted delivery of antigenic molecules into the antigen-presenting cells (APCs) using surface designed nanoparticles. (Mechanism described by Pati et al. 2018)

10.8 Types of Nano-Immuno Activators

10.8.1 Liposomes

Liposomes are spherical sac of phospholipid molecules ranging from 20 to 30 nm in size. It can imitate cell membranes and directly fuse with microbial membranes. They are the most broadly discovered vaccine and drug delivery tool in nanomedicine. Liposomes can be produced by lipid hydration resulting in the formation of lipid bilayer around the aqueous core (Sharma and Sharma 1997). They are predominantly composed of biodegradable phospholipids (e.g., cholesterol, phosphatidylcholine, and phosphatidylserine). Liposomes carry vaccines by combining them with the target cell membrane (Tyagi et al. 2012). They can encapsulate both hydrophilic and hydrophobic compounds. The aqueous core can accumulate hydrophilic particles, while the phospholipid bilayer encloses hydrophobic particles. According to some studies, the delivery of antigenic proteins trapped in multilamellar lipid vesicles produces active T- and B-cell responses (Moon et al. 2011). Antigenic peptides coupled to phosphatidylserine (PS) liposomes are internalized by APCs to develop T-helper cell-mediated immune responses (Ichihashi et al. 2013), and distribution of heat shock protein-encoding DNA vaccine using liposomes produced active immunity against fungal infections.

The most advantage of liposomes in nanomedicine is that they can not only deliver large amounts of drug payloads but also protect their encapsulated drugs from degradation, premature inactivation, and dilution in circulation and can be constructed in different forms for various pathways of administration.

10.8.2 VLPs (Virus like Particles)

VLPs are complexes between 30 and 90 nm consisting of self-assembled viral proteins with no nucleic acid genome or lipid envelope. They are more like empty viruses with intact protein hulls and, if desired, membrane envelopes but devoid of genetic material that makes them safe for human use (Naskalska and Pyrc 2015). VLPs contain repeated, high-density distributions of viral surface proteins that display conformational viral epitopes capable of producing robust T-cell and B-cell immune responses (Akahata et al. 2010). Nonprotein antigens and small organic molecules can also be chemically coupled to the viral surface to create VLP bioconjugates (Maurer et al. 2005). Consequently, VLPs can defend against viruses and heterologous antigens (Grgacic and Anderson 2006) and boost the immunogenicity of weak antigens as well. *Salmonella typhi* membrane antigen, influenza A M2 protein, and H1V1 Nef gonadotropin-releasing hormone (GnRH)-mounted VLPs produce boosted antigen-specific humoral and cellular immune responses (Gao et al. 2018).

10.8.3 Dendrimers

Dendrimers contain a blend of amines and amides and with nanostructures that are hyperbranched, three-dimensional, and monodispersed (Pati et al. 2018). Because of their features, such as hyperbranched, well-defined globular structures, excellent structural uniformity, multivalence, variable chemical composition, and robust biological compatibility, dendrimers hold enormous potential for biomedical applications (Kesharwani et al. 2014) as shown in Fig. 10.6. However, the application of dendrimers as scaffolds of prodrugs is chiefly interesting. Therefore, dendrimer has been considered as an efficient bioactive delivery vehicle because of their distinctive biological characteristic such as high drug payload, lipid bilayer interactions, targeting potential, blood plasma retention time, filtration, intracellular internalization, biodistribution, transfection, and strong colloidal and biological constancy (Hawker and Frechet 1990; Kumar et al. 2010).

A single dose of a dendrimer with many antigens produces strong antibody and T-cell responses against the Ebola virus, H1N1 influenza, and *Toxoplasma gondii* (Chahal et al. 2016). Irrespective of their toxicity, dendrimers have been referred as "smart" drug carriers because of their ability as intracellular drug delivery vehicles to cross biological barricades, to distribute in the body during the time required to

10 Therapeutic Applications of Nanotechnology in the Prevention of Infectious Diseases

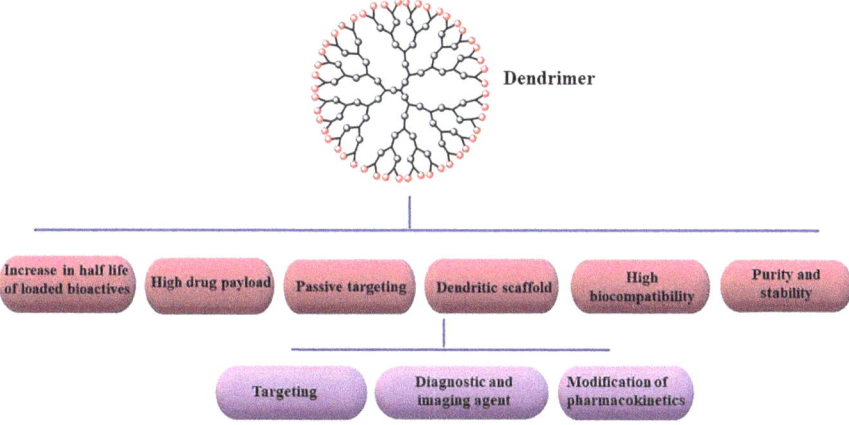

Fig. 10.6 Main features and applications derived from dendrimers

Table 10.2 List of antigens delivered via different nanocarriers for the medication of different infectious diseases

Antigen	Nanocarrier used	Disease	References
Against bacterial infection			
Antigenic protein	Poly (D,L-lactic-co-glycolic acid) nanospheres	Anthrax	Manish et al. (2013)
DNA encoding T cell epitopes of Esat-6 and FL	Chitosan nanoparticle	Tuberculosis	Feng et al. (2013)
Mycobacterium lipids	Chitosan nanoparticle	Tuberculosis	Das et al. (2017)
Polysaccharides	Liposomes	Pneumonia	Abraham (1992)
Bacterial toxic and parasitic protein	Liposomes	Cholera and malaria	Alving et al. (1986)
Antigenic protein	Nano emulsion	Cystic fibrosis	Makidon et al. (2010)
Antigenic protein	Nano emulsion	Anthrax	Bielinska et al. (2007)
Against viral infection			
Antigenic protein	Chitosan nanoparticles	Hepatitis B	Prego et al. (2010)
Viral protein	Gold nanoparticles	Foot and mouth disease	Chen et al. (2010)
Membrane protein	Gold nanoparticles	Influenza	Tao and Gill (2015)

Antigen	Nanocarrier used	Disease	References
Viral plasmid DNA	Gold nanoparticles	HIV	Xu et al. (2012)
Hepatitis B surface antigen	Poly(D,L-lactic-co-glycolic acid) nanospheres	Hepatitis B	Thomas et al. (2011)
Hepatitis B surface antigen	Alginate-coated chitosan nanoparticle	Hepatitis B	Borges et al. (2008)
Capsid protein	VLPs	Norwalk virus infection	Ball et al. (1998)
Nucleocapsid protein	VLPs	Hepatitis	Geldmacher et al. (2005)
Fusion protein	VLPs	Human papilloma virus	Oh et al. (2004)
Multiple proteins	VLPs	Rotavirus	Parez et al. (2006)
Virus proteins	VLPs	Bluetongue virus	Roy et al. (1994)
Viral protein	Polypeptide nanoparticles	Coronavirus for severe acute respiratory syndrome (SARS)	Pimentel et al. (2009)

exert a clinical effect, and to target specific complex (Kesharwani et al. 2014) (Table 10.2).

10.9 Conclusion

To tackle the growing number of antibiotic-resistant strains of pathogenic microorganisms, new antimicrobial agents are urgently required. In the field of medicine, nanoparticles are extensively examined because of their unique physicochemical properties. By providing fast healing of infection without destroying the surrounding cells, nanotechnology achieves impressive outcomes in the treatment of infectious diseases. The current study discusses recent advancements in nanomedicines, including technical innovations in targeted drug delivery and innovative diagnostic methodologies. By manipulating certain diagnostic and treatment criteria, such as controlled slow release of encapsulated medications, nanoparticles make a big leap from conventional diagnosis and treatment to more modern and upgraded one. The diverse types of nanomedicine that have been explored are very promising but still have far to go.

Acknowledgments The authors would like to thank the Director of the Institute of Advanced Study in Science and Technology (IASST), Assam, India, for providing facilities for this work. They also want to acknowledge the Department of Science and Technology (DST), Government of India, for supporting this work.

References

Abalkhil TA, Alharbi SA, Salmen SH, Wainwright M (2017) Bactericidal activity of biosynthesized silver nanoparticles against human pathogenic bacteria. Biotechnol Biotechnol Equip 31(2):411–417

Abraham E (1992) Intranasal immunization with bacterial polysaccharide containing liposomes enhances antigen-specific pulmonary secretory antibody response. Vaccine 10(7):461–468

Ahmad T, Wani IA, Lone IH, Ganguly A, Manzoor N, Ahmad A, Al-Shihri AS (2013) Antifungal activity of gold nanoparticles prepared by solvothermal method. Mater Res Bull 48(1):12–20

Akahata W, Yang ZY, Andersen H, Sun S, Holdaway HA, Kong WP et al (2010) A virus-like particle vaccine for epidemic Chikungunya virus protects nonhuman primates against infection. Nat Med 16(3):334–338

Allaker RP, Vargas-Reus MA, Ren GG (2012) Nanometals as antimicrobials. Antimicrob Polym 2011:327–350

Alving CR, Richards RL, Moss J, Alving LI, Clements JD, Shiba T, Hockmeyer WT (1986) Effectiveness of liposomes as potential carriers of vaccines: applications to cholera toxin and human malaria sporozoite antigen. Vaccine 4(3):166–172

Arias LR, Yang L (2009) Inactivation of bacterial pathogens by carbon nanotubes in suspensions. Langmuir 25(5):3003–3012

Aslan S, Loebick CZ, Kang S, Elimelech M, Pfefferle LD, Van Tassel PR (2010) Antimicrobial biomaterials based on carbon nanotubes dispersed in poly (lactic-co-glycolic acid). Nanoscale 2(9):1789–1794

Atkins HS, Morton M, Griffin KF, Stokes MG, Nataro JP, Titball RW (2006) Recombinant Salmonella vaccines for biodefence. Vaccine 24(15):2710–2717

Aziz N, Fatma T, Varma A, Prasad R (2014) Biogenic synthesis of silver nanoparticles using Scenedesmus abundans and evaluation of their antibacterial activity. Journal of Nanoparticles, Article ID 689419, https://doi.org/10.1155/2014/689419

Aziz N, Faraz M, Pandey R, Sakir M, Fatma T, Varma A, Barman I, Prasad R (2015) Facile algae-derived route to biogenic silver nanoparticles: Synthesis, antibacterial and photocatalytic properties. Langmuir 31:11605–11612. https://doi.org/10.1021/acs.langmuir.5b03081

Aziz N, Pandey R, Barman I, Prasad R (2016) Leveraging the attributes of Mucor hiemalis-derived silver nanoparticles for a synergistic broad-spectrum antimicrobial platform. Front Microbiol 7:1984. https://doi.org/10.3389/fmicb.2016.01984

Aziz N, Faraz M, Sherwani MA, Fatma T, Prasad R (2019) Illuminating the anticancerous efficacy of a new fungal chassis for silver nanoparticle synthesis. Front Chem 7:65. https://doi.org/10.3389/fchem.2019.00065

Ball JM, Hardy ME, Atmar RL, Conner ME, Estes MK (1998) Oral immunization with recombinant Norwalk virus-like particles induces a systemic and mucosal immune response in mice. J Virol 72(2):1345–1353

Bekele AZ, Gokulan K, Williams KM, Khare S (2016) Dose and size-dependent antiviral effects of silver nanoparticles on feline calicivirus, a human norovirus surrogate. Foodborne Pathog Dis 13(5):239–244

Beverley PC (2002) Immunology of vaccination. Br Med Bull 62(1):15–28

Bielinska AU, Janczak KW, Landers JJ, Makidon P, Sower LE, Peterson JW, Baker JR (2007) Mucosal immunization with a novel nanoemulsion-based recombinant anthrax protective antigen vaccine protects against Bacillus anthracis spore challenge. Infect Immun 75(8):4020–4029

Bitounis D, Ali-Boucetta H, Hong BH, Min DH, Kostarelos K (2013) Prospects and challenges of graphene in biomedical applications. Adv Mater 25(16):2258–2268

Borges O, Cordeiro-da-Silva A, Tavares J, Santarem N, de Sousa A, Borchard G, Junginger HE (2008) Immune response by nasal delivery of hepatitis B surface antigen and codelivery of a CpG ODN in alginate coated chitosan nanoparticles. Eur J Pharm Biopharm 69(2):405–416

Bortner CD, Cidlowski JA (2007) Cell shrinkage and monovalent cation fluxes: role in apoptosis. Arch Biochem Biophys 462(2):176–188

Buzea C, Pacheco II, Robbie K (2007) Nanomaterials and nanoparticles: sources and toxicity. Biointerphases 2(4):MR17–MR71

Carmona-Gutierrez D, Eisenberg T, Büttner S, Meisinger C, Kroemer G, Madeo F (2010) Apoptosis in yeast: triggers, pathways, subroutines. Cell Death Differentiat 17(5):763–773

Cataldo F, Da Ros T (eds) (2008) Medicinal chemistry and pharmacological potential of fullerenes and carbon nanotubes, vol Vol. 1. Springer Science & Business Media

Cataldo F. Solubility of fullerenes in fatty acids esters: a new way to deliver in vivo fullerenes. Theoretical calculations and experimental results. InMedicinal Chemistry and Pharmacological Potential of Fullerenes and Carbon Nanotubes 2008 (pp. 317-335). Springer, Dordrecht.

Chahal JS, Khan OF, Cooper CL, McPartlan JS, Tsosie JK, Tilley LD, Ploegh HL (2016) Dendrimer-RNA nanoparticles generate protective immunity against lethal Ebola, H1N1 influenza, and toxoplasma gondii challenges with a single dose. Proc Natl Acad Sci 113(29):E4133–E4142

Chakoli AN (2019) Poly (L-Lactide) Bionanocomposites. In: Peptide synthesis. IntechOpen

Chen YS, Hung YC, Lin WH, Huang GS (2010) Assessment of gold nanoparticles as a size-dependent vaccine carrier for enhancing the antibody response against synthetic foot-and-mouth disease virus peptide. Nanotechnology 21(19):195101

Dakal TC, Kumar A, Majumdar RS, Yadav V (2016) Mechanistic basis of antimicrobial actions of silver nanoparticles. Front Microbiol 7:1831

Das I, Padhi A, Mukherjee S, Dash DP, Kar S, Sonawane A (2017) Biocompatible chitosan nanoparticles as an efficient delivery vehicle for Mycobacterium tuberculosis lipids to induce potent cytokines and antibody response through activation of γδ T cells in mice. Nanotechnology 28(16):165101

Davachi SM, Kaffashi B (2015) Preparation and characterization of poly L-lactide/triclosan nanoparticles for specific antibacterial and medical applications. Int J Polym Mater Polym Biomater 64(10):497–508

Davies J, Davies D (2010) Origins and evolution of antibiotic resistance. Microbiol Mol Biol Rev 74(3):417–433

Ducheyne P, Healy K, Hutmacher DE, Grainger DW, Kirkpatrick CJ (eds) (2011) Comprehensive biomaterials. Elsevier, Amsterdam, p 229. ISBN 9780080552941

Duran N, Duran M, De Jesus MB, Seabra AB, Favaro WJ, Nakazato G (2016) Silver nanoparticles: a new view on mechanistic aspects on antimicrobial activity. Nanomedicine 12(3):789–799

Feng G, Jiang Q, Xia M, Lu Y, Qiu W, Zhao D, Wang Y (2013) Enhanced immune response and protective effects of nano-chitosan-based DNA vaccine encoding T cell epitopes of Esat-6 and FL against Mycobacterium tuberculosis infection. PLoS One 8(4):e61135

Gao Y, Wijewardhana C, Mann JF (2018) Virus-like particle, liposome, and polymeric particle-based vaccines against HIV-1. Front Immunol 9:345

Geldmacher A, Skrastina D, Borisova G, Petrovskis I, Kruger DH, Pumpens P, Ulrich R (2005) A hantavirus nucleocapsid protein segment exposed on hepatitis B virus core particles is highly immunogenic in mice when applied without adjuvants or in the presence of pre-existing anti-core antibodies. Vaccine 23(30):3973–3983

Grgacic EV, Anderson DA (2006) Virus-like particles: passport to immune recognition. Methods 40(1):60–65

Gupta B, Revagade N, Hilborn J (2007) Poly (lactic acid) fiber: an overview. Prog Polym Sci 32(4):455–482

Hawker CJ, Frechet JM (1990) Preparation of polymers with controlled molecular architecture. A new convergent approach to dendritic macromolecules. J Am Chem Soc 112(21):7638–7647

Hebeish A, El-Rafie H, El-Sheikh MA, Seleem AA, El-Naggar ME (2014) Antimicrobial wound dressing and anti-inflammatory efficacy of silver nanoparticles. Int J Biol Macromol 65:509–515

Hwang IS, Lee J, Hwang JH, Kim KJ, Lee DG (2012) Silver nanoparticles induce apoptotic cell death in Candida albicans through the increase of hydroxyl radicals. FEBS J 279(7):1327–1338

Ichihashi T, Satoh T, Sugimoto C, Kajino K (2013) Emulsified phosphatidylserine, simple and effective peptide carrier for induction of potent epitope-specific T cell responses. PLoS One 8(3):e60068

Iftach Yacoby, Itai Benhar, (2008) Antibacterial nanomedicine. Nanomedicine 3(3):329–341

Ignatova M, Starbova K, Markova N, Manolova N, Rashkov I (2006) Electrospun nano-fibre mats with antibacterial properties from quaternised chitosan and poly (vinyl alcohol). Carbohydr Res 341(12):2098–2107

Kang S, Herzberg M, Rodrigues DF, Elimelech M (2008) Antibacterial effects of carbon nanotubes: size does matter! Langmuir 24(13):6409–6413

Kesharwani P, Jain K, Jain NK (2014) Dendrimer as nanocarrier for drug delivery. Prog Polym Sci 39(2):268–307

Kolar M, Urbanek K, Latal T (2001) Antibiotic selective pressure and development of bacterial resistance. Int J Antimicrob Agents 17(5):357–363

Kugel AJ, Ebert SM, Stafslien SJ, Hevus I, Kohut A, Voronov A, Chisholm BJ (2012) Synthesis and characterization of novel antimicrobial polymers containing pendent triclosan moieties. React Funct Polym 72(1):69–76

Kumar P, Meena KP, Kumar P, Choudhary C, Thakur DS, Bajpayee P (2010) Dendrimer: a novel polymer for drug delivery. JITPS 1(6):252–269

Kumari Y, Kaur G, Kumar R, Singh SK, Gulati M, Khursheed R, Ghosh D (2019) Gold nanoparticles: new routes across old boundaries. Adv Colloid Interf Sci 274:102037

Liam Matthews, Rupinder K Kanwar, Shufeng Zhou, Vasu Punj, Jagat R. Kanwar, (2010) Applications of Nanomedicine in Antibacterial Medical Therapeutics and Diagnostics. The Open Tropical Medicine Journal 3(1):1–9

LinShu Liu, Tony Jin, Victoria Finkenstadt, Cheng-Kung Liu, Peter Cooke, David Coffin, Kevin Hicks, Charlie Samer, (2009) Antimicrobial Packaging Materials from Poly(Lactic Acid) Incorporated with Pectin-Nisaplin® Microparticles. Chemistry & Chemical Technology 3(3):221–230

Machado MC, Cheng D, Tarquinio KM, Webster TJ (2010) Nanotechnology: pediatric applications. Pediatr Res 67(5):500–504

Maddela NR, Chakraborty S, Prasad R (2021) Nanotechnology for Advances in Medical Microbiology. Springer Singapore (ISBN 978-981-15-9915-6) https://www.springer.com/gp/book/9789811599156

Makidon PE, Knowlton J, Groom JV, Blanco LP, LiPuma JJ, Bielinska AU, Baker JR (2010) Induction of immune response to the 17 kDa OMPA Burkholderia cenocepacia polypeptide and protection against pulmonary infection in mice after nasal vaccination with an OMP nanoemulsion-based vaccine. Med Microbiol Immunol 199(2):81–92

Manish M, Rahi A, Kaur M, Bhatnagar R, Singh S (2013) A single-dose PLGA encapsulated protective antigen domain 4 nanoformulation protects mice against Bacillus anthracis spore challenge. PLoS One 8(4):e61885

Manke A, Wang L, Rojanasakul Y (2013) Mechanisms of nanoparticle-induced oxidative stress and toxicity. BioMed Research International

Maurer P, Jennings GT, Willers J, Rohner F, Lindman Y, Roubicek K, Bachmann MF (2005) A therapeutic vaccine for nicotine dependence: preclinical efficacy, and Phase I safety and immunogenicity. Eur J Immunol 35(7):2031–2040

Medda S, Hajra A, Dey U, Bose P, Mondal NK (2015) Biosynthesis of silver nanoparticles from Aloe vera leaf extract and antifungal activity against Rhizopus sp. and Aspergillus sp. Appl Nanosci 5(7):875–880

Mizuno K, Zhiyentayev T, Huang L, Khalil S, Nasim F, Tegos GP, Hamblin MR (2011) Antimicrobial photodynamic therapy with functionalized fullerenes: quantitative structure-activity relationships. J Nanomed Nanotechnol 2(2):1

Mohammed MA, Syeda J, Wasan KM, Wasan EK (2017) An overview of chitosan nanoparticles and its application in non-parenteral drug delivery. Pharmaceutics 9(4):53

Mohammad J. Hajipour, Katharina M. Fromm, Ali Akbar Ashkarran, Dorleta Jimenez de Aberasturi, Idoia Ruiz de Larramendi, Teofilo Rojo, Vahid Serpooshan, Wolfgang J. Parak, Morteza Mahmoudi, (2012) Antibacterial properties of nanoparticles. Trends in Biotechnology 30(10):499–511

Moon JJ, Suh H, Bershteyn A, Stephan MT, Liu H, Huang B, Goodwin JT (2011) Interbilayer-crosslinked multilamellar vesicles as synthetic vaccines for potent humoral and cellular immune responses. Nat Mater 10(3):243–251

Nagpal K, Singh SK, Mishra DN (2010) Chitosan nanoparticles: a promising system in novel drug delivery. Chem Pharm Bull 58(11):1423–1430

Nakamura S, Mashino T (2009) Biological activities of water-soluble fullerene derivatives. In Journal of Physics: Conference series (Vol. 159, No. 1, p. 012003). IOP Publishing

Naskalska A, Pyrc K (2015) Virus like particles as immunogens and universal nanocarriers. Pol J Microbiol 64(1):3–13

Oh YK, Sohn T, Park JS, Kang MJ, Choi HG, Kim JA, Kim CK (2004) Enhanced mucosal and systemic immunogenicity of human papillomavirus-like particles encapsidating interleukin-2 gene adjuvant. Virology 328(2):266–273

Pacurari M, Qian Y, Fu W, Schwegler-Berry D, Ding M, Castranova V, Guo NL (2012) Cell permeability, migration, and reactive oxygen species induced by multiwalled carbon nanotubes in human microvascular endothelial cells. J Toxic Environ Health A 75(2):112–128

Panacek A, Kolar M, Vecerova R, Prucek R, Soukupova J, Krystof V, Kvitek L (2009) Antifungal activity of silver nanoparticles against Candida spp. Biomaterials 30(31):6333–6340

Parez N, Fourgeux C, Mohamed A, Dubuquoy C, Pillot M, Dehee A, Garbarg-Chenon A (2006) Rectal immunization with rotavirus virus-like particles induces systemic and mucosal humoral immune responses and protects mice against rotavirus infection. J Virol 80(4):1752–1761

Pati R, Shevtsov M, Sonawane A (2018) Nanoparticle vaccines against infectious diseases. Front Immunol 9:2224

Pimentel TA, Yan Z, Jeffers SA, Holmes KV, Hodges RS, Burkhard P (2009) Peptide nanoparticles as novel immunogens: design and analysis of a prototypic severe acute respiratory syndrome vaccine. Chem Biol Drug Des 73(1):53–61

Popelka A, Novak I, Lehocky M, Chodak I, Sedliacik J, Gajtanska M, Spirkova M (2012) Anti-bacterial treatment of polyethylene by cold plasma for medical purposes. Molecules 17(1):762–785

Prabaharan M (2008) Chitosan derivatives as promising materials for controlled drug delivery. J Biomater Appl 23(1):5–36

Prapruddivongs C, Sombatsompop N (2012) Compos Part B Eng 43:2730

Prasad R, Swamy VS (2013) Antibacterial activity of silver nanoparticles synthesized by bark extract of Syzygium cumini. Journal of Nanoparticles 2013, https://doi.org/10.1155/2013/431218

Prasad R (2014) Synthesis of silver nanoparticles in photosynthetic plants. Journal of Nanoparticles, Article ID 963961, 2014, https://doi.org/10.1155/2014/963961

Prasad R, Kumar V, Kumar M, Choudhary D (2019) Nanobiotechnology in Bioformulations. Springer International Publishing (ISBN 978-3-030-17061-5) https://www.springer.com/gp/book/9783030170608

Prego C, Paolicelli P, Diaz B, Vicente S, Sanchez A, Gonzalez-Fernandez A, Alonso MJ (2010) Chitosan-based nanoparticles for improving immunization against hepatitis B infection. Vaccine 28(14):2607–2614

Rampino A, Borgogna M, Blasi P, Bellich B, Cesàro A (2013) Chitosan nanoparticles: preparation, size evolution and stability. Int J Pharm 455(1-2):219–228

Riley RS, Day ES (2017) Gold nanoparticle-mediated photothermal therapy: applications and opportunities for multimodal cancer treatment. Wiley Interdiscip Rev Nanomed Nanobiotechnol 9(4):e1449

Roy P, Bishop DH, LeBlois H, Erasmus BJ (1994) Long-lasting protection of sheep against bluetongue challenge after vaccination with virus-like particles: evidence for homologous and partial heterologous protection. Vaccine 12(9):805–811

Saglam N, Korkusuz, F, Prasad R (2021) Nanotechnology Applications in Health and Environmental Sciences. Springer International Publishing (ISBN: 978-3-030-64410-9) https://www.springer.com/gp/book/9783030644093

Samia AC, Dayal S, Burda C (2006) Quantum dot-based energy transfer: perspectives and potential for applications in photodynamic therapy. Photochem Photobiol 82(3):617–625

Sharma A, Sharma US (1997) Liposomes in drug delivery: progress and limitations. Int J Pharm 154(2):123–140

Shvedova AA, Pietroiusti A, Fadeel B, Kagan VE (2012) Mechanisms of carbon nanotube-induced toxicity: focus on oxidative stress. Toxicol Appl Pharmacol 261(2):121–133

Tao W, Gill HS (2015) M2e-immobilized gold nanoparticles as influenza a vaccine: role of soluble M2e and longevity of protection. Vaccine 33(20):2307–2315

Tegos GP, Demidova TN, Arcila-Lopez D, Lee H, Wharton T, Gali H, Hamblin MR (2005) Cationic fullerenes are effective and selective antimicrobial photosensitizers. Chem Biol 12(10):1127–1135

Thomas C, Rawat A, Hope-Weeks L, Ahsan F (2011) Aerosolized PLA and PLGA nanoparticles enhance humoral, mucosal and cytokine responses to hepatitis B vaccine. Mol Pharm 8(2):405–415

Tyagi RK, Garg NK, Sahu T (2012) Vaccination strategies against malaria: novel carrier (s) more than a tour de force. J Control Release 162(1):242–254

Varkey J, editor. Peptide Synthesis. BoD–Books on Demand; 2019 Dec 18.

Vecitis CD, Zodrow KR, Kang S, Elimelech M (2010) Electronic-structure-dependent bacterial cytotoxicity of single-walled carbon nanotubes. ACS Nano 4(9):5471–5479

Wang X, Xuan S, Song L, Yang H, Lu H, Hu Y (2012) Synergistic effect of POSS on mechanical properties, flammability, and thermal degradation of intumescent flame retardant polylactide composites. J Macromol Sci Part B 51(2):255–268

World Health Organization (2016) Global Tuberculosis report. World Health Organization

Xu L, Liu Y, Chen Z, Li W, Liu Y, Wang L, Ma L (2012) Surface-engineered gold nanorods: promising DNA vaccine adjuvant for HIV-1 treatment. Nano Lett 12(4):2003–2012

Zhang YJ, Gao B, Liu XW (2015) Topical and effective hemostatic medicines in the battlefield. Int J Clin Exp Med 8(1):10

Chapter 11
Nanotechnology's Promising Role in the Control of Mosquito-Borne Disease

Gopalan Rajagopal, Shenbagamoorthy Sundarraj, Krishnan Anand, and Sakkanan Ilango

Contents

11.1	Introduction	346
	11.1.1 Nanotechnology	346
	11.1.2 Vector-Borne Diseases	346
	11.1.3 Nanotechnology in Mosquito Control	348
11.2	Drug Delivery System of Nanoparticles for Mosquitoes Borne Diseases	349
	11.2.1 Nanoliposomes	349
	11.2.2 Nanosuspensions	353
	11.2.3 Polymer-Based Nanoparticles	354
11.3	Nanoparticle Synthesis by Biological Methods	356
	11.3.1 Green-Based Nanoparticle	356
	11.3.2 Microorganism-Based Nanoparticle	358
11.4	Nanotechnology for Arbovirus Detection and Control	360
	11.4.1 Biosensor	360
	11.4.2 Insect Repellents	362
11.5	Conclusion	363
References		363

G. Rajagopal (✉) · S. Sundarraj · S. Ilango
Postgraduate and Research Department of Zoology, Ayya Nadar Janaki Ammal College, Sivakasi, India

K. Anand
Department of Chemical Pathology, School of Pathology, Faculty of Health Sciences and National Health Laboratory Service, University of the Free State, Bloemfontein, South Africa

11.1 Introduction

11.1.1 Nanotechnology

Nanotechnology can be described as the scientific and technological aspects of the architecture, synthesis, features, and functionality of substances at the nanometer scale. This technological know-how opens up innovative role in insect pest management and their opportunities in a blend with bio-based insecticides in order to improve stratigies for pest control. Nanoparticle formulations provide protection for degradation-prone agents, denaturation in areas of intense pH, and further prolong the exposure time of a compound by increasing the retention of the formula through bio-adhesion (Prasad et al. 2019; Sahoo et al. 2007). Nanobiotechnology is the combination of engineering and molecular biology, which focuses on the new classification and recognition charge of the multifunctional devices and structure for organic and chemical assessment with higher sensitivity, specificity, and a greater charge of recognition. Nano-objects with essential analytical purposes consist of nanotubes, nanochannels, nanoparticles, nano-blocks, nanopores and nanofibers (Fortina et al. 2005).

Recent literature suggested that the particular sectors of nanobiotechnology, consisting of liposomes nanoparticles for drug delivery, emulsions, imaging, biomaterials, food, optical, electronics, pathogens, biosensors and in vitro diagnostics. The special bodily and chemical houses of nanomaterials (small size, multiplied reactivity, excessive surface-to-volume ratio), whilst are probably to supply fitness advantages can also additionally be related with deleterious outcomes on cells and tissues. Nanomaterials have dimensions comparable to organelles observed in the cell and have the plausibility to intrude into necessary cell functions, ensuing in conceivable toxicity (Zhang et al. 2012). Advances in nanomaterial modification enable for the attachment of remarkable biomolecules including microbes, pathogens, proteins and nucleic acids (Crean et al. 2011). The development of eco-friendly nano-formulations with efficient delivery system and small quantities of nano-pesticides will be in great demand in the future (Bhattacharyya et al. 2016; Gupta et al. 2018).

11.1.2 Vector-Borne Diseases

Vector-borne diseases cause major public health tribulations, and their control is accomplished mainly with the usage of synthetic chemical insecticides. Since ancient times, mosquitoes consider as disease causing agent o human. They serve as vectors for pathogens that trigger life-threatening human diseases (Chandra et al. 2008). In 124 countries they place at risk up to 55% of the world's population (Beatty et al. 2007). Nowadays the control of biological vectors are prime importance for controlling the vector borne diseases (Kovendan et al. 2012). In worldwide, various methods were developed to control the vector-borne diseases such as dengue hemorrhagic fever, dengue syndrome, malaria, Japanese encephalitis, yellow fever, chikungunya fever and lymphatic filariasis (Ali et al. 1995). Amid the many diseases, filariasis and dengue are the eminent parasites to cause infections in tropical regions (Pandey et al. 2007). Mosquito-borne diseases can be controlled by targeting the causative parasites and pathogens (Hamaidia & Soltani 2014). There are different forms of mosquitoes and several of them are capable of spreading diseases, including *A. aegypti*, *Culex quinquefasciatus* and *Anopheles stephensi* (Diptera: Culicidae).

A. aegypti are arboviral vectors that are spreading tropical diseases and also becoming one of the greatest public health issues. They are the primary for the transmission of dengue, Chikungunya, Zika, Malaria, Filariasis, Japanese encephalitis, West Nile fever and yellow fever infection. Dengue is the most ubiquitous disease, with about 390 million infections per year (Bhatt et al. 2013). Chikungunya virus has caused over 2.5 million infections in the last decade and has more recently it spreads in the Americas (Staples and Fischer 2014) and emerging in Europe (Schaffner et al. 2013) Yellow fever has caused approximately 200,000 severe cases per year in Africa (Garske et al. 2014), Zika viruses have triggered about four million infections in America (Boeuf et al. 2016). Unfortunately, most mosquito-vectored arboviruses have no treatment accessible for dengue in particular.

Culex quinquefasciatus is the most prevalent type of mosquitoes in populated environments. It may also spread West Nile virus (WNV), SLEV (St. Louis Encephalitis Virus), Pathogenic Protozoa and Japanese Encephalitis virus (JEV) (Bhattacharya et al. 2016). Blood pathogens or the lymph system, muscles and connective tissues in the vertebrate hosts are source of filariasis. The WHO initiated its Global Lymphatic Filariasis Initiative in 2000 and reiterated the goal date by 2020 (WHO 2014).

Malaria is caused by plasmodium, a single-cell protozoan that is transmitted by the mosquito *Anopheles sp*. It mostly affected in developing region as resultant of millions of death occur by malaria (Khan et al. 2018, 2019). The WHO report of 2015 showed that nearly 438,000 people have died of malarial infection. Till now, more than 216 million people are affected per year. While the disease is prevalent, it spreads severely in tropical and subtropical regions (Ismail et al. 2018; Kolluri et al. 2018; Khan et al. 2019).

Various strategies for managing vector-borne diseases have been established worldwide (Baird 2000). Recently, there has been a resurgence in the onset of mosquito-borne diseases to improvements in public health policies and susceptibility changes in mosquitoes to pathogen transmission. Whilst the usage of chemical insecticides has been proven to control mosquitoes over decades, but the application of chemical pesticides in the long term causes a malevolent effect on human and environment. The usage of synthetic insecticide contributes to major pesticide health issues that range from stomach distress, nausea, vomiting, dizziness and depression to skin cancer, eye disorders and acquired defects (Lorenz 2007). Side effects from toxic insects, rodents, amphibians and fishes on the climate, to enhanced tolerance of mosquitoes (Denholm et al. 2002). Nature product such as plants has the potent active to control the mosquito. Still, they have various limitations such as seasonal variation and unavailability among the years and effectiveness on multiple metabolites (Suryawanshi et al. 2015). Recently, nanomaterials have to arise as a promising novel method to control mosquito-borne diseases.

11.1.3 Nanotechnology in Mosquito Control

Nanotechnology is a revolutionary field in the various areas of research, including medicine, biology and agriculture. Formulation of nano-emulsion preparation of smart nano-pesticide using nanomaterials as active pesticide agents or nanocarriers for their delivery. It can also be determined as valuable as compared to the chemical insecticide for targeting delivery of pesticides for controlling the mosquitoes larva and adult (Ahmed et al. 2019). Extensive research on nanopesticides is expected to deal with the restrictions of the accessible strategies used for pest control and endowed with nano-based novel formulations that enter into the target (pest), remain steady and active in the environment, without impact on non-target organisms, using cost-effective formulation. Beside this, nanoparticles are used in the formulation of insect repellents and also for the rapid diagnosis as biosensors. Nanoparticles-coated pesticides are used for controlled release, which makes an impact on bioavailability and pesticides stability properties (Fig. 11.1). Chitosan nanoparticles play a vital insect component includes mosquitoes which has the part as well as in exoskeleton cuticle and most tissues such as foregut, midgut, hindgut and trachea. Biosynthesis of chitin is mostly reliant on the enzyme chitin synthetase

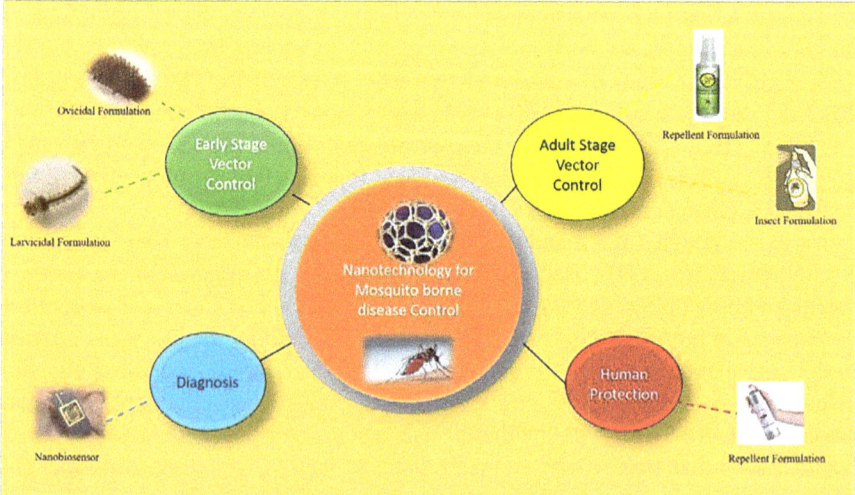

Fig. 11.1 Nanotechnology in Mosquito-Borne Disease Control. The usage of nanoparticles in many sectors in mosquito-borne disease control include, (i) Early-stage vector control, (ii) Adult-stage vector control, (iii) Diagnosis and (iv) Human protection

for the catalysing activity. In this process, they trigger the sugars for transmission to acceptors present in the mosquitoes. CHS A and CHS B are the chitin synthesis genes, where CHS A is present in cuticle and foregut, hindgut and trachea, and, CHS B is associated with the midgut of mosquitoes. The biosynthesis of epithelial cell chitin is linked to the peripheral matrix (PM) (Sen and Blau 2006; Kilama and Ntoumi 2009; Schwendener 2014).

Table 11.1 represents the key categories of nanoparticles used for the improvisation of potent actions that are conventionally done for the prevention of mosquito-borne diseases. It highlights the possible advantages of a variety of nanomaterials used, and some data have shown that they have been used thus far (encoded by accompanying references). Information of the various nanoparticles formed by the use of various nanoscale materials are provided in the following section.

11.2 Drug Delivery System of Nanoparticles for Mosquitoes Borne Diseases

Nanoparticles constitute versatile drug delivery system to control wide range of mosquiotes borne diseases. Nanoparticle mediated drug delivery systems are developed for the optimized release of sufficient and necessary amounts of drugs within a specific time, controlled and targeted delivery plays an important role in vector borne disease control. The large surface area also provides a high affinity to drug and tiny molecules, such as ligands or antibodies, which can be used for targeted and controlled release of therapeutic drugs. Possession of specific characteristics of nanoparticles, viz., high effective loading capacity, larger surface area, fast mass transfer to delivery drug, target delivery and the ability for easy attachment of various small therapeutic molecules, encouraged the use of NPs as nanocarriers. Nanoliposome also used to entrap anti-infectious drugs are active against infections due to facultative intracellular bacteria, parasites such as leishmania, protozoan, viruses such as the one causing mosquitoes borne diseases.The following types of nanocarriers have been used for the delivery of pesticides/drug such as, nanoliposomes, nanosuspension and polymer based nanoparticles.

11.2.1 Nanoliposomes

Liposomes are synthetic nano-sized vesicles comprising of phospholipids and the cholesterol layer surrounded by an aqueous core. Based on the number of layers and size of the nanoparticles, they are divided into three different types as small unilamellar, large unilamellar and multilamellar liposomes (Fig. 11.2). Affording to their physiochemical characterisation, the nanoparticles are either aqueously entrapped or might be inserted into the liposome layers (Shargh et al. 2012;

Table 11.1 Depiction the overview of types of nanoparticles used to control vector-borne diseases

S. No	Content	Types of Nanoparticles	Materials	Properties	References
1	Drug Delivery System of nanoparticles	Nanoliposomes	Lipids and micelles	Immunological Compatibility	Qiu et al. (2008), Fotoran et al. (2019), Shakeel et al. (2019)
		Nanosuspensions	Aqueous	Bioavailability and increase surface area	Omwoyo et al. (2014), Pessoa et al. (2018a)
		Polymer-Based Nanoparticles	Chitin and chitosan	Control drug release	Krishnaswamy and Orsat (2017), Tripathy et al. (2012), Ahmed and Aljaeid (2016)
2	Nanoparticle synthesis by biological methods	Green-based nanoparticle	Plant leaves, root, fruit and etc.,	Reduced size and the different shape and size of the particles	Parthiban et al. (2019), Murugan et al. (2016), Sowndarya et al. (2017), Mondal and Hajra (2016), Rajagopal et al. (2021)
		Microorganism-based nanoparticle	Bacteria and fungi	Unique nanostructures: bacterial nanocellulose, exopolysaccharides, bacterial nanowires, and biomineralised nanoscale materials	Soni and Prakash (2012), Salunkhe et al. (2011)

3	Nanotechnology for arbovirus detection and control	Biosensor	Nano-chips, nanowires and etc.,	Point-of-care (POC) and ASSURED (Affordable, Sensitive, Specific, User-friendly, Robust and rapid, Equipment-free, and Deliverable)	Campos et al. (2020), Pashchenko et al. (2018), Vinayagam et al. (2018)
		Insect Repellent	Insect formulation and Encapsulation	Drug Longevity, biocompatibility and biodegradability	Nogueira Barradas et al. (2016), Coelho et al. (2018), Gomes et al. (2019)

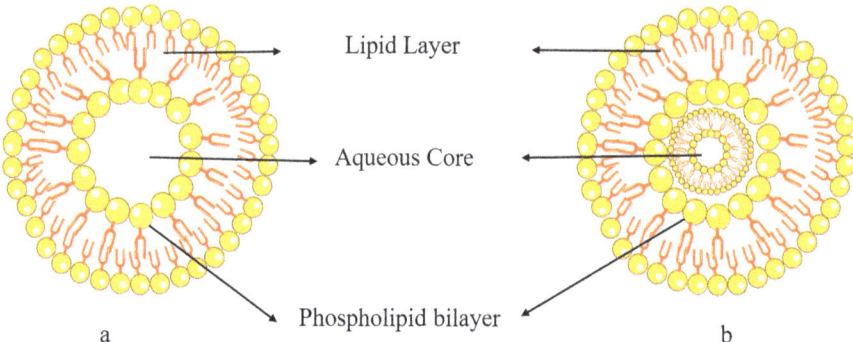

Fig. 11.2 Types of nanoliposomes synthesis. (**a**) Unilamellar liposomes, (**b**) Multilamellar liposomes

Firouzmand et al. 2013; Frezza et al. 2013; Eskandari et al. 2014; Kalat et al. 2014; Oliveira et al. 2014). These nanoliposomes compatability with immune system facilitated for targeted uptake by cells via receptor mediated endocytosis (Santos-Magalhães and Mosqueira 2010).

Nanoliposomes have been previously used for deliverying the drug or vaccine to control malaria. Despite that, the encapsulated therapeutic drug (Chloroquine) within the nanoliposomes shows potent activity by regulating the bioavailability and the dose-dependence inside the body. Pointedly, they reduce the toxic effect and also diminish the drug resistance risk (Qiu et al. 2008).

Recently, gel-core liposomes are used as a nanocarrier for the controlled delivery of the antimalarial drug. Gel core nanoliposomes are the composite of lipid-based carrier coated with the polymer for the controlled delivery of Pfs25 with immune-stimulatory adjuvant (CpGODN). In these gel-core liposomes, polymer integration into the liposomes interior aqueous core increases their stability, and they influence the controlled drug release via the slow release of the drug. The gradual diffusion enables the nanoliposome capability for long-lasting persistence of antigen without any need of stimulating (Baruah et al. 2017).

Lipid particles for drug delivery may be adjusted to form multi-layer vesicles with enhanced cargo interactions and stability. The in vitro treatment of *P. falciparum* NF54 with liposome encapsulated chloroquine. Lipids are capable of establishing hydrogen bonds rather than covalent connections and have been used to create stable vesicles-within-vesicles with a high capacity for entrapping antimalarial medicines such as chloroquine (hydrophilic) and artemisinin (hydrophilic) (lipophilic).While, vesicles deal with parasite-infected erythrocytes more directly than with normal red blood cells. A very interesting method for improved drug delivery is the hydrogen-bonded, multilamellar liposomes, since they are permissive not only to small compounds but also to bigger peptides (Fotoran et al. 2019).

Artemether and lumefantrine combination therapy has been well known for uncomplicated malaria treatment. Both drugs were formulated and freeze-dried with nanoliposomes. The initial bursting impact and the continued release trend over a duration of 30 h has been observed in *in vitro* drug-release research. Due to tissue distribution, the combination therapy of ART+LUM-NL's intake in reticuloendothelial system (RES) organs was high, particularly in liver and spleen. In vitro/in vivo toxicological examinations indicate that no erythrocytes are hemolyzed and that there is no indication of renal or hepatic damage in examined animals while treating the ART+LUM-NL's. It was predicted that nanoliposomes might increase the availability of artemether and lumefantrine by extending their in vivo retention. Thus, the nanocarriers are suitable candidate for controlling mosquitoes borne diseases. This type of formulation for malarial therapy does not show any symptoms of fibrosis, fats infiltration, core lobule necrosis and lymphocytic infiltration. (Shakeel et al. 2019).

For medicinal uses, liposomes are certified by the United States Food and Drug Administration. In the human and animal vaccine industries, vaccine components may be incorporated within the aquatic cavities of fibre recombinant vaccines. The liposome preparation laboratory and industrial techniques have been established and the possibility that they may be used in human vaccines is prospective (Adu-Bobie et al. 2003). A liposomal carrier's positive results are (1) their ability to ensure I antigen's protection and stability, including its native adherence, (2) enhanced allophycocyanin (APC) absorption by passive or active targeting and (3) strengthened or controlled antigen absorption.

11.2.2 Nanosuspensions

Nanosuspensions of drugs are nanosized, heterogeneous aqueous dispersions of insoluble drug particles stabilized by steric, electrostatic and surfactants. Thus, they prevent the agglomeration and confirming the stability of pharmaceutical nanosuspensions (Rabinow 2004). The preparation of nanosuspension is by the biphasic approach where the nanosized drug particles are dissolved in the aqueous solution via a top-down or bottom-up method to minimise the particle size (Fig. 11.3). The main advantage of nanosuspensions is the lowering of toxicity level by providing the bioavailability of the drug; besides, the diminished size increases the surface area.

Lumefantrine, a antimalarial drug, used to treat multi-drug resistance malaria. The mechanism of Lumefantrine is by converting the heme of *Plasmodium flaciparum* into hemozoin. The higher amount of accumulating toxic heme causes the death of the parasite and also control malarial infection. Nevertheless, Lumefantrine has a lesser water solubility and causes reduced bioavailability via oral and dietary way (Nakache et al. 2000; Baird and Hoffmann 2004). Nanosuspension-based drug delivery has the advantage to combat malarial infection. The enhanced mechanism of Lumefantrine-based nano-suspension shows a potent anti-malarial activity. In these methods, the size of the Lumefantrine may reduce from 72 μm to 0.251 μm via in vitro and in vivo experiments and tested against the *P. Yoelii nigeriensis* and *P. falciparum* (Omwoyo et al. 2014). In addition to these, the drug dihydro-

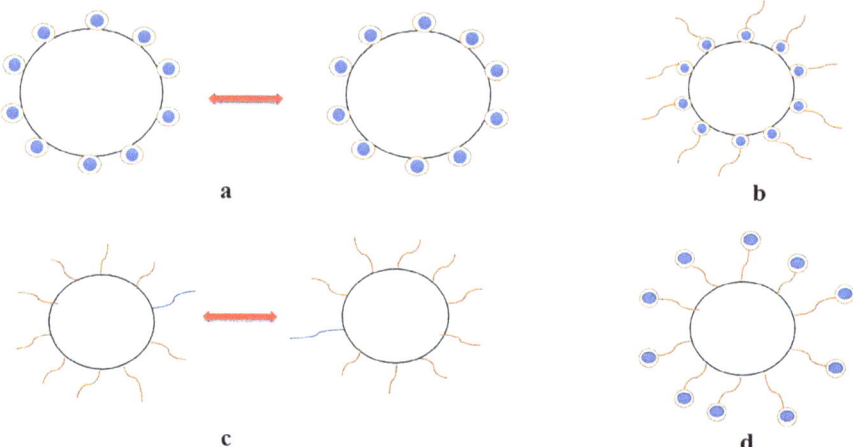

Fig. 11.3 Types of Nanosuspension, (**a**) Electrostatic stabilisation, (**b, d**) Electrosteric stabilisation, (**c**) Steric stabilisation

artemisinin (DHA) was also tested against *P. falciparum* in in vitro trials. The nanosuspension with the low amount of drug concentration reduced the risk of side effects rather than anti-malarial activity (Lee et al., 2014). Quercetin nanosuspension have been represented as a potential drug for controlling mosquitoes borne diseases (Pessoa et al. 2018).

Nanosuspensions are a distinctive and economically successful solution to addressing issues such as low solubility and poor bioavailability of hydrophobic medicines. Media milling and high-pressure homogenisation technologies have been successfully exploited for large-scale processing of nanosuspensions. The applications of nanosuspensions for different routes of administration have been extended by

11.2.3 Polymer-Based Nanoparticles

Polymeric nanoparticles are used as a carrier for the controlled release of drug in the pharmaceutical sectors. It may be used in two forms: nanocapsule and nanospheres (Fig. 11.4). The preparation of polymeric nanoparticles follows two strategies, namely top-down and bottom-up approach. In the top-down approach, a dispersion of polymer generates the polymer-based nanoparticles, while in the bottom-up approach, polymerisation of monomers forms the polymer-based nanoparticles. The main advantage of the polymer nanoparticles is biocompatibility and biodegradability, and their size vary from 1 to 1000 mm (Krishnaswamy and Orsat 2017; Prasad et al. 2017).

Primaquine (PQ) is a mostly used as anti-malarial drug against malarial parasite of *P. vivax* and *P. ovale*. Even though it has potent activity against the malarial parasites, it showed some side effect against the non-targeted organisms due to their

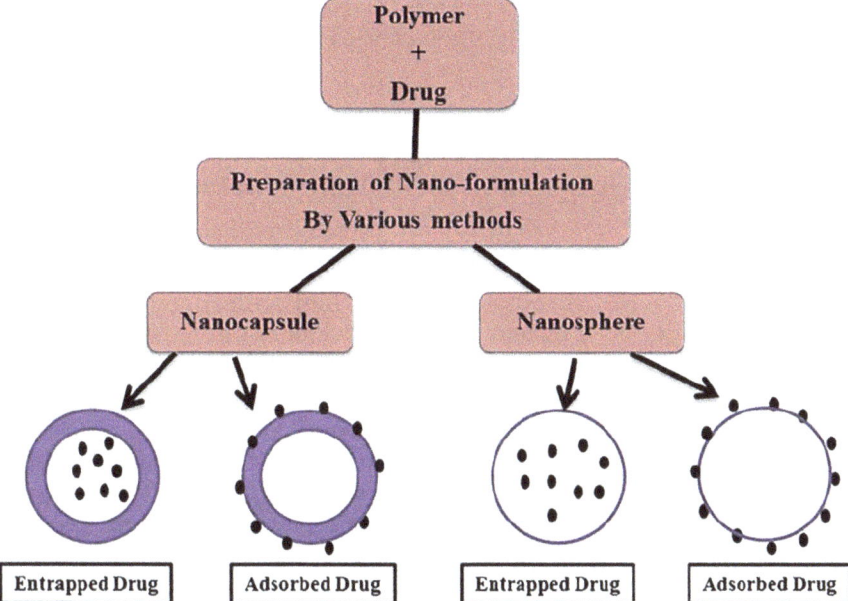

Fig. 11.4 Types of Polymer-based nanoparticles Preparation. (Source: Kumar et al. 2012)

short half-life and their lower bioavailability. The formation of PQ into a chitosan-based nanoparticle has a hope to overcome the risk of side-effects. Many researchers determine the PQ loaded solid lipid nanoparticles have the potential to reduce the multiplication of malarial parasite (Baird and Rieckmann 2003; Omwoyo et al. 2014; Dennis et al. 2015; Ahmed and Aljaeid 2016).

Chitin-based polymeric nanoparticles have more advantages owing to their lower toxicity, higher stability, water solubility, biocompatibility, biodegradability, sustainable release of the drug, quick fabrication and most importantly non-immunogenicity. All these make them an excellent agent for controlling mosquito-borne diseases (Baird and Rieckmann 2003; Dennis et al. 2015).

The encapsulation of Primaquine (PQ) with polymer-based nanoparticles are used for controlling malarial parasite (Moon et al. 2012; Bennet and Kim 2014). Scientists prepared nano-chloroquine (Nch) particles ranging from 150 to 300 nm in size, and they tested them against the spleen and the liver. The result shows reduction in the damage of spleen and liver (~37% and ~29%) as effective as conventional chloroquine (Tripathy et al. 2012).

Researchers have been exploring alternative delivery mechanisms in the last few decades to improve the effectiveness of multiple drugs. Nanotechnology is a promising novel field with aspirations of clinical science for developments in the wide range of applications in drug delivery. Polymeric nanoparticles provide a new path for recently developed disease site-specific drugs and current poorly soluble drugs to accomplish drug delivery and drug targeting. Overcoming the limitations in traditional drug delivery systems, polymeric nanoparticles are expected for easier implementation and efficient drug distribution, and potentially can increase safety and patient compliance.

11.3 Nanoparticle Synthesis by Biological Methods

11.3.1 Green-Based Nanoparticle

Chemical pesticides are commonly used to control the mosquito population, but it shows the adverse effect of the nontarget community also. Mosquito develops the resistance against synthetic pesticide pollutes the environmental (Milam et al., 2000). The use of chemical insecticide shows a negative impact on the environment and human health and generates resistance via mutation in mosquito population (Lees et al., 2014). In the past, the crude extract of the plant can be used for the larvicidal activity (Senthil-Nathan, 2015). In higher plants, lots of works have been done based on their biologically active material with antilarval properties (Ali et al., 2013). The green synthesis of silver nanoparticles exhibits more advantage than chemical synthesis. It does not produce any hazard to the environment, and it also shows that cost-effective as compared to the conventional method (Parthiban et al., 2019; Tripathy et al., 2013). Recently, the researchers focused on the nanoparticles coated with the natural compounds for prevention of mosquito vector without showing much toxicity (Benelli, 2016a; Kayalvizhi et al., 2016). The green engineered silver nanohybrid showed enhanced mortality range against mosquito because the metallic silver nanomaterial quickly penetrates insect cell through cuticle and mortality was induced by disturbing the normal physiological process (Benelli, 2016b). The green synthesised nanoparticles are well suitable for their reducing and the capping activity. Thus, they employed their ideal activity associated with their reduced size and the different shape and size of the particles (Tripathy et al., 2013). Green Nanotechnology is described as an environmentally safe, clean, nontoxic method to prepare nanoparticles. Beside this, the synthesis method is easy and cheap and does not need high energy, pressure, temperature and a toxic chemicals (Jeong et al. 2005; Benelli and Govindarajan 2017; Prasad et al. 2018b; Sarma et al. 2021). Some widely used nanoparticles include silver, gold, copper, titanium, zinc, silica, selenium, and chitin nanoparticles. In a biological method, the synthesis of nanomaterials consists of plant exact, which includes the whole part of the plant (Saxena et al. 2010; Prasad 2014; Joshi et al. 2018) (Fig. 11.5). Especially, the use of different botanicals act as both reducing and capping agents for the synthesis of nanoparticles with different biophysical features and the toxicity against the pathogens, parasites and the vectors (Benelli et al. 2016; Buhroo et al. 2017) (Fig. 11.6).

Effective green synthesised metallic silver nanoparticles were also obtained from many plant extracts including *Dicranopteris linearis*, *Chenopodium ambrosioides*, *Aristolochia indica*, *Gracilaria edulis*, *Couroupita guianensis*, and *Phyllanthus*

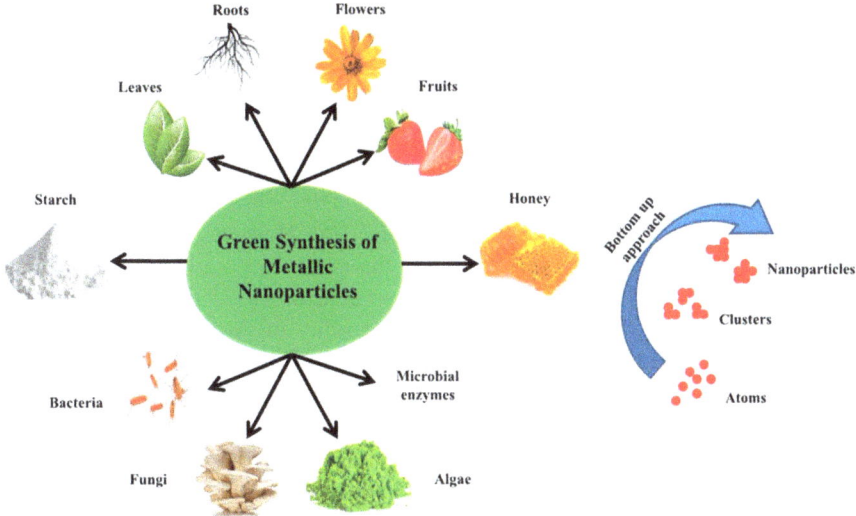

Fig. 11.5 Different types of green synthesis used for the preparation of metal nanoparticles. (Source: Kumar et al. 2020)

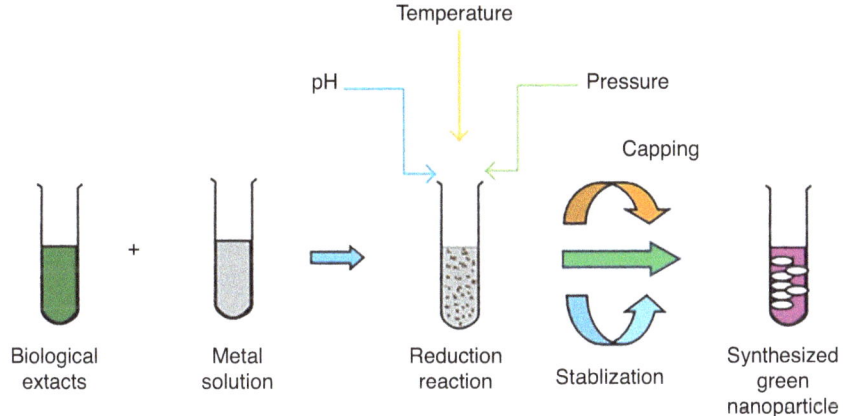

Fig. 11.6 Preparation of green synthesised nanoparticle. (Source: Patra and Back 2014)

niruri, against mosquito vectors such as *Anopheles, Aedes* and *Culex spp*. They showed lethal concentration in the range between 1 and 30 μg/mL, and the whole abolition of the larval species was achieved within 72 h of the first treatment (Benelli 2016b; Benelli ct al. 2018).

Green synthesised silver nanoparticles of *Sargassum muticum* and *D. linearis* are eradicating the 100% of egg hatchability of *A. aegypti* and the *Culex quinquefasciatus*. It showed potent activity within the range of 25 μg/mL and 20 μg/mL concentrations. Compared to *An. stephensi* and *Ae. aegypti, C. quinquefasciatus* showed resistance towards larvicidal activity. Silver nanoparticles from the plant extract of *Rubus ellipticus* showed there is no hatchability of *An. stephensi, Culex quinquefasciatus* and *Ae. aegypti* at the dose of 75 µg/mL, 90 µg/mL and 60 µg/mL, respectively (Azarudeen et al. 2017). Likewise, many silver nanoparticles derived from the leaf-extracts of various plants showed adulticidal activity against mosquito species such as *Ae. aegypti, Ae. albopictus, C. quinquefasciatus* and *An. stephensi*. Unluckily, the exact mechanism of the ovicidal and the adulticidal activity of green-based nanoparticles are unknown (Benelli et al. 2018). However, it has been hypothesised that the nanoparticles penetrate into the exoskeleton and bind with larval DNA. Finally, it alters the DNA function, causes rapid degradation of proteins and importantly decreases in the membrane permeability of insect. These all lead to affect the cell function and induce the cell death with the enormous production of ROS in the mosquito (Benelli 2016b; Azarudeen et al. 2017).

In addition to these, some previous studies revealed that the exposure of nanoparticles against *An. stephensi* and *Ae. aegypti* shows longevity in adult emergence and the fecundity of the females. Some of the examples are green fabricated nanoparticles from the plant extract of *Hypnea musciformis* and *P. aquilinum*, which reduced the longevity of *Ae. aegypti* and *An. Stephensi* and also female fecundity (Alto and Juliano 2001; Panneerselvam et al. 2016). Thus, the green fabricated nanoparticles act as an excellent candidate to control vector-borne diseases efficiently.

The synthesis of metal nanoparticles from plant-derived compounds are effective and showed with excellent anti-plasmodial potential as well as mosquitocidal properties. More than 100 studies have demonstrated a highly efficient mosquito toxicity to plant-based polydisperse metal nanoparticles. The use of various botanicals as reducers and stabilisers leads to different size, type and toxic properties of metal nanoparticles against mosquito vectors (Benelli et al. 2017).

11.3.2 Microorganism-Based Nanoparticle

Nowadays, microbial-based nanoparticles synthesis has a novel approach to control mosquito-borne diseases. Notably, the integration of nanoparticles by using microbes is well studied based on their interaction with metals, and they are also commercially employed in biotechnological processes such as bioremediation and bioleaching. Microbes are well recognised to produce inorganic compounds via intracellular or extracellular metabolites. Both metabolites are used for the synthesis of nanoparticles. Extracellular metabolites synthesised nanoparticles process are cheap and time-consuming compared to intracellular biosynthesis. Extracellular biosynthesis nanoparticles have potent medical applications. Thus, most of the studies are focused on this method to produce microbial nanoparticle synthesis (Durán et al. 2005; Gericke and Pinches 2006; Prasad et al. 2016, 2019a, b; Srivastava et al. 2021).

Most microorganisms are used for producing the nanoparticles like chemical synthesised nanoparticles (Fig. 11.7). In addition to these, the nanoparticles are synthesised with controlled size and determine the nanoparticle composition by

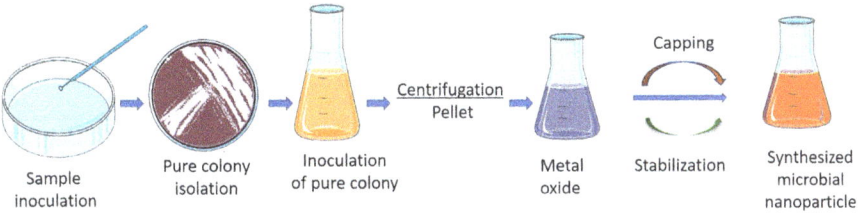

Fig. 11.7 Preparation of microbial synthesised nanoparticles. Metal nanoparticle mechanisms are revealed to be trapped on the surface of the cell membrane by metal nanoparticles. The trapped metal ions are then reduced to nanoparticles by reduction

microbial synthesis (Gericke and Pinches 2006). The mechanisms for toxicity of mosquito mortality have recently been studied in the treatment of nanoparticles. So, hypothetically suggested that the toxicity mechanism of nanoparticle aganist mosquito larva by the penetration of nanoparticle. In intracellular space, nanoparticles destroy enzymes and organelles resulting in loss of cell function and cell death (Selvan et al. 2018; Sowndarya et al. 2017).

The silver nanoparticles synthesised from the entomopathogenic bacteria *B. thuringiensis* (Bt) showed potent activity against *A. aegypti*, a dengue vector. The larvicidal activity of these nanomaterials showed potency at low concentration (Banu et al. 2014). This also reduces the xenobiotics chemical load in the environment (Banu and Balasubramanian 2014a, b). Likewise, other metals such as zinc, copper, selenium and gold are used to synthesise nanoparticles for controlling mosquito-borne diseases.

Fungi have numerous bioactive compounds compared to other microbial species. The filamentous fungi (Ascomycetes, *Fungi Imperfecti*, etc.) produced up to 6400 different compounds (Berdy 2005). Fungi are a promising source of secondary metabolites (Siddhardha et al. 2012). The types of fungi include *Cladosporium oxysporum*, *Cladosporium spherospermum*, *Gilmaniella subornata*, *Chaetomium indicum*, and *Penicillium purpurogenum* and their secondary metabolites are screened against the larvicidal activity (Siddhardha et al. 2012).

Fungal extracellular metabolite synthesised nanoparticles are stabilised by their proteins and they act as a reducing agent. Fungal proteins are associated with the nanoparticle synthesis (Prasad 2016, 2017; Prasad et al. 2018b; Abdel-Aziz et al. 2018). During synthesis, the strain produced specific NADH-dependent reductase (Kumar and McLendon 1997; Macdonald and Smith 1996).Synthesis of gold nanoparticles from the fungus *A. niger* are effective and eco-friendly approach against the mosquito (Soni and Prakash 2012). This group of researchers also reported that entomopathogenic fungi *Chrysosporium tropicum* acts against mosquitoes. They synthesised silver and gold nanoparticles from fungus *Chrysosporium tropicum* and studied for all the stages of larvae. Among these, the silver nanoparticles showed potent activity against the larvae of *C. quinquefasciatus*. While the gold nanoparticles were found to be very potent against *An. stephensi* (Soni and Prakash 2012, 2013).

The pathogenicity and nanoparticle synthesis ability of microbial isolates as *E. coli*, *A. bisporus*, *Pencillium* sp. and *Vibrio* sp. showed potent larvicidal activity within 24 h of post-treatment (Dhanasekaran and Thangaraj 2013). The synthesis of nanoparticle is large quantity for commercialization by this method. Then this particle is formulated and commercialised as the bio-larvicides for vector-borne disease and also for agriculture pest (Balasubramanian and Banu 2016).

11.4 Nanotechnology for Arbovirus Detection and Control

The development of nanotechnology-based biosensor is much attention for detection of arbovirus. On the other hand, there were under studied about the using nanotechnology to overcome the transmission of mosquito-borne diseases by preparing the formulation of larvicidal agent and the repellent (Campos et al. 2020).

11.4.1 Biosensor

In order to reduce and exhibit the further transmission of arbovirus causing diseases (Chikungunya, yellow fever, zika and dengue) the rapid detection method is a very effective tool (Patterson et al. 2016). Regarding this, the WHO has also accentuated the importance of developing point-of-care (POC) tests which are ASSURED (Affordable, Sensitive, Specific, User-friendly, Robust and rapid, Equipment-free, and Deliverable) (Pashchenko et al. 2018). A unique method for the detection of arboviruses must have those characteristics features and also facilitate the early diagnosis of diseases. For early diagnosis and protection from the mosquitoes borne diseases is major challenges in public health. There is a rapid and timely analysis plays a vital role at early confirming viral infection, therefore the affected candidate can be subsequently treated by clinically with the essential precautions (Rashid and Yusof 2018).

Biosensors are biological-based analytical devices, which are used to distinguish analysis in a complex sample matrix. They do not need a long sample treatment process. The main advantage of a biosensor is that the receptor specifically interacts with a diagnostic molecule. This interaction makes physicochemical changes including production of gases, mass, heat, ions, coloured moieties or light (Sethi 1994; Singh et al. 2020).

The multicoloured silver nanoplates are used to detect yellow fever and dengue, which was demonstrated by (Yen et al. 2015). Recently it was improved by using the LFA technology. The technology is associated with the nanotechnology tools of lab-on-a chip. The nanoparticle in the chip will change colour during aggregation. This technology improves the diagnosis selectivity and also sensitivity (Campos et al. 2020). For diagnosing serotype-specific DENV, a triangular silver nanoparticle coated biosensor was employed, based on the pH reduction method. It differentiates the DENV serotype from other serotypes. In this biosensor, they are identified based on colour response (Fig. 11.8). The colour response was accomplished by the interaction of TAg-DNA probe and the dengue virus RNA. Yet, it was not tested by using real samples from the dengue infected people, but it is a promising diagnosis tool for clinical POC diagnostic testing (Vinayagam et al. 2018).

Nano biosensor has a potential for early detection of arbovirus-borne diseases in an effective manner. Because it has specificity, sensitivity and stable recognition in complex matrix in real time analysis. Even though, the improved techniques and various protocols are needed and also in tested to improve the efficiency and production of nanobiosensor. For accomplishing the highly equipped nanobiosensor, we need a narrow scientific boundary amid other disciplines includeing biology, chemistry, sociology and nanotechnology. Nanotechnology based diagnosis device enough small quantity of blood sample and it takes only 40 mins for detection (Fig. 11.9) (Priye et al. 2017; Rong et al. 2019). In spite of these advances, there is a need to improve the devices for proper and accurate use of nanobiosensors in hospitals to early detection and prevention of arbovirus borne diseases (Campos et al. 2020).

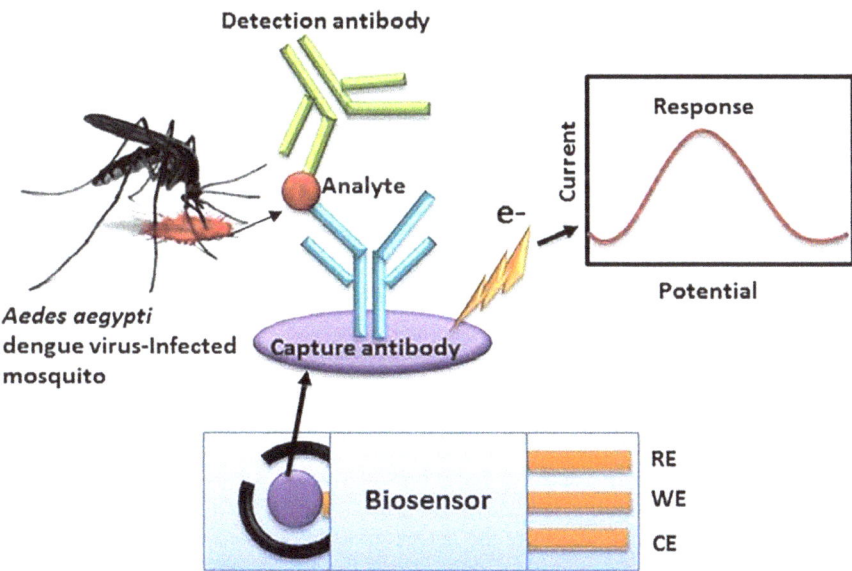

Fig. 11.8 Nanoparticles in Biosensor for detecting Dengue Fever. (Source: Anusha et al. 2019)

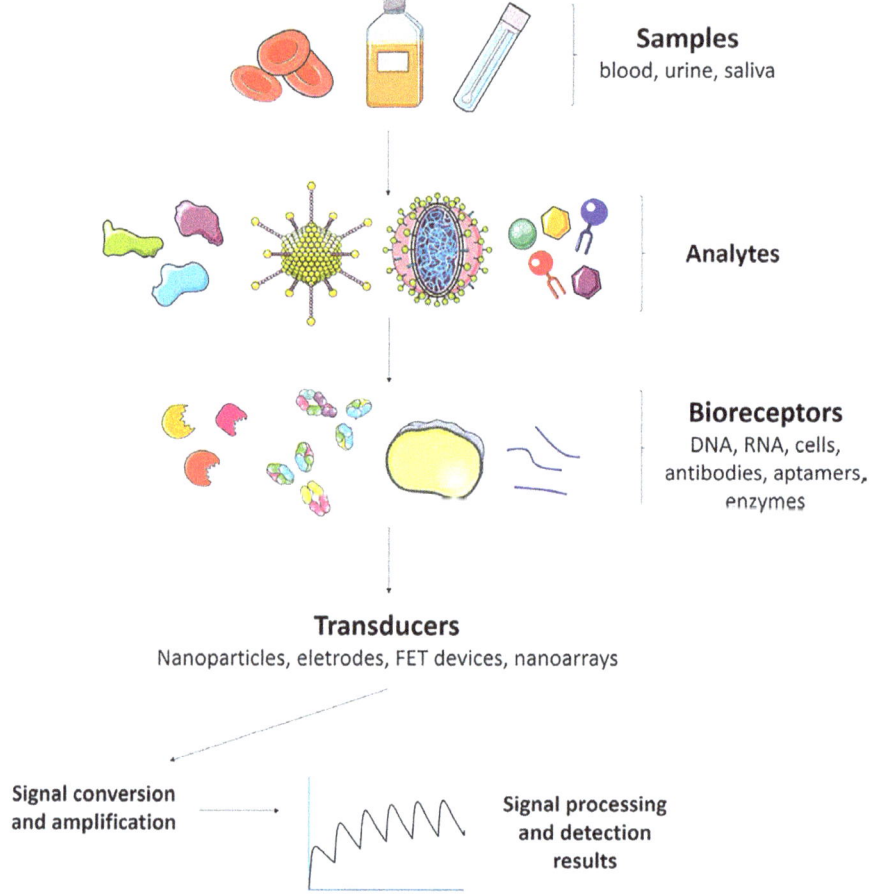

Fig. 11.9 The detection of NS1-anti-dengue and IgM antibodies, zika and/or chikungunya can involve different samples (blood, urine, saliva) In order to track arboviruses, analysts may communicate with bioreceptors. (Source: Campos et al. 2020)

11.4.2 Insect Repellents

The advancement of nanotechnology-based encapsulation of synthetic and natural repellent have an effective against the biological vectors. These constant releases of formulation deliver a controlled or slow release of potent agent to the environment. In addition to these, they increase the action duration and reduce the human exposure (for example, by permeation through the skin). Encapsulation also protects the active compound from premature degradation triggered by the effect of temperature, light, humidity and oxidation (Coelho et al. 2018). Various types of matrices including natural and synthetic are used for the preparation of carriers, for example, lipids, polysaccharides, proteins, polymers and others. Notably, the main characteristic feature of those matrices are biocompatibility and biodegradability and most importantly low cost (Nogueira Barradas et al. 2016).

Encapsulated DEET, a polymeric nanosphere with an average size of 114 ± 37 nm has high stability and low polydisperse index. The continuous release of nano-encapsulated DEET provides repellent activity against biological vectors up to 9 h. As compared to using DEET, the nano-encapsulated DEET has a longer time activity. It showed that the release mechanism was temperature dependent, thus it has high potent activity (Gomes et al. 2019).

11.5 Conclusion

Recently, control of mosquito-borne diseases facing a critical challenge, which deals with the scarcity of potent agents or detection tools against malaria and arboviruses transmitted diseases such as zika virus, dengue, and chikungunya. On this background the synthesis of nanoparticles from the biological product and their drug delivery system make an impact to control the vector-borne diseases. In addition to this, the rapid detection might be used to control and to take safety measures against vector-borne-diseases. The synthesis of nanoparticles is more advantageous than chemical insecticides owing to its target-specific mechanism. From an entomological point of view, nanoparticles have a potent activity to reduce the mosquito population, and persuade the egg mortality and oviposition pre-emption. As a conclusion, it is important to learn about nanoparticles preparation and to appreciate the new and more effective tools involving nanomaterials against vector-borne-diseases.

References

Abdel-Aziz SM, Prasad R, Hamed AA, Abdelraof M (2018) Fungal nanoparticles: A novel tool for a green biotechnology? In: Fungal Nanobionics: Principles and Applications (eds. Prasad R, Kumar V, Kumar M and Wang S), Springer Singapore Pte Ltd. 61–87

Adu-Bobie J et al (2003) Two years into reverse vaccinology. Vaccine. Elsevier, 21(7–8):605–610

Ahmed T et al (2019) Climatic conditions: conventional and nanotechnology-based methods for the control of mosquito vectors causing human health issues. Int J Environ Res Public Health. Multidisciplinary Digital Publishing Institute, 16(17):3165

Ahmed TA, Aljaeid BM (2016) Preparation, characterization, and potential application of chitosan, chitosan derivatives, and chitosan metal nanoparticles in pharmaceutical drug delivery. Drug Des Devel Ther. Dove Press, 10:483

Alto BW, Juliano SA (2001) Precipitation and temperature effects on populations of Aedes albopictus (Diptera: Culicidae): implications for range expansion. J Med Entomol. Oxford University Press Oxford, UK, 38(5):646–656

Anusha JR et al (2019) Electrochemical biosensing of mosquito-borne viral disease, dengue: a review. Biosens Bioelectron 142:111511

Azarudeen RMST et al (2017) Single-step biofabrication of silver nanocrystals using Naregamia alata: a cost effective and eco-friendly control tool in the fight against malaria, Zika virus and St. Louis encephalitis mosquito vectors. J Clust Sci. Springer, 28(1):179–203

Ali A, Nayar JK, Xue R-D (1995) Comparative toxicity of selected larvicides and insect growth regulators to a Florida laboratory population of Aedes albopictus. J Am Mosquito Control Assoc 11(1):72–76

Ali MYS, Ravikumar S, Beula JM (2013) Mosquito larvicidal activity of seaweeds extracts against Anopheles stephensi, Aedes aegypti and Culex quinquefasciatus. Asian Pac J Trop Dis 3(3):196–201

Baird JK (2000) Resurgent malaria at the millennium. Drugs. Springer, 59(4):719–743

Baird JK, Hoffman SL (2004) Primaquine therapy for malaria. Clin Infect Dis. The University of Chicago Press, 39(9):1336–1345

Baird JK, Rieckmann KH (2003) Can primaquine therapy for vivax malaria be improved? Trends Parasitol. Elsevier, 19(3):115–120

Balasubramanian C, Banu AN (2016) Microbial Nanoparticles as Mosquito Control Agents. In: Nanoparticles in the fight against parasites. Springer, pp 81–98

Banu AN, Balasubramanian C (2014a) Myco-synthesis of silver nanoparticles using Beauveria bassiana against dengue vector, Aedes aegypti (Diptera: Culicidae). Parasitol Res. Springer, 113(8):2869–2877

Banu AN, Balasubramanian C (2014b) Optimization and synthesis of silver nanoparticles using Isaria fumosorosea against human vector mosquitoes. Parasitol Res. Springer, 113(10):3843–3851

Banu AN, Balasubramanian C, Moorthi PV (2014) Biosynthesis of silver nanoparticles using Bacillus thuringiensis against dengue vector, Aedes aegypti (Diptera: Culicidae). Parasitol Res. Springer 113(1):311–316

Baruah UK et al (2017) Malaria treatment using novel nano-based drug delivery systems. J Drug Target. Taylor & Francis, 25(7):567–581

Beatty ME et al (2007) Estimating the total world population at risk for locally acquired dengue infection. In: American journal of tropical medicine and hygiene. Amer Soc Trop Med & Hygiene 8000 Westpark DR, STE 130, Mclean, VA 22101 USA, p 221

Benelli G (2016a) Green synthesized nanoparticles in the fight against mosquito-borne diseases and cancer—a brief review. Enzym Microb Technol. Elsevier, 95:58–68

Benelli G et al (2016) Mosquito vectors and the spread of cancer: an overlooked connection? Parasitol Res. Springer, 115(6):2131–2137

Benelli G (2016b) Plant-mediated biosynthesis of nanoparticles as an emerging tool against mosquitoes of medical and veterinary importance: a review. Parasitol Res. Springer, 115(1):23–34

Benelli G et al (2018) Mosquito control with green nanopesticides: towards the one health approach? A review of non-target effects. Environ Sci Pollut Res. Springer, 25(11):10184–10206

Benelli G, Caselli A, Canale A (2017) Nanoparticles for mosquito control: challenges and constraints. J King Saud Univ Sci. Elsevier, 29(4):424–435

Benelli G, Govindarajan M (2017) Green-synthesized mosquito oviposition attractants and ovicides: towards a nanoparticle-based "lure and kill" approach? J Clust Sci. Springer, 28(1):287–308

Bennet D, Kim S (2014) Polymer nanoparticles for smart drug delivery. Chapter

Berdy J (2005) Bioactive microbial metabolites. J Antibiot. Nature Publishing Group, 58(1):1–26

Bhatt S et al (2013) The global distribution and burden of dengue. Nature. Nature Publishing Group, 496(7446):504–507

Bhattacharya S, Basu P, Sajal Bhattacharya C (2016) The southern house mosquito, Culex quinquefasciatus: profile of a smart vector. J Entomol Zool Stud 4(2):73–81

Bhattacharyya A, Duraisamy P, Govindarajan M, Buhroo AA, Prasad R (2016) Nano-biofungicides: Emerging trend in insect pest control. In: Advances and Applications through Fungal Nanobiotechnology (ed. Prasad R), Springer International Publishing Switzerland 307–319

Boeuf P et al (2016) The global threat of Zika virus to pregnancy: epidemiology, clinical perspectives, mechanisms, and impact. BMC Med. BioMed Central, 14(1):1–9

Buhroo AA, Nisa G, Asrafuzzaman S, Prasad R, Rasheed R, Bhattacharyya A (2017) Biogenic silver nanoparticles from *Trichodesma indicum* aqueous leaf extract against *Mythimna separata* and evaluation of its larvicidal efficacy. J Plant Protect Res 57(2):194–200, https://doi.org/10.1515/jppr-2017-0026

Campos EVR et al (2020) Recent developments in nanotechnology for detection and control of Aedes aegypti-borne diseases. Front Bioeng Biotechnol. Frontiers, 8:102

Chandra G et al (2008) Mosquito control by larvivorous fish. Indian J Med Res 127(1):13

Coelho L et al (2018) Photostabilization strategies of photosensitive drugs. Int J Pharm. Elsevier, 541(1–2):19–25

Crean C et al (2011) Polyaniline nanofibres as templates for the covalent immobilisation of biomolecules. Synth Met. Elsevier, 161(3–4):285–292

Denholm I, Devine GJ, Williamson MS (2002) Insecticide resistance on the move. Science. American Association for the Advancement of Science, 297(5590):2222–2223

Dennis E et al (2015) Utilizing nanotechnology to combat malaria. J Infect Dis Ther. OMICS International

Dhanasekaran D, Thangaraj R (2013) Evaluation of larvicidal activity of biogenic nanoparticles against filariasis causing Culex mosquito vector. Asian Pac J Trop Dis. Elsevier, 3(3):174–179

Durán N et al (2005) Mechanistic aspects of biosynthesis of silver nanoparticles by several Fusarium oxysporum strains. J Nanobiotechnol. Springer, 3(1):8

Eskandari F et al (2014) Immunoliposomes containing Soluble Leishmania Antigens (SLA) as a novel antigen delivery system in murine model of leishmaniasis. Exp Parasitol. Elsevier, 146:78–86

Firouzmand H et al (2013) Induction of protection against leishmaniasis in susceptible BALB/c mice using simple DOTAP cationic nanoliposomes containing soluble Leishmania antigen (SLA). Acta Trop. Elsevier, 128(3):528–535

Fortina P et al (2005) Nanobiotechnology: the promise and reality of new approaches to molecular recognition. Trends Biotechnol. Elsevier, 23(4):168–173

Fotoran WL et al (2019) A multilamellar nanoliposome stabilized by interlayer hydrogen bonds increases antimalarial drug efficacy. Nanomedicine. Elsevier, 22:102099

Frezza TF et al (2013) Liposomal-praziquantel: efficacy against Schistosoma mansoni in a preclinical assay. Acta Trop. Elsevier, 128(1):70–75

Garske T et al (2014) Yellow fever in Africa: estimating the burden of disease and impact of mass vaccination from outbreak and serological data. PLoS Med. Public Library of Science, 11(5):e1001638

Gericke M, Pinches A (2006) Biological synthesis of metal nanoparticles. Hydrometallurgy. Elsevier, 83(1–4):132–140

Gomes GM et al (2019) Encapsulation of N, N-diethyl-meta-toluamide (DEET) via miniemulsion polymerization for temperature controlled release. J Appl Polym Sci. Wiley Online Library, 136(9):47139

Gupta N, Upadhyaya CP, Singh A, Abd-Elsalam KA, Prasad R (2018) Applications of silver nanoparticles in plant protection. In: Nanobiotechnology Applications in Plant Protection (eds. Abd-Elsalam K and Prasad R), Springer International Publishing AG 247–266

Hamaidia K, Soltani N (2014) Laboratory evaluation of a biorational insecticide, kinoprene, against Culex pipiens larvae: Effects on growth and development. Annual Res Rev Biol. 2263–2273

Ismail M et al (2018) Liposomes of dimeric artesunate phospholipid: a combination of dimerization and self-assembly to combat malaria. Biomaterials. Elsevier, 163:76–87

Jeong SH, Yeo SY, Yi SC (2005) The effect of filler particle size on the antibacterial properties of compounded polymer/silver fibers. J Mater Sci. Springer, 40(20):5407–5411

Joshi N, Jain N, Pathak A, Singh J, Prasad R, Upadhyaya CP (2018) Biosynthesis of silver nanoparticles using *Carissa carandas* berries and its potential antibacterial activities. J Sol-Gel Sci Techn 86(3):682–689. https://doi.org/10.1007/s10971-018-4666-2

Kalat SAM et al (2014) Use of topical liposomes containing meglumine antimoniate (Glucantime) for the treatment of L. major lesion in BALB/c mice. Exp Parasitol. Elsevier, 143:5–10

Kayalvizhi T, Ravikumar S, Venkatachalam P (2016) Green synthesis of metallic silver nanoparticles using Curculigo orchioides rhizome extracts and evaluation of its antibacterial, larvicidal, and anticancer activity. J Environ Eng. American Society of Civil Engineers, 142(9):C4016002

Khan SU et al (2018) Nanosilver: new ageless and versatile biomedical therapeutic scaffold. Int J Nanomedicine. Dove Press, 13:733

Khan SU et al (2019) Antimicrobial potentials of medicinal plant's extract and their derived silver nanoparticles: a focus on honey bee pathogen. Saudi J Biol Sci. Elsevier, 26(7):1815–1834

Kilama W, Ntoumi F (2009) Malaria: a research agenda for the eradication era. Lancet. Elsevier, 374(9700):1480–1482

Kolluri N, Klapperich CM, Cabodi M (2018) Towards lab-on-a-chip diagnostics for malaria elimination. Lab Chip. Royal Society of Chemistry, 18(1):75–94

Kovendan K et al (2012) Studies on larvicidal and pupicidal activity of Leucas aspera Willd. (Lamiaceae) and bacterial insecticide, Bacillus sphaericus, against malarial vector, Anopheles stephensi Liston.(Diptera: Culicidae). Parasitol Res. Springer, 110(1):195–203

Krishnaswamy K, Orsat V (2017) Sustainable delivery systems through green nanotechnology. In: Nano-and microscale drug delivery systems. Elsevier, pp 17–32

Kumar CV, McLendon GL (1997) Nanoencapsulation of cytochrome c and horseradish peroxidase at the galleries of α-zirconium phosphate. Chem Mater. ACS Publications, 9(3):863–870

Kumar H et al (2020) Flower-based green synthesis of metallic nanoparticles: applications beyond fragrance. Nanomaterials 10(4):766

Kumar S et al (2012) Nanotechnology as emerging tool for enhancing solubility of poorly water-soluble drugs. Bionano Sci 2(4):227–250

Lee S-W et al (2014) Effect of temperature on the growth of silver nanoparticles using plasmon-mediated method under the irradiation of green LEDs. Materials. Multidisciplinary Digital Publishing Institute, 7(12):7781–7798

Lees RS, Knols B, Bellini R, Benedict MQ, Bheecarry A, Bossin HC, Chadee DD, Charlwood J, Dabire RK, Djogbenou L (2014) Improving our knowledge of male mosquito biology in relation to genetic control programmes. Acta Trop. 132:S2–S11

Lorenz ES (2007) Potential health effects of pesticides. Pennsylvania State University, Pennsylvania

Macdonald IDG, Smith WE (1996) Orientation of cytochrome c adsorbed on a citrate-reduced silver colloid surface. Langmuir. ACS Publications, 12(3):706–713

Moon JJ et al (2012) Antigen-displaying lipid-enveloped PLGA nanoparticles as delivery agents for a Plasmodium vivax malaria vaccine. PLoS One. Public Library of Science, 7(2)

Milam CD, Farris JL, Wilhide JD (2000) Evaluating mosquito control pesticides for effect on target and nontarget organisms. Arch Environ Contam Toxicol 39(3):324–328

Nakache E et al (2000) Biopolymer and polymer nanoparticles and their biomedical applications. In: Handbook of nanostructured materials and nanotechnology. Elsevier, pp 577–635

Nogueira Barradas T et al (2016) Polymer-based drug delivery systems applied to insects repellents devices: a review. Curr Drug Deliv. Bentham Science Publishers, 13(2):221–235

Oliveira CB et al (2014) Liposomes produced by reverse phase evaporation: in vitro and in vivo efficacy of diminazene aceturate against Trypanosoma evansi. Parasitology. Cambridge University Press, 141(6):761–769

Omwoyo WN et al (2014) Preparation, characterization, and optimization of primaquine-loaded solid lipid nanoparticles. Int J Nanomedicine. Dove Press, 9:3865

Panneerselvam C et al (2016) Fern-synthesized nanoparticles in the fight against malaria: LC/MS analysis of Pteridium aquilinum leaf extract and biosynthesis of silver nanoparticles with high mosquitocidal and antiplasmodial activity. Parasitol Res. Springer, 115(3):997–1013

Pashchenko O et al (2018) A comparison of optical, electrochemical, magnetic, and colorimetric point-of-care biosensors for infectious disease diagnosis. ACS Infect Dis. ACS Publications, 4(8):1162–1178

Patra JK, Baek K-H (2014, 2014) Green nanobiotechnology: factors affecting synthesis and characterization techniques. J Nanomat

Patterson J, Sammon M, Garg M (2016) Dengue, Zika and chikungunya: emerging arboviruses in the New World. West J Emerg Med. California Chapter of the American Academy of Emergency Medicine (Cal/AAEM), 17(6):671

Pessoa LZ d S et al (2018) Nanosuspension of quercetin: preparation, characterization and effects against Aedes aegypti larvae. Rev Bras. SciELO Brasil, 28(5):618–625

Prasad R (2014) Synthesis of silver nanoparticles in photosynthetic plants. Journal of Nanoparticles, Article ID 963961, 2014, https://doi.org/10.1155/2014/963961

Prasad R (2016) Advances and Applications through Fungal Nanobiotechnology. Springer, International Publishing Switzerland (ISBN: 978-3-319-42989-2)

Prasad R (2017) Fungal Nanotechnology: Applications in Agriculture, Industry, and Medicine. Springer Nature Singapore Pte Ltd. (ISBN 978-3-319-68423-9)

Prasad R (2019a) Microbial Nanobionics: Basic Research and Applications. Springer International Publishing (ISBN 978-3-030-16534-5) https://www.springer.com/gp/book/9783030165338

Prasad R (2019b) Microbial Nanobionics: State of Art. Springer International Publishing (ISBN 978-3-030-16383-9) https://www.springer.com/gp/book/9783030163822

Prasad R, Pandey R, Barman I (2016) Engineering tailored nanoparticles with microbes: quo vadis. WIREs Nanomed Nanobiotechnol 8:316–330. https://doi.org/10.1002/wnan.1363

Prasad R, Pandey R, Varma A, Barman I (2017) Polymer based nanoparticles for drug delivery systems and cancer therapeutics. In: Natural Polymers for Drug Delivery (eds. Kharkwal H and Janaswamy S), CAB International, UK 53–70

Prasad R, Jha A, Prasad K (2018b) Exploring the Realms of Nature for Nanosynthesis. Springer International Publishing (ISBN 978-3-319-99570-0) https://www.springer.com/978-3-319-99570-0

Prasad R, Kumar V, Kumar M, Wang S (2018a) Fungal Nanobionics: Principles and Applications. Springer Nature Singapore Pte Ltd. (ISBN 978-981-10-8666-3) https://www.springer.com/gb/book/9789811086656

Prasad R, Kumar V, Kumar M, Choudhary D (2019) Nanobiotechnology in Bioformulations. Springer International Publishing (ISBN 978-3-030-17061-5) https://www.springer.com/gp/book/9783030170608

Priye A et al (2017) A smartphone-based diagnostic platform for rapid detection of Zika, chikungunya, and dengue viruses. Sci Rep. Nature Publishing Group, 7(1):1–11

Parthiban E, Manivannan N, Ramanibai R, Mathivanan N (2019) Green synthesis of silver-nanoparticles from Annona reticulata leaves aqueous extract and its mosquito larvicidal and anti-microbial activity on human pathogens. Biotechnol Rep 21:e00297.

Pandey V, Agrawal V, Raghavendra K, & Dash AP (2007) Strong larvicidal activity of three species of Spilanthes (Akarkara) against malaria (Anopheles stephensi Liston, Anopheles culicifacies, species C) and filaria vector (Culex quinquefasciatus Say). Parasitol Res 102(1):171–174

Qiu L, Jing N, Jin Y (2008) Preparation and in vitro evaluation of liposomal chloroquine diphosphate loaded by a transmembrane pH-gradient method. Int J Pharm. Elsevier 361(1–2):56–63

Rabinow BE (2004) Nanosuspensions in drug delivery. Nat Rev Drug Discov. Nature Publishing Group 3(9):785–796

Rashid JIA, Yusof NA (2018) Laboratory diagnosis and potential application of nucleic acid biosensor approach for early detection of dengue virus infections. Biosci Biotechnol Res Asia 15(2):245–255

Rong Z et al (2019) Smartphone-based fluorescent lateral flow immunoassay platform for highly sensitive point-of-care detection of Zika virus nonstructural protein 1. Anal Chim Acta. Elsevier, 1055:140–147

Rajagopal G, Nivetha A, Sundar M, Panneerselvam T, Murugesan S, Parasuraman P, Kumar S, Ilango S, Kunjiappan S (2021) Mixed phytochemicals mediated synthesis of copper nanoparticles for anticancer and larvicidal applications. Heliyon:e07360

Sahoo SK, Parveen S, Panda JJ (2007) The present and future of nanotechnology in human health care. Nanomedicine. Elsevier, 3(1):20–31

Santos-Magalhães NS, Mosqueira VCF (2010) Nanotechnology applied to the treatment of malaria. Adv Drug Deliv Rev. Elsevier, 62(4–5):560–575

Sarma H, Joshi S, Prasad R, Jampilek J (2021) Biobased Nanotechnology for Green Applications. Springer International Publishing (ISBN 978-3-030-61985-5) https://www.springer.com/gp/book/9783030619848

Saxena A, Tripathi RM, Singh RP (2010) Biological synthesis of silver nanoparticles by using onion (Allium cepa) extract and their antibacterial activity. Dig J Nanomater Bios 5(2):427–432

Schaffner F, Medlock JM, Bortel V, W. (2013) Public health significance of invasive mosquitoes in Europe. Clin Microbiol Infect. Elsevier, 19(8):685–692

Schwendener RA (2014) Liposomes as vaccine delivery systems: a review of the recent advances. Therap Adv Vaccines. SAGE Publications Sage UK: London, England, 2(6):159–182

Selvan SM et al (2018) Green synthesis of copper oxide nanoparticles and mosquito larvicidal activity against dengue, zika and chikungunya causing vector Aedes aegypti. IET Nanobiotechnol. IET, 12(8):1042–1046

Sen GL, Blau HM (2006) A brief history of RNAi: the silence of the genes. FASEB J. Federation of American Societies for Experimental Biology, 20(9):1293–1299

Sethi RS (1994) Transducer aspects of biosensors. Biosens Bioelectron. Elsevier, 9(3):243–264

Shakeel K et al (2019) Development and in vitro/in vivo evaluation of artemether and lumefantrine co-loaded nanoliposomes for parenteral delivery. J Liposome Res. Taylor & Francis, 29(1):35–43

Shargh VH et al (2012) Cationic liposomes containing soluble Leishmania antigens (SLA) plus CpG ODNs induce protection against murine model of leishmaniasis. Parasitol Res. Springer, 111(1):105–114

Siddhardha B et al (2012) Dubey and Basaveswara Rao MV2 in vitro antimicrobial and larvicidal spectrum of certain bioactive fungal extracts. Int J Res Pharm Biomedical Sci 3:115–155

Singh S, Kumar V, Dhanjal DS, Datta S, Prasad R, Singh J (2020) Biological Biosensors for Monitoring and Diagnosis. In: Singh J, Vyas A, Wang S, Prasad R (eds) Microbial Biotechnology: Basic Research and Applications. Springer Nature Singapore 317–336

Soni N, Prakash S (2012) Synthesis of gold nanoparticles by the fungus Aspergillus niger and its efficacy against mosquito larvae. Rep Parasitolo. Dove Press, 2:1–7

Soni N, Prakash S (2013) Possible mosquito control by silver nanoparticles synthesized by soil fungus (Aspergillus niger 2587). Scientific Research Publishing

Sowndarya P, Ramkumar G, Shivakumar MS (2017) Green synthesis of selenium nanoparticles conjugated Clausena dentata plant leaf extract and their insecticidal potential against mosquito vectors. Artif Cells Nanomed Biotechnol. Taylor & Francis, 45(8):1490–1495

Srivastava S, Usmani Z, Atanasov AG, Singh VK, Singh NP, Abdel-Azeem AM, Prasad R, Gupta G, Sharma M, Bhargava A (2021) Biological nanofactories: Using living forms for metal nanoparticle synthesis. Mini-Reviews in Medicinal Chemistry 21(2):245–265

Staples JE, Fischer M (2014) Chikungunya virus in the Americas—what a vectorborne pathogen can do. N Engl J Med. Mass Medical Soc, 371(10):887–889

Suryawanshi RK et al (2015) Towards an understanding of bacterial metabolites prodigiosin and violacein and their potential for use in commercial sunscreens. Int J Cosmet Sci. Wiley Online Library, 37(1):98–107

Salunkhe RB, Patil SV, Patil CD, Salunke BK (2011) Larvicidal potential of silver nanoparticles synthesized using fungus Cochliobolus lunatus against Aedes aegypti (Linnaeus, 1762) and Anopheles stephensi Liston (Diptera; Culicidae). Parasitology Res 109(3):823–831

Senthil-Nathan S (2015) A review of biopesticides and their mode of action against insect pests. In Environmental sustainability. Springer, (pp 49–63)

Tripathy S et al (2012) Synthesis, characterization of chitosan–tripolyphosphate conjugated chloroquine nanoparticle and its in vivo anti-malarial efficacy against rodent parasite: a dose and duration dependent approach. Int J Pharm. Elsevier, 434(1–2):292–305

Tripathy S et al (2013) The impact of nanochloroquine on restoration of hepatic and splenic mitochondrial damage against rodent malaria. J Nanopart. Hindawi Publishing Corporation, 2013

Vinayagam S et al (2018) DNA-triangular silver nanoparticles nanoprobe for the detection of dengue virus distinguishing serotype. Spectrochim Acta A Mol Biomol Spectrosc. Elsevier, 202:346–351

WHO (2014) Lymphatic Filariasis "Fact Sheet N 102." World Health Organization: Geneva, Switzerland

Yen CW, de Puig H, Tam JO, Gómez-Márquez J, Bosch I, Hamad-Schifferli K, Gehrke L (2015) Lab Chip

Yen CW, de Puig H, Tam JO, Gómez-Márquez J, Bosch I, Hamad-Schifferli K, Gehrke L (2015) Multicolored silver nanoparticles for multiplexed disease diagnostics: distinguishing dengue, yellow fever, and Ebola viruses. Lab on a Chip, 15(7):1638–1641

Zhang X-Q et al (2012) Interactions of nanomaterials and biological systems: implications to personalized nanomedicine. Adv Drug Deliv Rev. Elsevier, 64(13):1363–1384

Chapter 12
Phytosynthesized Metal Nanomaterials as an Effective Mosquitocidal Agent

M. Suresh, Satheeshkumar Balu, S. Cathy Jose, and Jaison Jeevanandam

Contents

12.1	Introduction	370
12.2	Synthesis Approaches of Metal Nanomaterials	371
	12.2.1 Physical Approach	372
	12.2.2 Chemical Approach	373
	12.2.3 Biological Approach	375
12.3	Limitations of Physicochemical Approach Compared to Biosynthesis Approaches	376
	12.3.1 Advantages and Disadvantages of Biological Approach	377
12.4	Phytosynthesis of Metal Nanomaterials	378
	12.4.1 Phytosynthesis of Common Metallic Nanoparticles	379
12.5	Mosquitocidal Activity of Metal Nanomaterials	381
	12.5.1 Gold Nanoparticles	382
	12.5.2 Silver Nanoparticles	383
	12.5.3 Copper Nanoparticles	384
	12.5.4 Other Metal Nanoparticles	384
12.6	Mosquitocidal Mechanism of Phytosynthesized Metal Nanoparticles	385
12.7	Future Perspective and Conclusion	387
References		389

M. Suresh (✉)
Loyola Institute of Frontier Energy (LIFE), School of Environmental Toxicology and Biotechnology, Postgraduate and Research Department of Advanced Zoology and Biotechnology, Loyola College, Chennai, Tamil Nadu, India

S. Balu
Department of Ceramic Technology, A.C. Tech Campus, Anna University, Chennai, Tamil Nadu, India

S. C. Jose
Postgraduate and Research Department of Advanced Zoology and Biotechnology, Loyola College, Chennai, Tamil Nadu, India

J. Jeevanandam
CQM - Centro de Química da Madeira, MMRG, Universidade da Madeira, Funchal, Portugal
e-mail: jaison.jeevanandam@staff.uma.pt

© The Author(s), under exclusive license to Springer Nature Switzerland AG 2022
A. Krishnan et al. (eds.), *Emerging Nanomaterials for Advanced Technologies*,
Nanotechnology in the Life Sciences, https://doi.org/10.1007/978-3-030-80371-1_12

12.1 Introduction

Mosquitoes are notable vectors of various diseases, including dengue, malaria, filariasis and yellow fever (Khater et al. 2019). According to the World Health Organization (WHO) in 2020, malaria is the most common parasitic infection transmitted by mosquitoes, which leads to 219 million cases globally, with more than 400,000 deaths, every year. Further, Dengue is a viral infection, which is transmitted by mosquitoes and affects 96 million symptomatic cases and 40,000 deaths globally, every year (WHO 2020). Thus, there is a quest to develop a novel and efficient mosquitocidal agent for a long time to eradicate the mosquito population. There are several conventional mosquito repellents or mosquitocidal agents, such as dichlorodiphenyltrichloroethane (DDT), lindane, malathion, fenitrothion, propoxur, chlorpyrifos-methyl, pirimiphos-methyl, bendiocarb, permethrin, cypermethrin, alpha-cypermethrin, cyfluthrin, deltamethrin, lambda-cyhalothrin, bifenthrin and etofenprox, that are synthesized via chemicals. These chemicals-based mosquitocidal agents have exhibited adverse toxicity towards humans and the environment with several limitations (Suresh et al. 2020). Thus, biological mosquitocides, such as plant extracts as mosquito repellents and mosquitocides, microbial mosquitocides, biological control agents, including fishes, sterile insect technique (SIT), boosted SIT, transgenic mosquitoes and symbiont-based methods with bacteria are used as alternatives to the chemical mosquitocides (Huang et al. 2017). However, the efficiency of these bio-mosquitocides are low, as compared to the chemical-based mosquitocidal agents (Benelli et al. 2016). Further, mosquitoes develop resistance towards most of the mosquitocidal agents, due to their ability to adapt to harsh environments and rapid evolution with less life cycle time (Senthil-Nathan 2019). Thus, there is a requirement for a novel mosquitocidal agent with high efficacy to inhibit the growth and development of mosquitoes and are safer for non-target organisms.

Nanotechnology is an emerging field of forefront innovation, which involves the contribution from various fields, including chemistry, physics, biology, material science and medicine (Quazi 2020). In recent times, nanomaterials are introduced as a potential alternative to conventional mosquitocidal chemicals and biological agents due to their high surface-to-volume ratio and ability to inhibit their growth at the cellular level (Borgheti-Cardoso et al. 2020). However, nanomaterials prepared via physical and chemical approaches are either costly or toxic to plants, animals, humans and the environment (Maroufpour et al. 2020; Sarma et al. 2021; Saglam et al. 2021). Thus, nanomaterials fabricated using plant extracts are widely used recently as an effective mosquitocidal agent (Rana et al. 2020). The synergistic mosquitocidal property of the phytochemicals from plants and nanomaterials is beneficial in inhibiting the population of mosquitoes by targeting their egg, pupa, larva and adult (Seetharaman et al. 2018). Moreover, the phytochemicals as surface functional groups in these nanomaterials are beneficial in reducing their adverse toxic effects (Rheder et al. 2018). Hence, this chapter is an overview of the various synthesis approaches to fabricate metal nanoparticles and the significance of phytosynthesis approach. In addition, the mosquitocidal property of the metal nanoparticles and their mechanism of action are also discussed.

12.2 Synthesis Approaches of Metal Nanomaterials

The synthesis approaches are highly significant in the fabrication of nanomaterials to yield them with exclusive and desired properties for specific applications. Various synthesis approaches, such as physical, chemical and biological methods, are available for the formation of distinct nanomaterials as listed in Table 12.1 (Gudikandula and Charya Maringanti 2016).

Table 12.1 Size and morphology of metal nanoparticles synthesized via various approaches

Nanomaterial	Synthesis method	Size (nm)	Morphology	Reference
Gold	Laser ablation	11–5	Spherical	Naharuddin et al. (2020)
Gold	Sputtering	1.6–7.4	Spherical	Ishida et al. (2016)
Gold	Sonochemical	92.37–112.3	Spherical	Usman et al. (2019)
Silver	Laser ablation	26–21	Spherical	Sadrolhosseini et al. (2019c)
Silver	Sputtering	5.9, 5.4 and 3.8	Spherical, worm-like network	Asanithi et al. (2012)
Silver	Cryomilling	4–8	Spherical	Kumar et al. (2016)
Silver	Ball milling	1–30	Spherical	Rak et al. (2016)
Silver	Sonochemical	3–8	Spherical	Vinoth et al. (2017)
Platinum	Sonochemical	3.5	–	Jameel et al. (2020)
Platinum	Microwave-assisted synthesis	2–8	Spherical	Inwati et al. (2016)
Gold	Chemical reduction	15–30	Spherical	Dong et al. (2020)
Copper	Chemical reduction	28.3	Cubic	Khan et al. (2016)
Platinum	Pulsed laser ablation	4.7	Spherical	Mendivil Palma et al. (2016)
Gold	Microbial synthesis	106	Flower shaped	Singh et al. (2016a)
Gold, silver	Microbial synthesis	102, 92.4	Spherical	Singh et al. (2016b)
Silver	Microbial Synthesis	15–25	Spherical	Gudikandula et al. (2017)
Platinum	Microbial synthesis	28.96	Cubic, spherical, truncated	Gupta and Chundawat (2019)
Silver, copper	Green alga synthesis	40–100, 10–70	Cubic, spherical, truncated	Arya et al. (2018)
Gold	Brown alga synthesis	8.4	Spherical	González-Ballesteros et al. (2017)
Silver	Marine green alga synthesis	25	Spherical	Edison et al. (2016)

12.2.1 Physical Approach

Fabrication of nanomaterials through physical methods involves high electrical or thermal energy and mechanical pressure that can cause material abrasion, melting and evaporation, which eventually leads to generation of nanoparticles (NPs). The swift reaction time to yield the final product, high purity, uniform size and shape of nanoparticles are the major advantages of physical methods (Jeyaraj et al. 2019). Several physical methods, such as microwave, ball milling, vapor deposition, laser ablation, arc discharge, flame pyrolysis and sputter deposition are available for the production of NPs (Jeyaraj et al. 2019). Recently, Shah and Zhang (2019) synthesized novel hexagonal gold nanoparticles (AuNPs) by microwave-assisted synthesis method. The study reported for the first time that the hydrolyzed organosilane (3-mercaptopropyl) trimethoxysilane (MPTMS) can be used to reduce the Au^+ ions under microwave irradiation. The results revealed that the hexagonal AuNPs were formed under different irradiation power, where the higher irradiation power can lead to the formation of larger amounts of AuNPs. In addition, the average size of the AuNPs was found to increase with an increment in the irradiation time, when the reaction was completed within 60s of the time period (Shah and Zheng 2019). In another study, Bayazit et al. (2016) combined microwave irradiation method with microflow chemistry to fabricate AuNPs with controllable size. The spherical AuNPs with the mean diameter between 4 and 15 nm were formed within 90s of irradiation time. Further, the modification ratio of citrate to Au were identified to be the influencing factor of the AuNPs morphology, which leads to the formation of Au nanowire-like structure with polygonal heads. The study explored that the fabrication of AuNPs with smaller particle size and size distribution is possible with ultrafast and effective microwave-assisted technique (Bayazit et al. 2016).

Recently, Rafique et al. (2019) utilized another novel physical approach to produce silver nanoparticles (AgNPs) via the laser ablation method using continuous wave (CW) diode laser and a pulsed neodymium-doped yttrium aluminium garnet (Nd:YAG) laser. The study revealed that utilization of two distinct laser types can produce spherical AgNPs with various sizes, which includes 20 and 9 nm, when using continuous wave (CW) laser and Nd:Yag laser, respectively. It can be noted that the clean and surface contaminant-free AgNPs were produced via simple laser ablation method. The advantage of this method is its simplicity, rapid reaction time, low cost and eco-friendliness to produce ultra-pure AgNPs, compared to other conventional methods (Rafique et al. 2019). Likewise, Nancy et al. (2018) synthesized AgNPs with controllable sizes in liquid medium via laser ablation for different laser energy densities. The results emphasized that nearly spherical-shaped AgNPs were produced. Further, the variation in the laser energy has been identified to produce AgNPs with various sizes. The smaller AgNPs with a size range of 5 nm were obtained with lower laser energy, whereas increment in the laser energy to high level has led to the growth of particle size to 35 nm (Nancy et al. 2018). Moreover, the mechanism of nanoparticle synthesis has been identified to involve the heat generated by the incidence of laser on the metal target followed by melting and vaporization. Later, the metal plate releases the metal nanoparticles depending on the absorbed energy and plasma plume expands (Sadrolhosseini et al. 2019a).

In another physical approach, Yadav and Vasu (2016) studied the synthesis of copper nanoparticles (CuNPs) by a wet milling process using a planetary ball milling equipment. The ball milling process involves the mechanical co-grinding of metal powders in a cylindrical container with metallic balls. The study revealed that an increment in the milling time will lead to decrement in the crystal size. Further, it is reported that the average crystal size of the milled CuNPs is 21 nm, after 40 h of milling time (Yadav and Vasu 2016). Furthermore, Ramesh et al. (2020) used waste copper chip material to produce CuNPs via high-energy ball milling process (HEBM). After 72 h of milling, hollow sphere-shaped CuNPs were obtained with the size range of 50 nm. This study demonstrated that ball milling can be an effective and economical method to produce CuNPs (Ramesh et al. 2020). Moreover, Wang et al. (2017) synthesized novel platinum nanoparticles (PtNPs) using atomic layer deposition (ALD) method. In this study, the effect of reaction temperature, number of ALD cycles and the type of substrate for the formation of PtNPs were studied. The results revealed that the growth orientation of Pt and its shape can be altered via various substrates. In addition, the particle size and density vary according to the type of ALD methods, number of cycles and reaction temperatures. The study demonstrated that the performance of PtNPs can be changed by varying the fabrication parameters of the ALD process (Wang et al. 2017). In another study, Sadrolhosseini et al. (2019b) synthesized PtNPs via novel laser ablation technique in graphene oxide solution. The study revealed the formation of platinum-graphene oxide nanocomposite with the size range of 12 to 22 nm, where the particle size further decreased with an increment in the concentration of PtNPs and ablation time (Sadrolhosseini et al. 2019b). Even though physical approaches are highly beneficial in yielding smaller sized nanoparticles with high purity, the cost of equipment required is high, which remains as a limitation of physical synthesis, compared to the other nanoparticle synthesis approaches (Jeevanandam et al. 2016).

12.2.2 Chemical Approach

Chemical synthesis approaches can be used to synthesize nanoparticles with desired size and shape. The significant factor to control the size and shape of nanoparticles is the nucleation and growth kinetics of the particle formation in a solution (Nikam et al. 2018). The main advantage of this method is its simplicity and easy scalability, under regular environmental conditions. It has been reported that even the minute changes in the reaction parameters, such as precursors, reducing agents, capping agents, solvent and pH, can lead to a substantial modification in the size, shape, distribution and self-assembly patterns of the nanoparticles (Iravani et al. 2014). The mechanism involved in chemical synthesis is the formation of atomic groups that are accomplished by the chemical reactions under mild reaction condition. Thus, the resultant atomic groups can undergo nucleation followed by growth process, which can lead to the formation of nanoparticles with a definite shape and size (Yu et al. 2008). Tyagi et al. (2016) synthesized citrate reduced gold nanoparticles (AuNPs) via the simplified chemical reduction method to yield fairly monodispersed gold nanoparticles. The study discussed the role of pH and the ratio of reactant concentration to explore the size control of AuNPs. The results demonstrated that the nanoparticles were uniform in morphology and monodispersed at optimal pH, whereas the particles were of non-uniform size and shape with a change in the optimal pH (Tyagi et al. 2016). In another study, Abkenar and Naderi (2016) synthesized AuNPs by the chemical reduction method using copper anode slime as a gold precursor with 0.1% of Au, which is an insoluble product that is deposited during the electrorefining of copper at the bottom of the electrorefining tank. The extraction of gold from the anode slime was performed using thiourea solution and the resultant cationic complex of gold-thiourea was used as gold solution. The reducing agents such as VenMet solution and sodium citrate were used to synthesize gold nanoparticles directly from the gold-thiourea solution. The results revealed that the reduction of gold-thiourea using two different reducing agents can yield various sizes and shapes of gold nanoparticles, including cubical and spherical nanoparticles with an average size range of 15–190 nm and 190–500 nm for VenMat solution and sodium citrate, respectively. This study suggested that the variation in the reducing agent and other synthesis parameters can lead to nanoparticles with distinct morphology (Abkenar and Naderi 2016). Therefore, chemical reduction is the simple and economical method to produce nanoparticles with a diverse range of size and shape.

Recently, Cheng et al. (2016) synthesized novel silver nanoparticles (AgNPs) with controllable size and morphology. The silver reduction was combined with citrate as reducing agent and tannic acid as both auxiliary reducing agent and stabilizer. It is worthy to note that the alteration of the reaction parameters and concentration of tannic acid in the reaction mixture has resulted in various AgNPs morphologies. Further, it has been revealed that the larger sized AgNPs will be yielded upon increasing the concentration of tannic acid. In addition, the reaction condition at 60 °C has led to the formation of particles with diverse shapes, such as spherical, rod-like, triangle and polygon NPs, with an increment in the tannic acid concentration (Cheng et al. 2016). Likewise, Gurusamy et al. (2017) synthesized multi-coloured silver nanoparticles of various sizes by varying the concentration precursor, capping and reducing agent, such as silver nitrate, poly-(N-vinylpyrrolidone) and hydrazine hydrate, respectively. The results strongly suggested that the variation in the morphology of AgNPs is dependent on both capping and reducing agents. Their findings also indicated that the successful synthesis of AgNPs with various sizes and shapes can be obtained by a simple and economical chemical reduction method (Gurusamy et al. 2017). In the same way, the copper nanoparticles were synthesized by several chemical methods and the most popular method is the chemical reduction method that utilizes exclusive reducing agents to reduce copper from its salts. As an evidence, Begletsova et al. (2017) produced copper nanoparticles via cetylpyridinium chloride as stabilizing agent and hydrazine hydrate (HH) as reducing agent by the simple chemical reduction method. The study demonstrated that the alterations in the volume of HH and pH can yield stable CuNPs with a size range of 40 to 80 nm under optimal reaction conditions

(Begletsova et al. 2017). In another study, novel platinum nanoparticles (PtNPs) were synthesized by Nagao et al. (2017) (Nagao et al. 2017) by utilizing sodium borohydride to reduce Pt ions and polyethyleneimine (PEI) as a protective agent. The results revealed that cubic and tetrahedral PtNPs are synthesized at lower pH, whereas spherical PtNPs were obtained at higher pH of the PEI solution. It is evident from this study that the one pot chemical synthesis approach possesses ability to yield PtNPs with narrow size distribution by simply adjusting the pH of the PEI. Even though, chemical approaches are simple and requires less cost, compared to physical methods, the utilization of toxic chemicals is a major limitation for utilizing these nanoparticles in biomedical applications (Jeevanandam et al. 2016).

12.2.3 Biological Approach

Nature consists of several organisms, both macroscopic and microscopic, such as bacteria, fungi, algae, yeast and plants. These organisms contain natural biomolecules which have been identified to be significant in the production of nanoparticles with distinct morphologies. They can act as a green fuel for the formation of safe and environment-friendly nanoparticles (Sharma et al. 2019). The use of nanoparticles in the field of biomedical and agriculture application has led to a need for producing nanoparticles with eco-friendly and reliable methods. The best option to achieve this goal is to utilize biological entities to reduce the metal ions to metal nanoparticles (Kaur 2018). Thus, the synthesis of metallic nanoparticles (AuNPs) via biological entities, such as microbes and plants, has gained the focus of researchers due to the current interest to produce green nanoparticles with unique properties, compared to their bulk counterparts (Sehgal et al. 2018). Camas et al. (2019) developed a novel biogenic approach for the synthesis of AuNPs using marine bacterial strains. This study is the first to report the use of actinobacterial strain (K1D109), which belongs to the *Cittricoccus sp.* isolated from *Petrosia ficiformis* sponge, to reduce the Au ions to AuNPs. The study showed that the optimum reaction conditions, such as 35 °C and reaction time of 24 h can lead to the formation of spherical AuNPs with the size range of 24–65 nm (Camas et al. 2019). In another study, Molnar et al. (2018) synthesized 6 to 40 nm-sized AuNPs using 29 thermophilic filamentous fungi. The results revealed that two steps were involved in the formation of fungi mediated AuNPs, such as reduction of Au^{3+} to Au^0 and stabilization of core AuNPs by proteins as capping ligands (Molnár et al. 2018). The mechanism involved in the formation of nanoparticles includes entrapment of metal ions inside or on the surface of the microbial cells, through the electrostatic attraction between the ions and the negatively charged cell wall as well as the reduction in the presence of enzymes (Li et al. 2011).

Other than gold nanoparticles, AgNPs are the most common nanoparticles that are used in various applications, such as solar fuel cell efficiency improvement, radio sensitizers and as antimicrobial agents. Shu et al. (2020) synthesized green AgNPs using yeast extract as reducing agent. The experimental results revealed that the yeast extracts reduced Ag ions and obtained well dispersed and uniform spherical-shaped AgNPs with an average size of 13.8 nm. Further, the study emphasized that the biomolecules of amino acids, alpha-linolenic acid and carbohydrates present in the yeast extract have an active role in the formation of AuNPs (Shu et al. 2020). In another study, Saravanan et al. (2017) synthesized AgNPs using *Leuconostoc lactis* bacterial strains that produce exopolysaccharide (EPS), which are isolated from idli batter. The EPS was precipitated through centrifugation, dissolved in Milli Q water and dialyzed at 4 °C for 48 h for purification. The extracted EPS from *Leuconostoc lactis* bacterial strain was used to reduce the Ag ions under constant stirring for 24 h. The results revealed that the bacterial EPS can act as reducing and stabilizing agent to yield 35 nm-sized spherical AgNPs (Saravanan et al. 2017). Recently, Arya et al. (2018) used green alga *Botryococcus braunii* to synthesize copper and silver nanoparticles. The green alga was isolated by the serial dilution method and incubated for three weeks and the algal biomass was used to reduce the Cu and Ag ions. The results revealed that the green algae can be successfully used to synthesize cubical morphology with an average size of copper and silver nanoparticles found to be 10–70 nm and 40–100 nm, respectively (Arya et al. 2018). Likewise, Gupta et al. (2019) used the biological approach to synthesize platinum nanoparticles via the fungus *Fusarium oxysporum*. The results revealed that the fungus-mediated green synthesis approach can produce cubical, spherical and truncated triangular-shaped platinum nanoparticles with an average size of 28.96 nm. The study stated that the presence of proteins, long chain fatty acids, amides and polysaccharides in the fungal extract as reducing and capping agents is responsible for the formation of PtNPs (Gupta and Chundawat 2019). Even though, microbial synthesis approaches of nanoparticles are less- or nontoxic to humans and the environment, the stability of the nanoparticle and the longer reaction time are the specific limitations which hurdle their large-scale synthesis, compared to conventional (physical and chemical) synthesis approaches. Thus, researchers prefer a hybrid synthesis approach to combine conventional and biosynthesis approaches for the fabrication of stable, monodispersed and less-toxic nanoparticles (Jeevanandam et al. 2016).

12.3 Limitations of Physicochemical Approach Compared to Biosynthesis Approaches

The production of toxic chemicals, salts and stabilizing agents as byproducts is enormous in physicochemical nanoparticle synthesis approaches, compared to green synthesis or biogenic techniques (Das et al. 2017). Both physical and chemical synthesis require high power consumption, high radiation, highly concentrated reducing and stabilizing agents as well as sophisticated apparatus, which will lead to higher costs than green synthesis methods. Further, the byproducts and the nanoparticles from the physical and chemical synthesis methods are often a great threat for the environment and the human health as it can lead to acute or chronic toxicity. Hence, green synthesis approach is highly recommended to synthesized nanoparticles in recent times, as it requires less energy and is eco-friendly in nature (Parveen et al. 2016a; Soshnikova et al. 2018a; Prasad 2019a, b).

12.3.1 Advantages and Disadvantages of Biological Approach

The most significant advantage of biological method is its ability to yield nanoparticles in a single-step bio-reduction process (Parveen et al. 2016b) without or less energy consumption. In addition, the nanoparticles from biogenic approaches are highly reproducible, polydisperse, non-toxic and eco-friendly with less health risk and are widely suggested to be beneficial in biomedical and clinical applications (Das et al. 2017; Aziz et al. 2016, 2019; Prasad et al. 2016).

12.3.1.1 Bacteria

Advantages Bacterial strains are cheap to procure, grow rapidly, easy to cultivate, compared to other biological systems, and can lead to high production of nanoparticles. Conditions such as oxygenation, temperature and incubation time can be controlled to modify the morphology of nanoparticles. The deftness of bacterial extracts in the manipulation of nanoparticle morphology provides supremacy over other biological organisms.

Disadvantages Intracellular synthesis of biogenic nanoparticles can yield nanoparticles slowly, after several hours or days, compared to physical and chemical methods. Further, the nanoparticles synthesized by microorganisms are prone to disintegration or agglomeration and deteriorate over a time period (Jamkhande et al. 2019).

12.3.1.2 Fungi

Advantages Fungi have high forbearance to metals and are easy to handle. The nanoparticles extracted from fungi are highly beneficial in agricultural pest control, due to their less toxicity towards plants and specific biomolecules as surface functional groups. Fungi produce several proteins, which eventually helps to increase the rate of nanoparticle synthesis (Pantidos and Horsfall 2014; Prasad 2016, 2017; Prasad et al. 2016, 2018a, b; Abdel-Aziz et al. 2018).

Disadvantages Most of the fungal species, that are utilized for nanoparticle synthesis, are highly toxic to humans or plants, which is a major obstacle for large scale production (Vahabi et al. 2011). Further, the fungal synthesized nanoparticles are unstable, polydispersed and are not pure, due to the presence of several biomolecules as surface functional groups (Rai et al. 2015).

12.3.1.3 Algae

Advantages Algae-synthesized nanoparticles are highly effective as anti-fungal agents, anti-bacterial agents, anti-biofilms and biosensors to detect diseases, such as cancer and diabetes (Aziz et al. 2014, 2015). Further, the synthesis requires less temperature, compared to other biosynthesis approaches, and are less toxic and eco-friendly in nature.

Disadvantages Nanoparticles synthesized are pernicious to eukaryotic cells (LewisOscar et al. 2016).

12.3.1.4 Virus

Advantages The virus-mediated nanoparticle synthesis or virus-like nanoparticles can be synthesized in large quantities in a short span. They are robust, polyvalent and dynamic. The viral nanoparticles obtained from plant viruses and bacteriophages are more beneficial in nature, as they are safe to humans with less side effects. Further, VNPs are beneficial in targeted cancer imaging and therapy (Steinmetz 2010).

Disadvantages The engineered viral nanoparticles may lead to mutations, which may cross-infect other species (Jeevanandam et al. 2018).

12.4 Phytosynthesis of Metal Nanomaterials

The limitations in the microbe-mediated nanoparticle fabrication have led to the emergence of phytosynthesis approaches, where biomolecules named phytochemicals extracted from plants are utilized as reducing and stabilizing agents for nanoparticle synthesis. Recently, the attraction towards phytosynthesis from plant biomass or extracts in increasing, due its eco-friendliness, affordability, rapidity, cost-effectiveness, non-toxicity and its ability to produce an abundance of nanoparticles. The phenomenon of phytosynthesis is simpler as it avoids multistep processes, such as cultivation and maintenance of microbes, preparation and extraction of enzymes (Prasad 2014; Andra et al. 2019). Further, synthesis of nanoparticles via plant extracts is rapid, compared to microbial synthesis approach, as plants serve as a natural source of metabolites, which can act as reducing, capping and stabilizing agents (Bamoharram et al. 2012). Moreover, the production of nanoparticles from plants is more stable with low contamination level. In recent times, the use of green synthesis with plant extracts is widely under research due to its essential requirement in various fields, especially biomedical and environmental applications (Zhu et al. 2019). Additionally, phytosynthesis of plant extract produces nanoparticles that can be used to increase the biomass in agriculture and improve fruit taste and crop yield of edible plants, such as carrot, radish, watermelon and tomato (Husen and Siddiqi 2014; Singhal et al. 2017).

12.4.1 Phytosynthesis of Common Metallic Nanoparticles

Nature provides an enormous amount of plant species, in which most of them can be utilized for the synthesis of novel nanoparticles. The phenomenon in which green plants are used for nanoparticle synthesis is termed as green route or phytosynthesis approach, which is a subclass of biosynthesis methods. Green chemistry or green route approach plays a major role in producing distinct types of metals from different parts of the plant. Recently, it has been reported that the common metal nanoparticles, such as gold, silver, palladium, platinum, copper and ferric iron can be synthesized via phytosynthesis approach (Kuppusamy et al. 2016).

12.4.1.1 Gold Nanoparticles

Gold is the most common metal to be synthesized via phytosynthesis, due to its wide applications in several fields ranging from biomedical to electronics. Patil et al. (2017) extracted phytochemicals from the galls of *Rhus chinensis* plant for the synthesis of spherical- and oval-shaped, 20–40 nm-sized gold nanoparticles. The study showed that the hydroxyl groups, ether and ester in gallotannins, tannins and polyphenols, are responsible for the gold nanoparticle formation. The resultant Au nanoparticles are demonstrated to possess dose-dependent cytotoxicity towards various types of cancer cell lines, such as human gastric adenocarcinoma (MKN-28), human liver (Hep-3B) and human bone osteosarcoma (MB-63) (Patil et al. 2017). Likewise, Soshnikova et al. (2018b) synthesized 5–15 nm-sized, spherical Au nanoparticles using distinct varieties of dried *Fructus Amomi* (cardamom) fruits, such as *Amomum villosum* and *Elettaria cardamomum*. This study also revealed that the existence of phenols with hydroxyl groups, tannins with precipitated alkaloids and proteins, and terpenoids with several isoprene units in the plant extract served as a reducing and capping agent for the nanoparticle formation (Soshnikova et al. 2018b). Similarly, Gupta et al. (2019) fabricated 10–25 nm-sized, spherical Au nanoparticles via the phytochemicals extracted from the medicinal plant named *Ocimum gratissimum* Linn. The Fourier transform infrared (FTIR) spectra revealed the presence of terpenoids, which have been identified as the prime reducing and capping agents to form gold nanoparticles (Gupta et al. 2019). Further, Balamurugan et al. (2016) synthesized 5–50 nm-sized, spherical gold nanoparticles via flower extracts of *Peltophorum pterocarpum* as capping and reducing agent. The reducing sugars, sterols, carbohydrates, glycosides and flavonoids present in the flower extract are proposed to be responsible for nanoparticle formation (Balamurugan et al. 2016).

12.4.1.2 Silver Nanoparticles

Silver is another metal that is phytosynthesized as nanosized particles, which has gained applicational importance in the field of biomedical and pharmaceuticals, next to gold nanoparticles, due to its excellent antimicrobial property (Srikar et al. 2016). Murugan et al. (2016) utilized aquatic plants, such as sponge-weed named *Codium tomentosum*, extract for the synthesis of 20–40 nm-sized, irregular-shaped silver nanoparticles. The FTIR results revealed that the presence of peptide linkages, polyphenols, proteins, enzymes or polysaccharides, flavones and terpenoids in the aqueous extract is the deciding factor in the formation of silver nanoparticles (Murugan et al. 2016). Further, Pilaquinga et al. (2019) fabricated 10–14 nm-sized, spherical-shaped silver nanoparticles via the fruit extract of *Solanum mammosum*. The presence of alkaloids, such as indol-alkaloids, alpha-chaconine, alpha-tomatine, alpha-solanine, pirrolizidin, tropane and glycoalkaloids are identified to be crucial for silver nanoparticle formation as well as their mosquitocidal property (Pilaquinga et al. 2019). Furthermore, Arjunan et al. (2012) extracted phytochemicals from *Annona squamosa* and used them for the synthesis of irregular-shaped silver nanoparticles, where the FTIR spectra indicated that the carbonyl groups in the amino acid residues play a crucial role in the nanoparticle formation. The study also showed that these phytosynthesized nanoparticles possess enhanced mosquitocidal property against *Anopheles stephensi*, *Aedes aegypti* and *Culex quinquefasciatus* to combat diseases such as filariasis, malaria and dengue (Arjunan et al. 2012). Moreover, Dinesh et al. (2015) utilized *Aloe vera* leaf extract for the synthesis of 35–55 nm-sized silver nanoparticles of spherical and cubical morphology. The study emphasized that the existence of carbonyl groups from polyphenols, such as epigallocatechin gallate, catechin gallate, theaflavin, epigallocatechin, epicatechin gallate and gallocatechin gallate, in the leaf extract leads to the formation of nanoparticles (Dinesh et al. 2015). Besides, Velmurugan et al. (2015) used the leaf extract of *Prunus* x *yedoensis* for the synthesis of 20–70 nm-sized, oval- and spherical-shaped silver nanoparticles. The FTIR spectra revealed that the existence of alkaloids, phytosterol, amino acids, flavonoids and triterpenoids as the surface functional biomolecules in plant extracts are responsible for the effective nanoparticle synthesis (Velmurugan et al. 2015).

12.4.1.3 Copper Nanoparticles

Copper nanoparticles are also widely used in biomedical applications (Yadav et al. 2017) and phytosynthesized CuNPs will be highly beneficial to enhance its biological properties. Tahvilian et al. (2019) fabricated 45–50 nm-sized, spherical-shaped CuNPs using the aqueous leaf extract of *Allium saralicum*. It has been identified from the FTIR spectra that the carbonyl and sp-2 carbon groups present in the leaf extract play a critical role in the synthesis of nanoparticles and reduce their cytotoxicity (Tahvilian et al. 2019). Similarly, Rajesh et al. (2018) fabricated ~15 nm-sized, monodispersed, spherical CuNPs with the help of bud extract from *Syzygium aromaticum*. The FTIR spectra revealed that the proteins along with carotenoids, tannins, alkaloids and flavonoids in the extract act as reducing and stabilizing agents for the nanoparticle formation (Rajesh et al. 2018). Likewise, Kuppusamy et al. (2017) synthesized 45–100 nm-sized, spherical-shaped CuNPs using aqueous *Commelina nudiflora* extract. The flavonoids, proteins, alkaloids and reducing sugars present in the plant extract are identified to be the reducing and capping agents that are responsible for the fabrication of nanoparticles (Kuppusamy et al. 2017). Moreover, Harne et al. (2012) utilized the aqueous latex extract of *Calotropis procera* for the fabrication of 5–30 nm-sized, spherical and monodispersed CuNPs. The study emphasized via FTIR spectra that the secondary amine, carboxylic acid, amide II, amide III, alcohol and C-N stretching of amines along with proteins act as a surface functional group for the nanoparticle formation (Harne et al. 2012).

12.4.1.4 Other Metal Nanoparticles

Recently, several metal nanoparticles have been synthesized via phytosynthesis approach. Pan et al. (2020) utilized the skin extract of red peanut named *Arachis hypogea* for the synthesis of iron nanoparticles. Biomolecules, such as flavonols, epicatechin, phenolic compounds and anthocyanins, present in the skin extract are responsible for the nanoparticle formation (Pan et al. 2020). Further, Cui et al. (2020) used polysaccharide from *Ginkgo biloba* leaves for synthesizing palladium nanoparticles and stabilizing them to form spherical, monodispersed and 7–14 nm sized nanoparticles (Cui et al. 2020). Furthermore, Sahin et al. (2020) synthesized 22 nm-sized, monodispersed and spherical-shaped palladium nanoparticles from the peel extract of *Punica granatum* (Pomegranate). This study showed that the presence of tannins, phenolics and flavonoids in the peel extract is responsible for the nanoparticle formation (Şahin Ün et al. 2020). Moreover, 1–6 nm-sized, spherical-shaped platinum nanoparticles were synthesized recently with the help of black cumin seed (*Nigella sativa*) extract. The synergistic effect of biomolecules present in the seed extract has been identified to be the responsible factor for the nanoparticle formation (Aygun et al. 2020). It can be noted from all these studies that the phytosynthesis approach predominantly yields spherical nanoparticles, and their shapes can be modified via alterations in the synthesis parameters (Jeevanandam et al. 2019).

12.5 Mosquitocidal Activity of Metal Nanomaterials

Mosquito vectors that belongs to the order *Diptera* and genera *Anopheles*, *Aedes* and *Culex* possess the ability to transmit various diseases, such as dngue, malaria, chikungunya, filariasis, Japanese encephalitis, yellow fever and Zika. These mosquito-borne diseases are responsible for millions of deaths every year, globally. In recent times, several nanomaterials are used to inhibit the growth and development of mosquitoes in all their life stages. In addition, the mosquitocidal activities of metal nanoparticles such as gold, silver or platinum are proven to be enhanced by phytosynthesis approaches, compared to physicochemical methods (Soni and Prakash 2012; Syed and Ahmad 2012; Castro et al. 2015; Lalitha et al. 2020; Suresh et al. 2020).

12.5.1 Gold Nanoparticles

In recent studies, broth extracts of *Cacumen platycladi, Coleus amboinicus, Artemisia nilagirica, Terminalia arjuna, Salicornia brachiata, Zingiber officinale* and *Phoenix dactylifera* have been used to synthesize gold nanoparticles. Hence, gold nanoparticles (AuNPs) that are biosynthesized using distinct plants, bacteria, algae, lichens, and fungi have been reported to be potent and successful insecticides against mosquito larvae even at less dosages (Murugan et al. 2015). Recent reports stated that the research on gold nanoparticles to demonstrate their insecticidal ability is insufficient, compared to that on AgNPs. The biosynthesized AuNPs from *Jatropha curcas* latex are reported to possess inhibitory effect against serum trypsin in different species of insects, including *A. Aegypti* (Patil et al. 2016). In another study, AuNPs are proven to disrupt the reproduction and development in German cockroaches named *Blattella germanica* (L.). Further, the larvicidal effects of AuNPs synthesized from the zein biopolymer (Ze-AuNPs) were also tested against *A. aegypti* vector. Histopathological results showed remarkable physiological changes, such as complete abdominal disintegration (midgut and caeca), caudal hair loss in antenna, lower, lateral, and upper head (Benelli 2018). Thus, AuNPs can be utilized as a potential insecticidal agent to inhibit the growth of a wide range of insects. A similar study of AuNPs synthesized from the leaf extracts of *Artemisia vulgaris L.* was found to exhibit a larvicidal effect against third and fourth instars of *A. aegypti*. This study also showed that the nanoparticle can lead to physiological damage in epithelial cells, cortex, and midgut along with AuNPs deposition in the midgut region (Saranya et al. 2020).

A recent study reported that the AuNPs synthesized using the aqueous extracts of lichens named *Parmelia sulcata* possess exclusive mosquitocidal property. The results revealed the inhibition of *A. stephensi and A. aegypti* larvae (I–IV instar), pupae and adult, and even affected the hatching of eggs, depending on the concentration of nanoparticles. Further, the comparative study showed that the nanoparticle inhibited *A. stephensi* more than *A. aegypti*, even at low concentration (Gandhi et al. 2019). Similarly, Murugan et al. (2015) emphasized the synthesis of AuNPs using *Cymbopogon citratus* to inhibit the growth of *A. stephensi* and *A. aegypti*. The study showed that the nanoparticles can lead to acute toxicity in the mosquito with a lethal concentration (LC50) of 18.8 to 41.5 parts per million (ppm) and boosted the predation efficiency of copepods named *Mesocyclops apsericornis* (Murugan et al. 2015). In another study, low doses of biosynthesized gold nanoparticles using *Couroupita guianensis* also showed high toxicity against *A. stephensi* (Subramaniam et al. 2016).

12.5.2 Silver Nanoparticles

Silver is the most common mosquitocidal agent among the metal nanoparticles. Khader et al. (2018) compared the larvicidal potential of several selected medicinal plants with phytosynthesized silver nanoparticles against third instar larvae of *A. aegypti* and *Culex tritaeniorhynchus*. In this study, alcoholic extracts of *Annona squamosa*, *Phyllanthus amarus*, *Eclipta prostrata* and *Cocconia grandis* are used for the synthesis of spherical-shaped, less than one-micron-sized silver nanoparticles. This study showed that the phytosynthesized silver nanoparticles are highly efficient in inhibiting the growth of mosquito larvae, compared to the alcoholic plant extracts (Khader et al. 2018). Likewise, Parthiban et al. (2019) recently fabricated AgNPs using the aqueous leaf extract of *Annona reticulata* and investigated its potential larvicidal activity against *A. aegypti* mosquito. The resultant nanoparticles were spherical in shape with sizes ranging from 7–8 nm and functional groups, such as amide groups and aromatic rings from triterpenoids, polyphenols and flavonoids. The study revealed that these phytosynthesized AgNPs are highly effective in inhibiting the growth of fourth instar larvae of *A. aegypti* mosquito with LC50 value of 4.43 µg/mL (Parthiban et al. 2019). Similarly, Shelar et al. (2019) synthesized silver nanoparticles via *Momordica charantia* plant and investigated its larvicidal and helminthicidal property. The study showed that the flavonoids, polyphenols and terpenoids present in the aqueous fruit peel extract is responsible for the formation of silver nanoparticles below 100 nm in size. These nanoparticles were also identified to possess the ability of inhibiting 85% of *A. albopictus* and *A. aegypti* larvae at 20 ppm of concentration (Shelar et al. 2019). Further, Kovendan et al. (2016) fabricated cubical and spherical shaped, 40–60 nm sized silver nanoparticles via leaf extract of *Psychotria nilgiriensis*. The study emphasized the nanoparticle formation is due to the presence of amine, carboxyl and hydroxyl groups in the leaf extract. These nanoparticles were identified to possess inhibitory effect against larva and pupa of *A. aegypti* mosquito at even low doses (3 ppm) (Kovendan et al. 2016). Furthermore, the essential oils extracted from *Curcuma zedoaria* were used to fabricate globular-shaped, ~100 nm-sized silver nanoparticles and their larvicidal activity against insecticide-susceptible and resistant *Culex quinquefasciatus* strains were evaluated. The study showed that the existence of eucalyptol, beta-tumerone, beta-sesquiphellandrene and alpha-zingiberene in the essential oils was responsible for nanoparticle formation. The resultant nanoparticles were identified to possess complete inhibitory activity against susceptible and resistant mosquito strain after 24 h of exposure with LC50 of 0.57 and 0.64 ppm, respectively (Sutthanont et al. 2019). It is evident from all these studies that the phytosynthesized silver nanoparticles are highly beneficial in inhibiting the growth of mosquitoes in all its life stages.

12.5.3 Copper Nanoparticles

Among metal nanoparticles, nanosized copper nanoparticles were also reported to be highly beneficial as a potential mosquitocidal agent. Abd El Hafiz Hassanain et al. (2019) utilized leaf extracts of *Lantana camara* to synthesize 11–17.8 nm-sized, spherical-shaped copper nanoparticles. These nanoparticles were identified to possess effective larvicidal activity against fourth instar larvae of *Anopholes multicolor*, that can cause malaria in humans, with LC50 value of 12.6 ppm at lower dose of 20 ppm (Abd El Hafiz Hassanain et al. 2019). Further, asymmetrical, dispersed and aggregated copper nanoparticles of ~100 nm in size were synthesized via aqueous leaf extract of *Artocarpus heterophyllus*, which contains phenolic and flavonoid compounds as the stabilization agent to form nanoparticles. The study also showed that 10 mg/litre of these nanoparticles can lead to 100% mortality against first to fourth instar larvae of *A. aegypti* mosquito (Sharon et al. 2018). Furthermore, Mondal and Hajra (2016) extracted the phytochemicals with alcoholic and phenolic groups from the petals of *Tagetes* species (Marigold) and *Helianthus* species (sunflower) and utilized them for the synthesis of spherical-shaped, 5–20 nm-sized copper nanoparticles. The resultant nanoparticles at 10 mg per litre of dose, possess 15% of enhanced larvicidal activity against *C. quinquifasciatus* after 24 h of incubation (Mondal and Hajra 2016). Moreover, Angajala et al. (2014) synthesized spherical, 50–100 nm sized copper nanoparticles using the aqueous leaf extract of *Aegle marmelos* correa, which contains phytochemicals, such as beta-sitosterol as reducing agent along with other phytocompounds for the nanoparticle formation. The study further emphasized the larvicidal efficacy of phytosynthesized copper nanoparticles against *Anopheles stephensi*, *Aedes aegypti* and *Culex quinquefasciatus*, with better LC50 of 500.06 ppm against *A. stephensi*, compared to the standard Temephos larvicide and crude plant extract (Angajala et al. 2014). Recently, Shehu et al. (2020) synthesized a novel copper-cobalt bimetallic nanoparticles via the fruit extract of the Palmyra palm with the majority of terpenoids as reducing agent to convert the precursors into nanosized particles. These nanoparticles are proven to possess moderate larvicidal activity against first, second and third instar larvae of *C. quinquefasciatus* with LC50 value of 12, 14.7 and 16 ppm (Shehu et al. 2020).

12.5.4 Other Metal Nanoparticles

Apart from gold, silver and copper, few other novel metal nanoparticles were also identified to possess enhanced mosquitocidal property. Sowndarya et al. (2017) synthesized novel 46–79 nm-sized selenium nanoparticles via the leaf extracts of *Clausena dentata*. The study showed that the formation of selenium nanoparticles is due to the conversion of the aldehyde group to carboxylic acid in the metal ion by terpenoids in the plant extract. Further, the study demonstrated that the phytosynthesized selenium nanoparticle possess a high mortality rate against the fourth instar larvae of *Aedes aegypti*, *Anopheles stephensi* and *Culex quinquefasciatus* at a low concentration of 104.13, 240.71 and 99.60 mg/L, respectively (Sowndarya et al. 2017). Furthermore, Elemike et al. (2017) fabricated silver nanocomposite using the aqueous leaf extract of *Achyranthes aspera*. The results revealed that the spherical and polydispersed nanoparticles of silver were embedded on the phytochemical extracts as a matrix. These nanocomposites were identified to possess enhanced larvicidal activity against *Aedes aegypti* mosquito without causing any toxic effect against non-toxic organisms, such as *Daphnia magna*, *Gambusia affinis* and *Moina macrocopa* (Sharma et al. 2020). Similar nanocomposite of silver with aggregated spherical morphology and size lower than 30 nm was synthesized via aqueous leaf extract of *Rubus ellipticus* and was investigated for their toxicity and oviposition deterrent activity against mosquito vectors, which can lead to malaria, filariasis and Zika virus. The FTIR spectral results emphasized that the terpenoids and flavonoids present in the leaf extract is responsible for nanoparticle formation and conjugate with them to form as bio-nanocomposite. These nanocomposites were identified to possess enhanced toxicity against the eggs, larvae and adults of *A. aegypti*, *A. stephensi* and *C. quinquefasciatus* with zero hatchability, reduced egg laying capability by gravid females and effective adulticidal potential. Besides, these nanocomposites were also found to be safer for non-target organisms, such as *Diplonychus indicus*, *Anisops bouvieri* and *Gambusia affinis* (AlQahtani et al. 2017). Furthermore, Elemike et al. (2017) stated that a novel silver-silver oxide nanoparticles with spherical morphology and average particle size of 23.6 nm can be synthesized using the aqueous *Eupatorium odoratum* leaf extract. The resultant nanoparticles were identified to possess enhanced inhibitory activity against the third and fourth instar larvae of *C. quinquefasciatus* after 12 h of exposure with LC50 value of 461.6 and 553.4 ppm (Elemike et al. 2017). Recently, Rajkumar et al. (2019) fabricated a novel chitosan-derived silver nanocomposites with 30–60 nm in size and spherical morphology using the aqueous leaf extract of *Carmona retusa* (Vahl) Masam. These bio-nanocomposites were also identified to possess enhanced larvicidal activity against *A. aegypti*, *A. stephensi* and *C. quinquefasciatus* at low LC50 and LC90 values (Rajkumar et al. 2019) (Table 12.2).

12.6 Mosquitocidal Mechanism of Phytosynthesized Metal Nanoparticles

In general, several works in the literature reported that the phytochemicals extracted from the plants possess enhanced mosquitocidal property (Senthil-Nathan 2019). The nanoparticles prepared with the help of plant extracts will possess these phytochemicals, which eventually increases the mosquitocidal property of the nanoparticles, compared to the crude phytochemicals via the synergistic effect (Gupta 2020). Figure 12.1 shows the probable mosquitocidal mechanism of the phytosynthesized metal nanoparticles. The metal nanoparticles synthesized via plant extracts

Table 12.2 Mosquitocidal activity of various phytosynthesized metal nanoparticles

Metal nanoparticle	Plant extract	Mosquitocidal activity	Reference
Gold	*Jatropha curcas* latex	*A. aegypti*	Patil et al. (2016)
Gold	*Artemisia vulgaris*	Third and fourth instars of *A. aegypti*	Saranya et al. (2020)
Gold	*Parmelia sulcate*	I-IV instar larvae of *A. stephensi* and *A. aegypti*	Gandhi et al. (2019)
Gold	*Cymbopogon citratus*	*A. stephensi* and *A. aegypti*	Murugan et al. (2015)
Gold	*Couroupita guianensis*	*A. stephensi*	Subramaniam et al. (2016)
Silver	*Annona squamosa*, *Phyllanthus amarus*, *Eclipta prostrata* and *Cocconia grandis*	*A. aegypti* and *Culex tritaeniorhynchus* third instar larvae	Khader et al. (2018)
Silver	*Annona reticulata*	*A. aegypti*	Parthiban et al. (2019)
Silver	*Momordica charantia*	*A. albopictus* and *A. aegypti*	Shelar et al. (2019)
Silver	*Psychotria nilgiriensis*	Larva and pupa of *A. aegypti*	Kovendan et al. (2016)
Silver	*Curcuma zedoaria* essential oil	Insecticide-susceptible and resistance *Culex quinquefasciatus*	Sutthanont et al. (2019)
Copper	*Lantana camara*	Fourth instar larvae of *Anopholes multicolor*	Abd El Hafiz Hassanain et al. (2019)
Copper	*Artocarpus heterophyllus*	Fourth instar larvae of *A. aegypti*	Sharon et al. (2018)
Copper	Marigold and sunflower	*C. quinquifasciatus* larvae	Mondal and Hajra (2016)
Copper	*Aegle marmelos* correa	*Anopheles stephensi*, *Aedes aegypti* and *Culex quinquefasciatus* larva	Angajala et al. (2014)
Copper-cobalt bimetallic	Palmyra palm fruit	I-III instar larvae of *C. quinquifasciatus*	Shehu et al. (2020)
Selenium	*Clausena dentata*	*Aedes aegypti*, *Anopheles stephensi* and *Culex quinquefasciatus*	Sowndarya et al. (2017)
Silver nanocomposite	*Achyranthes aspera*	*Aedes aegypti* larvae	Sharma et al. (2020)
Silver nanocomposite	*Rubus ellipticus*	Eggs, larvae and adults of *A. aegypti*, *A. stephensi* and *C. quinquefasciatus*	AlQahtani et al. (2017)
Silver-silver oxide	*Eupatorium odoratum*	Third and fourth instar larvae of *C. quinquefasciatus*	Elemike et al. (2017)

(continued)

Table 12.2 (continued)

Metal nanoparticle	Plant extract	Mosquitocidal activity	Reference
Chitosan-silver nanocomposite	*Carmona retusa*	*A. aegypti*, *A. stephensi* and *C. quinquefasciatus* larva	Rajkumar et al. (2019)

are usually suspended in the mixture of water and phytochemicals (Rana et al. 2020). These aqueous nanoparticles must be sprayed on the water surface, where the life stages of mosquito, such as eggs, larvae and adult, will grow and develop (Lourthuraj et al. 2020). The nanoparticles will slowly disintegrate into their ions, where they will either enter into the mosquito cells (egg, larvae or adult) or attach to the surface of the cell. In both these cases, the nanoparticles will lead to an increase in the cell membrane potential, produce reactive oxygen species and elevate lipid peroxidation (Buhroo et al. 2017; Suresh et al. 2020). Further, these toxic reactions of the nanoparticles will be synergistically supported by the phytochemicals that are toxic to the mosquitoes (Foko et al. 2019). Furthermore, the main significance of phytochemical-assisted metal nanoparticle is their ability to reduce the toxicity of nanoparticles towards other non-target organisms, compared to metal nanoparticles synthesized via chemical approaches (Lalitha et al. 2020). These toxic reactions of phytosynthesized metal nanoparticles will lead to the inhibition of the mosquito life stages, affect their growth and development, and mitigate their population.

12.7 Future Perspective and Conclusion

It is evident from the literature as mentioned in the previous sections that the phytosynthesized metal nanoparticles are highly beneficial as potential mosquitocidal agents, compared to the conventional physical and chemical synthesized nanosized metal particles (Barabadi et al. 2019). The high production cost in physical methods and the toxicity towards non-target organisms via chemical synthesized nanoparticles are the major limitations of conventional synthesis approaches (Shanmuganathan et al. 2019). These limitations can be addressed via phytosynthesis approach, where the plants are available in large quantities, which eventually reduces the production costs and the biomolecules in the plant extracts reduces the potential toxicity of nanoparticles towards non-target organisms (Ramanathan and Aqra 2019). However, it is not possible to attain nanoparticles with high stability via phytosynthesis approach, compared to physical and chemical methods (Granata et al. 2016). The challenge of stability in phytosynthesis can be overcome by using a biopolymer as a template or matrix (Jeevanandam et al. 2020). In future, there is a possibility of biopiracy to utilize several traditional or plants belong to keystone species for the large-scale fabrication of metal nanoparticles with mosquitocidal property (Dwivedy et al. n.d.). Thus, it is necessary to gain clarity on the actual mechanism of nanoparticle formation via phytochemicals. Further, it will be possible in the future that the

Fig. 12.1 Mosquitocidal mechanism of phytosynthesized metal nanoparticles

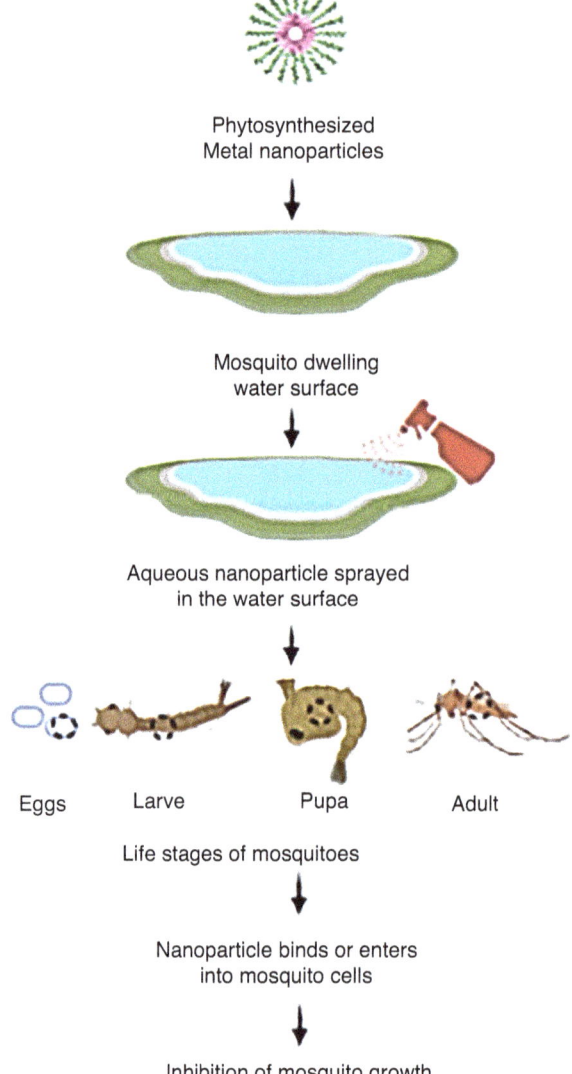

phytochemicals responsible for nanoparticle formation as well as with enhanced mosquitocidal property can be made to be expressed in the plants via genetic engineering and grow them via plant tissue culture approach (Shkryl et al. 2018). The incorporation of genetic engineering and tissue culture in the production and extraction of nanoparticles will reduce the biopiracy issues and can be utilized for the large-scale, commercial production of phytosynthesized metal nanoparticles with effective mosquitocidal property. Hence, phytosynthesized metal nanoparticles or nanocomposites can be a potential mosquitocidal agent to mitigate the population of mosquitoes in the future.

Acknowledgments The author (Dr. Jaison Jeevanandam) acknowledges the support of FCT-Fundação para a Ciência e a Tecnologia (Base Fund UIDB/00674/2020 and Programmatic Fund UIDP/00674/2020, Portuguese Government Funds), ARDITI-Agência Regional para o Desenvolvimento da Investigação Tecnologia e Inovação through the project M1420-01-0145-FEDER-000005-CQM+ (Madeira 14-20 Program). All the other authors would like to thank their respective department and university for the support during the preparation of this chapter.

References

Abd El Hafiz Hassanain N, Zeinhom Shehata A, Mohamed Mokhtar M, Mohamed Shaapan R, Abd El Hafiz Hassanain M, Zaky S (2019) Comparison between insecticidal activity of Lantana camara extract and its synthesized nanoparticles against Anopheline mosquitoes. Pak J Biol Sci 22(7):327–334. https://doi.org/10.3923/pjbs.2019.327.334

Abdel-Aziz SM, Prasad R, Hamed AA, Abdelraof M (2018) Fungal nanoparticles: A novel tool for a green biotechnology? In: Fungal Nanobionics: Principles and Applications (eds. Prasad R, Kumar V, Kumar M and Wang S), Springer Singapore Pte Ltd. 61–87

Abkenar AK, Naderi M (2016) Chemical synthesis of gold nanoparticles with different morphology from a secondary source. J Iran Chem Soc 13(12):2173–2184

AlQahtani FS, AlShebly MM, Govindarajan M, Senthilmurugan S, Vijayan P, Benelli G (2017) Green and facile biosynthesis of silver nanocomposites using the aqueous extract of Rubus ellipticus leaves: toxicity and oviposition deterrent activity against Zika virus, malaria and filariasis mosquito vectors. J Asia Pac Entomol 20(1):157–164

Andra S, Balu SK, Jeevanandham J, Muthalagu M, Vidyavathy M, San Chan Y, Danquah MK (2019) Phytosynthesized metal oxide nanoparticles for pharmaceutical applications. Naunyn Schmiedeberg's Arch Pharmacol 392(7):755–771

Angajala G, Pavan P, Subashini R (2014) One-step biofabrication of copper nanoparticles from Aegle marmelos correa aqueous leaf extract and evaluation of its anti-inflammatory and mosquito larvicidal efficacy. RSC Adv 4(93):51459–51470. https://doi.org/10.1039/C4RA10003D

Arjunan NK, Murugan K, Rejeeth C, Madhiyazhagan P, Barnard DR (2012) Green synthesis of silver nanoparticles for the control of mosquito vectors of malaria, filariasis, and dengue. Vector-Borne Zoonotic Dis 12(3):262–268

Arya A, Gupta K, Chundawat TS, Vaya D (2018) Biogenic synthesis of copper and silver nanoparticles using green alga Botryococcus braunii and its antimicrobial activity. Bioinorg Chem Appl 2018:1

Asanithi P, Chaiyakun S, Limsuwan P (2012) Growth of silver nanoparticles by DC magnetron sputtering. J Nanomater 2012:2

Aygun A, Gülbagca F, Ozer LY, Ustaoglu B, Altunoglu YC, Baloglu MC, Atalar MN, Alma MH, Sen F (2020) Biogenic platinum nanoparticles using black cumin seed and their potential usage as antimicrobial and anticancer agent. J Pharm Biomed Anal 179:112961

Aziz N, Faraz M, Pandey R, Sakir M, Fatma T, Varma A, Barman I, Prasad R (2015) Facile algae-derived route to biogenic silver nanoparticles: Synthesis, antibacterial and photocatalytic properties. Langmuir 31:11605–11612. https://doi.org/10.1021/acs.langmuir.5b03081

Aziz N, Fatma T, Varma A, Prasad R (2014) Biogenic synthesis of silver nanoparticles using Scenedesmus abundans and evaluation of their antibacterial activity. Journal of Nanoparticles, Article ID 689419. http://dx.doi.org/10.1155/2014/689419

Aziz N, Pandey R, Barman I, Prasad R (2016) Leveraging the attributes of *Mucor hiemalis*-derived silver nanoparticles for a synergistic broad-spectrum antimicrobial platform. Front Microbiol 7:1984. https://doi.org/10.3389/fmicb.2016.01984

Aziz N, Faraz M, Sherwani MA, Fatma T, Prasad R (2019) Illuminating the anticancerous efficacy of a new fungal chassis for silver nanoparticle synthesis. Front Chem 7:65. https://doi.org/10.3389/fchem.2019.00065

Balamurugan M, Kaushik S, Saravanan S (2016) Green synthesis of gold nanoparticles by using Peltophorum pterocarpum flower extracts. Nano Biomed Eng 8(4):213–218

Bamoharram FF, Ahmadpour A, Heravi MM, Ayati A, Rashidi H, Tanhaei B (2012) Recent advances in application of polyoxometalates for the synthesis of nanoparticles. Synthesis and Reactivity in Inorganic, Metal-Organic, and Nano-Metal Chemistry 42(2):209–230

Barabadi H, Alizadeh Z, Rahimi MT, Barac A, Maraolo AE, Robertson LJ, Masjedi A, Shahrivar F, Ahmadpour E (2019) Nanobiotechnology as an emerging approach to combat malaria: a systematic review. Nanomed Nanotechnol Biol Med 18:221–233

Bayazit MK, Yue J, Cao E, Gavriilidis A, Tang J (2016) Controllable synthesis of gold nanoparticles in aqueous solution by microwave assisted flow chemistry. ACS Sustain Chem Eng 4(12):6435–6442

Begletsova NN, Shinkarenko OA, Selifonova EI, Tsvetkova OY, Zakharevich A, Chernova RK, Kletsov AA, Glukhovskoy EG (2017) Synthesis of copper nanoparticles stabilized with cetylpyridinium chloride micelles. Adv Mater Let 8(4):404–409

Benelli G (2018) Mode of action of nanoparticles against insects. Environ Sci Pollut Res 25(13):12329–12341

Benelli G, Jeffries CL, Walker T (2016) Biological control of mosquito vectors: past, present, and future. Insects 7(4):52

Borgheti-Cardoso LN, San Anselmo M, Lantero E, Lancelot A, Serrano JL, Hernández-Ainsa S, Fernàndez-Busquets X, Sierra T (2020) Promising nanomaterials in the fight against malaria. J Mater Chem B 8(41):9428–9448

Buhroo AA, Nisa G, Asrafuzzaman S, Prasad R, Rasheed R, Bhattacharyya A (2017) Biogenic silver nanoparticles from *Trichodesma indicum* aqueous leaf extract against *Mythimna separata* and evaluation of its larvicidal efficacy. J Plant Protect Res 57(2):194–200. https://doi.org/10.1515/jppr-2017-0026

Camas M, Celik F, Sazak Camas A, Ozalp HB (2019) Biosynthesis of gold nanoparticles using marine bacteria and Box–Behnken design optimization. Part Sci Technol 37(1):31–38

Castro L, Blázquez ML, González F, Muñoz JÁ, Ballester A (2015) Biosynthesis of silver and platinum nanoparticles using orange peel extract: characterisation and applications. IET Nanobiotechnol 9(5):252–258

Cheng Y, Wang F, Fang C, Su J, Yang L (2016) Preparation and characterization of size and morphology controllable silver nanoparticles by citrate and tannic acid combined reduction at a low temperature. J Alloys Compd 658:684–688

Cui Y, Lai X, Liu K, Liang B, Ma G, Wang L (2020) Ginkgo Biloba leaf polysaccharide stabilized palladium nanoparticles with enhanced peroxidase-like property for the colorimetric detection of glucose. RSC Adv 10(12):7012–7018

Das RK, Pachapur VL, Lonappan L, Naghdi M, Pulicharla R, Maiti S, Cledon M, Dalila LMA, Sarma SJ, Brar SK (2017) Biological synthesis of metallic nanoparticles: plants, animals and microbial aspects. Nanotechnol Environ Eng 2(1):18

Dinesh D, Murugan K, Madhiyazhagan P, Panneerselvam C, Mahesh Kumar P, Nicoletti M, Jiang W, Benelli G, Chandramohan B, Suresh U (2015) Mosquitocidal and antibacterial activity of green-synthesized silver nanoparticles from Aloe vera extracts: towards an effective tool against the malaria vector Anopheles stephensi? Parasitol Res 114(4):1519–1529. https://doi.org/10.1007/s00436-015-4336-z

Dong J, Carpinone PL, Pyrgiotakis G, Demokritou P, Moudgil BM (2020) Synthesis of precision gold nanoparticles using Turkevich method. KONA Powder Particle J 37:224–232

Dwivedy AK, Singh VK, Kumar M, Upadhyay N, Das S, Chaudhari AK, Dubey NK. Bioprospection of traditionally used medicinal plants: an overview

Edison TNJI, Atchudan R, Kamal C, Lee YR (2016) Caulerpa racemosa: a marine green alga for eco-friendly synthesis of silver nanoparticles and its catalytic degradation of methylene blue. Bioprocess Biosyst Eng 39(9):1401–1408

Elemike EE, Onwudiwe DC, Ekennia AC, Sonde CU, Ehiri RC (2017) Green synthesis of Ag/Ag2O nanoparticles using aqueous leaf extract of Eupatorium odoratum and its antimicrobial and mosquito Larvicidal activities. Molecules 22(5). https://doi.org/10.3390/molecules22050674

Foko LPK, Meva FEA, Moukoko CEE, Ntoumba AA, Njila MIN, Kedi PBE, Ayong L, Lehman LG (2019) A systematic review on anti-malarial drug discovery and antiplasmodial potential of green synthesis mediated metal nanoparticles: overview, challenges and future perspectives. Malar J 18(1):337

Gandhi AD, Murugan K, Umamahesh K, Babujanarthanam R, Kavitha P, Selvi A (2019) Lichen Parmelia sulcata mediated synthesis of gold nanoparticles: an eco-friendly tool against Anopheles stephensi and Aedes aegypti. Environ Sci Pollut Res 26(23):23886–23898

González-Ballesteros N, Prado-López S, Rodríguez-González J, Lastra M, Rodríguez-Argüelles M (2017) Green synthesis of gold nanoparticles using brown algae Cystoseira baccata: its activity in colon cancer cells. Colloids Surf B Biointerfaces 153:190–198

Granata G, Yamaoka T, Pagnanelli F, Fuwa A (2016) Study of the synthesis of copper nanoparticles: the role of capping and kinetic towards control of particle size and stability. J Nanopart Res 18(5):133

Gudikandula K, Charya Maringanti S (2016) Synthesis of silver nanoparticles by chemical and biological methods and their antimicrobial properties. J Exp Nanosci 11(9):714–721

Gudikandula K, Vadapally P, Charya MS (2017) Biogenic synthesis of silver nanoparticles from white rot fungi: their characterization and antibacterial studies. Open Nano 2:64–78

Gupta P (2020) Development of mosquitocidal herbal nanoemulsions for controlling the dengue vector Aedes aegypti

Gupta K, Chundawat TS (2019) Bio-inspired synthesis of platinum nanoparticles from fungus Fusarium oxysporum: its characteristics, potential antimicrobial, antioxidant and photocatalytic activities. Mater Res Exp 6(10):1050d1056

Gupta A, Pandey S, Variya B, Shah S, Yadav JS (2019) Green synthesis of gold nanoparticles using different leaf extracts of Ocimum gratissimum Linn for anti-tubercular activity. Curr Nanomed (Formerly: Recent Patents on Nanomedicine) 9(2):146–157

Gurusamy V, Krishnamoorthy R, Gopal B, Veeravagan V (2017) Systematic investigation on hydrazine hydrate assisted reduction of silver nanoparticles and its antibacterial properties. Inorgan Nano-Metal Chem 47(5):761–767

Harne S, Sharma A, Dhaygude M, Joglekar S, Kodam K, Hudlikar M (2012) Novel route for rapid biosynthesis of copper nanoparticles using aqueous extract of Calotropis procera L. latex and their cytotoxicity on tumor cells. Colloids Surf B Biointerfaces 95:284–288

Huang Y-JS, Higgs S, Vanlandingham DL (2017) Biological control strategies for mosquito vectors of arboviruses. Insects 8(1):21

Husen A, Siddiqi KS (2014) Phytosynthesis of nanoparticles: concept, controversy and application. Nanoscale Res Lett 9(1):229

Inwati GK, Rao Y, Singh M (2016) In situ free radical growth mechanism of platinum nanoparticles by microwave irradiation and electrocatalytic properties. Nanoscale Res Lett 11(1):1–8

Iravani S, Korbekandi H, Mirmohammadi SV, Zolfaghari B (2014) Synthesis of silver nanoparticles: chemical, physical and biological methods. Res Pharm Sci 9(6):385

Ishida Y, Akita I, Sumi T, Matsubara M, Yonezawa T (2016) Thiolate–protected gold nanoparticles via physical approach: unusual structural and photophysical characteristics. Sci Rep 6:29928

Jameel MS, Aziz AA, Dheyab MA (2020) Comparative analysis of platinum nanoparticles synthesized using sonochemical-assisted and conventional green methods. Nano-Struct Nano-Objects 23:100484

Jamkhande PG, Ghule NW, Bamer AH, Kalaskar MG (2019) Metal nanoparticles synthesis: an overview on methods of preparation, advantages and disadvantages, and applications. J Drug Deliv Sci Technol 53:101174

Jeevanandam J, Chan YS, Danquah MK (2016) Biosynthesis of metal and metal oxide nanoparticles. Chem Bio Eng Rev 3(2):55–67

Jeevanandam J, Pal K, Danquah MK (2018) Virus-like nanoparticles as a novel delivery tool in gene therapy. Biochimie 157:38–47

Jeevanandam J, Chan YS, Danquah MK (2019) Effect of pH variations on morphological transformation of biosynthesized MgO nanoparticles. Part Sci Technol 38(5):573–586

Jeevanandam J, Sundaramurthy A, Sharma V, Murugan C, Pal K, Kodous MHA, Danquah MK (2020) Sustainability of one-dimensional nanostructures: fabrication and industrial applications. In: Sustainable nanoscale engineering. Elsevier, pp 83–113

Jeyaraj M, Gurunathan S, Qasim M, Kang M-H, Kim J-H (2019) A comprehensive review on the synthesis, characterization, and biomedical application of platinum nanoparticles. Nano 9(12):1719

Kaur P (2018) Biosynthesis of nanoparticles using eco-friendly factories and their role in plant pathogenicity: a review. Biotechnol Res Innov 2(1):63–73

Khader SZA, Syed Zameer Ahmed S, Sathyan J, Mahboob MR, Venkatesh P, Ramesh K (2018) A comparative study on larvicidal potential of selected medicinal plants over green synthesized silver nano particles. Egypt J Basic Appl Sci 5(1):54–62. https://doi.org/10.1016/j.ejbas.2018.01.002

Khan A, Rashid A, Younas R, Chong R (2016) A chemical reduction approach to the synthesis of copper nanoparticles. Int Nano Lett 6(1):21–26

Khater HF, Selim AM, Abouelella GA, Abouelella NA, Murugan K, Vaz NP, Govindarajan M (2019) Commercial mosquito repellents and their safety concerns. In: Malaria. IntechOpen

Kovendan K, Chandramohan B, Dinesh D, Abirami D, Vijayan P, Govindarajan M, Vincent S, Benelli G (2016) Green-synthesized silver nanoparticles using Psychotria nilgiriensis: toxicity against the dengue vector Aedes aegypti (Diptera: Culicidae) and impact on the predatory efficiency of the non-target organism Poecilia sphenops (Cyprinodontiformes: Poeciliidae). J Asia Pac Entomol 19(4):1001–1007

Kumar N, Biswas K, Gupta RK (2016) Green synthesis of Ag nanoparticles in large quantity by cryomilling. RSC Adv 6(112):111380–111388

Kuppusamy P, Yusoff MM, Maniam GP, Govindan N (2016) Biosynthesis of metallic nanoparticles using plant derivatives and their new avenues in pharmacological applications–an updated report. Saudi Pharm J 24(4):473–484

Kuppusamy P, Ilavenil S, Srigopalram S, Maniam GP, Yusoff MM, Govindan N, Choi KC (2017) Treating of palm oil mill effluent using Commelina nudiflora mediated copper nanoparticles as a novel bio-control agent. J Clean Prod 141:1023–1029

Lalitha K, Kalaimurgan D, Nithya K, Venkatesan S, Shivakumar MS (2020) Antibacterial, antifungal and Mosquitocidal efficacy of copper nanoparticles synthesized from Entomopathogenic nematode: insect–host relationship of bacteria in secondary metabolites of Morganella morganii sp.(PMA1). Arab J Sci Eng 45:4489

LewisOscar F, Vismaya S, Arunkumar M, Thajuddin N, Dhanasekaran D, Nithya C (2016) Algal nanoparticles: synthesis and biotechnological potentials. Algae–Organisms Imminent Biotechnol 7:157–182

Li X, Xu H, Chen Z-S, Chen G (2011) Biosynthesis of nanoparticles by microorganisms and their applications. J Nanomater 2011(8):1–16

Lourthuraj AA, Selvam MM, Hussain MS, Abdel-Warith A-WA, Younis EMI, Al-Asgah NA (2020) Dye degradation, antimicrobial and larvicidal activity of silver nanoparticles biosynthesized from Cleistanthus collinus. Saudi J Biol Sci 27(7):1753–1759

Maroufpour N, Mousavi M, Abbasi M, Ghorbanpour M (2020) Biogenic nanoparticles as novel sustainable approach for plant protection. In: Biogenic nano-particles and their use in agroecosystems. Springer, pp 161–172

Mendivil Palma MI, Krishnan B, Rodriguez GAC, Das Roy TK, Avellaneda DA, Shaji S (2016) Synthesis and properties of platinum nanoparticles by pulsed laser ablation in liquid. J Nanomater 2016:1

Molnár Z, Bódai V, Szakacs G, Erdélyi B, Fogarassy Z, Sáfrán G, Varga T, Kónya Z, Tóth-Szeles E, Szűcs R (2018) Green synthesis of gold nanoparticles by thermophilic filamentous fungi. Sci Rep 8(1):1–12

Mondal NK, Hajra A (2016) Synthesis of copper nanoparticles (CuNPs) from petal extracts of marigold (Tagetes sp.) and sunflower (Helianthus sp.) and their effective use as a control tool against mosquito vectors. J Mosquito Res 6(19):1–9

Murugan K, Benelli G, Panneerselvam C, Subramaniam J, Jeyalalitha T, Dinesh D, Nicoletti M, Hwang J-S, Suresh U, Madhiyazhagan P (2015) Cymbopogon citratus-synthesized gold nanoparticles boost the predation efficiency of copepod Mesocyclops aspericornis against malaria and dengue mosquitoes. Exp Parasitol 153:129–138

Murugan K, Panneerselvam C, Subramaniam J, Madhiyazhagan P, Hwang J-S, Wang L, Dinesh D, Suresh U, Roni M, Higuchi A, Nicoletti M, Benelli G (2016) Eco-friendly drugs from the marine environment: spongeweed-synthesized silver nanoparticles are highly effective on Plasmodium falciparum and its vector Anopheles stephensi, with little non-target effects on predatory copepods. Environ Sci Pollut Res 23(16):16671–16685. https://doi.org/10.1007/s11356-016-6832-9

Nagao H, Ichiji M, Hirasawa I (2017) Synthesis of platinum nanoparticles by reductive crystallization using polyethyleneimine. Chem Eng Technol 40(7):1242–1246

Naharuddin NZA, Sadrolhosseini AR, Bakar MHA, Tamchek N, Mahdi MA (2020) Laser ablation synthesis of gold nanoparticles in tetrahydrofuran. Opt Mater Exp 10(2):323–331

Nancy P, James J, Valluvadasan S, Kumar RA, Kalarikkal N (2018) Laser–plasma driven green synthesis of size controlled silver nanoparticles in ambient liquid. Nano-Struct Nano-Objects 16:337–346

Nikam A, Prasad B, Kulkarni A (2018) Wet chemical synthesis of metal oxide nanoparticles: a review. CrystEngComm 20(35):5091–5107

Pan Z, Lin Y, Sarkar B, Owens G, Chen Z (2020) Green synthesis of iron nanoparticles using red peanut skin extract: synthesis mechanism, characterization and effect of conditions on chromium removal. J Colloid Interface Sci 558:106–114

Pantidos N, Horsfall LE (2014) Biological synthesis of metallic nanoparticles by bacteria, fungi and plants. J Nanomed Nanotechnol 5(5):1

Parthiban E, Manivannan N, Ramanibai R, Mathivanan N (2019) Green synthesis of silvernanoparticles from Annona reticulata leaves aqueous extract and its mosquito larvicidal and anti-microbial activity on human pathogens. Biotechnol Rep 21:e00297

Parveen K, Banse V, Ledwani L (2016a) Green synthesis of nanoparticles: their advantages and disadvantages. AIP Publishing LLC, p 020048

Parveen K, Banse V, Ledwani L (2016b) Green synthesis of nanoparticles: their advantages and disadvantages. AIP Publishing LLC, p 020048

Patil CD, Borase HP, Suryawanshi RK, Patil SV (2016) Trypsin inactivation by latex fabricated gold nanoparticles: a new strategy towards insect control. Enzym Microb Technol 92:18–25

Patil MP, Ngabire D, Thi HHP, Kim M-D, Kim G-D (2017) Eco-friendly synthesis of gold nanoparticles and evaluation of their cytotoxic activity on cancer cells. J Clust Sci 28(1):119–132

Pilaquinga F, Morejón B, Ganchala D, Morey J, Piña N, Debut A, Neira M (2019) Green synthesis of silver nanoparticles using Solanum mammosum L. (Solanaceae) fruit extract and their larvicidal activity against Aedes aegypti L. (Diptera: Culicidae). PLoS One 14(10):e0224109. https://doi.org/10.1371/journal.pone.0224109

Prasad R (2014) Synthesis of silver nanoparticles in photosynthetic plants. Journal of Nanoparticles, Article ID 963961, 2014, http://dx.doi.org/10.1155/2014/963961

Prasad R (2016) Advances and Applications through Fungal Nanobiotechnology. Springer, International Publishing Switzerland (ISBN: 978-3-319-42989-2)

Prasad R (2017) Fungal Nanotechnology: Applications in Agriculture, Industry, and Medicine. Springer Nature Singapore Pte Ltd. (ISBN 978-3-319-68423-9)

Prasad R (2019a) Plant Nanobionics: Advances in the Understanding of Nanomaterials Research and Applications. Springer International Publishing (ISBN 978-3-030-12495-3) https://www.springer.com/gp/book/9783030124953

Prasad R (2019b) Plant Nanobionics: Approaches in Nanoparticles Biosynthesis and Toxicity. Springer International Publishing (ISBN 978-3-030-16379-2) https://www.springer.com/gp/book/9783030163785

Prasad R, Pandey R, Barman I (2016) Engineering tailored nanoparticles with microbes: quo vadis. WIREs Nanomed Nanobiotechnol 8:316–330. https://doi.org/10.1002/wnan.1363

Prasad R, Kumar V, Kumar M, Wang S (2018a) Fungal Nanobionics: Principles and Applications. Springer Nature Singapore Pte Ltd. (ISBN 978-981-10-8666-3) https://www.springer.com/gb/book/9789811086656

Prasad R, Jha A and Prasad K (2018b) Exploring the Realms of Nature for Nanosynthesis. Springer International Publishing (ISBN 978-3-319-99570-0) https://www.springer.com/978-3-319-99570-0

Quazi HA (2020) Commercializing nanotechnology: a roadmap to taking Nanoproducts from laboratory to market. CRC Press

Rafique M, Rafique MS, Kalsoom U, Afzal A, Butt SH, Usman A (2019) Laser ablation synthesis of silver nanoparticles in water and dependence on laser nature. Opt Quant Electron 51(6):179

Rai M, Maliszewska I, Ingle A, Gupta I, Yadav A (2015) Diversity of microbes in synthesis of metal nanoparticles: progress and limitations. In: Bio-nanoparticles: biosynthesis and sustainable biotechnological implications, pp 1–30

Rajesh KM, Ajitha B, Reddy YAK, Suneetha Y, Reddy PS (2018) Assisted green synthesis of copper nanoparticles using Syzygium aromaticum bud extract: physical, optical and antimicrobial properties. Optik 154:593–600

Rajkumar R, Shivakumar MS, Senthil Nathan S, Selvam K (2019) Preparation and characterization of Chitosan nanocomposites material using silver nanoparticle synthesized Carmona retusa (Vahl) Masam leaf extract for antioxidant, anti-cancerous and insecticidal application. J Clust Sci 30(4):1145–1155. https://doi.org/10.1007/s10876-019-01578-9

Rak MJ, Friščić T, Moores A (2016) One-step, solvent-free mechanosynthesis of silver nanoparticle-infused lignin composites for use as highly active multidrug resistant antibacterial filters. RSC Adv 6(63):58365–58370

Ramanathan AA, Aqra MW (2019) An overview of the green road to the synthesis of nanoparticles. J Mater Sci Res Rev 2(3):1–11

Ramesh S, Vetrivel S, Suresh P, Kaviarasan V (2020) Characterization techniques for nano particles: A practical top down approach to synthesize copper nano particles from copper chips and determination of its effect on planes. Mater Today Proc 33:2626

Rana A, Yadav K, Jagadevan S (2020) A comprehensive review on green synthesis of nature-inspired metal nanoparticles: mechanism, application and toxicity. J Clean Prod 272:122880

Rheder DT, Guilger M, Bilesky-José N, Germano-Costa T, Pasquoto-Stigliani T, Gallep TBB, Grillo R, dos Santos CC, Fraceto LF, Lima R (2018) Synthesis of biogenic silver nanoparticles using Althaea officinalis as reducing agent: evaluation of toxicity and ecotoxicity. Sci Rep 8(1):1–11

Sadrolhosseini AR, Mahdi MA, Alizadeh F, Rashid SA (2019a) Laser ablation technique for synthesis of metal nanoparticle in liquid. IntechOpen

Sadrolhosseini AR, Habibiasr M, Shafie S, Solaimani H, Lim HN (2019b) Optical and thermal properties of laser-ablated platinum nanoparticles graphene oxide composite. Int J Mol Sci 20(24):6153

Sadrolhosseini AR, Rashid SA, Shafie S, Soleimani H (2019c) Laser ablation synthesis of Ag nanoparticles in graphene quantum dots aqueous solution and optical properties of nanocomposite. Appl Phys A 125(2):82

Şahin Ün Ş, Ünlü A, Ün İ, Ok S (2020) Green synthesis, characterization and catalytic activity evaluation of palladium nanoparticles facilitated by Punica granatum peel extract. Inorganic Nano-Metal Chem:1–9. https://doi.org/10.1080/24701556.2020.1832118

Saglam N, Korkusuz, F, Prasad R (2021) Nanotechnology Applications in Health and Environmental Sciences. Springer International Publishing (ISBN: 978-3-030-64410-9) https://www.springer.com/gp/book/9783030644093

Saranya S, Selvi A, Babujanarthanam R, Rajasekar A, Madhavan J (2020) Insecticidal activity of nanoparticles and mechanism of action. In: Model organisms to study biological activities and toxicity of nanoparticles. Springer, pp 243–266

Saravanan C, Rajesh R, Kaviarasan T, Muthukumar K, Kavitake D, Shetty PH (2017) Synthesis of silver nanoparticles using bacterial exopolysaccharide and its application for degradation of azo-dyes. Biotechnol Rep 15:33–40

Sarma H, Joshi S, Prasad R, Jampilek J (2021) Biobased Nanotechnology for Green Applications. Springer International Publishing (ISBN 978-3-030-61985-5) https://www.springer.com/gp/book/9783030619848

Seetharaman PK, Chandrasekaran R, Gnanasekar S, Chandrakasan G, Gupta M, Manikandan DB, Sivaperumal S (2018) Antimicrobial and larvicidal activity of eco-friendly silver nanoparticles synthesized from endophytic fungi Phomopsis liquidambaris. Biocatal Agric Biotechnol 16:22–30

Sehgal N, Soni K, Gupta N, Kanchan K (2018) Microorganism assisted synthesis of gold nanoparticles: a review. Asian J Biomed Pharm Sci 8(64):22–29

Senthil-Nathan S (2019) A review of resistance mechanisms of synthetic insecticides and botanicals, phytochemicals, and essential oils as alternative Larvicidal agents against mosquitoes. Front Physiol 10:1591

Shah KW, Zheng L (2019) Microwave-assisted synthesis of hexagonal gold nanoparticles reduced by Organosilane (3-Mercaptopropyl) trimethoxysilane. Materials 12(10):1680

Shanmuganathan R, Karuppusamy I, Saravanan M, Muthukumar H, Ponnuchamy K, Ramkumar VS, Pugazhendhi A (2019) Synthesis of silver nanoparticles and their biomedical applications- a comprehensive review. Curr Pharm Des 25(24):2650–2660

Sharma D, Kanchi S, Bisetty K (2019) Biogenic synthesis of nanoparticles: a review. Arab J Chem 12(8):3576–3600

Sharma A, Tripathi P, Kumar S (2020) One-pot synthesis of silver nanocomposites from Achyranthes aspera: an eco-friendly larvicide against Aedes aegypti L. Asian Pac J Trop Biomed 10(2):54–64. https://doi.org/10.4103/2221-1691.275420

Sharon EA, Velayutham K, Ramanibai R (2018) Biosynthesis of copper nanoparticles using Artocarpus heterophyllus against dengue vector Aedes aegypti. Int J Life Sci Scienti Res eISSN 2455(1716):1716

Shehu Z, Danbature WL, Magaji B, Adam MM, Bunu MA, Mai AJ, Mela Y (2020) Green synthesis and nanotoxicity assay of copper-cobalt bimetallic nanoparticles as a novel nanolarvicide for mosquito larvae management. Int J Biotechnol 9(2):99–104

Shelar A, Sangshetti J, Chakraborti S, Singh AV, Patil R, Gosavi S (2019) Helminthicidal and Larvicidal potentials of biogenic silver nanoparticles synthesized from medicinal plant Momordica charantia. Med Chem 15(7):781–789. https://doi.org/10.2174/1573406415666190430142637

Shkryl YN, Veremeichik GN, Kamenev DG, Gorpenchenko TY, Yugay YA, Mashtalyar DV, Nepomnyaschiy AV, Avramenko TV, Karabtsov AA, Ivanov VV (2018) Green synthesis of silver nanoparticles using transgenic Nicotiana tabacum callus culture expressing silicatein gene from marine sponge Latrunculia oparinae. Artif Cells Nanomed Biotechnol 46(8):1646–1658

Shu M, He F, Li Z, Zhu X, Ma Y, Zhou Z, Yang Z, Gao F, Zeng M (2020) Biosynthesis and antibacterial activity of silver nanoparticles using yeast extract as reducing and capping agents. Nanoscale Res Lett 15(1):14

Singh P, Kim YJ, Wang C, Mathiyalagan R, Yang DC (2016a) Microbial synthesis of flower-shaped gold nanoparticles. Artificial Cells Nanomed Biotechnol 44(6):1469–1474

Singh P, Singh H, Kim YJ, Mathiyalagan R, Wang C, Yang DC (2016b) Extracellular synthesis of silver and gold nanoparticles by Sporosarcina koreensis DC4 and their biological applications. Enzym Microb Technol 86:75–83

Singhal U, Khanuja M, Prasad R, Varma A (2017) Impact of synergistic association of ZnO-nanorods and symbiotic fungus *Piriformospora indica* DSM 11827 on *Brassica oleracea* var. botrytis (Broccoli). Front Microbiol 8:1909. https://doi.org/10.3389/fmicb.2017.01909

Soni N, Prakash S (2012) Efficacy of fungus mediated silver and gold nanoparticles against Aedes aegypti larvae. Parasitol Res 110(1):175–184

Soshnikova V, Kim YJ, Singh P, Huo Y, Markus J, Ahn S, Castro-Aceituno V, Kang J, Chokkalingam M, Mathiyalagan R (2018a) Cardamom fruits as a green resource for facile synthesis of gold and silver nanoparticles and their biological applications. Artificial Cells Nanomed Biotechnol 46(1):108–117

Soshnikova V, Kim YJ, Singh P, Huo Y, Markus J, Ahn S, Castro-Aceituno V, Kang J, Chokkalingam M, Mathiyalagan R, Yang DC (2018b) Cardamom fruits as a green resource for facile synthesis of gold and silver nanoparticles and their biological applications. Artificial Cells Nanomed Biotechnol 46(1):108–117. https://doi.org/10.1080/21691401.2017.1296849

Sowndarya P, Ramkumar G, Shivakumar MS (2017) Green synthesis of selenium nanoparticles conjugated Clausena dentata plant leaf extract and their insecticidal potential against mosquito vectors. Artif Cells Nanomed Biotech 45(8):1490–1495. https://doi.org/10.1080/21691401.2016.1252383

Srikar SK, Giri DD, Pal DB, Mishra PK, Upadhyay SN (2016) Green synthesis of silver nanoparticles: a review. Green Sustain Chem 6(1):34–56

Steinmetz NF (2010) Viral nanoparticles as platforms for next-generation therapeutics and imaging devices. Nanomed Nanotechnol Biol Med 6(5):634–641

Subramaniam J, Murugan K, Panneerselvam C, Kovendan K, Madhiyazhagan P, Dinesh D, Kumar PM, Chandramohan B, Suresh U, Rajaganesh R (2016) Multipurpose effectiveness of Couroupita guianensis-synthesized gold nanoparticles: high antiplasmodial potential, field efficacy against malaria vectors and synergy with Aplocheilus lineatus predators. Environ Sci Pollut Res 23(8):7543–7558

Suresh M, Jeevanandam J, Chan YS, Danquah MK, Kalaiarasi JMV (2020) Opportunities for Metal Oxide Nanoparticles as a Potential Mosquitocide. Bio Nano Sci 10(1):292–310. https://doi.org/10.1007/s12668-019-00703-2

Sutthanont N, Attrapadung S, Nuchprayoon S (2019) Larvicidal activity of synthesized silver nanoparticles from Curcuma zedoaria essential oil against Culex quinquefasciatus. Insects 10(1). https://doi.org/10.3390/insects10010027

Syed A, Ahmad A (2012) Extracellular biosynthesis of platinum nanoparticles using the fungus Fusarium oxysporum. Colloids Surf B Biointerfaces 97:27–31

Tahvilian R, Zangeneh MM, Falahi H, Sadrjavadi K, Jalalvand AR, Zangeneh A (2019) Green synthesis and chemical characterization of copper nanoparticles using Allium saralicum leaves and assessment of their cytotoxicity, antioxidant, antimicrobial, and cutaneous wound healing properties. Appl Organomet Chem 33(12):e5234. https://doi.org/10.1002/aoc.5234

Tyagi H, Kushwaha A, Kumar A, Aslam M (2016) A facile pH controlled citrate-based reduction method for gold nanoparticle synthesis at room temperature. Nanoscale Res Lett 11(1):362

Usman AI, Aziz AA, Noqta OA (2019) Green sonochemical synthesis of gold nanoparticles using palm oil leaves extracts. Mater Today Proc 7:803–807

Vahabi K, Mansoori GA, Karimi S (2011) Biosynthesis of silver nanoparticles by fungus Trichoderma reesei (a route for large-scale production of AgNPs). Insciences J 1(1):65–79

Velmurugan P, Cho M, Lim S-S, Seo S-K, Myung H, Bang K-S, Sivakumar S, Cho K-M, Oh B-T (2015) Phytosynthesis of silver nanoparticles by Prunus yedoensis leaf extract and their antimicrobial activity. Mater Lett 138:272–275

Vinoth V, Wu JJ, Asiri AM, Anandan S (2017) Sonochemical synthesis of silver nanoparticles anchored reduced graphene oxide nanosheets for selective and sensitive detection of glutathione. Ultrason Sonochem 39:363–373

Wang C, Hu L, Lin Y, Poeppelmeier K, Stair P, Marks L (2017) Controllable ALD synthesis of platinum nanoparticles by tuning different synthesis parameters. J Phys D Appl Phys 50(41):415301

WHO (2020) Vector-borne diseases. WHO. https://www.who.int/en/news-room/fact-sheets/detail/vector-borne-diseases. Accessed 17th November 2020 2020

Yadav SK, Vasu V (2016) Synthesis and characterization of copper nanoparticles, using combination of two different sizes of balls in wet ball milling. Int J Emerg Trends Sci Technol 3(04):2348–9480

Yadav L, Tripathi RM, Prasad R, Pudake RN, Mittal J (2017) Antibacterial activity of Cu nanoparticles against *E. coli, Staphylococcus aureus* and *Pseudomonas aeruginosa*. Nano Biomed Eng 9(1):9–14. https://doi.org/10.5101/nbe.v9i1.p9-14

Yu C-H, Tam K, Tsang ES (2008) Chemical methods for preparation of nanoparticles in solution. Handbook of Metal Physics 5:113–141

Zhu X, Pathakoti K, Hwang H-M (2019) Green synthesis of titanium dioxide and zinc oxide nanoparticles and their usage for antimicrobial applications and environmental remediation. In: Green synthesis, characterization and applications of nanoparticles. Elsevier, pp 223–263

Chapter 13
Perspectives of Metals and Metal Oxide Nanoparticles for Antimicrobial Consequence – An Overview

R. L. Rengarajan, A. Rathinam, N. Suganthy, B. Balamuralikrishnan, A. Vijaya Anand, and S. Velayuthaprabhu

Contents

13.1	Introduction...	398
	13.1.1 What Are Nanostructures?...	398
13.2	Metal Nanoparticles as Antimicrobial Agents............................	399
	13.2.1 Mode of Action of Metal and Metal Oxide Nanoparticles..	400
13.3	Antimicrobial Action of Nanoparticles.......................................	402
	13.3.1 Silver Nanoparticles (Ag NPs)......................................	402
	13.3.2 Selenium Nanoparticles (Se NPs)................................	404
	13.3.3 Magnesium Oxide Nanoparticles (MgO NPs)...............	404
	13.3.4 Zinc Oxide Nanoparticles (ZnO_NPs)..........................	405

R. L. Rengarajan and A. Rathinam contributed equally with all other contributors.

R. L. Rengarajan
Department of Animal Science, Bharathidasan University, Tiruchirappalli, Tamil Nadu, India

A. Rathinam (✉)
Department of Animal Science, Bharathidasan University, Tiruchirappalli, Tamil Nadu, India

Key Laboratory for Genome Stability and Disease Prevention, Guangdong Province, Shenzhen University, Shenzhen, China

N. Suganthy
Department of Nanoscience and Technology, Alagappa University, Karaikudi, Tamil Nadu, India

B. Balamuralikrishnan
Department of Food Science and Biotechnology, College of Life Science, Sejong University, Seoul, South Korea

A. Vijaya Anand
Department of Human Genetics and Molecular Biology, Bharathiar University, Coimbatore, Tamil Nadu, India

S. Velayuthaprabhu
Department of Biotechnology, Bharathiar University, Coimbatore, Tamil Nadu, India

© The Author(s), under exclusive license to Springer Nature Switzerland AG 2022
A. Krishnan et al. (eds.), *Emerging Nanomaterials for Advanced Technologies*, Nanotechnology in the Life Sciences, https://doi.org/10.1007/978-3-030-80371-1_13

		13.3.5	Gold Nanoparticles (Au NPs)	405
		13.3.6	Titanium Oxide Nanoparticles (TiO$_2$ NPs)	405
		13.3.7	Calcium Carbonate and Magnesium Oxide composites (CaCO$_3$ and MgO)	406
		13.3.8	Aluminium Oxide Nanoparticles (Al$_2$O$_3$ NP^s)	406
		13.3.9	Copper and Copper Oxide Nanoparticles (Cu & CuO NPs)	407
13.4	Nanoantibiotics: An Option to Fight Against Antibiotic Resistance			407
		13.4.1	Silica Nanoparticles (SiO$_2$ NPs)	408
13.5	Nanoparticle and Biofilm Interaction			410
13.6	Application of Metallic Nanoparticles			410
		13.6.1	Dental Materials	410
		13.6.2	Antitumor Properties	410
		13.6.3	Textile Industry	411
		13.6.4	Food Management	411
		13.6.5	Wastewater Management	412
13.7	Conclusion			412
References				412

13.1 Introduction

13.1.1 What Are Nanostructures?

A nanoparticle that typically possesses 0.2–100 nm scale. On the other hand, it can be in the form of very fine particles, nanoclusters, and nanocrystals.

The emergence of multidrug-resistance microbes has turned into a global medical challenge, which prompted researchers towards the search for alternatives for the effective antimicrobial treatment for infection (Beyth et al. 2015). With the beginning of nanotechnology in biomedical research, scientists have attempted to use nanoparticles because of their unique physicochemical characters, including, smaller size with greater surface area, enhanced catalytic, magnetic and optical property, and their biocompatibility (Gao et al. 2014). The proposed mechanism of nanoparticles against microbes involves (i) free metal ion mediated toxicity released as of nanoparticle surface and (b) oxidative stress-mediated apoptosis due to the release of reactive oxygen species (Seil and Webster 2012; Prasad et al. 2016). Morphological and physicochemical properties of nanoparticles have great influence in the antimicrobial activities of nanoparticles. Nanoparticles with smaller size and positive charge exhibit enhanced bactericidal activity due to their stability, water-solubility, targeting capability of drugs, and bioavailability (Kumar et al. 2015; Prasad et al. 2019, 2020).

Nanoparticles have a remarkable potential to control disease-causing pathogens, especially multidrug-resistant microbes, through efficient release of the drug molecule (Sampath Kumar et al. 2015). Hence, nano-based treatment, including nano-medicine or nano-drug, are showing promising therapeutic efficiency to combat microbial diseases (Kumar et al. 2014). Figure 13.1 highlighs various general applications of nanoparticles. This chapter focuses on recent advancements in the

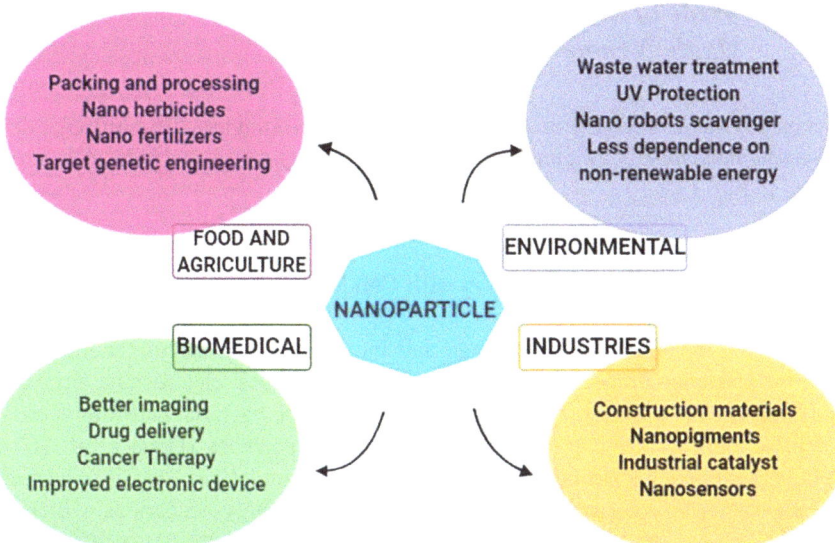

Fig. 13.1 Applications of nanoparticles in various industries

antimicrobial property and activity of the nanoparticles, with their mechanism of action and possibilities to reduce the nanoparticle-mediated toxicity.

13.2 Metal Nanoparticles as Antimicrobial Agents

Metals have been widely used as antimicrobial agents from the ancient times, around 1500 BP, where the Indians, Romans, Egyptians, and Greeks utilized silver and copper utensils to store food and water as a preservative and disinfectant (Galib et al. 2011). However, the medical applications of metals came to a standstill position after the discovery of antibiotics as early as 1920. In recent times, antimicrobial resistance or multidrug-resistance due to its biofilm-forming ability has turned into an epidemic problem globally, which might cause more than ten million deaths by 2050. Hence, researchers have turned their attention to alternative antimicrobial therapy (Prasad et al. 2020).

Metal has the ability to discriminate between bacterial and mammalian cells and interact with bacteria, interfering with the metabolic pathways leading to cellular death even in multidrug-resistant bacteria (Shenmchuk et al. 2010). The advancement in nanotechnology has increased the utilization of metal nanoparticles in the biomedicine field specifically as an antimicrobial agent because of their distinctive properties, including nano size and high surface-to-volume ratio with enhanced surface reactivity (Pelgrift and Friedman 2013).

13.2.1 Mode of Action of Metal and Metal Oxide Nanoparticles

Metals and metal oxide nanoparticles exhibit antimicrobial properties through three different mechanisms as mentioned below and the general mechanism is represented in Fig. 13.2.

13.2.1.1 Nanoparticle Interaction with Cell Membrane

The entry of the metal-based nanomaterials into the cell is based on the interaction with the transmembrane system via van der Waals forces, electrostatic attraction, and hydrophobic and receptor–ligand interactions. The bacterial cell wall acts as a defensive barrier which prevents entry of foreign particles (Dizaj et al. 2014). Cellular components differ between Gram-positive and Gram-negative bacteria. The Gram-negative bacteria have a rich source of lipopolysaccharides, which are

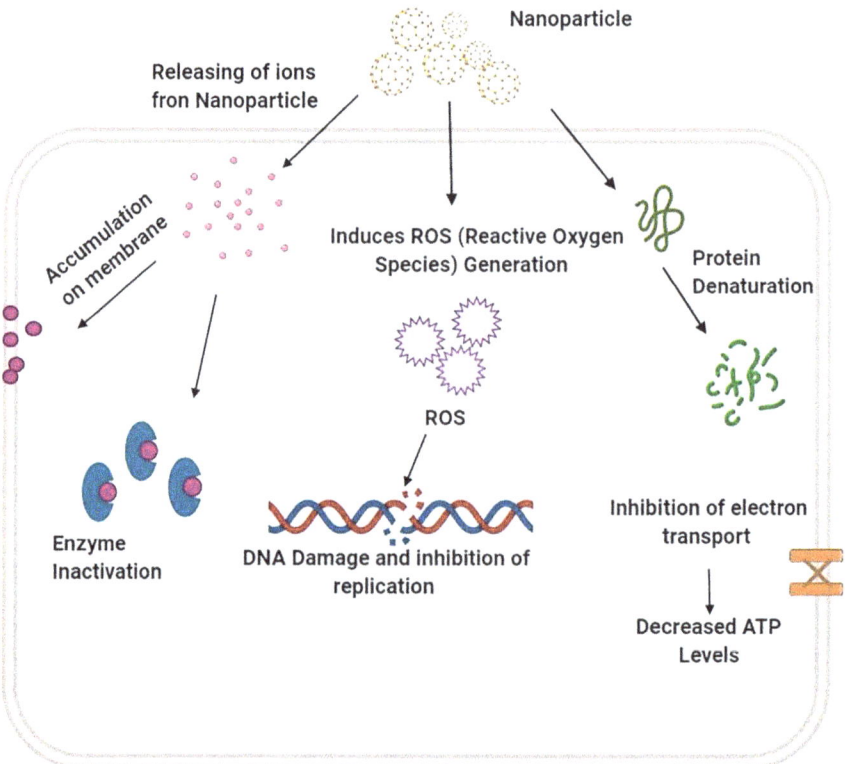

Fig. 13.2 Mechanism of action of antimicrobial activity of nanoparticles

negatively charged and interact quickly with positively charged nanoparticles based on the electrostatic force of attraction.

However, Gram-positive bacteria are composed of teichoic acid in which nanoparticles are distributed along with the phosphate group, which prevents its aggregation. Despite the charge, penetration of nanoparticles is quicker in Gram-positive bacteria, since the cell wall is thin, which is formed by peptidoglycan with abundant pores, while the thick lipopolysaccharides in Gram-negative bacteria act as a barrier to the entry of nanoparticles. Interaction of nanoparticles with the cell membrane alters the structural integrity of the membrane, disrupting cell membrane and leading to the efflux of water into the cytosol, which is compensated by the bacterial protein efflux pump and electron transport creating ionic imbalance (Shaikh et al. 2019). The penetrated nanoparticles interact with the components of the cell, including protein, DNA, enzymes, and lysosome, leading to oxidative stress, protein deactivation, enzyme inhibition, electrolyte imbalance, impaired respiration and disruption in energy transduction, ultimately leading to cell death (Prasad and Swamy 2013, Swamy and Prasad 2012; Prasad et al. 2011, 2012; Joshi et al. 2018; Gupta et al. 2018).

The shape, size, and charge of the nanoparticles influence the penetration into the cell membrane. Most of the metal nanoparticles such as gold (Au), silver (Ag), zinc oxide (ZnO), magnesium oxide (MgO), and titanium oxide (TiO_2) exhibit similar action (Gao et al. 2014).

13.2.1.2 Oxidative Stress-Mediated Cell Death

Nanoparticles induce the formation of reactive oxygen species, which includes hydrogen peroxide (H_2O_2), hydroxyl radicals (OH^-) or superoxide anions (O_2^-), and singlet oxygen, which interact with the cell membrane in the lipid core, leading to lipid peroxidation disrupting cell membrane, protein oxidation, inhibition of enzymes, and damage to DNA and RNA. Reactive oxygen species also inhibit the transcription and translation machinery and electron transport chain, leading to the death of the bacteria (Li et al. 2012).

13.2.1.3 Interaction of Dissolved Metal Ions with Protein and DNA

The metal ions released from nanoparticles are quickly absorbed into the cell through the cell membrane via ion transporters, which interact directly with the functional group of proteins and nucleic acids, like mercapto (–SH), carboxyl (–COOH), and amino (–NH) groups, inducing oxidative damage inhibiting its enzymatic activity affecting the normal physiological process culminating to death (Leung et al. 2014).

13.3 Antimicrobial Action of Nanoparticles

Scientific evidence on the antimicrobial activity of both the metal and metal oxide nanoparticles are highlighted in Table 13.1.

13.3.1 Silver Nanoparticles (Ag NPs)

Ag NPs are often used in biomedical devices and also widely employed in antimicrobial therapy (Durán et al. 2016; Zhu et al. 2015). Scientific studies have revealed that Ag NPs exhibit antibacterial activity by inducing pits and gaps in the cell membrane of bacteria leading to the destruction of the cell. Penetrated Ag NPs also interact with –SH groups of the enzymes, which may lead to the disruption of metabolic processes and causes cell death (Lok et al. 2007; Prasad and Swamy 2013; Prasad 2014; Aziz et al. 2014, 2015, 2016, 2019). Jo et al. (Jo et al. 2009) observed the antimicrobial potential of Ag NPs against foodborne fungi, which rarely produce spore. Ag NPs have also been reported for their antifungal property against plant pathogens, which controls the spore-forming ability and it was observed to be less toxic than synthetic fungicides.

Antimicrobial activities of Ag NPs have increased the attention towards nanoparticles in the food, pharmaceutical, and biomedical industries (Prasad 2014). The enhanced bactericidal activity of Ag NPs against *Streptococcus mutans*, when compared to other nanoparticles, suggests its application for treating dental caries, which is caused by *S. mutans* (Hernández-Sierra et al. 2008). Nanocomposite of Ag and ZnO NPs exhibit bactericidal activity against *L. plantarum* in orange juice indicating that Ag NPs can be used in food packaging industries (Emamifar et al. 2011). The possible antibacterial activity of the nanoparticles is represented in Fig. 13.3.

Mounting evidence on antibacterial potentials of Ag NPs against foodborne pathogens like *L. monocytogenes, S. typhimurium, E. coli*, and *V. parahaemolyticus* illustrated that Ag NPs may be used as an alternative cleansing and disinfectant agent on surfaces related to food environments (Zarei et al. 2014). The size and shape of Ag NPs influence their antimicrobial property. A small size with truncated triangular morphology showed high antibacterial activity due to its high-atom-density surfaces and ease penetration into the bacterial cell wall. Ag NPs exhibit antiviral properties by interrupting viral replication, inhibiting viral entry into host cells by attenuating fusion, infectivity, and CD-4-dependant virion binding, and also acts as a virucidal agent (Lara et al. 2010). Ag NPs also inhibit the post-entry stages of the life cycle in HIV-1 and Herpes Simplex Virus Type 1. Ag NPs encapsulated with mercaptoethane sulfonate compete with binding to the cellular heparan sulfate through its sulfonate end groups inhibiting its entry into the cell with no cytotoxic effect in mammalian cells (Baram-Pinto et al. 2009; Elechiguerra et al. 2005). Other studies also revealed the antiviral properties of Ag NPs against Hepatitis B virus, monkey-pox virus, and respiratory syncytial virus (Lu et al. 2008).

Table 13.1 Antimicrobial property of metal nanoparticles and their mode of action

Nanoparticles	Size (nm)	Pathogen	Mode of action	References
Ag Nps	20–30	*Bacillus* and *M. tuberculosis*	By interacting with the cell wall and the lysis of the cell	Zhou et al. (2012)
	25	Methicillin-resistant *S. aureus*	By interacting with the cell wall and the lysis of the cell	Panáček et al. (2006)
	20–30	*V. cholera*	By modifying the permeability of the cell membrane and the respiration	Krishnaraj et al. (2010)
Gold	25–30	*K. pneumoniae*	By changing the structure of the outer membrane and death of the cell	Murugan et al. (2015)
	43–79	*K. pneumoniae*	Not yet identified	Madhiyazhagan et al. (2015)
Copper Zinc oxide	43–79	*S. typhi*	By affecting the integrity of the cell membrane	Murugan et al. (2015)
Selenium nanoparticles	30–100	*B. anthracis*	Not yet identified	Singh et al. (2015)
Titanium Oxide nanoparticles	120	Drug-resistant *N. gonorrhoeae*	By affecting the integrity of the cell membrane	Li et al. (2013)
	5–15	*L. monocytogenes*	By separating the cytoplasmic membrane from the cell wall and changes in morphology of the cell	Tamayo et al. (2014)
	55	*E.coli* O157:H7	By interacting with the cell wall and the lysis of the cell	Paredes et al. (2014)
	20–30	BCG	By interacting with the cell wall and the lysis of the cell	Zhou et al. (2012)
	9–22	*K. pneumoniae*	By changing the structure of the outer membrane and death of the cell	Prema and Thangapandiyan (2013)
	—	Drug-resistant *N. gonorrhoeae*	By disrupting the cell and the leakage of intracellular contents	Addae et al. (2014)
	20–30	*S. typhi*	By affecting the cell wall of the bacteria may lead to cell death	Yallappa et al. (2015)
	5	*L. monocytogenes*	By separating the cytoplasmic membrane from the cell wall and changes in morphology of the cell	Jin et al. (2009)
	30–200	*Aspergillus fumigates*	Antifungal activity	Shakibaie et al. (2015)
	62–74	*Prevotella intermedia*	Photocatalytic activity	Vargas-Reus et al. (2012)

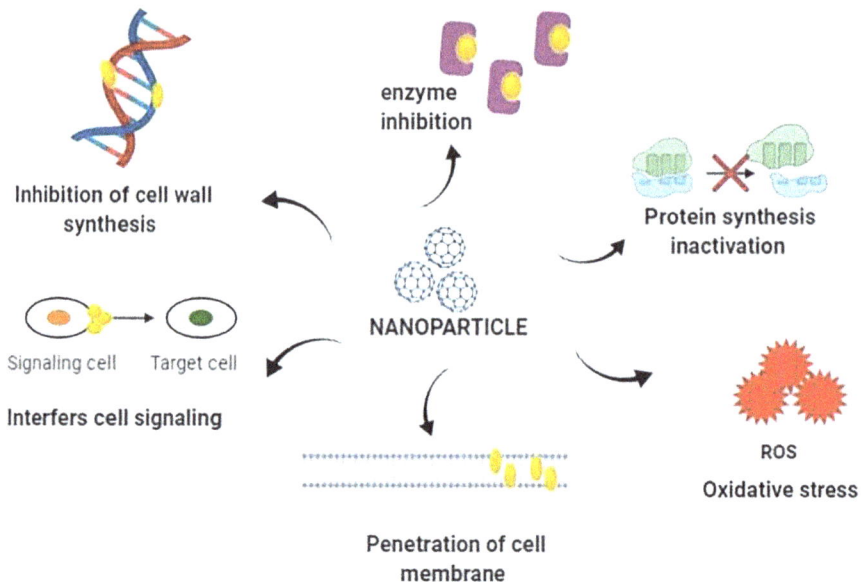

Fig. 13.3 Mechanism of action of antibacterial activity of nanoparticles

Fungus pathogenic species *Candida albicans* has been shown to cause severe bloodstream infection in humans. *C. albicans* has the potential for transformation between yeast and hyphal morphologies that tend to build up biofilm and also reflect its pathogenic factor. Ag NPs (1 nm) have been developed in order to fight against *C. albicans* biofilm. The results showed that Ag NPs greatly inhibit or reduce the biofilm formation (Lara et al. 2015).

13.3.2 Selenium Nanoparticles (Se NPs)

Se NPs showed potent antifungal activity upon *C. albicans* and *A. Fumigates*. Cerium oxide NPs significantly reduced the inflammatory response related to peritonitis, indicating the new therapeutic avenue for the management intra-abdominal infection (Shakibaie et al. 2015; Manne et al. 2015).

13.3.3 Magnesium Oxide Nanoparticles (MgO NPs)

MgO NPs revealed their potent antiviral properties against hand, foot and mouth disease-causing virus. It is reported that 250 μg/mL concentrations of MgO NPs did not cause any abnormality in Razi Bovine kidney cell line (Rafiei et al. 2015).

13.3.4 Zinc Oxide Nanoparticles (ZnO_NPs)

ZnO NPs possess an antioxidant and anticoccidial effect on *Eimeria papillata*-triggered infection in the jejunum (Dkhil et al. 2015). Besides, ZnO NPs have also been shown to possess antimicrobial properties against an array of pathogens such as *S. typhimurium*, *C. jejuni*, and other foodborne microbes such as *E. coli*, *Listeria monocytogenes*, *V. cholera*, *K. pneumoniae*, and *N. gonorrhea* (Tayel et al. 2011; Xie et al. 2011; Liu et al. 2009; Jin et al. 2009; Salem et al. 2015; Nagarajan and Kuppusamy 2013; Bhuyan et al. 2015). Khan et al. (Khan et al. 2015) highlighted that ZnO NPs are valuable nanoparticles in order to maintain and keep the hygiene of the oral cavity therein minimizing the bacterial colonies.

Approximately 550 tons of ZnO NPs have been produced annually, employed in a profit-making product and it is also showing a reduced amount of toxic as compared to CuO and Ag NPs to humans (Keller et al. 2013; Bondarenko et al. 2013). The antimicrobial property of ZnO NPs is nearly similar to Ag NPs (Adams et al. 2014).

13.3.5 Gold Nanoparticles (Au NPs)

Au NPs have many beneficial properties including their easy preparation, fluorescence, scattering effect, and biocompatibility which are widely used in the field of biomedicine (Huang et al. 2007). Au nanosphere could be made by reduction of auric acid in the presence of sodium citrate and the size of the Au NPs is also varied by the addition of different concentrations of reducing agent (Turkevich et al. 1951). Au NPs are stable and can interact (covalently and noncovalently) with appropriate ligands such as peptides, antibodies, and DNA (Han et al. 2007; Paciotti et al. 2006). Au NPs possess light-absorbing ability, and hence can conjugate with antibodies that may act against selected pathogens, *Staphylococcus aureus*, to kill them photothermally or hyperthermally (Zharov 2006; Norman et al. 2008).

13.3.6 Titanium Oxide Nanoparticles (TiO$_2$ NPs)

TiO$_2$ NPs are largely used in the food products such as candy and chewing gums. An adult human is exposed about 1 mg/kg B.WT/day of titanium (Weir et al. 2012). It is one of the most abundant nanoparticles in terms of production throughout the world, approximately 3000 tons/year (Keller et al. 2013). TiO$_2$ NPs have shown promising antimicrobial properties upon various microbes such as *E. coli*, *S. aureus*, *P. aeruginosa*, *E. faecium*, *B. subtilis*, and *K. pneumoniae* (Rajakumar et al. 2012). The nanoparticles (62–74 nm) with minimal inhibitory concentration value were tested for the above mentioned bacteria ranges between 40 and 80 µg/mL.

Similarly, another study showed the TiO_2 NPs' minimal inhibitory concentration value to be 1187.5 µg/mL on biofilm-forming bacteria *P. gingivalis, F. nucleatum, A. actinomycetemcomitans*, and *Prevotella intermedia* (Vargas-Reus et al. 2012). TiO_2 NPs has a unique characteristic that is photocatalytic activity, when exposed to photoactivation it notably increased the antimicrobial property on a number of species *P. aeruginosa, B. fragilis, E. hire, B. fragilis, S. aureus,* and *S. Typhimurium* (Maness et al. 1999).

13.3.7 Calcium Carbonate and Magnesium Oxide composites ($CaCO_3$ and MgO)

The $CaCO_3$ NPs' antimicrobial potential has been assessed on *S. typhimurium, S. aureusi E. coli,* and *B. subtilis*, it reveals the better antimicrobial activity of CaO on an aforementioned bacterial species. On the other hand, $CaCO_3$/MgO nanocomposites showed greater antibacterial potential on *E. coli* and *S. aureus*. The mechanism of antibacterial potential of CaO and MgO is generating the superoxide anion on their surface and besides augmenting in pH value because of the hydration of CaO and MgO (Yamamoto et al. 2010).

13.3.8 Aluminium Oxide Nanoparticles (Al_2O_3 NPs)

Aluminum oxide NPs (Al_2O_3 NPs) possesses several uses in industrial and home care products. It also evaluated for the antimicrobial property on *E. coli*, higher concentration of nanoparticles showed a moderate growth inhibitory effect when tested against bacteria. This attribution of the Al NPs is due to the surface charge interactions between the cell membrane and cell wall with nanoparticles, and also infiltration within the cells of the bacteria (Sadiq et al. 2009).

Basically, Al NPs are thermodynamically stable (Martınez-Flores et al. 2003). They possess close to neutral pH, which holds a + ve charge on their surface. The positively charged Al NPs tend to interact with negatively charged bacterial cells', *E. coli*, surface (Li and Logan 2004). This is due to the electrostatic force between bacteria and nanoparticle surface, in association with hydrophobic interactions and polymer cross-link. This antimicrobial activity of the nanoparticles produce reactive oxygen species that disrupt the cell wall as well as membrane, subsequently causing cell death (Ruparelia et al. 2008). Al NPs save the cells from death due to oxidative stress, thereby, act as free radical scavengers. Their action is based on nanostructure (Mohammad et al. 2008).

13.3.9 Copper and Copper Oxide Nanoparticles (Cu & CuO NPs)

Cu & CuO NPs revealed microbicidal activity against a variety of disease-causing bacterial species. For example, CuO NPs (50–100 nm) produced in a green chemistry manner pronounced a considerable antimicrobial activity upon various disease-causing species such as *Shigella dysenteriae, V. cholera*, and *K. pneumoniae* (Sutradhar et al. 2014). One more study showed that CuO NPs (20–95 nm) have antimicrobial activity on methicillin-resistant *S. aureus* and *E. coli* (Ren et al. 2009).

Oral pathogens *P. gingivalis, S. mutans,* and *F. nuceatum* could be inhibited in the range of minimal inhibitory concentration values 250–500 µg/mL of CuO NPs (10–50 nm). CuO NPs (23 nm) showed their antimicrobial activity on *Enterococcus faecalis*; however, they do not show antimicrobial activity on *K. pneumoniae* (Ahamed et al. 2014). The antimicrobial activity is based on the shape and size of the nanoparticles (Azam et al. 2012). It has been reported that Cu^{2+} ion release from the nanoparticles could trigger mutation in DNA and consequently produce the reactive oxygen species (Pan et al. 2010).

On the other hand, another study opposes that the release of Cu^{2+} ions from the nanoform is insignificant, so the less antimicrobial activity of the nanoparticles. Indeed, reactive oxygen species production is not because of the Cu^{2+} ions, but its nanoform. In the same manner, nanoparticles enter into the bacterial cell membrane and produce the reactive oxygen species, therein causing cell death (Applerot et al. 2012; Yadav et al. 2017). CuO NPs shows bactericidal activity via inhibiting *E. coli's* cellular respiration and also interacting with various biomolecules (Wahab et al. 2013). Ren et al. (Ren et al. 2009) reported that *B. subtilis* has an attraction towards amines and carboxylic groups present on the cell membrane. The nanoparticles may have the potential to inhibit the enzyme or degrade the protein through interaction with protein-SH group (Schrand et al. 2010). CuO tagged with linoleic acid inactivates certain enzymes (Das et al. 2010).

13.4 Nanoantibiotics: An Option to Fight Against Antibiotic Resistance

Frequent antibiotic treatment is commonly known to augment the resistance that develops in multidrug-resistant organisms, an increased dose of antibiotics also causes adverse side effects. So, an alternative is needed to overcome these problems; metallic nanoparticles are well characterized and possess antimicrobial properties. The functions of nanoparticles are structured in Fig. 13.4. Nanoparticles could be conjugated with antibiotics that can gradually reduce the dosage, toxicity and improve the antimicrobial activity against multidrug-resistant and great bioavailability (Allahverdiyev et al. 2011). Nanoantibiotics showed promising antimicrobial activity and also have a safe delivery of antibiotics (Abeylath and Turos

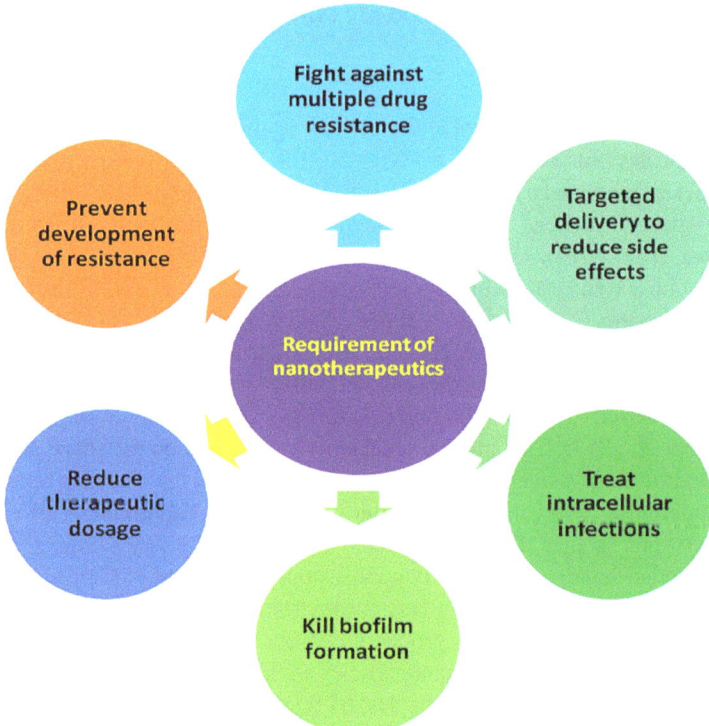

Fig. 13.4 Function of nanostructures in antimicrobial treatment

2008; Huh and Kwon 2011). A recent study revealed that nanomaterials could reduce the development of drug-resistant bacteria (Mühling et al. 2009).

13.4.1 Silica Nanoparticles (SiO₂ NPs)

The SiO_2 has antimicrobial activity at the nano-scale level because of the surface area enhancement (Dhapte et al. 2014). The Si NPs also inhibit the adherence of bacteria on an oral biofilm (Cousins et al. 2007). The combination treatment for antibacterial activities of Cu/SiO_2 properties had been studied well by disk diffusion method. Results demonstrated that the antibacterial (*C. albicans, P. citrinum, E. cloacae, E. coli,* and *S. aureus*) properties of $Cu-SiO_2$ nanocomposite were noticeably identified against bacteria (both Gram-positive and Gram-negative) and fungi because Cu NPs of $Cu-SiO_2$ were formed on the surface of SiO_2 NP (Kim et al. 2007).

Another combination of Ag-Si nanocomposite was also analyzed against various microbes and the results were compared with the conventional methods, including silver nitrate and silver zeolite. The Ag-Si nanocomposite showed improved antimicrobial activities (Egger et al. 2009). To determine the minimal inhibitory concentration and minimal bactericidal concentration the Ag-SiO$_2$ particles were tested against *S. aureus* and *E. coli* using the standard serial dilution method. The findings of the result suggest that Ag-SiO$_2$ NPs have greater antibacterial property (Xu et al. 2009). Literature showed that Si nanowires combined with the living cells and bacteria disrupted the functions of cell-like cell differentiation, adhesion and spreading. The Ag NPs Si nanowires revealed the antibacterial properties and also showed biocompatibility with epithelial cells of the lung adenocarcinoma (Li and Logan 2004). These results showed that the combination of Si with Ag and/or nanocomposites has the antimicrobial potential and usefulness in the biomedical field.

13.4.1.1 Silica–Antibiotics Combination

Gentamicin-Loaded Silica Nanoparticles

Gentamicin-loaded with silica xerogel is showed potential antimicrobial against *Salmonella, Mycobacterium* species, and *Brucella* which severely causes chronic infections and also a big challenge to eradicate. Silica nanoparticles could effectively deliver the antibiotic with minimizing the dosage, increasing longevity and reducing the toxicity level of the antibiotic (Seleem et al. 2009).

Silica NPs Conjugated with Tetracycline Antibiotic

The antibiotic tetracycline encapsulated with silica NPs was demonstrated by Capeletti et al. (Capeletti et al. 2014). The prepared nanostructures were tested against vulnerable *E. coli* and the results were compared to pristine tetracycline alone and a mixture of tetracycline + ampicillin for its antimicrobial efficiency. The Si-based NPs have interacted with the lipopolysaccharides of the peptidoglycan layer of the outer membrane of the bacteria.

The formation of the hydrogen bond between the saccharides and hydroxyl groups is identified on the silica surface. This complex can weaken the peptidoglycan layer and it may lead to the disruption of the cells. It is also noted that there is no toxicity effect in the mammalian cells. The antibiotics can disrupt the cell wall of the bacteria, and also they do not affect the mammalian cells. This is because of the differences in the structure of the bacterial cell walls and mammals. It is also evident that the bactericidal actions of encapsulated tetracycline may be due to the hydrolyzing effect of nanoantibiotic on sugar molecules of peptidoglycan.

13.5 Nanoparticle and Biofilm Interaction

Biofilm is the key factor which makes the microbes resistant to antibiotics and the immune system. Scientific reports revealed that nanoparticles disrupt the structural integrity of biofilm by attaching with exopolysacharides. Ag NPs inhibits the production of exopolysacharides, which might be due to the mechanism of antibiofilm property of Ag NPs against *Klebsiella pneumoniae* and *E. coli* (Su et al. 2009).

Nanoparticles influence the rate of biofilm formation and bacterial adhesion, whose mechanism of action is not yet completely elucidated. MgO NPs adhere and diffuse into biofilms leading to the disruption of the membrane potential, promote lipid peroxidation and binds with DNA and disrupts its function, altering the normal functioning of bacteria inhibiting its ability to form biofilm (Lellouche et al. 2012). Potassium ion channels play a vital role in long-distance electrical signal conduction within the bacterial biofilm and also influences the metabolic activity of bacteria (Lundberg et al. 2013). Most of the nanoparticles like Ag NPs and ZnO NPs have been identified to interrupt the ion channels and inhibit the metabolic activity thereby attenuating the biofilm formation.

13.6 Application of Metallic Nanoparticles

The general applications of metallic nanoparticles are highlighted in Fig. 13.5.

13.6.1 Dental Materials

Bacteria accumulate in the plaque and spreads over the mouth, which causes damage to teeth. The groups coated with CuO and ZnO NPs have the possibility to restrain the growth of *S. mutans* (Ramazanzadeh et al. 2015). Addition of titanium dioxide NPs (TiO_2 NPs) showed the antibacterial effects following light exposure (Aboelzahab et al. 2012).

13.6.2 Antitumor Properties

Although medicines are available in plenty, many people die from cancer. Chemotherapeutic drugs are often associated with severe side effects. Hence, NP-based drugs recently grab great interest in terms of nano-size, efficient delivery, biocompatibility, site targeting, and fewer side effects (Tourinho et al. 2012). Various types of nano-carriers have been reported so far, including polymeric micelles, dendrimers, liposomes, and inorganic in anti-cancer treatment in order to

Fig. 13.5 Applications of metallic nanoparticles

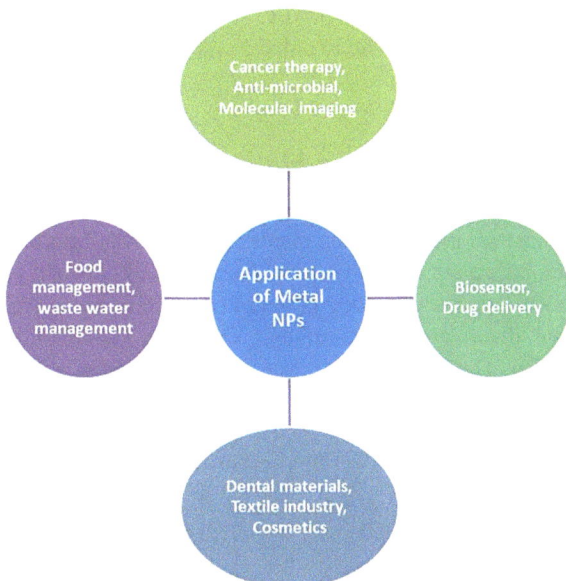

reduce the side effects (Liang et al. 2020). Inorganic nanomaterials such as metal, including silver, gold, and nickel as well as metal oxides of iron, zinc, and titanium show potential influence in medicine, including cancer therapy, cell imaging, and gene or drug delivery (Sunderam et al. 2019).

13.6.3 Textile Industry

Recently, the blending of nanoparticles with textiles during fabrication has improved greatly. Ag NPs are employed widely for the improvement of properties such as antibacterial, self-cleaning, and UV blocking of finished fabrics (Fouda et al. 2017). Besides, ZnO-NPs are included in textile manufacturing in order to increase UV locking and for the antibacterial properties (Mohamed et al. 2019). Inorganic nanoparticles are preferred over organic nanoparticles in the textile industry as UV blockers (Riva et al. 2006).

13.6.4 Food Management

The nanotechnology is a useful tool in two main areas in the food industry including food packaging and food additives/ingredients (Sharaf et al. 2019; Chausal et al. 2021). Nanoparticles are used in the food industry in a variety of applications such as nano-particulate delivery systems, wrapping, and food safety (Prasad et al. 2017).

Nanometal oxides, ZnO NPs, are employed in polymeric materials in the production of packing tissue to enhance the antimicrobial properties (Espitia et al. 2012).

13.6.5 Wastewater Management

Industries are one of the major causes of water pollution. The effluents released by the industries contain several hazardous chemicals. Nanotechnology has offered a novel approach in the treatment of wastewater in terms of removing toxic metals and disinfection (Prasad and Thirugnanasanbandham 2019; Uddandarao et al. 2019). The photocatalytic activity of Palladium with ZnO NPs allows the elimination of disease-causing microbes from the wastewater (Mishra et al. 2020). An array of metals in nanoscales, including Ag, ZnO, CuO, TiO_2, and carbon nanotubes, has high potential disinfection properties in the polluted water (Rafique et al. 2020).

13.7 Conclusion

In an epoch of increasing multidrug resistance where the microbes are developing resistance against several antibiotics, treatment against infectious disease has turned into a serious global issue due to increased mortality and morbidity. The emergence of nanotechnology has introduced nanoparticles as a *via*ble alternative to antibiotics specifically against multidrug-resistance bacteria, through their multifaceted activity such as cell wall penetration, oxidative stress-mediated damage to biomolecules altering the gene regulation and metabolism thereby blocking the bacteria defence and survival. Despite several reports on the antimicrobial efficiency of nanoparticles, their toxicity in mammalian cells still remains a debate. Hence, future research should focus on understanding the molecular mechanism behind the antibacterial activity of nanoparticles and fabrication of biocompatible engineered nanocomposite to attenuate human and environmental toxicity.

Acknowledgement Dr. Rathinam Ayyasamy acknowledges DST-Science and Engineering Research Board (SERB), India, for the award of National Post-Doctoral Fellowship (NPDF) (PDF/2016/001777) and the Royal Society of Chemistry (RSC) for the award of Research Mobility Grant (RM1802-8094).

References

Abeylath SC, Turos E (2008) Drug delivery approaches to overcome bacterial resistance to β-lactam antibiotics. Expert Opin Drug Del 5(9):931–949

Aboelzahab A, Azad AM, Dolan S, Goel V (2012) Mitigation of Staphylococcus aureus-mediated surgical site infections with Ir photoactivated TiO_2 coatings on Ti implants. Adv Healthc Mater 1(3):285–291

Adams CP, Walker KA, Obare SO, Docherty KM (2014) Size-dependent antimicrobial effects of novel palladium nanoparticles. PLoS One 9(1):e85981

Addae E, Dong X, McCoy E, Yang C, Chen W, Yang L (2014) Investigation of antimicrobial activity of photothermal therapeutic gold/copper sulfide core/shell nanoparticles to bacterial spores and cells. J Biol Eng 8(1):11

Ahamed M, Alhadlaq HA, Khan MA, Karuppiah P, Al-Dhabi NA (2014) Synthesis, characterization, and antimicrobial activity of copper oxide nanoparticles. J Nanomater:1–4

Allahverdiyev AM, Abamor ES, Bagirova M, Rafailovich M (2011) Antimicrobial effects of TiO_2 and Ag_2O nanoparticles against drug-resistant bacteria and *leishmania* parasites. Future Microbiol 6:933–940

Applerot G, Lellouche J, Lipovsky A, NitzanY LR, Gedanken A, Banin E (2012) Understanding the antibacterial mechanism of CuO nanoparticles: revealing the route of induced oxidative stress. Small 8(21):3326–3333

Azam A, Ahmed SA, Oves M, Khan MS, Memic A (2012) Size-dependent antimicrobial properties of CuO nanoparticles against gram-positive and gram-negative bacterial strains. Int J Nanomedicine 7:3527–3535

Aziz N, Fatma T, Varma A, Prasad R (2014) Biogenic synthesis of silver nanoparticles using Scenedesmus abundans and evaluation of their antibacterial activity. Journal of Nanoparticles, Article ID 689419, https://doi.org/10.1155/2014/689419

Aziz N, Faraz M, Pandey R, Sakir M, Fatma T, Varma A, Barman I, Prasad R (2015) Facile algae-derived route to biogenic silver nanoparticles: Synthesis, antibacterial and photocatalytic properties. Langmuir 31: 11605–11612. https://doi.org/10.1021/acs.langmuir.5b03081

Aziz N, Pandey R, Barman I, Prasad R (2016) Leveraging the attributes of Mucor hiemalis-derived silver nanoparticles for a synergistic broad-spectrum antimicrobial platform. Front Microbiol 7:1984. https://doi.org/10.3389/fmicb.2016.01984

Aziz N, Faraz M, Sherwani MA, Fatma T, Prasad R (2019) Illuminating the anticancerous efficacy of a new fungal chassis for silver nanoparticle synthesis. Front Chem 7:65. https://doi.org/10.3389/fchem.2019.00065

Baram-Pinto D, Shukla S, Perkas N, Gedanken A, Sarid R (2009) Inhibition of herpes simplex virus type 1 infection by silver nanoparticles capped with mercaptoethane sulfonate. Bioconjug Chem 20(8):1497–1502

Beyth N, Houri-Haddad Y, Domb A, Khan W, Hazan R (2015) Alternative antimicrobial approach: nano-antimicrobial materials. Evid Based Complement Alternat Med

Bhuyan T, Mishra K, Khanuja M, Prasad R, Varma A (2015) Biosynthesis of zinc oxide nanoparticles from *Azadirachta indica* for antibacterial and photocatalytic applications. Mater Sci Semicond Process 32:55–61

Bondarenko O, Juganson K, Ivask A, Kasemets K, Mortimer M, Kahru A (2013) Toxicity of Ag, CuO and ZnO nanoparticles to selected environmentally relevant test organisms and mammalian cells in vitro: a critical review. Arch Toxicol 87(7):1181–1200

Capeletti LB, de Oliveira LF, Goncalves KDA, de Oliveira JFA, Saito A, Kobarg J, Santos JHZD, Cardoso M (2014) Tailored silica-antibiotic nanoparticles: overcoming bacterial resistance with low cytotoxicity. Langmuir 30(25):7456–7464

Chausali N, Jyoti Saxena J, Prasad R (2021) Recent trends in nanotechnology applications of bio-based packaging. Journal of Agriculture and Food Research, https://doi.org/10.1016/j.jafr.2021.100257

Cousins BG, Allison HE, Doherty PJ, Edwards C, Garvey MJ, Martin DS, Williams RL (2007) Effects of a nanoparticulate silica substrate on cell attachment of Candida albicans. J Appl Microbiol 102(3):757–765

Das R, Gang S, Nath SS, Bhattacharjee R (2010) Linoleic acid capped copper nanoparticles for antibacterial activity. J Bionanosci 4(1–2):82–86

Dhapte V, Kadam S, Pokharkar V, Khanna K, Dhapte V (2014) Versatile SiO_2 nanoparticles polymer composites with pragmatic properties. ISRN Inorg Chem:1–8

Dizaj SM, Lotfipour F, Barzegar-Jalali M, Zarrintan MH, Adibkia K (2014) Antimicrobial activity of the metals and metal oxide nanoparticles. Mater Sci Eng C 44:278–284

Dkhil MA, Al-Quraishy S, Wahab R (2015) Anticoccidial and antioxidant activities of zinc oxide nanoparticles on Eimeria papillata-induced infection in the jejunum. Int J Nanomedicine 10:1961–1968

Durán N, Durán M, De Jesus MB, Seabra AB, Fávaro WJ, Nakazato G (2016) Silver nanoparticles: a new view on mechanistic aspects on antimicrobial activity. Nanomedicine 12(3):789–799

Egger S, Lehmann RP, Height MJ, Loessner MJ, Schuppler M (2009) Antimicrobial properties of a novel silver-silica nanocomposite material. Appl Environ Microbiol 75(9):2973–2976

Elechiguerra JL, Burt JL, Morones JR, Camacho-Bragado A, Gao X, Lara HH, Yacaman MJ (2005) Interaction of silver nanoparticles with HIV-1. J Nanobiotechnol 3(1):1–10

Emamifar A, Kadivar M, Shahedi M, Soleimanian-Zad S (2011) Effect of nanocomposite packaging containing Ag and ZnO on inactivation of Lactobacillus plantarum in orange juice. Food Control 22(3–4):408–413

Espitia PJP, Soares NFF, dos Reis Coimbra JS, de Andrade NJ, Cruz RS, Medeiros EAA (2012) Zinc oxide nanoparticles: synthesis, antimicrobial activity and food packaging applications. Food Bioprocess Technol 5(5):1447–1464

Fouda A, Mohamed A, Elgamal MS, El-Din Hassan S, Salem Salem S, Shaheen TI (2017) Facile approach towards medical textiles *via* myco-synthesis of silver nanoparticles. Der Pharma Chemica 9(13):11–18

Frens G (1973) Controlled nucleation for the regulation of the particle size in monodisperse gold suspensions. Nat Phys Sci 241(105):20–22

Galib MB, Mashru M, Jagtap C, Patgiri BJ, Prajapati PK (2011) Therapeutic potentials of metals in ancient India: a review through Charaka Samhita. J Ayurveda Integr Med 2(2):55

Gao W, Thamphiwatana S, Angsantikul P, Zhang L (2014) Nanoparticle approaches against bacterial infections. Wiley Interdiscip Rev Nanomed Nanobiotechnol 6(6):532–547

Gupta N, Upadhyaya CP, Singh A, Abd-Elsalam KA, Prasad R (2018) Applications of silver nanoparticles in plant protection. In: Nanobiotechnology Applications in Plant Protection (eds. Abd-Elsalam K and Prasad R), Springer International Publishing AG 247–266

Han G, Ghosh P, Rotello VM (2007) Functionalized gold nanoparticles for drug delivery. Nanomedicine (Lond) 2(1):113–123

Hernández-Sierra JF, Ruiz F, Pen DCC, Martínez-Gutiérrez F, Martínez AE, Guillén ADJP, Tapia-Pérez H, Castañón GM (2008) The antimicrobial sensitivity of Streptococcus mutans to nanoparticles of silver, zinc oxide, and gold. Nanomedicine 4(3):237–240

Huang WC, Tsai PJ, Chen YC (2007) Functional gold nanoparticles as photothermal agents for selective-killing of pathogenic bacteria. Nanomedicine (Lond) 6:777–787

Huh AJ, Kwon YJ (2011) "Nanoantibiotics": a new paradigm for treating infectious diseases using nanomaterials in the antibiotics resistant era. J Control Release 156(2):128–145

Jin T, Sun D, Su JY, Zhang H, Sue HJ (2009) Antimicrobial efficacy of zinc oxide quantum dots against Listeria monocytogenes, Salmonella enteritidis, and Escherichia coli O157: H7. J Food Sci 74(1):46–52

Jo YK, Kim BH, Jung G (2009) Antifungal activity of silver ions and nanoparticles on phytopathogenic fungi. Plant Dis 93(10):1037–1043

Joshi N, Jain N, Pathak A, Singh J, Prasad R, Upadhyaya CP (2018) Biosynthesis of silver nanoparticles using Carissa carandas berries and its potential antibacterial activities. J Sol-Gel Sci Techn 86(3):682–689. https://doi.org/10.1007/s10971-018-4666-2

Keller AA, McFerran S, Lazareva A, Suh S (2013) Global life cycle releases of engineered nanomaterials. J Nanopart Res 15(6):1692

Khan ST, Wahab R, Ahmad J, Al-Khedhairy AA, Siddiqui MA, Saquib Q, Ali BA, Musarrat J (2015) Coo thin nanosheets exhibit higher antimicrobial activity against tested gram-positive bacteria than gram-negative bacteria. Korean Chem Eng Res 53(5):565–569

Kim YH, Lee DK, Cha HG, Kim CW, Kang YS (2007) Synthesis and characterization of antibacterial ag- SiO_2 nanocomposite. J Phys Chem C 111(9):3629–3635

Krishnaraj C, Jagan EG, Rajasekar S, Selvakumar P, Kalaichelvan PT, Mohan N (2010) Synthesis of silver nanoparticles using Acalypha indica leaf extracts and its antibacterial activity against waterborne pathogens. Colloids Surf B Biointerfaces 76(1):50–56

Kumar L, Verma S, Bhardwaj A, Vaidya S, Vaidya B (2014) Eradication of superficial fungal infections by conventional and novel approaches: a comprehensive review. Artif Cells Nanomed Biotechnol 42(1):32–46

Kumar L, Verma S, Prasad DN, Bhardwaj A, Vaidya B, Jain AK (2015) Nanotechnology: a magic bullet for HIV AIDS treatment. Artif Cells Nanomed Biotechnol 43(2):71–86

Lara HH, Ayala-Nuñez NV, Ixtepan-Turrent L, Rodriguez-Padilla C (2010) Mode of antiviral action of silver nanoparticles against HIV-1. J Nanobiotechnol 8(1):1–10

Lara HH, Romero-Urbina DG, Pierc C, Lopez-Ribot JL, Arellano-Jimenez MJ, Jose-Yacaman M (2015) Effect of silver nanoparticles on Candida albicans biofilms: an ultrastructural study. J Nanobiotechnol 13(1):1–12

Lellouche J, Friedman A, Lellouche JP, Gedanken A, Banin E (2012) Improved antibacterial and antibiofilm activity of magnesium fluoride nanoparticles obtained by water-based ultrasound chemistry. Nanomedicine 8(5):702–711

Leung YH, Ng AM, Xu X, Shen Z, Gethings LA, Wong MT, Chan CM, Guo MY, Ng YH, Djurišić AB, Lee PK (2014) Mechanisms of antibacterial activity of MgO: non-ROS mediated toxicity of MgO nanoparticles towards Escherichia coli. Small 10(6):1171–1183

Li B, Logan BE (2004) Bacterial adhesion to glass and metal-oxide surfaces. Colloids Surf B Biointerfaces 36(2):81–90

Li Y, Zhang W, Niu J, Chen Y (2012) Mechanism of photogenerated reactive oxygen species and correlation with the antibacterial properties of engineered metal-oxide nanoparticles. ACS Nano 6(6):5164–5173

Li LH, Yen MY, Ho CC, Wu P, Wang CC, Maurya PK, Chen PS, Chen W, Hsieh WY, Chen HW (2013) Non-cytotoxic nanomaterials enhance antimicrobial activities of cefmetazole against multidrug-resistant Neisseria gonorrhoeae. PLoS One 8(5):e64794

Liang T, Qiu X, Ye X, Liu Y, Li Z, Tian B, Ya D (2020) Biosynthesis of selenium nanoparticles and their effect on changes in urinary nanocrystallites in calcium oxalate stone formation. *3.* Biotech 10(1):23

Liu YJ, He LL, Mustapha A, Li H, Hu ZQ, Lin MS (2009) Antibacterial activities of zinc oxide nanoparticles against Escherichia coli O157: H7. J Appl Microbiol 107(4):1193–1201

Lok CN, Ho CM, Chen R, He QY, Yu WY, Sun H et al (2007) Silver nanoparticles: partial oxidation and antibacterial activities. J Biol Inorg Chem 12:527–534

Lu L, Sun RW, Chen R, Hui CK, Ho CM, Luk JM, Lau GK, Che CM (2008) Silver nanoparticles inhibit hepatitis B virus replication. Antivir Ther 13(2):253

Lundberg ME, Becker EC, Choe S (2013) MstX and a putative potassium channel facilitate biofilm formation in *Bacillus subtilis*. PLoS One 8(5):e60993

Madhiyazhagan P, Murugan K, Kumar AN, Nataraj T, Dinesh D, Panneerselvam C, Subramaniam J, Kumar PM, Suresh U, Roni M, Nicoletti M (2015) Sargassum muticum-synthesized silver nanoparticles: an effective control tool against mosquito vectors and bacterial pathogens. Parasitol Res 114(11):4305–4317

Maness PC, Smolinski S, Blake DM, Huang Z, Wolfrum EJ, Jacoby WA (1999) Bactericidal activity of photocatalytic TiO2 reaction: toward an understanding of its killing mechanism. Appl Environ Microbiol 65(9):4094–4098

Manne ND, Arvapalli R, Nepal N, Thulluri S, Selvaraj V, Shokuhfar T, He K, Rice KM, Asano S, Maheshwari M, Blough ER (2015) Therapeutic potential of cerium oxide nanoparticles for the treatment of peritonitis induced by polymicrobial insult in Sprague-Dawley rats. Crit Care Med 43(11):477–489

Martınez-Flores E, Negrete J, Villasenor GT (2003) Structure and properties of Zn-Al-Cu alloy reinforced with alumina particles. Mater Des 24(4):281–286

Mishra V, Arya A, Chundawat TS (2020) High catalytic activity of Pd nanoparticles synthesized from green alga Chlorella vulgaris in Buchwald-hartwig synthesis of N-aryl Piperazines. Curr Organocatal 7(1):23–33

Mohamed AA, Fouda A, Abdel-Rahman MA, Hassan SED, El-Gamal MS, Salem SS, Shaheen TI (2019) Fungal strain impacts the shape, bioactivity and multifunctional properties of green synthesized zinc oxide nanoparticles. Biocatal Agric Biotechnol 19:101103

Mohammad G, Mishra VK, Pandey HP (2008) Antioxidant properties of some nanoparticle may enhance wound healing in T2DM patient. Dig J Nanomater Biostruct 3(4):159–162

Mühling M, Bradford A, Readma JW, Somerfield PJ, Handy RD (2009) An investigation into the effects of silver nanoparticles on antibiotic resistance of naturally occurring bacteria in an estuarine sediment. Mar Environ Res 68(5):278–283

Murugan K, Venus JSE, Panneerselvam C, Bedini S, Conti B, Nicoletti M, Sarkar SK, Hwang JS, Subramaniam J, Madhiyazhagan P, Kumar PM (2015) Biosynthesis, mosquitocidal and antibacterial properties of Toddalia asiatica-synthesized silver nanoparticles: do they impact predation of guppy Poecilia reticulata against the filariasis mosquito Culex quinquefasciatus? Environ Sci Pollut Res Int 22(21):17053–17064

Nagarajan S, Kuppusamy KA (2013) Extracellular synthesis of zinc oxide nanoparticle using seaweeds of gulf of Mannar, India. J Nanobiotechnol 11(1):39

Norman RS, Stone JW, Gole A, Murphy CJ, Sabo-Attwood TL (2008) Targeted photothermal lysis of the pathogenic bacteria, Pseudomonas aeruginosa, with gold nanorods. Nano Lett 8(1):302–306

Paciotti GF, Kingston DG, Tamarkin L (2006) Colloidal gold nanoparticles: a novel nanoparticle platform for developing multifunctional tumor-targeted drug delivery vectors. Drug Dev Res 67(1):47–54

Pan X, Redding JE, Wiley PA, Wen L, McConnell J, Zhang B (2010) Mutagenicity evaluation of metal oxide nanoparticles by the bacterial reverse mutation assay. Chemosphere 79(1):113–116

Panáček A, Kvítek L, Prucek R, Kolář M, Večeřová R, Pizúrová N, Sharma VK, Nevěčná TJ, Zbořil R (2006) Silver colloid nanoparticles: synthesis, characterization, and their antibacterial activity. J Phys Chem B 110(33):16248–16253

Paredes D, Ortiz C, Torres R (2014) Synthesis, characterization, and evaluation of antibacterial effect of ag nanoparticles against Escherichia coli O157: H7 and methicillin-resistant Staphylococcus aureus (MRSA). Int J Nanomedicine 9:1717–1729

Pelgrift RY, Friedman AJ (2013) Nanotechnology as a therapeutic tool to combat microbial resistance. Adv Drug Deliv Rev 65(13–14):1803–1815

Prasad R (2014) Synthesis of silver nanoparticles in photosynthetic plants. Journal of Nanoparticles, Article ID 963961, 2014, https://doi.org/10.1155/2014/963961

Prasad KS, Pathak D, Patel A, Dalwadi P, Prasad R, Patel P, Kaliaperumal SK (2011) Biogenic synthesis of silver nanoparticles using Nicotiana tobaccum leaf extract and study of their antibacterial effect. Afr J Biotechnol 9 (54): 8122–8130

Prasad R, Swamy VS, Varma A (2012) Biogenic synthesis of silver nanoparticles from the leaf extract of Syzygium cumini (L.) and its antibacterial activity. Int J Pharma Bio Sci 3(4):745–752

Prasad R, Swamy VS (2013) Antibacterial activity of silver nanoparticles synthesized by bark extract of Syzygium cumini. Journal of Nanoparticles 2013, https://doi.org/10.1155/2013/431218

Prasad R, Pandey R, Barman I (2016) Engineering tailored nanoparticles with microbes: quo vadis. WIREs Nanomed Nanobiotechnol 8:316–330. https://doi.org/10.1002/wnan.1363

Prasad R, Kumar V, Kumar M (2017) Nanotechnology: Food and Environmental Paradigm. Springer Nature Singapore Pte Ltd. (ISBN 978-981-10-4678-0)

Prasad R, Thirugnanasanbandham K (2019) Advances Research on Nanotechnology for Water Technology. Springer International Publishing https://www.springer.com/us/book/9783030023805

Prasad R, Kumar V, Kumar M, Choudhary D (2019) Nanobiotechnology in Bioformulations. Springer International Publishing (ISBN 978-3-030-17061-5) https://www.springer.com/gp/book/9783030170608

Prasad R, Siddhardha B, Dyavaiah M (2020) Nanostructures for Antimicrobial and Antibiofilm Applications. Springer International Publishing (ISBN 978-3-030-40336-2) https://www.springer.com/gp/book/9783030403362

Prema P, Thangapandiyan S (2013) In-vitro antibacterial activity of gold nanoparticles capped with polysaccharide stabilizing agents. Int J Phar Pharma Sci 5:310

Rafiei S, Rezatofighi SE, Ardakani MR, Madadgar O (2015) In vitro anti-foot-and-mouth disease virus activity of magnesium oxide nanoparticles. IET Nanobiotechnol 9(5):247–251

Rafique M, Shafiq F, Gillani SSA, Shakil M, Tahir MB, Sadaf I (2020) Eco-friendly green and biosynthesis of copper oxide nanoparticles using *Citrofortunella microcarpa* leaves extract for efficient photocatalytic degradation of Rhodamin B dye form textile wastewater. Optik 208:164053

Rajakumar G, Abdul Rahuman A, Mohana Roopan S, Gopiesh Khanna V, Elango G, Kamaraj C, Abduz Zahir A, Velayutham K (2012) Fungus-mediated biosynthesis and characterization of TiO_2 nanoparticles and their activity against pathogenic bacteria. Spectrochim Acta A 91:23–29

Ramazanzadeh B, Jahanbin A, Yaghoubi M, Shahtahmassbi N, Ghazvini K, Shakeri M, Shafaee H (2015) Comparison of antibacterial effects of ZnO and CuO nanoparticles coated brackets against Streptococcus mutans. J Dent (Shiraz) 16(3):200–205

Ren G, Hu D, Cheng EW, Vargas-Reus MA, Reip P, Allaker RP (2009) Characterisation of copper oxide nanoparticles for antimicrobial applications. Int J Antimicrob Agents 33(6):587–590

Riva A, Algaba IM, Pepió M (2006) Action of a finishing product in the improvement of the ultraviolet protection provided by cotton fabrics. Modelisation of the effect. Cellulose 13(6):697–704

Ruparelia JP, Chatterjee AK, Duttagupta SP, Mukherji S (2008) Strain specificity in antimicrobial activity of silver and copper nanoparticles. Acta Biomater 4(3):707–716

Sadiq IM, Chowdhury B, Chandrasekaran N, Mukherjee A (2009) Antimicrobial sensitivity of Escherichia coli to alumina nanoparticles. Nanomedicine 5(3):282–286

Salem W, Leitner DR, Zing FG, Schratter G, Prassl R, Goessler W, Reidl J, Schild S (2015) Antibacterial activity of silver and zinc nanoparticles against *Vibrio cholerae* and *enterotoxic Escherichia coli*. Int J Med Microbiol 305(1):85–95

Sampath Kumar TS, Madhumathi K, Rubaiya Y, Doble M (2015) Dual mode antibacterial activity of ion substituted calcium phosphate nanocarriers for bone infections. Front Bioeng Biotechnol 3:59

Schrand AM, Rahman MF, Hussain SM, Schlager JJ, Smith DA, Syed AF (2010) Metal-based nanoparticles and their toxicity assessment. Wiley Interdiscip Rev Nanomed Nanobiotechnol 2(5):554–568

Seil JT, Webster TJ (2012) Antimicrobial applications of nanotechnology: methods and literature. Int J Nanomedicine 7:2767

Seleem MN, Munusamy P, Ranjan A, Alqublan H (2009) Silica-antibiotic hybrid nanoparticles for targeting intracellular pathogens. Antimicrob Agents Chemother 53(10):4270–4274

Shaikh S, Nazam N, Rizvi SMD, Ahmad K, Baig MH, Lee EJ, Choi I (2019) Mechanistic insights into the antimicrobial actions of metallic nanoparticles and their implications for multidrug resistance. Int J Mol Sci 20(10):2468

Shakibaie M, Salari Mohazab N, Ayatollahi Mousavi SA (2015) Antifungal activity of Selenium nanoparticles synthesized by bacillus species Msh-1 against *Aspergillus fumigatus* and *Candida albicans*. Jundishapur J Microbiol 8(9):263–281

Sharaf OM, Al-Gamal MS, Ibrahim GA, Dabiza NM, Salem SS, El-ssayad MF, Youssef AM (2019) Evaluation and characterization of some protective culture metabolites in free and nanochitosan-loaded forms against common contaminants of Egyptian cheese. Carbohydr Polym 223:115094

Shenmchuk O, Braga D, Grepioni F, Turner RJ (2010) Co-crystallization of antibacterials with inorganic salts: paving the way to activity enhancement. RSC Adv 10:2146–2149

Singh P, Kim YJ, Singh H, Wang C, Hwang KH, Farh MEA, Yang DC (2015) Biosynthesis, characterization, and antimicrobial applications of silver nanoparticles. Int J Nanomedicine 10:2567–2577

Su HL, Chou CC, Hung DJ, Lin SH, Pao IC, Lin JH, Huang FL, Dong RX, Lin JJ (2009) The disruption of bacterial membrane integrity through ROS generation induced by nanohybrids of silver and clay. Biomaterials 30(30):5979–5987

Sunderam V, Thiyagarajan D, Lawrence AV, Mohammed SSS, Selvaraj A (2019) In-vitro antimicrobial and anticancer properties of green synthesized gold nanoparticles using Anacardium occidentale leaves extract. Saudi J Biol Sci 26(3):455–459

Sutradhar P, Saha M, Maiti D (2014) Microwave synthesis of copper oxide nanoparticles using tea leaf and coffee powder extracts and its antibacterial activity. J Nanostruct Chem 4:86

Swamy VS, Prasad R (2012) Green synthesis of silver nanoparticles from the leaf extract of Santalum album and its antimicrobial activity. J Optoelectronic and Biomedical Materials 4(3): 53–59

Tamayo LA, Zapata PA, Vejar ND, Azocar MI, Gulppi MA, Zhou X, Thompson GE, Rabagliati FM, Paez MA (2014) Release of silver and copper nanoparticles from polyethylene nanocomposites and their penetration into Listeria monocytogenes. Mater Sci Eng C Mater Biol Appl 40:24–31

Tayel AA, El-Tras WF, Moussa S, El-Baz AF, Mahrous H, Salem MF, Brimer L (2011) Antibacterial action of zinc oxide nanoparticles against foodborne pathogens. J Food Saf 31(2):211–218

Tourinho PS, Van Gestel CA, Lofts S, Svendsen C, Soares AM, Loureiro S (2012) Metal-based nanoparticles in soil: fate, behavior, and effects on soil invertebrates. Environ Toxicol Chem 31(8):1679–1692

Turkevich J, Stvenson PC, Hillier J (1951) A study of the nucleation and growth processes in the synthesis of colloidal gold. Disc Farad Soc 11:55–75

Uddandarao P, Balakrishnan RM, Ashok A, Swarup S, Sinha P (2019) Bioinspired ZnS: Gd nanoparticles synthesized from an endophytic fungi *Aspergillus flavus* for fluorescence-based metal detection. Biomimetics 4(1):11

Vargas-Reus MA, Memarzadeh K, Huang J, Ren GG, Allaker RP (2012) Antimicrobial activity of nanoparticulate metal oxides against peri-implantitis pathogens. Int J Antimicro Agents 40(2):135–139

Wahab R, Khan ST, Dwivedi S, Ahamed M, Musarrat J, Al-Khedhairy AA (2013) Effective inhibition of bacterial respiration and growth by CuO microspheres composed of thin nanosheets. Colloids Surf B Biointerfaces 111:211–217

Weir A, Westerhoff P, Fabricius L, von Goetz N (2012) Titanium dioxide nanoparticles in food and personal care products. Environ Sci Technol 46(4):2242–2250

Xie Y, He Y, Irwin PL, Jin T, Shi X (2011) Antibacterial activity and mechanism of action of zinc oxide nanoparticles against *Campylobacter jejuni*. Appl Environ Microbiol 77(7):2325–2331

Xu K, Wang JX, Kang XL, Chen JF (2009) Fabrication of antibacterial monodispersed ag-SiO2 core-shell nanoparticles with high concentration. Mater Lett 63(1):31–33

Yadav L, Tripathi RM, Prasad R, Pudake RN, Mittal J (2017) Antibacterial activity of Cu nanoparticles against *E. coli, Staphylococcus aureus* and *Pseudomonas aeruginosa*. Nano Biomed Eng. 9(1): 9–14. https://doi.org/10.5101/nbe.v9i1.p9-14

Yallappa J, Manjanna BL, Dhananjaya U, Vishwanatha B, Ravishankar H, Gururaj P, Niranjana HBS (2015) Phytochemically functionalized Cu and Ag nanoparticles embedded in MWCNTs for enhanced antimicrobial and anticancer properties. Nano Micro Lett 8:120–130

Yamamoto O, Ohira T, Alvarez K, Fukuda M (2010) Antibacterial characteristics of CaCO3-MgO composites. Mat Sci Eng B 173(1–3):208–212

Zarei M, Jamnejad A, Khajehali E (2014) Antibacterial effect of silver nanoparticles against four foodborne pathogens. Jundishapur J Microbiol 7(1)

Zharov V (2006) Photothermal nanotherapeutics and nanodiagnostics for selective killing of bacteria targeted with gold nanoparticles. Biophys J 90:619–627

Zhou Y, Kong Y, Kundu S, Cirillo JD, Liang P (2012) Antibacterial activities of gold and silver nanoparticles against *Escherichia coli* and bacillus Calmette-Guérin. J Nanobiotechnol 10:19

Zhu Y, Cao QS, Wang M, Gu Y, Luo H, Meng F, Liu X, Lai H (2015) Hierarchical micro/nano-structured titanium with balanced actions to bacterial and mammalian cells for dental implants. Int J Nanomedicine 10:6659–6674

Chapter 14
Advancement in Nanomaterial Synthesis and its Biomedical Applications

Benil P. Bharathan, Rajakrishnan Rajagopal, Ahmed Alfarhan, Mariadhas Valan Arasu, and Naif Abdullah Al-Dhabi

Contents

14.1	Introduction	420
14.2	Historical Development of Nanomaterials in Relevance to Biology and Medicine	420
	14.2.1 History of Nanobiology	423
14.3	Infection and Their Treatment – Current Knowledge	425
	14.3.1 Antibiotics Targeting Cell Wall	425
	14.3.2 Inhibition of Protein Biosynthesis	427
	14.3.3 Inhibitors of DNA Replication	428
	14.3.4 Folic Acid Metabolism Inhibitors	429
14.4	Mechanisms of Antimicrobial Resistance	429
	14.4.1 Prevention of Accumulation of Antimicrobials	429
14.5	Non-antibiotic Treatments for Infections	431
	14.5.1 Phage Therapy	431
	14.5.2 Bacteriocins	433
	14.5.3 Killing Factors	434
	14.5.4 Antibacterial Activities of Non-antibiotic Drugs	434
	14.5.5 Quorum Quenching	435
14.6	Mechanisms of Antibacterial Activity of Nanoparticles	435
	14.6.1 Direct Absorption of Nanoparticles	435
	14.6.2 Reactive Oxygen Species (ROS) Production	436
	14.6.3 Cell Wall Damage	436
14.7	Nanomaterials in Controlling Infections	436
	14.7.1 Nanotechnology-Based Drug Delivery Systems	437
	14.7.2 Nanotechnology-Based Vaccines and Immunostimulatory Adjuvants	442
14.8	Role of Nanomaterials in Other Diseases	445
	14.8.1 Neurodegeneration	445
	14.8.2 Cancer Therapy	447

B. P. Bharathan
Department of Agadatantra, Vaidyaratnam P.S Varier Ayurveda College,
Kottakkal, Kerala, India

R. Rajagopal (✉) · A. Alfarhan · M. V. Arasu (✉) · N. A. Al-Dhabi
Department of Botany and Microbiology, College of Science, King Saud University,
Riyadh, Saudi Arabia
e-mail: mvalanarasu@ksu.edu.sa

© The Author(s), under exclusive license to Springer Nature Switzerland AG 2022
A. Krishnan et al. (eds.), *Emerging Nanomaterials for Advanced Technologies*,
Nanotechnology in the Life Sciences, https://doi.org/10.1007/978-3-030-80371-1_14

14.8.3 Nanotechnology in Diabetes Mellitus.. 452
14.9 Future Perspective of Nanomedicine and Biology...................................... 454
14.10 Conclusion... 455
References.. 455

14.1 Introduction

Health has been the most important criteria to measure the development of a country. Health indices have always been dwindling throughout the world due to various factors that emerged time to time. It was infectious diseases that claimed the lives of millions during the initial stages of human settlement (Dobson and Carper 1996). With the discovery of antibiotics and the advancement of health care, we have succeeded in overcoming infections to a greater extent initially. Then came an era of non-communicable diseases with a steep rise in the death rate among different countries. The determinants of health have been redefined in the light of escalating rates of diseases like coronary artery disease, stroke, cancers etc. At the dawn of the twenty-first century we were facing a grave situation of highly prevalent non-communicable diseases with a soaring rate of infectious diseases contributed by newly emerging pandemics (Heinrich et al. 2020) of Ebola, SARS-CoV-2 etc. Making the matters more complicated is the resurgence of bacterial strains that are resistant to multiple antibiotics. The situation is warranting to shift the concentration from developing newer antibiotics to adopt newer strategies to make the available antibiotics more efficacious. For this, medical sciences embraced the newly emerged science of nanotechnology. The application of nanoscience in medical sciences have already developed different speciality streams like nanovaccines, nanodiagnostics, nanopharmaceutics so on and so forth (Islan et al. 2017; Saglam et al. 2021). In this chapter we discuss about the research and developments in the field of nanomedicine in controlling infections and other diseases.

14.2 Historical Development of Nanomaterials in Relevance to Biology and Medicine

Evolution of nanoscience can be traced back to oldest civilizations that existed on earth. In the fifth century B.C. philosophers debated on the continuity of matter and ascertained that any matter can be divided into infinitesimally smaller subunits culminating in indivisible and invisible matter (Bayda et al. 2020). This smallest subunit of matter is equivalent to the modern entity of 'atom', and can be considered as a recognition and acceptance of materials in micro- and nanoscales in those days. In the fourth century AD Romans created the Lycurgus cup, a dichroic glass cup which appears green on direct light and reddish purple in transmitted light, is considered as the oldest manmade nanomaterial. The dichroic property of Lycurgus cup was established in 1990 through transmission electron microscopy. It is due to the presence of silver-gold alloy nanoparticles in 7:3 proportion of 50–100 nm size containing dispersed copper of about 10% (Elsner 2013). The vast array of ceramic glazes used by the Islamic world and Europe contained silver and copper nanoparticles (Padovani et al. 2003; Barber and Freestone 1990). They were extensively used from the ninth to seventeenth centuries. Later Italians utilized similar nanoparticles during the Renaissance period for pottery making. They were influenced by the Ottoman techniques prevalent from the thirteenth to eighteenth centuries using 'Damascus' saber blades, cementite nanowires and carbon nanotubes which provided strength, resilience and keen edges respectively to produce 'Damascus' saber blades strength, made from cementite nanowires for resilience and carbon nanotubes for keen edges (Sciau 2012). Michael Faraday studied colloidal gold in suspension, its optical and electronic properties and demonstrated how the gold nanoparticles changed their colour in solutions under specific lighting conditions (Lin et al. 1986). The ancient Indian treatment systems like Ayurveda and Siddha also utilized metals and minerals in the nanoscale for the treatment of various ailments. Detailed procedures of calcining the metals and minerals into "Bhasma' of nano-proportions were utilized in the treatment of various ailments at small doses. The science detailing these methods of purification and reduction along with their modes of use in the treatment branch termed 'Rasasastra' had enormous patronage from Buddhism (Ranade and Acharya 2015). The famous Indian Alchemy principles also lay hidden in these scriptures. Calcined gold, silver, copper, lead, tin, antimony, sulphur, arsenic etc. were utilized for the treatment of various ailments with the aid of appropriate vehicles. Even biological materials like amber, stag horn, conch shell and elephant tusk were also calcined and used as medicine in the form of nanoparticles. The qualities of certain metals like silver and gold were known to prevent infections from the tenth to thirteenth centuries. This knowledge was inculcated while making confectionaries prepared in India and its adjoining provinces by covering with finely thin silver or gold foils beaten up for several days to make them reach nano-proportions, which were called 'Chandi ka warq' and 'Sone ka warq' respectively. This practice is still in vogue in parts of Indian subcontinent.

The modern evolution of nanotechnology took place in 1959 when Richard Feynman, Nobel laureate and famous American physicist introduced it conceptually with the famous lecture *"There's Plenty of Room at the Bottom"* given at California Institute of Technology (Richard 1960). The concept laid by Feynman helped in constructing machines of molecular proportions for performing humongous task of handling large volumes of data in a tiniest of space. The materialization of the concepts laid down by Feynman during the latter half of the twentieth century earned him the title 'Father of modern nanotechnology'. The spark created by Feynman was carried forward by many other eminent personalities like Norio Taniguchi from Japan who coined the term "nanotechnology" in 1974 (Taniguchi 1974).

Towards the end of the twentieth century several more scientists were attracted to the opportunities hidden in the field of nanotechnology and came up with increased momentum of research activities. Several opportunities were suggested for the synthesis of nanomaterials and broadly they were classified as two categories: top-down and bottom-up approaches (Khan et al. 2011).

The breaking down of a material into nano-sized particles from a relatively large material by the use of precision engineering and lithography is termed top-down approach (Madou 2011). Majority of micro-electronics industry utilizes top-down approach with precision engineering while lithography utilizes patterning of surfaces through deposition of materials or through exposure ions, electrons or light (Biswas et al. 2012).Bottom-up approach on the other hand utilizes self-assembly of atoms or molecules by physical or chemical interactions to form materials in nanoscale range (1–100 nm) (Luby et al. 2015).

The first published book on nanotechnology was by K. Eric Drexler in the year 1986 entitled "Engines of Creation: The Coming Era of Nanotechnology", which introduced the term 'Molecular engineering' (Bayda et al. 2020; Drexler 1981). He described the build-up of complex machines from self-assembly of individual atoms to form nanostructures. Later in 1991 he associated with Peterson and Pergamit to publish yet another milestone publication titled "Unbounding the Future: The Nanotechnology Revolution" in which they introduced terms like 'nanobots' or 'assemblers' for medical applications, and the term 'nanomedicine' was introduced henceforth (Drexler et al. 1991). The evolution of nanotechnology is subdivided into four distinct generations and as per Mihail Roco, we are presently in the fourth generation of nanotechnology era (Roco 2007) (Fig. 14.1). Nanomaterials and their uses in the field of medicine are presented in Table 14.1.

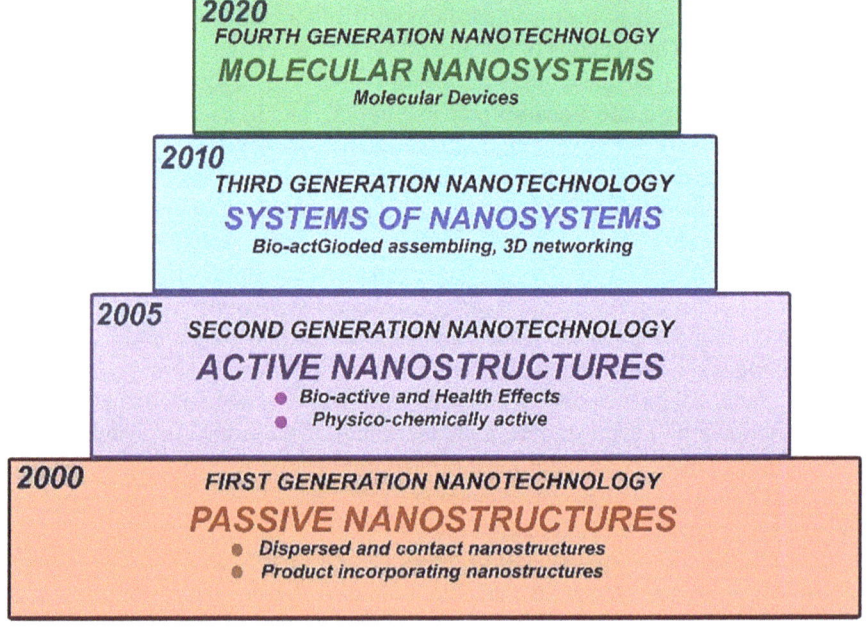

Fig. 14.1 Generations of Nanotechnology

Table 14.1 Nanomaterials used in Medicine

Nanomaterials	Description	Uses
Liposomes	Small spherical artificial vesicles produced from natural nontoxic phospholipids and cholesterol	Gene therapy, drug delivery, Targeted therapy
Nanoparticles	Colloidal particles ranging from 10 to 100 nm size which are biocompatible and biodegradable	Gene delivery, protein delivery, drug delivery, gene expression vector, gene transfection
Dendrimers	Macromolecular compounds with a central core with radiating branches of less than 10 nm	Gene delivery, intravascular drug delivery, intrabronchial drug delivery,
Carbon nano tubes	0.5–3 nm diameter tubes with 200–1000 nm length	Detection of DNA mutation, detection of disease biomarkers
Nano crystals	2–9.5 nm size materials smaller than 100 nm composed of atoms in a polycrystalline arrangement	Carrier for poorly soluble drugs, labelling of cancer markers in diagnostics
Nano shells	Spherical nanoparticle consisting of a dielectric core covered by a thin metallic shell	Tumour-specific imaging, deep tissue thermal ablation
Nano wires	Tiny wires of nanoscale dimensions	Disease protein biomarker detection, DNA mutation detection, gene expression detection
Quantum dots	2–9.5 nm semiconductor crystal having optical and electrical properties	Optical detection of genes, detection of proteins in cell assays, tumour and lymph node visualization

14.2.1 History of Nanobiology

The history of use of nanotechnology in the fields of biology and medicine extends from diagnosis, drug delivery to molecular imaging. Biological systems including the human body are an assembly of nanoscaled structures undergoing self-assembly and self-organization to form higher-order structure of biological units and organisms ranging from micro-, meso- and macroscale. Life processes involving elementary biological units like cell-membrane, DNA, proteins or lipids are of nano-dimensions (Fig. 14.2). These biological units and their functions are better comprehended, guided and manipulated with the help of nanotechnology (Logothetidis 2006). The application of nanoscience in the field of biology is henceforth called bio-nanotechnology. Miniaturization is the essential feature of nanomedicine wherein the nanometer scale is similar in size to the macromolecules like enzymes, receptors, and carrier proteins. The nanoscale devices in the range of 20–50 nm can easily enter cells as well as exit the vascular circulation and enter tissues swiftly. The birth of a new stream of nanotechnology called "nano-pharmaceuticals" developed and marketed products using nanotechnology for drug delivery of regenerative medicine, nanoparticles with antimicrobial activities and nanochips, nanoelectrodes and nano-biosensors for the detection of biomarkers. In medicine, nanotechnology coupled with Biotechnology, Information technology

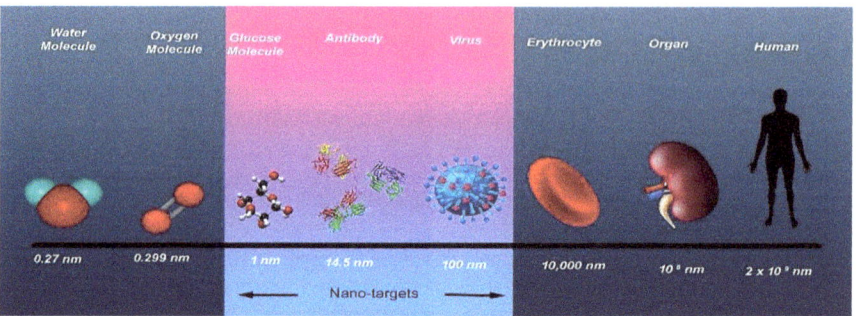

Fig. 14.2 Biological systems in Nanoscale

and Cognitive science (NBIC developments) are thought to contribute immensely in the fields of in vitro detection, in vivo diagnosis, multimodal imaging, chemotherapy, phototherapy, gene therapy, immunotherapy, theranostics and their clinical translation (Logothetidis 2006; McGinn 2012).

Profound advancements are made in the field of nano-oncology to improve the efficacy of traditional chemotherapy by incorporating nanomaterials to target the tumour site (Kumar et al. 2016a, b; Palazzolo et al. 2019a, b). Researches are advancing in using nanomaterials to modulate essential biological process for cancer therapy like autophagy, quenching of oxidative stress and to exert cytotoxic activity against the cancer cells (Sharma et al. 2019). Other than cancers, diabetes mellitus, neurodegenerative diseases as well as detecting and curing bacteria, fungi and viruses associated with infections are also being investigated with the help of nanotechnology. Another instrumental advancement in the field of bio-nanotechnology is the development of "Scaffolded DNA origami" by Paul Rothemund in 2006 by self-assembling DNA nanostructures in a "one-pot" reaction (Rothemund 2006). This forms the first application of DNA nanotechnology, which was conceptualized by Nadrian Seeman way back in 1982 (Seeman 1982). DNA nanotechnology is the hot seat of interdisciplinary research presently with inputs from physics, chemistry, materials science, computer science and medicine. Nano-informatics incorporating the vast opportunities of computer science is another field which is not being used to its fullest potential in the field of medicine (Sharma et al. 2019). Predictive analysis of nanocarriers by employing powerful machine-learning algorithms predicts their cellular uptake, activity and cytotoxicity. Nano-informatics can utilize powerful tools like data mining, network analysis, quantitative structure-property relationship (QSPR), quantitative structure-activity relationship (QSAR) and ADMET (absorption, distribution, metabolism, excretion, and toxicity) for predictions. The opportunities and possibilities for nanoscience are many and they are being utilized elaborately during the twenty-first century and hence we aptly termed as the "next industrial revolution".

The following properties of nanomaterials created tremendous attraction for them to be utilized as nanomedicine. (1) Biomolecules with nano-size dimensions are capable of regulating cellular biochemical pathways and in turn cellular homoeostasis. Properly engineered nanoparticles can interfere with the biomolecules at any stage in the molecular processes giving us the capability to hinder or promote any biochemical process in the system biology of humans. (2) Nanomaterials have fairly good solubility due to the advantage of their size, and this can be further enhanced by modifying their surface properties. (3) Owing to their higher surface-to-volume ratio and greater surface area, they are capable of carrying a higher therapeutic payload to their target site. (4) Due to their property of selective targeting, nanoparticles can deliver therapeutic dosages to their specific target, reducing the untoward effects on the nearby healthy tissues. (5) Nanoparticles can now be utilized for personalized diagnosis and therapy (Bisht and Rayamajhi 2016).

14.3 Infection and Their Treatment – Current Knowledge

Infectious diseases posed the greatest threat to the establishment of human race on earth. The human population of present day is to a greater extent due to the victory over infectious diseases that started with the discovery of antibiotics. From the discovery of penicillin in 1928 by Alexander Fleming, the number of antibiotics in treatment have increased at a sudden pace, now reaching to a sizeable population amounting to hundreds. Antibiotic resistance also emerged simultaneously leading to the development of 'superbugs' which pose a real challenge to researchers. Antibiotic resistance has provided an evolutionary advantage to the microbes to survive the newly emerging antibiotics. Methicillin-resistant *Staphylococcus aureus*, fluoroquinolone-resistant *S. aureus*, erythromycin-resistant *Streptococcus pyogenes* and *S. pneumoniae* and vancomycin-resistant enterococci are few of the resistant strains that gain medical attention on a public health purview recently (Kapoor et al. 2017).

Antibiotics evolved after 1928 are classified and dealt based on their mechanism of action (Fig. 14.3);

14.3.1 Antibiotics Targeting Cell Wall

Gram-positive bacteria are protected by an outer cell wall which is tough, rigid and mesh-like. Meanwhile, the Gram-negative bacteria is covered by a thin cell wall surrounded by a second outer membrane. The space enclosed between the cell wall and the outer membrane is termed periplasm. The outer membrane is an additional protective layer which protects the bacteria from the entry of external substances. However, they are provided with channels called porins. which allow the entry of molecules into the cell (Hauser 2015). The tough cell wall helps in maintaining their shape and protects them from the osmotic and mechanical stressors. The cytoplasmic membrane prevents the entry and exit of ions and maintains the cytoplasmic components.

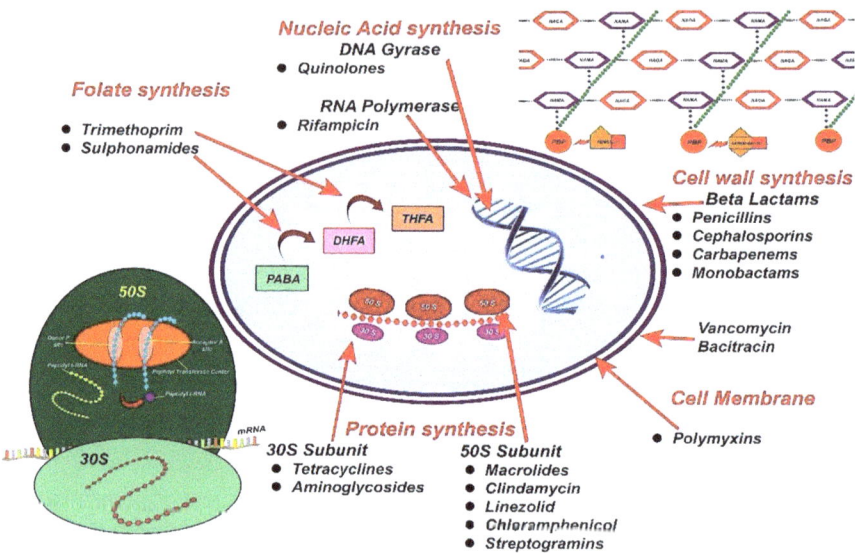

Fig. 14.3 Site of action of antibiotics in bacteria

Bacterial cell wall is made up of peptidoglycan, which is a polysaccharide layer cross-linked to glycan strands facilitated by the action of transglycosidases. The cross-linking extends from the sugars in the polymers to the peptides on the other strand. Precisely, the D-alanyl-alanine portion of peptide chain is cross-linked by glycine in the presence of penicillin binding proteins (PBPs). This cross-linking makes the cell walls stronger, and this biochemical pathway is targeted by a group of antibiotics like β-lactams and glycopeptides (Kahne et al. 2005; Reynolds 1989; Strohl 1997; Benton et al. 2007; Leach et al. 2007).

14.3.1.1 Beta-Lactam Antibiotics

They are the oldest and a broad class of antibiotics with a β-lactam ring in their molecular structure. They irreversibly inhibit the enzyme transpeptidase, an enzyme required for bacterial cell wall synthesis. The final step in transpeptidation in the synthesis of cell wall is facilitated by a transpeptidase called penicillin binding protein (PBP). PBP binds to D-alanyl-D-alanine attached at the end of the peptidoglycan precursors, muropeptides, to cross link the peptidoglycan. β-lactam antibiotics target the PBPs, as the β-lactam ring mimics the D-alanyl D-alanine portion of the peptide chain (Džidić et al. 2008). As the PBP links to the β-lactam ring, they will not be available to cross-link the peptidoglycan and the synthesis of the new cell wall abruptly ends. The disruption of peptidoglycan layer damages the permeability barrier of the bacterium. Penicillin derivatives, cephalosporins, monobactams and carbapenems belong to β-lactam antibiotics. In spite of escalating antimicrobial resistance, these antibiotics are still clinically important due to their short $t_{1/2}$, low volume of distribution with significant kidney tubular secretion (MacDougall 2017).

14.3.1.2 Glycopeptides

Glycopeptides are cyclic or polycyclic glycosylated non-ribosomal peptides produced by filamentous actinomycetes belonging to various groups. The glycopeptides inhibit the cross-linking of the peptidoglycan by binding on to the D-alanyl D-alanine portion of peptide side chain. They in turn cross-link peptides within and between peptidoglycan on the surface of the cytoplasmic membrane. Vancomycin, a glycopeptide antibiotic prevents the binding of PBP with the D-alanyl subunit and binds noncovalently with the terminal carbohydrate, thus inhibiting the formation of cell wall (Grundmann et al. 2006).

14.3.2 Inhibition of Protein Biosynthesis

The genetic information in the bacterial DNA is first transcribed on to the m-RNA followed by translation of the triplet codons on the m-RNA by the ribosomes. The entire process of protein synthesis is catalysed by ribosomes and myriads of cytoplasmic factors. The bacterial 70S ribosomes are consisting of two sub-units, the 30S and 50S subunits which are integral to the functioning of the ribosomes (Yoneyama and Katsumata 2006; Vannuffel and Cocito 1996; Johnston et al. 2002). Antimicrobials target either the 30S or 50S sub-units of the bacterial ribosomes.

14.3.2.1 Inhibitors of 30S Subunits

Aminoglycosides

Aminoglycosides (AG) are positively charged molecules attaching on to the negatively charged outer membrane of the bacterial cell. This process leads to the formation of large pores, which allow the penetration of the antibiotic into the bacterial cell. The passage of the antibiotic through the cell membrane requires activation of energy-dependent transport channels which requires oxygen and an active proton motive force. Due to this peculiar requirement, AG are less active against anaerobic organisms, but they are synergistic to cell wall disrupting antibiotics like β-lactams and glycopeptides as they provide easier access to the AG into the cell. On their entry, AGs form hydrogen bonds with the conserved portions in 16S r-RNA of the 30S subunit near the A site, causing misreading and abrupt termination of translation of m-RNA. Tetracycline, chlortetracycline, doxycycline and minocycline exhibit antimicrobial effect following this mechanism (Wise 1999).

14.3.2.2 Inhibition of 50S Subunits

Chloramphenicol

Chloramphenicol is a broad-spectrum antibiotic isolated from *Salmonella venezuelae*. It is characterised by the presence of a nitrobenzene moiety derived from dichloroacetic acid. This antibiotic binds with the conserved sequences of the 23S r-RNA of the 50S subunit at the *peptidyl transferase* cavity and prevents the binding of t-RNA to the A site of the ribosome (Yoneyama and Katsumata 2006; Vannuffel and Cocito 1996).

Macrolides

Macrolides contain macrocyclic lactone ring and hence the name. Almost all members of this group are isolated from Streptomyces, and erythromycin is the best-known member of this group. Macrolides interferes with the early stage of protein synthesis, translocation, by targeting the conserved sequences of 23S r-RNA of 50S subunit and binds within the nascent peptide exit tunnel (NPET) adjacent to the peptidyl transferase centre and results in premature detachment of the newly formed incomplete peptide chains (Strohl 1997; Leach et al. 2007). Along with macrolides, linocosamides and streptogramins B also exhibit similar mechanism (Yoneyama and Katsumata 2006; Wise 1999).

Oxazolidinones

Oxazolidinones are compounds containing 2-oxazolidine, exhibiting good activity against Gram-positive bacteria. Oxazolidinones are a relatively newer group of antibiotics which are completely synthetic and act by interfering with protein synthesis at multiple stages like binding to 23S r-RNA of the 50S subunit and suppressing 70S inhibition and interact with peptidyl-t-RNA (Lambert 2005; Bozdogan and Appelbaum 2004; Strohl 1997; Leach et al. 2007).

14.3.3 Inhibitors of DNA Replication

14.3.3.1 Quinolones

Fluoroquinolones (FQ) inhibit DNA gyrase, the enzyme that nicks the double-stranded DNA, and introduces negative supercoils and reseals the nicked ends. They thus prevent the positive supercoiling essential for the transcription or replication of the strands. DNA gyrase enzyme consists of four chains – two A subunits and two B subunits. The A subunit is responsible for creating the nick in the DNA strand, while B subunit adds the negative supercoils and the other A subunit reseal the strands. FQs have high affinity to A subunit and perform strand cutting and resealing. In Gram-positive bacteria, topoisomerase IV is the enzyme responsible for nicking the DNA strands, and it shows high affinity towards FQs. Hence, FQs have more potency against Gram-positive bacteria. In mammals, the homologous enzyme for DNA gyrase is topoisomerase II, which has very low affinity for FQ (Yoneyama and Katsumata 2006; Wise 1999; Higgins et al. 2003; Strohl 1997).

14.3.4 Folic Acid Metabolism Inhibitors

14.3.4.1 Sulfonamides and Trimethoprim

These antibiotics interfere with the folic acid metabolism pathways. Sulpha drugs and trimethoprim act on the folic acid metabolic pathway at two different steps. Sulphonamides have high affinity towards dihydropteroate synthase and inhibit it competitively compared to its natural substrate p-amino benzoic acid. Trimethoprim on the other hand inhibits dihydrofolate reductase, which catalyses a step much farther to that of dihydropteroate synthase (Yoneyama and Katsumata 2006; Tenover 2006; Straus and Hancock 2006).

14.4 Mechanisms of Antimicrobial Resistance

14.4.1 Prevention of Accumulation of Antimicrobials

This is achieved by either decreasing the uptake of the drug or increasing the efflux from the cell by changing the permeability of the outer membrane. Bacterial cells uptake materials from the outside by diffusion through porins, diffusion through the plasma membrane or by self-uptake. Porins are located on the outer membrane of Gram-negative bacteria and admit small hydrophilic molecules to pass through it. As the number of porins on the outer membrane decreases, the organism develops resistance against antibiotics that gain entry into the cell through them. β-lactam antibiotics and FQ resistance to Gram-negative bacteria is due to this process (Fig. 14.4).

14.4.1.1 Efflux Pumps

Efflux pumps are the membrane proteins that export the antibiotics from the interior of the cell to outside. The resistance of the organism against an antibiotic is determined by the efficiency of the efflux pumps (Džidić et al. 2008). Some resistant strains have developed speedy efflux pumps, which pump out the antibiotics at the

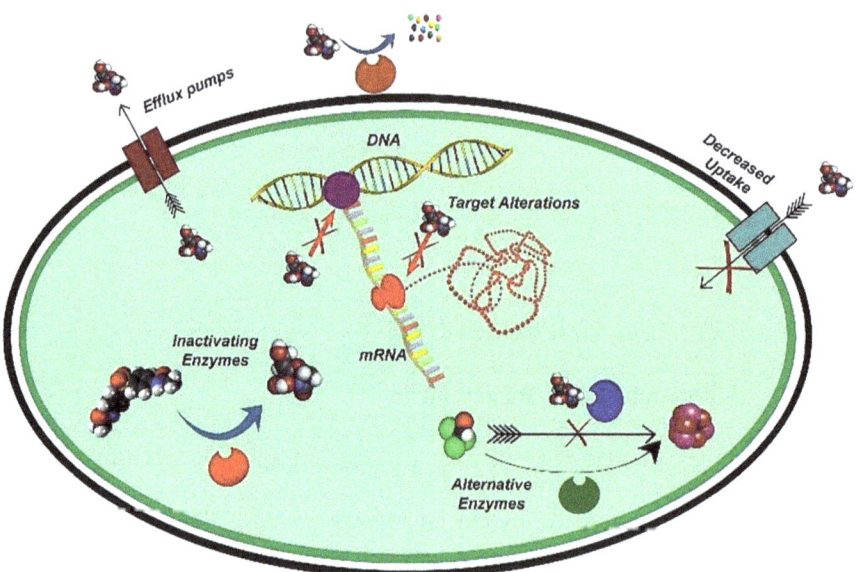

Fig. 14.4 Mechanism of antibiotic resistance in bacteria

same rate as they enter the cells and prevent the drugs from reaching their targets (Nikaido 1994; Kumar and Schweizer 2005). These pumps are located on the cytoplasmic membrane as compared to porins which are present on the outer membrane. Efflux pumps can be specific to a particular antibiotic, but most of them are multidrug transporters capable of pumping out wide range of unrelated antibiotics (Nikaido and Zgurskaya 1999; Webber and Piddock 2003).

14.4.1.2 Modification of Target Molecule

The binding of the antibiotic molecule to its target site is governed by the symmetry and dynamics of the interaction. Any natural or acquired change in the target molecule could hamper the effective drug interaction. A bacterium can acquire the change in target molecule through mutations in bacterial genes. Alterations in the 30S and 50S subunits can lead to resistance to antibiotics that affect protein synthesis like macrolides, chloramphenicol etc. Modification of PBP is another mechanism for resistance favoured by Gram-positive bacteria. It creates decreased affinity of the β -lactam antibiotics to the PBP (Mobashery and Azucena 1999; Lambert 2005). β -lactamase production is the complementary mechanism adopted by Gram-negative microbes for resistance. *E. faecium* resistance to ampicillin, Streptococcus pneumoniae resistance to penicillin etc., are mediated through this mechanism. Cell wall precursors are changed by mutations by certain bacteria. Antibiotics that inhibit cell wall synthesis are performed by binding to the D-alanyl-D-alanine moieties of peptidoglycan. Changing the D-alanyl-D-alanine to D-alanyl-lactate prevents glycopeptides cross linking (Džidić et al. 2008; Grundmann et al. 2006). *E.faecium* and *E. faecalis* develop resistance through this mechanism. Mutation to DNA gyrase and topoisomerase IV genes produces defective enzymes that leads to replication failure and prevents antibiotics like FQ to bind. Ribosomal protection mechanisms and RNA polymerase mutations also impart resistance to bacterial strains (Kapoor et al. 2017).

14.4.1.3 Antibiotic Inactivation

Chiefly there are three enzymes that inactivate antibiotics – β-lactamases, aminoglycoside-modifying enzymes and chloramphenicol acetyltransferases (Alekshun and Levy 2007). β-lactamases hydrolyse β-lactam ring containing ester and amide bond like penicillin, cephalosporins, monobactams etc. Aminoglycoside modifying enzymes like phosphoryl-transferases, nucleotidyl-transferases and adenylyl-transferases reduce affinity to bind to 30S ribosomal subunit and generate resistance to multiple organisms (Maurice et al. 2008; Strateva and Yordanov 2009). Chloramphenicol-acetyl-transferases is another mechanism to build resistance to the antibiotic chloramphenicol (Tolmasky 2000). These enzymes present in some Gram-positive and some Gram-negative organisms along with few *Haemophilus influenza* strains acetylates hydroxyl groups of chloramphenicol making them unable to bind to the 50S subunits of ribosomes.

WHO has warned that several of the infectious diseases will become incurable due to resistance.For overcoming these several other treatment strategies have been scientifically developed. Here we list the major classes of alternatives for antibiotic treatment through non-antibiotic treatments to control infections.

14.5 Non-antibiotic Treatments for Infections

- Phage therapy
- Bacteriocins
- Killing factors in microbes
- Antibacterial activities of non-antibiotic drugs
- Quorum quenching

14.5.1 Phage Therapy

Phages are a group of viruses that have the capability to infect and kill bacteria. The first mentioning on phages was done by Ernest Hankin in 1896 and later the name bacteriophage was coined by Felix d'Herelle. Among all the different methods of non-antibiotic treatment of infections above, phage therapy is the only treatment that entered clinical trials and was produced in large scale for the purpose of treatment during the 1940s. Humans are administered with phages orally, rectally, locally, parenterally as intravenous injections or through respiratory passages as aerosols or intrapleural injections (Sulakvelidze et al. 2001). Today, phage therapy is not considered as an armamentarium against bacterial infections in almost all parts of the world. One reason for its declined used is attributed to the early introduction of phage therapy before understanding its purpose. The advent of stronger and newer antibiotics pushed phage therapy to the verge of extinction (Kutter et al. 2010).

Bacteriophages replicate inside the bacterial cells following two major pathways:

1. Lytic module
    Lytic pathway involves the following steps:
    (a) Attachment of the phage on to the bacterial wall
    (b) Injection of its DNA inside the host cell
    (c) Termination of the synthesis of bacterial cell components
    (d) Phage DNA replication and formation of new phage capsids
    (e) Assembly and release of phage components by lysis of the host cell.

2. Lysogenic module

In this pathway the initial steps of attachment and inoculation of the phage DNA into the bacterial host cell are similar to that of the lytic pathway. In the third step, the phage DNA will anneal on to the host chromosome and gets integrated (lysogenization), and the replication of the phage DNA occurs along with the replication of bacterial DNA for subsequent several multiplications (Prophage). The prophages after several cycles of multiplication free themselves from the bacterial genome and induce the bacterial genome to synthesize phage components, which are released after the lysis of the bacterial cell. Phages inhibit bacterial restriction enzyme through genome modification (Andriashvili et al. 1986). As this infection cycle can go on for several lifecycles, lysogenic phages are not ideally used for phage therapy (Lorch 1999).

Phage therapy had not been successful in combating bacterial infections as the scientists were more concerned with the application of the phages while the clinical data pertaining to their use were completely ignored. As the phages replicates inside the bacterium host, they exhibit the interesting phenomena of self-dosing. Phages are highly specific about their host and due to this reason, its usage was relatively safe as the friendly micro flora of the body is spared from the attack of the phages. But this also has a disadvantage that the causative agent of the disease needs to be identified properly before dosing. This specificity is lacking in the case of antibiotics and they became more popular than the phage therapy in the years that followed. The most feared complication from phage therapy is the release of endotoxins by the bacteria lysed by the phages. Even though phage therapy took up gradually crossing many geographical restrictions, none of its advocates recognized the role bacterial immunity. The bacterial immune responses are of two kinds – the innate and adaptive responses. The distinction between and the self and non-self-DNA through the restriction modification forms the innate response. This is achieved on the basis of the DNA methylation pattern and on the lack of phage replication machinery. The adaptive immunity in bacteria was discovered with the identification of CRISPR (Clustered regularly interspaced short palindromic repeats) sequences. CRISPR-mediated immune response involves complex processes like gene silencing mechanism (Brouns et al. 2008; Barrangou et al. 2007; Hale et al. 2009).

14.5.2 Bacteriocins

Bacteriocins are peptides which are bactericidal, produced by bacteria belonging to several groups which is evolved for eliminating competition from other strains. *E. coli* strain V produces a dialyzable and heat-stable compound called colicin V that inhibits the growth of coliphages (*E. coliφ*) at very low concentration. The use of bacteriocins is not extensive in clinical infections but limited to food preservation (Cleveland et al. 2001; O'sullivan et al. 2002). Application of bacteriocins are seen in food industry, where the probiotic microorganisms used in packaged foods secretes bacteriocins and prevents the growth of harmful bacteria. Bacteriocins are of four major types; Class I (antibiotics) are <5 kDa small heat-resistant peptides; class II are <15 kDa small heat-stable, membrane-active and unmodified peptides; class III consists of heat-labile proteins with size >15 kDa and class IV bacteriocins are lipid or carbohydrate moieties bound to it (Sablon et al. 2000; Garneau et al. 2002; Klaenhammer 1993). Bacteriocins exert inhibitory action mainly on Gram positive bacteria than Gram-negative, as they possess an outer membrane made mainly out of lipopolysaccharides (LPS). This outer membrane provides an evolutionary advantage of preventing free diffusion of molecules heavier than 0.6 kDa. The smallest known bacteriocin is of about 3 kDa size. Bacteriocins act by binding to the cell membrane and alter their osmotic stability leading to cell death (Cotter et al. 2013). Even though Gram-negative bacteria evades the action of bacteriocins, some of them gain entry to the cell interior through specific receptors over the outer membrane like OmpF, FhuA or over the inner membrane like SbmA, YejABEF and TonB. Absence of the outer protective LPS layer makes Gram-negative bacteria more susceptible to bacteriocins leading to cell death (Stevens et al. 1991). The mode of mechanism of bacteriocin is doubtful yet it is postulated that they bind to the lipid-II, which is essential for the transport of peptidoglycan subunit from cytoplasm to cell wall, thus interrupting the synthesis of cell wall leading to cell death. Several other mechanisms have also been postulated regarding the mode of action of bacteriocins; they include pore formation in cell wall using lipid-II as a docking molecule, de-energizing membrane and dissipating proton motive force, preventing uptake of amino acids and triggering their release from the cell, excluding potassium ions, depolarizing cytoplasmic membrane, hydrolysis and partial efflux of cellular ATP (Cotter et al. 2005; Abee et al. 1994). The advantage of bacteriocin use is that they do not harm the beneficial microbiota of the body, but they are effective only against a small group of bacteria. Bacteriocins are short-acting than antibiotics, due to action of proteolytic enzymes on them which easily converts them to non-toxic amino acids. This can be overcome by the using dendrimers that prevent their degradation and deliver them at the site of infection (Tam et al. 2002; Bracci et al. 2003). Resistance development is another negative aspect of bacteriocin that limits its activity. Bacteriocin resistance is acquired through the immunity gene present in bacterial strain or through alteration in gene expression.

14.5.3 Killing Factors

Killing factors are the factors released by bacterial cells to kill sibling cells during starvation. This phenomenon was noticed in *B. subtilis* and is named as "cannibalism". "Cannibalism" is exhibited through a set of genes that induce lysis of their sister cells in their milieu during nutrient scarcity (Nandy et al. 2007). The lysed cells provide the nutrients for the killer cells for their survival and spore formation. *B. subtilis* exhibits predation rather than cannibalism, that is, it lyses the bacterial cells of other species. The feature of cannibalism is due to two peptides – sporulation delaying protein and sporulation killing factor. Killer *B. subtilis* cells preferentially target non-*B. subtilis* cells. This shows the antibiotic action of killer peptides (Burbulys et al. 1991).

14.5.4 Antibacterial Activities of Non-antibiotic Drugs

The drugs that are developed to treat non-infectious diseases but having antimicrobial activities are called non-antibiotics (Williams 1995; Lind and Kristiansen et al. 1990). They are effective against some Gram-negative bacteria, Gram positive bacteria, viruses, fungi, protozoa etc. (Jones 1996). Barbiturates, diuretic drugs, beta-adrenergic receptor antagonists, antihistamines, mucolytic agents, non-steroid anti-inflammatory drugs, proton pump inhibitors and psychotherapeutic drugs are studied in this regard. Alteration of cell permeability is considered as their mode of action in this regard. Alternative mechanisms like affection of efflux pump of microbes, cross-membrane ion transport, cell energy transport and activity of membrane-bound enzymes are also studied (Cederlund and Mårdh 1993). The concentration at which these drugs produce their activity is much higher than the physiologically observed dose. The combination of nonantibiotic drug with antibiotics may make the resistant bacteria susceptible to the non-effective drug. The combination of beta-lactam with phenothiazines when administered to beta-lactam-resistant microorganisms like MRSA makes them sensitive (Amaral and Kristiansen 2000). The advantage of such an approach is that the physiological impacts of the drugs involved are well established.

14.5.5 Quorum Quenching

Quorum quenching is the process of communication between bacterial cells through messenger molecules. It is established that such communication plays a significant part in the beginning of virulence mechanism. Two mechanisms in this regard have been recognized and analyzed (Nigam et al. 2014). First type involves recognition of signal by means of cytosolic transcription factor while in the second type it is facilitatedvia an auto inducing recognized through a membrane receptor. Any microbe can own either type of quorum sensing system or both. The former form of quorum sensing is facilitated by derivatives of acyl homoserine lactone while the auto inducing system involves peptides. It is assumed that if this signalling is weakened, the virulence of microbes can be controlled without selection pressure. It is known that selection pressure leads to the development of new drug-resistant microbes. Therefore, this method that hinders the spread of microbes without imposing selection pressure may be crucial to prevent evolution of drug-resistant microbes. Quorum quenching is achieved either through mimicking of quorum sensing molecules that compete with analogous quorum sensing molecules or through inhibiting enzymes involved in synthesis of quorum sensing molecules. The report of Triclosan as an anti-bacterial agent inhibiting enoyl-ACP reductase which produces an important intermediate in AHL biosynthesissupport this mechanism (Hoang and Schweizer 1999). The important advantages of quorum sensing inhibitors are that it limits evolution of new drug-resistant forms as well as virulence and bio-film formation (Czajkowski and Jafra 2009).

14.6 Mechanisms of Antibacterial Activity of Nanoparticles

Mainly three mechanisms have been identified as the modes of action of nanoparticles in killing the microbes. They are as follows: (1) due to direct intake of nanoparticles, (2) reactive oxygen species (ROS) production and (3) disruption of cell wall by nanoparticles.

14.6.1 Direct Absorption of Nanoparticles

Studies have shown that nanoparticles of silver (AgNPs) leach out silver ions when it comes in contact with the body fluids due to variation in the pH. These silver ions have high affinity towards enzymes containing thiol group rich amino acids like Cysteine. Thus, they have high affinity towards respiratory (NADH dehydrogenase) and electron transport group of enzymes (Prasad and Swamy 2013; Swamy and Prasad 2012). This interaction results in the uncoupling of the ATP molecule from the respiratory chain. They create a phosphate deficient environment inside the bacterial cell by actively efflux phosphate ions and at the same time reduces their reuptake. They reduce proton motive force resulting from the loss of protons due to the binding of silver ions to the various transport proteins. They are also shown to increase the frequency of DNA mutations during the translation process and in leakage of intracellular contents from cytoplasm retrenchment and cell membrane degradation.

14.6.2 Reactive Oxygen Species (ROS) Production

ROS are formed as by-products of all metabolic pathways inside the cells of all respiring organisms. There are several inbuilt mechanisms in organisms to tackle this life-threatening escalation of ROS. Antioxidant defence mechanisms like glutathione, glutathione disulphide, reduced glutathione, glutathione peroxidases and peroxiredoxins play an important role in quenching the surge in oxidative stress. Excessive production of ROS brings in damages to the lipid bilayer of membranes, dysfunction of mitochondria and DNA damage. Metals in general and silver ions in particular catalyses ROS generation in the presence of oxygen dissolved in milieu. The generation of ROS is done either by inactivating the respiratory chain enzymes or the superoxide dismutase which scavenge the superoxide radicals or by a combined mechanism of both. Thiol group binding and inactivating metal ions thus provide significant antibacterial effect.

14.6.3 Cell Wall Damage

The low size of the metal ions and their electrostatic attraction to the negatively charged cell membrane allows them to easily adhere to them and penetrate to create pits by releasing lipopolysaccharides and proteins from them. This varies the permeability of the membrane by releasing muramic acid by binding to the N-acetylmuramic acid and N-acetylglucosamine of the peptidoglycan strands (Feng et al. 2000; Sondi and Salopek-Sondi 2004; Morones et al. 2005; Song et al. 2006).

14.7 Nanomaterials in Controlling Infections

Reducing the size of the particles to nanoscale has the advantage of easily surpassing the natural biological barriers inside living systems owing to their high surface-to-volume ratio. Nanosize of particles having high surface-to-volume ratio permits them to surpass the barriers to biological systems and molecules. Molecular interactions could be enhanced by manipulating the size, shape and chemical characteristics of the nanomaterials (Kim et al. 2010) and can be utilized as therapeutic and diagnostic agents in the form of vehicles. Nanotechnology could be used to overcome the bacterial resistance by creating new and improved antimicrobial agents. The use of nanomaterials in infectious diseases can be discussed under the following headings (Blecher et al. 2011);

- Drug delivery systems
- Drug infused nanoparticles
- Immunomodulation

14.7.1 Nanotechnology-Based Drug Delivery Systems

14.7.1.1 Chitosan

Chitosan is a biopolymer made of natural polysaccharide exhibiting polycationic interactions with negatively charged microbial cell wall and other cytoplasmic membranes leading to the disruption of the cell membrane and subsequent leakage of intracellular elements from deranged osmotic stability.They also have the capability to enter the nucleus of the bacterial cell and other microorganisms like fungi and bind to the DNA and thus inhibits the mRNA and protein synthesis (Ma et al. 2008; Qi et al. 2004). Chitosan molecules have high affinity towards bacteria and fungi due to their relatively higher surface-to-charge ratio and surface charge density. Hence, they exhibit greater antimicrobial activity against Gram-positive as well as Gram-negative organisms. Thus, they exhibit antimicrobial activity against the most notorious pathogens like *E. coli* and *Staphylococcus aureus*. (Banerjee et al. 2010; Sanpui et al. 2008). Chitosan is derived from a natural substance, chitin, which is the structural ingredient of the exoskeleton of crustaceans. Studies have shown that the chitosan nanoparticles are more efficacious than chitosan alone or antibiotics like doxycycline. Its polycationic characteristic and high affinity towards metals are utilized in engaging it along with several other nanoparticles like metallic NPs (copper and silver), nitric oxide releasing nanoparticles and drug containing NPs used for targeted drug delivery and as carriers (Qi et al. 2004). Chitosan also escalates the antimicrobial property of these nanomaterials. In experimental models, silver nanoparticles-embedded membranes increased their zone of inhibition when incorporated with 70% of chitosan. Similarly, incorporation of chitosan in silver nanoparticles decreased the mean inhibitory concentration (MIC) against *S. aureus* (Ma et al. 2008).

14.7.1.2 Metallic Nanoparticles

Silver

Silver is traditionally used as an antimicrobial agent in treating conditions like burns and wounds. It is believed that silver ions on gaining entry into the bacterial cell wall and membranes target the DNA, respiratory enzymes and other proteins containing sulphur or thiol groups resulting in loss of capability to replicate and finally cell death results (Aziz et al. 2014, 2015, 2016, 2019). Silver nanoparticles (Ag-NP) are attributed with small size and large surface area, and this makes them capable of easily penetrating the bacterial cell wall and other biological membranes (Rai et al. 2009; Pal et al. 2007; Ruparelia et al. 2008; Prasad 2014). Thus, the nanoparticle size is proportionated with its antimicrobial activity; smaller the size, greater the effect. Shape of the nanoparticles also influences its activity. Small triangular and truncated nanoparticles are more efficacious than round or rod-shaped particles. Hence, small and triangular nanoparticles (<10 nm) exhibit more antimicrobial efficacy than large round or rod-shaped nanoparticles (Pal et al. 2007). Ag-NPs thus exhibit a very varied antimicrobial property against viruses, bacteria and fungi due to their surface interaction. Several bacterial species belonging to both Gram-negative and Gram-positive strains succumb to them. This wide range of activity is achieved at relatively lower concentration of Ag-NPs than the conventional silver preparations. This paves way to lower dosing and lesser toxicity from silver. These conclusions are relatively theoretical and studies pertaining to the toxicity of Ag-NPs are yet to be unravelled. Ag-NPs also exhibit synergism with other antibiotics. This is another area of interest in Ag-NP research. Activities of penicillin G, amoxicillin, erythromycin, clindamycin and vancomycin increased against organisms like *S.aureus* and *E. coli*. Among these agents, erythromycin showed the greatest inhibition. As Ag-NPs exhibit wide range of targets, microorganisms should develop multiple, simultaneous compensatory mutations to develop resistance. Hence, Ag-NPs can overcome bacterial resistance against antibiotics and at the same time enhance their efficacy (Sanpui et al. 2008). Owing to these qualities Ag-NPs find its application in medical devises that constantly come in contact with body fluids and poses a threat to infection by serving as acoatings on them to prevent microbial colonization, in wound dressings and in enhancing the potency of antibiotics.

Copper

Copper has been less engaged as an antimicrobial agent when compared to silver. Yet its use as an antifungal agent is well documented in history as early as the nineteenth century (Cioffi et al. 2005). Copper oxide as a nanomaterial has recently gained importance due to its cost effectiveness and compatibility with other polymers. Yet copper oxide nanoparticles (CuO-NPs) are inferior to Ag-NP in their antimicrobial property against *E. coli* as well as methicillin-resistant *S.aureus* (MRSA) (Ruparelia et al. 2008; Ren et al. 2009). But they are found to be more efficacious against *B. subtilis*, which is attributed to the copper's affinity to amine and carboxyl groups on the cell surface of these pathogens (Ruparelia et al. 2008; Yadav et al. 2017). Compared to Ag-NPs, CuO-NP exhibit a broader range of activity, especially against fungi. Copper loaded nanoparticle laden polymer thin films demonstrated significant antifungal activity against *S. cervisiae* yeast, moulds and bacteria including *E. coli, S. aureus* and *Listeria monocytogenes*. They significantly reduced the number of colony forming units (CFUs) especially in *S. cervisiae*, where no CFUs were noted (Cioffi et al. 2005). These results project out the biostatic property of CuO-NPs. On comparison with silver, CuO-NPs exhibit broader range of antimicrobial activity and weaker activity against most of the bacteria, they are strong antifungal agents and are capable of preventing surface microbial colonization especially on medical instruments.

Titanium

Titanium dioxide (TiO_2) has gained importance as a nanomaterial recently due to its activation on exposure to ultraviolet light forming active oxygen species, a process termed as photocatalysis. The family of active oxygen species generated includes hydrogen peroxide and hydroxyl radicals. They are responsible for obliterating the bacterial cell membranes resulting in cell death (Kim et al. 2003). This property of TiO_2 nanoparticles (TiO_2-NP) has been engaged in water and air purification and their activity against pathogenic opportunistic microorganisms (Martinez-Gutierrez et al. 2010). TiO_2-NP infused thin film composite membranes has been recently shown to have significant antimicrobial activity and prevents bacterial attachment to membrane surface and formation of *E. coli* biofilm on medical instruments by disrupting the bacterial cell membrane (Martinez-Gutierrez et al. 2010). Researches has been conducted by coupling TiO_2-NP with Ag-NP for their antimicrobial property. Even though TiO_2-AgNPs proved less efficacious than TiO_2 alone or Ag-NP alone, against their activities against Gram-positive bacteria, Gram-negative bacteria and various fungi responsible for opportunistic infections and colonization of medical devices, they showed more pronounced activity against several strains of fungi of medical importance. These results highlight that it might be more beneficial to combine metal nanoparticles to augment the antimicrobial activity.

Magnesium

Several non-metals like chlorine, bromine and iodine has been traditionally used as antibacterial agents, but their toxicities limit their use especially in medical conditions. Their antimicrobial property is due to the formation of covalent metal-halogen complexes that interact with specific cellular enzymes or through oxidative stress leading to lipid peroxidation ending in the leakage of intracellular contents culminating in cell death (Lellouche et al. 2009). Magnesium- halogen nanoparticle also exhibit antimicrobial activity through the same pathway. Magnesium oxide has a unique place among them as they easily adsorb and retain halogens and at nanoparticle level, these activities increase fivefold. Other advantages of this compound are that, as they bind with halogens, the compound gets converted to powder form which helps in easy handling. Magnesium – halogen nanoparticles are found to be more effective against endospores of different bacterial strains like *E. coli*, *B.*

megaterium and *B. subtilis*. Among them *E. coli* and *B. megaterium* are highly susceptible to MgO-halogen nanoparticle with complete destruction of entire endospores in less than 20 minutes. Several combinations of Mg-halogen combinations have been investigated for their antimicrobial activity. MgF_2-NPs also showed similar antibacterial effect comparable to MgO-halogen nanoparticles. MgF_2 – NPs exhibited dose-dependent inhibition of growth of *E. coli* and *S. aureus* and also inhibited biofilm formation on medical devices.

Zinc

Zinc oxide (ZnO) is another compound of interest in the field of antibacterial effect. It is the compound which is approved by FDA with respect to its antibacterial as well as safety profiles. Zinc oxide nanoparticles (ZnO-NPs) are demonstrated to be effective against major food-borne pathogens like *E. coli* O157:H7, *Listeria monocytogenes* and *Salmonella* spp. (Jin et al. 2009). ZnO-NPs are found to inhibit the growth of *E. coli* O157:H7 in a dose-dependent manner (Liu et al. 2009) Their antimicrobial action is attributed to membrane binding leading to lipid and protein destruction on membranes. This leads to altered membrane permeability and leakage of intracellular contents. These pathological changes are initiated by the generation of reactive oxygen species (Bhuyan et al. 2015).

14.7.1.3 Nitric Oxide – Releasing Nanoparticles

Nitric oxide (NO) has been identified as a molecule with varied physiological functions in the body. Several phagocytic cells like macrophages enhances the production and release of NO on stimulation through the transcription of inducible nitric oxide synthase (iNOS) (Englander and Friedman 2010). The NO thus released demonstrate antimicrobial activity by either of the several mechanisms identified like direct microbial DNA damage through generation of peroxynitrite, inactivation of zinc metalloproteins and interfering cellular respiration or by stimulating innate antimicrobial pathways that enhances host immune response. Also, there are research reports on the activity of NO as a wound healing agent. Considering these properties, a nanomaterial system has been created that donates or delivers NO to the site. NO – releasing nanoparticles (NO-NPs) entraps NO in a dry matrix and releases gaseous NO free radicals on exposure to moisture. This is utilized as a topical agent for promoting wound healing and preventing infections. NO–NPs can easily be applied on skin and ensures sustained delivery to the affected areas over prolonged period of time. These nanomaterials have been tested against *S. aureus*, *Acinetobacter* and MRSA – infected wounds and showed accelerated wound closure and decreased microbe burden (Martinez et al. 2009).

14.7.1.4 Drug – Infused Nanoparticles

Intracellular infections from organisms like *Salmonella*, *Listeria* and *M. tuberculosis* are difficult to eliminate due to the advantage they achieved from highly evolved evolutionary mechanisms like escaping phagosomes, phagosome–lysosome fusion inhibition and their ability to stay dormant (Pinto-Alphandary et al. 2000). These mechanisms downregulate several drug targets. Complicating the scenario is the fact that several antibiotics cannot penetrate intracellularly to act on these microorganisms. In such conditions, nanoparticles and liposomes are devised as potential drug carriers as they get endocytosed into phagocytic cells carrying intracellular pathogens. Several antibiotic-encapsulated liposomes are developed to deliver antibiotics like β-lactams like penicillin, ampicillin and cephalosporins, macrolides, aminoglycosides and fluoroquinolones for enhanced bacterial killing. Ampicillin-encapsulated liposomes against *Salmonella*, liposome-encapsulated tobramycin against *P. aeruginosa* and amphotericin B-encapsulated liposome against moulds and yeasts have been studied (Fattal et al. 1991). In spite of their success as a promising drug delivery system, there are limiting factors that impede their application in medicine. Some of the highly discussed limitations are their size, charge, purity, solubility of contents, stability, antigenicity, biocompatibility etc. There are two ways to deliver the drug to its target site by the nanocarriers – passive and active targeting. In passive targeting, nanocarriers are transported through the inter-endothelial cell spaces on the neovascularized vessels of the tumour mass or through the fenestrations on the vessels at the sites of inflammation. Passive transport is dependent on the plasma permeability and their time of retention. In the case of active targeting, nanocarriers are transported based on the receptor–ligand interaction, and this requires fenestrae or specific receptors. Change in the pH of the medium or a sudden oxidative burst can facilitate the delivery of the drug to its required target. Locally they can be transported with the aid of magnetic guidance and radio frequency-mediated delivery.

14.7.1.5 Immunomodulatory Effects

The immune system helps the body to fight against foreign substances including pathogens. It is broadly classified into innate and adaptive systems. The nonspecific immune response is termed innate immune response, which involves the recognition of the pathogen-specific molecular patterns (PAMPs) of the invading pathogens (Plummer and Manchester 2011). Antigen presenting cells (APCs) engulf the pathogen after the PAMPs have been recognized by the pattern recognition receptors (PRPs) present on cells. The cascade of events following the APCs engulfing and presenting the antigens to the cells of the adaptive immune system culminates in the activation of CD4 and CD8 lymphocytes subsequent to the induction of T and B cells (Plummer and Manchester 2011). The entire machinery works on an integrated and coordinated network of events mediated by cytokines which decides on the type of response to each pathogen. For example, Interferon gamma activation promotes T helper type 1 (Th1) response while Interleukin-4 (IL-4) and IL-5 activation results in a Th2 response (Sun et al. 2009). A synchronized interaction between APCs, T and B cells and inflammatory cytokines is essential for a functionally robust immune system.

Vaccines play an integral part of infectious disease control and require a balanced stimulation of both innate and acquired immune systems. Vaccines basically belong to either of the three categories viz., live attenuated, killed or fragmented (subunit) vaccines. With respect to the fairness of immunogenic response generated, live attenuated vaccines are the best, but they pose a threat of reactivation and attaining virulence inside the host. While killed and subunit vaccines are devoid of this threat, their immunogenicity is very poor, requiring repeated dosing and thereby increasing the cost. Since a trained personnel is essential to administer a vaccine in the form of injection, vaccine coverage is decreasing in many countries in spite of a global intervention. Mucosal application with the help of nanotechnology is one alternative to overcome this hurdle. There are several barriers to cross for the vaccine to reach the APC and induce a strong enough immune response. In intranasal or inhalational vaccines, the vaccine components should be smaller (< 5um) and not to be cleared by the exhalation (Chadwick et al. 2010). As the size specified is in accordance with that of a nanomaterial, nanoparticles are capable enough to modulate and influence the immune system at various levels of interaction. Nanoparticles can thus augment the efficacy of both oral and injected vaccines by enhancing the vaccine exposure time to the immune system as well as an increase in the uptake of antigens by APCs (Bal et al. 2010). Thus they improve the immune response to microbes like viruses and bacterial components and also lead to a more controlled and modified cytokine response (Huang et al. 2010).

14.7.2 Nanotechnology-Based Vaccines and Immunostimulatory Adjuvants

14.7.2.1 Synthetic Polymers

Polymeric nanoparticles are utilized as carriers for different vaccines like DNA vaccines and for proteins as well. Choice of nanomaterial is based on the requirements of the vaccine. Poly ε-caprolactone polymers are used to overcome the physicochemical differences in the digestive tract in the case of oral vaccines. Multicomponent particles are created encapsulating DNA in a PCL microparticle and this system is named nanoparticle-in-microsphere hybrid oral delivery system (NiMOS). The central core of NiMOS is susceptible to proteolytic degradation while the exterior is susceptible to lipase digestion. As the particles transit through the gut, the outer coat survives the proteolytic degradation and as they reach the intestines, got acted on by lipases and release the inner gelatinous core containing the vaccine which gets absorbed from the intestines and produce the antigenic effect. Many of the synthetic polymers used for vaccine delivery are found to be non-immunostimulatory. Poly(lactide-co-glycolide) (PLGA) particles does not lead to the rise in pro-inflammatory cytokine levels even after being taken up by the macrophages. But they produced stronger and sustained IgG response when encapsulating plasmid DNA owing to the stability of PLGA nanoparticles during their delivery into appropriate cells. This highlight the fact that nano polymers alone cannot be used as immunostimulants, but they can enhance immunization when they cross the mucosal barriers. A polymeric nanoparticle, Polymethyl methyl methacrylate (PMMA) is highly immunostimulant when administered as a vaccine adjuvant. It increased the antibody titre to 100-fold in HIV2 virus vaccine in animal models and also exhibited enhanced IgG and IgM antibody production against ovalbumin. Several other nanomaterials have also shown immunomodulatory activity for example; Carboxyfullerene nanoparticles increased immunologic activity by neutrophil activation resulting in microbial death in a *S. pyogenes* infection model in mouse. Some nanoparticles enhanced immune responsiveness in conjunction with Toll-like receptor agonists. PLGA NP modified with tetanus toxoid and MUCI lipopeptide (a TLR ligand) showed significant T cell activation and responsiveness in comparison to PLGA NP and tetanus toxoid or MUCI lipopeptide alone (Diwan et al. 2003).

Polymeric oligonucleotides of DNA or RNA termed Aptamers, which folds in 3-dimensional configuration with high affinity to target proteins, peptides or small molecules of drug or vitamin showed pronounced antimicrobial activity, inhibition of HIV reverse transcriptase and vaccinia virus replication and to overcome β-lactamase resistance in Gram-positive and Gram-negative bacteria. Aptamers are selected from random nucleic acid libraries and technologies are now available to prepare them with highest degree of purity, stability and selectivity.

14.7.2.2 Nanoemulsions

Nanoemulsions constitute lipophilic or hydrophilic substances dispersed as either water-in-oil or oil-in-water forms. Such carrier systems are engaged in vaccine delivery through mucosal barriers as they are easily endocytosed by the surface cells of the mucosa and easily presented to the APCs. Nanoemulions have also showed good immunostimulatory effect. The hepatitis B vaccine currently in use is given as an IM injection of recombinant hepatitis B surface antigen (HBsAg), containing aluminium salt (alum) as an adjuvant. Alum stimulates a Th2 immune response in the host with an ineffective CD8 response to the virus infected hepatocytes. Alum is also responsible for other complications like the formation of erythema and nodules over the injection site. On the contrary, recombinant HBsAg nanoemulsion (HBsAg – NE)-based intranasal vaccine produces an effective Th1 response in the host without any chance for local inflammation (Muttil et al. 2010). They are also capable of producing comparable levels of IgG antibody body level and significant levels of mucosal IgA antibodies as well. Nanoemulsion vaccines are also being tried for inactivated influenza and vaccinia virus infections.

14.7.2.3 Immune Stimulating Complexes

Immune-stimulating complexes (ISCOMs) are carriers and immunostimulatory adjuvants, which are nanosized spherical micelles containing saponin-derived components like Quil A, derived from the tree bark Quillaja (Helgeby et al. 2006). They are efficient carriers easily taken up by APCs and can activate and upregulate the expression of MHC I and II on APCs, even in the absence of an antigen. The overall result of this is the induction of pro-inflammatory cytokines like IL1, IL6, IL8 and IFNχ. Carriers containing saponin-derived components produce Th1 type response, but this may change depending on the adjuvant carried (Sun et al. 2009). ISCOMs carrying Leishmania antigen induce a Th2 response. Studies on various antigens incorporated on ISCOMs are available including influenza, hepatitis B, herpes virus, *Helicobacter pylori* and *Corynebacterium* (Sun et al. 2009).

14.7.2.4 Cytidine-Phosphate-Guanosine (CpG) Motifs

Bacterial oligodeoxynucleotides containing unmethylated CpG motifs are used as vaccine adjuvants with good immunostimulatory activity. They lead to enhanced secretion of IL12, surface expression of MHC, Th1 recruitment and activation and secretion of IgG antibodies as they are recognized by the APCs. They are recognized by the APCs through the activation of Toll-like-receptor 9 (TLR-9). CpG motifs incorporated into a nanoemulsion of water-oil-water has inactivated influenza virus and induced strong and sustained immune response when compared to conventional vaccines.

14.7.2.5 Chitosan

Chitosan improved vaccine delivery in the case of oral vaccines, as it proved to be an efficient carrier system to transport vaccines across mucous membrane. It also find its use as an immune adjuvant as it promotes antigen uptake and cytokine production. When incorporated with the glycoproteins present over the surface of influenza virus like HA and NA, a pronounced immunological response in the form of high titres of serum IgG and mucosal IgA antibodies were observed after intranasal administration (Brunner et al. 2000). Escalated IgA levels are attributed to the relatively strong interaction of chitosan with the sialic acid residues over the mucin. This interaction results in the opening up of the mucosal tight junctions enhancing the mucosal membrane transport and vaccine retention time (Florindo et al. 2009). Chitosan has gained importance as vaccine adjuvants with immunomodulatory stimulation capability. N-trimethyl chitosan (TMC) nanoparticles coupled with ovalbumin and diphtheria toxoid where swiftly internalized by the APC of the skin, Langerhans cells, and induced a profound expression of CD83, CD86 and MHC-II in murine models on intradermal administration. Once they are internalized, the antigens set free from the chitosan after lysosomal degradation, the immunological processes unfurl inside the body and the antigen will be presented to the T cells followed by a Th2-mediated antibody response. The antibody titre is higher than either of the antigen alone and this establishes its immunostimulatory effect as an adjuvant.

14.7.2.6 Metallic Nanoparticles

Every drug or gene transported in a nanoparticle should be well protected against the immune recognition by the host's body because of the activation of the innate immune response and subsequent inactivation of the drug. To a great extend this property is well exhibited by metal nanoparticles (Massich et al. 2009). Polyvalent oligonucleotide coupled with gold nanoparticle produced only 25% less macrophage activation evidenced by IFNβ when compared to lipid-complexed DNA.

14.8 Role of Nanomaterials in Other Diseases

14.8.1 Neurodegeneration

Neurodegeneration is defined as the destruction in the structure and function of neurons leading to their loss. This basic event cascades to a plethora of incurable pathological anomalies to central nervous system collectively called as neurodegenerative disorders which include Alzheimer's disease, Parkinson's disease, Prion disease, amyotrophic lateral sclerosis and Huntington's disease (Rubinsztein 2006). Neurodegenerative changes occur at different levels in the nervous system ranging from molecular level to systemic level. Distorted synaptic functions, a higher prevalence of intra-neuronal deposits resulting from misfolded proteins are some of the common events related with many neurodegenerative disorders, which helps in devising therapeutic modalities. Many more pathological events like genetic mutations, protein misfolding, protein aggregation, mitochondrial dysfunction, damage to the nucleic acid (DNA), structural and functional disruption of organelle membranes, neuronal and microglial apoptosis, autophagy and transglutaminase binding strongly influence the manifestation of neurodegenerative diseases. As neurodegenerative diseases occur during the later decades of life, the events leading to its manifestations ae attributed to mitochondrial DNA mutations and oxidative stress. Progression of neurodegenerative disease with advancing age is proportionated with the amount of neuronal loss and contributes a great deal of stress on the families emotionally, socially and financially. Considering the therapeutic challenges in neurodegenerative diseases, blood-brain-barrier (BBB) poses the greatest challenge. BBB represents a selectively semipermeable barrier between the nervous tissue and the blood compartments. It helps in the maintenance of neuronal cell functions, regulation of transport of nutrients and metabolites and the overall protection of the brain tissue. It avoids the entry of large therapeutic molecules as well a small molecule drugs from entering brain. Carrier-mediated transporters like large neutral amino-acid transporter, glucose transporter (GLUT1), cationic amino-acid transporter (CAT1), adenosine transporter (CNT2) and monocarboxylic acid transport small molecules to brain (Pardridge 2003). Certain circulating blood elements like leukocytes, erythrocytes, neutrophils, and other cells like exosomes can cross BBB at a faster rate. Drug delivery utilizing these living cells and exosomes paved way to the discovery of novel chemotherapeutic drug delivery to treat disorders of brain. Curcumin loaded exosomes experimentally showed improved cognitive function in mouse. Receptor-mediated transcytosis (RMT) is the common modality now used to deliver chemotherapeutic drugs to brain tissue across BBB. Endogenous macromolecular neuropeptides including transferrin, hormones, lipoproteins and insulin reaches brain tissue through RMT from blood through specific receptors. RMT are expressed on the luminal side of the endothelial cells and they favour endocytosis and transcytosis of molecules across the BBB (Abdul Razzak 2019). In RMT, nanoparticles are utilized after surface modification to bind to the transmembrane receptor associated with its transport. Macromolecular drug delivery to brainrelated diseases can also be increased tremendously by delivering drugs directly into the brain tissue with the help of electrostatic interactions between nanoparticle, which bears a positive charge, and the negatively charged BBB membrane through a process termed adsorptive-mediated transcytosis (AMT). AMT does not interfere with the normal cellular physiology as other drug delivery methods do. Nanocarriers used to cross BBB are;

14.8.1.1 Metal Nanoparticles

Transition metals like Zn, Cu and Fe are present in a sufficiently large amount in the brain and they play a crucial role in their cellular physiology including the metalloenzyme function. They act as a therapeutic agent as well as an agent that carries diagnostic agent due to its specific physicochemical properties. These metal NPs are also used for the detection of biomarkers. They are used in various shapes and sizes, surface charge and by coupling with various surface ligands for targeting for effective drug delivery for NDs. Functionalized super-paramagnetic iron oxide (SPION) is presently used for theranostic applications in imaging techniques like MRI and targeted therapy in NDs (Luo et al. 2020).

14.8.1.2 Lipid-Based Nanoparticles

Lipid-based NPs are developed as theranostic agents in the treatment of various NDs, and they are unique in the aspect of negligible side effects when compared to other technologies. This is achieved through modifying their properties by enabling drugs and ligands to bind on to their surface (Niu et al. 2019). The well-known phytoconstituents in NDs is curcumin, which has many limitations due to its poor solubility, low bioavailability and instability. Soft lipid-based nanocurcumin has provided a new hope in inhibiting neuronal loss in Parkinson's, Alzheimer's, ALS and Huntington's diseases (Rakotoarisoa and Angelova 2018).

14.8.1.3 Hydrogels

Hydrogels are 3D polymeric mesh-works capable of holding water. They find extensive use as a neuroprotective agent. Hydrogels are capable of systemic delivery of drugs directly into the brain tissue for targeted action in NDs. Activin B-loaded hydrogels have been developed, capable of slowly releasing activin-B over a period of several weeks for the treatment of Parkinson's disease (Albani et al. 2013).

14.8.1.4 Dendrimers

Dendrimers are credited with the position of one of the smallest nano-formulations used in the treatment of NDs. Dendrimers are manipulated by changing their size, core-shell or the surface functional groups to be used as nanocarriers for drug or gene delivery to the brain tissue. Moreover, they have strong anti-amyloidogenic activity that makes them ideal for the treatment of PD, Prion diseases and Alzheimer's. They also find their application in other areas like sterilization of medical instruments by phosphorus-containing dendrimers specially to prevent transmission of Prion disease (Šebestík et al. 2012).

14.8.1.5 Polymeric Nanoparticles

Polymer nanoparticles have the advantage of very low toxicity due to their fast excretion rate. They are block co-polymeric molecules that are biocompatible and biodegradable, and made up of lactic-co-glycolic acid (PLGA), polylactic acid (PLA), PLGA-PEG etc. (Calzoni et al. 2019; Prasad et al. 2017). Promising results have been shown by curcumin-loaded PLGA nanoparticles in the treatment of Alzheimer's disease, showing an enhanced drug delivery and reduced oxidative stress and inflammation (Barbara et al. 2017; Mukherjee et al. 2020).

14.8.2 Cancer Therapy

Cancer therapy protocols of current era are restricted to surgery, radiation and chemotherapy. These treatment regimens, though successful to some extent, will not provide complete eradication of disease and on the contrary damage healthy tissues also. The limited success rate of the conventionally used chemotherapy is due to the lack of water-soluble chemotherapeutic agents, lack of drug sensitivity of the cancer cells and resistance to multiple chemotherapeutic drugs developed due to their repetitive administration. The modern-day cancer therapeutics aims at boosting the natural capacity of the body to identify and destroy abnormal cells. Cancer cells have also adapted to evade the body's immune response by downregulating tumour surface antigen expression, extrusion of certain proteins to deactivate the immune cells or altering the cells in the surrounding microenvironment in order to suppress immune response. Immunotherapy acts by either stimulating the activities of components of immune system or by inhibiting the signals produced by the cancer cells that suppresses the immune responses.

14.8.2.1 Treatment Modalities in cancer

Immune Checkpoint Modulators

Natural proteins secreted by the cancer cells to prevent the damage of the normal and tumour cells by triggering the immunological responses are called immune checkpoint modulators. Suppressing the stimulation of immune response proteins will result in the activation of immune responses and their capability to destroy cancer cells. Activated cytotoxic T lymphocytes expressing CTLA4 on their surface is an FDA approved immune checkpoint inhibitor.

Adoptive Cell Transfer

Adoptive cell transfer is a promising treatment modality in cancer therapy where infiltrated T cells are isolated from the tumour samples of the patient and they are segregated based on their highest response shown to recognize patient's tumour cells. These cells are isolated in a laboratory, cloned and cultured to generate a large population, which are later activated by immune signalling proteins, cytokines, and are reinfused into the patient's body (Perica et al. 2015).

Therapeutic Antibodies

Therapeutic antibodies are lab-designed therapeutic molecules like antibody-drug conjugates (ADCs) that selectively destroy the cancer cells. In ADCs, a cytotoxic substance like bacterial toxin or cytotoxic small molecule drug or a radioactive compound is coated with antibodies or fragment antibodies by chemically linking them on to their surface. The target molecule expressed over the surface of the tumour cells will bind with the antibodies over the ADCs. The ADCs will be subsequently internalized by the tumour cells and the cytotoxic substance will destroy the cell (Scott et al. 2012).

Cancer Treatment Vaccines

Cancer vaccines are prepared from patients' own tumour cells or substances produced by the tumour cells. They are designed to treat already developed cancers by strengthening the body's natural defences against the cancer (Guo et al. 2013). Several new approaches have been developed in this novel treatment strategy. Contrary to the vaccines developed for other diseases, which prevent the occurrence of the disease, cancer vaccines are effectively utilized for their curative aspects. Vaccines targeting the tumour antigens form a major group of vaccines. As the tumour antigens vary drastically with the types of tumour as well as with individuals, two categories of cancer vaccines are identified; vaccines against tumour antigens that are specific to a particular tumour (Molecular vaccines) and those that are non-specific (Cellular vaccines) (Cheever et al. 2009). The efficacy of a vaccine relies on effectiveness of the adjuvant. Different adjuvants like incomplete Freund's adjuvant in emulsified form, particulates and saponins are extensively researched. Many cancer vaccines act by modulating T cell immunity. Generation of CD8+ T effector and memory cells from CD4+ T cell help response is crucial in the effectiveness of cancer vaccines. Several other strategies including patient-derived immune cell vaccines, expression of tumour antigens with the help of recombinant viral vaccines, peptide vaccines, DNA vaccines and whole cell vaccines have been developed against cancers (Hearnden et al. 2013).

Chimeric Antigen Receptor T-Cell Therapy (CAR-T Cell Therapy)

CAR-T cell therapy which is an advanced Adoptive cell transfer technology involving harvesting of T cells which are then genetically engineered to express specific chimeric antigen receptors (CARs) on their surface with which the tumour cells are identified. The CAR-T cells are then proliferated to a population of billions or more in laboratory and are then introduced to the bloodstream of the patient where they proliferate and identify the tumour cells with the help of the genetically engineered receptor and destroys them. Though they are being used clinically, they have some untoward effects like toxicity related to uncontrolled T cell activity, cytokine release syndrome, shared expression of tumour-specific antigens by healthy cells and "on-target, off-tumour" toxicity (Zhang and Xu 2017).

Stem Cell Transplant

Stem cell transplant is another area of interest that provided positive results. Autologous, allogenic and syngeneic transplantation of marrow cells collected from healthy subjects are transfused through a vein of the patient which gets engrafted in the bone marrow produces a new lineage of blood cells. Autologous transfusion represents self stem cell transplantation before chemotherapy, while allogenic transplantation is the stem cell transplantation from a suitable donor; whereas syngeneic transplantation is done using stem cells from an identical sibling. In all the three types of transplantations, the marrow stem cells are harvested in aseptic conditions and are stored in a specific nutrient medium at particular freezing temperature after filtration.

14.8.2.2 Nanotechnology Used in cancer Treatment

Nanocarriers

Chemotherapy utilizes the drugs that kill cancer cells, but many of these agents also destroys healthy cells also leading to adverse events. Nanoparticles are utilized as the vehicles to deliver chemotherapeutic medications directly to the tumour site sparing the healthy tissue. Nanomaterials provide an effective size of the particles between 10 and 100 nm which enables them to reach the cells more efficiently. Secondly, nanoparticles can easily adsorb onto the cell matrix and thus becomes easily integrated into the cell and thus increases the effectiveness of the drug. This is partly contributed by the high dissolution rate of the nanoparticles owing to its larger surface-to-volume ratio. Thirdly, they exhibit drug delivery to specific target site and thus overcomes the hurdle of poor solubility and drug uptake. Fourthly, they can prevent dose related adverse events effectively by controlled drug release. Various pharmaceutical techniques like nanosphere encapsulation has provided efficient methods to prolong exposure to drug by slow and controlled drug release which is the most desirable property for a chemotherapeutic agent. Thus, utilization of nanotechnology in pharmaceutical industry improves therapeutic index, solves problems in drug delivery, protects drug from degradation prior to their binding to the target site, enhances bioavailability, facilitates drug absorption by the tumours cells and prevents interaction with normal cells.

Passive Targeting

Nanoparticles can easily pervade to tissues from the blood stream due to their nanoscale size and surface properties. This property is utilized in cancer therapy as many tumours contain blood vessels with leaky walls which help the drug to concentrate in the tumour tissue and exert its cytotoxicity only to the tumour cells, sparing the healthy normal cells. This property is termed as enhanced permeability and retention (Jasim 2017; Greish 2010).

Active Targeting

Active targeting refers to the properties of a nanoparticle to target cancer cells selectively based on the chemical affinity to the molecules expressed on their surface. Molecules attached to a nanoparticle can target and interact with the receptor expressed over the cancer cells. Thus, the drug can be delivered into the cancerous cells letting free the normal cells (Kumar 2012).

Destruction from inside the Cell (Photothermal Targeting)

Photothermal targeting is utilizing the capability of nanomaterials to convert specific wavelengths of light to thermal energy which can be used for the purpose of treatment in conditions like cancers. Photothermal therapy (PTT) are highly efficacious yet with minimal side effects when compared to traditional therapies (O'Neal et al. 2004. They precisely exterminated the tumour tissue without damaging any of the normal tissues. PTT is attributed with the advantage of producing minimal trauma, less toxicity, precise targeting, wide applicability, repeatability in targeting, usefulness in palliative treatment, improvising disease curability in association with surgical procedures, elimination of occult cancers, preservation of aesthetic appeal of the patients and protection of organ functions. They utilize the near-infrared region of the electromagnetic spectrum (700–1000 nm), which has the highest penetrating power in biological tissue. Nanomaterials with photothermal properties are injected into the patient which then accumulates in the cancer cells due to its enhanced permeability and retention (EPR) effect (Yavuz et al. 2009). On reaching the target site, they absorb wavelengths in the near-infrared region and generate heat energy from them sufficient to kill the tumour cells. It is explained that the light energy is absorbed by the nanomaterial in the form of photons, a part of which is then transmitted as photons and part of it is converted into heat energy which raises the temperature of the nanomaterial. The local rise in temperature around the tumour cells causes dilatation of vessels around the tumour mass leading to accelerated blood flow into the tumour tissue dissipating the enhanced temperature into the cancerous cells. As tumour mass is attributed with poor blood flow rate (less than 5% of normal tissues), the hyperpyrexia induced by the photothermal agent will not be dissipated out of the tumour mass leading to a swift rise in temperature of the tumour of the order of 42 degrees, which is sufficient to damage and kill the tumour (Shah et al. 2008). The nanosized photothermal agent injected into the body are excreted through two major routes – through kidneys (when their diameter is around 5 nm) or through liver (when their diameter is around 20–25 nm). Based on the nature of the nanomaterial used, there are four broad categories of photothermal agents: (1) Noble metal photothermal agents: gold, silver, platinum etc. (2) Photothermal agents made of carbon materials: graphene oxide, carbon nanotubes etc. (3) Photothermal agents from transition metal dichalcogenides: copper sulphide, zinc sulphide, bismuth sulphide, tungsten sulphide, bismuth selenide etc. (4) Organic- and dye-based photothermal agents: prussian blue, indocyanine green, polypyrrole, polyaniline, dopamine, melanin, thiophene etc. Noble metals like gold, silver and platinum exhibit a phenomenon called surface plasmon resonance (SPR) effect due to their highly ordered electromagnetic field and a greater electromagnetic absorption owing to their electronic structure in the conductive band. (Zhang et al. 2018). Gold is considered as an ideal material for photothermal effect due to its strong surface plasmon resonance effect. Gold nanoparticles of varied shapes like nanorods, nanoshells, nanostars, nanocages and nanobranches are used as photothermal agents owing to its extended stability, biocompatibility and non-toxic nature (Cobley et al. 2010). Silver nanoparticles, being one of the most widely used and cheapest of all noble metal in nanoscale, is possessing a unique property of surface enhanced Raman scattering (SERS) effect which is considered as an abnormal surface optical phenomenon due to which it has an immense absorptive capability at near-infrared wavelengths. Silver also exhibits strong localized surface plasmon resonance quality and as the near-infrared wavelengths fall on them, causes surface free electron resonance leading to the formation of an exothermic electron gas which transfers its energy to the surrounding environment through the medium of the nanoparticle resulting in rise of temperature. Platinum is also sharing properties similar to gold and silver yet will lower toxicity and high biocompatibility. They are being used in cancer therapy to scavenge oxygen free radicals generated in cancer cells by enzyme mimetics. Carbon materials like Graphene has greater absorption coefficient in near-infrared wavelengths with additional electrical, mechanical, optical and thermal properties making it more advantageous in PTT. They have added advantages of having large functional groups containing oxygen, low toxicity and production cost and highly biocompatible. It is being considered as the most promising and reputed carbon nanomaterial of the twenty-first century. Nanotubes made of carbon also show tremendous photothermal properties and can be utilized for thermal destruction of tumour cells. They can also be used as a drug carrier and can load chemotherapeutic agents on their surface and helps in combining phototherapy with chemotherapy synergistically. Transition metal chalcogenides attracted tremendous research interest recently due to its excellent thermal effect. They are being used as layered structures like tungsten disulphide and molybdenum disulphide or bismuth selenide, bismuth sulphide, silver sulphide and copper sulphide exhibiting high extinction coefficient and enhanced absorption in the near-infrared region and visible light.

Simultaneous Delivery of Two Drugs

Multidrug delivery using nanoparticles is utilized in resistant cancers with the tendency to relapse frequently. In the most resistant form of cancers like the triple-negative breast cancer treatment, they have been successfully implemented. Simultaneous delivery of multiple interventions like small interfering RNA (siRNA) along with chemotherapeutic agent doxorubicin in the treatment of breast cancer has been carried out successfully (Jiang et al. 2014; Sebastian 2017). These researches promise a new ray of hope even to the most resistant type of cancers.

14.8.3 Nanotechnology in Diabetes Mellitus

With the rising prevalence of obesity and aberrant lifestyle habits, the diabetes disease burden is rising globally. The global prevalence of diabetes is expected to touch 600 million mark by the turn of 2035 (Forouhi and Wareham 2010). Diabetes mellitus (DM) is a metabolic disorder caused by insufficient or absent secretion of insulin, which is secreted by a group cells in the pancreas termed islets of Langerhans. DM is classified mainly into two – Type 1 and Type 2 DM. Type 1 DM also called insulin-dependent DM is caused by the extensive destruction of the beta cells of islets of Langerhans by an autoimmune response involving the T-cells. Type 2 DM on the other hand is caused due to insulin resistance or sensitivity associated with decreased production (Subramani et al. 2012).

14.8.3.1 Nanoparticles for Insulin Delivery

Some of the common drug delivery systems utilizing nanotechnology are listed below; (Yih and Al-Fandi 2006; Attivi et al. 2005)

- Polymeric biodegradable nanoparticles
- Ceramic nanoparticles
- Polymeric nanoparticles
- Dendrimer
- Liposomes

Polymeric nanoparticles are manufactured as colloidal solid particles (with dimensions ranging from 10 to 1000 nm) and based on the difference in their preparation they are of two types – Nanospheres and Nano-capsules. Nanospheres are polymeric nanoparticles having the drugs uniformly dispersed. Nano-capsules on the other hand have drugs dispersed in polymer membrane bound cavities that form vesicular systems (Tsapis et al. 2002; Si et al. 2003). These nanomaterials degrade on hydrolysis and delivers the medication to the targeted tissue.Polymeric nanoparticles are utilized for insulin transport utilizing a nonporous membrane bound polymer-insulin matrix containing glucose oxidase grafted on to their surface. An escalation in blood glucose level results in release of insulin after the biodegradation of the membrane of these nanoparticles (Morishita et al. 1992). Successful trials for oral insulin regimens with casein combined polymeric calcium phosphate-polyethylene glycol containing insulin have been undertaken (Sarmento et al. 2007). Casein prevents inactivation of insulin by gastric enzymes and the nanoparticles remain in small intestine for a longer duration favouring prolonged slower absorption and bioavailability due to its muco-adhesive property. Ceramic nanoparticles have high biocompatibility, low size (50 nm) and dimensional stability and are made up of materials such as calcium phosphate, alumina, titanium or silica (Rai et al. 2016). Porous hydroxyapatite nanoparticles have been used for delivering insulin into the intestines. Similarly, insulin-loaded poly nanoparticles are developed as nasally administered insulin. Bio Micro Electro Mechanical Systems (BioMEMS) are developed to implant them for controlled and prolonged release of insulin for long-term blood glucose control. Apart from these systems, gold nanoparticles along with chitosan as a reducing agent are being developed to carry insulin (Rai et al. 2016). Gold nanoparticles served a dual purpose in this case as a reducing agent as well as enhancing the penetration and uptake of insulin.

DM is also associated with lot of systemic complications like retinopathy, coronary artery disease (CAD), nephropathy, inflammatory gum diseases, delayed wound healing, inflammatory skin diseases and neuropathy. Polyacrylic acid, polylactide and chitosan have been used as carriers for ophthalmic drug delivery in DM.

Though nanomaterials find their application in almost all types of diseases, this section has been devoted to the most 'notorious' and most prevalent diseases of public health importance. Hence, we discussed the use of nanomaterials in cancers and neurodegenerative diseases. This does not in any way cut down the importance of research and use of nanomaterials in other diseases like cardiovascular, respiratory, reproductive and endocrine disorders. It is also worthwhile to discuss here the rising concern regarding the use of nanomaterials in medicine as diagnostics and therapeutics, that is, their role as an endocrine disruptor is gaining momentum in many spheres of health. Though they appear as a 'sole curse to a billion boons', yet researches in this area must go on and, in its run, should also solve this 'blotch on its face' permanently.

14.9 Future Perspective of Nanomedicine and Biology

The possibilities of nanotechnologies are immense and beyond every known boundary. As we have been discussing the positive aspects of nanotechnology so far, it is time to see some of vices that this technology brings to humans. When it comes to its application in the field of medicine and biology, the dose and toxicity are of major concern. Even with the first medically used nanoparticle, nanosilver, itself toxicities have been noticed in the form of permanent pigmentation over the skin at higher doses which has been termed 'Argyrosis' (McShan et al. 2014; Korani et al. 2011). Similarly, titanium dioxide nanoparticles in inhalation and ingestion have produced pathological lesions on lung, liver, kidneys, spleen and brain in animal models. The first and foremost concern in the future perspective is to bring down the toxicity. As we are facing the global transmission from SARS-CoV-2, the world is still lurking in the dark for a permanent cure in the form of a vaccine (Liu et al. 2020; Singh 2020). The percentage of susceptible population is very high and the infection can take a serious turn at any point in the form of cytokine storm, septic shock, metabolic acidosis, coagulation dysfunction, acute respiratory distress syndrome (ARDS) finally leading to death. An effective treatment is still lacking due to drug resistance and non-specific targeting. This highlights the need for a vaccine immediately to put a stop to this scourge. Nanotechnology is the only solution to design new strategies to design vaccines against this virus and the preparedness to oversee the emerging outbreaks yet to come in the future. Cancer treatment is yet another field requiring improvisation. We are still not in a position to get rid of the cytotoxic drugs out of our cancer armamentarium. Targeted therapies are to be refined to spare the healthy tissue from the toxicity of the chemotherapeutic agents. As we are moving back to the era of high mortality from infectious diseases, it is high time to move to the phytoremedies coupled with nanotechnologies in the cure of infectious diseases. In-depth researches are needed to ascertain effective drug delivery systems in this regard. Nanotechnology still stands the forerunner to advance all technologies to the next millennium.

14.10 Conclusion

Nanotechnology has been in existence from the time humans started to live in communities in this planet. Every aspect of life has been touched by nanoscience. Application of nanotechnology in biology and medicine revolutionized health care. Nanoscience gained entry in prevention, diagnosis and treatment of diseases. The spectrum of diseases ranging from the non-communicable diseases to communicable diseases equally shared the benefits of nanoscience. Though the world has been benefited from nanotechnology and its applications, the inherent toxicity is masquerading its benefits. The future is for the technologies that address the toxicities arising from the nanomaterials used for medical purposes.

References

Abdul Razzak R, Florence GJ, Gunn-Moore FJ (2019) Approaches to CNS drug delivery with a focus on transporter-mediated transcytosis. Int J Mol Sci 20(12):3108

Abee T, Rombouts FM, Hugenholtz J, Guihard G, Letellier L (1994) Mode of action of nisin Z against Listeria monocytogenes Scott A grown at high and low temperatures. Appl Environ Microbiol 60(6):1962–1968

Albani D, Gloria A, Giordano C, Rodilossi S, Russo T, D'Amora U, Tunesi M, Cigada A, Ambrosio L, Forloni G (2013) Hydrogel-based nanocomposites and mesenchymal stem cells: a promising synergistic strategy for neurodegenerative disorders therapy. Sci World J. https://doi.org/10.1155/2013/270260

Alekshun MN, Levy SB (2007) Molecular mechanisms of antibacterial multidrug resistance. Cell 128(6):1037–1050

Amaral L, Kristiansen JE (2000) Phenothiazines: an alternative to conventional therapy for the initial management of suspected multidrug resistant tuberculosis. A call for studies. Int J Antimicrob Agents 14(3):173–176

Andriashvili IA, Kvachadze LI, Bashakidze RP, Adamiia R, Chanishvili TG (1986) Molecular mechanism of phage DNA protection from the restriction endonucleases of Staphylococcus aureus cells. Mol Gen Mikrobiol Virusol (8):43

Attivi D, Wehrle P, Ubrich N, Damge C, Hoffman M, Maincent P (2005) Formulation of insulin-loaded polymeric nanoparticles using response surface methodology. Drug Dev Ind Pharm 31(2):179–189

Aziz N, Fatma T, Varma A, Prasad R (2014) Biogenic synthesis of silver nanoparticles using *Scenedesmus abundans* and evaluation of their antibacterial activity. Journal of Nanoparticles, Article ID 689419. http://dx.doi.org/10.1155/2014/689419

Aziz N, Faraz M, Pandey R, Sakir M, Fatma T, Varma A, Barman I, Prasad R (2015) Facile algae-derived route to biogenic silver nanoparticles: Synthesis, antibacterial and photocatalytic properties. Langmuir 31:11605–11612. https://doi.org/10.1021/acs.langmuir.5b03081

Aziz N, Pandey R, Barman I, Prasad R (2016) Leveraging the attributes of Mucor hiemalis-derived silver nanoparticles for a synergistic broad-spectrum antimicrobial platform. Front Microbiol 7:1984. https://doi.org/10.3389/fmicb.2016.01984

Aziz N, Faraz M, Sherwani MA, Fatma T, Prasad R (2019) Illuminating the anticancerous efficacy of a new fungal chassis for silver nanoparticle synthesis. Front Chem 7:65. https://doi.org/10.3389/fchem.2019.00065

Bal SM, Slütter B, van Riet E, Kruithof AC, Ding Z, Kersten GF, Jiskoot W, Bouwstra JA (2010) Efficient induction of immune responses through intradermal vaccination with N-trimethyl chitosan containing antigen formulations. J Control Release 142(3):374–383

Banerjee M, Mallick S, Paul A, Chattopadhyay A, Ghosh SS (2010) Heightened reactive oxygen species generation in the antimicrobial activity of a three component iodinated chitosan− silver nanoparticle composite. Langmuir 26(8):5901–5908

Barbara R, Belletti D, Pederzoli F, Masoni M, Keller J, Ballestrazzi A, Vandelli MA, Tosi G, Grabrucker AM (2017) Novel Curcumin loaded nanoparticles engineered for Blood-Brain barrier crossing and able to disrupt Abeta aggregates. Int J Pharm 526(1-2):413–424

Barber DJ, Freestone IC (1990) An investigation of the origin of the colour of the Lycurgus cup by analytical transmission electron microscopy. Archaeometry 32(1):33–45

Barrangou R, Fremaux C, Deveau H, Richards M, Boyaval P, Moineau S, Romero DA, Horvath P (2007) CRISPR provides acquired resistance against viruses in prokaryotes. Science 315(5819):1709–1712

Bayda S, Adeel M, Tuccinardi T, Cordani M, Rizzolio F (2020) The history of nanoscience and nanotechnology: from chemical–physical applications to nanomedicine. Molecules 25(1):112

Benton B, Breukink E, Visscher I, Debabov D, Lunde C, Janc J, Mammen M, Humphrey P (2007) Telavancin inhibits peptidoglycan biosynthesis through preferential targeting of transglycosylation: evidence for a multivalent interaction between telavancin and lipid II. Int J Antimicrob Agents 29:S51–S52

Bhuyan T, Mishra K, Khanuja M, Prasad R, Varma A (2015) Biosynthesis of zinc oxide nanoparticles from *Azadirachta indica* for antibacterial and photocatalytic applications. Mater Sci Semicond Process 32:55–61

Bisht G, Rayamajhi S (2016) ZnO nanoparticles: a promising anticancer agent. Nano 3(Godište 2016):3–9

Biswas A, Bayer IS, Biris AS, Wang T, Dervishi E, Faupel F (2012) Advances in top–down and bottom–up surface nanofabrication: techniques, applications & future prospects. Adv Colloid Interf Sci 170(1-2):2–7

Blecher K, Nasir A, Friedman A (2011) The growing role of nanotechnology in combating infectious disease. Virulence 2(5):395–401

Bozdogan B, Appelbaum PC (2004) Oxazolidinones: activity, mode of action, and mechanism of resistance. Int J Antimicrob Agents 23(2):113–119

Bracci L, Falciani C, Lelli B, Lozzi L, Runci Y, Pini A, De Montis MG, Tagliamonte A, Neri P (2003) Synthetic peptides in the form of dendrimers become resistant to protease activity. J Biol Chem 278(47):46590–46595

Brouns SJ, Jore MM, Lundgren M, Westra ER, Slijkhuis RJ, Snijders AP, Dickman MJ, Makarova KS, Koonin EV, Van Der Oost J (2008) Small CRISPR RNAs guide antiviral defense in prokaryotes. Science 321(5891):960–964

Brunner C, Seiderer J, Schlamp A, Bidlingmaier M, Eigler A, Haimerl W, Lehr HA, Krieg AM, Hartmann G, Endres S (2000) Enhanced dendritic cell maturation by TNF-α or cytidine-phosphate-guanosine DNA drives T cell activation in vitro and therapeutic anti-tumor immune responses in vivo. J Immunol 165(11):6278–6286

Burbulys D, Trach KA, Hoch JA (1991) Initiation of sporulation in B. subtilis is controlled by a multicomponent phosphorelay. Cell 64(3):545–552

Calzoni E, Cesaretti A, Polchi A, Di Michele A, Tancini B, Emiliani C (2019) Biocompatible polymer nanoparticles for drug delivery applications in cancer and neurodegenerative disorder therapies. J Funct Biomater 10(1):4

Cederlund H, Mårdh PA (1993) Antibacterial activities of non-antibiotic drugs. J Antimicrob Chemother 32(3):355–365

Chadwick S, Kriegel C, Amiji M (2010) Nanotechnology solutions for mucosal immunization. Adv Drug Deliv Rev 62(4-5):394–407

Cheever MA, Allison JP, Ferris AS et al. (2009). The prioritization of cancer antigens: a national cancer institute pilot project for the acceleration of translational research. Clinical cancer research 15(17):5323–5337.

Cioffi N, Torsi L, Ditaranto N, Tantillo G, Ghibelli L, Sabbatini L, Bleve-Zacheo T, D'Alessio M, Zambonin PG, Traversa E (2005) Copper nanoparticle/polymer composites with antifungal and bacteriostatic properties. Chem Mater 17(21):5255–5262

Cleveland J, Montville TJ, Nes IF, Chikindas ML (2001) Bacteriocins: safe, natural antimicrobials for food preservation. Int J Food Microbiol 71(1):1–20

Cobley CM, Au L, Chen J, Xia Y (2010) Targeting gold nanocages to cancer cells for photothermal destruction and drug delivery. Expert Opin Drug Deliv 7(5):577–587

Cotter PD, Hill C, Ross RP (2005) Bacteriocins: developing innate immunity for food. Nature Reviews Microbiology 3(10):777–788.

Cotter PD, Ross RP, Hill C (2013) Bacteriocins—a viable alternative to antibiotics? Nat Rev Microbiol 11(2):95–105

Czajkowski R, Jafra S (2009) Quenching of acyl-homoserine lactone-dependent quorum sensing by enzymatic disruption of signal molecules. Acta Biochim Pol 56(1)

Diwan M, Elamanchili P, Lane H, Gainer A, Samuel J (2003) Biodegradable nanoparticle mediated antigen delivery to human cord blood derived dendritic cells for induction of primary T cell responses. J Drug Target 11(8–10):495–507

Dobson AP, Carper ER (1996) Infectious diseases and human population history. Bioscience 46(2):115–126

Drexler EK (1981) Molecular engineering: an approach to the development of general capabilities for molecular manipulation. Proc Natl Acad Sci U S A 78:5275–5278

Drexler EK, Peterson C, Pergamit G (1991) Unbounding the future: the nanotechnology revolution. William Morrow and Company, Inc., New York, NY

Džidić S, Šušković J, Kos B (2008) Antibiotic resistance mechanisms in bacteria: biochemical and genetic aspects. Food Technol Biotechnol 46(1)

Elsner J (2013) The Lycurgus cup. New light on old glass: recent research on byzantine mosaics and glass. The British Museum, London, UK, pp 103–111

Englander L, Friedman A (2010) Nitric oxide nanoparticle technology: a novel antimicrobial agent in the context of current treatment of skin and soft tissue infection. J Clin Aesthet Dermatol 3(6):45

Fattal E, Rojas J, Youssef M, Couvreur P, Andremont A (1991) Liposome-entrapped ampicillin in the treatment of experimental murine listeriosis and salmonellosis. Antimicrob Agents Chemother 35(4):770–772

Feng QL, Wu J, Chen GQ, Cui FZ, Kim TN, Kim JO (2000) A mechanistic study of the antibacterial effect of silver ions on Escherichia coli and Staphylococcus aureus. J Biomed Mater Res 52(4):662–668

Florindo HF, Pandit S, Lacerda L, Gonçalves LM, Alpar HO, Almeida AJ (2009) The enhancement of the immune response against S. equi antigens through the intranasal administration of poly-ε-caprolactone-based nanoparticles. Biomaterials 30(5):879–891

Forouhi NG, Wareham NJ (2010) Epidemiology of diabetes. Medicine 38(11):602–606

Garneau S, Martin NI, Vederas JC (2002) Two-peptide bacteriocins produced by lactic acid bacteria. Biochimie 84(5–6):577–592

Greish K (2010) Enhanced permeability and retention (EPR) effect for anticancer nanomedicine drug targeting. In: Cancer Nanotechnol. Humana Press, pp 25–37

Grundmann H, Aires-de-Sousa M, Boyce J, Tiemersma E (2006) Emergence and resurgence of meticillin-resistant Staphylococcus aureus as a public-health threat. Lancet 368(9538):874–885

Guo C, Manjili MH, Subjeck JR, Sarkar D, Fisher PB, Wang XY (2013) Therapeutic cancer vaccines: past, present, and future. In: Advances in cancer research, vol 119. Academic Press, pp 421–475

Hale CR, Zhao P, Olson S, Duff MO, Graveley BR, Wells L, Terns RM, Terns MP (2009) RNA-guided RNA cleavage by a CRISPR RNA-Cas protein complex. Cell 139(5):945–956

Hauser AR (2015) Cell envelope. In: Antibiotic basic for clinicians, 2nd edn. Wolters Kluwer (India) Pvt. Ltd., New Delhi, pp 3–5

Heinrich MA, Martina B, Prakash J (2020) Nanomedicine strategies to target coronavirus. Nano Today:100961

Hearnden C, Lavelle EC (2013) Adjuvant strategies for vaccines: the use of adjuvants within the cancer vaccine setting. In: Prendergast GC, Jaffee EM (eds) Cancer immunotherapy: immune suppression and tumor growth, 2nd edn. Elsevier, New York, 655

Helgeby A, Robson NC, Donachie AM, Beackock-Sharp H, Lövgren K, Schön K, Mowat A, Lycke NY (2006) The combined CTA1-DD/ISCOM adjuvant vector promotes priming of mucosal and systemic immunity to incorporated antigens by specific targeting of B cells. J Immunol 176(6):3697–3706

Higgins PG, Fluit AC, Schmitz FJ (2003) Fluoroquinolones: structure and target sites. Curr Drug Targets 4(2):181–190

Hoang TT, Schweizer HP (1999) Characterization of Pseudomonas aeruginosa enoyl-acyl carrier protein reductase (FabI): a target for the antimicrobial triclosan and its role in acylated homoserine lactone synthesis. J Bacteriol 181(17):5489–5497

Huang MH, Lin SC, Hsiao CH, Chao HJ, Yang HR, Liao CC, Chuang PW, Wu HP, Huang CY, Leng CH, Liu SJ (2010) Emulsified nanoparticles containing inactivated influenza virus and CpG oligodeoxynucleotides critically influences the host immune responses in mice. PLoS One 5(8):e12279

Islan GA, Durán M, Cacicedo ML, Nakazato G, Kobayashi RK, Martinez DS, Castro GR, Durán N (2017) Nanopharmaceuticals as a solution to neglected diseases: is it possible? Acta Trop 170:16–42

Jasim A, Abdelghany S, Greish K (2017) Current update on the role of enhanced permeability and retention effect in cancer nanomedicine. In Nanotechnology-based approaches for targeting and delivery of drugs and genes. Academic Press, 62–109.

Jiang T, Mo R, Bellotti A, Zhou J, Gu Z (2014) Gel–liposome mediated co-delivery of anticancer membrane-associated proteins and small-molecule drugs for enhanced therapeutic efficacy. Adv Funct Mater 24(16):2295–2304

Jin T, Sun D, Su JY, Zhang H, Sue HJ (2009) Antimicrobial efficacy of zinc oxide quantum dots against Listeria monocytogenes, Salmonella enteritidis, and Escherichia coli O157: H7. J Food Sci 74(1):M46–M52

Johnston NJ, Mukhtar TA, Wright GD (2002) Streptogramin antibiotics: mode of action and resistance. Curr Drug Targets 3(4):335–344

Jones GR (1996) Successful cancer therapy with promethazine: the rationale. Med Hypotheses 46(1):25–29

Kahne D, Leimkuhler C, Lu W, Walsh C (2005) Glycopeptide and lipoglycopeptide antibiotics. Chem Rev 105(2):425–448

Kapoor G, Saigal S, Elongavan A (2017) Action and resistance mechanisms of antibiotics: a guide for clinicians. J Anaesthesiol Clin Pharmacol 33(3):300

Khan F, Khattak NS, Khan US, Rahman A (2011) Historical development of magnetite nanoparticles synthesis. J Chem Soc Pak 33(6):793

Kim BY, Rutka JT, Chan WC (2010) Nanomedicine. N Engl J Med 363(25):2434–2443

Kim SH, Kwak SY, Sohn BH, Park TH (2003) Design of TiO2 nanoparticle self-assembled aromatic polyamide thin-film-composite (TFC) membrane as an approach to solve biofouling problem. J Membr Sci 211(1):157–165

Klaenhammer TR (1993) Genetics of bacteriocins produced by lactic acid bacteria. FEMS Microbiol Rev 12(1–3):39–85

Korani M, Rezayat SM, Gilani K, Bidgoli SA, Adeli S (2011) Acute and subchronic dermal toxicity of nanosilver in guinea pig. Int J Nanomedicine 6:855

Kumar A, Schweizer HP (2005) Bacterial resistance to antibiotics: active efflux and reduced uptake. Adv Drug Deliv Rev 57(10):1486–1513

Kumar Khanna V (2012) Targeted delivery of nanomedicines. ISRN Pharmacol 2012

Kumar V, Bayda S, Hadla M, Caligiuri I, Russo Spena C, Palazzolo S, Kempter S, Corona G, Tooli G, Rizzolio F (2016a) Enhanced chemotherapeutic behavior of open-caged DNA@doxorubicin nanostructures for cancer cells. J Cell Physiol 231:106–110

Kumar V, Palazzolo S, Bayda S, Corona G, Tooli G, Rizzolio F (2016b) DNA nanotechnology for cancer therapy. Theranostics 6:710–725

Kutter E, De Vos D, Gvasalia G, Alavidze Z, Gogokhia L, Kuhl S, Abedon ST (2010) Phage therapy in clinical practice: treatment of human infections. Curr Pharm Biotechnol 11(1):69–86

Lambert PA (2005) Bacterial resistance to antibiotics: modified target sites. Adv Drug Deliv Rev 57(10):1471–1485

Leach KL, Swaney SM, Colca JR, McDonald WG, Blinn JR, Thomasco LM, Gadwood RC, Shinabarger D, Xiong L, Mankin AS (2007) The site of action of oxazolidinone antibiotics in living bacteria and in human mitochondria. Mol Cell 26(3):393–402

Lellouche J, Kahana E, Elias S, Gedanken A, Banin E (2009) Antibiofilm activity of nanosized magnesium fluoride. Biomaterials 30(30):5969–5978

Lin ST, Franklin MT, Klabunde KJ (1986) Nonaqueous colloidal gold. Clustering of metal atoms in organic media. 12. Langmuir 2(2):259–260

Lind K, Kristiansen JE (2000) Effect of some psychotropic drugs and a barbiturate on mycoplasmas. Int J Antimicrob Agents 14(3):235–238

Liu L, Liu Z, Chen H, Liu H, Gao Q, Cong F, Gao G, Chen Y (2020) Subunit nanovaccine with potent cellular and mucosal immunity for COVID-19. ACS Appl Biol Mater

Liu YJ, He LL, Mustapha A, Li H, Hu ZQ, Lin MS (2009) Antibacterial activities of zinc oxide nanoparticles against Escherichia coli O157: H7. J Appl Microbiol 107(4):1193–1201

Lorch A (1999) Bacteriophages: an alternative to antibiotics. Biotechnol Dev Monit 39:14–17

Logothetidis S (2006) Nanotechnology in medicine: The medicine of tomorrow and nanomedicine. Hippokratia 10:7–21

Luby Š, Lubyová M, Šiffalovič P, Jergel M, Majková E (2015) A brief history of nanoscience and foresight in nanotechnology. In: Nanomaterials and nanoarchitectures. Springer, Dordrecht, pp 63–86

Luo S, Ma C, Zhu MQ, Ju WN, Yang Y, Wang X (2020) Application of Iron oxide nanoparticles in the diagnosis and treatment of neurodegenerative diseases with emphasis on Alzheimer's disease. Front Cell Neurosci 14

Ma Y, Zhou T, Zhao C (2008) Preparation of chitosan–nylon-6 blended membranes containing silver ions as antibacterial materials. Carbohydr Res 343(2):230–237

MacDougall C (2017) Penicillins, cephalosporins, and other β-lactam antibiotics. In: Brunton LL, Hilal-Dandan R, Knollmann BC (eds) Goodman & Gilman's: The Pharmacological Basis of Therapeutics, 13th edn. McGraw-Hill Education, New York, 1023–1038.

Madou MJ (2011) Manufacturing techniques for microfabrication and nanotechnology. CRC press

Martinez LR, Han G, Chacko M, Mihu MR, Jacobson M, Gialanella P, Friedman AJ, Nosanchuk JD, Friedman JM (2009) Antimicrobial and healing efficacy of sustained release nitric oxide nanoparticles against Staphylococcus aureus skin infection. J Investig Dermatol 129(10):2463–2469

Martinez-Gutierrez F, Olive PL, Banuelos A, Orrantia E, Nino N, Sanchez EM, Ruiz F, Bach H, Av-Gay Y (2010) Synthesis, characterization, and evaluation of antimicrobial and cytotoxic effect of silver and titanium nanoparticles. Nanomedicine 6(5):681–688

Massich MD, Giljohann DA, Seferos DS, Ludlow LE, Horvath CM, Mirkin CA (2009) Regulating immune response using polyvalent nucleic acid– gold nanoparticle conjugates. Mol Pharm 6(6):1934–1940

Maurice F, Broutin I, Podglajen I, Benas P, Collatz E, Dardel F (2008) Enzyme structural plasticity and the emergence of broad-spectrum antibiotic resistance. EMBO Rep 9(4):344–349

McGinn RE (2012) What's Different, Ethically, About Nanotechnology? Foundational Questions and Answers. In A. S. Khan (Ed.), Nanotechnology: ethical and Social Implications. CRC Press, Boca Raton, Florida, pp. 67–89

McShan D, Ray PC, Yu H (2014) Molecular toxicity mechanism of nanosilver. J Food Drug Anal 22(1):116–127

Morishita M, Morishita I, Takayama K, Machida Y, Nagai T (1992) Novel oral microspheres of insulin with protease inhibitor protecting from enzymatic degradation. Int J Pharm 78(1–3):1–7

Morones JR, Elechiguerra JL, Camacho A, Holt K, Kouri JB, Ramírez JT, Yacaman MJ (2005) The bactericidal effect of silver nanoparticles. Nanotechnology 16(10):2346

Mukherjee S, Madamsetty VS, Bhattacharya D, Roy Chowdhury S, Paul MK, Mukherjee A (2020) Recent advancements of nanomedicine in neurodegenerative disorders theranostics. Adv Funct Mater 2003054

Muttil P, Prego C, Garcia-Contreras L, Pulliam B, Fallon JK, Wang C, Hickey AJ, Edwards D (2010) Immunization of guinea pigs with novel hepatitis B antigen as nanoparticle aggregate powders administered by the pulmonary route. AAPS J 12(3):330–337

Nandy SK, Bapat PM, Venkatesh KV (2007) Sporulating bacteria prefers predation to cannibalism in mixed cultures. FEBS Lett 581(1):151–156

Nigam A, Gupta D, Sharma A (2014) Treatment of infectious disease: beyond antibiotics. Microbiol Res 169(9-10):643–651

Nikaido H, Zgurskaya HI (1999) Antibiotic efflux mechanisms. Curr Opin Infect Dis 12(6):529–536

Nikaido H (1994) Prevention of drug access to bacterial targets: permeability barriers and active efflux. Science 264(5157):382–388

Niu X, Chen J, Gao J (2019) Nanocarriers as a powerful vehicle to overcome blood-brain barrier in treating neurodegenerative diseases: focus on recent advances. Asian J Pharm Sci 14(5):480–496

O'sullivan L, Ross RP, Hill C (2002) Potential of bacteriocin-producing lactic acid bacteria for improvements in food safety and quality. Biochimie 84(5–6):593–604

O'Neal DP, Hirsch LR, Halas NJ, Payne JD, West JL (2004) Photo-thermal tumor ablation in mice using near infrared-absorbing nanoparticles. Cancer Lett 209(2):171–176

Padovani S, Sada C, Mazzoldi P, Brunetti B, Borgia I, Sgamellotti A, Giulivi A, d'Acapito F, Battaglin G (2003) Copper in glazes of renaissance luster pottery: nanoparticles, ions, and local environment. J Appl Phys 93(12):10058–10063

Pal S, Tak YK, Song JM (2007) Does the antibacterial activity of silver nanoparticles depend on the shape of the nanoparticle? A study of the gram-negative bacterium Escherichia coli. Appl Environ Microbiol 73(6):1712–1720

Palazzolo S, Hadla M, Spena CR, Bayda S, Kumar V, Lo Re F, Adeel M, Caligiuri I, Romano F, Corona G et al (2019a) Proof-of-concept multistage biomimetic liposomal DNA origami nanosystem for the remote loading of doxorubicin. ACS Med Chem Lett 10:517–521

Palazzolo S, Hadla M, Russo Spena C, Caligiuri I, Rotondo R, Adeel M, Kumar V, Corona G, Canzonieri V, Tooli G et al (2019b) An effective multi-stage liposomal DNA origami nanosystem for in vivo cancer therapy. Cancers 11:1997

Pardridge WM (2003) Blood-brain barrier drug targeting: the future of brain drug development. Mol Interv 3(2):90

Perica K, Varela JC, Oelke M, Schneck J (2015) Adoptive T cell immunotherapy for cancer. Rambam Maimonides Med J 6(1):e0004

Pinto-Alphandary H, Andremont A, Couvreur P (2000) Targeted delivery of antibiotics using liposomes and nanoparticles: research and applications. Int J Antimicrob Agents 13(3):155–168

Plummer EM, Manchester M (2011) Viral nanoparticles and virus-like particles: platforms for contemporary vaccine design. Wiley Interdiscip Rev Nanomed Nanobiotechnol 2:174–196

Prasad R (2014) Synthesis of silver nanoparticles in photosynthetic plants. Journal of Nanoparticles, Article ID 963961, 2014, http://dx.doi.org/10.1155/2014/963961

Prasad R, Pandey R, Varma A, Barman I (2017) Polymer based nanoparticles for drug delivery systems and cancer therapeutics. In: Natural Polymers for Drug Delivery (eds. Kharkwal H and Janaswamy S), CAB International, UK 53–70

Prasad R, Swamy VS (2013) Antibacterial activity of silver nanoparticles synthesized by bark extract of *Syzygium cumini*. Journal of Nanoparticles 2013, http://dx.doi.org/10.1155/2013/431218

Qi L, Xu Z, Jiang X, Hu C, Zou X (2004) Preparation and antibacterial activity of chitosan nanoparticles. Carbohydr Res 339(16):2693–2700

Rai M, Yadav A, Gade A (2009) Silver nanoparticles as a new generation of antimicrobials. Biotechnol Adv 27(1):76–83

Rai VK, Mishra N, Agrawal AK, Jain S, Yadav NP (2016) Novel drug delivery system: an immense hope for diabetics. Drug Deliv 23(7):2371–2390

Rakotoarisoa M, Angelova A (2018) Amphiphilic nanocarrier systems for curcumin delivery in neurodegenerative disorders. Medicines 5(4):126

Ranade AV, Acharya R (2015) Arka and its pharmaceutical attributes in Indian alchemy (Rasashastra): a comprehensive review. Int J Ayurvedic Med 6(4):280–288

Ren G, Hu D, Cheng EW, Vargas-Reus MA, Reip P, Allaker RP (2009) Characterisation of copper oxide nanoparticles for antimicrobial applications. Int J Antimicrob Agents 33(6):587–590

Reynolds PE (1989) Structure, biochemistry and mechanism of action of glycopeptide antibiotics. Eur J Clin Microbiol Infect Dis 8(11):943–950

Richard F (1960) There's plenty of space at the bottom. Caltech Eng Sci 23(5):22–36

Rothemund PWK (2006) Folding DNA to create nanoscale shapes and patterns. Nature 440:297–302

Roco MC (2007). National nanotechnology initiative-past, present, future. In Goddard III WA, Brenner D, Lyshevski SE and Iafrate GJ (eds) Handbook on nanoscience, engineering and technology, 2nd ed, CRC Press, Boca Raton, Florida, pp. 3.1–3.26.

Rubinsztein DC (2006) The roles of intracellular protein-degradation pathways in neurodegeneration. Nature 443(7113):780–786

Ruparelia JP, Chatterjee AK, Duttagupta SP, Mukherji S (2008) Strain specificity in antimicrobial activity of silver and copper nanoparticles. Acta Biomater 4(3):707–716

Mobashery S, Azucena EF (1999) Bacterial antibiotic resistance. In: Encyclopedia of life sciences. Nature Publishing Group, London, UK

Sablon E, Contreras B, Vandamme E (2000) Antimicrobial peptides of lactic acid bacteria: mode of action, genetics and biosynthesis. In: New products and new areas of bioprocess engineering. Springer, Berlin, Heidelberg, pp 21–60

Saglam N, Korkusuz, F, Prasad R (2021) Nanotechnology Applications in Health and Environmental Sciences. Springer International Publishing (ISBN: 978-3-030-64410-9) https://www.springer.com/gp/book/9783030644093

Sanpui P, Murugadoss A, Prasad PD, Ghosh SS, Chattopadhyay A (2008) The antibacterial properties of a novel chitosan–Ag-nanoparticle composite. Int J Food Microbiol 124(2):142–146

Sarmento B, Ribeiro A, Veiga F, Sampaio P, Neufeld R, Ferreira D (2007) Alginate/chitosan nanoparticles are effective for oral insulin delivery. Pharm Res 24(12):2198–2206

Sciau P (2012) Nanoparticles in ancient materials: the metallic lustre decorations of medieval ceramics. INTECH Open Access Publisher

Scott AM, Allison JP, Wolchok JD (2012) Monoclonal antibodies in cancer therapy. Cancer Immun Arch 12(1)

Sebastian R (2017) Nanomedicine-the future of cancer treatment: a review. J Cancer Prev Curr Res 8(1):00–265

Šebestík J, Reiniš M, Ježek J (2012) Dendrimers in Neurodegenerative Diseases. In: Biomedical applications of peptide-, glyco-and glycopeptide dendrimers, and analogous dendrimeric structures. Springer, Vienna, pp 209–221

Seeman NC (1982) Nucleic acid junctions and lattices. J Theor Biol 99:237–247

Shah J, Park S, Aglyamov SR, Larson T, Ma L, Sokolov KV, Johnston KP, Milner TE, Emelianov SY (2008) Photoacoustic imaging and temperature measurement for photothermal cancer therapy. J Biomed Opt 13(3):034024

Sharma N, Sharma M, Sajid Jamal QM, Kamal MA, Akhtar S (2019) Nanoinformatics and biomolecular nanomodeling: a novel move en route for e_ective cancer treatment. Environ Sci Pollut Res Int:1–15

Si PZ, Zhang ZD, Geng DY, You CY, Zhao XG, Zhang WS (2003) Synthesis and characteristics of carbon-coated iron and nickel nanocapsules produced by arc discharge in ethanol vapor. Carbon 41(2):247–251

Singh B (2020) Biomimetic nanovaccines for COVID-19. Appl Sci Technol Ann 1(1):176–182

Sondi I, Salopek-Sondi B (2004) Silver nanoparticles as antimicrobial agent: a case study on E. coli as a model for Gram-negative bacteria. J Colloid Interface Sci 275(1):177–182

Song HY, Ko KK, Oh LH, Lee BT (2006) Fabrication of silver nanoparticles and their antimicrobial mechanisms. Eur Cells Mater 11(Suppl 1):58

Stevens KA, Sheldon BW, Klapes NA, Klaenhammer TR (1991) Nisin treatment for inactivation of Salmonella species and other gram-negative bacteria. Appl Environ Microbiol 57(12):3613–3615

Strateva T, Yordanov D (2009) Pseudomonas aeruginosa–a phenomenon of bacterial resistance. J Med Microbiol 58(9):1133–1148

Straus SK, Hancock RE (2006) Mode of action of the new antibiotic for Gram-positive pathogens daptomycin: comparison with cationic antimicrobial peptides and lipopeptides. Biochim Biophys Acta (BBA) 1758(9):1215–1223

Subramani K, Pathak S, Hosseinkhani H (2012) Recent trends in diabetes treatment using nanotechnology. Dig J Nanomater Bios 7(1)

Sulakvelidze A, Alavidze Z, Morris JG (2001) Bacteriophage therapy. Antimicrob Agents Chemother 45(3):649–659

Sun HX, Xie Y, Ye YP (2009) ISCOMs and ISCOMATRIX™. Vaccine 27(33):4388–4401

Swamy VS, Prasad R (2012) Green synthesis of silver nanoparticles from the leaf extract of *Santalum album* and its antimicrobial activity. J Optoelectronic and Biomedical Materials 4(3):53–59

Tam JP, Lu YA, Yang JL (2002) Antimicrobial dendrimeric peptides. Eur J Biochem 269(3):923–932

Taniguchi N (1974) On the basic concept of "nano-technology". In: Proceedings of international conference on production engineering. Part II, Japan Society of Precision Engineering, Tokyo

Tenover FC (2006) Mechanisms of antimicrobial resistance in bacteria. Am J Med 119(6):S3–10

Tolmasky ME (2000) Bacterial resistance to aminoglycosides and beta-lactams: the Tn1331 transposon paradigm. RNA 5(10):11

Tsapis N, Bennett D, Jackson B, Weitz DA, Edwards DA (2002) Trojan particles: large porous carriers of nanoparticles for drug delivery. Proc Natl Acad Sci 99(19):12001–12005

Vannuffel P, Cocito C (1996) Mechanism of action of streptogramins and macrolides. Drugs 51(1):20–30

Strohl WR (1997) Biotechnology of antibiotics. Marcel Dekker Inc., New York, NY

Webber MA, Piddock LJ (2003) The importance of efflux pumps in bacterial antibiotic resistance. J Antimicrob Chemother 51(1):9–11

Williams JD (1995) Selective toxicity and concordant pharmacodynamics of antibiotics and other drugs. J Antimicrob Chemother 35(6):721–737

Wise R (1999) A review of the mechanisms of action and resistance of antimicrobial agents. Can Respir J 6:20A-A

Yadav L, Tripathi RM, Prasad R, Pudake RN, Mittal J (2017) Antibacterial activity of Cu nanoparticles against *E. coli, Staphylococcus aureus* and *Pseudomonas aeruginosa*. Nano Biomed Eng 9(1):9–14. https://doi.org/10.5101/nbe.v9i1.p9-14

Yavuz MS, Cheng Y, Chen J, Cobley CM, Zhang Q, Rycenga M, Xie J, Kim C, Song KH, Schwartz AG, Wang LV (2009) Gold nanocages covered by smart polymers for controlled release with near-infrared light. Nat Mater 8(12):935–939

Yih TC, Al-Fandi M (2006) Engineered nanoparticles as precise drug delivery systems. J Cell Biochem 97(6):1184–1190

Yoneyama H, Katsumata R (2006) Antibiotic resistance in bacteria and its future for novel antibiotic development. Biosci Biotechnol Biochem 70(5):1060–1075

Zhang E, Xu H (2017) A new insight in chimeric antigen receptor-engineered T cells for cancer immunotherapy. J Hematol Oncol 10(1):1–1

Zhang H, Chen G, Yu B, Cong H (2018) Emerging advanced nanomaterials for cancer photothermal therapy. Rev Adv Mater Sci 53(2):131–146

Chapter 15
Perspectives of Nanotechnology in Aquaculture: Fish Nutrition, Disease, and Water Treatment

Ndakalimwe Naftal Gabriel, Habte-Michael Habte-Tsion, and Mayday Haulofu

Contents

15.1	Introduction...	464
15.2	Nanotechnology Application in Fish Nutrition...	464
	15.2.1 Nanoparticles' Role in Fish Nutrition...	466
	15.2.2 Nanotechnology Application in the Aquafeed Industry...	471
15.3	Nanotechnology Application in Aquaculture Disease Control...	472
	15.3.1 Nanoparticles as Antibacterial Agents in Aquaculture...	473
	15.3.2 Nanoparticles as Vaccine/Drug Delivery Vector...	476
15.4	Nanotechnology Application for Water Quality Management in Aquaculture...	477
	15.4.1 Nanocatalysts and Nanoadsorbents in Aquaculture...	478
15.5	Conclusion and Future Perspectives...	478
References...		479

N. N. Gabriel (✉)
Department of Fisheries and Aquatic Sciences, Sam Nujoma Campus, University of Namibia, Henties Bay, Namibia
e-mail: ngabriel@unam.na

H.-M. Habte-Tsion
Cooperative Extension-Aquaculture Research Institute, University of Maine, Orono, Maine, United States

M. Haulofu
Sam Nujoma Coastal and Marine Resources Research Center, University of Namibia, Henties Bay, Namibia

© The Author(s), under exclusive license to Springer Nature Switzerland AG 2022
A. Krishnan et al. (eds.), *Emerging Nanomaterials for Advanced Technologies*, Nanotechnology in the Life Sciences, https://doi.org/10.1007/978-3-030-80371-1_15

15.1 Introduction

Globally, food fish demand has been on the rise for the past seven decades (annual consumption rate at 3.1%), at a rate nearly double that of the annual global human population growth (1.6%) (FAO 2020). The sad reality is that the fishing sector (capture fisheries) which has been the main supplier of food fish over the years is alone unable to meet the current and future global food fish demand. To meet the global food fish demand, aquaculture, which is one of the fastest growing food-producing sectors, is believed to be a good opportunity to complement capture fisheries. The stress on aquaculture to close the supply and demand gap of food fish has led to a shift from extensive to intensive methods such as the recirculatory aquaculture systems (RAS). In the intensive systems, fish are stocked at high density, and this has been shown to cause stress in farmed fish, thereby affecting fish performance and welfare (Sneddon et al. 2016; Hoseini et al. 2019). In addition, these systems could also be accumulation grounds for pollutants either from water sources or fish feeds (Wang and Wang 2012; Boonanuntanasarn et al. 2014) and diseases (Romero et al. 2012; Culot et al. 2019). Thus, since the emerging of intensive farming systems, the sustainability of aquaculture has been predominantly criticized.

The benefits associated with aquaculture such as the provision of accessible food, income generation, and community empowerment could have led to the radical search for effective strategies to mitigate negative impacts, instead of discouraging the practice. One of the emerging technologies is nanotechnology, which is defined as the "science and engineering concerned with the design, synthesis, characterization, and application of materials that possess a functional organization on the nano-metric scale (10^{-9} m) (Silva 2010)". Nanoparticles are characterized by higher reactivity and can change the pharmacological properties of active principles (Jiang et al. 2019). This technology is widely researched in aquaculture for various purposes such as vaccine delivery (Rajeshkumar et al. 2009), gene transfer (Murata et al. 1998), drug delivery (Lavertu et al. 2006; Wei et al. 2007), delivery of nutrients (Ashouri et al. 2015), nutraceuticals (Aklakur et al. 2016), and water filtration and remediation (Khosravi-Katuli et al. 2017). Therefore, this chapter reviewed the application of nanotechnology in aquaculture with specific focus on fish nutrition, diseases, and water quality management (Fig. 15.1), presenting trends and perspectives.

15.2 Nanotechnology Application in Fish Nutrition

Traditionally, feeding fish has relied on providing fish with food in the form of a pellet/ bycatch/ fish-offal. The pellet is chiefly formulated based on the daily nutritional fish requirements for major components such as proteins, carbohydrates, fats, minerals, and vitamins. Recently, nutritionists utilized nanotechnology to create various delivery systems such as encapsulation, protection, and controlled release of

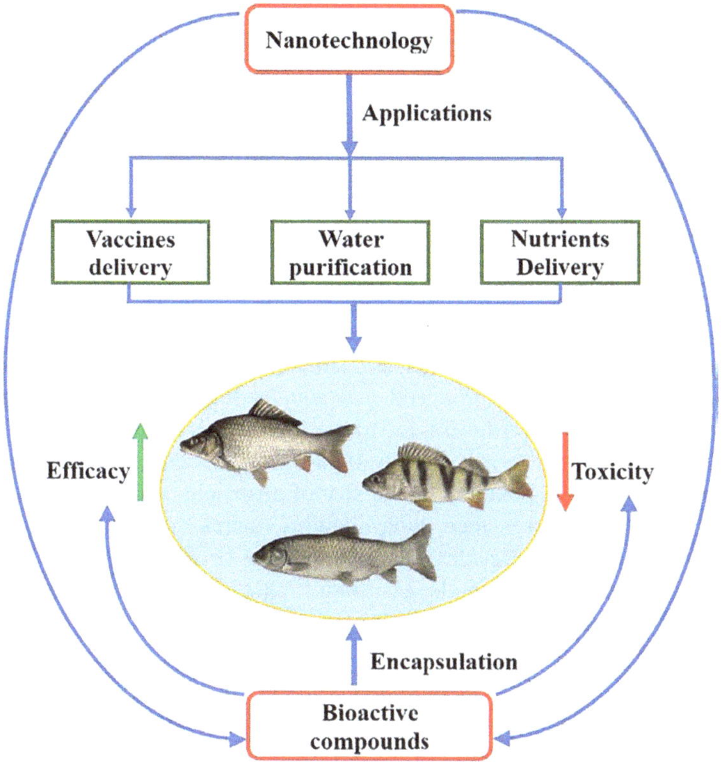

Fig. 15.1 Schematic representation of nanotechnology applications in aquaculture. (Adapted from Shah and Mraz (2020))

micronutrients. Hence, nanotechnology has an important potential to boost nutritional assessment and measures of bioavailability. For instance, ultrasensitive detection of nutrients and metabolites increases the understanding of nutrient and biomolecular interactions in specific tissues.

In nutritional research, gastrointestinal tract has always been the preferred and most important route of feed/food delivery principles including for nanoparticles. Nanoparticles can route to the gastrointestinal tract in many ways such as (1) ingestion or swallow pathway: ingestion directly from food and water and from therapeutic nano-drugs administration; (2) inhalation pathway: inhaled nanoparticles can be swallowed and enter to the gastrointestinal tract following clearance from the respiratory tract; and (3) oral pathway: oral or smart delivery into gastrointestinal tract, in which particle uptake in the gastrointestinal tract depends on diffusion and accessibility through mucus and contact with the cells of the gastrointestinal tract (Hoet et al. 2004). The smaller the particle diameter the faster is the diffusion through gastrointestinal tract mucus to reach the cells of intestinal lining, followed by uptake through gastrointestinal tract barrier to reach the blood (Hoet et al. 2004).

In fish, one important idea is that nanoparticles will enhance aquafeeds by increasing the proportion of fish feed nutrients that pass across the gut tissue and into the fish, rather than passing directly through the fish digestive system unused (Handy 2012). Specifically, the delivery systems of nanoparticles are aimed to improve the bioavailability, bioaccessibility and hence efficacy of the nutrients by improving their solubility and protection of fish gut. The delivery systems specifically consist of micronutrients trapped within nanoparticles that may be fabricated from surfactants, lipids, proteins, and/or carbohydrates (Joye et al. 2014). The small particle size in these systems has several advantages over the conventional delivery systems including improved bioavailability, higher stability to aggregation and gravitational separation, and higher optical clarity (Joye et al. 2014). For instance, immuno-modulatory ingredients such as phenolic compounds, vitamins, and minerals are being increasingly introduced into aquafeeds to improve fish health and growth performance. Nevertheless, incorporating these nutraceuticals into feeds is often challenging due to their low bioavailability, which can be solved by encapsulating the bioactive components. As the size of a particle containing encapsulated bioactive agents decreases their bioavailability increases, enabling their faster digestion and absorption. Besides, nanoparticles can be formulated to survive passage through specific regions of the gastrointestinal tract and then release their payload at a specified point, thus maximizing their potential immune-nutritional benefits (Jafari and McClements 2017).

15.2.1 Nanoparticles' Role in Fish Nutrition

In fish nutrition, nanoparticles are playing an important role in improving growth performance and immuno-biochemical (health) status of fish. They are usually incorporated in little amount, however, at a higher cost. Therefore, intensive care should be taken in their usage to maximize their utilization and avoid wastage (Friends of the Earth 2008). Consequently, many studies have been reported to address the functions and levels of nanoparticles and various methods have been adopted in aquatic animals (Table 15.1):

Sahu et al. (2008) conducted a 60 days experiment to investigate the effect of dietary *Curcuma longa* nanoparticles (0.1, 0.5, 1.0 and 5.0 g kg^{-1} of orally supplemented feed) on enzymatic and immunological profiles of rohu, *Labeo rohita* (Ham.), infected with *Aeromonas hydrophila*. Dietary *C. longa* nanoparticle significantly enhanced lysozyme activity, superoxide anion production, and serum bactericidal activity; and promoted protection against *Aeromonas hydrophila* (Sahu et al. 2008). In vitro and in vivo studies on the effects of dietary curcumin nanoparticles (0.5 & 1%, orally supplemented feed) reported that dietary curcumin nanoparticles (1) significantly enhanced growth, survival rates, and disease resistance; (2) decreased lipid peroxidation product; (3) promoted antioxidant status and protein content; (4) improved liver proactive effects; (5) increased haemoglobin content, RBC count and haematocrit; and (6) enhanced overall growth performance and

Table 15.1 Application of nanotechnology/ nanoparticles role in fish nutrition

Nanoparticles	Function	Fish Species	References
Aloe vera	A diet supplemented with 1% *Aloe vera* nanoparticles significantly promoted the growth parameters of fish in contrast to a control group.	Siberian sturgeon (*Acipenser baerii*)	Sharif Rohani et al. (2017)
Ginger	Fish fed with 1 and 0.5 g ginger nanoparticles per kg feed showed 100% relative percentage survival, whereas fish fed with 0.5 g ginger per kg feed showed 20% mortality rate and 71% relative percentage survival. These findings confirmed that ginger nanoparticles as a successful formulation in the prevention of motile *Aeromonas septicaemia* in common carp fingerlings compared to ginger.	Common carp (*Cyprinus carpio*)	Korni and Khalil (2017)
Azolla microphylla	Significantly ameliorated the levels of metabolic enzymes, hepatotoxic markers, oxidative stress markers, altered tissue enzymes, reduced hepatic ions, abnormal liver histology, etc. Based on those results, it was suggested that *Azolla microphylla* phytochemically synthesized gold nanoparticles as an effective protector against acetaminophen-induced hepatic damage in fresh water common carp.	Common carp fish (*Cyprinus carpio* L.)	Kunjiappan et al. (2015)
Azadirachta indica (neem)	Significantly elevated functional activity of immunological parameters in fish treated with these nanoparticles. It was concluded that they have a potential immunomodulatory and antibacterial activity.	Mrigala carp (*Cirrhinus mrigala*)	Rather et al. (2017)
Curcuma longa	Significantly enhanced lysozyme activity, superoxide anion production, and serum bactericidal activity; and improved protection against *Aeromonas hydrophila*.	Rohu (*Labeo Rohita*)	Sahu et al. (2008)

(continued)

Table 15.1 (continued)

Nanoparticles	Function	Fish Species	References
Curcumin	Enhanced growth, survival rates, and disease resistance of *A. testudineus* (Bloch). Promoted antioxidant status and protein content of the fish. Decreased lipid peroxidation product. Increased haemoglobin content, RBC count, and haematocrit in the fish. Improved over all health status of the fish.	*Anabas testudineus* (Bloch)	Manju et al. (2009, 2012, 2013)
Curcumin	Remarkably minimized CCl4-induced liver damage by upregulating hepatocyte antioxidative capacity and inhibiting NF-kB, IL-1b, TNF-α, and IL-12 expression in Jian carp.	Jian carp (*Cyprinus carpio* var. Jian)	Cao et al. (2015)
Curcumin	Notably improved growth performance, feed utilization, oxidative status, immune responses, and disease resistance of fish. Promoted non-specific immune defense mechanisms against *Vibrio alginolyticus*. Enhanced hepatic lesions in aflatoxin B infected fish.	Nile tilapia (*Oreochromis niloticus*)	Elgendy et al. (2016); Mahmoud et al. (2017); Manal (2018)
Curcumin	Enhanced growth performance and increased disease resistance against *Edwardsiella tarda* infection.	Mrigala carp (*Cirrhinus mrigala*)	Leya et al. (2017)
Curcumin	Promoted performance of catfish and increased their disease resistance, reducing use of antimicrobials in fish farming.	Channel catfish (*Ictalurus punctatus*)	Hafiz et al. (2017)
Curcumin	Enhanced the activities of digestive enzymes. Modulated the expression of GH in brain and growth factors such as IGF-1 and IGF-2 in muscle of *O. mossambicus*.	Mozambique tilapia (*Oreochromis mossambicus*)	Midhun et al. (2016)
Selenium (Se)	Nano-Selenium (Se, 1 mg kg^{-1} diet) showed significant improvement in the growth and antioxidant defense system of common carp in contrast to a control group.	Common carp (*Cyprinus carpio*)	Ashouri et al. (2015)
Selenium (Se), zinc (Zn) and manganese (Mn)	Dietary nanoparticles such as nano-selenium (Se), zinc (Zn), and manganese (Mn) in early weaning diets enhanced stress resistance and bone mineralization of gilthead seabream.	Gilthead seabream (*Sparus aurata*)	Izquierdo et al. (2017)

(continued)

Table 15.1 (continued)

Nanoparticles	Function	Fish Species	References
Iron (Fe)	A diet supplemented with iron (Fe) nanoparticles and *Lactobacillus casei* as a probiotic significantly promoted growth performance of rainbow trout.	Rainbow trout (*Oncorhynchus mykiss*)	Mohammadi and Tukmechi (2015)
Manganese (Mn)	Dietary MnO nanoparticles (16 mg kg^{-1} diet) significantly elevated the growth performance and antioxidant defense system of freshwater prawn.	Freshwater prawn (*Macrobrachium rosenbergii*)	Asaikkutti et al. (2016)
Copper (Cu)	Supplementation of dietary copper (Cu) nanoparticle (20 mg kg^{-1} diet) significantly improved the growth, biochemical status, digestive and metabolic enzyme activities, antioxidant, and non-specific immune response of aquatic animals.	Freshwater prawn (*M. rosenbergii*) and Red sea bream (*Pagrus major*)	Muralisankar et al. (2016); El Basuini et al. (2017)

health status of *A. testudineus* (Bloch) (Manju et al. 2009, 2012, 2013). Cao et al. (2015) studied the effects of curcumin nanoparticles (0.1%, 0.5%, or 1.0% of orally supplemented feed) on antioxidative activities and cytokine production in Jian carp (*Cyprinus carpio* var. Jian) with CCl$_4$-induced liver damage. Dietary curcumin nanoparticles significantly reduced CCl$_4$-induced liver damage in Jian carp by upregulating hepatocyte antioxidative capacity and inhibiting NF-kB, IL-1b, TNF-a, and IL-12 expression (Cao et al. 2015). Supplementation of curcumin nanoparticles (0.5, 1, or 2% of diet) significantly improved non-specific immune defense mechanisms of fish against *Vibrio alginolyticus*; promoted hepatic lesions in aflatoxin B infected fish; improved hepatosomatic index (HIS) values; and enhanced growth performance, feed utilization, oxidative status, immune responses, and disease resistance of tilapia, *Oreochromis niloticus* (Elgendy et al. 2016; Mahmoud et al. 2017; Manal 2018). Leya et al. (2017) evaluated the effects of curcumin nanoparticles supplemented diet (0.25, 0.5, 1, 1.5 and 2% of orally supplemented diet) on growth and non-specific immune parameters of mrigala carp (*Cirrhinus mrigala*) against *Edwardsiella tarda* infection. Dietary curcumin nanoparticles improved growth performance and increased disease resistance against *Edwardsiella tarda* infection in *C. mrigala* (Leya et al. 2017). Curcumin nanoparticle supplementation (0.5 & 1% of orally supplemented diet) significantly enhanced performance and increased disease resistance of catfish, *Ictalurus punctatus* (Hafiz et al. 2017). Midhun et al. (2016) evaluated modulation of digestive enzymes, GH, IGF-1, and IGF-2 genes in the teleost, Tilapia (*Oreochromis mossambicus*) by dietary curcumin nanoparticles (0.5 & 1%). Dietary curcumin nanoparticles significantly improved the activities of digestive enzymes and modulated the expression of GH in brain and growth factors such as IGF-1 and IGF-2 in the muscles of tilapia, *O. mossambicus* (Midhun et al. 2016).

Studies in other nanoparticles, Sharif Rohani et al. (2017), evaluated the effects of three different levels (0.5, 1.0, and 1.5% of the diet) of *Aloe vera* nanoparticles on the growth performance, survival rate, and body composition of Siberian sturgeon (*Acipenser baerii*). This study reported that a diet supplemented with 1% *Aloe vera* nanoparticles significantly promoted the growth factors of fish in contrast to a control group but did not find significant difference in body composition of fish (Sharif Rohani et al. 2017). Korni and Khalil (2017) studied the effect of ginger and its nanoparticles on growth performance, cognition capability, immunity, and prevention of motile *Aeromonas septicaemia* in common carp (*Cyprinus carpio*) fingerlings. Fish fed with ginger nanoparticles (1 and 0.5 g kg^{-1} of diet) showed better growth performance, and significantly increased total protein, globulin, and lysozyme of fish; and showed 100% relative percentage survival (RPS) compared to control group (Korni and Khalil 2017). Kunjiappan et al. (2015) investigated the hepatoprotective and antioxidant effects of *Azolla microphylla*-based gold nanoparticles against acetaminophen-induced toxicity in a fresh water common carp fish (*Cyprinus carpio* L.) Their results showed that gold nanoparticles significantly ameliorated the levels of metabolic enzymes, hepatotoxic markers, oxidative stress markers, altered tissue enzymes, reduced hepatic ions, abnormal liver histology etc. Based on those results, it was suggested that *A. microphylla* phytochemically synthesized gold nanoparticles as an effective protector against acetaminophen-induced hepatic damage in fresh water common carp (Kunjiappan et al. 2015). Rather et al. (2017) evaluated the immunomodulatory potential of green synthesis of silver nanoparticles (G-AgNPs) using *Azadirachta indica* (neem) in *Cirrhinus mrigala* fingerlings challenged with *Aeromonas hydrophila*. This study reported that dietary G-AgNPs significantly increased the functional activity of immunological parameters (nitro-blue tetrazolium assay, myeloperoxidase activity, phagocytic activity, anti-protease, and lysozyme activity), enhanced disease resistance and improved survival rate; and it was concluded that biosynthesized silver nanoparticles have immunomodulatory and antibacterial activity (Rather et al. 2017). Formulation of solid lipid nanoparticles-encapsulated 6-coumarin-loaded pectin microparticles showed improved uptake of the compound by two gilthead seabream (*Sparus aurata* L.) cell types compared to a competitor 6-coumarin-loaded pectin microparticles, which makes solid lipid nanoparticles as suitable nanocarriers for the delivery of biologically active substances in fish (Trapani et al. 2015).

Furthermore, dietary nano-minerals or dietary minerals at the nanoscale size may pass into cells more readily than their larger counterparts, and this accelerates their assimilation process into the fish. For example, dietary selenium (Se, 1 mg kg^{-1} of diet) nanoparticles significantly promoted growth and antioxidant defense system of common carp (*Cyprinus carpio*) in contrast to a control group (Ashouri et al. 2015). In rainbow trout, a dietary iron (Fe) nanoparticles and *Lactobacillus casei* as a probiotic significantly improved growth performance and feed utilization, such as weight gain, specific growth rate, daily growth rate, condition factor, and food conversion rate (Mohammadi and Tukmechi 2015). Nanoparticles such as nanoselenium (Se), zinc (Zn) and manganese (Mn) in early weaning diets for gilthead seabream (*Sparus aurata;* Linnaeus, 1758) enhanced stress resistance and bone mineralization (Izquierdo et al., 2017). Dietary copper (Cu) nanoparticle (20 mg kg^{-1} of diet) significantly improved the growth, biochemical status, digestive and metabolic enzyme activities, antioxidant, and non-specific immune response of red sea bream, *Pagrus major* (El Basuini et al. 2017) and freshwater prawn, *M. rosenbergii* (Muralisankar et al. 2016). Supplementation of manganese oxide (MnO) nanoparticles (16 mg kg^{-1} diet) significantly elevated the growth performance and antioxidant defense system of freshwater prawn (*Macrobrachium rosenbergii*) (Asaikkutti et al. 2016).

15.2.2 Nanotechnology Application in the Aquafeed Industry

There are numerous potential applications of nanotechnology in feed industry, including: (i) minor modifications of natural ingredients to enhance taste, palatability and sensory improvement such as flavor, color, and texture; (ii) enhancing nutrition quality of foods by stabilizing active ingredients such as nutraceuticals in feed matrices, packaging, and product innovation to extend shelf-life, (iii) increasing bioavailability of essential nutrients (Food Safety Authority of Ireland 2008). Nano-delivery of bioactive/nutrient in feedstuffs or in vivo in fish is enabled through improved knowledge of feed materials at the nanoscale. The different nanomaterials that have the potential to be used for this purpose are nanocomposites, nanoclays, and nanotubes. The nanoproducts that would find applications are nanosensors, nanoimaging, and nanochips and nanofilters. Similarly, the potential nano-delivery systems are nanocapsules, nanocochleates, nanoballs, nanodevices, nanomachines, and nanorobots (Thulasi et al. 2013).

In aquafeed, nanotechnology may also play significant roles in the delivery of micronutrients to aquatic animals. For instance, nanomaterials can be used to coat nutrients that could normally degrade, such as fatty acids, or have limited assimilation efficiency across the gut of fishes, because they are poorly soluble (i.e. fat-soluble vitamins) (Handy 2012). Nanoencapsulation technology has been suggested for vitamins, minerals, carotenoids, and fatty acids, with increasing bioavailability being the main goal (Acosta 2009; Bouwmeester et al. 2009).

Several vitamins and their precursors, such as carotenoids, are insoluble in water. Nevertheless, nanotechnology helps to address these problems. Specifically, when prepared as nanoparticles, these vitamins and their precursors can easily be homogenized with cold water, which enables to increase their bioavailability. For example, Vitamin B$_{12}$ absorption from the gut under physiological conditions occurs via receptor-mediated endocytosis; and the ability to increase oral bioavailability of various peptides (granulocyte colony stimulating factor, erythropoietin) and particles by covalent coupling to vitamin B$_{12}$ has been reported by Russell Jones (2001) and Russell Jones et al. (1999). Vitamin E is a term describing all tocopherol and tocotrienol derivatives, which exhibit the biological activity of alpha tocopherol. Its structure is sensitive to light, heat, and oxygen; consequently, synthetic versions of vitamin E are less expensive, but have lower biological activity (Thulasi et al. 2013).

Nano-micelles made from casein can be used as a vehicle for hydrophobic ingredients such as vitamin D_2 (Semo et al. 2007).

Nanoscale mineral supplements might provide a source of trace metals, without the extensive faecal losses normally associated with mineral salts (e.g. Fe salts; Carriquiriborde et al. 2004). Nanomaterials may also offer an alternative to organic forms of food supplements, where antinutritional factors (incidental pesticides, toxic metals, etc.) in the ingredient can sometimes be a problem (Berntssen et al. 2010).

Nanomaterials can be used to change the physical properties of aquafeed in addition to enhancing the bioavailability and stability of aquafeed. For example, feed wastage and pollution in aquaculture due to poor feed quality (stability, texture or inappropriate buoyancy of the pellet) is a continuing problem (Handy and Poxton 1993); and small supplementations of nanomaterials can significantly alter the physical properties of these pellets. Specifically, the additions of single-walled carbon nanotubes to trout feed can result in a hard pellet that does not fragment easily in water (Handy 2012). Rainbow trout readily eat feed containing nanomaterials up to 100 mg kg^{-1} TiO$_2$ nanoparticles (Ramsden et al. 2009) and/or 500 mg kg^{-1} C60 and 500 mg kg^{-1} single-walled carbon nanotubes (Fraser et al., 2010) without loss of appetite or growth rate. Therefore, adding a few milligrams of nanomaterial/nanoparticles to aquafeed modify the physical properties of pellets, which could play important roles in the development of aquafeed industry, ultimately sustainable growth of aquaculture industry.

15.3 Nanotechnology Application in Aquaculture Disease Control

Aquaculture sector (especially intensive and super-intensive commercial farms) has grieved major economic losses because of disease outbreaks caused by several pathogenic agents (i.e. bacteria, viruses, and parasites) (Huang et al. 2015; Shinn et al. 2015; Tandel et al. 2017). Traditionally, these pathogens could be treated with chemical disinfectants and antibiotics either through feed, immersion, or injection. However, the use of these chemicals in aquaculture has been criticised, because, they are no longer effective i.e. several pathogenic bacteria i.e. *Aeromonas hydrophila, A. salmonicida, Yersinia ruckeri, Vibrio, Listeria, Pseudomonas,* and *Edwardsiella* species have been reported to be insensitive against most common antibiotics used in aquaculture (Sørum 2008; Swain et al. 2014). In addition, the excessive use of these chemicals in aquaculture could be toxic to other organisms including humans and the environment (Shah and Mraz 2019; Malheiros et al. 2020). This could have paved ways to the search for better alternative technology to control bacteria, viruses, and parasites in aquaculture. Today, nanotechnology has become the new alternative with potential to be used as antimicrobial agents, vaccines, and diagnosis tools for disease causing agents in fish farming (Shaalan et al. 2016).

15.3.1 Nanoparticles as Antibacterial Agents in Aquaculture

Different metal nanoparticles (biologically or chemically synthesized) have been recommended as alternative antibacterial agents, with potential to eradicate or reduce the use of traditional antibiotics in aquaculture (Gunalan et al. 2012; Shaalan et al. 2016) (Table 15.2). Biologically synthesized metal nanoparticles (derivatives of plants, bacteria and fungi) are more advocated over chemically synthesized ones because of their high antimicrobial activity, environmental friendliness, simplicity and affordability (Kalishwaralal et al. 2008; Gunalan et al. 2012; Prasad 2014; Prasad et al. 2016, 2018; Srivastava et al. 2021; Sarma et al. 2021). Some of the antibacterial metal nanoparticles studied in aquaculture include zinc nanoparticles (ZnNPs), silver nanoparticles (AgNPs), copper oxide (CuONPs), gold nanoparticles (AuNPs), and titanium dioxide (TiO_2NPs) (Swain et al. 2014), with AgNPs, ZnNPs, and AuNPs being the widely studied nanoparticles. These nanoparticles could be used either alone or in combination with each other (Venegas et al. 2018). This section reviews the commonly reported metal nanoparticles as antibacterial agents in aquaculture (Table 15.2).

Zinc Nanoparticles (ZnNPs) These nanoparticles are gaining popularity due to their multifunctional properties; antibacterial and antifungal properties (Wang et al. 2008; Di Cesare et al. 2012). Zinc-oxide (chemically synthesized) reportedly showed broad spectrum antibacterial activity against *Aeromonas hydrophila*, *Edwardsiella tarda*, *Flavobacterium branchiophilum*, *Vibrio* sp., *Staphylococcus aureus*, *Bacillus cereus*, and *Citrobacter* sp. (Swain et al. 2014), which are some of the important pathogenic bacteria in aquaculture. Remarkably, ZnO-NPs synthesized with aloe extracts showed high broad antibacterial activity when compared to the chemically synthesized ZnO nanoparticles (Gunalan et al. 2012). Similarly, ZnO-NPs synthesized with *A. hydrophila* showed antibacterial activity against *Enterococcus faecalis*, *Pseudomonas aeruginosa*, *Candida albicans*, *Escherichia coli*, and *Aspergillus flavus* (Jayaseelan et al. 2012). In addition, dietary supplementation of ZnO-NPs reportedly enhanced resistance of *Labeo rohita* (Swain et al. 2019), and *Oreochromis mossambicus* (Anjugam et al. 2018) against *A. hydrophila*. Generally, the antibacterial mechanisms of nanoparticles are fairly understood. The active oxygen species generated by the metal oxide particles is considered the main mode of action, thereby these particles inhibit bacterial proliferation by disrupting the bacterial cell membrane, hence destroying the cell content (Liu et al. 2009; Gunalan et al. 2012; Bhuyan et al. 2015).

Silver Nanoparticles (AgNPs) They have been widely reported to elicit antibacterial activity against a broad spectrum of pathogenic bacteria of economic importance in aquaculture. Silver nanoparticles are reported to inhibit bacterial growth through different mechanisms: Ag+ binds to the bacterial cell membrane proteins resulting in the distraction of the membrane (Lara et al. 2010; Aziz et al. 2014, 2015, 2016, 2019), and by disrupting the cell division or bacteria reproduction process (Huang et al. 2011). A study by Elayaraja et al. (2017) demonstrated that silver nanoparticles synthesised using bacterial cellulose (Ag-NPs-BC) had high

Table 15.2 Some nanoparticles studied as antibacterial agents in aquaculture

Nanoparticles (NPs)	Screened bacteria	Antibacterial activity (Yes/No)	References
Zinc nanoparticles (ZnNPs) Zinc oxide (ZnO) (chemical)	*Aeromonas hydrophila; Edwardsiella tarda; Flavobacterium branchiophilum; Vibrio* sp.; *Staphylococcus aureus; Bacillus cereus; Citrobacter* sp.;	Yes	Swain et al. (2014)
ZnO (chemical) Bulk-ZnO (chemical)	*Vibrio harveyi*	Yes No	Ramomoorthy et al. (2013)
ZnO (Chemical) ZnO-*Aloe vera* (biological)	*Serratia marcescens; S. aureus; Proteus mirabilis; C. freundii*	Yes	Gunalan et al. (2012)
ZnO-*A. hydrophila* (biological)	*Enterococcus faecalis; Pseudomonas aeruginosa; Candida albicans; Escherichia coli; Aspergillus flavus*	Yes	Jayaseelan et al. (2012)
Dietary ZnONPs	*A. hydrophila*	Yes	Anjungam et al. (2018)
ZnO (chemical)	*A. salmonicida; Yersinia ruckeri; Aphanomyces invada*	Yes	Shaalan et al. (2017)
ZnO-Ag (mixture) (chemical)	*Pseudomonas* spp.	Yes	Venegas et al. (2018)
Silver nanoparticles (AgNPs) AgNPs (chemical)	*A. salmonicida* subsp. *Salmonicida S. aureus; E. coli* O157:H$_7$; *Streptococcus pyogenes; V. fluvialis*	Yes Yes Yes Yes	Shaalan et al. (2018) Ayala-Nunes et al. (2009) Lara et al. (2010) Meneses-Marquez et al. (2019)
Ag-TiO$_2$(chemical)	*A. hydrophila; E. tarda; F. branchiophilum; Vibrio sp; S. aureus; Citrobacter* sp.	Yes	Swain et al. (2014)
AgNPs-*Citrus limon* (biological)	*E. tarda; S. aureus*	Yes	Swain et al. (2014)
AgNPs-Tea leaf	*V. harveyi*	Yes	Vaseeharan et al. (2010)
AgNPs-*Calotropis gigantea* extracts (biological)	*V. alginolyticus*	Yes	Baskaralingam et al. (2012)
AgNPs-BC (biological)	*V. harveyi; V. parahaemolyticus*	Yes	Elayaraja et al. (2017)
AgNPs-red algae (biological)	*V. harveyi; V. parahaemolyticus; V. alginolyticus; V. anguillarum*	Yes	Fatima et al. (2020)

(continued)

Table 15.2 (continued)

Nanoparticles (NPs)	Screened bacteria	Antibacterial activity (Yes/No)	References
Gold nanoparticles (AuNPs) Fucoidan-AuNPs (biological)	*A. hydrophila*	Yes	Vijayakumar et al. (2017)
Acanthophora spicifera-AuNPs (biological)	*V. harveyi*	Yes	Babu et al. (2020)
	S. aureus	No	
Herbal extracts-AuNPs (biological)	*A. hydrophila*	Yes	Fernando and Cruz (2020)
	S. agalactiae	Yes	
Anacardium occidentale-AuNPs (biological)	*A. hydrophila*	No	
	A. bestiarum	Yes	Velmurugan et al. (2014)
	P. fluorescens	Yes	
	E. tarda	No	
AuNPs-zeolites	*E. coli; Salmonella typhi*	Yes	Lima et al. (2013)
Nigella sativa essential oil-AuNPs (NsEO-AuNPs)	*S. aureus; V. harveyi*	Yes	Manju et al. (2016)
AuNPs (chemical)	*V. parahaemolyticus*	Yes	Tello-Olea et al. (2019)

Yes = inhibited bacterial growth; No = did not inhibit bacterial growth

bactericidal activity against *V. parahaemolyticus* and *V. harveyi*, which are some of the deadliest bacterial pathogens in shrimp aquaculture. Similarly, biologically synthesized Ag-NPs were recommended as alternative antibiotics in controlling *S. aureus* and *E. tarda* (Swain et al. 2014), and *V. harveyi* infection in *Feneropenaeus indicus* (Vaseeharan et al. 2010). Interestingly, AgNPs demonstrated effectiveness against multi-drug resistant bacteria such as methicillin-resistant *S. aureus* (MRSA) (Ayala-Nunez et al. 2009), ampicillin-resistant *E. coli* O157:H_7, and erythromycin-resistant *Streptococcus pyogenes* (Lara et al. 2010). This is indeed an indication that these nanoparticles have the ability to eradicate the use of ineffective antibiotics to fight bacterial diseases in aquaculture.

Gold Nanoparticles (AuNPs) They are one of the emerging nanoparticles, and they can be more preferred mainly because of their less toxicity to animals (Li et al. 2014). Different gold nanoparticles have been reported to possess antibacterial properties, with the potential to eliminate bacteria responsible for huge production and economic losses in aquaculture (Table 15.2). A study by Vijayakumar et al. (2017) demonstrated that fucoidan (marine polysaccharide)-coated gold nanoparticles (Fu-AuNPs) inhibited the biofilm of *A. hydrophila*, and reduced mortality in *A. hydrophila*-infected *Oreochromis mossambicus* juveniles. *Acanthophora spicifera* (marine red algae)-mediated gold particles (As-AuNPs) exhibited the highest antibacterial activity against *V. harveyi* than *S. aureus* (Babu et al. 2020). Gold nanoparticles reportedly act against bacterial pathogens via a number of pathways such as their ability to collapse the bacterial membrane potential, inhibit ATPase activities, and subsequently the ATP level; and inhibit the subunit of ribosome from binding tRNA (Cui et al. 2012). In addition, AuNPs synthesized with crude herbal extracts reportedly inhibited *A. hydrophila* biofilm formation via the disruption of their quorum sensing ability (communication between cells) (Fernando and Cruz 2020). The communication between bacterial cells has been the target to control bacterial virulence for promising antibacterial agents (Rasmussen et al. 2005).

15.3.2 Nanoparticles as Vaccine/Drug Delivery Vector

In aquaculture, drugs are traditionally administered through feed, injection, or immersion. The traditional drug delivery methods are considered to be ineffective for several reasons such as poor bioavailability and absorption of the drugs to the targeted cells (Moges et al. 2020). Recently, the use of nanoparticles in drug formulation and delivery has gained attention in the fight against pathogens in aquaculture (Table 15.3). With this technology, the compound of interest (i.e. antibiotics, vitamins, vaccines, probiotics) is encapsulated into a compound of the nanoscale, thereby increasing absorption of the compound to targeted region, because nanoparticles are able to penetrate through cellular barriers (Sivakumar 2016; Moges et al. 2020); hence better protection against pathogens compared to traditional drug delivery methods.

Chitosan (Chit.) (polysaccharides) and poly-lactic glycolipids acid (PLGA) (copolymer) nanoparticles are the widely studied nanoparticles for drug delivery. These nanoparticles are commonly used due to their outstanding physiochemical properties such as biocompatibility, bioactivity, non-toxicity, and biodegradability (De Jong and Borm 2008; Lü et al. 2009). Chitosan nanoparticles combined with infectious salmon anaemia virus (ISAV) gene as an adjuvant were used to develop a DNA vaccine to control ISAV in Atlantic salmon culture (Rivas-Aravena et al. 2015). Chitosan nanoparticles-based vaccine was developed for *Lates calcarifer* against *V. anguillarum* (Rajesh Kumar et al. 2008). In addition, PLGA nanoparticles loaded with rifampicin were reported to show efficacy against *Mycobacterium marinum* in zebra fish larvae (Fenaroli et al. 2014). The use of chitosan and PLGA nanoparticles in combination were also reported in aquaculture. For instance, a plasmid DNA vaccine (pDNA) combined with PLGA and chitosan nanoparticles complex (pDNA-PLGA-Chit-NPs) significantly activated immune parameters in *Labeo rohita* and increased their survival after *Edwardsiella tarda* infection (Leya et al. 2020). This is said to be attributed to the ability of the complex vaccine to act synergistically to provide the host with amplified protective immunity against pathogens (Leya et al. 2020).

Table 15.3 Some chitosan and Poly lactic-co-glycotic acid nanoparticles studied as drug delivery agents in aquaculture

Nanoparticles (Chit-PLGA)	Pathogens	Fish species	References
pDNA-PLGA-Chit-NPs pDNA-PLGA-NPs PLGA-NPs Chit-NPs	*Edwardsiella tarda*	*Labeo rohita*	Leya et al. (2020)
Chit-ISAV	Alphavirus	Atlantic salmon	Rivas-Aravena et al. (2015)
Chit-DNA (pVAOMP38)	*Vibrio anguillarum*	*Lates calcarifer*	Rajesh Kumar et al. (2008)
Chit-DNA (pEGFP-N2OMPK, pDNA)	*V. parahaemolyticus*	*Acanthopagrus schlegelii* Bleeker	Li et al. (2013)
Chit-inactivated *E. ictaluri* and infectious spleen and kidney necrosis virus.	*E. ictalurid*	*Pelteobagrus fulvidraco; Siniperca chuasi*	Zhang et al. (2019) Zhu et al. (2019)
Chit-*Piscirickettsia salmonis* membrane	*Piscirickettsia salmonis*	*Dario rerio*	Tandberg et al. (2018)
PLGA-rifampicin	*Mycobacterium marinum*	*Dario rerio*	Fenaroli et al. (2014)

Chit Chitosan, *PLGA* Poly lactic-co-glycotic acid, *NPs* Nanoparticles, *p* plasmid; *ISAV* Infectious Atlantic salmon anaemia virus

15.4 Nanotechnology Application for Water Quality Management in Aquaculture

In aquaculture, animals are fed with high-protein feeds, and fertilizers, especially in semi-intensive systems, are used to stimulate natural feeds to sustain the growth of farmed animals and stimulate production. However, the challenge is in the handling/management of uneaten feed and waste products, which often contribute to the culture water quality (Ninh et al. 2016). Consequently, water in poorly managed aquaculture systems may be enriched with nutrients and organic and suspended matter (Boyd 2001; Sikder et al. 2016) which are associated with negative effects on fish growth, increased fish stress, and high risks of infectious diseases (Boyd and Tucker 1998; Boyd 2001). Water quality management is, therefore, vital for aquaculture operations.

Contrary to conventional wastewater treatment methods such as chemical treatment, filtration, and ion exchange (Muzammil et al. 2016), aquaculture effluents are treated via sedimentation, constructed wetlands, and water treatment reservoirs (Boyd 2001; Kerepeczki et al. 2011). However, these techniques are said to be ineffective in the complete removal of contaminants (Le et al. 2019). Therefore, the use of nanomaterials has been recommended as the best alternative technology in the purification of water either for human consumption (Gehrke et al. 2015) or fish culture (Sichula et al. 2011). This section outlines the use of nano-catalysts and nano-adsorbents for wastewater treatment in aquaculture.

15.4.1 Nanocatalysts and Nanoadsorbents in Aquaculture

Nano-catalysts are being employed in wastewater treatment for the chemical oxidation of organic and inorganic pollutants (Muzammil et al. 2016). Titanium oxide (TiO_2) and Zinc oxide (ZnO) are some of the widely used nanoparticles in photocatalysis. Their efficiency depends on the interaction with light energy and presence of metallic nanoparticles/semi-conductor metals (Acheampong and Antwi 2016). For instance, titanium oxide (TiO_2) was reported to remove bacterial cells (Litter 2015). This is because, TiO_2 possess high antimicrobial abilities that permits its use in inactivating pathogenic organisms such as bacteria found in wastewater (Wu et al. 2014; Amin et al. 2014). In another study, TiO_2 was reported to reduce the viability of several waterborne pathogens such as protozoa, fungi, *E. coli*, and *P. aeruginosa,* after 8 hours of simulated solar exposure (Amin et al. 2014). In addition, titanium was able to remove heavy metals such as chromium and arsenic from wastewater (Litter 2015).

A study by Le et al. (2019) tested the removal of heavy metal ions using rod-shaped ZnO particles under utraviolet light and visible light. This study observed that ZnO nanoparticles could remove heavy metal ions such as Cu(II), Ag(I) and Pb(II) at an efficiancy rate greater than 85%, but not very efficient at removing Cr(VI), Mn(II), Cd(II), and Ni(II) ions, regardless of the light source used. Similar to TiO_2, ZnO nanoparticle produced by solution combustion method (SCM) has also demonstrated effectiveness in removing *E. coli* from water (Masoumbaigi et al. 2015). Another important nano-catalyst in wastewater treatment is nanosilver. These nanoparticles synthesized with fungal species have been reported to remove *E. coli, Staphylococcus* sps, and *Pseudomonas* sps in wastewater (Moustafa 2017), which are some of the pathogenic bacteria in aqauculture as demostrated above in Sect. 15.3.

In addition to nanocatalysts, nanoadsorbents are likewise impressive water treatment methods, used to remove heavy metals, nutrients, and microbes from water (Thines et al. 2017). One such nanoadsorbent is activated charcoal (AC). A study by Aly et al. (2016) and Sichula et al. (2011) indicated that AC successfully removed ammonia from aquaculture production systems and reduced unionized ammonia concentrations in *O. niloticus* culture respectively. The application of nanocatalysts and nanoadsorbents is, therefore, a promising approach for the management of water quality in aquaculture production systems. By adopting nanotechnologies, microbial and heavy metal contamination can be adressed in aqaculture and in so doing, manage the challenges of nutrient accumulation, ultimately disease proliferation.

15.5 Conclusion and Future Perspectives

Challenges associated with increased intensification in aquaculture such as poor feed quality and utilization, increased disease outbreaks, and poor water quality cannot be overemphasised. This chapter has provided substantial evidence that nanomaterials have the potential to enhance feed quality (nutritional and physical properties), feed utilization, drug formulation and delivery, disease treatment, and water quality management in aquaculture; hence improving fish growth and better economic return. Despite promising research findings, the speed of implementation of this technology in aquaculture is still limited. One of the limitations is the complex manufacturing process of nanomaterials, which requires expensive equipments and services, and this could directly influence the cost of nanoproducts. Therefore, small-scale fish farmers may be financially limited to participate in the manufacturing process of the nanomaterials. Another limitation is that some nanoparticles, particularly the chemically synthesized ones, are toxic to animals at higher dosages and may negatively affect the development of animals (Verma et al. 2017, 2018).

Moving forward, there is a need to adopt nanoparticles manufacturing approaches such as biological methods, which are described to be simple and produce non-toxic, environmentally friendly, and affordable products (Kalishwaralal et al. 2008; Prabhu and Poulose 2012; Thangadurai et al. 2020, 2021; Maddela et al. 2021). This way, aquaculture farmers at all levels (from small scale to commercial) would be able to harness the benefits associated with nanotechnology. All in all, nanotechnology is playing important roles in the sustainable development of aquaculture.

References

Acheampong MA, Antwi DMB (2016) Modification of titanium dioxide for wastewater treatment application and its recovery for reuse. J Environ Sci Eng Technol 5:498–510

Acosta E (2009) Bioavailability of nanoparticles in nutrient and nutraceutical delivery. Curr Opin Colloid Interface Sci 14:3–15

Aklakur M, Asharf Rather M, Kumar N (2016) Nanodelivery: an emerging avenue for nutraceuticals and drug delivery. Crit Rev Food Sci Nutr 56:2352–2361

Aly HA, Abdel Rahim MM, Lotfy AM et al (2016) The applicability of activated carbon, natural zeolites, and probiotics (EM®) and its effects on ammonia removal efficiency and fry performance of european seabass *Dicentrarchus labrax*. J Aquac Res Dev 7:11. https://doi.org/10.4172/2155-9546.100045

Amin MT, Alazba AA, Manzoor U (2014) A review of removal of pollutants from water/wastewater using different types of nanomaterials. Adv Mater Sci Eng. https://doi.org/10.1155/2014/825910

Anjugam M, Vaseeharan B, Iswarya A et al (2018) Effect of β-1, 3 glucan binding protein based zinc oxide nanoparticles supplemented diet on immune response and disease resistance in *Oreochromis mossambicus* against *Aeromonas hydrophila*. Fish Shellfish Immunol 76:247–259

Asaikkutti A, Bhavan PS, Vimala K et al (2016) Dietary supplementation of green synthesized manganese-oxide nanoparticles and its effect on growth performance, muscle composition and digestive enzyme activities of the giant freshwater prawn *Macrobrachium rosenbergii*. J Trace Elem Med Biol 35:7–17

Ashouri S, Keyvanshokooh S, Salati AP et al (2015) Effects of different levels of dietary selenium nanoparticles on growth performance, muscle composition, blood biochemical profiles and antioxidant status of common carp (*Cyprinus carpio*). Aquaculture 446:25–29

Ayala-Núñez NV, Lara HH, Turrent LDCI et al (2009) Silver nanoparticles toxicity and bactericidal effect against methicillin- resistant *Staphylococcus aureus*: nanoscale does matter. Nanobiotechnology 5:2–9

Aziz N, Fatma T, Varma A, Prasad R (2014) Biogenic synthesis of silver nanoparticles using Scenedesmus abundans and evaluation of their antibacterial activity. Journal of Nanoparticles, Article ID 689419, https://doi.org/10.1155/2014/689419

Aziz N, Faraz M, Pandey R, Sakir M, Fatma T, Varma A, Barman I, Prasad R (2015) Facile algae-derived route to biogenic silver nanoparticles: Synthesis, antibacterial and photocatalytic properties. Langmuir 31:11605–11612. https://doi.org/10.1021/acs.langmuir.5b03081

Aziz N, Pandey R, Barman I, Prasad R (2016) Leveraging the attributes of Mucor hiemalis-derived silver nanoparticles for a synergistic broad-spectrum antimicrobial platform. Front Microbiol 7:1984. https://doi.org/10.3389/fmicb.2016.01984

Aziz N, Faraz M, Sherwani MA, Fatma T, Prasad R (2019) Illuminating the anticancerous efficacy of a new fungal chassis for silver nanoparticle synthesis. Front Chem 7:65. https://doi.org/10.3389/fchem.2019.00065

Babu B, Palanisamy S, Vinosha M et al (2020) Bioengineered gold nanoparticles from marine seaweed *Acanthophora spicifera* for pharmaceutical uses: antioxidant, antibacterial, and anticancer activities. Bioprocess Biosyst Eng. https://doi.org/10.1007/s00449-020-02408-3

Baskaralingam V, Sargunar CG, Lin YC et al (2012) Green synthesis of silver nanoparticles through *Calotropis gigantea* leaf extracts and evaluation of antibacterial activity against *Vibrio alginolyticus*. Nanotechnol Dev 2:e3. https://doi.org/10.4081/nd.2012.e3

Berntssen MHG, Julshamn K, Lundebye AK (2010) Chemical contaminants in aquafeeds and Atlantic salmon (*Salmo salar*) following the use of traditional- versus alternative feed ingredients. Chemosphere 78:637–646

Bhuyan T, Mishra K, Khanuja M, Prasad R, Varma A (2015) Biosynthesis of zinc oxide nanoparticles from *Azadirachta indica* for antibacterial and photocatalytic applications. Mater Sci Semicond Process 32:55–61

Boonanuntanasarn S, Khaomek P, Pitaksong T et al (2014) The effects of the supplementation of activated charcoal on the growth, health status and fillet composition- odor of Nile tilapia (*Oreochromis niloticus*) before harvesting. Aquac Int 22:1417–1436

Bouwmeester H, Dekkers S, Noordam MY et al (2009) Review of health safety aspects of nanotechnologies in food production. Regul Toxicol Pharmacol 53:52–62

Boyd CE (2001) Decision support systems for water resources management. In: AWRA/UCOWR summer specialty conference, Snowbird, Utah, , 27–30 June 2001, p 153

Boyd CE, Tucker CS (1998) Ecology of aquaculture ponds. In: Pond aquaculture water quality management. Springer, Boston, MA. https://doi.org/10.1007/978-1-4615-5407-3_2

Cao L, Ding W, Du J et al (2015) Effects of curcumin on antioxidative activities and cytokine production in Jian carp (*Cyprinus carpio* var. Jian) with CCl4-induced liver damage. Fish Shellfish Immunol 43:150–157

Carriquiriborde P, Handy RD, Davies et al (2004) Physiological modulation of iron metabolism in rainbow trout (*Oncorhynchus mykiss*) fed low and high iron diets. J Exp Biol 207:75–86

Cui Y, Zhao Y, Tian Y et al (2012) The molecular mechanism of action of bactericidal gold nanoparticles on *Escherichia coli*. Biomaterials 33:2327–2333

Culot A, Grosset N, Gautier M (2019) Overcoming the challenges of phage therapy for industrial aquaculture: a review. Aquaculture 513:734423

De Jong WH, Borm PJ (2008) Drug delivery and nanoparticles: applications and hazards. Int J Nanomedicine 3:133–149

Di Cesare A, Vignaroli C, Luna GM et al (2012) Antibiotic-resistant enterococci in seawater and sediments from a coastal fish farm. Microb Drug Resist 18:502–509

Elayaraja S, Zagorsek K, Li F et al (2017) In situ synthesis of silver nanoparticles into TEMPO-mediated oxidized bacterial cellulose and their antivibriocidal activity against shrimp pathogens. Carbohydr Polym 166:329–337

ElBasuini MF, El-Hais AM, Dawood MAO et al (2017) Effects of dietary copper nanoparticles and vitamin C supplementations on growth performance, immune response and stress resistance of red sea bream, *Pagrus major*. Aquacult Nutr 23:1329–1340

Elgendy M, Hakim A, Ibrahim et al (2016) Immunomodulatory effects of curcumin on nile tilapia, *Oreochromis niloticus* and its antimicrobial properties against *Vibrio alginolyticus*. J Fish Aquat Sci 11:206–215

FAO (2020) The state of the world fisheries and aquaculture 2020. FAO, Rome

Fatima R, Priya M, Indurthi L et al (2020) Biosynthesis of silver nanoparticles using red algae *Portieria hornemannii* and its antibacterial activity against fish pathogens. Microb Pathog 138:103780. https://doi.org/10.1016/j.micpath.2019.103780

Fenaroli F, Westmoreland D, Benjaminsen J et al (2014) Nanoparticles as drug delivery system against tuberculosis in zebrafish embryos: direct visualization and treatment. ACS Nano 8:7014–7026

Fernando SID, Cruz KGJ (2020) Ethnobotanical biosynthesis of gold nanoparticles and its down-regulation of Quorum Sensing-linked AhyR gene in *Aeromonas hydrophila*. SN Appl Sci 2:1–8

Food Safety Authority of Ireland (2008) The relevance of food safety of applications of nanotechnology in the food and feed industries. Court lower abbey street, Abbey, Dublin 1. www.fsai.ie

Fraser TWK, Reinardy HC, Shaw BJ et al (2010) Dietary toxicity of single-walled carbon nanotubes and fullerenes (C60) in rainbow trout (*Oncorhynchus mykiss*). Nanotoxicology 5:98–108

Friends of the Earth (2008) Out of the laboratory and onto our plates: nanotechnology in food and agriculture, 2nd edn A report prepared for Friends of the Earth

Gehrke I, Geiser A, Somborn-Schulz A (2015) Innovations in nanotechnology for water treatment. Nanotechnol Sci Appl 8:1–17

Gunalan S, Sivaraj R, Rajendran V (2012) Green synthesized ZnO nanoparticles against bacterial and fungal pathogens. Prog Nat Sci 22:693–700

Hafiz S, Srivastava KK, Newton JC et al (2017) Efficacy of curcumin as an immunostimulatory dietary supplement for channel catfish. Am J Anim Vet Sci 12:1–7

Handy RD (2012) FSBI briefing paper: nanotechnology in fisheries and aquaculture. Fisheries Society of the British Isles, pp 1–29

Handy RD, Poxton MG (1993) Nitrogen pollution in mariculture – toxicity and excretion of nitrogenous compounds by marine fish. Rev Fish Biol Fish 3:205–241

Hoet P, Bruske-Hohlfeld I, Salata O (2004) Nanoparticles-known and unknown health risks. J Nanobiotechnol 2:12–27

Hoseini SM, Yousefi M, Hoseinifar SH et al (2019) Effects of dietary arginine supplementation on growth, biochemical, and immunological responses of common carp (*Cyprinus carpio* L.), stressed by stocking density. Aquaculture 503:452–459

Huang CM, Chen CH, Pornpattananangkul D et al (2011) Eradication of drug resistant *Staphylococcus aureus* by liposomal oleic acids. Biomaterials 32:214–221

Huang S, Wang L, Liu L et al (2015) Nanotechnology in agriculture, livestock, and aquaculture in China. A review. Agron Sustain Dev 35:369–400

Izquierdo MS, Ghrab W, Roo J et al (2017) Organic, inorganic and nanoparticles of Se, Zn and Mn in early weaning diets for gilthead seabream (*Sparus aurata*; Linnaeus, 1758). Aquac Res 48:2852–2867

Jafari SM, McClements DJ (2017) Chapter one - Nanotechnology approaches for increasing nutrient bioavailability. Adv Food Nutr Res 81:1–30

Jayaseelan C, Rahuman AA, Kirthi AV et al (2012) Novel microbial route to synthesize ZnO nanoparticles using *Aeromonas hydrophila* and their activity against pathogenic bacteria and fungi. Spectrochim Acta Part A Mol Biomol Spectrosc 90:78–84

Jiang Y, Chekuri S, Fang RH et al (2019) Engineering biological interactions on the nanoscale. Curr Opin Biotechnol 55:1–8

Joye IJ, Davidov-Pardo G, McClements DJ (2014) Nanotechnology for increased micronutrient bioavailability. Trends Food Sci Technol 40:168–182

Kalishwaralal K, Deepak V, Ramkumarpandian S et al (2008) Extracellular biosynthesis of silver nanoparticles by the culture supernatant of *Bacillus licheniformis*. Mater Lett 62(29):4411–4413

Kerepeczki É, Gál D, Kosáros T et al (2011) Natural water treatment method for intensive aquaculture effluent purification. Studia Universitatis" Vasile Goldis" Arad. Seria Stiintele Vietii (Life Sciences Series) 21: 827

Khosravi-Katuli K, Prato E, Lofrano G et al (2017) Effects of nanoparticles in species of aquaculture interest. Environ Sci Pollut Res 24:17326–17346

Korni FMM, Khalil F (2017) Effect of ginger and its nanoparticles on growth performance, cognition capability, immunity and prevention of motile *Aeromonas septicaemia* in *Cyprinus carpio* fingerlings. Aquac Nutr 23:1492–1499

Kunjiappan S, Bhattacharjee C, Chowdhury R (2015) Hepatoprotective and antioxidant effects of *Azolla microphylla* based gold nanoparticles against acetaminophen induced toxicity in a fresh water common carp fish (*Cyprinus carpio* L.). Nanomedicine 2:88–110

Lara HH, Ayala-Núnez NV, Turrent LDCI et al (2010) Bactericidal effect of silver nanoparticles against multidrug-resistant bacteria. World J Microbiol Biotechnol 26:615–621

Lavertu M, Methot S, Tran-Khanh N et al (2006) High efficiency gene transfer using chitosan/DNA nanoparti-cles with specific combinations of molecular weight anddegree of deacetylation. Biomaterials 27:4815–4824

Le AT, Pung SY, Sreekantan S et al (2019) Mechanisms of removal of heavy metal ions by ZnO particles. Heliyon 5:e01440. https://doi.org/10.1016/j.heliyon.2019.e01440

Leya T, Ahmad I, Sharma R et al (2020) Bicistronic DNA vaccine macromolecule complexed with poly lactic-co-glycolic acid-chitosan nanoparticles enhanced the mucosal immunity of *Labeo rohita* against *Edwardsiella tarda* infection. Int J Biol Macromol. https://doi.org/10.1016/j.ijbiomac.2020.04.048

Leya T, Raman RP, Srivastava PP et al (2017) Effects of curcumin supplemented diet on growth and non-specific immune parameters of *Cirrhinus mrigala* against *Edwardsiella tarda* infection. Int J Curr Microbiol Appl Sci 6:1230–1243

Li L, Lin SL, Deng L et al (2013) Potential use of chitosan nanoparticles for oral delivery of DNA vaccine in black seabream *Acanthopagrus schlegelii* Bleeker to protect from *Vibrio parahaemolyticus*. J Fish Dis 36:987–995

Li X, Robinson SM, Gupta A et al (2014) Functional gold nanoparticles as potent antimicrobial agents against multi-drug-resistant bacteria. ACS Nano 8:10682–10686

Lima E, Guerra R, Lara V et al (2013) Gold nanoparticles as efficient antimicrobial agents for *Escherichia coli* and *Salmonella typhi*. Chem Cent J 7:11. https://doi.org/10.1186/1752-153X-7-11

Litter MI (2015) Mechanisms of removal of heavy metals and arsenic from water by TiO2-heterogeneous photocatalysis. Pure Appl Chem 87:557–567

Liu YJ, He LL, Mustapha A et al (2009) Antibacterial activities of zinc oxide nanoparticles against *Escherichia coli* O157: H7. J Appl Microbiol 107:1193–1201

Lü JM, Wang X, Marin-Muller C et al (2009) Current advances in research and clinical applications of PLGA-based nanotechnology. Expert Rev Mol Diagn 9:325–341

Maddela NR, Chakraborty S, Prasad R (2021) Nanotechnology for Advances in Medical Microbiology. Springer Singapore (ISBN 978-981-15-9915-6) https://www.springer.com/gp/book/9789811599156

Mahmoud HK, Al-Sagheer AA, Reda FM et al (2017) Dietary curcumin supplement influence on growth, immunity, antioxidant status, and resistance to *Aeromonas hydrophila* in *Oreochromis niloticus*. Aquaculture 475:16–23

Malheiros DF, Sarquis IR, Ferreira IM et al (2020) Nanoemulsions with oleoresin of *Copaifera reticulata* (Leguminosae) improve anthelmintic efficacy in the control of monogenean parasites when compared to oleoresin without nanoformulation. J Fish Dis 43:687–695

Manal I (2018) Impact of garlic and curcumin on the hepatic histology and cytochrome P450 gene expression of aflatoxicosis *Oreochromis niloticus* using RT-PCR. Turkish J Fish Aquat Sci 18:405–415

Manju M, Sherin TG, Rajasekharan KN et al (2009) Curcumin analogue inhibits lipid peroxidation in a freshwater teleost, *Anabas testudineus* (Bloch) -an *in vitro* and *in vivo* study. Fish Physio Biochem 35:413–420

Manju M, Vijayasree AS, Akbarsha MA et al (2013) Protective effect of dietary curcumin in *Anabas testudineus* (Bloch) with a special note on DNA fragmentation assay on hepatocytes and micronucleus assay on erythrocytes *in vivo*. Fish Physiol Biochem 39:1323–1330

Manju S, Malaikozhundan B, Vijayakumar S et al (2016) Antibacterial, antibiofilm and cytotoxic effects of *Nigella sativa* essential oil coated gold nanoparticles. Microb Pathog 91:129–135

Manju M, Akbarsha MA, Oommen OV (2012) *In vivo* protective effect of dietary curcumin in fish *Anabas testudineus* (Bloch). Fish Physiol Biochem 38:309–318

Masoumbaigi H, Rezaee A, Hosseini H et al (2015) Water disinfection by zinc oxide nanoparticle prepared with solution combustion method. Desalin Water Treat 56:2376–2381

Meneses-Márquez JC, Hamdan-Partida A, del Carmen M-DM et al (2019) Use of silver nanoparticles to control *Vibrio fluvialis* in cultured angelfish *Pterophyllum scalare*. Dis Aquat Org 137:65–72

Midhun SJ, Arun D, Edatt L et al (2016) Modulation of digestive enzymes, GH, IGF-1 and IGF-2 genes in the teleost, Tilapia (*Oreochromis mossambicus*) by dietary curcumin. Aquac Int 24:1277–1286

Moges FD, Patel P, Parashar SKS et al (2020) Mechanistic insights into diverse nano-based strategies for aquaculture enhancement: a holistic review. Aquaculture 519:734770. https://doi.org/10.1016/j.aquaculture.2019.734770

Mohammadi N, Tukmechi A (2015) The effects of iron nanoparticles in combination with *Lactobacillus casei* on growth parameters and probiotic counts in rainbow trout (*Oncorhynchus mykiss*) intestine. J Vet Res 70:47–53

Moustafa MT (2017) Removal of pathogenic bacteria from wastewater using silver nanoparticles synthesized by two fungal species. Water Sci 31:164–176

Muralisankar T, Saravana Bhavan P, Radhakrishnan S et al (2016) The effect of copper nanoparticles supplementation on freshwater prawn *Macrobrachium rosenbergii* post larvae. J Trace Elem Med Biol 34:39–49

Murata J, Ohya Y, Ouchi T (1998) Design of quaternary chitosan conjugate having antennary galactose residues as a gene delivery tool. Carbohydr Polym 32:105–109

Muzammil A, Miandad R, Muhammad W et al (2016) Remediation of wastewater using various nano-materials. Arab J Chem. https://doi.org/10.1016/j.arabjc.2016.10.004

Ninh NTH, Dung NM, Cuong HN (2016) Water quality management for sustainable aquaculture production in the Mekong Delta. In: International conference of the Mekong, Salween and red river: sharing knowledge and perspectives across borders, Faculty of Political Science, 12 Nov 2016. Chulalongkorn University, Bangkok

Prabhu S, Poulose EK (2012) Silver nanoparticles: mechanism of antimicrobial action, synthesis, medical applications, and toxicity effects. Int Nano Lett 2:32. https://doi.org/10.1186/2228-5326-2-32

Prasad R (2014) Synthesis of silver nanoparticles in photosynthetic plants. Journal of Nanoparticles, Article ID 963961, 2014, https://doi.org/10.1155/2014/963961

Prasad R, Pandey R, Barman I (2016) Engineering tailored nanoparticles with microbes: quo vadis. WIREs Nanomed Nanobiotechnol 8:316–330. https://doi.org/10.1002/wnan

Prasad R, Jha A, Prasad K (2018) Exploring the Realms of Nature for Nanosynthesis. Springer International Publishing (ISBN 978-3-319-99570-0 https://www.springer.com/978-3-319-99570-0

Rajeshkumar S, Ishaq AVP, Parameswaran V et al (2008) Potential use of chitosan nanoparticles for oral delivery of DNA vaccine in Asian sea bass (*Lates calcarifer*) to protect from *Vibrio (Listonella) anguillarum*. Fish Shellfish Immunol 25:47–56

Rajeshkumar S, Venkatesan C, Sarathi M et al (2009) Oral delivery of DNA construct using chitosan nanoparticles to protect the shrimp from white spot syndrome virus (WSSV). Fish Shellfish Immun 26:429–437

Ramamoorthy S, Kannaiyan P, Moturi M et al (2013) Antibacterial activity of zinc oxide nanoparticles against *Vibrio harveyi*. Indian J Fish 60:107–112

Ramsden CS, Smith TJ, Shaw BS, Handy RD (2009) Dietary exposure to titanium dioxide nanoparticles in rainbow trout, (*Oncorhynchus mykiss*): no effect on growth, but subtle biochemical disturbances in the brain. Ecotoxicology 18:939–951

Rasmussen TB, Bjarnsholt T, Skindersoe ME et al (2005) Screening for quorum-sensing inhibitors (QSI) by use of a novel genetic system, the QSI selector. J Bacteriol 187:1799–1814

Rather MA, Bhat IA, Sharma N et al (2017) Synthesis and characterization of *Azadirachta indica* constructed silver nanoparticles and their immunomodulatory activity in fish. Aquac Res 48:3742–3754

Rivas-Aravena A, Fuentes Y, Cartagena J et al (2015) Development of a nano- particle-based oral vaccine for Atlantic salmon against ISAV using an alphavirus re-plicon as adjuvant. Fish Shellfish Immunol 45:157–166

Romero J, Feijoo CG, Navarrete P (2012) Antibiotics in aquaculture – use, abuse and alternatives. In: Carvalho D, David SG, Silva R (eds) Health and environment in aquaculture. InTech, Croatia

Russell-Jones GJ (2001) The potential use of receptormediated endocytosis for oral drug delivery. Adv Drug Deliv Rev 46:59–73

Russell-Jones GJ, Arthur L, Walker H (1999) Vitamin B_{12} mediated transport of nanoparticles across Caco-2 cells. Int J Pharm 179:247–255

Sahu S, Das BK, Mishra BK et al (2008) Effect of dietary Curcuma longa on enzymatic and immunological profiles of rohu, *Labeo rohita* (Ham.), infected with *Aeromonas hydrophila*. Aquac Res 39:1720–1730

Sarma H, Joshi S, Prasad R, Jampilek J (2021) Biobased Nanotechnology for Green Applications. Springer International Publishing (ISBN 978-3-030-61985-5) https://www.springer.com/gp/book/9783030619848

Semo E, Kesselman E, Danino D, Livney YD (2007) Casein micelle as a natural nano-capsular vehicle for nutraceuticals. Food Hydrocoll 21:936–942

Shaalan M, El-Mahdy M, Theiner S et al (2018) Silver nanoparticles: their role as antibacterial agent against *Aeromonas salmonicida* subsp. salmonicida in rainbow trout (*Oncorhynchus mykiss*). Res Vet Sci 119:196–204

Shaalan M, Saleh M, El-Mahdy M et al (2016) Recent progress in applications of nanoparticles in fish medicine: a review. Nanomed Nanotechnol 12:701–710

Shaalan MI, El-Mahdy MM, Theiner S et al (2017) In vitro assessment of the antimicrobial activity of silver and zinc oxide nanoparticles against fish pathogens. Acta Vet Scand 59:49

Shah BR, Mraz J (2019) Advances in nanotechnology for sustainable aquaculture and fisheries. Rev Aquacult 12:925–942

Shah BR, Mraz J (2020) Advances in nanotechnology for sustainable aquaculture and fisheries. Rev Aquacult 12:925–942

Sharif Rohani M, Haghighi M, Bazari Moghaddam S (2017) Study on nanoparticles of *Aloe vera* extract on growth performance, survival rate and body composition in Siberian sturgeon (*Acipenser baerii*). Iran J Fish Sci 16:457–468

Shinn AP, Pratoomyot J, Bron JE et al (2015) Economic costs of protistan and metazoan parasites to global mariculture. Parasitology 142:196

Sichula J, Makasa ML, Nkonde GK et al (2011) Removal of ammonia from aquaculture water using maize cob activated carbon. Malawi J Aquac Fish 1:10–15

Sikder MNA, Min WW, Ziyad AO et al (2016) Sustainable treatment of aquaculture effluents in future-a review. Int Res J Adv Eng Sci 1:190–193

Silva AA (2010) Nanotechnology applications and approaches for neuroregeneration and drug delivery to the central nervous system. Ann N Y Acad Sci 1199:221–230

Sivakumar SM (2016) Therapeutic potential of chitosan nanoparticles as antibiotic delivery system: challenges to treat multiple drug resistance. Asian J Pharm 10:S63. https://doi.org/10.22377/ajp.v10i2.624

Sneddon LU, Wolfenden DCC, Thomson JS (2016) Stress management and welfare. Fish Physiol 35:463–539

Sørum H (2008) Antibiotic resistance associated with veterinary drug use in fish farms. In: Lie Ø (ed) Improving farmed fish quality and safety. Food science, technology, and nutrition. Woodhead Publishing, pp 157–182

Srivastava S, Usmani Z, Atanasov AG, Singh VK, Singh NP, Abdel-Azeem AM, Prasad R, Gupta G, Sharma M, Bhargava A (2021) Biological nanofactories: Using living forms for metal nanoparticle synthesis. Mini-Reviews in Medicinal Chemistry 21:245–265

Swain P, Das R, Das A et al (2019) Effects of dietary zinc oxide and selenium nanoparticles on growth performance, immune responses and enzyme activity in rohu, *Labeo rohita* (Hamilton). Aquac Nutr 25:486–494

Swain P, Nayak SK, Sasmal A et al (2014) Antimicrobial activity of metal based nanoparticles against microbes associated with diseases in aquaculture. World J Microbiol Biotechnol 30:2491–2502

Tandberg J, Lagos L, Ropstad E et al (2018) The use of chitosan-coated membrane vesicles for immunization against salmonid rickettsial septicemia in an adult zebrafish model. Zebrafish 15:372–381

Tandel GM, John KR, George MR et al (2017) Current status of viral diseases in Indian shrimp aquaculture. Acta Virol 61:131–137

Tello-Olea M, Rosales-Mendoza S, Campa-Córdova AI et al (2019) Gold nanoparticles (AuNP) exert immunostimulatory and protective effects in shrimp (*Litopenaeus vannamei*) against *Vibrio parahaemolyticus*. Fish Shellfish Immunol 84:756–767

Thangadurai D, Sangeetha J, Prasad R (2020) Functional Bionanomaterials. Springer International Publishing (ISBN 978-3-030-41464-1) https://www.springer.com/gp/book/9783030414634

Thangadurai D, Sangeetha J, Prasad R (2021) Nanotechnology for Food, Agriculture, and Environment. Springer International Publishing (ISBN 978-3-030-31937-3) https://www.springer.com/gp/book/9783030319373

Thines RK, Mubarak NM, Nizamuddin S et al (2017) Application potential of carbon nanomaterials in water and wastewater treatment: a review. J Taiwan Inst Chem Eng 72:116–133

Thulasi A, Rajendran D, Jash S et al (2013) Nanobiotechnology in animal nutrition. In: Sampath KT, Ghosh J, Bhatta R (eds) Animal nutrition and reproductive physiology. Satish Serial Publishing House, New Delhi

Trapani A, Mandracchia D, Di Franco C, Cordero H, Morcillo P, Comparelli R, et al (2015) *In vitro* characterization of 6-Coumarin loaded solid lipid nanoparticles and their uptake by immunocompetent fish cells. Colloids Surf B: Biointerfaces 127:79–88

Vaseeharan B, Ramasamy P, Chen JC (2010) Antibacterial activity of silver nanoparticles (AgNps) synthesized by tea leaf extracts against pathogenic Vibrio harveyi and its protective efficacy on juvenile *Feneropenaeus indicus*. Lett Appl Microbiol 50:352–356

Velmurugan P, Iydroose M, Lee SM et al (2014) Synthesis of silver and gold nanoparticles using cashew nutshell liquid and its antibacterial activity against fish pathogens. Indian J Microbiol 54:196–202

Venegas MA, Bollaert MD, Jafari A et al (2018) Nanoparticles against resistant *Pseudomonas* spp. Microb Pathog 118:115–117

Verma SK, Jha E, Panda PK et al (2018) Rapid novel facile biosynthesized silver nanoparticles from bacterial release induce biogenicity and concentration dependent in vivo cytotoxicity with embryonic zebrafish-a mechanistic insight. Toxicol Sci 161:125–138

Verma SK, Jha E, Sahoo B et al (2017) Mechanistic insight into the rapid one-step facile biofabrication of antibacterial silver nanoparticles from bacterial release and their biogenicity and. RSC Adv 7:40034–40045

Vijayakumar S, Vaseeharan B, Malaikozhundan B et al (2017) A novel antimicrobial therapy for the control of *Aeromonas hydrophila* infection in aquaculture using marine polysaccharide coated gold nanoparticle. Microb Pathog 110:140–151

Wang R, Wang WX (2012) Contrasting mercury accumulation patterns in tilapia (*Oreochromis niloticus*) and implications on somatic growth dilution. Aquat Toxicol 114:23–30

Wang X, Lu J, Xu M et al (2008) Sorption of pyrene by regular and nanoscaled metal oxide particles: influence of adsorbed organic matter. Environ Sci Technol 42:7267–7272

Wei H, Zhang XZ, Chen WQ et al (2007) Self-assembled thermosensitive micelles based on poly (L-lac-tide-star block-N-isopropylacrylamide) for drug delivery. J Biomed Mater Res A 83:980–989

Wu MJ, Bak T, O'Doherty et al (2014) Photocatalysis of titanium dioxide for water disinfection: challenges and future perspectives. Int J Photochem 2014:1–9

Zhang J, Fu X, Zhang Y et al (2019) Chitosan andanisodamine improve the immune efficacy of inactivated infectious spleen and kidney necrosis virus vaccinein *Siniperca chuatsi*. Fish Shellfish Immunol 89:52–60

Zhu W, Zhang Y, Zhang J et al (2019) Astragalus polysaccharides, chitosan andpoly(I:C) obviously enhance inactivated *Edwardsiella ictaluri* vaccine potency in yellow catfish *Pelteobagrus fulvidraco*. Fish Shellfish Immunol 87:379–385

Chapter 16
Nanomaterials in Electrochemical Biosensors and Their Applications

J. R. Anusha, Mariadhas Valan Arasu, Naif Abdullah Al-Dhabi, and C. Justin Raj

Contents

16.1	Introduction.	487
16.2	Nanomaterials in Biosensors.	489
16.3	Types of Nanomaterials in Biosensor.	490
	16.3.1 Nanosized Metal Compounds.	490
	16.3.2 Carbon Nanomaterials.	496
	16.3.3 Polymeric Nanomaterials.	499
	16.3.4 Nanosized Biomaterials.	502
16.4	Applications of Nanobiosensors.	505
	16.4.1 In the Medical Field.	506
	16.4.2 In Environmental Monitoring.	508
	16.4.3 In Food Industry.	509
	16.4.4 In Agriculture.	509
16.5	Challenges and Future Perspectives.	510
References.		511

16.1 Introduction

Recent advances in nanotechnology have highly influenced our lifestyle in the fields of medicine, environment, energy, engineering, construction, and telecommunication (Ozin et al. 2009). Moreover, its application in various disciplines also transforms the economy of countries and change the production process of industries. Nanoscale deals with the object having a size range from 1 to 100 nm. It is a

---

J. R. Anusha (✉) · C. J. Raj
Department of Chemistry, Dongguk University, Seoul, Republic of Korea

M. V. Arasu · N. A. Al-Dhabi
Department of Botany and Microbiology, College of Science, King Saud University, Riyadh, Saudi Arabia

© The Author(s), under exclusive license to Springer Nature Switzerland AG 2022
A. Krishnan et al. (eds.), *Emerging Nanomaterials for Advanced Technologies*, Nanotechnology in the Life Sciences, https://doi.org/10.1007/978-3-030-80371-1_16

complex challenge and achievement, with great effort from the scientific community, including both theoretical and experimental science. Initially, the physicist and Nobel laureate Richard Feynman was the first to introduce the concept of nanotechnology and later Prof. Norio Taniguchi of Tokyo University of Science (1974) created the term "nanotechnology" to describe the precision manufacture of materials at the nanoscale (Drexler 2004). In later years, nanotechnology research started to resolve many problems in science and technology, but the real wave of applications is only beginning to break in the last three decades.

Nanotechnology starts within the panel of the green innovation technologies, which impact the potential application and excellent vision towards social and economic development (Colvin 2003; Sarma et al. 2021). The scientists have dedicated years of research to understand the phenomena of atoms and molecules and discovered various nanostructured materials. For example, the discovery of quantum dots, nanowires, nanosheets, proteins, liposomes, viruses, antibodies, deoxyribonucleic acid, and antibodies. (Ramos et al. 2017). Moreover, based on the size and dimensions, nanomaterials are classified as zero dimension (0D), one dimension (1D), two dimensions (2D), and three dimensions (3D). Recently, the manipulation of nanomaterials and utilization with the technologies such as the controlled size, dimension, morphology and characterization, lead to a better understanding of the relationship between nanomaterials and their properties. The bulk materials engineered to the nanoscale can relatively contributes to a large surface area and makes the material more chemically reactive, mechanically stable, tune the optical, magnetic and electrical properties (Parikha 2016). Moreover, based on the size, dimension, and peculiar properties, nanomaterials are subjected to day-to-day applications in various commercial areas.

In environmental science and technology, nanomaterials have a considerable role in the development of nonpolluting materials to make the chemical reactions more perfect and effective during the manufacturing processes. Moreover, nanosensors are used to detect harmful or hazardous gases and chemicals. Also the nanomaterials can optimize the remediation of water, soil, and air contaminations (Karn et al. 2004). Conductive and semiconductor nanomaterials have used in energy conversion and energy storage systems. The nanomaterials such as quantum dots, graphene, carbon nanotubes, and nanowires are considerably increasing the processing speed in computers and other electronic devices (Zhang 2017). In the construction field, nanomaterials used in cement, mortar, concrete, paints, insulating materials and glass help to improve the durability, strength, fire resistance, and heat stability of the constructions. Besides, the fire-resistant, heat-controlling, and water-resistant textiles made utilizing nanomaterials have a significant interest in the commercial market (Oke et al. 2017). Similarly, in automobile industries, nanomaterials, carbon-based microfiber and nanofibers, are used to fabricate heat-resistant windows, reinforcement materials in tires, anti-rust coatings. and even help to make auto bodies lighter (Mathew et al. 2019).

In the medical field, nanomaterials and nano-based sensors are considerable for diagnosis and treatment of various diseases in their early stage without any side effects. For example, most of the glucose detectors in our daily life used to measure blood sugar level are built based on nanomaterials. They are also been used to develop implants, prosthetics, and chip-based drug delivery systems (Srinivasan et al. 2015). Meanwhile, nanobiosensors are capable of detecting single cancer cells in blood or other regions of the patient's body enabling focused treatment (Tansil and Gao 2015; Kargozar and Mozafari 2018). Thus, the advancement of nanomaterials leads to the development of various emerging new smart biosensor devices. These smart biosensors can detect a low concentration of the desired analyte within a fraction of seconds. Moreover, the modified nanomaterials are predicted to increase the performance of the biosensor with high sensitivity, selectivity, and low limit of detection.

16.2 Nanomaterials in Biosensors

The pioneers Clark and Lyons first introduced biosensors in 1960, and in 1967 the scientists Updike and Hicks reported the first enzyme-based sensor. Followed by their inventions variety of research works have described the primary function and feature of biosensors, that sense biological and chemical materials. A biosensor is a smart analytical device that generates an electronic signal by receptor-target analyte interactions. It consists of a bioreceptor, a transducer, a signal processor for converting electronic to the desired signal and an interface. This sensing has been accomplished by another biologically active material called bioreceptor. The second component transducer system converts the analyte and its respective bioreceptor interactions into an electrical signal. The electric signals from the transducer are received by the third component called a signal processor, which amplifies to read and interpret the data. Besides these components, another essential requirement is the immobilization of bioreceptor which enhances the reaction with bioanalyte feasibly and efficiently. For the immobilization of biorecognition elements, various nanostructured materials like metal/metal oxides, carbonaceous materials, polymer, biomaterials and hybrid composites are used to enhance the electroanalytical performances of the biosensors (Fig. 16.1) (Pandit et al. 2016).

Recently, numerous biosensor devices utilize nanomaterials for improvising upon the sensing mechanism. However, the physicochemical properties of the nanomaterial control or determine its purpose in biosensor applications. For example, quantum dots are used as fluorescent sensing platforms for biomolecular

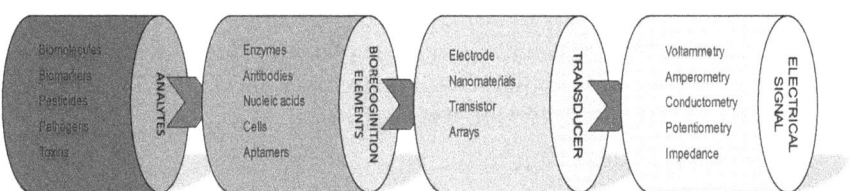

Fig. 16.1 Schematic illustration of electrochemical biosensor

detection, the colloidal nanoparticles conjugated with antibodies are used in immunolabeling and immunosensing applications. Moreover, metal compound nanomaterials are very effective in the detection of nucleic acid sequencing. Continuing along the same direction, metal oxides nanomaterials have been used for biochemical sensors, like glucose, cholesterol, urea, uric acid, ascorbic acid, and other metabolic intermediates. Similarly, carbon nanotubes have been widely used for glucose and insulin biosensor due to their rapid detection and biocompatibility (Gerwen et al. 1998).

In biosensor, the selection of nanomaterials for electrode fabrication depends upon a multitude of factors and their optimization processes. The most important factors are the physicochemical properties in addition to their selectivity and sensitivity. Foremost, the main focus is the nanofabrication process, which involves two critical procedures, namely the construction and design of the sensitive nanoscale surface. Four basic techniques are used for developing nanomaterial-based biosensor electrodes, namely lithography, surface etching strategies, thin-film strategies, and chemical bonding. Among these techniques, the lithography technique provides large surface areas for effective immobilization with high precision and sensing accuracy (Pak et al. 2001).

16.3 Types of Nanomaterials in Biosensor

16.3.1 Nanosized Metal Compounds

Metal and metal oxides nanoparticles (MNPs and MONPs) are widely used in highly sensitive novel electroanalytical devices. These nanostructures can enhance the selectivity of the device when conjugated with biorecognition molecules. Both MNPs and MONPs exhibit significant electrical and electrochemical properties because of their tunable bandgap, size, stability, and large surface area. In biosensor fabrications, the techniques like physical adsorption, electrodeposition, and chemical covalent bonding are employed to fix MNPs and MONPs on the surface of the working electrode. Generally, MNPs such as gold (Au), platinum (Pt), silver (Ag), and palladium (Pd) have been extensively used in biosensors. These MNPs are used alone or in combination with other nanostructures in biosensor electrodes. On the other hand, the transition MONPs are mainly used in biosensors, which include oxides of iron, silver, copper, zinc, cobalt, manganese, nickel, vanadium, zirconium, titanium, and tungsten. Biomolecular sensors fabricated using these metallic nanostructures can contribute to significant signal amplification, higher sensitivity, selectivity, and great advances in sensing as well as quantification of various biomolecules and ions (For example Table 16.1) (Lia et al. 2010).

Table 16.1 Metal and metal oxide nanomaterials-based biosensor and performances

S No.	Nanomaterial	bioreceptor	Sensing agent	Deduction limit	Reference
1	Au NPs-MoS$_2$	Laccuse	Catechol	2 µM	Zhang et al. (2020)
2	Ag-GQDs	ssDNA	*Legionella Pneumophila*	1 zM	Mobed et al. (2020)
3	Au-Pt NPs	—	Daptamycin	0.161 pM	Ozcelikay et al. (2020)
4	Pd NPs	A549 aptamer	Cancer cells	8 cells mL^{-1}	Cui et al. (2020)
5	Antimone (Sb) QDs	Catalase	H$_2$O$_2$ from ovarian cancer samples	4.4 µM	Fatima et al. (2020)
6	Fe$_2$O$_3$	—	Carcinoembryonic antigen	—	Kumar et al. (2019)
7	ZnO	Lectin	Arboviruses	—	Simão et al. (2020)

16.3.1.1 Gold Nanoparticles (AuNPs)

The unique properties of gold nanoparticles (AuNPs) like easy conjugation to biomolecules of interest and low toxicity increase their interest in the biological applications. The potential properties of AuNPs such as quantum scale dimension, extreme conductivity, high catalytic activity, high surface-to-volume ratio and excellent biocompatibility facilitate their wide application in biosensors. The AuNPs are transducers in a biosensor, which produce a detectable signal on biological recognition of the primary event originating on their surface or interfaces. Moreover, the adsorption of biomolecules on the AuNPs surfaces can retain their bioactivity and stability because of the biocompatibility and the high surface free energy of the AuNPs. Apart from the pristine AuNPs, the AuNPs coupled with or dispersed in polymeric compounds like polyaniline, polypyrrole, and chitosan are also potential for the fabrication of electrochemical biosensors. These AuNPs/polymer composite electrodes have considerable stability in different solvents, better biocompatibility, high processability, and are even retainable or reusable. For example, the carboxymethyl chitosan-AuNPs nanocomposite designed by Xu and coworkers is potential for H$_2$O$_2$ electrochemical sensing (Xu et al. 2006). This nanocomposite exhibited good hydrophilic nature with biocompatibility for enzyme immobilization and showed excellent biosensor performances with good stability and sensitivity.

The electrochemical glucose biosensor constructed by surface-immobilized periplasmic glucose receptors on AuNPs displayed a sensitive detection of glucose with a low detection limit (0.18 µM). In this, the genetically engineered cysteine is crucial for the immobilization of the receptor (protein) to the AuNPs by a direct sulfur-gold bond (Andreescu and Luck 2008). Similar to the polymer matrix, the AuNPs revealed excellent performance when worked in combination with other metal nanoparticles. For instance, the copper-nanoflower decorated

AuNPs-graphene oxide nanofiber-based electrochemical biosensor fabricated by Baek et al. showed high sensitivity and selectivity in glucose detection (Baek et al. 2020). The large surface-to-volume ratio, catalytic, and interface-dominant properties of AuNPs can decrease the overpotentials of many analytically important electrochemicals and even contribute to some reversible redox reactions. These advances appropriate the fabrication of enzyme-free AuNPs biosensors. Moreover, the AuNPs contributes to the oxidation and reduction of H_2O_2 and leads to the development of biosensors for the detection of various kinds of analytes. For example, Pekmez and coworkers reported an enzyme-free H_2O_2 electrochemical biosensor utilizing a disposable electrode fabricated by the poly(2-aminophenylbenzimidazole)/AuNPs-coated pencil graphite (Teker et al. 2019).

16.3.1.2 Silver Nanoparticles (AgNPs)

Silver nanoparticles (AgNPs) are another significant MNPs widely used in biomedical applications owing to their high conductivity, amplified electrochemical signal, high sensitivity, antibacterial activity, and excellent biocompatibility (Prasad 2014; Aziz et al. 2014, 2015, 2016, 2019). Recently, AgNPs and their nanocomposites have been extensively used for the design and construction of novel analytical techniques for various analytes like the early-stage diagnosis of diseases by disease markers, biosensors, and other disinfection agents. The silver nanoparticles composited with metal oxides, CNTs, silicates, graphene, polymers, dendrimers etc., considerably enhance the performance of biosensor. Since, the AgNP composites synergize the conductivity, electrocatalytic activity and biocompatibility of the materials. Researchers developed silver nanoparticles-based biosensor for the detection of H_2O_2, glucose, dopamine, uric acid, ascorbic acid, etc. For example, Xu et al. developed a high-performance enzymatic biosensor with polymeric nanoparticles and conductive silver nanoparticles for H_2O_2 detection. Here, horseradish peroxidase as a model enzyme was co-assembled with an amphiphilic and photo-cross-linked with polypeptides. This biosensor demonstrated high sensitivity and stability with low detection limit and wide detection range for H_2O_2 in milk and human urine samples. This promising feature of its application in the real samples can provide inspiring thoughts for the development of new biosensing systems (Xu et al. 2019).

16.3.1.3 Platinum Nanoparticles (PtNPs)

Recently, platinum nanomaterials (PtNPs) have engrossed in electrochemical biosensors for disease diagnosis and other biomedical applications, because these noble metal NPs have unique electronic properties and electrocatalytic activities for many chemical reactions. These properties completely depend upon size, shape, compositions, crystal orientation, surface reactive site etc., of the NPs. The PtNPs also ease the electron transfer and also easily modified with a wide range of biomolecules and compounds. The PtNPs and nanocomposites exhibit fast, reliable, and precise bioanalytical methods. The electrochemical sensors based on platinum nanoparticles revealed enhanced sensitivity and selectivity towards the detection of various biomolecules. Various research reports are available for platinum nanoparticles and their nanocomposites-based biosensors. For instance, a novel glucose biosensor was constructed based on platinum nanoparticles into polyaniline-montmorillonite hybrid nanocomposites for glucose detection in human serum. This platinum nanocomposite biosensor exhibited excellent stability over two months and with high sensitivity and selectivity (Zheng et al. 2020).

16.3.1.4 Palladium Nanoparticles (PdNPs)

Similarly, the palladium nanoparticles are also an interesting candidate for the construction of biomedical devices due to their high catalytic activity and sensing properties. For the facile selective catalytic and sensing property of the PdNPs, the size and shape of the nanoparticles are very important, which can be attained during the synthesis process. Due to the unique catalytic performance, electronic properties and sensing behavior, different forms of palladium-based nanostructures like composites, bimetallic compositions, metal oxides, and carbon composites were used in biosensing devices. Most commonly, PdNPs and their composite-based biosensors have been utilized for the detection of glucose, H_2O_2, dopamine, cholesterol, etc. (Phan et al. 2020). Dopamine is a significant catecholamine neurotransmitter, which is used for the early diagnosis and treatment of diseases like Parkinson, Alzheimer's, and Schizophrenia. In a dopamine biosensor, the PdNPs are incorporated with the nanoporous gold for the construction of the sensor electrode. This combination of PdNPs enhances the electrocatalytic effects on the dopamine with the broad detection range, high sensitivity, and excellent selectivity (Yi et al. 2017).

16.3.1.5 Quantum Dots

Quantum dots (QDs) are another prominent 0D-nanomaterial used for the bioanalytics and biomedical applications. The QDs are luminescent semiconducting nanocrystals, mostly the metal chalcogenides (MX; M = Cd, Pb, Cu, etc., X = S, Se, Te). These QDs can provide a high absorption spectrum with a size-dependent narrow emission spectrum. Small-sized quantum dots have unique optical and electronic properties, which make them a promising candidate in biomedical, chemical, and physical applications. Especially in biosensors, the QDs are a construction element for efficient detection of various biomolecules. The surface of QDs can easily modify with biomolecules such as peptides, antibodies, enzymes, and DNAs for the construction of sensing devices. QDs with high photostability, high quantum yield, and long life makes them ideal for fluorescence-based biosensor. Interestingly, the fluorescence resonance energy transfer (FRET) method has widely used for QDs-biosensors (Ma et al. 2018). Here, fluorescence signals can be detected by a single molecule detection method, as this technique has distinct advantages such as high signal-to-noise ratio, low sample consumption, rapid analysis time, and high sensitivity. This analytical method is used to detect various biomolecules including enzymes, microRNAs, DNA methylation, and DNAs.

For disease diagnosis, the target enzymes such as caspases, DNA glycosylase, DNA methyltransferase, terminal deoxynucleotide transferase (TdT), O-GlcNAc transferase, protease are detected by QD-based FERT biosensors. For instance, Petryayeva and Algar designed a QDs (CdSeS/ZnS) immobilized paper-based protease biosensor. They demonstrated that the QDs are viable probes that serve as alternatives to AuNPs for the next-generation paper-type diagnosing kits. This paper-based biosensor detected the analyte within 5–60 min at a low level of 1–2 nM protease (Petryayeva and Algar 2013). Similarly, glycoproteins, prostate-specific antigen, and carcinoembryonic antigen are also targeted for early diagnosis and treatment using QDs-based sensors.

Glycoproteins are essential large family proteins, and their abnormal expressions represent the presence of some dangerous diseases. So, glycoproteins are significant biomarkers for biomedical diagnosis and research. In recent work, a fluorescent FRET probe between the glucosamine-Mn-doped ZnS QDs and mercaptophenylboronic acid (MBA)-capped AuNPs is demonstrated for the detection of glycoproteins. The designed sensor probe was accurate to determine 10^{-9} M glycoproteins and showed wide linear detection range for glycoproteins like α1-acid glycoprotein (AGP; 0–0.5 µM) and immunoglobulin G (IgG; 0–2.4 µM). Further, the QDs-based FRET probe exhibited high selectivity and anti-interference ability in the detection of ACP from serum (Chang et al. 2017).

MicroRNAs are short noncoding RNAs, which are responsible for gene expression by binding to the 3′-untranslated regions of target mRNA. The microRNA dysregulation may lead to a variety of human diseases such as cardiovascular diseases and cancers. Ho and Willner fabricated a QD-based biosensor for the detection of miR-141 (microRNA), a promising biomarker for prostate cancer. This biosensor achieved the lowest of 1 pM detection limit and implemented for the analysis of miR-141 in serum samples. And the analysis showed impressive result between the serum samples of healthy individuals and prostate cancer patients (Jou et al. 2015).

16.3.1.6 Iron Oxide Nanoparticles

Magnetite (Fe_3O_4) and Hematite (Fe_2O_3) nanoparticles are widely used phases of iron oxides in electrochemical sensors. Because, the iron nanoparticles are excellent immobilization matrix due to its various interesting catalytic and physicochemical properties. Moreover, iron oxide nanostructures are used for the fabrications of biosensor electrodes, which are potential for sensing various analytes like glucose, H_2O_2, heavy metals, and organic entities (Kaushik et al. 2009; Lee et al. 2016;

Absalan et al. 2015; Fang et al. 2005; Zhang et al. 2011). Further, the combination of iron oxide nanostructures with various nanostructures like carbon, metal compounds, and polymer has considerably improved their electrocatalytic properties. Moreover, these compositions lead to the enhancement of electron-transfer kinetics, surface-to-volume ratio, biocompatibility, electrical conductivity, mechanical strength, etc. Besides, the iron oxide-polymer composites are used for the fabrications of lightweight, flexible disposable biosensors due to the synergetic effects rendered by the combination of metal oxide and polymer (Zhu et al. 2015; Wen et al. 2014). For example, Kumar et al. devised an iron oxide/poly(3,4-ethylenedioxythiophene): poly(styrenesulfonate) (PEDOT:PSS) modified conducting paper-based biosensor. The resultant biosensor was promising for the detection of cancer biomarker (carcinoembryonic antigen) and achieved a low detection rate of 4–25 ng mL^{-1} with high sensitivity (10.2 µA ng^{-1} mL cm^{-2}) (Kumar et al. 2019).

In addition, other magnetic nanoparticles, such as cobalt oxide, nickel oxide,- based biosensors have been broadly used to detect a wide range of analyte targets like proteins, enzymes, drugs, DNA/mRNA, pathogens, and tumor cells. Furthermore, these magnetic nanoparticles can be used as labels or integrated into transducer materials, which effectively enhance the sensitivity and stability of the biosensors. Moreover, high accessible active surface and superior electron-transfer behavior of magnetic nanoparticles are the bonus advantages of these magnetic materials for electrochemical biosensor (Rocha-Santos 2014). Furthermore, the noble metal nanoparticles supported with these magnetic metal oxide nanoparticles display much higher electrocatalytic activity. Recently, Lang et al., developed nanoporous supported cobalt oxide hybrid microelectrodes as a nonenzymatic electrochemical glucose biosensor. This amperometric glucose biosensor exhibited a multi-linear detection with high sensitivity (12.5 mA mM^{-1} cm^{-2}) and very low limit of detection 5 nM (Lang et al. 2013).

16.3.1.7 Zinc Oxide Nanoparticles

Zinc oxide is another important metal oxide widely used in biosensors. Its promising properties like wide bandgap, high exciton, better electrochemical activities, low cost, nontoxicity, chemical and photochemical stability, biocompatibility, and high-electron communication features attract the material for designing biosensors (Bhuyan et al. 2015). Different crystallite size and morphologies of ZnO can be synthesized by controlling various factors like pH, reaction temperature, capping agents and surfactants. Nanoflowers, nanorods, nanoplates, nanorods, nanocubes, and nanospheres are the versatile morphologies of ZnO nanomaterials. ZnO nanostructure exhibits good sensitivity and stability in electrochemical biosensors. The ZnO acts as an active platform with particular binding affinity for the immobilization of biological recognition component. The 1D ZnO nanostructures like nanorods, nanowires, and nanotubes are interesting due to their large surface area and can provide a direct and rapid electron transport pathway. In field-effect transistor biosensors, the vertical and lateral 1D ZnO have exhibited long-term monitoring potential, large surface area, high enzyme immobilization efficiency, long-term stability, and simple fabrication techniques (Zhao et al. 2010). Further, the 3D ZnO nanoarchitecture is also demonstrated as a promising candidate, and this structure highly influences the biosensing process. Mostly, these nanostructures are synthesized by the bottom-up approach and are potential for developing amperometric, potentiometric, and impedimetric biosensors (Napi et al. 2019). Anusha et al., fabricated glucose biosensor with the aid of nanoporous ZnO with glucose oxidase enzyme. The cyclic voltammetry and impedance spectroscopic analysis showed enhanced glucose sensing property with good analytical performance and high sensitivity (Anusha et al. 2014).

16.3.1.8 Other Metal Oxide Nanoparticles

Manganese oxide nanoparticle is another beneficial and considerably studied material for electrochemical biosensor. Different crystalline phases like MnO, MnO_2, and Mn_3O_4 of the manganese oxide nanomaterials are significantly utilized in biosensors, since manganese oxide is a low-cost, nontoxic, environmentally friendly, and natural abundant material. In biosensor, various one-dimensional manganese oxide morphologies like nanowires, nanorods, nanobelts, and nanoneedles are widely used and showed promising sensing properties (Majd et al. 2016). In addition, the copper oxide (CuO or Cu_2O) is a p-type semiconductor which is an extensively studied metal oxide nanostructures for the electrochemical sensors. It can be easily synthesized with different morphologies, integrated with other nanostructures like CNTs, graphene, activated carbon, conducting polymers, and have potential applications in highly selective and sensitive electrochemical sensors (Li et al. 2015).

Similarly, the rare earth oxide, cerium dioxide has a considerable interest in the field of biosensors owing to its high catalytic activity and a better immobilization matrix. The electrochemical H_2O_2 sensor developed using single-walled carbon nanohorns/cerium oxide nanoparticles showed excellent H_2O_2 sensing with a low limit of detection 0.1 mM. Besides, it showed high stability, excellent reproducibility and increase sensitivity. The research team also examined the performance of biosensor in milk and cleaning liquid. The results showed a substantial selectivity towards H_2O_2 even such complex matrices and confirmed it as a highly promising material for the development of numerous biosensor (Bracamonte et al. 2017).

16.3.2 Carbon Nanomaterials

Carbon nanostructures are very popular due to their advanced physicochemical properties and specific structures. Carbon nanomaterials exist in a variety of dimensions: 0D, 1D, 2D, and 3D nanostructures. The graphene quantum dots (GQDs) are a zero dimensional carbon nanostructure consists of single or few layer of *sp2* bonded carbon atoms (graphene). The GQDs exhibits excellent chemical, physical, and biological properties that allow them in numerous potential medical applications. The 1D carbon nanotubes have a peculiar tubular structure, high mechanical stability, good biocompatibility, and excellent electron transfer properties. Graphene is a 2D structure, which has many advantages such as large active surface area, excellent conductivity, high carrier capacity, and stability. Nanocrystalline diamond and fullerite are the best examples of three-dimensional structures. These carbon nanostructures have been potentially utilized for the development of biosensors, and they displayed sensing properties regarding their various structures. The biosensors fabricated using some technologically important carbon nanostructures are briefly discussed below.

16.3.2.1 Carbon Nanotubes (CNTs)

The carbon nanotubes (CNTs) have a unique one dimensional structure, a high electrical conductivity, large surface area, good chemical stability and excellent electron transfer properties arouse growing interest in modern electronic and biomedical devices. In electrochemical biosensors, CNTs are employed as transducers, which significantly enhance the sensitivity and detection properties. The high surface-to-volume ratio of CNT alters immobilization of greater concentrations of bioreceptor. The high electrical conductivity of CNTs and their feasible surface functionalization properties are highly appropriate for the recognition of a target and transduction of their signals (Clancy et al. 2018). Moreover, their ability to penetrate within the biological membranes makes them relevant for in vivo photoacoustic imaging. Based on the mechanism, target recognition, and transduction, the CNT-based biosensors are classified as electrochemical biosensors, immunosensors, and optical biosensors.

CNTs are highly appreciable for the development of sensitive biosensors, which enable the easy and early diagnosis of various diseases including cancer. Figure 16.2 schematically illustrates CNT biosensors for the detection of various cancer biomarkers coupled with DNA, enzymes, antibodies, proteins, aptamers, and peptides (Shobha and Muniraj 2015; Choi et al. 2010). CNT-based immunosensor is still in the incipient stage, and many challenges are there to overcome for successful commercialization. The multiwall carbon nanotubes (MWCNTs)-coated paper-based disposable bipolar electrode is used for the diagnosis of prostate-specific antigen, and this electrochemiluminescent detection showed high sensitivity and specificity in electrochemical biosensing (Feng et al. 2014). The multi-array sensor was fabricated using chitosan/MWNTs via electrodeposition technique. The resultant sensor array demonstrated the simultaneous detection of endogenous metabolites, drugs, pH, and temperature (Baj-Rossi et al. 2014). In another work, the early-stage diagnosis of prostate cancer has been achieved through a biosensor designed using DNA strands functionalized SWNTs and MWNTs electrodes for the effective detection of PSA in blood samples. Zheng and coworkers devised the electrochemical HeLa and HL60 cancer cell sensors using folic acid functionalized polydopamine-coated

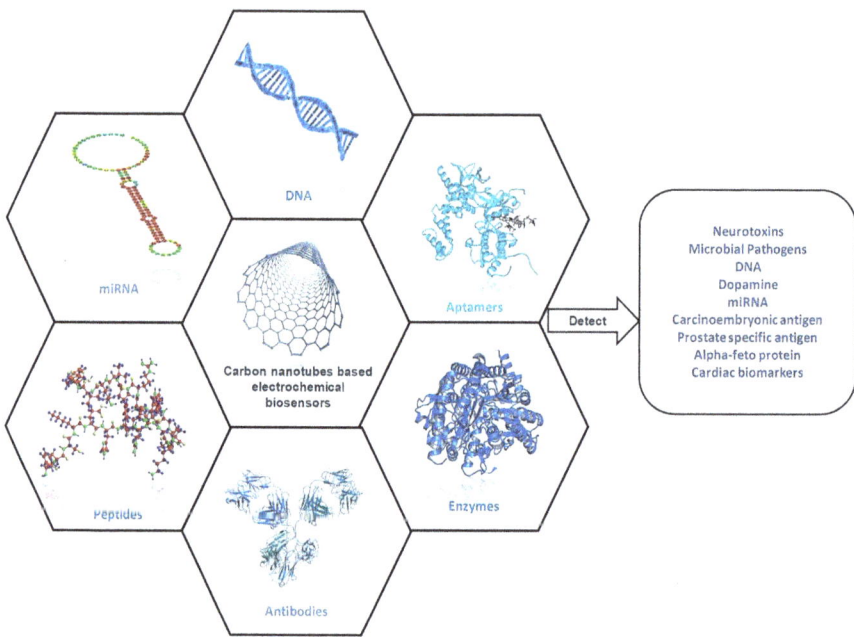

Fig. 16.2 Carbon nanotubes (CNTs)-based bio-recognition elements for the detection of analytes

CNTs. Here, the cancer cells were detected by evaluating the overexpressing folate receptors (Zheng et al. 2012).

CNTs-based biosensor electrodes are applicable for the detection of cancer biomarkers such as microRNA, DNA, alpha-fetoprotein, PSA, or CEA and quantified by the clinically significant metabolites like lactate, cholesterol, glucose, and glutamate (Dey et al. 2013). In recent years, a wide variety of CNTs-based amperometric sensors, field-effect transistors (FET) biosensors and impedimetric sensors have been designed and reported for the detection of various cancer biomarkers. For example, CNT-based impedimetric biosensor was fabricated using vertically aligned carbon nanotube on Ni/SiO$_2$/Si layers by photolithography process for the efficient sensing of SW48 cells (Abdolahad et al. 2012). Similarly, Barekat et al. developed a screen-printed amperometric biosensor using CNT to detect the formaldehyde released from U251 human glioblastoma cells. The biosensor is comparatively cheap and displayed sensitive, selective, and rapid detection of the target (Bareket et al. 2010). Moreover, the CNT-based electrochemical biosensors are potential towards the detection of H_2O_2, ions, metabolites, and protein biomarkers (Tilmaciu and Morris 2015).

16.3.2.2 Graphene

Next to CNTs in the field of the biosensor, graphene is widely used 2D carbon nanomaterial with one atom to a few nanometers thickness. Graphene is the basic building block of all other dimensional graphite materials. Various forms of graphene such as graphene oxide, reduced graphene oxide, and graphene nanoribbons are prominently used in electroanalytical detection applications towards various disease diagnosis. The peculiar properties of graphene, like excellent electrical conductivity, high optical and mechanical properties, biocompatibility, rich active sites, and large specific surface area, make them a potential material for the construction of biosensors with the outstanding performance (Krishnan et al. 2019). Graphene with high water solubility and biocompatibility was rendered by an optimized chemical functionalization process. This graphene structure can be much easier to alter with some chemical and biological functionalities for better sensing properties. Similarly, the graphene oxide reduced to form functionalized graphene have a wrinkled structure due to the presence of lattice and edge defects, and this is different than the rippled structure observed in pristine graphene. These lattice or edge defects have a better affinity towards the biomolecules or other functionalities to determine the performance of biosensors. Among various graphene structures, the porous graphene (3D graphene) structure is recently confirmed as a promising candidate for immobilizing enzyme and enhancing performance in biomolecule sensing (Taniselass et al. 2019).

Recent reports establish that the graphene and its composite electrodes have extraordinary electron-transport property, high surface area, and anchoring site for the effective immobilization of enzymes. Liu et al. fabricated dopamine biosensor using PEDOT: PSS/graphene composites on fluorine doped tin oxide (FTO) electrode. The resultant biosensor detected dopamine from the aqueous medium in the presence of uric and ascorbic acid. The device showed an interference-free detection of dopamine with a detection limit of ~105 nM and sensitivity of ~27.7 $\mu A \; \mu M^{-1} \; cm^{-2}$ (Liu et al. 2017). In another study, the $Ni(OH)_2$ nanoflakes/graphene oxide (GO) nanosheets were used for the fabrication of the dopamine electrochemical sensor, which exhibit an excellent selectivity under the interference of uric acid, also showed high repeatability and stability (Yue et al. 2019).

16.3.3 Polymeric Nanomaterials

Recently, polymeric nanomaterials were widely used in biosensor platforms due to their biocompatibility, long-term stability, tunable surface functionalities for the attachment of biomolecules, and flexibility (Prasad et al. 2017). For instance, polymeric structures with planned structure, homo-polymers, copolymers, and molecular shape recognition structures are utilized for the fabrication of biosensors. The most important factors that influence the performance of polymer-based sensor were biofunctionalization of the exposed active surface and biomolecules, durability by the type of binding with biomolecules, high specificity towards bioanalytes, and increased electrochemical signal transduction. Generally, polymer-based biosensors are used for the detection of DNA, protein, antigens, carbohydrates, enzymes, and metabolites due to their high sensitivity, selectivity, and linearity.

16.3.3.1 Conducting Polymers

Conducting polymers have an immense interest in biosensor devices due to their special physical and chemical properties and scalable and easy processing. The considerable electrical conductivity, biocompatibility, possible surface modification, low cost, large surface area, etc., of the polymers added special advantage for developing sensing systems. Moreover, the lightweight, mechanical durability, and flexibility of the polymer nanostructures are viable for the construction of flexible biosensor devices. Conducting polymers nanomaterial-based biosensors, especially field-effect transistor biosensor, DNA chips, aptasensors, and immunosensors have exhibited high sensing performance towards biomolecules or biological species. Conducting polymeric nanostructures like polypyrrole, polyaniline, and poly(3,4-ethylene dioxythiophene) are commonly used to fabricate sensors.

Polypyrrole (PPy) is a well-known and widely used conductive copolymer formed by polymerization of pyrrole. Recently, PPy nanostructures such as nanoparticles, nanotubes, hollow nanospheres, and core-shell nanomaterials have been used in biosensor devices due to facile functionalization, high conductivity, and environmental stability and friendliness. The PPy can be electrochemically polymerized on any required electrode surface in a controlled and optimized manner, which is promising for designing biosensors. Moreover, the selected biological compounds or aptamers can easily dope or incorporate with PPy layer via electrodeposition to enrich the sensing properties. For a microfluidic aptasensor, the aptamer and PPy nanowire were integrated directly over gold electrode via one-step electrodeposition process by Huang and coworkers. The device detected IgE protein solutions in a linear range between 0.1 and 100 nM with excellent stability and specificity. Moreover, the microfluidic aptasensor exhibited a low detection limit for a cancer biomarker MUC1, which was found comparatively precise than the commercial MUC1 diagnosis assay (Huang et al. 2011). Yuan et al. fabricated an ultrasensitive immunosensor on a disposable indium doped tin oxide (ITO) glass for the detection of gypican-3 (GPC3) in human serum (GPC3 is a tumor marker for hepatocellular carcinoma). In this sensor, the pyrrole-α-carboxylic layer was electrochemically polymerized on ITO and the GPC3 antibody was directly covalently bonded with the carboxyl group. The resultant immunosensor demonstrated high sensitivity towards the detection of GPC3 analyte with considerable stability and reproducibility (Yuan et al. 2015).

Polyaniline (PANI) nanomaterials are another widely used conducting polymers in sensors due to their high conductivity, good environmental stability, and diverse color change in response to different redox states. PANI has been used in biosensors, neural probes, tissue engineering, and controlled drug delivery applications. Zhai et al. fabricated an extremely sensitive enzymatic glucose biosensor with hydrogel heterostructure composed of platinum nanoparticles and PANI. The enzymes were immobilized onto the PANI porous hydrogel structure, and it effectively catalyzes the glucose oxidation reaction. Moreover, the platinum nanoparticles catalyzed the reduction of H_2O_2 produced during the enzymatic reaction. This glucose sensor exhibited enhanced performance with ultrahigh sensitivity, a low limit of detection, and a fast response (Zhai et al. 2013).

Poly(3,4-ethylenedioxythiophene) (PEDOT) is a polythiophene derivative synthesized by conventional oxidative or electropolymerization techniques. PEDOT is a transparent thin film, which exhibits high electrical conductivity, excellent stability in the oxidized state, high charge mobility, and serves as a suitable matrix for enzyme immobilization. The PEDOT 1D structures like nanorods, nanowires, and nanotubes have a unique mechanism of direct electron transfer during electrochemical detection. For example, Vasantha et al. modified glassy carbon electrode with PEDOT nanostructure and reported as a dual sensor for the detection of dopamine and ascorbic acid. From the analysis, they observed a hydrophobic interaction between the aromatic groups of dopamine and the PEDOT film. Meanwhile, ascorbic acid was detected using the electrostatic interaction between the heteroatom of PEDOT and ascorbate ions (Vasantha and Chen 2006).

16.3.3.2 Molecularly Imprinted Polymers

In recent times, a widely developing technique called molecular imprinting allows the formulation of specific recognition sites in polymer matrices. This molecular imprinted polymer is potential for the development of the efficient biosensor for sensing various chemicals and biological molecules (Behera et al. 2020). Molecularly imprinted polymers (MIP) has achieved by simple wet chemical techniques precipitation, dispersion and emulsion seed polymerization techniques. In biological sensors, MIPs have to couple with appropriate transducers for quantitative detection. The combination of metal nanoparticles and polymer provides excellent functional nanomaterials for biomedical applications. The molecularly imprinted electroactive nanoparticles of water dispersible quality and a macromolecular self-assembly technique were employed for the development of paracetamol sensor. The resultant sensor exhibited good sensitivity towards the detection of paracetamol with a low detection limit of 0.3 µM (Luo et al. 2016).

16.3.3.3 Dendrimers

Dendrimers are 3D, hyper-branched, monodispersed, nanoscale polymeric materials with a high density of surface functional groups with unique molecular weight, shape, and size. These dendrimers structures have a single chemically addressable group called core or focal point, the dendron, and the surface functional groups. Various techniques have been developed for the synthesis of dendrimers such as divergent approach and convergent approach. In biosensors, the high mechanical and chemical stability, presence of excess surface functional groups and hydrophilicity established dendrimers as a viable immobilization matrix for biomolecules. Moreover, these properties enhanced the overall performance of the sensor and increased the sensitivity, stability, reproducibility and reusability. Dendrimers are poor conductors, but metallic nanoparticles can easily be coupled with functional groups to improve conductivity. Besides, the dendrimers have a considerable amount of amine groups on the surface, which is a much favorable site for the conjugation of numerous bioreceptors (Satija et al. 2011). Shende and Sahu fabricated enzyme-conjugated PAMAM dendrimers for the estimation of glucose content in saliva. In this work, the Fe_3O_4 nanoparticles encapsulated in PAMAM dendrimers to immobilize glucose oxidase in the presence glutaraldehyde a cross-linking agent. The biosensor showed a rapid response and high reproducibility (Shende and Sahu 2021). A surface plasma resonance (SPR) sensor was reported based on self-assembled monolayer/reduced graphene oxide-polyamidoamine dendrimer thin film for the effective detection of DENV-2 E-proteins (a dengue virus envelop protein). The sensor revealed high specificity, sensitivity, binding affinity, and selectivity towards DENV-2E proteins for the diagnosis of dengue virus (Omar et al. 2020).

16.3.4 Nanosized Biomaterials

Over the last two decades, the nanoscale biomaterials like biopolymers, proteins, nucleic acids, enzymes, and their nanocomposites are widely used in the biosensor, drug delivery, bio-imaging, biocatalysts, and cell targeting applications. By the combination of biotechnology and nanotechnology, nanocomposite materials were synthesized with extreme catalytic and recognition ability for electrochemical biosensor platform. Recently, electrochemical biosensor designed utilizing bio-nanomaterials have great attention in research and clinical field owing to its excellent performance like high selectivity, reproducibility, sensitivity and biocompatibility. Few of these nano-sized biomaterials utilized for sensor applications are discussed in the following section.

16.3.4.1 Chitosan

Among numerous commonly available natural polymers, chitosan is a marine-based biopolymer widely applied in the biomedical field. Chitosan is a natural polysaccharide, the most important derivative of chitin and derived using the N-deacetylated method. Chitin is obtained from the hard outer skeleton of arthropods including crab, shrimp, lobster, and beaks of cephalopods. It is also present in the cell wall of insects and some microorganisms like fungi. Due to its biocompatibility, environmental stability, good mechanic strength, hydrophilicity, low cost, and easy-processability, it is used as a substrate material or immobilization matrix in electrochemical biosensors. The chitosan has some amino and hydroxyl groups in its structure which can easily crosslink with various nanomaterials and biomolecules. Recently, chitosan nanocomposites formed in combination with inorganic metal compounds, carbon nanostructures, and bio-complexes are used for biosensors and biomedical devices. The chitosan nanomaterials-based biosensors revealed good sensitivity, selectivity, and stability for the detection of an analyte such as H_2O_2, proteins, glucose, DNAs, uric acid, biomolecules, microbial pathogens (bacteria to viruses).

Chitosan also acts as a mediator material between the biological recognition element and electrode surfaces. Zhang et al. reported a single-wall CNT/chitosan composite electrode as an aflatoxin B1 biosensor, where the chitosan is entrapped and attached with the SWCNTs on the glassy electrode surface. The differential pulse voltammetry achieved 3.5 pg/mL limit of detection (Zhang et al. 2016). Similarly, an amperometric biosensor was developed based on glucose oxide immobilized chitosan nanoparticles/gold electrode. Interestingly, these chitosan nanoparticles were synthesized from gladius of squid, *U. duvauceli*, by ionic gelation process. The biosensor possessed good amperometric response with high sensitivity. In addition, the Michaelis Menten kinetics revealed that a low value represents excellent substrate affinity towards the enzyme due to the presence of chitosan nanoparticles (Fig. 16.3) (Anusha et al. 2015).

16.3.4.2 Aptamers

Aptamers are artificial oligonucleotide ligands that bind to a highly specific target with high affinity. Aptamers can be isolated from a large random sequence pool using an in vitro process called systematic evolution of ligands (SELEX) process. Currently, numerous highly specific and high-affinity aptamers have been produced for various target molecules like proteins, peptides, and whole-cell. These aptamers are employed in biosensors as a recognition element known as aptasensors. In aptasensors, the aptamers can be immobilized without influencing its affinity and stability during denaturation and renaturation process. In an electrochemical aptasensor, the aptamers are immobilized onto the electrode surface, and the analyte binding

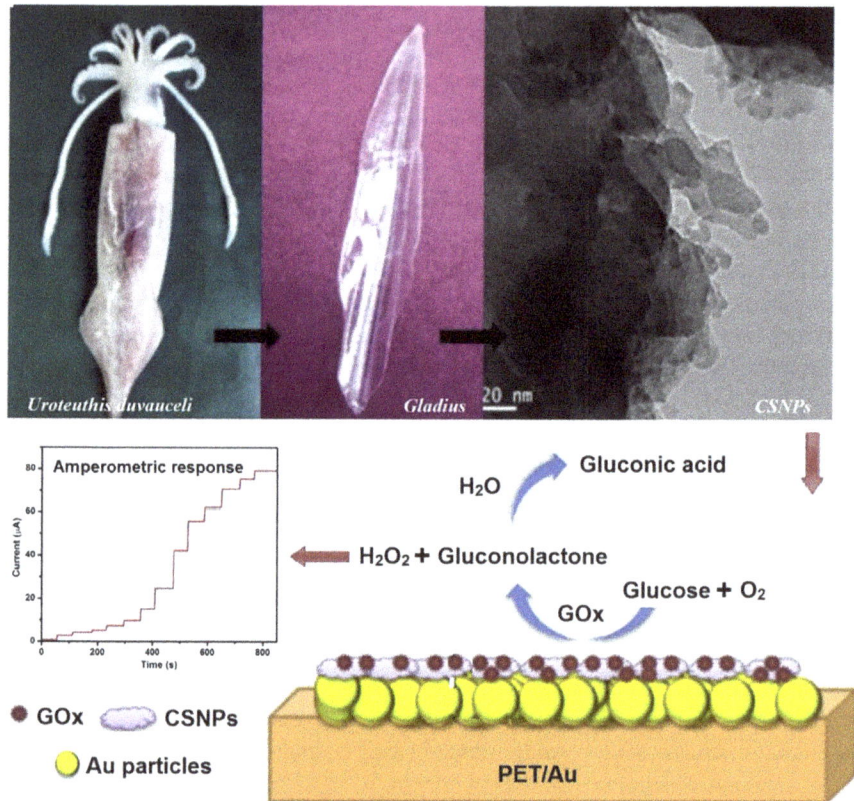

Fig. 16.3 Schematic illustration of electrochemical enzymatic glucose biosensor fabricated using chitosan nanoparticles from gladius of squid, *U. duvauceli* on PET/gold electrode. (Adopted from Anusha et al. 2015)

process is detected via the change in electrochemical signals. The main advantages of aptasensor are high sensitivity, miniaturized, and low fabrication cost (Hong et al. 2012).

Aptasensor is use for the diagnosis of cancer, and detection of various biomarkers and infectious disease-causing microorganisms. Generally, the aptasensors are used to detect biomarkers like thrombin, immunoglobulin E (Ig E), retinol binding protein (RBP4), C-reactive protein (CRP), N-terminal pro-brain natriuretic peptide (NT-proBNP), and interferon (IFN) in urine, blood, saliva, serum etc., for the early diagnosis of diseases. Zhang et al. developed an electrochemical aptasensor using functionalized mesoporous silica with MWCNT nanocomposites for thrombin detection. This device exhibited high sensitivity and wider linearity for the detection of thrombin in the range between 0.0001 and 80 nm with a low detection limit of 50 fM (Zhang et al. 2013). Using aptamer AS1411 and graphene modified electrode, a label-free cancer cell detection biosensor was developed by Feng et al. This cancer aptasensor can distinguish cancer cells from the normal one and detect as low as 1000 cells (Feng et al. 2011).

Microbial infections caused by pathogenic bacteria and viruses can be detected and identified for public health protection. Mostly, the virus-infected cells can be detected by investigating the particular targets found on the host cell surface (Yao et al. 2020). For example, Tombelli et al. fabricated an aptasensor using RNA aptamers immobilized over piezoelectric crystal (quartz) for the detection of HIV-1 Tat protein based on QCM and SPR sensing. Both sensing platforms showed similar reproducibility, sensitivity, and specificity (Tombelli et al. 2005).

16.3.4.3 DNA Nanomaterials

DNA is a perfect material for biosensing applications due to its unique functionalities and structural versatility. The DNA biosensors are very sensitive, cost-effective and potential point-of-care diagnostic tools. Moreover, robust sensing and biocompatibility provide multiple readout strategies. Nanoscale DNA structures are used in devices such as fluorescence, FERT, electrochemical signaling, Raman spectroscopy and nanoparticle-based color change. DNA-based biosensor work based on the specific recognition events happens between a substrate and the target analyte. In particular, DNA tetrahedron-based biosensors combined with the surface-based assays are considerable for electrochemical detection (Arun 2017). For example, detection of Mucin 1 (MUC1) a tumor marker model, very significant for the early diagnosis of patients with tumor or carcinomas (Deng et al. 2017). A novel combination of aptamer and DNA nanostructure were utilized for the fabrication of biosensor. For the detection of acetamiprid, a partial paired complementary DNA of acetamiprid aptamer and four single-stranded DNA sequences were inserted at each side of tetrahedral DNA and modified with gold nanoparticles. The developed DNA-based aptasensor showed high performance with a low limit of detection and assists the selective detection of pesticide residues (Yao et al. 2020). Interestingly, electrochemical genosensors and immunosensors based on functional nanomaterials were used for new electroanalytical techniques, developing a perfect point-of-care diagnosis, DNA/enzyme amplification techniques, and miniaturization of devices.

16.4 Applications of Nanobiosensors

The highly versatile and multifunctional behavior of nanobiosensors leads to many and perhaps endless applications, from disease diagnosis to environmental monitoring, food industry, agriculture, marine sector, etc.

16.4.1 In the Medical Field

In medical science, nanobiosensors play a lead role in diagnosing various disease. Biosensors are applicable for the detection of metabolic disorders, serum antigens, and carcinogens (Table 16.2). Serum analysis is applied in some of the routine applications in diagnosis of diseases like cancer, diabetes, allergy, and many other disorders. The nanomaterials make the diagnosis of diseases further sensitive, rapid, and more precise. For the diagnosis of diabetes mellitus, glucose biosensors are being

Table 16.2 Summary of advantages and limitations of biorecognition elements used in biosensors for the detection of various analytes in the biomedical field

S No.	Biorecognition Elements	Analyte	Advantages	Limitations
1	Enzymes	Substrate	Strong binding capacity; excellent catalytic activity; ability to inhibit or catalyze reactions; significant stability over years	Less stability at harsh conditions; interfere with endogenous enzymes, and require multiple assays
2	Antibody	Antigen	Permit variable detection of bacterial cells and metabolite toxins; antibody-antigen binding specific interactions; noninvasive capability and direct recognition ability.	Unable to differentiate live from dead cells; highly challenging production; with high cost and long time.
3	Nucleic acid	Microbial pathogens, oligonucleotides, etc.	Promptly produced and regenerated in comparison with antibodies, and enzymes; and recognition is based on nucleic acid sequence identification base-pairing attribute.	Unstable at extreme conditions; poor solubility in aqueous media; high production cost
4	Whole cells	Antibiotics, proteins, vitamins, pollutants, etc.	Low production cost; highly stable; very few purification procedures.	Specific environmental conditions required; limited self-life; interfere with multiple biochemical pathways and create similar cellular response which produces false-positive results.
5	Aptamers	Pathogenic microorganisms, cancer cells, etc.	Easily prepared; highly stable; recognition and detection based on shape.	Produce false results when interacting with large molecules

used pervasively in the medical field, as they provide precise control over blood glucose level. Before the development of biosensors, the detection of blood glucose was complicated, time-consuming, and expensive. But the vast advances in blood glucose biosensors directed the usage at home in routine. The incorporation of nanomaterials in biosensors allows the enzyme immobilization easy, and this allowed the reuse and recycle of enzymes. In glucose biosensor, glucose oxidase enzyme is immobilized in polymer nanomaterials which improve the sensitivity and accuracy of the device. Moreover, the biosensors are widely used in hospitals to diagnose infectious diseases, particularly to diagnose urinary tract infections and to identify pathogens.

At present, more than one million people are suffering from heart diseases, so early phase identification is very important. A human interleukin (IL)-10 biosensor is used for the early diagnosis of heart failure (Lee et al. 2012). Various other methods have used for the diagnosis of cardiovascular diseases, including immunoaffinity column assay, enzyme linked immunosorbent assay and fluorometric assay require skilled personnel, laborious, expensive, and time-consuming. For cancer detection, immunosensor array for clinical immunophenotyping of acute leukemia, histone deacetylase inhibitor assay from resonance energy transfer, hormone-based biochip for a quick and accurate diagnosis. Dengue is a mosquito-borne virus, cause dengue fever. The early detection of this virus infection is very important for the control of the disease. The conventional methods used for the diagnosis such as immunological and molecular techniques are time consuming, expensive and require skilled professional. But the construction of electrochemical biosensors helps for the early diagnosis of dengue virus serotypes (Fig. 16.4) (Anusha et al. 2019).

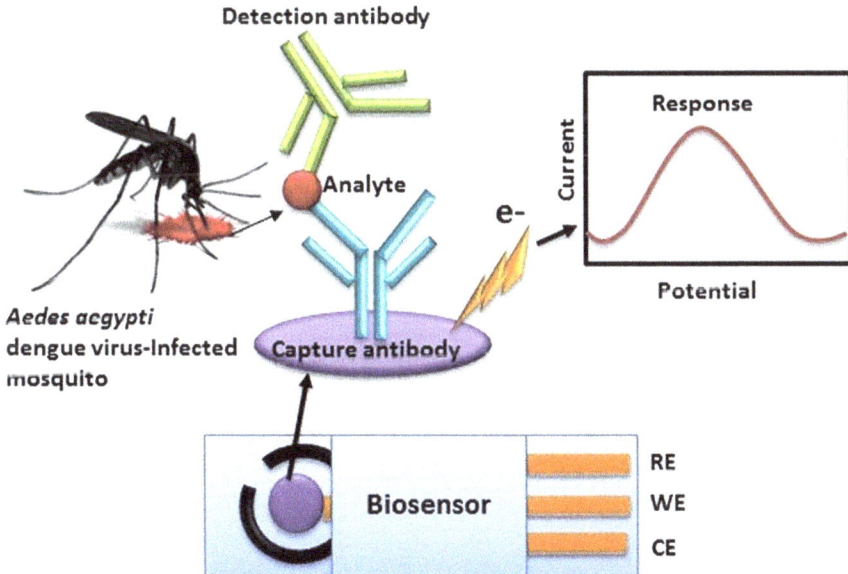

Fig. 16.4 The construction of electrochemical biosensor for the detection of dengue virus. (Adopted from Anusha et al. 2019)

16.4.2 In Environmental Monitoring

Our environment undergoes rapid change almost every second due to many factors, especially human activities. The detection of environmental pollutants, toxins, heavy metals, climate change, and many features are crucial to keeping the environment safe and healthy to live. The sensors play a vital role in the monitoring of the environment to rectify the problems. The nanomaterials-based devices with electronic probes require a small amount of analyte sample for analysis. Pesticide residue found in the soil, water, air and food cause major public health issues. Conventional methods used for the detection of pollutants involve liquid or gas chromatography (Zhao et al. 2007). This method is highly sensitive and reliable but requires skilled technicians, long process and procedure for meticulous sample preparation and very expensive. In contrast, electrochemical and immunoassay techniques are inexpensive, with simple fabrication and ease of miniaturization. Despite these advantages, the biosensors failed to conquer the market place. Hence nanotechnology is incorporated in biosensors to enhance sensitivity and reliability. Nanomaterials such as carbon nanotubes, graphene, quantum dots, and metal nanoparticles have been effectively used for novel biosensors (Fig. 16.5). For instance, acetylcholinesterase (AChE) enzyme is responsible for the destruction of neurotransmitter acetylcholine with nerve tissue, causing neurotoxicity and leads to death. The organophosphorus pesticides have an affinity for binding and inhibiting AChE. The AChE enzyme-biosensors have been used for the detection of pesticides and nerve agents, where AChE has been immobilized over nanomaterials to improve

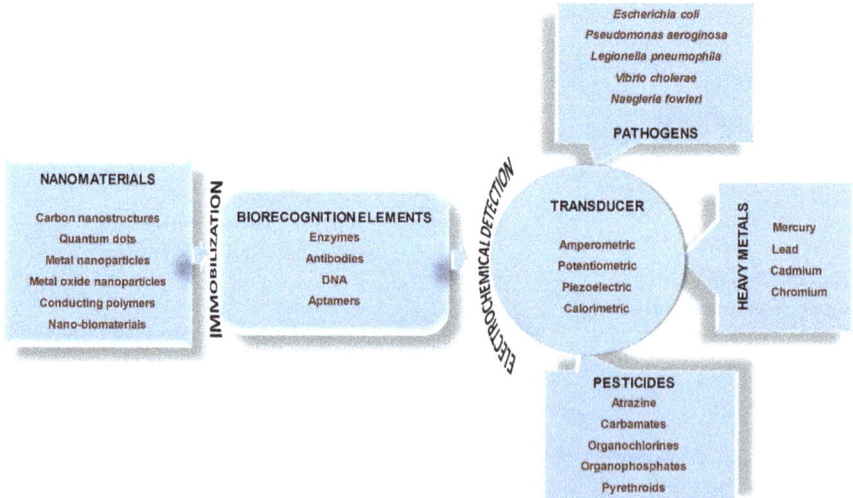

Fig. 16.5 Various nanomaterials and biorecognition elements utilized for the construction of electrochemical biosensor and analytical techniques used for the detection of environmental polluting agents such as heavy metals, pesticides, and pathogens

electrochemical response and stability. A self-assembled AChE on CNT- modified electrode was used for the detection of paraoxon (Liu and Lin 2006). Similarly, the substrate-specific detection mechanism has been developed for the detection of nitrates, inorganic phosphates, different contaminants, and parameters of biological oxygen demand. The application of nanobiosensors in environmental monitoring is highly time saving, energy saving, and economical.

16.4.3 In Food Industry

In the food industry, quality and safety are analyzed using traditional chemical experiments and spectroscopic analysis. These conventional methods are time-consuming, expensive, and require skilled laboratory technicians. The implementation of novel biosensors is a breakthrough in the food industry. The nanomaterials like CNTs, nanowires, MNPs, and other nanostructured materials are used as catalytic tools, immobilization platform to enhance biosensing. In most cases, the integration of enzymes, antibodies, and DNA sequences with nanomaterials provides a novel hybrid system for food industry applications (Chausali et al. 2021). Few examples that are applied in food analysis are pathogens, insecticides, and sugars. The main food poison or illnesses are caused due to having food contaminated by microbial pathogens like bacteria, fungi, or viruses. Microorganisms like *Salmonella*, *E. coli*, and *L. monocytogenes* are most examined pathogens regarding their detection and quantification in food. Biosensing platforms with nanomaterials are used to detect or identify these pathogens. For example, Dungchaia et al. developed a highly sensitive biosensor based on gold nanoparticles immobilized with monoclonal antibodies on polystyrene microwells for the detection of *S. typhi* (Dungchaia et al. 2008). The monitoring of carbohydrate level during the fermentation process is crucial in the food industry. Zhao and coworkers reported glucose detection in the food products using the Pt and Pd nanoparticles decorated graphite-based amperometric biosensor. This hybrid electrode increases the performance of glucose detection with high sensitivity and selectivity (Zhao et al. 2007).

16.4.4 In Agriculture

Nanobiosensors have a vital role in the detection of plant infections, and thus subsequent prevention of diseases is a considerable measure to sustain the availability of food and food products. The advances in the field of biosensor increase the growing global demands for higher food production, control diseases, food quality control and alarm climate conditions, which are very important in agriculture (Prasad et al. 2014, 2017a, b, c). Nanomaterial-based biosensors have a significant role in the detection of plant infections, phytohormones, metabolic content, abiotic stress, mi RNAs, etc., in a very short time of time. Traditionally, plant diseases have been detected through direct observation of change in the plant morphology. Later, some laboratory techniques such as serological, electron microscope, and polymerase chain reaction have been used for the early detection. In general, the plant viruses are detected using immunoassays and optical DNA hybridization biosensor (Khot et al. 2012; Mufamadi and Sekhejane 2017). Gold nanoparticles are highly used in sensor functionalization systems for the detection of pathogens. Gold nanorods were utilized for the fabrication of SPR immunosensor to monitor dual viruses of an orchid plant, *Cymbidium mosaic virus* or *Odontoglossum ringspot virus,* which achieved LODs of 48 and 42 pg/mL respectively (Lin et al. 2014). Apart from gold nanoparticles, nanomaterials such as graphene, CNTs, metal oxide nanowires were also used for the construction of a biosensor for pathogen and mycotoxin detection. A DNA biosensor was successfully developed with ZnO nanoparticles and chitosan nanocomposite modified in the gold electrode for the identification of the soil-borne fungi, *Trichoderma harzianum.* The electrochemical analysis revealed good sensitivity and selectivity towards the target analyte (Siddiquee et al. 2014). In agriculture, soil humidity is an important factor for the effective yield of the crop. Various materials such as polymers, ceramics, and composites are utilized for the development of humidity biosensor. For example, $Na_2Ti_3O_7$ nanotubes coated on Al_2O_3 ceramic substrate in Ag-Pd as interdigitated electrodes are used for the fabrication of impedance-based humidity sensor (Zhang et al. 2008).

16.5 Challenges and Future Perspectives

Nanomaterials-based biosensors have been developed to answer many toughest analytical challenges. An ideal biosensor can detect a feeble quantity of analyte with more accuracy, precision, and rapidity. They must possess high mechanical, thermal, and chemical stability, and could be produced at low cost with easy procedures. In recent sensors, nanomaterials improve the features like biocompatibility, nontoxicity, and powerless operation, however, the desired properties and efficiency of the biosensors remain a challenge. But in electrochemical biosensors, the incorporation of nanomaterials leads to numerous challenges respective to the material of choice. Especially, the nanomaterials used in the fabrications of biosensors need high stability and operate under various conditions like strong ionic buffers, high temperature, and humidity and must be durable for even weeks and months. Nevertheless, these discrepancies can be overcome with surface modification or engineering the properties of nanomaterials in respective to their applications.

Another major advantage of nanomaterials is their surface-to-volume ratio. The main advantage of nanomaterials over bulk is their large specific surface area capable for the immobilization and better transduction. However, it is important to control and optimize the nonspecific binding over a large active surface, since it leads to the surface fouling. This surface fouling affects the signal intensity with a drop in relevant signal-to-noise ratio. This challenge can be overcome by real-world matrixes, that is, combining the nanomaterials with the antifouling materials. For example, polymers with nanofiber structures have excellent antifouling property, and the surface charges can optimize the appropriate binding process. Further, the synthesis of nanomaterials-associated biosensors requires a cost-effective, reproducible, and scalable process. To achieve this a commercially viable product with reliable functionalization of materials is very important. Some detailed research is crucial to optimize some feasible methods like printing techniques, electrodeposition, and other solution-based processes like drop-casting, spray coating, and dip-coating.

In addition, the technology must focus on the development of novel materials with promising properties to solve the biocompatibility and immobilization problems. And need high concentration on developing new materials, design of multiarrays, miniaturization and flexible/wearable sensors. Moreover, future research direction and development appreciates more new findings; to avoid nonspecific reactions, to increase the signal-to-noise ratio, improve electron transfer rate, and better studies on the interface chemistry.

References

Abdolahad M, Taghinejad M, Taghinejad H, Janmaleki M, Mohajerzadeh S (2012) A vertically aligned carbon nanotube-based impedance sensing biosensor for rapid and high sensitive detection of cancer cells. Lab Chip 12:1183–1190

Absalan G, Akhond M, Bananejad A, Ershadifar H (2015) Highly sensitive determination of nitrite using a carbon ionic liquid electrode modified with Fe_3O_4 magnetic nanoparticle. J Iran Chem Soc 12:1293–1301

Andreescu S, Luck LA (2008) Studies of the binding and signaling of surface-immobilized periplasmic glucose receptors on gold nanoparticles: a glucose biosensor application. Anal Biochem 375:282

Anusha JR, Kim HJ, Fleming AT, Das SJ, Yu KH, Kim BC, Raj CJ (2014) Simple fabrication of ZnO/Pt/chitosan electrode for enzymatic glucose biosensor. Sens Actuators B Chem 202:827–833

Anusha JR, Raj CJ, Cho BB, Fleming AT, Yu KH, Kim BC (2015) Amperometric glucose biosensor based on glucose oxidase immobilized over chitosan nanoparticles from gladius of Uroteuthis duvauceli. Sens Actuators B Chem 215:536–543

Anusha JR, Kim BC, Yu KH, Raj CJ (2019) Electrochemical biosensing of mosquito-borne viral disease, dengue: A review. Biosens Bioelectron 142:111511

Arun RC (2017) DNA Nanobiosensors: An Outlook on Signal Readout Strategies. J Nanomater 2820619:1–9

Aziz N, Fatma T, Varma A, Prasad R (2014) Biogenic synthesis of silver nanoparticles using Scenedesmus abundans and evaluation of their antibacterial activity. Journal of Nanoparticles, Article ID 689419, https://doi.org/10.1155/2014/689419

Aziz N, Faraz M, Pandey R, Sakir M, Fatma T, Varma A, Barman I, Prasad R (2015) Facile algae-derived route to biogenic silver nanoparticles: Synthesis, antibacterial and photocatalytic properties. Langmuir 31:11605–11612. https://doi.org/10.1021/acs.langmuir.5b03081

Aziz N, Pandey R, Barman I, Prasad R (2016) Leveraging the attributes of Mucor hiemalis-derived silver nanoparticles for a synergistic broad-spectrum antimicrobial platform. Front Microbiol 7:1984. https://doi.org/10.3389/fmicb.2016.01984

Aziz N, Faraz M, Sherwani MA, Fatma T, Prasad R (2019) Illuminating the anticancerous efficacy of a new fungal chassis for silver nanoparticle synthesis. Front Chem 7:65. https://doi.org/10.3389/fchem.2019.00065

Baek SH, Roh J, Park CY, Kim MW, Shi R, Kailasa SK, Park TJ (2020) Cu-nanoflower decorated gold nanoparticles-graphene oxide nanofiber as electrochemical biosensor for glucose detection. Mater Sci Eng C 107:110273

Baj-Rossi C, De Micheli G, Carrara S (2014) Electrochemical biochip for applications to wireless and batteryless monitoring of free-moving mice. Conf Proc IEEE Eng Med Biol Soc 2014:2020–2023

Bareket L, Rephaeli A, Berkovitch G, Nudelman A, Rishpon J (2010) Carbon nanotubes based electrochemical biosensor for detection of formaldehyde released from a cancer cell line treated with formaldehyde-releasing anticancer prodrugs. Bioelectrochemistry 77:94–99

Behera BK, Prasad R, Behera S (2020) Bioprinting. In: Behera BK, Prasad R, Behera S (eds) Competitive Strategies in Life Sciences, Springer 137–156

Bhuyan T, Mishra K, Khanuja M, Prasad R, Varma A (2015) Biosynthesis of zinc oxide nanoparticles from Azadirachta indica for antibacterial and photocatalytic applications. Mater Sci Semicond Process 32:55–61

Bracamonte MV, Melchionna M, Giuliani A, Nasi L, Tavagnacco C, Prato M, Fornasiero P (2017) H_2O_2 sensing enhancement by mutual integration of single walled carbon nanohorns with metal oxide catalysts: the CeO_2 case. Sens Actuators B Chem 239:923–932

Chang L, He X, Chen L, Zhang Y (2017) A fluorescent sensing for glycoproteins based on the FRET between quantum dots and Au nanoparticles. Sens Actuators B Chem 250:17–23

Chausali N, Jyoti Saxena J, Prasad R (2021) Recent trends in nanotechnology applications of bio-based packaging. Journal of Agriculture and Food Research, https://doi.org/10.1016/J.jafr.2021.100257

Choi YE, Kwak JW, Park JW (2010) Nanotechnology for early cancer detection. Sensors 10:428

Clancy AJ, Bayazit MK, Hodge SA, Skipper NT, Howard CA, Shaffer MS (2018) Charged Carbon Nanomaterials: Redox Chemistries of Fullerenes, Carbon Nanotubes, and Graphenes. Chem Rev 118:7363–7408

Colvin VL (2003) The potential environmental impact of engineered nanomaterials. Nat Biotechnol 21:1166–1170

Cui L, Shen J, Ai S, Wang X, Zhang CY (2020) In-situ synthesis of covalent organic polymer thin film integrates with palladium nanoparticles for the construction of a cathodic photoelectrochemical cytosensor. Biosens Bioelectron 168:112545

Deng C, Pi X, Qian P, Chen X, Wu W, Xiang J (2017) High-Performance Ratiometric Electrochemical Method Based on the Combination of Signal Probe and Inner Reference Probe in One Hairpin-Structured DNA. Anal Chem 89:966–973

Dey RS, Bera RK, Raj CR (2013) Nanomaterial-based functional scaffolds for amperometricsensing of bioanalytes. Anal Bioanal Chem 405:3431–3448

Drexler KE (2004) Nanotechnology: From Feynman to Funding. B Sci Technol Soc 24:21–27

Dungchai W, Siangproh W, Chaicumpac W, Tongtawed P, Chailapakula O (2008) Salmonella typhi determination using voltammetric amplification of nanoparticles: a highly sensitive strategy for metalloimmunoassay based on a copper-enhanced gold label. Talanta 77:727–732

Fang B, Wang G, Zhang W, Li M, Kan X (2005) Fabrication of Fe_3O_4 nanoparticles modified electrode and its application for voltammetric sensing of dopamine. Electroanalysis 17:744–748

Fatima B, Hussain D, Bashir S, Hussain HT, Aslam R, Nawaz R, Rashid HN, Bashir N, Majeed S, Ashiq MN, Najam-Ul-Haq M (2020) Catalase immobilized antimonene quantum dots used as an electrochemical biosensor for quantitative determination of H_2O_2 from CA-125 diagnosed ovarian cancer samples. Mater Sci Eng C 117:111296

Feng L, Chen Y, Ren J, Qu X (2011) A graphene functionalized electrochemical aptasensor for selective label-free detection of cancer cells. Biomaterials 32:2930–2937

Feng QM, Pan JB, Zhang HR, Xu JJ, Chen HY (2014) Disposable paper-based bipolar electrode for sensitive electrochemiluminescence detection of a cancer biomarker. Chem Commun 50:10949–10951

Gerwen PV, Laureyn W, Laureys W, Huyberechts G, Beeck MOD, Baert K, Suls J, Sansen W, Jacobs P, Hermans L, Mertens R (1998) Nanoscaled interdigitated electrode arrays for biochemical sensors. Sens Actuators B Chem B 49:73–80

Hong P, Li W, Li J (2012) Applications of aptasensors in clinical diagnostics. Sensors 12:1181–1193

Huang J, Luo X, Lee I, Hu Y, Cui XT, Yun M (2011) Rapid real-time electrical detection of proteins using single conducting polymer nanowire-based microfluidic aptasensor. Biosens Bioelectron 30:306–309

Jou AF, Lu CH, Ou YC, Wang SS, Hsu SL, Willner I, Ho JA (2015) Diagnosing the miR-141 prostate cancer biomarker using nucleic acid-functionalized CdSe/ZnS QDs and telomerase. Chem Sci 6:659–665

Kargozar S, Mozafari M (2018) Nanotechnology and Nanomedicine: Start small, think big. Mater Today: Proceedings 5(7):15492–15500

Karn B, Masciangioli T, Zhang WX, Masciangioli TM (2004) Nanotechnology and the environment. American Chemical Society

Kaushik A, Solanki PR, Ansari AA, Sumana G, Ahmad S, Malhotra BD (2009) Iron oxide-chitosan nanobiocomposite for urea sensor. Sens Actuators B Chem 138:572–580

Khot LR, Sankaran S, Maja JM, Ehsani R, Schuster EW (2012) Application of nanomaterials in agricultural production and crop protection. Crop Prot 35:64–70

Krishnan SK, Singh E, Singh P, Meyyappan M, Nalwa HS (2019) A review on graphene-based nanocomposites for electrochemical and fluorescent biosensors. RSC Adv 9:8778–8881

Kumar S, Umar M, Saifi A, Kumar S, Augustine S, Srivastava S, Malhotra BD (2019) Electrochemical paper based cancer biosensor using iron oxide nanoparticles decorated PEDOT:PSS. Anal Chim Acta 1056:135–145

Lang XY, Fu HY, Hou C, Han GF, Yang P, Liu YB, Jiang Q (2013) Nanoporous gold supported cobalt oxide microelectrodes as high-performance electrochemical biosensors. Nat Commun 4:2169

Lee M, Zine N, Baraket A, Zabala M, Campabadal F, Caruso R, Trivella MG, Renault NJ, Errachid A (2012) A Novel Biosensor Based on Hafnium Oxide: Application for Early Stage Detection of Human Interleukin-10. Sensors Actuators B Chem 175:201–207

Lee S, Oh J, Kim D, Piao Y (2016) A sensitive electrochemical sensor using an iron oxide/graphene composite for the simultaneous detection of heavy metal ions. Talanta 160:528–536

Li B, Zhou Y, Wu W, Liu M, Mei S, Zhou Y, Jing T (2015) Highly selective and sensitive determination of dopamine by the novel molecularly imprinted poly(nicotinamide)/CuO nanoparticles modified electrode. Biosens Bioelectron 67:121–128

Lia Y-Y, Schluesenerb HJ, Xua S (2010) Gold nanoparticle-based biosensors. Gold Bull 43:29–41

Lin HY, Huang CH, Lu SH, Kuo IT, Chau LK (2014) Direct detection of orchid viruses using nanorod-based fiber optic particle plasmon resonance immunosensor. Biosens Bioelectron 51:371–378

Liu G, Lin Y (2006) Biosensor based on self-assembling acetylcholinesterase on carbon nanotubes for flow injection/amperometric detection of organophosphate pesticides and nerve agents. Anal Chem 78:835–843

Liu D, Rahman MM, Ge C, Kim J, Lee JJ (2017) Highly stable and conductive PEDOT:PSS/graphene nanocomposites for biosensor applications in aqueous medium. New J Chem 41:15458–15465

Luo J, Ma Q, Wei W, Zhu Y, Liu R, Liu X (2016) Synthesis of Water-Dispersible Molecularly Imprinted Electroactive Nanoparticles for the Sensitive and Selective Paracetamol Detection. ACS Appl Mater Interfaces 8:21028–21038

Ma F, Li C, Zhang J (2018) Development of quantum dot-based biosensors: principles and applications. J Mater Chem B 6:6173–6180

Majd SM, Salimi A, Astinchap B (2016) Manganese Oxide Nanoparticles/Reduced Graphene Oxide as Novel Electrochemical Platform for Immobilization of FAD and its Application as Highly Sensitive Persulfate Sensor. Electroanalysis 28:493–502

Mathew J, Joy J, George SC (2019) Potential applications of nanotechnology in transportation: A review. J King Saud Uni-Sci 31(4):586–594

Mobed A, Hasanzadeh M, Shadjou N, Hassanpour S, Saadati A, Agazadeh M (2020) Immobilization of ssDNA on the surface of silver nanoparticles-graphene quantum dots modified by gold nanoparticles towards biosensing of microorganism. Microchem J 152:104286

Mufamadi MS, Sekhejane PR (2017) Nanomaterial-based biosensors in agriculture application and accessibility in rural smallholding farms: food security. In: Prasad R, Kumar M, Kumar V (eds) Nanotechnology. Springer, Singapore

Napi MLM, Sultan SM, Ismail R, How KW, Ahmad MK (2019) Electrochemical-Based Biosensors on Different Zinc Oxide Nanostructures: A Review. Materials 12:2985

Oke AE, Aigbavboa CO, Semenya K (2017) Energy Savings and Sustainable Construction: Examining the Advantages of Nanotechnology. Energy Procedia 142:3839–3843

Omar NAS, Fen YW, Abdullah J, Kamil YM, Daniyal WMEMMD, Sadrolhosseini AR, Mahdi MA (2020) Sensitive Detection of Dengue Virus Type 2 E-Proteins Signals Using Self-Assembled Monolayers/Reduced Graphene Oxide-PAMAM Dendrimer Thin Film-SPR Optical Sensor. Sci Rep 10:2374

Ozcelikay G, Kurbanoglu S, Yarman A, Scheller FW, Ozkan SA (2020) Au-Pt nanoparticles based molecularly imprinted nanosensor for electrochemical detection of the lipopeptide antibiotic drug Daptomycin. Sens Actuators B Chem 320:128285

Ozin GA, Arsenault AC, Cademartiri L (2009) Nanochemistry: A chemical approach to nanomaterials, 2nd edn. Royal Society of Chemistry

Pak SC, Penrose W, Hesketh PJ (2001) An ultrathin platinum film sensor to measure biomolecular binding. Biosens Bioelectron 16.371–379

Pandit S, Dasgupta D, Dewan N, Prince A (2016) Nanotechnology based biosensors and its application. Pharma Innov 5(6):18–25

Parikha M (2016) Biosensors and their applications – A review. J Oral Biol Craniofacial Res 6:153–159

Petryayeva E, Algar WR (2013) Proteolytic assays on quantum-dot-modified paper substrates using simple optical readout platforms. Anal Chem 85:8817–8825

Phan TTV, Huynh T-C, Manivasagan P, Mondal S, Oh J (2020) An Up-To-Date Review on Biomedical Applications of Palladium Nanoparticles. Nanomaterials 10:66

Prasad R (2014) Synthesis of silver nanoparticles in photosynthetic plants. Journal of Nanoparticles, Article ID 963961, 2014, https://doi.org/10.1155/2014/963961

Prasad R, Kumar V, Prasad KS (2014) Nanotechnology in sustainable agriculture: present concerns and future aspects. Afr J Biotechnol 13(6):705–713

Prasad R, Pandey R, Varma A, Barman I (2017) Polymer based nanoparticles for drug delivery systems and cancer therapeutics. In: Natural Polymers for Drug Delivery (eds. Kharkwal H and Janaswamy S), CAB International, UK 53–70

Prasad R, Kumar M, Kumar V (2017a) Nanotechnology: An Agriculture paradigm. Springer Nature Singapore Pte Ltd. (ISBN: 978-981-10-4573-8)

Prasad R, Kumar V, Kumar M (2017b) Nanotechnology: Food and Environmental Paradigm. Springer Nature Singapore Pte Ltd. (ISBN 978-981-10-4678-0)

Prasad R, Bhattacharyya A, Nguyen QD (2017c) Nanotechnology in sustainable agriculture: Recent developments, challenges, and perspectives. Front Microbiol 8:1014. https://doi.org/10.3389/fmicb.2017.01014

Ramos AP, Cruz MAE, Tovani CB, Ciancaglini P (2017) Biomedical applications of nanotechnology. Biophys Rev 9:79–89

Rocha-Santos TA (2014) Sensors and biosensors based on magnetic nanoparticles. TrAC Trends Anal Chem 62:28–36

Sarma H, Joshi S, Prasad R, Jampilek J (2021) Biobased Nanotechnology for Green Applications. Springer International Publishing (ISBN 978-3-030-61985-5) https://www.springer.com/gp/book/9783030619848

Satija J, Sai VVR, Mukherji S (2011) Dendrimers in biosensors: Concept and applications. J Mater Chem 21:14367–14386

Shende P, Sahu P (2021) Enzyme bioconjugated PAMAM dendrimers for estimation of glucose in saliva. Inter J Polymeric Mater Polymeric Biomater 70:469–475

Shobha BN, Muniraj NJR (2015) Design, modeling and performance analysis of carbon nanotube with DNA strands as biosensor for prostate cancer. Microsyst Technol 21:791–800

Siddiquee S, Rovina K, Yusof NA, Rodrigues KF, Suryani S (2014) Nanoparticle-enhanced electrochemical biosensor with DNA immobilization and hybridization of Trichoderma harzianum gene. Sens BioSensing Res 2:16–22

Simão EP, Silva DB, Cordeiro MT, Gil LH, Andrade CA, Oliveira MD (2020) Nanostructured impedimetric lectin-based biosensor for arboviruses detection. Talanta 208:120338

Srinivasan M, Rajabi M, Mousa SA (2015) Multifunctional Nanomaterials and Their Applications in Drug Delivery and Cancer Therapy. Nanomaterials 5:1690–1703

Taniselass S, Arshad MM, Gopinath SC (2019) Graphene-based electrochemical biosensors for monitoring noncommunicable disease biomarkers. Biosens Bioelectron 130:276–292

Tansil SN, Gao Z (2015) Nanoparticles in biomolecular detection. Nano Today 1:28–37

Teker MS, Karaca E, Pekmez NQ, Tamer U, Pekmez K (2019) An Enzyme-free H_2O_2 Sensor Based on Poly(2-Aminophenylbenzimidazole)/Gold Nanoparticles Coated Pencil Graphite Electrode. Electroanalysis 31:75–82

Tilmaciu CM, Morris MC (2015) Carbon nanotube biosensors. Front Chem 3:59

Tombelli S, Minunni M, Luzi E, Mascini M (2005) Aptamer-based biosensors for the detection of HIV-1 Tat protein. Bioelectrochem 67:135–141

Vasantha V, Chen SMJ (2006) Electrocatalysis and simultaneous detection of dopamine and ascorbic acid using poly (3, 4-ethylenedioxy) thiophene film modified electrodes. J Electroanal Chem 592:77–87

Wen T, Zhu W, Xue C, Wu J, Han Q, Wang X, Zhou X, Jiang H (2014) Novel electrochemical sensing platform based on magnetic field-induced self-assembly of Fe_3O_4@Polyaniline nanoparticles for clinical detection of creatinine. Biosens Bioelectron 56:180–185

Xu Q, Mao C, Liu NN, Zhu JJ, Sheng J (2006) Direct electrochemistry of horseradish peroxidase based on biocompatible carboxymethyl chitosan-gold nanoparticle nanocomposite. Biosens Bioelectron 22:768–773

Xu S, Huang X, Chen Y, Liu Y, Zhao W, Sun Z, Wong CP (2019) Silver Nanoparticle-Enzyme Composite Films for Hydrogen Peroxide Detection. ACS Appl Nano Mater 2(9):5910–5921

Yao Y, Wang GX, Shi XJ, Li JS, Yang FZ, Cheng ST, Zhang H, Dong HW, Guo YM, Sun X, Wu YX (2020) Ultrasensitive aptamer-based biosensor for acetamiprid using tetrahedral DNA nanostructures. J Mater Sci 55:15975–15987

Yi X, Wu Y, Tan G, Yu P, Zhou L, Zhou Z, Chen J, Wang Z, Pang J, Ning C (2017) Palladium nanoparticles entrapped in a self-supporting nanoporous gold wire as sensitive dopamine biosensor. Sci Rep 7:7941

Yuan G, He J, Li Y, Xu W, Gao L, Yu C (2015) A novel ultrasensitive electrochemical immunosensor based on carboxy-endcapped conductive polypyrrole for the detection of gypican-3 in human serum. Anal Methods 7:1745–1750

Yue HY, Zhang HJ, Huang S, Gao X, Song SS, Wang Z, Wang WQ, Guan EH (2019) A novel non-enzymatic dopamine sensors based on NiO-reduced graphene oxide hybrid nanosheets. J Mater Sci 30(5):5000–5007

Zhai D, Liu B, Shi Y, Pan L, Wang Y, Li W, Zhang R, Yu G (2013) Highly sensitive glucose sensor based on pt nanoparticle/polyaniline hydrogel heterostructures. ACS Nano 7:3540–3546

Zhang F (2017) Grand Challenges for Nanoscience and Nanotechnology in Energy and Health. Front Chem 5:80

Zhang Y, Fu W, Yang H, Li M, Li Y, Zhao W, Sun P, Yuan M, Ma D, Liu B, Zou G (2008) A novel humidity sensor based on Na2Ti3O7 nanowires with rapid response-recovery. Sens Actuators B Chem 135:317–321

Zhang Z, Zhu H, Wang X, Yang X (2011) Sensitive electrochemical sensor for hydrogen peroxide using Fe 3 O 4 magnetic nanoparticles as a mimic for peroxidase. Microchim Acta 174:183–189

Zhang J, Chai Y, Yuan R, Yuan Y, Bai L, Xie SA (2013) A highly sensitive electrochemical aptasensor for thrombin detection using functionalized mesoporous silica@multiwalled carbon nanotubes as signal tags and DNAzyme signal amplification. Analyst 138:6938–6945

Zhang X, Li CR, Wang WC, Xue J, Huang YL, Yang XX, Tan B, Zhou XP, Shao C, Ding SJ, Qiu JF (2016) A novel electrochemical immunosensor for highly sensitive detection of aflatoxin B1 in corn using single-walled carbon nanotubes/chitosan. Food Chem 192:197–202

Zhang Y, Li X, Li D, Wei Q (2020) A laccase based biosensor on AuNPs-MoS2 modified glassy carbon electrode for catechol detection. Colloids Surf B 186:110683

Zhao ZW, Chen XJ, Tay BK, Chen JS, Han ZJ, Khor KA (2007) A novel amperometric biosensor based on ZnO:Co nanoclusters for biosensing glucose. Biosens Bioelectron 23:135–139

Zhao Z, Lei W, Zhang X, Wang B, Jiang H (2010) ZnO-Based Amperometric Enzyme Biosensors. Sensors 10(2):1216–1231

Zheng TT, Zhang R, Zou L, Zhu JJ (2012) A label-free cytosensor for the enhanced electrochemical detection of cancer cells using polydopamine-coated carbon nanotubes. Analyst 137:1316–1318

Zheng H, Liu M, Yan Z, Chen J (2020) Highly selective and stable glucose biosensor based on incorporation of platinum nanoparticles into polyaniline-montmorillonite hybrid composites. Microchem J 152:104266

Zhu W, Jiang G, Xu L, Li BZ, Cai QZ, Jiang HJ, Zhou XM (2015) Facile and controllable one-step fabrication of molecularly imprinted polymer membrane by magnetic field directed self-assembly for electrochemical sensing of glutathione. Anal Chim Acta 886:37–47

Chapter 17
Nano-Adsorbents and Nano-Catalysts for Wastewater Treatment

Zeenat Sheerazi and Maqsood Ahmed

Contents

17.1	Introduction	517
17.2	Synthesis Approaches for Nano-Catalysts/Adsorbents	519
	17.2.1 Sol-Gel Method	521
	17.2.2 Micro-Emulsion Method	522
	17.2.3 Hydrothermal Synthesis	523
	17.2.4 Co-precipitation	523
	17.2.5 Polyol Synthesis	523
17.3	Types of Nano-Adsorbents and Nano-Catalysts	524
	17.3.1 Metal-Based Nano Adsorbents	524
	17.3.2 Polymer-Based Nano-Adsorbents	530
	17.3.3 Silica- and Carbon-Based Nano-Adsorbents	531
	17.3.4 Nano-Catalysts for Wastewater Treatment	531
17.4	Conclusion	532
References		533

17.1 Introduction

Environmental conditions are very alarming all around us due to the anthropogenic activities disturbing the water bodies (Ray and Shipley 2015; Schwarzenbach et al. 2010). Even though anthropogenic processes are responsible for polluting water, natural processes also introduce some toxic metals into the water bodies due to weathering conditions, erosion of rock and soil, and rainwater (Wang and Mulligan

---

Z. Sheerazi
Department of Chemistry, Jamia Millia Islamia, New Delhi, India
e-mail: maqsood@csiriict.in

M. Ahmed (✉)
Indian Institute of Chemical Technology, Hyderabad, India

2006). Water is the nectar for our life, but its quality is disturbed due to the contamination by industrial wastes, pesticides, harmful synthetic dyes and organic pollutants (pharmaceuticals, pesticide, phenols, fertilizer, plasticizer, detergent, oils, hydrocarbons etc.). Around the world more than 0.78 billion people face shortage of fresh and hygienic water. This causes serious infections and death of more than 200 million people every year, with 5000 to 6000 children deaths (Amin et al. 2014). Pollutant-removal techniques including biological treatment system, physicochemical, chlorination, ozonation, UV photolysis and ion exchange are not more effective for polluted water treatment (Amin et al. 2014). Due to enhancing nanotechnologies the role of nanomaterials in water purification attracted the attentions due to their nano size, large surface area, reusability, stability, electrical and optical properties (Lu et al. 2016; Prasad and Thirugnanasanbandham 2019). Various shapes of nanomaterials like nanowires, single and multi-walled carbon nanotubes (CNT), nano-colloids, carbon quantum dots (CQD), nano-membrane and films have been studied for diverse applications including wastewater treatment, due to their small size and large surface area (Fig. 17.1). These materials can be used in the form of absorbents for the capture of pollutants and catalytic degradation of larger organic molecules from polluted water (Khajeh et al. 2013). Metallic oxide nanoparticles including mono-metallic, bi-metallic and tri-metallic oxides are used in the form of absorbents and catalysts for wastewater remediation; the magnetic behaviour and presence of variable oxidation states make them efficient absorbents as well as catalysts (Ray and Shipley 2015; Khajeh et al. 2013). Nanostructural silica- and alumina-based materials were also studied in the form of nano-porous materials which provide accessible active sites. (Jadhav et al. 2019; Banerjee et al. 2019; Afkhami et al. 2011; Kalfa et al. 2009a). Polymer nanocomposite with hierarchical surface porosity is also an interesting material for the treatment of waste water (Zhao et al. 2018a). Moreover the nanomaterials can be functionalized with different types of

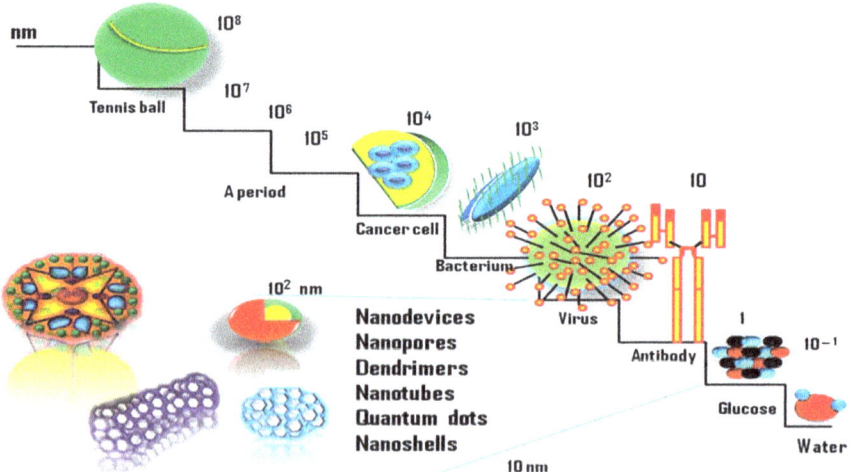

Fig. 17.1 Different morphologies of nanostructural materials

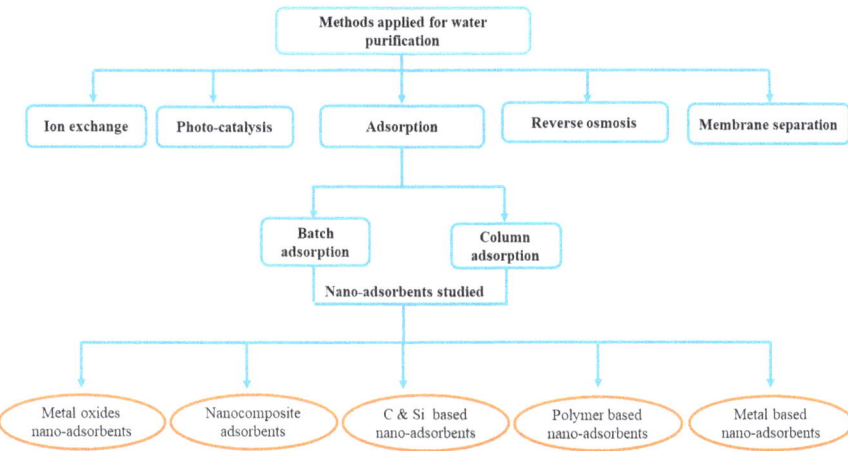

Fig. 17.2 Various chemical methods for waste water treatment and nano-adsorbents

organic functionalities possessing higher affinity for toxic metal ions (As, Cd, Hg, Cr etc.). Such materials include organo-modified nanostructural materials which can be functionalized via ex situ as well as in situ synthesis. The common methods used for the synthesis of metallic based nanoparticle involves co-precipitation, sol-gel and hydrothermal techniques. The nanoporous materials are commonly prepared through soft templating routes by using some surfactant and block copolymers as templates.

The most common methods applied for water purification are ion exchange (Akieh et al. 2008; Ismail et al. 2010; Qiu and Zheng 2009), reverse osmosis (Dialynas and Diamadopoulos 2009; Mohsen-Nia et al. 2007), electrochemical treatment (Hunsom et al. 2005; Deng and Englehardt 2007; Rana et al. 2004), membrane filtration (Qdais and Moussa 2004), photo catalysis and adsorption (Fu and Wang 2011). But the adsorption technique is the most recognised and convenient, as it does not require high amount of energy. The adsorption methods involve two different procedures viz. batch adsorption and column adsorption. The batch adsorption is carried out by adding the adsorbate into the sorbent solution under continuous agitation or stirring. The latter will be carried out through the continuous flow of contaminated water over fixed bed of adsorbent and is typically applied at industrial scale. Various chemical methods for waste water treatment and nano-adsorbents are depicted in Fig. 17.2.

17.2 Synthesis Approaches for Nano-Catalysts/Adsorbents

Most commonly the nanomaterials can be synthesised in two major categories viz. top to bottom and bottom to up approaches (Fig. 17.3). In top to down approach the bulk material is breakdown to nanoscale by mechanical process with uniform

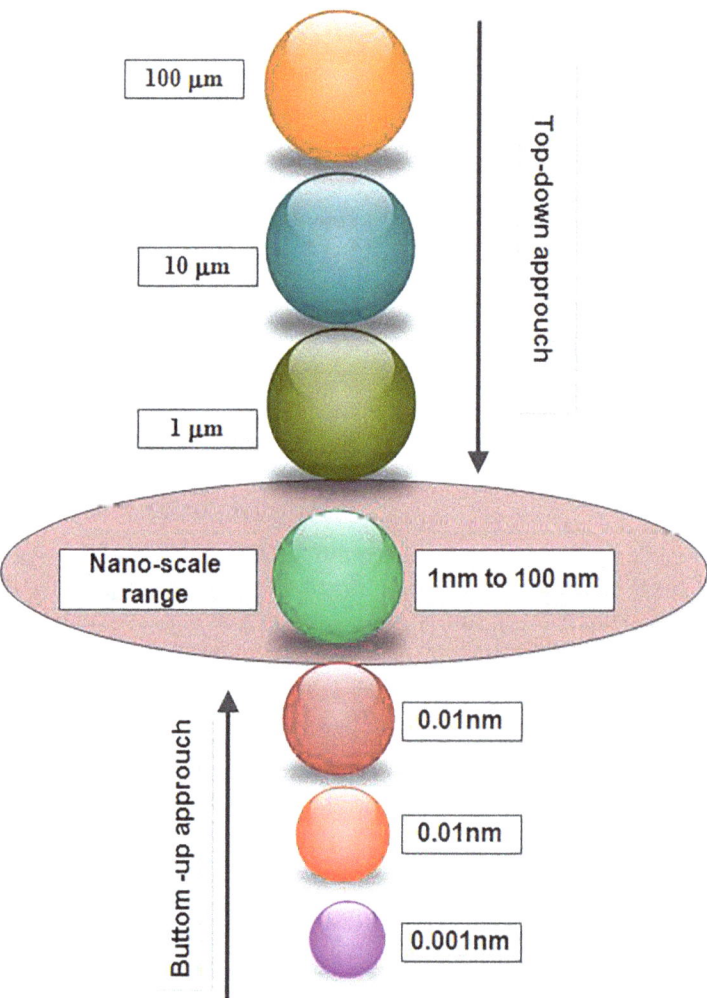

Fig. 17.3 Top-down bottom-up approaches for synthesis of nanomaterials

particle size and morphology. The latter involves the assembling of molecules and atoms to create various ranges of particles in nanoscales (Prasad et al. 2016). It is further divided into chemical, physical and biological methods. Among the three, the chemical method is more simplistic, low cost and quick in action. The chemical method is well established and more common, in which the nanomaterials are prepared through the interaction of molecules and atoms. It enables the easy functionalization of nanomaterials, which can be applied as adsorbents and catalysts in wastewater treatment. These chemical methods involve different synthetic chemistry (depicted in Fig. 17.5) viz. sol-gel, micro-emuslion, hydrothermal synthesis, Co-precipitation method, chemical vapour deposition (CVD) & Chemical vapour

17 Nano-Adsorbents and Nano-Catalysts for Wastewater Treatment

synthesis (CVC), microwave and ultrasound-assisted synthesis (Devatha and Thalla 2018; Rane et al. 2018).

17.2.1 Sol-Gel Method

In sol-gel technique the colloidal solution called sol of the metal precursor or any inorganic species is formed through hydrolysis, followed by the condensation or formation of gel-like diphasic system containing both liquid and solid phases. The morphologies of these materials range from discrete particles to the polymer network range. Nanomaterials of different inorganic metals and their oxides like TiO_2, ZnO, SnO_2, CdSe, Fe_2O_3, ZrB_2, $GdVO_4$, Ta_2O_5, CeO_2 and nanoalumina are synthesised through sol-gel techniques and some of them are studied in the field of waste water treatment (Xiao et al. 2009; Sharma et al. 2008; Bayal and Jeevanandam 2012; Nautiyal et al. 2015; Reda 2010; Zhang et al. 2011; Chumha et al. 2014; Sreethawong et al. 2013). Variety of mesoporous silica nanoparticles (MSN) are synthesised through sol gel methods and studied for adsorption processes (Chen et al. 2018; Qin et al. 2018). Jie Chen et al. demonstrated the study of different morphologies of MSN synthesised through sol-gel and micro-emulsion method (Chen et al. 2018). The synthesis of MSN involves the soft templating route, in which surfactants are used as templates (Fig. 17.4). Mostly the silica

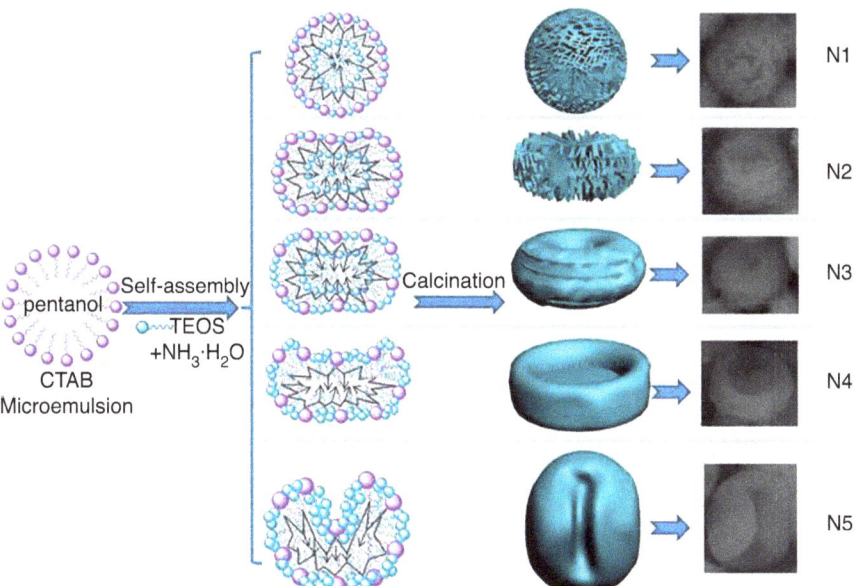

Fig. 17.4 Illustration of the synthetic procedure of MSNs with different morphologies. (Reprinted with permission from Ref. (Chen et al. 2018) Copyright (2020) American Chemical Society)

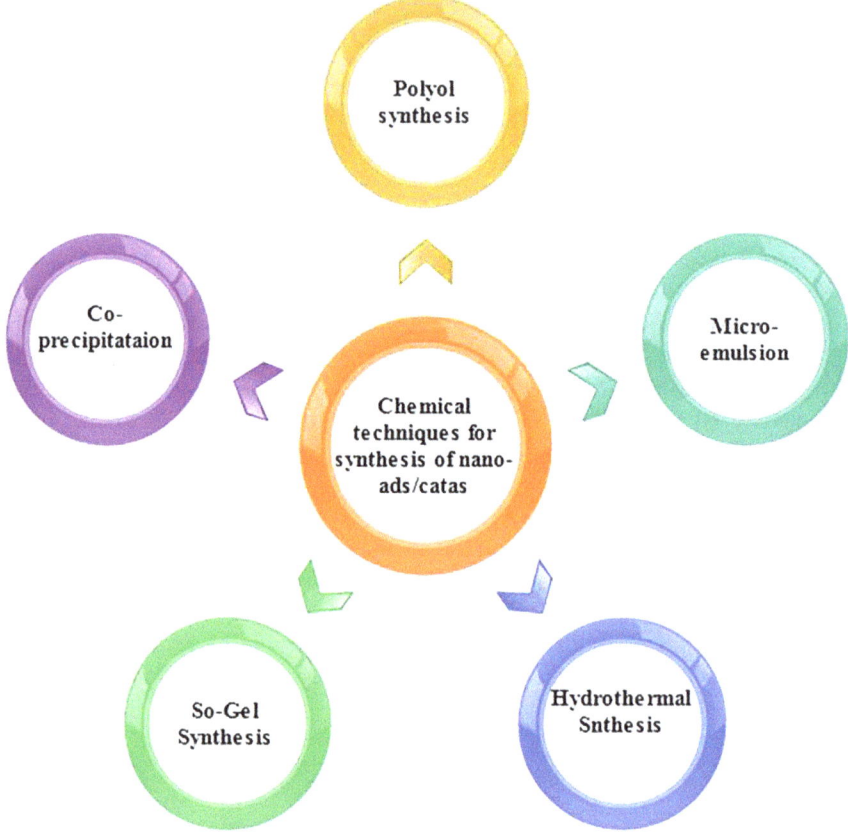

Fig. 17.5 Schematic representation of different techniques used for synthesis of nano-adsorbents and nano-catalysts through chemical method

precursors-tetraethylorthosilicate (TEOS) first hydrolysed and gets start condensing at different pH ranges. The surfactant/template is removed through calcination or solvent extraction methods. The same approach can be applied to other mesoporous materials by using their salt precursors with modified procedures.

17.2.2 Micro-Emulsion Method

Diameter range of 600 nm to 800 nm monodispersed spherical droplets of oil in water (o/w) or water in oil (w/o) depending upon the type of surfactant forming emulsion which is called as micro-emulsion system. Water in oil (w/o), also called as reverse micellar system, acts as a best reaction site for nanomaterial synthesis like Ag, Cu, CdS, Ni, Au, Rh, silica-CdS, Ag-Au, Pd-Au and Pd-Ag nanoparticles. Other non-metallic nanomaterials of polystyrene (20–30 nm), cholesterol, rhodiarome, retinol, polyaniline, etc. can be synthesized through micro-emulsion technique. (Joshi and Kumar 2018; Wang et al. 2010a; Dhand et al. 2010).

17.2.3 Hydrothermal Synthesis

The process of crystallization under autogenesis pressure produced at higher temperature in the presence of water vapours is called hydrothermal synthesis. The synthesis is usually carried out in a stain less steel autoclave hydrothermal rectors, or any closed autoclavable container. Hydrothermal synthesis is a fast and facile process for the synthesis of different nanostructural materials such as Fe_3O_4, NiO, CuO, $CoFe_2O_4$ and ZnO (Zhao et al. 2007; Yang and Pan 2012; Guo et al. 2012; Choi et al. 2013; Cao et al. 2013; Behbahani et al. 2012; Maryanti et al. 2014). Guohong Qiu et al. demonstrated the microwave-assisted hydrothermal synthesis of α-Fe_2O_3 and studied the applications for As (III) removal from water (Qiu et al. 2011). Synthesis of various nanomaterials like Cu nanoparticles, hydroxyapatite/ biochar nanocomposite, hydroxyl sodalite zeolite nanoparticles, Mn_2O_3 and $CoFe_2O_4$ were carried out through hydrothermal synthesis (Kumar et al. 2014a; Kumar et al. 2014b; Nassar et al. 2016; Abdelrahman et al. 2019; Hermosilla et al. 2020). These materials were studied for the removal of different contaminating agents from water. Microwave-assisted hydrothermal synthesis is also important for nanomaterial synthesis, which can be used for adsorption process and in catalysis.

17.2.4 Co-precipitation

In co-precipitation, continuous occurrence of nucleation, growth coarsening and agglomeration will take place during synthesis. Nucleation is the key step to take place and the product formed is usually the insoluble species under the condition of super saturation. Super saturation condition is mandatory to start precipitation, which are usually the result of following chemical reaction (Rane et al. 2018)

$$XAy + (aqueous) + YBx - ((aqueous)) \rightarrow AxBy (solid)$$

17.2.5 Polyol Synthesis

Polyol methods is the synthesis route for the preparation of a wide range of metal-based nanoparticles (Ag, Pr, Pt, Pd, Cu), metal oxides nanoparticles (ZnO, indium-tin-oxide; ITO, Gd_2O_3), magnetic nanoparticles and mixed metal nanoparticles (Dhand et al. 2015). Polyethylene glycol is used as a solvent and a complexing and reducing agent at the same time, so the process is called as polyol process. Sharif Ahmad et al. (Ghosal et al. 2013) demonstrated the formation of nickel nanostructure using natural polyol and studied their dye adsorption properties. Nanoparticles of cuprous oxides were obtained in the forms of nanoboxes, nanocubes and nanospheres through polyol process by Lei Huang et al. (Huang et al. 2008).

17.3 Types of Nano-Adsorbents and Nano-Catalysts

A majority of inorganic adsorbents and catalysts belong to the transition metals and rare earth metals and their oxide forms. A few elements of main groups (like Si, Al, Mg) also play a role in the formation of active adsorbents and catalysts. The metallic oxides are well-studied nano-adsorbents in the form of mixed metal oxides (monometallic, bi-metallic, or tri-metallic oxides). In addition to metal oxides, pure metallic nano-adsorbents are also studied for their applications. Other adsorbents like surface-modified metal oxide, alumina, zeolites, and silica-based mesoporous materials are also good adsorbents for waste water treatment. Different forms of nano-adsorbents and nano-catalysts are described below.

17.3.1 Metal-Based Nano Adsorbents

Nanomaterials of transition metal are well studied in the various applications of catalysis, adsorption, and in the formation of semi-conductors and other devices. Due to the demands of efficient catalysts and adsorbents with paramagnetic behaviour for removal of harmful toxic metals and synthetic dyes from textile water, researchers focused on the transition-metal based nanomaterials for this application (Ge et al. 2012). Although number of transition metal ions are toxic in their ionic form like As(III), Cr (VI), Pb (II) and Hg (II) which contaminate water through their contamination but well design material is not soluble into water and work as adsorbents or catalyst. The introduction of organo-functionality onto the surface of nanomaterials also enhances the desirable activity of the material (Ge et al. 2012). Different types of metal-based nano-adsorbents with different structures and properties are discussed in the sections that follow.

17.3.1.1 Iron Oxide Nano-Adsorbents

Separation of adsorbents along with adsorbate from aqueous solution is a challenging task. The magnetically active forms of iron oxides nanomaterials (Fe_3O_4 and γ-Fe_2O_3) act as excellent adsorbents for the capture of toxic elements, which can be easily separated from solution. These nanomaterials used for the removal of different toxic heavy metals (viz. Cr, Co, Pb, Cu, As and Ni from water (Badruddoza et al. 2013; Lei et al. 2014; Tan et al. 2014; Shipley et al. 2010). Such materials can be synthesised in different forms like nanorods, nanoparticles, and nanotubes. Nanoparticles of magnetite (Fe_3O_4) are ideal nano-adsorbents for the capture of As (III) and As (V) from water (Shipley et al. 2010). Arsenic adsorption studies along with the effect of common containments present in water were carried out by Shipley et al. in (Shipley et al. 2010) through batch adsorption study. About 83 mg L^{-1} of arsenates were adsorbed within 1 hour by using 0.5 g L^{-1} magnetite nanoparticles as adsorbents at optimized conditions. In 2013, Roy et al. studied maghemite nanotubes (MHNT) as effective nano-adsorbents for Cu(II), Pb(II) and Zn(II) ion capture from water (Roy and Bhattacharya 2012). The nanotubes of γ-Fe_2O_3 were synthesised with microwave and were studied for adsorption kinetics (Roy and Bhattacharya 2013). The adsorption isotherm model (Langmuir and Freundlich) is used to study the adsorption capacity of adsorbents, and the Langmuir model showed good agreement to the observed data of the adsorption of metal ion onto the MHNT then Freundlich model (Roy and Bhattacharya 2012). Favourable adsorption study was done for different metal ions (Cu (II), Zn (II), and Pb (II) studied which are depicted in Table 17.1. Magnetite (Fe_3O_4) nano-rods were also studied as nano adsorbents by Karami in 2013 for the capture of Fe (II), Pb(II), Zn(II), Ni (II), Cd(II) and Cu(II) ions at lab scale with best fitted adsorption data with the theoretical models as shown in Table 17.1 (Karami 2013). When magnetite nanorods are compared with

Table 17.1 Various iron oxide nano-adsorbents studied for removal of toxic metal ions

Iron oxide nano-adsorbents	Shape and size of material (nm)	BET surface area (m^2g^{-1})	Targeted metal ions	Isothermal models	Sorption capacity	Ref.
Fe_3O_4	Nano-sphere, 19.3	60	As(V) As(III)	Langmuir q_m (mgg^{-1})	1.19 1.13	(Shipley et al. 2010)
Fe_3O_4, nano-rods	Rods, 55–65 Length 900–1000	–	Fe(II) Pb(II) Zn(II) Ni(II) Cd(II) Cu(II)	Langmuir q_m (mgg^{-1})	127.01 112.86 107.27 95.42 88.39 79.1	(Karami 2013)
γ-Fe_2O_3	Tubes, 10–15 Length 150–250	321	Cu(II) Zn(II) Pb(II)	Langmuir q_m (mgg^{-1})	111.11 84.95 71.42	(Roy and Bhattacharya 2012)
α-Fe_2O_3 (3D flower like)	Flower shape, 5000–7000	–	As(V) Cr(VI)	Langmuir q_m (mgg^{-1})	41.46 33.82	(Liang et al. 2013)
α-Fe_2O_3 (3D sphere)	Spheres, 37	31	Pb(II) Cd(II) Cu(II)	Freundlich qe mgg^{-1}	3.11 0.51 0.051 0.31	(Shipley et al. 2013)

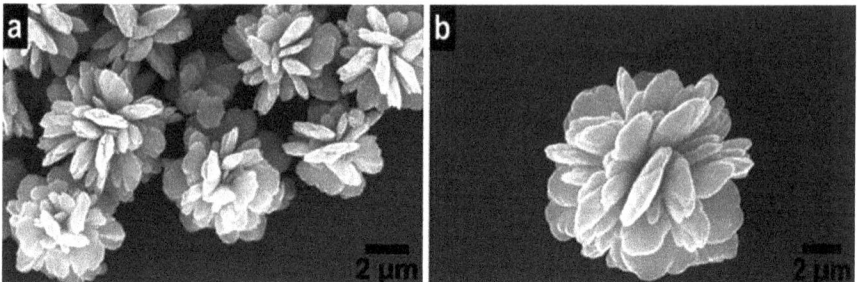

Fig. 17.6 SEM images of α-Fe$_2$O$_3$ micro-flowers after hydrothermal treatment at 150 °C for 12 h. (Reprinted from Ref. (Liang et al. 2013), Copyright 2020, with permission from Elsevier)

maghemite nanotubes, the former showed better adsorption capacity for Zn (II) and Pb (II) but exception for Cu, which might be due to the morphology of the nanorods (Karami 2013).

Nano-size hematite (α-Fe$_2$O$_3$) phase of iron oxide is a non-magnetic which is used in catalysis and environmental applications (Liang et al. 2013). Shipley et al. study the role of nono α-Fe$_2$O$_3$ for the removal of Pb (II), Cd ((II), Cu (II) and Zn(II) from aqueous solution (Shipley et al. 2013). The different adsorption parameters like pH, which affects the charge of adsorbent and temperature, were studied to elucidate the best absorption capacity. The surface chemistry of nano α-Fe$_2$O$_3$ depends upon optimized adsorption pH, which is attributed to the presence of –OH groups on the exterior surface of nano-adsorbents responsible for binding with toxic metals (Shipley et al. 2013). The isotherms models applied to the experimental data are in agreement with the observed conditions and are depicted in the Table 17.1. Hierarchical nano-structures of α-Fe$_2$O$_3$ with more surface area were studied by Liang et al. as adsorbents in order to achieve maximum adsorption capacity (Liang et al. 2013). The prepared structures are self-assembled and flower-like (SEM image shown Fig. 17.6), synthesised through hydrothermal treatment (Liang et al. 2013). Both the adsorption models, viz. Langmuir and Freundlich models, were studied in order to investigate the adsorption capacity of the flower-shaped nano α-Fe$_2$O$_3$ adsorbents.

17.3.1.2 Titanium Oxide Nano-Adsorbents

Titanium dioxide (TiO$_2$) has many applications of photo catalysis and photovoltaic, H$_2$ sensing, coatings and environmental applications for removal of pollutants (Bavykin et al. 2006a). Lao et at study the adsorption of arsenic by using TiO$_2$ at optimized parameters with 21 successive treatment adsorption cycles with regenerated TiO$_2$ (Luo et al. 2010). They study the adsorption kinetics through batch adsorption study which is fitted to pseudo second-order kinetic model (R value >0.999) and rate constant of 0.84 g mg^{-1} h^{-1}. Commercial TiO$_2$ nanoparticles were studied by Engates and Shipley et al. in 2011 for the adsorption of Pb(II), Cd(II) and

Table 17.2 Various iron oxide nano-adsorbents studied for removal of toxic metal ions

TiO$_2$ Nano-adsorbents	Shape and size of material (nm)	BET surface area (m^{2}g^{-1})	Targeted metal ions	Isothermal models	Sorption capacity	Ref.
TiO$_2$ nanoparticles	Nano-sphere, 8.3	185	Pb(II), Cd(II), Ni(II)	Langmuir q$_m$ (mgg^{-1})	83.04 15.19 6.75	(Engates and Shipley 2011)
Layered protonated titanate sheets	Nano-sheets 2-15 nm thickness 0.78 nm interlayer distance	379	Pb(II)	Langmuir q$_m$ (mgg^{-1})	366 mgg^{-1}	(Yang et al. 2008a)
Na$_2$Ti$_3$O$_7$ -T3	Nano fibres	321	Ba(II), Sr(II), Pb(II)	Sorption saturate capacity (mgg^{-1})	160.64 55.20 279.45	(Yang et al. 2008a)
Na$_{1.5}$H$_{0.5}$Ti$_3$O$_7$ -T3(H)	Nano fibres	—	Ba(II), Sr(II), Pb(II)	Sorption saturate capacity (mgg^{-1})	130.44 49.94 244.26	(Yang et al. 2008a)
Titanate nano-flower	Flower shape, 600–100 nm	290	Pb(II), Cd(II), Ni(II), Zn(II)	Langmuir q$_m$ (mgg^{-1})	304.3 168.6 88.05 98.1	(Huang et al. 2012)
Titanate nano-tubes	Tubes. 200 nm length, 7–10 nm outer dia.	230	Pb(II), Cd(II), Ni(II), Zn(II)	Langmuir q$_m$ (mgg^{-1})	147.4 76.76 40.09 44.67	(Huang et al. 2012)
Titanate nano-wires	Wires shape 10 μm length, 40–240 nm dia	30	Pb(II), Cd(II), Ni(II), Zn(II)	Langmuir q$_m$ (mgg^{-1})	106.09 47.55 24.83 27.66	(Huang et al. 2012)

Ni(II), and the results are depicted in Table 17.2 (Engates and Shipley 2011). The Langmuir adsorption model is well fitted to the process and indicates the monolayer adsorption on to the surface of TiO$_2$ nano-adsorbents, the adsorption efficiency of TiO$_2$ nanoparticles is more in comparison to bulk anantase TiO$_2$ (Engates and Shipley 2011). In addition to TiO$_2$, titanates are also useful as adsorbents for the removal of heavy metals from water (Kasap et al. 2012). Titanates in different forms like nano-sheets, nano-fibers fibers were also reported for adsorption process (Bavykin et al. 2006b; Lin et al. 2014; Yang et al. 2008a; Yang et al. 2008b; Bancroft et al. 1982). Huang et al. studied the titanate nanoflower, titanate nanotubes and titanate nanowires for the removal of Zn^{2+}, Ni^{2+} and Cd^{2+} using a ternary system (Huang et al. 2012). These three nanomaterials were synthesized through hydrothermal methods in alkaline condition which is followed by protonation in acidic media (Huang et al. 2012). The adsorption study suggested the strong adsorption capacity of titanate nanotubes.

17.3.1.3 Cobalt Oxide Nano-Adsorbents

Nanostructural cobalt oxide with various structural morphologies has been synthesised by hydrothermal and solvothermal techniques (Nassar and Ahmed 2012; Ribeiro et al. 2018). The Co-based nanomaterials are found to be good agents for waste water remediation. M.Y. Nassar et al. prepared the cobalt oxide nanomaterials with different morphologies and studied the application for removal of organic dye (methylene blue dye). The material showed maximum adsorption of 99.19% for MB in 24 h (Nassar and Ahmed 2012). Surface functionalized nano-adsorbents with various organo-functionalities like amine and thiol is more interesting due to the grafting of proper and desirable active sites on to the surface of nanomaterials. Qurrat-ul-Ain et al. synthesised magnetic Co-Fe oxide nanoparticles (CoFeNp) and functionalized their surface with two separate amine functionalities (hydrazine and dodecyl amine) (Khurshid et al. 2020). After complete characterization, the material was carried out for adsorption of six different negatively charged azoic dyes, which include acid Orange 7, reactive Red-P2B, naphthol Blue Black, Acid Orange 52, reactive Orange 16 and amaranth. The experimental data showed the pseudo-second order kinetics, in which film diffusion was the dominant phenomenon compared to intra-particle diffusion. The composite of $CoFe_2O_4$ modified with tragacanth gum was prepared and studied for methyl orange (MO) and methyl red (MR) from waste water (Moghaddam et al. 2020).

17.3.1.4 Zinc Oxide Nano-Adsorbents

Zn oxide is not a much-studied material for adsorption; however it has more applications in photocatalysis and gas sensing. Its nontoxic nature and availability of surface hydroxyl groups makes it as good adsorbent for the removal of Zn(II), Cd(II), and Hg(II) ions from aqueous solution (Sheela et al. 2012). Nanomaterials are prepared through the precipitation method and calcined at 400 °C and carried out for batch adsorption study. The adsorption study is well fitted with the theoretical isotherm models; it is suggested that due to small hydrated ionic radii of Hg (II) and more electronegativity adsorption efficiency for Hg is more as compare to Zn(II), and Cd(II) ions (Sheela et al. 2012). ZnO hollow microspheres were prepared by Wang et al., and their adsorption study was compared with ZnO nanopowder and nano-plates (Wang et al. 2010b). Hollow microspheres showed better adsorption performance over ZnO nanopowder and nano-plates.

17.3.1.5 Mixed Metal Oxides Nano-Adsorbents

Synthesis of nanoscale mixed metal oxides like spinel (Wang and Kang 2012; Giri et al. 2002; Khedr et al. 2006), Ti-based bimetallic and trimetallic oxides (Galindo et al. 2007), In (III)-Sn (IV) oxides have many applications in electrical, magnetic and conducting properties. These bi-metallic and tri-metallic oxide nano-adsorbents were also synthesised through similar methods and studied for adsorption of heavy metal ions and synthetic dyes. Gupta et al. studied Fe-Ti mixed oxides for arsenic removal from ground water in West Bengal (India) and Bangladesh. Iron doped titanium oxide adsorbent was prepared by Lin Chen et al. in 2012 through precipitation method using $Ti(SO_4)_2$ and $FeSO_4$ salts (Chen et al. 2012). Adsorption study of the materials was carried out for removal of fluoride from drinking water which showed the adsorption capacity of 53.22 mg/g, obtained by fitting experimental data to the Langmuir isotherm model. It is suggested that Fe doped into the titanium oxide increases the –OH groups on adsorbent surface, which enhance the adsorption efficiency for fluoride (Chen et al. 2012). Mesoporous Ce-Zr mixed oxides were synthesised through salvo-thermal synthesis by Qi Li et al. (Su et al. 2015). These nanomaterials were studied for the removal of phosphate ions from water through batch adsorption study. The phosphate adsorption capacity is ~112.23 mg/g. The material can be regenerated after desorption by NaOH solution. Yaswanth K. Penke and co-workers studied the tri-metallic oxides (Mn-Al-Fe and Cu-Al-Fe) as nano-adsorbent for the removal of arsenic. XPS studies showed the redox behaviour of adsorbents and showed 75-90% adsorption of As (III) (Penke et al. 2019). Co-Fe oxide and many other mixed ferrites like $MnFe_2O_4$, $ZnFe_2O_4$, $MgFe_2O_4$, $NiFe_2O_4$, $CuFe_2O_4$, and $CoFe_2O_4$ have been studied by Hu et al. (2007) for the removal of Cr((VI) (Hu et al. 2007). Different forms of bi-metallic, tri-metallic and mixed oxides nano-adsorbents are summarized in Table 17.3.

17.3.1.6 Aluminium Oxide Nano-Adsorbents

Aluminium trioxide nanoparticles, Al_2O_3, are well-studied and efficient nano-adsorbents due to their large surface area. Afkhami et al. studied the adsorption of Pb(II) and Cr(III) ions on the surface of modified alumina nanomaterials (Afkhami et al. 2011). Kalfa et al. used nanoscale Al_2O_3 on single-walled carbon nanotubes for adsorption study of Cd ions. The material was prepared through sol-gel technique and mentioned to be a better absorbent for Cd in caparison to the single-walled carbon nanotubes (Kalfa et al. 2009b).

17.3.1.7 Magnesium Oxide Nano-Adsorbents

Nanoparticles of magnesium oxide (MgO) are useful for the removal different heavy metal ions. Different morphologies like microsphere were studied by Gupta et al. in 2015 for removal of heavy metal. Various nanostructures of MgO like

Table 17.3 Mixed metal oxide nano-adsorbents and catalysts studied for water purification

Nano-catalyst/ nano-adsorbent	Targeted metal ion	Method of preparation	Sorption capacity (%) Qe(mg/g)	Process	Ref.
$Ag-Sc_{0.01}Ti_{0.99}O_{1.99}$	RhB	PPM	90%	Photo catalytic	(da Silva et al. 2014)
TiO_2-Flakes	RhB	Sol-gel (Dip-coating)	73.2%	Photo catalytic	(Li et al. 2015)
FAP-TiO_2	Pb^{2+},Cd^{2+} Cr^{3+},Fe^{3+}	Solvothermal	99%	Adsorption	(Wang et al. 2020)
Nano-TiO_2	Cr(VI), Phenol	–	~99%	Photo catalytic	(Chi et al. 2019)
Ti_3C_2/$SrTiO_3$	U(VI)	Hydrothermal	77%	Photo catalytic	(Deng et al. 2019)
V_2O_5	MB	Hydrothermal	437	Adsorption	(Avansi et al. 2015)
CS-VTM	CR	Hydrothermal	99.1%	Adsorption	(Zhang et al. 2020a)
Ni-V_2O_5	RhB	Sol-gel	100%	Photo catalytic	(Rafique et al. 2020)
$MnFe_2O_4$	RhB	Sol-gel	90%	Photo catalytic	(Zhang et al. 2020b)
MnO_2@PmPD	Pb^{2+}	Oxidation	104.88	Adsorption	(Xiong et al. 2020)
OMS-2	U(VI) Eu(III)	Hydrothermal	348 106	Adsorption	(Yin et al. 2020)
FPL	M_O	Hydrothermal	833.33	Adsorption	(Natarajan et al. 2020)
Fe_2O_4/COP	AO, RhB	Solvothermal	107.11, 131.23	Adsorption	(Shakeri et al. 2020)
HFOR	p-ASA As(V)	Co-ppt	22 60	Adsorption	(Liu et al. 2020)
$CoFe_2O_4$@γ-Fe_2O_3	Cr(VI)	Hydrothermal coprecipitation	50	Adsorption	(Campos et al. 2019)

nanorods, nanotubes, and nanocubes are reported as nano-adsorbents for heavy metals. MgO nanomaterials were prepared and used for the removal of azo and anthraquinone reactive dyes from water by Gholamreza Moussavi et al. (Moussavi and Mahmoudi 2009).

17.3.2 Polymer-Based Nano-Adsorbents

In addition to the inorganic adsorbents, polymer-based nanocomposites are also well-studied nano-adsorbents due to their macromolecular structure and the variety of functional groups. These materials possess good physical properties, large surface area, mechanical rigidity, and can be regenerated. These materials usually include, polyaniline, polystyrene and polyacrylic ester matrix. Compared to single polymers nano-adsorbents, dual polymers are more efficient for adsorption, due to their abundant surface functional groups (Wu et al. 2016). The combination of Polypyrrole (PPy) with PANI and polyacrylonitrile is found to be a suitable adsorbent for Co (II) (Javadian 2014a; Javadian 2014b; Wang et al. 2013). The adsorption study Co (II) on PANI/PPy polymer nanaofiber showed efficiency of 99.68% in 11 mins. The data are well fitted with Freundlich model and followed the pseudo-second-order kinetic model. Checkol et al. studied an efficient material consisting of poly (3,4 ethylenedioxythiophene)/polystyrene sulfonate (PEDOT/PSS) and the lignin (LG) for adsorption Pb(II) (Checkol et al. 2018). PPy/polyacrylonitrile core-shell nanostructures were also prepared and studied for Cr(VI) removal from aqueous solution (Wang et al. 2013). A variety of polymer-based nano-adsorbents including dual polymers, polymer-carbon composites, polymer silica composites, and polymer-metal composites are well studied and described in detail by Xiangke Wang et al. (Zhao et al. 2018b).

17.3.3 Silica- and Carbon-Based Nano-Adsorbents

Carbon nanotube in combination with other metals as support enhances the adsorption behaviour of adsorbents. Di et al. reported supported Ce in the form of CeO_2 on to the carbon nanotube and study its adsorptive behaviour for wastewater treatment (Di et al. 2006). The hydrated forms of these rare earth metals have high affinity for anions like fluoride, arsenate and phosphate (Zhang et al. 2003; Tokunaga et al. 1995). Carbon nanotubes were also used for supporting magnetic iron oxide by Gupta et al. (Gupta et al. 2011). The combined adsorptive behaviour of both the components in nanocomposite enhances the adsorption behaviour for Cr removal.

Silica nanoparticles are well-studied adsorbents for waste water treatment. They are also used for coating of metal oxides nanoparticles before their functionalization, with some organic functionality to improve adsorption. Bulk mesoporous silica like SBA-15, MCM-41, MCM-48, and HMS are found to be more superior adsorbents due to their large surface area, uniform pore size distribution and abundant hydroxyl groups. Their surface can be modified with various functional groups (like amines, thiol) to adsorb metal ions.

17.3.4 Nano-Catalysts for Wastewater Treatment

In addition to the adsorption of the toxic pollutants from water by nano-adsorbents, nano-catalysts are also play a role in polluted water remediation. Both the materials (nano-adsorbents & nano-catalysts) mostly have similar structure, composition, stability, surface area, and particle size, but differ only in their mode of action for purification. Nano-absorbents capture the whole polluting agents (adsorbate) on to their surface through physiosorption or chemisorption. On the other hand nano-catalysts simply act as catalysts to make more toxic ions into less toxic ones, like reduction of nitrite and catalytic degradation of dye molecules into smaller fractions. Recently nano-sized Pd-Ag alloy was studied as nano-catalyst by J. P. Troutman et al. in 2020 and used for the reduction of nitrite in drinking water (Troutman et al. 2020). Nitrate is one of the most common pollutants present in ground water. It is harmful after ingestion and causes methemoglobinemia due to the formation of nitrite after the reduction of nitrate (blue-baby syndrome). Catalytic reduction of nitrite to N_2 gas or NH_3 is a promising route for the detoxification of nitrite ions. Titanium-based nano-catalysts are well used photocatalysts for the degradation of dyes. Polycarpos Falaras et al. in 2013 studied anion-doped mesoporous titania materials for the degradation of a hazardous material microcystin-LR (MC-LR) cyanotoxin pollutant in the presence of visible light (Likodimos et al. 2013). The materials were developed through sol-gel technique and co-doping of N and F anions (Likodimos et al. 2013). Rajender S. Varma and his co-workers studied and compiled the literature about green synthesis of nano-catalysts and their application in wastewater treatment (Nasrollahzadeh et al. 2020). There are many routes for the fabrication of bio-based chemicals to the nanotubes, nanowires etc. which are found to be excellent catalysts for reduction and degradation of water pollutant (Nasrollahzadeh et al. 2019a; Nasrollahzadeh et al. 2019b; Singh et al. 2016). Usually the metal salts precursors and plants or microorganisms are mixed and fabricated to develop nanostructures. The terpenoids, phenolic acid, carbohydrates, proteins, vitamins and alkaloids will work as capping reducing agents for the development of sustainable nanostructure (Prasad 2014; Bhuyan et al. 2015; Prasad et al. 2016, 2018; Srivastava et al. 2021). Metal oxide-based photocatalysts like mesoporous TiO_2 are also well-studied catalysts for the degradation of dyes present in water (Ahirwar et al. 2016). Most of the 3D metal-based materials are used as adsorbents and catalysts for wastewater treatment through adsorption and photo-degradation respectively. TiO_2 is a well-known photocatalyst used in the degradation of dyes present in water, due to their strong oxidizing properties (Lian et al. 2020). Most of the functionalized nonporous zeolites type materials studied for other catalytic activities can also be studied for adsorption as well as catalytic activities for wastewater treatment (Ahmed and Sakthivel 2017; Yadav et al. 2016; Ahmed et al. 2016). Bimetallic oxide-based nano-catalysts are summarized in Table 17.3.

17.4 Conclusion

The whole chapter presents the detailed study of nanomaterials' applications in waste water treatment. Varity of nanomaterials in different morphology, shape and size we studied as nano-adsorbent and nano-catalyst for the adsorption of various effluents from waste water are summarized in the current chapter. The different chemical techniques (viz. hydrothermal, co-precipitation and sol-gel) used for the synthesis of nano-adsorbents and catalysts are stated with examples. Metal-based nano-adsorbents and nano-catalysts in the form of oxides, and bimetallic, and trimetallic oxides are discussed in details. Considerable work has been done to develop magnetic metallic oxides which were studied for the removal of toxic metal ions and are easy to separate from the water. The nano-types of such materials are very interesting due to the small size and large surface area. Moreover the porous silica, alumina and TiO_2 nanomaterials are also included in the chapter due to their potential application in metal ion capture and photocatalytic degradation.

Acknowledgements MA thanks CSIR-IICT and communication number IICT/Pubs./2020/247. ZS thanks JMI for her fellowship.

References

Abdelrahman EA, Tolan DA, Nassar MY (2019) A tunable template-assisted hydrothermal synthesis of hydroxysodalite zeolite nanoparticles using various aliphatic organic acids for the removal of zinc (II) ions from aqueous media. J Inorg Organomet Polym Mater 29:229–247

Afkhami A, Saber-Tehrani M, Bagheri H, Madrakian T (2011) Flame atomic absorption spectrometric determination of trace amounts of Pb (II) and Cr (III) in biological, food and environmental samples after preconcentration by modified nano-alumina. Microchim Acta 172(1-2):125–136

Ahirwar D, Bano M, Khan F (2016) Synthesis of mesoporous TiO 2 and its role as a photocatalyst in degradation of indigo carmine dye. J Sol-Gel Sci Technol 79:228–237

Ahmed M, Sakthivel A (2017) Preparation of cyclic carbonate via cycloaddition of CO_2 on epoxide using amine-functionalized SAPO-34 as catalyst. J CO2 Util 22:392–399

Ahmed M, Yadav R, Sakthivel A (2016) In situ preparation, characterization, and catalytic application of various amine functionalized microporous SAPO-37. J Nanosci Nanotechnol 16:9298–9306

Akieh MN, Lahtinen M, Väisänen A, Sillanpää M (2008) Preparation and characterization of sodium iron titanate ion exchanger and its application in heavy metal removal from waste waters. J Hazard Mater 152:640–647

Amin MT, Alazba AA, Manzoor U (2014) A review of removal of pollutants from water/wastewater using different types of nanomaterials. Adv Mater Sci Eng 2014

Avansi W, de Mendonça VR, Lopes OF, Ribeiro C (2015) Vanadium pentoxide 1-D nanostructures applied to dye removal from aqueous systems by coupling adsorption and visible-light photodegradation. RSC Adv 5:12000–12006

Badruddoza AZM, Shawon ZBZ, Rahman MT, Hao KW, Hidajat K, Uddin MS (2013) Ionically modified magnetic nanomaterials for arsenic and chromium removal from water. Chem Eng J 225:607–615

Bancroft GM, Metson JB, Kanetkar SM, Brown JD (1982) Surface studies on a leached sphene glass. Nature 299:708–710

Banerjee S, Dubey S, Gautam RK, Chattopadhyaya MC, Sharma YC (2019) Adsorption characteristics of alumina nanoparticles for the removal of hazardous dye, Orange G from aqueous solutions. Arab J Chem 12:5339–5354

Bavykin DV, Friedrich JM, Walsh FC (2006a) Protonated titanates and TiO2 nanostructured materials: synthesis, properties, and applications. Adv Mater 18:2807–2824

Bavykin DV, Friedrich JM, Walsh FC (2006b) Protonated titanates and TiO2 nanostructured materials: synthesis, properties, and applications. Adv Mater 18:2807–2824

Bayal N, Jeevanandam P (2012) Synthesis of metal aluminate nanoparticles by sol–gel method and studies on their reactivity. J Alloys Compd 516:27–32

Behbahani A, Rowshanzamir S, Esmaeilifar A (2012) Hydrothermal synthesis of zirconia nanoparticles from commercial zirconia. Proc Eng 42:908–917

Bhuyan T, Mishra K, Khanuja M, Prasad R, Varma A (2015) Biosynthesis of zinc oxide nanoparticles from *Azadirachta indica* for antibacterial and photocatalytic applications. Mater Sci Semicond Process 32:55–61

Campos AFC, de Oliveira HAL, da Silva FN, da Silva FG, Coppola P, Aquino R, Mezzi A, Depeyrot J (2019) Core-shell bimagnetic nanoadsorbents for hexavalent chromium removal from aqueous solutions. J Hazard Mater 362:82–91

Cao Y, Hu P, Jia D (2013) Phase-and shape-controlled hydrothermal synthesis of CdS nanoparticles, and oriented attachment growth of its hierarchical architectures. Appl Surf Sci 265:771–777

Checkol F, Elfwing A, Greczynski G, Mehretie S, Inganäs O, Admassie S (2018) Highly stable and efficient lignin-PEDOT/PSS composites for removal of toxic metals. Adv Sustain Syst 2(1):1700114

Chen L, He S, He B-Y, Wang T-J, Chao-Li S, Zhang C, Jin Y (2012) Synthesis of iron-doped titanium oxide nanoadsorbent and its adsorption characteristics for fluoride in drinking water. Ind Eng Chem Res 51:13150–13156

Chen J, Sheng Y, Song Y, Chang M, Zhang X, Cui L, Meng D, He Z, Shi Z, Zou H (2018) Multimorphology mesoporous silica nanoparticles for dye adsorption and multicolor luminescence applications. ACS Sustain Chem Eng 6:3533–3545

Chi Y, Tian C, Li H, Zhao Y (2019) Polymerized titanium salts for algae-laden surface water treatment and the algae-rich sludge recycle toward chromium and phenol degradation from aqueous solution. ACS Sustain Chem Eng 7:12964–12972

Choi BH, Park S-A, Park BK, Chun HH, Kim Y-T (2013) Controlled synthesis of La1−xSr$_x$CrO$_3$ nanoparticles by hydrothermal method with nonionic surfactant and their ORR activity in alkaline medium. Mater Res Bull 48:3651–3656

Chumha N, Kittiwachana S, Thongtem T, Thongtem S, Kaowphong S (2014) Synthesis and characterization of GdVO$_4$ nanoparticles by a malic acid-assisted sol–gel method. Mater Lett 136:18–21

da Silva DW, Manfroi DC, Teixeira GF, Perazolli LA, Zaghete MA, Cavalheiro AA (2014) Photocatalytic decomposition of rhodamine-B using scandium and silver-modified TiO2 powders. In: Advanced materials research, vol 975. Trans Tech Publications Ltd, pp 213–218

Deng Y, Englehardt JD (2007) Electrochemical oxidation for landfill leachate treatment. Waste Manag 27:380–388

Deng H, Li Z-j, Lin W, Yuan L-y, Lan J-h, Chang Z-y, Chai Z-f, Shi W-q (2019) Nanolayered Ti3C2 and SrTiO3 composites for photocatalytic reduction and removal of uranium (VI). ACS Appl Nano Mater 2:2283–2294

Devatha CP, Thalla AK (2018) Green synthesis of nanomaterials. In: Synthesis of Inorganic nanomaterials. Woodhead Publishing, pp 169–184

Dhand C, Das M, Sumana G, Srivastava AK, Pandey MK, Kim CG, Datta M, Malhotra BD (2010) Preparation, characterization and application of polyaniline nanospheres to biosensing. Nanoscale 2:747–754

Dhand C, Dwivedi N, Loh XJ, Ying ANJ, Verma NK, Beuerman RW, Lakshminarayanan R, Ramakrishna S (2015) Methods and strategies for the synthesis of diverse nanoparticles and their applications: a comprehensive overview. RSC Adv 127:105003–105037

Di Z-C, Ding J, Peng X-J, Li Y-H, Luan Z-K, Liang J (2006) Chromium adsorption by aligned carbon nanotubes supported ceria nanoparticles. Chemosphere 62:861–865

Dialynas E, Diamadopoulos E (2009) Integration of a membrane bioreactor coupled with reverse osmosis for advanced treatment of municipal wastewater. Desalination 238:302–311

Engates KE, Shipley HJ (2011) Adsorption of Pb, Cd, Cu, Zn, and Ni to titanium dioxide nanoparticles: effect of particle size, solid concentration, and exhaustion. Environ Sci Pollut Res 18:386–395

Fu F, Wang Q (2011) Removal of heavy metal ions from wastewaters: a review. J Environ Manag 92:407–418

Galindo IR, Viveros T, Chadwick D (2007) Synthesis and characterization of titania-based ternary and binary mixed oxides prepared by the sol– gel method and their activity in 2-propanol dehydration. Ind Eng Chem Res 46:1138–1147

Ge F, Li M-M, Ye H, Zhao B-X (2012) Effective removal of heavy metal ions Cd^{2+}, Zn^{2+}, Pb^{2+}, Cu^{2+} from aqueous solution by polymer-modified magnetic nanoparticles. J Hazard Mater 211:366–372

Ghosal A, Shah J, Kotnala RK, Ahmad S (2013) Facile green synthesis of nickel nanostructures using natural polyol and morphology dependent dye adsorption properties. J Mater Chem A 1:12868–12878

Giri AK, Kirkpatrick EM, Moongkhamklang P, Majetich SA, Harris VG (2002) Photomagnetism and structure in cobalt ferrite nanoparticles. Appl Phys Lett 80:2341–2343

Guo J, Zhou X, Lu Y, Zhang X, Kuang S, Hou W (2012) Monodisperse spindle-like FeWO4 nanoparticles: controlled hydrothermal synthesis and enhanced optical properties. J Solid State Chem 196:550–556

Gupta VK, Agarwal S, Saleh TA (2011) Chromium removal by combining the magnetic properties of iron oxide with adsorption properties of carbon nanotubes. Water Res 45:2207–2212

Hermosilla D, Han C, Nadagouda MN, Machala L, Gascó A, Campo P, Dionysiou DD (2020) Environmentally friendly synthesized and magnetically recoverable designed ferrite photocatalysts for wastewater treatment applications. J Hazard Mater 381:121200

Hu J, Lo IMC, Chen G (2007) Comparative study of various magnetic nanoparticles for Cr (VI) removal. Sep Purif Technol 56:249–256

Huang L, Peng F, Yu H, Wang H (2008) Synthesis of Cu2O nanoboxes, nanocubes and nanospheres by polyol process and their adsorption characteristic. Mater Res Bull 43:3047–3053

Huang J, Cao Y, Liu Z, Deng Z, Tang F, Wang W (2012) Efficient removal of heavy metal ions from water system by titanate nanoflowers. Chem Eng J 180:75–80

Hunsom M, Pruksathorn K, Damronglerd S, Vergnes H, Duverneuil P (2005) Electrochemical treatment of heavy metals (Cu^{2+}, Cr^{6+}, Ni^{2+}) from industrial effluent and modeling of copper reduction. Water Res 39:610–616

Ismail AA, Mohamed RM, Ibrahim IA, Kini G, Koopman B (2010) Synthesis, optimization and characterization of zeolite A and its ion-exchange properties. Colloids Surf A Physicochem Eng Asp 366:80–87

Jadhav SA, Garud HB, Patil AH, Patil GD, Patil CR, Dongale TD, Patil PS (2019) Recent advancements in silica nanoparticles based technologies for removal of dyes from water. Colloid Interface Sci Commun 30:100181

Javadian H (2014a) Application of kinetic, isotherm and thermodynamic models for the adsorption of Co (II) ions on polyaniline/polypyrrole copolymer nanofibers from aqueous solution. J Ind Eng Chem 20:4233–4241

Javadian H (2014b) Application of kinetic, isotherm and thermodynamic models for the adsorption of Co (II) ions on polyaniline/polypyrrole copolymer nanofibers from aqueous solution. J Ind Eng Chem 20:4233–4241

Joshi P, Kumar D (2018) Nanosensors: from chemical to green synthesis for wastewater remediation. Nanotechnol Sustain Water Resour:301–328

Kalfa OM, Yalçinkaya Ö, Türker AR (2009a) Synthesis and characterization of nano-scale alumina on single walled carbon nanotube. Inorg Mater 45(9):988–992

Kalfa OM, Yalçinkaya Ö, Türker AR (2009b) Synthesis and characterization of nano-scale alumina on single walled carbon nanotube. Inorg Mater 45(9):988–992

Karami H (2013) Heavy metal removal from water by magnetite nanorods. Chem Eng J 219:209–216

Kasap S, Piskin S, Tel H (2012) Titanate nanotubes: preparation, characterization and application in adsorption of strontium ion from aqueous solution. Radiochimica Acta 100:925

Khajeh M, Laurent S, Dastafkan K (2013) Nanoadsorbents: classification, preparation, and applications (with emphasis on aqueous media). Chem Rev 113:7728–7768

Khedr MH, Omar AA, Abdel-Moaty SA (2006) Magnetic nanocomposites: preparation and characterization of Co-ferrite nanoparticles. Colloids Surf A Physicochem Eng Asp 281:8–14

Khurshid S, Gul Z, Khatoon J, Shah MR, Hamid I, Khan IAT, Aslam F (2020) Anionic azo dyes removal from water using amine-functionalized cobalt–iron oxide nanoparticles: a comparative time-dependent study and structural optimization towards the removal mechanism. RSC Adv 10:1021–1041

Kumar K, Yogesh H, Muralidhara B, Nayaka YA, Hanumanthappa H, Veena MS, Kiran Kumar SR (2014a) Hydrothermal synthesis of hierarchical copper oxide nanoparticles and its potential application as adsorbent for Pb (II) with high removal capacity. Sep Sci Technol 49:2389–2399

Kumar K, Yogesh H, Muralidhara B, Nayaka YA, Hanumanthappa H, Veena MS, Kiran Kumar SR (2014b) Hydrothermal synthesis of hierarchical copper oxide nanoparticles and its potential application as adsorbent for Pb (II) with high removal capacity. Sep Sci Technol 49:2389–2399

Lei Y, Chen F, Luo Y, Zhang L (2014) Three-dimensional magnetic graphene oxide foam/Fe 3 O 4 nanocomposite as an efficient absorbent for Cr (VI) removal. J Mater Sci 49:4236–4245

Li B, Zhao J, Liu J, Shen X, Mo S, Tong H (2015) Bio-templated synthesis of hierarchically ordered macro-mesoporous anatase titanium dioxide flakes with high photocatalytic activity. RSC Adv 5(20):15572–15578

Lian Z, Wei C, Gao B, Yang X, Chan Y, Wang J, Chen GZ et al (2020) Synergetic treatment of dye contaminated wastewater using microparticles functionalized with carbon nanotubes/titanium dioxide nanocomposites. RSC Adv 10:9210–9225

Liang H, Xu B, Wang Z (2013) Self-assembled 3D flower-like α-Fe2O3 microstructures and their superior capability for heavy metal ion removal. Mater Chem Phys 141:727–734

Likodimos V, Han C, Pelaez M, Kontos AG, Liu G, Zhu D, Liao S et al (2013) Anion-doped TiO2 nanocatalysts for water purification under visible light. Ind Eng Chem Res 52(39):13957–13964

Lin C-H, Wong DS-H, Shih-Yuan L (2014) Layered protonated titanate nanosheets synthesized with a simple one-step, low-temperature, urea-modulated method as an effective pollutant adsorbent. ACS Appl Mater Interfaces 6:16669–16678

Liu B, Liu Z, Wu H, Pan S, Cheng X, Sun Y, Yanhua X (2020) Effective and simultaneous removal of organic/inorganic arsenic using polymer-based hydrated iron oxide adsorbent: capacity evaluation and mechanism. Sci Total Environ 140508

Lu H, Wang J, Stoller M, Wang T, Bao Y, Hao H (2016) An overview of nanomaterials for water and wastewater treatment. Adv Mater Sci Eng 2016

Luo T, Cui J, Hu S, Huang Y, Jing C (2010) Arsenic removal and recovery from copper smelting wastewater using TiO2. Environ Sci Technol 44:9094–9098

Maryanti E, Damayanti D, Gustian I (2014) Synthesis of ZnO nanoparticles by hydrothermal method in aqueous rinds extracts of Sapindus rarak DC. Mater Lett 118:96–98

Moghaddam AZ, Jazi ME, Allahrasani A, Ganjali MR, Badiei A (2020) Removal of acid dyes from aqueous solutions using a new eco-friendly nanocomposite of CoFe2O4 modified with Tragacanth gum. J Appl Poly Sci 137:48605

Mohsen-Nia M, Montazeri P, Modarress H (2007) Removal of Cu2+ and Ni2+ from wastewater with a chelating agent and reverse osmosis processes. Desalination 217:276–281

Moussavi G, Mahmoudi M (2009) Removal of azo and anthraquinone reactive dyes from industrial wastewaters using MgO nanoparticles. J Hazard Mater 168:806–812

Nasrollahzadeh M, Atarod M, Sajjadi M, Sajadi SM, Issaabadi Z (2019a) Plant-mediated green synthesis of nanostructures: mechanisms, characterization, and applications. In: Interface science and technology, vol 28. Elsevier, pp 199–322

Nasrollahzadeh M, Sajadi SM, Issaabadi Z, Sajjadi M (2019b) Biological sources used in green nanotechnology. In: Interface science and technology, vol 28. Elsevier, pp 81–111

Nasrollahzadeh M, Sajjadi M, Iravani S, Varma RS (2020) Green-synthesized nanocatalysts and nanomaterials for water treatment: current challenges and future perspectives. J Hazard Mater 123401

Nassar MY, Ahmed IS (2012) Template-free hydrothermal derived cobalt oxide nanopowders: synthesis, characterization, and removal of organic dyes. Mater Res Bull 47:2638–2645

Nassar MY, Amin AS, Ahmed IS, Abdallah S (2016) Sphere-like Mn2O3 nanoparticles: facile hydrothermal synthesis and adsorption properties. J Taiwan Inst Chem Eng 64:79–88

Natarajan S, Anitha V, Gajula GP, Thiagarajan V (2020) Synthesis and characterization of magnetic superadsorbent Fe3O4-PEG-Mg-Al-LDH nanocomposites for ultrahigh removal of organic dyes. ACS Omega 5(7):3181–3193

Nautiyal P, Seikh MM, Lebedev OI, Kundu AK (2015) Sol–gel synthesis of Fe–Co nanoparticles and magnetization study. J Magn Magn Mater 377:402–405

Prasad R (2014) Synthesis of silver nanoparticles in photosynthetic plants. Journal of Nanoparticles, Article ID 963961, 2014, http://dx.doi.org/10.1155/2014/963961

Prasad R, Pandey R, Barman I (2016) Engineering tailored nanoparticles with microbes: quo vadis. WIREs Nanomed Nanobiotechnol 8:316–330. https://doi.org/10.1002/wnan.1363

Prasad R, Jha A, Prasad K (2018) Exploring the Realms of Nature for Nanosynthesis. Springer International Publishing (ISBN 978-3-319-99570-0) https://www.springer.com/978-3-319-99570-0

Prasad R, Thirugnanasanbandham K (2019) Advances Research on Nanotechnology for Water Technology. Springer International Publishing https://www.springer.com/us/book/9783030023805

Penke YK, Anantharaman G, Ramkumar J, Kar KK (2019) Redox synergistic Mn-Al-Fe and Cu-Al-Fe ternary metal oxide nano adsorbents for arsenic remediation with environmentally stable As (0) formation. J Hazard Mater 364:519–530

Qdais HA, Moussa H (2004) Removal of heavy metals from wastewater by membrane processes: a comparative study. Desalination 164:105–110

Qin P, Yang Y, Zhang X, Niu J, Yang H, Tian S, Zhu J, Minghua L (2018) Highly efficient, rapid, and simultaneous removal of cationic dyes from aqueous solution using monodispersed mesoporous silica nanoparticles as the adsorbent. Nano 8:4

Qiu W, Zheng Y (2009) Removal of lead, copper, nickel, cobalt, and zinc from water by a cancrinite-type zeolite synthesized from fly ash. Chem Eng J 145:483–488

Qiu G, Huang H, Genuino H, Opembe N, Stafford L, Dharmarathna S, Suib SL (2011) Microwave-assisted hydrothermal synthesis of nanosized α-Fe2O3 for catalysts and adsorbents. J Phys Chem C 115:19626–19631

Rafique M, Hamza M, Shakil M, Irshad M, Tahir MB, Kabli MR (2020) Highly efficient and visible light-driven nickel-doped vanadium oxide photocatalyst for degradation of Rhodamine B Dye. Appl Nanosci

Rana P, Mohan N, Rajagopal C (2004) Electrochemical removal of chromium from wastewater by using carbon aerogel electrodes. Water Res 38:2811–2820

Rane AV, Kanny K, Abitha VK, Thomas S (2018) Methods for synthesis of nanoparticles and fabrication of nanocomposites. In: Synthesis of inorganic nanomaterials. Woodhead Publishing, pp 121–139

Ray PZ, Shipley HJ (2015) Inorganic nano-adsorbents for the removal of heavy metals and arsenic: a review. RSC Adv 5(38):29885–29907

Reda SM (2010) Synthesis of ZnO and Fe_2O_3 nanoparticles by sol–gel method and their application in dye-sensitized solar cells. Mater Sci Semicond Process 13:417–425

Ribeiro RAP, de Lazaro SR, Gracia L, Longo E, Andrés J (2018) Theoretical approach for determining the relation between the morphology and surface magnetism of Co3O4. J Magn Magn Mater 453:262–267

Roy A, Bhattacharya J (2012) Removal of Cu (II), Zn (II) and Pb (II) from water using microwave-assisted synthesized maghemite nanotubes. Chem Eng J 211:493–500

Roy A, Bhattacharya J (2013) A binary and ternary adsorption study of wastewater Cd (II), Ni (II) and Co (II) by γ-Fe2O3 nanotubes. Sep Purif Technol 115:172–179

Schwarzenbach RP et al (2010) Global water pollution and human health. Ann Rev Environ Res 35:109–136

Shakeri S, Rafiee Z, Dashtian K (2020) Fe3O4-based melamine-rich covalent organic polymer for simultaneous removal of auramine O and rhodamine B. J Chem Eng Data 65(2):696–705

Sharma YC, Srivastava V, Upadhyay SN, Weng CH (2008) Alumina nanoparticles for the removal of Ni (II) from aqueous solutions. Ind Eng Chem Res 47:8095–8100

Sheela T, Arthoba Nayaka Y, Viswanatha R, Basavanna S, Venkatesha TG (2012) Kinetics and thermodynamics studies on the adsorption of Zn (II), Cd (II) and Hg (II) from aqueous solution using zinc oxide nanoparticles. Powder Technol 217:163–170

Shipley HJ, Yean S, Kan AT, Tomson MB (2010) A sorption kinetics model for arsenic adsorption to magnetite nanoparticles. Environ Sci Pollut Res 17:1053–1062

Shipley HJ, Engates KE, Grover VA (2013) Removal of Pb (II), Cd (II), Cu (II), and Zn (II) by hematite nanoparticles: effect of sorbent concentration, pH, temperature, and exhaustion. Environ Sci Pollut Res 20:1727–1736

Singh P, Kim Y-J, Zhang D, Yang D-C (2016) Biological synthesis of nanoparticles from plants and microorganisms. Trends Biotechnol 34:588–599

Sreethawong T, Ngamsinlapasathian S, Yoshikawa S (2013) Facile surfactant-aided sol–gel synthesis of mesoporous-assembled Ta2O5 nanoparticles with enhanced photocatalytic H_2 production. J Mol Catal A Chem 374:94–101

Srivastava S, Usmani Z, Atanasov AG, Singh VK, Singh NP, Abdel-Azeem AM, Prasad R, Gupta G, Sharma M, Bhargava A (2021) Biological nanofactories: Using living forms for metal nanoparticle synthesis. Mini-Reviews in Medicinal Chemistry 21(2):245–265

Tan L, Xu J, Xue X, Lou Z, Zhu J, Baig SA, Xinhua X (2014) Multifunctional nanocomposite Fe 3 O 4@ SiO 2-mPD/SP for selective removal of Pb (ii) and Cr (vi) from aqueous solutions. RSC Adv 4:45920–45929

Tokunaga S, Haron MJ, Wasay SA, Wong KF, Laosangthum K, Uchiumi A (1995) Removal of fluoride ions from aqueous solutions by multivalent metal compounds. Int J Environ Stud 48:17–28

Troutman JP, Li H, Haddix AM, Kienzle BA, Henkelman G, Humphrey SM, Werth CJ (2020) PdAg alloy nanocatalysts: toward economically viable nitrite reduction in drinking water. ACS Catal 10:7979–7989

Wang Z-l, Kang ZC (2012) Functional and smart materials: structural evolution and structure analysis. Springer Science & Business Media

Wang S, Mulligan CN (2006) Effect of natural organic matter on arsenic release from soils and sediments into groundwater. Environ Geochem Health 28:197–214

Wang Z, Wang Y, Xu D, Kong ES-W, Zhang Y (2010a) Facile synthesis of dispersible spherical polythiophene nanoparticles by copper (II) catalyzed oxidative polymerization in aqueous medium. Synth Met 160:921–926

Wang X, Cai W, Lin Y, Wang G, Liang C (2010b) Mass production of micro/nanostructured porous ZnO plates and their strong structurally enhanced and selective adsorption performance for environmental remediation. J Mater Chem 20:8582–8590

Wang J, Pan K, He Q, Cao B (2013) Polyacrylonitrile/polypyrrole core/shell nanofiber mat for the removal of hexavalent chromium from aqueous solution. J Hazard Mater 244:121–129

Wang P, Sun D, Deng M, Zhang S, Bi Q, Zhao W, Huang F (2020) Amorphous phosphated titanium oxide with amino and hydroxyl bifunctional groups for highly efficient heavy metal removal. Environ Sci Nano 7:1266–1274

Wu M-T, Tsai Y-L, Chiu C-W, Cheng C-C (2016) Synthesis, characterization, and highly acid-resistant properties of crosslinking β-chitosan with polyamines for heavy metal ion adsorption. RSC Adv 6(106):104754–104762

Xiao H, Ai Z, Zhang L (2009) Nonaqueous sol– gel synthesized hierarchical CeO2 nanocrystal microspheres as novel adsorbents for wastewater treatment. J Phys Chem C 113:16625–16630

Xiong T, Yuan X, Wang H, Jiang L, Wu Z, Wang H, Cao X (2020) Integrating the (311) facet of MnO_2 and the fuctional groups of poly (m-phenylenediamine) in core–shell MnO_2@ poly (m-phenylenediamine) adsorbent to remove Pb ions from water. J Hazard Mater 389:122154

Yadav R, Ahmed M, Singh AK, Sakthivel A (2016) In-situ preparation of functionalized molecular sieve material and a methodology to remove template. Sci Rep 6:1–9

Yang J, Pan J (2012) Hydrothermal synthesis of silver nanoparticles by sodium alginate and their applications in surface-enhanced Raman scattering and catalysis. Acta Mater 60:4753–4758

Yang D, Zheng Z, Liu H, Zhu H, Ke X, Xu Y, Wu D, Sun Y (2008a) Layered titanate nanofibers as efficient adsorbents for removal of toxic radioactive and heavy metal ions from water. J Phys Chem C 112:16275–16280

Yang DJ, Zheng ZF, Zhu HY, Liu HW, Gao XP (2008b) Titanate nanofibers as intelligent absorbents for the removal of radioactive ions from water. Adv Mater 20:2777–2781

Su Y, Yang W, Sun W, Li Q, Shang JK (2015) Synthesis of mesoporous cerium–zirconium binary oxide nanoadsorbents by a solvothermal process and their effective adsorption of phosphate from water. Chem Eng J 268:270–279

Yin L, Hu B, Zhuang L, Dong F, Li J, Hayat T, Alsaedi A, Wang X (2020) Synthesis of flexible cross-linked cryptomelane-type manganese oxide nanowire membranes and their application for U (VI) and Eu (III) elimination from solutions. Chem Eng J 381:122744

Zhang Y, Yang M, Huang X (2003) Arsenic (V) removal with a Ce (IV)-doped iron oxide adsorbent. Chemosphere 51:945–952

Zhang Y, Li R, Jiang Y, Zhao B, Duan H, Li J, Feng Z (2011) Morphology evolution of ZrB_2 nanoparticles synthesized by sol–gel method. J Solid State Chem 184:2047–2052

Zhang W, Lan Y, Ma M, Chai S, Zuo Q, Kim K-H, Gao Y (2020a) A novel chitosan–vanadium-titanium-magnetite composite as a superior adsorbent for organic dyes in wastewater. Environ Int 142:105798

Zhang H, Guan W, Zhang L, Guan X, Wang S (2020b) Degradation of an organic dye by bisulfite catalytically activated with iron manganese oxides: the role of superoxide radicals. ACS Omega

Zhao D, Wu X, Guan H, Han E (2007) Study on supercritical hydrothermal synthesis of $CoFe_2O_4$ nanoparticles. J Supercrit Fluids 42(2):226–233

Zhao G, Huang X, Tang Z, Huang Q, Niu F, Wang X (2018a) Polymer-based nanocomposites for heavy metal ions removal from aqueous solution: a review. Polym Chem 9:3562–3582

Zhao G, Huang X, Tang Z, Huang Q, Niu F, Wang X (2018b) Polymer-based nanocomposites for heavy metal ions removal from aqueous solution: a review. Polymer Chemistry 9:3562–3582. M34

Chapter 18
Nano-Bioremediation Using Biologically Synthesized Intelligent Nanomaterials

S. Sakthinarendran, M. Ravi, and G. Mirunalini

Contents

18.1 Introduction.. 541
18.2 Conventional Technology of Soil Remediation................................ 543
    18.2.1 Physical Methods.. 543
    18.2.2 Chemical Methods.. 544
    18.2.3 Biological Methods... 545
18.3 Knowledge of Nanotechnological Application in Soil Remediation.... 545
    18.3.1 Nanomaterials Used in Soil Remediation.............................. 545
    18.3.2 Nano-Bioremediation of Organic Pollutants.......................... 546
    18.3.3 Nano-Bioremediation of Inorganic Pollutants....................... 547
    18.3.4 The Fate of Nanoparticles Used in Soil System.................... 548
18.4 Green Synthesis of Nanoparticle... 548
18.5 Intelligent Nano-Biosensors for Soil Remediation: An Innovative Approach..... 549
18.6 Conclusion and Future Perspective... 549
References... 550

18.1 Introduction

The total landmass present in the world accounts for about 13,003 million hectares. 37.6% of the total landmass is classified as an "agriculture area" by FAO (Marklund and Batello 2008). The use of synthetic fertilizer and pesticides in agriculture contaminate the soil affecting its health and fertility. For instance, urbanization and industrialization in China lead to the contamination of 19% agricultural soil (Zhao et al. 2014). Toxic elements like cadmium (Cd), copper (Cu), nickel (Ni), Zinc (Zn), etc. contaminate the soils, sediments, and groundwaters, posing a high threat to the

---

S. Sakthinarendran · M. Ravi (✉) · G. Mirunalini
Centre for Ocean Research, DST-FIST Sponsored Centre, ESTC Cell – Marine Biotechnology, Sathyabama Institute of Science and Technology, Chennai, India

environment and human health (Antoniadis et al. 2017; Sarkar et al. 2017; Niazi et al. 2018). The contaminants enter the soil system through various anthropogenic activities like spillages of pesticides and herbicides, industrial discharges, and discharges from service industries (solvent use, cleaning, and paint removal). Moreover, an organic compound such as trichloroethane (TCA), trichloroethylene (TCE), perchloroethane (PCA), etc.

According to WHO, survey data of 2015 reports about 494,550 deaths and 9.3 disability life due to long-term exposure to Pb. Even many young children's deaths have occurred when exposed to Pb-contaminated soil in countries like Nigeria, Senegal, and other countries (WHO 2018). Similarly, 35–77 million people got poisoned in Bangladesh due to soil contamination (Smith et al. 2000). These incidents show the importance and severity of the impact of soil contamination.

The focus on remediation of soil is the severity of risk based on different soil and human health contaminants. Remediation is done to preserve the limiting source (soil) for the future generation. Depending on the country, region, state, and local (community), the cleanup strategy must be employed. Soil contamination can also occur in nature, depending on the geochemical properties of source rocks, weathering process, volcanic eruption, etc. (Cui et al. 2018). Anthropogenic activities like agricultural practices, industrial production, military practices, mining, smelting operation, etc. add up toxic element concentration in soil. The toxic elements are collectively called as potential toxic element (PTE) (Hou and Li 2017).

The conventional methods of soil remediation could be categorized into physical and chemical methods. Physical remediation methods include excavation and removal, barrier system that prevents entry of contaminants to the soil, etc. Chemical methods include stabilization and solidification using chemical reaction agents. Similarly, biological remediation includes employing microbes for degradation or converting toxic elements to non-toxic ones Prasad and Aranda (2018). However, physical and chemical methods are not feasible and produces toxic residues like toxic sludge.

On the other hand, biological treatment takes its own time of action (Khan et al. 2018). To overcome these limitations, an urge for new sustainable technology is required. Nanotechnology is a promising field of science at the nanoscale level. It provides a sustainable technology for removing contamination of soil, thereby enhancing its health and maintaining soil fertility (Prasad et al. 2014, 2017). Nanomaterials are highly reactive, have high surface-to-volume ratio, and are smaller in size. These characteristics made these materials useful in situ remediations of soil compared to other traditional methods (Panpatte et al. 2016). The remediation mechanism is based on sorption, reduction, or chemical oxidation (Guerra et al. 2018). The remediation is of two types in situ and ex situ. The former treats the soil in the contaminated site, whereas the latter removes soil from the contaminated site and treats it externally outside its environment. Out of which in situ remediation was found to be feasible and effective.

18.2 Conventional Technology of Soil Remediation

18.2.1 Physical Methods

It includes soil washing, vitrification, encapsulation, electrokinesis, and permeable barrier system. We will see in brief about each technique.

18.2.1.1 Vitrification

Vitrification is a process of converting materials into a glass or glass-like substances. It could be applied as both in situ and ex situ methods. It employs heat to destroy organic compounds through pyrolysis or combustion and fusing inorganic metals into glass-like materials. These glass structures will be composed of oxides of silicon, boron, and alkaline earth metals. There are three heat treatment stages called first, second, and third heat generation (Reddi and Inyang 2000).

18.2.1.2 Electrokinetic Technique

This technique is suitable for an adequate grain soil system and effective in situ solutions. Electrodes are placed into the contaminated site, and a direct electrical current is applied that induces the movement of ions present in the soil towards the electrodes. Three principles are applied simultaneously: electro-osmosis, electromigration, and electrophoresis (Czurda et al. 2002). It can be used to remove organic as well as inorganic contaminants.

18.2.1.3 Permeable Barrier System

Usually, it is called pump-and-treat technology wherein groundwater is taken out of the aquifer, treating it in a water treatment plant, then back to the aquifer, or discharging it into the ground. This method was found inefficient with organic pollutants in groundwater. So, as an alternative method, the permeable wall was developed. Lower-density nonaqueous liquids will float on the water surface, and nonaqueous dense particles will settle down at the aquifer (Starr and Cherry 1994).

18.2.1.4 Encapsulation

It is a preventive measure taken to avoid further spreading of contaminants from the actual site of occurrence. For instance, bentonite is usually used as supporting slurry walls for the trench. Moreover, thin walls are a cost-effective way of encapsulation. A heavy steel beam is placed into the ground, which is vibrated with a high-pressure jet. Similar advancements in techniques include sheet pile walls, bored pile walls, injection walls, artificial ground freezing, etc. (Philip 2001).

18.2.1.5 Soil Washing

It is a widely utilized technique for removing heavy metals and organic contaminants from the soil system. The main principle is selective categorizing fine contaminants, followed by solid/liquid phase separation of the remaining suspension. It does not directly remove contaminants but separates soil fraction containing high pollutants from low pollutant soil. The separation could be done using magnetic separation. The two primary steps are wet liberation and classification unit (Wilichowski 2001).

18.2.2 Chemical Methods

The chemical method includes precipitation, ion exchange, and membrane filter process.

18.2.2.1 Precipitation

In this technique, metal ions are dissolved with precipitant resulting in the formation of insoluble compounds. Further, these solid sediments could be removed using solid or liquid filtration techniques. Several materials are used as precipitating agents includes digested sludge, iron salts, calcium hydroxide, and aluminum iodide salts. It was found very effective against metal oxides (Bradl and Xenidis 2005).

18.2.2.2 Ion Exchange

It is a ubiquitous method for the removal of heavy metals. The basic principle behind this technique is an ion exchanger matrix with dissociable counter ions. The most common materials employ as matrices are polystyrene or polyacrylate, whereas condensation resins were made up of phenol and formaldehyde (Hahn 1987).

18.2.2.3 Flocculation

This method transforms the suspended colloidal particle into an easily separating form. Further, it can be removed using any mechanical means from supernatant or using flocculant. The main inorganic flocculation chemicals are ferric and ferrous salts, aluminum iodide salts, and calcium hydroxide (Lagaly 1986).

18.2.2.4 Stabilization

This is very effective in situ application, and it immobilizes or stabilizes, thereby reducing the mobility of contaminants. It is done by chemical/physical means. The stabilizing agents are directly injected into the contaminated site. These agents convert the toxic substance into less soluble, immobile, and less toxic (US EPA 1989).

18.2.3 Biological Methods

The most common biological approach is microbial remediation and phytoremediation of heavy metals contaminants in soil. However, the only limitation is that it takes its course of time to come into effect.

18.2.3.1 Microbial Degradation

Microbes like bacteria, fungi, actinomycetes, etc. In one way, the rhizosphere bacterial community has a close relationship with the root system, thereby forming a sheath, thus preventing toxic heavy metals (Inamuddin et al. 2021). Similarly, vesicular-arbuscular mycorrhizal (VAM) limits outside contaminants' uptake by plants (Paul and Clark 1996).

18.2.3.2 Phytoremediation

Plants have several mechanisms to sequester or stabilize the elements and prevent translocation into sensitive terrestrial portion. The plant takes up non-essential elements such as As, Cd, Na, Se, and Pb. Plants uptake of water and transpiration is an essential process (Ensley 2000). Simultaneously, photovolatilization of a volatile organic compound and certain metalloids is achieved through translocation and transpiration (Fig. 18.1).

18.3 Knowledge of Nanotechnological Application in Soil Remediation

18.3.1 Nanomaterials Used in Soil Remediation

Several types of nanomaterials could be employed in the remediation of soil. They are nanoscale: zeolite, zero-valent iron, iron oxide, phosphate, iron sulfide, carbon nanotubes, etc. Zeolite is employed as an adsorbent and catalyst for different

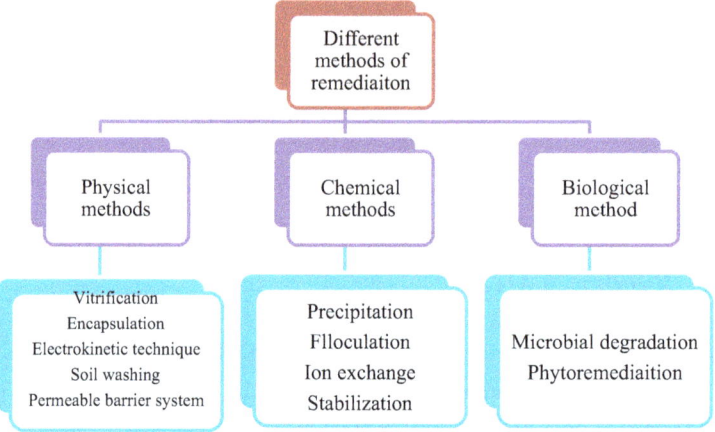

Fig. 18.1 Various methods employed in soil remediation

pollutants. These materials have a porous structure containing many cations making it readily exchangeable to other solutions. Zeolite application has provided a reduction in Hg uptake by some plants (Haidouti 1997). Then nanoiron oxide and nano-zero-valent iron oxide provides effective remediation while not having any secondary contamination. Since iron is already present in the soil, it is cost-effective and very effective against stabilizing heavy metals. This is due to their very high adsorbing capacity, which is being studied in a different context (Hua et al. 2012). Phosphate-based nanoparticles have a similar effect on pollutants and produce highly insoluble phosphorous compounds for absorbing heavy metal pollution. These particles were utilized in the soil amendment. Figure 18.2 depicts the various nanomaterials used in soil remediation.

18.3.2 Nano-Bioremediation of Organic Pollutants

Bioremediation is a practical, eco-friendly method of soil remediation using biological organisms as a tool for remediation (Kumar et al. 2021). It will be of double benefit when nanotechnology could be coupled with bioremediation. Nano remediation was utilized for chemical decontamination over the last two decades. However, integration in bioremediation is a new development, still at its infantry stage. Singh et al. studied the effect of stabilized Pb/Fe bimetallic nanoparticles on lindane contamination in soil followed by treatment using *Sphingomonas* sp. strain. It showed better efficacy in combining both techniques (Singh et al. 2013).

Nanomaterials enhance the availability of organic contaminants to biological agents. Similarly, altered membrane selectivity phytotoxic nanomaterials increases organic pollutants (Gong et al. 2018). Moreover, Le et al. 2015 studied the efficacy of bimetallic Pb/nFe on chemical oxidation of hexachlorinated biphenyls; further, it was degraded using *Burkholderia xenovorans* (Le et al. 2015). Similarly, De la

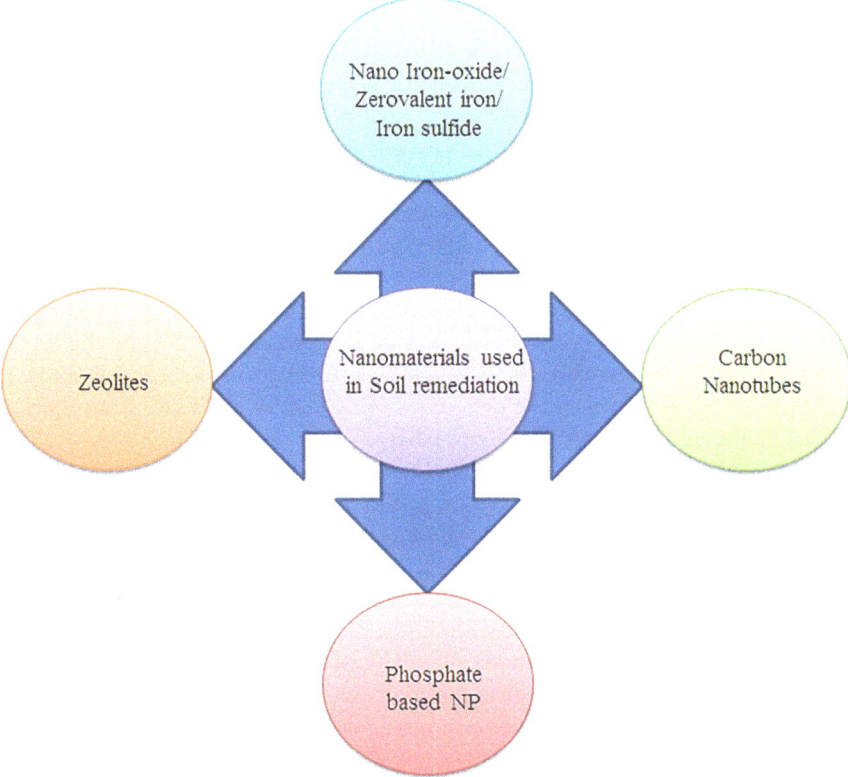

Fig. 18.2 Different nanomaterials utilized in soil remediation

Torre-Roche et al. also investigated DDT's accumulation by fullerene nanoparticles has increased the uptake of DDE significantly (De la Torre-Roche et al. 2012). Wu et al. also investigated the reduction of toxicity and translocation of polybrominated diphenyl ethers to Chinese cabbage by the application of Ni/Fe bimetallic nanoparticles. On the other hand, materials like carbon nanotubes harm *Chlorella vulgaris* grown in diuron-contaminated soil. Much work must be established to use this potential technique effectively.

18.3.3 *Nano-Bioremediation of Inorganic Pollutants*

Remediation of inorganic pollutants like heavy metals could be achieved by nano-bioremediation. Liang et al. showed a significant impact by nano-hydroxyapatite and nano- carbon black on lead phytoextraction by ryegrass (*Lolium temulentum*) (Liang et al. 2017). Hu et al. suggested that the accumulation of heavy metals in plants nanomaterials brings about a change in cell wall permeability (Hu et al. 2015). Different nanomaterials respond differently to various heavy metals and uptake of the same in plants (Gong et al. 2018).

18.3.4 The Fate of Nanoparticles Used in Soil System

Reports related to the fate of nanoparticles in water systems are more, whereas much work has not been done in the soil system. Nanomaterials deployed in the soil for various purposes interact first with soil components (organic or inorganic), and then depending on their nature, it undergoes physical, chemical, or biological changes (Darlington et al. 2009; Ben-Moshe et al. 2010). The most common physical changes in aggregation with the same type nanomaterials (homoaggregation) or aggregation with other soil constituents or pollutants (heteroaggregation). As a result, it reduces nanoparticles' mobility and behavior (Lowry et al. 2012; Batley et al. 2013). Soil organic matter also plays a vital role in behavior and the fate of nanomaterials by adsorption and stabilization. Even at a low concentration of 0.05 mg, L-1 of HS revoked the toxicity of nC60 (Lei et al. 2018). It also has an impact on the solubility and stability of NMs.

Since nanoparticles size is the minimal range, it can enter plants through osmotic pressure, cell wall pores, and capillary force. In most cases, the application of NMs over plants shows a positive result, but some plants also show a phytotoxic effect against NMs (Mazaheri-Tirani and Dayani 2020). The toxic effects of NPs could be observed in germination, biomass, and root elongation (Lin and Xing 2007; Racuciu and Creanga 2007; Lee et al. 2010). Similarly, it has a toxic effect on soil microbes too. Wu et al. (2020) showed carbon nanotubes' effect on functional genes and pathways of soil microbial communities, especially on carbon and nitrogen cycles. The toxicity of NMs is based on the concentration, nature, and synthesis process (Chen et al. 2019).

18.4 Green Synthesis of Nanoparticle

The very first essential step in nanotechnology is the synthesis of desired nanoparticles according to its target function. Nanoparticles could be synthesized through physical, chemical, and biological methods. There are numerous reports on the techniques of synthesizing nanoparticles. The most used physical approach includes evaporation-condensation, thermal decomposition, sputtering and sonication, etc. In comparison, chemical approaches include the sol-gel method, colloidal method, and chemical reducing technique using reducing agents.

These synthesis techniques could be categorized into the top-down method and bottom up methods. The former approach is made by etching nanoparticle from a substrate i.e., scaling down a bulk material to nanoparticles. In contrast, the other method is based on engraving particles onto a substrate, i.e., atoms are stacked to get a crystal plane, which is further arranged to get nanostructures since these methods use inorganic reagents that make them toxic to the environment and human health.

Therefore, an alternative method using bio-organism (plants extract, microbes, algae, secondary by-products like protein, lipids, etc.) was adopted for NP synthesis

(Prasad et al. 2016, 2018; Srivastava et al. 2021; Sarma et al. 2021). Green synthesis of nanoparticles makes use of eco-friendly, non-toxic, cost-effective reagents. So, the biological method of synthesis undergoes a bottom-up approach using reducing and stabilizing agents (Singh et al. 2011; Aziz et al. 2014, 2015, 2016, 2019; Joshi et al. 2018). Synthesis of NPs by using agro-waste should be employed to reduce the cost (Sangeetha et al. 2017). The feasibility of scaling it up to mass production will lead to waste utilization and reducing the production cost.

18.5 Intelligent Nano-Biosensors for Soil Remediation: An Innovative Approach

A biosensor is an analytical device that senses the biological changes and provides data into readable form. It comprises three crucial components, namely, detector, transducer, and bioreceptor (Dhole and Pitambara 2019; Singh et al. 2020). A biosensor at the nanoscale is called a nano-biosensor.

For the detection of heavy metals in the soil system, microbial cells can react to an available fraction of heavy metal ions, developed like luminescent bacterial sensors (Ivask et al. 2004). The application of intelligent nano-biosensors for environmental remediation is at the infantry stage. The concept behind intelligent nano-biosensors is they analyze the contaminated site with their biosensor capability and procure data, analyze it, and provide an apt solution to be employed in the site. For such a high-end device, more research must be taken to understand the soil system's pollutants. Then pollutant mediated changes in soil composition must be determined—similarly, nanomaterials' effect in various soil systems, its effect on soil microbes, and associated plants.

18.6 Conclusion and Future Perspective

Despite the promising potential of nanomaterials in application over environment and soil remediation, extensive research has been done in the development of new innovative technology for soil remediation. The limitation present in the current technology of remediation stresses nanotechnology shows higher results than conventional techniques. Since different nanomaterials react differently to pollutants, more research has to be done in understanding such effect at the same time to know about the fate of nanomaterials in the soil system. Extensive research should be done in understanding the fate of nanomaterials in the soil system and its toxic effects. Similarly, government bodies should implement regulations and guidelines for nanomaterials used in soil remediation. The effect of different nanomaterials in different soil systems should also be analyzed. So, green synthesized nanomaterials can be employed as non-toxic to the environment and a sustainable one. Combining nanotechnology strategies with bioremediation and biosensor in soil remediation is beneficial and could be used for future remediation. Nanotechnology will provide effective remediation of toxic pollutants in a cost-effective, sustainable, and without much disturbance in ecosystem balance.

References

Antoniadis V, Levizou E, Shaheen SM, Ok YS, Sebastian A, Baum C, Prasad MN, Wenzel WW, Rinklebe J (2017) Trace elements in the soil-plant interface: phytoavailability, translocation and phytoremediation – a review. Earth Sci Rev 171:621–645

Aziz N, Faraz M, Pandey R, Sakir M, Fatma T, Varma A, Barman I, Prasad R (2015) Facile algae-derived route to biogenic silver nanoparticles: Synthesis, antibacterial and photocatalytic properties. Langmuir 31:11605–11612. https://doi.org/10.1021/acs.langmuir.5b03081

Aziz N, Fatma T, Varma A, Prasad R (2014) Biogenic synthesis of silver nanoparticles using Scenedesmus abundans and evaluation of their antibacterial activity. Journal of Nanoparticles, Article ID 689419, https://doi.org/10.1155/2014/689419

Aziz N, Faraz M, Sherwani MA, Fatma T, Prasad R (2019) Illuminating the anticancerous efficacy of a new fungal chassis for silver nanoparticle synthesis. Front Chem 7.65. https://doi.org/10.3389/fchem.2019.00065

Aziz N, Pandey R, Barman I, Prasad R (2016) Leveraging the attributes of Mucor hiemalis-derived silver nanoparticles for a synergistic broad-spectrum antimicrobial platform. Front Microbiol 7:1984. https://doi.org/10.3389/fmicb.2016.01984

Batley GE, Kirby JK, McLaughlin MJ (2013) Fate and risks of nanomaterials in aquatic and terrestrial environments. Acc Chem Res 46(3):854–862. https://doi.org/10.1021/ar2003368

Ben-Moshe T, Dror I, Berkowitz B (2010) Transport of metal oxide nanoparticles in saturated porous media. Chemosphere 81(3):387–393. https://doi.org/10.1016/j.chemosphere.2010.07.007

Bradl H, Xenidis A (2005) Chapter 3 Remediation techniques. In: Interface science and technology, pp 165–261. https://doi.org/10.1016/s1573-4285(05)80022-5

Czurda K, Huttenloch P, Gregolec G, Roehl KE (2002) In: Simon FG, Meggyes T, McDonald C (eds) Advanced groundwater remediation: active and passive technologies. Thomas Telford, London, pp. 173–192. [31] Acar YB, Alshawabkeh AN. Environ. Sci. Technol.

Chen M, Sun Y, Liang J, Zeng G, Lee Z, Tang L, Zhu Y, Jiang D, Song B (2019) Understanding the influence of carbon nanomaterials on microbial communities. Environ Int 126:690–698

Cui J-L, Zhao Y-P, Li J-S, Beiyuan J-Z, Tsang DC, Poon C-S, Chan T-S, Wang W-X, Li X-D (2018) Speciation, mobilization, and bioaccessibility of arsenic in geogenic soil profile from Hong Kong. Environ Pollut 232:375–384

Darlington TK, Neigh AM, Spencer MT, Guyen OTN, Oldenburg SJ (2009) Nanoparticle characteristics affecting environmental fate and transport through soil. Environ Toxicol Chem 28:1191–1199. https://doi.org/10.1897/08-341.1

De La Torre-Roche R, Hawthorne J, Deng Y, Xing B, Cai W, Newman LA, Wang C, Ma X, White JC (2012) Fullerene-Enhanced Accumulation of p,p′-DDE in Agricultural Crop Species. Environ Sci Technol 46(17):9315–9323. https://doi.org/10.1021/es301982w

Dhole A, Pitambara M (2019) Nanobiosensors: a novel approach in precision agriculture. In: Panpatte DG, Jhala YK (eds) Nanotechnology for agriculture. Springer Nature Singapore Pvt. Ltd.

Ensley BD (2000) In: Raskin J, Ensley BD (eds) Phytoremediation of toxic metals – using plants to clean up the environment. Wiley, New York, pp. 3–31

Gong X, Huang D, Liu Y, Peng Z, Zeng G, Xu P, Cheng M, Wang R, Wan J (2018) Remediation of contaminated soils by biotechnology with nanomaterials: bio-behavior, applications, and perspectives. Crit Rev Biotechnol 38(3):455–468. https://doi.org/10.1080/07388551.2017.1368446

Guerra FD, Attia MF, Whitehead DC, Alexis F (2018) Nanotechnology for environmental remediation: materials and applications. Molecules 23:1760

Hahn HH (1987) Wassertechnologie. Springer, Berlin

Haidouti C (1997) Inactivation of mercury in contaminated soils using natural zeolites. Sci Total Environ 208(1–2):105–109. https://doi.org/10.1016/S0048-9697(97)00284-2

Hou D, Li F (2017) Complexities surrounding China's soil action plan. Land Degrad Dev 28(7):2315–2320

Hu Z, Xie Y, Jin G et al. (2015) Growth responses of two tall fescue cultivars to Pb stress and their metal accumulation characteristics. Ecotoxicology 24:563–572. https://doi.org/10.1007/s10646-014-1404-6

Hua M, Zhang S, Pan B, Zhang W, Lv L, Zhang Q (2012) Heavy metal removal from water/wastewater by nanosized metal oxides: a review. J Hazard Mater 211–212:317–331. https://doi.org/10.1016/j.jhazmat.2011.10.016

Inamuddin, Ahamed MI, Prasad R (2021) Recent Advances in Microbial Degradation. Springer Singapore (ISBN: 978-981-16-0518-5) https://www.springer.com/gp/book/9789811605178

Ivask A, François M, Kahru A, Dubourguier HC, Virta M, Douay F (2004) Recombinant luminescent bacterial sensors for the measurement of bioavailability of cadmium and lead in soils polluted by metal smelters. Chemosphere 55(2):147–156. https://doi.org/10.1016/j.chemosphere.2003.10.064

Joshi N, Jain N, Pathak A, Singh J, Prasad R, Upadhyaya CP (2018) Biosynthesis of silver nanoparticles using *Carissa carandas* berries and its potential antibacterial activities. J Sol-Gel Sci Techn 86(3):682–689. https://doi.org/10.1007/s10971-018-4666-2

Khan NT, Jameel N, Khan MJ (2018) A brief overview of contaminated soil remediation methods. Biotechnol Ind J 14(4):171

Kumar V, Prasad R, Kumar M (2021) Rhizobiont in Bioremediation of Hazardous Waste. Springer Singapore (ISBN 978-981-16-0601-4) https://www.springer.com/gp/book/9789811606014

Marklund LG, Batello C (2008) FAO datasets on land use, land use change, agriculture and forestry and their applicability for National Greenhouse Gas reporting

Mazaheri-Tirani M, Dayani S (2020) In vitro effect of zinc oxide nanoparticles on Nicotiana tabacum callus compared to ZnO micro particles and zinc sulfate ($ZnSO_4$). Plant Cell Tiss Organ Cult 140:279–289. https://doi.org/10.1007/s11240-019-01725-0

Lagaly G (1986) In Campbell FT, Pfefferkom R, Rounsaville JR (eds) Ullmann's encyclopedia of industrial chemistry, vol A7. Verlag Chemie, Weinheim, pp. 341–367

Le TT, Nguyen K-H, Jeon J-R, Francis AJ, Chang Y-S (2015) Nano/bio treatment of polychlorinated biphenyls with evaluation of comparative toxicity. J Hazard Mater 287:335–341. https://doi.org/10.1016/j.jhazmat.2015.02.001

Lei C, Sun Y, Tsang DCW, Lin D (2018) Environmental transformations and ecological effects of iron-based nanoparticles. Environ Pollut 232:10–30. https://doi.org/10.1016/j.envpol.2017.09.052

Lee CW, Mahendra S, Zodrow K, Li D, Tsai YC, Braam J, Alvarez PJ (2010) Developmental phytotoxicity of metal oxide nanoparticles to Arabidopsis thaliana. Environ Toxicol Chem 29:669–675

Liang S-x, Jin Y, Liu W, Li X, Shen S-g, Ding L (2017) Feasibility of Pb phytoextraction using nano-materials assisted ryegrass: Results of a one-year field-scale experiment. J Environ Manage 190:170–175

Lin D, Xing B (2007) Phototoxicity of nanoparticles: inhibition of seed germination and root growth. Environ Pollut 150:243–250

Lowry GV, Gregory KB, Apte SC, Lead JR (2012) Transformations of Nanomaterials in the Environment. Environ Sci Technol 46(13):6893–6899. https://doi.org/10.1021/es300839e

Niazi NK, Bibi I, Shahid M, Ok YS, Burtonc ED, Wang H, Shaheen SM, Rinklebe J, Lüttge A (2018) Arsenic sorption to perilla leaf biochar in aqueous environments: an advanced spectroscopic and microscopic examination. Environ Pollut 232:31–41

Philip LK (2001) Eng Geol 60:209

Paul EA, Clark FE (1996) Soil microbiology and biochemistry. Academic Press, San Diego

Panpatte DG, Jhala YK, Shelat HN, Vyas RV (2016) Nanoparticles – the next generation technology for sustainable agriculture. In: Singh DP, Singh HB, Prabha R (eds) Microbial inoculants in sustainable agricultural productivity, functional applications, vol 2. Springer, New Delhi, pp 289–300

Prasad R, Aranda E (2018) Approaches in Bioremediation. Springer International Publishing https://www.springer.com/de/book/9783030023683

Prasad R, Bhattacharyya A, Nguyen QD (2017) Nanotechnology in sustainable agriculture: Recent developments, challenges, and perspectives. Front Microbiol 8:1014. https://doi.org/10.3389/fmicb.2017.01014

Prasad R, Kumar V, Prasad KS (2014) Nanotechnology in sustainable agriculture: present concerns and future aspects. Afr J Biotechnol 13(6):705–713

Prasad R, Jha A, Prasad K (2018) Exploring the Realms of Nature for Nanosynthesis. Springer International Publishing (ISBN 978-3-319-99570-0) https://www.springer.com/978-3-319-99570-0

Prasad R, Pandey R, Barman I (2016) Engineering tailored nanoparticles with microbes: quo vadis. WIREs Nanomed Nanobiotechnol 8:316–330. https://doi.org/10.1002/wnan.1363

Reddi LN, Inyang HI (2000) Geoenvironmental engineering- principles and applications. Marcel Dekker, New York

Racuciu M, Creanga DE (2007) TMA-OH coated magnetic nanoparticles internalized in vegetal tissue. Rom J Phys 52:395–402

Sangeetha J, Thangadurai D, Hospet R, Purushotham P, Manowade KR, Mujeeb MA, Mundaragi AC, Jogaiah S, David M, Thimmappa SC, Prasad R, Harish ER (2017) Production of bionanomaterials from agricultural wastes. In: Nanotechnology (eds. Prasad R, Kumar M, Kumar V), Springer Nature Singapore Pte Ltd. 33–58

Sarkar SK, Mondal P, Biswas JK, Kwon EE, Ok YS, Rinklebe J (2017) Trace elements in surface sediments of the Hooghly (Ganges) estuary: distribution and contamination risk assessment. Environ Geochem Health:1–14

Sarma H, Joshi S, Prasad R, Jampilek J (2021) Biobased Nanotechnology for Green Applications. Springer International Publishing (ISBN 978-3-030-61985-5) https://www.springer.com/gp/book/9783030619848

Smith AH, Lingas EO, Rahman M (2000) Contamination of drinking-water by arsenic in Bangladesh: a public health emergency. Bull World Health Organ 78:1093–1103

Singh S, Kumar V, Dhanjal DS, Datta S, Prasad R, Singh J (2020) Biological Biosensors for Monitoring and Diagnosis. In: Singh J, Vyas A, Wang S, Prasad R (eds) Microbial Biotechnology: Basic Research and Applications. Springer Nature Singapore 317–336

Singh M, Manikandan S, Kumaraguru AK (2011) Nanoparticles: a new technology with wide applications. Res J Nanosci Nanotechnol 1(1):1–11

Singh R, Manickam N, Mudiam MKR, Murthy RC, Virendra Misra (2013) An integrated (nano-bio) technique for degradation of γ-HCH contaminated soil. J Hazard Mater 258–259:35–41. https://doi.org/10.1016/j.jhazmat.2013.04.016

Srivastava S, Usmani Z, Atanasov AG, Singh VK, Singh NP, Abdel-Azeem AM, Prasad R, Gupta G, Sharma M, Bhargava A (2021) Biological nanofactories: Using living forms for metal nanoparticle synthesis. Mini-Reviews in Medicinal Chemistry 21(2):245–265

Starr RC, Cherry JA (1994) Ground water. 32:465

US Environmental Protection Agency (1989) Stabilization/solidification of CERCLA and RCRA wastes, physical tests, chemical testing procedures, technology screening and field activities, EPA/625/6-89/022. Office of Research and Development, Cincinnati, OH

WHO (2018) Lead Poisoning and Health. http://www.who.int/en/news-room/fact-sheets/detail/lead-poisoning-and-health

Wilichowski M (2001) In: Stegmann R, Brunner G, Calmano W, Matz G (eds) Treatment of contaminated soil. Springer, Berlin, Heidelberg, pp. 417–433

Wu F, You Y, Wemer D, Jiao S, Hi J, Zhang X, Wan Y, Liu J, Wang B, Wang X (2020) Carbon nanomaterials affect carbon cycle-related functions of the soil microbial community and coupling of nutrient cycles. J Hazard Mater 390:122–144

Zhao FJ, Ma Y, Zhu YG, Tang Z, McGrath SP (2014) Soil contamination in China: current status and mitigation strategies. Environ Sci Technol 49(2):750–759

Chapter 19
Recent Developments in Nanotechnological Interventions for Pesticide Remediation

Rictika Das and Debajit Thakur

Contents

19.1	Introduction.	554
19.2	Background of Nanotechnology.	554
19.3	Nanobiotechnology.	555
19.4	Nanomaterials.	556
	19.4.1 Nanoparticles in Pesticide Remediation.	556
	19.4.2 Green Synthesis of Nanoparticles.	556
19.5	Mechanism Behind Nanomaterial-Based for Pesticide Sensing and Remediation.	559
	19.5.1 Homogeneous Chemistry.	559
	19.5.2 Heterogeneous Chemistry.	559
19.6	Various Types of Nanoparticles for Pesticide Sensing, Remediation, and Elimination.	560
	19.6.1 Metal Nanoparticles.	560
	19.6.2 Bimetallic Nanoparticles.	563
	19.6.3 Metal Oxide Nanoparticles.	564
19.7	Nanocomposites.	567
19.8	Nanobiocomposites.	568
19.9	Nanotubes.	568
	19.9.1 Carbon Nanotubes.	569
	19.9.2 Halloysite Nanotubes (HNTs).	570
19.10	Nanobioremediation.	570
19.11	Biosensors for the Detection of Pesticides.	571
	19.11.1 Nanoparticle-Based Biosensors.	571
	19.11.2 Nanoparticle-Based Optical Biosensors.	573
	19.11.3 Nanotube-Based Electrochemical Biosensors.	573

R. Das
Microbial Biotechnology Laboratory, Life Sciences Division, Institute of Advanced Study in Science and Technology (IASST), Paschim Boragaon Garchuk,
Guwahati, Assam, India

Department of Molecular Biology and Biotechnology, Cotton University,
Guwahati, Assam, India

D. Thakur (✉)
Microbial Biotechnology Laboratory, Life Sciences Division, Institute of Advanced Study in Science and Technology (IASST), Paschim Boragaon Garchuk,
Guwahati, Assam, India

© The Author(s), under exclusive license to Springer Nature Switzerland AG 2022
A. Krishnan et al. (eds.), *Emerging Nanomaterials for Advanced Technologies*,
Nanotechnology in the Life Sciences, https://doi.org/10.1007/978-3-030-80371-1_19

19.12 Future Perspectives... 573
19.13 Concluding Remarks.. 574
References.. 575

19.1 Introduction

In modern agriculture, pesticides are considered as an inescapable part for suppressing various flora and fauna pests in addition to prompt progression in urbanization and heavy industrialization that has ultimately urged to meet the demand for increase in agriculture yield to fulfill the needs of steady growth of population rate. Use of agrochemicals such as insecticides, herbicides, and fungicides kills the undesirable organisms (pests) along with some beneficial organisms existing in the ecosystem and also degrades the soil quality to a larger extent (Bhattacharyya et al. 2016). Moreover, pesticides tend to persist in the habitat for longer period causing serious issues like accumulation of undesirable residues, leading toxicity into the earth's stratum surface and also raising issues over food security using by the animals as well as the public. Worldwide, environmental protection and safety of public health has been considered as a pivotal issue and need to be addressed before time. Earlier classical methods have been used for removal of toxic wastes by chemical oxidation, adsorption, and biological oxidation, but these methods are time-consuming, with least cost effective. The emergence of nanotechnology in environmental sector has attained a great deal of interest and thus will be helpful for overall remediation through the application of nanoparticles (Prasad and Aranda 2018; Shash et al. 2019; Thangadurai et al. 2020; Saglam et al. 2021). The use of desired nanomaterials in environmental remediation with large surface-to-volume ratios act as sterling adsorbents, catalysts, and sensors thus in course increases the reactivity. Hence, there was utmost necessary for less time constraint as well as cost-effective method to remediate contaminated soil and groundwater at the hazardous sites and also breakdown to less hazardous products into the surroundings.

19.2 Background of Nanotechnology

During the period of 1980s, nanotechnology has come to the fore and gained popularity both in scientific and public domain having dimension sized ranging from 1 to 100 nm, where its unique phenomena enable novel applications. It basically deals with the use of nanomaterials in various scientific fields of biology, medicine, chemistry, physics, material science, engineering, etc. Currently the increasing efforts of using nanotechnology in the environmental sectors have improved the overall effectiveness of classical based remediation methods through the application of nanoparticles. With the constant development in technological tools, the implementation of nanotechnology in the field of detection of pollutants through techniques like

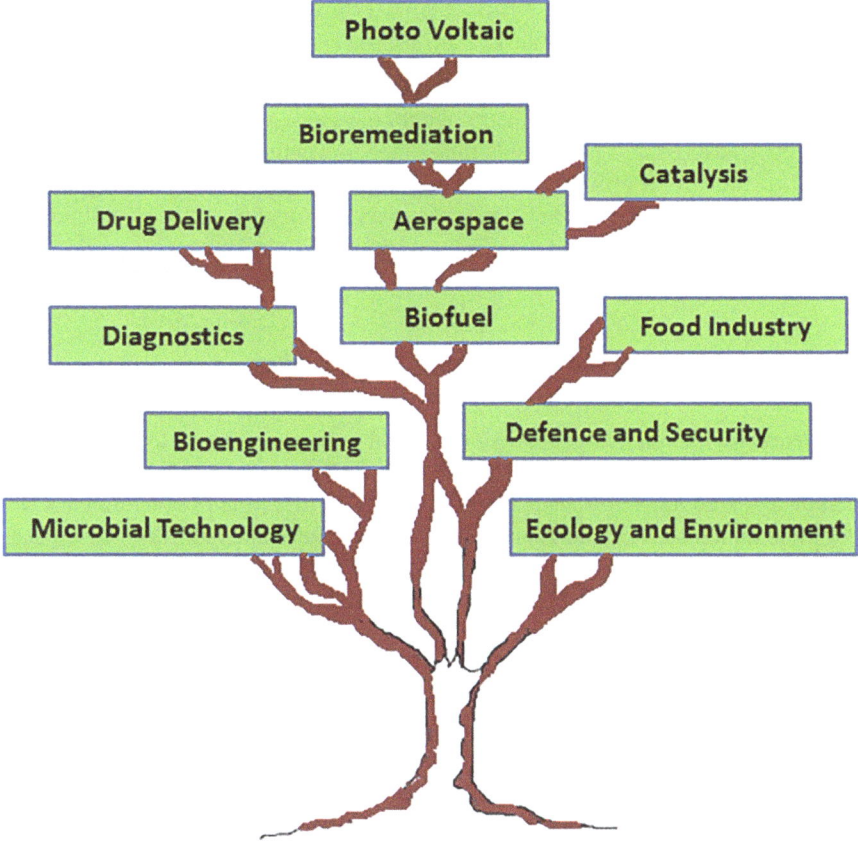

Fig. 19.1 Various branches of nanotechnology

surface-enhanced Raman scattering and electrochemical or optical detection is the need of the hour (Fulekar 2010; Tanwar et al. 2021) (Fig. 19.1).

19.3 Nanobiotechnology

In biotechnology, the knowledge and techniques of biology are applied to alter the molecular, genetic, and cellular processes to generate products and services that are being used in diverse fields from medicine to agriculture. Thus, nanobiotechnology is considered to be the unique fusion of two most progressive fields: biotechnology and nanotechnology where nanobiotechnology uses (nanoscale) biological starting materials. It has substantial potential for the programmed nano-/microfabrication of structured materials, to build tools for studying biological systems with specificity, better sensitivity, and a higher degree of recognition.

19.4 Nanomaterials

Nanomaterials has great potential in environmental remediation because of their large surface areas compared to their volumes (surface-to-volume ratio), thus acting as superior adsorbents and catalysts than other conventional tools within the range from 1 to 100 nm for efficient removal of hazardous chemicals and biological contaminants from the habitat. They can lead to very sensitive detection of pollutants to remediate the contaminants at a quicker rate with lesser hazardous by-products.

19.4.1 Nanoparticles in Pesticide Remediation

Nanoparticles exhibit a large number of special properties relative to bulk material; because of their specific size > than 100 nm; large surface area with exceptional parameters predominantly leads to a higher rate of detection of contaminants that allows to remediate these contaminants at a faster rate (Sun et al. 2006; Tosco et al. 2014). The utilization of zero-valent metals such as nickel, iron, and palladium has proved to be effective and better results in decontamination of toxic substances. Nanoparticles are synthesized by two approaches: The first approach is top-down synthesis thus involving the breakdown of bulk materials to nanoscale for obtaining nanoparticles, while the second approach is bottom-up synthesis involving the stacking-up of atoms and molecules of the bulk material mainly for the fabrication of nanoparticles. Now-a-days, the use of nanoparticles has been increased significantly by using countless nanoparticles for detecting, degrading, and removing contaminants and turn-up to be the most used in situ approach for remediation purpose (Ding et al. 2008). The major groups of NPs used for detection and degradation of pesticides are metal NPs, bimetallic NPs, and metal oxide NPs that have been frequently studied by the researchers.

19.4.2 Green Synthesis of Nanoparticles

In the past numerous years, nanoparticles synthesized by physicochemical techniques thus increase the accumulation of toxic, hazardous, and non-ecofriendly chemicals into the environment which impart non-lethal impacts on non-target organisms as well as on human population. With advancement of technology, the earnest need in synthesizing eco-friendly nanoparticles using non-toxic precursors having mild reactions and cost-effectiveness with environmental sustainability has help to emerge "green technology" by combining nanotechnology with green chemistry to exploit the potential of biological entities over physicochemical methods. Green synthesis approach provides a fast, easy, and eco-friendly nanoparticle production with environmental sustainability, simple, and reproducible approach with less hazards in the environment.

19.4.2.1 Bacterial Synthesis of Nanoparticles

Prokaryotes are comprised of single cell organisms such as bacteria that are considered as first choice for biosynthesis of nanoparticles due to their simplest structure and easy metabolism. The strong affinity for metals and its metal binding property by bacteria have helped in the synthesis of Au, Ag, Pt, Pd, Ti, nanoparticles, and so forth. The development of resistance mechanism like suppression and enhanced of influx system and efflux system, respectively, extracellular complexation, intracellular chelation, or precipitation and enzyme detoxification of metals (Silver 2003; Prasad et al. 2016) by bacteria after exposure to harsh metals and their metal ions have evolved in the large-scale synthesis of nanoparticles. Bacteria under the genus *Bacillus, Klebsiella, Lactobacillus,* and *Pseudomonas* fall under the category of nanoparticles by applying green technology. For instance, extracellular synthesis of nanoparticles by the member of Enterobacteriaceae (*Klebsiella pneumonia, Escherichia coli,* and *Enterobacter cloacae*) was first reported by Shahverdi et al. (2007). Similarly, silver nanoparticles were first synthesized by *Bacillus thuringiensis*.

19.4.2.2 Phytosynthesis of Nanoparticles

Plants possess the basic biological molecules such as carbohydrates, protein, and enzymes that have the immense potential to reduce metal salts for synthesis of nanoparticles. It is truly a one-step biosynthesis process where different plant extracts are employed due to their cost-effectiveness, easily scalable, safe to handle, less toxicity to overcome the drawbacks possessed by the conventional methods in synthesizing nanoparticles (Gurunathan et al. 2009; Prasad 2014; Srivastava et al. 2021). Biofabrication of nanoparticles by plant-based method can be easily available for large-scale production as compared to nanoparticles synthesized by microbe-based method since the latter rely more on the preservation of microbial culture that might generate toxic moieties which can be responsible for threatening both the environment and human population (Anuradha et al. 2015). Plant extracts are usually prepared from various parts such as extracts from plant leaves, juices of different medicinal plants, etc. are basically involved in mixing of plant extract with that of metal ions in a fixed ratio for synthesizing of nanoparticles. The nanoparticles are characterized by UV, XRD, and FTIR data analysis finally once they are synthesized (Table 19.1).

19.4.2.3 Nanoparticles Synthesized by Fungi and Yeast

Eukaryotic organisms such as fungi and yeast also come under the green synthesis approach of nanoparticles. Fungi has the potential to produce large-scale production of nanoparticles compared to prokaryotes (bacteria) because if the presence of various intracellular enzymes (Chen et al. 2009; Mohanpuria et al. 2008; Aziz et al. 2016, 2019; Prasad 2016, 2017; Prasad et al. 2018; Abdel-Aziz et al. 2018). Apart from monodispersity, fungi also help in the synthesis of well-defined dimensions of nanoparticles. The use of specific enzymes or metabolites; use of isolated

Table 19.1 Biosynthesis of nanoparticles from microbes and plants

Name of NP	Microorganism	References	Plant	References
Silver NPs	*Staphylococcus aureus* *Streptomyces* sp.	Kumar et al. (2011a, b), Alani et al. (2012)	*Sinapis arvensis* *Trigonella foenum-graecum*	Lam et al. (2018), Kavitha et al. (2013)
	Streptomyces naganishii *Brevibacterium casei*	Duran et al. (2011), Tripathi et al. (2015)	*Artemisia nilagirica* *Lantana camara*	Rasheed et al. (2017), Dimitrov (2006)
Gold NPs	*Rhodococcus* sp.	Yadav (2017)	*Abelmoschus esculentus*	Chaturvedi and Verma (2015)
	Klebsiella pneumonia	Balaji et al. (2009)	*Angelica, Hypericum, Hamamelis Eucalyptus, Ocimum, Mentha*	Subbaiya et al. (2014)
	Rhodopseudomonas capsulate	Park et al. (2011)	*Stevia rebaudiana*	Manivasagan et al. (2016)
	Rhodococcus sp., *Streptomyces* sp., *Streptomyces viridogens*	Ahmad et al. (2003a, b), Balagurunathan et al. (2011)	*Zingiber officinale*	Sinha et al. (2015) Pasca et al. (2014)
Iron NPs	*Shewanella oneidensis*, *Klebsiella oxytoca*	Narayanan and Sakthivel (2011), Binupriya et al. (2010)	*Aloe vera, Eucalyptus tereticornis*	Kumar et al. (2011a, b) Mishra et al. (2015)
	C. globosum	Elcey et al. (2014)	*Rosemarinus officinalis* Green tea	Kumar et al. (2012)
	E. coli, *Plerotus* sp.	Arcon et al. (2012) Kaul et al. (2012)	*Dodonaea viscose*	Kumar et al. (2012) Phumying et al. (2013)
Zinc NPs	*Lactobacillus*	Zhang (2003); Lee et al. (2008)	*Aloe vera*	Laokul and Maensiri (2009); Phumying et al. (2013)
	Streptomyces sp.	Raliya and Tarafdar (2014); Raliya and Tarafdar (2013)	*Nyctanthes arbor-tristis*	Taranath and Patil (2016)
	Candida albicans	Xu et al. (2005); Mazumdar and Haloi (2011)	*Nyctanthes arbor-tristis*	Jamdagni et al. (2016)

proteins instead of fungi culture has shown promising results in nanoparticle production. There are many reports regarding the ability of fungi to produce metal and metal oxides nanoparticles. The fungus *Verticillium* sp. is reported to reduce $AuCl_4$ ions and synthesis of Au nanoparticles on both the outer surface and the inner fungal cells with negligible reduction in the solution (Mukherjee et al. 2001; Shankar et al. 2004). *Fusarium oxysporum* have turn up for the extracellular synthesis of highly stable Au and Ag nanoparticles within the dimension range of 2–50 nm (Mukherjee et al. 2002; Ahmad et al. 2003a, b). The fungus *Aspergillus flavus* employed for the synthesizing of Ag nanoparticles wherein two proteins of 32 and 35 kDa are involved in the synthesizing process as well as gaining stability of synthesized Ag nanoparticles. Yeast (single-celled microorganisms) is also engaged in the synthesis of metallic nanoparticles (Ag and Au) by *Saccharomyces cerevisiae* regarded as more advantageous as compared to bacteria.

19.5 Mechanism Behind Nanomaterial-Based for Pesticide Sensing and Remediation

Pesticide detection, degradation, and finally removal from the environment basically involve two different type of chemistry: (a) homogeneous and (b) heterogeneous (Bond 1997). Thus, it is regarded as the fundamental method behind the chemistries for nanomaterials-based pesticide detection and removal, especially from water bodies.

19.5.1 Homogeneous Chemistry

This method first involves the nanoparticles diffused in water sample in presence of pesticide. The diffused nanoparticles then helps in the degradation or detection of pesticides present in water. The usage of most of the surface area, presented by the nanoparticles, is the fringe benefit of using this method. However, these nanoparticles are difficult to be removed from the water system, once they get diffused into it and might release toxic effects into the water system which can be a major concern of this method.

19.5.2 Heterogeneous Chemistry

This method involves the immobilized nanoparticles used on varied support materials before their use for detection of pesticides and remediation. Then these support materials are diffused into the water samples in presence of pesticide. The presence

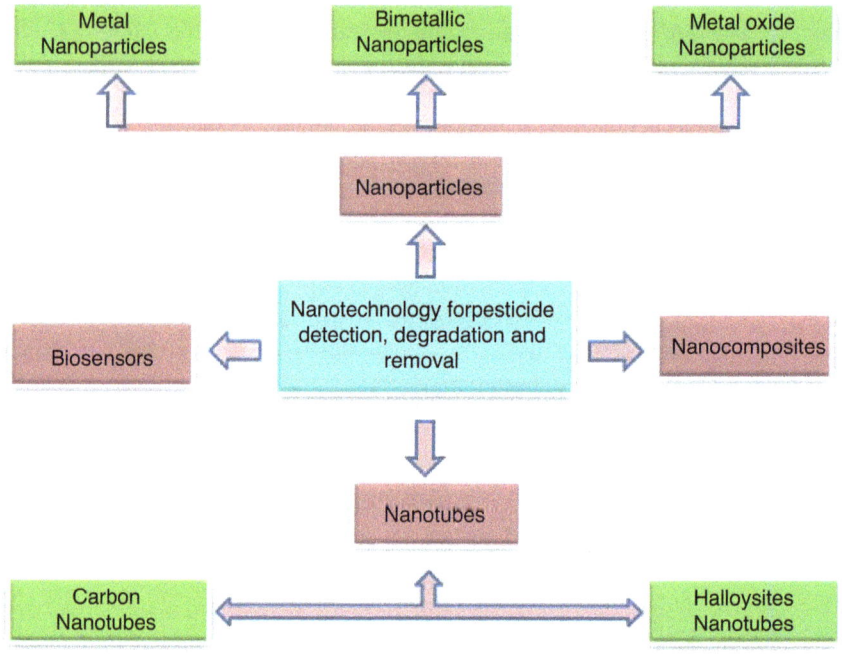

Fig. 19.2 Classification of nanotechnology-based approaches for pesticide sensing and detection

of immobilized nanoparticles into the support material helps in the detection and degradation of pesticides in the water system. The boon for heterogenous chemistry is the reuse of support systems for different water samples while detecting and degrading pesticides. The clump formation of nanoparticles is prevented through immobilization on solid supports are thus regarded as another benefit of using this method (Figs. 19.2 and 19.3).

19.6 Various Types of Nanoparticles for Pesticide Sensing, Remediation, and Elimination

19.6.1 Metal Nanoparticles

The NPs under the category of noble metals, namely, gold (Au), silver (Ag), platinum (Pt), and palladium (Pd) along with the transition metal nanoparticles like iron (Fe), copper (Cu), and zinc (Zn) have found limitless applications in environmental remediation. The incomparable surface chemistry of these NPs allows the redox reaction that takes place at the exterior of the nanoparticle thus playing a significant part in the decomposition of toxic pesticides to small-scale size and less hazardous pollutants (Street et al. 2014).

19 Recent Developments in Nanotechnological Interventions for Pesticide Remediation

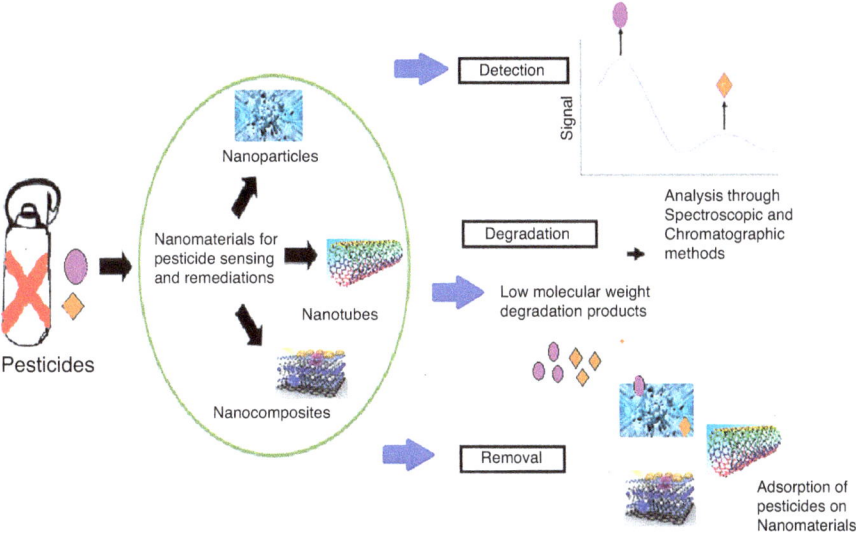

Fig. 19.3 Graphical representation of pesticide sensing, degradation, and elimination using nanomaterials

19.6.1.1 Gold Nanoparticles

From ancient times, gold has been regarded as a precious metal, and among the nanoparticles, gold is leading and has revolutionized the daily life. The AuNPs are familiar to have some properties like exhibiting various colors with change in size. The color change of AuNPs at different stages of agglomeration provides them to be a suitable visible material for analyte (Tsai et al. 2005). The sensitivity and specificity of AuNPs can be increased by their surface modification toward different pesticides. Sensors conjugated with AuNPs also help in the detection of organophosphates and organochlorines in the environment. The detection of DDT (organochlorine), by using AuNPs, is basically a colorimetric assay, where anti-DDT antibodies conjugate with AuNPs in the presence of DDT with various concentrations. The tested sample along with DDT blocks the anti-DDT antibodies followed by reduction of color intensity which is inversely proportional to the amount of DDT present in the tested sample.

19.6.1.2 Silver Nanoparticles

The AgNPs possess some specific properties of shape and size that enables their use in different fiber composites, biosensors and as antimicrobial properties (Rawtani et al. 2013; Prasad and Swamy 2013; Joshi et al. 2018). The optical properties of AgNPs with different sizes help in the sensing of pesticides. The surface modification of AgNPs similarly to AuNPs thus helps in increasing the sensitivity and specificity for the pesticides detection. Generally, dipterex an organophosphate insecticide found to be contaminated in aquatic bodies is detected by using AgNPs capped with citrate. When immobilized acetylcholinesterase enzyme is present along with citrate-capped AgNPs, it forms pink color due to the occurrence of thiocholine from acetylthiocholine in the presence of the enzyme acetylcholinesterase. But, no occurrence of thiocholine or pink color development takes place when the organophosphate dipterex is present since it inhibits the acetylcholinesterase enzyme (Lia et al. 2014). Due to the presence of acetylcholinesterase enzyme, it allows the formation of thiocholine on the substrate. This enzymatic action helps in the conversion of yellow color solution of RB-AgNPs to grey color. At the same time, the fluorescence of rhodamine B dye was also unquenched. Whereas, due to the presence of pesticides, no color formation or fluorescence occurs in the tested sample (Luo et al. 2017). Thus, this method of detection of pesticides is based on both colorimetric as well as fluorescence assay.Similarly, a herbicide known as paraquat has also been detected by using citrate-capped AgNPs. It also helps in the detection of carbaryl, an insecticide present in the vegetables, fruits, and river water with the help of modified rhodamine B dye. Various pesticides such as paraoxon and thiram (fungicide) are detected by using AgNPs as active substrates in surface-enhanced Raman spectroscopy (SERS). The emission of Raman signals from the target pesticide that gets deposited on the nanostructure surface (in this case AgNP) is increased by using SERS, thereby helping to detect very low limits of pesticides in the tested sample (Wang et al. 2014; Tanwar et al. 2021).

19.6.1.3 Iron Nanoparticles

The importance of iron is well understood in environmental remediation due to its contaminant portability, adsorption property, and mainly breakdown of iron into two valence states, i.e., ferrous iron Fe(II) (water soluble) and ferric iron (III) (water insoluble). The rusting of iron is well known in presence of oxygen helps in the formation of iron oxide. Thus both iron and iron oxide nanoparticles are considered as convenient nanoparticles for environmental remediation. For instance, the degradation of the pesticide 2,4-dichlorophenol is being degraded by biosynthesizing FeNPs by means of adsorption. The split of the benzene ring on 2,4-dichlorophenol during the degradation process emerge in the formation of acetone and acetic acid as the starting products which are then detected on GC-MS (Guo et al. 2017).

The utilization of zero-valent iron nanoparticles (ZV-FeNPs) is considered as a useful technique for remediation of organochlorine as well as micropollutant from the environment. It has been reported that lindane, an organochlorine, has been degraded by using ZV-FeNPs. It was found that the organochlorine was degraded to a large extent by the particles post polymer stabilization thus raising the exposure period of the pesticide. Lindane degradation mainly involves dichloroelimination and dehydrohalogenation, and during this degradation phase benzene, chlorobenzene, and dichlorobenzene are regarded as the main products (San Roman et al. 2013) (Table 19.2).

Table 19.2 Nanomaterials used for detection of pesticides

Nanomaterial	Type of nanomaterial	Modification	Pesticide	Matrix	Detection limit	References
Metal nanoparticle	AuNPs	Conjugation with anti-DDT antibodies	DDT	Grapes, Cauliflower	27 ng/mL 0.65–	Lisa et al. (2009)
		Conjugation with IgG antibody	Kitazine	Tomato, Cucumber	2.44 mL/mL	Malarkodi et al. (2017)
	AgNPs	Capping with citrate	Dipterex	Water	0.18 ng/mL	Lia et al. (2014)
		Conjugation with Rhodamine B	Carbaryl	Tomato, Apple, River water	0.023 ng/L	Luo et al. (2017)
Metal oxide nanoparticle	SiO$_2$NPs	Immobilization of AChE and AuNPs	Paraoxon	Spiked pesticide solutions	500 nM	Luckham and Brennan (2010)
		Conjugation with anti-FNT and anti-CLT antibodies	FNT and CLT	Spiked pesticide solutions	0.25 ng/mL	Wang et al. (2013)
Nanotube	Carbon nanotubes	Coating with silica and immobilization of phthalocyanine ruthenium	FNT	Orange juice	0.45 mg/mL	Canevari et al. (2016)
		Immobilization of AgNPs	Dimethoate	Orange and lake water	0.01 mg/mL	Hsu et al. (2017)
	Halloysite nanotubes	Immobilization of TiO$_2$NPs	Parathion	Strawberry, celery, apple		Saraji et al. (2016)

AuNPs gold nanoparticles, *AgNPs* silver nanoparticles, *DDT* dichlorodiphenyl trichloroethane, *SiO$_2$ NPs* silica nanoparticles, *AChE* acetylcholinesterase, *FNT* Feni-trothion, *CLT* chlorpyrifos methyl, *TiO$_2$NPs* titanium oxide nanoparticles

19.6.2 Bimetallic Nanoparticles

These nanoparticles consist of some interesting characteristics of combination of two metal nanoparticles in the interior of a single nanoparticle (Zaleska-Medynska et al. 2016). The utilization of bimetallic nanoparticles (BNPs) in the field of pesticides removal and degradation takes place by means of reduction by using Fe/Ni NPs (Liu et al. 2014). The degradation of the organophosphate, profenofos, takes place by using Fe/Ni BNPs as a catalyst. Here, nanoscale zero-valent iron (nZVI) particles act as a reducing factor for the pesticide degradation wherein Ni safeguards the surface of nZVI particles from corrosion. Various studies have found the utilization of Fe/NiNPs, having nZVI particles for dechlorination of the herbicide sulfentrazone (Nascimento et al. 2016). In addition to this, the degradation of 4-chlorophenol takes place where superoxide radicals provide an effective mechanism for its degradation via bimetallic system having nZVI/Ni particles (Shen et al. 2017). Recently, chlorpyrifos, an organophosphate, where degradation is being carried out by green synthesized of Ag/CuBNPs where BNPs act as nano-catalyst that provides an environmental friendly route for water purification from pesticide contamination (Rosbero and Camacho 2017) (Table 19.3).

19.6.3 Metal Oxide Nanoparticles

This type of nanoparticle due to their superconducting nature is widely used for environmental remediation. These superconducting properties of metal oxide NPs come up with an effective and specific photocatalysis activity that has been applied in diverse research work for sensing and remediation of pesticides. Various types of metal oxide NPs are involved for pesticides sensing, degradation, and removal from diverse sources. These nanoparticles (NPs) are mainly silicon oxide (SiO_2), zinc oxide (ZnO), titanium oxide (TiO_2), and iron oxide (Fe_2O_3 or Fe_3O_4) have been taken into account.

19.6.3.1 Titanium Oxide Nanoparticles

TiO_2NPs have proved to be a promising candidate for metal oxide NPs due to their exceptional features, like photocatalysis, cost-effective, non-toxicity, and stability in connection with chemicals. They possess huge surface area for photocatalytic activity, thus increasing their consumption for pollutant remediation from the surrounding. It was reported that a study was carried out at Dindigul district in Tamil Nadu, where TiO_2NPs have been applied for chlorpyrifos and monocrotophos degradation present in pond and deep well water. The degradation of these pesticides was triggered by irradiation of the photocatalyst in presence of UV light. It was found that with increase in glow time, there was an increase in the photodegradation efficiency (Amalraj and Pius 2015).

Mesoporous TiO_2NPs is regarded as the most widely used mesoporous material due to their large surface area and intrinsic property. This NP was first synthesized by Antonelli and Ying using modified sol-gel method (Antonelli and Ying 1995). These NPs are used for the microextraction of six organochlorine pesticides, namely, hexachlororbenzene (HCB), *trans*-chlordane, *cis*-chlordane, o,p-DDT, p,p-DDT, and mirex. Fabrication of solid-phase microextraction fiber by using TiO_2NPs helps in the removal of these pesticides. TiO_2NPs helps in the degradation of carbendazim, a widely used fungicide, by doping with Fe and Si ions. Thus by doping, the photocatalytic activity of NPs has been increased resulting to a larger extent of degradation (98%) of the fungicide in presence of UV light (Kaur et al. 2016).

19 Recent Developments in Nanotechnological Interventions for Pesticide Remediation 565

Table 19.3 Nanomaterials used for degradation of pesticides

Nanomaterial	Type of nanomaterial	Modification	Pesticide	Matrix	Mechanism of degradation	Degradation efficiency (%)	Reference
Metal nanoparticle	FeNPs	Immobilization with laccase	Chlorpyrifos	Spiked pesticide solution	Enzyme-based catalysis	99	Das et al. (2017)
		Coating with carboxymethyl cellulose	Lindane	Water	Dichloroelimination and dehydrohalogenation	95	San Roman et al. (2013)
Bimetallic nanoparticle	Fe/NiNPs	e	Profenofos	Spiked pesticide solution	Catalytic reduction	94.5	Mansourieh et al. (2019)
		e	Sulfentrazone	Spiked pesticide solution	Dechlorination	100	Nascimento et al. (2016)
	Ag/CuNPs	e	Chlorpyrifos	Water	Catalytic reduction	e	Rosbero and Camacho (2017)
Metal oxide nanoparticle	TiO$_2$NPs	e	Chlorpyrifos and Monocrotophos	Pond and bore well water	Photocatalysis	>95	Amalraj and Pius (2015)
		Doping with Fe and Si ions	Carbendazim	Spiked pesticide solution	Photocatalysis	98	Kaur et al. (2016)
	ZnONPs	e	Methylparathion and parathion	Water	Photocatalysis	93	Sharma et al. (2016)
Nanocomposite	Graphene oxide and AG NPs	e	Chlorpyrifos, Endosulfan and DDE	Water	Catalytic dehalogenation	95	Koushik et al. (2016)

FeNPs iron nanoparticles, *Fe/Ni NPs* iron/nickel nanoparticles, *Ag/Cu NPs* silver/copper nanoparticles, *TiO$_2$NPs* titanium oxide nanoparticles, *ZnO NPs* zinc oxide nanoparticles, *AgNPs* silver nanoparticles, *DDE* dichlorodiphenyl-dichloroethylene, *Fe$_3$O$_4$ NPs* ferric oxide nanoparticles

19.6.3.2 Zinc Oxide Nanoparticles

ZnO NPs possess both distinctive chemico-physical nature, due to their specific size and high density at the edge of the surface. The surface functionalization of ZnO NPs increases their detection and catalytic properties and thereby engaged in pesticide remediation from varied samples. ZnO NPs act as nano-photocatalyst that helps in the degradation of methyl parathion and parathion present in water samples. These pesticides were basically degraded after irradiation of the photocatalyst with the UV light. It showed around 93% degradation of the target pesticides when carried out under optimum condition (Sharma et al. 2016).

ZnO NPs helped in the removal of permethrin, a broadly used neurotoxic pesticide in agriculture field that was removed from nearby water samples. This NP along with chitosan helps in bead formation for the effective removal of the pesticide. Highest removal efficiency was found to be 99% of the pesticide at neutral pH. The beads formation has proved to be a convenient one for water purification with 56% recovery after three cycles (Dehaghi et al. 2014).

19.6.3.3 Iron Oxide Nanoparticles

These nanoparticles mainly consist of maghemite (γ-Fe_2O_3) and magnetite (Fe_3O_4) particles having wide applications in pesticides remediation from varied samples. The basic concept of nanoparticles is the increase in the surface area to volume ratio remarkably thus the immobilization of iron oxide nanoparticles in different matrices helps in the sensing and degradation of the target pesticides. This perspective of involvement of iron oxide NPs for pesticide remediation thus increased by controlling the size, shape and surface properties of these NPs with both efficiency and specificity. Thus it plays an important role in the detection of various kinds of agro-inputs used in different sources. Glyphosate, a broad-spectrum systemic herbicide, is being removed from water system by using iron oxide NPs, entrapped in mesoporous silica where the immobilization increases greatly the surface area and the porous nature of the magnetic adsorbent (Fiorilli et al. 2017). The systemic fungicide, fenarimol, is being removed of post immobilization in palygorskite, a type of clay mineral by using iron oxide (Fe_2O_3) NPs thereby increasing the holding capacity of palygorskite for removal of the fungicide fenarimol. This method showed 70% adsorption rate for fenarimol, thus suggesting the broader use of iron oxide NPs for sustained removal of fungicide (Ouali et al. 2015).

19.6.3.4 Silica Oxide Nanoparticles

These are also known as silicon dioxide nanoparticles or nano-silica particles. They have some special characteristics of adsorbent; spherical, porous nature; as well as the increase in surface area which allow the extraction efficiency of varied pesticides from different sources. Sulfonylurea found in water samples are removed by using silica oxide NPs post functionalization with N-methylimidazole. The process of functionalization helps in the enhancement of adsorption of polar pesticide on the surface of silica oxide NPs. Different types of organophosphates, namely, chlorpyrifos, methidathion, dicrotophos diazinon, mathamidophos, and malathion, have been extracted by using silica oxide NPs after co-functionalization with polar cyanopropyltriethoxysilane (CNPrTEOS) and non-polar methyltrimethoxysilane (MTMOS) and are finally analyzed by using HPLC or GC-MS (Ibrahim et al. 2013). Further, silica oxide NPs have been used by using various methods such as electrochemical, optical, and surface-enhanced Raman spectroscopy (SERS) for pesticides detection (Bapat et al. 2016). The enzyme acetylcholinesterase (AChE) binds with the silica oxide NPs that have been used for the detection of the pesticide paraoxon by using a colorimetric assay particularly known as "dipstick" assay. The detection of the pesticide at very low limits is thus increased by entrapment of AuNPs in silica oxide NPs along with the enzyme (Luckham and Brennan 2010). Similarly paraoxon can also be degraded by using enzymes such as organophosphate hydrolase and carboxyesterase that have been immobilized on mesoporous SiO2 NPs (Boubbou et al. 2012). Chlorpyrifos methyl (CLT) and fenitrothion (FNT) are detected by using antibody-tagged silica oxide NPs where monoclonal antibodies (anti-FNT and anti-CLT) isolated from mouse was found to link covalently with the silica oxide NPs for detection of pesticides (Wang et al. 2013). For detection of pesticides, SiO_2 NPs have been utilized with the help of optical, electrochemical, SERS, or fluorescence methods (Bapat et al. 2016); AChE immobilized SiO_2NPs used for detection of the paraoxon. Again, AuNPs entrapped in SiO_2 NPs along with the enzyme to increase the detection of the pesticide at very low levels (Luckham and Brennan 2010). Moreover, antibody-tagged SiO_2 NPs have been used for detecting fenitrothion (FNT) and chlorpyrifos methyl (CLT) where monoclonal antibodies from mouse (anti-FNT and anti-CLT) were covalently linked with the NPs for the pesticide detection (Wang et al. 2013). Recently, these type of procedures are regarded as a promising approach where silica oxide NPs are involved for degradation of broad range of pesticides using biotic system due to cost-effectiveness and eco-friendly nature which have attracted the attention among the researchers.

19.7 Nanocomposites

A nanocomposite falls under a broad range of materials that consist of a multiphase solid material that incorporates nanosized particles (i.e., metals, semiconductors) into a matrix having at least one dimension in the regime of nanoscopic size. Recently, several nanocomposite materials that include nanoparticles of metals, metal oxides, carbon nanotubes, plant-based nanocomposites, etc. with specific properties have played a major role for environmental remediation of pesticides more effectively from contaminated sites. Generally, nanocomposites express great surface area to large surface and volume ratio compared to normal adsorbents (Kamigaito 1991). Nowadays, graphene oxide (GO) and reduced grapheme oxide (rGO) has been widely used for the production of nanocomposites by using various metal and metal oxide NPs for pollutants remediation. Nanocomposite with Fe_3O_4 NPs has been developed by using rGO for elimination of triazine (broad-spectrum herbicide). The presence of electrostatic interlinkage between nanocomposite and analyte is thus capable of pesticide removal with high adsorption (Boruah et al. 2016). Apart from this, rGO along with AgNPs are also used for the degradation of organophosphates and organochlorine pesticides. This involves a two-step mechanism where AgNP induce removal of halogen from pesticide followed on adsorption of the degradation product of the target pesticide (Koushik et al. 2016). Nanocomposites are also used to prepare montmorillonite clay (a very soft phyllosilicate group of minerals) by using hexadimethrine, a cationic polymer, to increase the efficiency of transduction for removal of a commonly used herbicide, 2-methyl-4-chlorophenoxy acetic acid (MCPA) (Gamiz et al. 2015). Thus, the utilization of nanocomposites works as a boon when it is combined with different materials, which signifies the improvement for the detection and degradation of various pesticides.

19.8 Nanobiocomposites

These are the noble class of composite materials which have great potential where nanofillers are used in biopolymer matrix of the nanocomposite system. Nowadays, bio-based products have shown promising results due to their sustainability in the environment. Biopolymer possesses hydrophilic nature that makes them efficient adsorbents for pollutant remediation from aquatic system. Recently, chitosan-based nanobiocomposite have grabbed more attention among the researchers due to their phenomenal characteristics of biodegradable and biocompatible. The intercalation process helped in the synthesis of Ag/chitosan nanobiocomposite where both chitosan and silver nitrate solution are mix together and followed by microwave irradiation (Saifuddin et al. 2011). Microwave irradiation is a method that uses "one-pot" synthesizing of metal NPs by using metal salts and solutions of polymer surfactants. When it is exposed to this method, silver nitrate reduction takes place for the formation of Ag NPs. Finally, synthesis of Ag/chitosan nanobiocomposite is obtained for the removal of atrazine (herbicide) from drinking water.

19.9 Nanotubes

A nanomaterial with long and hollow cylindrical shaped with length varying from nm to mm is usually defined as a nanotube. The diameter of this tube-form nanotubes ranges in nanometers of (~1–100 nm). Nanotubes functions as good adsorbents due to its different parameters such as big surface area, surface modification with high aspect ratio. Thus these parameters have enabled the use of nanotubes for detection and degradation of pesticides from the contaminated sites. Based on the above based criteria, nanotubes are classified into carbon nanotubes (CNTs) and halloysite nanotubes (HNTs) that have attracted considerable attention among the researchers because of their unique properties and big surface area.

19.9.1 Carbon Nanotubes

Carbon nanotubes (CNTs) are believed to be promising material as building blocks that have better efficiency as compared to traditional adsorbents (e.g., activated carbon) due to the presence of mesoporous structure and high surface area consisting of different functional groups like phenol, carboxyl, and hydroxyl. They have higher efficiency in adsorbing large number of organic compounds and thus the process of adsorption takes place between the electrostatic attraction and formation of chemical bonds with an outer diameter ranging from 4 to 30 nm. The structure of CNTs depicts with hollow, ordered, graphene-based nanomaterial and bonded with sp^2 hybridization and acts as an exceptionally strong interaction. Some ideal properties of CNTs like thermal conductivity, high tensile strength, less weight, and high aspect ratio have not only limited their applications to electrical, electronics, sensors, and thermal devices but also attracted the researchers in the field of environmental nanotechnology (E-nanotechnology) for removal of recalcitrant from various contaminated sites. They are classified into two types:

(i) *Single-Walled CNTs (SWCNTs)*
    Nanotubes with single sheet of grapheme shell is rotated up, to form a tube-form structure are known as single-walled CNTs (SWCNTs). It was first reported in the year 1993 (Iijima et al. 1993). The diameter of SWCNTs ranges less than 1 nanometer.
(ii) *Multi-walled CNTs (MWCNTs)*
    Nanotubes with multi-walled sheets of grapheme shell consisting of concentric SWCNTs having an outer diameter and inner diameter of (50–80) nm and (5–15) nm, respectively, and spacings between the adjacent layers is of 3.4 Å. In the recent years, adsorption is considered as the most efficient and feasible technology for pollutants removal by transferring the required pollutant from water phase to solid phase (adsorbent) to collect the removal. It was found that MWCNTs are utilized as good nano-adsorbents to eliminate pesticide residues through solid phase extraction technique from tea and later on analyzed by using GC-MS/MS. The spiked pesticide residues in the tea samples are being removed and detected in GC-MS/MS and hence proved to be a skilled technique for pesticides removal.

Fenitrothion, an organophosphate insecticide, is being used in fruits, vegetables, rice, cereals, stored grains, etc. where the insecticide is determined by electrochemical detection with phthalocyanine ruthenium (RuPc); RuPc is being used as the catalyst for the redox reaction by using silica-coated MWCNT. The method behind the detection of dimethoate, an organophosphate insecticide to kill insects and mites with oxidized MWCNTs capped with AgNPs, catalyzes the oxidation of ample red (AR)-hydrogen peroxide system into resorufin, a crystalline dye. So, in the presence of dimethoate, this oxidation reaction is inhibited (Hsu et al. 2017).

19.9.2 Halloysite Nanotubes (HNTs)

HNTs are naturally viable clay nanomaterial having different morphologies, such as tube-form, spheroidal with elongated tubes, the last being the most common among all the three. They exhibit large surface area with both positively and negatively charged from inner and outer surface respectively thus providing greater adsorption properties (Rawtani and Agrawal 2012). This adsorption capacity of HNT plays a major role in detection and degradation of toxic organic compounds from the environment. Due to their increase biocompatibility and lower cytotoxicity, HNT plays a major part in recent applications such as tumor cell isolation, as scaffolds for tissue engineering, novel drug, and gene delivery. Apart from all these, most importantly, the non-toxic effects exhibited by HNTs have been regarded as a boon into the environmental remediation by the replacement of toxic and expensive carbon nanotubes.

19.10 Nanobioremediation

Nowadays, bio-based sustainable remediation has gained a lot of attention due to its low risks by minimizing the subsidiary impacts of waste generation, consumption of natural resource, etc. But it was observed that bioremediation techniques are time-consuming and might have negative impact on the existing microorganisms if present at higher concentration. Basically, a single technology may not be sufficient for removal of recalcitrant from the polluted site that might be expensive and may not be efficient, high specificity, non-hazardous, and viable. This led to the combination of multiple technologies along with their applications for a single as well as effective remediation technology with less cost effective, high specificity, and better efficiency. Therefore, a promising integration of nanotechnology and biotechnology could overcome this limitation and emerged a more strong and sustainable remediation method known as nanobioremediation. It basically uses the applications of physico-chemical (fast, but expensive) and biological methods (cheap, but relatively slow) for biodegradation of soil and water contaminants to a low risk level and less toxic environment. It was found that nanoparticles used by plants, fungi, and microbes could enhance the microbial activity by removal of pollutants such as both organic and inorganic toxins and heavy metals from the surrounding (Singh and Walker 2006).

19.11 Biosensors for the Detection of Pesticides

Biosensor was first introduced by Cammann in the year 1977. Earlier pesticide residues are being determined in soil and water by using some sophisticated instruments like high-performance liquid chromatography and mass spectroscopy, liquid/gas chromatography. Since these approaches are regarded as highly efficient but at the same time require hardcore sample preparation, along with some highly qualified technicians for analyzing the samples. Therefore, great efforts are devoted for replacement of conventional methods with better sensitivity screening, low cost, and stability for detecting low levels of pesticides. Researchers have carried out rigorous efforts for the development of efficient, cost-effective, eco-friendly nanomaterial-based biosensors, which can detect the presence as well as concentration of organic compounds when present in limited or low amounts. A biosensor basically comprises of three parts: a component that recognizes the analyte and produces a signal, a signal transducer, and a reader device (Fig. 19.4).

19.11.1 Nanoparticle-Based Biosensors

19.11.1.1 Enzyme Biosensors

Here, enzymes are being used as identification component for detection of hazardous substances from both stratum and beneath of earth's surface with high accuracy and precision, high specificity and sensitivity, robustness, and safety. Through enzyme biosensor, information is not obtained regarding a particular pesticide; rather they provide detection of broad categories of pesticides. Since acetylcholinesterase is mostly inhibited by organophosphorous pesticides, hence it is regarded as the basis for enzyme biosensor. Acetylcholinesterase (AChE), butyrylcholinesterase (BChE), or urease is being used as biological receptors where they act as catalytic activity reducers for inhibition of enzyme-based biosensors for detecting pesticides is regarded as the basic mechanism behind the enzyme biosensor. In these reactions, several methods like amperometric, conductometric, and optical are employed for choline detection, the main reaction end-product. Hence, researchers have developed AChE-based biosensors where acetylcholine (ACh) is converted by

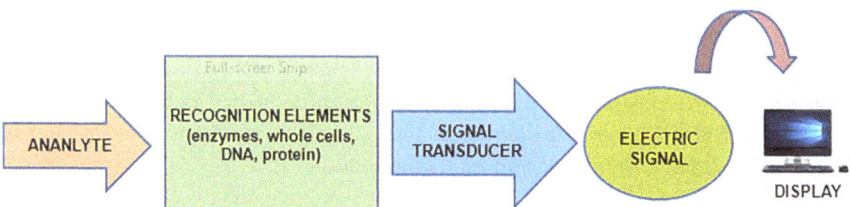

Fig. 19.4 Various parts of a biosensor

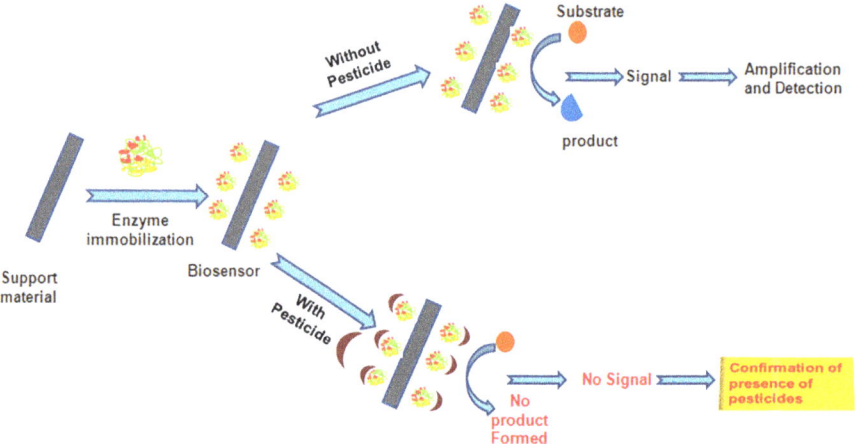

Fig. 19.5 Graphical representation of the mechanism behind pesticide detection using enzyme-based biosensor

AChE into acetic acid and choline (Ch) in presence of H_2O. The intensity of this reaction gives information about the detection of pesticides.

$$Acetylcholine + H_2O \xrightarrow{AChE} Choline + Acetic\,acid$$

Enzyme-based biosensors customize with different nanoparticles like quantum dots (QDs) and gold nanoparticles (AuNPs) are basically used for organophosphates, organochlorines detection in the environment. For instance, monocrotophos is being able to detect by this type of customized-based enzyme biosensors (Fig. 19.5).

19.11.1.2 Immunosensor

This type of sensor basically provides information regarding a specific pesticide, having high selectivity and sensitivity of antibody-antigen reaction. Conductimetric immunosensor helped in the detection of atrazine, by using antibodies labeled with NPs (Valera et al. 2008). Diuron, a substituted phenyl urea herbicide, was developed by an electrochemical immunosensor for its fast screening (Sharma et al. 2011). Fabrication of polystyrene substrate helps in the removal of low cost electrodes thus by modifying with Prussian Blue (PB)-AuNP film that helped in the enhancement of transfer of electrons in the domain of the gold electrode thereby increasing its sensitivity compared to unmodified gold electrodes.

19.11.2 Nanoparticle-Based Optical Biosensors

The use of nanoparticles has an important role in developing efficient optical biosensors for pesticide detection. Nanoparticles such as semiconductor QDs are frequently used in fluorescent sensing. These QDs or polymer nanoparticles are considered as highly photostable than a conventional fluorophore and thus allow high fluorescence quantum yields and also exhibit high sensitivity. Monocrotophos (organophosphate insecticide) can be detected through optical biosensor by CdTe as fluorescence probe (Sun et al. 2011). Recently, it was found that QDs-based fluorescence assays are able to detect several organophosphates and the activity of AChE (Saa et al. 2010; Chen et al. 2013; Yu et al. 2014; Zheng et al. 2011; Buiculescu et al. 2010; Garai-Ibabe et al. 2014).

19.11.3 Nanotube-Based Electrochemical Biosensors

The interaction between an analyte with an electrode (e.g., platinum, gold, silver, graphite) is usually measured in terms of potential or current for detection of electrochemical wherein various changes like chronoamperometry and chronopotentiometry cyclic voltammetry are observed by using abundant of techniques (Grieshaber et al. 2008). The use of nanoparticles such as modified carbon nanotubes have basically helped in enzyme electrodes, specifically electrochemical biosensor in various fields such as biosensing, biomedical engineering, nanoelectronics, and bioanalysis. Recently electrochemical biosensor has been developed by using CNTs for inhibiting the activity of AChE (Du et al. 2007; Oliveira and Mascaro 2011; Firdoz et al. 2010; Qu et al. 2010). This biosensor has shown promising results in terms of sensitivity and stability regarding pesticides monitoring in aquatic system.

19.12 Future Perspectives

The field of nanotechnology has great perspective in retransforming the previous used conventional techniques with high specificity, cost-effective, small-scale size, low detection limits and high sustainability in environmental remediation. In addition to these, utilization of various nanotechnology-based nanomaterials such as metals and metal oxides helps in increasing the removal of organic pollutants by means of reducing or oxidizing of metals along with functional groupings of chemical groups that can selectively detect the target pesticides from the contaminated sites. The use of nanoparticles such as AuNPs, TiO_2 NPs, ZnO NPs, AgNPs, and SiO_2 NPs together with nanocomposites and nanotubes like CNTs and HNTs helps to detect or sense pesticides when they are present at very negligible level. The surface modifications of these nanoparticles help in the enhancement of both sensitivity and specificity for pesticide detection. For instance, ZnO NPs and nanotubes like CNTs and HNTs have proved to increase the removal efficiency by 99–100%. The practical applications of CNTs or magnetic composites should be explored more in the future. Biosensors are considered as superior candidate for pesticides recognition in complex samples. Enzyme inhibition-based biosensors have shown better results for detection of pesticides. The large-scale use of polymeric adsorbents has shown promising results in adsorption of metals and organic pollutants. Moreover, the capability of reuse and increasing the lifespan of these adsorbents should be investigated further to reduce the cost for pesticide remediation. Further, the understandings of nanotechnological applications in the bioremediation and nano-bio-interactions accelerate the development of nanopesticide. The wide use of the pesticide for the management of tea pest aggravates various concerns such as non-target toxicity to parasitoids and insect predators, development of resistance, and upsetting the ecological balance and heavy load of pesticide residue in tea leaves. Exploiting advantages of nanotechnological interventions and aligning it with green chemistry and environmental sustainability principles hold tremendous potential in combating tea plant pest. Recent report demonstrated alteration of the nonsystemic behavior of the pesticide ferbam on tea leaves by engineered gold nanoparticles (Hou et al. 2015). These finding open up new avenue of research and translational development in nanopesticide for an integral part of Integrated Pest Management (IPM) practices of tea plant.

19.13 Concluding Remarks

Nanoremediation has emerged as a new scope for environmental remediation from the field of nanobiotechnology that has immense potential for diminishing recalcitrants from the habitat to a more greenish environment. The effectiveness of nanoremediation has helped in the overall reduction of xenobiotics that has approximately reach to zero level in in situ nanoremediation due to cost effective, high competency as well as large-scale application compared to ex situ nanoremediation which needs to be workout for better results. The applications of various nanoparticles and biosensors having some specific characteristics of large surface area, high specificity, and small-scale size with quick response have helped to overcome the drawbacks for detection of pesticides at very low levels of detection in the environment over the conventional techniques. However, the toxicity of nanoparticles (carbon nanotubes, metals, and metal oxide NPs) cannot be denied completely because of their lethal effect on the significant proliferation over the microorganisms. Hence, more meticulous research is utmost necessary in this regard so that nanoremediation emerged as a promising tool in the future for contaminant remediation and also for better environmental sustainability.

Acknowledgments The authors gratefully thank the Director of IASST for all the support. This work is funded by DST Women Scientist-A scheme under the Government of India, bearing the reference No. (SR/WOS-A/LS-127/2018(G).

References

Abdel-Aziz SM, Prasad R, Hamed AA, Abdelraof M (2018) Fungal nanoparticles: A novel tool for a green biotechnology? In: Fungal Nanobionics: Principles and Applications (eds. Prasad R, Kumar V, Kumar M and Wang S), Springer Singapore Pte Ltd. 61–87

Ahmad A, Senapati S, Khan MI, Kumar R, Ramani R, Srinivas V, Sastry M (2003a) Intracellular synthesis of gold nanoparticles by a novel alkalotolerant actinomycete, Rhodococcus species. Nanotechnology 14(7):824

Ahmad PM, Senapati S, Mandal D, Khan MI, Kumar R, Sastry M (2003b) Colloids Surf B 28:313

Alani F, Moo-Young M, Anderson W (2012) World J Microbiol Biotechnol 28:1081

Amalraj A, Pius A (2015) Photocatalytic degradation of monocrotophos and chlorpyrifos in aqueous solution using TiO2 under UV radiation. J Water Process Eng 7:94–101

Antonelli DM, Ying JY (1995) Synthesis of hexagonally packed mesoporous TiO_2 by a modified sol–gel method. Angew Chem Int Ed 34:2014–2017

Anuradha J, Abbasi T, Abbasi SA (2015) An eco-friendly method of synthesizing gold nanoparticles using an otherwise worthless weed pistia (*Pistiastratiotes* L.). J Adv Res 6(5):711–720

Arcon I, Piccolo O, Paganelli S, Baldi F (2012) Biometals 25(5):875

Aziz N, Pandey R, Barman I, Prasad R (2016) Leveraging the attributes of *Mucor hiemalis*-derived silver nanoparticles for a synergistic broad-spectrum antimicrobial platform. Front Microbiol 7:1984. https://doi.org/10.3389/fmicb.2016.01984

Aziz N, Faraz M, Sherwani MA, Fatma T, Prasad R (2019) Illuminating the anticancerous efficacy of a new fungal chassis for silver nanoparticle synthesis. Front Chem 7:65. https://doi.org/10.3389/fchem.2019.00065

Balagurunathan R, Radhakrishnan M, Rajendran RB, Velmurugan D (2011) Indian J Biochem Biophys 48:331

Balaji DS, Basavaraja S, Deshpande R, Mahesh DB, Prabhakar BK, Venkataraman A (2009) Colloids Surf B Biointerfaces 68:88

Bapat G, Labade C, Chaudhari A, Zinjarde S (2016) Silica nanoparticle based techniques for extraction, detection, and degradation of pesticides. Adv Colloid Interf Sci 237:1–14

Bhattacharyya A, Duraisamy P, Govindarajan M, Buhroo AA, Prasad R (2016) Nano-biofungicides: Emerging trend in insect pest control. In: Advances and Applications through Fungal Nanobiotechnology (ed. Prasad R), Springer International Publishing Switzerland 307–319

Binupriya AR, Sathishkumar M, Vijayaraghavan K, Yun SI (2010) J Hazard Mater 177:539

Bond GC (1997) In: Ertl G, Knozinger H, Weitkamp J (eds) Handbook of heterogeneous catalysis. VCH, Weinheim, pp 752–770

Boruah PK, Sharma B, Hussain N, Das MR (2016) Magnetically recoverable Fe3O4/graphene nanocomposite towards efficient removal of triazine pesticides from aqueous solution: investigation of the adsorption phenomenon and specific ion effect. Chemosphere 168:1–10

Boubbou KE, Schofield DA, Landry CC (2012) Enhanced enzymatic activity of OPH in ammonium-functionalized mesoporous silica: surface modification and pore effects. J Phys Chem C 116:17501–17506

Buiculescu R, Hatzimarinaki M, Chaniotakis NA (2010) Biosilicated CdSe/ZnS quantum dots as photoluminescent transducers for acetylcholinesterase-based biosensors. Anal Bioanal Chem 398:3015–3021

Canevari TC, Prado TM, Cincotto FH, Machado SAS (2016) Immobilization of ruthenium phthalocyanine on silica-coated multi-wall partially oriented carbon nanotubes: electrochemical detection of fenitrothion pesticide. Mater Res Bull 76:41–47
Chaturvedi V, Verma P (2015) Bioresour Bioproc 2:18
Chen YL, Tuan HY, Tien CW (2009) Augmented biosynthesis of cadmium sulfide nanoparticles by genetically engineered *Escherichia coli*. Biotechnol Prog 25:1260–1266
Chen Z, Ren X, Tang F (2013) Optical detection of acetylcholine esterase based on CdTe quantum dots. Chin Sci Bull 58:2622–2627
Das A, Singh J, Yogalakshmi KN (2017) Laccase immobilized magnetic iron nanoparticles: fabrication and its performance evaluation in chlorpyrifos degradation. Int Biodeterior Biodegr 117:183–189
Dehaghi SM, Rahmanifar B, Moradi AM, Azar PA (2014) Removal of permethrin pesticide from water by chitosanezinc oxide nanoparticles composite as an adsorbent. J Saudi Chem Soc 18:348–355
Dimitrov D (2006) Colloids Surf A Physicochem Eng Asp 8:282
Ding YS, Zhang TL (2008) Using Chou's pseudo amino acid composition to predict subcellular localization of apoptosis proteins: an approach with immune genetic algorithm-based ensemble classifier. *Pattern Recogn Lett* 29(13):1887–1892
Du D, Huang X, Cai J, Zhang A (2007) Comparison of pesticide sensitivity by electrochemical test based on acetylcholinesterase biosensor. Biosens Bioelectron 23, No. 2:285–289
Durán N, Marcato PD, Durán M, Yadav A, Gade A, Rai M (2011) Appl Microbiol Biotechnol 90:1609
Elcey C, Kuruvilla AT, Thomas D (2014) Int J Curr Microbiol Appl Sci 3:408
Fiorilli S, Rivoira L, Calì G, Appendini M, Bruzzoniti MC, Coïsson M, Onida B (2017) Iron oxide inside SBA-15 modified with amino groups as reusable adsorbent for highly efficient removal of glyphosate from water. Appl Surf Sci 411:457–465
Firdoz S, Ma F, Yue X, Dai Z, Kumar A, Jiang B (2010) A novel Amperometric biosensor based on single walled carbon nanotubes with acetylcholine esterase for the detection of Carbaryl Pestcide in water. Talanta 83(1):269–273
Fulekar MH (2010) Nanotechnology: importance and applications. IK International Pvt Ltd
Gamiz B, Hermosin MC, Cornejo J, Celis R (2015) Hexadimethrine-montmorillonite nanocomposite: characterization and application as a pesticide adsorbent. Appl Surf Sci 332:606–613
Garai-Ibabe G, Saa L, Pavlov V (2014) Thiocholine mediated stabilization of *in situ* produced CdS quantum dots: application for the detection of acetylcholinesterase activity and inhibitors. Analyst 139:280–284
Grieshaber D, MacKenzie R, Voros J, Reimhult E (2008) Electrochemical biosensors-sensor principles and architectures. Sensors 8(3):1400–1458
Guo M, Weng X, Wang X, Chen Z (2017) Biosynthesized iron-based nanoparticles used as a heterogeneous catalyst for the removal of 2,4-dichlorophenol. Sep Purif Technol 175:222–228
Gurunathan S, Kalishwaralal K, Vaidyanathan R, Venkataraman D, Pandian SR, Muniyandi J, Hariharan N, Eom SH (2009) Colloids Surfaces B: Biointerfaces 74:328–335
Hou R, Pang S, He L (2015) In situ SERS detection of multi-class insecticides on plant surfaces. Analytical Methods 7(15):6325–6330
Hsu CW, Lin ZY, Chan TY, Chiu TC, Hu CC (2017) Oxidized multiwalled carbon nanotubes decorated with silver nanoparticles for fluorometric detection of dimethoate. Food Chem 224:353–358
Ibrahim WAW, Ismail WNW, Sanagi MM (2013) Selective and simultaneous solid phase extraction of polar and non-polar organophosphorus pesticides using sol-gel hybrid silica-based sorbent. J Teknologi Sci Eng 62:83–87
Iijima S, Ichihashi T (1993) Single-shell carbon nanotubes of 1-nm diameter. Nature 363.6430:603–605
Jamdagni P, Khatri P, Rana JS (2016) J King Saud Univ-Sci 30:168. https://doi.org/10.1016/j.jksus.2016.10.002

Joshi N, Jain N, Pathak A, Singh J, Prasad R, Upadhyaya CP (2018) Biosynthesis of silver nanoparticles using *Carissa carandas* berries and its potential antibacterial activities. J Sol-Gel Sci Techn 86(3):682–689. https://doi.org/10.1007/s10971-018-4666-2

Kamigaito O (1991) What can be improved by nanometer composites? J Jpn Soc Powder Powder Metall 38(3):315–321

Kaul R, Kumar P, Burman U, Joshi P, Agrawal A, Raliya R, Tarafdar (2012) J, Mater Sci-Poland 30:254

Kaur T, Sraw A, Wanchoo RK, Toor AP (2016) Visible elight induced photocatalytic degradation of fungicide with Fe and Si doped TiO2 nanoparticles. Mater Today Proc 3:354–361

Kavitha KS, Baker S, Rakshith D, Kavitha HU, Rao HCY, Harini BP, Satish S (2013) Int Res J Bio Sci 2:66

Koushik D, Gupta SS, Maliyekkal SM, Pradeep T (2016) Rapid dehalogenation of pesticides and organics at the interface of reduced graphene oxide-silver nanocomposite. J Hazard Mater 308:192–198

Kumar D, Karthik L, Kumar G, Roa KB (2011a) Pharmacologyonline 3:31100–31111

Kumar KP, Paul W, Sharma CP (2011b) Process Biochem 46:2007

Kumar KM, Mandal BK, Sinha M, Krishnakumar V (2012) Spectrochim Acta A Mol Biomol Spectrosc 86:490

Lam SJ, Wong EHH, Boyer C, Qiao GG (2018) Prog Polym Sci 76:40. https://doi.org/10.1016/j.progpolymsci.2017.07.007

Laokul P, Maensiri S (2009) J Optoelecron Adv Mat 11:857

Lee C, Kim JY, Lee WI, Nelson KL, Yoon J, Sedlak DL (2008) Environ Sci Technol 42:4927

Lia Z, Wang Y, Ni Y, Kokot S (2014) Unmodified silver nanoparticles for rapid analysis of the organophosphorus pesticide, dipterex, often found in different waters. Sens Actuator B 193:205–211

Lisa M, Chouhan RS, Vinayaka AC, Manonmani HK, Thakur MS (2009) Gold nanoparticles based dipstick immunoassay for the rapid detection of dichlorodiphenyltrichloroethane: an organochlorine pesticide. Biosens Bioelectron 25:224–227

Liu WJ, Qian TT, Jiang H (2014) Bimetallic Fe nanoparticles: recent advances in synthesis and application in catalytic elimination of environmental pollutants. Chem Eng J 236:448–463

Luckham RE, Brennan JD (2010) Bioactive paper dipstick sensors for acetylcholinesterase inhibitors based on solegel/enzyme/gold nanoparticle composites. Analyst 135:2028–2035

Luo Q, Li Y, Zhang M, Qiu P, Deng Y (2017) A highly sensitive, dual-signal assay based on rhodamine B covered silver nanoparticles for carbamate pesticides. Chin Chem Lett 28:345–349

Malarkodi C, Rajeshkumar S, Annadurai G (2017) Detection of environmentally hazardous pesticide in fruit and vegetable samples using gold nanoparticles. Food Control 80:11–18

Manivasagan P, Venkatesan J, Sivakumar K, Kim SK (2016) Crit Rev Microbiol 42:209

Mansouriieh N, Sohrabi MR, Khosravi M (2019) Optimization of profenofos organophosphorus pesticide degradation by zero-valent bimetallic nanoparticles using response surface methodology. Arab J Chem 12(8):2524–2532

Mazumdar H, Haloi N (2011) J Microbiol Biotechnol Res 1:39

Mishra S, Dixit S, Soni S (2015) Bio-Nanoparticles: Biosynthesis and Sustainable Biotechnological Implications, vol 20, p 141. https://doi.org/10.1002/9781118677629.ch7

Mohanpuria P, Rana NK, Yadav SK (2008) Biosynthesis of nanoparticles: technological concepts and future applications. J Nanopart Res 10:507–517

Mukherjee P, Ahmad A, Mandal D, Senapati S, Sainkar SR, Khan MI, Ramani R, Parischa R, Ajaykmat PV, Alam M, Sastry M, Kumar R (2001) Angew Chem Int Ed 40:3585

Mukherjee P, Senapati S, Mandal D, Ahmad A, Khan MI, Kumar R, Sastry M (2002) Extracellular synthesis of gold nanoparticles by the fungus Fusarium oxysporum. Chembiochem 3(5):461–463

Narayanan KB, Sakthivel N (2011) J Hazard Mater 189:519

Nascimento MA, Lopes RP, Cruz JC, Silva AA, Lima CF (2016) Sulfentrazone dechlorination by iron-nickel bimetallic nanoparticles. Environ Pollut 211:406–413

Oliveira AC, Mascaro LH (2011) Evaluation of Ace- tylcholinesterase biosensor based on carbon nanotube paste in the determination of Chlorphenvinphos. Inter J Anal Chem:Article ID 974216

Ouali A, Belaroui LS, Bengueddach A, Galindo AL, Pena A (2015) Fe2O3epalygorskite nanoparticles, efficient adsorbates for pesticide removal. Appl Clay Sci 115:67–75

Park Y, Hong YN, Weyers A, Kim YS, Linhardt RJ (2011) IET Nanobiotechnol 5:69

Pasca RD, Mocanu A, Cobzac SC, Petean I, Horovitz O, Tomoaia-Cotisel M (2014) Biogenic Syntheses of Gold Nanoparticles Using Plant Extracts. Part Sci Technoly 32(2):131–137

Phumying S, Labuayai S, Swatsitang E, Amornkitbamrung V, Maensiri S (2013) Mat Res Bull 48:2060

Prasad R (2014) Synthesis of silver nanoparticles in photosynthetic plants. Journal of Nanoparticles, Article ID 963961, 2014, https://doi.org/10.1155/2014/963961

Prasad R (2016) Advances and Applications through Fungal Nanobiotechnology. Springer, International Publishing Switzerland (ISBN: 978-3-319-42989-2)

Prasad R (2017) Fungal Nanotechnology: Applications in Agriculture, Industry, and Medicine. Springer Nature Singapore Pte Ltd. (ISBN 978-3-319-68423-9)

Prasad R, Aranda E (2018) Approaches in Bioremediation. Springer International Publishing. https://www.springer.com/de/book/9783030023683

Prasad R, Kumar V, Kumar M, Wang S (2018) Fungal Nanobionics: Principles and Applications. Springer Nature Singapore Pte Ltd. (ISBN 978-981-10-8666-3). https://www.springer.com/gb/book/9789811086656

Prasad R, Pandey R, Barman I (2016) Engineering tailored nanoparticles with microbes: quo vadis. WIREs Nanomed Nanobiotechnol 8:316–330. https://doi.org/10.1002/wnan.1363

Prasad R, Swamy VS (2013) Antibacterial activity of silver nanoparticles synthesized by bark extract of *Syzygium cumini*. Journal of Nanoparticles https://doi.org/10.1155/2013/431218

Qu Y, Sun Q, Xiao F, Shi G, Jin L (2010) Layer- by-layer self-assembled Acetylcholienesterase/ PAMAM- Au on CNTs modified electrode for sensing pesticides. Bioelectrochemistry 77(2):139–144

Raliya R, Tarafdar JC (2013) ZnO Nanoparticle Biosynthesis and Its Effect on Phosphorous- Mobilizing Enzyme Secretion and Gum Contents in Clusterbean (Cyamopsis tetragonoloba L.). Agribiol Res 2(1):48–57

Raliya R, Tarafdar JC (2014) Biosynthesis and characterization of zinc, magnesium and titanium nanoparticles: an eco-friendly approach. Int Nano Lett 4:1

Rasheed T, Bilal M, Iqbal HMN, Li C (2017) Colloids Surf B 158:408. https://doi.org/10.1016/j.colsurfb.2017.07.020

Rawtani D, Agrawal YK (2012) Halloysite as support matrices: a review. Emerg Mater Res 1(4):212–220

Rawtani D, Agrawal YK, Prajapati P (2013) Interaction behavior of DNA with halloysite nanotubeesilver nanoparticle-based composite. BioNano Sci 3:73–78

Rosbero TMS, Camacho DH (2017) Green preparation and characterization of tentacle-like silver/ copper nanoparticles for catalytic degradation of toxic chlorpyrifos in water. J Environ Chem Eng 5:2524–2532

Saa L, Virel A, Sanchez-Lopez J, (2010) Pavlov V. Analytical applications of enzymatic growth of quantum dots. Chem Eur J 16:6187–6192

Saifuddin N, Nian CY, Zhan LW, Ning KX (2011) Chitosan-silver nanoparticles composite as point-of-use drinking water filtration system for household to remove pesticides in water. Asian J Biochem 6(2):142–159

Saglam N, Korkusuz, F, Prasad R (2021) Nanotechnology Applications in Health and Environmental Sciences. Springer International Publishing (ISBN: 978-3-030-64410-9). https://www.springer.com/gp/book/9783030644093

San Roman I, Alonso ML, Bartolom EL, Galdames A, Goiti E, Ocejo M, Moragues M, Alonso RM, Vilas JL (2013) Relevance study of bare and coated zero valent iron nanoparticles for lindane degradation from its by-product monitorization. Chemosphere 93:1324–1332

Saraji M, Jafari M, Mossaddegh M (2016) Halloysite nanotubes-titanium dioxide as a solid-phase microextraction coating combined with negative corona discharge-ion mobility spectrometry for the determination of parathion. Anal Chim Acta 926:55–62

Shahverdi AR, Fakhimi A, Shahverdi HR, Minaian S (2007) Synthesis and effect of silver nanoparticles on the antibacterial activity of different antibiotics against *Staphylococcus aureus and Escherichia coli*. Nanomedicine 3(2):168–171

Shankar SS, Rai A, Ahmad A, Sastry M (2004) J Colloid Interface Sci 2:496

Sharma AK, Tiwari RK, Gaur MS (2016) Nanophotocatalytic UV degradation system for organophosphorus pesticides in water samples and analysis by Kubista model. Arab J Chem 9:1755–1764

Sharma P, Sablok K, Bhalla V, Suri CR (2011) A novel disposable electrochemical immunosensor for phenyl urea herbicide diuron. Biosensors and Bioelectronics 26(10):4209–4212

Shash S, Ramanan VV, Prasad R (2019) Sustainable Green Technologies for Environmental Management. Springer Singapore (ISBN: 978-981-13-2772-8) https://www.springer.com/la/book/9789811327711

Shen W, Mu Y, Wang B, Ai Z, Zhang L (2017) Enhanced aerobic degradation of 4-chlorophenol with iron-nickel nanoparticles. Appl Surf Sci 393:316–324

Silver S (2003) FEMS Microbiol Rev 27(2–3):341–353

Singh BK, Walker A (2006) Microbial degradation of organophosphorus compounds. FEMS Microbiol Rev 30(3):428–471

Sinha S, Paul ND, Halder N, Sengupta D, Patra SK (2015) Appl Nanosci 5:703

Srivastava S, Usmani Z, Atanasov AG, Singh VK, Singh NP, Abdel-Azeem AM, Prasad R, Gupta G, Sharma M, Bhargava A (2021) Biological nanofactories: Using living forms for metal nanoparticle synthesis. Mini-Reviews in Medicinal Chemistry 21(2):245–265

Street A, Sustich R, Duncan J, Savage N (2014) Nanotechnology applications for clean water: solutions for improving water quality, 2nd edn. Elsevier, Waltham

Subbaiya R, Shiyamala M, Revathi K, Pushpalatha R, Selvam MM (2014) Int J Curr Microbiol App Sci 3:83

Sun H, Zhang QF, Wu JL (2006) Electroluminescence from ZnO nanorods with an n-ZnO/p-Si heterojunction structure. Nanotechnology 17(9):2271

Sun X, Liu B, Xia K (2011) A sensitive and regenerable biosensor for organophosphate pesticide based on self-assembled multilayer film with CdTe as fluorescence probe. Luminescence 26(6):616–621

Tanwar S, Paidi SK, Prasad R, Pandey R, Barman I (2021) Advancing Raman spectroscopy from research to clinic: Translational potential and challenges. Spectrochimica Acta Part A: Molecular and Biomolecular Spectroscopy. https://doi.org/10.1016/j.saa.2021.119957

Taranath TC, Patil BN (2016) Int J Mycobacteriol 5:197. https://doi.org/10.1016/j.ijmyco.2016.03.0041

Thangadurai D, Sangeetha J, Prasad R (2020) Nanotechnology for Food, Agriculture, and Environment. Springer International Publishing (ISBN 978-3-030-31937-3). https://www.springer.com/gp/book/9783030319373

Tosco T, Papini MP, Viggi CC, Sethi R (2014) Nanoscale zerovalent iron particles for groundwater remediation: a review. J Clean Prod 77:10–21

Tripathi V, Fraceto LF, Abhilash PC (2015) Sustainable clean-up technologies for soils contaminated with multiple pollutants: Plant-microbe-pollutant and climate nexus. Ecol Eng 82:330–335

Tsai CS, Yu TB, Chen CT (2005) Gold nanoparticle-based competitive colorimetric assay for detection of protcineprotein interactions. Chem Commun 0:4273–4275

Valera E, Ramon-Azcon J, Sanchez FJ, Marco MP, Rodriguez A (2008) Conductimetric Immunosensor for Atrazine detection based on antibodies labelled with gold nanoparticles. Sensors Actuators B 134(1):95–103

Wang X, Mu Z, Shangguan F, Liu R, Pu Y, Yin L (2013) Simultaneous detection of Fenitrothion and Chlorpyrifos-Methyl with a photonic suspension array. PLOS One 8 (6):e66703. https://doi.org/10.1371/journal.pone.0066703

Wang B, Zhang L, Zhou X (2014) Synthesis of silver nanocubes as a SERS substrate for the determination of pesticide paraoxon and thiram. Spectrochim Acta Mol Biomol Spectrosc 121:63–69

Wells M (2007) Vanishing bees threaten U.S. Crops. BBC News, London

Xu J-C, Mei L, Guo X-Y, Li H-U (2005) J Mol Catal A Chem 226:123

Yadav KK (2017) J Mater Environ Sci 8:740

Yu T, Ying T-Y, Song Y-Y, Li Y-J, Wu F-H, Dong X-Q, Shen J-S (2014) A highly sensitive sensing system based on photoluminescent quantum dots for highly toxic organophosphorus compounds. RSC Adv 4:8321–8327

Zaleska-Medynska A, Marchelek M, Diak M, Grabowska E (2016) Noble metalbased bimetallic nanoparticles: the effect of the structure on the optical, catalytic and photocatalytic properties. Adv Colloid Interf Sci 229:80–107

Zhang W-X (2003) J Nanopart Res 5:323

Zheng Z, Zhou Y, Li X, Liua S (2011) Tang Z, Highly-sensitive organophosphorous pesticide biosensors based on nanostructured films of acetylcholinesterase and CdTe quantum dots. Biosens Bioelectron 26:3081–3085

Chapter 20
Potential Applications of Nanomaterials in Agronomy: An African Insight

Hupenyu A. Mupambwa, Adornis D. Nciizah, Patrick Nyambo, Ernest Dube, Binganidzo Muchara, Morris Fanadzo, and Martha K. Hausiku

Contents

20.1	Introduction	582
20.2	African Smallholder Crop Production Challenges	582
	20.2.1 Soil Fertility Management	582
	20.2.2 Plant Pest Management	584
	20.2.3 Drought/Water Shortages and Management	585
20.3	Current Nanotechnologies in Agriculture	586
	20.3.1 Nano-fertilizers	587
	20.3.2 Nanopesticides	588
20.4	Mycosynthesis of Nanomaterials	589
20.5	Potential of Nanotechnologies in African Smallholder Agriculture	590
	20.5.1 The Soil	591
	20.5.2 The Crop Yield	594
20.6	Socioeconomic Implications of Nanotechnology on Agriculture	594
20.7	Conclusion	596
References		596

H. A. Mupambwa (✉) · M. K. Hausiku
Sam Nujoma Marine and Coastal Resources Research Centre, Sam Nujoma Campus, University of Namibia, Henties Bay, Namibia

A. D. Nciizah
Agricultural Research Council – Natural Resources & Engineering, Pretoria, South Africa

P. Nyambo
Risk and Vulnerability Science Centre, Faculty of Science and Agriculture, University of Fort Hare, Alice, South Africa

E. Dube
School of Natural Resources Management, Nelson Mandela Metropolitan University, George Campus, South Africa

B. Muchara
Graduate School of Business Leadership, University of South Africa, Midrand, South Africa

M. Fanadzo
Department of Agriculture, Faculty of Applied Sciences, Cape Peninsula University of Technology, Wellington, South Africa

© The Author(s), under exclusive license to Springer Nature Switzerland AG 2022
A. Krishnan et al. (eds.), *Emerging Nanomaterials for Advanced Technologies*, Nanotechnology in the Life Sciences, https://doi.org/10.1007/978-3-030-80371-1_20

20.1 Introduction

The green revolution which took place in the twentieth century saw the introduction of huge machinery in crop production such as tractors, ploughs, planters, and harvesters, as well as the significant use of inorganic nutrient sources in the soil. Notwithstanding the environmental and soil degradation consequences of the green revolution, as it relied on chemicals like inorganic fertilizers which only fed the crop and not the soil, it also resulted in significant improvement in crop yields and food security which was critical in feeding the growing world population. Due to the need for continued increase in food production per unit area to accommodate the ever-increasing world population, the twenty-first century has now seen a move towards sustainable intensification. The introduction of sustainable intensification is a concept that is related to increasing crop productivity in agriculture with limited or reduced impacts on the environment while maintaining the same or even higher productivity within the same area (Fraceto et al. 2016). As part of sustainable intensification, biotechnology is gaining momentum as the way forward in driving this concept where living organisms and their derivatives are utilized in increasing the efficiency of various processes linked to agriculture. From realizing that all matter is made up of atoms to the discovery of how all living organisms are made up of cells with various organelles inside, science has been inspired by nature, with most research driving towards the manipulation of matter and living organisms at their basic fundamental building structures. Through the exploitation of the various fundamental principles of nature, researchers are now able to control different physical, biological, and chemical properties at a nanoscale (Ghasemmezhad et al. 2019). This has led to the development of a new branch of science called nanotechnology, and this is seeing wide applications even in agriculture where nano-materials are now being used to power sustainable intensification and increase farmer's resilience and adaptation to climate change (Prasad et al. 2014, 2017a, b). With nano-technology being an emerging science, its applicability to resource poor African farmers is still limited. Our chapter presents some of the production challenges that are faced by these resource poor farmers, the nanotechnologies that are available in agriculture and provides a guide to where nanotechnology can be utilized under the African farming scenario.

20.2 African Smallholder Crop Production Challenges

20.2.1 Soil Fertility Management

Soils in much of Africa, where smallholder farming is practiced, are poor in quality and prone to rapid degradation. This setttlement of smallholder farmers on poor quality soils can be traced back to the colonial past, whereby Africans were forced to settle on marginal lands. Sandy soils, that are inherently acidic and low in phosphorus (P) and zinc (Zn), predominate most smallholder farm lands in Africa (Mafongoya et al. 2006; Vanlauwe and Giller 2006). In commercial crop production, chemical fertilizers and composts play a major role in increasing yield. However, such is not generally the case in smallholder systems of Africa, where most farmers are resource poor. Thus, the poor soil fertility situation of resource poor farmers is exacerbated by continuous mono-cropping and removal of nutrients without any replenishment from either inorganic fertilizers or manures. Additionally, conventional farming practices such as overgrazing, intensive tillage, short-to-no fallow periods, lack of organic matter input, and limited crop rotation have been identified as further causes of soil degradation.

Poor soil fertility is long known to be a major reason why resource poor smallholder farmers abandon farming as a means of livelihood. It is also a component of the vicious cycle of poverty in African smallholder farmer communities. Barret and Bevis (2015) demonstrated a strong, self-reinforcing link between poor soil fertility and poverty, whereby, "poor soil constrained agricultural production and household capital, and low household capital constrained investments in improving soils." They recommended that interventions that provide poor families with inputs for improving soil fertility may all help break the soil–poverty cycle. It has often been suggested by various researchers that agronomical research on soil fertility improvement in African smallholder farmer systems may need to focus on alternative sources of nutrients to inorganic fertilizers for soil fertility improvement. Proposed nutrient sources include livestock manures, human waste, biomass transfer, cover crops, nitrogen (N)-fixing legumes, composts, and rock phosphates, among others. However, numerous challenges have also been observed regarding practical application of these alternative soil amendments.

Malnourished animals grazing on depleted veldts as those in resource poor smallholder communities of Africa will produce manures of poor quality, which are also difficult to collect as the animal's graze over vast areas. Similarly, organic sources for production of nutrient rich composts are scarce in resource poor communities and have at times been shown to stimulate harmful pests and diseases, especially under conservation agriculture (CA) (Chiduza and Dube 2013). For various ethical reasons, the acceptability of human wastes as soil fertility ameliorant may be contentious. It is accepted that not all legumes will fix N and at times require inoculation. Depending on the harvest index, legumes grown on poor soils may remove net amounts of N from the soil (Vanlauwe and Giller 2006). Since legumes require land, labor, and P fertilizer, the cost of legume N on poor soils may easily exceed that of purchasing N fertilizer. Rock phosphates are typically poor soluble, and this limits their applicability as they have to composted first before application (Vanlauwe and Giller 2006).

In the past two decades, there has been a research drive towards CA practices that encourage accumulation of SOM as well as prevent soil erosion for ameliorating soil fertility problems. The major principles of CA are minimum soil disturbance, crop rotation, and permanent soil cover through cover crops (Dube et al. 2012). Despite numerous efforts aimed at promoting CA by both government and nongovernmental organizations, uptake of the CA technology in Africa remains low, owing to various challenges in its implementation. As highlighted by Giller et al. (2009) in the heretic's review of CA, the lack of suitable no-till farming tools, pests and diseases, competing uses of crop residues in mulching, and animal feeding are some of the barriers to wide-scale CA adoption in resource poor smallholder farming systems of South Africa. Integrated soil fertility management (ISFM), which aims to increase food production through strategic combination of traditional and new technologies is the new approach being advocated for improving soil fertility in resource poor smallholder communities (Bekunda et al. 2010, Vanlauwe et al. 2015). It has been defined as soil fertility management practices/principles that include the use of fertilizer, organic inputs, improved cultivars/germplasm combined with various other crop production technologies, and the knowledge on how to adapt these practices to local conditions, aiming at maximizing agronomic use efficiency of the applied nutrients and improving crop productivity (Vanlauwe et al. 2010). To this regard, nanotechnologies present new opportunities yet to be explored for ISFM in resource poor smallholder farming systems.

20.2.2 Plant Pest Management

Apart from poor soil fertility, pests have been singled out by several researchers as another major cause of poor crop yields in smallholder farmer systems of Africa, including conservation agriculture systems (Chiduza and Dube 2013; Fanadzo et al. 2018). Pest and soil fertility management complement each other. Thus, the benefits of improved soil fertility may not be realized in the absence of good pest management and vice versa. Without adequate pest management, it may be wasteful to invest in soil fertility improvement. A plant pest is any organism that reduces the quantity, quality, or value of a plant grown for food, fiber, or recreation. The three major groups of plant pests are animal pests (such as insects, mites, other arthropods, and vertebrates such as birds and rodents), plant pathogens (such as fungi, bacteria, viruses and nematodes), and weeds. Of these, weeds, insects and plant pathogens have been identified are the main biological constraints faced by smallholder farmers. Economic losses caused by these pests do not only include the direct action of these organisms as they damage crops, but also indirect economic losses related to the costs of controlling the pest and environmental damage caused by pesticides.

In South Africa, weeding is identified as one of the activities demanding the most labor in smallholder farming (van Averbeke et al. 1998; Fanadzo 2007, 2010; Fanadzo et al. 2010). Most smallholder farmers are aware of the detrimental effects of weeds, but do not have the time or the means to control them especially where tractor mechanization has resulted in an increased area of land being cultivated. Owing to a history of poor weed management, soils of production lands of the resource poor smallholder farming systems typically have reservoirs of problematic weed seeds. In severe cases, weed problems have been known to cause abandonment of cropped fields (Fanadzo et al. 2010). Using smallholder irrigation schemes in South Africa as a case study, Fanadzo et al. (2010) reported that there were no responses to N fertilization and irrigation owing to poor weed management. In the absence of adequate weed management, N fertilizer promotes weed growth more than crop growth. Because resource-poor smallholder farmers cannot afford herbicides and lack technical expertise for their use, they typically rely on hand hoes for weeding. Literacy rates are low, such that even where pesticides are made freely available, there is a general lack of technical knowledge on how to use them efficiently. There is drudgery through the many hours spend on hoe weeding and, in most cases, weeding is carried out by women and children (Giller et al. 2009). The move towards CA has presented more challenges with regard to pest management. Conventional tillage was useful for controlling not just weeds, but animal pests as well through exposing the pests to their natural enemies or directly by physical damage inflicted during the tillage process. Tillage was also used to bury crop residues that harbor plant pathogens into deeper layers of the soil where they cause less or no disease.

Labor shortages largely caused by rural–urban migration and shift from farming by younger community members into other activities means that hand weeding can be a daunting task. Thus, in many cases, labor is mostly family labor, comprising mainly of older people. The effect of HIV/AIDs has weakened the labor force, and the COVID-19 pandemic is likely to exacerbate the situation. Under irrigated crop production in South Africa, poor crop stands observed in farmlands have mainly been attributed to late weeding caused by shortage of labor (Fanadzo et al. 2010). Due to competing demands on family labor and the overlapping of the optimal times for weeding with other crop activities, there is a shortage of labor for hand-weeding at the optimal times. As a result, weeding is either not done at all or is performed inadequately or too late after the weeds have already reduced the crop yield potential. It is clear that herbicides have a major role to play in increasing agricultural production and improvement of rural welfare. The integration of herbicides into small-scale farmer pest management systems can minimize labor requirements and increase profitability. Recent developments in herbicide nanotechnology mean that herbicides may be applied in smaller proportions at a nanoscale, thus potentially reducing the costs and hazards associated with herbicide use (Bhattacharyya et al. 2016). These and other potential benefits of nano-technologies in pest management at the resource poor smallholder farmer level will be discussed in this book chapter.

20.2.3 Drought/Water Shortages and Management

Africa is experiencing extreme weather conditions due to climate change, and these include high annual rainfall variability and unreliability. Shifts in seasonal rainfall and temperature patterns have seen frequent droughts and major disruption of smallholder farmer livelihoods. For example, during the period 2015 to 2018, South Africa experienced one of the worst droughts in history, with a rainfall average of

403 mm in 2015, the lowest recorded since 1904. The extent of its severity was qualified as greater than the 1992 to 1993 drought, experienced by the entire region of the Southern Africa (Baudoin et al. 2017). Droughts can have devastating effects on smallholder farmer livelihoods and adapting to them is complicated because of the absence of a solution that can be suitable for a wide range of situations or problems. Compared to their large-scale commercial counterparts, smallholder farmers are more vulnerable to drought impacts, and they struggle to cope and adapt. Their capacity is restricted because of their limited or lack of access to water resources, finance, reliable markets, land, knowledge, managerial skills, and extension support.

It is suspected that as the continent is becoming drier overall, atmospheric temperatures are also rising. Hence, there has been a recent drive toward much research aimed at drought mitigation in smallholder farmer systems of Africa. Options proposed include rainwater harvesting, mulching, reduced tillage practices, and drought-tolerant cultivars, among others. These are referred to as "climate smart" agricultural technologies and include conservation agriculture (Ramanan et al. 2020). Rockstrom (2000) identified two broad strategies for increasing yields in dryland farming when water availability in the root zone constrains crop growth: (1) capturing more water and allowing it to infiltrate into the root zone and (2) using the available water more efficiently (increasing water productivity) by increasing the plant water uptake capacity and/or reducing non-productive soil evaporation. Practices used to achieve this include water harvesting, supplemental irrigation, deficit irrigation, precision irrigation techniques, and soil–water conservation practices. Generally, infield water management in smallholder farming is weak, partly due to the infield irrigation equipment which is inefficient, as inefficient, as well as the absence of proper irrigation scheduling. Lack of scheduling results in over- or under-irrigation depending on crop type and growth stage. There is a need for simple irrigation technologies that can improve water use efficiency in resource poor smallholder farming systems of Africa, where water is increasingly becoming a scarce resource.

20.3 Current Nanotechnologies in Agriculture

Production challenges such as extreme weather conditions, poor soil fertility, and plant pests and pathogens which cause extensive yield losses of between 20% and 40% per annum require innovative solutions that are not only cheaper by highly effective as well. Combining nanotechnology with current agricultural technologies such as climate smart agriculture could potentially significantly reduce production costs and boost yield especially for resource poor smallholder farmers. Already, there is a huge concern on the overuse of pesticides, which results in significant amounts being lost in runoff, drift, as well as causing pesticide resistance (Hayles et al. 2017). Nanotechnology can potentially play a significant role in building agricultural resilience to climate change, improved productivity, and food security (Venkatramanan et al. 2020). More specifically nanotechnology reduces the need

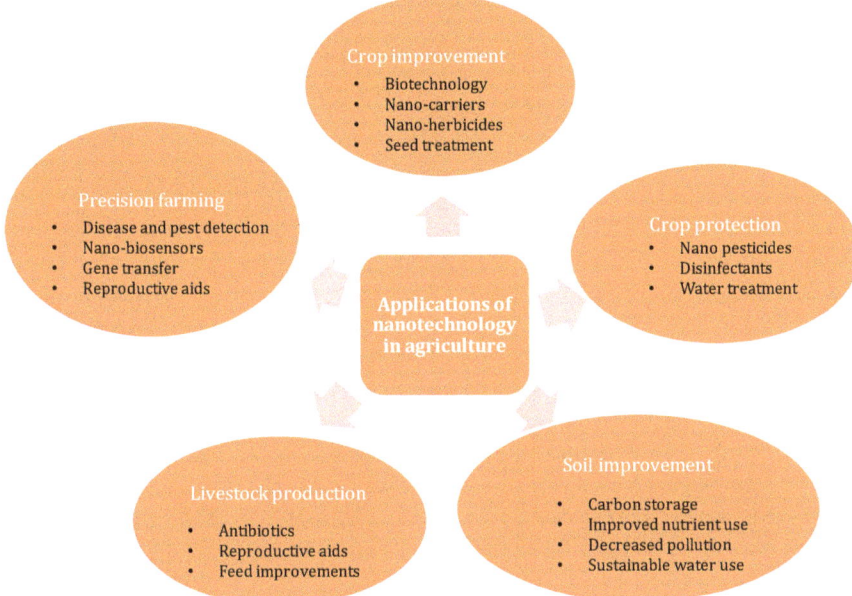

Fig. 20.1 Schematic diagram showing the potential applications of nanotechnology in agriculture. (Adopted from Dasgupta et al. 2014 and Shang et al. 2019)

and frequency for plant protection (Iqbal 2019) and improves crop and soil quality as well as nutrient and water use efficiency. Potential agricultural applications of nanotechnology are summarized in Fig. 20.1.

20.3.1 Nano-fertilizers

Any product with nanomaterials that are used to enhance nutrient use efficiency can be referred to as nano-fertilizers (Mikkelsen 2018). Nano-fertilizers are made by manipulating nutrients and minerals on a molecular level, typically amounts of mineral that is smaller than 100 nanometers (1 mm = 1,000,000 nm). Nano-fertilizers minimize environmental contamination and wastage by reducing the frequency and target application of fertilizer (El-Ramady et al. 2018). Compared to traditional fertilizers, nano-fertilizers have a higher surface area that increases their solubility and reaction capacity within the soil and can be designed to slowly release nutrients which limit leaching and result in more efficient plant nutrient uptake Cui et al. (2010). The small size of the particles increases their concentration per unit area, thus boosting chances of uptake/penetration into plants through roots and leaf surfaces. Furthermore, minimum to no energy is needed to move the nutrients from the soil into the plant's vascular system; therefore, the plant has more energy for other growth processes. Other properties of nano-fertilizers are high solubility, controlled and timely release, stability, effectiveness, improved targeted activity by delivering desired concentration, and reduced toxicity with easy, safe distribution, and disposal (Pramanik et al. 2020). Researchers have put nano-fertilizers into three classes, (1) nanoscale fertilizer (nanoparticles that contain nutrients), (2) nanoscale additives (traditional fertilizers with nanoscale additives), and (3) nanoscale coating (traditional fertilizers coated with nanoparticles) (Rai et al. 2012, Mikkelsen 2018). Even though the nano-fertilizers technology is still in its infancy, various studies done show the potential of nano-fertilizers in agriculture. Most of the studies reported that nano-fertilizers improved crop growth, yield, and quality parameters of the crop, which resulted better food products for human and animals (Mahanta et al. 2019). Examples of nano-fertilizers include hydroxyapatite, ammonium-charged zeolites, nano-calcite ($CaCO_3$-40%) with nano-SiO_2 (4%), MgO (1%), and Fe_2O_3 (1%) and ZnO nanoparticles (Adhikari et al. 2015, Kah et al. 2018). For example, Marzouk et al. (2019) reported that applying zinc nano-fertilizer as a foliar spray improved the vegetative growth, fresh pod yield, pods physical quality, and nutritional value of two snap bean cultivars. Table 20.1 highlights some of the research that has been done on nano-fertilizers.

20.3.2 Nanopesticides

The green revolution indicated above resulted in the use of various chemical compounds for use in the control of pests and diseases in order to increase crop yields. This saw the introduction of chemicals like 2,4 D and glyphosate for the control of weeds, together with various insecticides, which all resulted in reduced crop competition and increased yields. However, the introduction of these pesticides also resulted in various environmental challenges such as death of bees and fish due to their excessive use. One of the critical drivers for the application of nano-technology is a reduction in the amount of pesticide needed to assure crop protection, which may be achieved by several ways such as by improved apparent solubility, controlled release, targeted delivery, enhanced bioavailability, increased leaf adhesion, and improved stability of the active ingredient in the environment (Kah et al. 2013). Nanopesticides are more effective than conventional pesticides because they have a more targeted delivery and effect. In addition, they pose less risk to non-target organisms and use less water and energy during the application of nanopesticides since they are required in smaller amounts and less frequently than the conventional pesticides. Consequently, nanopesticides reduce wastage and labor costs while sustainably increasing crop productivity (Sahu 2020). The application of nanopesticides involves smart delivery methods (nanoencapsulates and nanocontainers); controlled pesticide release; physical and biological changes in plant structures; and improved wettability and efficacy at reduced dosages (Khot et al. 2012). The use of Nanopesticides could complement current efforts to promote sustainable agriculture which entails minimized use of agro-chemicals in order to protect the environment (Bhattacharyya et al. 2016).

Table 20.1 Summary of research undertaken on nano-fertilizers in agriculture

Nanomaterial	Observations	References
Ferrous nano-oxide particles	Lipid and protein levels, fatty acids, Fe, Mg, Ca, and P, chlorophyll contents of Soybean (*Glycine max L.*) Seed were improved with an increasing in concentration of ferrous nano-oxide from 0 to 0.75 g L^{-1}, but the increase from 0.75 to 1 g L^{-1} resulted in a reduction all the parameters above	Sheykhbaglou et al. (2018)
Nano-KH$_2$PO$_4$	Shoot and root biomass increased under nano-KH$_2$PO$_4$ due to the high physiological efficiency of P in these organs. However, P uptake was lower in nano-KH$_2$PO$_4$ compared to KH$_2$PO$_4$.	Miranda-Villagomez et al. (2019)
Banana peel nano-fertilizer blend	Germination increased from 14% (control without nano) to 97% and from was enhanced from 25% (control without nano) to 93.14% for tomato plant and fenugreek, respectively	Hussein et al. (2019)
Nano chitosan-NPK fertilizer	Nano chitosan-NPK fertilizer compared to the control, significantly increased the harvest index, crop index, and mobilization index of the determined wheat yield variables, as compared with control. The number of days to maturity was reduced from 170 to 130 after treatment with nano chitosan-NPK fertilizer	Abdel-Aziz et al. (2016)
ZnO nanoparticles	The delayed and low germination of lentil seed was proportional with an increase in concentration of ZnO nanoparticles from 25 µg mL^{-1} to 200 µg mL^{-1}	Mahanta et al. (2019)
Iron oxide nanoparticles	The study showed that significant increment of plant height (37 cm), leaf area (45.4 cm^2/plant), number of symbodial branches per plant (15.1), seed cotton yield (14.9 g/pot), and boll weight (3.5 g/boll) were increased due to foliar application of magnetite nanoparticles	Kanjana (2019)
Nano-TiO$_2$	An increase in nanoTiO$_2$ treatment from 0.25–4% increased spinach the plant dry weight, chlorophyll formation, ribulose bisphosphate carboxylase/oxygenase activity, and the photosynthetic rate	Zheng et al. (2005)
Fe nano-oxide (Fe$_3$O$_4$, N1) and Mg nano-oxide (MgO, N$_2$)	Bulk density increased with increase in Fe from 0% to 5% while it decreased increase in Mg levels	Bayat et al. (2018)

20.4 Mycosynthesis of Nanomaterials

Mushrooms are the group of fungi with characteristic fruiting bodies that are large enough to be seen by naked eyes. They have been used as food and medicine in many parts of the world for a long time (Kamalebo 2018). Besides their use as food and medicine, mushrooms have also been used in bioremediation (using mushrooms to detoxify contaminated soil) and myco-filtration (using fungal mycelium to remove biological contaminants from surface water passing directly into sensitive watersheds). Nanotechnology has of late been growing significantly with promising applications in the medical, electronics, photonics, and catalyst industries. Metal nanoparticles (NPs) can be synthesized by physical, chemical, hybrid, and biological techniques (Khanna et al. 2019). Biological techniques have been of interest due to the potential of the NPs produced through these techniques in the production of antimicrobial materials, catalysts in biological labeling, gene therapy, and DNA among others uses (Adeeyo and Odiyo 2018). Biological agents such as plants and fungi including mushrooms have been used successfully for synthesis of metal NPs (Balakumaran et al. 2016; Prasad et al. 2016, 2018; Srivastava et al. 2021; Sarma et al. 2021).

The use of edible and medicinal mushrooms to synthesize metal NPs, a process known as mycosynthesis, has however demonstrated to be advantageous over other biological techniques due to their high tolerance toward heavy metals and their potential to secrete large amount of enzymes that are essential for reduction of metal ions into their nano-forms (Prasad 2016, 2017; Prasad et al. 2018; Abdel-Aziz et al. 2018). Furthermore, fungi provide relatively quick and ecologically "clean" metallic NPs compared to the use of other biological entities (Shah et al. 2015). The process of mycosynthesis involves the extraction of various phytochemicals which include polyphenols, flavonoids, terpenoids, and high-low molecular weight proteins participate in the formation of metallic nanoparticles upon reduction of their precursor salts and stabilization of the NPs in a complex redox-mediated process. The basic principle behind mycosynthesis is the reduction of metal ions into their NPs, and as the reduction takes place, the biomolecules aggregate with the metal ions to form nano-sized corona (Patil et al. 2019; Aziz et al. 2016, 2019). Due to the complexity of the process of mycosynthesis, there is extensive research being undertaken in this field biological synthesis of nanoparticles by evaluating various fungi as highlighted in Table 20.2.

20.5 Potential of Nanotechnologies in African Smallholder Agriculture

As highlighted above, most of the African resource poor farmers' soils are now degraded with poor physical properties, reduced organic matter content, and limited biological activity which greatly reduces the resilience of these farmers in the face of a changing climate. The main challenge that these farmers often face is that of poor soil quality measured by their very low organic matter, reduced water holding capacity, and poor nutrient retention capacity. Furthermore, due to the monocropping introduced to most of these farmers, African farmers tend to face serious yield-reducing competition from pests and diseases. Nanotechnology is quite novel to propose for the resource poor smallholder African farmers, but we propose some of the technologies that can be critical to research on for adoption by these farmers in the near future.

20.5.1 The Soil

Nanotechnology promises to offer quicker and cheaper solutions to reverse soil degradation under the African soils and soil improvement through use of nanotechnology can be at a macro- and micro-level. At a macro-level, the nanoparticles strengthen soil structure and modify the pore fluid, while at a microscale the nanoparticles influence the colloidal properties and nature of the soil (Ozin et al. 2009). A study by Aminiyan et al. (2015) showed significant improvement in soil structural stability after applying nanozeolite, zeolite, and plant residues. In

Table 20.2 Selected current research focusing on the use of mushroom species in the mycosynthesis of metal nanoparticles (NPs)

Mushroom species used and nanomaterial synthesized	How nanomaterial was synthesized	Nanomaterial characterization	Nanoparticle use	References
Ramaria botryti – **Silver–gold composite nanoparticle**	Aqueous solution of polysaccharide extracted from fruit bodies mixed with chloroauric acid tetra hydrate and silver nitrate solutions	Change in color of the solution from yellow to reddish violet and Ag@AuCNPs was confirmed by UV–vis spectroscopy, where a peak at 506 nm was observed	Antioxidants and antibacterial assays as well as catalytic activity toward the reduction of 4-nitrophenol	Bhanja et al. (2020)
Ganoderma lucidum - **Silver nanoparticles**	Ethanol extract of powdered mushroom fruit bodies was obtained by using a microwave process. The extract was diluted using distilled water then 15 mg of $AgNO_3$ salt was mixed with the mushroom extract then placed on a magnetic stirrer system for the reduction of Ag+ ions to Ag0	Transparent solution changed to a brown-reddish color. UV-vis spectrum exhibited a broad absorption peak between 400–460 nm which indicates the existence of Ag NPs TEM images showed Ag NPs are spherical with a diameter range of 15–22 nm	Antioxidant, DNA cleavage, and antibacterial activities	Aygün et al. (2020)

(continued)

Table 20.2 (continued)

Mushroom species used and nanomaterial synthesized	How nanomaterial was synthesized	Nanomaterial characterization	Nanoparticle use	References
Fomitopsis pinicola - **Titanium oxide and silver nanoparticles**	Aqueous solution of powdered fruit bodies were mixed with $AgNO_3$ and TiO_2, respectively.	UV-visible spectroscopy showing diffraction lines corresponding to (111), (200) and (220) planes indicating the presence of Ag NPs. FTIR spectroscopy determined the presence of mushroom biomolecules. SEM used to analyzed synthesized TiO_2 NPs and Ag NPs. TEM used to determine size and shape of the synthesized TiO_2 NPs and Ag NPs	Antibacterial and anti-proliferative properties	Rehman et al. (2020)
Pleurotus djamor - **Titanium dioxide**	Aqueous extract of fresh fruit bodies mixed with TiO_2	UV–Vis spectrum of TiO_2 NPs showed maximum absorption at 345 nm. EDX pattern confirmed the purity of TiO_2 NPs	Antibacterial, anticancer, and mosquito larvicidal activity	Manimaran et al. (2020)
Flammulina velutipes - **Silver nanoparticles**	Aqueous extract of fresh fruiting bodies mixed with mixed with chloroauric acid	Change in color from light yellow purple color UV–vis spectrometer	Catalyst for decolorization of methylene blue	Rabeea et al. (2020)
Pleurotus giganteus - **Silver nanoparticles**	Aqueous solution of fresh fruit bodies were mixed with $AgNO_3$	UV–Vis spectroscopy surface plasmon resonance peak around at 420 nm	Antimicrobial and α-amylase inhibitory activity	Debnath et al. (2020)

(continued)

Table 20.2 (continued)

Mushroom species used and nanomaterial synthesized	How nanomaterial was synthesized	Nanomaterial characterization	Nanoparticle use	References
Agaricus bisporus - **Gold nanoparticles**	Chloroauric acid heated on a magnetic stirrer hotplate from 80 °C to 100 °C in a 250-ml flask then aqueous solution from fresh fruit bodies was added.	The EDX spectrum confirmed the identity of the nanoparticles as being that of gold	Degradation of Azo dye	Dheyab et al. (2020)

addition to soil structural stability, application of nanozeolite significantly increased soil organic carbon in soil aggregates. This finding indicated the potential of nanozeolite in soil carbon sequestration, hence reducing loss of carbon to the atmosphere. Under African systems, one of the main limitations to improved soil quality is their lack of soil organic matter, and this tends to reduce soil biological, chemical, and physical properties and thus their productivity. Several researchers have proposed technologies such as conservation agriculture and application of animal manures as a way of increasing soil organic carbon. However, all these technologies proposed require addition of huge quantities of organic matter to the soil over several years to accrue positive soil quality benefits. For example, Nyambo et al. (2018) determined that addition of high rates of up to 200 t ha^{-1} biochar is required for improved soil bulk density, soil organic carbon, aggregate stability, and microbial biomass carbon, and this may not be feasible for full field application especially under smallholder farmer setup. In another study, Aminiyan et al. (2015) showed that the application of nanozeolite improved the concentration of soil organic carbon in light and heavy fractions, which also play significant roles in soil stability. In addition, nanozeolites play critical roles in improving the water-holding capacity of soil which could be very critical in dry areas. Padidar et al. (2015) reported a significant reduction in wind erosion with use of nanoclay. This was attributed to improved dry aggregate stability after applying nanoclay. What is critical to observe is that nanotechnology offers a much more feasible and faster method of improving African farmers' soil quality relative to the current technologies being promoted. However, since this technology is high tech and very new, there is need for research that can practically evaluate the feasibility of these nanomaterials in actually increasing soil quality, as such information is lacking. With other researchers reporting that some nanoparticles do not contribute to the soil organic carbon themselves, but rather affect the organic matter decomposition rates, research also evaluate the combination of nanomaterials, and the current organic agriculture-based systems will also be important.

20.5.2 The Crop Yield

To African farmers, weeds and pests present their greatest nemesis to achieving higher yields even with increased fertilization. What is interesting is that most of the African farmers have not adopted the use of pesticides in their conventional substance farming systems. Though still far-fetched, nanotechnologies still offer a twenty-first century opportunity for increased yields among these farmers. The use of nanotechnology is proposed to have the ability to reduce the dosages of herbicides and insecticides that are needed to achieve similar results, which has cost-saving effects to these farmers whilst reducing the environmental effects of these insecticides. This research on the potential of such improved technologies is still very limited from an African perspective, and if the resilience of these farmers is to be improved in the face of a changing climate, there is a need for research in this area.

20.6 Socioeconomic Implications of Nanotechnology on Agriculture

There are many goods (nanoproducts) on the market that are produced using nanotechnology capabilities, or that are nanotechnology-based, ranging from foods, cosmetics, household appliances, computers, cellular phones, medicines, textiles, ceramics, construction materials, sports equipment, and military weapons. In food, nanotechnology is used in product packaging, nutritional supplements, and agricultural production. As such, the types of nanotechnologies are diverse; hence their application in agriculture is wide. However, researchers in nanotechnology and agriculture emphasize that the broad application of nanotechnology is aimed at resolving the current challenges of sustainability, food insecurity, and climate change (Pasiri et al. 2014). As such, nanotechnologies are being promoted as new source of key improvements for the agricultural sector through reduction of the amount of sprayed chemical products by smart delivery of active ingredients, minimize nutrient losses in fertilization, and increase yields through optimized water and nutrient management (Pasiri et al. 2014). Furthermore, nanotechnology-derived devices are also being explored in the field of plant breeding and genetic transformation.

Despite these potential advantages, nanotechnology applications in the agricultural sector are still comparably marginal and have not yet made it to the market to any large extent in comparison with other industrial sectors. This has led to deficiencies in data sets related to adoption of nanotechnology thereby posing challenges on credible economic evaluation of the economic impact of nanotechnology to the economies and the society. Some economic researchers have suggested that nanotechnology could be a new post-industrial economy emerging, which is shaped by nanoscience (Canton 2015). The NanoEconomy is based on manipulating matter on demand, thus fundamentally changing the products, services, markets, channels, jobs, and supply chains that we know today (Canton undated).

The limited availability of nanotechnology data in the market has weakened the economic research output in this area. As such, Canton (undated) points a gap in knowledge by highlighting that there is no economic theory sufficiently well developed to understand the impact of nanotechnology and science on the economy. There is reliance on technology adoption theories, which may fail to give credible information in situations where the technology is new, and data is limited. However, evolutionary economic models might provide a useful context for explaining the impact of core technologies and sciences that create economic change. In this instance, economic change may be described as the movement of capital, the invention of new products and services, the emergence of new markets, and the production of wealth (Canton undated; Musee et al. 2010). In addition, the formulation of new economic opportunity such as job creation and investment potential would be characteristic of these economic changes as well. Competitive advantage may also be an outcome. The basis of this approach to evolutionary economics is the recognition that core technologies represent economic shifts impacting on agricultural markets, customers, as well as both upstream and downstream industries. There is often a convergence of technologies that end affecting communities. It is the convergence of these technologies that create the most significant economic changes. It is becoming clear that as nanoscience becomes the next core scientific advancement, it also may represent the next evolutionary economic shift in the societies.

Although there is need for continuous research on actual as well as potential risks associated with nanotechnology, some research highlights some concerns of the technology. For instance, Musee et al. (2012) argued that the toxicity of nanoparticles affects both human health and other forms of biological organisms in the environment. This is because the increasing accumulation of nanoparticles in ecosystems with potential for transfer to higher organisms through the food chain may result in exposure of humans through foods such as fish, bacteria, earthworms, and snails and vegetables, among others (Musee et al. 2012). The fact that nanoparticles can move along the food chain, the extent to which consumers should be vigilant for possible toxic effects of engineered nanomaterials should be emphasized. Although potential health and environmental risks of nanomaterials are scientifically documented and numerous uncertainties remain, the public funds dedicated to evaluating these risks are extremely low, especially in Africa. Furthermore, the African context is unique in that there is a challenge to balance the need to use technology to fight hunger and poverty while are at the same time mitigating the adverse effects of such technology to the society The current lack of information and supervision of nanotechnology, in the African economies has also posed a serious regulatory challenge of the technology in Africa and some parts of the globe; hence some activists call for precautionary regulatory frameworks (Musee et al. 2012). It is important to note that the uncertainty about nano-related risks has not impeded the introduction of nanotechnology products into the market. Globally, most nanotechnology research and policies put in place have been largely geared toward accelerating nanotechnology introduction into the markets with only very limited consideration of precautionary approaches to address the potential risks of this emerging technology (Royal Society and the Royal Academy of Engineering 2004).

20.7 Conclusion

Nanotechnology is an emerging field from an African perspective, when it comes to its applications in agriculture where most subsistence farmers are still using green revolution technologies. Our chapter however gives an insight on technologies that are being developed elsewhere in other developed countries and indicates where these technologies can benefit the resource poor African farmers. What is critical for a start is to promote research that looks at the soil quality improving capabilities and functionality under the African context. Furthermore, nanotechnology techniques that improve the efficacy of current pesticides can be critical in reducing their cost and environmental effects, making the aspect of sustainable intensification a reality. From an African perspective, there is a need for intensive research if nanotechnology is to be adopted by the resource poor farmers in this twenty-first century; otherwise, this may remain a pipe-dream.

References

Abdel-Aziz HMM, Hasaneen MNA, Omer AM (2016) Nano chitosan-NPK fertilizer enhances the growth and productivity of wheat plants grown in sandy soil. Span J Agric Res 14(1):e0902. https://doi.org/10.5424/sjar/2016141-8205

Abdel-Aziz SM, Prasad R, Hamed AA, Abdelraof M (2018) Fungal nanoparticles: A novel tool for a green biotechnology? In: Fungal Nanobionics: Principles and Applications (eds. Prasad R, Kumar V, Kumar M and Wang S), Springer Singapore Pte Ltd. 61–87

Adeeyo AO, Odiyo JO (2018) Biogenic synthesis of silver nanoparticle from mushroom exopolysaccharides and its potentials in water purification. Open Chem J 5(1):64–75. https://doi.org/10.2174/1874842201805010064

Adhikari T, Kundu S, Biswas AK, Tarafdar JC, Rao AS (2015) Characterization of zinc oxide nano particles and their effect on growth of maize (Zea mays L.) plant. J Plant Nutr 38:1505–1515

Aminiyan MM, Sinegani AAS, Sheklabadi M (2015) Assessment of changes in different fractions of the organic carbon in a soil amended by nanozeolite and some plant residues: incubation study. Int J Recycl Org Waste Agricult 4:239–247

Aygün A, Özdemir S, Gülcan M, Cellat K, Şena F (2020) Synthesis and characterization of Reishi mushroom-mediated green synthesis of silver nanoparticles for the biochemical applications. J Pharm Biomed Anal 178. https://doi.org/10.1016/j.jpba.2019.112970

Aziz N, Pandey R, Barman I, Prasad R (2016) Leveraging the attributes of *Mucor hiemalis*-derived silver nanoparticles for a synergistic broad-spectrum antimicrobial platform. Front Microbiol 7:1984. https://doi.org/10.3389/fmicb.2016.01984

Aziz N, Faraz M, Sherwani MA, Fatma T, Prasad R (2019) Illuminating the anticancerous efficacy of a new fungal chassis for silver nanoparticle synthesis. Front Chem 7:65. https://doi.org/10.3389/fchem.2019.00065

Balakumaran MD, Ramachandran R, Balashanmugam P, Mukeshkumar DJ, Puthupalayam K (2016) Mycosynthesis of silver and gold nanoparticles: optimization, characterization and antimicrobial activity against human pathogens. Microbiol Res 182:8–20. https://doi.org/10.1016/j.micres.2015.09.009

Barrett CB, Bevis LE (2015) The self-reinforcing feedback between low soil fertility and chronic poverty. Nat Geosci 8:907

Baudoin, M.A., Vogel C., Nortjea, K., & Naika, M. (2017). Living with drought in South Africa:

Bayat H, Kolahchi Z, Valaey S, Rastgou M, Mahdavi S (2018) Novel impacts of nanoparticles on soil properties: tensile strength of aggregates and compression characteristics of soil. Arch Agron Soil Sci 64:6, 776–789. https://doi.org/10.1080/03650340.2017.1393527

Bekunda M, Sanginga N, Woomer PL (2010) Restoring soil fertility in sub-Sahara Africa. In: Advances in agronomy, vol 108. Academic Press, pp 183–236

Bhanja SK, Samanta SK, Mondal B, Jana S, Ray J, Pandey A, Tripathy T (2020) Green synthesis of Ag@Au bimetallic composite nanoparticles using a polysaccharide extracted from Ramaria botrytis mushroom and performance in catalytic reduction of 4-nitrophenol and antioxidant, antibacterial activity. Environ Nanotechnol Monitor Manag 14. https://doi.org/10.1016/j.enmm.2020.100341

Bhattacharyya A, Duraisamy P, Govindarajan M, Buhroo AA, Prasad R (2016) Nano-biofungicides: Emerging trend in insect pest control. In: Advances and Applications through Fungal Nanobiotechnology (ed. Prasad R), Springer International Publishing Switzerland 307–319

Canton J. undated. The Emerging NanoEconomy: Key Drivers, Challenges and Opportunities. In Nanotechnology: Societal Implications—Individual Perspectives (Ed: Roco M.C.) National Science Foundation. https://citeseerx.ist.psu.edu/viewdoc/download?doi=10.1.1.74.5080&rep=rep1&type=pdf

Chiduza C, Dube E (2013) Maize production challenges in high biomass input smallholder farmer conservation agriculture systems: a practical research experience from South Africa. In: African crop science conference proceedings 11:23–27

Chiduza C, Dube E (2017) Maize production challenges in high biomass input smallholder farmer conservation agriculture systems: a practical research experience from South Africa. African Crop Science Conference Proceedings, 11:23–27

Cui HX, Sun CJ, Liu Q, Jiang J, Gu W (2010) Applications of nanotechnology in agrochemical formulation, perspectives, challenges and strategies. In: International conference on Nanoagri, Sao Pedro, Brazil, 20–25 June 2010, pp 28–33

Dasgupta N, Ranjan S, Deepa M, Chidambaram R, Ashutosh K, Rishi S (2014) Nanotechnology in agro-food: from field to plate. Food Res Int 69:381–400

Debnath G, Das P, Saha AK (2020) Green synthesis of silver nanoparticles using mushroom extract of Pleurotus giganteus: characterization, antimicrobial, and α-amylase inhibitory activity. Bionanoscience 9:611–619. https://doi.org/10.1007/s12668-019-00650-y

Dheya MA, Owaid MN, Rabeea MA, Aziz AA, Jameel MS (2020) Mycosynthesis of gold nanoparticles by the Portabello mushroom extract, Agaricaceae, and their efficacy for decolorization of Azo dye. Environ Nanotechnol Monitor Manag 14. https://doi.org/10.1016/j.enmm.2020.100312

Dube E, Chiduza C, Muchaonyerwa P (2012) Conservation agriculture effects on soil organic matter on a Haplic Cambisol after four years of maize–oat and maize–grazing vetch rotations in South Africa. Soil Tillage Res 123:21–28

El-Ramady H, El-Ghamry A, Mosa A, Alshaal T (2018) Nanofertilizers vs. biofertilizers: new insights. Environ Biodivers Soil Secur 2:40 50

Fanadzo M (2007) Weed management by small-scale irrigation farmers – the story of Zanyokwe, SA Irrig 29(6):20–24

Fanadzo M (2010) Improving productivity of maize-based smallholder irrigated cropping systems: a case study of Zanyokwe Irrigation Scheme, Eastern Cape, South Africa. PhD Thesis, University of Fort Hare, South Africa

Fanadzo M, Chiduza C, Mnkeni PNS, Van der Stoep L, Steven J (2010) Crop production management practices as a cause for low water productivity at Zanyokwe Irrigation Scheme. Water SA 36(1):27–36

Fanadzo M, Dalicuba M, Dube E (2018) Application of conservation agriculture principles for the management of field crops pests. In: Sustainable Agriculture Reviews, vol 28. Springer, Cham, pp 125–152

Fraceto LF, Grillo R, de Medeiros GA, Scognamiglio V, Rea G, Bartolucci C (2016) Nanotechnology in agriculture: which innovation potential does it have? Front Environ Sci 4:20. https://doi.org/10.3389/fenvs.2016.00020

Ghasemmezhad A, Ghorbanpour M, Sohrabi O, Ashnavar M (2019) A general overview on application of nanoparticles to agriculture and plant science, Comprehens. Anal Chem 87:85–110

Giller KE, Witter E, Corbeels M, Tittonell P (2009) Conservation agriculture and smallholder farming in Africa: the heretics' view. Field Crop Res 114(1):23–34

Hayles J, Johnson L, Worthley C, Losic D (2017) Nanopesticides: a review of current research and perspectives, Editor(s): Alexandru Mihai Grumezescu, New Pesticides and Soil Sensors, Academic Press 193–225. https://doi.org/10.1016/B978-0-12-804299-1.00006-0

Hussein HS, Shaarawy HH, Hussien NH, Hawash SI (2019) Preparation of nano-fertilizer blend from banana peels. Bullet Nat Res Centre 43(1):26

Kah M, Kookana RS, Gogos A, Bucheli TD (2018) A critical evaluation of nanopesticides and nanofertilizers against their conventional analogues. Nat Nanotechnol 13:677–684

Kah M, Beulke S, Tiede K, Hofmann T (2013) Nanopesticides: state of knowledge, environmental fate, and exposure modeling. Crit Rev Environ Sci Technol 43:1823–1867

Kamalebo HM, Wa Malale HMS, Ndabaga CM, Degreef J, De Kesel A (2018) Uses and importance of wild fungi: traditional knowledge from the Tshopo province in the Democratic Republic of the Congo. J Ethnobiol Ethnomed 1214(1):13. https://doi.org/10.1186/s13002-017-0203-6. PMID: 29433575; PMCID: PMC5809825.

Kanjana D (2019) Foliar study on effect of iron oxide nanoparticles as an alternate source of iron fertilizer to cotton. Int J Chem Stud 7:4374–4379

Khanna P, Kaur A, Goyal D. (2019) Algae based metallic nanoparticles: Synthesis, characterization and applications. J Microbiol Methods. 163:105656. https://doi.org/10.1016/j.mimet.2019.105656. Epub. PMID: 31220512

Khot LR, Sankaran S, Maja JM, Ehsani R, Schuster EW (2012) Applications of nanomaterials in agricultural production and crop protection: A review. Crop Prot 35:64–70

Mafongoya PL, Bationo A, Kihara J, Waswa BS (2006) Appropriate technologies to replenish soil fertility in southern Africa. Nutr Cycl Agroecosyst 76(2-3):137–151

Mahanta N, Dambale A, Rajkhowa M (2019) Nutrient use efficiency through nano fertilizers. Int J Chem Stud 7(3):2839–2842

Manimaran K, Murugesan S, Ragavendran C, Balasubramani G, Natarajan D, Ganesan A, Seedevi P (2020) Biosynthesis of TiO_2 nanoparticles using edible mushroom (*Pleurotus djamor*) extract: mosquito Larvicidal, histopathological, antibacterial and anticancer effect. J Clust Sci. https://doi-org.ezproxy.unam.edu.na/10.1007/s10876-020-01888-3

Marzouk NM, Abd-Alrahman HA, EL-Tanahy AMM, Mhmoud SH (2019) Impact of foliar spraying of nano micronutrient fertilizers on the growth, yield, physical quality, and nutritional value of two snap bean cultivars in sandy soils. Bull Natl Res Cent 43:84. https://doi.org/10.1186/s42269-019-0127-5

Mikkelsen R (2018) Nanofertilizer and nanotechnology: a quick look. Better Crops 102. https://doi.org/10.24047/BC102318

Miranda-Villagómez E, Trejo-Téllez LI, Gómez-Merino FC, Sandoval-Villa M, Sánchez-García P, Aguilar-Méndez MÁ (2019) Nanophosphorus fertilizer stimulates growth and photosynthetic activity and improves P status in Rice. J Nanomater 2019:5368027

Iqbal MA (2019) Nano-fertilizers for sustainable crop production under changing climate: a global perspective, Sustainable Crop Production, Mirza Hasanuzzaman, Marcelo Carvalho Minhoto Teixeira Filho, Masayuki Fujita and Thiago Assis Rodrigues Nogueira, IntechOpen, https://doi.org/10.5772/intechopen.89089. Available from: https://www.intechopen.com/books/sustainable-crop-production/nano-fertilizers-for-sustainable-crop-production-under-changing-climate-a-global-perspective

Musee N, Foladori G, Azoulay D (2012) Social and environmental implications of nanotechnology development in Africa. CSIR (Nanotechnology Environmental Impacts Research Group, South Africa) / ReLANS (Latin American Nanotechnology and Society Network) / IPEN (International POPs Elimination Network)

Musee N, Obersholster PJ, Sikhwivhilu L, Botha A-M (2010) The effects of engineered nanoparticles on survival, reproduction, and behaviour of freshwater snail, Physa acuta (Draparnaud, 1805). Chemosphere 81:1196–1203

Nyambo P, Taeni T, Chiduza C, Araya T (2018) Effects of maize residue biochar amendments on soil properties and soil loss on acidic Hutton soil. Agronomy 8(11):256

Ozin GA, Arsenault AC, Cademartiri L (2009) Nanochemistry: a chemical approach to nanomaterials. Royal Society of Chemistry, London

Padidar M, Jalalian A, Abdouss M, Najafi P, Honarjoo N, Fallahzade J (2015) Effects of Nanoclay on Some Physical Properties of Sandy Soil and Wind Erosion. Int J Soil Sci 11(1):9–13

Patil HBV, Nithin KS, Sachhidananda S, Hatna S, Siddaramaiah H, Chandrashekarad KT, Kumara BYS (2019) Mycofabrication of bioactive silver nanoparticle: photo catalysed synthesis and characterization to attest its augmented bio-efficacy. Arab J Chem 12(8):4596–4611

Pramanik P, Krishnan P, Maity A, Mridha N, Mukherjee A, Rai V (2020) In: Dasgupta et al (eds) Environmental nanotechnology Volume 4, Environmental chemistry for a sustainable world 32. https://doi.org/10.1007/978-3-030-26668-4_9

Prasad R (2016) Advances and Applications through Fungal Nanobiotechnology. Springer, International Publishing Switzerland (ISBN: 978-3-319-42989-2)

Prasad R (2017) Fungal Nanotechnology: Applications in Agriculture, Industry, and Medicine. Springer Nature Singapore Pte Ltd. (ISBN 978-3-319-68423-9)

Prasad R, Kumar V, Prasad KS (2014) Nanotechnology in sustainable agriculture: present concerns and future aspects. Afr J Biotechnol 13(6):705–713

Prasad R, Kumar V, Kumar M, Wang S (2018) Fungal Nanobionics: Principles and Applications. Springer Nature Singapore Pte Ltd. (ISBN 978-981-10-8666-3). https://www.springer.com/gb/book/9789811086656

Prasad R, Bhattacharyya A, Nguyen QD (2017a) Nanotechnology in sustainable agriculture: Recent developments, challenges, and perspectives. Front Microbiol 8:1014. doi: 10.3389/fmicb.2017.01014

Prasad R, Kumar M, Kumar V (2017b) Nanotechnology: An Agriculture paradigm. Springer Nature Singapore Pte Ltd. (ISBN: 978-981-10-4573-8)

Prasad R, Pandey R, Barman I (2016) Engineering tailored nanoparticles with microbes: quo vadis. WIREs Nanomed Nanobiotechnol 8:316–330. https://doi.org/10.1002/wnan.1363

Prasad R, Jha A, Prasad K (2018) Exploring the Realms of Nature for Nanosynthesis. Springer International Publishing (ISBN 978-3-319-99570-0) https://www.springer.com/978-3-319-99570-0

Rabeea MA, Owaid MN, Aziz AA, Jameel MS, Dheya MA (2020) Mycosynthesis of gold nanoparticles using the extract of Flammulina velutipes, Physalacriaceae, and their efficacy for decolonization of methylene blue. J Environ Chem Eng 8. https://doi.org/10.1016/j.jece.2020.103841

Rai V, Acharya S, Dey N (2012) Implications of nanobiosensors in agriculture. J Biomat Nanobiotechnol 3:315–324

Ramanan V, Shah S, Prasad R (2020) Global Climate Change and Environmental Policy: Agriculture Perspectives. Springer Singapore (ISBN: 978-981-13-9570-3) https://www.springer.com/gp/book/9789811395697

Rehman S, Jermy R, Asiri SM, Shah MA, Farooq R, Ravinayagam V, Ansaria MA, Alsalem Z, Jindan RA, Reshi Z, Khan FA (2020) Using Fomitopsis pinicola for bioinspired synthesis of titanium dioxide and silver nanoparticles, targeting biomedical applications. Royal Soc Chem Adv 10:32137–32147. https://doi.org/10.1039/D0RA02637A

Rockstrom J (2000) Water resources management in smallholder farms in Eastern and Southern Africa: an overview. Phys Chem Earth 25:275–283

Royal Society and the Royal Academy of Engineering (2004) Nanoscience and nanotechnologies: opportunities and uncertainties. Royal Society, United Kingdom

Sahu G (2020) Role and prospects of nanotechnology in agriculture. Agric Food E-Newslett 2(8):377–379, Article ID: 31139. Available at www.agrifoodmagazine.co.in. Accessed on 28/09/20

Sarma H, Joshi S, Prasad R, Jampilek J (2021) Biobased Nanotechnology for Green Applications. Springer International Publishing (ISBN 978-3-030-61985-5) https://www.springer.com/gp/book/9783030619848

Shang Y, Hasan MK, Ahammed GJ, Li M, Yin H, Zhou J (2019) Applications of nanotechnology in plant growth and crop protection: a review. Molecules 24:2558

Sheykhbaglou R, Sedghi M, Fathi-Achachlouie B (2018) The effect of ferrous nano-oxide particles on physiological traits and nutritional compounds of soybean (Glycine max L.) seed. An Acad Bras Cienc 90(1):485–494

Shah M, Fawcett D, Sharma S, Tripathy SK, Poinern G (2015) Green synthesis of metallic nanoparticles via biological entities. Materials (Basel, Switzerland), 8(11):7278–7308. https://doi.org/10.3390/ma8115377

Srivastava S, Usmani Z, Atanasov AG, Singh VK, Singh NP, Abdel-Azeem AM, Prasad R, Gupta G, Sharma M, Bhargava A (2021) Biological nanofactories: Using living forms for metal nanoparticle synthesis. Mini-Reviews in Medicinal Chemistry 21(2):245–265

Van Averbeke W, M'marete CK, Igodan CO, Belete A (1998) An investigation into food plot production at irrigation schemes in central Eastern Cape. WRC Report 719/1/98. Water Research Commission, Pretoria

Vanlauwe B, Giller KE (2006) Popular myths around soil fertility management in sub-Saharan Africa. Agric Ecosyst Environ 116(1-2):34–46

Vanlauwe B, Bationo A, Chianu J, Giller KE, Merckx R, Mokwunye U et al (2010) Integrated soil fertility management: operational definition and consequences for implementation and dissemination. Outlook Agric 39(1):17–24

Vanlauwe B, Descheemaeker K, Giller KE, Huising J, Merckx R, Nziguheba G et al (2015) Integrated soil fertility management in sub-Saharan Africa: unravelling local adaptation. Soil 1(1):491–508

Venkatramanan V, Shah S, Prasad R (2020) Global Climate Change and Environmental Policy: Resilient and Smart Agriculture. Springer Singapore (ISBN: 978-981-329-855-2) https://www.springer.com/gp/book/9789813298552

Zheng L, Hong F, Lu S, Liu C (2005) Effect of nano-TiO_2 on strength of naturally aged seeds and growth of spinach. Biol Trace Elem Res 104(1):83–91

Chapter 21
Nanomaterials for Wastewater Remediation: Resolving Huge Problems with Tiny Particles

Ambikapathi Ramya ⓘ, Periyasamy Dhevagi, and S. S. Rakesh

Contents

21.1	Introduction	601
21.2	Sources of Wastewater	602
21.3	Mechanism	604
21.4	Remediation	605
	21.4.1 Magnetic Nanomaterials	606
	21.4.2 Transition Metal Oxide NPs	611
	21.4.3 Carbon-Based Nanoparticles	612
	21.4.4 Nanomembranes	614
21.5	Conclusion and Future Prospect	614
References		615

21.1 Introduction

Water is the prime requirement for the endurance of animals and human beings. The earth is covered by 70% of water, and only 2.5% is clean water. Water is utilized for domestic purpose and numerous industrial activities being discharged enormous quantity of untreated wastewater into water bodies. Besides, farming activities consume a range of fertilizers, pesticides and insecticides which reach water bodies through run-off.

In general, the microorganisms, inorganic and organic materials, cause water contamination. The inorganic pollutants consist of metals like Pb, Hg, Cd, Cr, etc., and the organic pollutants consist of agricultural pesticides, pharmaceuticals, textile wastes such as dyes, personal care products, household wastes like detergents,

---

A. Ramya (✉) · S. S. Rakesh · P. Dhevagi
Department of Environmental Sciences, Tamil Nadu Agricultural University, Coimbatore, Tamil Nadu, India

phenolic compounds and halogenated and aromatic compounds. Discharge of these pollutants in the lakes, rivers and/or drinking water resources leads to serious environmental issues like water pollution, water scarcity and health risks (Saleh 2015). According to the World Health Organization (WHO), up to 12 million people are affected by consumption of polluted water every year.

In present decades, water treatment techniques have turn into a considerable attention worldwide in increasing fast growth of industries and ecological contamination (Geise et al. 2010; Sharma et al. 2019; Yahya et al. 2018). Removal of pollutants from water is most important for the developing countries and some of the techniques are employed for the water treatment. Among, various wastewater treatment technologies, nanotechnology acquire several advantages owing to its unique physicochemical properties. Oxidation, adsorption and degradation are the nano-techniques used for the removal of contaminants which persist in the polluted water. Chemical oxidation and advanced oxidation are efficient processes for the removal of organic contaminants in water. The oxidants used in the chemical oxidation process are ozone, chlorine, chlorite, hydrogen peroxide, etc. Fenton process is one of the advanced oxidation process (AOP) which degrades the organic contaminants from water by redox reactions, dehydrogenation and electrophilic addition. AOP does not eliminate the byproducts completely during the treatment which leads to health risk to humans (Feng et al. 2013; Celiz et al. 2009).

Adsorption is a critically significant method to remediate pollutants from the polluted water, and it is an economic method. The efficiency of the adsorption depends on the nature of adsorbate, adsorbent and condition of the operation. The mechanism of adsorption generally attributes to electrostatic, π–π and Van der Waals interactions. Activated carbon, carbon nanomaterials, metal organic frameworks, clays and zeolites are used as the adsorbent (Yu et al. 2016). This remediation mechanism can be applied in water treatment nanotechnology to eliminate a number of contaminants. Hence, this chapter highlights sources of wastewater, properties of various nanomaterials, mechanism of nanoremediation and efficiency of nanomaterials used for target pollutants in wastewater.

21.2 Sources of Wastewater

Organic contaminants in waster ecosystem cause serious issues in the environment. In addition to the organic contaminants, heavy metals in water also lead to the health issue to the living organism. The pharmaceutical products are also identified in surface and groundwater ecosystem. The personal care products and endocrine disrupting compounds are high volatile and high polar in nature. Due to this intractable and persistent nature in the environment, these compounds does not undergo the degradation process (Archer et al. 2017; Tijani et al. 2019; Snyder et al. 2003).

The world population is increasing day by day, and so the world faces the lack of food. Pesticides play a significant part in the production of crops and vegetables, and it includes fungicides, herbicides, insecticides, etc. To control the pest in farming, more than a million tons of pesticides have been prepared each year. Usage of pesticides in farming leads to fast transfer of pesticide residue to the water bodies and penetrated into food chain. In farming countries, water pollution due to pesticides are incredibly frequent, and it is hazardous to human beings. Consumption of water affected by pesticide causes organ damage, cancer, reproduction effects, nervous system damage and also birth defects. The famous pesticides are organophosphate, organochlorine, N-methyl carbamate, arsenic-containing fungicides, chlorophenoxy, nitrophenol, pentachlorophenol, rodenticides and fumigants (Milne 2018).

Organic dyes are the most common organic pollutants obtained from textile, plastic, construction, leather, food industries, cosmetics and paper industries. It was anticipated that above 15% of dyes (approximately 400 ton/day) were liberated into the water body during the dyeing process (Garcia et al. 2007; Vanhulle et al. 2008). These are synthetic organic compounds which have the mutagenic and carcinogenic effects (Tang et al. 2016). The azo dyes, for instance, methylene blue and rhodamine B, are commonly used in the recent decades. These organic dyes are nondegradable naturally (Chen et al. 2011; Mirzazadeh and Lashanizadegan 2018; Han et al. 2015). Methylene blue causes eye and skin irritation and also respiratory tract irritation upon contact during the consumption (Jain et al. 2007; Kumar and Kumaran 2005). In addition, usage of detergents has increased in our daily life, and it affects the water ecosystem. These detergents are classified as cationic detergents like quaternary ammonium cations; anionic detergents like linear alkylbenzene sulfonates; and non-ionic and zwitterionic detergents. The general industrial chemicals like aromatic compounds and bisphenol A cause many health issues during consumption (Sui et al. 2011).

The water contaminated with heavy metals shows high density and atomic weight as high as five times than pure water, and these metals naturally occur via different sources. Heavy metal sources including soils, rocks and volcanic explosion and other anthropogenic sources including mining actions contaminates the water bodies. Arsenic (As), chromium (Cr), lead (Pb), mercury (Hg), cadmium (Cd), copper (Cu) and nickel (Ni) are some of the common metals which cause heavy metal contamination in water. Amongst all the heavy metals, arsenic and chromium are the principal reasons of water contamination by both natural and anthropogenic sources. According to the World Health Organization (WHO), arsenic is classified as group 1 human carcinogenic substance. Consumption of arsenic-polluted water leads to various types of skin diseases such as hyperpigmentation, hyperkeratosis and cancers like kidney, skin, lung and bladder (Khare 2016; Singh et al. 2015; Halem et al. 2009).

21.3 Mechanism

Elimination of contaminants present in wastewater can be done by photocatalysis, and the catalyst used for the degradation process in photocatalysis is known as photocatalyst which alters rate of the reaction only. After the completion of degradation process, the catalyst can be recovered and reused. During the photocatalytic activity, the following reactions are taking place in the remediation of contaminants in the water body:

- The contaminants transferred are adsorbed on the surface of the catalyst.
- The photonic activation and decomposition of the adsorbed contaminants.
- Desorption of the reaction product.
- Reaction products are eliminated from the surface of the catalyst.

$$e^- + O_2 \rightarrow O_2^{\cdot -}$$
$$O_2^{\cdot -} + 2H^+ + 2e^- \rightarrow OH^{\cdot} + OH^-$$
$$h^+ + H_2O \rightarrow OH^{\cdot} + OH^+ / OH^{\cdot}$$
$$OH^{\cdot} + OC \rightarrow CO_2(g) + H_2O(l) + Other$$
$$e^- + OC \rightarrow OC^- \rightarrow Degradation\ products$$
$$h^+ + OC \rightarrow OC^+ \rightarrow Degradation\ products$$

where OC organic contaminants.

The photocatalytic degradation of pollutants present in water can be carried out with the use of photons, electrons and catalyst (Fig. 21.1). The photocatalytic degradation of materials such as dyes, heavy metals, organic pollutants, microplastics, etc., engages the photons, catalyst and electrons. These were engaged the individual energy stage in the atom. Depending upon the atoms, every individual energy state was divided into several energy levels. The energy bands are formed depends on close energy state. Atom filled with electrons by donation of the energy and

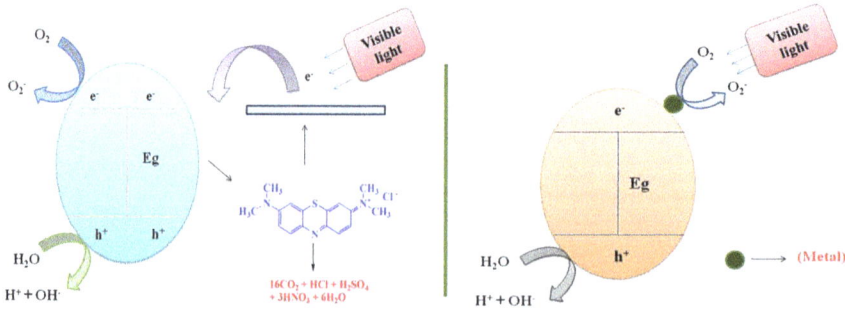

Fig. 21.1 Photocatalytic mechanism of nanomaterials

conductivity band was created. This action can be done by photocatalyst, and then the different factors were contributed in the important parts of photocatalytic activity. The electrons are energized to conductance band from valence band in the presence of sunlight. The assortment of the photogenerated n-type and p-type from nanoparticles prohibited beneath sunlight gives enhanced time of charge transporters and additional proficient redox properties (Nivetha et al. 2019).

21.4 Remediation

Nowadays, wastewater obtained from industrial sources are treated by different processes, and they have the capability to remediate the contaminants. There are numerous methods that have been adopted for the wastewater treatment such as phytoremediation, bioremediation, etc. But these methods have some disadvantages like inefficiency, regeneration, waste products, weak selectivity, etc. To overcome these demerits, alternate method must be needed. Hence, nanotechnologies play a pivotal role in wastewater treatment processes. Adsorption and degradation process is considered as the most significant process in the water purification. The degradation efficiency of contaminants relies on several factors like source of irradiation, adsorbent dosage, contact time, effect of pH, temperature and nature and type of the

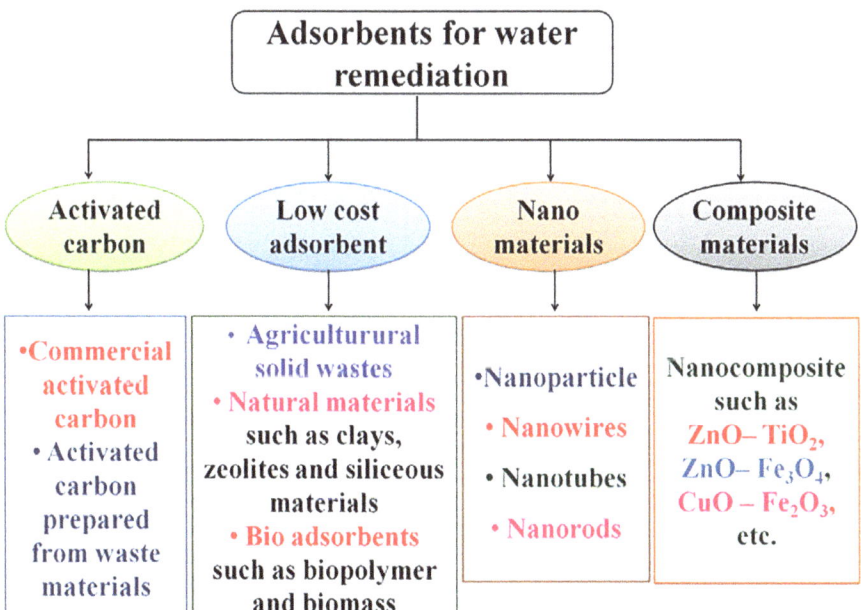

Fig. 21.2 Nano-adsorbents for wastewater remediation

catalyst used in the wastewater remediation. Some of the important nano-based materials are depicted in Fig. 21.2 for wastewater remediation.

21.4.1 Magnetic Nanomaterials

The application of various nanomaterials and its degradation efficiency in pollutant remediation were depicted in Tables 21.1 and 21.2.

21.4.1.1 Zero-Valent Iron Nanoparticles

Zero-valent iron (ZVI) nanoparticles have been widely considered and applied for the wastewater treatment (Fig. 21.3). Owing to the ease of oxidation, high reactivity, non-hazardous nature, economic, plenty, synthesize easily, bioavailabilities and large surface area, it is used in wastewater treatment processes (Zhu and Chen 2019; Zhang 2003). The ZVI is used to eliminate the organic pollutants, polychlorinated biphenyl, organic chlorinated solvent and heavy metals present in wastewater. It is used as catalyst in the degradation and oxidation of organic pollutants transferred to hydrogen peroxide in presence of dissolved oxygen. From ZVI, two electrons are transported repeatedly to reduce the hydrogen peroxide into the water. In addition, the hydroxyl radicals are produced, and it has the ability to oxidize the organic pollutants during the permutation of hydrogen peroxide and Fe^{2+} (Fenton reaction). It has the standard redox potential value 0.440 V and shows that it can eliminate the halogenated compounds via reductive dehalogenation (Fu et al. 2014; He and Zhao 2007). It quickly removes hazardous non-aqueous phase liquids like halogenated partially volatile compound, halogenated volatile material and non-halogenated partially volatile materials to non-hazardous materials in polluted water.

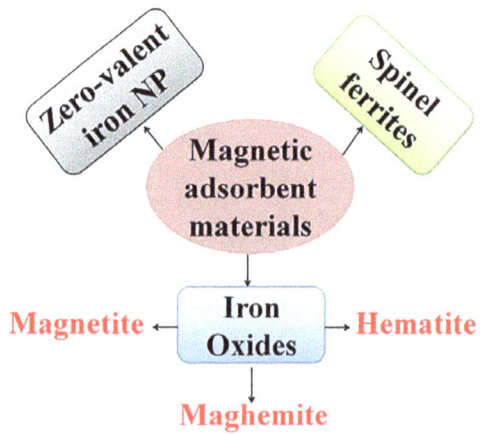

Fig. 21.3 Iron-based nanomaterials for wastewater remediation

Table 21.1 Degradation efficiency of various nanomaterials in pollutant removal

S. no	Material	Pollutants	Degradation efficiency and time	Reference
1	ZVI NPs	Amoxicillin	86.5% & 25 min	Zha et al. (2014)
2	ZVI NPs	Methylene blue	72.1% & 30 min	Hamdy et al. (2018)
3	ZVI NPs	Pb(II)	95% & 1 h	Ahmed et al. (2017)
4	ZVI NPs	Cd(II) Cu(II) Pb(II) Ni(II)	71.4% 100% 99.9% 96.6%	Danila et al. (2018)
5	ZVI/kaolinite	Acid black 1	98% & 2 h	Kakavandi et al. (2019)
6	ZVI/peroxymonosulfate	Tetracycline	88.5% & 5 min	Cao et al. (2019)
7	Cu	Safranin Carbol fuchsin Malachite green Methylene blue	92% 94% 97% 85%	Dlamini et al. (2019)
8	$CuO\text{-}Fe_2O_3$	Rhodamine B	100% & 1 h	Alp et al. (2019)
9	ZnO	Ag(I) Pb(II) Cr(VI)	97.92% 85.18% 43.34%	Le et al. (2019)
10	ZnO	Cd Cu Fe Pb	99.03% & 1 h 97.39% & 1 h 100% & 1 h 97.64% & 1 h	El-Dafrawy et al. (2017)
11	ZnO	Crystal violet dye	90% & 2 h	Franco et al. (2019)
12	ZnO	Dibenzothiophene	97% & 3 h	Khalafi et al. (2019)
13	ZnO MWCNT ZnO/MWCNT	Reactive blue 203	85.4% & 20 min 19% & 20 min 99.1% & 20 min	Bagheri et al. (2020)
14	$TiO_2/CoFe_2O_4$	4-nitrophenol	94% & 35 min	Ibrahim et al. (2019)
15	$Fe_3O_4/TiO_2/SiO_2$	Methylene blue	98% & 2 h	Abbas et al. (2016)
16	$Fe_2O_3/BiVO_4$	Methylene blue and rhodamine B	100% 20 min	Wen et al. (2019)
17	$MgFe_2O_4\text{-}TiO_2@GO$	Methylene blue	100% & 5 h	Kaur and Kaur (2019)

(continued)

Table 21.1 (continued)

S. no	Material	Pollutants	Degradation efficiency and time	Reference
18	Cu Fe_2O_4/graphene oxide	Acid orange 7	95% & 50 min	Ayazi et al. (2016)
19	$ZnFe_2O_4$/AgI	E. coli	100% & 1 h 20 min	Xu et al. (2018)
20	TiO_2-CdS/reduced graphene oxide	Methylene blue	97.5% & 20 min	Kassaee et al. 2011
21	Reduced graphene oxide/ TiO_2	Methylene blue	92% & 2 h	Kireeti et al. (2016)
22	TiO_2/graphene oxide	Phenol	99.3% & 8 h	Pizarro et al. (2015)
23	Ag–TiO_2	Chloramphenicol	100% & 30 min	Shokri et al. (2013)
24	Fe–TiO_2	Metronidazole	97% & 2 h	Malakootian et al. (2019)
25	Zr–TiO_2	Bisphenol A	100% & 1 h 20 min	Gao et al. (2010)
26	Cu–TiO_2	Naproxen sodium	87% & 6 h	Hinojosa-Reyes et al. (2019)
27	Multiwalled carbon nanotubes (MWCNT)	Phenol, anti-inflammatory and nonsteroidal drugs, polychlorinated biphenyls	80–99%	Hu et al. (2015)
28	CNT	As(V)	80%	Peng et al. (2005)

In aqueous solution, zero-valent iron nanoparticles undergo oxidation readily either by oxygen or react with subsurface component which is occurring naturally leading to low reactivity and surface passivation. The reaction circumstances like pH, strength of ions, initial concentration of pollutant, method of synthesis, stabilizing agent used during the synthesis, medium for oxidation either air or water, and time play a crucial role in the degradation of organic contaminants. ZVI nanoparticles have some disadvantages as follows:

(i) Passivation – Owing to the less active iron hydroxides formation on ZVI nanoparticles surface during the reaction
(ii) Aggregation – Owing to the Van der Walls and magnetic force and also formation of less active small particles
(iii) Poor retrievability
(iv) Potential health and environmental risk – Because of bioaccumulation (Lu and Astruc 2020)

Table 21.2 Various nanomaterials and its adsorption capacity in pollutant removal

S. no	Material	Pollutant	Adsorption capacity (mg/g)	Reference
1	EDTA-GO	Pb(II)	479 ± 46	Clemonne et al. (2012)
2	MWCNT	Magnetic carbon Methylene blue	149 399	Ma et al. (2012)
3	MWCNT	Pb(II) Cd(II)	97.08 10.86	Li et al. (2003)
4	Chitosan	Pb(II)	398	Qi et al. (2004)
5	Graphene sand composite	Cr(VI)	2859.38	Dubey et al. (2015)
6	Nitrogen-doped magnetic CNT	Cr(III)	638.56	Shin et al. (2011)
7	MnO_2-MWCNT	Cr(III)	99.01	Tian et al. (2014)
8	TiO_2	Cd(II)	29.28	Sharaf El-Deen and Zhang (2016)
9	TiO_2	Cd Cu Ni Pb	120.1 50.2 39.3 21.7	Mahdavi et al. (2013)

21.4.1.2 Iron Oxide

Iron oxide nanoparticles have existed in different structures like maghemite, hematite and magnetite (Fig. 21.3). These iron oxide nanoparticles acquire polymorphism which includes temperature-induced phase transition (Cornell and Schwertmann 2003). Owing to the strong magnetic property, porosity and precise surface area, iron nanoparticles displayed tremendous properties in adsorption process (Huang and Chen 2009; Nizamuddin et al. 2019).

Hematite (α-Fe_2O_3) is an n-type semiconductor, and it has the band gap value of 2.1–2.2 eV, and it has paramagnetic phase at Curie temperature (T_c = 682.85 °C). According to JCPDS file No. 33-0664, it has trigonal crystal structure which belongs to R-3c space group and unit cell parameter $a = b$ = 4.9865 Å, c = 13.5016 Å (Rozenberg et al. 2002). Crystallinity, particle size, cation doping, exchange interactions and subparticle structure influence the magnetic properties of the hematite. Not only the size and shape of hematite but also the nature of the dopant plays a vital role to increase the adsorption properties and lead to the efficient catalyst in the wastewater treatment (Tadic et al. 2019). Owing to poor conductivity and low efficiency of separation, its photocatalytic activity is controlled (Zhang et al. 2017). During visible light irradiation, it absorbs approximately 43% of light which helps to degrade the effluent from the polluted water under light source (Santhosh et al. 2019). Kang et al. (2019) reported that hematite-based material can enhance the degradation efficiency of rhodamine B in the presence of photocatalyst. Chen et al. (2019) reported the admirable degradation kinetics of antibiotics such as ciprofloxacin, norfloxacin, sulfadiazine and tetracycline under solar light irradiation in the presence of AgBr/Ag$_3$PO$_4$@natural hematite as a photocatalyst. The rate constants of antibiotics are 0.16, 0.19, 0.34 and 0.10 min^{-1} for ciprofloxacin, norfloxacin, sulfadiazine and tetracycline, respectively.

Magnetite (Fe$_3$O$_4$) is nothing but a combination of ferrous and ferric ions, different to all other metal oxides. Magnetite divulges both p-type and n-type semiconductor with low band gap energy of 0.1 eV. According to JCPDS file No. 19-0629, it has a cubic inverse spinel structure which belongs to Fd3m space group (Okube et al. 2012). Owing to high surface energy, it is not stable in the aqueous environment. Surface functionalization generates the stability of the magnetite, and it increases the efficiency of elimination of pollutant. It can be used as adsorbent to remove the heavy metals from the polluted water compared with other adsorbents. Fan et al. (2019) tried to eliminate the lead from lead-containing solution using carboxymethyl-cellulose-immobilized magnetite nanoparticles, and it shows that the utmost adsorption capability of lead ion was attained at 152 mg/g.

Maghemite (γ-Fe$_2$O$_3$) has an analogous crystal arrangement to magnetite, and it has cubic structure. Because of high magnetization saturation, it is broadly employed as an appropriate adsorbent in wastewater treatment (Wu et al. 2015; Leone et al. 2018). It reveals ferromagnetic nature and more stable in aqueous environment. It is more superior adsorbent for heavy metal than magnetite due to small size and high specific surface area (Martinez-Boubeta and Simeonidis 2019). Rajput et al. (2017) reported the removal efficiency was found at 59.2 and 25 for lead (II) and copper (II), respectively, in the presence of maghemite.

21.4.1.3 Spinel Ferrites

Spinel ferrites have the general formula MFe$_2$O$_4$, where M represents the divalent metal ions with ionic radius ranging between 0.6 and 1 Å. Examples for divalent ions are Cu, Ni, Mg, Mn, Co, Zn, Cd, etc. (Ashour et al. 2014). These are magnetic semiconductors which are widely used in the field of water treatment. Depends on nature, sharing of cations and synthesis method, the properties and application of spinel ferrites varies (Fig. 21.3). The attractive band gap makes the spinel ferrite a more efficient catalyst in heavy metal removal and increases the ability of photodegradation. Owing to biocompatibility, magnetic behaviour and chemical stability, spinel ferrites and its composites are utilized in wastewater treatment. Aromatic nitro compounds are identified as common pollutant in agricultural and industrial wastewater owing to their stability and solubility in water which can be efficiently degraded by photocatalytic activity of spinel ferrites.

21.4.2 Transition Metal Oxide NPs

21.4.2.1 Titania

Titanium oxide (TiO_2) is a semiconductor material which acts as photocatalyst under ultraviolet light. The electron present on the surface of substrates is transported to the conduction band. The various effluents present in the water can be degraded using titania because of its stability, abundance and less hazardous (Shakeel et al. 2016). Owing to the intrinsic properties like wide band gap (3.2 eV) and low quantum yield, the use of titania in water treatment was limited under visible light (Upadhyay et al. 2014; Qin et al. 2015). To induce the solar efficiency the nanomaterials are undergoing modification. During contact of catalyst with light sources, it produces electron and hole pairs. The electrons and hole pairs move around to the surface of the catalyst, and the redox reaction occurs for absorbing the pollutant (Fujishima et al. 2008; Di Paola et al. 2012). Transition metal or non-metal supporting is one of the tactics to decrease the band gap of titania which stimulates the photocatalyst under different light sources or direct sunlight radiation. Titania thin films and titania-decorated alumina showed high efficiency in the remediation of creatinine and methylene blue dye. Titania nanoparticles have the ability to degrade the highly hazardous materials such as antibiotic and chemotherapeutic doxorubicin present in water. Titania nanorods, nanobelts, nanowires, nanotubes, nanomembranes and nanofibers have been used to remediate the wastewater. Titania doped with other metals such as silver, iron, copper, zirconium, etc. improved electrical, catalytic and optical properties.

21.4.2.2 Copper Oxide

Copper oxide nanoparticles have been employed for the wastewater treatment owing to its incredible optical, superconductive, electrical, magnetic and thermal properties and also its low cost, low toxicity and abundance. The monovalent copper oxide nanoparticle is a p-type semiconductor with the narrow band gap of 2.0–2.5 eV. Copper oxide nanoparticles have the affinity to adsorb molecular oxygen to proliferate photogenerated electrons. Yadav et al. (2021) tried to remove the organic dyes such as Congo red (CR), methylene blue (MB), methyl orange (MO) and methyl red (MR) from the water using copper oxide nanoparticles. The increasing order of degradation efficiency was found to be MO > MB > CR > MR. Husein et al. (2019) synthesized copper nano-adsorbent for the removal of pharmaceutical pollutants from real wastewater samples. Ibuprofen, naproxen and diclofenac were identified as pollutants in real water samples. The removal capacities were calculated as 36.0, 33.9 and 33.9 mg/g for ibuprofen, naproxen and diclofenac. Dlamini et al. (2019) collected three different samples such as coal mine water, domestic wastewater and Mzingazi river water. Copper nanoparticle showed 85 and 76% removal efficiency of phosphate and sulphate from coal mine water, respectively.

The removal efficiency was found at 80, 89, 63, 62 and 64% for phosphate, total nitrogen, nitrate, aluminium and sulphate, respectively, in domestic water sample. The removal efficiency was found at 92 and 52% for phosphate and total nitrogen in Mzingazi river water, respectively.

21.4.2.3 Zinc Oxide

Zinc oxide nanoparticles are a semiconductor material which acts as better photocatalyst in the removal of organic dyes. It is an n-type semiconductor with the band gap value of 3.37 eV. Owing to the non-hazardous nature, effective adsorption properties, good thermal, mechanical and chemical properties and zinc oxide nanoparticles can be employed for the elimination of organic and inorganic nanomaterials (Mustapha et al. 2020). It has the admirable UV and visible light adsorption and reflective properties, and it has high surface activity due to its large number of active adsorption sites. The wavelength and intensity of light source is more important, owing to the essential characteristic of the material in photocatalytic reaction. The catalyst used for the wastewater treatment not only eliminates the chemicals but also eliminates the microbial contaminants present in water. Photocatalytic inactivation of microbes was a tedious process, and the process differs with concentration, physiological state and kind of microbes. The morphology, nature and concentration of the catalyst influence the rate of microbial inactivation. Anusa et al. (2017) removed the heavy metals such as Cu(II), Pb(II) and Cd(II) using zinc oxide nanoparticles at different pH from simulated industrial wastewater. The removal efficiency of Cu(II) at pH = 2, 4, 6 and 8 was 99.15, 99.25, 100 and 100%. The removal efficiency of Pb(II) at pH = 2 was 63.61 and at pH = 4, 6 and 8 was 77.47%. The removal efficiency of Cd(II) at pH = 2, 4, 6 and 8 was 87.05, 96.50, 98.05 and 97.85%.

21.4.3 Carbon-Based Nanoparticles

21.4.3.1 Carbon Nanotubes (CNTs)

Currently, carbon-enriched materials such as carbon nanotubes, graphene oxides, activated carbon, carbon fibres and biochar are used as adsorbents in water purification. CNTs have π–π conjugative structure with hexagonal arrays, and every carbon atom has sp^2 hybridization. It is hydrophobic in nature (Gupta et al. 2013). Carbon has the capability to form carbon to carbon long chains due to its binding ability in both straight and complex branching which facilitates double or triple bond formation and collection of atoms in different geometrical arrangements (Mubarak et al. 2014). Because of non-hazardous and high adsorption property, carbon-based nanomaterials have been broadly used for the elimination of heavy metals present in water. Electrostatic communication, ligand replacement, surface complex formation and adsorption–precipitation within metal ion and functional groups present in surface of carbon nanotubes (CNTs) are the steps followed in the mechanism of elimination of heavy metals in water. Due to the higher surface area to volume ratio of CNTs, the absorption property will be increased (Ruthiraan et al. 2015).

Multiwalled carbon nanotubes (MWCNTs) were synthesized by chemical vapour deposition, and the synthesized MWCNTs which has been found in between the width 60 and 70 nm range revealed the 100% efficiency in the removal of Cd(II) at pH = 10 and 12 with 0.5 mg/mL from 100 ppm metallic solution. The pH plays a vital role in the elimination of heavy metals. The heavy metals can be removed easily in the acidic condition. In the case of basic pH, the metals form precipitate because of the hydroxide formation. The MWCNTs shows the higher efficiency in the elimination of heavy metals in water, but its efficiency depends upon the pH of the reaction (Bhanjana et al. 2017).

Carbon microspheres can be utilized for the elimination of chromium, nickel and copper. Magnetic carbon nanomaterials have been employed for the elimination of heavy metals. MWCNTs-incorporated ZVI nanoparticles have been used in the removal of arsenic in the pH ranging between 6 and 7. During the oxidation in water, ZVI forms Fe^{2+} and Fe^{3+} hydroxides and leads to the formation of arsenic complexes, and these complexes can easily be eliminated from the water because the precipitation is taking place (Alijani and Shariatinia 2017).

21.4.3.2 Graphene Nanomaterials

Graphene have been obtained from graphite, and it acquires good electrical and mechanical properties and also it shows better thermal conductivity with honeycomb network structure (Aghigh et al. 2015; Chatterjee et al. 2015). It exists in various forms such as graphene oxide and reduced graphene oxide which are used for the removal of heavy metals (Gao et al. 2011). Graphene oxide (GO) is an oxidative product of graphene, and reduced graphene oxide (r-GO) is a reduction product of graphene oxide. GO have different oxygen functional group, while r-GO can be changed by functional groups, for instance, hydroxyl, amine and carboxylic acid group with more structural imperfection than graphene. To enhance the adsorption nature of GO, different kinds of functional groups are added to modify the GO. Due to the high surface area with fine chemical constancy, GO and r-GO are used for the wastewater remediation. The mechanism of heavy metal elimination from water using graphene-based nanomaterials depends on the electrostatic interactions and surface metal hydroxide. Based upon the surface area and surface charge, the interaction is taking place. Surface area is directly proportional to adsorption ability which is straightly with respect to particularly tunable morphology of the GO-supported nanomaterials. GO-supported metal oxides increase the ability of elimination of heavy metals in the water because of the increase of metal oxide's electronegative charge over the GO (Ghorbani et al. 2020).

21.4.4 Nanomembranes

Filtration and membranes are extremely efficient methodologies for purification water and remediation of wastewater. The remediation includes elimination of heavy metals, inorganic ions, organic pollutants such as dyes, pesticides, pharmaceutical products, etc. and bio-based products like microorganisms (Zhang et al. 2018). Reverse osmosis is the process which is used to purify the water and desalination of seawater till now. In between ultrafiltration and reverse osmosis, there is a process called nano-filtration, using membranes that have been employed for the wastewater treatment and desalination of seawater. Nanomembrane filtration is a very efficient method for wastewater treatment. It can be divided into inorganic and organic membranes in which zeolite, silicon dioxide and 2D graphene-based nanomaterials are inorganic membranes and organic polymer-based nanomaterials belong to organic nanomembranes (Liu et al. 2014; Pedrosa et al. 2019; Huang et al. 2014). The organic polymer membranes consist of natural polymer such as chitosan, cellulose acetate and synthetic polymeric membrane such as polyacrylonitrile, polyurethane, polyamidoamine, polysulfone, polyethersulfone, polyvinyl alcohol and polyamide. The efficiency of membranes mainly depends on the structure and weight of the molecule, pore size and volume, polarity, hydrophilicity and hydrophobicity (Cyna et al. 2002; Ahmad et al. 2008; Bonne et al. 2000; Tepus et al. 2009).

21.5 Conclusion and Future Prospect

The universe is in need of advanced water treatment technologies to get freshwater for drinking and agricultural purposes. Nanotechnology mutiny will play a crucial part in resolving the difficulty of increasing demands of freshwater and disseminated water recycle. Nanomaterials are attractive material which is used for water treatment because of its fascinating physicochemical properties. Engineering nanomaterials like nanoadsorbents, photocatalysts, nanomembranes, etc. provide the prospective for new water treatment technologies, and it can be adapted to precise applications in the removal of pollutants from contaminated water. Owing to their distinctive properties such as high reaction rate, high surface area-to-volume ratio, increased surface associated behaviour (antimicrobial properties and catalysis), high conductivity and self-assembling property on substrate, nanomaterials show high efficiency in the removal of pollutants. Nanomaterials can act as catalyst to purify the water under ultraviolet light source and freely existing sun irradiation. Nanomaterials can be used to eliminate harmful organic pollutants, microplastics and microbes via catalysis using ultraviolet and solar irradiation. A nanomembrane (semi porous membranes) is used to convert hard water into soft water by blocking monovalent and bivalent ions present in water body. In future, nanotechnology plays a fascinating role in water treatment, water monitoring, etc. that can effectively stop an extensive assortment of contaminant present in water together with affordability and ease of operation. There is no debate that nanotechnologies play a vital role in the field of wastewater treatment because of its unique nature.

References

Abbas N, Shao GN, Imran SM, Haider MS, Kim HT (2016) Inexpensive synthesis of a high performance Fe3O4-SiO2-TiO2 photocatalyst: magnetic recovery and reuse. Front Chem Sci Eng 10:405–416

Aghigh A, Alizadeh V, Wong HY, Islam MS, Amin N, Zaman M (2015) Recent advances in utilization of graphene for filtration and desalination of water: a review. Desalination 365:389–397

Ahmad AL, Tan LS, Shukor SRA (2008) Dimethoate and atrazine retention from aqueous solution by nanofiltration membranes. J Hazard Mater 151:71–77

Ahmed MA, Bishay ST, Ahmed FM, El-Dek SI (2017) Effective Pb^{2+} removal from water using nanozerovalent iron stored 10 months. Appl Nanosci 7:407–416

Alijani H, Shariatinia Z (2017) Effective aqueous arsenic removal using zero valent iron doped MWCNT synthesized by in situ CVD method using natural α-Fe_2O_3 as a precursor. Chemosphere 171:502–511

Alp E, Esgin H, Kazmanli MK, Genç A (2019) Synergetic activity enhancement in 2D CuO-Fe_2O_3 nanocomposites for the photodegradation of rhodamine B. Ceram Int 45:9174–9178

Anusa R, Ravichandran C, Sivakumar EKT (2017) Removal of heavy metal ions from industrial waste water by nano-ZnO in presence of electrogenerated Fenton's reagent. Int J ChemTech Res 10(7):501–508

Archer E, Petrie B, Hordern BK, Wolfaardt GM (2017) The fate of pharmaceuticals and personal care products (PPCPs), endocrine disrupting contaminants (EDCs), metabolites and illicit drugs in a WWTW and environmental waters. Chemosphere 174:437–446

Ashour AH, Hemeda OM, Heiba ZK, Al-Zahrani SM (2014) Electrical and thermal behavior of PS/ferrite composite. J Magn Magn Mater 369:260–267

Ayazi Z, Khoshhesab ZM, Norouzi S (2016) Modeling and optimizing of adsorption removal of Reactive Blue 19 on the magnetite/graphene oxide nanocomposite via response surface methodology. Desalination Water Treat 57:25301–25316

Bagheri M, Najafabadi NR, Borna E (2020) Removal of reactive blue 203 dye photocatalytic using ZnO nanoparticles stabilized on functionalized MWCNTs. J King Saud Univ Sci 32(1):799–804

Bhanjana G, Dilbaghi N, Kim KH, Kumar S (2017) Carbon nanotubes as sorbent material for removal of cadmium. J Mol Liq 242:966–970

Bonne PAC, Beerendonk EF, Van der Hoek JP, Hofman JAMH (2000) Retention of herbicides and pesticides in relation to aging of RO membranes. Desalination 132:189–193

Cao J, Lai L, Lai B, Yao G, Chen X, Song L (2019) Degradation of tetracycline by peroxymonosulfate activated with zero valent iron: performance, intermediates, toxicity and mechanism. Chem Eng J 364:45–56

Celiz MD, Tso J, Aga DS (2009) Pharmaceutical metabolites in the environment: analytical challenges and ecological risks. Environ Toxicol Chem 28(12):2473–2484

Chatterjee SG, Chatterjee S, Ray AK, Chakraborty AK (2015) Graphene–metal oxide nanohybrids for toxic gas sensor: a review. Sens Actuators B Chem 221:1170–1181

Chen C, Liu J, Liu P, Yu B (2011) Investigation of photocatalytic degradation of methyl orange by using nano-sized ZnO catalysts. Adv Chem Eng Sci 1:9–14

Chen L, Yang S, Huang Y, Zhang B, Kang F, Ding D, Cai T (2019) Degradation of antibiotics in multi component systems with novel ternary $AgBr/Ag_3PO_4$@ natural hematite heterojunction photocatalyst under simulated solar light. J Hazard Mater 371:566–575

Clemonne J, Madadrang, Kim HY, Gao G, Wang N, Zhu J, Feng H, Gorring M, Kasner ML, Hou S (2012) Adsorption behavior of EDTA-graphene oxide for Pb(II) removal. ACS Appl Mater Interfaces 4:1186–1193

Cornell RM, Schwertmann U (2003) The iron oxides: structure, properties, reactions, occurrences and uses. John Wiley & Sons. ISBN: 3527302743

Cyna B, Chagneaub G, Bablon G, Tanghe N (2002) Two years of nanofiltration at the Méry-sur-Oise plant, France. Desalination 147:69–75

Danila V, Vasarevicius S, Valskys V (2018) Batch removal of Cd(II), Cu(II), Ni(II), and Pb(II) ions using stabilized zero-valent iron nanoparticles. Energy Procedia 147:214–219

Di Paola A, García-López E, Marcì G, Palmisano L (2012) A survey of photocatalytic materials for environmental remediation. J Hazard Mater 211–212:3–29

Dlamini NG, Basson AK, Pullabhotla VSR (2019) Optimization and application of bioflocculant passivated copper nanoparticles in the wastewater treatment. Int J Environ Res Public Health 16(12):2185

Dubey R, Bajpai J, Bajpai AK (2015) Green synthesis of graphene sand composite (GSC) as novel adsorbent for efficient removal of Cr (VI) ions from aqueous solution. J Water Process Eng 5:83–94

El-Dafrawy SM, Fawzy S, Hassan SM (2017) Preparation of modified nanoparticles of zinc oxide for removal of organic and inorganic pollutant. Trends Appl Sci Res 12:1–9

Fan H, Ma X, Zhou S, Huang J, Liu Y, Liu Y (2019) Highly efficient removal of heavy metal ions by carboxymethyl cellulose immobilized Fe3O4 nanoparticles prepared via high gravity technology. Carbohydr Polym 213:39–49

Feng L, Van Hullebusch ED, Rodrigo MA, Esposito G, Oturan MA (2013) Removal of residual anti-inflammatory and analgesic pharmaceuticals from aqueous systems by electrochemical advanced oxidation processes, a review. Chem Eng J 228:944–964

Franco P, Sacco O, De Marco I, Vaiano V (2019) Zinc oxide nanoparticles obtained by supercritical antisolvent precipitation for the photocatalytic degradation of crystal violet dye. Catalysts 9:346

Fu F, Dionysiou DD, Liu H (2014) The use of zero-valent iron for groundwater remediation and wastewater treatment: a review. J Hazard Mater 267:194–205

Fujishima A, Zhang XT, Tryk DA (2008) TiO_2 photocatalysis and related surface phenomena. Surf Sci Rep 63:515–582

Gao B, Lim TM, Subagio DP, Lim TT (2010) Zr-doped TiO_2 for enhanced photocatalytic degradation of bisphenol A. Appl Catal A 375:107–115

Gao W, Majumder M, Alemany LB, Narayanan TN, Ibarra MA, Pradhan BK, Ajayan PM (2011) Engineered graphite oxide materials for application in water purification. ACS Appl Mater Interfaces 3:1821–1826

Garcia JC, Oliveira U, Silva AEC, Oliveira CC, Nozaki J, de Souza NE (2007) Comparative study of the degradation of real textile effluents by photocatalytic reactions involving UV/TiO2/H2O2 and UV/Fe2+/H2O2 systems. J Hazard Mater 147:105–110

Geise GM, Lee HS, Miller DJ, Freeman BD, McGrath JE, Paul DR (2010) Water purification by membranes: the role of polymer science. J Polym Sci B Polym Phys 48(15):1685–1718

Ghorbani M, Seyedin O, Aghamohammadhassan M (2020) Adsorptive removal of lead (II) ion from water and wastewater media using carbon-based nanomaterials as unique sorbents: a review. J Environ Manag 254:109814

Gupta VK, Kumar R, Nayak A, Saleh TA, Barakat M (2013) Adsorptive removal of dyes from aqueous solution onto carbon nanotubes: a review. Adv Colloid Interf Sci 193:24–34

Halem DV, Basker S, Amy G, Van Dijk J (2009) Arsenic in drinking water: a worldwide water quality concern for water supply companies. Drink Water Eng Sci 2:29–34

Hamdy A, Mostafa MK, Nasr M (2018) Zero-valent iron nanoparticles for methylene blue removal from aqueous solutions and textile wastewater treatment, with cost estimation. Water Sci Technol 78(2):367–378

Han PLJ, Zhu G, Hojamberdev M, Peng J, Zhang X, Lu Y, Ge B (2015) Rapid adsorption and photocatalytic activity for Rhodamine B and Cr(VI) by ultrathin BiOI nanosheets with high exposed {001} facets. New J Chem 39:1874–1882

He F, Zhao D (2007) Manipulating the size and dispersibility of zerovalent iron nanoparticles by use of carboxymethyl cellulose stabilizers. Environ Sci Technol 41(17):6216–6221

Hinojosa-Reyes M, Camposeco-Solis R, Ruiz F, Rodríguez-González V, Moctezuma E (2019) Promotional effect of metal doping on nanostructured TiO_2 during the photocatalytic degradation of 4-chlorophenol and naproxen sodium as pollutants. Mater Sci Semicond Process 100:130–139

Hu C, He M, Chen B, Hu B (2015) Simultaneous determination of polar and apolar compounds in environmental samples by a polyaniline/hydroxyl multi-walled carbon nanotubes composite-coated stir bar sorptive extraction coupled with high performance liquid chromatography. J Chromatogr A 1394:36–45

Huang SH, Chen DH (2009) Rapid removal of heavy metal cations and anions from aqueous solutions by an amino-functionalized magnetic nano-adsorbent. J Hazard Mater 163(1):174–179

Huang HB, Ying YL, Peng XS (2014) Graphene oxide nanosheet: an emerging star material for novel separation membranes. J Mater Chem A 2:13772–13782

Husein DZ, Hassanien R, Al-Hakkani MF (2019) Green-synthesized copper nano-adsorbent for the removal of pharmaceutical pollutants from real wastewater samples. Heliyon 5:e02339

Ibrahim I, Athanasekou C, Manolis G, Kaltzoglou A, Nasikas NK, Katsaros F, Devlin E, Kontos AG, Falaras P (2019) Photocatalysis as an advanced reduction process (ARP): the reduction of 4-nitrophenol using titania nanotubes ferrite nanocomposites. J Hazard Mater 372:37–44

Jain R, Mathur M, Sikarwar S, Mittal A (2007) Removal of the hazardous dye rhodamine B through photocatalytic and adsorption treatments. J Environ Manag 85(4):956–964

Kakavandi B, Takdastan A, Pourfadakari S, Ahmadmoazzam M, Jorfi S (2019) Heterogeneous catalytic degradation of organic compounds using nanoscale zero valent iron supported on kaolinite: mechanism, kinetic and feasibility studies. J Taiwan Inst Chem Eng 96:329–340

Kang MJ, Yu H, Lee W, Cha HG (2019) Efficient $Fe_2O_3/Cg-C_3N_4$ Z-scheme heterojunction photocatalyst prepared by facile one step carbonizing process. J Phys Chem Solids 130:93–99

Kassaee M, Motamedi E, Majdi M (2011) Magnetic Fe_3O_4-graphene oxide/polystyrene: fabrication and characterization of a promising nanocomposite. Chem Eng J 172:540–549

Kaur J, Kaur M (2019) Facile fabrication of ternary nanocomposite of $MgFe_2O_4 - TiO_2$@ GO for synergistic adsorption and photocatalytic degradation studies. Ceram Int 45(7):8646–8659

Khalafi T, Buazar F, Ghanemi K (2019) Phycosynthesis and enhanced photocatalytic activity of zinc oxide nanoparticles toward organosulfur pollutants. Sci Rep 9:6866

Khare JSSS (2016) Heavy metal toxicity in the ecosystem and its impacts. Global J Eng Sci Soc Sci Stud 2. ISSN: 2394-3084

Kireeti KV, Chandrakanth G, Kadam MM, Jha N (2016) A sodium modified reduced graphene oxide-Fe3O4 nanocomposite for efficient lead (II) adsorption. RSC Adv 6:84825–84836

Kumar KV, Kumaran A (2005) A removal of methylene blue by mango seed kernel powder. Biochem Eng J 27(1):83–93

Le AT, Pung SY, Sreekantan S, Matsuda A (2019) Mechanisms of heavy metal ions removal by ZnO particles. Heliyon 5:e01440

Leone VO, Pereira MC, Aquino SF, Oliveira LCA, Correa S, Ramalho TC, Gurgel LVA, Silva AC (2018) Adsorption of diclofenac on a magnetic adsorbent based on maghemite: experimental and theoretical studies. New J Chem 42:437–449

Li YH, Ding J, Luan Z, Di Z, Zhu Y, Xu C, Wu D, Wei B (2003) Competitive adsorption of Pb^{2+}, Cu^{2+} and Cd^{2+} ions from aqueous solutions by multiwalled carbon nanotubes. Carbon 41(14):2787–2792

Liu T, Li B, Hao Y, Yao Z (2014) MoO_3-nanowire membrane and $Bi_2Mo_3O_{12}/MoO_3$ nano-heterostructural photocatalyst for wastewater treatment. Chem Eng J 244:382–390

Lu F, Astruc D (2020) Nanocatalysts and other nanomaterials for water remediation from organic pollutants. Coord Chem Rev 408:213180

Ma J, Yu F, Zhou L, Jin L, Yang M, Luan J, Tang Y, Fan H, Yuan Z, Chen J (2012) Enhanced adsorptive removal of methyl orange and methylene blue from aqueous solution by alkali-activated multiwalled carbon nanotubes. ACS Appl Mater Interfaces 4(11):5749–5760

Mahdavi S, Jalali M, Afkhami A (2013) Heavy metals removal from aqueous solutions using TiO_2, MgO, and Al_2O_3 nanoparticles. Chem Eng Commun 200:448–470

Malakootian M, Olama N, Malakootian M, Nasiri A (2019) Photocatalytic degradation of metronidazole from aquatic solution by TiO_2-doped Fe^{3+} nano-photocatalyst. Int J Environ Sci Technol 16:4275–4284

Martinez-Boubeta C, Simeonidis K (2019) Chapter 20 – magnetic nanoparticles for water purification, in: S. Thomas, D. Pasquini, S.-.Y. Leu, D.A. Gopakumar (Eds.), Nanoscale Materials in Water Purification, Elsevier, pp. 521–552.

Milne GWA (ed) (2018) The Ashgate handbook of pesticides and agricultural chemicals. Routledge

Mirzazadeh H, Lashanizadegan M (2018) ZnO/CdO/reduced graphene oxide and its high catalytic performance towards degradation of the organic pollutants. J Serbian Chem Soc 83(2):221–236

Mubarak N, Sahu J, Abdullah E, Jayakumar N, Ganesan P (2014) Single stage production of carbon nanotubes using microwave technology. Diam Relat Mater 48:52–59

Mustapha S, Ndamitso MM, Abdulkareem AS, Tijani JO, Shuaib DT, Ajala AO, Mohammed AK (2020) Application of TiO_2 and ZnO nanoparticles immobilized on clay in wastewater treatment: a review. Appl Water Sci 10:49

Nivetha A, Devi SM, Prabha I (2019) Fascinating physic-chemical properties and resourceful applications of selected cadmium nanomaterials. J Inorg Organomet Polym Mater 29:1423–1438

Nizamuddin S, Siddiqui MTH, Mubarak NM, Baloch HA, Abdullah EC, Mazari SA, Griffin GJ, Srinivasan MP, Tanksale A (2019) Chapter 17 – Iron oxide nanomaterials for the removal of heavy metals and dyes from wastewater. In: Thomas S, Pasquini D, Leu SY, Gopakumar DA (eds) Nanoscale materials in water purification. Elsevier, pp 447–472

Okube M, Yasue T, Sasaki S (2012) Residual density mapping and site selective determination of anomalous scattering factors to examine the origin of the Fe K pre-edge peak of magnetite. J Synchrotron Radiat 19:759–767

Pedrosa M, Drazic G, Tavares PB, Figueiredo JL, Silva AMT (2019) Metal-free graphene-based catalytic membrane for degradation of organic contaminants by persulfate activation. Chem Eng J 369:223–232

Peng X, Luan Z, Ding J, Di Z, Li Y, Tian B (2005) Ceria nanoparticles supported on carbon nanotubes for the removal of arsenate from water. Mater Lett 59:399–403

Pizarro C, Rubio MA, Escudey M, Albornoz MF, Muñoz D, Denardin J, Fabris JD (2015) Nanomagnetite-zeolite composites in the removal of arsenate from aqueous systems. J Braz Chem Soc 26:1887–1896

Qi L, Xu Z, Jiang X, Hu C, Zou X (2004) Preparation and antibacterial activity of chitosan nanoparticles. Carbohydr Res 339:2693–2700

Qin C, Li Z, Chen G, Zhao Y, Lin T (2015) Fabrication and visible-light photocatalytic behavior of perovskite praseodymium ferrite porous nanotubes. J Power Sources 285:178–184

Rajput S, Singh LP, Pittman CU Jr, Mohan D (2017) Lead (Pb2+) and copper (Cu2+) remediation from water using superparamagnetic maghemite (γ-Fe2O3) nanoparticles synthesized by Flame Spray Pyrolysis (FSP). J Colloid Interface Sci 492:176–190

Rozenberg GK, Dubrovinsky LS, Pasternak MP, Naaman O, Le Bihan T, Ahuja R (2002) High pressure structural studies of hematite Fe_2O_3. Phys Rev B 65:064112

Ruthiraan M, Mubarak NM, Thines RK, Abdullah EC, Sahu JN, Jayakumar NS, Ganesan P (2015) Comparative kinetic study of functionalized carbon nanotubes and magnetic biochar for removal of Cd^{2+} ions from wastewater. Korean J Chem Eng 32:446–457

Saleh TA (2015) Mercury sorption by silica/carbon nanotubes and silica/activated carbon: a comparison study. J Water Supply Res Technol AQUA 64(8):892

Santhosh C, Malathi A, Dhaneshvar E, Bhatnagar A, Grace AN, Madhavan J (2019) Chapter 16 – Iron oxide nanomaterials for water purification. In: Thomas S, Pasquini D, Leu SY, Gopakumar DA (eds) Nanoscale materials in water purification. Elsevier, pp 431–446

Shakeel M, Jabeen F, Shabbir S, Asghar MS, Khan MS, Chaudhry AS (2016) Toxicity of nanotitanium dioxide (TiO_2-NP) through various routes of exposure: a review. Biol Trace Elem Res 172:1–36

Sharma S, Dutta V, Singh P, Raizada P, Sani AR, Bandegharaei AH, Thakur VK (2019) Carbon quantum dot supported semiconductor photocatalysts for efficient degradation of organic pollutants in water: a review. J Clean Prod 228:755–769

Sharaf El-Deen SEA, Zhang FS (2016) Immobilisation of TiO2-nanoparticles on sewage sludge and their adsorption for cadmium removal from aqueous solutions. J Exp Nanosci 11:239–58

Shin KY, Hong JY, Jang J (2011) Heavy metal ion adsorption behavior in nitrogen-doped magnetic carbon nanoparticles: isotherms and kinetic study. J Hazard Mater 190:36–44

Shokri M, Jodat A, Modirshahla N, Behnajady M (2013) Photocatalytic degradation of chloramphenicol in an aqueous suspension of silver-doped TiO_2 nanoparticles. Environ Technol 34:1161–1166

Singh R, Singh S, Parihar P, Singh VP, Prasad SM (2015) Arsenic contamination, consequences and remediation techniques a review. Ecotoxicol Environ Saf 112:247–270

Snyder SA, Westerhoff P, Yoon Y, Sedlak DL (2003) Pharmaceuticals, personal care products, and endocrine disruptors in water: implications for the water industry. Environ Eng Sci 20:449–469.

Sui Q, Huang J, Liu Y, Chang X, Ji G, Deng S, Xie T, Yu G (2011) Rapid removal of bisphenol A on highly ordered mesoporous carbon. J Environ Sci 23:177–182

Tadic M, Trpkov D, Kopanja L, Vojnovic S, Panjan M (2019) Hydrothermal synthesis of hematite (α-Fe_2O_3) nanoparticle forms: synthesis conditions, structure, particle shape analysis, cytotoxicity and magnetic properties. J Alloys Compd 792:599–609

Tang L, Wang J, Wang L, Jia C, Lv G, Liu N, Wu M (2016) Facile synthesis of silver bromide based nanomaterials and their efficient and rapid selective adsorption mechanisms toward anionic dyes. ACS Sustain Chem Eng 4:4617–4625

Tepus B, Simonic M, Petrinic I (2009) Comparison between nitrate and pesticide removal from ground water using adsorbents and NF and RO membranes. J Hazard Mater 170:1210–1217

Tian Z, Yang B, Cui G, Zhang L, Guo Y, Yan S (2014) Synthesis of poly(m-phenylenediamine)/iron oxide/acid oxidized multi-wall carbon nanotube for removal of hexavalent chromium. RSC Adv 5:2266–2275

Tijani JO, Fatoba OO, Babajide OO, Petrik LF, Zhu K, Chen C (2019) Chapter 6 – Application of nZVI and its composites into the treatment of toxic/radioactive metal ions. In: Chen C (ed) Interface science and technology. Elsevier, pp 281–330

Upadhyay RK, Soin N, Roy SS (2014) Role of graphene/metal oxide composites as photocatalysts, adsorbents and disinfectants in water treatment: a review. RSC Adv 4(8):3823–3851

Vanhulle S, Trovaslet M, Enaud E, Lucas M, Taghavi S, Van Der Lelie D, Van Aken B, Foret M, Onderwater RCA, Wesenberg D, Agathos SN, Schneider YJ, Corbisier AM (2008) Decolorization, cytotoxicity, and genotoxicity reduction during a combined ozonation/fungal treatment of dye-contaminated wastewater. Environ Sci Technol 42:584–589

Wen Y, Zhao Y, Guo M, Xu Y (2019) Synergetic effect of Fe_2O_3 and $BiVO_4$ as photocatalyst nanocomposites for improved photo Fenton catalytic activity. J Mater Sci 54:8236–8246

Wu W, Wu Z, Yu T, Jiang C, Kim WS (2015) Recent progress on magnetic iron oxide nanoparticles: synthesis, surface functional strategies and biomedical applications. Sci Technol Adv Mater 16:023501

Xu Y, Liu Q, Xie M, Huang S, He M, Huang L, Xu H, Li H (2018) Synthesis of zinc ferrite/silver iodide composite with enhanced photocatalytic antibacterial and pollutant degradation ability. J Colloid Interface Sci 528:70–81

Yadav S, Chauhan M, Mathur D, Jain A, Malhotra P (2021) Sugarcane bagasse-facilitated benign synthesis of Cu_2O nanoparticles and its role in photocatalytic degradation of toxic dyes: a trash to treasure approach. Environ Dev Sustain 23(2):2071–2091. https://doi.org/10.1007/s10668-020-00664-7

Yahya N, Aziz F, Jamaludin NA, Mutalib MA, Ismail AF, Salleh WNW, Jaafar J, Yusof N, Ludin NA (2018) A review of integrated photocatalyst adsorbents for wastewater treatment. J Environ Chem Eng 6(6):7411–7425

Yu F, Li Y, Han S, Ma J (2016) Adsorptive removal of antibiotics from aqueous solution using carbon materials. Chemosphere 153:365–385

Zha S, Cheng Y, Gao Y, Chen Z, Megharaj M, Naidu R (2014) Nanoscale zero valent iron as a catalyst for heterogeneous Fenton oxidation of amoxicillin. Chem Eng J 255:141–148

Zhang WX (2003) Nanoscale iron nanoparticles for environmental remediation: an overview. J Nanopart Res 5:323–332

Zhang K, Liu Y, Deng J, Xie S, Lin H, Zhao X, Yang J, Han Z, Dai H (2017) Fe_2O_3/3DOM $BiVO_4$: high performance photocatalysts for the visible light-driven degradation of 4-nitrophenol. Appl Catal B 202:569–579

Zhang Y, Wei S, Hu Y, Sun S (2018) Membrane technology in wastewater treatment enhanced by functional nanomaterials. J Clean Prod 197:339–348

Zhu K, Chen C (2019) Chapter 6: Application of nZVI and its composites into the treatment of toxic/radioactive metal ions, in: C. Chen (Ed.), Interface Science and Technology, Elsevier, pp. 281–330.

Chapter 22
Impact of Nanomaterials on Waste Management: An Insight to the Modern Concept of Waste Abatement

Ram Kumar Ganguly, Susanta Kumar Chakraborty, Sujoy Midya, and Balasubramani Ravindran

Contents

22.1	Introduction..	622
22.2	Nanomaterials in the Remediation of Toxic Chemicals and Heavy Metals..............	623
22.3	Nanotechnology in Wastewater Treatment..	625
22.4	Nanomaterials in Solid Waste Management...	630
22.5	Integration of Nanotechnology with Bioremediation..	631
22.6	Risk Assessment..	633
22.7	Recent Trends and Future Outlook...	634
22.8	Conclusion...	635
References..		635

Abbreviation

CNT   Carbon nanotube
NP   Nanoparticle
nZVI   Nonzerovalent iron

R. K. Ganguly (✉) · S. K. Chakraborty · S. Midya
Department of Zoology, Vidyasagar University, Midnapore, West Bengal, India

B. Ravindran
Department of Environmental Energy and Engineering, Kyonggi University, Suwon, South Korea

22.1 Introduction

Wastes stance a solemn menace to society. In order to meet the over increasing economic demands, all the countries of the world have been trying to increase their production for both industrial and domestic sectors, generating huge amount of wastes which are the prime reason for contaminating and polluting the natural environment (Ganguly and Chakraborty 2020a). However, such environmental encumbrance can be reduced through less conservation of resources, reduction of chemical use and energy. Wastes emerged from different industrial and agricultural sectors include solid wastes, wastewaters and air contaminants. The harmful impact of such pollutants can be reduced through control of production, emission and effective management of wastes (Boldrin et al. 2014). Therefore, it has appeared to become a challenge for the entrepreneur to ensure a green environment despite huge production demand of commodity. Nanotechnology has proven to be a most emerging trend in the field of green engineering as it involves the utilization of minuscule particles having distinct structural, magnetic and electrokinetic properties that enable better removal of contaminants through adsorption (Vázquez-Núñez et al. 2020; Rizwan et al. 2014). Such functionalities reduce energy consumption and facilitate a sustainable environment (Bhushan 2017; Poole Jr and Owens 2003; Paul and Robeson 2008).

Nanotechnology is apprehensive with the realm of imperceptible diminutive particles and is governed by different underlying principles of physics and surface chemistry. It involves the production of materials in the form of a nanometer or one billionth of a metre (Dasgupta et al. 2017). The technology gains its attention after the development of AFM (atomic force microscope) which allows the manipulation and relative comparison at the atomic scale (Brar et al. 2010). Such particles pose a high surface area to the mass ratio which impacts significantly over physical, chemical, biological and mechanical characteristics. Bolyard et al. (2013) estimated a huge rise in the production of nanomaterials from 1000 tons (2011) to 58,000 tons (2020). The presence of such unique properties led to better adsorption with the different types of contaminants. Nanotechnology can be employed as both in form of in situ or ex situ waste management strategies. In situ strategies implies the development of reactive barrier utilizing non zero-valent iron nanoparticles (nZVI), magnetic nanoparticles, etc. in the flow of contaminants generated from ordinary waste management techniques (Thomé et al. 2015; Singh et al. 2020). Ex situ strategies involve the utilization of photocatalytic degradation and other filtration processes such as ultrafiltration, reverse osmosis, nanofiltration, etc. which are based on adsorption property of nanoparticles (Andrade et al. 2015; Crane et al. 2015). Nanomaterials are often used in the preparation of membrane matrix which helps in greater removal of contaminants through adsorption (Huang et al. 2019; Jiang et al. 2017). Such methodologies are also used in the removal of heavy metals from the canals or streams carrying wastewater and thereby help in cleaning of water as a part of sustainable environment through the reduction, recycling, reusing and other hazardous materials (Madhura et al. 2019; Mubarak et al. 2014). Different engineered nanoparticles proved their efficiency by displaying different biophysical properties such as adsorption, photocatalytic degradation, nanofiltration, etc. which limits the consumption of reactants and imparts better conversion of waste into value-added products. Therefore, nanoparticles-supplemented waste management technologies can mitigate a wide range of environmental problems through the reduction of different organic, inorganic and nonpoint source pollutants through the development of nanocoated membranes, environmental sensors, etc. (Fan et al. 2016). The unique properties of nanoparticles lead to huge implementation in the context of waste management which has urged expansion along with modification of existing nanotechnology such as the development of different types of carbon nanotubes (single-walled CNTs, multiwalled CNTs, hybrid CNTs), graphenes and magnetic NPs alongside the improvement of technologies such as nanobioremediation, nanofiltration, etc. (Kim and Kwak 2007). Nowadays nanoparticles like nZVI, silver, CNTs, etc. are used in food products, cosmetics, textiles, etc. and therefore get impinged with our daily life. Recently, increased application of different engineered nanoparticles in the management of wastewater streams and industrial wastes further leads to contamination of the environment. Therefore, there is a multitude of sources by which the minuscule particles get contaminated with the environment (Boldrin et al. 2014; Le et al. 2015). Owing to their physical and chemical properties, these substances can get easily intermixed with the environment. Such disposal to the natural environment causes severe toxicity to the environment and explicitly damage life forms. Several types of research have been done concerning the impact of nanomaterials on human health and the environment. All of these ecotoxicological studies have generated information for the prevention of the overproduction and overuse of nanoparticles (Lee et al. 2008). Therefore, the present study has expposed different ways and means for the application of different nanoparticles in light of remediation of contaminants in different areas of waste management and also highlight the risk imposed by such implementation towards a sustainable green environment.

22.2 Nanomaterials in the Remediation of Toxic Chemicals and Heavy Metals

Industrial discharge has been estimated to around 10 million tons per year across the world which includes the discharge of several toxic chemicals such as dibenzofurans, dibenzo-p-dioxins, etc. posing severe cytotoxicity and interact with different biotic and abiotic environmental factors (Ganguly and Chakraborty 2018, 2020a). Several research methodologies had been proposed in context to sludge abatement, but the application of NPs has appeared to be most effective among other methodologies as it increases the rate of degradation, reduction of heavy metal content, etc. in order to maintain ecosystem health (Fig. 22.1). Different parameters of nanomaterials such as size, shape, surface area, surface coating, etc. play a pivotal role in

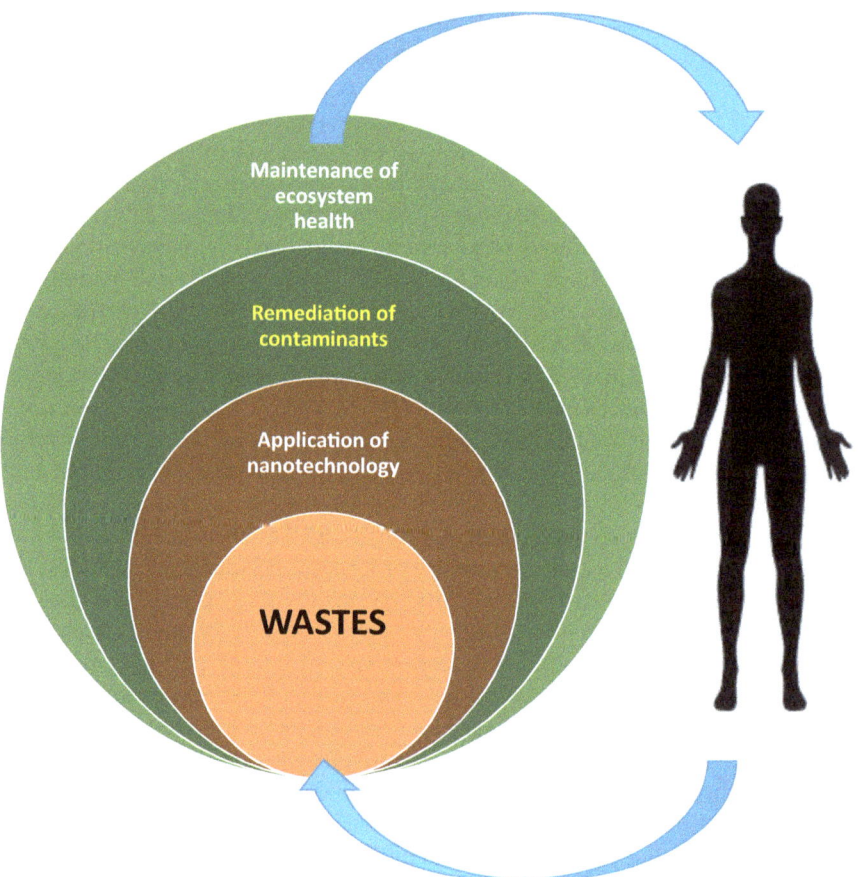

Fig. 22.1 Schematic representation of human-manipulated waste management using nanotechnology

bioremediation depending upon some external features such as media, pH, temperature, etc. of the contaminated environment. During the last couple of decades, several forms of nanomaterials have gained importance such as (i) dendrimers, (ii) carbon-based nanomaterials, (iii) single enzyme-linked nanoparticle, etc. (Crespilho et al. 2006; Rizwan et al. 2014). Nanomaterials are often conjugated with the membrane and are used in contaminated air or water streams for remediation of different hazardous materials using the process of adsorption or absorption. For example, activated carbon plays an important role in the treatment of non-point source pollution; different fullerenes, carbon nanotubes (CNTs), are found to adsorb benzene compounds and polycyclic aromatic hydrocarbons (PAHs) from the contaminated environment (Shan et al. 2009).

However, adsorption properties of different types of CNTs depend upon morphology, structure, etc. and also upon different physiochemical properties such as hydrophobicity, polarizability, etc. However, an adsorption affinity depends upon π-π (donor-acceptor) electron interactions (Vázquez-Núñez et al. 2020). Similar interactions are also found for graphene sheets. Such properties contribute for the development of maximum adsorption affinity for nitroaromatics. Multiwalled CNTs play an immense role in the adsorption of different aromatics such as benzene, cyclohexane, toluene, nitrobenzene and metal ions such as Pb^{+2}, atrazine, etc. from the contaminated environment (Lu et al. 2005; Wang et al. 2000; Yan et al. 2008). Similar evidences are also found in the case of iron nanomaterials for remediation of organics and inorganics. nZVI is widely used in the removal of Ba^{+2}, As^{+3} and different types of humic acid compounds (Celebi et al. 2007). A number of steps have so far been undertaken such as lignocellulose-based anion removal media which entails nanocoating technology for better removal of halocarbons, sulphur, phosphorus and different environmental toxins. Micellar-enhanced filtration (MEF) proved its potential role in the removal of Cu^{+2} from groundwater (Thomé et al. 2015). Due to the specific surface characterization, nanomaterials can reach a certain depth and help in the removal of organochlorine pesticides, polychlorinated benzene organics (Table 22.1). nZVI nanoparticles are found to be effective in remediation of arsenic, chromium, lead, etc. Ferragels are much more efficient for removal of heavy metal ions. Poly acrylic-supported nZVI has proved its efficacy in the removal of chlorinated hydrocarbons from soil and groundwater. The name "Dendrimers" derived from the words "dendri" meaning the branch of the tree and "meros" which means part of a tree. These are branched polymers made of several smaller subunits consisting of the central core, interior radial symmetry and terminal branch cell. These macromolecules have void spaces that allow them to interact with several metal ions which help in wastewater and dye treatment industries.

22.3 Nanotechnology in Wastewater Treatment

Water pollution has posed real threat to the very existence of mankind as water alongside playing supportive roles, represents the most essential components for human survival. Therefore, the supply of high-quality water with low-cost technology appears to be a global concern. Several innovations had been made for better treatment of wastewater. Owing to better adsorption, magnetic and photocatalytic properties, amendment of nanotechnology proved to be a promising tool of modern wastewater management strategies (Baruah et al. 2016; Madhura et al. 2018; Prasad and Thirugnanasanbandham 2019).

The application of nanoparticles helps in the reuse of water which results in the regular supply of recycled water in agricultural and other industrial sectors. Such infrastructural developments are also associated with proper distribution networks which can reduce energy consumption especially for developing countries which suffer from trepidations about the hasty worsening of water quality and immense pressure for pure production of water to meet environmental standards and good

Table 22.1 Commonly used nanoparticles in remediation of organic contaminants

Organic contaminants	Use of ENM	Removal ability (mg/g)	Citation
Chlorobenzene	TiO_2 combined with bentonite nanocomposite	0.2	Mishra et al. (2017)
Trichloroethane	Biochar hybridized Fe^{+2} or Ni^{+2} nanoparticles	20	Li et al. (2017)
	Multiwalled CNTs	0.35	Ma et al. (2011)
Dichlorobenzene	CNTs	30.8	Peng et al. 2003
	Graphitized CNTs	28.7	Peng et al. 2003
	TiO_2 nanocomposite		Salamat et al. (2017)
Hexachlorocyclohexane	nZVI	0.23	Li et al. (2007)
Rhodamine B	NiO_2 nanoparticles	2.5	Suo et al. (2013)
	nZVI nanoparticles	87.72	Shi et al. (2017)
	CNT membrane		Wei et al. (2014)
	Gold nanoparticles	40.3	Xiong et al. (2010)
	TiO_2 nanocomposite	6	Wang et al. (2000)
Methyl orange	TiO_2 nanoparticles hybridized with silver	135	Yang et al. (2013)
	TiO_2 nanoparticles hybridized with tin oxide	38	Yang et al. (2013)
Azo dye black	nZVI nanoparticles	299	Shu et al. (2007)
Methylene blue	Hybrid of graphene oxide with iron oxide nanocomposite		Deng et al. (2013)
	Graphene oxide nanoparticles	84	Fan et al. (2013)
	TiO_2 nanoparticles	50	Kim and Kwak (2007)
	Manganese oxide nanoparticles	68	Chen and He (2008)
	Graphene oxide filters		Hou et al. (2013)
	Silver hybrid TiO_2 nanoparticles	135	Yang et al. (2013)
Orange II	Gold/zinc nanoparticles	35	Cho et al. (2012)
Naphthalene	Single-walled CNTs	8	Moradi et al. (2012)
Guaiacol	TiO_2 nanoparticles	49.7	Peiro et al. (2001)

(continued)

Table 22.1 (continued)

Organic contaminants	Use of ENM	Removal ability (mg/g)	Citation
Phenol	TiO$_2$ nanoparticles	37.6	Peiro et al. (2001)
	MnO hybridized peroxymonosulphate nanoparticles	62.5	Saputra et al. (2013)
	Ruthenium oxide hybridized peroxymonosulphate nanocomposite	125	Muhammad et al. (2012)

ENM represent engineered nanomaterials used in different treatment

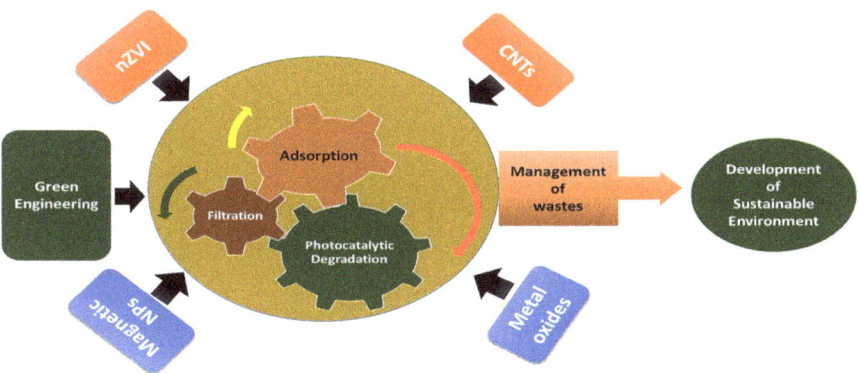

Fig. 22.2 Represents the role of nanoparticles in green engineering for the development of sustainable environment

public health (Han et al. 2009; Girginova et al. 2010). Such a situation caters to the need for development of high-performance water treatment technologies. The use of nanotechnology proves to be an easy method as it can target a wide array of pollutants. Such innovations are mainly based on four concepts such as adsorptive elimination of pollutants; decontamination of microbes; photocatalytic degradation; and filtration technology (Fig. 22.2). For example, nanoparticles prepared from silver are widely used as filters for wastewater treatment (Zhao et al. 2017). Furthermore, nanoparticles prepared from titanium oxide and fullerenes are used as photocatalytic degradation which disinfects different microbial populaces (Han et al. 2009; Goutam et al. 2018). Such technology favours substantial qualitative enhancement of wastewater.

Adsorbents in nanoscale have played an important role in adsorbing different organic and inorganic complexes from the aqueous environment. Nanoparticles with small size with large active surface area functionalize the property of adsorbing contaminants from the environment. Several carbon-based nanotubes, fullerenes and graphenes have good thermal stability demonstrate excellent adsorbing capacity of several organic contaminants from the environment (Deng et al. 2013; Jiang

Table 22.2 Commonly used nanoparticles in remediation of inorganic contaminants

Inorganic contaminants	Use of ENM	Removal ability (mg/g)	Citation
Lead	nZVI	1667	Zhang et al. (2013)
	Magnetite nanoparticles	77	Fan et al. (2013)
	Zirconium oxide nanoparticles	320	Hua et al. (2013)
	Multiwalled CNTs	66	Zhou et al. (2014)
	Manganese ferrite nanoparticles	69	Ren et al. (2012)
Cadmium	ZnO nanoparticles	217	Khezamia et al. (2017)
	nZVI	67	Zhang et al. (2014)
	Graphene oxide	91	Deng et al. (2013)
	Zirconium dioxide nanoparticles	215	Hua et al. (2013)
	Multiwalled CNTs	22	Vuković et al. (2010)
Copper	nZVI	340	Zhang et al. (2014)
	Manganese hybridized ferrite nanoparticles	61	Ren et al. (2012)
	Multiwalled CNTs	39	
Cobalt	nZVI	172	Uzum et al. (2008)
	Bentonite magnetic composite	22.73	Chen et al. (2011)
Mercury	Magnetite nanoparticles	17	Girginova et al. (2010)
	nZVI	80	Yan et al. (2010)
Arsenic	Cerium oxide nanoparticles	14	Zhong et al. (2007)
	Aluminium oxide nanocomposites	140	Wu et al. (2012)
	nZVI	As(III) = 36; As(V) = 29	Wang et al. (2014)
	Zirconia dioxide nanocomposites	As(III) = 95; As(V) = 85	Luo et al. (2013)
	Multiwalled CNTs	As(III) = 1.7; As(V) = 0.2	Ntim and Mitra (2012)
Nitrate	Carbon nanosheets	140	Tofighy and Mohammadi (2012)
Phosphate	Graphene oxide nanocomposite	18	Zong et al. (2013)
	Aluminium oxide nanocomposite	60	Wu et al. (2014)

ENM represent engineered nanomaterials used in different treatment

et al. 2017). The integration of magnetic nanoparticles and CNTs reflects the removal of inorganic metal ions (Huang et al. 2019; Cao et al. 2020) (Table 22.2).

The adsorption of CNT depends upon the functional nature of adsorbate such as the presence of phenolic, carboxylic and lactone groups which increase the rate of adsorption and are proved to be efficient in wastewater treatment (Ma et al. 2011).

The CNT has proved their efficiency in the removal of hydrocarbon components from petroleum and water in several higher magnitudes than commercial polycarbonate membranes and therefore widely used in wastewater purification. Hybridization of calcium alginate with carbon nanotubes had increased the removal of Cu^{+2} from solution (Singh et al. 2012; Mubarak et al. 2014).

Earlier studies had revealed the role of carbon-based nanoparticles in the removal of organic pollutants (Yu et al. 2014; Matsumura et al. 2018; Jabbari et al. 2016). Different nano network polymers had been developed to increase the rate of mineralization such as poly (ethylene) glycol-modified urethane acrylate (PUMA) which had increased bioavailability of phenanthrene molecule in water (Riaz and Park 2020).

The 3D sponge-like CNT played a significant role in the adsorption of dichlorobenzene and showed an easy recovery from the contaminated aqueous environment (Camilli et al. 2014). It was demonstrated that the presence of chlorine groups favoured the adsorption of chlorobenzene by both CNTs and graphenes (Balamurugan and Subramanian 2013; Pasti et al. 2018). However, graphene has been proved to be much more effective in comparison to CNTs owing to economic feasibility, and it provides two basal planes for adsorption of contaminants (Shan et al. 2009).

Another aspect of wastewater treatment is the removal of heavy metals in which different magnetic nanoparticles such as magnetite, maghemite and iron oxide particles play the significant task of remediation (Chen et al. 2011).

Recently, innovations have been made over the development of membranes impinged with nanomaterials for the treatment of wastewater coming out from industrial and agricultural wastes. Different types of filtration processes such as nanofiltration, ultrafiltration and reverse osmosis are performed for the treatment of wastewater which involves purification of water and removal of the organic contaminants (Braeken et al. 2006; Frank et al. 2002). Several fabricated alumina-coated ultrafiltration membranes had been developed for the removal of synthetic dyes. Ericsson et al. (1996) demonstrated that nanofiltration could effectively remove organic matter and different microbiota such as viruses, bacteria, etc. (Ericsson et al. 1996; ven der Bruggen and Vandecasteele 2003). The ultralow pressure reverse osmosis process coupled nanofiltered membranes endowed with hydrophilic and hydrophobic ultra-filtration membranes and is highly used for desalination of brackish water. Such innovations also proved to be useful for the removal of organic and inorganic contaminants (Hassan et al. 2000; Ozaki et al. 2000). However, supplementation with dendrimers enhanced the removal of different metal ions along with complex organics at low cost in comparison to conventional ion exchangers (Christen 2004). Nanofilter membrane is widely used to filter the outlet stream of the food processing industry. Such industries release dense slurry-type water enriched with starch materials. Nanofiltration can filter out the dextrose from other oligosaccharides and thereby prepare purified glucose solution.

Although different opinions exist on the remediation mechanism of CNT and nanofiltration which are mainly based on electrostatic interactions, concentration gradient and pressure difference over charged membrane, both the technologies work hand in hand for qualitative enrichment of wastewater (Andrade et al. 2015; Frank et al. 2002). The elimination of contaminants (both organic and inorganic) is determined by properties of membranes which include porosity, membrane charges, etc. along with properties of contaminants such as hydrophobicity, ionization potential, etc. which altogether strengthen the operating conditions of membrane system (Riaz and Park 2020; Wei et al. 2014).

22.4 Nanomaterials in Solid Waste Management

Valourization of wastes plays an immense role in the conversion of wastes into value-added products. Several technological innovations have been made to undermine wastes and their toxicity to the surrounding environment (Ganguly and Chakraborty 2020b). Therefore, several bioconversion methods have been adopted to decrease the environmental burden of several organic wastes and to increase potentially value-added products.

Among different mechanical processes, enzymatic or catalytic breakdown of wastes are supposed to have played an immense role as a process of waste abatement (Ezeilo et al. 2017; Ganguly and Chakraborty 2018). But it suffers from shorter lifetime owing to their higher rate of oxidation; rendering catalyst inactive after a certain time.

Therefore, enzymes are immobilized over inert support to improve enzyme stability, recovery and inhibition (Cipolatti et al. 2016; Chen et al. 2017). In such a context, nanomaterials are thought to render better roles such as gold nanoparticles-silica nanocomposite serves as a novel enzyme immobilization matrix owing to its higher surface area, biocompatibility and electrical conductivity (Pingarrón et al. 2008; Thangaraj and Solomon 2019). Significance relies on the development of polyamidoamide (PMAM) dendrimers with cobalt hexacyanoferrate-modified gold nanoparticles which alternate with polyvinyl sulfonic acid layers on indium tin oxide electrodes and immobilize glucose oxidase which plays a contributing role in wastewater treatment (Crespilho et al. 2006; Akin et al. 2010). Such innovations face the question of the reusability of an enzyme. Therefore, magnetic iron-based nanoparticles are used with the enzyme for better separation of reactants and products (Yazid et al. 2016; Vaghari et al. 2016). Such nanoparticles proved to be very useful in facillitating the functioning of enzymes such as trypsin and peroxides which render them much more stable, economical and efficient.

Another important aspect involves the removal of toxic and carcinogenic dye pollutants using several technologies such as photocatalytic degradation, adsorption, etc. which help remediation of environmental contaminants. Owing to the large surface area, efficiency of nanoparticles such as Fe_3O_4 can be enhanced in adsorption process and thereby can undertake in better way for removal of toxic dye which pose a high adsorption range for acid green, crocein orange G, etc. (Salamat et al. 2017; Ahmed et al. 2013). Nanoscale zerovalent iron (NZVI), with doses ranging between 0.16 and 0.33 g/L, has revealed an adsorptive capability of

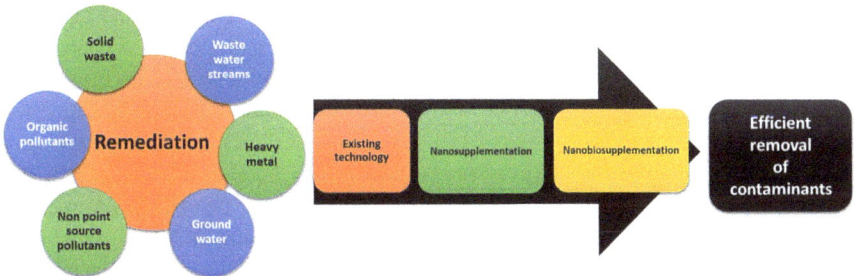

Fig. 22.3 Represent the applicative role of nanotechnology and its advances in remediation of environmental contaminants

609.4 mg/g for C.I. Acid Black 24 (pH 4–9; Initial conc. 25–100 mg/L; time 15–30 min) (Shu et al. 2007). Adsorbent prepared with nano-chitosan can be utilized to remove Acid Green 27 (Ragab et al. 2019).

The incineration process plays an important role in the debasement of solid wastes as it involves a greater rate of degradation and minimal landfilling and performed as a potential source of energy (Ganguly and Chakraborty 2020a). But such techniques release toxic dioxin and severely contaminate the surrounding environment. Therefore, the admixture of wastes along with nanoparticles proved to be much more efficient and eco-friendly (Fig. 22.3).

Nanoparticles exhibit a property of photocatalytic degradation and by the process can be a part of effective measure in the remediation of dyes (Adeleke et al. 2018; Raliya et al. 2017; Ahmed et al. 2013; Roy et al. 2022). As for example, the roles relating to the photocatalytic degradation of Acid Orange 7, Reactive Orange 16, Malachite green, etc. using nanoparticles prepared from TiO_2.

Upon UV stimulation, nano-TiO_2 harvests electron-hole pairs which led to the development of photodegradable TiO_2 nanocomposite (Lin et al. 2017; Nazarpour Laghani and Ebrahimian Pirbazari 2017). For the augmentation of work efficiency, modifications of TiO_2 have been installed with the addition of polystyrene or polyvinyl chloride support which enhanced the adequate dispersion and support eco-friendly management of wastes. Such utilization of polystyrene and polyvinyl chloride also reduces the detrimental effect of such pollutant or "white pollution." The photocatalytic degradation of PVC/TiO_2 nanohybrid proved to be a good alternative to landfilling (Yang et al. 2010).

22.5 Integration of Nanotechnology with Bioremediation

Another aspect of remediation is the utilization of different plants, animals or microbiota for the eradication of contaminants which is termed as bioremediation. Manipulation of wastes with biological organisms results in a greater reduction of both types of organic and inorganic pollutants (Murakami et al. 2006; Prasad et al. 2021). Nowadays, several attempts have been made with the integration of nanomaterials and bioremediation technologies for the effective elimination of contaminants. Nanomaterials enhance the process of bioremediation as they facilitate microbial growth or upregulate different microbial enzymes which promotes the process of elimination of different organic pollutants (Prasad and Aranda 2018; Borah et al. 2022).

However, such integration gives promising results upon proper environmental conditions, nature of pollutants and type of nanoparticles. Nano-bio-treatment of nZVI with *Sphingomonas* sp. was able to show good results in the elimination of polybrominated diphenyl ethers (Kim et al. 2012). The bacterial group showed promising growth in nanomaterial solution and therefore is very helpful in the remediation of contaminants from polluted sites. Such preparation along with the addition of biosurfactant and electrokinetics elements have been successful in the remediation of nitrate ions, heavy metals, polychlorinated biphenyls and organic chlorines (Fan et al. 2016).

Bioremediation studies have demonstrated that bacteria and plants are capable of immobilizing metals which transformed both organic and inorganic contaminants. During recent years, there are some promising positive results of the combined use of NMs and bioremediation technologies to eliminate contaminants from the environment. Nanoparticles prepared from Pd/nFe and in conjunction with *Burkholderia xenovorans* were used in dehalogenation of Aroclor 1248 (Le et al. 2015). Earlier studies had revealed its efficiency in terms of reduction of toxicity of polychlorinated benzene compounds (Bhattacharya et al. 2016; De Lima et al. 2012). However, the efficiency of such nano-compounds suffers some limitations as several humic acids undergo competition with the pollutants for surface binding sites of nZVI. Similar degradation of chlorophenols was noticed using *R. rhodochrous* immobilized with magnetic nanoparticles (Hou et al. 2013). It involves the bioconversion of compounds into chlorocatechols through the upregulation of microbial genes Cat A, Cat B and Cat C, and it also enhances the degradation of aniline compounds (Matsumura et al. 2018).

Such a nano-combined bioremediation of polychlorinated benzenes had been noticed for CNTs with *Arthrobacter* sp. However, a high concentration of CNT prevents the rate of biodegradation and the growth of microbiota, but low concentration favours microbial growth and expression of different bioremediation genes. The polyvinylpyrrolidone-coated iron oxide NPs in the combination of *Halomonas* sp. demonstrate remediation of heavy metals such as Cu^{+2} and Pb^{+2} (Cao et al. 2020). CNTs also favour the breakdown of azo dyes in the presence of electron-withdrawing groups such as sulphanilamide which facilitate the breakdown of the azo bond.

Although nano-bio-remediation plays an important role in the context of elimination of contaminants, the application of such materials in the natural environment should be made under proper surveillance. There are several shreds of evidence depicting that nanomaterials do not support bioaugmentation; it decreases the diversity of natural microbes and reduces the natural enzymes present in the environment. However, several instances support such change in the early phase of treatment which again regains in later phases due to the resilience property of the ecosystem. Thus inadvertent utilization of nanomaterials should be checked to maintain a sustainable environment and ecosystem health.

22.6 Risk Assessment

Several regulations based on the monitoring of the scope of applicabilities and efficacies of nanoparticles by different authorities such as Registration, Evaluation, Authorisation and Restriction of Chemicals (REACH; Europe) and Toxic Substances Control Act (TSCA; USA) have been undertaken to prevent unintentionally or overuse of nanomaterials (Brar et al. 2010). For example, the production of CNTs has been limited or under termination because of their acute toxicity to the environment. The toxic potential of nanomaterials depends on the nature of nanoproducts. Many research studies have advocated with regard to the reduction of the persistence of nanomaterials in the process of any biological or physical transformation in the natural environment. Such release of nanoparticles takes place during a different phase of waste management processes such as the collection and recycling of solid waste, waste incineration and land disposal (Ganguly and Chakraborty 2020a). Therefore, several recommendations have been made regarding risk assessments of nanoparticles based on the rate of exposure and hazard potential to prevent the overproduction of nanoproducts as they impart severe harm to flora and fauna of the environment (Fig. 22.4). Nanoparticles such as multiwalled CNTs, zinc oxide, zinc, etc. inhibit root development and germination of seeds in lettuce, cucumber, radish, etc. (Lin and Xing 2008). Such phytotoxic effects were much more prominent for nonfunctional CNTs. The bioavailable concentration of Cu^{+2} nanoparticles shows toxicity to wheat plants (*Triticum* sp.) and mung bean plants (*Phaseolus* sp.) (Lee et al. 2008). However, earlier studies had revealed a shift in soil microbiome structure upon contamination with silver, gold and Cu nanoparticles. Owing to the nano-dimension, such particles become immobilize as they enter the minute spaces

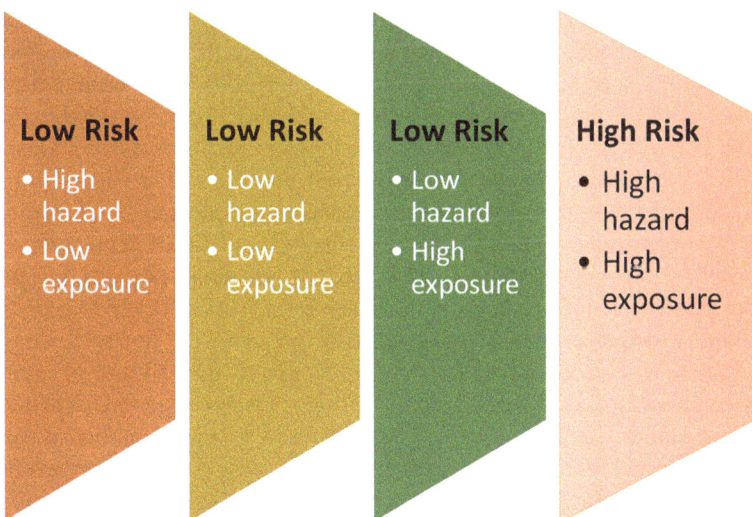

Fig. 22.4 Risk assessment of nanomaterials upon exposure to environment

between soil particles and get embedded within the soil matrix. The extent of sorption depends upon the nature of nanoparticles and soil conditions. Other instances such as contamination of graphenes in soil diminish the number of fast-growing bacteria and soil arthropod resulting to alteration of the ecosystem. Organic matter such as the presence of different humic acids interact with the nanoparticles and modify the kinetics of sorption. Similarly, the presence of high organic load also favours the sorption kinetics of nanomaterials among sludge water. The earlier studies had reported the role of organic carbon in the sorption potential of nanoparticles (Ahmad et al. 2001; Gunasekara and Xing 2003).

Nowadays, nanomaterials are engineered to produce different products in order to enhance the qualitative enrichment of the products and are termed as engineered nanomaterials (ENM). Such high usefulness of products results in the generation of contaminants in the nano-dimension and is often termed as "nanowaste". Such leaching of nanoparticles occurs from the manufacturing process of nanoproducts, nonfunctional or end-of-life nanoproducts and waste materials contaminated with NPs. Owing to their dimension, surface area, electrokinetic properties, solubility, etc., their exposure poses toxicity to the natural environment, and the future consequences are presently impulsive. Therefore, different ecotoxicological researches are required to predict the fate of such molecules for the development of a sustainable environment. Hence, production and reuse of nanomaterials should be done concerning ecosystem health so that hazardous effects can be mitigated.

22.7 Recent Trends and Future Outlook

Nanotechnology including application of nanomaterials along with some innovations played a pivotal role in context of waste management. Several considerable research developments have been amended which are based upon integration of nanotechnology specifically nanobiotechnology for effective remediation of contaminants with an aim to maintain ecosystem health. Nanobiotechnology is an evolving technology which involves the development of different nanofluids, nanoproteomics, nanobiomembrane, nanobiosensor, etc. For better functionalization, special efforts are constantly being made over coating procedures for better impregnation of nanoparticles (NPs). Progress is observed in field of application of biomolecules for the improvement of nanoproduct functionalization. Utilization of amyloid protein or milk proteins in fibrillary form help in better remediation of metal ions. Membrane coating using cellulose polymers are prepared along with implementation of bacterial species proved to be a better tool in removal of environmental contaminants. Different features of living organisms are used for better functioning of nano-assisted membranes. Several forms of mixed matrix membranes such as nanoparticle entrapped membranes are now being regularly used in effective removal of industrial contaminants. For example, utilization of different types of lipid biomembranes for effective removal of metal ions and different organic contaminants. Development of aquaporin-based biomimetic membranes functions better in comparison to conventional membranes through enhanced reverse osmosis and elaborately used in desalination applications. Furthermore, development of nanofibrous webs through electrospinning technology along with microbial supplementation plays a pivotal role in removal of wastewater contaminants (Wendorff et al. 2012). Such technology is highly efficient over existing nanotechnology owing to better porosity, surface area and economic feasibility. Engineering of microbial fuel cell (MFC) technology with microbial population and nanoparticles such as CNTs are also playing a promising role in remediation of organic or inorganic contaminants and help in derivation of energy (Hou et al. 2016). Therefore, different innovations have been made for better removal of environmental pollutants through nanotechnology. Although development of nanobiotechnology alongwith the use of nanofibers have been rendering valuable services to remediate pollutants, disposal of such nanocompounds further increases the concern towards "nanowaste". However, adequate ecotoxicological research should be done for the utilization of such products, and their adequate disposal through proper planning and execution will lead to the maintenance of ecological integrity.

22.8 Conclusion

Different types of nanoparticles along with its differential utilization in different forms such as nanocoated membrane, photocatalytic degradation, etc. play a very significant role in the context of waste management. These minuscule particles play a major role in the remediation of different organic and inorganic contaminants from the environment. However, in the milieu of waste treatment such as water streams or in solid waste treatment, these particles get admixed to our natural environment and sometimes become detrimental roles threatening the structure and functioning of the ecosystem health. Therefore, overproduction and overutilization can impart negative effects which marked it as a "necessary evil". Therefore, utilization of nanoparticles demands the development of some comprehensive ideas and ecotoxicological researches so that one can pursue research for the generation of qualitative products to achieve the target of having a greener environment in the foreseeable future.

Acknowledgement The study acknowledges Vidyasagar University in supporting all sorts of library facilities.

References

Adeleke JT, Theivasanthi T, Thiruppathi M, Swaminathan M, Akomolafe T, Alabi AB (2018) Photocatalytic degradation of methylene blue by ZnO/NiFe$_2$O$_4$ nanoparticles. Appl Surf Sci 455:195–200

Ahmad R, Kookana RS, Alston AM, Skjemstad JO (2001) The nature of soil organic matter affects sorption of pesticides. 1. Relationships with carbon chemistry as determined by 13C CPMAS NMR spectroscopy. Environ Sci Technol 35:878–884

Ahmed MA, El-Katori EE, Gharni ZH (2013) Photocatalytic degradation of methylene blue dye using Fe_2O_3/TiO_2 nanoparticles prepared by sol–gel method. J Alloys Compd 553:19–29

Akin M, Yuksel M, Geyik C, Odaci D, Bluma A, Höpfner T, Beutel S, Scheper T, Timur S (2010) Alcohol biosensing by polyamidoamine (PAMAM)/cysteamine/alcohol oxidase-modified gold electrode. Biotechnol Prog 26(3):896–906

Andrade PF, de Faria AF, Oliveira SR, Arruda MAZ, do Carmo Gonçalves M (2015) Improved antibacterial activity of nanofiltration polysulfone membranes modified with silver nanoparticles. Water Res 81:333–342

Balamurugan K, Subramanian V (2013) Adsorption of chlorobenzene onto (5,5) armchair single-walled carbon nanotube and graphene sheet: toxicity versus adsorption strength. J Phys Chem C 117(41):21217–21227

Baruah S, Najam Khan M, Dutta J (2016) Perspectives and applications of nanotechnology in water treatment. Environ Chem Lett 14(1):1–14

Bhattacharya K, Mukherjee SP, Gallud A, Burkert SC, Bistarelli S, Bellucci S, Bottini M, Star A, Fadeel B (2016) Biological interactions of carbon-based nanomaterials: from coronation to degradation. Nanomedicine 12:333–351

Bhushan B (ed) (2017) Springer handbook of nanotechnology. Springer

Boldrin A, Hansen SF, Baun A, Hartmann NIB, Astrup TF (2014) Environmental exposure assessment framework for nanoparticles in solid waste. J Nanopart Res 16(6):2394

Bolyard S, Reinhart D, Santra S (2013) Behavior of engineered nanoparticles in landfill leachate. Environ Sci Technol 47(15):8114–8122. https://doi.org/10.1021/es305175e

Borah SN, Koch N, Sen S, Prasad R, Sarma H (2022) Novel nanomaterials for nanobioremediation of polyaromatic hydrocarbons. In: Sharma H, Dominguez DC, Lee W-Y (eds) Emerging Contaminants in the Environment. Elsevier 643–667

Braeken L, Bettens B, Boussu K, Van der Meeren P, Cocquyt J, Vermant J, van der Bruggen B (2006) Transport mechanisms of dissolved organic compounds in aqueous solution during nanofiltration. J Membr Sci 279:311–319

Brar SK, Verma M, Tyagi RD, Surampalli RY (2010) Engineered nanoparticles in wastewater and wastewater sludge–evidence and impacts. Waste Manag 30(3):504–520

Camilli L, Pisani C, Gautron E, Scarselli M, Castrucci P, D'Orazio F, Passacantando M, Moscone D, De Crescenzi M (2014) A three-dimensional carbon nanotube network for water treatment. Nanotechnology 25(6):065701

Cao X, Alabresm A, Chen YP, Decho AW, Lead J (2020) Improved metal remediation using a combined bacterial and nanoscience approach. Sci Total Environ 704:135378

Celebi O, Uezuem C, Shahwan T, Erten HN (2007) A radiotracer study of the adsorption behavior of aqueous Ba^{2+} ions on nanoparticles of zero-valent iron. J Hazard Mater 148(3):761–767

Chen H, He J (2008) Facile synthesis of monodisperse manganese oxide nanostructures and their application in water treatment. J Phys Chem C 112(45):17540–17545

Chen L, Huang Y, Huang L, Liu B, Wang G, Yu S (2011) Characterization of Co(II) removal from aqueous solution using bentonite/iron oxide magnetic composites. J Radioanal Nucl Chem 290(3):675–684

Chen M, Zeng G, Xu P, Lai C, Tang L (2017) How do enzymes "meet" nanoparticles and nanomaterials? Trends Biochem Sci 42(11):914–930

Cho S, Jang J-W, Hwang S, Lee JS, Kim S (2012) Self-assembled gold nanoparticle–mixed metal oxide nanocomposites for self-sensitized dye degradation under visible light irradiation. Langmuir 28(50):17530–17536

Christen K (2004) Novel nanomaterial strips contaminants from waste streams. Environ Sci Technol 38(23):453A–454A

Cipolatti EP, Valerio A, Henriques RO, Moritz DE, Ninow JL, Freire DM, Manoel EA, Fernandez-Lafuente R, de Oliveira D (2016) Nanomaterials for biocatalyst immobilization–state of the art and future trends. RSC Adv 6(106):104675–104692

Crane RA, Pullin H, Macfarlane J, Silion M, Popescu IC, Andersen M, Calen V, Scott TB (2015) Field application of iron and iron–nickel nanoparticles for the ex situ remediation of a uranium-bearing mine water effluent. J Environ Eng 141(8):04015011

Crespilho FN, Ghica ME, Florescu M, Nart FC, Oliveira ON Jr, Brett CM (2006) A strategy for enzyme immobilization on layer-by-layer dendrimer–gold nanoparticle electrocatalytic membrane incorporating redox mediator. Electrochem Commun 8(10):1665–1670

Dasgupta N, Ranjan S, Ramalingam C (2017) Applications of nanotechnology in agriculture and water quality management. Environ Chem Lett 15(4):591–605

De Lima R, Seabra AB, Durán N (2012) Silver nanoparticles: a brief review of cytotoxicity and genotoxicity of chemically and biogenically synthesized nanoparticles. J Appl Toxicol 32:867–879

Deng J-H, Zhang X-R, Zeng G-M, Gong J-L, Niu Q-Y, Liang J (2013) Simultaneous removal of Cd(II) and ionic dyes from aqueous solution using magnetic graphene oxide nanocomposite as an adsorbent. Chem Eng J 226:189–200

Ericsson B, Hallberg M, Wachenfeldt J (1996) Nanofiltration of highly colored raw water for drinking water production. Desalination 108(1–3):129–141

Ezeilo UR, Zakaria II, Huyop F, Wahab RA (2017) Enzymatic breakdown of lignocellulosic biomass: the role of glycosyl hydrolases and lytic polysaccharide monooxygenases. Biotechnol Biotechnol Equip 31(4):647–662

Fan L, Luo C, Sun M, Qiu H, Li X (2013) Synthesis of magnetic β-cyclodextrin–chitosan/graphene oxide as nanoadsorbent and its application in dye adsorption and removal. Colloids Surf B Biointerfaces 103:601–607

Fan G, Wang Y, Fang G, Zhu X, Zhou D (2016) Review of chemical and electrokinetic remediation of PCBs contaminated soils and sediments. Environ Sci Process Impacts 18:1140–1156

Frank MJW, Westerink JB, Schokker A (2002) Recycling of industrial waste water by using a two-step nanofiltration process for the removal of colour. Desalination 145:69–74

Ganguly RK, Chakraborty SK (2018) Assessment of microbial roles in the bioconversion of paper mill sludge through vermicomposting. J Environ Health Sci Eng 16:205–212

Ganguly RK, Chakraborty SK (2020a) Paper pulp mill wastes: a curse or boon – modern approach of recycling. In: Hussain C (ed) Handbook of environmental materials management. Springer, Cham, p 2020. https://doi.org/10.1007/978-3-319-58538-3_216-1

Ganguly RK, Chakraborty SK (2020b) Eco-management of industrial organic wastes through the modified innovative vermicomposting process: a sustainable approach in tropical countries. In: Bhat S, Vig A, Li F, Ravindran B (eds) Earthworm assisted remediation of effluents and wastes. Springer, Singapore, pp 161–177. https://doi.org/10.1007/978-981-15-4522-1_10

Girginova PI, Daniel-da-Silva AL, Lopes CB, Figueira P, Otero M, Amaral VS, Pereira E, Trindade T (2010) Silica coated magnetite particles for magnetic removal of Hg2+ from water. J Colloid Interface Sci 345(2):234–240

Goutam SP, Saxena G, Singh V, Yadav AK, Bharagava RN, Thapa KB (2018) Green synthesis of TiO2 nanoparticles using leaf extract of Jatropha curcas L for photocatalytic degradation of tannery wastewater. Chem Eng J 336:386–396

Gunasekara AS, Xing B (2003) Sorption and desorption of naphthalene by soil organic matter: importance of aromatic and aliphatic components. J Environ Qual 32:240–246

Han F, Kambala VSR, Srinivasan M, Rajarathnam D, Naidu R (2009) Tailored titanium dioxide photocatalysts for the degradation of organic dyes in wastewater treatment: a review. Appl Catal A Gen 359(1–2):25–40

Hassan AM, Farooque AM, Jamaluddin ATM, Al-Amoudi AS, Al-Sofi MAK, Al-Rubaian AF, Kither NM, Al-Tisan IAR, Rowaili A (2000) A demonstration plant based on the new NF-SWRO process. Desalination 131:157–171

Hou L, Zhu D, Wang X, Wang L, Zhang C, Chen W (2013) Adsorption of phenanthrene, 2-naphthol, and 1-naphthylamine to colloidal oxidized multiwalled carbon nanotubes: effects of humic acid and surfactant modification. Environ Toxicol Chem 32(3):493–500

Hou J, Liu F, Wu N, Ju J, Yu B (2016) Efficient biodegradation of chlorophenols in aqueous phase by magnetically immobilized aniline-degrading *Rhodococcus rhodochrous* strain. J Nanobiotechnol 14:5

Hua M, Jiang Y, Wu B, Pan B, Zhao X, Zhang Q (2013) Fabrication of a new hydrous Zr(IV) oxide-based nanocomposite for enhanced Pb(II) and Cd(II) removal from waters. ACS Appl Mater Interfaces 5(22):12135–12142

Huang D, Wu J, Wang L, Liu X, Meng J, Tang X, Tang C, Xu J (2019) Novel insight into adsorption and co-adsorption of heavy metal ions and an organic pollutant by magnetic graphene nanomaterials in water. Chem Eng J 358:1399–1409

Jabbari V, Veleta JM, Zarei-Chaleshtori M, Gardea-Torresdey J, Villagrán D (2016) Green synthesis of magnetic MOF@ GO and MOF@ CNT hybrid nanocomposites with high adsorption capacity towards organic pollutants. Chem Eng J 304:774–783

Jiang L, Liu Y, Liu S, Zeng G, Hu X, Hu X, Guo Z, Tan X, Wang L, Wu Z (2017) Adsorption of estrogen contaminants by graphene nanomaterials under natural organic matter preloading: comparison to carbon nanotube, biochar, and activated carbon. Environ Sci Technol 51(11):6352–6359

Khezamia L, Tahaa KK, Amamic E, Ghiloufid I, El Mird L (2017) Removal of cadmium(II) from aqueous solution by zinc oxide nanoparticles: kinetic and thermodynamic studies. Desalin Water Treat 62:346–354

Kim DS, Kwak S-Y (2007) The hydrothermal synthesis of mesoporous TiO2 with high crystallinity, thermal stability, large surface area, and enhanced photocatalytic activity. Appl Catal A Gen 323:110–118

Kim YM, Murugesan K, Chang YY, Kim EJ, Chang YS (2012) Degradation of polybrominated diphenyl ethers by a sequential treatment with nanoscale zero valent iron and aerobic biodegradation. J Chem Technol Biotechnol 240:525–532

Le TT, Nguyen KH, Jeon JR, Francis AJ, Chang YS (2015) Nano/bio treatment of polychlorinated biphenyls with evaluation of comparative toxicity. J Hazard Mater 287:335–341

Lee W-M, An Y-J, Yoon H, Kweon H-S (2008) Toxicity and bioavailability of copper nanoparticles to the terrestrial plants mung bean (Phaseolus radiatus) and wheat (Triticum aestivum): plant agar test for water-insoluble nanoparticles. Environ Toxicol Chem 27(9):1915–1921

Li X-Q, Brown DG, Zhang WX (2007) Stabilization of biosolids with nanoscale zero-valent iron (nZVI). J Nanopart Res 9(2):233–243

Li H, Qiu Y-F, Wang X-L, Yang J, Yu Y-J, Chen Y-Q (2017) Biochar supported Ni/Fe bimetallic nanoparticles to remove 1, 1,1-trichloroethane under various reaction conditions. Chemosphere 169:534–541

Lin D, Xing B (2008) Root uptake and phytotoxicity of ZnO nanoparticles. Environ Sci Technol 42:5580–5585

Lin LY, Nie Y, Kavadiya S, Soundappan T, Biswas P (2017) N-doped reduced graphene oxide promoted nano TiO_2 as a bifunctional adsorbent/photocatalyst for CO2 photoreduction: effect of N species. Chem Eng J 316:449–460

Lu CS, Chung YL, Chang KF (2005) Adsorption of trihalomethanes from water with carbon nanotubes. Water Res 39(6):1183–1189

Luo X, Wang C, Wang L, Deng F, Luo S, Tu X, Au C (2013) Nanocomposites of graphene oxide-hydrated zirconium oxide for simultaneous removal of As (III) and As (V) from water. Chem Eng J 220:98–106

Ma X, Anand D, Zhang X, Talapatra S (2011) Adsorption and desorption of chlorinated compounds from pristine and thermally treated multiwalled carbon nanotubes. J Phys Chem C 115(11):4552–4557

Madhura L, Kanchi S, Sabela MI, Singh S, Bisetty K, Inamuddin (2018) Membrane technology for water purification. Environ Chem Lett 16(2):343–365

Madhura L, Singh S, Kanchi S, Sabela M, Bisetty K (2019) Nanotechnology-based water quality management for wastewater treatment. Environ Chem Lett 17(1):65–121

Matsumura E, Sakai M, Hayashi K, Mavukkandy MO, Zaib Q, Arafat HA (2018) CNT/PVP blend PVDF membranes for the removal of organic pollutants from simulated treated wastewater effluent. J Environ Chem Eng 6(5):6733–6740

Mishra A, Mehta A, Sharma M, Basu S (2017) Impact of Ag nanoparticles on photomineralization of chlorobenzene by TiO_2/bentonite nanocomposite. J Environ Chem Eng 5(1):644–651

Moradi O, Yari M, Moaveni P, Norouzi M (2012) Removal of p-nitrophenol and naphthalene from petrochemical wastewater using SWCNTs and SWCNT-COOH surfaces. Fuller Nanotube Carbon Nanostruct 20(1):85–98

Mubarak NM, Sahu JN, Abdullah EC, Jayakumar NS (2014) Removal of heavy metals from wastewater using carbon nanotubes. Sep Purif Rev 43(4):311–338

Muhammad S, Shukla PR, Tadé MO, Wang S (2012) Heterogeneous activation of peroxymonosulphate by supported ruthenium catalysts for phenol degradation in water. J Hazard Mater 215:183–190

Murakami S, Takenaka S, Aoki K (2006) Constitutive expression of cat ABC genes in the aniline-assimilating bacterium *Rhodococcus* species AN-22: production, purification, characterization and gene analysis of CatA, CatB and CatC. Biochem J 393:219–226

Nazarpour Laghani S, Ebrahimian Pirbazari A (2017) Photocatalytic treatment of synthetic wastewater containing 2, 4 dichlorophenol by ternary MWCNTs/Co-TiO2 nanocomposite under visible light. J Water Environ Nanotechnol 2(4):290–301

Ntim SA, Mitra S (2012) Adsorption of arsenic on multiwall carbon nanotube–zirconia nanohybrid for potential drinking water purification. J Colloid Interface Sci 375(1):154–159

Ozaki H, Sharma K, Saktaywin W, Wang D, Yu Y (2000) Application of ultra low pressure reverse osmoisi membrane to water and wastewater. Water Sci Technol 42(12):123–135

Pasti L, Rodeghero E, Beltrami G, Ardit M, Sarti E, Chenet T, Stevanin C, Martucci A (2018) Insights into adsorption of chlorobenzene in high silica MFI and FAU zeolites gained from chromatographic and diffractometric techniques. Fortschr Mineral 8(3):80

Paul DR, Robeson LM (2008) Polymer nanotechnology: nanocomposites. Polymer 49(15):3187–3204

Peiro AM, Ayllon JA, Peral J, Domenech X (2001) TiO2- photocatalyzed degradation of phenol and ortho-substituted phenolic compounds. Appl Catal B Environ 30(3–4):359–373

Peng X, Li Y, Luan Z, Di Z, Wang H, Tian B, Jia Z (2003) Adsorption of 1,2-dichlorobenzene from water to carbon nanotubes. Chem Phys Lett 376(1–2):154–158

Pingarrón JM, Yanez-Sedeno P, González-Cortés A (2008) Gold nanoparticle-based electrochemical biosensors. Electrochim Acta 53(19):5848–5866

Poole CP Jr, Owens FJ (2003) Introduction to nanotechnology. John Wiley & Sons

Prasad R, Aranda E (2018) Approaches in Bioremediation. Springer International Publishing https://www.springer.com/de/book/9783030023683

Prasad R, Thirugnanasanbandham K (2019) Advances Research on Nanotechnology for Water Technology. Springer International Publishing. https://www.springer.com/us/book/9783030023805

Prasad R, Nayak SC, Kharwar RN, Dubey NK (2021) Mycoremediation and Environmental Sustainability, Volume 3. Springer International Publishing (ISBN: 978-3-030-54421-8) https://www.springer.com/gp/book/9783030544218

Ragab A, Ahmed I, Bader D (2019) The removal of brilliant green dye from aqueous solution using nano hydroxyapatite/chitosan composite as a sorbent. Molecules 24(5):847

Raliya R, Avery C, Chakrabarti S, Biswas P (2017) Photocatalytic degradation of methyl orange dye by pristine titanium dioxide, zinc oxide, and graphene oxide nanostructures and their composites under visible light irradiation. Appl Nanosci 7(5):253–259

Ren Y, Li N, Feng J, Luan T, Wen Q, Li Z, Zhang M (2012) Adsorption of Pb(II) and Cu(II) from aqueous solution on magnetic porous ferrospinel MnFe2O4. J Colloid Interface Sci 367(1):415–421

Riaz S, Park SJ (2020) An overview of TiO_2-based photocatalytic membrane reactors for water and wastewater treatments. J Ind Eng Chem 84:23–41

Rizwan M, Singh M, Mitra CK, Morve RK (2014) Ecofriendly application of nanomaterials: nanobioremediation. J Nanoparticles 2014:431787

Roy A, Ananda Murthy HC, Ahmed H, Islam MN, Prasad R (2022) Phytogenic nanoparticles for degradation of dyes. Journal of Renewable Materials. https://doi.org/10.32604/jrm.2022.019410

Salamat S, Younesi H, Bahramifar N (2017) Synthesis of magnetic core–shell Fe3O4@ TiO2 nanoparticles from electric arc furnace dust for photocatalytic degradation of steel mill wastewater. RSC Adv 7(31):19391–19405

Saputra E, Muhammad S, Sun HQ, Ang HM, Tade MO, Wang SB (2013) A comparative study of spinel structured Mn3O4, Co3O4 and Fe3O4 nanoparticles in catalytic oxidation of phenolic contaminants in aqueous solutions. J Colloid Interface Sci 407:467–473

Shan G, Surampalli RY, Tyagi RD, Zhang TC (2009) Nanomaterials for environmental burden reduction, waste treatment, and nonpoint source pollution control: a review. Front Environ Sci Eng China 3(3):249–264

Shi X, Ruan W, Hu J, Fan M, Cao R, Wei X (2017) Optimizing the removal of rhodamine B in aqueous solutions by reduced graphene oxide-supported nanoscale zerovalent iron (NZVI/RGO) using an artificial neural network-genetic algorithm (ANN-GA). Nanomaterials 7(6):134

Shu HY, Chang MC, Yu HH, Chen WH (2007) Reduction of an azo dye Acid Black 24 solution using synthesized nanoscale zerovalent iron particles. J Colloid Interface Sci 314(1):89–97

Singh L, Pavankumar AR, Lakshmanan R, Rajarao GK (2012) Effective removal of Cu^{2+} ions from aqueous medium using alginate as biosorbent. Ecol Eng 38(1):119–124

Singh R, Behera M, Kumar S (2020) Nano-bioremediation: an innovative remediation technology for treatment and management of contaminated sites. In: Bioremediation of industrial waste for environmental safety. Springer, Singapore, pp 165–182

Suo Z, Dong X, Liu H (2013) Single-crystal-like NiO colloidal nanocrystal-aggregated microspheres with mesoporous structure: synthesis and enhanced electrochemistry, photocatalysis and water treatment properties. J Solid State Chem 206:1–8

Thangaraj B, Solomon PR (2019) Immobilization of lipases–a review. Part II: carrier materials. ChemBioEng Rev 6(5):167–194

Thomé A, Reddy KR, Reginatto C, Cecchin I (2015) Review of nanotechnology for soil and groundwater remediation: Brazilian perspectives. Water Air Soil Pollut 226(4):121

Tofighy MA, Mohammadi T (2012) Nitrate removal from water using functionalized carbon nanotube sheets. Chem Eng Res Des 90(11):1815–1822

Uzum C, Shahwan T, Eroglu AE, Lieberwirth I, Scott TB, Hallam KR (2008) Application of zerovalent iron nanoparticles for the removal of aqueous Co2+ ions under various experimental conditions. Chem Eng J 144(2):213–220

Vaghari H, Jafarizadeh-Malmiri H, Mohammadlou M, Berenjian A, Anarjan N, Jafari N, Nasiri S (2016) Application of magnetic nanoparticles in smart enzyme immobilization. Biotechnol Lett 38(2):223–233

Vázquez-Núñez E, Molina-Guerrero CE, Peña-Castro JM, Fernández-Luqueño F, de la Rosa-Álvarez M (2020) Use of nanotechnology for the bioremediation of contaminants: a review. PRO 8(7):826

ven der Bruggen B, Vandecasteele C (2003) Removal of pollutants from surface water and groundwater by nanofiltration: overview of possible applications in the drinking water industry. Environ Pollut 122:435–445

Vuković GD, Marinković AD, Čolić M, Ristić MĐ, Aleksić R, Perić-Grujić AA, Uskoković PS (2010) Removal of cadmium from aqueous solutions by oxidized and ethylenediamine-functionalized multi-walled carbon nanotubes. Chem Eng J 157(1):238–248

Wang Y, Cheng H, Zhang L, Hao Y, Ma J, Xu B, Li W (2000) The preparation, characterization, photoelectrochemical and photocatalytic properties of lanthanide metal-ion-doped TiO2 nanoparticles. J Mol Catal A Chem 151(1–2):205–216

Wang C, Luo H, Zhang Z, Wu Y, Zhang J, Chen S (2014) Removal of As (III) and As (V) from aqueous solutions using nanoscale zero valent iron-reduced graphite oxide modified composites. J Hazard Mater 268:124–131

Wei G, Yu H, Quan X, Chen S, Zhao H, Fan X (2014) Constructing all carbon nanotube hollow fiber membranes with improved performance in separation and antifouling for water treatment. Environ Sci Technol 48(14):8062–8068

Wendorff JH, Agarwal S, Greiner A (2012) Electrospinning: materials, processing, and applications, 1st edn. John Wiley & Sons, Weinheim

Wu K, Liu T, Xue W, Wang X (2012) Arsenic(III) oxidation/adsorption behaviors on a new bimetal adsorbent of Mn-oxide-doped Al oxide. Chem Eng J 192:343–349

Wu K, Liu T, Ma C, Chang B, Chen R, Wang X (2014) The role of Mn oxide doping in phosphate removal by Al-based bimetal oxides: adsorption behaviors and mechanisms. Environ Sci Pollut Res Int 21(1):620–630

Xiong Z, Zhang LL, Ma J, Zhao X (2010) Photocatalytic degradation of dyes over graphene–gold nanocomposites under visible light irradiation. Chem Commun 46(33):6099–6101

Yan XM, Shi BY, Lu JJ, Feng CH, Wang DS, Tang HX (2008) Adsorption and desorption of atrazine on carbon nanotubes. J Colloid Interface Sci 321(1):30–38

Yan W, Herzing AA, Kiely CJ, Zhang W-X (2010) Nanoscale zerovalent iron (nZVI): aspects of the core-shell structure and reactions with inorganic species in water. J Contam Hydrol 118(3–4):96–104

Yang C, Gong C, Peng T, Deng K, Zan L (2010) High photocatalytic degradation activity of the polyvinyl chloride (PVC)–vitamin C (VC)–TiO_2 nano-composite film. J Hazard Mater 178(1–3):152–156

Yang X, Fu H, Wong K, Jiang X, Yu A (2013) Hybrid Ag@ TiO2 core–shell nanostructures with highly enhanced photocatalytic performance. Nanotechnology 24(41):415601

Yazid NA, Barrena R, Sánchez A (2016) The immobilisation of proteases produced by SSF onto functionalized magnetic nanoparticles: application in the hydrolysis of different protein sources. J Mol Catal B Enzym 133:S230–S242

Yu JG, Zhao XH, Yang H, Chen XH, Yang Q, Yu LY, Jiang JH, Chen XQ (2014) Aqueous adsorption and removal of organic contaminants by carbon nanotubes. Sci Total Environ 482:241–251

Zhang Y, Su Y, Zhou X, Dai C, Keller AA (2013) A new insight on the core–shell structure of zerovalent iron nanoparticles and its application for Pb(II) sequestration. J Hazard Mater 263:685–693

Zhang Y, Li Y, Dai C, Zhou X, Zhang W (2014) Sequestration of Cd(II) with nanoscale zero-valent iron (nZVI): characterization and test in a two-stage system. Chem Eng J 244:218–226

Zhao F, Chen S, Hu Q, Xue G, Ni Q, Jiang Q, Qiu Y (2017) Antimicrobial three dimensional woven filters containing silver nanoparticle doped nanofibers in a membrane bioreactor for wastewater treatment. Separation and Purification Technol 175:130–139

Zhong L-S, Hu J-S, Cao A-M, Liu Q, Song W-G, Wan L-J (2007) 3D flowerlike ceria micro/nanocomposite structure and its application for water treatment and CO removal. Chem Mater 19(7):1648–1655

Zhou L, Ji L, Ma P-C, Shao Y, Zhang H, Gao W, Li Y (2014) Development of carbon nanotubes/$CoFe_2O_4$ magnetic hybrid material for removal of tetrabromobisphenol A and Pb(II). J Hazard Mater 265:104–114

Zong E, Wei D, Wan H, Zheng S, Xu Z, Zhu D (2013) Adsorptive removal of phosphate ions from aqueous solution using zirconia functionalized graphite oxide. Chem Eng J 221:193–203

Chapter 23
Applicability of Emerging Nanomaterials in Microbial Fuel Cells as Cathode Catalysts

Vikash Kumar, Prasanta Pattanayak, and Subrata Hait

Contents

23.1	Introduction..	644
23.2	Basic Principle and Architecture of Microbial Fuel Cell (MFC).........................	645
23.3	Cathode Catalysis..	647
	23.3.1 Mechanism of Oxygen Reduction Reaction (ORR).............................	647
	23.3.2 Catalyst Materials Used in MFCs..	647
23.4	Emerging Nanomaterials in Cathode Catalysis...	649
	23.4.1 Transition Metal/Metal Oxide and Alloy-Based Cathode Catalysts.....	649
	23.4.2 Metal-Carbon Hybrid Catalysts..	650
	23.4.3 Metal-Activated Carbon Hybrids..	650
	23.4.4 Metal-Carbon Nanofibers (CNFs) and Nanotubes (CNTs)...................	651
	23.4.5 Metal-Graphene-Based Nanocomposites...	651
	23.4.6 Metal-Conducting Polymer-Based Nanocomposites...........................	652
23.5	Potentiodynamic Effects of Nanomaterials...	653
23.6	Anode Modifications...	656
23.7	Conclusions...	659
References..		659

V. Kumar · S. Hait (✉)
Department of Civil and Environmental Engineering, Indian Institute of Technology Patna, Patna, Bihar, India
e-mail: shait@iitp.ac.in

P. Pattanayak
Advanced Polymer Laboratory, Department of Polymer Science & Technology, University of Calcutta, Kolkata, West Bengal, India

23.1 Introduction

Innovation in this technological era has amalgamated different streams of science as an answer to the future world. With depleting fuel resources, the urge for alternative energies, particularly in the areas of renewable energies, is continuously looking for different sustainable technologies as an alternative. One such "green" approach is microbial fuel cell (MFC), which is a microbial powered electrochemical device for bio-electricity generation. The employed microbes oxidize organic feeds at the anode and release electrons and protons that eventually travel towards the cathode (Hosseini and Ahadzadeh 2012; Han et al. 2018; Pattanayak et al. 2019). At the cathode, these incoming electrons and protons get accepted by a particular substrate of interest that finally reduces to form a product. In total, the microbes generate electrons and protons by organic feed oxidation, where cathode catalysis is used to capture the liberated electrons and protons to form a specific product from reduction. For example, oxygen reduction into water is a commonly catalysed process at the cathode that captures the incoming electron and protons in the reaction (Logan et al. 2006).

This process of catalysis is however quite complex. It deals with various aspects of recombining substrates and their feasibility at the reaction site. Substantial investigations in the areas of cathode catalysis have been conducted, where platinum (Pt) is the most commercialized catalysts used in oxygen reduction reaction (ORR). However, it lags in many aspects of prolonged applicability. Major limitations like poisoning, limited availability and high cost are some of its major setbacks (Pattanayak et al. 2020a; Kumar et al. 2018). As a substitute, wider studies on low-cost cathode catalysts have also been conducted that can efficiently conduct the oxygen reduction reactions (ORR) in the process.

In the rigor, several transition metal oxide-based catalysts such as V_2O_5, carbon-supported nickel-phthalocyanine/MnOx, GO-Zn/Co, Spinel-type Cu/Co, and Ni/Co-oxides, α-Fe2O3/polyaniline, Cu_2O/RGO and polyaniline/β-MnO_2 materials have been tested as cathode catalysts in MFCs (Noori et al. 2018; Tiwari et al. 2017; Tang and Ng 2017; Papiya et al. 2019; Li and Zhou 2018; Zhou et al. 2018; Ayyaru et al. 2019). Titanium dioxide (TiO_2) is another widely studied nanomaterial because of its ample availability, higher electro-activity, chemical durability and eco-friendliness (Pattanayak et al. 2020a). The particle shape and size of metal oxides are generally regulated by the conventional hydrothermal process, which is widely employed for their preparation. Modified carbon electrodes, such as carbon blacks, carbon fibre, carbon felt, carbon nanotubes (CNT) and carbon papers have also been duly applied in bio-electrochemical systems for wider applications (Yang et al. 2019). Recently, graphene oxide (GO) has been proved as a high-performance material because of its enhanced specific surface area and mechanical stability (Pattanayak et al. 2019; Papiya et al. 2019). Apart from that, modifications like doping metals in the electrodes, providing conductive polymer (CP) matrix to nanoparticles, etc., have also shown higher electron transfer efficiency in fuel cell operation. Therefore, it has been suggested that the incorporation of specific nanocomposites in the electrode structure or coating with efficient electro-active nanomaterials can enhance the electrochemical process in the system.

The chapter here encompasses the oxygen reduction efficiency of these nanomaterials in MFCs, majorly as a cathode catalyst. The larger surface area of nanocatalyst serves as an advantage for higher electro-catalytical activity, which ultimately improves the power generation and stability in the system (Khurmi and Sedha 2008; Kirubakaran et al. 2009). Having a prospective role in treating wastewater and generating bio-electricity in MFCs, there are several factors that require serious outlook, primarily in the domains of hydrogen yield, purity and overall efficiency of the system. There are ample opportunities particularly in the areas of electrode and catalyst development that can largely influence the electrical conductivity, stability, power density and overall cost economics of the system. In effect, designing this nanomaterial-based electrode setup offers a promising tool for enhancing the systemic efficiency, where a comprehensive illustration about how influential these nanomaterials are in practical usage is discussed in detail in this chapter. Hopefully, this will endow a broad insight into the recent advances and ongoing alterations in the areas of oxygen reduction and cathode catalysis.

23.2 Basic Principle and Architecture of Microbial Fuel Cell (MFC)

In principle, MFC can be a single- or dual-chambered system, comprising anode and cathode areas, where a proton exchange membrane (PEM) separates these two segments for individual operation. At the anode, microbial oxidation of organic feed occurs, where the employed electrogenic microbes finally liberate electrons and protons in the process (Bosch-Jimenez et al. 2017; Gnana et al. 2014). The released protons get internally transferred towards the cathode via PEM, where it meets the externally transported electron through an external circuit. A reducing agent (oxygen) takes up the incoming electron and proton in presence of a cathode catalyst and finally gets reduced into water. The overall process results in organic feed degradation and bio-electricity generation from the system (Nandy et al. 2016). This can be a single-chambered MFC with an air cathode or a dual-chambered system with anodic and cathodic compartments separated by a polymer electrolyte membrane (PEM) as a barrier (Fig. 23.1).

The whole circuit gets completed by oxygen reduction into water molecules as represented by the reaction (Peera et al. 2021):

$$\text{At anode}: C_6H_{12}O_6 + 6H_2O \xrightarrow{\text{Microbes}} 6CO_2 + 24H^+ + 24e^- \quad (E^\circ = -0.428V/NHE) \tag{23.1}$$

$$\text{At cathode}: O_2 + 4H^+ + 4e^- \rightarrow 2H_2O \quad (E^\circ = 0.805V/NHE) \tag{23.2}$$

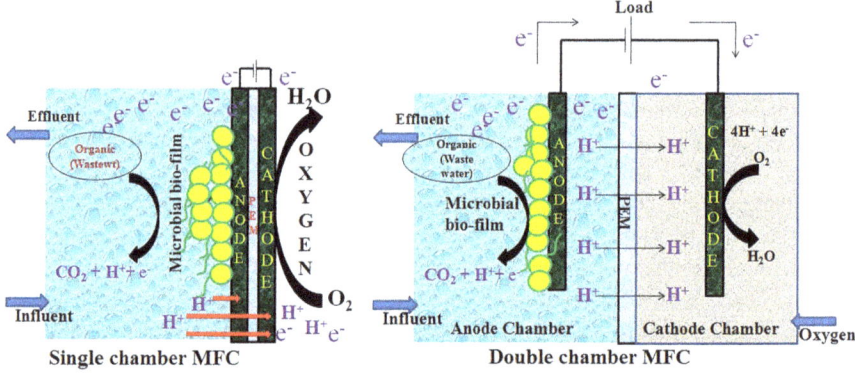

Fig. 23.1 A schematic illustration of different types of MFCs

$$\text{Overall reaction}: C_6H_{12}O_6 + 6O_2 \rightarrow 6CO_2 + 6H_2O \quad (E_{cell} = 1.233\text{V}) \quad (23.3)$$

Where NHE stands for normal hydrogen electrode and a cathode, at which oxygen (usually air) is reduced to form water to complete the circuit. In effect, MFC performance is dependent on the different types of parameters, such as microbial strains involved in the process, anodic feed, type of electrodes, separating barrier/membrane, cathode catalyst and reaction kinetics of the cathodic oxygen reduction reaction (ORR). The overall system efficiency can be expressed in different ways, where it is generally obtained as systemic power density. Power (P) in watt (W) is calculated as:

$$P = I \times E_{cell}$$

Where I (A) and E_{cell} represent the respective current and cell potential (V) of the system. This is calculated from the polarization graph (I–V curve) of the system. Power is calculated and normalized by the anode surface area, where microbial action for substrate degradation occurs. Moreover, in several cases, cathodic ORR is also taken as the controlling factor in systemic power generation, where power density is normalized by the given cathodic area. In addition to these, volumetric power density also contributes in describing the obtained power density of the system, by normalizing it with the specific volume of the reactor, as:

$$P = (E \times I)/U$$

Where P depicts volumetric power density (W/m³), E denotes cell potential (V), I represents current (A) and U is given as the total reactor volume (m³).

The major setback in MFC is the lagged oxygen reduction reaction (ORR) that plays a crucial role in limiting the overall performance of the system. Platinum (Pt) has been widely used as a cathode catalyst for ORR, although it lags with severe drawbacks of fouling, lowered performance over time, expensiveness and limited availability (Pattanayak et al. 2020a; Papiya et al. 2019). Thus, in order to minimize the overall systemic cost in MFCs, novel attempts have been made with low-cost alternatives for higher stability and electro-catalytical activities.

23.3 Cathode Catalysis

Reduction occurs at the cathode, where cathode catalysis influences the overall performance of the system. In a catalyst devoid MFC, a slower oxygen reduction reaction (ORR) becomes the prime concern in the system. The high activation energy barrier limits the basic reductive process at the cathode, which eventually gets aided with the lower accessibility of H^+ and hydroxyl/hydroxide ions (OH^-). This, in turn, augments the higher over potential in the system. To lower that, one needs to look closer at the mechanism involved in the cathodic reduction reaction (Ou and Chen 2013).

23.3.1 Mechanism of Oxygen Reduction Reaction (ORR)

Oxygen reduction reaction (ORR) is a multi-electron transfer reaction that is proposed with two diverse mechanisms, namely, four-electron ($4e^-$) and two-electron pathway ($2e^-$). In $4e^-$ pathway, O_2 is directly converted into H_2O molecules, whereas in $2e^-$ route, O_2 goes through a two-electron reduction process and H_2O_2 is produced as an intermediate (Table 23.1).

23.3.2 Catalyst Materials Used in MFCs

In order to avoid the slower reaction kinetics of oxygen reduction at the cathode, catalysts are generally employed to lower the required activation energy of the reaction. This, in turn, increases the overall ORR kinetics, where the role of catalysts

Table 23.1 Oxygen reduction reaction (ORR) in alkaline and acidic electrolyte (Pattanayak et al. 2020a, b; Noori et al. 2018)

Aqueous medium	Process	Oxygen reduction reaction (ORR)
Alkaline electrolyte	$4e^-$	$O_2 + H_2O + 4e^- \longrightarrow 4OH^-$
	$2e^-$	$O_2 + H_2O + 2e^- \longrightarrow HO_2^- + OH^-$ $HO_2^- + H_2O + 2e^- \longrightarrow 3OH^-$
Acidic electrolyte	$4e^-$	$O_2 + 4H^+ + 4e^- \longrightarrow 4OH^-$
	$2e^-$	$O_2 + 2H^+ + 2e^- \longrightarrow H_2O_2$ $H_2O_2 + 2H^+ + 2e^- \longrightarrow 2H_2O$

becomes quintessential for enhanced systemic performance (Sonawane et al. 2018). The most widely used ORR catalyst is platinum (Pt), which has been majorly addressed in different system designs (Mishra and Jain 2016; Lu et al. 2013). Being a cost-effective approach, the economic viability of Pt hinders its commercial usage in the long run (Yuan et al. 2010). To avoid that, numerous efforts have been put forth to find an alternative low-cost cathode catalyst for MFC usage. This includes the role of different carbonaceous materials such as activated carbon, modified carbon blacks, carbon nanofibers, CNTs and graphene as cathode catalyst (Fig. 23.2). These graphitized nanostructures possess several cutting-edge advantages such as increased surface area, higher stability and conductivity over other conventional materials (Lu et al. 2013; Papiya et al. 2018; Cui et al. 2015). As shown by Ghasemi et al. (2016), nanostructured activated carbons were employed in MFCs, owing to their enhanced surface area that qualifies as a potent cathode catalyst with low cost and high electrical conductivity. Further, their electrical conductivity has been augmented by mixing activated carbon with additional carbon blacks during fabrication (Yuan et al. 2011; Ansari et al. 2016). In effect, numerous metals like [cobalt (Co), iron (Fe), nickel (Ni), manganese (Mn), vanadium (V)], metal oxide (M_xO_y) e.g., [cuprous oxide (Cu_2O), vanadium oxide (V_2O_5), MnO_2, etc.], metal oxide- carbon hybrids such as [Cu_2O-activated carbon, MnO_2-activated carbon, etc.], metal-CNTs such as [Ni-CNT, Mn-CNT, etc.), and metal nitrogen carbon (M-N-C) complexes such as [Fe-N-carbon, Ni/N-CNTs, etc.] have also been demonstrated as possible catalysts for cathodic reactions (Lu et al. 2013; Yuan et al. 2010; Ansari et al. 2016;

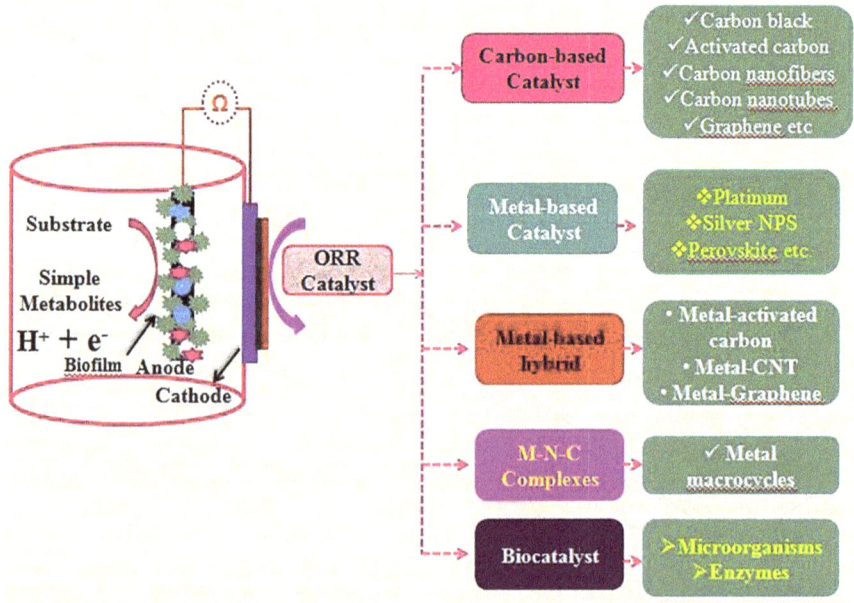

Fig. 23.2 Schematic representation of distinct ORR catalysts employed in MFCs

Khilari et al. 2014; Esmaeili et al. 2014; Kirubaharan et al. 2016). These will be dealt in detail in the next segments.

23.4 Emerging Nanomaterials in Cathode Catalysis

Several novel nanomaterials have been tried and tested in MFCs as the catalyst for augmenting the overall reaction kinetics. These can be distinguished based on incorporated nanosized fillers into the nanocomposite matrix, which plays a vital role in improving systemic performance. Here, the prime concern becomes the incorporation of specific novel properties within the designed nanocomposite structure (such as durability, electrical conductivity and chemical reactivity) that consists of single or several nanoparticle materials with one or more constituents of dimension less than 100 nm. These are grouped as follows.

23.4.1 Transition Metal/Metal Oxide and Alloy-Based Cathode Catalysts

Platinum (Pt), being the widely studied oxygen-reducing catalyst, has severe limitations in terms of high cost, limited availability and fouling tendencies (Pattanayak et al. 2019, 2020a). This brings upon other cheap substrates such as transition metals (Fe, Co, Ni, Mn) and metal-free heteroatom doped (N, P, S, F) catalysts that can also lower the activation energy required for an efficient ORR (Peera et al. 2021; Li et al. 2019). As a consequence, binary metal catalysts (like Pt-Fe, Pt-Cu, Pt-Mn) (Coleman et al. 2015; Ammam and Easton 2013) and multicomponent Pt-based catalysts have also been studied as efficient catalysts to bring economic feasibility in the system utility (Santoro et al. 2015; Liu et al. 2017). Although this brings the cost down, the durability of these catalysts is still a prime concern in its widespread usage. As an instance, Liu et al. (2017) used glycerol for stabilizing the prepared Pt-Fe alloy catalyst. This resulted in an approximate 18% higher power density over the used Pt/C catalyst. The results showed Fe leaching after some time from the alloy composition, even though its maximum power density was relatively more stable than that of Pt/C under neutral environments.

Similarly, different metal oxide (MO_x) nanoparticles such as CuO_x, NiO, VO_x and MnO_x have also been introduced as effective, affordable materials for enhancing the conductivity of the designed catalysts (Noori et al. 2016; Liu et al. 2015a; Yuan et al. 2016a). For several years, these MO_x have been known in the areas of energy conversion and fuel cell designs. However, at present, more emphasis is given on their nanocomposite usage with different base substrates. As an example, a nanocomposite cathode catalyst matrix of Ni–NiO/PPy–rGO revealed an enhanced power density of 678.79 ± 34 mW/m^2 with ~81.52% chemical oxygen demand

(COD) removal from single-chambered MFC. The results indicated this relatively inexpensive nanocomposite better over the used Pt/C catalyst (Pattanayak et al. 2019). Similarly, manganese and vanadium oxide nanocomposites in combination with carbon material have been tested as cathode catalysts in MFCs (Ghoreishi et al. 2014). In distinct attempts, investigations with V_2O_5 catalyst in lithium-ion batteries have been tested; however, quite limited studies are there in its catalytical usage in MFCs. As an example, Noori et al. (2016) have shown MnO_2 nanotubes and V_2O_5 microflowers as cathode catalysts in MFCs. The results indicated ~31% higher efficiency of V_2O_5 microflowers over employed MnO_2 nanotubes. In another instance, the group compared the performance of V_2O_5/Vulcan XC with V_2O_5 nanorods based on reduced graphene oxide in MFCs as cathode catalyst (Noori et al. 2017). Here too, V_2O_5 nanorods revealed a higher power density of 533 mW/m^2 over V_2O_5/Vulcan XC (384 mW/m^2). In comparison, the calculated overall systemic cost of V_2O_5/reduced graphene was way more economical than the commercially used Pt/C catalyst (Noori et al. 2017). Similarly, several other MO_x nanocomposites, e.g. calcium titanium oxide, CoO_x, PbO_2 and ZrO_2, have been used in MFCs as potent catalysts for ORR.

23.4.2 Metal-Carbon Hybrid Catalysts

The segment here will enumerate different metal-carbon oxygen-reducing hybrids (e.g. activated carbon-metal, CNTs-metal, graphene-metal hybrids, etc.) as cathode catalysts for enhanced ORR activity in MFCs.

23.4.3 Metal-Activated Carbon Hybrids

The excellent electrical conductivity of activated carbon has put it on top of carbon black, as an efficient alternative base support material for other metal catalysts. For instance, Ge et al. (2015) prepared nano-cobalt (II, III) oxide (nano-Co_3O_4) via a hydrothermal process, where mixing it with activated carbon as support yielded an enhanced power density from the system. The effects were apparently visible in the system, where lower internal resistances were particularly observed, especially with the coated electrodes (Ge et al. 2015). Similarly, a nano urchin-like nickel cobaltite ($NiCo_2O_4$) was constructed via a hydrothermal process, where again its mixing with activated carbon depicted increased power density from the system (Ge et al. 2016). Initial experimentations on activated carbon were done on silk fibroins, which were used to fabricate activated carbon from it (Zhang et al. 2009a). Later, it was thought to augment the catalytical activity of this activated carbon. This was primarily achieved by incorporating heteroatom (e.g. N) into the structural backbone of carbon materials. In an instance, Huang et al. (2017) showed cobalt (II) oxide (CoO) nanosheets layered with N-incorporated activated carbon, as an efficient MFC cathode catalyst for oxygen reduction reactions. This in situ formulated composite indicated an enhanced performance of ~1650 mW/m^2, with ~122.5% more efficiency over the control system. The effect of nanocomposite depicted higher ORR activity from the employed MFC cathode catalyst (Huang et al. 2017).

23.4.4 Metal-Carbon Nanofibers (CNFs) and Nanotubes (CNTs)

Numerous investigations on CNTs and CNFs have been focused as cathode-based catalysts, for analysing their performance in MFCs (Teng et al. 2019; Karra et al. 2013; Liu et al. 2010a). Primarily, CNTs and CNFs exhibit high surface area, anti-corrosive property and superior electro-activity because of its increased graphitic nature of composed sp^2 carbons. KOH activation is commonly done to attain a high surface area and porosity in CNTs and CNFs. Similarly, activated carbon, carbon black and graphite also undergo acid and alkaline treatment for activation and surface area increment of the material. Another route of transformation is heteroatom incorporation, which eventually enhances the ORR activity in CNTs and CNFs. Besides, incorporating various transition metals, e.g. Co, Fe, Mn on CNTs and CNFs, has also shown superior oxygen reductive property in MFC setups (Yaping et al. 2011; Liu et al. 2010b; Yong et al. 2011; Jung-Chen et al. 2019). CNTs are mainly categorized as single-walled CNTs (SWCNTs) and multiwalled CNTs (MWCNTs). Both have been evaluated as cathode catalysts in MFC operation. As mentioned earlier, MnO$_2$ is an efficient low-cost metal catalyst, whose combined effect with CNTs has been equally studied in MFC application. For instance, in an in situ hydrothermal process, MnO$_2$ was deposited over CNTs that showed marked increments in ORR electron transfer and thereby resulted in enhanced performance in MFC setups (Zhang et al. 2011). In another study, novel manganese-polypyrrole-carbon nanotube (Mn-PPY-CNT) composites were formulated by a solvothermal method that exhibited higher ORR activity with a catalyst loading of 2 mg/cm^2 in the system (Lu et al. 2013). This, in effect, showed an increased power output and durability in MFCs in comparison to the conventionally used Pt/C catalyst.

23.4.5 Metal-Graphene-Based Nanocomposites

In general, pristine graphene is considered inert, as it lacks the defects that are used as electroactive centres for catalytic activity and higher conduction. To avert that, one atom thick graphene is projected as an electroactive substrate that allows easier physiochemical modifications in its graphitic structure. This, in turn, improves the overall nature of the material with a larger surface area and enhanced electro-conductivity (Kannan and Gnanakumar 2016). In effect, both graphene and graphite

Table 23.2 Graphene/graphite-based nanocomposites as cathode catalysts in MFCs

Catalyst	Cathode material	Type of MFC	Maximum power density (mW/m^2)	References
NG/Co-N	Carbon cloth	Single chambered	713.6 (571.3) Pt/C catalyst (JM 40%)	Cao et al. (2016)
GO/MgO	Carbon cloth	Single chambered	755.63 (870.75)	Li et al. (2017)
Ni-NiO/PPy-rGO	Carbon cloth	Single chambered	678.79 ± 34 (481.02)	Pattanayak et al. (2019)
V$_2$O$_5$/rGO	SS wire mesh	Single chambered	533 ± 37 (512 ± 51)	Noori et al. (2017)
Ni-Co/MGO	Carbon cloth	Single chambered	1003.2 (483.48)	Papiya et al. (2019)
Ni-Co (1:1)/SPAni	Carbon cloth	Single chambered	659.79 ± 20 (483.48)	Papiya et al. (2018)
K-(PPy-Co-PANI)-rGO	Carbon cloth	Single chambered	~763 ± 38 (483.48)	Pattanayak et al. (2020b)

have actively been studied as catalyst support, with other substrates such as metal catalysts (Table 23.2) (Yuan et al. 2016a; Quan et al. 2015). In a study, Wen et al. (2012) revealed a hybrid low-cost cathode catalyst for MFC, fabricating MnO$_2$-graphene nanosheets via microwave irradiation method. This resulted in enhanced cathode catalysis over simple MnO$_2$ with an improved power output of 2.08 W/m^2. Similarly, Khilari et al. (2013) prepared nanotubular MO$_2$/graphene oxide nanocomposite as an oxygen-reducing catalyst by a hydrothermal technique that again provided an overall increased performance of 3.35 W/m^2 in MFC usage. In comparison, the study revealed the better durability and shorter lag period of nanocomposites in activating the oxygen reduction reaction over other used catalysts.

23.4.6 Metal-Conducting Polymer-Based Nanocomposites

Another promising material for higher electron conduction is conducting polymers (CP) that endows specific properties like metal-like conductivity and reversible doping/de-doping characteristics that can efficiently serve in various catalytical processes (Fig. 23.3) (Wang et al. 2017). Apart from that, it plays a crucial role in augmenting the surface area for other substrates, to attach and form an active nucleation site for substrates like electron and air, for cathodic oxygen reduction.

The majority of these conducting polymers have aromatic sites, consisting of conjugated double bonds, such as polyacetylene (PA), polypyrrole (PPy), polythiophene (PTh), poly(p-phenylene vinylene) (PPV), polyaniline (PANI) and poly(3,4-ethylene dioxythiophene) (PEDOT). Some of the most widely studied CPs are polypyrrole (PPy), polyaniline (PAni), polythiophene (PTh) and poly(aniline-co-pyrrole) (Pattanayak et al. 2018). With such variations, these CPs exhibit brilliant

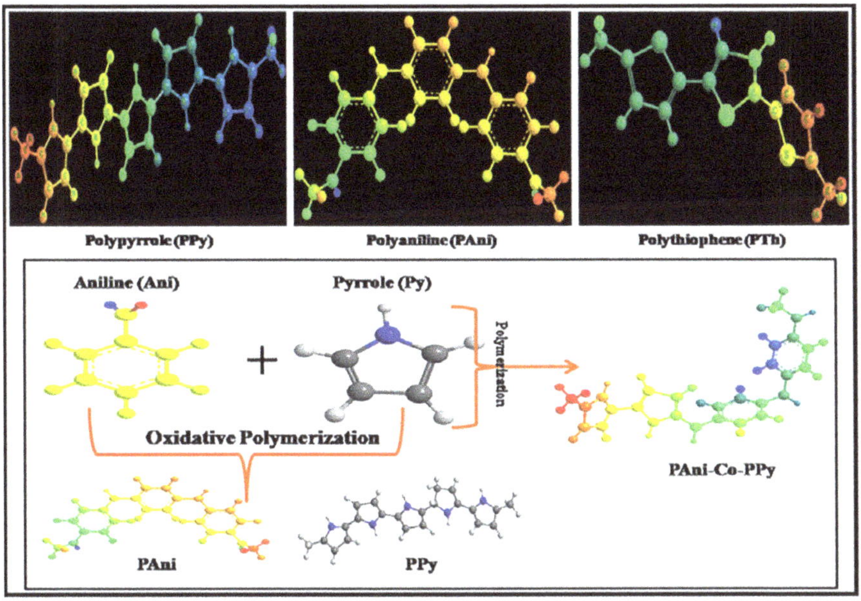

Fig. 23.3 Chemical structure and synthesis of three common CPs and copolymers such as polypyrrole (PPy), polyaniline (PAni), polythiophene (PTh) and poly(aniline-co-pyrrole)

electrochemical properties, with their conductivity ranging between 10^{-6} and 10^3 Scm^{-1}. These can further be optimized depending on the doping level that can increase the conductivity level in a numeric span of 10^{-10} to 10^4 Scm^{-1} (Wang et al. 2017; Pattanayak et al. 2018; Rudra et al. 2019). The chemical structures and stability characteristics of some common conducting polymers (CPs) are shown in Table 23.3.

CPs and their corresponding composites serve as the best supporting candidates for nanoparticle deposition, which, in turn, alters the electro-activity of the formed composite. With superior electrical conductivity, tunable specific capacitance and low fabrication cost, these CPs serve as a brilliant platform for different electrochemical reactions, including cathodic-ORR and anodic-microbial oxidation (Table 23.4). In effect, these also aid in increasing thermal stability within catalytical supports, which is done by infusing oxidative chemical or electrochemical sensitive nanomaterials.

23.5 Potentiodynamic Effects of Nanomaterials

The high catalytic activity and cost-effective nature are the two most desired criteria in cathode catalysis in MFCs. A broad range of nanomaterials has been investigated and tested for their efficacy and durability in the long run. Combined studies of Pt

Table 23.3 Chemical structures and characteristics of some common conducting polymers (CPs)

CPs	Chemical structure	Maximum conductivity (Scm^{-1})	Stability
Polyaniline (PAni)		9–10	Stable
Polypyrrole (PPy)		1000–2000	Reasonably stable
Polythiophene (PTh)		90–100	Stable
Polyacetylene		1.5×10^5	Reacts with air
Poly(p-phenylene vinylene)		1000	Stable un-doped form

Table 23.4 MFC performance with different nanomaterial-conducting polymer-based cathode catalysts

S. no.	Catalysts	Electrode type	MFC type	Maximum power density (mW/m²)	References
1	PANI/C/FePc	Wet-proofed carbon cloth	Air-cathode	630.5	Yuan et al. (2011)
2	75% wt PANI/MWNT	Graphite felt	Single chambered	488	Cui et al. (2015)
3	PPy/C	Non-wet proofing carbon cloth	Single chambered	402	Yuan et al. (2010)
4	Mn–PPY–CNT	Glass carbon electrode (GCE)	Single chambered	213	Lu et al. (2013)
5	CNT/PPy nanocomposite	Carbon cloth	Dual chambered	113.5	Ghasemi et al. (2016)
6	MnCo$_2$O$_4$ NRs/PPy	Carbon paper	Air-cathode	6.11 W/m³	Khilari et al. (2014)
7	Ni–NiO/PPy–rGO	Carbon cloth	Single chambered	678.79	Pattanayak et al. (2019)
8	Pani–MnO$_2$	Carbon paper	Dual chambered	0.0588 W/m²	Ansari et al. (2016)
9	Polypyrrole (PPy)/kappa-carrageenan(KC)	Carbon paper	Dual chambered	72.1	Esmaeili et al. (2015)
10	MnO$_2$/PPy/MnO$_2$ nanotubes	Carbon cloth	Single chambered	721	Yuan et al. (2015)
11	(PANI/C/FePc)	Carbon cloth	Single chambered	630.5	Yuan et al. (2011)

alloyed with other transition metals (e.g. Pt-M, where M = Mn, Co, Ni, Fe, V, Cu, Ti) have been used to minimize the amount of Pt and assay its overall catalytic activity in the system (Gupta et al. 2009). Being low cost, transition metal oxides nanocomposites, like manganese (IV) dioxide (MnO_2), have been widely analysed as ORR catalysts in MFCs. It is regarded as a promising cathode catalyst in MFC, because of its eco-friendly nature, low price and high chemical stability. In a study, Zhang et al. (2009b) showed three manganese dioxide materials, α–MnO_2, β–MnO_2 and γ–MnO_2 as an alternative ORR catalyst over the commercially used platinum (Pt) catalyst (Zhang et al. 2009b). It was observed that all the catalysts, especially β–MnO_2, indicated much higher catalytic activity over the corresponding α–MnO_2 and γ–MnO_2. The reason was attributed to the higher surface area of nanocomposite β–MnO_2 over other variants used. In other experiments, anchoring different carbon supports along MnO_2, such as graphite, activated carbon (AC), carbon nanotube (CNT) and graphite oxide (GO), was shown to improve the overall conductivity of the catalyst in the system. The mode of recombining ingredients also plays a crucial role in the overall performance of the catalyst. For example, in situ preparation of carbon nanotubes (CNTs) coated with manganese dioxide (MnO_2/CNTs) by hydrothermal methods revealed much higher electro-activity with increased ORR than the mechanically mixed MnO_2/CNTs (Zhang et al. 2011). Similarly, the hydrothermal preparation of manganese dioxide-graphene nanosheet (MnO_2/GNS) catalyst indicated higher ORR activity in the system. The cyclic voltammetric (CV) studies revealed the oxygen reduction peak at -0.43 V, which was higher than that of normal MnO_2 (-0.71 V), and almost equivalent to Pt/C catalyst (-0.44 V) (Wen et al. 2012). The reason was attributed to the close attachment of MnO_2 nanoparticles to graphene nanosheets, which was responsible for the excellent catalytic activity of MnO_2/GNS composite.

A study conducted by Dessie et al. (2020) revealed Mn_2O_3/C nanopowder as an efficient cathode catalyst with improved ORR and better stability over commercial Pt/C catalyst in the system. Current findings revealed that MnO_x performance can be augmented by increasing the oxygen oxidation state or incorporating materials (e.g. nanomaterials) that basically modify their electronic structure and electron transfer behaviour in catalytic reactions. For example, the nest-like oxygen-deficient $Cu_{1.5}Co_{1.5}O_4$ contains higher micropores and active sites for O_2 access that ultimately enhances the electro-activity and conduction at the catalyst surface. In an instance, Roche et al. (2010) reported MnOx/C as a cathode catalyst with a systemic power density of 161 mW/m^2, which was slightly lower than the obtained power output from Pt/C (193 mW/m^2). In order to increase the catalyst surface area with MnOx substrates, an alternative bi-metallic hybrid approach has also been employed. For example, the bimetal catalyst h-Co_3O_4@$MnCo_2O_{4.5}$ was shown to exhibit a higher ORR activity that was attributed to its increased specific surface area (130.4 m^2/g) in the system (Liu et al. 2019). Similarly, Hu et al. (2015) showed hydrothermal in situ preparation of $MnCo_2O_4$/C catalyst with improved ORR in dual-chambered MFC. It was highlighted that the four electron pathway was more favourable for ORR via $MnCo_2O_4$/C catalyst with a maximum power density of 545 mW/m², which was relatively advanced over the plain cathode (Pmax = 214 mW/m²) used in the system (Hu et al. 2015).

A study conducted by Zhang et al. (2014) showed an N-incorporated activated carbon catalyst with increased N content (5.56% pyridinic N and 8.65% total N), activated by an acid/alkaline pre-treatment using cyanamide as the nitrogen precursor. The modified activated carbon showed a 44% higher power density over the conventionally used Pt/C catalyst. Furthermore, N-doping and nitrogenous chemicals, such as ammonium bicarbonate, were found to enhance the porosity and pore size distribution in the matrix. This, in turn, reduced the charge transfer resistance in the system and subsequently improved the overall performance of ORR and MFC. Chronoamperometric results revealed around 73% decrement in the current density of Pt/C after 7 h, whereas the nitrogen-doped activated carbon cathode showed only about 30% decrement. Moreover, linear sweep voltammetric (LSV) analysis revealed an onset potential of 180 mV with nitrogen-doped activated carbon cathode with noticeable positive shifts in ORR (Zhang et al. 2014).

Other combined effects of nanomaterials with noble metals such as Au, Ag and Pd have been duly studied. A bio-electrochemical system containing Pd nanoparticle-modified with p-type Si nanowire (Pd-SiNW) photocathode was tested by Han et al. (2017) in MFC. Here, minimal internal resistance (670 Ω) was observed from Pd-SiNW photocathode under illumination, whereas it was found highly increased in dark experiments (2184 Ω). As a control, carbon paper was studied as a photocathode that indicated respective high internal resistances, e.g. 2372 and 2376 Ω, under illumination and dark experiments. The chronoamperometric results indicated an initial current density of ~0.8 A/m² under illumination that gradually decreased to ~0.4 A/m² after 36 h of operation in Pd-SiNW electrodes. This in turn was much higher than the control system, which was attributed to the catalytical effect of used nanoparticle in the framework. A few case studies of the potentiodynamic effects of used nanoparticles in MFCs as cathode catalyst is shown in Table 23.5.

23.6 Anode Modifications

Improvisations in electrode fabrications have gathered increased attention in the last few years, where carbon materials have been widely used because of their high availability, porosity and better electrical property. To further augment the efficacy of electrodes, different nanocomposites with better characteristics such as increased durability, electro-activity and thermal endurance have been aimed and designed. For example, modified stainless steel has been used to improve the performance output in the system. This showed an enhanced power density of 2880 mW/m² with stainless steel wool/PAni/polypyrrole nanocomposite (Nitisoravut et al. 2017; Zhou et al. 2011). Another widely used substrate is graphene, which is known for its larger surface area and superior conductivity. In combination with different conducting polymers and nanocomposites, these have also been tested in MFCs as anode. Some examples of these modified anodes are given in Table 23.6.

Table 23.5 Electrochemical performance of nanomaterials in MFC analysis

Catalyst used	Reduction voltage (V) by cyclic voltammetry	Reduction current (mA) by cyclic voltammetry	Electrolyte used for ORR	Electrochemical impedance spectroscopy (EIS) (Ω)	Maximum current density (mA/m^2)	Maximum power density (mW/m^2)	COD removal efficiency (%)	References
In situ MnCo$_2$O$_4$ NRs/PPy	~0.3	–	O$_2$-saturated 0.1 M KOH	14.86	–	420	89.8	Khilari et al. (2014)
Ni-Co (1:1)/SPAni	~0.715	0.049	PBS solution	43.46	1778.4	659.79 ± 20	91.5	Papiya et al. (2018)
β-MnO$_2$	–0.16	–0.27	O$_2$-saturated 0.2 M NaCl solution	200	2341	172 ± 7	–	Zhang et al. (2009b)
Fe-AAPyr	0.307	–	K-PB (0.1 M) solution	–	–	482 ± 5 mW/cm^2	–	Kodali et al. (2017)
N-G@CoNi/BCNT	0.06	–	1.0 M PBS	~10	11.2 A/m^2	2.0 ± 0.1 W/m^2	–	Hou et al. (2016)
CNT/PPy	–0.5	–0.17	NaCl solution	1540	226.25	~113	96	Ghasemi et al. (2016)

Table 23.6 Comparative performance of the modified nanomaterial-based anode in MFCs

Anode materials	MFC assembly	Bacteria/seed	Power density (mW/m^2)	References
Tartaric acid doped PANI/carbon cloth	Dual chambered	S. oneidensis MR-1	490	Liao et al. (2015)
PANI/SSFF	Dual chambered	Domestic wastewater	360	Sonawane et al. (2018)
PANI/CNT/GF	Dual chambered	S. putrefaciens	257	Huang et al. (2016)
MnFe$_2$O$_4$/PANI/CC	Single chambered	S. putrefaciens	11.2 W/m^3	Khilari et al. (2015)
PANI/CF	Dual chambered	S. cerevisiae	460	Hidalgo et al. (2016)
Poly (3,4-ethylenedioxythiophene) (PEDOT) modified carbon cloth	Dual chambered	S. loihica strain PV-4	140	Liu et al. (2015b)
(PANI+G+CC)	Single chambered	Lake sediment	884	Huang et al. (2016)
Conductive polypyrrole hydrogels and carbon nanotubes composite (CPHs/CNTs)	Dual chambered	Mixed bacterial culture	1898	Tang et al. (2015)
MnO$_2$/polypyrrole/MnO$_2$ nanotubes (NT-MPMs)	Single chambered	Mixed bacterial culture	934.8	Yuan et al. (2016b)
BC/PANI nano-biocomposite	Dual chambered	Anaerobic sludge	616	Mashkour et al. (2016)
PANI-LMC	Dual chambered	Mixed bacterial culture	1280	Zou et al. (2017)
PPy-ACNF/CNT	Dual chambered	Shewanella oneidensis	598	Jung and Roh (2017)
PU/graph/PPy	Dual chambered	Municipal wastewater	305.5 mW/m^3	Pérez-Rodríguez et al. (2016)
PPy/SS	Single chambered	Anaerobic granular sludge	1190.94	Pu et al. (2018)
Ppy/SAC/SS	Dual chambered	Mixed bacterial culture	45.2 W/m^3	Wu et al. (2018)
PANI/rGO	Single chambered	Mixed bacterial culture	862	Zhao et al. (2018)
SS-P/PANi	Single chambered	Synthetic wastewater	0.078 mW/cm^2	Sonawane et al. (2018)
MWCNT-MnO$_2$/PPy	Dual chambered	Sewage	1125.4	Mishra and Jain (2016)

In effect, these modifications have shown better microbial biofilm adhesion and growth at the anode, which in turn have aided in improving the Coulombic efficiency (CE) and overall performance of the system. Whether it's the larger surface area, increased conductivity or enhancing the durability of the electrodes, nanomaterials and their hybrid modifications have proven their substantial applicability in the long run. In the last few years, such endeavours have shown an approximate 10,000-fold improvement in the overall current density of the system. To further augment the system, the multidimensional role of nanomaterials would find specificity in different areas of hydrogen evolution, microbial adhesion, anodic-feed oxidation, electronic/ionic conductivity, ORR and increased stability of the system. This will not only benefit MFCs but also aid in other areas of separation and bio-electrochemical technologies such as microbial electrosynthesis (MES), desalination and microbial electrolysis systems. It can well be predicted that the nanocomposite-based ORR catalysts would play a pivotal role in determining the fate of such existing technologies in large-scale field application.

23.7 Conclusions

The chapter here covers the implications of emerging nanomaterials as improved cathode catalysts in MFCs. This involves oxygen reduction into water, known as oxygen reduction reaction (ORR) that uses the incoming electrons and protons from the anode. Cathode catalysts lower the required activation energy for the process and thereby increases the overall oxygen reduction rate at cathode. The commercially used Pt/C has major drawbacks in terms of their expensiveness and long-term stability. As a substitute, different low-cost nanomaterials/nanocomposites, comprising of transition metals oxides, and carbon materials have been fabricated and tested in MFCs that have shown improved ORR and power efficiency in the system. The chapter here enumerates a comprehensive list of these emerging nanomaterials that have been gaining viability with electrode modifications and in other areas of catalysis. In hope to deliver better efficacy, more such nanomaterial designs and alterations are desired to achieve major adaptations in bio-electrochemical and separation technologies with practical field applications.

References

Ammam M, Easton EB (2013) Oxygen reduction activity of binary PtMn/C, ternary PtMnX/C (X = Fe, Co, Ni, Cu, Mo and, Sn) and quaternary PtMnCuX/C (X = Fe, Co, Ni, and Sn) and PtMnMoX/C (X= Fe, Co, Ni, Cu and Sn) alloy catalysts. J Power Sources 236:311–320

Ansari SA, Parveen N, Han TH, Ansari MO, Cho MH (2016) Fibrous polyaniline@ manganese oxide nanocomposites as supercapacitor electrode materials and cathode catalysts for improved power production in microbial fuel cells. Phys Chem Chem Phys 18(13):9053–9060

Ayyaru S, Mahalingam S, Ahn YH (2019) A non-noble V$_2$O$_5$ nanorods as an alternative cathode catalyst for microbial fuel cell applications. Int J Hydrog Energy 44:4974–4984

Bosch-Jimenez P, Martinez-Crespiera S, Amantia D, Della Pirriera M, Forns I, Shechter R, Borràs E (2017) Non-precious metal doped carbon nanofiber air-cathode for Microbial Fuel Cells application: oxygen reduction reaction characterization and long-term validation. Electrochim Acta 228:380–388

Cao C, Wei L, Su M, Wang G, Shen J (2016) Enhanced power generation using nano cobalt oxide anchored nitrogen-decorated reduced graphene oxide as a high-performance air-cathode electrocatalyst in biofuel cells. RSC Adv 6:52556–52563

Coleman EJ, Chowdhury MH, Co AC (2015) Insights into the oxygen reduction reaction activity of Pt/C and PtCu/C catalysts. ACS Catal 5:1245–1253

Cui HF, Du L, Guo PB, Zhu B, Luong JH (2015) Controlled modification of carbon nanotubes and polyaniline on macroporous graphite felt for high-performance microbial fuel cell anode. J Power Sources 283:46–53

Dessie Y, Tadesse S, Eswaramoorthy R (2020) Review on manganese oxide based biocatalyst in microbial fuel cell: nanocomposite approach. Mater Sci Energy Technol 3:136–149

Esmaeili C, Ghasemi M, Heng LY, Hassan SH, Abdi MM, Daud WRW, Ismail AF (2014) Synthesis and application of polypyrrole/carrageenan nano-bio composite as a cathode catalyst in microbial fuel cells. Carbohydr Polym 114:253–259

Esmaeili C, Abdi MM, Mathew AP, Ionoobi M, Oksman K, Rezayi M (2015) Synergy effect of nanocrystalline cellulose for the biosensing detection of glucose. Sensors 15:24681–24697

Ge B, Li K, Fu Z, Pu L, Zhang X (2015) The addition of ortho-hexagon nano spinel Co$_3$O$_4$ to improve the performance of activated carbon air cathode microbial fuel cell. Bioresour Technol 195:180–187

Ge B, Li K, Fu Z, Pu L, Zhang X, Liu Z, Huang K (2016) The performance of nano urchin-like NiCo2O4 modified activated carbon as air cathode for microbial fuel cell. J Power Sources 303:325–332

Ghasemi M, Daud WRW, Hassan SH, Jafary T, Rahimnejad M, Ahmad A, Yazdi MH (2016) Carbon nanotube/polypyrrole nanocomposite as a novel cathode catalyst and proper alternative for Pt in microbial fuel cell. Int J Hydrog Energy 41:4872–4878

Ghoreishi KB, Ghasemi M, Rahimnejad M, Yarmo MA, Daud WRW, Asim N, Ismail M (2014) Development and application of vanadium oxide/polyaniline composite as a novel cathode catalyst in microbial fuel cell. Int J Energy Res 38:70–77

Gnana KG, Kirubaharan CJ, Udhayakumar S, Karthikeyan C, Nahm KS (2014) Conductive polymer/graphene supported platinum nanoparticles as anode catalysts for the extended power generation of microbial fuel cells. Ind Eng Chem Res 53:16883–16893

Gupta G, Slanac DA, Kumar P, Wiggins-Camacho JD, Wang X, Swinnea S, Johnston KP (2009) Highly stable and active Pt–Cu oxygen reduction electrocatalysts based on mesoporous graphitic carbon supports. Chem Mater 21:4515–4526

Han HX, Shi C, Yuan L, Sheng GP (2017) Enhancement of methyl orange degradation and power generation in a photoelectrocatalytic microbial fuel cell. Appl Energy 204:382–389

Han TH, Parveen N, Shim JH, Nguyen ATN, Mahato N, Cho MH (2018) Ternary composite of polyaniline graphene and TiO$_2$ as a bifunctional catalyst to enhance the performance of both the bioanode and cathode of a microbial fuel cell. Ind Eng Chem Res 57:6705–6713

Hidalgo D, Tommasi T, Bocchini S, Chiolerio A, Chiodoni A, Mazzarino I, Ruggeri B (2016) Surface modification of commercial carbon felt used as anode for microbial fuel cells. Energy 99:193–201

Hosseini MG, Ahadzadeh I (2012) A dual-chambered microbial fuel cell with Ti/nano-TiO$_2$/Pd nano-structure cathode. J Power Sources 220:292–297

Hou Y, Yuan H, Wen Z, Cui S, Guo X, He Z, Chen J (2016) Nitrogen-doped graphene/CoNi alloy encased within bamboo-like carbon nanotube hybrids as cathode catalysts in microbial fuel cells. J Power Sources 307:561–568

Hu D, Zhang G, Wang J, Zhong Q (2015) Carbon-supported spinel nanoparticle MnCo$_2$O$_4$ as a cathode catalyst towards oxygen reduction reaction in dual-chamber microbial fuel cell. Aust J Chem 68:987–994

Huang L, Li X, Ren Y, Wang X (2016) In-situ modified carbon cloth with polyaniline/graphene as anode to enhance performance of microbial fuel cell. Int J Hydrog Energy 41:11369–11379

Huang Q, Zhou P, Yang H, Zhu L, Wu H (2017) CoO nanosheets in situ grown on nitrogen-doped activated carbon as an effective cathodic electrocatalyst for oxygen reduction reaction in microbial fuel cells. Electrochim Acta 232:339–347

Jung HY, Roh SH (2017) Carbon nanofiber/polypyrrole nanocomposite as anode material in microbial fuel cells. J Nanosci Nanotechnol 17:5830–5833

Jung-Chen W, Wei-Mon Y, Wei-Hung C, Sangeetha T, Chen-Hao W, Chin-Tsan W (2019) Innovative multi-processed N-doped carbon and Fe$_3$O$_4$ cathode for enhanced bioelectro-Fenton microbial fuel cell performance. Int J Energy Res 43:7594–7603

Kannan MV, Gnanakumar G (2016) Current status, key challenges and its solutions in the design and development of graphene based ORR catalysts for the microbial fuel cell applications. Biosens Bioelectron 77:1208–1220

Karra U, Manickam SS, McCutcheon JR, Patel N, Li B (2013) Power generation and organics removal from wastewater using activated carbon nanofiber (ACNF) microbial fuel cells (MFCs). Int J Hydrog Energy 38:1588–1597

Khilari S, Pandit S, Ghangrekar MM, Das D, Pradhan D (2013) Graphene supported α-MnO2 nanotubes as a cathode catalyst for improved power generation and wastewater treatment in single-chambered microbial fuel cells. RSC Adv 3:7902–7911

Khilari S, Pandit S, Das D, Pradhan D (2014) Manganese cobaltite/polypyrrole nanocomposite-based air-cathode for sustainable power generation in the single-chambered microbial fuel cells. Biosens Bioelectron 54:534–540

Khilari S, Pandit S, Varanasi JL, Das D, Pradhan D (2015) Bifunctional manganese ferrite/polyaniline hybrid as electrode material for enhanced energy recovery in microbial fuel cell. ACS Appl Mater Interfaces 7:20657–20666

Khurmi RS, Sedha RS (2008) Materials science. S. Chand

Kirubaharan CJ, Yoo DJ, Kim AR (2016) Graphene/poly (3, 4-ethylenedioxythiophene)/Fe$_3$O$_4$ nanocomposite–an efficient oxygen reduction catalyst for the continuous electricity production from wastewater treatment microbial fuel cells. Int J Hydrog Energy 41:13208–13219

Kirubakaran A, Jain S, Nema RK (2009) A review on fuel cell technologies and power electronic interface. Renew Sust Energ Rev 13:2430–2440

Kodali M, Santoro C, Serov A, Kabir S, Artyushkova K, Matanovic I, Atanassov P (2017) Air breathing cathodes for microbial fuel cell using Mn-, Fe-, Co-and Ni-containing platinum group metal-free catalysts. Electrochim Acta 231:115–124

Kumar R, Singh L, Ab Wahid Z, Mahapatra DM, Liu H (2018) Novel mesoporous MnCo$_2$O$_4$ nanorods as oxygen reduction catalyst at neutral pH in microbial fuel cells. Bioresour Technol 254:1–6

Li M, Zhou S (2018) α-Fe2O3/polyaniline nanocomposites as an effective catalyst for improving the electrochemical performance of microbial fuel cell. Chem Eng J 339:539–546

Li M, Zhou S, Xu M (2017) Graphene oxide supported magnesium oxide as an efficient cathode catalyst for power generation and wastewater treatment in single chamber microbial fuel cells. Chem Eng J 328:106–116

Li Y, Li Q, Wang H, Zhang L, Wilkinson DP, Zhang J (2019) Recent progresses in oxygen reduction reaction electrocatalysts for electrochemical energy applications. Electrochem Energy Rev 2:518–538

Liao ZH, Sun JZ, Sun DZ, Si RW, Yong YC (2015) Enhancement of power production with tartaric acid doped polyaniline nanowire network modified anode in microbial fuel cells. Bioresour Technol 192:831–834

Liu D, Ming Z, Chang L, Ling L, Dong S (2010a) Development of high performance of Co/Fe/N/CNT nanocatalyst for oxygen reduction in microbial fuel cells. Talanta 81:444–448

Liu XW, Sun XF, Huang YX, Sheng GP, Zhou K, Zeng RJ, Dong F, Wang SG, Xu AW, Tong ZH, Yu HQ (2010b) Nano-structured manganese oxide as a cathodic catalyst for enhanced oxygen reduction in a microbial fuel cell fed with a synthetic wastewater. Water Res 44:5298–5305

Liu Z, Li K, Zhang X, Ge B, Pu L (2015a) Influence of different morphology of three-dimensional CuxO with mixed facets modified air-cathodes on microbial fuel cell. Bioresour Technol 195:154–161

Liu X, Wu W, Gu Z (2015b) Poly (3,4-ethylenedioxythiophene) promotes direct electron transfer at the interface between Shewanella loihica and the anode in a microbial fuel cell. J Power Sources 277:110–115

Liu D, Mo X, Li K, Liu Y, Wang J, Yang T (2017) The performance of spinel bulk-like oxygen-deficient CoGa2O4 as an air-cathode catalyst in microbial fuel cell. J Power Sources 359:355–362

Liu Y, Chi X, Han Q, Du Y, Huang J, Lin X, Liu Y (2019) Metal–organic framework-derived hierarchical Co_3O_4@$MnCo_2O_{4.5}$ nanocubes with enhanced electrocatalytic activity for Na–O_2 batteries. Nanoscale 11:5285–5294

Logan BE, Hamelers B, Rozendal R, Schröder U, Keller J, Freguia S et al (2006) Microbial fuel cells: methodology and technology. Environ Sci Technol 40(17):5181–5192

Lu M, Guo L, Kharkwal S, Ng HY, Li SFY (2013) Manganese-polypyrrole-carbon nanotube, a new oxygen reduction catalyst for air-cathode microbial fuel cells. J Power Sources 221:381–386

Mashkour M, Rahimnejad M, Mashkour M (2016) Bacterial cellulose-polyaniline nano-biocomposite: a porous media hydrogel bioanode enhancing the performance of microbial fuel cell. J Power Sources 325:322–328

Mishra P, Jain R (2016) Electrochemical deposition of MWCNT-MnO_2/PPy nano-composite application for microbial fuel cells. Int J Hydrog Energy 41:22394–22405

Nandy A, Kumar V, Kundu PP (2016) Effect of electric impulse for improved energy generation in mediatorless dual chamber microbial fuel cell through electroevolution of Escherichia coli. Biosens Bioelectron 79:796–801

Nitisoravut R, Thanh CN, Regmi R (2017) Microbial fuel cells: advances in electrode modifications for improvement of system performance. Int J Green Energy 14:712–723

Noori MT, Ghangrekar MM, Mukherjee CK (2016) V2O5 microflower decorated cathode for enhancing power generation in air-cathode microbial fuel cell treating fish market wastewater. Int J Hydrog Energy 41:3638–3645

Noori MT, Mukherjee CK, Ghangrekar MM (2017) Enhancing performance of microbial fuel cell by using graphene supported V2O5-nanorod catalytic cathode. Electrochim Acta 228:513–521

Noori MT, Bhowmick GD, Tiwari BR, Ghangrekar OM, Ghangrekar MM, Mukherjee CK (2018) Carbon supported Cu-Sn bimetallic alloy as an excellent low-cost cathode catalyst for enhancing oxygen reduction reaction in microbial fuel cell. J Electrochem Soc 165:F621

Ou L, Chen S (2013) Comparative study of oxygen reduction reaction mechanisms on the Pd (111) and Pt (111) surfaces in acid medium by DFT. J Phys Chem C 117:1342–1349

Papiya F, Pattanayak P, Kumar P, Kumar V, Kundu PP (2018) Development of highly efficient bimetallic nanocomposite cathode catalyst, composed of Ni:Co supported sulfonated polyaniline for application in microbial fuel cells. Electrochim Acta 282:931–945

Papiya F, Das S, Pattanayak P, Kundu PP (2019) The fabrication of silane modified graphene oxide supported Ni–Co bimetallic electrocatalysts: a catalytic system for superior oxygen reduction in microbial fuel cells. Int J Hydrog Energy 44:25874–25893

Pattanayak P, Pramanik N, Kumar P, Kundu PP (2018) Fabrication of cost-effective non-noble metal supported on conducting polymer composite such as copper/polypyrrole graphene oxide (Cu_2O/PPy–GO) as an anode catalyst for methanol oxidation in DMFC. Int J Hydrog Energy 43:11505–11519

Pattanayak P, Papiya F, Pramanik N, Kundu PP (2019) Deposition of Ni-NiO nanoparticles on the reduced graphene oxide filled polypyrrole: evaluation as cathode catalyst in microbial fuel cells. Sustain Energy Fuels 3:1808–1826

Pattanayak P, Papiya F, Kumar V, Singh A, Kundu PP (2020a) Performance evaluation of poly (aniline-co-pyrrole) wrapped titanium dioxide nanocomposite as an air-cathode catalyst material for microbial fuel cell. Mater Sci Eng C 118:111492

Pattanayak P, Pramanik N, Papiya F, Kumar V, Kundu PP (2020b) Metal-free keratin modified poly (pyrrole-co-aniline)-reduced graphene oxide based nanocomposite materials: a promising cathode catalyst in microbial fuel cell application. J Environ Chem Eng 8:103813

Peera SG, Maiyalagan T, Liu C, Ashmath S, Lee TG, Jiang Z, Mao S (2021) A review on carbon and non-precious metal based cathode catalysts in microbial fuel cells. Int J Hydrog Energy 46(4):3056–3089. https://doi.org/10.1016/j.ijhydene.2020.07.252

Pérez-Rodríguez P, Ovando-Medina VM, Martínez-Amador SY, Rodríguez-de la Garza JA (2016) Bioanode of polyurethane/graphite/polypyrrole composite in microbial fuel cells. Biotechnol Bioprocess Eng 21:305–313

Pu KB, Ma Q, Cai WF, Chen QY, Wang YH, Li FJ (2018) Polypyrrole modified stainless steel as high performance anode of microbial fuel cell. Biochem Eng J 132:255–261

Quan X, Mei Y, Xu H, Sun B, Zhang X (2015) Optimization of Pt-Pd alloy catalyst and supporting materials for oxygen reduction in air-cathode microbial fuel cells. Electrochim Acta 165:72–77

Roche I, Katuri K, Scott K (2010) A microbial fuel cell using manganese oxide oxygen reduction catalysts. J Appl Electrochem 40:13

Rudra R, Pattanayak P, Kundu PP (2019) Conducting polymer-based microbial fuel cells. In: Enzymatic fuel cells: materials and applications, vol 44. Materials Research Foundations, pp 173–187

Santoro C, Serov A, Narvaez Villarrubia CW, Stariha S, Babanova S, Schuler AJ, Atanassov P (2015) Double-chamber microbial fuel cell with a non-platinum-group metal Fe-N-C cathode catalyst. ChemSusChem 8:828–834

Sonawane JM, Al-Saadi S, Raman RS, Ghosh PC, Adeloju SB (2018) Exploring the use of polyaniline-modified stainless steel plates as low-cost, high-performance anodes for microbial fuel cells. Electrochim Acta 268:484–493

Tang X, Ng HY (2017) Cobalt and nitrogen-doped carbon catalysts for enhanced oxygen reduction and power production in microbial fuel cells. Electrochim Acta 247:193–199

Tang X, Li H, Du Z, Wang W, Ng HY (2015) Conductive polypyrrole hydrogels and carbon nanotubes composite as an anode for microbial fuel cells. RSC Adv 5:50968–50974

Teng C, Manhong H, Yuxuan H, Wei Z (2019) Enhanced performance of microbial fuel cells by electrospinning carbon nanofibers hybrid carbon nanotubes composite anode. Int J Hydrog Energy 44:3088–3098

Tiwari BR, Noori MT, Ghangrekar MM (2017) Carbon supported nickel-phthalocyanine/MnOx as novel cathode catalyst for microbial fuel cell application. Int J Hydrog Energy 42:23085–23094

Wang J, Wang J, Kong Z, Lv K, Teng C, Zhu Y (2017) Conducting-polymer-based materials for electrochemical energy conversion and storage. Adv Mater 29:1703044

Wen Q, Wang S, Yan J, Cong L, Pan Z, Ren Y, Fan Z (2012) MnO_2–graphene hybrid as an alternative cathodic catalyst to platinum in microbial fuel cells. J Power Sources 216:187–191

Wu G, Bao H, Xia Z, Yang B, Lei L, Li Z, Liu C (2018) Polypyrrole/sargassum activated carbon modified stainless-steel sponge as high-performance and low-cost bioanode for microbial fuel cells. J Power Sources 384:86–92

Yang W, Lu JE, Zhang Y, Peng Y, Mercado R, Li J, Chen S (2019) Cobalt oxides nanoparticles supported on nitrogen-doped carbon nanotubes as high-efficiency cathode catalysts for microbial fuel cells. Inorg Chem Commun 105:69–75

Yaping Z, Yongyou H, Sizhe L, Jian S, Bin H (2011) Manganese dioxide-coated carbon nanotubes as an improved cathodic catalyst for oxygen reduction in a microbial fuel cell. J Power Sources 196:9284–9289

Yong Y, Bo Z, Yongwon J, Shengkui Z, Shungui Z, Sunghyun K (2011) Iron phthalocyanine supported on aminofunctionalized multi-walled carbon nanotube as an alternative cathodic oxygen catalyst in microbial fuel cells. Bioresour Technol 102:5849–5854

Yuan Y, Zhou S, Zhuang L (2010) Polypyrrole/carbon black composite as a novel oxygen reduction catalyst for microbial fuel cells. J Power Sources 195:3490–3493

Yuan Y, Ahmed J, Kim S (2011) Polyaniline/carbon black composite-supported iron phthalocyanine as an oxygen reduction catalyst for microbial fuel cells. J Power Sources 196:1103–1106

Yuan H, Deng L, Tang J, Zhou S, Chen Y, Yuan Y (2015) Facile synthesis of MnO2/polypyrrole/MnO2 multiwalled nanotubes as advanced electrocatalysts for the oxygen reduction reaction. ChemElectroChem 2:1152–1158

Yuan H, Hou Y, Abu-Reesh IM, Chen J, He Z (2016a) Oxygen reduction reaction catalysts used in microbial fuel cells for energy-efficient wastewater treatment: a review. Mater Horiz 3:382–401

Yuan H, Deng L, Chen Y, Yuan Y (2016b) MnO$_2$/Polypyrrole/MnO$_2$ multi-walled-nanotube-modified anode for high-performance microbial fuel cells. Electrochim Acta 196:280–285

Zhang F, Cheng S, Pant D, Van Bogaert G, Logan BE (2009a) Power generation using an activated carbon and metal mesh cathode in a microbial fuel cell. Electrochem Commun 11:2177–2179

Zhang L, Liu C, Zhuang L, Li W, Zhou S, Zhang J (2009b) Manganese dioxide as an alternative cathodic catalyst to platinum in microbial fuel cells. Biosens Bioelectron 24:2825–2829

Zhang Y, Hu Y, Li S, Sun J, Hou B (2011) Manganese dioxide-coated carbon nanotubes as an improved cathodic catalyst for oxygen reduction in a microbial fuel cell. J Power Sources 196:9284–9289

Zhang B, Wen Z, Ci S, Mao S, Chen J, He Z (2014) Synthesizing nitrogen-doped activated carbon and probing its active sites for oxygen reduction reaction in microbial fuel cells. ACS Appl Mater Interfaces 6:7464–7470

Zhao N, Ma Z, Song H, Wang D, Xie Y (2018) Polyaniline/reduced graphene oxide-modified carbon fiber brush anode for high-performance microbial fuel cells. Int J Hydrog Energy 43:17867–17872

Zhou M, Chi M, Luo J, He H, Jin T (2011) An overview of electrode materials in microbial fuel cells. J Power Sources 196:4427–4435

Zhou X, Xu Y, Mei X, Du N, Jv R, Hu Z, Chen S (2018) Polyaniline/β-MnO2 nanocomposites as cathode electrocatalyst for oxygen reduction reaction in microbial fuel cells. Chemosphere 198:482–491

Zou L, Qiao Y, Zhong C, Li CM (2017) Enabling fast electron transfer through both bacterial outer-membrane redox centers and endogenous electron mediators by polyaniline hybridized large-mesoporous carbon anode for high-performance microbial fuel cells. Electrochim Acta 229:31–38

Chapter 24
Metal Oxide Nanostructured Materials for Photocatalytic Hydrogen Generation

Bishal Kumar Nahak, Lucky Kumar Pradhan, T. Suraj Kumar Subudhi, Arveen Panigrahi, Biranchi Narayan Patra, Satya Sopan Mahato, and Shrabani Mahata

Contents

24.1	Introduction		666
	24.1.1	Nanotechnology in Energy Systems	666
	24.1.2	Present Scenario of Conventional and Potential Methods of Energy Generation	668
24.2	Fundamental Photocatalytic Hydrogen Generation Process		669
	24.2.1	Thermochemical Water Splitting	670
	24.2.2	Photobiological Water Splitting	671
	24.2.3	Photoelectrochemical Water Splitting	672
	24.2.4	Photocatalytic Water Splitting	674
24.3	Heterojunction Architecture		678
	24.3.1	Type 1	679
	24.3.2	Type 2	680
	24.3.3	Type 3	681
	24.3.4	Z-Scheme	681
	24.3.5	Binary Semiconductor Photocatalyst	681
	24.3.6	Ternary Oxide Photocatalysts	683
24.4	Metal-Semiconductor Heterojunction Photocatalysis		684
24.5	Semiconductor-Semiconductor Heterojunction Photocatalyst		685
24.6	Modifications in Photocatalysts		686
	24.6.1	Metal and Non-metal Implantation	686
	24.6.2	Effect of Co-catalyst Loading	687
	24.6.3	Dye Sensitization	689

B. K. Nahak · L. K. Pradhan · T. S. K. Subudhi · A. Panigrahi · S. S. Mahato
Department of Electronics and Communication Engineering, National Institute of Science and Technology, Berhampur, India

B. N. Patra
Department of Electrical and Electronics Engineering, National Institute of Science and Technology, Berhampur, India

S. Mahata (✉)
Department of Chemistry, National Institute of Science and Technology, Brahmpur, India
e-mail: shrabani.mahata@nist.edu

© The Author(s), under exclusive license to Springer Nature Switzerland AG 2022
A. Krishnan et al. (eds.), *Emerging Nanomaterials for Advanced Technologies*, Nanotechnology in the Life Sciences, https://doi.org/10.1007/978-3-030-80371-1_24

24.7 Operating Conditions Affecting the Photocatalytic Hydrogen Generation............ 690
  24.7.1 Particle Size... 690
  24.7.2 Surface Area.. 690
  24.7.3 Reaction Temperature... 690
  24.7.4 Catalyst... 691
  24.7.5 Effect of pH.. 691
24.8 Photocatalytic Reactors.. 691
  24.8.1 Thin-Film-Type Photocatalytic Reactors..................................... 692
  24.8.2 Slurry-Type Photocatalytic Reactors.. 692
24.9 Challenges and Future Perspective... 693
24.10 Applications of Hydrogen Generation.. 694
  24.10.1 Hydrogen as a Feedstock... 694
  24.10.2 Hydrogen in Fertilizers Industries.. 694
  24.10.3 Fuel Industry... 695
  24.10.4 Methanol Production... 695
  24.10.5 Hydrogen in Electronics Industry.. 696
  24.10.6 Hydrogen as a Fuel.. 696
24.11 Conclusion... 696
References... 697

24.1 Introduction

24.1.1 Nanotechnology in Energy Systems

In this fast-moving and automated world, rise in world's demand for energy in a continuous manner made an obvious step for improvement of sustainable and efficient technology which helps in generation and storage of energy (Zang 2011). For this reason, nanoscience and technology advancement is the potential means for increasing energy efficiency across all sectors and balancing renewable energy output economically through new technologies and optimized manufacturing techniques (Steyn 2009).

Nanotechnology is used in different fields like medical and health care (Boisseau and Loubaton 2011), electronics and IT sectors, energy and many more (Wang and Wu 2012). Nanotechnology includes fabrication, designing and creation of device (Thomas et al. 2014). This vast field of attraction is due to its size (below 100 nanometres) (Roduner 2006). This feature helps in creation, storage and transfer of energy (Zhang et al. 2013). For the nanomaterials, a good advantage is the ratio in surface area to volume (Lahann 2008), which offers a reduction in weight with improved stability and mechanical characteristics. Generation of traditional energy using fossil fuels (Serrano et al. 2009) can be replaced with the renewable energy such as solar, tidal, geothermal, wind, biomass, or hydro energy, which provides important improvement in increasing environmental security (Krstić and Wells 2010). Nano-coated (Nguyen-Tri et al. 2018), wear-resistant drill samples, nano-membranes can expand the idea of carbon dioxide separation and environment neutral storage possibilities for power generation in coal power plants because the selectivity of molecules enables (Okeke and Iloanusi 2014; Lambauer et al. 2012; Choudhary et al. n.d.) and also extends the storage life and helps protect against various corrosion agents, in order to carry out this process for generation of power in an eco-friendly manner for long term (Fulekar et al. 2014). Nanotechnologies may make a decisive contribution to the realization of this vision by, among other things, nano-sensor devices and power-electronic components capable of solving extremely complex problems (Lin et al. 2014; Zeb et al. 2019).

The usage of nanotechnology in energy conversion and energy generation in the following areas:

Batteries: When the battery is not in operation, nanomaterials can be used as a coating to isolate the electrodes from any fluids in the battery. The liquids and solids interact in the present battery technology, producing a low-level discharge. The shelf life of a battery is shortened by this (Abu-Lebdeh and Davidson 2012; Wong and Dia 2017).

Geothermal: By allowing efficient energy generation at lower temperature, nanoscience is now helping to make geothermal power more practical. It enables the lifetime and performance optimization of systems for the production of deposits of petroleum and natural gas or geothermal energy, thereby saving costs. The fluid's heat-retaining properties are also strengthened by nanoparticles (Wiesner and Bottero 2007; Serrano et al. 2009).

Nano-optimization for fuel cell and solar cell: The use of nanomaterials such as lead selenide results in the release of more electrons (and thus more electricity) when struck by a light photon. In addition, nanotechnology modifies the structural properties of PV cells (Ghasemi et al. 2013; Gong et al. 2009; Wang et al. 2011; Tsuchiya et al. 2011; Sethi et al. 2011; Wong et al. 2014; Sharma et al. 2015).

Nanomaterial in water technology: The use of nanomaterials has been considered in point-of-use water purification systems. Nanostructured materials, like greater relative surface areas, display many advantages over traditional microstructured materials for water purification. These techniques of chlorine-free water treatment are of major relevance as carcinogenic disinfection by products may arise when natural water constituents combine with chloramines or chlorine (Carpenter et al. 2015; Savage and Diallo 2005).

Nanomembrane and nano-sieve: A type of membrane which can handle high temperature and useful in petrochemical industry for the preplacement of conventional separation technique and applications which lies in the area of removing water from solvents and bio-fuels (Rogers et al. 2011; Technip n.d.; Wang et al. 2018).

Nuclear reactions: Low-energy nuclear reactions (LENRs) are produced from hydrogen (or deuterium)-loaded metal alloy nanoparticles by pressurizing the particle-containing vessel. The key finding to date is that the excess energy found in experiments to date is well above the highest estimate of what could be applied to known chemical reactions (Sanchez and Sobolev 2010; Keller 2007).

Nanotechnology in energy applications: This technology is used mostly for energy efficiency, eco-friendly in the areas of hydrogen generation, membranes and other energy areas (Hussein 2015; Serrano et al. 2009; Hu et al. 2010).

24.1.2 Present Scenario of Conventional and Potential Methods of Energy Generation

As of now, there are different methods of energy generation and conversion in the world with the help of different machineries. The methods are as follows:

Coal Power Generation: Coal-based power plants produce steam by burning coal in a boiler. This steam under high pressure helps in spinning the turbine and creating electricity. Around 41% of energy consumption in the world is based on coal. Bituminous coal is used for energy production (Mokhatab et al. 2006; Diallo and Brinker 2011; Spliethoff 2010).

Nuclear Power Generation: This also works in the concept of transformation of thermal to electrical power, while the nuclear reactor acts as a heat source. Heat is used to produce steam which can be used to run the turbine and hence produces the required electricity. 11% of the world's energy consumption is produced from nuclear energy. U-235 is basically used for nuclear energy production (Häfele 1990; Bang and Jeong 2011; Serrano et al. 2009; Lenzen 2008; Abu-Khader 2009).

Hydropower Generation: This basically converts the potential energy to kinetic energy which results in rotation of shafts in a turbine which in return produces electrical energy. Around 16% of the world's energy consumption is based on this form of energy. This is basically built across the rivers in the form of dams (Singh and Singal 2017; Chen and Chen 2013; Serrano et al. 2010; Madhavan 2020).

Natural Gas Power Generation: This also works in the mechanism of thermal power plants where the natural gas is used for steam generation which makes the shaft rotate and thus produce electricity. 11% of the world's energy depends on this (Gür 2016; Heppenstall 1998; Kumar et al. 2011; Elcock 2007; Esmaeili 2011).

Geothermal Power Generation: This is mainly based on geothermal energy which is derived from the Earth's subsurface in the form of steam or water for production of clean electricity (Bayer et al. 2013; Zarrouk and Moon 2014).

Wind Power Generation: The main source of this is wind energy which helps rotate wind turbines which then helps in conversion of mechanical to electrical energy. Basically, this is used in high altitudes as there is chances of high wind flow (Vargas et al. 2019; Li et al. 2011; Archer and Jacobson 2005; Lu et al. 2009; Foley et al. 2012; Sivaramakrishna and Reddy 2013).

Wastewater and Biogas Energy Generation: In most instances, the biogas is utilized in combustible engines which converts into mechanical energy powering the electric generators, which probably produce electricity directly by fuel cell. But this method involves relatively clean gas and costly fuel cells. This gas has carbon dioxide, methane and traces of hydrogen, nitrogen and carbon monoxide (Hahn et al. 2014; Rasi et al. 2011; Wiley et al. 2011; Arjuna et al. 2017; Shen et al. 2015).

Tidal Power Generation: The tides coming in a seabed is basically used for this, but at present scenario, the commercial use of this energy is quite less. But this can be potential energy for energy generation. At present, this is converted into different forms of power but mainly electricity (Adcock et al. 2015; Khan et al. 2017;

Westwood 2004; Khan et al. 2009; Schiermeier et al. 2008; Rourke et al. 2010; Ahmadi et al. 2018).

Solar Power Generation: In this case, the sun is the source of energy with the help of photovoltaic cells which converts the sun radiation into electric, and it is called solar thermal power generation. Using solar panels, we can directly transform solar radiation into electricity. In the recent days, there are different methods coming up which uses solar energy into different energies like hydrogen energy, etc., with the help of modified photovoltaic cells (Khan and Arsalan 2016; Qi et al. 2018; Veziroglu 2007; Singh 2013; Hosenuzzaman et al. 2015; Dincer 2011; Müller-Steinhagen and Trieb 2004; Abdin et al. 2013; Vayssieres 2010; Reddy et al. 2014).

In present scenario, the total world production of energy is produced by fossil fuels (80.6%), renewable sources (16.7%) and nuclear (2.7%) as of 2010 (Fridleifsson 2001; Boyle 2004). The major renewable sources are biomass energy (11.44%), hydro energy (3.34%), wind energy (0.51%), etc. (Fronk et al. 2010; Holm and Arch 2013; Goldemberg 2000).

The upcoming renewable energy technologies are solar power, floating wind turbines, printable organic cells, biomass gasification, tidal energy, microbial fuel cell and different starting stage technologies like graphene-based super caps, monolithic microscale heat pumps, hydrogen energy, etc. which are nanotech-based projects which have futuristic scope (Ellabban et al. 2014; Dincer 2000; Sun et al. 2011; Rabaey and Verstraete 2005; Santoro et al. 2017; Huang et al. 2012; Vivekchand et al. 2008; Iverson and Garimella 2008). The usage of these technologies is less in number because the installation cost of these projects are higher compared to present form of fossil fuel-based projects. In this past few years, the cost of production is decreasing, and efficiency is increasing day by day. Thus, the dependency of world market in this green technology is becoming greater day by day (Jhariya et al. 2018; Zerta et al. 2008).

24.2 Fundamental Photocatalytic Hydrogen Generation Process

In the growing world of the twenty-first century, finding an energy source which can fulfil all the demands to being a sustainable energy and environment friendly is still a major challenge for the world (Acar et al. 2016). Due to significant advantages of renewable energies, they are considered as a sustainable alternative over the fossil fuel (Bilgen et al. 2004). Fossil fuels are limited and release toxic gases into the atmosphere. With so many drawbacks, demand of fossil fuels is still at peak since its discovery and is used in diverse applications. The consumption of the highly in-demand fossil fuels leads to increasing rate of releasing CO_2 into the atmosphere. Due to human activities, the rate of net increase of CO_2 is 3×10^{12} kg yr^{-1}, which consequences the annual increase in CO_2 concentration by 0.4%. These kind of growth of CO_2 gases impacts on the world very negatively by increasing the intensity and frequency of weather events like floods, hurricanes, heat waves, etc. which put a direct impact on glacier retreat and agricultural yield and decrease the rate of growth in human development. After overlooking all the consequences of using fossil fuels, hydrogen is considered as a clean source of energy to overcome all drawbacks of fossil fuel and cover the demand of energy in future. It's carbon-free impacts on environment can be an advantage for the future generation to view the nature and utilize it properly (Züttel et al. 1923).

24.2.1 Thermochemical Water Splitting

Among the various ways for splitting of water, thermochemical water splitting process is a very old and impactful process. It requires high temperature, which can be taken by concentrated solar power (Abanades et al. 2006). Another way to produce the high temperature is from the waste heat of nuclear power reactions and chemical reactions, which acts as a very feasible solution to generate hydrogen and oxygen from water through thermochemical water splitting process (Brown et al. 2002).

In the thermochemical water splitting technique, the major energy source is high temperature, which separate the hydrogen from the mixture which was earlier in equilibrium condition. Gibbs function (ΔG or free energy) needs to be zero for this process; otherwise, the hydrogen and oxygen will mix together to form water. To make sure the Gibbs function to be zero, the temperature needs to be about 4700 K. Apart from the temperature, there are also material separation at high temperature, and STP thermodynamics of water are also some difficulties in this process. However, after that project of the 1960s, many attempts were done to improve the efficiency and feasibility of the process (Funk 2001).

Attempts to increase the efficiency of the cycle (process) was taken into good result in the late 1960s. Hydrogen generation rate to the amount of heat supplied shows efficiency of this process. Some cycles were done using different elements like Cl, S, Li, Ni, Fe, Ca, I, Ba, Sb, etc. The Fe-Cl family cycles among others show some good results (Funk 2001). After all this experiments, some projects were done in international level at Ispra, Italy, through a program named Direct Production of Hydrogen with Nuclear Heat. The summary of the research performed was published by Beghi (Beghi 1986). The main outcome of this program is a process which is known as "GA process". That process was on the development of Mark 16 process, which is also known as the "three closures for sulphuric acid decomposition" shown in Fig. 24.1. The attractive thing in GA process is the determination of Bunsen reaction of iodine, sulphur dioxide and water. The overall efficiency of this process was estimated to be about 47% (O'keefe et al. 1982).

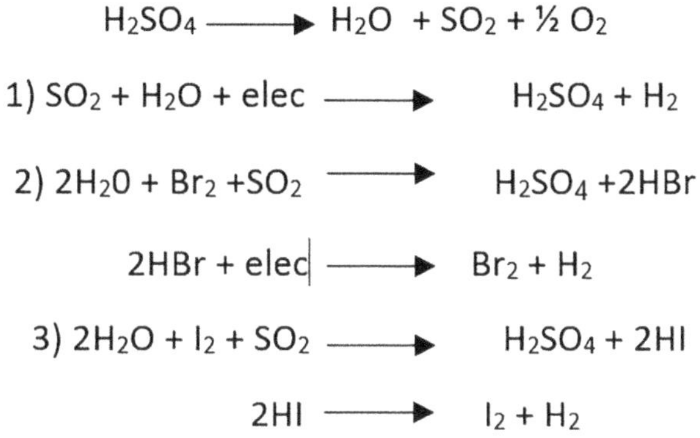

Fig. 24.1 Three closures for sulphuric acid decomposition (Abas et al. 2015)

$$H_2SO_4 \longrightarrow H_2O + SO_2 + \tfrac{1}{2} O_2$$

1) $SO_2 + H_2O + elec \longrightarrow H_2SO_4 + H_2$

2) $2H_2O + Br_2 + SO_2 \longrightarrow H_2SO_4 + 2HBr$

$2HBr + elec \longrightarrow Br_2 + H_2$

3) $2H_2O + I_2 + SO_2 \longrightarrow H_2SO_4 + 2HI$

$2HI \longrightarrow I_2 + H_2$

24.2.2 Photobiological Water Splitting

In the present world, the consumption of fossil fuel only covers 86% demand of required energy. Photobiological water splitting is an effective way to generate hydrogen molecules by using photoautotrophic microorganisms from water and sunlight (Ghirardi et al. 2009). These microorganisms such as algae, dark fermentative bacteria, sulphur bacteria and cyanobacteria are used to generate biohydrogen by consuming the carbon dioxide gas (CO_2) (Benemann 1997). In this whole process, the sunlight and the microorganisms' main role is to produce biohydrogen. Basically, these microorganisms synthesized the hydrogen using CO_2 and H_2O. There are two techniques for this process: (i) direct photobiological hydrogen generation and (ii) indirect photobiological hydrogen generation. In direct process, hydrogen is directly generated by hydrogenase process with the help of molecules such as carbohydrates, whereas in indirect method, the gas is produced after storing glycogen or carbohydrates (Dasgupta et al. 2010). Hence, here is the described reaction of indirect method:

$$6H_2O + 6CO_2 + light \rightarrow C_6H_{12}O_6 + 6O_2 \tag{24.1}$$

$$C_6H_{12}O_6 + 2H_2O \rightarrow 4H_2 + 2CH_3COOH + 2CO_2 \tag{24.2}$$

$$2CH_3COOH + 4H_2O + light \rightarrow 8H_2 + 4CO_2 \tag{24.3}$$

In the process of photobiological hydrogen production, the microalgae donate e^- for which it requires sunlight. This process can produce oxygen or not depending upon the microorganisms carried out in this process. Oxygen is generated when the process was carried out by eukaryotic microalgae and cyanobacteria (Melis et al. 2000), whereas the process does not generate oxygen when purple non-sulphur bacteria takes part in this process (McKinlay 2014). There are two enzymes available, i.e. hydrogenase and nitrogenase, which activities concluded the production of biohydrogen depending upon the presence of molecular oxygen. However, the molecular oxygen produced by photosynthesis creates a burden for the microorganisms; the production of biohydrogen is shown in Fig. 24.2 for which the researches are going to overcome this problem (Prince and Kheshgi 2005).

Earlier there are only two methods for this process, i.e. aerobic and anaerobic. However, these reactions can give a short description about the two methods (Kruse et al. 2005):

$$H_2O \rightarrow 2H^+ + 2e^- + \tfrac{1}{2}O_2 \; (\text{Aerobic}) \tag{24.4}$$

$$2H^+ + 2e^- \rightarrow H_2 \; (\text{Anaerobic}) \tag{25.5}$$

24.2.3 Photoelectrochemical Water Splitting

Direct use of light radiation for splitting water to oxygen and hydrogen, when light energy interacts with the surface of semiconductor (SC), is called photoelectrochemical (PEC) water splitting. Producing of hydrogen with this process is very acceptable for current world and future world. In this process, the first step includes choosing a suitable photoelectrochemical material or SC, which has suitable band

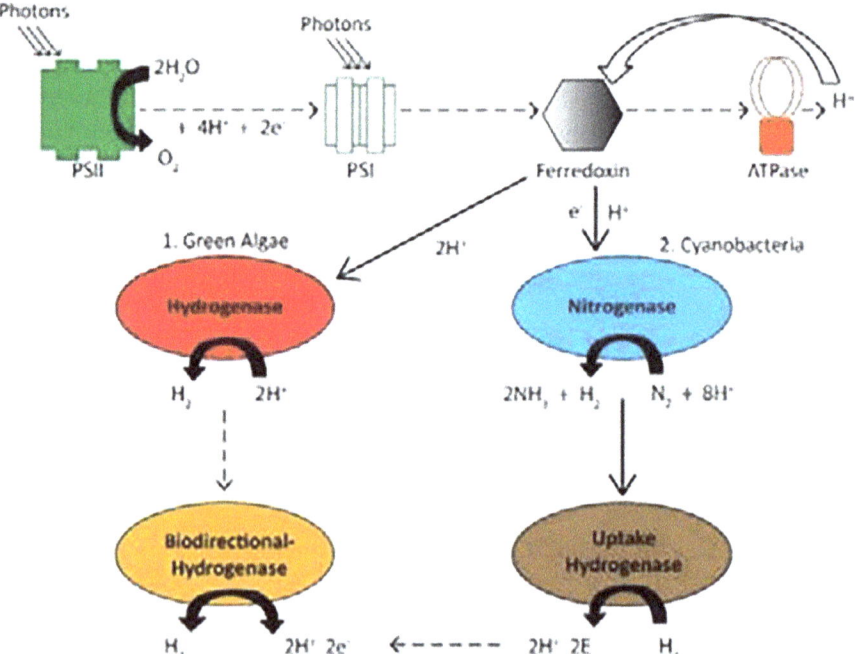

Fig. 24.2 Mechanism of photobiological hydrogen production

gap, efficient performance in visible light, photostable, resists to photo corrosion, etc. However, in a single material, it is very difficult to find all these properties as well as at affordable cost for large scale implementation. Because some material has small band gap but may not have good stability, so in this way different materials have different types of advantages and disadvantages. To overcome this burden, many ways have been founded to develop the material properties, which have put a positive factor on the efficiency of the substrate for production of hydrogen and for industrial-level purpose also.

As choosing of material is an important step of this process, Fig. 24.3 shows the band gap of different materials and the position of CB and VB according to E vs NHE bar. For water splitting action, the CB and VB need to be at least 1.23 eV band gap, so that both the reaction, i.e. photo reduction at CB and photo oxidation in VB, can be processed smoothly. We can consider TiO_2 for this process due to its number of advantages, which can clarify the reasons of each material properties required for hydrogen production BY photolysis of water. If we consider the minimum required band gap and position of CB and VB for the photoreduction and photooxidation, respectively, then we may find many materials. But some have large band gap, and there might not be any positive future modifications that can be done with the material for this disadvantage. However, in TiO_2, some modifications have been done for its large band gap issue.

Mechanism has been shown in Fig. 24.4. The light energy having adequate energy to penetrate the material can excite an electron to jump from VB to CB. Further, a hole is generated at VB which interacts with the water and generates oxygen molecule and H^+ ions, which will interact with the free electron and form hydrogen molecule.

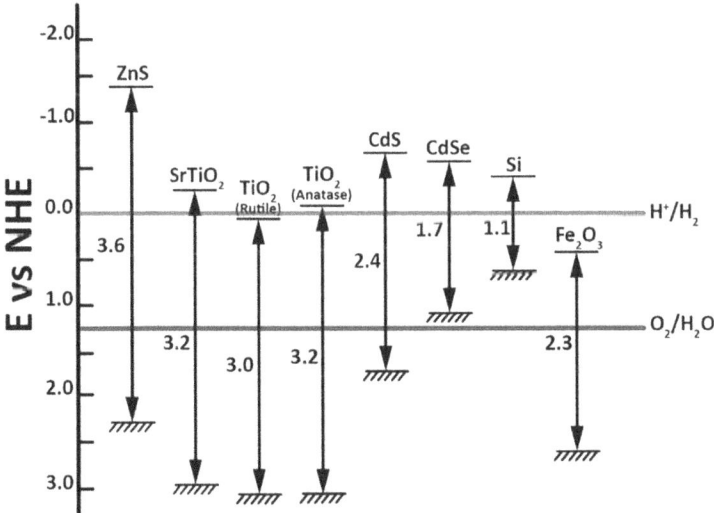

Fig. 24.3 Band gap of different semiconductors

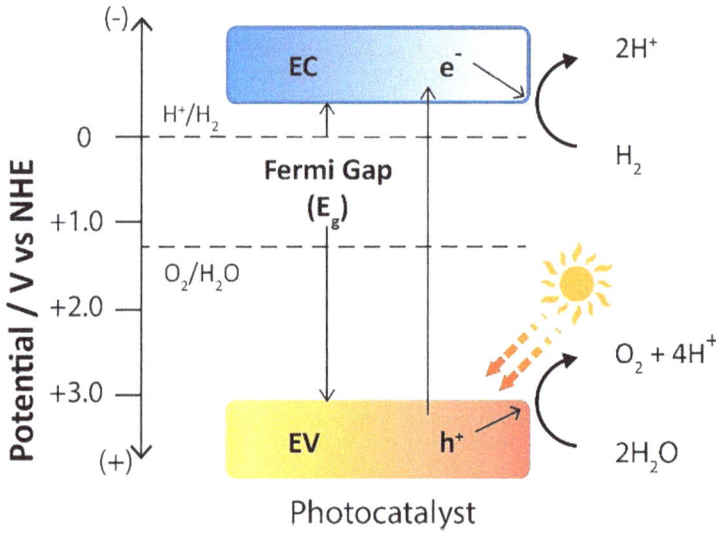

Fig. 24.4 Mechanism of photocatalytic hydrogen generation

However, apart from all these processes, there are some limitations like not able to utilize longer-wavelength photons or light energy from sun for hydrogen production. Some recent development was done to overcome these burdens up to some instant, and they are doping of metals in semiconductor to lower its band gap and make it functionable at the frequency of visible light or little bit longer wavelength. This metal doping also helps in decreasing the rate of recombination of electron and holes which are generated to interact with water, so that the efficiency can increase. Apart from all these developments, there are also heterogeneous structures formed for better efficiency.

24.2.4 Photocatalytic Water Splitting

The process in which water splits using semiconductor as a catalyst is called photocatalytic water splitting. It was believed that this process was discovered in the early 1980s, and from that moment till now, the research is going on for the development of its catalyst, as it is acting as a main role in the process of photocatalysis; however, the cocatalyst's improvement is also important (Maeda 2011). Titanium dioxide or titania (TiO_2) has acquired a major role in this field and was studied over more than quarter of a century because of its suitable material properties which are very much reliable for this application. Those reliable qualities are its robust architecture, which makes it more stable than other materials; low cost, which makes it more suitable for commercial purpose; and environment friendly, a very much required advantage to replace fossil fuel in the field of energy. However, it has one major drawback, which lags its efficiency and purpose of application. That major drawback is its large energy band gap (~3.3 eV), which ensures its efficiency is maximum only in the UV radiation than the visible light, which makes it very much unrealistic to be implemented for practical purpose (Luo et al. 2004; Dholam et al. 2009; Selcuk et al. 2012). No doubt the researches are going on regarding its improvement, and many ways have been founded (Fujishima and Zhang 2006). Some ways to improve the material properties of TiO_2 are metal oxide doping, dye sensitization, heterogeneous structure, etc. These ways not only make TiO_2 work under visible light but also improve its efficiency by increasing some other aspects which ensure its growth in production of hydrogen from water splitting (Ni et al. 2007). Apart from TiO_2, there are also many materials that have been founded performing good performance like TiO_2; they are ZrO_2, $SrTiO_3$, $KTaO_3$ and $BiVO_4$ (Sakata et al. 1983; Serpone and Pelizzetti 1989), but there are some SCs available which are quite less compatible for this application due to their drawbacks, and those drawbacks are not environment friendly, not stable with its properties for this particular application and photo corrosion occurs while application is going on, which means anion generated from photocatalyst is oxidized by the photogenerated holes instead of water. Those semiconductors are SiC, CdS and ZnO (Kudo and Miseki 2009; Linsebigler et al. 1995; Zou et al. 2011).

24.2.4.1 Thermodynamics and Kinetics of Photocatalytic Water Splitting

Photocatalysis is an endothermic and a multi-electron process (Colón 2016). This process requires an initial energy to fulfil Gibbs free energy change which is $\Delta H° = 238$ kJ mol^{-1} and can be used for rearranging valence electrons of water for the production of hydrogen and oxygen molecules (Maeda 2011; Acar et al. 2014). In general, water splitting process can be explained by Eqs. (24.6) and (24.7). As shown in Eq. (24.6), to generate a single hydrogen molecule, the energy necessary by H_2O to decompose and produce hydrogen molecule is 2.458 eV (Zamfirescu et al. 2012). Since complete water splitting process requires four hydrogen molecules as shown in Eq. (24.7), required input energy will be 4.915 eV, which could be done using UV radiation of smaller wavelength than 252.3 nm or by photons from visible wavelength (<504.5 nm) (Zamfirescu et al. 2011). This process involves two half reactions, oxidation of water to form O_2 molecule and reduction of protons to form H_2 molecule.

$$\text{Half reactions} \quad H_2O \xrightarrow{TiO_2/h\nu} \frac{1}{2}O_2 + 2H^+; \quad 2H^+ + 2e^- \rightarrow H_2 + \frac{1}{2}O_2 \quad (24.6)$$

$$\text{Complete water splitting reaction} \quad 2H_2O \xrightarrow{TiO_2/h\nu} O_2 + 4H^+; \\ 4H^+ + 4e^- \rightarrow 2H_2 + O_2 \quad (24.7)$$

When the photocatalyst is subjected to radiation having larger energy than E_g, it excites the e$^-$ to CB and leaves a hole behind at VB. These excited electrons and holes can move freely so they have chances of being delocalized within the semiconductor. As a result, the electrons may quickly attain internal equilibrium within energy level rather than moving out beyond band gap as shown in Fig. 24.5. Electron state at equilibrium known as "quasi-equilibrium states" and potential of e⁻ and h⁺ in these states are shown in Eqs. (24.8) and (24.9) (Liu et al. 2014; Shehzad et al. 2018). In radiation energy more than E_g, e⁻ and h⁺ dependent reactions can be explained from Eq. (24.10). Although when photocatalyst is at thermal equilibrium $\Delta H = 0$, i.e. $\Delta G = 0$, as a result, zero net force requires to carry out the photocatalysis reaction. This shows heat is not responsible force for generating e⁻ and h⁺ pairs. Thus, for photocatalysis:

$$F_n = E_c + k_B T \ln \frac{n}{N_c} \tag{24.8}$$

$$F_p = E_v + k_B T \ln \frac{n}{N_v} \tag{24.9}$$

$$\Delta G = -\#F_n - F_p\# = -E_g - k_B T \ln \frac{np}{N_c N_v} \tag{24.10}$$

The efficiency of hydrogen generation is affected by band gap and intensity of light. Mainly, there are two types of light being considered from the electromagnetic spectrum: UV light whose wavelength range lies in between 200 and 400 nm, while the other one is visible light whose wavelength lies in between 400 and 800 nm. Light absorption gets very less in semiconductors having wider band gap. Semiconductor having band gap higher than 3.15 eV can be activated in ultraviolet light; semiconductor having band gap lower than 3.15 eV visible light can be used to drive reaction. Thus, a semiconductor is suitable for being a photocatalyst having band gap 1.23 eV < Eg < 3.15 eV (Yan et al. 2010; Chun et al. 2003).

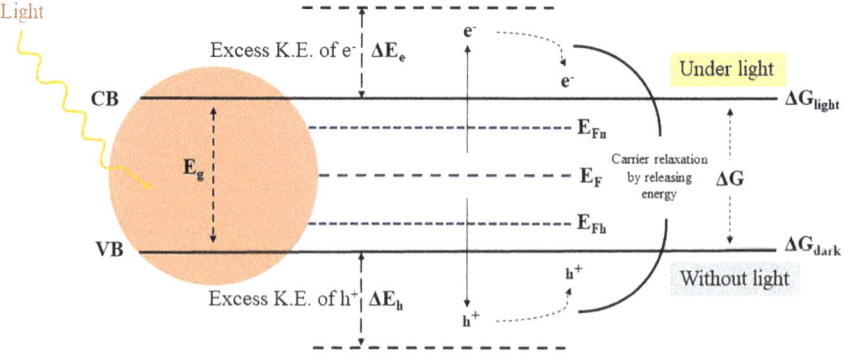

Fig. 24.5 Thermodynamics and Gibbs free energy in presence and absence of light

24.2.4.2 Hydrogen Generation Efficiency

Comparing the efficiency of photocatalysis has always been difficult to judge, as different researchers used different methods or chemical tests for their respective experiment. These chemical tests mainly include the finding of solar-to-hydrogen conversion and quantum efficiencies or apparent quantum yields of the photocatalytic activities. Basically, at particular wavelength, the quantum yield is calculated, similarly at particular spectrum of light the solar-to-hydrogen conversion calculated. However, some universal principles or standards have been taken into consideration for finding all these efficiencies. Solar-to-hydrogen conversion is the very genuine term used for finding the efficiency of hydrogen production from any process.

That term is given here by Eq. (24.11):

$$STH = \left[\frac{\left| j_{sc} \left(\frac{mA}{Cm^2} \right) \right| \times (1.23\,V) \times \eta F}{P_{total} \left(\frac{mW}{cm^2} \right)} \right] AM1.5G \qquad (24.11)$$

where power density of incident light (AM 1.5G) is represented by P_{total} and the product of faradic efficiency (η_F), the thermodynamic voltage necessary for decomposition of water (1.23 V) and photocurrent density (j_{sc}) at zero potential (short-circuit photocurrent) are in the numerator. Similarly, SHE can be represented as shown in Eq. (24.12):

$$STH = \left[\frac{(mmole\,H_2/s)(237\,KJ/mole)}{P_{total}\left(\frac{mW}{cm^2}\right) \times Area(cm^2)} \right] AM1.5G \qquad (24.12)$$

where the product of area in contact with incident light and total power the incident sunlight (AM 1.5G) are represented in denominator section. In numerator, the H_2 evolution rate is multiplied by the Gibbs free energy required for the generation of one mole of hydrogen from water.

There are some conditions and methods upon which the observable quantum yields depend, i.e. the reaction conditions, measuring methods, etc. So, Eq. (24.13) shown here is the general equation for calculating apparent quantum yields. This equation is only valid for catalysts radiated at source of monochromatic wavelength:

$$\text{AQE\%} = \frac{\text{No. of reacted electrons}}{\text{No. of incident photons}} \times 100$$
$$= \frac{2 \times \text{No. of evolved H}_2 \text{ molecules}}{\text{No. of incident photons}} \times 100 \quad (24.13)$$

However, hydrogen generation performance could be evaluated by the photocatalysts radiated under sunrays. This is shown in Eq. (24.14):

$$\text{SHE\%} = \frac{\text{Output energy as H2 evolved}}{\text{Energy og incident solar light}} \times 100 \quad (24.14)$$

After all these equations, one thing can be concluded that depending upon the light source, different equations can be applied for calculating the efficiency, but for one particular system or source of light, all the conversation cannot be applied.

24.3 Heterojunction Architecture

Till today, various semiconductor photocatalyst (e.g. Fe_2O_3 (Sivula et al. 2010), WO_3 (Hodes et al. 1976; Bignozzi et al. 2013), $BiVO_4$ (Sayama et al. 2003; Li et al. 2013)) have been reported for water splitting purpose. Out of all these, very less amount of materials is able to split water in visible light conditions. This poses a major downside and decreases the efficiency of solar-to-hydrogen conversion, which eventually affects the rate of generation of hydrogen. Thus, there is an urgent need of new materials and strategies for increasing efficiency in energy conversion. Lowering the rate of recombination of the generated e^- and h^+ pairs is one of the main factors in rising the hydrogen gas generation. When light falls on the top of SC, e^- and h^+ pairs are produced, the e^- from the VB promoted to the CB leaving behind a h^+ at VB.

There are different techniques reported for lowering the recombination speed of e^- and h^+ pairs: (i) using sacrificial agents, it removes either of the holes or electrons in the system and making the other charge carriers isolated which helps to carry out one half reaction (oxidation/reduction) (Kudo et al. 1998); (ii) alternating the morphology of the photocatalyst, this significantly increases the efficiency as it increases the surface region and shortens the length of diffusion of the charge carriers to surface (Moniz et al. 2015); (iii) creating heterojunction structure, charge carriers generated are transferred to other photocatalyst thereby increasing the lifetime of electro-hole pairs (Rhee et al. 1998; Martin et al. 2014).

Different types of heterojunctions could be achieved by adding a SC with a metal forming a Schottky junction, whereas the assembling of two SC materials can give rise to a semiconductor heterojunction. These SC materials are of two types, p-type (majority holes carriers) and n-type (majority electron carriers). Based on this,

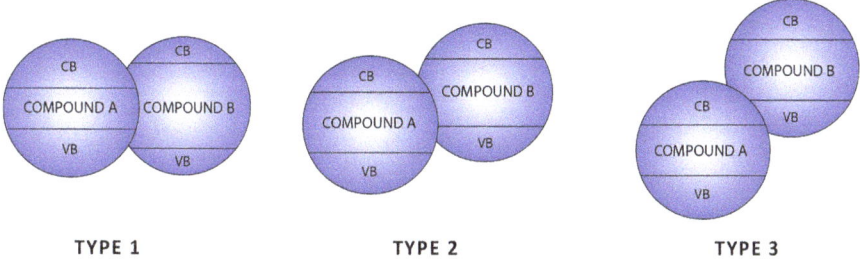

Fig. 24.6 Band alignment in type 1, type 2 and type 3 heterojunctions

heterojunction can be classified into three categories: (i) p-p type, two p-types of SCs; (ii) n-n type, two n-types SCs; and (iii) p-n type in which one p-type semiconductor and one n-type semiconductor is used. There exist three forms of heterojunctions for band alignment as shown in Fig. 24.6.

24.3.1 Type 1

In type 1 heterojunctions, first material E_{CB} is more −ve compared to that of second, and it's E_{VB} more +ve than second. It has been reported that electron and holes tend to flow to less −ve E_{CB} and less +ve E_{VB} in SC-SC heterojunction, respectively. In this alignment, electrons start accumulating at SC having lower band gap (in this case second material, SC-2); therefore, the probability of recombination of these electrons is more which may not bring any significant improvement in photocatalysis. Further, this leads to creating very weak internal electric field, which helps in better charge separation. When first material has higher E_F level compared to second, until E_F levels are synchronized, electrons will start flowing towards second material as shown in Fig. 24.7(a). This contributes to formation of the space charge at the junction. Since first material is more +vely charged, the path of the internal electric field would be from first to second material (Zhang and Yates Jr 2012). This means after incident of photon the transfer of electron is not allowed but at the same time holes can migrate from SC-1 to SC-2. When we reverse the conditions, i.e. SC-2 becomes more positive than SC-1, electrons start migrating from SC-2 to SC-1, whereas holes are prohibited from migration and get accumulated at the interface shown in Fig. 24.7(b).

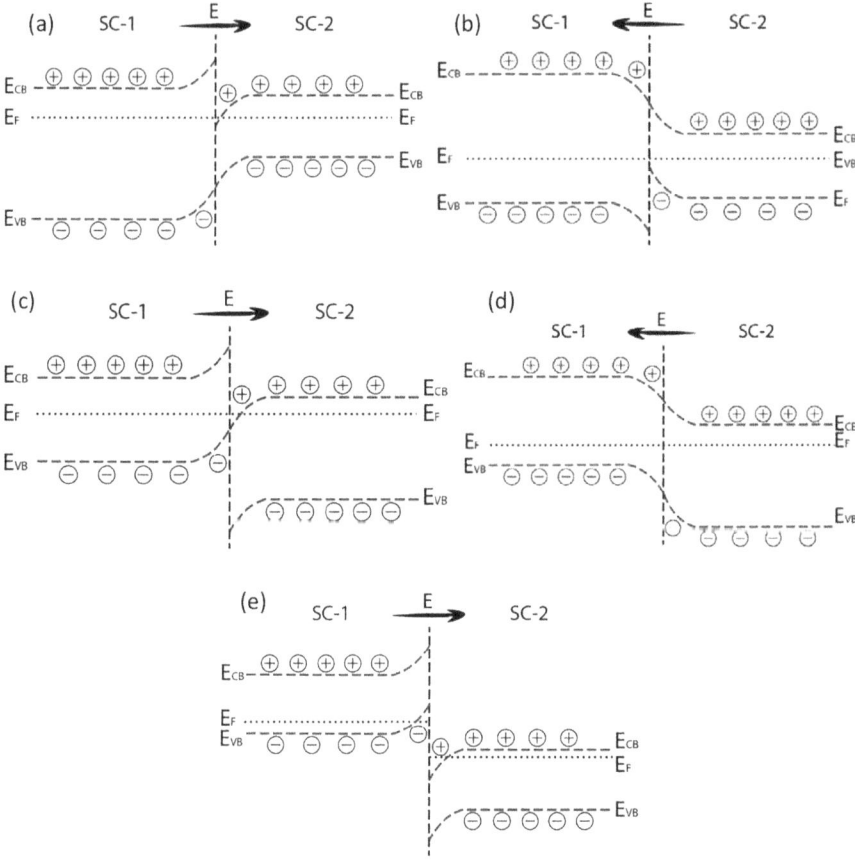

Fig. 24.7 Types of heterojunctions and their charge transfer due to existing internal electric field

24.3.2 Type 2

In type 2 heterojunction, first material E_{CB} is more −ve, while E_{VB} is less positive compared to second material. This alignment supports better charge separation. First material has a higher E_F level than second, and an internal electrical field has occurred in the direction from material 1 to material 2 shown in Fig. 24.7(c). The generated e^- in space charge area of first material will migrate towards interior, and h^+ will travel near to interface at the instant of time the charge carriers of SC-2 will oppose the flow of SC-1. Thus, recombination will occur at interface due to transfer of h^+ of material 1 and e^- from material 2 (QingáLu 2009), but material 1 e^- and material 2 h^+ are retained for catalytic reactions. All these process takes form of Z-shape therefore it is referred as Z-scheme (Liu et al. 2016; Jin et al. 2015; Zhou et al. 2014). When E_F of material 1 is lower than material 2 and material 1 is p-type SC and material 2 is n-type SC, then the internal electric field (E) will flow from n-type to p-type shown in Fig. 24.7(d). As the E_{CB} of n-type is less −ve than p-type, e^- of p-type will start migrating towards interface and will transfer to E_{CB} of n-type, and similarly, holes will move to E_{VB} of n-type.

24.3.3 Type 3

In type 3 heterojunction, both E_{CB} and E_{VB} of material 1 are higher than that of material 2. In this type of alignment, both of the SCs across the junction can be treated as individual SCs, because e^- and h^+ transfer doesn't take place in between them shown in Fig. 24.7(e). Both h^+ of material 1 and e^- of material 2 will migrate towards junction; there they get intervened by energy levels barrier. A twisted Z-scheme can be formed by introducing an appropriate charging mediator as a bridge connecting these two semiconductors (Gholipour et al. 2015).

24.3.4 Z-Scheme

The groundbreaking idea of developing a Z-scheme photocatalysis was initiated by Bard in the year 1979. His idea was to introduce an electron mediator as a charge carrier channel between two photocatalytic systems which can be further utilized for oxidation and reduction process (Bard 1979). This proposed idea was first coming into demonstration in the year 2001, where IO_3^-/I^- shuttle redox couple was placed as an ionic mediator to monitor the charge transfer between two photosystems (Sayama et al. 2001). This became the first-generation Z-scheme photocatalytic systems, which brought a significant change in hydrogen generation process. Moreover, recently, Wang et al. proposed the idea of Z-scheme photocatalyst sheets, which confirms to be a potential candidate for STH conversion (Wang et al. 2016). Different structures of z-scheme have been studied such as powder photocatalysis, thin film form of Z-scheme which have dual layer of particulate sheet. Research on the evolutionary idea of Z-scheme photocatalysis was started since 1979, and till today extensive efforts and numerous studies have been devoted to develop a photocatalytic system which can fully utilize solar radiation.

24.3.5 Binary Semiconductor Photocatalyst

24.3.5.1 Oxide-Based Binary Photocatalyst

Teoh et al. synthesized TiO_2 nanoparticles (NPs) via the flame spray pyrolysis (FSP) process, which showed higher activity of photomineralization of sucrose shared with commercially available Degussa P-25 (Teoh et al. 2005). FSP techniques yields

NPs which have high crystallinity, regulated surface area and morphology without involving any post-treatment process. He also published a comprehensive analysis of his successful attempt for using FSP to produce different binary photocatalysts (Teoh 2013). Similarly, Zhanxia et al. developed nanoflower made up of zirconia (ZrO_2). They used zirconium oxide sulphate and sodium acetate as precursor materials and adapted hydrothermal process to furnish the required 3D nanoflower. They found that using ZrO_2 nanoflower enhances photocatalytic activity for dye degradation and hydrogen generation which could be due to superior absorbance of photons. Many researchers have explored ZnO and CeO_2 nanomaterials synthesized in hydrothermal process for hydrogen generation and found to have well-defined properties for photocatalysis (Roig et al. 2011; Slostowski et al. 2013).

24.3.5.2 Nitride-Based Binary Photocatalyst

In comparison to binary chalcogenides/oxides, binary nitrides are being recently explored for photocatalytic purpose. Gallium nitride was chosen as a parent semiconductor photocatalyst after having seen its excellent properties in the field of sensing because of its stability and mechanical power (Ye et al. 2018). Kida et al. was first to report the use of GaN for water splitting, for which he used GaN in powdered form (Hu et al. 2005). Apart from GaN, tantalum nitride is an important transition binary nitride compound used for photolysis of water, due to its narrow band gap and preferable band gap potentials (Luo et al. 2012; Jing and Guo 2006).

24.3.5.3 Chalcogenide-Based Binary Photocatalyst

Heterojunction SCs binary chalcogenides such as zinc sulphide (ZnS) and cadmium sulphide (CdS) photocatalyst were studied in vast scale in the last four decades because of its potential applications in environment purification, hydrogen production and CO_2 reduction. Out of these ZnS chalcogenides, one is most commonly studied due to its 3.2–4.4 eV band difference (Yonenaga 2001). Hu et al. reported a simple approach for ZnS nanoporous nanoparticles via solution phase thermal decomposition technique. He found that prepared samples outperformed the commercial TiO_2 in dye degradation by degrading eosin B dye (Kida et al. 2006). Further, he confirmed the enhancement due to spherical shape and high monodispersity.

Similarly, the synthesis of hollow spherical CdS was reported by Luo et al. (Tabata et al. 2010). Jing et al. developed a new technique to replace the conventional synthetic precipitation process for the generation of nanostructured CdS photocatalysts. Its thermal sulphidation process reduces the CdS phase transition and crystallization temperature. The synthesized CdS by this process enhanced their stability against air and photo corrosion. CdS-based photocatalyst displayed a significant performance due to its narrow band gap and higher conduction potential as compared to H_2O/H_2 reduction potential (Chun et al. 2003).

24.3.6 Ternary Oxide Photocatalysts

Ternary oxide SCs comprise of two different metallic cation and single anion. In recent decades, these materials are being studied more due to their stable nature.

24.3.6.1 Perovskite Photocatalysts (ABO_3 Type)

Perovskite oxide materials having general formula ABO_3 as SC photocatalyst has been used in wide scale due to phenomenal properties of converting STH. Here, A is a cation either rare earth metal or alkali, and B is cation transition metal element having smaller size than A. Its ideal structure is in cubic lattice form. Strontium titanate ($SrTiO_3$) has immense application in heterojunction photocatalysis because of its thermal stability and photo-corrosion resistance, and its chemical physical properties can be altered by changing its composition (Fujinami et al. 2010). $SrTiO_3$ have band gap of 3.2 eV which helps to absolute mineralize or degrade organic dye and enhance water splitting reaction (Ohno et al. 2005; Ahuja and Kutty 1996; Kato and Kudo 2001). Unlike titanates, tantalates also exhibit photocatalytic properties, in which they can split water without any addition of reduction and oxidation catalyst, due to their high CB and VB edge than redox potential of water. Similarly, lithium tantalate ($LiTaO_3$) is also being utilized for decomposing water in the presence of UV light without any addition of catalyst due to wide forbidden energy gap of 4.6–4.7 eV (AkilaKesavan 2014; Kato and Kudo 2001).

24.3.6.2 Delafossite Photocatalyst (ABO_2 Type)

This type of oxides has general formula of ABO_2, where A and B are metals with +1 and +3 oxidation states, respectively. They possess a layered structure with a sheet of linearly coordinated A cations and BO_6 octahedral edges sharing its boundaries. These set of oxides have Ag containing group-3 elements of $AgMO_2$ (M = Al, Ga, In) (Maruyama et al. 2006; Ouyang et al. 2006; Maruyama et al. 2006; Ouyang et al. 2008). These oxides also include α-$AgInO_2$ (1.92 eV), α-$AgGaO_2$ (2.38 eV), β-$AgAlO2$ (2.95 eV) and C (2.18 eV), where α refers to delafossite and β refers to cristobalite structure. Ouyang et al. performed a photocatalytic experiment using these SCs and found their efficiency was in the order of α-$AgGaO_2$ > β-$AgAlO_2$ > β-$AgAlO_2$ > α-$AgInO_2$ (Ouyang et al. 2008; Ouyang et al. 2009).

24.3.6.3 Spinel Photocatalyst (AB_2O_4 Type)

These compounds have a general formula of AB_2O_4, where A and B are metals with oxidation state of +2 and +3, respectively. They have a cubic with close-packed lattice. Compounds such as calcium indate ($CaIn_2O_4$), containing p-block element indium with d_{10} configuration comes under this division. $CaIn_2O_4$ shows photodegradation of organic dyes under viewable light (Tang et al. 2003; Tang et al. 2004a). The process of $CaIn_2O_4$ via solid state reaction involves high temperature (>1000 °C) and long calcination time (12 h) (Tang et al. 2003; Tang et al. 2004a; Tang et al. 2004b). In order to address this, Ding et al. has implemented a combustion synthetic route method to replace the conventional SSR method for the production of high crystalline and surface area $CaIn_2O_4$ nanotubes. He found that the synthesized $CaIn_2O_4$ photocatalyst shows 66% of degradation of toluene gas, and when dispersed in water in presence of Pt, it was able to produce hydrogen 1.23 $\mu mol\ h^{-1}\ g^{-1}$, which is 24 times higher than traditional SSR method (Ding et al. 2009).

24.3.6.4 ABO_4 Type Photocatalyst

Bismuth vanadate is a viewable light-responsive SC that has acquired more popularity due to its ferroelastic, polymorphic and electronic properties. Such properties are highly dependent on crystalline structure of material. There are three polymorphic kinds of $BiVO_4$, which are scheelite-monoclinic (SM), scheelite-tetragonal (ST) and zircon tetragonal (ZT) (Tokunaga et al. 2001; Bierlein and Sleight 1975). These three SM are found to be more effective to produce O_2 from aqueous $AgNO_3$ and degradation of endocrine compounds. Compared to conventional aqueous process routes, Yu and Kudo found hydrothermal routes to synthesize $SM-BiVO_4$ with improved morphology and surface texture to boost the photocatalytic hydrogen evolution for O_2 from $AgNO_3$ by changing the pH of the medium (Yu and Kudo 2006).

24.4 Metal-Semiconductor Heterojunction Photocatalysis

Recent developments in plasma noble metals (Linic et al. 2011; Duan et al. 2014; Pradhan et al. 2001) and semiconductive nanomaterials (Trotochaud et al. 2013; Ma et al. 2012; Zandi and Hamann 2014) remain the most important inorganic material for photocatalysis. These can be utilized for removing organic pollutant, bacterial detoxifications and producing hydrogen (Trotochaud et al. 2013; Zandi and Hamann 2014). The recent studies show that the combining plasmonic metals and semiconductors give rise to a better catalyst for solar-to-energy conversion. These combination of metal-semiconductor forms a heterojunction which may generate new properties when both are placed close to each other shown in Fig. 24.8(a). This can aid in the rapid transfer of charge carriers produced from one carrier to another; it will remove the electrons from metal and semiconductor which will further hinder the process of recombination (Yu et al. 2014; Costi et al. 2010). Also, these materials give different amalgamation of facets on its surfaces, giving an opportunity to substrate molecules getting absorbed. All mentioned advantages make these kinds

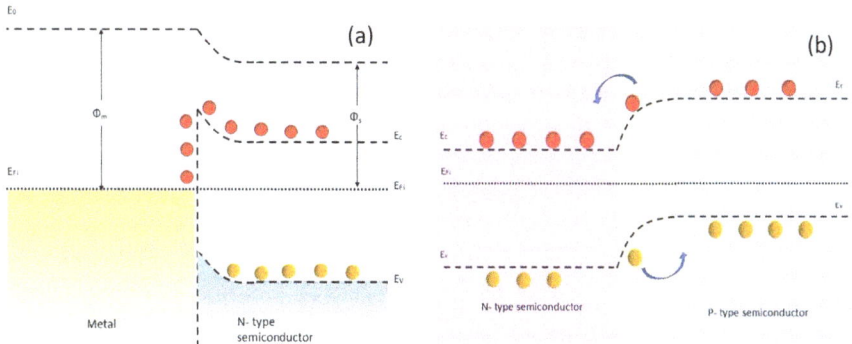

Fig. 24.8 (**a**) M-S junction and (**b**) S-S heterojunction

of structures more favourable to use. These structures can be further categorized into: (i) In first case materials are photoactive which means on excitation charges will get transfer from one to other material. For instance, plasmonic gold is combined with high band gap semiconductor materials such as TiO_2; charge will start flowing from gold to TiO_2 to initiate the catalytic reaction (Linic et al. 2011; Duan et al. 2014); (ii) the second one is metal and SC; both of them are photoresponsive and absorb solar radiation such as Au-CdS and Au-CdSe (Costi et al. 2010). However, very few reports are available till date for these materials, their promising advantages encourage the researchers to work more on it, and proper band alignment may help more facile electron transfer.

24.5 Semiconductor-Semiconductor Heterojunction Photocatalyst

Recently, heterojunctions formed by combining two SCs are most common; generally p- and n- SC are chosen for this. Figure 24.8(b) shows the junction of S-S type heterojunction. As a result, space charge region is created at the interface and electric field is produced due to diffusion of charge carriers. This type heterojunction is related to band alignment of SC. Having knowledge of band structure of semiconductor is crucial as the band alignment could be easily drawn by assigning band edge positions in energy diagram. Further, band structure tells about the redox ability of semiconductor heterojunction. The more −ve E_{CB} represents strong reducing strength, while more +ve E_{VB} means more strong oxidizing strength.

24.6 Modifications in Photocatalysts

24.6.1 Metal and Non-metal Implantation

The effectiveness of photocatalytic hydrogenation is governed by how effectively the rate of recombination is decreased and charge carriers get separated to take part in oxidation and reduction reaction (Hoffmann et al. 1995). It has been a continuous point of discussion for materials scientists to find more active materials which can improve the performance of photocatalytic process for industrial-scale applications (Ollis and Al-Ekabi 1993; Parent et al. 1996). One of the best approaches to resolve this issue by doping of different metal or non-metal ions in to the host photocatalyst. Plenty of potential dopants have been found and studied for photocatalytic dye-degradation rate. Various metals such as rhodium (Rh), silver (Ag), platinum (Pt), copper (Cu), iron (Fe), vanadium (V), chromium (Cr), nickel (Ni), aluminium (Al), gold (Au), Lithium (Li), palladium (Pd), Magnesium (Mg), etc. are some of the most common dopants that are being used (Choi et al. 2002; Serpone et al. 2010; Siemon et al. 2002; Litter 1999; Tan et al. 2003; Su et al. 2004; Vamathevan et al. 2002; Tayade et al. 2006; Tayade et al. 2011). Doping involves, the guest metals/non-metals is incorporated into the parent semiconductor material which create vacant sites and defects in atomic structure. It improves the base properties of the host material by modifying its material properties. Doping of metals in TiO_2 stops the electrons to move away from the surface which prevents it from recombining with the valance band holes (Gerischer and Heller 1992; Gerischer and Heller 1991; Wang et al. 1992). It also facilitates electron transfer at TiO_2 surface; thus, an effective e^- transfer occurs to e^- acceptors than donors in case of the undoped TiO_2. Oxygen traps the electron and produces superoxide ion, whereas hole oxidizes the hydroxyl ion to hydroxyl radicals. Various chemical and physical methods are used for the metal doping in semiconductor such as chemical vapour deposition, sol-gel, water in oil microemulsion, wet impregnation, hydrothermal, ion-assisted sputtering.

Ghasemi et al. investigated photocatalytic activity of different transitional metal ions (Fe, Cr, Co, Fe, Ni, Mn and Zn)-loaded TiO_2 film for acid blue 92 dye degradation. He claimed that with the presence of metal ions, the photocatalytic behaviour of TiO_2 was dramatically increased and the most prominent and active metal was Fe. Further, he confirmed that this was due to its and reduced band gap, high surface area and small crystallite size (Ghasemi et al. 2009). Sobana et al. explored the degrading Direct blue 53 (DB 53) and Direct red 23 (DR 23) dyes using Ag-doped TiO_2 through wet impregnated process in UV radiation. This change was due to e^--h^+ segregation by e^- trapping of Ag nanoparticles (Sobana et al. 2006). In addition, similar findings were found by Whang et al., highest degradation rate of methylene blue (MB) was at 2.0 wt% of Ag-doped TiO_2 (Whang et al. 2009). Choi et al. analysed chloroform degradation by doping various valance metal ions (Fe^{3+}, Os^{3+}, Rh^{3+}, Al^{3+}, Co^{3+}, V^{4+}, Re^{5+} and Mo^{5+}) into TiO_2 photocatalyst. In all cases, they used 0.5% of dopant concentration. He found that Fe^{3+}-doped TiO_2 showed maximum efficiency, while Al^{3+} and Co^{3+} doping caused decrease in efficiency. For the degradation of rhodamine B in aqueous solution under visible light and UV, the photocatalytic activity of cerium-doped TiO_2 was investigated (Choi et al. 2002). Cerium doping improved performance with a maximum photocatalytic behaviour at a doping concentration of 0.2–0.4%. This facilitated formation of super oxide anion radicals and enhances the charge separation. However, overdoping of Ce causes adverse effect on photocatalytic activity (Tong et al. 2007). Similarly, using V^{5+}-doped TiO_2 synthesized by coprecipitation method showed decrease in photocatalytic activity, while using V^{4+} showed increase in photocatalytic activity (Martin et al. 1994).

Beside using metal dopants, many non-metal dopants also showed very good efficiency photocatalytic activity. Asahi et al. are considered to be the first people who reported the use of non-metal as a dopant and its significance of efficient overlap between dopant and band states of titania oxide (Asahi et al. 2001). In addition, Zheng et al. reported enhanced the photocatalytic activity by using N-doped TiO_2 nano-fibers and removing surface bonded N species via post treatment in air (Zheng et al. 2013). This makes top of nano-fibres to become efficient for absorbing organic molecule and improve activation for oxygen molecules. Boron was first documented by Zhao et al.; in this he prepared B-doped TiO_2 via sol-gel method with boric acid (Zhao et al. 2004). Band gap was shifted 2.93 eV through boron doping and lowered by the use of Ni-B-doping to 2.85 eV. Both B-doped and Ni-B-doped TiO_2 showed a large increase in activity relative to uncoupled TiO_2. Tailoring of TiO_2 with several other halogen elements are also referred in some articles. Iodine doped of TiO_2 was first reported by Hong et al. through tetrabutyl titanate hydrolysis in the presence of HIO_3 (Wu et al. 2018). Prepared samples are being used for photocatalytic phenol degradation, exhibiting photocatalytic activity under visible irradiation, 59% degradation after 2 h.

24.6.2 Effect of Co-catalyst Loading

Nobel metals and transition metals are most commonly used co-catalyst for photocatalytic hydrogen generation. When loading of metal co-catalyst is done in to photocatalyst, generated e^- start moving towards the surface of photocatalyst which get interrupted by the co-catalyst due to low E_F level of noble metal than parent photocatalyst (shown in Fig. 24.9). Meanwhile the generated h^+ remains in the parent photocatalyst and moves towards the surface. Subsequently, this process creates two separate half reactions oxidation and reduction. Loading of co-catalyst on semiconductor is necessary as it reduces the overvoltage for oxygen and hydrogen evolution systems. Different transition metals are being employed to enhance the photoreduction of the proton (Matsumura et al. 1983). These co-catalysts lower down the activation energy for the hydrogen generation and capture the generated electron to suppress the charge carrier's recombination rate (Maeda et al. 2007a; Jang et al. 2008). Till date most used co-catalyst are transition metal oxides (Lu and Hwang 2011; Husin et al. 2011), Pt (Zou and Arakawa 2003), Ru (Hara et al. 2003), NiO

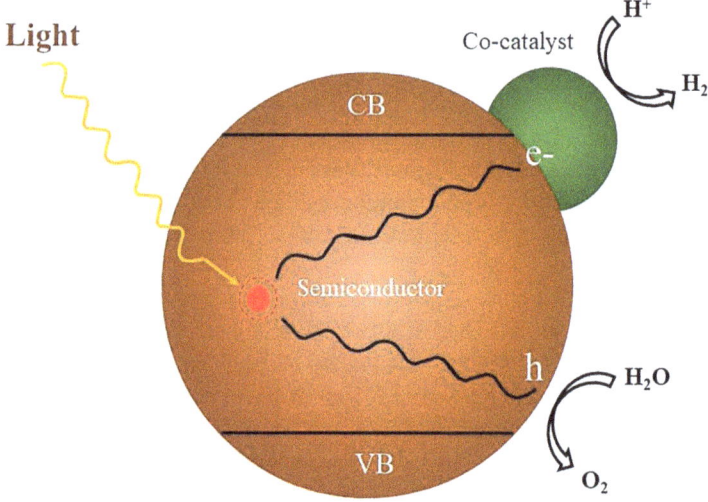

Fig. 24.9 Charge transfer between co-catalyst and photocatalyst

(Hu and Teng 2010), Co_3O_4 (Long et al. 2006), Rh (Long et al. 2006), Pd (Sayama et al. 1998) and Au (Jin et al. 2006). It has been seen that without co-catalyst many of the photocatalyst doesn't shows hydrogen generation. Compared to the photocatalyst/solution interface, the barrier height of the photocatalyst/noble metal interface is thin, enclosing effective charge separation of photogenerated electrons and holes (Daskalaki et al. 2010; Prousek 1996). Nozik designed a system as Schottky-type photochemical diode (Nozik 1977). Later, Sathish et al. stated that there exists a dependency in transition metal properties and the rate of development of hydrogen (Sathish et al. 2006). The most promising candidates to drive the reaction are the co-catalyst metal having more redox potential and work function and weak metal-H bond strength. Jang et al. developed platinized CdS/TiO_2 photocatalyst using different methods such as photo deposition, wet impregnation and chemical reduction. He found that Pt fabricated by PD showed very less photocatalytic activity because part of Pt resides inside CdS, whereas all other Pt located at surface of TiO_2 nanoparticles particles in case of WI and CR. Domen et al. were first to perform water splitting using NiO-loaded $SrTiO_3$ powder (Domen et al. 1982; Domen et al. 1986). NiO for hydrogen evolution is activated by reducing hydrogen and oxidizing oxygen to make NiO/Ni dual layer which supports e^- movement from parent photocatalyst to co-catalyst (Maeda et al. 2006a). They also introduced a new co-catalyst, Rh-Cr metal oxide ($Rh_{2-y}Cr_yO_3$) nanoparticles, that improves the decomposition of water (Maeda et al. 2006b; Maeda et al. 2007b). Dispersion of $Rh_{2-y}Cr_yO_3$ nanoparticles on $(Ga_{1-x}Zn_x)(N_{1-x}O_x)$ ($x = 0.12$; GaN:ZnO) produced an effective catalyst for H_2 production (Maeda et al. 2006a; Maeda et al. 2007b).

24.6.3 Dye Sensitization

The dye-based sensitization is utilized variably because of its visible light utilization which would be helpful for conversion. Some unique property-based dye which has redox and viewable light responsivity property could be utilized widely in solar cells and photo splitting activity. Whenever light is illuminated, the dye gets excited and sends the e^- to E_{CB} of the semiconductor as illustrated in Fig. 24.10 which in return helps in starting the catalytic reaction (Ni et al. 2007). In this process, the dyes potential energy level should be higher than that of semiconductor so as to inject electron to that particular semiconductor (Nahak et al. 2021; Mondal et al. 2011; Mahata et al. 2015). The significance of dye is that it utilizes the visible light of sun which has more effect compared to that of UV light (Mahata and Kundu 2009).

There is low production in hydrogen if the semiconductor is absent as this acts as charge separators. The excited electrons can transfer to noble metals whose prime target is to start the water reduction reaction. To continue the reaction cycle electron-rich components like iodine pair, EDTA can be used in the solution (Mahshid et al. 2007). It follows a following reaction which is expressed:

$$dye \xrightarrow{h\nu} dye^*$$

$$dye^* \xrightarrow{TiO_2} dye^+ + e^-$$

$$dye^+ + e^- \rightarrow dye$$

For increasing the efficiency of converting into electrical energy or hydrogen energy, flow of electron injection should be high, and backward reaction should be low, which perfectly combines the electron and hole ratio (O'regan and Grätzel 1991; Dhanalakshmi et al. 2001). The fast injection and slow backward reaction make the dye-sensitization method suitable for the generation of hydrogen. The dye

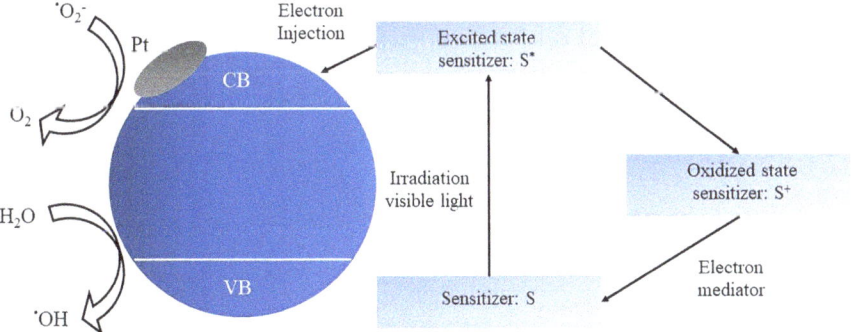

Fig. 24.10 Photo splitting activity of dye sensitized hydrogen generation process

molecules which are absorbed by the semiconductor could effectively inject electrons to the semiconductor for photocatalytic activity. By the help of maxima wavelength, it is not possible to say which dye will be helpful for energy generation which may be due to electron injection into the semiconductor (Gurunathan et al. 1997). Further research is required to understand charge injection dynamics of different dyes for better understandings.

24.7 Operating Conditions Affecting the Photocatalytic Hydrogen Generation

24.7.1 Particle Size

Morphology of photocatalyst such as agglomerate and particle dimension is one of the key factors affecting the process of photocatalysis. In photocatalytic dye-degradation, there exists a relation between photocatalyst morphology and the organic compound; thus, particle size of the synthesized photocatalyst plays a vital role in it. Particles having smaller size possess many active sites at their surface. These active sites absorb molecules of water, that is further reduced by photogenerated electrons. Smaller particles minimize diffusion length from their core to the bulk of surface photocatalyst for photogenerated electron and hole pairs (Liu et al. 2008; Kumar and Pandey 2017).

24.7.2 Surface Area

Many studies show that anatase TiO_2 having higher crystallinity and large surface area is more suitable for hydrogen generation (Jitputti et al. 2007). It becomes important to control thermal treatment of the photoelectrode to maintain the proper crystalline of anatase phase. It is preferable to use anatase phase of TiO_2 with high surface area and crystallinity (Sreethawong et al. 2005; Sreethawong and Yoshikawa 2005; Jitputti et al. 2008; Leung et al. 2010).

24.7.3 Reaction Temperature

The reactor temperature is considered as important factor while studying on photocatalytic hydrogen generation. It is expected that temperature inside the reactor gradually increases due to continuous radiation of lamp. Rodriguez et al. reported that as they increased the reaction temperature, hydrogen production also increased. Photocatalytic efficiency improves with an increase in reaction temperature, but if we increase temperature more than 80 °C, it starts promoting recombination of electron and hole pairs (Hashimoto et al. 2005).

24.7.4 Catalyst

Co-catalyst addition increases the overall hydrogen processing performance and accelerates the chemical reaction of the surface. The most widely used metals are Pd, Ru, Pt and Rh and meal oxides such as NiO and RuO_2. Amount of catalyst loaded at a certain solution volume will influence and improve the efficiency of hydrogeneration, and for different photocatalysis, the amount of loading varies. For instance, some catalyst shows optimum activity at 100 mg of catalyst, while others may show at 50 mg of catalyst Transition metals, especially noble metals, are considered as effective co-catalyst for photocatalytic reactions. When these metals are doped onto the surface of photocatalyst, the electrons (generated due to photon incident) starts migrating towards the surface of the parent photocatalyst. The noble metal is impeded here, as the noble metals' fermi level is often lower than the parent photocatalyst (Kumar et al. 2013). Meanwhile the h^+ stay at parent semiconductor photocatalyst, and then it starts migrating towards surface. Thus, a separation of charge carriers takes place, and separated e^- and h^+ will get involved in reduction and oxidation, respectively.

24.7.5 Effect of pH

pH plays a vital role in hydrogen generation in photocatalytic solutions. By adding NaOH, KOH, HCl, CH_3COOH, H_2SO_4 and HNO_3 to the solution, the solution's pH can be adjusted to a specific concentration. In order to explore more how pH affects the photocatalytic action, several experiments have been performed at different concentration of pH. It has been reported that hydrogen production gets increased at pH = 6–8 but decreases for pH = 9–10. This occurs due to the presence of sacrificial ions like HS^- to S^{2-} ions, etc. (Preethi and Kanmani 2014).

24.8 Photocatalytic Reactors

Apart from all other factors discussed in previous sections of the chapter, photocatalytic reactors also influence the rate of hydrogen production. An ideal photoreactor must distribute the light uniformly throughout the chamber so that each particle gets sufficient light radiation which results higher hydrogen generation efficiency. For water splitting applications and photocatalytic degradation by light irradiation,

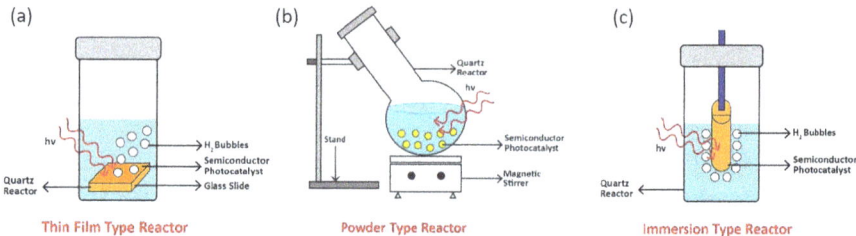

Fig. 24.11 Different types of reactors: (**a**) thin-film-type photocatalytic chambers, (**b**) immersion-type photocatalytic chambers, (**c**) powder-type photocatalytic chambers

various types of reactors may be used. These are usually categorized into two types: (1) thin-film-type photoreactors and (2) slurry-type photoreactors shown in Fig. 24.11 (Reddy et al. 2018).

24.8.1 Thin-Film-Type Photocatalytic Reactors

In reactors of these types, thin layer of the photocatalyst material is to be coated on the glass/metal slide and light is being irradiated on it. These reactors are not much effective and commercially used as compared to powder-type reactors, as only side of the glass slide interact with the light, while the other side gets very less amount of light on its surface. Hence, these reactor systems are so popularly used.

24.8.2 Slurry-Type Photocatalytic Reactors

Reactors of these types are much advantageous and also mostly used than thin film photocatalytic reactors. Further, slurry-type reactors are divided as (1) dispersion-type and (2) immersion-type reactors. Out of these two, dispersion types are very famous, since it contains the photocatalysts in powder form due to which each particle will come in contact with light and can actively take part in photocatalytic reaction, whereas in immersion-type reactors, photocatalyst is coated on a road-like shape and light is being irradiated from all the side, so that each surface will come in contact with light and can give maximum efficiency. These kinds of photoreactors are commonly used when both gas and liquid phase reactants come in, while the catalyst is solid phase.

24.9 Challenges and Future Perspective

In this chapter, we have discussed hydrogen as a good alternative for existing conventional energy source in near future. However, apart from these advantages, it also has some demerits; the nature of hydrogen is highly destructive, low density and highly reactive. Efficient generation and storage of clean hydrogen are commonly adopted as two key technical problems in extending the use of hydrogen. Today, renewable power-generating techniques such as biomass, tidal, hydro, solar and wind account for a very less percent of the world's overall power consumption (Guo et al. 2008). Further, there are some problem arises for refilling, evaporation loss occurs when hydrogen is transported from the reservoir to the tank through a transfer line which is cooled up to −253 °C. Due to difference of temperature gradient across the valves, there is evaporation loss which cannot be supressed fully (Schüth 2009). However, there are some solutions that can minimize the loss during filling by using coaxial cryogenic connectors (Topsøe et al. 2004). Transportation of hydrogen is very expensive because the weight and cost of hydrogen storage container is very high and the containers are vacuum isolated and are coated with radiation shield up to 40 layers, which is quite difficult to assemble (Guo et al. 2008).

Hydrogen needs to be stored safely and effectively to use as an energy carrier, particularly for vehicle applications. Hydrogen storage on board a vehicle can be seen as a key to the fuel cell's commercial success. The key challenge is the storage of adequate hydrogen on board and also provide equivalent driving range of 300 miles without sacrificing passenger or cargo space with regard to vehicle safety, weight, length, performance and cost constraints (Chalk and Miller 2006). It can be stored as gaseous or liquid or an atomic form as of a hydride. Another key challenge for hydrogen generation is the separation of hydrogen processes like hydrolysis and reforming. The cost of generation of hydrogen is too high.

Semiconductor-based photocatalysts has been considered as a potential material for renewable hydrogen generation (Topsøe et al. 2004). The studies upon this sector had taken many decades, in which the development of the photocatalysts and the efficiency has been focused. Those studies concluded regarding the metal doping, efficiency study in visible light, changes of band gap taken through some methods, capping and many more (Liao et al. 2012; Dvoranova et al. 2002; Ohno et al. 2003; Yanagida et al. 1995). Apart from all these, some conclusions were found, which can help this sector to implement future purpose at a large scale. Some assumptions are also taken which attracted interests to keep on the development on this.

One of them is large-scale implementation of photocatalytic water splitting over an area of 5 Km × 5 Km which can produce 570 tons of hydrogen a day, which can cover up to one-third of the energy requirements of future generations (Soltani et al. 2013). Some studies are going on for more development in heterogeneous photocatalysts which efficiently increases the production of hydrogen; some are trying to develop an inexpensive water splitting reactor in which a thin film with solutions and photocatalysts will generate hydrogen (Maeda and Domen 2010; Wang et al. 2019). The consequences of all these studies had come one by one, and this impacts on the future studies upon this sector, which may ensure an industrial-level acceptable way to hydrogen generation for various of applications (James et al. 2009; Takata and Domen 2019; Chowdhury et al. 2017).

24.10 Applications of Hydrogen Generation

In these chapter, we discuss about the hydrogen generation; here a question arises in everyone's mind, why only hydrogen? As we have studied from our childhood that our earth's atmosphere consists of direct hydrogen of 0.000055% and indirectly consist of 99.96% in the form of water, compounds, etc. Due to its abundant availability, we choose hydrogen for various applications. The "G20 Karuizawa Innovation Action Plan on Energy Transitions" and "the Global Environment for Sustainable Growth" was launched on 16 June 2019 by the International Renewable Energy Agency (IRENA) to build future sustainable pathways to hydrogen-enabled clean energy (Du and Eisenberg 2012). Hydrogen is a rising star. Hydrogen enables the greater integration of renewable energy. The environment-friendly nature of hydrogen is taken into consideration; when it combusts, no CO_2 is exhausted as it releases heat and water as byproduct (Chalk and Miller 2006). It is used in vehicles, industry, electricity and transports for its decarbonizing nature. As a clean energy vector, hydrogen can easily transport and stored.

Hydrogen is versatile in nature and can be utilized in different ways. On the basis of their use, it is classified into two categories: one is feedstock and another as an energy vector (Deng et al. 2008).

24.10.1 Hydrogen as a Feedstock

In the current era of development, the attention is towards the renewable energy; that's why different industries used hydrogen in different industrial processes. The basic building block of manufacturing NH_3 is hydrogen; thus, around 55% of hydrogen produced in the world is used for the synthesis of ammonia (Balat and Balat 2009), 25% is used in refineries where it is used to process intermediate oil product, 10% is used for the methanol production, and rest of it is used for different worldwide application (Cheekatamarla and Finnerty 2006).

24.10.2 Hydrogen in Fertilizers Industries

Above we have studied that the majority of hydrogen produced is used in the production of ammonia or also known as azane, which is obtained by artificial nitrogen fixation process called Haber-Bosch process (Bogdanović and Schwickardi 1997).

This process employs high pressure to force chemical reactions; it fixes atmospheric nitrogen with natural gas hydrogen to create ammonia. The method uses large amount of pressure, since the N_2 are bound together by triple bonds (Weber et al. 2006). Haber-Bosch method uses a catalyst made of ruthenium or iron with an internal temperature of more than 800° F (426° C) and 200 atm pressure to unite nitrogen with hydrogen.

Hence, the outcome of these process is ammonia, which is used as a refrigerant in plants. The below image is the experimentation setup of ammonia by Haber-Bosch process (Asadullah et al. 2002).

24.10.3 Fuel Industry

The demand of crude oil is increasing drastically; to maintain the demand, more crude oil is extracted. The more the crude oil, the more the hydrogen for the refining of crude oil (Soria et al. 2006). Hydrogen is used as a catalyst to transfer crude oil to refined fuel like gasoline and diesel (Kalamaras and Efstathiou 2013). Hydrogen is used to remove contaminated element like sulphur. There are two processes of using hydrogen:

- Hydrocracking
- Hydroprocessing

Hydrocracking is the process of breaking and the hydrogenation of hydrocarbons to produce refined fuel with higher H/C ratio, whereas in hydroprocessing, the sulphur and the nitrogen compounds are hydrogenated and produce H_2S and NH_3 as a waste (Weber et al. 2006).

24.10.4 Methanol Production

Methanol is the widely used chemical, it is widely used in different field like production of acetic acid and formaldehyde, and it is also used as an antifreeze element (Ni et al. 2006); it is also used in the production of hydrocarbons (Suelves et al. 2005). In the presence of high pressure and temperature, hydrogen reacts with carbon dioxide to form methanol (Asadullah et al. 2002):

$$CO_2 + 3H_2 \rightarrow CH_3OH + H_2O$$

24.10.5 Hydrogen in Electronics Industry

Hydrogen is used as a scavenger to reduce residual oxygen and to protect the device from the deposition of oxygen in the top layer so no damage occurs (Hohn and Schmidt 2001). It is also used in the epitaxial growth of polysilicon to reduce the silicon tetrachloride to silicon to obtain a pure silicon water (Ogden et al. 1999).

24.10.6 Hydrogen as a Fuel

Fuel is a basic demand in our day-to-day lives, without fuel we can't imagine the world. Fuel plays a vital role in the development of the country. It is used in different fields like logistic, transportation, electricity, etc., but there is a problem of using crude oil products (Ogden et al. 1999). The petroleum products release polluted gas when combusted (Zaluski et al. 2003). So, in order to overcome these circumstances, people are looking towards the alternative renewable energy (Asadullah et al. 2002). For this, hydrogen is considered as a fuel due to its abundant availability and ease of transportation and storage (Farrauto et al. 2003). Hydrogen can be utilized directly and indirectly as fuel. Directly means without further converting the hydrogen into different form, and it is utilized in combustion engine and fuel cell (Kordesch 1967). Indirectly means the molecular hydrogen is converted to gas, liquid or energy, and it is used like liquid hydrogen is used in the rockets and aircrafts.

24.11 Conclusion

This chapter addresses the main aspects of generation of hydrogen for the future society for increasing sustainability in the development of energy system. These approaches are discussed in the preceding sections; there are many other options which are explored for hydrogen generation, such as thermochemical water splitting, photochemical water splitting, photobiological water splitting, etc. While the key photocatalytic activity monitoring factors in semiconductor photocatalyst have been established, several aspects of the role of inorganic photocatalysts remain unknown. For the photocatalytic water splitting, compounds having electrons in d_0 cell (Ti, Zr, Nb and Ta) and d_{10} cell (Ga, In, Ge, Sn and Sb) both showed significant enhancement in performance. Oxides are dominant, but the response has also been shown to be catalysed by nitrides and chalcogenides. Most notably, the underlying mechanism for decomposing water into reduction and oxidation on the surface of the semiconductor has not been fully elucidated yet. There has been no complete analysis of the impact of variable material preparations and surface impurities on the catalytic activity of semiconductors. The growth of better photocatalysts will also benefit from recent developments in nanoscience. Quantum size effects can now be used to refine both the electronic structure and nanostructure reactivity, while synthetic techniques can be used to monitor catalyst morphology down to the nanoscale in order to further improve the efficiency of photochemical water splitting systems. Further efforts to develop novel co-catalysts could provide additional breakthroughs in the acquisition of highly effective photocatalysts. Such scientific progress is currently underway alongside new technological directions for the preparation of photocatalysts and innovative processes for system development.

References

Abanades S, Charvin P, Flamant G, Neveu P (2006) Screening of water-splitting thermochemical cycles potentially attractive for hydrogen production by concentrated solar energy. Energy 31(14):2805–2822

Abas N, Kalair A, Khan N (2015) Review of fossil fuels and future energy technologies. Futures 69:31–49

Abdin Z, Alim MA, Saidur R, Islam MR, Rashmi W, Mekhilef S, Wadi A (2013) Solar energy harvesting with the application of nanotechnology. Renew Sust Energ Rev 26:837–852

Abu-Khader MM (2009) Recent advances in nuclear power: a review. Prog Nucl Energy 51(2):225–235

Abu-Lebdeh Y, Davidson I (eds) (2012) Nanotechnology for lithium-ion batteries. Springer Science & Business Media

Acar C, Dincer I, Zamfirescu C (2014) A review on selected heterogeneous photocatalysts for hydrogen production. Int J Energy Res 38(15):1903–1920

Acar C, Dincer I, Naterer GF (2016) Review of photocatalytic water-splitting methods for sustainable hydrogen production. Int J Energy Res 40(11):1449–1473

Adcock TA, Draper S, Nishino T (2015) Tidal power generation–a review of hydrodynamic modelling. Proc Inst Mech Eng Part A J Power Energy 229(7):755–771

Ahmadi MH, Ghazvini M, Sadeghzadeh M, Alhuyi Nazari M, Kumar R, Naeimi A, Ming T (2018) Solar power technology for electricity generation: a critical review. Energy Sci Eng 6(5):340–361

Ahuja S, Kutty TRN (1996) Nanoparticles of $SrTiO_3$ prepared by gel to crystallite conversion and their photocatalytic activity in the mineralization of phenol. J Photochem Photobiol A Chem 97(1-2):99–107

AkilaKesavan G (2014) Nanotechnology and its applications. Scitech J 1(06):12–13

Archer CL, Jacobson MZ (2005) Evaluation of global wind power. J Geophys Res Atmos 110(D12)

Arjuna J, Sitorus TB, Hazwi M, Sitio A (2017) Performansi mesin otto yang menggunakan bahan bakar biogas dari limbah cair sawit. Jurnal Energi dan Manufaktur 10(1):17–22

Asadullah M, Ito SI, Kunimori K, Yamada M, Tomishige K (2002) Energy efficient production of hydrogen and syngas from biomass: development of low-temperature catalytic process for cellulose gasification. Environ Sci Technol 36(20):4476–4481

Asahi RYOJI, Morikawa TAKESHI, Ohwaki T, Aoki K, Taga Y (2001) Visible-light photocatalysis in nitrogen-doped titanium oxides. Science 293(5528):269–271

Balat M, Balat M (2009) Political, economic and environmental impacts of biomass-based hydrogen. Int J Hydrog Energy 34(9):3589–3603

Bang IC, Jeong JH (2011) Nanotechnology for advanced nuclear thermal-hydraulics and safety: boiling and condensation. Nucl Eng Technol 43(3):217–242

Bard AJ (1979) Photoelectrochemistry and heterogeneous photo-catalysis at semiconductors. J Photochem 10(1):59–75

Bayer P, Rybach L, Blum P, Brauchler R (2013) Review on life cycle environmental effects of geothermal power generation. Renew Sust Energ Rev 26:446–463

Beghi GE (1986) A decade of research on thermochemical hydrogen at the Joint Research Centre-Ispra. In: Hydrogen systems. Pergamon, pp 153–171

Benemann JR (1997) Feasibility analysis of photobiological hydrogen production. Int J Hydrog Energy 22(10-11):979–987

Bierlein JD, Sleight AW (1975) Ferroelasticity in $BiVO_4$. Solid State Commun 16(1):69–70

Bignozzi CA, Caramori S, Cristino V, Argazzi R, Meda L, Tacca A (2013) Nanostructured photoelectrodes based on WO 3: applications to photooxidation of aqueous electrolytes. Chem Soc Rev 42(6):2228–2246

Bilgen S, Kaygusuz K, Sari A (2004) Renewable energy for a clean and sustainable future. Energy Sources 26(12):1119–1129

Bogdanović B, Schwickardi M (1997) Ti-doped alkali metal aluminium hydrides as potential novel reversible hydrogen storage materials. J Alloys Compd 253:1–9

Boisseau P, Loubaton B (2011) Nanomedicine, nanotechnology in medicine. Comptes Rendus Physique 12(7):620–636

Boyle G (2004) Renewable energy. 456

Brown LC, Besenbruch GE, Schultz KR, Marshall AC, Showalter SK, Pickard PS, Funk JF (2002) Nuclear production of hydrogen using thermochemical water-splitting cycles. In: Proc Int congress on advanced nuclear power plants, vol 9, pp 475–485

Carpenter AW, de Lannoy CF, Wiesner MR (2015) Cellulose nanomaterials in water treatment technologies. Environ Sci Technol 49(9):5277–5287

Chalk SG, Miller JF (2006) Key challenges and recent progress in batteries, fuel cells, and hydrogen storage for clean energy systems. J Power Sources 159(1):73–80

Cheekatamarla PK, Finnerty CM (2006) Reforming catalysts for hydrogen generation in fuel cell applications. J Power Sources 160(1):490–499

Chen ZM, Chen GQ (2013) Demand-driven energy requirement of world economy 2007: a multi-region input–output network simulation. Commun Nonlinear Sci Numer Simul 18(7):1757–1774

Choi W, Termin A, Hoffmann MR (2002) The role of metal ion dopants in quantum-sized TiO_2: correlation between photoreactivity and charge carrier recombination dynamics. J Phys Chem 98(51):13669–13679

Choudhary BK, Majumdar K, Deb S An overview of application of nanotechnology in environmental restoration

Chowdhury P, Malekshoar G, Ray AK (2017) Dye-sensitized photocatalytic water splitting and sacrificial hydrogen generation: current status and future prospects. Inorganics 5(2):34

Chun WJ, Ishikawa A, Fujisawa H, Takata T, Kondo JN, Hara M, Kawai M, Matsumoto Y, Domen K (2003) Conduction and valence band positions of Ta_2O_5, TaON, and Ta_3N_5 by UPS and electrochemical methods. J Phys Chem B 107(8):1798–1803

Colón G (2016) Towards the hydrogen production by photocatalysis. Appl Catal A Gen 518:48–59

Costi R, Saunders AE, Banin U (2010) Colloidal hybrid nanostructures: a new type of functional materials. Angew Chem Int Ed 49(29):4878–4897

Dasgupta CN, Gilbert JJ, Lindblad P, Heidorn T, Borgvang SA, Skjanes K, Das D (2010) Recent trends on the development of photobiological processes and photobioreactors for the improvement of hydrogen production. Int J Hydrog Energy 35(19):10218–10238

Daskalaki VM, Antoniadou M, Li Puma G, Kondarides DI, Lianos P (2010) Solar light-responsive $Pt/CdS/TiO_2$ photocatalysts for hydrogen production and simultaneous degradation of inorganic or organic sacrificial agents in wastewater. Environ Sci Technol 44(19):7200–7205

Deng ZY, Ferreira JM, Sakka Y (2008) Hydrogen-generation materials for portable applications. J Am Ceram Soc 91(12):3825–3834

Dhanalakshmi KB, Latha S, Anandan S, Maruthamuthu P (2001) Dye sensitized hydrogen evolution from water. Int J Hydrog Energy 26(7):669–674

Dholam R, Patel N, Adami M, Miotello A (2009) Hydrogen production by photocatalytic water-splitting using Cr-or Fe-doped TiO_2 composite thin films photocatalyst. Int J Hydrog Energy 34(13):5337–5346

Diallo M, Brinker CJ (2011) Nanotechnology for sustainability: environment, water, food, minerals, and climate. In: Nanotechnology research directions for societal needs in 2020. Springer, Dordrecht, pp 221–259

Dincer I (2000) Renewable energy and sustainable development: a crucial review. Renew Sust Energ Rev 4(2):157–175

Dincer F (2011) The analysis on photovoltaic electricity generation status, potential and policies of the leading countries in solar energy. Renew Sust Energ Rev 15(1):713–720

Ding J, Sun S, Bao J, Luo Z, Gao C (2009) Synthesis of $CaIn_2O_4$ rods and its photocatalytic performance under visible-light irradiation. Catal Lett 130(1-2):147–153

Domen K, Naito S, Onishi T, Tamaru K (1982) Photocatalytic decomposition of liquid water on a NiO-$SrTiO_3$ catalyst. Chem Phys Lett 92(4):433–434

Domen K, Kudo A, Onishi T (1986) Mechanism of photocatalytic decomposition of water into H_2 and O_2 over NiO-$SrTiO_3$. J Catal 102(1):92–98

Du P, Eisenberg R (2012) Catalysts made of earth-abundant elements (Co, Ni, Fe) for water splitting: recent progress and future challenges. Energy Environ Sci 5(3):6012–6021

Duan C, Wang H, Ou X, Li F, Zhang X (2014) Efficient visible light photocatalyst fabricated by depositing plasmonic Ag nanoparticles on conductive polymer-protected Si nanowire arrays for photoelectrochemical hydrogen generation. ACS Appl Mater Interfaces 6(12):9742–9750

Dvoranova D, Brezova V, Mazúr M, Malati MA (2002) Investigations of metal-doped titanium dioxide photocatalysts. Appl Catal B Environ 37(2):91–105

Elcock D (2007) Potential impacts of nanotechnology on energy transmission applications and needs (No. ANL/EVS/TM/08-3). Argonne National Lab (ANL), Argonne, IL

Ellabban O, Abu-Rub H, Blaabjerg F (2014) Renewable energy resources: current status, future prospects and their enabling technology. Renew Sust Energ Rev 39:748–764

Esmaeili A (2011) Applications of nanotechnology in oil and gas industry. In: AIP conference proceedings, vol 1414. American Institute of Physics, pp 133–136

Farrauto R, Hwang S, Shore L, Ruettinger W, Lampert J, Giroux T, Liu Y, Ilinich O (2003) New material needs for hydrocarbon fuel processing: generating hydrogen for the PEM fuel cell. Annu Rev Mater Res 33(1):1–27

Foley AM, Leahy PG, Marvuglia A, McKeogh EJ (2012) Current methods and advances in forecasting of wind power generation. Renew Energy 37(1):1–8

Fridleifsson IB (2001) Geothermal energy for the benefit of the people. Renew Sust Energ Rev 5(3):299–312

Fronk BM, Neal R, Garimella S (2010) Evolution of the transition to a world driven by renewable energy. J Energy Resour Technol 132(2)

Fujinami K, Katagiri K, Kamiya J, Hamanaka T, Koumoto K (2010) Sub-10 nm strontium titanate nanocubes highly dispersed in non-polar organic solvents. Nanoscale 2(10):2080–2083

Fujishima A, Zhang X (2006) Titanium dioxide photocatalysis: present situation and future approaches. C R Chim 9(5-6):750–760

Fulekar MH, Pathak B, Kale RK (2014) Nanotechnology: perspective for environmental sustainability. In: Environment and sustainable development. Springer, New Delhi, pp 87–114

Funk JE (2001) Thermochemical hydrogen production: past and present. Int J Hydrog Energy 26(3):185–190

Gerischer H, Heller A (1991) The role of oxygen in photooxidation of organic molecules on semiconductor particles. J Phys Chem 95(13):5261–5267

Gerischer H, Heller A (1992) Photocatalytic oxidation of organic molecules at TiO_2 particles by sunlight in aerated water. J Electrochem Soc 139(1):113

Ghasemi S, Rahimnejad S, Setayesh SR, Rohani S, Gholami MR (2009) Transition metal ions effect on the properties and photocatalytic activity of nanocrystalline TiO_2 prepared in an ionic liquid. J Hazard Mater 172(2-3):1573–1578

Ghasemi M, Daud WRW, Hassan SH, Oh SE, Ismail M, Rahimnejad M, Jahim JM (2013) Nanostructured carbon as electrode material in microbial fuel cells: a comprehensive review. J Alloys Compd 580:245–255

Ghirardi ML, Dubini A, Yu J, Maness PC (2009) Photobiological hydrogen-producing systems. Chem Soc Rev 38(1):52–61

Gholipour MR, Dinh CT, Béland F, Do TO (2015) Nanocomposite heterojunctions as sunlight-driven photocatalysts for hydrogen production from water splitting. Nanoscale 7(18):8187–8208

Goldemberg J (ed) (2000) World energy assessment: energy and the challenge of sustainability. United Nations Development Programme, New York, pp 1–29

Gong K, Du F, Xia Z, Durstock M, Dai L (2009) Nitrogen-doped carbon nanotube arrays with high electrocatalytic activity for oxygen reduction. Science 323(5915):760–764

Guo ZX, Shang C, Aguey-Zinsou KF (2008) Materials challenges for hydrogen storage. J Eur Ceram Soc 28(7):1467–1473

Gür TM (2016) Comprehensive review of methane conversion in solid oxide fuel cells: prospects for efficient electricity generation from natural gas. Prog Energy Combust Sci 54:1–64

Gurunathan K, Maruthamuthu P, Sastri MVC (1997) Photocatalytic hydrogen production by dye-sensitized Pt/SnO$_2$ and Pt/SnO$_2$/RuO$_2$ in aqueous methyl viologen solution. Int J Hydrog Energy 22(1):57–62

Häfele W (1990) Energy from nuclear power. Sci Am 263(3):136–145

Hahn H, Krautkremer B, Hartmann K, Wachendorf M (2014) Review of concepts for a demand-driven biogas supply for flexible power generation. Renew Sust Energ Rev 29:383–393

Hara M, Nunoshige J, Takata T, Kondo JN, Domen K (2003) Unusual enhancement of H$_2$ evolution by Ru on TaON photocatalyst under visible light irradiation. Chem Commun 24:3000–3001

Hashimoto K, Irie H, Fujishima A (2005) TiO$_2$ photocatalysis: a historical overview and future prospects. Jpn J Appl Phys 44(12R):8269

Heppenstall T (1998) Advanced gas turbine cycles for power generation: a critical review. Appl Therm Eng 18(9-10):837–846

Hodes G, Cahen D, Manassen J (1976) Tungsten trioxide as a photoanode for a photoelectrochemical cell (PEC). Nature 260(5549):312–313

Hoffmann MR, Martin ST, Choi W, Bahnemann DW (1995) Environmental applications of semiconductor photocatalysis. Chem Rev 95(1):69–96

Hohn KL, Schmidt LD (2001) Partial oxidation of methane to syngas at high space velocities over Rh-coated spheres. Appl Catal A Gen 211(1):53–68

Holm D, Arch D (2013) Renewable energy future for the developing world. Trans Renew Energy Syst

Hosenuzzaman M, Rahim NA, Selvaraj J, Hasanuzzaman M, Malek AA, Nahar A (2015) Global prospects, progress, policies, and environmental impact of solar photovoltaic power generation. Renew Sust Energ Rev 41:284–297

Hu C-C, Teng H (2010) Structural features of p-type semiconducting NiO as a co-catalyst for photocatalytic water splitting. J Catal 272(1):1–8

Hu JS, Ren LL, Guo YG, Liang HP, Cao AM, Wan LJ, Bai CL (2005) Mass production and high photocatalytic activity of ZnS nanoporous nanoparticles. Angew Chem Int Ed 44(8):1269–1273

Hu X, Li G, Yu JC (2010) Design, fabrication, and modification of nanostructured semiconductor materials for environmental and energy applications. Langmuir 26(5):3031–3039

Huang Y, Liang J, Chen Y (2012) An overview of the applications of graphene-based materials in supercapacitors. Small 8(12):1805–1834

Husin H, Su WN, Chen HM, Pan CJ, Chang SH, Rick J, Chuang WT, Sheu HS, Hwang BJ (2011) Photocatalytic hydrogen production on nickel-loaded La x Na 1−x TaO 3 prepared by hydrogen peroxide-water based process. Green Chem 13(7):1745–1754

Hussein AK (2015) Applications of nanotechnology in renewable energies—a comprehensive overview and understanding. Renew Sust Energ Rev 42:460–476

Iverson BD, Garimella SV (2008) Recent advances in microscale pumping technologies: a review and evaluation. Microfluid Nanofluid 5(2):145–174

James BD, Baum GN, Perez J, Baum KN (2009) Technoeconomic analysis of photoelectrochemical (PEC) hydrogen production. DOE report

Jang JS, Ham DJ, Lakshminarasimhan N, yong Choi, W. and Lee, J.S. (2008) Role of platinum-like tungsten carbide as cocatalyst of CdS photocatalyst for hydrogen production under visible light irradiation. Appl Catal A Gen 346(1-2):149–154

Jhariya N, Roshan R, Mahato SS, Mahata S (2018) Hydrothermal synthesis of surface functionalized semiconducting nano crystals and study of their photo induced interaction with natural dye. IRJET 5(5):3892–3895

Jin Z, Zhang X, Lu G, Li S (2006) Improved quantum yield for photocatalytic hydrogen generation under visible light irradiation over eosin sensitized TiO_2—investigation of different noble metal loading. J Mol Catal A Chem 259(1-2):275–280

Jin J, Yu J, Guo D, Cui C, Ho W (2015) A hierarchical Z-scheme $CdS–WO_3$ photocatalyst with enhanced CO_2 reduction activity. Small 11(39):5262–5271

Jing D, Guo L (2006) A novel method for the preparation of a highly stable and active CdS photocatalyst with a special surface nanostructure. J Phys Chem B 110(23):11139–11145

Jitputti J, Pavasupree S, Suzuki Y, Yoshikawa S (2007) Synthesis and photocatalytic activity for water-splitting reaction of nanocrystalline mesoporous titania prepared by hydrothermal method. J Solid State Chem 180(5):1743–1749

Jitputti J, Pavasupree S, Suzuki Y, Yoshikawa S (2008) Synthesis of TiO_2 nanotubes and its photocatalytic activity for H_2 evolution. Jpn J Appl Phys 47(1S):751

Kalamaras CM, Efstathiou AM (2013) Hydrogen production technologies: current state and future developments. In: Conference papers in energy. Hindawi

Kato H, Kudo A (2001) Water splitting into H_2 and O_2 on alkali tantalate photocatalysts $ATaO_3$ (A = Li, Na, and K). J Phys Chem B 105(19):4285–4292

Keller KH (2007) Nanotechnology and society. J Nanopart Res 9(1):5–10

Khan J, Arsalan MH (2016) Solar power technologies for sustainable electricity generation–a review. Renew Sust Energ Rev 55:414–425

Khan MJ, Bhuyan G, Iqbal MT, Quaicoe JE (2009) Hydrokinetic energy conversion systems and assessment of horizontal and vertical axis turbines for river and tidal applications: a technology status review. Appl Energy 86(10):1823–1835

Khan N, Kalair A, Abas N, Haider A (2017) Review of ocean tidal, wave and thermal energy technologies. Renew Sust Energ Rev 72:590–604

Kida T, Minami Y, Guan G, Nagano M, Akiyama M, Yoshida A (2006) Photocatalytic activity of gallium nitride for producing hydrogen from water under light irradiation. J Mater Sci 41(11):3527–3534

Kordesch KV, Union Carbide Corp (1967) Apparatus for hydrogen generation. U.S. Patent 3,338,681

Krstić PS, Wells JC (eds) (2010) Nanotechnology for electronics, photonics, and renewable energy, vol 78. Springer, New York

Kruse O, Rupprecht J, Bader KP, Thomas-Hall S, Schenk PM, Finazzi G, Hankamer B (2005) Improved photobiological H_2 production in engineered green algal cells. J Biol Chem 280(40):34170–34177

Kudo A, Miseki Y (2009) Heterogeneous photocatalyst materials for water splitting. Chem Soc Rev 38:253–278

Kudo A, Ueda K, Kato H, Mikami I (1998) Photocatalytic O_2 evolution under visible light irradiation on BiVO 4 in aqueous $AgNO_3$ solution. Catal Lett 53(3-4):229–230

Kumar A, Pandey G (2017) A review on the factors affecting the photocatalytic degradation of hazardous materials. Mater Sci Eng Int J 1(3):1–10

Kumar S, Kwon HT, Choi KH, Cho JH, Lim W, Moon I (2011) Current status and future projections of LNG demand and supplies: a global prospective. Energy Policy 39(7):4097–4104

Kumar DP, Shankar MV, Kumari MM, Sadanandam G, Srinivas B, Durgakumari V (2013) Nano-size effects on CuO/TiO_2 catalysts for highly efficient H_2 production under solar light irradiation. Chem Commun 49(82):9443–9445

Lahann J (2008) Nanomaterials clean up. Nat Nanotechnol 3(6):320–321

Lambauer J, Voss A, Fahl U (eds) (2012) Nanotechnology and energy: science, promises, and limits. CRC Press

Lenzen M (2008) Life cycle energy and greenhouse gas emissions of nuclear energy: a review. Energy Convers Manag 49(8):2178–2199

Leung DY, Fu X, Wang C, Ni M, Leung MK, Wang X, Fu X (2010) Hydrogen production over titania-based photocatalysts. ChemSusChem 3(6):681–694

Li J, Wu Z, Tan X, Chen B (2011) Review of wind power generation and relative technology development. Electric Power Construction/Dianli Jianshe 32(8):64–72

Li R, Zhang F, Wang D, Yang J, Li M, Zhu J, Zhou X, Han H, Li C (2013) Spatial separation of photogenerated electrons and holes among {010} and {110} crystal facets of BiVO 4. Nat Commun 4(1):1–7

Liao CH, Huang CW, Wu J (2012) Hydrogen production from semiconductor-based photocatalysis via water splitting. Catalysts 2(4):490–516

Lin ZH, Cheng G, Wu W, Pradel KC, Wang ZL (2014) Dual-mode triboelectric nanogenerator for harvesting water energy and as a self-powered ethanol nanosensor. ACS Nano 8(6):6440–6448

Linic S, Christopher P, Ingram DB (2011) Plasmonic-metal nanostructures for efficient conversion of solar to chemical energy. Nat Mater 10(12):911–921

Linsebigler AL, Lu G, Yates JT Jr (1995) Photocatalysis on TiO_2 surfaces: principles, mechanisms, and selected results. Chem Rev 95(3):735–758

Litter MI (1999) Heterogeneous photocatalysis: transition metal ions in photocatalytic systems. Appl Catal B Environ 23(2-3):89–114

Liu Y, Xie L, Li Y, Yang R, Qu J, Li Y, Li X (2008) Synthesis and high photocatalytic hydrogen production of $SrTiO_3$ nanoparticles from water splitting under UV irradiation. J Power Sources 183(2):701–707

Liu B, Zhao X, Terashima C, Fujishima A, Nakata K (2014) Thermodynamic and kinetic analysis of heterogeneous photocatalysis for semiconductor systems. Phys Chem Chem Phys 16(19):8751–8760

Liu J, Cheng B, Yu J (2016) A new understanding of the photocatalytic mechanism of the direct Z-scheme gC_3N_4/TiO_2 heterostructure. Phys Chem Chem Phys 18(45):31175–31183

Long M, Cai W, Cai J, Zhou B, Chai X, Wu Y (2006) Efficient photocatalytic degradation of phenol over $Co_3O_4/BiVO_4$ composite under visible light irradiation. J Phys Chem B 110(41):20211–20216

Lu SY, Hwang BJ (2011) Hydrogen energy and fuel cell technologies Foreword

Lu X, McElroy MB, Kiviluoma J (2009) Global potential for wind-generated electricity. Proc Natl Acad Sci 106(27):10933–10938

Luo H, Takata T, Lee Y, Zhao J, Domen K, Yan Y (2004) Photocatalytic activity enhancing for titanium dioxide by co-doping with bromine and chlorine. Chem Mater 16(5):846–849

Luo M, Liu Y, Hu J, Liu H, Li J (2012) One-pot synthesis of CdS and Ni-doped CdS hollow spheres with enhanced photocatalytic activity and durability. ACS Appl Mater Interfaces 4(3):1813–1821

Ma SSK, Hisatomi T, Maeda K, Moriya Y, Domen K (2012) Enhanced water oxidation on Ta_3N_5 photocatalysts by modification with alkaline metal salts. J Am Chem Soc 134(49):19993–19996

Madhavan AA (2020) Nanotechnology: advances and real-life applications. CRC Press, p 71

Maeda K (2011) Photocatalytic water splitting using semiconductor particles: history and recent developments. J Photochem Photobiol C: Photochem Rev 12(4):237–268

Maeda K, Domen K (2010) Photocatalytic water splitting: recent progress and future challenges. J Phys Chem Lett 1(18):2655–2661

Maeda K, Teramura K, Lu D, Takata T, Saito N, Inoue Y, Domen K (2006a) Photocatalyst releasing hydrogen from water. Nature 440(7082):295–295

Maeda K, Teramura K, Masuda H, Takata T, Saito N, Inoue Y, Domen K (2006b) Efficient overall water splitting under visible-light irradiation on $(Ga_{1-x}Zn_x)(N_{1-x}O_x)$ dispersed with Rh– Cr mixed-oxide nanoparticles: effect of reaction conditions on photocatalytic activity. J Phys Chem B 110(26):13107–13112

Maeda K, Teramura K, Domen K (2007a) Development of cocatalysts for photocatalytic overall water splitting on (Ga 1−x Zn x)(N 1−x O x) solid solution. Catal Surv Jpn 11(4):145–157

Maeda K, Teramura K, Lu D, Saito N, Inoue Y, Domen K (2007b) Roles of Rh/Cr_2O_3 (core/shell) nanoparticles photodeposited on visible-light-responsive $(Ga_{1-x}Zn_x)(N_{1-x}O_x)$ solid solutions in photocatalytic overall water splitting. J Phys Chem C 111(20):7554–7560

Mahata S, Kundu D (2009) Hydrothermal synthesis of aqueous nano-TiO_2 sols. Mater Sci-Pol 27(2):463–470

Mahata S, Mondal B, Mahata SS, Usha K, Mandal N, Mukherjee K (2015) Chemical modification of titanium isopropoxide for producing stable dispersion of titania nano-particles. Mater Chem Phys 151:267–274

Mahshid S, Askari M, Ghamsari MS (2007) Synthesis of TiO_2 nanoparticles by hydrolysis and peptization of titanium isopropoxide solution. J Mater Process Technol 189(1-3):296–300

Martin ST, Morrison CL, Hoffmann MR (1994) Photochemical mechanism of size-quantized vanadium-doped TiO_2 particles. J Phys Chem 98(51):13695–13704

Martin DJ, Reardon PJT, Moniz SJ, Tang J (2014) Visible light-driven pure water splitting by a nature-inspired organic semiconductor-based system. J Am Chem Soc 136(36):12568–12571

Maruyama Y, Irie H, Hashimoto K (2006) Visible light sensitive photocatalyst, delafossite structured α-$AgGaO_2$. J Phys Chem B 110(46):23274–23278

Matsumura M, Saho Y, Tsubomura H (1983) Photocatalytic hydrogen production from solutions of sulfite using platinized cadmium sulfide powder. J Phys Chem 87(20):3807–3808

McKinlay JB (2014) Systems biology of photobiological hydrogen production by purple non-sulfur bacteria. In: Microbial bioenergy: hydrogen production. Springer, Dordrecht, pp 155–176

Melis A, Zhang L, Forestier M, Ghirardi ML, Seibert M (2000) Sustained photobiological hydrogen gas production upon reversible inactivation of oxygen evolution in the green Alga Chlamydomonas reinhardtii. Plant Physiol 122(1):127–136

Mokhatab S, Fresky MA, Islam MR (2006) Applications of nanotechnology in oil and gas E&P. J Pet Technol 58(04):48–51

Mondal B, Usha K, Mahata S, Kumbhakar P, Nandi MM (2011) Synthesis and characterization of nanocrystalline TiO_2 thin films for use as photoelectrodes in dye sensitized solar cell application. Transactions of the Indian Ceramic Society 70(3):173–177

Moniz SJ, Shevlin SA, Martin DJ, Guo ZX, Tang J (2015) Visible-light driven heterojunction photocatalysts for water splitting–a critical review. Energy Environ Sci 8(3):731–759

Müller-Steinhagen H, Trieb F (2004) Concentrating solar power. A review of the technology. Ingenia Inform QR Acad Eng 18:43–50

Nahak BK, Subudhi TSK, Pradhan LK, Panigrahi A, Roshan R, Mahato SS, Mahata S (2021) An investigation on photocatalytic dye degradation of rhodamine 6G dye with Fe-and Ag-doped TiO_2 thin films. In: Proceedings of the fourth international conference on microelectronics, computing and communication systems. Springer, Singapore, pp 295–307

Nguyen-Tri P, Nguyen TA, Carriere P, Ngo Xuan C (2018) Nanocomposite coatings: preparation, characterization, properties, and applications. Int J Corros 2018:4749501

Ni M, Leung DY, Leung MK, Sumathy K (2006) An overview of hydrogen production from biomass. Fuel Process Technol 87(5):461–472. Konieczny A, Mondal K, Wiltowski T, Dydo P. Catalyst development for thermocatalytic decomposition of methane to hydrogen. Int J Hydrogen Energy. 2008;33(1):264–272

Ni M, Leung MK, Leung DY, Sumathy K (2007) A review and recent developments in photocatalytic water-splitting using TiO_2 for hydrogen production. Renew Sust Energ Rev 11(3):401–425

Nozik AJ (1977) Photochemical diodes. Appl Phys Lett 30(11):567–569

O'keefe D, Allen C, Besenbruch G, Brown L, Norman J, Sharp R, McCorkle K (1982) Preliminary results from bench-scale testing of a sulfur-iodine thermochemical water-splitting cycle. Int J Hydrog Energy 7(5):381–392

O'regan B, Grätzel M (1991) A low-cost, high-efficiency solar cell based on dye-sensitized colloidal TiO$_2$ films. Nature 353(6346):737–740

Ogden JM, Steinbugler MM, Kreutz TG (1999) A comparison of hydrogen, methanol and gasoline as fuels for fuel cell vehicles: implications for vehicle design and infrastructure development. J Power Sources 79(2):143–168

Ohno T, Mitsui T, Matsumura M (2003) Photocatalytic activity of S-doped TiO$_2$ photocatalyst under visible light. Chem Lett 32(4):364–365

Ohno T, Tsubota T, Nakamura Y, Sayama K (2005) Preparation of S, C cation-codoped SrTiO$_3$ and its photocatalytic activity under visible light. Appl Catal A Gen 288(1-2):74–79

Okeke I, Iloanusi O (2014) Application of nanotechnology for improving energy security. In: Proceedings of the 1st African international conference/workshop on applications of nanotechnology to energy, health and environment. UNN, pp 237–243

Ollis DF, Al-Ekabi H (1993) Photocatalytic purification and treatment of water and air: proceedings of the 1st international conference on TiO$_2$ photocatalytic purification and treatment of water and air, London, Ontario, Canada, 8-13 November, 1992. Elsevier Science Ltd.

Ouyang S, Zhang H, Li D, Yu T, Ye J, Zou Z (2006) Electronic structure and photocatalytic characterization of a novel photocatalyst AgAlO$_2$. J Phys Chem B 110(24):11677–11682

Ouyang S, Li Z, Ouyang Z, Yu T, Ye J, Zou Z (2008) Correlation of crystal structures, electronic structures, and photocatalytic properties in a series of Ag-based oxides: AgAlO$_2$, AgCrO$_2$, and Ag$_2$CrO$_4$. J Phys Chem C 112(8):3134–3141

Ouyang S, Kikugawa N, Chen D, Zou Z, Ye J (2009) A systematical study on photocatalytic properties of AgMO$_2$ (M = Al, Ga, In): effects of chemical compositions, crystal structures, and electronic structures. J Phys Chem C 113(4):1560–1566

Parent Y, Blake D, Magrini-Bair K, Lyons C, Turchi C, Watt A, Wolfrum E, Prairie M (1996) Solar photocatalytic processes for the purification of water: state of development and barriers to commercialization. Sol Energy 56(5):429–437

Pradhan N, Pal A, Pal T (2001) Catalytic reduction of aromatic nitro compounds by coinage metal nanoparticles. Langmuir 17(5):1800–1802

Preethi V, Kanmani S (2014) Photocatalytic hydrogen production using Fe$_2$O$_3$-based core shell nano particles with ZnS and CdS. Int J Hydrog Energy 39(4):1613–1622

Prince RC, Kheshgi HS (2005) The photobiological production of hydrogen: potential efficiency and effectiveness as a renewable fuel. Crit Rev Microbiol 31(1):19–31

Prousek JOSEF (1996) Advanced oxidation processes for water teatment. Photochemical processes. Chemické listy 90(5):307–315

Qi J, Zhang W, Cao R (2018) Solar-to-hydrogen energy conversion based on water splitting. Adv Energy Mater 8(5):1701620

QingáLu G (2009) Enhanced photocatalytic hydrogen evolution by prolonging the lifetime of carriers in ZnO/CdS heterostructures. Chem Commun 23:3452–3454

Rabaey K, Verstraete W (2005) Microbial fuel cells: novel biotechnology for energy generation. Trends Biotechnol 23(6):291–298

Rasi S, Läntelä J, Rintala J (2011) Trace compounds affecting biogas energy utilisation–a review. Energy Convers Manag 52(12):3369–3375

Reddy KG, Deepak TG, Anjusree GS, Thomas S, Vadukumpully S, Subramanian KRV, Nair SV, Nair AS (2014) On global energy scenario, dye-sensitized solar cells and the promise of nanotechnology. Phys Chem Chem Phys 16(15):6838–6858

Reddy NL, Rao VN, Kumari MM, Kakarla RR, Ravi P, Sathish M, Karthik M, Venkatakrishnan SM (2018) Nanostructured semiconducting materials for efficient hydrogen generation. Environ Chem Lett 16(3):765–796

Rhee KH, Morris EP, Barber J, Kühlbrandt W (1998) Three-dimensional structure of the plant photosystem II reaction centre at 8 Å resolution. Nature 396(6708):283–286

Roduner E (2006) Size matters: why nanomaterials are different. Chem Soc Rev 35(7):583–592

Rogers JA, Lagally MG, Nuzzo RG (2011) Synthesis, assembly and applications of semiconductor nanomembranes. Nature 477(7362):45–53

Roig Y, Marre S, Cardinal T, Aymonier C (2011) Synthesis of exciton luminescent ZnO nanocrystals using continuous supercritical microfluidics. Angew Chem Int Ed 50(50):12071–12074

Rourke FO, Boyle F, Reynolds A (2010) Tidal energy update 2009. Appl Energy 87(2):398–409

Sakata T, Kawai T, Gratzel M (1983) In: Gratzel M (ed) Energy resources through photochemistry and catalysis. Elsevier

Sanchez F, Sobolev K (2010) Nanotechnology in concrete–a review. Constr Build Mater 24(11):2060–2071

Santoro C, Arbizzani C, Erable B, Ieropoulos I (2017) Microbial fuel cells: from fundamentals to applications. A review. J Power Sour 356:225–244

Sathish M, Viswanathan B, Viswanath RP (2006) Alternate synthetic strategy for the preparation of CdS nanoparticles and its exploitation for water splitting. Int J Hydrog Energy 31(7):891–898

Savage N, Diallo MS (2005) Nanomaterials and water purification: opportunities and challenges. J Nanopart Res 7(4-5):331–342

Sayama K, Yase K, Arakawa H, Asakura K, Tanaka A, Domen K, Onishi T (1998) Photocatalytic activity and reaction mechanism of Pt-intercalated $K_4Nb_6O_{17}$ catalyst on the water splitting in carbonate salt aqueous solution. J Photochem Photobiol A Chem 114(2):125–135

Sayama K, Mukasa K, Abe R, Abe Y, Arakawa H (2001) Stoichiometric water splitting into H_2 and O_2 using a mixture of two different photocatalysts and an IO_3^-/I^- shuttle redox mediator under visible light irradiation. Chem Commun 23:2416–2417

Sayama K, Nomura A, Zou Z, Abe R, Abe Y, Arakawa H (2003) Photoelectrochemical decomposition of water on nanocrystalline $BiVO_4$ film electrodes under visible light. Chem Commun 23:2908–2909

Schiermeier Q, Tollefson J, Scully T, Witze A, Morton O (2008) Energy alternatives: electricity without carbon. Nat News 454(7206):816–823

Schüth F (2009) Challenges in hydrogen storage. Eur Phys J Spec Top 176(1):155–166

Selcuk MZ, Boroglu MS, Boz I (2012) Hydrogen production by photocatalytic water-splitting using nitrogen and metal co-doped TiO_2 powder photocatalyst. React Kinet Mech Catal 106(2):313–324

Serpone N, Pelizzetti E (1989) Photocatalysis: fundamentals and applications. Wiley, New York

Serpone N, Horikoshi S, Emeline AV (2010) Solar engineering solar engineering. J Photochem Photobiol, C 11(2):114–131

Serrano E, Rus G, Garcia-Martinez J (2009) Nanotechnology for sustainable energy. Renew Sust Energ Rev 13(9):2373–2384

Serrano E, Li K, Rus G, García-Martínez J (2010) Nanotechnology for energy production. In: Nanotechnology for the energy challenge. Wiley-VCH Verlag GmbH & Co, KGaA Weinheim, pp 3–32

Sethi VK, Pandey M, Shukla MP (2011) Use of nanotechnology in solar PV cell. Int J Chem Eng Appl 2(2):77

Sharma S, Jain KK, Sharma A (2015) Solar cells: in research and applications—a review. Mater Sci Appl 6(12):1145

Shehzad N, Tahir M, Johari K, Murugesan T, Hussain M (2018) A critical review on TiO_2 based photocatalytic CO_2 reduction system: strategies to improve efficiency. J CO2 Util 26:98–122

Shen Y, Linville JL, Urgun-Demirtas M, Mintz MM, Snyder SW (2015) An overview of biogas production and utilization at full-scale wastewater treatment plants (WWTPs) in the United States: challenges and opportunities towards energy-neutral WWTPs. Renew Sust Energ Rev 50:346–362

Siemon U, Bahnemann D, Testa JJ, Rodríguez D, Litter MI, Bruno N (2002) Heterogeneous photocatalytic reactions comparing TiO_2 and Pt/TiO_2. J Photochem Photobiol A Chem 148(1-3):247–255

Singh GK (2013) Solar power generation by PV (photovoltaic) technology: a review. Energy 53:1–13

Singh VK, Singal SK (2017) Operation of hydro power plants-a review. Renew Sust Energ Rev 69:610–619

Sivaramakrishna N, Reddy CKR (2013) Hybrid power generation through combined solar–wind power and modified solar panel. Int J Eng Trends Technol 4(5):1414–1417

Sivula K, Zboril R, Le Formal F, Robert R, Weidenkaff A, Tucek J, Frydrych J, Gratzel M (2010) Photoelectrochemical water splitting with mesoporous hematite prepared by a solution-based colloidal approach. J Am Chem Soc 132(21):7436–7444

Slostowski C, Marre S, Bassat JM, Aymonier C (2013) Synthesis of cerium oxide-based nanostructures in near-and supercritical fluids. J Supercrit Fluids 84:89–97

Sobana N, Muruganadham M, Swaminathan M (2006) Nano-Ag particles doped TiO_2 for efficient photodegradation of direct azo dyes. J Mol Catal A Chem 258(1-2):124–132

Soltani N, Saion E, Yunus WMM, Navasery M, Bahmanrokh G, Erfani M, Zare MR, Gharibshahi E (2013) Photocatalytic degradation of methylene blue under visible light using PVP-capped ZnS and CdS nanoparticles. Sol Energy 97:147–154

Soria A, Szabo L, Russ P, Suwala W, Hidalgo I, Purwanto A (2006) World energy technology outlook 2050. European Commission, Joint Research Centre 21

Spliethoff H (2010) Power generation from solid fuels. Springer Science & Business Media

Sreethawong T, Yoshikawa S (2005) Comparative investigation on photocatalytic hydrogen evolution over Cu-, Pd-, and Au-loaded mesoporous TiO_2 photocatalysts. Catal Commun 6(10):661–668

Sreethawong T, Suzuki Y, Yoshikawa S (2005) Synthesis, characterization, and photocatalytic activity for hydrogen evolution of nanocrystalline mesoporous titania prepared by surfactant-assisted templating sol–gel process. J Solid State Chem 178(1):329–338

Steyn WJ (2009) Potential applications of nanotechnology in pavement engineering. J Transp Eng 135(10):764–772

Su C, Liao CH, Wang JD, Chiu CM, Chen BJ (2004) The adsorption and reactions of methyl iodide on powdered Ag/TiO_2. Catal Today 97(1):71–79

Suelves I, Lázaro MJ, Moliner R, Corbella BM, Palacios JM (2005) Hydrogen production by thermo catalytic decomposition of methane on Ni-based catalysts: influence of operating conditions on catalyst deactivation and carbon characteristics. Int J Hydrog Energy 30(15):1555–1567

Sun Y, Wu Q, Shi G (2011) Graphene based new energy materials. Energy Environ Sci 4(4):1113–1132

Tabata M, Maeda K, Higashi M, Lu D, Takata T, Abe R, Domen K (2010) Modified Ta_3N_5 powder as a photocatalyst for O_2 evolution in a two-step water splitting system with an iodate/iodide shuttle redox mediator under visible light. Langmuir 26(12):9161–9165

Takata T, Domen K (2019) Particulate photocatalysts for water splitting: recent advances and future prospects. ACS Energy Lett 4(2):542–549

Tan TTY, Yip CK, Beydoun D, Amal R (2003) Effects of nano-Ag particles loading on TiO_2 photocatalytic reduction of selenate ions. Chem Eng J 95(1-3):179–186

Tang J, Zou Z, Yin J, Ye J (2003) Photocatalytic degradation of methylene blue on $CaIn_2O_4$ under visible light irradiation. Chem Phys Lett 382(1-2):175–179

Tang J, Zou Z, Katagiri M, Kako T, Ye J (2004a) Photocatalytic degradation of MB on MIn_2O_4 (M = alkali earth metal) under visible light: effects of crystal and electronic structure on the photocatalytic activity. Catal Today 93:885–889

Tang J, Zou Z, Ye J (2004b) Effects of substituting Sr^{2+} and Ba^{2+} for Ca^{2+} on the structural properties and photocatalytic behaviors of $CaIn_2O_4$. Chem Mater 16(9):1644–1649

Tayade RJ, Kulkarni RG, Jasra RV (2006) Transition metal ion impregnated mesoporous TiO_2 for photocatalytic degradation of organic contaminants in water. Ind Eng Chem Res 45(15):5231–5238

Tayade RJ, Bajaj HC, Jasra RV (2011) Photocatalytic removal of organic contaminants from water exploiting tuned bandgap photocatalysts. Desalination 275(1-3):160–165

Technip KTI Nano-sieve offers alternative to conventional separation techniques used in the petrochemical industry

Teoh WY (2013) A perspective on the flame spray synthesis of photocatalyst nanoparticles. Materials 6(8):3194–3212

Teoh WY, Mädler L, Beydoun D, Pratsinis SE, Amal R (2005) Direct (one-step) synthesis of TiO_2 and Pt/TiO_2 nanoparticles for photocatalytic mineralisation of sucrose. Chem Eng Sci 60(21):5852–5861

Thomas S, Rafiei S, Maghsoodlou S, Afzali A (2014) Foundations of nanotechnology, volume two: nanoelements formation and interaction. CRC Press

Tokunaga S, Kato H, Kudo A (2001) Selective preparation of monoclinic and tetragonal $BiVO_4$ with scheelite structure and their photocatalytic properties. Chem Mater 13(12):4624–4628

Tong T, Zhang J, Tian B, Chen F, He D, Anpo M (2007) Preparation of Ce–TiO_2 catalysts by controlled hydrolysis of titanium alkoxide based on esterification reaction and study on its photocatalytic activity. J Colloid Interface Sci 315(1):382–388

Topsøe H, Egeberg RG, Knudsen KG (2004) Future challenges of hydrotreating catalyst technology. Prepr Pap-Am Chem Soc Div Fuel Chem 49(2):568

Trotochaud L, Mills TJ, Boettcher SW (2013) An optocatalytic model for semiconductor–catalyst water-splitting photoelectrodes based on in situ optical measurements on operational catalysts. J Phys Chem Lett 4(6):931–935

Tsuchiya M, Lai BK, Ramanathan S (2011) Scalable nanostructured membranes for solid-oxide fuel cells. Nat Nanotechnol 6(5):282–286

Vamathevan V, Amal R, Beydoun D, Low G, McEvoy S (2002) Photocatalytic oxidation of organics in water using pure and silver-modified titanium dioxide particles. J Photochem Photobiol A Chem 148(1-3):233–245

Vargas SA, Esteves GRT, Maçaira PM, Bastos BQ, Oliveira FLC, Souza RC (2019) Wind power generation: a review and a research agenda. J Clean Prod 218:850–870

Vayssieres L (ed) (2010) On solar hydrogen and nanotechnology. John Wiley & Sons

Veziroglu TN (2007) 21st Century's energy: Hydrogen energy system. In: Assessment of hydrogen energy for sustainable development. Springer, Dordrecht, pp 9–31

Vivekchand SRC, Rout CS, Subrahmanyam KS, Govindaraj A, Rao CNR (2008) Graphene-based electrochemical supercapacitors. J Chem Sci 120(1):9–13

Wang ZL, Wu W (2012) Nanotechnology-enabled energy harvesting for self-powered micro-/nanosystems. Angew Chem Int Ed 51(47):11700–11721

Wang CM, Heller A, Gerischer H (1992) Palladium catalysis of O_2 reduction by electrons accumulated on TiO_2 particles during photoassisted oxidation of organic compounds. J Am Chem Soc 114(13):5230–5234

Wang S, Yu D, Dai L (2011) Polyelectrolyte functionalized carbon nanotubes as efficient metal-free electrocatalysts for oxygen reduction. J Am Chem Soc 133(14):5182–5185

Wang Q, Hisatomi T, Jia Q, Tokudome H, Zhong M, Wang C, Pan Z, Takata T, Nakabayashi M, Shibata N, Li Y (2016) Scalable water splitting on particulate photocatalyst sheets with a solar-to-hydrogen energy conversion efficiency exceeding 1%. Nat Mater 15(6):611–615

Wang S, Wang C, Peng Z, Chen S (2018) A new technique for nanoparticle transport and its application in a novel nano-sieve. Sci Rep 8(1):1–10

Wang Z, Li C, Domen K (2019) Recent developments in heterogeneous photocatalysts for solar-driven overall water splitting. Chem Soc Rev 48(7):2109–2125

Weber G, Fu Q, Wu H (2006) Energy efficiency of an integrated process based on gasification for hydrogen production from biomass. Dev Chem Eng Miner Process 14(1-2):33–48

Westwood A (2004) Ocean power: wave and tidal energy review. Refocus 5(5):50–55

Whang TJ, Huang HY, Hsieh MT, Chen JJ (2009) Laser-induced silver nanoparticles on titanium oxide for photocatalytic degradation of methylene blue. Int J Mol Sci 10(11):4707–4718

Wiesner M, Bottero JY (2007) Environmental nanotechnology. McGraw-Hill Professional Publishing, New York

Wiley PE, Campbell JE, McKuin B (2011) Production of biodiesel and biogas from algae: a review of process train options. Water Environ Res 83(4):326–338

Wong K, Dia S (2017) Nanotechnology in batteries. J Energy Resour Technol 139(1)

Wong KV, Perilla N, Paddon A (2014) Nanoscience and nanotechnology in solar cells. J Energy Resour Technol 136(1)

Wu XF, Sun Y, Li H, Wang YJ, Zhang CX, Zhang JR, Su JZ, Wang YW, Zhang Y, Wang C, Zhang M (2018) In-situ synthesis of novel pn heterojunction of Ag_2CrO_4-$Bi_2Sn_2O_7$ hybrids for visible-light-driven photocatalysis. J Alloys Compd 740:1197–1203

Yan SC, Lv SB, Li ZS, Zou ZG (2010) Organic-inorganic composite photocatalyst of g-C_3N_4 and TaON with improved visible light photocatalytic activities. Dalton Trans 39:1488, e91

Yanagida S, Ogata T, Shindo A, Hosokawa H, Mori H, Sakata T, Wada Y (1995) Semiconductor photocatalysis: size control of surface-capped CdS nanocrystallites and the quantum size effect in their photocatalysis. Bull Chem Soc Jpn 68(3):752–758

Ye Z, Kong L, Chen F, Chen Z, Lin Y, Liu C (2018) A comparative study of photocatalytic activity of ZnS photocatalyst for degradation of various dyes. Optik 164:345–354

Yonenaga I (2001) Thermo-mechanical stability of wide-bandgap semiconductors: high temperature hardness of SiC, AlN, GaN, ZnO and ZnSe. Phys B Condens Matter 308:1150–1152

Yu J, Kudo A (2006) Effects of structural variation on the photocatalytic performance of hydrothermally synthesized $BiVO_4$. Adv Funct Mater 16(16):2163–2169

Yu X, Shavel A, An X, Luo Z, Ibanez M, Cabot A (2014) Cu_2ZnSnS_4-Pt and Cu_2ZnSnS_4-Au heterostructured nanoparticles for photocatalytic water splitting and pollutant degradation. J Am Chem Soc 136(26):9236–9239

Zaluski L, Zaluska A, Strom-Olsen J MCGILL INIVERSITY (2003) Method of hydrogen generation for fuel cell applications and a hydrogen-generating system. U.S. Patent Application 10/257,943

Zamfirescu C, Dincer I, Naterer GF (2011) Analysis of a photochemical water splitting reactor with supramolecular catalysts and a proton exchange membrane. Int J Hydrog Energy 36(17):11273–11281

Zamfirescu C, Naterer GF, Dincer I (2012) Photo-electro-chemical chlorination of cuprous chloride with hydrochloric acid for hydrogen production. Int J Hydrog Energy 37(12):9529–9536

Zandi O, Hamann TW (2014) Enhanced water splitting efficiency through selective surface state removal. J Phys Chem Lett 5(9):1522–1526

Zang L (ed) (2011) Energy efficiency and renewable energy through nanotechnology. Springer, Berlin, p 451

Zarrouk SJ, Moon H (2014) Efficiency of geothermal power plants: a worldwide review. Geothermics 51:142–153

Zeb S, Ullah I, Karim A, Muhammad W, Ullah N, Khan M, Komal W (2019) A review on nanotechnology applications in electric components. Nanotechnology 2(2)

Zerta M, Schmidt PR, Stiller C, Landinger H (2008) Alternative World Energy Outlook (AWEO) and the role of hydrogen in a changing energy landscape. Int J Hydrog Energy 33(12):3021–3025

Zhang Z, Yates JT Jr (2012) Band bending in semiconductors: chemical and physical consequences at surfaces and interfaces. Chem Rev 112(10):5520–5551

Zhang Q, Uchaker E, Candelaria SL, Cao G (2013) Nanomaterials for energy conversion and storage. Chem Soc Rev 42(7):3127–3171

Zhao W, Ma W, Chen C, Zhao J, Shuai Z (2004) Efficient degradation of toxic organic pollutants with $Ni_2O_3/TiO_{2-x}B_x$ under visible irradiation. J Am Chem Soc 126(15):4782–4783

Zheng Z, Zhao J, Yuan Y, Liu H, Yang D, Sarina S, Zhang H, Waclawika ER, Zhu H (2013) Tuning the surface structure of nitrogen-doped TiO_2 nanofibres—An effective method to enhance photocatalytic activities of visible-light-driven green synthesis and degradation. Chem Eur J 19(18):5731–5741

Zhou P, Yu J, Jaroniec M (2014) All-solid-state Z-scheme photocatalytic systems. Adv Mater 26(29):4920–4935

Zou Z, Arakawa H (2003) Direct water splitting into H_2 and O_2 under visible light irradiation with a new series of mixed oxide semiconductor photocatalysts. J Photochem Photobiol A Chem 158(2-3):145–162

Zou Z, Ye J, Sayama K, Arakawa H (2011) Direct splitting of water under visible light irradiation with an oxide semiconductor photocatalyst. In: Materials for sustainable energy: a collection of peer-reviewed research and review articles from nature publishing group, pp 293–295

Züttel A, Remhof A, Borgschulte A, Friedrichs O (1923) Hydrogen: the future energy carrier. Philos Trans R Soc London Ser A 368:3329–3342

Chapter 25
Recent Advances in the Synthesis of Heterocycles Over Heterogeneous Cerium-Based Nanocatalysts

Cong Chien Truong, Dinesh Kumar Mishra, and Hoang Long Ngo

Contents

25.1	Introduction..	710
25.2	Applications of Cerium-Based Catalysts in the Synthesis and Functionalization of Heterocycles...	712
	25.2.1 Commercial CeO_2..	712
	25.2.2 Synthetic Nano-CeO_2...	717
	25.2.3 Cerium Mixed Oxides..	729
	25.2.4 Cerium-Solid Material Composite..	737
	25.2.5 CeO_2 as Solid Support..	742
25.3	Cerium-Based Catalysts for the Vapour-Phase Synthesis of Heterocycles...............	746
25.4	Cerium-Based Catalysts for the Synthesis of CO_2-Derived Heterocycles...............	746
25.5	Summary and Outlook...	749
References...		751

Abbreviations

$(NH_4)_2Ce(NO_3)_6$	Ceric ammonium nitrate
$(NH_4)_2CO_3$	Ammonium carbonate
Brij35	Polyoxyethylene (23) lauryl ether
$Ce(NO_3)_3 \cdot 6H_2O$	Cerium (III) nitrate hexahydrate
$CeCl_3$	Cerium (III) chloride
CeO_2/ceria	Cerium (IV) oxide

C. C. Truong (✉)
Department of Bio-functional Molecular Engineering, Graduate School of Science and Engineering, University of Toyama, Toyama, Japan

D. K. Mishra
Department of Chemical Engineering and Research Institute of Industrial Science (RIST), Hanyang University, Seoul, South Korea

H. L. Ngo
NTT Hi-Tech Institute, Nguyen Tat Thanh University, Ho Chi Minh City, Vietnam

CH$_3$CN	Acetonitrile
CH$_4$	Methane
CO	Carbon monoxide
CO$_2$	Carbon dioxide
CTAB	Cetyltrimethylammonium bromide
Cu(NO$_3$)$_2$	Copper (II) nitrate
DMF	N,N-Dimethylformamide
DMSO	Dimethyl sulfoxide
Eu(NO$_3$)$_2$	Europium (III) nitrate
Luperox 101	2,5-Bis(*tert*-butylperoxy)-2,5-dimethylhexane
Lupersol TAEC	2-Ethylhexyl 2-methylbutan-2-yloxy carbonate
Mg (NO$_3$)$_2$	Magnesium (II) nitrate
MIL-101	Chromium terephthalate metal organic framework
NH$_3$	Ammonia solution
NO	Nitrogen (II) oxide
NPs	Nanoparticles
PEG 400	Polyethylene glycol 400
Pluronic 17R4	Poly(propylene glycol)-*block*-poly(ethylene glycol)-*block*-poly(propylene glycol)
Pluronic P123	Poly(ethylene glycol)-*block*-poly(propylene glycol)-*block*-poly(ethylene glycol)
PVA	Polyvinyl alcohol
PVP	Polyvinylpyrrolidone
Zr(NO$_3$)$_2$	Zirconium (II) nitrate

25.1 Introduction

Heterocycle is an important class of compounds in organic chemistry, which can be found in wide applications from natural to man-made products. In nature, numerous heterocyclic skeletons can be found in plant/marine metabolites, chlorophyll, genetic building blocks, vitamins, essential oils, enzymes and so on (Walsh 2015). Alternatively, novel synthetic compounds containing various heteroatoms and/or fused ring systems have been successfully constructed over the years (Taylor et al. 2016). For the assembly of complex molecules, these privileged structures turned out to be versatile and valuable building blocks in the synthesis of natural products (Carson and Kerr 2009; Majumdar and Chattopadhyay 2011), organic semiconductors (Zhao et al. 2017), high-density energy materials (Yin and Shreeve 2017), agrochemicals (Lamberth 2013), polymers (Lu 1998), etc. Due to the diverseness in architectural complexity, molecular functionality and bioactivity, the exploration of heterocycles is considered of great significance in medicinal chemistry (Fig. 25.1) (Gomtsyan 2012). For instance, the US FDA databases show that 59% of small-molecule drugs are composed of *N*-heterocyclic fragments (Vitaku et al. 2014). In addition, other top-selling heterocyclic pharmaceuticals are currently exploited as

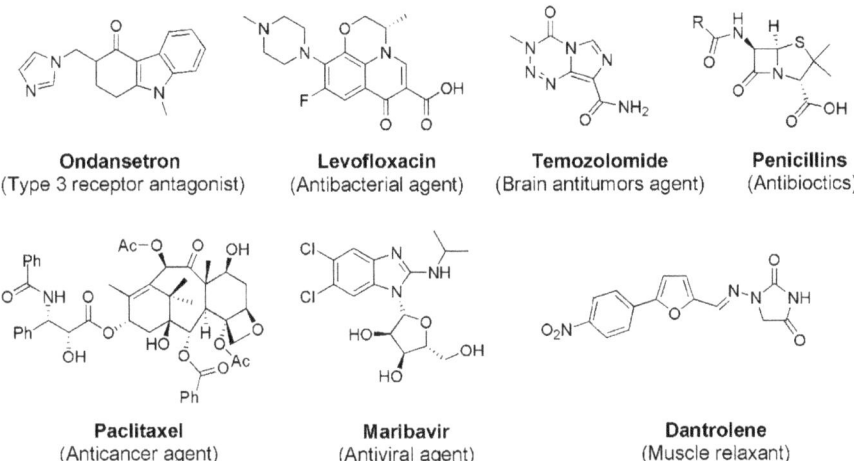

Fig. 25.1 Heterocyclic pharmaceuticals

anticancer, antibiotic, antiviral, antibacterial, diuretic and antineoplastic agents (Baumann et al. 2011; Baumann and Baxendale 2013; Ali et al. 2015; Feng et al. 2016; Delost et al. 2018). From these reasons, seeking simple but effective processes for the eco-friendly production of heterocycles has been considered as formidable challenges in both academia and industry throughout the years.

Recently, nanocatalysts have been widely acknowledged as powerful tools in the domain of heterogeneous catalysis, where nanostructured metal oxides and their hybrid materials attracted significant attention due to their superb catalytic efficiency in many chemical transformations (Wang et al. 2009; Guo et al. 2014; Gadipelly and Mannepalli 2019). In this manner, simple preparation, excellent thermal/chemical stability, high surface area, tunable control of acidity/basicity, low cost and recyclability are conducive to their versatility. Among the rare earth metal-based nanoparticles, most of the researches focused on the application of cerium-based materials as both catalyst and support due to the abundant, unique and tunable features of cerium (Sun et al. 2012; Zhang et al. 2012; Paier et al. 2013; Huang and Gao 2014). For example, the oxygen vacancies and reversible valence change (Ce^{4+} and Ce^{3+}) in CeO_2 allowed this nanostructure to participate in copious reactions such as oxidation, hydrogenation, methane reforming, water-gas shift, CO_2 conversion and others (Chang et al. 2019; Rodriguez et al. 2017). Moreover, the outstanding catalytic performance of ceria-supported transitional metals (e.g., Pd, Pt, Rh and Au), cerium mixed oxides, or cerium-doped solid materials in CH_4/CO/NO oxidation (Cargnello et al. 2012; Colussi et al. 2009; Spezzati et al. 2017; Qi and Li 2015), ozonation (Orge et al. 2012; Xu et al. 2016), hydrogenation (Akbayrak 2018; Hu et al. 2018) and photochemical reactions (Channei et al. 2014; Fiorenza et al. 2016; Shi et al. 2011) was also realized. Prompted by aforementioned reasons, several research groups have recently turned their keen eyes on the utility of cerium-based solids in organic chemistry (Vivier and Duprez 2010; Naaz et al. 2019), where the acid/base, redox or dual (acid/base-redox) sites on these heterogeneous catalysts served the crucial roles in determining the activity. To the best of our knowledge, a holistic overview on the practicality of cerium-based nanocatalysts in the construction and functionalization of heterocycles has not been reported. In this book chapter, numerous examples on the green and sustainable assembly of heterocyclic frameworks over well-defined cerium oxide/mixed oxides, cerium composites, cerium-doped solids and ceria-supported metals are introduced. In particular, the deployment of nanoceria in the chemical fixation of CO_2 towards valuable cyclic products is also explored. Furthermore, mechanistic description on each transformation is discussed in detail to give further insight on the activity of cerium-based nanocatalyst.

25.2 Applications of Cerium-Based Catalysts in the Synthesis and Functionalization of Heterocycles

25.2.1 Commercial CeO_2

In 2014, Edayadulla and Lee (2014) explored the catalysis of commercial CeO_2 NPs in the divergent synthesis of quinoxalin-2-amines and 3,4-dihydroquinoxalin-2-amines. By using 5 mol% of CeO_2, the one-pot condensation of 1,2-diamines, isocyanides with aldehydes or ketones could undergo smoothly in water to render a multiple of quinoxalin-2-amine and 3,4-dihydroquinoxalin-2-amine derivatives, respectively. Furthermore, the utility of CeO_2 NPs was also successfully attempted in the construction of indophenazine derivatives from the coupling of 1,2-phenylenediamine, isatins with *tert*-butyl isocyanide. The model mechanistic concourse towards quinoxalin-2-amine starting from 1,2-phenylenediamine, aldehyde and isocyanide is described to follow a cascade of imine formation/addition of isocyanide/annulation/isomerization/oxidation (Scheme 25.1), where CeO_2 NPs are demonstrated to facilitate the generation of imine and the insertion of isocyanide into imine.

Later, Shrestha et al. (2016) expanded the utility of CeO_2 NPs for the eco-friendly assembly of spiro[indoline-3,4-pyrano[2,3-*c*]pyrazole] derivatives. Under assistance of 30 mol% of CeO_2, a plenty of fused spirooxindoles could be afforded in the range yields of 75–93% from the aqueous-phase condensation of β-ketoesters with phenylhydrazines, malononitrile and isatins (Scheme 25.2). Particularly, a number of designed spirooxindole derivatives showed promising results on the potent antioxidant and antibacterial activities.

In another case, Sharma et al. (2018) established a novel synthetic strategy for fused tetrahydroisoquinolines and pyrrolo[3,4-c]quinoline-1,3-diones by coupling *N*,*N*-dimethylanilines **1** with *N*-substituted maleimides **2** over CeO_2 NPs. As shown in Scheme 25.3, tetrahydroisoquinoline derivatives with a high tolerance of functionality were obtainable upon performing the oxidative annulation of **1** and **2** with

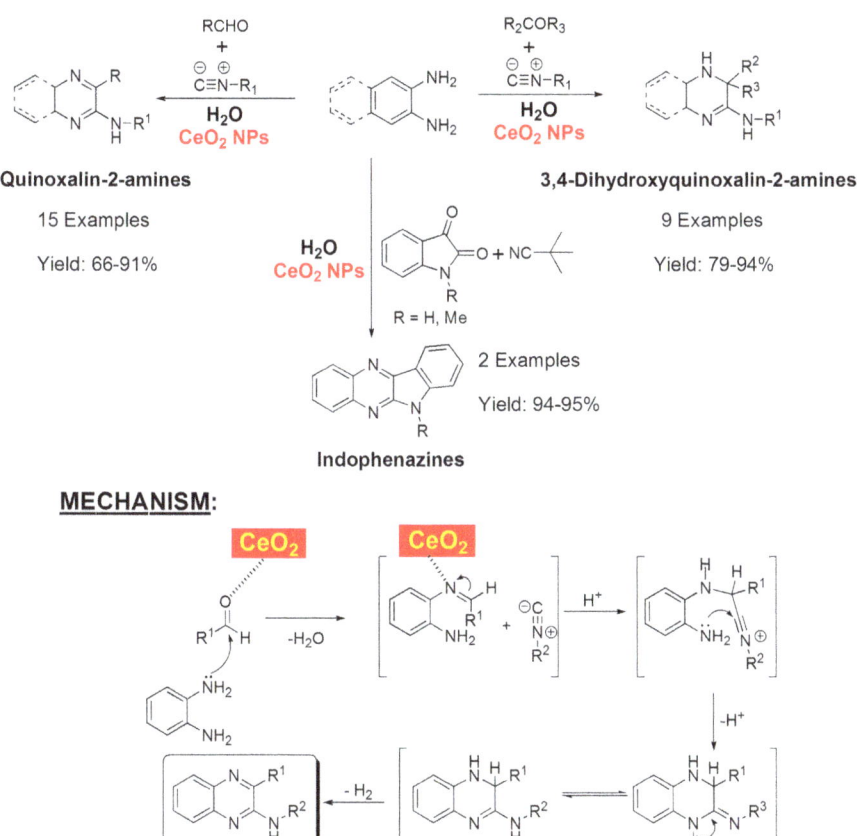

Scheme 25.1 Divergent synthesis of quinoxalin-2-amines, 3,4-dihydroquinoxalin-2-amines and indophenazines over CeO_2 NPs

20% mol of CeO_2 in air under optimal conditions. Afterwards, these resulting tetrahydroisoquinolines were efficiently transformed into quinoline-1,3-diones through the dehydrogenative/N-demethylative cascade in the presence of 2,3-dichloro-5,6-dicyano-1,4-benzoquinone (DDQ). Unluckily, the activity of recovered CeO_2 NPs was found to gradually drop after four recycles. In the mechanistic proposal, the model assembly of pyrrolo[3,4-c]quinoline-1,3 diones is proposed to follow the sequential stage of oxidative annulation/dehydrogenation/N-demethylation.

Besides, CeO_2 was also exploited as a robust catalyst in functionalizing the heterocyclic skeletons. For example, CeO_2 NPs was effective in promoting the aerobic cross-dehydrogenative coupling (CDC) of N-aryl tetrahydroisoquinolines with either nitroalkanes or acetone, which delivered a collection of corresponding 1-substituted-2-aryl-1,2,3,4-tetrahydroisoquinoline derivatives (Sharma et al. 2016a). Through a set of control experiments, the model mechanism for the oxidative CDC of N-phenyl tetrahydroisoquinoline and nitromethane via radical pathway is

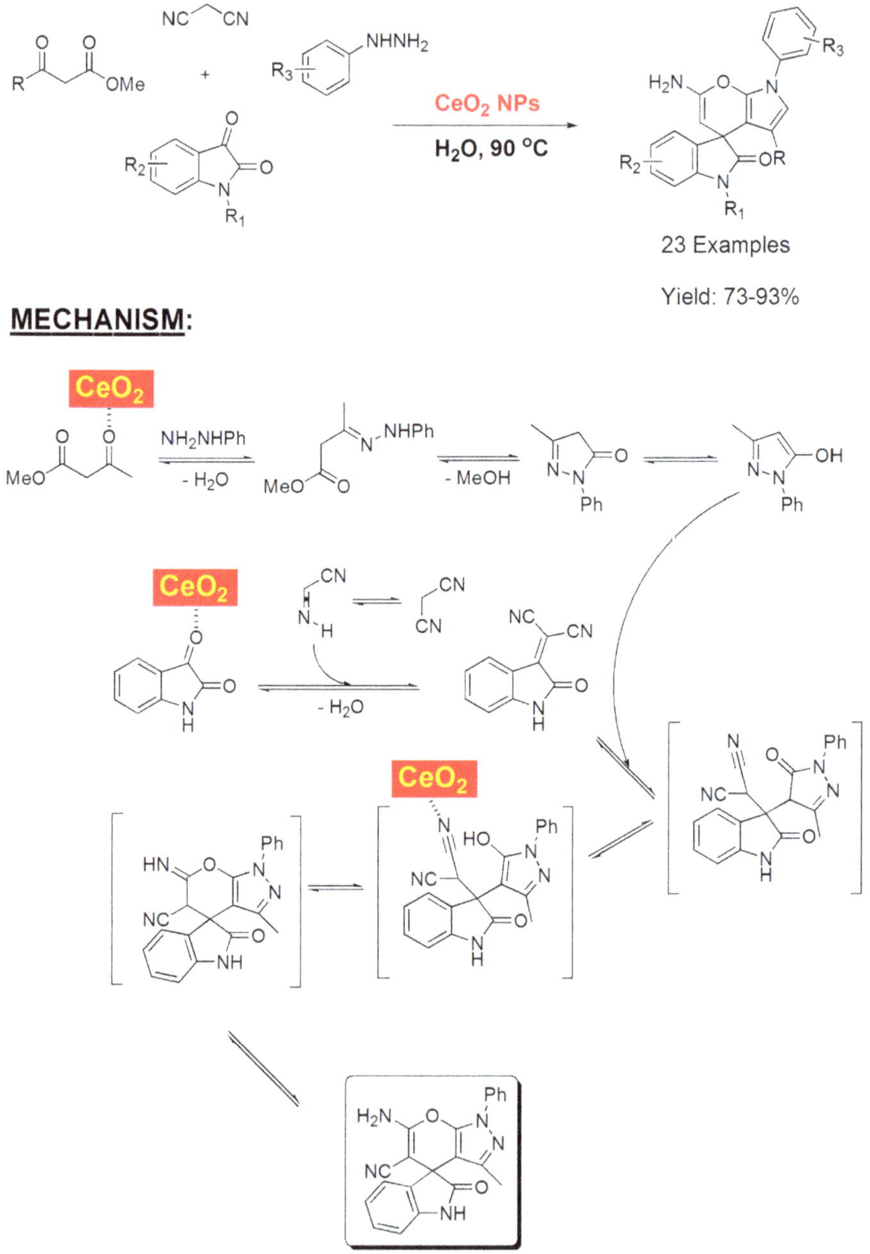

Scheme 25.2 CeO$_2$-mediated assembly of spiro[indoline-3,4-pyrano[2,3-c]pyrazole] derivatives in water

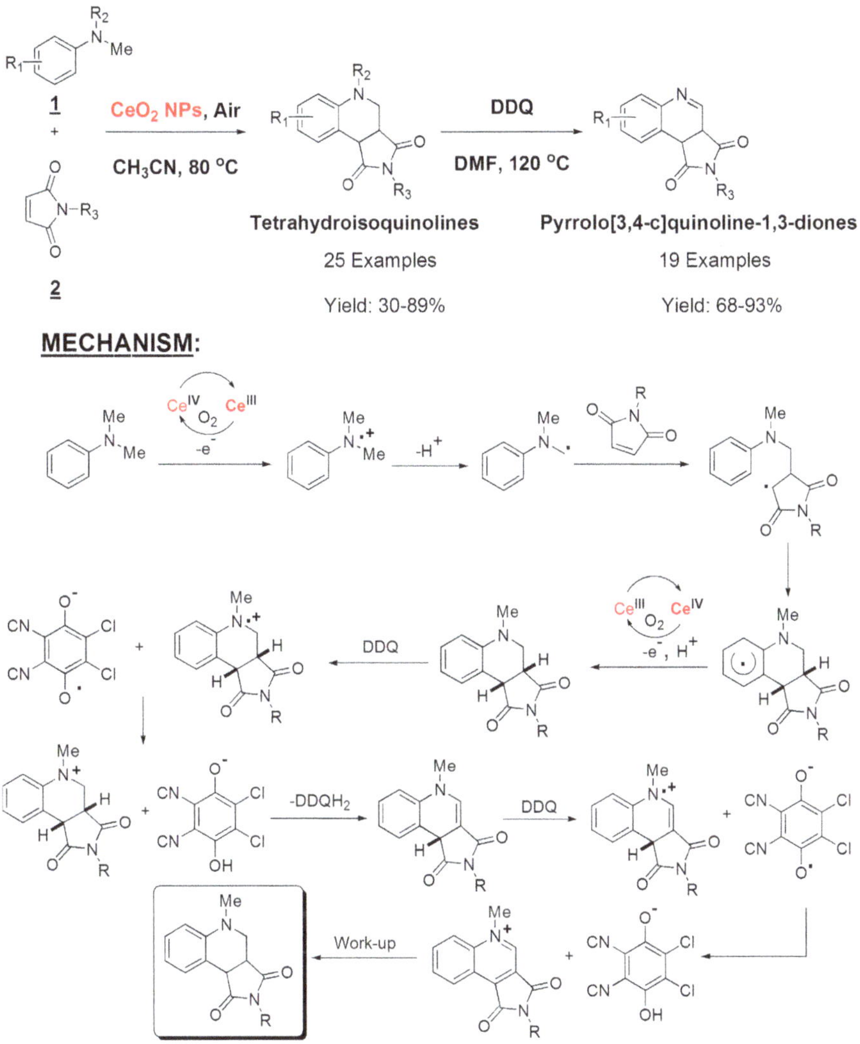

Scheme 25.3 Synthesis of tetrahydroisoquinolines and pyrrolo[3,4-c]quinoline-1,3-diones from the CeO_2/DDQ-mediated coupling of N,N-dimethylanilines and N-substituted maleimides

established in Scheme 25.4. In this context, Ce^{4+} would be transformed into Ce^{3+} and vice versa in the presence of O_2 during the single-electron transfer (SET) to facilitate the formation of iminium intermediate. Significantly, only a minor diminution in the yields of N-aryl tetrahydroisoquinoline was observed after four circulations of spent CeO_2.

In addition, Rashed et al. (2020) demonstrated that the commercial CeO_2 (JRC-CEO-1, 185.3 m²/g) could stimulate the solvent-free alkenylation of oxindole with aldehydes (Scheme 25.5). Specifically, this synthetic protocol was applicable to

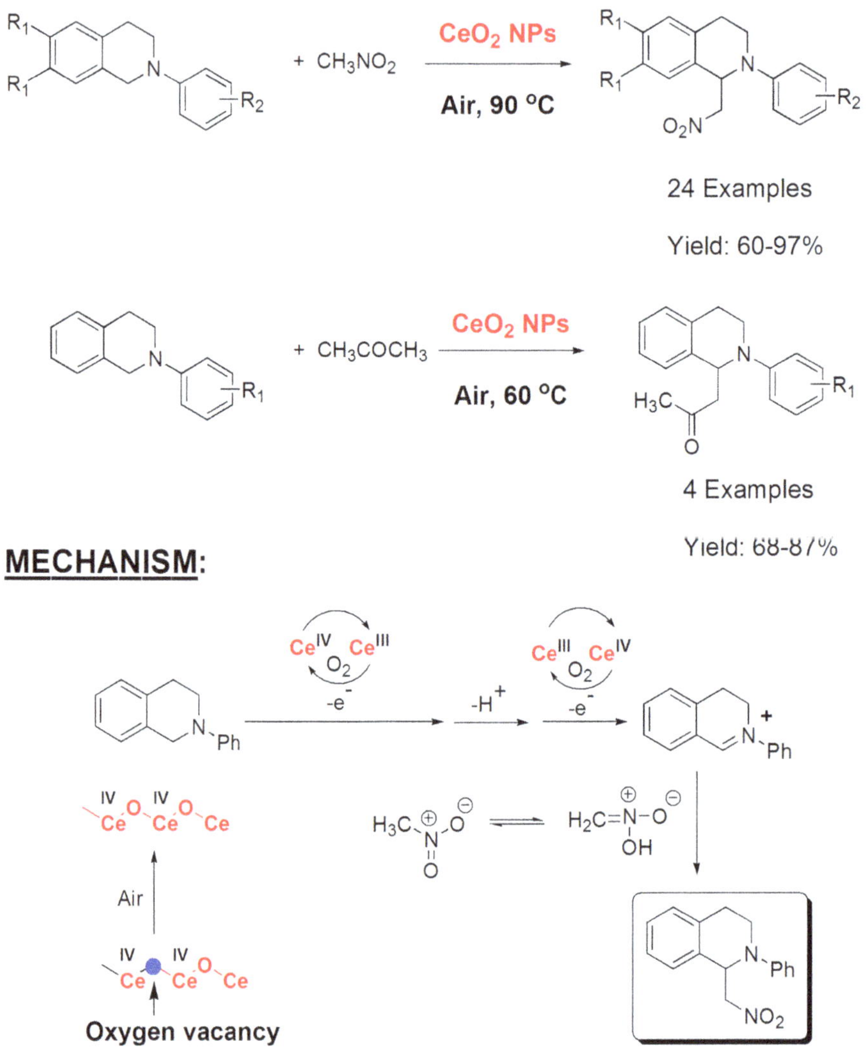

Scheme 25.4 Functionalization of *N*-aryltetrahydroisoquinolines with nitroalkanes and acetone over nanostructured CeO_2

both aliphatic and aromatic aldehydes, furnishing 87–99% yields of C_3-alkenylated oxindole products with high selectivity in *E* isomers. In this study, a close relationship between catalytic activity and morphology of CeO_2 calcined at different temperatures (i.e. 300, 500, 600, 800 and 1000 °C) was described. Surprisingly, the ceria with increasing calcination temperature would display higher catalytic activity despite their lower specific area, which might be attributed to the presence of non-defect (111) surface as active sites for the alkenylation reaction. Another reason came from the assumption that elevating the calcination temperature in the

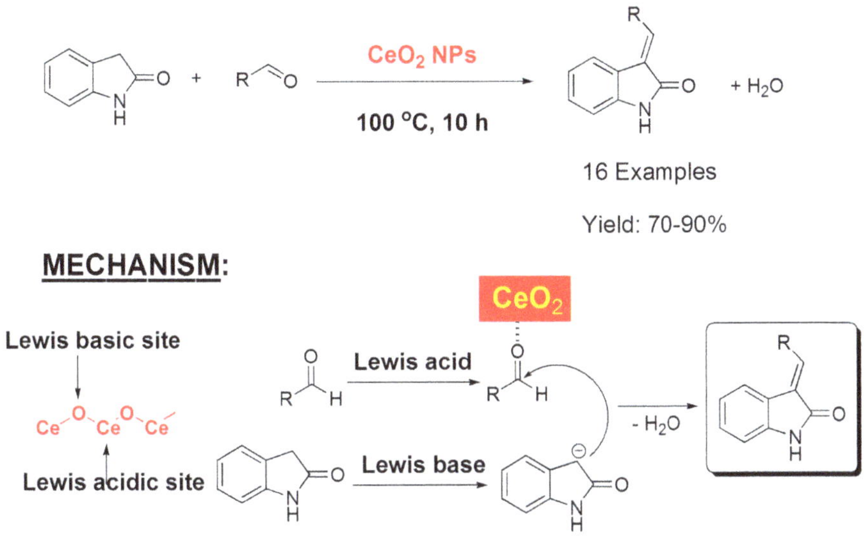

Scheme 25.5 C_3-alkenylation of oxindole with aldehydes towards 3-alkyledene oxindoles over CeO_2 NPs

pretreatment stage led to a higher density of Lewis active sites. Indeed, the outstanding catalytic activity of nanostructured CeO_2 in this alkenylation was accredited to the bifunctional Lewis acid-base property, in which the basic sites (oxygen atom) would deprotonate the C_α-H bond of oxindole to trigger the corresponding enolate ion. Meanwhile, the acidic sites (cerium atom) would activate the carbonyl group of aldehyde, thereby enhancing the reactivity of C=O bond towards the nucleophilic attack of enolate.

25.2.2 Synthetic Nano-CeO_2

In nanotechnology, a plethora of techniques have been developed to fabricate the metal oxide nanoparticles (Table 25.1) (Rane et al. 2018). With each type of synthetic mode, the nanostructured oxides with different physical-chemical properties (e.g. particle size, porosity, defect, crystal structure, polarity and acidity/basicity) can be selectively controlled. In this regard, the reaction conditions such as starting precursors, capping agents, pH, ageing time/temperature and calcination temperature are key factors governing the outcome of final nanostructures. For instance, copious exemplars showing the impact of synthetic procedures on the specific morphology of CeO_2 NPs are illustrated in Table 25.2.

In this chapter, all of the reported nanostructured CeO_2 could be prepared from four main synthetic categories of co-precipitation, template, biological and sol-gel pattern.

Table 25.1 Synthetic techniques of nanoparticles

Synthetic modes of nanoparticles			
Co-precipitation synthesis	Sol-gel synthesis	Ultrasound synthesis	Laser ablation synthesis
Hydrothermal synthesis	Template synthesis	Microwave-assisted synthesis	Sputtering synthesis
Inert gas condensation synthesis	Microemulsion synthesis	Spark discharge synthesis	Biological synthesis

Table 25.2 Impact of synthetic methods on the morphology of CeO_2

Method	Cerium precursor	Capping agent	Particle size (nm)	Morphology	References
Precipitation	Cerium (III) nitrate	–	9–18	Cubic hexagonal	Chen and Chen (1993)
		PVP	27	Spherical	
Microemulsion	Cerium (III) nitrate	Hexamethylenetetramine	7–10	Spherical	Arya et al. (2014)
	Cerium (III) nitrate-Cerium (III) chloride	Brij35	6–13	Cubic	Bumajdad et al. (2004)
Hydrothermal	Cerium (III) nitrate	–	8–16	Cubes, rods	Arya et al. (2014)
	Cerium (III) chloride	Citric acid	<5	Spherical	López et al. (2015)
Biological	Cerium (III) nitrate	*Hibiscus sabdariffa*	3.9	Amorphous	Thovhogi et al. (2015)
	Cerium (IV) ammonium nitrate	Fructose/glucose/lactose	2–6	Spherical/agglomerate	Kargar et al. (2015)
Sol-gel	Cerium (III) nitrate	Oleylamine	1.2–35	Spherical, tadpole, wire	Yu et al. (2005)

Co-precipitation (Guo et al. 2015) This is the most facile and convenient strategy to fabricate metal oxide nanoparticles by adding a precipitating agent (organic or inorganic bases) into the aqueous solution of metal salts at room or elevated temperature. As soon as the concentration of species present in the solution reaches the critical point, a cascade of nucleation/growth/agglomeration reaction will take place. In some cases, employing the surfactants and capping agents is necessary to selectively manipulate the physiochemical and catalytic features of the final metal oxides. Undoubtedly, multiple factors such as precursors, nature of bases, pH of the reaction medium, temperature, and stirring rates strongly influence the property of designed metal nanoparticles. For instance, Chen and Chang (2005) disclosed that increasing the temperature in the co-precipitation of $Ce(NO_3)_3 \cdot 6H_2O$ with NH_3 led to a morphological change of CeO_2 particles from cubic to hexagonal, whilst lowering the temperature induced the smaller size of ceria particles. On the other hand, the elevation of pH towards 12 in the reaction medium helped to decrease the crystallite size of CeO_2 (Ramachandran et al. 2019). Other influential factors in the co-precipitation for CeO_2 such as cerium precursors and precipitating agents are illustrated in the Table 25.3 as well. In fact, rapid, safe, low-cost, facile and organic solvent-free aspects are acknowledged as remarkable merits of this synthetic strategy.

Sol-gel synthesis (Parashar et al. 2020; Laberty-Robert et al. 2006) This model is associated with the rapid hydrolysis of metal-organic precursors in water and/or organic solvents to generate the corresponding metal oxo-hydroxides, which subsequently undergo the condensation to form an extended matrix of metal hydroxides. Next, the polymerization of these hydroxides will lead to the establishment of a dense network porous gel. Afterwards, the ultrafine porous metal oxides can be obtained upon drying and heating the gel at high temperatures. In this situation, the nature of both metal precursors and solvents considerably determines the morphology and particle size of final metal oxides. As an example, Yu et al. (2005) revealed that spherical CeO_2 could be triggered from the sol-gel treatment of $Ce(NO_3)_3.6H_2O$, diphenyl ether with oleylamine. On the other hand, the addition of oleic acid in this mixture resulted in wired or tadpole-like CeO_2 regarding to the amount of oleic acid. More examples on the sol-gel approach towards different CeO_2 NPs are depicted in Table 25.4.

Template-assisted synthesis (Yu et al. 2013) This technique mainly concerns with the deployment of hard/soft materials (e.g. carbon nanotube, alumina, zeolites, silica and polymers) as a host, where the nanoparticles will be fabricated and confined within the pores or channels of the template after calcination. By applying a proper choice of starting precursors, loading amounts and type of templates, it is able to render controlled-sized nanostructures with various morphologies (Table 25.5).

Table 25.3 Different types of nano-CeO_2 obtained from the co-precipitation

Cerium precursor	Precipitating agents	Particle size (nm)	Morphology	References
Cerium (III) nitrate	Ammoniac-ammonium bicarbonate	120–460	Spherical	Zhang et al. (2009)
	Sodium hydroxide	5	Rod	Du et al. (2007)
	Ammonia-hydrogen peroxide-hexamethylenetetramine	6	Cubic	Kamruddin et al. (2004)
Cerium (IV) ammonium nitrate	Urea	~8	Cubic	Tsai (2004)

Table 25.4 Different types of nano-CeO$_2$ obtained from the sol-gel strategy

Cerium Precursor	Medium	Particle size (nm)	Morphology	References
Cerium (IV) nitrate	Oleylamine-trioctylamine-diphenyl ether	1.2–3.5	Spherical, tadpole, wire	Yu et al. (2005)
Cerium (III) salts	PVA-sucrose	6–9	Cubic	Soni and Biswas (2013)
Cerium (IV) ammonium nitrate	CTAB-methanol-aniline	3.4–10.4	Sponge-like	Tillirou and Theocharis (2008)

Table 25.5 Different types of nano-CeO$_2$ from the template-directed synthetic pattern

Cerium precursor	Template	Particle size (nm)	Morphology	References
Cerium (III) nitrate	Carbon spheres	300	Hollow spherical	Xu et al. (2014)
Ammonium cerium (IV) nitrate	Polymethyl methacrylate	5	Tubular	Schneider et al. (2011)
Cerium (III) nitrate	Chitosan	~4	Cubic	Sifontes et al. (2011)

Biological synthesis (Malik et al. 2017) This synthetic mode involves the application of biological materials such as microorganisms (e.g. bacteria, fungi, yeast and algae), plant parts (e.g. leaves, fruit, flower, bark and seed) or sugars as natural reducing agents to assist the fabrication of nanoparticles. In the presence of these biochemical reductants, the metal ions from precursor salts are initially reduced to atoms which subsequently nucleate into small clusters. Originating from these metal clusters, the nanoparticles will grow in different manners depending on the concentration of metal ions, pH, reaction time, temperatures, and types of reducing agents. For example, the plate-like CeO$_2$ could be fabricated by employing fresh egg white (Kargar et al. 2015; Maensiri et al. 2007), in which ovalbumin/lysozyme (egg proteins) were demonstrated to serve the function of bio-capping/stabilizing agent. Alternatively, several investigations on the plant-mediated synthesis of CeO$_2$ NPs using the extract of *Hibiscus sabdariffa* flower, *Petroselinum crispum* leaf and *Olea europaea* leaf as phyto-chelating/capping agents were also reported (Thovhogi et al. 2015; Korotkova et al. 2019; Maqbool 2017). Additionally, Thakur et al. (2019) were able to produce spherical CeO$_2$ (5–20 nm) by using the culture filtrate of *Curvularia lunata*. As depicted in Table 25.6, various exemplars on the bio-directed fabrication of CeO$_2$ NPs are also introduced.

Table 25.6 Different types of CeO$_2$ NPs obtained from the biological synthetic pattern

Capping agent	Cerium precursor	Particle size (nm)	Morphology of NPs	References
Egg white	Cerium(III) acetate	6–30	Plate-like	Maensiri et al. (2007)
Gloriosa superba	Cerium(III) chloride	5	Spherical	Arumugam et al. (2015)
Ricinus communis leaf extract	Cerium(III) chloride	34	Irregular	Suvetha Rani (2020)
Honey	Cerium(III) nitrate	23	Cubic	Darroudi et al. (2014)
Aspergillus niger culture filtrate	Cerium (III) chloride	5–20	Cubic-spherical	Gopinath et al. (2015)

25.2.2.1 Nanostructured CeO$_2$ from the Co-precipitation Method

In 2015, Safaei-Ghomi et al. (2015a) reported the application of CeO$_2$ derived from the co-precipitation of Ce(NO$_3$)$_3$.6H$_2$O with NH$_3$ as an effective nanocatalyst for the assembly of 2-aminocyclohex-1-ene-1-carboxylic esters (Scheme 25.6).

Later, the co-precipitated CeO$_2$ NPs was also deployed to facilitate the room-temperature synthesis of polysubstituted dihydropyridines from the four-component coupling of aromatic aldehydes, ethyl cyanoacetate, arylamines and dimethyl acetylenedicarboxylate (Safaei-Ghomi et al. 2015b). In this study, the CeO$_2$ with particles size of 11 nm showed the superior activity over other nanosized catalysts such as CaO (35 nm), ZnO (24 nm), CuO (40 nm), MgO (18 nm) and SnO (28 nm), therefore enabling for high yields of polysubstituted dihydropyridines. As shown in the Scheme 25.7, the CeO$_2$-mediated coupling followed a set of sequential reactions of Knoevenagel condensation/Michael addition/annulation/tautomerization.

Subsequently, a high-yielding process of C-tethered bispyrazol-5-ols from the CeO$_2$-mediated multicomponent condensation of dimethyl acetylenedicarboxylate, phenylhydrazine and aromatic aldehydes in water was described by Safaei-Ghomi et al. (2015c). In this setting, the excellent activity of lab-prepared CeO$_2$ NPs was attributed to the high surface area (33.2 m^2/g) with respect to that of bulk CeO$_2$ (5.2 m^2/g), CaO (1.2 m^2/g) and ZrO$_2$ (4.9 m^2/g). Another reason came from the high distribution of oxygen vacancies as Lewis acidic sites on the surface of lab-designed CeO$_2$ NPs. On account of these factors, the CeO$_2$ NPs was able to produce derivatives of C-tethered bispyrazol-5-ol in high isolated yields (Scheme 25.8).

Likewise, Safaei-Ghomi et al. (2016) also introduced CeO$_2$ as a recyclable nanocatalyst for the *mechanochemical* synthesis of 2-amino-4,6-diarylbenzene-1,3-dicarbonitriles. As depicted in Scheme 25.9, the CeO$_2$-mediated reaction is suggested to undergo a mechanistic sequence of Knoevenagel condensation/Michael addition/annulation/aromatization at room temperature.

Later, D'Alessandro et al. (2015) described the usefulness of CeO$_2$ in triggering the solvent-free multicomponent Hantzsch reaction. Remarkably, it is revealed that a switchable construction of 1,4-dihydropyridine and 2-phenylpyridine could be

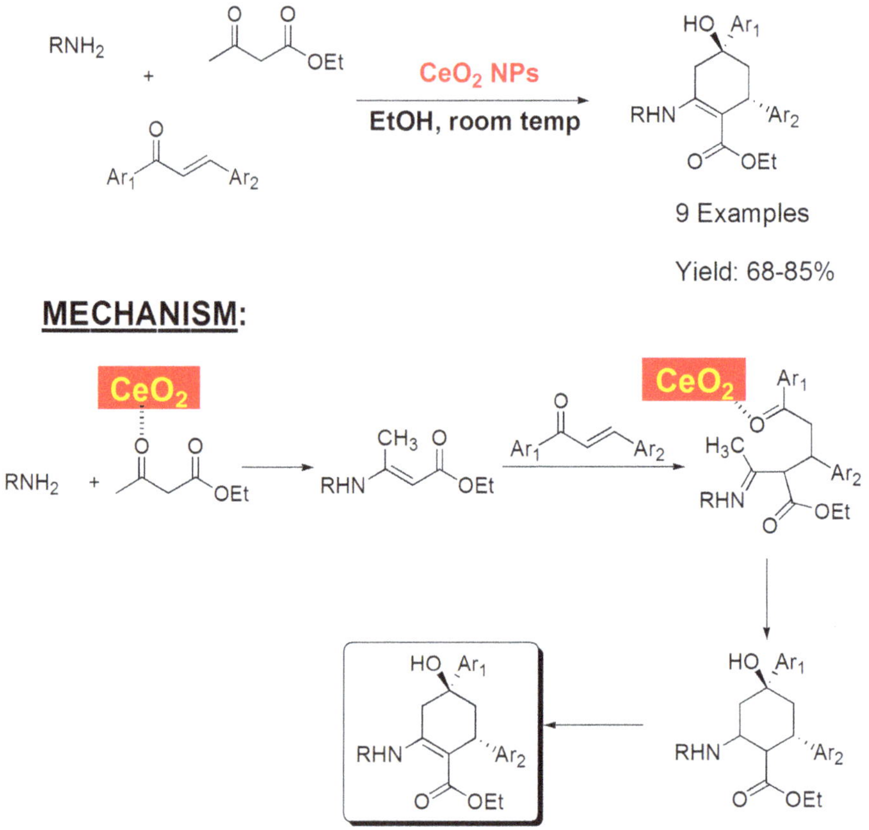

Scheme 25.6 Room-temperature synthesis of 2-aminocyclohex-1-ene-1-carboxylic esters over CeO_2 NPs

accomplished from the coupling of benzaldehyde, methyl acetoacetate and ammonium acetate under different temperatures. Remarkably, it is found that 97% yield of phenylpyridine was generated at 25 °C, while elevating the reaction temperature to 80 °C offered 75% yield of 1,4-dihydropyridine. In both cases, the recovered CeO_2 NPs could maintain the original activity after four consecutive trials. Similarly, Suresh et al. (2016) disclosed that a novel scaffold of fused triazolo/tetrazolo[1,5-*a*]pyrimidine could be assembled under the catalysis of CeO_2 NPs. In this manner, the CeO_2-mediated condensation of substituted aromatic aldehydes, benzoylacetonitrile with 5-aminotriazole/5-aminotetrazole, took place smoothly in water to generate two types of fused pyrimidine products. The catalytic role of CeO_2 NPs in this tandem Knoevenagel/Michael addition/intermolecular cyclization/intermolecular dehydrogenation reaction is clearly clarified in Scheme 25.10.

In another example, Gharib et al. (2013) fabricated the nanostructured CeO_2 by precipitating the aqueous solution of $(NH_4)_2Ce(NO_3)_6$ with NH_3. Thanks to the high surface area, the lab-designed CeO_2 was capable of promoting the *aqueous-phase*

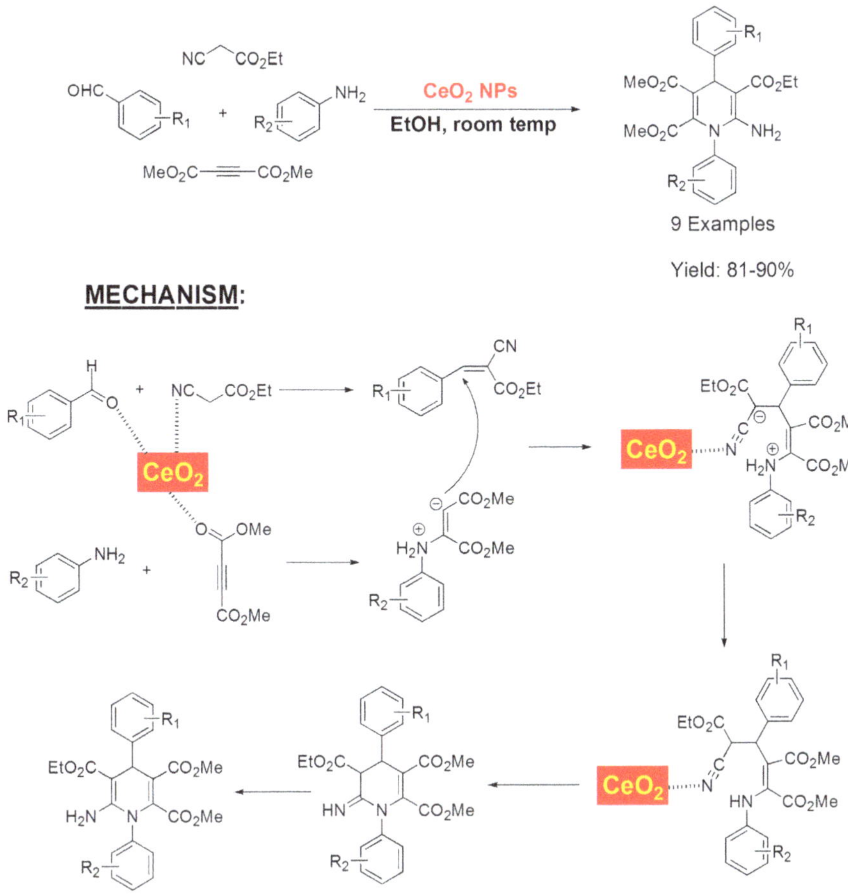

Scheme 25.7 Construction of polysubstituted dihydropyridines over CeO_2 NPs

coupling of Lawsone reagent with 3-methyl-1-phenyl-1*H*-pyrazol-5-amine and substituted benzaldehydes under reflux condition. Accordingly, seven derivatives of 3-methyl-1-phenyl-1*H*-benzo[*g*]pyrazolo[3,4-*b*]quinoline-5,10-diones could be furnished in good to excellent yields (66–94.5%).

To construct the multiple heterocyclic scaffold of imino-pyrrolidine-thione, Wang et al. (2016) applied the porous CeO_2 nanorods obtained from the hydrothermal treatment of $Ce(NO_3)_3 \cdot 6H_2O$ with $(NH_4)_2CO_3$ to mediate the coupling of 2-mercaptobenzoxazole/2-mercaptobenzothiazole with a mixture of substituted benzaldehydes, malononitrile and isocyanide. As illustrated in Scheme 25.11, the Ugi four-component condensation could run smoothly in a binary mixture of CH_3CN-H_2O (3:1, *v/v*) with 5 mol% of nanoporous CeO_2 to deliver a broad library of imino-pyrrolidine-thiones. Under identical condition, commercial and other synthetic CeO_2 NPs with different morphologies (i.e. linear, granular and fusiform) were found to give lower yield of coupling product with respect to the titled

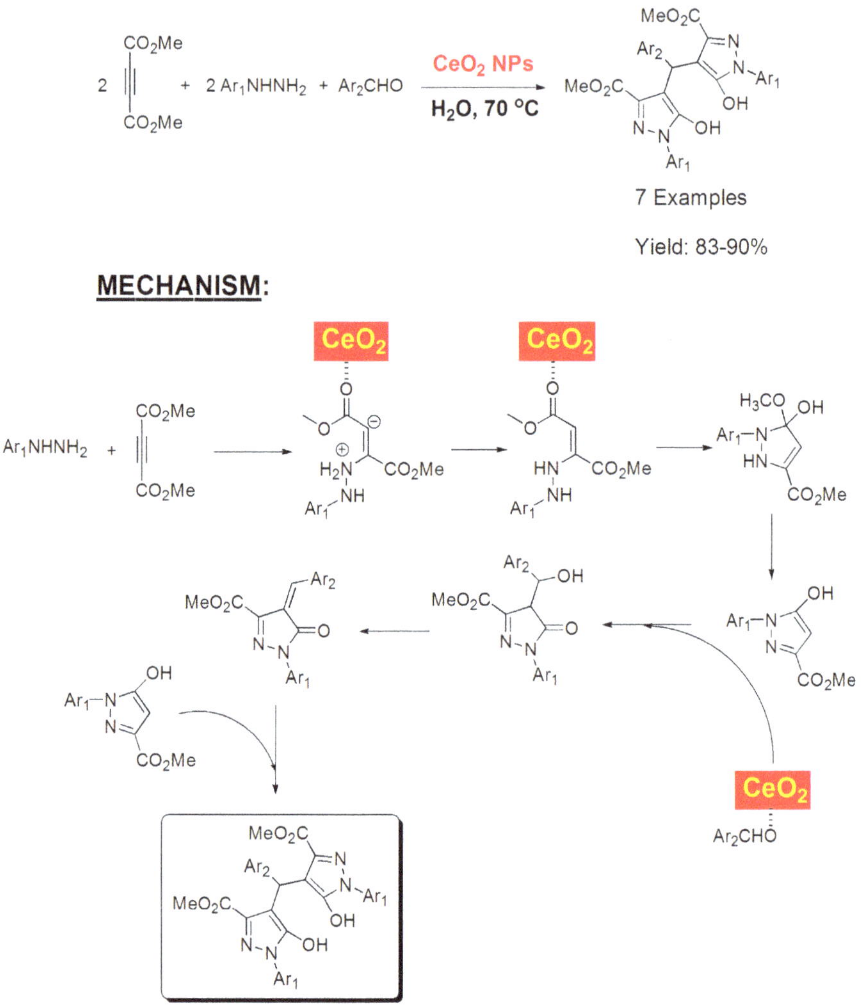

Scheme 25.8 Construction of C-tethered bispyrazol-5-ols over CeO_2 NPs

nanoporous CeO_2. Unfortunately, the loss of oxygen storage in the spent CeO_2 was assumed to take place, thereby leading to a significant drop in the catalytic performance after the third recycle.

To address intrinsic drawbacks in the current manufacture of azole compounds (benzimidazoles, benzothiazoles and benzoxazoles), Shelkar et al. (2013) established a facile and eco-friendly strategy to construct these privileged skeletons upon employing CeO_2 nanocatalyst prepared from the surfactant-assisted co-precipitation under ultrasonic irradiation (Terribile et al. 1998). In comparison with other tested metal oxides (i.e. ZnO, TiO_2, MnO_2, SiO_2, Al_2O_3, La_2O_3 and Cu_2O NPs), the robust CeO_2 NPs displayed the preeminence in fostering high yields of benzimidazoles,

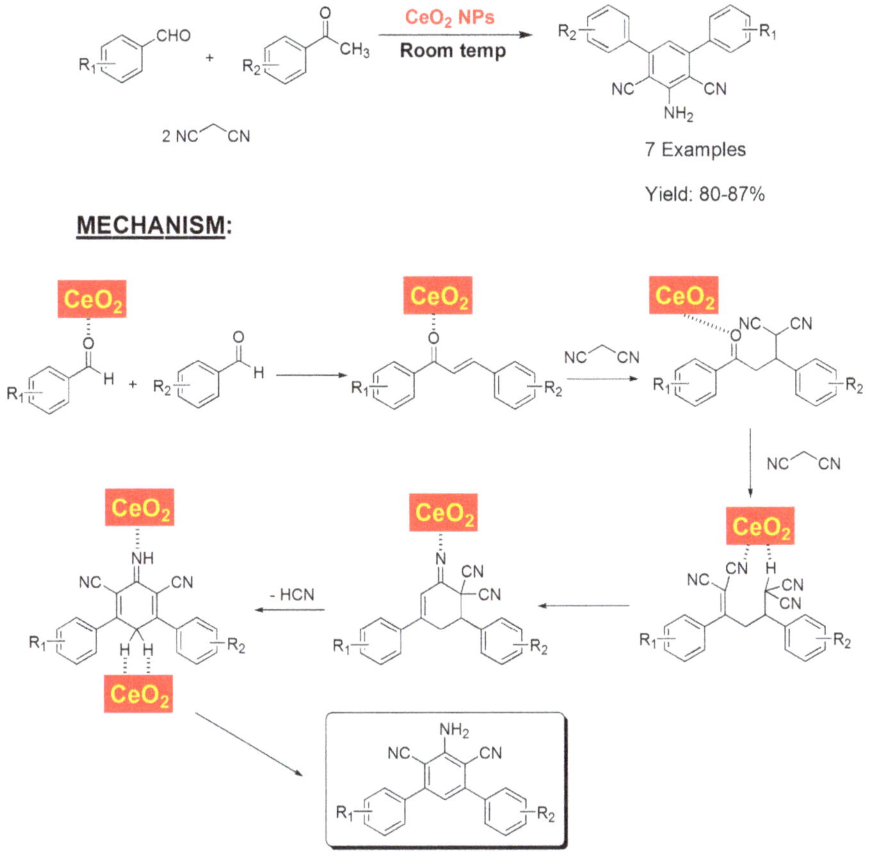

Scheme 25.9 Solvent-free access of 2-amino-4,6-diarylbenzene-1,3-dicarbonitriles over CeO_2 NPs

benzothiazoles and benzoxazoles from the aqueous-phase coupling of 1,2-phenylenediamine/2-aminothiophenol/2-aminophenol with aldehydes, respectively (Scheme 25.12).

25.2.2.2 Nanostructured CeO_2 from the Polymer-Directed Method

In 2011, Girija et al. (2011) fabricated the polymer-directed CeO_2 nanoparticles by treating the mixture of $(NH_4)_2Ce(NO_3)_6$, hexylamine and polyethylene glycol 6000 (PEG-6000) under microwave irradiation, which was then examined for the catalytic assembly of polyhydroquinolines. In this context, the solvent-free multicomponent condensation of aldehydes, ethyl acetoacetate, dimedone and ammonium acetate was carried out under the assistance of both microwave radiation and CeO_2 NPs, finally providing 88–97% yields of target polyhydroquinolines. However, a

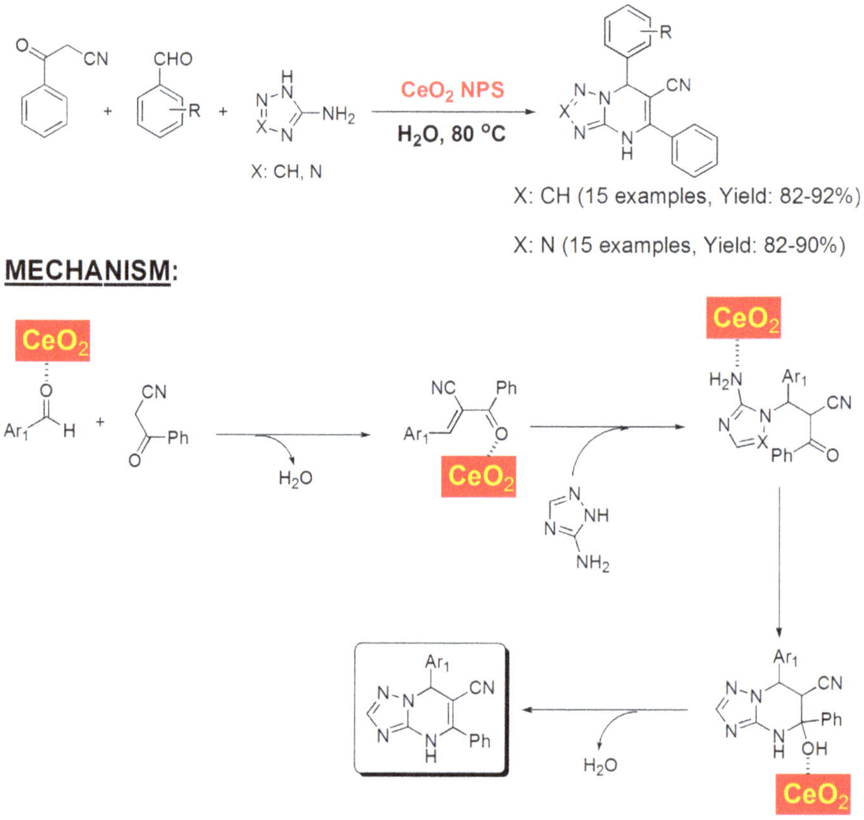

Scheme 25.10 Construction of fused triazolo/tetrazolo[1,5-*a*]pyrimidines over CeO_2 nanocatalyst

gradual loss in the catalytic activity of recovered CeO_2 was observed due to the slow oxidation of Ce NPs during the recycling trials.

By combining the reverse microemulsion system of bis(2-ethylhexyl) sulfosuccinate-lecithin-isooctane-water with different polymers of polyvinylpyrrolidone (PVP), block copolymer P123 or reverse block copolymer 17R4 as structural controller during the preparative procedure, Samai et al. (2016) were able to prepare a set of CeO_2 (i.e. CeO_2-PVP; CeO_2-P123; and CeO_2-17R4) with controlled nanoparticle sizes. Noticeably, it is uncovered that the relationship between the morphology and the catalytic performance of these titled nano-CeO_2 was intimately correlated with the directing polymeric agents. In this aspect, CeO_2-PVP with the largest surface area (58.0 m^2/g) displayed superior results in comparison with CeO_2-P123 (45 m^2/g) and CeO_2-17R4 (40.96 m^2/g) upon coupling nitrostyrene, 1,3-dicarbonyl compounds and aromatic primary amines. Accordingly, a collection of *N*-aryl pyrroles in the range yields of 59–77% was successfully produced over recyclable CeO_2-PVP nanocatalyst.

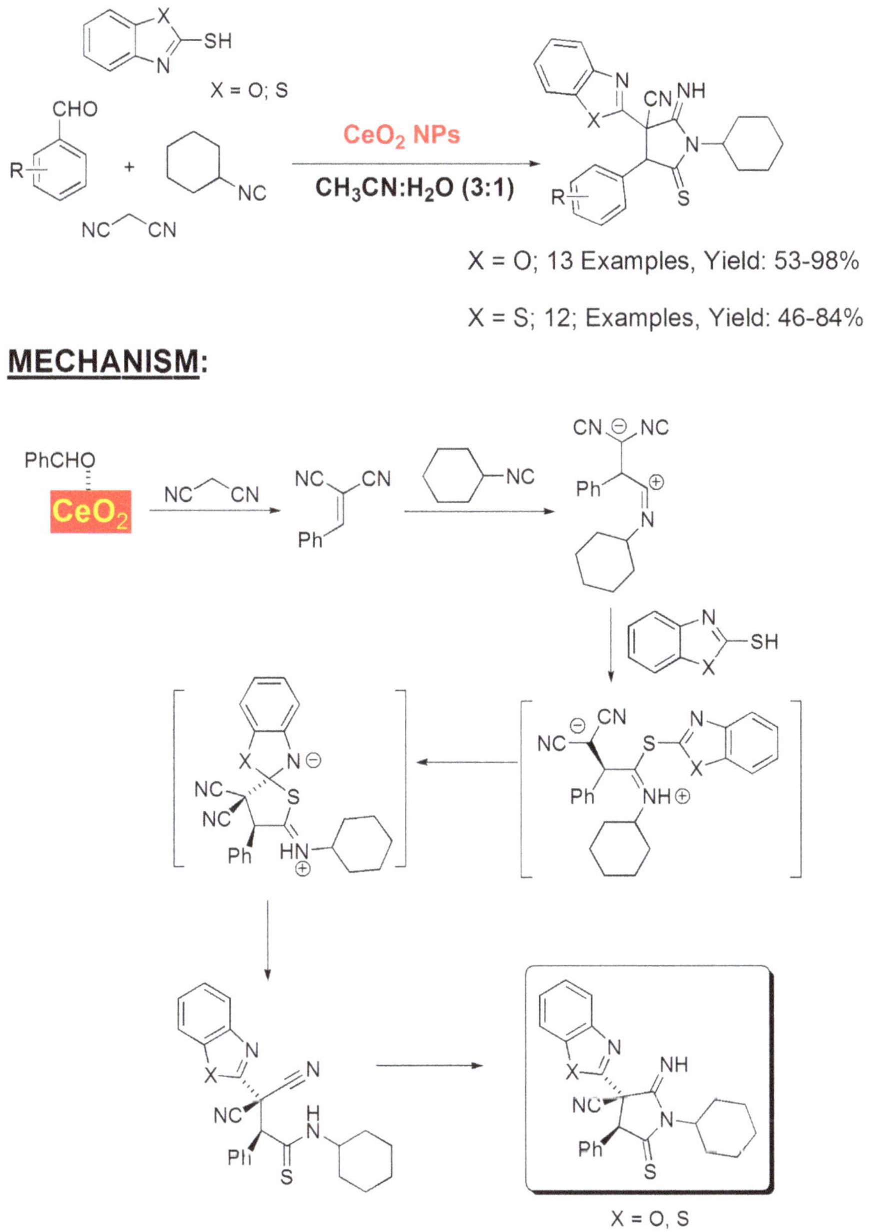
X = O; S
CHO
R
NC
NC CN
CeO₂ NPs
CH₃CN:H₂O (3:1)
N
X
CN NH
N
S
X = O; 13 Examples, Yield: 53-98%
X = S; 12; Examples, Yield: 46-84%
MECHANISM:
PhCHO
CeO₂
Ph
X = O, S

Scheme 25.11 Multicomponent synthesis of imino-pyrrolidine-thiones over CeO_2 nanoparticles

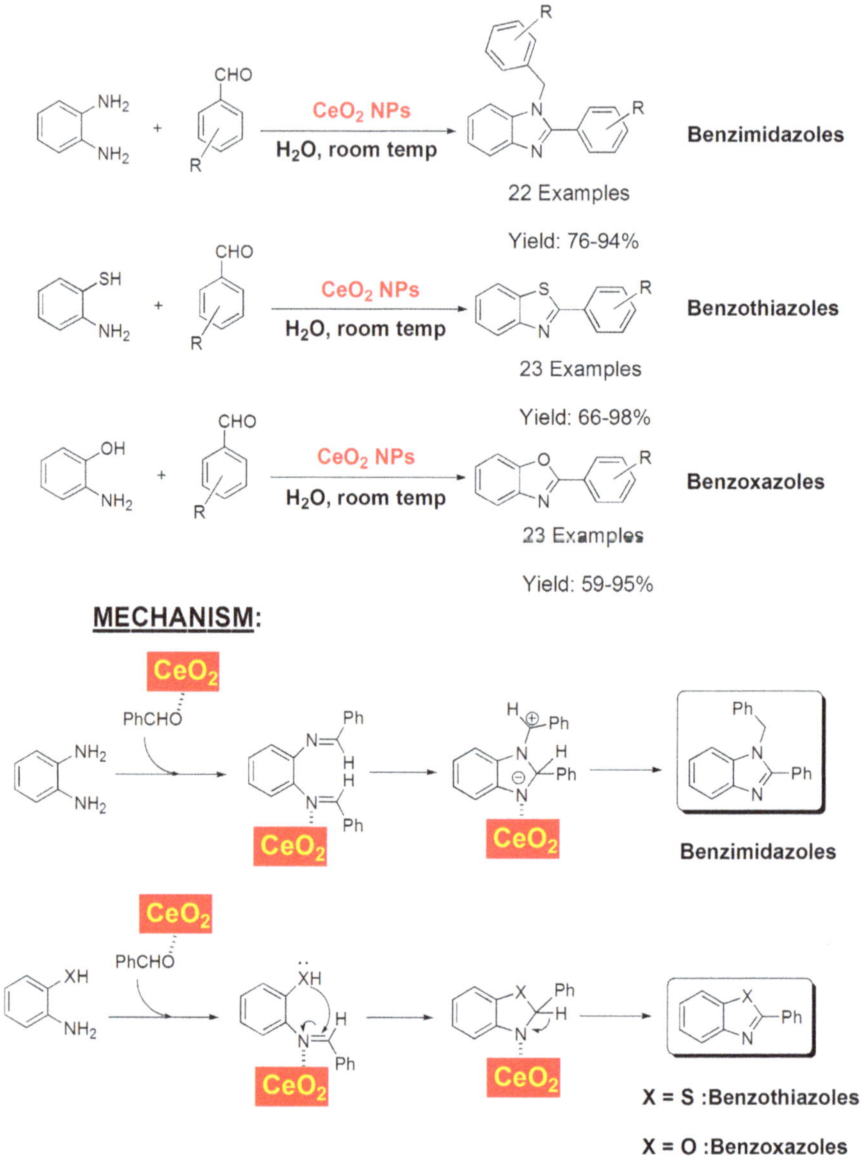

Scheme 25.12 Nano-CeO$_2$-mediated synthesis of benzimidazoles, benzothiazoles and benzoxazoles

25.2.2.3 Nanostructured CeO$_2$ from the Biology-Directed Method

Recently, plant extracts or bio-based materials have been deployed as greener alternatives to chemical reductants/oxidants/precipitating agents (e.g. cetyltrimethylammonium bromide, polyethylene glycols, monoethanolamine, ammonium hydroxide and polyvinylpyrrolidone) in the fabrication of ceria nanoparticles (Arumugam et al. 2015; Ferreira et al. 2016). With these nature-derived compounds, the preparative procedure can circumvent a complicated and tedious purification process (washing, calcination, Soxhlet extraction, etc.) to deliver organic-free CeO_2 NPs. Prompted by these examples, Zamani et al. (2018) explored the walnut shell powder to assist the fabrication of nano-CeO_2 in the absence of any surfactants or precipitating agents, in which the particle size could be tuned by controlling the ratio of Ce source/biomass. In this case, it is found that the presence of walnut shell as a cheap and green template is necessary to trigger smaller size of ceria, where the optimal ratio of Ce source/biomass was established at the ratio of 6.9:10. Hence, the resulting CeO_2 with a particle size of 9 nm was able to stimulate the aqueous-phase coupling of *o*-phenylenediamine and acetone with *tert*-butyl isocyanide at 80 °C to give 93% yield of 3,4-dihydroquinoxalin-2-amine.

25.2.3 Cerium Mixed Oxides

There are several documented methods for fabricating mixed metal oxides such as co-precipitation, wet impregnation, sol-gel, hydrothermal treatment, etc. (Courty and Marcilly 1976; Cousin and Ross 1990). In such cases, various true mixed oxides or solid solutions with the deposition of different metals can be readily composed to render a set of binary, ternary, quaternary or multiple-component mixed metal oxides, respectively. Undoubtedly, the mixed metal oxides display distinctive properties of acidity-basicity, oxidation-reduction, morphology (e.g. particle size, pore volume, surface area and defect) and thermal/chemical stability in comparison to pure metal oxides (Grzybowska-Swierkosz 1987; Wang et al. 2017). In addition, the bonding network between metals in mixed oxides allows the reagents to approach the active sites in an effective and selective manner, therefore increasing the yield and selectivity of the target products (Gawande et al. 2012; Burange and Gawande 2016). Thanks to these prominent features, the cerium-based mixed oxides have been widely deployed in the production of chemicals, organic synthesis, combustion of pollutants and energy applications (Orge et al. 2012; Shen et al. 2009; Zhang et al. 2018; Liu et al. 2019; Melchionna and Fornasiero 2014). For example, the nanocomposite of CeO_2-ZrO_2 obtained by the co-precipitation gave 90% yield of acetophenone from the deprotection of acetophenone oxime, whilst the pure CeO_2 only delivered 60% yield under identical condition (Deshpande et al. 2008). In another case, the catalytic activity of $Mn_3Gd_{7-x}Ce_x(SiO_4)_6O_{1.5}$ in the degradation of tetracycline was improved by the introduction of cerium in the structure, ascribable to the generation of active sites, the redox potential and an increase in the oxygen storage capacity (Fu et al. 2019). Likewise, Albadi et al. reported the practicality of CuO@CeO_2 nanocomposite for the construction of various heterocyclic structures through the multicomponent patterns (Scheme 25.13). Towards this end, the CuO@CeO_2 catalyst was composed from the co-precipitation of KOH with an aqueous mixture of $Ce(NO_3)_3$ and $Cu(NO_3)_2$. In the presence of CuO@CeO_2 nanocatalyst,

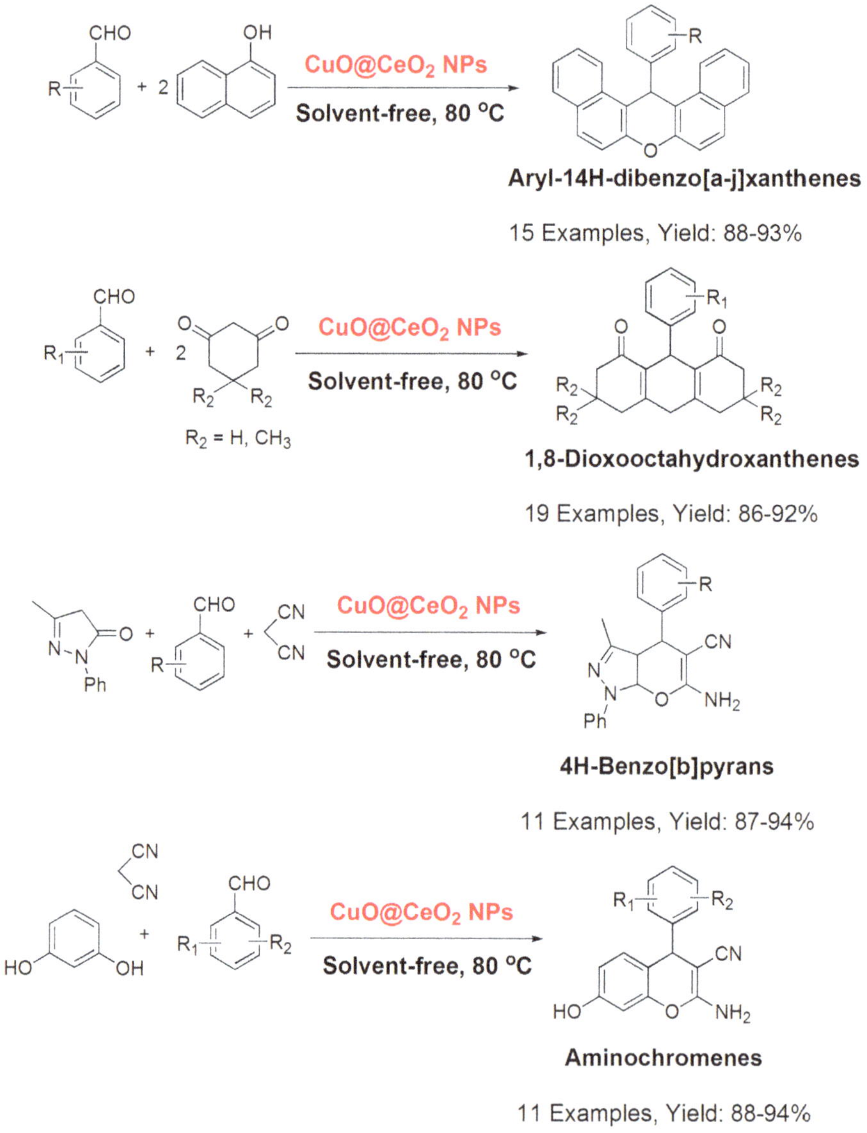

Scheme 25.13 Solvent-free synthesis of various heterocycles over CuO@CeO$_2$ nanocomposite

the solvent-free assembly of aryl-14*H*-dibenzo[a-j]xanthenes (Albadi et al. 2013a), 1,8-dioxooctahydroxanthenes (Albadi et al. 2013b), 4*H*-benzo[b]pyrans (Albadi et al. 2013c) and aminochromenes (Albadi et al. 2013d) was achievable with no difficulty.

Besides, Albadi et al. (2014a) also applied the nanostructured CuO@CeO$_2$ as a heterogeneous Lewis acid to induce the assembly of biscoumarins from benzaldehydes and 4-hydroxycoumarin in water (Scheme 25.14).

To develop a benign protocol for 1,4-disubstituted-1,2,3-triazoles, Albadi et al. (2014b) deployed the amberlite-supported azide as an alternative source of azide ion and CuO@CeO$_2$ as a heterogeneous copper catalyst. In this regard, the CuO@CeO$_2$-mediated click synthesis of functionalized triazoles by refluxing a mixture of aryl terminal alkynes and α-bromo ketones/ benzyl bromides with amberlite-supported azide in ethanol could provide excellent isolated yields of various triazoles in an eco-friendly manner (13 examples, 88–92%). In such examples, it is verified that the robust CuO@CeO$_2$ with no leeching of Cu could retain the outstanding catalytic activity after several recycling trials.

Furthermore, the practicality of nanostructured MgO@CeO$_2$ as an active solid catalyst in the construction of heterocyclic skeletons was also recognized (Scheme 25.15). In this setting, a collection of imidazo[4,5-c]pyrazoles (Moydeen et al. 2017), 2-amino-4-arylthiophene-3-carboxamides and thieno[2,3-d]pyrimidin-4(3H)-one-s (Shafighi et al. 2018) could be furnished in a high efficacy. After several recycles, no significant loss in the performance of recovered MgO@CeO$_2$ was observed, indicating the robustness of this titled nanocatalyst during the transformation.

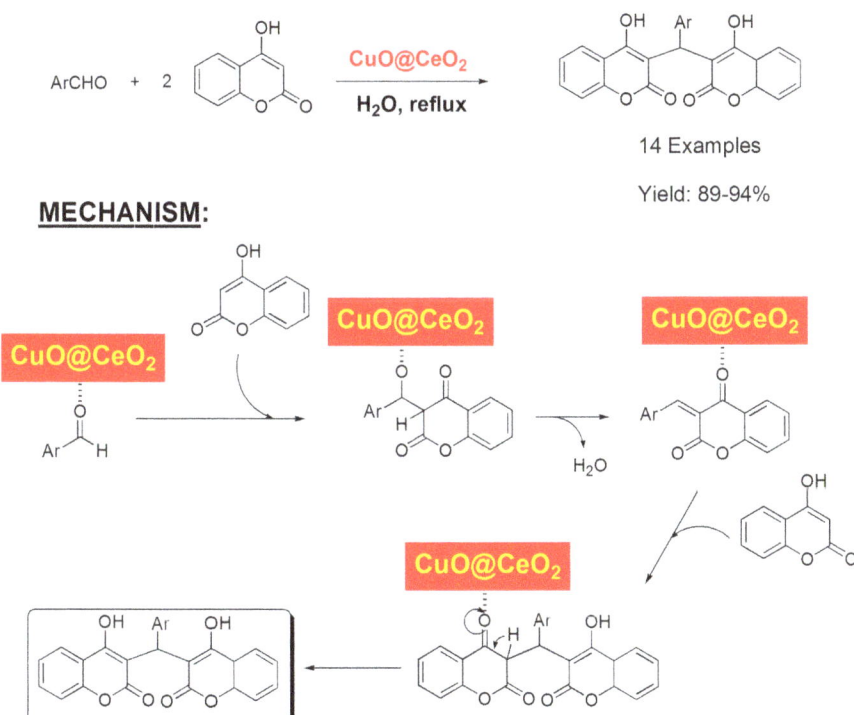

Scheme 25.14 Synthesis of biscoumarin derivatives over CuO@CeO$_2$ nanocatalyst

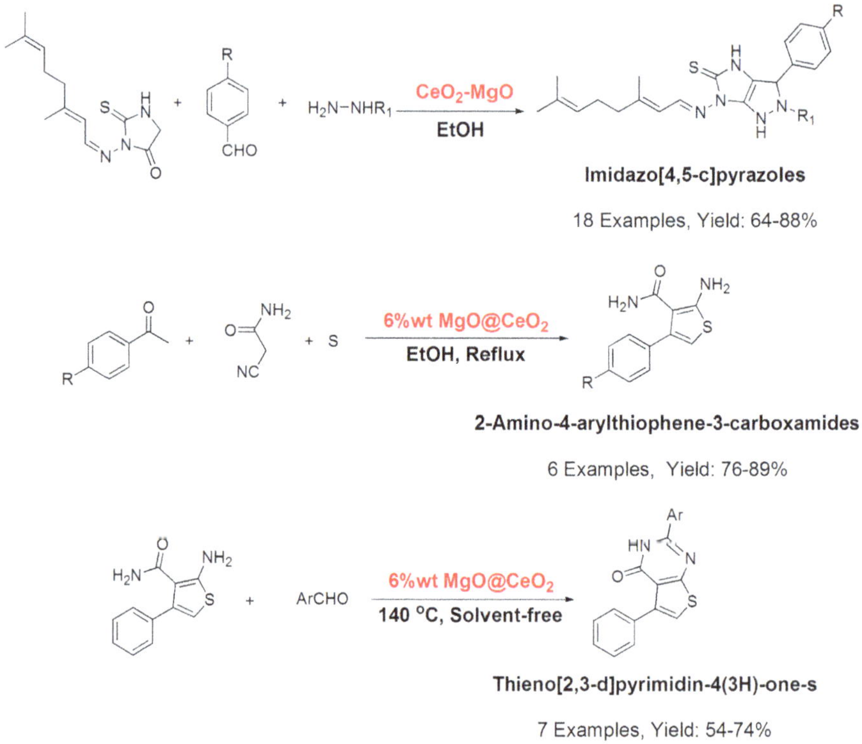

Scheme 25.15 Preparation of diversified heterocycles *over* MgO@CeO$_2$ nanocatalyst

In 2016, Vijay Kumar et al. (2016) designed the Eu$_2$O$_3$@CeO$_2$ nanocomposite for the multicomponent synthesis of phenyldiazenylacridinedione-carboxylic acids. For that objective, the binary oxide was prepared from the co-precipitation of Ce(NO$_3$)$_3$ and Eu(NO$_3$)$_3$ with NH$_3$ solution upon setting the optimal molar of Ce/Eu at a value of 8:2. The structural analysis indicated that the introduction of Eu on CeO$_2$ helped to induce the oxygen defects and to increase the surface area, leading to the superior catalytic activity of Eu$_2$O$_3$@CeO$_2$ over pure CeO$_2$. Hence, the catalysis of Eu$_2$O$_3$@CeO$_2$ in the aqueous-phase coupling of 1,3-dicarbonyl compounds, 4-hydroxy-3-methoxy-5-(substituted-phenyl-diazenyl)-benzaldehydes with glycine, enabled high yielding of (4-hydroxy-3-methoxy-5-(substituted-phenyldiazenyl)-dihydropyridineacetic acids. As described in the Scheme 25.16, Eu$_2$O$_3$@CeO$_2$ served as a heterogeneous Lewis acid in activating the C=O bonds during the multicomponent synthesis.

In another study, Ghayour et al. (2018) introduced ZnO@CeO$_2$ with 30.1 wt% of ZnO for the solvent-free coupling of aldehydes with 2-amino-4,5,6,7-tetrahydrobenzo[*b*]thiophene-3-carboxamide, where 62–92% yields of the thieno[2,3-*d*]pyrimidin-4(3*H*)-ones were achievable. To stimulate the construction of novel chromene derivatives bearing azo segment, Sagar Vijay Kumar et al. (2016)

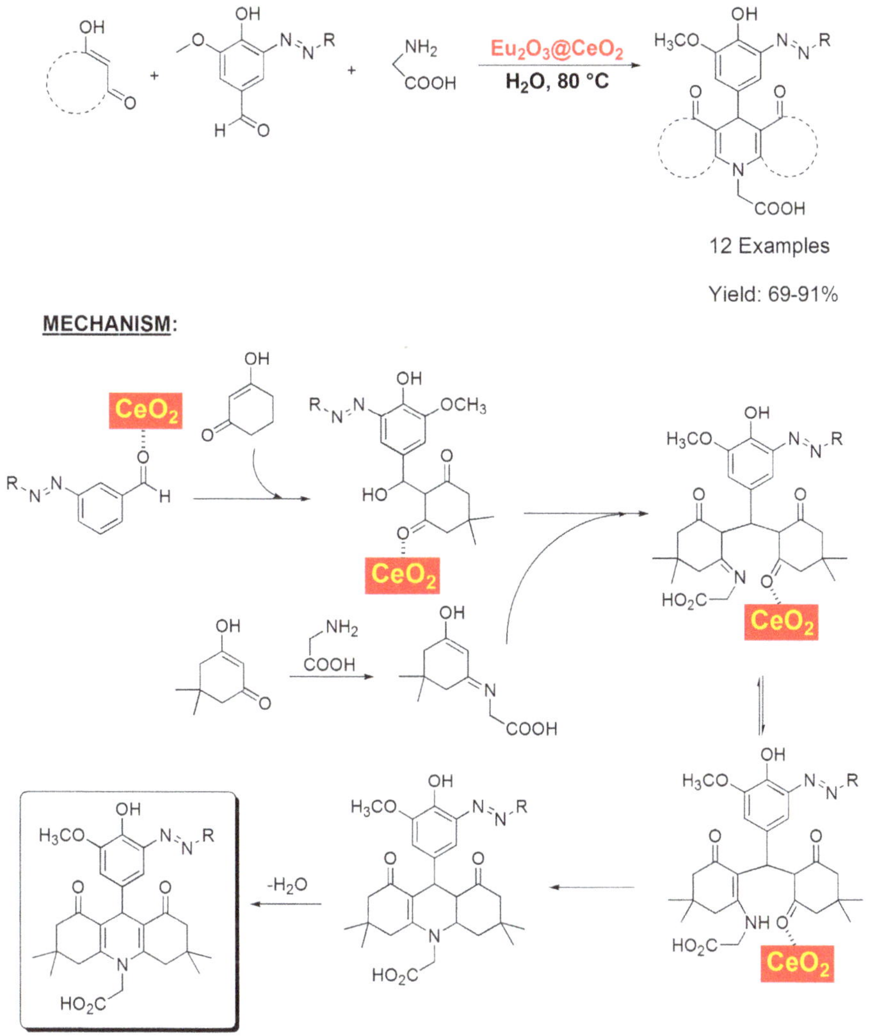

Scheme 25.16 Preparation of phenyldiazenyl-acridinedione-carboxylic acid derivatives from the Eu$_2$O$_3$@CeO$_2$ mediated multicomponent reaction

devised the co-precipitated nanocomposite of ZrO$_2$@CeO$_2$ (ratio Zr/Ce = 1:1) as a potential candidate for the room-temperature condensation of malononitrile, 4-hydroxy-3-methoxy-5-(substituted-phenyl-diazenyl) benzaldehydes with different compounds of 1,3-dicarbonyls. The mechanistic pathway leading to the formation of 2-amino-4-(4-hydroxy-3-methoxy-5-(substituted-phenyl-diazenyl)-chromene-3-carbonitriles is assumed to follow a set of Knoevenagel condensation/Michael addition/tautomerization/annulation reaction, in which ZrO$_2$@CeO$_2$ helped to activate the C=O and C≡N bond (Scheme 25.17).

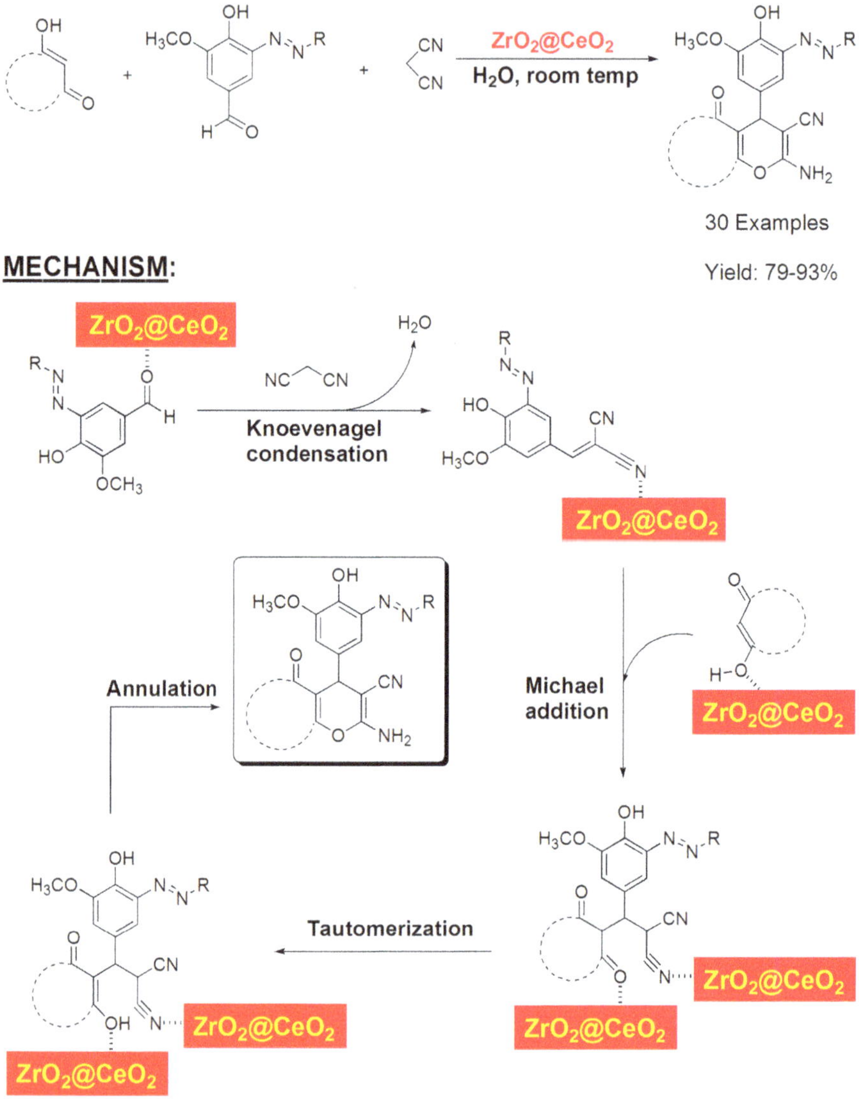

Scheme 25.17 Synthesis of novel azo chromenes over $ZrO_2@CeO_2$

Another exemplar of cerium-based mixed metal oxides comes from the preparation of $Ce_1Mg_{0.6}Zr_{0.4}O_2$ composite as reported by Rathod et al. (2010). In this scenario, the aqueous mixture of $(NH_4)_2Ce(NO_3)_6$, $Mg(NO_3)_2$ and $Zr(NO_3)_2$ was co-precipitated with NH_3 and PEG-400, followed by calcination at 500 °C to render the titled nanocomposite. Through the structural characterization, the authors claimed that all three metals (Ce, Mg and Zr) in $Ce_1Mg_{0.6}Zr_{0.4}O_2$ had a strong mutual interaction and were highly dispersed on the surface. Besides, the insertion of magnesium into the lattice of cerium-zirconium led to a decrease in the size of particles along with an enhancement in the acidic-basic active sites, thereby enhancing the efficiency of $Ce_1Mg_{0.6}Zr_{0.4}O_2$ NPs in promoting the construction of tetrahydrobenzo[*b*]pyrans. By refluxing the mixture of substituted benzaldehydes, malononitrile and dimedone with $Ce_1Mg_{0.6}Zr_{0.4}O_2$ in ethanol, the authors were able to obtain excellent yields of corresponding pyran derivatives (10 examples, yield: 90–94%).

Besides, cerium is also treated as a metal dopant in some metal oxides to improve the catalytic performance of these materials in various transformations (Fu et al. 2019; Fayaz et al. 2016; Do et al. 2018). For instance, 5 wt% CeO_2 doped on NiMnO (calcined at 400 °C) could induce 100% conversion of benzyl alcohol to benzaldehyde (Sultana et al. 2015). Doping 2.5 wt% of ceria on $CuMnO_x$ helped to improve the efficacy of the catalyst in the low temperature oxidation of CO (Dey and Dhal 2020). Likewise, Samantaray et al. (2012) prepared a set of CeO_2@CaO nanocomposites by the citrate method and introduced them as main catalysts for the access of aminochromenes. It is revealed that the amorphous citrate template enabled the generation of macropores on the surface of resulting porous materials, in which the phase of binary oxide with the particle size of 5–25 nm was well dispersed in the phase of calcia. Furthermore, the incorporation of Ce^{4+} into the lattice of CaO might also increase the active basic sites on the surface of CeO_2@CaO composites, thereby improving their catalytic capability with respect to that of pure CaO. In this study, the authors stated that the CeO_2@CaO with 20 mol% of CeO_2 displayed the supreme performance in providing a structural diversity of 2-amino-2-chromenes (10 examples, yield: 76–85%) upon treating a mixture of substituted benzaldehydes and malononitrile with α-napthol in water at 80 °C. Meanwhile, Maddila et al. (2016) explored the recyclable cerium-vanadium-loaded alumina catalyst (Ce-V@Al_2O_3) for the solvent-free synthesis of multisubstituted pyridines. Herein, setting the total loading of Ce-V on the Al_2O_3 support at 2.5 wt% was verified to offer the best result thanks to the optimal distribution of acidic-basic sites on the surface of hybrid catalyst. Accordingly, the room-temperature manufacture of functional pyridines from aromatic aldehydes, malononitrile and ethanol was accomplished in a facile and selective manner (11 examples, yield: 86–94%). Subsequently, CeO_2@ZrO_2 was developed as an effective catalyst to induce the four-component annulation of substituted benzaldehydes, malononitrile and hydrazine hydrate with ethyl acetoacetate at room temperature, where a broad library of pyrano[2,3-c]pyrazole was rendered in the range yields of 89–98% (Maddila et al. 2017a). Alternatively, Khan et al. (2019) reported the high-yielding formation of quinolines from the CeO_2@TiO_2-mediated coupling of anilines, aldehydes with acetophenone in solvent-free condition.

In most cases, a proper choice of solvent to dissolve the product, suction filtration or centrifugation must be employed to separate the heterogeneous catalyst from the reaction mixture, causing great annoyances during the catalyst recovery. To overcome these barriers, magnetically recoverable nanocatalysts would become more ideal in terms of "green chemistry" viewpoint (Polshettiwar et al. 2011). In this aspect, superparamagnetic Fe_3O_4 (magnetite) which is considered as a cheap, stable and easy-to-prepare support has been widely implemented to immobilize active catalysts in many reactions (Gawande et al. 2013a; Sharma et al. 2016b), since the active catalyst@Fe_3O_4 composite would be easy to recover from the reaction medium by an external magnet. Motivated by these works, Gawande et al. (2013b) designed a magnetic nanocatalyst of magnetite-ceria (CeO_2@Fe_3O_4) for the room-temperature construction of dihydropyridines and tetrahydropyridine (Scheme 25.18). Similarly, Shelkar et al. (2015) designed CeO_2@Fe_3O_4 with 7.44 wt% of Ce as a cheap and active nanocatalyst for the C-H functionalization of heteroarenes (Scheme 25.19). In this approach, the arylation was implemented by heating the mixture of benzoxazole/benzothiazole (1 equiv.) and aryl halides (1 equiv.) with K_2CO_3 (2 equiv.) in DMSO under the assistance of 5 mol% of CeO_2@Fe_3O_4, which led to a myriad of 2-aryl-substituted derivatives of benzoxazole and benzothiazole.

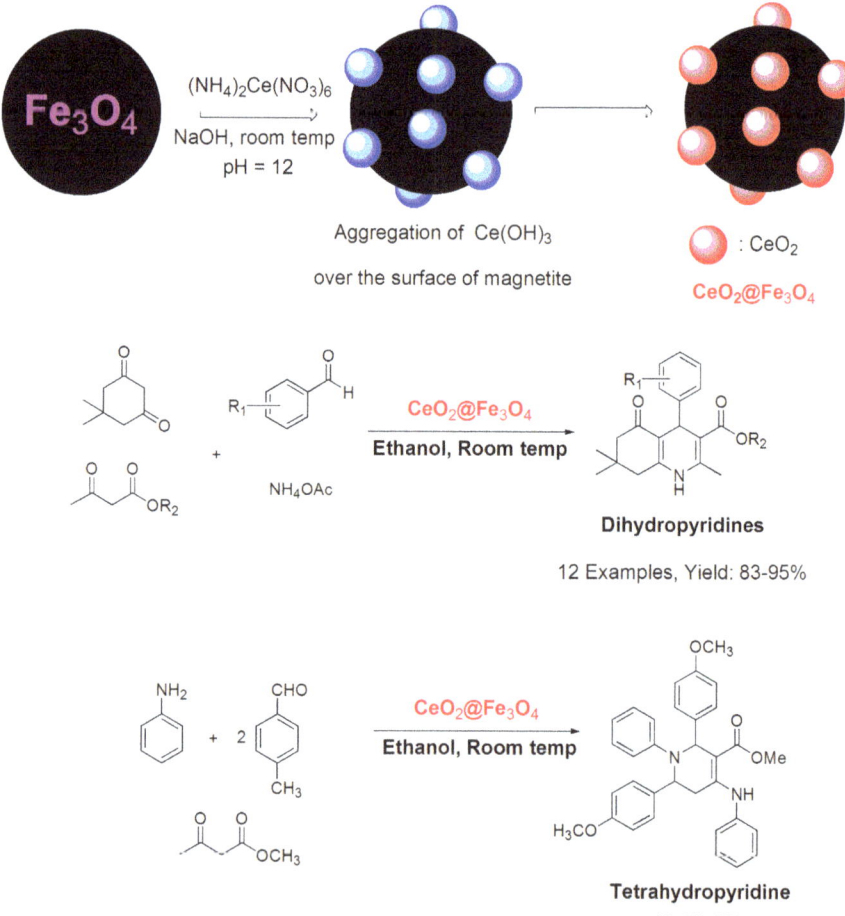

Scheme 25.18 Fabrication and utility of CeO_2@Fe_3O_4 in the manufacture of dihydropyridines and tetrahydropyridine

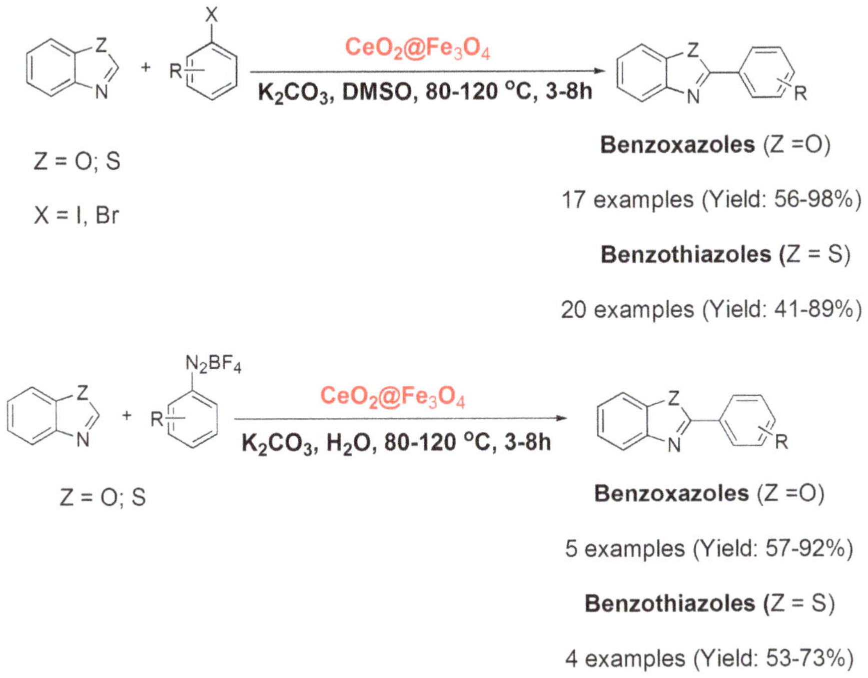

Scheme 25.19 C-H functionalization of benzoxazoles and benzothiazoles over magnetic CeO_2@Fe_3O_4 nanocatalyst

Strikingly, the CeO_2@Fe_3O_4-mediated *N*-arylation could be accomplished in greener condition by replacing the mixture of aryl halides-DMSO with a cheap combination of arenediazonium salts and water. In those examples, the magnetic CeO_2@Fe_3O_4 nanocatalyst could be readily recovered and recyclable for several batches with a negligible deactivation.

25.2.4 Cerium-Solid Material Composite

25.2.4.1 CeO_2-Polymer

In 2005, Sabitha and Shailaja (2005, 2008) designed a hybrid catalyst composed of CeO_2 NPs and polymer to promote the assembly of heterocycles. For this target, the titled composite, CeO_2@(VP-co-DVB), was readily prepared from the suspension copolymerization of 4-vinylpyridine (VP), 1,4-divinylbenzene (DVB) and $CeCl_3$ in basic condition upon using polyvinylpyrrolidone K30 as a removable template and an initiator mixture of Lupersol TAEC/Luperox 101 (Scheme 25.20). Strikingly, it is indicated that the robust CeO_2@(VP-co-DVB) catalyst could induce the synthesis

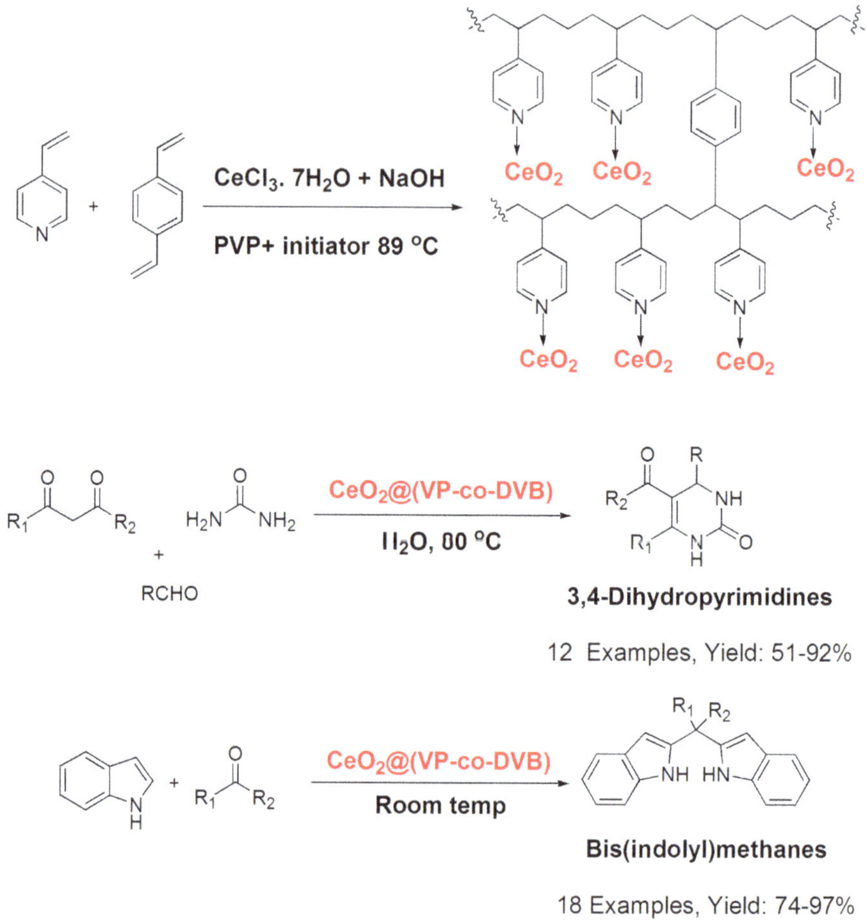

Scheme 25.20 Preparation and practicality of CeO$_2$@(VP-co-DVB) nanocatalyst in the synthesis of 3,4-dihydropyrimidines and bis(indolyl)methanes

of both 3,4-dihydropyrimidines (12 examples, yield: 51–92%) and bis(indolyl) methanes (18 examples, yield: 74–97%) in high efficacy after multiple recycles.

25.2.4.2 CeO$_2$-Silica

Generally, silica is acknowledged as a versatile solid material owing to its own specific properties of high surface area, high thermal stability and a great flexibility in pore sizes and acidic-basic sites (Agotegaray and Lassalle 2017). Accordingly, silica has been widely employed as an exceptional template to immobilize active species in a well-dispersed manner, therefore providing a great volume of powerful silica-supported catalysts in the domain of heterogeneous catalysis (Akelah 1981).

In this context, MCM41 and SBA15 are two exemplary mesoporous silica which have widespread applications as either heterogeneous catalysts or solid supports in various transformations (Bhattacharyya et al. 2006; Rahmat 2010). Owing to a large specific surface area along with a well-defined pore structure of the mesoporous template, active species (metals or metal complexes) can be incorporated and uniformly dispersed on the wall of mesopores of MCM41/SBA15 to deliver a plenty of active heterogeneous catalysts (Liang et al. 2017). For example, Akondi et al. (2012) successfully fabricated CeO_2@MCM-41 with 15 wt% of Ce (CeO_2@MCM-41) by the wet impregnation to facilitate the oxidative coupling of 2-naphthol with substituted anilines. Similarly, the excellent catalytic activity of CeO_2@MCM-41 in the manufacture of mono- and bis-dihydropyrimidin-2(1H)-ones (Vadivel et al. 2013), benzoxanthenones/benzochromenones (Akondi et al. 2014) and caprolactam (Babu et al. 2016) was also recorded (Scheme 25.21). In these cases, the immobilization of cerium on the inner surface of mesopores of parent MCM41 is accountable for the improvement in the stability and catalytic performance with respect to CeO_2. Besides, Saadati-Moshtaghin and Zonoz (2019) developed a novel three-component composite of Fe_3O_4-MCM41-CeO_2 as a new hybrid solid catalyst for the solvent-free manufacture of tetrahydobenzo[b]pyrans from the condensation of aromatic aldehydes, malononitrile and dimedone (15 examples, yield: 69–96%).

Apart from MCM41, silica (SiO_2) is also regarded as a versatile solid support in heterogeneous catalysis (Ramazani et al. 2017). From the perspective of economical metrics, no template-directed SiO_2 is considered cheaper and more

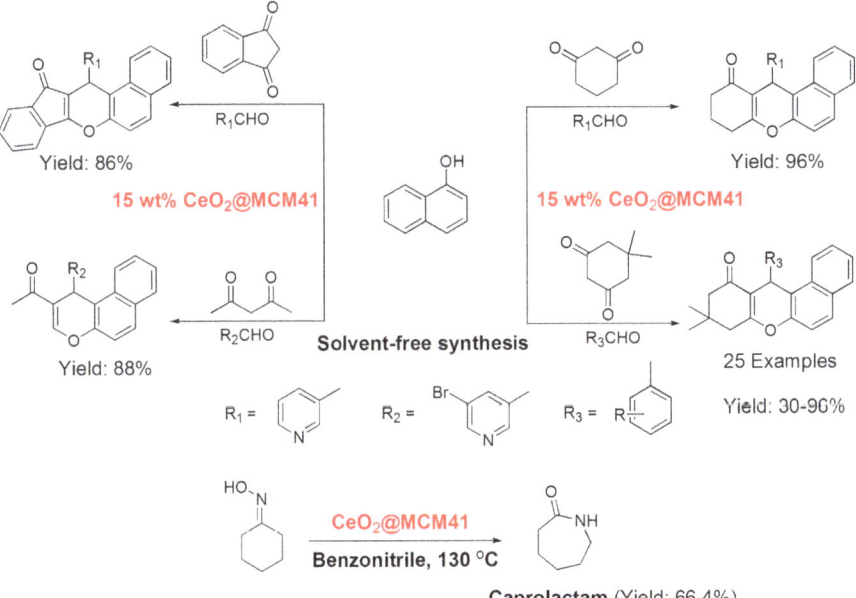

Scheme 25.21 Utility of CeO_2@MCM41 in the assembly of benzoxanthenones, benzochromenones and caprolactam

easy-to-prepare than MCM41. Moreover, the "sol-gel chemistry" is also acknowledged as a powerful tool in preparing metal oxides and oxide-supported metal catalysts (Esposito 2019). Prompted by these reasons, Akondi et al. (2016) later deployed sol-gel-derived SiO_2 in place of MCM41 to immobilize CeO_2 for the fabrication of nanostructured $CeO_2@SiO_2$. The textural analysis indicated that Ce^{4+} species from the oxidation of Ce^{3+} were successfully incorporated and tightly bound inside the mesoporous silica framework during the preparative procedure, thereby hampering the possible leaching of cerium to the reaction media. With only 0.9 mol% of mesoporous $CeO_2@SiO_2$ as main catalyst, the multicomponent condensation of aliphatic/aromatic aldehydes, 2-napthol and phenyl hydrazine with ethyl acetoacetate in water was induced to trigger a library of substituted pyrazolones in high yield and selectivity. The mechanism leading to the formation of substituted pyrazolones over $CeO_2@SiO_2$ is suggested to follow a sequential reaction of Knoevenagel condensation/Michael addition through two different pathways (Scheme 25.22).

25.2.4.3 CeO$_2$-Clay Composite

Another noticeable class of biomaterial is associated with hydroxyapatite [HAP; $Ca_{10}(PO_4)_6(OH)_2$] (Lu et al. 2019). This functional solid is highly recognized due to its outstanding properties such as high thermal stability, strong adsorption capability and tunable acidity/basicity (Pokhrel 2018). Accordingly, several investigations on the application of HAP as a solid catalyst or support in heterogeneous catalysis have been well executed (Fihri et al. 2017; Dobosz et al. 2016; Yan et al. 2016). In a typical study, Maddila et al. (2017b) doped ceria nanoparticles on hydroxyapatite ($CeO_2@HAP$) to induce the high-yielding assembly of pyrido[2,3-d]pyrimidine derivatives from the room-temperature coupling of benzaldehydes, dimethylbarbituric acid and ammonium acetate.

Honeycomb monolith (HM) is a type of solid material containing an extended matrix of long parallel and straight channels which are separated by thin walls (Govender and Friedrich 2017). This unique structural property generates a large number of void fractions and a large surface area to volume ratio. Furthermore, other major merits of honeycomb monolithic material encompass the high thermal conductivities, low pressure drops and ease of manufacturing and recyclability (Sungkono et al. 1997; Boger et al. 2004; Hosseini et al. 2020). Due to these reasons, HMs coated with metals/metal oxides are currently explored as heterogeneous catalysts in the NO_x reduction, N_2O decomposition, removal of SO_2-NO_x, syngas production, CO oxidation (Russo et al. 2007; Rico-Pérez et al. 2013; Vita et al. 2018a; Vita et al. 2018b; Davo-Quinonero et al. 2019) and organic synthesis (Gatica et al. 2016; Pratap et al. 2020). Recently, Venkatesh et al. (2015) prepared and applied the synthetic cordierite HM ($Mg_2Al_4Si_5O_{18}$) as a support to immobilize a set of cerium-based solid acids (i.e. sulphated CeO_2, CeO_2-ZrO_2 and sulphated CeO_2-ZrO_2) for the assembly of quinoxaline framework. It is disclosed that these cerium-based solid acids after coating with cordierite HM could display their supremacy over corresponding powder solid acids for the assembly of quinoxalines. In this

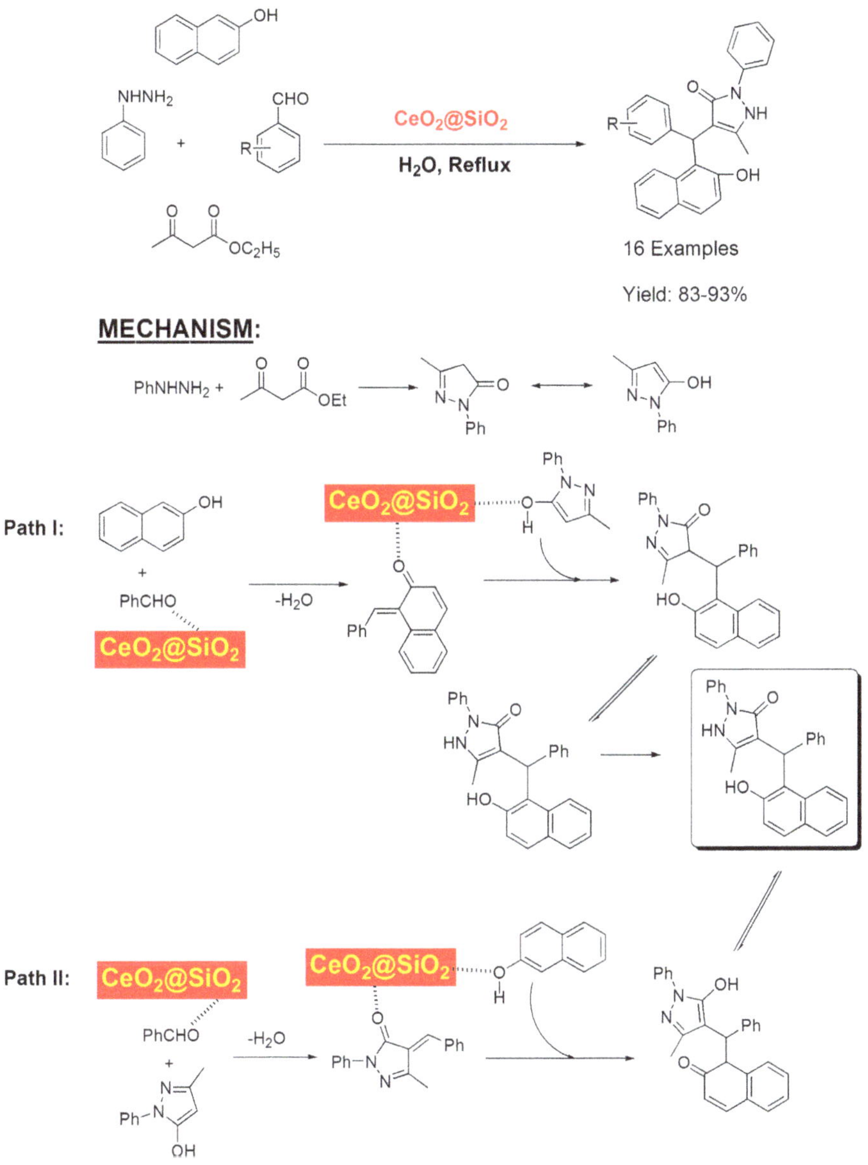

Scheme 25.22 Four-component assembly of substituted pyrazolones over $CeO_2@SiO_2$ nanocomposite

setting, high surface area, good dispersion of active sites and strong acidity of these acid-coated cordierites were acknowledged as main factors accounting for the high yield and selectivity of final products. Owing to high numbers of moderate and strong acid sites, HM-coated CeO_2-ZrO_2 was selected as the potential candidate for this synthetic paradigm, finally providing a group of quinoxalines in the range yields of 72–89%. Strikingly, the spent catalyst could undergo six circulations with a negligible drop in the catalytic activity.

25.2.4.4 CeO$_2$-Carbon Template

Thanks to the unique properties (e.g. large surface area, excellent crystallinity, high physical/chemical/thermal stability and well-defined porosity), multi-walled carbon nanotubes (MWCNTs) have been exploited as versatile support in the fabrication of heterogeneous catalysts for the divergent synthesis of heterocycles (Safari and Gandomi-Ravandi 2014a, b, c; Zarnegar et al. 2015). By following this trend, Harikrishna et al. (2020) recently developed ceria-doped MWCNTs (CeO$_2$/MWCNTs) with 2.5 wt% of CeO$_2$ as a heterogeneous catalyst for the one-pot synthesis of pyridine-3-carboxamides. Under the promotion of recyclable CeO$_2$/MWCNTs nanocatalyst, the four-component coupling of acetoacetanilide, ammonium acetate and substituted aromatic aldehydes with ethyl cyanoacetate took place with no difficulty at room temperature. Accordingly, excellent isolated yields of pyridine-3-carboxamides (90–97%) could be delivered within a short period of time.

25.2.5 CeO$_2$ as Solid Support

In addition to being exploited as effective catalysts in the construction of diversified heterocyclic frameworks, several research groups also attempted to utilize the versatility of CeO$_2$ as a solid support to immobilize palladium metal (Pd@CeO$_2$) to facilitate the assembly of heterocycles (Scheme 25.23). For instance, Chen et al. (2014) reported the remarkable activity of Pd@CeO$_2$ in the oxidative synthesis of N-(2-pyridyl)indole derivatives. In this study, the Pd@CeO$_2$ was demonstrated to outperform other commercial catalysts (e.g. Rh@C, Ru@C and Pd@C) for this oxidative C-H activation. Unfortunately, the spent Pd@CeO$_2$ could not be recycled well, delivering a sharp drop in the yield of annulated products after two recycling tests. Later, Zhang et al. (2017) developed a facile one-pot redox strategy to fabricate self-assembled Pd/CeO$_2$ hybrid catalyst with 5.82 wt% of Pd, where the high-temperature stage of calcination and reduction was avoided in the pretreatment. Thanks to the high surface area and defect sites of CeO$_2$, the Pd/CeO$_2$ catalyst was able to trigger a quantitative yield of gamma-valerolactone (GVL) from the hydrogenation of levulinic acid (LA) under mild condition (90 °C, 4 bar of H$_2$), which showed the catalytic superiority over commercial Pd/C (yield: 7.5%) and conventional Pd/CeO$_2$ derived from the precipitation-reduction method (yield: 45.3%). In another case, Ge et al. (2018) applied Pd/CeO$_2$ (3 wt% Pd) as an effective nanophotocatalyst to trigger the photochemical synthesis of asymmetrical heterobiaryls.

With the aim of improving the isolated yields of benzimidazoylquinoxalines from current protocols, Climent et al. (2013) established an alternative synthetic pattern where the Au@CeO$_2$ with 2.33 wt% of Au was exploited as a potential

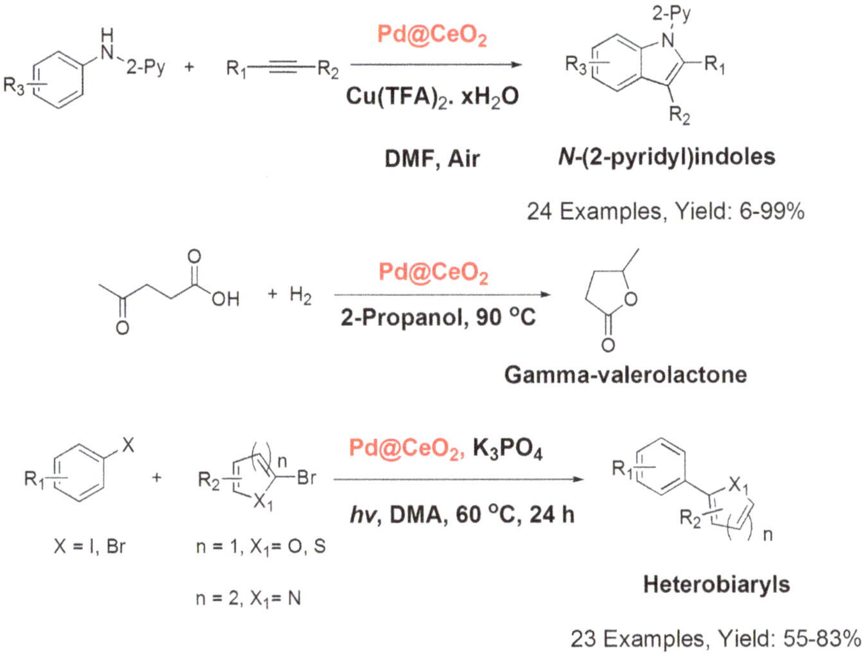

Scheme 25.23 Construction of various heterocyclic structures over Pd@CeO_2 nanocomposite

catalyst candidate. With the assistance of titled material, the trial for benzimidazoyl-quinoxalines could be attempted via two different manners (Scheme 25.24). In the straightforward approach, the oxidative coupling of the biomass-derived glycerol with 1,2-phenylene diamine was carried out in diglyme at 140 °C. In this case, two intermediates of quinoxalin-2-ylmethanol **A** and 1-(1H-benzo-[d]imidazol-2-yl)ethane-1,2-diol **B** were simultaneously generated from the coupling of 1,2-phenylene diamine with glyceraldehyde derived from the oxidation of glycerol. Afterwards, these intermediates would slowly undergo the oxidative condensation with 1,2-phenylene diamine to render the final 2-(1H-benzo[d]imidazol-2-yl)quinoxaline products (11 examples, yield: 24–80%). To expand the synthetic scope for constructing the benzimidazoylquinoxaline derivatives containing different substituents on both heteroaromatic moieties, the authors turned to deploy the one-pot two-step strategy upon starting with glyceraldehyde. In such case, 1-(1H-benzo-[d]imidazol-2-yl)ethane-1,2-diol **B** would be generated in water as the sole intermediate under the catalysis of Au@CeO_2 at room temperature, which was subsequently converted into a variety of substituted 2-(1H-benzo[d]imidazol-2-yl)quinoxaline upon oxidative coupling with substituted 1,2-phenylene diamines in diglyme at 140 °C (12 examples, yield: 63–79%). In each recycling trial, the recovered Au@CeO_2 was introduced to the calcination in O_2 at 250 °C prior to use, showing no significant loss in the original catalytic activity.

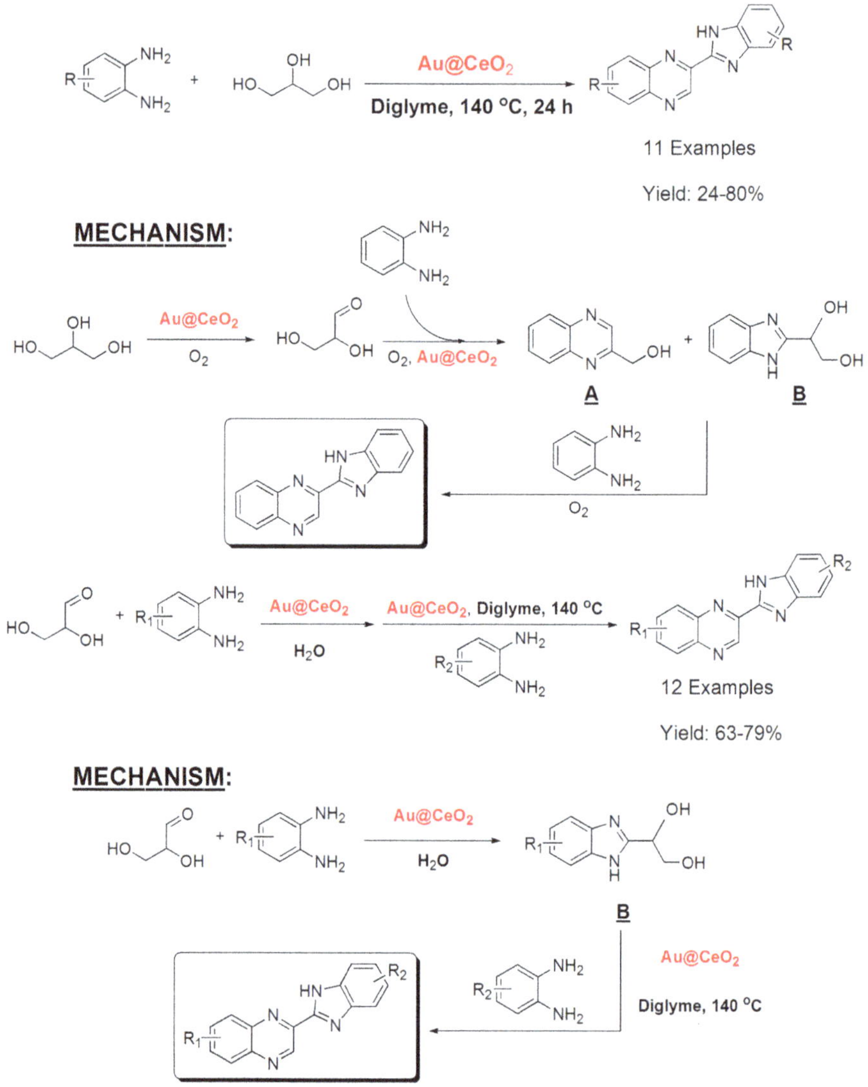

Scheme 25.24 Au@CeO$_2$-catalyzed synthesis of benzimidazoylquinoxalines

Furthermore, the practicality and efficiency of ceria-supported ruthenium (Ru@CeO$_2$) or ceria-supported platinum (Pt@CeO$_2$) as recyclable catalysts in the construction and functionalization of heterocycles such as indoles (Shimura et al. 2011), quinazolinones (An et al. 2018), γ-valerolactone (Gao et al. 2020), oxindoles (Chaudhari et al. 2014a) and quinazolines (Chaudhari et al. 2014b) have been realized in recent years (Scheme 25.25).

Another typical implementation of ceria-supported metal in the construction and functionalization of heteroarenes was introduced by Amadine et al. (2014), where

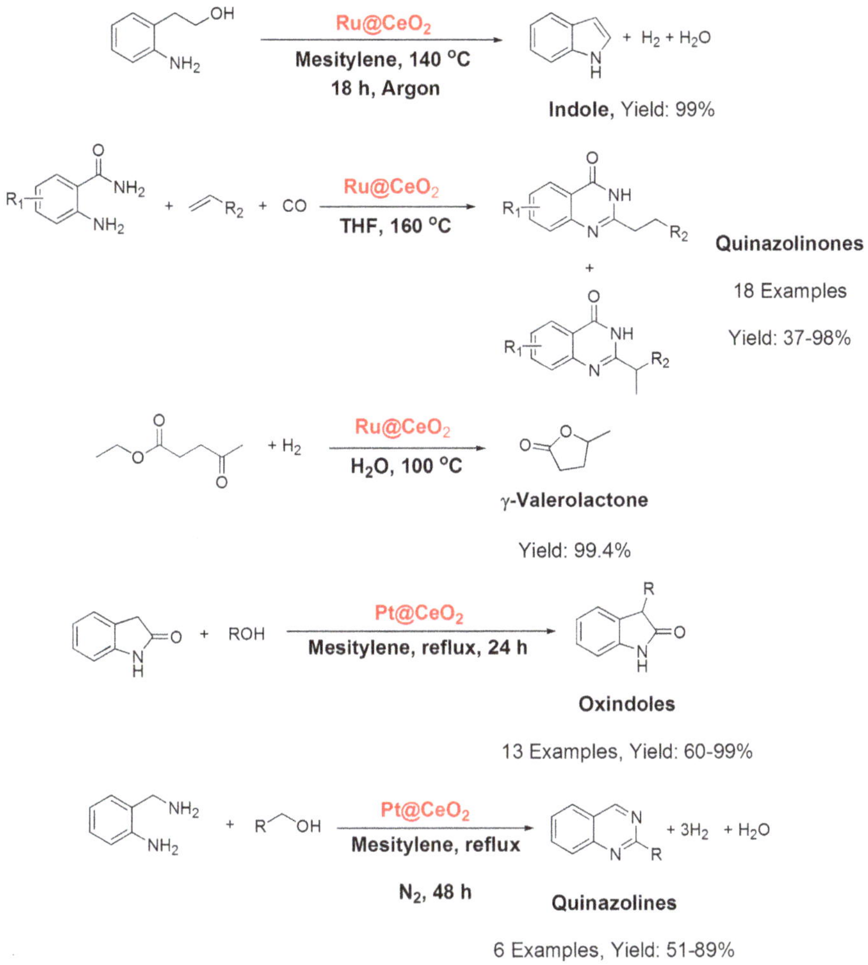

Scheme 25.25 Assembly of functionalized heterocycles over Ru@CeO$_2$ and Pt/CeO$_2$ catalyst

the ceria-supported copper nanoparticles (Cu@CeO$_2$) displayed the excellent catalytic activity in the *N*-arylation of indole with various aryl bromides. Although 82–89% isolated yields of *N*-arylated indoles could be achieved under optimal conditions, a considerable drop in the activity of spent Cu@CeO$_2$ was observed after three cycles. In this context, the reasons were likely attributed to the unavoidable oxidation of Cu^0 to Cu^{2+} and the poisonous deposition of in situ generated KBr on the surface of Cu@CeO$_2$. Later, Amini et al. (2016) reported the utility of robust Cu@CeO$_2$ nanocomposite (10 wt% Cu) to formulate a collection of 1,2,3-triazole derivatives (yield: 62–96%) from the 1,3-dipolar cycloaddition of terminal alkynes with sodium azide and benzyl halide derivatives in water.

25.3 Cerium-Based Catalysts for the Vapour-Phase Synthesis of Heterocycles

Recently, the catalysis of ceria-supported metal oxides for the vapour-phase synthesis of γ-butyrolactone (GBL) has been investigated due to the great importance and high output demand of GBL in industry (Schwarz et al. 2019). For example, Bhanushali et al. (2019) introduced the ceria supported copper (CuO@CeO$_2$) with 10 wt.% of Cu as an effective catalyst for the fixed-bed dehydrogenation of 1,4-butanediol (1,4-BDO) at 240 °C. Thanks to the high surface area, good dispersion of cooper on the ceria support and an enhancement in basicity, the CuO@CeO$_2$ could induce the dehydrogenation in an effective manner to trigger 93% conversion of 1,4-BDO and 98% selectivity of GBL. Subsequently, 10 wt% of Cu supported on CeO$_2$-Al$_2$O$_3$ (3:1 ratio) catalyst was able to promote the one-pot synthesis of GBL and benzyl alcohol from the simultaneous 1,4-BDO dehydrogenation and benzaldehyde hydrogenation, in which 90% conversion of 1,4-BDO and 95% selectivity of GBL were accomplished (Bhanushali et al. 2020a). Lately, 99% yield and 99% selectivity of GBL from the direct dehydrogenation of 1,4-BDO at 240 °C could be reachable in the presence of mesoporous 10 wt% CuO@CeO$_2$-Al$_2$O$_3$ (3:1 ratio) (Bhanushali et al. 2020b). In these examples, a remarkable decrease in conversion of 1,4-BDO up to 45% was unavoidable due to the coke deposition and agglomeration of copper nanoparticles after a long-time span on stream at high temperature.

Thanks to the high atom-economic, low cost and benign aspects, the vapour-phase synthesis of 3-methylindole from glycerol and aniline has drawn much interest over the past few years. In such transformation, the heterogeneous catalysts containing a large specific area along with a great number of weak acidic sites are strongly required to offer high yield and selectivity of the target product (Sun et al. 2010; Cui et al. 2013). Recently, Ke et al. (2020) applied the Cu/MIL-101 modified with CeO$_2$ (0.03 mmol/g) to prepare 59% yield of 3-methylindole from this synthetic paradigm. In this study, the authors stated that the addition of CeO$_2$ was conducive to the catalytic activity for the sake of (i) enhancing the mutual interaction of Cu and MIL-101; (ii) inhibiting the sintering of active components during the transformation; and (iii) increasing the number of weak acid sites on the surface of catalyst. Later, Qu et al. (2020) successfully fabricated mesoporous catalyst of Ag/SBA-15 modified with ZnO-CeO$_2$ (1 mmol/g of Ag, 1 mmol/g of ZnO and 0.05 mmol/g of CeO$_2$) to upgrade the yield of 3-methylindole up to 62%.

25.4 Cerium-Based Catalysts for the Synthesis of CO$_2$-Derived Heterocycles

Apart from being employed as heterogeneous catalysts in the manufacture of CO$_2$-based products such as ureas (Tamura et al. 2016a), carbamates (Tomishige et al. 2019), carbonates (Tomishige et al. 2020) and polycarbonates (Gu et al. 2019;

Tamura et al. 2016b), cerium-based materials have been widely applied in the catalytic fixation of CO_2 towards heterocycles as well. For example, Tamura et al. (2013a) established a novel catalyst system composed of CeO_2 and 2-propanol to promote the synthesis of cyclic ureas from CO_2 and diamines. In this study, it is revealed that the presence of 2-propanol was essential to suppress the competitive formation of N-alkylated amines. Through the kinetic and FTIR investigations, the CeO_2-mediated cyclization is proposed to follow a cascade reaction of (i) simultaneous adsorption of diamine with CO_2 and CeO_2 to generate carbamic acid and carbamate adspecies on ceria; (ii) decomposition of carbamate species to a free amino group; (iii) annulation to cyclic urea by the intramolecular attack of amino group to the activated carbamate part; and (iv) desorption of the cyclic urea product and regeneration of CeO_2 (Scheme 25.26).

For the manufacture of CO_2-based cyclic carbonates such as ethylene carbonate (EC) and propylene carbonate (PC), Tomishige et al. (2004) reported the coupling pattern of ethylene glycol (EG) or propylene glycol (PG) with CO_2 over $Ce_xZr_{1-x}O_2$ solid solution. The authors stated that the acetonitrile as solvent helped to improve the catalytic activity and the equilibrium yield of carbonates in this reaction. Additionally, the maximal yield of both EC and PC could be obtained with CeO_2–ZrO_2 (Ce/[Ce+Zr] = 0.5) calcined at 800 °C or CeO_2–ZrO_2 (Ce/[Ce+Zr] = 0.2 and 0.33) calcined at 1000 °C. Later, Honda et al. (2014) examined the convenience of CeO_2 for the approach to five-/six-membered cyclic carbonates from diols and CO_2. It is verified that the introduction of excessive 2-cyanopyridine (2-CP) as a dehydrating agent was indispensable to overcome the equilibrium limitation, where the in situ generated water was effectively trapped by 2-CP. In the mechanistic description (Scheme 25.27), CeO_2 served as a Lewis acid to deprotonate the O-H bond of diol, thereby generating the cerium alkoxide **I** at the first stage. Afterwards, this alkoxide would allow the insertion of CO_2 to form the carbonate specie **II**, followed by the intramolecular cyclization and dehydration to result in the final cyclic carbonate.

In the production of glycerol carbonate, Liu et al. (2016) carried out the CeO_2-mediated carbonylation of glycerol with CO_2 in the presence of DMF and dehydrating agent (2-cyanopyridine). For this purpose, the authors attempted three types of CeO_2 derived from the traditional precipitation (TP), hydrothermal (HT) and citrate sol-gel (SG) method for the CO_2 carbonylation. From the CO_2-TPD and H_2-TPR analysis, the basicity and oxygen vacancy density of these designed CeO_2 followed the order of nano-rod CeO_2 (HT) > nanoparticulate CeO_2 (TP) > sponge-like CeO_2 (SG). Hence, the highest yield of glycerol carbonate (78.9%) was provided under the mediation of nano-rod CeO_2 upon heating glycerol with CO_2 at 150 °C. By following the same strategy, Liu et al. (2018) applied the $Ce_{098}Zr_{002}O_2$ derived from the hydrothermal method to deliver 36.3% yield of glycerol carbonate.

To investigate the conversion of CO_2 into 2-oxazolidinones, Juarez et al. (2010) examined CeO_2 NP (5 nm) and commercial CeO_2 (40 nm) to promote the coupling of CO_2 and ω-aminoalcohols. Due to the high density of defects on the surface, the CeO_2 NPs (5 nm) displayed the best results in converting CO_2, and N-alkyl substituted aminoethanols into corresponding N-alkyl 1,3-oxazolidin-2-ones at

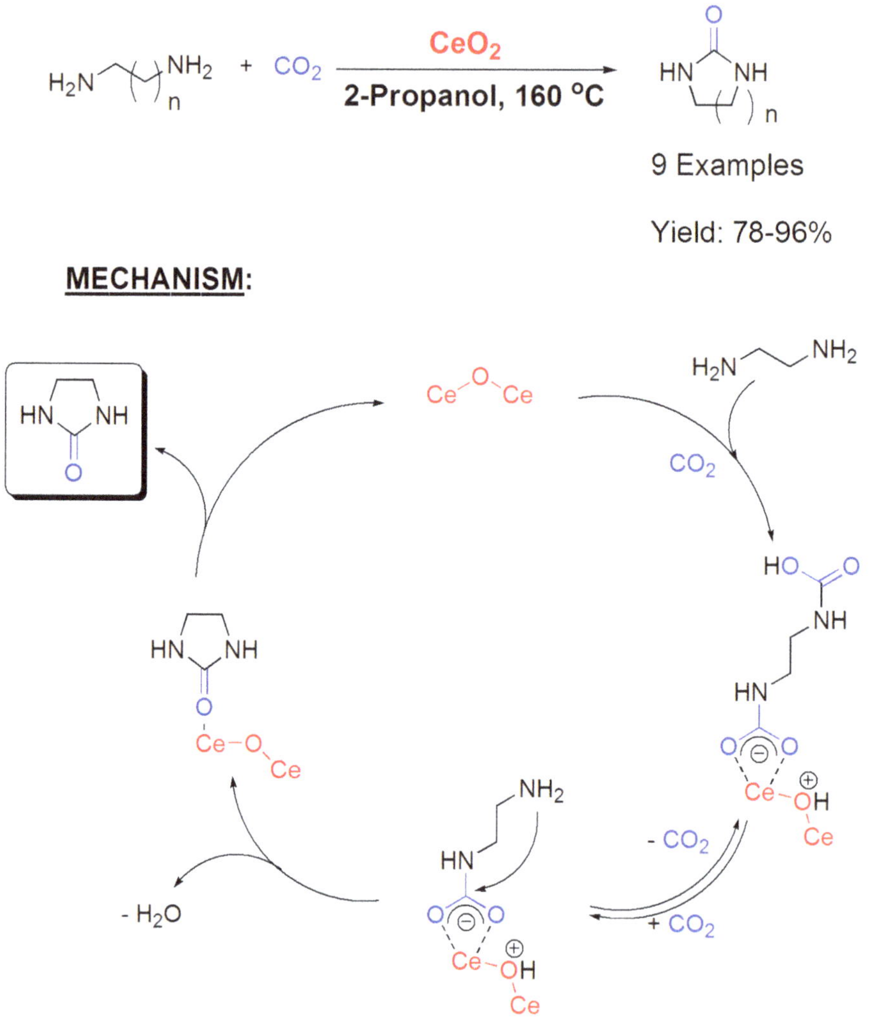

Scheme 25.26 Synthesis of cyclic ureas from CO_2 and diamines over CeO_2 in 2-propanol

160 °C. Later, Tamura et al. (2013b) stated that a variety of aminoalcohols could be selectively converted into corresponding cyclic carbamates in high yields (88–99%) over the catalytic system of CH_3CN-CeO_2. In particular, the chiral configuration of chiral centre at the α-position of the hydroxyl group of starting aminoalcohols was kept intact after the reaction. From the kinetic studies and FTIR analyses, the mechanistic pathway leading to the generation of 2-oxazolidinones over CeO_2 is suggested to follow four consecutive steps as shown in Scheme 25.28.

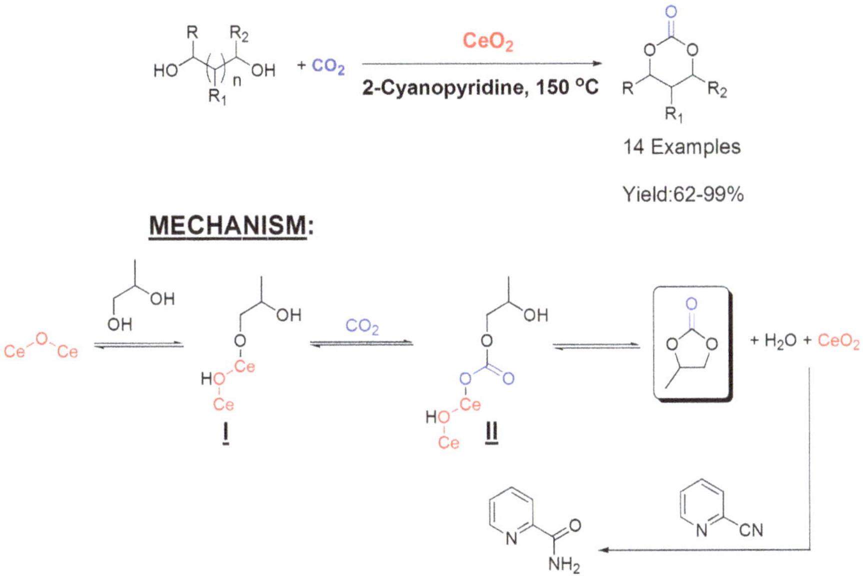

Scheme 25.27 Synthesis of cyclic carbonates from CO_2 and diols under the catalysis of CeO_2 and 2-cyanopyridine

25.5 Summary and Outlook

The great importance and omnipresence of heterocyclic frameworks in natural products, bioactive molecules, pharmaceuticals and key building blocks have continuously raised the interest in developing novel, eco-friendly and sustainable protocols to improve the time/energy consumption, atom economy and selectivity during the manufacture. In this chapter, the broad practicality of cerium-based nanomaterials as both catalysts and solid supports in the synthesis and functionalization of heterocycles was summarized, where a diversity of heterocyclic skeletons containing nitrogen, oxygen and/or sulphur atom was constructed successfully under heterogeneous conditions. For this objective, well-defined cerium-based materials such as cerium oxide/mixed oxides, cerium composites and ceria-supported metals were fabricated over different procedures including co-precipitation, sol-gel, hydrothermal, wet impregnation and so on. With tunable modifications of morphology, acidity-basicity, redox properties and oxygen storage capacity, these nanocatalysts turned out to be potential heterogeneous candidates for the divergent synthesis of heterocyclic compounds. Particularly, their excellent catalytic performances were also recognized in the chemical fixation of CO_2, where valuable cyclic products of ureas, carbonates and carbamates were accomplished in great success. In such transformations, cheap, robust, easy-to-handle and recoverable nature are noticeable merits of these cerium-based nanomaterials in comparison with other benchmark catalysts. However, further improvements in this synthetic

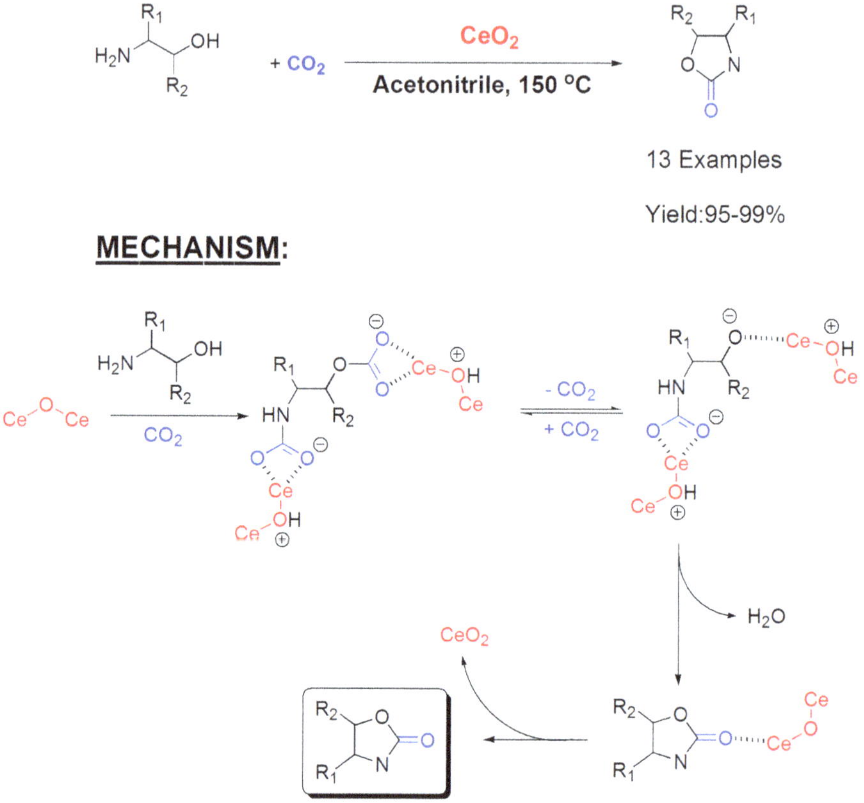

Scheme 25.28 CeO$_2$-mediated coupling of CO$_2$ and aminoalcohols towards N-alkyl 1,3-oxazolidin-2-ones

strategy need to be carried out, where the development of simple, cheap and effective processes for the fabrication of cerium-based nanomaterials is strongly desired. In this manner, utilizing the latest advent of nanotechnology in controlling the size, shape and composition of final cerium oxide/mixed oxides is considered of great importance to maximize the number and strength of active sites for specific reactions. Moreover, the architecture of the solid supports and their interaction with cerium nanoparticles need to be carefully considered as well. In another aspect, the catalysis of cerium-based nanostructures in the MCR patterns should be further explored to expand the molecular complexity of heterocyclic frameworks. Last but not least, solvent-free or aqueous-phase paradigms are highly appreciated for the eco-friendly and sustainable synthesis of heterocycles.

References

Agotegaray MA, Lassalle VL (2017) Silica: chemical properties and biological features. In: Silica-coated magnetic nanoparticles. Springer, Cham, pp 27–37

Akbayrak S (2018) Rhodium(0) nanoparticles supported on ceria as catalysts in hydrogenation of neat benzene at room temperature. J Colloid Interface Sci 530:459–464

Akelah A (1981) Use of functionalised silica in catalysis and organic synthesis. Br Polym J 13:107–110

Akondi AM, Trivedi R, Sreedhar B, Kantam ML, Bhargava S (2012) Cerium-containing MCM-41 catalyst for selective oxidative arene cross-dehydrogenative coupling reactions. Catal Today 198:35–44

Akondi AM, Kantam ML, Trivedi R, Sreedhar B, Buddana SK, Prakasham RS, Bhargava S (2014) Formation of benzoxanthenones and benzochromenones via cerium-impregnated-MCM-41 catalyzed, solvent-free, three-component reaction and their biological evaluation as antimicrobial agents. J Mol Catal A Chem 386:49–60

Akondi AM, Kantam ML, Trivedi R, Bharatam J, Vemulapalli SPB, Bhargava SK, Buddana SK, Prakasham RS (2016) Ce/SiO2 composite as an efficient catalyst for the multicomponent one-pot synthesis of substituted pyrazolones in aqueous media and their antimicrobial activities. J Mol Catal A Chem 411:325–336

Albadi J, Razeghi A, Abbaszadeh H, Mansournezhad A (2013a) CuO-CeO2 nanocomposite: an efficient recyclable catalyst for the synthesis of Aryl-14H-dibenzo[a-j]xanthenes. J Nanoparticles 2013:1–5

Albadi J, Mansournezhad A, Abbaszadeh H (2013b) CuO-CeO2Nanocomposite: a highly efficient recyclable catalyst for the green synthesis of 1,8-Dioxooctahydroxanthenes in water. J Chin Chem Soc:n/a–n/a

Albadi J, Mansournezhad A, Derakhshandeh Z (2013c) CuO–CeO2 nanocomposite: a highly efficient recyclable catalyst for the multicomponent synthesis of 4H-benzo[b]pyran derivatives. Chin Chem Lett 24:821–824

Albadi J, Razeghi A, Mansournezhad A, Azarian Z (2013d) CuO-CeO2 nanocomposite catalyzed efficient synthesis of aminochromenes. J Nanostruct Chem 3:85

Albadi J, Mansournezhad A, Salehnasab S (2014a) Green synthesis of biscoumarin derivatives catalyzed by recyclable CuO–CeO2 nanocomposite catalyst in water. Res Chem Intermed 41:5713–5721

Albadi J, Shiran JA, Mansournezhad A (2014b) Click synthesis of 1,4-disubstituted-1,2,3-triazoles catalysed by CuO–CeO2 nanocomposite in the presence of amberlite-supported azide. J Chem Sci 126:147–150

Ali I, Lone MN, Al-Othman ZA, Al-Warthan A, Sanagi MM (2015) Heterocyclic Scaffolds: centrality in anticancer drug development. Curr Drug Targets 16:711–734

Amadine O, Maati H, Abdelouhadi K, Fihri A, El Kazzouli S, Len C, El Bouari A, Solhy A (2014) Ceria-supported copper nanoparticles: a highly efficient and recyclable catalyst for N-arylation of indole. J Mol Catal A Chem 395:409–419

Amini M, Hassandoost R, Bagherzadeh M, Gautam S, Chae KH (2016) Copper nanoparticles supported on CeO2 as an efficient catalyst for click reactions of azides with alkynes. Catal Commun 85:13–16

An J, Wang Y, Zhang Z, Zhao Z, Zhang J, Wang F (2018) The synthesis of Quinazolinones from Olefins, CO, and Amines over a heterogeneous Ru-clusters/Ceria catalyst. Angew Chem Int Ed Engl 57:12308–12312

Arumugam A, Karthikeyan C, Haja Hameed AS, Gopinath K, Gowri S, Karthika V (2015) Synthesis of cerium oxide nanoparticles using Gloriosa superba L. leaf extract and their structural, optical and antibacterial properties. Mater Sci Eng C Mater Biol Appl 49:408–415

Arya A, Sethy NK, Das M, Singh SK, Das A, Ujjain SK, Sharma RK, Sharma M, Bhargava K (2014) Cerium oxide nanoparticles prevent apoptosis in primary cortical culture by stabilizing mitochondrial membrane potential. Free Radic Res 48:784–793

Babu CM, Vinodh R, Abidov A, Ravikumar R, Peng MM, Cha WS, Jang HT (2016) Caprolactam synthesis using Ce-MCM-41Catalysts. Int J Bio-Sci Bio-Technol 8:171–182

Baumann M, Baxendale IR (2013) An overview of the synthetic routes to the best selling drugs containing 6-membered heterocycles. Beilstein J Org Chem 9:2265–2319

Baumann M, Baxendale IR, Ley SV, Nikbin N (2011) An overview of the key routes to the best selling 5-membered ring heterocyclic pharmaceuticals. Beilstein J Org Chem 7:442–495

Bhanushali JT, Prasad D, Patil KN, Babu GVR, Kainthla I, Rao KSR, Jadhav AH, Nagaraja BM (2019) The selectively regulated vapour phase dehydrogenation of 1,4-butanediol to γ-butyrolactone employing a copper-based ceria catalyst. New J Chem 43:11968–11983

Bhanushali JT, Prasad D, Patil KN, Reddy KS, Rama Rao KS, Jadhav AH, Nagaraja BM (2020a) Simultaneous dehydrogenation of 1,4- butanediol to γ-butyrolactone and hydrogenation of benzaldehyde to benzyl alcohol mediated over competent CeO_2–Al_2O_3 supported Cu as catalyst. Int J Hydrog Energy 45:12874–12888

Bhanushali JT, Prasad D, Patil KN, Reddy KS, Kainthla I, Rao KSR, Jadhav AH, Nagaraja BM (2020b) Tailoring the catalytic activity of basic mesoporous cu/CeO_2 catalyst by Al_2O_3 for selective lactonization and dehydrogenation of 1,4-butanediol to γ-butyrolactone. Catal Commun 143:106049

Bhattacharyya S, Lelong G, Saboungi ML (2006) Recent progress in the synthesis and selected applications of MCM-41: a short review. J Exp Nanosci 1:375–395

Boger T, Heibel AK, Sorensen CM (2004) Monolithic catalysts for the chemical industry. Ind Eng Chem Res 43:4602–4611

Bumajdad A, Zaki MI, Eastoe J, Pasupulety L (2004) Microemulsion-based synthesis of CeO(2) powders with high surface area and high-temperature stabilities. Langmuir 20:11223–11233

Burange AS, Gawande MB (2016) Role of mixed metal oxides in heterogeneous catalysis. In: Encyclopedia of inorganic and bioinorganic chemistry. Wiley, Hoboken, pp 1–19

Cargnello M, Jaen JJD, Garrido JCH, Bakhmutsky K, Montini T, Gamez JJC, Gorte RJ, Fornasiero P (2012) Exceptional activity for methane combustion over modular Pd@CeO_2 subunits on functionalized Al_2O_3. Science 337:713–717

Carson CA, Kerr MA (2009) Heterocycles from cyclopropanes: applications in natural product synthesis. Chem Soc Rev 38:3051–3060

Chang K, Zhang H, Cheng M-j, Lu Q (2019) Application of Ceria in CO_2 conversion catalysis. ACS Catal 10:613–631

Channei D, Inceesungvorn B, Wetchakun N, Ukritnukun S, Nattestad A, Chen J, Phanichphant S (2014) Photocatalytic degradation of methyl orange by CeO_2 and Fe–doped CeO_2 films under visible light irradiation. Sci Rep 4:5757

Chaudhari C, Siddiki SMAH, Kon K, Tomita A, Tai Y, Shimizu K-i (2014a) C-3 alkylation of oxindole with alcohols by Pt/CeO_2 catalyst in additive-free conditions. Cat Sci Technol 4:1064–1069

Chaudhari C, Hakim Siddiki SMA, Tamura M, Shimizu K-i (2014b) Acceptorless dehydrogenative synthesis of 2-substituted quinazolines from 2-aminobenzylamine with primary alcohols or aldehydes by heterogeneous Pt catalysts. RSC Adv 4:53374–53379

Chen H-I, Chang H-Y (2005) Synthesis of nanocrystalline cerium oxide particles by the precipitation method. Ceram Int 31:795–802

Chen P-L, Chen IW (1993) Reactive cerium(IV) oxide powders by the homogeneous precipitation method. J Am Ceram Soc 76:1577–1583

Chen J, He L, Natte K, Neumann H, Beller M, Wu X-F (2014) Palladium@Cerium(IV) oxide-catalyzed oxidative synthesis of N-(2-Pyridyl)indoles via C–H activation reaction. Adv Synth Catal 356:2955–2959

Climent MJ, Corma A, Iborra S, Martínez-Silvestre S (2013) Gold catalysis opens up a new route for the synthesis of Benzimidazoylquinoxaline derivatives from biomass-derived products (glycerol). ChemCatChem 5:3866–3874

Colussi S, Gayen A, Farnesi Camellone M, Boaro M, Llorca J, Fabris S, Trovarelli A (2009) Nanofaceted Pd–O sites in Pd–Ce surface superstructures: enhanced activity in catalytic combustion of methane. Angew Chem Int Ed 48:8481–8484

Courty P, Marcilly C (1976) General synthesis methods for mixed oxide catalysts. In: Studies in surface science and catalysis, vol 1. Elsevier, Amsterdam, pp 119–145

Cousin P, Ross RA (1990) Preparation of mixed oxides: a review. Mater Sci Eng A 130:119–125

Cui Y, Zhou X, Sun Q, Shi L (2013) Vapor-phase synthesis of 3-methylindole from glycerol and aniline over zeolites-supported Cu-based catalysts. J Mol Catal A Chem 378:238–245

D'Alessandro O, Sathicq ÁG, Sambeth JE, Thomas HJ, Romanelli GP (2015) A study of the temperature effect on Hantzsch reaction selectivity using Mn and Ce oxides under solvent-free conditions. Catal Commun 60:65–69

Darroudi M, Hoseini SJ, Kazemi Oskuee R, Hosseini HA, Gholami L, Gerayli S (2014) Food-directed synthesis of cerium oxide nanoparticles and their neurotoxicity effects. Ceram Int 40:7425–7430

Davo-Quinonero A, Sorolla-Rosario D, Bailon-Garcia E, Lozano-Castello D, Bueno-Lopez A (2019) Improved asymmetrical honeycomb monolith catalyst prepared using a 3D printed template. J Hazard Mater 368:638–643

Delost MD, Smith DT, Anderson BJ, Njardarson JT (2018) From Oxiranes to oligomers: architectures of U.S. FDA approved pharmaceuticals containing oxygen heterocycles. J Med Chem 61:10996–11020

Deshpande SS, Sonavane SU, Jayaram RV (2008) A facile deprotection of oximes over mixed metal oxides under solvent-free conditions. Catal Commun 9:639–644

Dey S, Dhal GC (2020) Ceria doped CuMnOx as carbon monoxide oxidation catalysts: synthesis and their characterization. Surfaces Interfaces 18:100456

Do J, Chava R, Son N, Kim J, Park N-K, Lee D, Seo M, Ryu H-J, Chi J, Kang M (2018) Effect of Ce doping of a Co/Al2O3 catalyst on hydrogen production via propane steam reforming. Catalysts 8:413

Dobosz J, Hull S, Zawadzki M (2016) Catalytic activity of cobalt and cerium catalysts supported on calcium hydroxyapatite in ethanol steam reforming. Pol J Chem Technol 18:59–67

Du N, Zhang H, Chen B, Ma X, Yang D (2007) Ligand-free self-assembly of Ceria nanocrystals into nanorods by oriented attachment at low temperature. J Phys Chem C 111:12677–12680

Edayadulla N, Lee YR (2014) Cerium oxide nanoparticle-catalyzed three-component protocol for the synthesis of highly substituted novel quinoxalin-2-amine derivatives and 3,4-dihydroquinoxalin-2-amines in water. RSC Adv 4:11459

Esposito S (2019) "Traditional" Sol-Gel chemistry as a powerful tool for the preparation of supported metal and metal oxide catalysts. Materials (Basel) 12:668

Fayaz F, Danh HT, Nguyen-Huy C, Vu KB, Abdullah B, Vo D-VN (2016) Promotional effect of Ce-dopant on Al2O3-supported Co catalysts for syngas production via CO2 reforming of ethanol. Proc Eng 148:646–653

Feng M, Tang B, Liang SH, Jiang X (2016) Sulfur containing scaffolds in drugs: synthesis and application in medicinal chemistry. Curr Top Med Chem 16:1200–1216

Ferreira NS, Angélica RS, Marques VB, de Lima CCO, Silva MS (2016) Cassava-starch-assisted sol–gel synthesis of CeO 2 nanoparticles. Mater Lett 165:139–142

Fihri A, Len C, Varma RS, Solhy A (2017) Hydroxyapatite: a review of syntheses, structure and applications in heterogeneous catalysis. Coord Chem Rev 347:48–76

Fiorenza R, Bellardita M, D'Urso L, Compagnini G, Palmisano L, Scirè S (2016) Au/TiO2-CeO2 catalysts for photocatalytic water splitting and VOCs oxidation reactions. Catalysts 6:121

Fu J, Liu N, Mei L, Liao L, Deyneko D, Wang J, Bai Y, Lv G (2019) Synthesis of Ce-doped Mn3Gd7-xCex(SiO4)6O1.5 for the enhanced catalytic ozonation of tetracycline. Sci Rep 9:18734

Gadipelly C, Mannepalli LK (2019) Nano-metal oxides for organic transformations. Curr Opin Green Sustain Chem 15:20–26

Gao X, Zhu S, Dong M, Wang J, Fan W (2020) Ru/CeO2 catalyst with optimized CeO2 morphology and surface facet for efficient hydrogenation of ethyl levulinate to γ-valerolactone. J Catal 389:60–70

Gatica JM, García-Cabeza AL, Yeste MP, Marín-Barrios R, González-Leal JM, Blanco G, Cifredo GA, Guerra FM, Vidal H (2016) Carbon integral honeycomb monoliths as support of copper catalysts in the Kharasch–Sosnovsky oxidation of cyclohexene. Chem Eng J 290:174–184

Gawande MB, Pandey RK, Jayaram RV (2012) Role of mixed metal oxides in catalysis science—versatile applications in organic synthesis. Cat Sci Technol 2:1113

Gawande MB, Branco PS, Varma RS (2013a) Nano-magnetite (Fe3O4) as a support for recyclable catalysts in the development of sustainable methodologies. Chem Soc Rev 42:3371–3393

Gawande MB, Bonifácio VDB, Varma RS, Nogueira ID, Bundaleski N, Ghumman CAA, Teodoro OMND, Branco PS (2013b) Magnetically recyclable magnetite–ceria (Nanocat-Fe-Ce) nanocatalyst – applications in multicomponent reactions under benign conditions. Green Chem 15:1226

Ge Y, Diao P, Xu C, Zhang N, Guo C (2018) Visible light induced cross-coupling synthesis of asymmetrical heterobiaryls using Pd/CeO2 nanocomposite photocatalyst. Chin Chem Lett 29:903–906

Gharib A, Hashemipour Khorasani BR, Jahangir M, Roshani M, Safaee R (2013) Catalytic synthesis of 3-Methyl-1-phenyl-1H-benzo[g]pyrazolo[3,4-b]quinoline-5,10-dione derivatives using cerium oxide nanoparticles as heterogeneous catalyst in green conditions. Organic Chem Int 2013:1–5

Ghayour F, Mohammad Shafiee MR, Ghashang M (2018) ZnO-CeO2 nanocomposite: efficient catalyst for the preparation of thieno[2,3-d]pyrimidin-4(3H)-one derivatives. Main Group Metal Chem 41:1

Girija DK, Naik H, Sudhamani C, Kumar BV (2011) Cerium oxide nanoparticles-a green, reusable, and highly efficient heterogeneous catalyst for the synthesis of Polyhydroquinolines under solvent-free conditions. Arch Appl Sci Res 3:373

Gomtsyan A (2012) Heterocycles in drugs and drug discovery. Chem Heterocycl Compd 48:7–10

Gopinath K, Karthika V, Sundaravadivelan C, Gowri S, Arumugam A (2015) Mycogenesis of cerium oxide nanoparticles using Aspergillus niger culture filtrate and their applications for antibacterial and larvicidal activities. J Nanostruct Chem 5:295–303

Govender S, Friedrich H (2017) Monoliths: a review of the basics, preparation methods and their relevance to oxidation. Catalysts 7:62

Grzybowska-Swierkosz B (1987) Acidic properties of mixed transition metal oxides. Mater Chem Phys 17:121–144

Gu Y, Matsuda K, Nakayama A, Tamura M, Nakagawa Y, Tomishige K (2019) Direct synthesis of alternating polycarbonates from CO2 and diols by using a catalyst system of CeO2 and 2-Furonitrile. ACS Sustain Chem Eng 7:6304–6315

Guo Z, Liu B, Zhang Q, Deng W, Wang Y, Yang Y (2014) Recent advances in heterogeneous selective oxidation catalysis for sustainable chemistry. Chem Soc Rev 43:3480–3524

Guo T, Yao M-S, Lin Y-H, Nan C-W (2015) A comprehensive review on synthesis methods for transition-metal oxide nanostructures. CrystEngComm 17:3551–3585

Harikrishna S, Robert AR, Ganja H, Maddila S, Jonnalagadda SB (2020) A green, efficient and recoverable CeO2/MWCNT nanocomposite catalyzed click synthesis of pyridine-3-carboxamides. Appl Organomet Chem 34:e5796

Honda M, Tamura M, Nakao K, Suzuki K, Nakagawa Y, Tomishige K (2014) Direct cyclic carbonate synthesis from CO2 and Diol over carboxylation/hydration cascade catalyst of CeO2 with 2-Cyanopyridine. ACS Catal 4:1893–1896

Hosseini S, Moghaddas H, Masoudi Soltani S, Kheawhom S (2020) Technological applications of Honeycomb Monoliths in environmental processes: a review. Process Saf Environ Prot 133:286–300

Hu Z, Tan S, Mi R, Li X, Li D, Yang B (2018) Solvent-controlled reactivity of Au/CeO2 towards hydrogenation of p-Chloronitrobenzene. Catal Lett 148:1490–1498

Huang W, Gao Y (2014) Morphology-dependent surface chemistry and catalysis of CeO2nanocrystals. Cat Sci Technol 4:3772–3784

Juarez R, Concepcion P, Corma A, Garcia H (2010) Ceria nanoparticles as heterogeneous catalyst for CO2 fixation by omega-aminoalcohols. Chem Commun (Camb) 46:4181–4183

Kamruddin M, Ajikumar PK, Nithya R, Tyagi AK, Raj B (2004) Synthesis of nanocrystalline ceria by thermal decomposition and soft-chemistry methods. Scr Mater 50:417–422

Kargar H, Ghazavi H, Darroudi M (2015) Size-controlled and bio-directed synthesis of ceria nanopowders and their in vitro cytotoxicity effects. Ceram Int 41:4123–4128

Ke K, Wu F, Ren L, Jiao Y, Xing N, Shi L (2020) An efficient catalyst of Cu/MIL-101 modified with CeO2 for the conversion of biomass-derived glycerol with aniline to 3-methylindole. Catal Commun 136:105896

Khan S, Agasar M, Ghosh A, Keri RS (2019) A novel, multi-component method of preparation of Quinolines using recyclable CeO2-TiO2 nanocomposite catalyst under solvent-free conditions. Org Prep Proced Int 51:153–160

Korotkova AM, Borisovna PO, Aleksandrovna GI, Bagdasarovna KD, Vladimirovich BD, Vladimirovich KD, Alexandrovich FA, Yurievna KM, Nikolaevna BE, Aleksandrovich KD, Yurievich CM, Valerievich LS (2019) "Green" synthesis of cerium oxide particles in water extracts Petroselinum crispum. Curr Nanomater 4:176–190

Kumar PSV, Suresh L, Vinodkumar T, Reddy BM, Chandramouli GVP (2016) Zirconium doped Ceria nanoparticles: an efficient and reusable catalyst for a green multicomponent synthesis of novel Phenyldiazenyl–Chromene derivatives using aqueous medium. ACS Sustain Chem Eng 4:2376–2386

Laberty-Robert C, Long JW, Lucas EM, Pettigrew KA, Stroud RM, Doescher MS, Rolison DR (2006) Sol–gel-derived Ceria nanoarchitectures: synthesis, characterization, and electrical properties. Chem Mater 18:50–58

Lamberth C (2013) Heterocyclic chemistry in crop protection. Pest Manag Sci 69:1106–1114

Liang J, Liang Z, Zou R, Zhao Y (2017) Heterogeneous catalysis in Zeolites, Mesoporous Silica, and metal-organic frameworks. Adv Mater 29

Liu J, Li Y, Zhang J, He D (2016) Glycerol carbonylation with CO2 to glycerol carbonate over CeO2 catalyst and the influence of CeO2 preparation methods and reaction parameters. Appl Catal A Gen 513:9–18

Liu J, Li Y, Liu H, He D (2018) Transformation of CO2 and glycerol to glycerol carbonate over CeO2ZrO2 solid solution —— effect of Zr doping. Biomass Bioenergy 118:74–83

Liu J, Zhao Z, Xu C, Liu J (2019) Structure, synthesis, and catalytic properties of nanosize cerium-zirconium-based solid solutions in environmental catalysis. Chin J Catal 40:1438–1487

López JM, Gilbank AL, García T, Solsona B, Agouram S, Torrente-Murciano L (2015) The prevalence of surface oxygen vacancies over the mobility of bulk oxygen in nanostructured ceria for the total toluene oxidation. Appl Catal B Environ 174-175:403–412

Lu F (1998) Some heterocyclic polymers and polysiloxanes. J Macromol Sci Polym Rev 38:143–205

Lu Y, Dong W, Ding J, Wang W, Wang A (2019) Hydroxyapatite nanomaterials: synthesis, properties, and functional applications. In: Nanomaterials from clay minerals. Elsevier, Amsterdam/Cambridge, MA, pp 485–536

Maddila S, Maddila SN, Jonnalagadda SB, Lavanya P (2016) Reusable Ce-V loaded alumina catalyst for multicomponent synthesis of substituted pyridines in green media. J Heterocyclic Chem 53:658–664

Maddila SN, Maddila S, van Zyl WE, Jonnalagadda SB (2017a) CeO2/ZrO2 as green catalyst for one-pot synthesis of new pyrano[2,3-c]-pyrazoles. Res Chem Intermed 43:4313–4325

Maddila S, Gangu KK, Maddila SN, Jonnalagadda SB (2017b) A viable and efficacious catalyst, CeO2/HAp, for green synthesis of novel pyrido[2,3-d]pyrimidine derivatives. Res Chem Intermed 44:1397–1409

Maensiri S, Masingboon C, Laokul P, Jareonboon W, Promarak V, Anderson PL, Seraphin S (2007) Egg white synthesis and photoluminescence of Platelike clusters of CeO2Nanoparticles. Cryst Growth Des 7:950–955

Majumdar KC, Chattopadhyay SK (2011) Heterocycles in natural product synthesis. Wiley-VCH, Weinheim

Malik B, Pirzadah TB, Kumar M, Rehman RU (2017) Biosynthesis of nanoparticles and their application in pharmaceutical industry. In: Nanotexhnology, pp 235–252

Maqbool Q (2017) Green-synthesised cerium oxide nanostructures (CeO2-NS) show excellent biocompatibility for phyto-cultures as compared to silver nanostructures (Ag-NS). RSC Adv 7:56575–56585

Melchionna M, Fornasiero P (2014) The role of ceria-based nanostructured materials in energy applications. Mater Today 17:349–357

Moydeen M, Al-Deyab SS, Kumar RS, Idhayadhulla A (2017) Efficient synthesis of novel 3-Phenyl-5-thioxo-3,4,5,6-tetrahydroimidazo[4,5-c]pyrazole-2(1H)-carbothioamide derivatives using a CeO2-MgO catalyst and evaluation of antimicrobial activity. J Heterocyclic Chem 54:3208–3219

Naaz F, Farooq U, Ahmad T (2019) Ceria as an efficient Nanocatalyst for organic transformations. In: Nanocatalysts. IntechOpen, London

Orge CA, Órfão JJM, Pereira MFR, Duarte de Farias AM, Fraga MA (2012) Ceria and cerium-based mixed oxides as ozonation catalysts. Chem Eng J 200-202:499–505

Paier J, Penschke C, Sauer J (2013) Oxygen defects and surface chemistry of ceria: quantum chemical studies compared to experiment. Chem Rev 113:3949–3985

Parashar M, Shukla VK, Singh R (2020) Metal oxides nanoparticles via sol–gel method: a review on synthesis, characterization and applications. J Mater Sci Mater Electron 31:3729–3749

Pokhrel S (2018) Hydroxyapatite: preparation, properties and its biomedical applications. Adv Chem Eng Sci 08:225–240

Polshettiwar V, Luque R, Fihri A, Zhu H, Bouhrara M, Basset JM (2011) Magnetically recoverable nanocatalysts. Chem Rev 111:3036–3075

Pratap SR, Shamshuddin SZM, Thimmaraju N, Shyamsundar M (2020) Cordierite honeycomb monoliths coated with Al(III)/ZrO2 as an efficient and reusable catalyst for the Knoevenagel condensation: a faster kinetics. Arab J Chem 13:2734–2749

Qi G, Li W (2015) NO oxidation to NO2 over manganese-cerium mixed oxides. Catal Today 258:205–213

Qu Y, Gao Y, Lin S, Shi L (2020) Efficient synthesis of 3-methylindole using biomass-derived glycerol and aniline over ZnO and CeO2 modified Ag/SBA-15 catalysts. Mol Catal 493:111038

Rahmat (2010) A review: Mesoporous Santa Barbara Amorphous-15, types, synthesis and its applications towards biorefinery production. Am J Appl Sci 7:1579–1586

Ramachandran M, Subadevi R, Sivakumar M (2019) Role of pH on synthesis and characterization of cerium oxide (CeO2) nano particles by modified co-precipitation method. Vacuum 161:220–224

Ramazani A, Asiabi P, Aghahosseini H, Gouranlou F (2017) Review on the synthesis and functionalization of SiO2 nanoparticles as solid supported catalysts. Curr Org Chem 21:908–922

Rane AV, Kanny K, Abitha VK, Thomas S (2018) Methods for synthesis of nanoparticles and fabrication of nanocomposites. In: Synthesis of inorganic nanomaterials. Wiley, Weinheim, pp 121–139

Rashed MN, Touchy AS, Chaudhari C, Jeon J, Siddiki SMAH, Toyao T, Shimizu K-i (2020) Selective C3-alkenylation of oxindole with aldehydes using heterogeneous CeO2 catalyst. Chin J Catal 41:970–976

Rathod S, Arbad B, Lande M (2010) Preparation, characterization, and catalytic application of a Nanosized Ce1MgxZr1-xO2Solid heterogeneous catalyst for the synthesis of Tetrahydrobenzo[b]pyran derivatives. Chin J Catal 31:631–636

Rico-Pérez V, García-Cortés JM, de Lecea CS-M, Bueno-López A (2013) NOx reduction to N2 with commercial fuel in a real diesel engine exhaust using a dual bed of Pt/beta zeolite and RhOx/ceria monolith catalysts. Chem Eng Sci 104:557–564

Rodriguez JA, Grinter DC, Liu Z, Palomino RM, Senanayake SD (2017) Ceria-based model catalysts: fundamental studies on the importance of the metal-ceria interface in CO oxidation, the water-gas shift, CO2 hydrogenation, and methane and alcohol reforming. Chem Soc Rev 46:1824–1841

Russo N, Mescia D, Fino D, Saracco G, Specchia V (2007) N2O decomposition over perovskite catalysts. Ind Eng Chem Res 46:4226–4231

Saadati-Moshtaghin HR, Zonoz FM (2019) In situ preparation of CeO2 nanoparticles on the MCM-41 with magnetic core as a novel and efficient catalyst for the synthesis of substituted pyran derivatives. Inorg Chem Commun 99:44–51

Sabitha G, Reddy KB, Yadav JS, Shailaja D, Sivudu KS (2005) Ceria/vinylpyridine polymer nanocomposite: an ecofriendly catalyst for the synthesis of 3,4-dihydropyrimidin-2(1H)-ones. Tetrahedron Lett 46:8221–8224

Sabitha G, Reddy N, Prasad M, Yadav J, Sivudu K, Shailaja D (2008) Efficient synthesis of Bis(indolyl)methanes using nano Ceria supported on vinyl pyridine polymer at ambient temperature. Lett Org Chem 5:300–303

Safaei-Ghomi J, Kalhor S, Shahbazi-Alavi H, Asgari-Kheirabadi M (2015a) Three-component synthesis of cyclic β-aminoesters using CeO2 nanoparticles as an efficient and reusable catalyst. Turk J Chem 39:843–849

Safaei-Ghomi J, Heidari-Baghbahadorani E, Shahbazi-Alavi H, Asgari-Kheirabadi M (2015b) A comparative study of the catalytic activity of nanosized oxides in the one-pot synthesis of highly substituted dihydropyridines. RSC Adv 5:18145–18152

Safaei-Ghomi J, Asgari-Keirabadi M, Khojastehbakht-Koopaei B, Shahbazi-Alavi H (2015c) Multicomponent synthesis of C-tethered bispyrazol-5-ols using CeO2 nanoparticles as an efficient and green catalyst. Res Chem Intermed 42:827–837

Safaei-Ghomi J, Shahbazi-Alavi H, Kalhor S (2016) CeO2 nanoparticles: an efficient and robust catalyst for the synthesis of 2-amino-4,6-diarylbenzene-1,3-dicarbonitriles. Monatshefte für Chemie - Chemical Monthly 147:1933–1937

Safari J, Gandomi-Ravandi S (2014a) Application of the ultrasound in the mild synthesis of substituted 2,3-dihydroquinazolin-4(1H)-ones catalyzed by heterogeneous metal–MWCNTs nanocomposites. J Mol Struct 1072:173–178

Safari J, Gandomi-Ravandi S (2014b) Silver decorated multi-walled carbon nanotubes as a heterogeneous catalyst in the sonication of 2-aryl-2,3-dihydroquinazolin-4(1H)-ones. RSC Adv 4:11654–11660

Safari J, Gandomi-Ravandi S (2014c) Titanium dioxide supported on MWCNTs as an eco-friendly catalyst in the synthesis of 3,4-dihydropyrimidin-2-(1H)-ones accelerated under microwave irradiation. New J Chem 38:3514–3521

Samai B, Sarkar S, Chall S, Rakshit S, Bhattacharya SC (2016) Polymer-fabricated synthesis of cerium oxide nanoparticles and applications as a green catalyst towards multicomponent transformation with size-dependent activity studies. CrystEngComm 18:7873–7882

Samantaray S, Pradhan DK, Hota G, Mishra BG (2012) Catalytic application of CeO2–CaO nanocomposite oxide synthesized using amorphous citrate process toward the aqueous phase one pot synthesis of 2-amino-2-chromenes. Chem Eng J 193-194:1–9

Schneider JJ, Naumann M, Schafer C, Brandner A, Hofmann HJ, Claus P (2011) Template-assisted formation of microsized nanocrystalline CeO(2) tubes and their catalytic performance in the carboxylation of methanol. Beilstein J Nanotechnol 2:776–784

Schwarz W, Schossig J, Rossbacher R, Pinkos R, Höke H (2019) Butyrolactone. In: Ullmann's Encyclopedia of industrial chemistry. Weinheim, Wiley, pp 1–7

Shafighi S, Mohammad Shafiee MR, Ghashang M (2018) MgO-CeO2nanocomposite: efficient catalyst for the preparation of 2-aminothiophenes and thieno[2,3-d]pyrimidin-4(3H)-one derivatives. J Sulfur Chem 39:402–413

Sharma K, Borah A, Neog K, Gogoi DP (2016a) CeO2-catalyzed C-H functionalization ofN-Aryltetrahydroisoquinolines: an aerobic cross-Dehydrogenative coupling reaction between two sp3C-H bonds. Chem Select 1:4620–4623

Sharma RK, Dutta S, Sharma S, Zboril R, Varma RS, Gawande MB (2016b) Fe3O4(iron oxide)-supported nanocatalysts: synthesis, characterization and applications in coupling reactions. Green Chem 18:3184–3209

Sharma K, Das B, Gogoi P (2018) Synthesis of pyrrolo[3,4-c]quinoline-1,3-diones: a sequential oxidative annulation followed by dehydrogenation and N-demethylation strategy. New J Chem 42:18894–18905

Shelkar R, Sarode S, Nagarkar J (2013) Nano ceria catalyzed synthesis of substituted benzimidazole, benzothiazole, and benzoxazole in aqueous media. Tetrahedron Lett 54:6986–6990

Shelkar RS, Balsane KE, Nagarkar JM (2015) Magnetically separable nano CeO2: a highly efficient catalyst for ligand free direct C–H arylation of heterocycles. Tetrahedron Lett 56:693–699

Shen M, Wang J, Shang J, An Y, Wang J, Wang W (2009) Modification Ceria-Zirconia mixed oxides by doping Sr using the reversed microemulsion for improved Pd-only three-way catalytic performance. J Phys Chem C 113:1543–1551

Shi Z-L, Du C, Yao S-H (2011) Preparation and photocatalytic activity of cerium doped anatase titanium dioxide coated magnetite composite. J Taiwan Inst Chem Eng 42:652–657

Shimura S, Miura H, Wada K, Hosokawa S, Yamazoe S, Inoue M (2011) Ceria-supported ruthenium catalysts for the synthesis of indole via dehydrogenative N-heterocyclization. Cat Sci Technol 1:1340

Shrestha R, Sharma K, Lee YR, Wee YJ (2016) Cerium oxide-catalyzed multicomponent condensation approach to spirooxindoles in water. Mol Divers 20:847–858

Sifontes AB, Gonzalez G, Ochoa JL, Tovar LM, Zoltan T, Cañizales E (2011) Chitosan as template for the synthesis of ceria nanoparticles. Mater Res Bull 46:1794–1799

Soni B, Biswas S (2013) Ceria nanoparticles synthesized by a polymer precursor method. Int J Modern Phys Conf Ser 22:169–172

Spezzati G, Su Y, Hofmann JP, Benavidez AD, DeLaRiva AT, McCabe J, Datye AK, Hensen EJM (2017) Atomically dispersed Pd–O species on CeO2(111) as highly active sites for low-temperature CO oxidation. ACS Catal 7:6887–6891

Sultana SSP, Kishore DHV, Kuniyil M, Khan M, Alwarthan A, Prasad KRS, Labis JP, Adil SF (2015) Ceria doped mixed metal oxide nanoparticles as oxidation catalysts: synthesis and their characterization. Arab J Chem 8:766–770

Sun W, Liu D-Y, Zhu H-Y, Shi L, Sun Q (2010) A new efficient approach to 3-methylindole: vapor-phase synthesis from aniline and glycerol over Cu-based catalyst. Catal Commun 12:147–150

Sun C, Li H, Chen L (2012) Nanostructured ceria-based materials: synthesis, properties, and applications. Energy Environ Sci 5:8475

Sungkono IE, Kameyama H, Koya T (1997) Development of catalytic combustion technology of VOC materials by anodic oxidation catalyst. Appl Surf Sci 121-122:425–428

Suresh L, Vijay Kumar PS, Vinodkumar T, Chandramouli GVP (2016) Heterogeneous recyclable nano-CeO2 catalyst: efficient and eco-friendly synthesis of novel fused triazolo and tetrazolo pyrimidine derivatives in aqueous medium. RSC Adv 6:68788–68797

Suvetha Rani J (2020) Green synthesis and characterization of Ceria nanoparticles using Ricinus communis leaf extract. Int J Sci Res Publ 10:9743

Tamura M, Noro K, Honda M, Nakagawa Y, Tomishige K (2013a) Highly efficient synthesis of cyclic ureas from CO2 and diamines by a pure CeO2 catalyst using a 2-propanol solvent. Green Chem 15:1567

Tamura M, Honda M, Noro K, Nakagawa Y, Tomishige K (2013b) Heterogeneous CeO2-catalyzed selective synthesis of cyclic carbamates from CO2 and aminoalcohols in acetonitrile solvent. J Catal 305:191–203

Tamura M, Ito K, Nakagawa Y, Tomishige K (2016a) CeO2-catalyzed direct synthesis of dialkyl-ureas from CO2 and amines. J Catal 343:75–85

Tamura M, Ito K, Honda M, Nakagawa Y, Sugimoto H, Tomishige K (2016b) Direct copolymerization of CO2 and diols. Sci Rep 6:24038

Taylor AP, Robinson RP, Fobian YM, Blakemore DC, Jones LH, Fadeyi O (2016) Modern advances in heterocyclic chemistry in drug discovery. Org Biomol Chem 14:6611–6637

Terribile D, Trovarelli A, Llorca J, de Leitenburg C, Dolcetti G (1998) The synthesis and characterization of mesoporous high-surface area ceria prepared using a hybrid organic/inorganic route. J Catal 178:299–308

Thakur N, Manna P, Das J (2019) Synthesis and biomedical applications of nanoceria, a redox active nanoparticle. J Nanobiotechnol 17:84

Thovhogi N, Diallo A, Gurib-Fakim A, Maaza M (2015) Nanoparticles green synthesis by Hibiscus Sabdariffa flower extract: Main physical properties. J Alloys Compd 647:392–396

Tillirou AA, Theocharis CR (2008) Synthesis and characterization of mesoporous cerium oxide prepared using an organic base and a templating agent. Adsorp Sci Technol 26:687–692

Tomishige K, Yasuda H, Yoshida Y, Nurunnabi M, Li B, Kunimori K (2004) Catalytic performance and properties of ceria based catalysts for cyclic carbonate synthesis from glycol and carbon dioxide. Green Chem 6:206

Tomishige K, Tamura M, Nakagawa Y (2019) CO2 conversion with alcohols and amines into carbonates, ureas, and carbamates over CeO2 catalyst in the presence and absence of 2-Cyanopyridine. Chem Rec 19:1354–1379

Tomishige K, Gu Y, Nakagawa Y, Tamura M (2020) Reaction of CO2 with alcohols to linear-, cyclic-, and poly-carbonates using CeO2-based catalysts. Front Energy Res 8:117

Tsai M-S (2004) Powder synthesis of nano grade cerium oxide via homogenous precipitation and its polishing performance. Mater Sci Eng B 110:132–134

Vadivel P, Ramesh R, Lalitha A (2013) Ceric ion loaded MCM-41 catalyzed synthesis of substituted mono- and Bis-dihydropyrimidin-2(1H)-ones. J Catal 2013:1–8

Venkatesh SZM, Shamshuddin NM (2015) Mubarak, effective synthesis of quinoxalines over ceria based solid acids coated on honeycomb monoliths. Indian J Chem 54A:843–850

Vijay Kumar PS, Suresh L, Vinodkumar T, Chandramouli GVP (2016) Eu2O3 modified CeO2 nanoparticles as a heterogeneous catalyst for an efficient green multicomponent synthesis of novel phenyldiazenyl-acridinedione-carboxylic acid derivatives in aqueous medium. RSC Adv 6:91133–91140

Vita A, Italiano C, Ashraf MA, Pino L, Specchia S (2018a) Syngas production by steam and oxysteam reforming of biogas on monolith-supported CeO 2 -based catalysts. Int J Hydrog Energy 43:11731–11744

Vita A, Italiano C, Pino L, Frontera P, Ferraro M, Antonucci V (2018b) Activity and stability of powder and monolith-coated Ni/GDC catalysts for CO2 methanation. Appl Catal B Environ 226:384–395

Vitaku E, Smith DT, Njardarson JT (2014) Analysis of the structural diversity, substitution patterns, and frequency of nitrogen heterocycles among U.S. FDA approved pharmaceuticals. J Med Chem 57:10257–10274

Vivier L, Duprez D (2010) Ceria-based solid catalysts for organic chemistry. ChemSusChem 3:654–678

Walsh CT (2015) Nature loves nitrogen heterocycles. Tetrahedron Lett 56:3075–3081

Wang S, Wang Z, Zha Z (2009) Metal nanoparticles or metal oxide nanoparticles, an efficient and promising family of novel heterogeneous catalysts in organic synthesis. Dalton Trans 2009:9363–9373

Wang Y, Ge W, Fang Y, Ren X, Cao S, Liu G, Li M, Xu J, Wan Y, Han X, Wu H (2016) Porous CeO2 nanorod-catalyzed synthesis of poly-substituted imino-pyrrolidine-thiones. Res Chem Intermed 43:631–640

Wang Y, Arandiyan H, Scott J, Bagheri A, Dai H, Amal R (2017) Recent advances in ordered meso/macroporous metal oxides for heterogeneous catalysis: a review. J Mater Chem A 5:8825–8846

Xu P, Yu R, Ren H, Zong L, Chen J, Xing X (2014) Hierarchical nanoscale multi-shell Au/CeO2 hollow spheres. Chem Sci 5:4221–4226

Xu B, Qi F, Sun D, Chen Z, Robert D (2016) Cerium doped red mud catalytic ozonation for bezafibrate degradation in wastewater: efficiency, intermediates, and toxicity. Chemosphere 146:22–31

Yan B, Zhang Y, Chen G, Shan R, Ma W, Liu C (2016) The utilization of hydroxyapatite-supported CaO-CeO2 catalyst for biodiesel production. Energy Convers Manag 130:156–164

Yin P, Shreeve JNM (2017) Nitrogen-rich azoles as high density energy materials. In: Advances in heterocyclic chemistry, vol 121, pp 89–131

Yu T, Joo J, Park YI, Hyeon T (2005) Large-scale nonhydrolytic sol-gel synthesis of uniform-sized ceria nanocrystals with spherical, wire, and tadpole shapes. Angew Chem Int Ed Engl 44:7411–7414

Yu B, Zhang H, Zhao Y, Chen S, Xu J, Hao L, Liu Z (2013) DBU-based ionic-liquid-catalyzed carbonylation of o-Phenylenediamines with CO2 to 2-Benzimidazolones under solvent-free conditions. ACS Catal 3:2076–2082

Zamani A, Marjani AP, Alimoradlu K (2018) Walnut Shell-templated ceria nanoparticles: green synthesis, characterization and catalytic application. Int J Nanosci 17:1850008

Zarnegar Z, Safari J, Kafroudi ZM (2015) Co3O4–CNT nanocomposites: a powerful, reusable, and stable catalyst for sonochemical synthesis of polyhydroquinolines. New J Chem 39:1445–1451

Zhang QL, Yang ZM, Ding BJ (2009) Synthesis of cerium oxide nanoparticles by the precipitation method. Mater Sci Forum 610-613:233–238

Zhang D, Du X, Shi L, Gao R (2012) Shape-controlled synthesis and catalytic application of ceria nanomaterials. Dalton Trans 41:14455–14475

Zhang Y, Chen C, Gong W, Song J, Zhang H, Zhang Y, Wang G, Zhao H (2017) Self-assembled Pd/CeO 2 catalysts by a facile redox approach for high-efficiency hydrogenation of levulinic acid into gamma-valerolactone. Catal Commun 93:10–14

Zhang X, Song Y, Guan F, Zhou Y, Lv H, Wang G, Bao X (2018) Enhancing electrocatalytic CO2 reduction in solid oxide electrolysis cell with Ce0.9Mn0.1O2−δ nanoparticles-modified LSCM-GDC cathode. J Catal 359:8–16

Zhao X, Chaudhry ST, Mei J (2017) Heterocyclic building blocks for organic semiconductors. In: Advances in heterocyclic chemistry, vol 121, pp 133–171

Index

A
Acetylcholinesterase (AChE), 508, 571
Acinetobacter baumannii, 191
Actinic keratosis, 221, 223
Activated charcoal (AC), 478
Activated carbon, 624
Active targeting, 450
Active tumor targeting, 91
Activin B-loaded hydrogels, 447
Acute respiratory distress syndrome (ARDS), 454
Acyl homoserine lactone, 435
Adoptive cell transfer, 448
Adriamycin, 121
Adsorbents
 with nano-chitosan, 631
 in nanoscale, 627
Adsorption, 602
 adsorption–precipitation, 612
 CNTs, 624
 dichlorobenzene, 629
 efficiency, 602
 engineered nanoparticles, 623
 GO, 613
 hematite, 609
 mechanism, 602
Adsorptive-mediated transcytosis (AMT), 446
Advanced oxidation, 602
Advanced oxidation process (AOP), 602
African farmers' soil quality, 593
African smallholder agriculture, nanotechnologies
 crop yields, 594
 poor physical properties, 590
 poor soil quality, 590
 soils and soil improvement, 591, 593
 yield-reducing competition, 590
African smallholder crop production challenges
 drought/water shortages and management, 585–586
 plant pest management, 584–585
 soil fertility management, 582–584
Agarose nanoparticles (ANPs), 66, 67
Ag NPs Si nanowires, 409
AgNps synthesis, 227
Agriculture area, 541
Albumin nanospheres, 65
Alginate, 40, 43, 65, 66
Alkalyting specialists, 220
Allium saralicum extract, 380
Allyl isothiocyanate, 143
Aloe vera nanoparticles, 470
Aluminum oxide NPs (Al_2O_3 NPs), 406
Amberlite-supported azide, 731
Aminoglycosides (AG), 427, 431
Amperometric biosensor, 503
Amphiphilic macromolecules, 36
Ampicillin-encapsulated liposomes, 441
Ancient Indian treatment systems, 421
Animal pests, 584
Annona squamosa, 380
Antibacterial activity (AB), 165, 166
Antibacterial mechanisms, 473
Antibacterial metal nanoparticles, aquaculture
 AgNPs, 473, 475
 AuNPs, 475, 476
 types, 473
 ZnNPs, 473
Antibiotic resistance, 164, 425

Antibiotics, 420
Antibiotics targeting cell wall
  bacterial cell wall, 426
  β-lactam, 426, 427
  glycopeptides, 427
  gram-positive bacteria, 425
  PBPs, 426
  prions, 445
Antibiotic treatment, 407
Antibody-drug conjugates (ADCs), 448
Anticancer drugs, 119
Antifouling coatings, 172, 180, 182, 194
Antifungal agents, 245
Antigen presenting cells (APCs), 334, 441, 442
Antimalarial herbals, 282
Anti-metabolites, 220
Antimicrobial activity, 164, 165
Antimicrobial agents, 244, 245
Antimicrobial agents, metal NPs
  bacterial and mammalian cells, 399
  biofilm-forming ability, 399
  dissolved metal interaction, 401
  interaction with cell membrane, 400, 401
  oxidative stress-mediated cell death, 401
  silver and copper utensils, 399
Antimicrobial compounds, 169
Antimicrobial resistance mechanism
  accumulation prevention
    antibiotic inactivation, 431
    efflux pumps, 429, 430
    target molecule modification, 430, 431
  in bacteria, 430
Antimicrobial treatment, 398
Antimicrobials, 427
Antioxidant defence mechanisms, 436
Antitumor anti-infection agents, 220
Aptamers, 443, 491, 497, 500, 503–506
Aptasensors, 503, 504
Aquaculture
  advantages, 464
  antibacterial agents, 474–475
  chemicals, 472
  drug delivery agents, 477
  economic losses, 472
  food-producing sectors, 464
  sustainability, 464
Aquafeeds, 466
Aqueous leaf extract, 383
Arboviruses, 347, 360–363
Archeosome, 268
Argyrosis, 454
Armamentarium, 432

Arsenic-polluted water, 603
Artemisinin, 246
Atomic force microscope (AFM), 622
Atomic layer deposition (ALD), 373
Atoms, 164, 420
ATP molecule, 435
Azane, 694
Azolla microphylla-based gold nanoparticles, 470

B

Bacillus cereus, 183
Bacteria, 243
Bacterial cell imparting toxicity, 164
Bacterial cell surface properties, 167
Bacterial cells uptake materials, 429
Bacterial immune responses, 432
Bacterial oligodeoxynucleotides, 444
Bacterial strains, 164
Bacteriocin resistance, 434
Bacteriocins, 433, 434
Bacterium, 430
Bacteroides ferment dietary fiber, 287
Basal cell carcinoma, 223
Benzimidazoylquinoxalines, 742–744
Beta-lactam antibiotics, 426, 427
Beta-lactam-resistant microorganisms, 434
Bimetallic nanoparticles (BNPs), 166, 563–565
Bioactive compounds, 13, 22
  delivering and targeting, 17
  delivery, 7
  functional, 6
  incorporation, 8
  natural, 10
  stability, 22
Bioactive nutraceuticals, 275
Bioactive phytochemicals, 266
Bioaugmentation, 632
Bioavailability, 6, 7
Bio-based sustainable remediation, 570
Biochemical sensors, 490
Biocompatibility, 325, 333
Biocompatible engineered nanocomposite, 412
Bioconversion methods, 630
Biodegradable polymers, 17
Bio-electrochemical systems, 644, 656
Biofabrication, 557
Biofilm inhibitory concentration (BIC), 165
Biofilm-induced infections, 168
Biofilm matrix, 169
Biofilms, 410

Index   763

aluminum oxide NP, 191
antibiofilm fabrics, 190
antibiotic resistance, 188
antibiotic-free hydrogels, 190
antimicrobial coatings, 189
antimicrobial property, 190
bacterial cell membranes, 191
bacterial cells, 166
biofilm theory, 166
biomaterial surfaces, 189
calcium oxide (CaO), 190
coating approaches, 188
detrimental effects, 168
development, 166, 167, 189
drug-resistant bacterial strains, 189
electrospinning, 189
formation, 166
glycolytic pathway, 190
implant materials, 189
limb prosthesis, 188
magnesium oxide (MgO), 190
metal fluoride complexes, 190
microcapsules, 189
microorganisms, 189
microspheres, 189
morphology, 189
nanoparticles, 190
NIR carbon dots (CD), 190
orthopedic implant materials, 188
pathogenesis, 188
pH values, 190
polymer nanocomposites, 191
resistance/tolerance, 168–170
smart polymer surfaces, 190
zinc oxide nanoparticles, 189
Biogenic synthesis, 312, 313
Biological materials, 421
Biological method
    algae
        advantages, 378
        disadvantages, 378
    bacteria
        advantages, 377
        disadvantages, 377
    fungi
        advantages, 377
        disadvantages, 377
    virus
        advantages, 378
        disadvantages, 378
Biological mosquitocides, 370
Biological NPs synthesis approaches
    AgNps, 375, 376

AuNps, 375
    disadvantages, 376
    microbes and plants, 375
    microbial synthesis, 376
Biological organisms, 595
Biological systems, 423
Biological techniques, 590
Biomaterial-associated infection (BAI), 188
Bio-nanocomposites, 385
Bio-organism, 548
Biopharmaceutical classification systems
        (BCS), 88
Biopolymeric nanocarriers
    dendrimers, 265–266
    PMs, 265
    protein-based, 265
Biopolymers, 103
Bioreceptor, 489, 491, 497, 502
Biorecognition elements, 489, 506, 508
Bioremediation, 546, 589, 605
    with nanotechnology, 631, 632
Biosensors, 240, 346, 348, 351, 360, 361,
        549, 571
    description, 489
    nanofabrication process, 490
    for pesticides detection, 571
        enzyme biosensors, 571, 572
        immunosensor, 572
        nanoparticle-based optical
            biosensors, 573
        nanotube-based electrochemical
            biosensors, 573
    sensing mechanism, 489
Biosynthesized AgNPs, 226, 227
Biotechnology, 241, 254, 582
Biotransformation, 275
Biscoumarins, 731
Bismuth nanoparticles, 221
Bismuth vanadate, 684
Blattella germanica (L.), 382
Blood-brain-barrier (BBB), 445
Bottom-up approach, 422
Butyrylcholinesterase (BChE), 571

C

Calcium oxide nanoparticles (CaO NPs), 406
Calotropis procera extract, 381
Cancer
    homeostatic imbalance, 119
    as neoplasm, 119
Cancer cell lines, 379
Cancer chemotherapy, 277

Cancer mortality rates, 85
Cancer therapy
  body's immune response, 448
  chemotherapy, 447
  nanotechnology (*see* Nanotechnology-based cancer treatment)
  protocols, 447
  treatment modalities (*see* Cancer treatment modalities)
Cancer treatment modalities
  adoptive cell transfer, 448
  CAR-T cell therapy, 449
  immune checkpoint modulators, 448
  stem cell transplant, 449
  therapeutic antibodies, 448
  vaccines, 449
Cancer treatments
  chemotherapy, 85, 86
  conventional, 86
  hormone therapy, 85
  immunotherapy, 85
  nano DDS (*see* Nanomaterials-based drug delivery systems (nano DDS))
  nanomaterials as drug delivery system, 87
  nonsurgical, 85
  organic nanomaterials (*see* Organic nanomaterials)
  radiation therapy, 85
  surgical intervention, 85
Cancer vaccines, 449
Cannibalism, 434
Capsaicin, 275
Carbon-based drug delivery systems
  CNTs, 46–48
  fullerenes, 48, 49
  graphenes, 49, 50
Carbon-based nanodevices, 46
Carbon-based nanoparticles, 328, 329
Carbon nanomaterials
  CNTs, 497, 498
  dimensions, 496
  graphene, 497, 499
Carbon nanotubes (CNTs), 46–48, 71, 72, 102, 230, 329, 330, 497–498, 623, 624
  applications, 139
  adsorption, 628
  in cancer therapy, 139–141
    and drug delivery, 140
  cisplatin-loaded, 141
  curcumin-loaded, 140
  inner and outer surfaces, 139
  methotrexate-conjugated, 139
  nanotubes, 139

  types, 139
  wastewater remediation, 612, 613
Carboxyfullerene nanoparticles, 443
Carcinogenesis, 219
Carcinoma, 217, 218
Cardiotoxicity, 308
Carrier-mediated transporters, 446
Catalysis, 216
Cathode catalysis, 644
  ORR, 647
Cathode catalysts
  materials in MFCs, 647, 648
  mechanism of ORR, 647
Cationic detergents, 603
Cationic nanogels, 44
CD-4-dependant virion binding, 402
Cell permeability, 434
Cellulose nanomaterials, 19
Cell wall precursors, 430
Ceria-supported transitional metals, 711
Cerium-based catalysts
  commercial CeO_2 NPs, 712–715, 717
  nanostructured CeO_2
    from biology-directed method, 728, 729
    from co-precipitation method, 721–728
    from polymer-directed method, 725, 726
  for synthesis of CO_2-derived heterocycles, 746–750
  synthetic nano-CeO_2, 717
    biological synthesis, 720, 721
    co-precipitation, 718, 719
    morphology, 717, 718
    sol-gel synthesis, 719, 720
    template-assisted synthesis, 719, 720
  vapour-phase synthesis, heterocycles, 746
Cerium-based materials, 711
Cerium-based nanocatalysts, 712
Cerium mixed oxides
  Ce1Mg0.6Zr0.4O_2 composite, 734
  co-precipitated nanocomposite, ZrO_2@CeO_2, 733
  magnetic CeO2@Fe_3O_4 nanocatalyst, 737
  mixed metal oxides, 729, 734
  nanocomposite, CeO_2-ZrO_2, 729
  nanostructured MgO@CeO_2, 731
  superparamagnetic Fe_3O_4, 735
Cerium-solid material composite
  CeO_2-carbon template, 742
  CeO_2-clay composite, 740
  CeO_2-polymer, 737, 738
  CeO_2-silica, 738–741

Index   765

Cetylpyridinium chloride, 374
Chalcogenide-based binary photocatalyst, 682
Chemical integrity of drug, 75
Chemical NPs synthesis approaches
　advantage, 373
　atomic groups formation, 373
　cost, 375
　electrorefining, 374
　monodispersed gold nanoparticles, 374
　PtNps, 375
　reducing agents, 374
　silver reduction, 374
　size and shape, 373, 374
　VenMat solution, 374
Chemical oxidation, 602
Chemical synthesized nanoparticles, 387
Chemical vapor deposition (CVD)
　method, 139
Chemically synthesized ZnO
　nanoparticles, 473
Chemicals-based mosquitocidal agents, 370
Chemotherapeutic drugs, 86–88, 90, 92–95,
　101, 102
Chemotherapy, 86, 217, 220, 225, 308, 450
Chikungunya, 362, 363
Chimeric antigen receptor T-cell therapy
　(CAR-T cell therapy), 449
Chimeric antigen receptors (CARs), 449
Chitin, 503
Chitin-based polymeric nanoparticles, 355
Chitosan (CS), 66, 89, 98, 102, 437, 444,
　476, 503
Chitosan nanoparticles (CNPs), 65
Chitosan polymer, 40
Chloramphenicol, 428
Chloramphenicol-acetyl-transferases, 431
Chloroquine-loaded gold nanoparticle, 136
Choriocarcinoma, 218
Circulating blood elements, 446
Climate change resilience, 582, 590, 594
Clustered regularly interspaced short
　palindromic repeats (CRISPR), 433
CNT-based systems
　for cancer treatment, 103
CNTs-based biosensor electrodes, 498
CO_2 fixation, 712
Co-drug delivery, 132
Co-Fe oxide nanoparticles (CoFeNp), 528
Colony forming units (CFUs), 439
Combination therapy with nanomaterials, 104
Combined chemotherapy, 86
Commelina nudiflora extract, 381
Commercial CeO_2 NPs, 712–715, 717

Competitive advantage, 595
Complementary metal–oxide–semiconductors
　(CMOS), 252
Computed tomography (CT), 303
Conductimetric immunosensor, 572
Conducting polymers (CP), 500, 501,
　644, 652–654
Conservation agriculture (CA), 583, 584
Conservative drug carrier systems, 76
Continuous aqueous phase technique, 59
Continuous organic phase technique, 59
Continuous wave (CW), 372
Controlled and modified cytokine
　response, 442
Conventional cancer treatments, 86
Conventional chemotherapy, 86
Conventional drugs, 120, 149
Conventional farming practices, 583
Conventional mosquitocidal chemicals, 370
Conventional therapies, 303, 308
Conventional tillage, 585
Conventional waste water treatment
　methods, 477
Copper, 287
Copper and Copper oxide nanoparticles (Cu &
　CuO NPs), 407
Copper nanoparticles (CuNps), 373
Copper oxide nanoparticles (CuO NPs), 313,
　438, 439, 611, 612
　antifungal agents, 185
　broad-spectrum antibacterial activity, 185
　gram-positive bacteria, 185
　hydroxyl radicals, 185
　microbial cells, 185
Coronary artery disease (CAD), 453
Coulombic efficiency (CE), 659
Covalent conjugation method, 69
COVID-19
　nanonutraceuticals role, 284–286
CRISPR-mediated immune response, 433
Critical micelle concentration (CMC), 41
Crop production machinery, 582
Crop production technologies, 584
$CuO@CeO_2$ nanocatalyst, 731
Curcumin loaded exosomes, 446
Curcumin nanoparticle supplementation, 469
Cyclic-Arg-Gly-Asp (cRGD) peptide-
　conjugated polymeric micelles,
　122, 123
Cynodon dactylon, 314
Cysteine, 435
Cytidine-phosphate-guanosine (CpG), 444
Cytosol, 401

D

D-alanyl D-alanine portion, 427
Damascus' saber blades, 421
Degradation, 602
Dendrimer-doxorubicin conjugates, 129
Dendrimers, 44, 45, 70, 71, 95, 103, 229, 246, 265, 447, 502
   astramol dendrimers, 128
   cationic chlorambucil, 130
   dendrimer-surcumin conjugate, 129
   DHATX, 130
   as drug delivery agents, 129
   hydrophobic/hydrogen-bond and electrostatic, 128
   hyperbranched/brush polymers, 128
   J591 antibody-dendrimer conjugate, 130
   methotrexate-loaded polyether-copolyester, 130
   molecular structure, 128
   nanocapsules, 265
   nanoconjugate drug molecules, 266
   nanoparticles, 265
   PAMAM, 129
   polyamidoamine, 130
   trastuzumab-grafted PAMAM, 129
   virosomes, 266
Dendritic cells (DCs), 334
Dendronized polymers, 229
Dendropanax morbifera, 311
Dengue, 347, 359–361, 363, 370
Detoxification process, 275
Diabetes mellitus (DM), 452
Dietary copper (Cu), 471
Dietary *Curcuma longa* nanoparticles, 466
Dietary curcumin nanoparticles, 466, 469
Diets, 2
Differential scanning calorimetry (DSC), 128
Dihydroartemisinin (DHA), 353–354
Dimocarpus longan, 311
Disease transmission, 218
Distinctive biosensors, 252
Distorted synaptic functions, 445
Diuron, 572
DM treatment, nanotechnology insulin delivery, 453, 454
DNA biosensors, 505
DNA damage, 217, 219
DNA gyrase, 431
DNA mutations, 436
DNA nanotechnology, 424
DNA replication inhibitors, 428, 429
Docetaxel (DTX), 95, 97, 98, 103
Doxorubicin (DOX), 43, 94–102
Doxorubicin-loaded anti-HER2immunoliposomes, 127
Doxorubicin-loaded polylactide-poly(ethylene glycol) aptamer micelles, 121
Doxorubicin-loaded silica nanoparticles, 134
Doxycycline, 437
Drought/water shortages and management, 586
Drug delivery in skin cancer
   CNTs, 230
   dendrimers, 229
   liposomes, 228
   QDs, 230
   SLNs, 229
Drug delivery systems
   disadvantages of conventional, 30
   organic nanocarriers (*see* Organic nanocarriers)
   requirements, 30
Drug formulation and delivery, NPs, 476
Drug loading
   to Au NPs, 72–74
   to CNTs, 71, 72
   to dendrimers, 70, 71
   entrapment/encapsulation efficiency, 75
   to fullerenes, 72
   MSNs, 74
   to nanogels, 69, 70
   to polymeric nanoparticles, 58
      nanoencapsulation/entrapment techniques, 58
      synthesised by polymerization, 59–61
      synthesised from natural polymers, 65–67
      synthesised from synthetic polymers, 61, 63
   to SLNs, 68, 69
   synthesized NPs, 75
Drug–infused nanoparticles, 441
D-Tocopherol polyethylene glycol1000 succinate (TPGS), 96, 98
Dye-based sensitization, 689, 690

E

Ebola analysis, 252
Ebola infection, 245, 252
Ecologically "clean" metallic NPs, 590
Economical chemical reduction method, 374
Edwardsiella tarda infection, 476
Efflux pumps, 429, 430
Eimeria papillata-triggered infection, 405
Electrochemical analysis, 510

Index

Electrochemical biosensor, 507
Electrode fabrication, 490
Electrodeposition technique, 497
Electrodes, 543
Electrospinning technology, 635
Emulsification/solidification technique, 63
Emulsification/solvent diffusion (ESD), 63, 64
Emulsification/solvent evaporation method, 61, 62
Emulsification-evaporation method, 62
Emulsifiers, 267
Emulsion polymerization-continuous aqueous phase technique, 60
Emulsion polymerization-continuous organic phase technique, 59
Emulsion types, 263
Encapsulated DEET, 363
Encapsulation, 351, 362, 543
Endogenous macromolecular neuropeptides, 446
Endosomal pathway, 48
Endosome escape pathway, 45–46
Endotoxins, 432
Energy generation
  conventional and potential methods
    coal power generation, 668
    geothermal power generation, 668
    hydropower generation, 668
    natural gas power generation, 668
    nuclear power generation, 668
    solar power generation, 669
    tidal power generation, 668
    wastewater and biogas energy generation, 668
    wind power generation, 668
Engineered nanomaterials (ENM), 595, 626–628, 634
Engineering nanomaterials, 614
Enhanced immune responsiveness, 443
Enhanced mosquitocidal property, 388
Enhanced permeability and retention (EPR), 90, 91, 100, 451
Environmental challenges, 588
Environmental encumbrance, 622
Enzyme-based biosensors, 571, 572
Enzyme-linked immunosorbent assay (ELISA), 253
Enzymes, 169, 630
Epidermal growth factor (EGF), 129
Epidermal growth factor receptor (EGFR), 307
Epidermoid carcinoma (skin cancer)
  basal cell carcinoma, 223
  biosynthesized AgNPs, 226, 227
  curettage and drying up, 224
  effects of UV light, 224
  efficacy of nano treatment, 230–231
  malignant growths, 221
  melanoma, 221, 223
  metastasized, 221
  MTT test, 225–227
  nano drug delivery (*see* Drug delivery in skin cancer)
  nanoparticles, 221
  phytochemicals, 221
  radiation medicines, 225
  SCC, 221, 223
  signs and side effects, 223–224
  skin disease, 221
  sorts, skin growth, 222
  therapeutic treatment, 225
  treatment, 224
Epigenetics, 274
Erythromycin, 428
Erythromycin-resistant streptococcus pyogenes, 425
Escalated IgA levels, 444
Escherichia coli, 168, 170, 181, 185, 186, 189, 190, 333
Evolutionary economic models, 595
Ex situ strategies, 622
Exopolymeric substances (EPS), 167, 181
Extracellular polymer matrix (EPS), 169

F
Fenitrothion, 569
Fenton process, 602
Ferragels, 625
Filtration processes, 629
Fish nutrition, nanoparticles
  Aloe vera, 470
  Azolla microphylla-based, 470
  dietary copper, 471
  dietary curcumin, 466
  enhanced disease resistance, 470
  experiments, 466
  ginger, 470
  L. casei, 470
  MnO, 471
  nano-minerals/dietary minerals, 470
  non-specific immune defense mechanisms, 469
  role, 466–469
Flocculation, 544
Fluconazole, 245
Fluidity, 37

Fluorescence reverberation vitality transfer (FRET) silica nanoparticles, 250–251
Fluoroquinolone-resistant S. aureus, 425
Fluoroquinolones (FQ), 428, 429
Folic acid (FA), 91, 96, 98
Folic acid metabolism inhibitors, 429
Folic acid-graphene oxide-polyvinylpyrrolidone (FA-NGO-PVP), 146
Food fish demand, 464
Food-grade nanomaterials, 10
  lipids, 10
  nutraceuticals, 10
  SLNs, 10
Foods for special medical purposes (FSMPs), 3
Fourier transform infrared (FTIR), 379
FQ resistance, 429
Free radicals, 119
French medical procedure, 217
Fructus Amomi (cardamom) fruits, 379
FTIR spectra, 380
Fucoidan (marine polysaccharide)-coated gold nanoparticles (Fu-AuNPs), 475
Fucoidan mimetic glycopolymer-coated gold nanoparticle, 136
Fullerenes, 48, 49, 72
Functionalized CNTs (f-CNT), 47
Fundamental building structures, 582
Fungi, 243, 244, 359
Fungicides, 245
Fungus pathogenic species Candida albicans, 404
Fusarium oxysporum, 376

G

GA-CdTe nanocomposites, 143
Gambogic acid (GA), 143
Gel core nanoliposomes, 352
Gelatin nanoparticles, 65
Gemcitabine (GEM), 98, 99, 101, 132
Gemcitabine-loaded MSNs, 99
Genetic engineering, 388
Genetic tendency, 224
Germinoma, 218
Ginger nanoparticles, 470
Glioma, 218
Global food fish demand, 464
Glutathione, 280
Glycopeptides, 427
Glycoproteins, 494

Gold (Au) nanoparticles, 50–52, 72, 73, 101, 249, 250, 311, 331, 332, 405, 475, 476, 491, 492
  administration, 136
  advantages, 136
  biomedical applications, 136
  bio-nanotechnology research, 136
  biosensor device, 136
  chloroquine-loaded gold nanoparticle, 136
  combinational therapy, phytochemicals, 138–139
  covalent attachments and supramolecular assembly, 135
  dendrimer encapsulated, 136
  DNA labeled, 136
  fucoidan mimetic glycopolymer-coated, 136
  porphyran encapsulated, 136
  synthesis, 135
  in targeted cancer drug delivery, 137–138
  toward cancer therapy, 136–138
Gonadotropin-releasing hormone (GnRH), 336
Gram-negative microorganisms, 248
Gram-positive bacteria, 401, 425
Graphene, 49, 50, 497, 499
  wastewater remediation, 613
Graphene/graphite-based nanocomposites, 652
Graphene oxide (GO), 613, 644
Green-based nanoparticle, 350, 356, 357
Green chemistry, 735
Green chemistry/green route approach, 379
Green fabricated nanoparticles, 358
Green innovation technologies, 488
Green nanotechnology, 213, 215
Green revolution, 582, 596
Green synthesis approach, 376
Green synthesis of silver nanoparticles (G-AgNPs), 470
Green synthesis, NPs, 556
  bacterial synthesis, 557
  by fungi and yeast, 557–559
  phytosynthesis, 557, 558
Green synthesised silver nanoparticles, 357
Green technology, 556

H

Haber-Bosch process, 694
HA-CPT/CUR-NP nanoparticles, 125
Halloysite nanotubes (HNTs), 570
Hazardous materials, 622, 624
Health indices, 420
Heavy metals

sources, 603
water contamination, 603
Hematite (Fe_2O_3), 494
Heme oxygenase-1 (HO-1) inducible cytoprotective isoform, 280
Hepatitis B surface antigen (HBsAg), 443
Hepatitis B vaccine, 443
Hepatocellular carcinoma, 218
HER2-positive breast cancer cells, 128, 129
HER2-RQD nanoprobes, 144
Heterocycle, 710
   cerium-based catalysts
      commercial CeO_2 NPs, 712–715, 717
      nanostructured CeO_2 from biology-directed method, 728, 729
      nanostructured CeO_2 from co-precipitation method, 721–728
      nanostructured CeO_2 from polymer-directed method, 725, 726
      synthetic nano-CeO_2, 717–721
   CuO@CeO_2 nanocomposite, 730
   MgO@CeO_2 nanocatalyst, 732
Heterocyclic pharmaceuticals, 711
Heterogeneous catalysis
   CeO_2 as solid support, 742–745
Heterojunctions
   categories, 679
   metal-semiconductor forms, 684
   SC materials, 678
   SCs binary chalcogenides, 682
   SrTiO3, 683
   S-S type heterojunction, 685
   type 1 heterojunctions, 679
   type 2 heterojunctions, 680
   type 3 heterojunctions, 681
   types, 678
   Z-scheme photocatalysis, 681
High-energy ball milling process (HEBM), 373
Histone deacetylases (HDAC), 306
Hollow magnetic nanoparticle (HMNPs), 147, 148
Honeycomb monolith (HM), 740
Hormone therapy, 86
Hospital Associated/Acquired Infections (HAI), 246
Hot homogenization process, 68
Human epidermal growth factor receptor-2 (HER2) genes, 127, 128
Human interleukin (IL)-10 biosensor, 507
Human-manipulated waste management, 624
Humidity biosensor, 510
Hyaluronic acid (HA), 91

Hyaluronic acid (HA)-functionalized polymeric nanoparticle (HA-CPT/CUR-NPs), 124
Hydrocracking, 695
Hydrogels, 42, 447
Hydrogen generation
   applications
      in electronics industry, 696
      as feedstock, 694
      in fertilizers industries, 694, 695
      as a fuel, 696
      fuel industry, 695
      methanol production, 695
   earth's atmosphere, 694
   PEC water splitting, 672–674
   photobiological water splitting, 671, 672
   photocatalytic water splitting, 674, 675
   thermochemical water splitting, 670
Hydrophobic interactions, 9
Hydroprocessing, 695
Hydrothermal synthesis, 523
Hydrothermal treatment, 526
Hydroxypropyltrimethyl ammonium chloride chitosan (HACC), 189

I

IgA antibodies, 444
Immune checkpoint modulators, 448
Immune-stimulating complexes (ISCOMs), 444
Immunoassays, 252
Immuno-modulatory ingredients, 466
Immunosensor, 572
Immunotherapy, 85, 220, 448
In Gram-positive bacteria, 429
Incineration process, 631
Indian Alchemy principles, 421
Inducible nitric oxide synthase (iNOS), 440
Infections control, nanomaterials
   drug delivery systems (*see* Nanotechnology-based drug delivery systems)
   natural biological barriers, 436
   size manipulation, 436
   vaccines and immunostimulatory adjuvants (*see* Nanotechnology-based vaccines and immunostimulatory adjuvants)
Infectious agent, *see* Microbial pathogens
Infectious disease, 241–242, 425
Infectious salmon anaemia virus (ISAV), 476
Infield water management, 586

Inorganic nanocarriers
  Au NPs, 50–52
  core, 50
  MSNs, 56–58
  nanoshells, 52, 53
  QDs, 53–55
  shell, 50
  SPIONs (*see* Superparamagnetic iron-oxide nanoparticles (SPIONs))
Inorganic nanomaterials, 98, 411
  CNTs (*see* Carbon nanotubes (CNTs))
  GNPs (*see* Gold nanoparticles (GNPs))
  mesoporous silica nanoparticles, 98
  MSNs, 90–92
  QDs (*see* Quantum dots (QDs))
  types, 100
Inorganic nanoparticles, 149, 216
  as anticancer drug delivery vehicle, 120
  as nanocarriers (*see* Inorganic nanocarriers)
Inorganic pollutants, 601
Inrganic nanocarriers, 35
Insect repellents, 348, 351, 362, 363
Instrumental techniques, 269
Integrated soil fertility management (ISFM), 584
Intelligent nano-biosensors, 549
Intensive farming systems, 464
Interfacial deposition (ID), 63
Interfacial polymerization technique, 61
Interferon gamma activation, 441
International Agency for Research on Cancer, 302
Intracellular infections, 441
Intuitive immunological resistance system, 251
Iron oxide nanoparticles, 494, 495
  wastewater remediation
    hematite (α-Fe_2O_3), 609
    maghemite (γ-Fe_2O_3), 610
    magnetite (Fe_3O_4), 610
Irresistible illnesses, 246

K
Killing factors, 434
Kirsten rat sarcoma viral oncogene (KRAS), 307

L
Lactic-co-glycolic acid (PLGA), 447
Lactobacillus casei, 470
Lactoferrin-PLGA, 98
Large-scale commercial counterparts, 586

Lectin-conjugated silica nanoparticle, 135
Lethal concentration (LC50), 382
Leuconostoc lactis bacterial strains, 376
Leukemia, 218
Life processes, 423
Lifestyle-associated diseases, 6
Lignin (LG), 531
Limited surface plasmon (LSP), 250
Lipid-based amphiphilic drug delivery systems
  nanoliposomes, 37–38
  SLNs (*see* Solid lipid nanoparticles (SLNs))
Lipid-based NPs, 446
Lipopolysaccharides (LPS), 433
Liposoluble drugs, 61
Liposomal chemotherapeutic medications, 228
Liposomal delivery, 16
Liposomes, 15, 16, 37, 92, 94, 95, 228, 246, 264, 268, 272, 335, 336
  advantages, 15
  in cancer treatment, 127–128
  drug delivery system, 127
  drug distributing vesicles, 126
  drug formulations, 127
  large unilamellar vesicle, 126
  multilamellar vesicles, 126
  small unilamellar vesicle, 126
Liquid nitrogen, 225
Listeria monocytogenes, 189
Lithography, 422
Long-distance electrical signal conduction, 410
Lower critical solution temperature (LCST), 41
Lung cancer theranostics
  Ag NPs, 311
  anxiety, 302
  Au NPs, 311
  biogenic nanoparticles, 316, 317
  biogenic synthesis, 312, 313
  biological applications, 317
  cancer deaths, 302
  challenges, 307, 308
  chitosan nanoparticles, 315
  classification, 304
  conventional therapies, 303
  CuONPs, 313
  depression, 302
  diagnosis technologies, 302
  implementation, 303
  limitations, 307, 308
  long-term tobacco smoking and exposure, 302
  magnetic nanoparticles, 312, 313
  metastases, 303

Index   771

molecular biology, 304
MPM, 306, 307
nanoparticle-mediated gene delivery, 304
nanoparticles (NPs), 303
nanoscale formulations, 303
nanoscale materials, 303
nanotechnology, 303, 309, 310
NP synthesis, 303
NSCLC, 305, 306
polyherbal nanoparticles, 315, 316
SCLC, 304
stress, 302
TiO_2NPs, 314
treatment strategies, 302
Lycurgus cup, 420
Lymphoma, 218
Lysogenic module, 432
Lytic module, 432

M

Macrolides, 428
Macromolecular drug delivery, 446
Maghemite (γ-Fe_2O_3), 610
Maghemite nanotubes (MHNT), 525
Magnesium, 286, 439, 440
Magnesium oxide (MgO), 529, 530
Magnetic hyperthermia, 147, 148
Magnetic nanoparticles (MNPs), 87, 98, 102
    in cancer treatments, 147–149
    for drug delivery, 148
    magnetic elements, 147
    magnetic hyperthermia, 147
    photothermal impact, 147
Magnetic resonance imaging (MRI), 102, 103, 302–303
Magnetite (Fe_3O_4), 494, 610
Magnolia officinalis, 311
Malignancy, 218
Malignant growth, 217, 218, 220
    disease transmission, 218
Malignant pleural mesothelioma (MPM), 306, 307
Malnourished animals grazing, 583
Mammography, 217
Manganese oxide (MnO), 471, 496
Marsdenia tenacissima, 311
Materialization, 421
Materials science, 214, 215
Matrix Metallo Proteinases (MMP), 127
Matrix-assisted laser-desorption ionization-mass spectrometry (MALDI-MS), 128

Mean inhibitory concentration (MIC), 437
Medical procedure, 217, 220, 225
Medicinal chemistry, 710
Medicinal services practice, 253
Melanoma, 222, 223
Membrane coating, 634
Mesoporous silica nanoparticles (MSN), 56–58, 74, 87, 98–100, 521
Mesoporous TiO_2NPs, 564
Mesothelioma, 218
Metal compound nanomaterials, 490
Metallic nanoparticles, 165
    Ag-NP, 438
    copper, 438, 439
    magnesium, 439, 440
    TiO_2, 439
    zinc, 440
Metallic NPs applications
    antitumor properties, 410
    dental materials, 410
    food management, 411
    wastewater management, 412
Metallic silver, 214
Metal nanoparticles (MNPs), 490, 492, 509
    large-scale fabrication, 387
    size and morphology, 371
    synthesis, 590
Metal NPs synthesis approaches
    biological, 375–376
    chemical methods, 373–375
    fabrication, 371
    physical methods, 372–373
Metal oxide nanoparticles (MONs), 164, 169
    antibiotic efflux pumps, 193
    antifouling applications, 172
    antimicrobial activities, 172
    applications, 179–181
    bacterial cell membrane, 192
    bacterial strains, 192
    biofouling and biofilm formation, 172
    biological applications, 172
    biomedical applications, 172
    biosynthesis, 171, 172
    bottom-up approach, 173
    cell wall, 191
    chemical and hydrothermal synthesis, 170
    chloride/nitrate, 173
    coprecipitation synthesis, 171
    drying process, 174
    electrochemical synthesis, 171
    electron-hole pairs, 193
    Fenton reaction, 193
    hydroxyl radicals, 193

Metal oxide nanoparticles (MONs) (*cont.*)
  influence
    ammonia solution, 178
    annealing process, 180
    antibacterial activity, 175
    antifouling, 179
    antimicrobial studies, 175
    catalytic activity, 175
    copper oxides, 177
    crystal structure, 174
    crystallization process, 174
    CuO nanostructures, 178
    CuO structures, 177, 178
    DRS analysis, 175
    dry crystals, 175
    electronic structures, 179
    FE-SEM images, 177
    formation of crystal, 174
    growth process, 174
    hexamine, 177
    hydrothermal method, 175
    liquid/gas phase, 174
    mechanism, 175
    morphology, 175, 178
    nanomaterials, 175
    nanoparticles, 179
    nanorods, 175
    nanoscale, 179
    particles, 176, 179
    photocatalysis, 175
    physical and chemical properties, 179
    physical grinding method, 175, 176
    physiochemical properties, 178
    polymorphs, 178
    precipitation-decomposition method, 176
    p-type semiconductor, 179
    quantum dots (QDs), 179
    SnO_2 nanoparticles, 175
    sol-gel process, 175
    solvothermal approaches, 175
    solvothermal reaction, 178
    TiO_2 nanoparticles, 174, 176
    zinc nitrate and oxalic acid, 175
    ZnO NPs, 175
  intracellular enzymes, 193
  lipid peroxidation, 193
  mechanical and chemical processes, 172
  mechanism of action, 191, 192
  nanomaterials, 172
  nanoparticles, 174
  nanotechnology, 172
  nucleic acids, 193
  peroxyl radical interacts, 193
  photocatalytic action, 193
  physiochemical and biological properties, 172
  smaller-sized particles, 193
  sol-gel synthesis, 170, 174
  sonochemical synthesis, 170
  top-down process, 173
  toxicity, 192
  wet chemical synthesis, 171
Metal oxide-based nanocarriers, 35
Metal oxides nanoparticles (MONPs), 490, 491, 495
Metal-semiconductor heterojunction photocatalysis, 684
Metastasis, 217
Methanol, 695
Methicillin-resistant *S. aureus* (MRSA), 425, 438, 475
Methotrexate conjugate gold nanoparticle, 136
Methyl orange (MO), 528
Methyl red (MR), 528
MgO-halogen nanoparticle, 440
Micellar-enhanced filtration (MEF), 625
Micelle-like nanocarriers, 40
Micelles, 264
Microbial cells, 165
Microbial electrosynthesis (MES), 659
Microbial fuel cell (MFC) technology, 635
  catalyst materials, 647, 648
  oxygen reduction efficiency, 645
  PEM, 645
  performance, 646
  principle and architecture, 645
  types, 646
Microbial infections, 505
Microbial pathogens, 242
  bacteria, 243
  fungi, 243, 244
  prions, 244
  virus, 244
Microcystin-LR (MC-LR), 532
Microemulsion, 264, 521, 522
Microorganism-based nanoparticle, 358, 359
Microorganisms, 267, 509
MicroRNAs, 494
Microwave irradiation, 568
Minimum inhibitory concentration (MIC), 165
Mixed matrix membranes, 634
Mixed metal oxides, 729, 734
Modified carbon electrodes, 644
Modified nanomaterial-based anode, 658
Molecular engineering, 422

Index

Molecular imprinting, 501
Molecularly imprinted polymers (MIP), 501
Monometallic nanoparticles, 166
Mononuclear phagocytic system (MPS), 101
Mosquito-borne diseases, 381
Mosquitocidal activity, metal NPs
   AgNps, 383
   AuNps, 382, 383
   CuNps, 384
   enhanced inhibitory activity, 385
   FTIR spectra, 385
   phytosynthesized, 386–387
   selenium, 384
   spherical and polydispersed, 385
Mosquitocidal activity, phytosynthesized metal NPs
   aqueous nanoparticles, 387
   phytochemicals, 385
   synergistic effect, 385
   toxic reactions, 387
Mosquitocidal agents, 370
Mosquitoes, 370
Mucosal application, 442
Multicoloured silver nanoplates, 360
Multicomponent particles, 442
Multidrug delivery, 452
Multidrug-resistance bacteria, 412
Multidrug-resistance microbes, 398
Multidrug-resistant (MDR), 324, 412
Multidrug-resistant organisms, 407
Multidrug-resistant strains (MDR), 164
Multiwalled carbon nanotubes (MWCNTs), 329, 613
Multi-walled CNTs (MWCNTs), 569, 625
Myco-filtration, 589
Mycosynthesis, 589–593
Mycotoxins, 242, 245

N
Nanoadsorbents, 478
   aluminium oxide, 529
   catalysts, 524
   cobalt oxide, 527, 528
   inorganic adsorbents, 524
   iron oxides, 524–526
   metal-based, 524
   metallic oxides, 524
   MgO, 529, 530
   mixed metal oxides, 529, 530
   polymer-based, 530, 531
   silica- and carbon-based, 531
   titanium dioxide (TiO_2), 526, 527
   Zn oxide, 528
Nanoantibiotics
   antimicrobial activity, 407
   SiO_2 NPs, 408–409
Nano-assisted membranes, 634
Nano-based sensors, 488
Nano-based treatment, 398
Nanobiology
   instrumental advancement, 424
   nanocarriers predictive analysis, 424
   nanomaterials, 424
   nano-oncology, 424
   nanoscale, 423, 424
   NBIC, 424
Nanobiomembrane, 634
Nanobioremediation (NBR), 570, 632
   chemical decontamination, 546
   inorganic pollutants, 547
   organic contaminants, 546
Nanobiosensor, 361, 634
   applications
      in agriculture, 509–510
      in environmental monitoring, 508–509
      in food industry, 509
      in medical field, 506, 507
Nanobiotechnology, 555, 634, 635
   engineering and molecular biology, 346
   liposomes nanoparticles, 346
   nanomaterials, 346
Nano-blocks, 346
Nano-capsules, 453
Nanocarrier-mediated nutraceutical delivery, 269
Nanocarriers, 35, 119, 263, 326, 446
   anticancer activity against HeLa cells, 142
   biophysicochemical properties, 120
   and biosynthetic process, 268
   biopolymeric (see Biopolymeric nanocarriers)
   core-shell structured, 144
   drug camptothecin, 145
   liquid lipid preparations
      cubosomes, 264, 265
      hexosomes, 264, 265
      liposomes, 264
      micelles, 264
      microemulsions, 264
      nanoemulsions, 264
   polyvinyl caprolactam conjugate, 145
   ZnO QDs-conjugated gold nanoparticles, 142
Nano-catalysts, 522, 524, 531, 532
   heterogeneous catalysis, 711

Nanochannels, 346
Nano-combined bioremediation, 632
Nanocomposites, 166
    pesticides, 567, 568
Nano-crystalline silver particles, 214
NanoEconomy, 594
Nanoemulsions, 14, 15, 246, 254, 264, 443
    size, 14
Nano-encapsulated DEET, 363
Nanoencapsulated delivery system, 268
Nanoencapsulation, 267, 284
    minerals, 272
    probiotics, 272
    vitamins, 272
Nanoencapsulation technology, 471
Nanoengineered materials, 241
Nano-fertilizers
    classes, 588
    definition, 587
    environmental contamination and
        wastage, 587
    foliar spray, 588
    higher surface area, 587
    properties, 588
    quality parameters, 588
    research, 588, 589
Nanofibers, 346
Nanofibrous webs, 635
Nanofiltration, 622, 623, 629
Nanofluids, 634
Nanoformulated drugs, 270
Nanoformulations
    vitamins, 268
Nanogels, 69, 70
Nanographene
    conjugated inhibitors in cancer
        treatment, 145–147
    target-specific delivery, anticancer
        drug, 146
    2Dcarbon nanomaterial, 145
Nano-immuno activators
    dendrimers, 336–338
    liposomes, 335, 336
    VLPs, 336
Nano-informatics, 424
Nanoliposomes, 17, 37, 349, 352, 353
    advantages, 17
    benefit, 17
Nanolithography, 253
Nanomaterial-based biosensor electrodes, 490
Nanomaterial-based therapeutics, 232
Nanomaterials, 18, 19, 238, 545, 548, 556
    advantage, 510
    application, 19

automobile industries, 488
in bacterial detection
    AgNPs, 247–249
    Au nanoparticles, 249, 250
    fluorescent nanoparticles, 250–251
    LSP, 250
    magnetic nanoparticles, 246–247
    physicochemical and immunological
        techniques, 246
and biorecognition elements, 508
biological barriers, 326
in biosensors
    AgNPs, 492
    AuNPs, 491, 492
    iron oxide nanoparticles, 494, 495
    MnO, 496
    MNPs and MONPs, 490
    PdNPs, 493
    PtNPs, 492, 493
    QDs, 493, 494
    ZnO, 495, 496
cellulose, 19
ceramic glazes, 421
challenges, 510
complex manufacturing process, 479
conductive and semiconductor, 488
drug delivery carriers, 326
environmental science and technology, 488
in MFCs as cathode catalysts, 649
    anode modifications, 656–659
    graphene/graphite-based, 652
    metal-activated carbon hybrids, 650
    metal-carbon hybrid catalysts, 650
    metal-CNFs and CNTs, 651
    metal-CP-based, 652–654
    metal-graphene-based
        nanocomposite, 651–652
    potentiodynamic effects, 653, 655–657
    transition metal/metal oxide and
        alloy-based, 649–650
nano vaccine, 254
nanocarriers, 326
nanomedicine, 326, 327 (see also
    Nanoparticles (NPs))
nano-sized particles, 422
pathogenic microorganisms, 325
physical and chemical properties, 325
properties, 424
QS, 255, 256
quorum sensing, 255
research direction and development, 511
risk assessment, 633, 634
in scientific fields, 554
in solid waste management, 630, 631

Index

  in toxic chemicals and heavy metals
      remediation, 623, 624
  in viral detection
      approaches, 251
      biosensing methods, 253
      electrochemical biosensing, 252
      pathogens, 251
      SERS, 252
      unfavorable analysis, 251
      VLPs, 251
Nanomaterials role
  cancer therapy, 447–450
  DM treatment, nanotechnology, 452–454
  neurodegeneration (see
      Neurodegeneration)
Nanomedicine, 232, 279, 326, 327
  active targeting approach, 32
  conventional medicine, 30
  description, 30
  nanoparticle-based drug delivery, 32
  QD nanocarriers, 54
Nanomembranes
  wastewater remediation, 614
Nanometer
  size, 238
Nanometer-sized thin material, 262
Nano-micelles, 472
Nanominerals, 270
Nanonetwork polymers, 629
Nanonutraceutical carriers, 268
Nanonutraceutical formulations, 288
Nanonutraceutical research, 262
Nanonutraceuticals, 282, 283
  advantages, 271
  and anti-inflammatory activity, 278
  as antibacterial agents, 282–283
  in antiviral therapy, 283, 284
  in cancer, infectious diseases and
      inflammation, 274
  in chemotherapy of cancer, 274–277
  in COVID-19 therapy, 284–286
  in healthcare, 270
  in medical imaging, 279, 280
  in medicine, 270
  in prophylaxis, diagnosis and treatment of
      infectious diseases, 280–282
  nanophytochemicals, 273, 274
  pharmacokinetics, 270
  target-based drug delivery, 270
  toxicities, 287, 288
  vitamins and minerals, 270
Nanoparticle-based optical biosensors, 573
Nanoparticle-in-microsphere hybrid oral
    delivery system (NiMOS), 442
Nanoparticle-mediated toxicity, 399

Nanoparticles, 164, 309, 346, 548
  advantages, 262
  amalgamation, 213
  amalgamation, synthetic concoctions, 215
  antimicrobial property and activity, 399
  application industries, 399
  application, 8
  bio functionalization, 240
  biodegradable natural biopolymers, 8
  bio-electrochemical system, 656
  biofilm interaction, 410
  captivating application, 239
  characterization, 214
  chitosan, 332, 333
  classification, 35
  colossal specific surface zone, 216
  control disease-causing pathogens, 398
  deposition, 653
  ecological non-harmful manufactured
      conventions, 213
  extraordinary physicochemical
      characteristics, 214
  green synthesis, 120
  green-based nanoparticle, 350, 356, 357
  history, 240
  in industry frameworks, 239
  infectious diseases, 334, 335
  mechanism, 398
  metal NPs, 239
  microorganism-based, 358, 359
  MnO_2, 655
  morphological and physicochemical
      properties, 398
  MO_x, 649
  organic/inorganic materials, 262
  PEO-PCL, 124–126
  pharmacological properties, 464
  physical, compound and organic, 213
  physicochemical properties, 239
  physiochemical characteristics, 30
  PLLA, 333, 334
  properties, 216
  quantum dots, 7
  in remediation
      of inorganic contaminants, 628
      of organic contaminants, 626–627
  shapes, 31
  size, 30
  smaller-size, 120
  stability and storage, 75–76
  synthesis, 7
  top-down methods, 8
  TRC-NP, 124, 125
  types, 216
Nanoparticles manufacturing approaches, 479

Nanopesticides
  application, 588
  conventional pesticides, 588
  sustainable agriculture, 588
Nano-pharmaceuticals, 423
Nanophytochemicals, 273, 274
Nanopores, 346
Nanoproducts, 213
Nanoproteomics, 634
Nanoremediation, 602
Nanoscale biomaterials, 502
  aptamers, 503–505
  chitosan, 503
  DNA biosensors, 505
Nanoscale devices, 423
Nanoscale drug delivery strategies
  cellular internalization, drug
    nanocarriers, 32
  particles, 31
  passive and active targeting, 32, 33
  release of drug, 34
  SDDS, 34
Nanoscience, 212, 240, 241, 262, 595
  biology application, 423
  evolution, 420
  medical sciences, 420
  opportunities and possibilities, 424
Nanosensors, 241, 488
Nanoshells, 52, 53
Nano-sized biomaterials, 502
Nanosized liposomes, 37
Nanospheres, 453
Nanostructured channels, 241
Nanostructured lipid carriers (NLCs), 13, 22, 263, 264
  advantages, 13
  organic solvents, 13
  production, 13
  SLNs, 13
  synthesis, 14
  types, 13
Nanostructured materials, 240, 488, 489
Nanosuspensions, 350, 353, 354
Nanosystems, 240
Nanotechnologies in agriculture
  agricultural resilience, 586
  agricultural technologies, 586
  nano-fertilizers, 587, 588
  nanopesticides, 588
  schematic representation, 587
  socioeconomic implications, 594–595
Nanotechnology, 164, 212, 238, 241, 277, 309, 310, 487, 488
  AgNPs, 326, 328, 329
  antimicrobial agents, 324
  aspects, 454
  AuNPs, 331, 332
  biology and medicine applications, 455
  biomedical applications, 271
  biomedical tools, 324
  branches, 555
  carbon-based nanoparticles, 328, 329
  CNTs, 329, 330
  definition, 464
  in energy conversion and energy generation
    batteries, 667
    fuel and solar cell, 667
    geothermal, 667
    in energy applications, 667
    nanomembrane and nano-sieve, 667
    nuclear reactions, 667
    water technology, 667
  evolution, 421
  formulations, 324
  fullerenes, 330, 331
  generations, 422
  green engineering, 622
  green innovation technologies, 488
  human-manipulated waste
    management, 624
  levels, 238
  malignant growth, 220
  materials science, 215
  MDR, 324
  microorganisms, 324
  multidrug-resistant pathogens, 325
  nanometer, 324
  nanoparticles, 324
  nanoscience, 324
  opportunities, 421
  pathogenic microorganisms, 324
  size comparison, 239
  size of nanometer, 239
  vaccine deisgn, 454
  wastewater treatment technologies, 602
  with bioremediation, 631, 632
Nanotechnology application in aquaculture
  antibacterial metal nanoparticles, 473–476
  drug formulation and delivery, NPs, 476
  schematic representation, 465
  water quality management, 477–478
Nanotechnology applications in aquafeed industry
  bioactive/nutrient nano-delivry, 471
  bioavailability and stability, 472
  micronutrients delivery, 471

Index   777

nanomaterials, 471, 472
nanoscale mineral supplements, 472
precursors, 471
types, 471
vitamin B12 absorption, 471
Nanotechnology applications in fish nutrition
   aquafeed industry, 471–472
   bioactive agents, 466
   conventional delivery systems, 466
   delivery systems, 464
   feed/food delivery principles, 465
   feeding fish, 464
   nanoparticles, 466–471
Nanotechnology-based cancer treatment
   active targtieng, 450
   multidrug delivery, 452
   nanocarriers, 450
   passive targtieng, 450
   photothermal targeting, 451–452
Nanotechnology-based drug delivery systems
   chitosan, 437
   drug–infused NPs, 441
   immunomodulatory effects, 441, 442
   metallic nanoparticles, 437–440
   NO–NPs, 440
Nanotechnology-based nutraceuticals, 10
Nanotechnology-based vaccines and immunostimulatory adjuvants
   chitosan, 444, 445
   CpG, 444
   ISCOMs, 444
   metallic nanoparticles, 445
   nanoemulsions, 443
   synthetic polymers, 442, 443
Nanotechnology-derived devices, 594
Nanotherapuetics, 220–221
Nanotube-based electrochemical biosensors, 573
Nanotubes, 568, 569
Nanovitamins, 270
Nanowaste, 634, 635
Nanowires, 488, 495, 496, 500, 501, 509
Nascent peptide exit tunnel (NPET), 428
Near-infrared (NIR), 188
Neodymium-doped yttrium aluminium garnet (Nd:YAG), 372
Nephrotoxicity, 308
Neurodegeneration
   BBB, 446
   definition, 445
   dendrimers, 447
   hydrogels, 447
   lipid-based NPs, 446
   macromolecular drug delivery, 446

metal NPs, 446
NDs, 445
pathological events, 445
polymeric NPs, 447
RMT, 446
Neurodegenerative diseases (NDs), 445
Neutraceuticals
   induction of metabolism, 282
N-isopropylacrylamide (NIPAAM), 124, 125
Nitric oxide (NO), 440
Nitride-based binary photocatalyst, 682
NO–releasing nanoparticles (NO–NPs), 440
Nomaterials-based drug delivery systems (nano DDS)
   angiogenesis and tumor vasculatures, 90
   application, 87
   barticle size distribution, 88
   controlled drug release, 89
   drug loading, 88
   high drug payload, 88
   nanoparticles, 88
   surface modification/coating, 89
   tumor-targeted drug delivery, 90–92
Non zero-valent iron (nZVI) nanoparticles, 622, 623, 625, 630, 632
Non-antibiotic drug, 434
Non-antibiotic treatments, infections
   antibacterial activities, 434
   bacteriocins, 433, 434
   killing factors, 434
   phage therapy, 431–433
   quorum quenching, 435
Non-antibiotics, 434
Non-communicable diseases, 420, 455
Non-small cell lung cancer (NSCLC), 302, 305, 306
Nonsurgical cancer treatment, 85
Nosocomial diseases (NI), 246
Novel nanotechnology-based drug delivery systems, 31
NP-based drugs, 410
NPs antibacterial activity mechanisms
   cell wall damage, 436
   direct absorption, 435
   ROS production, 436
NPs antimicrobial activity
   AgNPs, 402
   Al_2O_3 NPs, 406
   AuNPs, 405
   CaO NPs, 406
   Cu & CuO NPs, 407
   mode of action, 403
   ZnO NPs, 405
N-trimethyl chitosan (TMC), 444

Nuclear factor kappa B (NFκB), 121, 124, 125
Nuclear magnetic resonance (NMR), 128
Nutraceutical delivery systems, 9
Nutraceuticals, 3, 5, 266, 267, 278, 280
   advantages, nanosizing, 269
   animal sources, 6
   beverages, 266
   bioactive compounds, 7
   bioactive phytochemicals, 266
   bioavailability, 6
   classification, 4, 266
   consumption, 3
   daily consumption, 280
   dietary supplements, 266
   FSMPs, 4
   functional foods, 266
   global nutraceutical market, 6
   immunomodulating effect, 285
   limitations, 269
   materials for encapsulation, 267
   metabolites, 5
   microbials, 6
   nanoformulation, 269
   phytochemicals possess, 5
   primary food elements, 266
   role, 3
   selection of nanocarriers, 267
   sources, 5
   vitamins and minerals, 5
N-vinyl-2-pyrrolidone (VP), 124, 125

O

Organic agriculture-based systems, 593
Organic contaminants, 602
Organic dyes, 603, 611, 612
Organic matter, 629, 634
Organic nanocarriers, 35
   amphiphilic macromolecules, 36
   amphiphilic systems, 36
   carbon-based, 35
   drug delivery systems
      carbon-based (*see* Carbon-based drug delivery systems)
      lipid-based amphiphilic (*see* Lipid-based amphiphilic drug delivery systems)
      polymer-based (*see* Polymer-based drug delivery systems)
   hydrophilic regions, 36
   lipophilic regions, 36
   metal oxide-based, 35
   self-assembly processes, 36

Organic nanomaterials, 92
   dendrimers, 95, 97
   liposomes, 92, 94, 95
   polymeric micelles, 94–96
   polymeric nanoparticles, 96–98
   SLNs (*see* Solid lipid nanoparticles (SLNs))
   types, 93
Organic nanoparticles, 36
   as anticancer drug delivery vehicle, 120
   as nanocarriers (*see* Organic nanocarriers)
Organic pollutants, 601, 606, 614
Organic polymer membranes, 614
Oxazolidinones, 428
Oxidation, 602
Oxidative stress-mediated cell death, 401
Oxide-based binary photocatalyst, 681
Oxygen reduction into water, 644
Oxygen reduction reaction (ORR), 644
   in alkaline and acidic electrolyte, 647
   cathodic ORR, 646
   chronoamperometric results, 656
   in CNTs and CNFs, 651
   in dual-chambered MFC, 655
   multi-electron transfer reaction, 647

P

Paclitaxel (PTX), 94–96, 98, 99, 101, 132
Paclitaxel co-drug delivery, 132
Palladium nanoparticles (PdNPs), 493
PAMAM-doxorubicin conjugate, 128
PAMAM-paclitaxel-conjugated omega-3 fatty acid, 130
Parkinson's disease, 15
Passive drug targeting, 91
Passive targeting, 450
Passive transport, 441
Passive tumor targeting, 91
Pathogenic bacteria, 472
Pathogenic diseases, 241
Pathogenic organisms, 478
Pathogenic viral maladies, 244
Pathogen-specific molecular patterns (PAMPs), 441
Pathophysiological mechanism, 274
Pattern recognition receptors (PRPs), 441
PBP modification, 430
PEDOT:PSS-PEG nanoparticles, 125
Peltophorum pterocarpum, 379
Penicillin binding proteins (PBPs), 426
Persuasive anti-malarial drugs, 246
Pesticide detection, 559

Index

Pesticides, 602
  biosensors for detection (*see* Biosensors)
  in modern agriculture, 554
  nanobiocomposites, 568
  nanocomposites, 567, 568
  nanotubes
    CNTs, 569
    fenitrothion, 569
    HNTs, 570
    MWCNTs, 569
    SWCNTs, 569
  NPs in pesticide remediation, 556
Pesticides bioremediation
  bimetallic NPs, 563–565
  metal NPs, 560
    AgNPs, 561, 562
    AuNPs, 561
    FeNPs, 562, 563
  metal oxide NPs, 564
    iron oxide (Fe_2O_3), 566
    silicon oxide, 566, 567
    TiO_2NPs, 564
    ZnO NPs, 566
Phage DNA, 432
Phage therapy, 431
  antibiotics, 432
  armamentarium, 432
  bacterial infections, 432
  bacteriophages, 432
  clinical trials, 432
  disadvantage, 432
  DNA methylation pattern, 433
  endotoxins, 432
  inhibit bacterial restriction enzyme, 432
  multiplication free cycles, 432
Pharmacodynamics
  probiotics, 273
Phenyldiazenyl-acridinedione-carboxylic acid derivatives, 733
Phosphatidylinositol 3-kinase (P13K), 307
Photobiological water splitting, 671, 672
Photocatalysis, 604
  agglomerate and particle dimension, 690
  endothermic and multi-electron process, 675
  hydrogen generation efficiency, 677
  zero net force, 676
  Z-scheme photocatalysis, 681
Photocatalysts, 604, 605, 609, 611, 612, 614
  modifications
    co-catalyst loading, 687–688
    dye-based sensitization, 689, 690
    metal and non-metal implantation, 686, 687

Photocatalytic degradation, 622, 623, 627, 630, 631, 635
Photocatalytic dye-degradation, 690
Photocatalytic hydrogen generation
  operating conditions
    co-catalyst addition, 691
    particle size, 690
    pH effect, 691
    reactor temperature, 690
    surface area, 690
Photocatalytic reactors
  slurry-type, 692
  thin-film-type, 692
  types of reactors, 692
Photocatalytic water splitting, 674, 675, 693
Photodegradable TiO_2 nanocomposite, 631
Photodynamic treatment, 220
Photoelectrochemical (PEC) water splitting, 672–674
Photosensitizer, 37
Photosensitizing agents, 220
Photo-stimulated nanoliposomes (PNLs), 37
Photothermal agents, 451
Photothermal targeting
  carbon materials, 452
  description, 451
  electromagnetic spectrum, 451
  EPR effect, 451
  nanotubes, 452
  photothermal agents, 451
  PTT, 451
  SERS, 451
  SPR, 451
  surface free electron resonance, 452
  tumour mass, 451
Photothermal therapy (PTT), 95, 96, 104, 145, 331, 451
Photovolatilization, 545
Phthalocyanine ruthenium (RuPc), 569
Physical NPs synthesis methods
  ALD methods, 373
  AuNps morphology, 372
  electrical/thermal energy pressure, 372
  laser ablation method, 372
  microwave irradiation method, 372
  PtNps, 373
  surface contaminant-free AgNps, 372
  swift reaction time, 372
  synthesized novel hexagonal AuNps, 372
  types, 372
  wet milling process, 373
Physicochemical nanoparticle synthesis approaches, 376
Phytochemical constituents, 287

Phytochemicals, 273, 276, 277, 370
Phytonutrients, 4
Phytoremediation, 545, 605
Phytoremedies, 454
Phytosynthesis approaches, metallic NPs
   AgNPs, 380
   AuNPs, 379
   CuNPs, 380, 381
   green chemistry, 379
   green synthesis, 378
   iron nanoparticles, 381
   palladium nanoparticles, 381
   phytochemicals, 378
   plant biomass/extracts, 378
Phytosynthesized metal nanoparticles, 388
Plant alkaloids, 220
Plant extracts, 213
Plant illnesses, 242
Plant pathogens, 584
Plant pest management
   animal pests, 584
   conservation agriculture systems, 584
   conventional tillage, 585
   economic losses, 584
   herbicides, 585
   labor shortages, 585
   late weeding, 585
   N fertilizer, 585
   nanotechnologies, 585
   plant pathogens, 584
   smallholder farming, 584
   soil fertility improvement, 584
Plasmid DNA vaccine (pDNA), 476
Platinum (Pt), 646
Platinum nanomaterials (PtNPs), 492, 493
Platinum nanoparticles (PtNps), 375
Pluronic123, 57
Pollutant, 518, 531, 532, 607, 609
Poly (3,4 ethylenedioxythiophene)/polystyrene sulfonate (PEDOT/PSS), 531
Poly (carboxybetaine methacrylate) (pCBMA), 189
Poly (D, L-lactide-co- glycolide) (PLGA), 40, 41
Poly (D, L-lactide-co-glycolide) (PLGA), 98
Poly (D,L-lactide-co-glycolide) (PLGA) nanoparticles, 126
Poly (ethylene glycol)monoacrylate (PEG-A), 124, 125
Poly (ethylene oxide)-modified poly (epsilon-caprolactone) (PEO-PCL), 124–126
Poly (ethylene) glycol-modified urethane acrylate (PUMA), 629

Poly (glycolic acid) (PGA), 40
Poly (L-lactic acid), 333, 334
Poly (N-isopropylacrylamide) (PNIPAm), 41
Poly (propylamine) (PPI), 95
Poly (sulfobetaine methacrylate) (pSBMA), 189
Poly (ε-caprolactone) (PCL), 98
Poly 3,4-ethylenedioxythiophene-poly styrenesulfonate (PEDOT:PSS), 124
Poly acrylic-supported nZVI, 625
Poly ε-caprolactone polymers, 442
Poly(3,4-ethylenedioxythiophene) (PEDOT), 501
Poly(amidoamine) PAMAM dendrimers, 71
Poly(lactic-co-glycolic acid) (PLGA) acid, 149
Poly(lactide-co-glycolide) (PLGA), 443
Polyamidoamine (PAMAM), 44, 45, 71, 95, 97
Polyamidoamine dendrimers, 129, 130, 148
Polyaniline (PANI) nanomaterials, 501
Polycyclic aromatic hydrocarbons (PAHs), 624
Polyelectrolytes, 22
Polyethylene (glycol) (PEG), 40, 41, 47, 73, 89, 92, 95, 97–99, 101, 103, 189
Polyethylene glycol-phosphatidyl ethanolamine (PEG-PE), 121, 122
Polyethyleneimine (PEI), 375
Polylactic acid (PLA), 98, 447
   composition, 18
   properties, 18
   synthetic polymers, 18
Poly-lactic glycolipids acid (PLGA), 476
Polylactide (PLA), 40, 41
Polylactide-co-glycolide (PLGA), 132
Poly-l-glutamic acid-gemcitabine conjugate, 132–133
Poly-l-lysine (PLL), 95, 97
Polymer-based drug delivery systems
   alginate, 40
   dendrimer-polymeric nanocarriers, 44, 45
   micelle-like nanocarriers, 40, 41
   nanogels, 42–44
   polymersomes, 41
   synthetic polymers, 40
Polymer-based vesicles, 41
Polymer-drug conjugates (PDCs), 87, 92, 97, 99
   cancer drugs, 132
   for drug delivery, 131
   nano medicines, 131
   types, 131

Index

Polymer electrolyte membrane (PEM), 645
Polymer nanocomposites, 170
　antibiofilm/antifouling coatings, 182
　antimicrobial surfaces, 181
　antimicrobial/antifouling coatings, 182, 183
　biopolymers, 181
　hydrophilic nanoparticles, 182
　metal oxide nanoparticles, 181
　nanofillers (NPs), 181
　nanoparticles, 181
　physical and chemical reactions, 181
　polymer matrix, 182
Polymer nanoparticles, 447
Polymer polylactide-co-glycolide, 133
Polymeric calcium phosphate-polyethylene glycol, 453
Polymeric micelles (PM), 94, 96, 265
　anticancer delivery vehicles, 121
　blood-stable, 123
　cisplatin-conjugated, 121
　combinational anticancer effect, 121
　cRGD peptide-conjugated, 123
　genexol, 121
　glucose transporter-1 and glutathione, 121
　naringin, 121
　non-covalent interactions, 121
　PEG-PE, 121
　pH-responsive, 123
　SN-38-incorporating polymeric micelles, 121
　vitamin E tumor-targeted immunomicelles, 121
　with mitoxantrone, 123
Polymeric nanogels, 42–44
Polymeric nanomaterials
　biofunctionalization, 499
　conducting polymers, 500, 501
　dendrimers, 502
　molecular imprinting, 501
Polymeric nanoparticles, 96–98, 354, 355, 442, 453
　approaches, 123
　in cancer treatments, 124–126
　potent applications, 123
　synthetic polymers, 123
Polymeric oligonucleotides, 443
Polymer-insulin matrix, 453
Polymers
　natural polymers, 130
　PDCs (*see* Polymer drug conjugates (PDCs))
　synthetic polymers, 130
Polymersomes, 41

Polymethyl methacrylate (PMMA), 59, 191, 443
Polyol synthesis, 522–524
Polypropylene imine (PPI), 128
Polypropylene-grafted polyacrylic acid (PP-gPAAc), 190
Polypyrrole (PPy), 500, 531
Polysaccharides, 21, 267
Polyvinylpyrrolidone-coated iron oxide NPs, 632
Poor soil fertility, 583
Positron emission tomography (PET), 103
Povel platinum nanoparticles (PtNPs), 373
PPI dendrimers, 128, 129
PPy nanostructures, 500
Precautionary regulatory frameworks, 595
Precision engineering, 422
Primaquine (PQ), 355
Primary food elements, 266
Prions, 244, 445
Pristine graphene, 651
Probiotics, 277
　in COVID therapy, 287
Production challenges, 586
Protein biosynthesis inhibition
　AG, 427
　bacterial DNA, 427
　chloramphenicol, 428
　macrolides, 428
　oxazolidinones, 428
　ribosomes and myriads, 427
Protein food, 267
Proteins, 20, 21
Proton exchange membrane (PEM), 645
Proto-oncogenes, 219
Punica granatum, 312
Purpose-of-care (POC) tests, 240, 252
Pytosynthesized metal nanoparticles, 387

Q
QD-Apt(Dox) complex, 55
Quantum dots (QDs), 53–55, 73, 74, 87, 99, 103, 230, 489, 493, 494
　as high-energy photons, 141
　bio labeling and biosensing, 141
　conjugated inhibitors in cancer treatment, 143–145
　cost-effective method, 144
　drug delivery and diagnostic systems, 142, 143
　semiconductor core, 141
　ZnO quantum dot-conjugated, 144
Quantum effects, 216

Quaternized chitosan (QCh), 333
Quercetin-loaded mesoporous silica
  nanoparticles, 135
Quinolones, 428
Quinoxalines, 740, 741
Quorum quenching (QQ), 255, 435
Quorum sensing (QS), 240, 255, 256, 435

R

Radiation therapy, 85, 308
Radio frequency-mediated delivery, 441
Rasasastra, 421
Reactive oxygen species (ROS), 190, 272,
  273, 278, 328, 331, 436
Receptor–ligand interactions, 400
Receptor-mediated transcytosis (RMT), 446
Recirculatory aquaculture systems (RAS), 464
Recombinant HBsAg–NE-based intranasal
  vaccine, 443
Registration, Evaluation, Authorisation and
  Restriction of Chemicals
  (REACH), 633
Remediation, 542
  wastewater (see Wastewater remediation)
Remediation mechanism, 542, 602
Renewable energies, 644, 669
Renewable power-generating techniques, 693
Restorative plants, 213
Reticuloendothelial system (RES), 92, 101
Reverse osmosis, 622, 629, 634
Risk assessment, 633, 634
RNA aptamers, 505
Rock phosphates, 583
ROS production, 436

S

Salting-out technique, 64
Sarcoma, 218
Scaffolded DNA origami, 424
Scanning electron microscope (SEM), 128
Screening techniques, 217
SEDDS (nanoemulsion/microemulsion/lipid
  concentrate), 264
Selenium, 286
Selenium nanoparticles, 384
Self and non-self-DNA, 433
Self-assembling DNA nanostructures, 424
Semiconductor photocatalyst, 678
Semiconductor-semiconductor heterojunction
  photocatalyst, 685
Shewanella oneidensis, 331
Siberian sturgeon (*Acipenser baerii*), 470

Siddha medicine aphorism, 2
Signal processor, 489
Silica nanoparticles (SiO_2 NPs), 99
  Ag-Si nanocomposite, 408
  antibacterial property, 409
  antimicrobial activity, 408
  as drug delivery systems, 134
  as silicon dioxide nanoparticles/nano
    silica, 133
  Stober method, 133
  synthesized particle, 133
  toward cancer therapy, 133–135
  with snake venom, 135
Silica NPs conjugated with tetracycline
  antibiotic, 409
Silver, 214
  dietary confirmation, 214
  inhibitory effect, 215
  nanoparticles, 213, 227
  solicitations, 214
  unadulterated silver, 214
Silver nanoparticle (SNP/Ag_2O), 311, 326,
  328, 329, 374, 438, 473, 475, 492
  antibacterial agent, 247
  antibacterial systems, 248
  antibiofilm activity, 184
  antibiofilm disturbance, 249
  antimicrobial action, 248
  antimicrobial activity, 184
  antimicrobial effects, 183
  communication, 249
  functionalization, 248
  glass ionomer cements, 184
  mechanism of action, 184
  microbial cells, 183
  multidrug-resistant strains, 182
  nano silver toxicity, 183
  polymer matrix, 184
  polysaccharides, 184
  prophylactic properties, 184
  self-assembled monolayers, 183
  small-sized/nano silver, 184
Single-walled carbon nanotube (SWCNT),
  181, 329, 569
Site-specific/targeted drug delivery, 76
Skin cancer detection, 227, 228
Skin malignancies, 222
Skin medicines, 215
Small cell lung cancer (SCLC), 302, 304
Small interfering RNA (siRNA), 228, 452
Smallholder farmer livelihoods, 585
Smart drug delivery, 9
Smart stimuli-responsive drug delivery
  systems (SDDS), 34

SMEDDS (nanoemulsion/microemulsion/lipid concentrate), 264
Soil fertility management
 abandon farming, 583
 CA, 583
 commercial crop production, 583
 conventional farming practices, 583
 cultivars/germplasm, 584
 inorganic fertilizers, 583
 interventions, 583
 ISFM, 584
 legumes, 583
 malnourished animals grazing, 583
 organic sources, 583
Soil humidity, 510
Soil organic carbon, 593
Soil remediation
 bentonite, 543
 biological approach, 545
 biological remediation, 542
 biological treatment, 542
 conventional methods, 542
 electrokinetic technique, 543
 flocculation, 544
 ion exchange, 544
 microbes, 545
 nanomaterials, 546
 osmotic pressure, 548
 physical methods, 543
 precipitant, 544
 pump-and-treat technology, 543
 solid/liquid phase, 544
 stabilization, 545
 vitrification, 543
Soil structural stability, 591
Soil–water conservation practices, 586
Solanum mammosum, 380
Sol-gel method, 521, 522, 548
Solid lipid nanoparticles (SLNs), 35, 68, 69, 87, 88, 92, 93, 96, 229, 264
 anticancer drug doxorubicin, 40
 drug loading models, 38–39
 innovative colloidal lipospheres, 38
 lipospheres, 38
 models of drug loading, 39
 pH-responsive cholesterol-polyethylene glycol-coated, 40
 physiological lipids, 38
 stabilizing surfactants, 38
 stimulus-responsive, 38
 thermoresponsive, 39
Solid lipid nanoparticles-encapsulated 6-coumarin-loaded pectin microparticles, 470
Solid-lipid nanoparticles (SLNs), 10, 12, 22

Solvent displacement (SD), 63
Solvothermal/hydrothermal process, 174
Spinel ferrites, 606, 610
Squamous cell carcinoma (SCC), 221, 223, 224
Staphylococcus aureus, 189, 333
Stem cell transplant, 449
Stimuli-responsive nanomaterials, 104
Stimulus-responsive drug release, 37, 43, 45, 47, 49, 56, 76
Stimulus-responsive nanogels, 42, 43
Stimulus-responsive polymersomes, 41
Stimulus-responsive prototypes, 34
Stober method, 133
Streptococcus aureus, 166, 168, 170, 183–187, 190, 193
Streptomyces, 428
Sulpha drugs, 429
Sulphonamides, 429
Superbugs, 425
Super-paramagnetic iron oxide (SPION), 446
Superparamagnetic iron-oxide nanoparticles (SPIONs), 56
Superparamagnetic nanoparticles, 147
Surface enhanced Raman scattering (SERS), 451
Surface modification/coating, 89
Surface plasma resonance (SPR) sensor, 50, 451, 502, 505, 510
Sustainable environment, 622, 627, 632, 634
Sustainable intensification, 582, 596
Synchronized interaction, 442
Synergistic mosquitocidal property, 370
Synthesized multi-coloured AgNPs, 374
Synthetic polymers, 18, 40
Synthetic polymers-based nano DDS, 98
Systematic evolution of ligands (SELEX) process, 503
Systemic complications, 453
Syzygium aromaticum extract, 381

T

Target-based nanonutraceutical drug delivery, 270
Targeted drug delivery, 437
T-cell receptors (TCR), 334
Temephos larvicide, 384
Temperature-sensitive polymeric nanocarriers, 41
Ternary oxide photocatalysts, 683
 ABO4 Type Photocatalyst, 684
 delafossite, 683
 perovskite oxide materials, 683
 spinel, 683

Th2-mediated antibody response, 445
Theranostic nanoparticles, 30
Theranostic polymeric nanoparticle (TPTN), 125, 126
Theranostics, 231
Therapeutic antibodies, 448
Thermochemical water splitting, 670
ThermoDox®, 94
Thermo-responsive chitosan-g-poly (N-vinylcaprolactam) nanoparticles (TRC-NPs), 124, 125
Thermosensitive micelles, 41
Thiol group binding, 436
TiO_2 nanoparticles (TiO_2-NP), 439
TiO_2 NPs' minimal inhibitory concentration value, 406
Tissue culture, 388
Titanium dioxide nanoparticles (TiO_2NPs), 314
Titanium dioxide (TiO_2), 187, 188, 439, 478, 611, 644
Toll-like receptor, 443
Toll-like-receptor 9 (TLR-9), 444
Top-down approach, 422
Topoisomerase IV genes, 431
Toxic chemicals, 623
Toxic Substances Control Act (TSCA), 633
Toxin-antitoxin (TA) system, 169
Transglycosidases, 426
Transition metal chalcogenides, 452
Transition metal oxide-based catalysts, 644
Transition metals, 446
Transmission electron microscopy (TEM), 128
Triclosan, 435
Trimethoprim, 429
Tuberculosis, 241, 246, 247
Turbidimetry technique, 39

U
UV light, 224

V
V_2O_5 nanorods, 650
Vaccine delivery, 464
Vaccines, 442
Valourization of wastes, 630
Vancomycin, 427
Vantimicrobial activity, 186
Vector-borne disease
  Anopheles sp., 347
  Culex quinquefasciatus, 347
nanodrug delivery
  nanoliposomes, 349, 352, 353
  nanosuspensions, 353, 354
  polymer-based nanoparticles, 354, 355
Vesicular-arbuscular mycorrhizal (VAM), 545
Virosomes, 266
Virulence mechanism, 435
Virus, 244
Viruslike particles (VLPs), 336
Vitamin A, 285
Vitamin C deficiency, 285
Vitamin D, 285
Vitamin E, 471
Vitamins, 277
Vitrification, 543

W
Warburg impact, 219
Waste abatement, 630
Waste management, 630, 631, 634
Wastewater remediation
  adsorption and degradation process, 605
  CNTs, 612, 613
  graphene, 613
  magnetic nanomaterials
    iron oxide, 609–610
    spinel ferrites, 606, 610
    ZVI nanoparticles, 606, 608
  nano-based materials, 605, 606
  nanomembranes, 614
  transition metal oxide NPs
    copper oxide, 611, 612
    TiO_2, 611
    zinc oxide, 607–608, 612
Wastewater treatment, 605, 629
  adsorption technique, 519
  anthropogenic processes, 517
  batch adsorption, 519
  catalysts, 518
  chemical method, 519, 520
  co-precipitation, 523
  environmental conditions, 517
  hydrothermal synthesis, 523
  metallic oxide nanoparticles, 518
  micro-emulsion method, 522
  nano-catalysts, 531, 532
  nanomaterials, 518–520
  nanoporous materials, 519
  nanoscale, 519
  nanostructural materials, 518
  pollutant-removal techniques, 518
  polymer nanocomposite, 518

Index   785

polyol synthesis, 522–524
sol-gel method, 521, 522
toxic metal ions, 519
Wastewaters
  and air contaminants, 622
  bioremediation, 605
  consumption of water, 603
  contaminants, 635
  detergents, 603
  elimination of contaminants, 604
  heavy metals, 603
  organic contaminants, 602
  organic dyes, 603
  pesticides, 602
  pharmaceutical products, 602
  photocatalytic activity, 604
  phytoremediation, 605
  remediation (*see* Wastewater remediation)
Water, 601
  photocatalytic degradation, pollutants, 604
  removal of pollutants, 602
Water pollution, 625
Water quality management
  aquaculture operations, 477
  heavy metal ions removal, 478
  nanoadsorbents, 478
  nano-catalysts, 478
  purification technology, 477
  TiO_2, 478
  waste water treatment methods, 477
Water technology, 667
Water treatment techniques, 602

X
X-beam photoelectron spectroscopy (XPS), 214

Z
Zein biopolymer (Ze-AuNPs), 382
Zeolite, 545
Zero dimension (0D), 488
Zero-valent iron (ZVI) nanoparticles, 606, 608
Zero-valent iron nanoparticles (ZV-FeNPs), 562
Zika viruses, 347, 362, 363
Zinc, 286
Zinc nanoparticles (ZnNPs), 440, 473, 478
Zinc oxide (ZnO) nanoparticles, 405, 440, 495, 496
  antibacterial activity, 186
  antibiofilm activity, 186
  antimicrobial activity, 186
  bacterial cell membrane, 186
  biofilms, 186
  biomolecules, 186
  chitosan, 187
  enzymes, 186
  mechanism of toxicity, 185
  montmorillonite, 187
  photoactivation, 186
  photocatalytic activity, 185
  polyaniline, 187
Zinc oxide NPs, 607–608, 612
ZnO nanoparticle, 478
Z-scheme, 680, 681